D1226339

Charles Darwin's Notebooks, 1836–1844

Notebook pages B126e and D134e (see pp. 201, 374–75 for a transcription). B126e was originally written c. September 1837, while the note in the bottom corner—transcribed in bold—was added in grey ink between 29 July and 20 October 1838. The additions [in brown crayon] were made in December 1856, when pages were excised and distributed to topical portfolios: 11 was the divergence portfolio (see Table of Location of Excised Pages, pp. 643–52). D134e was written in September 1838, when grey ink was Darwin's standard writing medium. The page was crossed in pencil, presumably after *the note was of no further use.*

Charles Darwin's Notebooks, 1836–1844

Geology, Transmutation of Species, Metaphysical Enquiries

TRANSCRIBED AND EDITED BY

Paul H. Barrett Michigan State University
Peter J. Gautrey Cambridge University Library
Sandra Herbert University of Maryland Baltimore County
David Kohn Drew University
Sydney Smith St Catharine's College Cambridge

BRITISH MUSEUM (NATURAL HISTORY)

CORNELL UNIVERSITY PRESS
Ithaca, New York

First published 1987 by Cornell University Press.

Library of Congress Cataloging-in-Publication Data

Darwin, Charles, 1809–1882.
Charles Darwin's notebooks, 1836–1844.

Bibliography: p.
Includes index.
Partial contents: Red notebook 1836–1837 / transcribed and edited by Sandra Herbert—Geology. Notebook A, 1837–1839 / transcribed and edited by Sandra Herbert. Glen Roy notebook / transcribed and edited by Sydney Smith, Paul H. Barrett, and Peter J. Gautrey—[etc.]
1. Evolution. 2. Geology. 3. Metaphysics. 4. Darwin, Charles, 1809–1882. I. Barrett, Paul H. II. Title.
QH365.Z9B37 1987 575.01'62 87-47593
ISBN 0-8014-1660-4

Printed in the United States of America

Contents

Acknowledgements vii

Historical Preface 1
Sydney Smith

Introduction 7
Sandra Herbert & David Kohn

Red Notebook [1836–1837] 17
Transcribed and edited by Sandra Herbert

Geology

Notebook A [1837–1839] 83
Transcribed and edited by Sandra Herbert

Glen Roy Notebook [1838] 141
Transcribed and edited by Sydney Smith, Paul H. Barrett, & Peter J. Gautrey

Transmutation of Species

Notebook B [1837–1838] 167
Transcribed and edited by David Kohn

Notebook C [1838] 237
Transcribed and edited by David Kohn

Notebook D [1838] 329
Transcribed and edited by David Kohn

Notebook E [1838–1839] 395
Transcribed and edited by David Kohn

Torn Apart Notebook [1839–1841] 457
Transcribed and edited by Sydney Smith & David Kohn

Summer 1842 472
Transcribed and edited by David Kohn

Zoology Notes, Edinburgh Notebook [1837–1839] 475
Transcribed and edited by Paul H. Barrett

Questions & Experiments [1839–1844] 487
Transcribed and edited by Paul H. Barrett

Metaphysical Enquiries

Notebook M [1838] 517
Transcribed and edited by Paul H. Barrett

Notebook N [1838–1839] 561
Transcribed and edited by Paul H. Barrett

Old & Useless Notes [1838–1840] 597
Transcribed and edited by Paul H. Barrett

Abstract of Macculloch [1838] 631
Transcribed and edited by Paul H. Barrett

Table of Location of Excised Pages 643

Bibliography 653

Biographical Index 693

Subject Index 701

Symbols used in the transcriptions of Darwin's notebooks

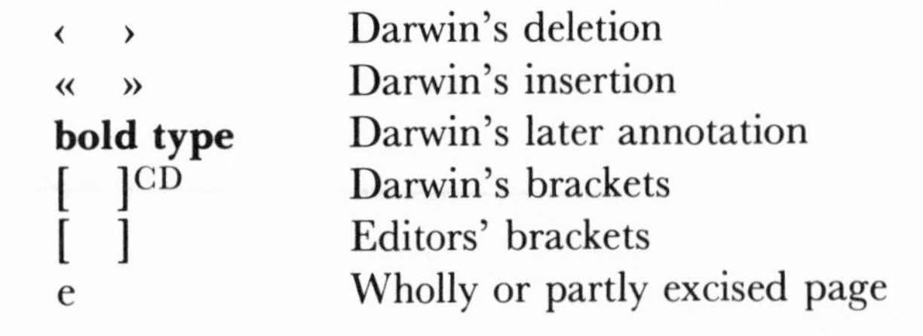

‹ ›	Darwin's deletion
« »	Darwin's insertion
bold type	Darwin's later annotation
[ ]CD	Darwin's brackets
[ ]	Editors' brackets
e	Wholly or partly excised page

Acknowledgements

The editors wish to thank George Pember Darwin on behalf of the family for kindly extending permission to publish these manuscripts. We also wish to thank the Royal College of Surgeons of England, the present owners of Down House, for granting access to the Red Notebook, and the Syndics of Cambridge University Library for making available the other manuscripts.

The British Museum (Natural History) has contributed more to this book than is usual for a publisher, and to all of the following the editors express their gratitude and appreciation. Maldwyn J. Rowlands, then Librarian, gave early encouragement. Robert Cross, formerly Head of Publications, carried the project forward from its inception and served as a firm anchor until its near-completion, when, on his retirement, responsibility passed to Clive Reynard, currently Head of Publications, who has seen the work through to publication. Myra Givans sub-edited the manuscripts with patience, authority, and enthusiasm, skilfully weaving together a text of many parts from five individual authors. Anthony P. Harvey, formerly Head of Library Services, aided the editors in setting an editorial standard; Dorothy Norman from the Library provided bibliographic assistance. Hugh Tempest-Radford designed the book's format. Others from the Museum providing assistance included Rex E. R. Banks, Ann Datta, A. W. Gentry, Susan Goodman, Ray Ingle, and David Snow. We would also like to thank the representatives of our co-publishers for their assistance: Richard Ziemacki at Cambridge University Press, and David Gilbert, Robb Reavill, and, formerly, Daniel Snodderly at Cornell University Press.

No library is of greater value for Darwin scholarship than the Cambridge University Library, and no group of librarians more helpful. We would like to thank Arthur Owen, Jayne Ringrose, and Margaret Pamplin of the Manuscripts Department; Godfrey Waller, Louise Aldridge, and Noni Hyde-Smith of the Manuscripts Reading Room; Anne Darvall and staff of the Reading Room; maps librarian Roger Fairclough; Gerry Bye and staff of the Photography Department and, for the colour microfilm, Carol Papworth; and Janet Coleby of the Conservation Department. We wish also to thank the following librarians and institutions: Reginald Fish, Zoological Society of London; N. H. Robinson, Royal Society of London; David Wileman and Francis Herbert, Royal Geographical Society; Ian Fraser, Local Studies Library, Shrewsbury; M. I. Williams, National Library of Wales; Claude Douault, Muséum National d'Histoire Naturelle; Andrée Lheritier and Claude Philippon, Bibliothèque Nationale; Brigitte Le Liepvre, Institut de France; Shirley Humphries, State Library of New South Wales; Constance Carter, Gary Fitzpatrick, and James Flatness of the Library of Congress; Martha Shirky, University of Missouri-Columbia; Marcia Goodman, University of Oklahoma; Vanessa Stubban, Kansas State University; Edwina Pancake, Douglas Hurd, and Peter Farrell, University of Virginia; Alan E. Fusonie, National Agricultural Library; and Carol Duane Jones, Michigan State University.

To members of the *Correspondence of Charles Darwin* project we owe the intellectual debt due to the closest of colleagues. In particular we wish to thank the project for permission to draw on its great store of biographical sketches, many of which appear in an altered format in the biographical index to this volume. Frederick Burkhardt and Stephen V. Pocock have given us particular assistance in this biographical research. In addition, Anne Secord was especially helpful in identifying linkages between letters and notebooks.

Philip Titheradge of the Darwin Museum at Down House, Downe, Kent has graciously assisted us with information on the manuscripts and books kept at the Museum.

Fellow scholars whose knowledge and judgement have benefited this edition include:

Stephen G. Brush, P. Thomas Carroll, Jane R. Camerini, Gordon Chancellor, Alain Corcos, John Cornell, Elizabeth Ermarth, Margaret W. Grimes, John C. Greene, M. J. S. Hodge, Nicholas Hotton III, Virginia Imig, G. A. Lindeboom, R. G. Mazzolini, Andrea Medacco, Franklin Mendels, Peter Maxwell-Stuart, Stan P. Rachootin, Frank H. T. Rhodes, Martin J. S. Rudwick, Simon Schaffer, James A. Secord, Phillip R. Sloan, Frank J. Sulloway, J. C. Thackray, Thomas Vargish, Magda Vasillov, Victor G. Wexler, Charlotte Watkins, Charles Webster, Leonard G. Wilson, and Penelope Wilson. To them we extend our appreciation.

For research assistance we would like to thank Janet Browne for creating the index and for comments as an informed and helpful reader. We would also like to thank Barbara Newell-Berghage and Nick Gil for substantial work on the bibliography. Others who have assisted us in important and timely ways include Saundra Barker, Mary Bartley, Jenny Fellowes, Virginia Imig, Elizabeth Kight, Stephen S. Kim, Chris MacLeod, Mindy Schwartz, Ronald Settle, Dale Simmons, and Jean Wilkins. To them we express our thanks.

The following university officers have assisted this project at critical junctures: Bard Thompson and Paolo Cucci at Drew University; Richard U. Byerrum, Richard Howe, Richard Seltin, and Diana Marinez at Michigan State University; and R. K. Webb at the University of Maryland Baltimore County. For assistance with computers we wish to thank Sam Motherwell and Chris Sendall at Cambridge University Library; Donald J. Weinshank and Stephan J. Ozminsky at Michigan State University; and David Czar, Natalie Stone, and Les Lloyd at Drew University. For typing we are indebted to Martha McAllister of Michigan State University, who produced the bibliography and texts of several of the notebooks, and to Carol Warner and Linda Hatmaker of the University of Maryland Baltimore County.

For hospitality the editors would like to express their gratitude to the Masters and Fellows of Churchill College, Darwin College, and St Catharine's College; the Warden and Fellows of Robinson College; the President and Fellows of Wolfson College; and the Program in History of Science at Princeton University.

For financial support of the project the editors are indebted to the National Science Foundation. Additional travel assistance was provided by the American Philosophical Society. Individual editors also wish to thank the following institutions, foundations, and agencies: Paul H. Barrett, Michigan State University; Peter J. Gautrey, Cambridge University Library; Sandra Herbert, the John Simon Guggenheim Memorial Foundation, the National Endowment for the Humanities, and the University of Maryland Baltimore County; and David Kohn, the American Council of Learned Societies and Drew University. The editors also wish to thank the Alfred P. Sloan Foundation for a publication subsidy.

While the editors have worked collaboratively, individual responsibilities being assigned as noted on the table of contents, recognition must also be given to Peter J. Gautrey who has supplied information as needed from the Darwin Archive, has checked all transcriptions against the original manuscripts, and has served as Secretary to the committee. Recognition must also be given to Sydney Smith for contributing his transcription notes of Notebooks B–E for use in editing these notebooks, and to Paul H. Barrett for identifying and assembling photocopies of many of the extracts from books and articles that were used in writing footnotes for Notebooks B–E and the Torn Apart Notebook.

Finally, the editors would like to thank members of their families who contributed their encouragement, and, in several instances, research to this edition. They are: Wilma Barrett, Margaret Anne Rathert, Paul H. Barrett, Jr., Thomas E. Barrett; Maureen Gautrey, Kerrin Gautrey, Sean Gautrey, Astell Gautrey; James Herbert, Kristen Herbert, Sonja Herbert, Charles Herbert, Helen Herbert, Emrick C. Swanson, Joann Swanson; Terry Kohn, Robert Kohn, Noah Benjamin Kohn, Jesse Kohn, Herbert M. Kidner; and Kate Smith.

Charles Darwin's Notebooks, 1836–1844

Historical Preface

SYDNEY SMITH

The documents here newly transcribed with the benefit of contemporary insight sustained by computer technology, are survivors from Darwin's period of maximal diversification of interests which drove him to make abstracts of everything available in print. This was, moreover, about the last time when such an activity was within the capacity of a single man. The results accumulated in the family home, extended to house the growing family and essential staff, record forty years of writing and experimenting, mostly written on high-quality paper made from linen rags.

Down House, Downe, Kent is today the store of material covering Darwin's life up to the return from the *Beagle* voyage. Papers relating to domestic matters, Darwin's health and activities in the garden, poultry and pigeon houses and so on are also at Down. The major store and site for accumulation of most of Darwin's working papers and significantly important manuscripts, and of the contents of this book, is the Darwin Archive of Cambridge University Library. It is my concern to give in outline the occasionally chaotic history and to record and give thanks for the recent and continuing generosity of contemporary members of the Darwin family and others. The Archive at Cambridge is incomparable in richness and as yet incompletely exploited treasure.

This volume aims to cover the years 1836 to 1844 when the theory of transmutation was conceived and was drafted in pencil in 1842 (DAR 6) to the second version of 1844 (DAR 7). Following the death of Emma in 1896, the manuscripts of the original sketches of 1842 and 1844 were found in a cupboard under the staircase at Down House. Supplementing the manuscript versions Darwin, plagued by ill-health and fearing premature death, had a fair copy (DAR 113) made in 1844 by Mr Fletcher, the Downe schoolmaster, which was interleaved for additions and corrections. This copy was returned to Darwin in September, 1844 and corrected by him against the original manuscript. On the 21st of the month Mr Fletcher, for his 'Species theory copying' was paid the sum of £2.0.0. In the event of his death, Darwin's wife Emma was charged with revealing the contents with the help of individual friends.

Public knowledge of the extent and quality of manuscripts, furniture, photographs and portraits in the family became clear when the Linnean Society of London celebrated the fiftieth anniversary of the meeting on 1 July 1858, at which the Darwin–Wallace communications were read by the Secretary. In 1909, the centenary of Darwin's birth and the fiftieth year after the 'Origin' was published, was made an occasion for celebration in Cambridge 22 to 24 June. At the same time an exhibition was mounted in the Old Library of Christ's College. The catalogue and also the Easter Term issue of the College Magazine should not be neglected. Most of the material in the Cambridge exhibition was also exhibited with many items from the British Museum (Natural History) in a further display in London in July. The Director of the Museum, Sir Sidney F. Harmer and Dr W. G. Ridewood, published a document of great significance to scholars, as some of the loans from the Royal College of Surgeons' Museum in Lincoln's Inn were lost in air raids on London during the War of 1939–45. The catalogue *Memorials of Charles Darwin, 1909*, was re-issued in 1910 and might well be reprinted, but not I fear for sixpence a copy. As Harmer says in the Preface:

> *it seemed best to illustrate some of Darwin's arguments by means of specimens, using as far as possible the species to which he himself referred in his writings, and in some cases the material which actually passed through his hands.*

At that time it is interesting to see that apart from notebooks of the *Beagle* period, only Notebook M, dealing chiefly with expression (item 42) was exhibited. Every delegate at the Cambridge celebrations had been given a printing of the pencil manuscript of the first (1842) version of the 'Origin'. Francis re-issued this and also the version of 1844 entitled 'The foundations of the Origin of Species'.

After the celebrations in 1909 Francis Darwin resided at 'Wychfield', his house in Cambridge with a few favourites from among his father's books. He also kept on his research room in the Botany School where his old assistant would set up plants for him to confirm or perhaps extend older measurements. He wrote an introduction to the Collected Works of his brother George, the Plumian Professor of Astronomy at Cambridge who died in 1913. He wrote on natural history and continued his lifelong practice at music—playing the bassoon in chamber works. He died in 1925 and left the residue of his father's library to the Botany School where they were distributed as reading copies on the open shelves and where, amongst other works by Herbert Spencer, Darwin's annotated numbers of his *Principles of Biology* were stumbled on. In 1925 Darwin himself was underrated. Francis left the great accumulation of documents assembled for *Life and Letters*, published in 1887, and with A. C. Seward, the two-volume *More Letters*, published in 1903 which weighted most essentially the scientific content of the published correspondence. Difficult subjects as Cirripedes and Pigeon breeding were still left on one side. Bernard, the inheritor of this accumulation, moved into 'Gorringes', a former dower house on the Lubbock estate in Downe village. Following the death of his wife in 1954 he moved to Kensington, where he died on 16 October 1963.

The more esteemed of the manuscripts belonged to Sir Charles Galton Darwin at the National Physical Laboratory: the *Beagle* Diary, and the manuscript 'Recollections of the Development of my Mind and Character', now called his 'Autobiography'. Owing to characteristic outspokenness in this work, complete publication was restricted during the lifetime of Leonard Darwin. He died in 1943.

On 4 September 1942, at one of the most troubled and difficult moments of World War II, Sir Alan Barlow, with his wife Nora at hand, wrote to the Librarian of Cambridge University as follows:

> *Dear Mr Scholfield,*
>
> *The Pilgrim Trust have decided to buy certain MSS of Charles Darwin, with the intention that the main part should be given to the Cambridge University Library, and the rest to Down House. I am writing to ask whether the Library would be willing to accept the gift . . . the greater number belong to Bernard Darwin & Mrs Cornford, and the rest to Sir Charles Darwin. I am co-executor with Bernard Darwin of the late Sir Francis Darwin, his and Mrs Cornford's father, and am writing on behalf of all three.*
>
> *The proposal is to give to Down House, the diary of the 'Beagle' (the property of Sir Charles Darwin), which is at present deposited with you, the field notebooks from which it was compiled; certain smaller items relating particularly to Down; and Charles Darwin's personal account books; and to give the rest to Cambridge. The principle of the division is to let Down have a popular exhibit, & items specially relating to Down, but to keep together in the University Library the rest of the material, in order that it may be available for any future student of Darwin & his work. The material throws a good deal of light on his methods of work & the growth of his theory of Evolution & Natural Selection.*

After specifying the location of Sir Charles (Galton) Darwin's MSS Sir Alan proceeds:

> *The rest of the material is in Sir Bernard Darwin's House-Gorringes, Downe, Kent. The Down*

House Trustees have accepted the items offered to them, [they were transferred by October 1942] *and Sir Charles Darwin will be writing to you to ask you to send the 'Beagle' Diary to them.*

He concludes

We all feel the documents should be in a public library rather than in private ownership, & though sale in the U.S.A. would have produced more cash, we would like them to remain in this country. We should be glad that through the munificence of the Pilgrim Trust, they should find a home at Cambridge.

Yours very truly,

Alan Barlow.

On 14th October 1942 the University accepted the gift, but in spite of appeals from Bernard that the Library collect their property, the excuse of staff shortage to catalogue and process accessions, petrol rationing for carriage allowed the Librarian to remain inert and unresponsive. Somewhat exasperated, Bernard Darwin wrote 9 May 1946: *May I remind you that yr Darwin documents are still here awaiting you, unblitzed and unburgled so far.* Scholfield was invited for lunch. He replied that he could not commit himself and suggested carriage be undertaken by rail or by Pickfords. Bernard answered 22 May 1946: *I am disinclined to send them off by rail or by Pickfords.* Meanwhile a request to the Library from Sweden for permission to consult the Darwin material had been forwarded to Bernard by the staff for him to attend to. This produced this expostulation 31 August 1948:

I really do think it is time you sent for your property here . . . I am most anxious to be quit of them and this sort of thing is not very encouraging to those who give. Do please take some step on this matter—I wish you would.

Mr Creswick, then Secretary of the Library, replied for Scholfield who was ill, and action to some purpose was at long last begun. Creswick was able to report to Sir Charles Darwin on 13 October 1948: *Those MSS destined for Cambridge have been collected from Downe & Barclays Bank.* There remained errors in assigning items between Down House and Cambridge to be set right, but there also remained items which were still missing. On 4 June 1949 when thanking Bernard for the gift of the 'Gorringes' 1932 catalogue, compiled by Miss Catherine Ritchie, Creswick concluded:

. . . all our efforts are being directed to completing the toll of known Darwin manuscripts and their present whereabouts in the interest of posterity and the fame of your great ancestor.

The 'Edinburgh Notebook' was missing. It was known to have been borrowed by the late Professor Ashworth who had Professed Zoology at Edinburgh University. Sir Charles Darwin as well as Bernard were approached, and the Librarian of the National Library of Scotland was asked to help sort things out. The widow of Professor Ashworth was able to supply the receipt of the registered parcel conveying the loan back to Bernard on 28 August, 1935. Moreover, she still had the note from him acknowledging its return. He accordingly searched at 'Gorringes' with more care, to reveal the notebook *tucked away in a locked drawer where I ought to have looked before.* (Letter 7 February 1949)

Meanwhile his wife's own search turned out the massive and important 'Diary of observations on Zoology of the places visited during the voyage'. Both items were sent to the

Library 7 February 1949. Reporting these events to Mrs Ashworth in a letter 8 February, Mr Creswick concluded:

> *. . . my enquiries about the Edinburgh notebook is complete success. This is to large extent due to your kindness in allowing me to see the papers relating to the use of the Notebook, and its return in 1935. My letter to Mr Darwin was so convincing, that he made a further search and found it together with a great bundle of other papers all in Charles Darwin's hand. We have therefore, as good reason to be grateful to you as if you had presented us with a valuable manuscript for the Library.*

It seems fair to conclude that Bernard did not know where the items still required to complete the gift to Cambridge may be in his house. The list of undelivered material was sent to Bernard on two occasions, without any response. In preparation for the move to Kensington, miscellaneous treasures which were outside the 'Gorringes' catalogue were placed in Box 'B' and the Box put into store. These items included Erasmus Darwin's correspondence with his contemporary Richard Lovell Edgeworth, author and inventor, together with part of a letter from Benjamin Franklin while he was U.S. Ambassador in Paris, as well as a great treasure of family letters which have proved invaluable for giving continuity to the Correspondence. Some items missing from Boxes A, C, D and E, reappeared over a period, but there were a small number still missing and not traced. Box 'B' had labels tied to a handle; an old dirty one inscribed 'Box "B" CD', seemingly dating from the division of the manuscripts between the five Boxes A to E. The other label was newer, 'B. Darwin 26/9/56, John Barker & Co. Ltd. Depository. Cromwell Crescent. W.' It seems that Bernard had the Box in his flat for about two years.

Box 'B' reappeared when Barker's ceased trading as a separate store and merged with Derry & Toms. The Box was returned to Bernard's custody, but was housed in the basement of the Science Museum, close to the Royal College of Art where Sir Robin Darwin, Bernard's son, was Principal. There was already other Darwin material deposited there: notably the letters sent to Darwin from the 1860s when he had to assume responsibilities congruent with his public notoriety; and in addition, observations shared by both Charles and his son Francis which hovered on the edge of Charles' own work and continued by Francis after his father died. When arranging his father's manuscripts, Francis overlooked this joint work, so much of the material on the Power of Movement in Plants appeared.

The storage in the Science Museum seems to have been arranged while Sir Terence Morrison Scott was Director. When he became Director of the British Museum (Natural History) the boxes moved with him. Sir Robin Darwin, informed of this mislaid property now assembled for inspection, wrote to Lady Barlow suggesting she and I call for an appraisal. This we did on the morning of 22 March 1962, initially meeting with Miss Skramovsky, secretary to the former Director, Sir Gavin de Beer, who still retained working space in the library of the Museum. The black metal deed box, its lock burst open by force, was filled with a confusion of manuscripts, amongst which was a small sealed envelope inscribed by Francis Darwin: 'Box C. D5. Darwin's Journal'. This is the description in the 'Gorringes' list. 'C.' implies the Journal had been removed from Box 'C', and moreover Francis signed the sealed envelope before 1925. It seemed probable that the arrangement of the documents was done by Francis, possibly with his father. I handed the envelope to Lady Barlow saying *you should open this*. The long-lost, original Journal was inside and we were released from the tidied up, distorted travesty from which Sir Gavin de Beer produced the reprint. After lunch with Sir Robin, we adjourned to his office at the top of the Royal College of Art where we were shown the books from Box 'H'. These consisted of Darwin's copies of his works with corrections and additions, together with copies formerly belonging to the family. These books were for sale and our advice was sought as to their possible value. We told Robin of the high significance of the volumes and suggested an expert should be consulted about their value. In the end, all the books reached Cambridge University Library a few months later *the gift of an anonymous donor*.

The envelopes listed in the 'Gorringes' catalogue under C.40 were found to contain most of the excised pages from the notebooks transcribed on the pages that follow. Marked with the number of envelope in crayon, the pages were cut out by Darwin, 'B' notebook, 7 Dec. 1856; 'C', 13 Dec. 1856; 'D', 14 Dec. 1856, and finally 'E', 15 Dec. 1856. Some pages are to be found in other places in the Darwin Archive at Cambridge, but the majority have been located in the C.40 envelopes in Box 'B' and now mounted in volumes DAR 205.1–11.

Shortly after this great day in London, Sir Robin agreed to the deposit of Box 'B' with the University Library in Cambridge but, while access was possible, the contents of 'B' were so chaotic that it took a good time to decide which of the materials were still Robin's property. Things were not finally agreed upon when he died suddenly in January 1974. He bequeathed to his two sisters those manuscripts in the black Box 'B' which had been placed therein entirely independent of the 'Gorringes' list. This was an exacting task, but by agreement with Sir Robin's lawyers, funds already given to the Library together with a matching Government Grant aimed at safeguarding for the Nation manuscripts of great importance, money was made available to provide annuities for the two sisters. The greater part of the excised pages were later transcribed by de Beer, Rowlands and Skramovsky and published in *Bulletin of the British Museum (Natural History)*, Historical Series. Issued in 1967, this work was read by me in proof. There has been difficulty in studying the notebooks in scattered publications, which is why the present attempt to bring them together in a seemly fashion was an urgent task. During many years of my serious indisposition the colleagues whose labours complete the work have earned my warm gratitude and I am sure the respect of future readers.

NOTE: All correspondence relating to the original gift is to be found *arranged in chronological order* in DAR 156.

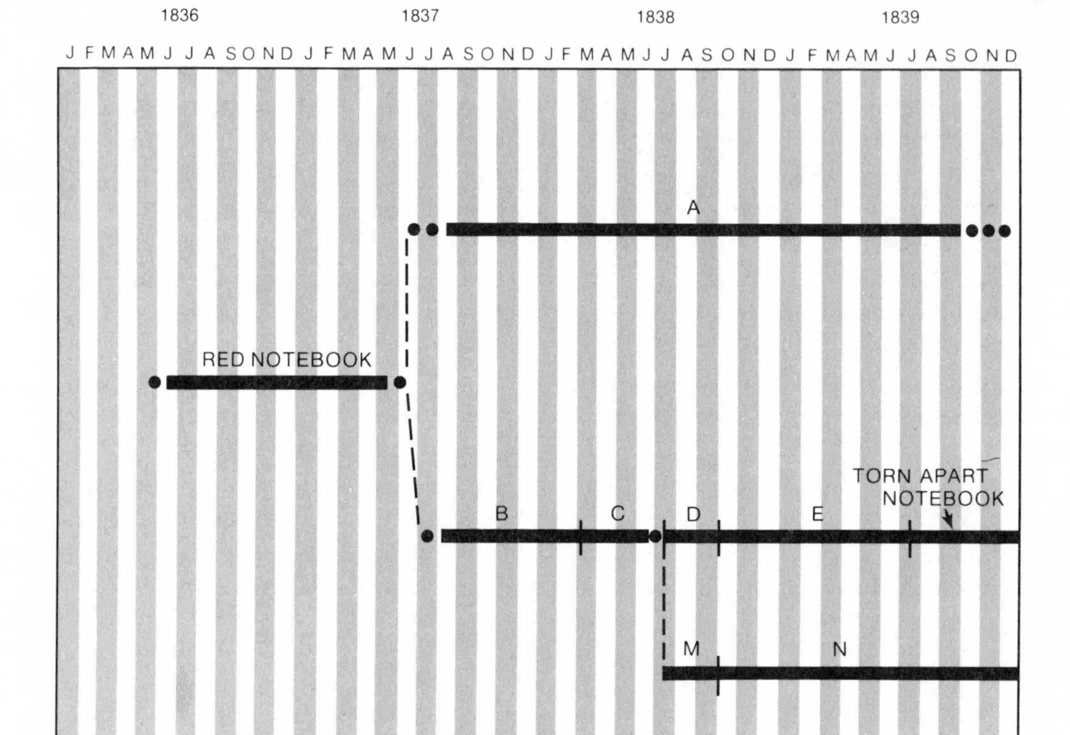

Fig. 1 Nine Darwin notebooks from 1836 to 1839. Solid lines represent the notebooks, dotted lines uncertainties in dating, and broken lines divisions of subject matter among members of the set.

Introduction

SANDRA HERBERT & DAVID KOHN

The English naturalist Charles Darwin lived from 1809 to 1882. Through his writing he affected the content and growth of science to a degree rarely matched in its history. This volume contains eleven notebooks and four related manuscripts from his most creative years as a theorist. In the notes contained in this volume he developed many of the major ideas contained in several of his geological publications, in *On the Origin of Species* (1859), *The Variation of Animals and Plants under Domestication* (1868), *The Descent of Man, and Selection in Relation to Sex* (1871) and *The Expression of the Emotions in Man and Animals* (1872).[1] These notes thus contain in outline the main programme of research and publication he was to follow during his life. They illumine his intellectual development and display the power and direction of his mind as scientist and as author.

The subtitle of this work 'Geology, Transmutation of Species, Metaphysical Enquiries', is drawn from Darwin's own words. He labelled one notebook 'Geology', one group of notebooks he characterized as being 'on Transmutation of Species"—', and another notebook he described as being 'connected with Metaphysical Enquiries'.[2] The earliest notes in this collection date from 1836, when Darwin was 27 years old and returning home from his service as naturalist aboard H.M.S. *Beagle* during the years 1831–36; the latest explicitly dated notes made in regular sequence are from 1844, though later entries exist.[3]

As an edition, this work is simultaneously a work of fresh publication, of restoration and, within limits, of interpretation. As a work of fresh publication it provides texts of six previously unpublished manuscripts: Notebook A; the Glen Roy Notebook; the extant pages of what we have characterized as the Torn Apart Notebook; the Summer 1842 Notes; the Zoology Notes, Edinburgh Notebook; and a notebook labelled Questions & Experiments. As a work of restoration this edition offers an integrated version of Notebooks B, C, D and E. When these were first published by Gavin de Beer in 1960, only pages still extant in the notebooks were included.[4] Darwin had excised numerous pages from his notebooks for use in later writing, and the location of these was then unknown. After the original publication of the notebooks, the majority of the pages excised by Darwin were found and published in 1961 and 1967.[5] While that publication was an appropriate solution at the time, it was not a lasting one since it is clearly an advantage to be able to read the notebooks straight through, as they were written, without having to consult separate publications. In the present edition that defect is corrected, the excised pages from Notebooks B, C, D and E having been replaced in their original positions. Other improvements have also been made in this edition. Most especially, care has been taken to note the layering of the manuscripts with respect to date.

[1] The standard guide to Darwin's publications is Freeman 1977. For concordances to *Origin*, *Descent*, and *Expression* see Barrett, Weinshank and Gottleber (1981); Barrett, Weinshank, Ruhlen and Ozminski (1987, in press); and Barrett, Weinshank, Ruhlen, Ozminski and Newell-Berghage (1986).

[2] Notebook A is labelled 'Geology'. In his 'Journal' begun in August 1838, Darwin referred to Notebook B as follows—'In July [1837] opened first note Book on "transmutation of Species".—' —'and to Notebook M as follows—'opened note book connected with Metaphysical Enquiries.' De Beer 1959:7–8, and for a text corrected against the original manuscript, *Correspondence* 2:431–32.

[3] See the Introduction to Questions & Experiments. The bulk of the notebook was completed by 1844, its latest explicit date. However, 1845–46 references do exist, and the date of last use of the notebook remains uncertain.

[4] De Beer 1960. The unexcised pages of Notebook B were also published in Barrett 1960.

[5] De Beer and Rowlands 1961; De Beer, Rowlands, and Skramovsky 1967.

In addition to being a work of fresh publication and restoration, the present edition is also, within limits, a work of interpretation. It is so in its grouping of manuscripts. In publishing the geological and the metaphysical notebooks alongside the transmutation notebooks, the editors are taking a synthetic view of Darwin's endeavours. His work as a geologist and his enquiries concerning man and behaviour are presented as integral to the formation of the species theory and to the development of his scientific outlook generally. Reading his geological notebooks alongside his transmutationist notebooks allows one to observe the connections between his views on species and his questions regarding the origin and structure of the earth. For a somewhat different reason, reading the metaphysical notebooks alongside the transmutation notebooks one perceives more readily what gains accrued to Darwin's theory from his own reading in traditions outside natural history. Were one to read more narrowly, these perspectives, and with them a sense of the integrity of Darwin's work, might be lost.

The seven notebooks lettered alphabetically—Notebooks A, B, C, D, E, M and N—form the core of this edition. They define the centre of the notebook period, summer 1837 to summer 1839, and elaborate its key topics: geology (Notebook A), transmutation of species (Notebooks B to E), and metaphysical enquiries (Notebooks M and N). The other eight manuscripts relate directly to this core. The Red Notebook contains both geological and transmutationist themes and was the predecessor to the later notebooks. The Glen Roy Notebook, while primarily a geological field notebook, contains occasional observations on breeding and instinct. The Torn Apart Notebook was a direct continuation of Notebook E, the brief Summer 1842 Notes being on related themes but later. The Zoology Notes, Edinburgh Notebook were a series of notes running roughly parallel in time and subject matter to the transmutation notebooks. Questions & Experiments carried forward transmutationist themes by a set of questions on breeding and inheritance and by a record of experiments to be tried on these subjects. The Old & Useless Notes and the Abstract of Macculloch complemented the major themes of Notebooks M and N.

In addition to forming the core of this volume, the alphabetically lettered notebooks bear a generative relationship to one another, the subject of transmutation serving as the stimulus for the differentiation of the group. During the years Darwin kept these notebooks he moved from asserting transmutation as an hypothesis to constructing a full theory of its operation. Before opening these notebooks he knew of the state of opinion on the subject and, while on the *Beagle* voyage, commented on it, often tangentially. In March 1837 after London zoologists had examined a number of the specimens he had collected on the voyage, he was ready to take up the transmutationist hypothesis. Darwin's earliest known explicitly transmutationist statements occur in the Red Notebook, scattered amongst observations and reading notes, primarily on geology. When this notebook was filled, Darwin began Notebooks A and B. Notebook A, devoted primarily to geology, he filled gradually over two years. Notebook B, devoted entirely to the subject of the transmutation of species, he filled relatively rapidly. It was succeeded in turn by Notebooks C, D, E and the Torn Apart Notebook all devoted to transmutation. In addition, in the course of keeping Notebook C, Darwin recorded an increasing number of observations on man, behaviour, and the metaphysical and epistomological implications of transmutation. When Notebook C was filled, Darwin opened a new parallel series of notebooks, labelled sequentially 'M' (possibly for 'metaphysics') and 'N', devoted to exploring his new interests. Meanwhile, in Notebook D Darwin first formulated the concept of natural selection, which he elaborated in subsequent notebooks. Over time, the scheme of these notebooks can be represented as illustrated in Fig. 1, page 6. Although Darwin stopped using notebooks to record his views in the 1840s, he continued to make notes on transmutation, which he organized into subject portfolios, through the rest of his career. The formal writing of the *Origin* went through three stages before publication: two drafts—the *1842 Sketch* and the *1844 Essay*; a long version of the argument, *Natural Selection*, never published in his lifetime; and, finally, the *Origin* itself in its first and subsequent five editions. Darwin's geological notes in the Red Notebook and in Notebook A were incorporated into various of his later publications. Material from Notebooks M and N appeared primarily in *The Descent of Man* and *The Expression of the Emotions in Man and Animals*.

In addition to the fifteen manuscripts included in this volume are others bearing on its subject that were excluded only for practical reasons. The Red Notebook itself has a predecessor in the form of 34 numbered pages at the back of a field notebook from the *Beagle* voyage labelled 'Santiago Book'. Begun in 1835, these pages contain entries of a theoretical nature and are directed towards future publication.[6] From the post-voyage period the 'St. Helena Model' notebook, begun in 1838, records Darwin's notes on his observations of Robert Seale's topographical model of that island, as well as some entries directed towards the species question.[7] In addition to notebooks, there are in the Darwin Archive at Cambridge University Library a considerable number of loose notes from the 1836–44 period that bear on the subject matter of the notebooks published here. Among these are notes, catalogued in the library as DAR 29, regarding specimens from the *Beagle* voyage. Also of interest are Darwin's collection of abstracts of books and scientific periodicals similar to the Abstract of Macculloch.

Entries in the majority of the fifteen manuscripts in the present collection are directed towards the construction of theory. They represent a series of brief expositions, memoranda and reading notes: theory in the process of gestation. Since the arguments presented in the notebooks were in the early stages of formation, they display the probing and discursive logic of discovery rather than the coherent and fully articulated logic of final exposition. In addition, to the reader familiar with *On the Origin of Species*, the notebooks may initially seem foreign, for where the *Origin*, at least in its first edition, offered few clues to its antecedents, the notebooks served as Darwin's record of his sources. Also in these manuscripts Darwin frequently reflected on the subject of scientific method with regard to his own work, as, for example, in his comment in [Notebook] D117: 'The line of argument «often» pursued throughout my theory is to establish a point as a probability by induction, & to apply it as hypothesis to other points. & see whether it will solve them.—'

The notebooks are also revealing of Darwin's day-to-day practice as a scholar. In the notebook texts, reading notes are abundant and useful for assaying questions of influence, but one can also find connections between the texts and Darwin's marginal comments in books from his own library, sometimes, as in the John Phillips' references in A147, transferred

[6] The notebook labelled 'Santiago Book' is stored at Down House, Darwin's former home in Kent, with the other field notebooks from the voyage. It is catalogued as 1.18. Darwin's first statement of his theory of coral reef formation is contained in the set of numbered pages at the rear of the notebook. For a transcription of the coral reef entries see *Correspondence* 1:567–71. Also see Barlow 1945:243–44.

[7] The notebook labelled 'St. Helena Model' is catalogued as 1.5 and stored at Down House, no doubt because it was believed to have been a product of the voyage. However, the notebook appears rather to be post-voyage in date, possibly having been opened in September 1838. On 12 September 1838 Darwin wrote a letter requesting permission to study the large plaster model of St Helena constructed by Robert F. Seale and housed, from at least as early as 1826, at the East India Company's Military College at Addiscombe (*Correspondence* 2:103, Seale 1834:8). Entries in the first fifteen pages of the notebook appear to correspond with a trip by Darwin to see the model and pertain to craters, which suggests a relation with Darwin's comment that in September 1838 he 'Began craters of Elevation Theory' (*Correspondence* 2:432, *VI*:93–96). The notebook continues into late 1838 and had some use in 1839. For quotations from it see Barlow 1945:257–60. Gordon Chancellor is presently preparing for publication an edition of the complete notebook. He has generously shared his work on the notebook with the editors of this volume.

directly from book margin to notebook page. With regard to Darwin's note-taking, the organization of the notebook occasionally changes in such a way that reveals a new category of meaning emerging, as in the notes on generation that begin a separate section of Notebook D at page 152. The notebooks also bear evidence of recurrent use, for Darwin repeatedly culled them for ideas and references. For an early instance of Darwin's re-reading of and commentary on his own work see B153 quoting the Red Notebook; for a much later example of re-reading see A33.

The circumstances under which Darwin kept the majority of notes in this collection were conducive to concentrated intellectual effort. With an income provided for him by his father, he set his own course. The prospect of his becoming a clergyman had evaporated during the voyage.[8] For a time he believed a Cambridge professorship would be required to provide a living, but even that proved unnecessary.[9] A handsome marriage settlement further secured independence.[10] In addition to leisure, Darwin also benefited from a change in venue, for during most of the notebooks' period he lived in London. His places of residence were as follows: 16 December 1836 to 6 March 1837 lodgings in Fitzwilliam Street in Cambridge, a week's stay at his brother's house at 43 Great Marlborough Street in London, then from 13 March to 31 December 1837 furnished rooms at 36 Great Marlborough Street, from 1839 to September 1842 a house for himself and his bride at 12 Upper Gower Street after which the family moved to Down House in Kent.[11] (Darwin married his cousin Emma Wedgwood on 29 January 1839.) It was the presence in London of the major scientific societies and institutions that had drawn him to the city, and, once there, he found peers. In turn he was valued. In 1841, as Darwin was contemplating departure from London, the geologist Charles Lyell wrote to him:

> *It will not happen easily that twice in ones life even in the large world of London a congenial soul so occupied with precisely the same pursuits & with an independance enabling him to pursue them will fall so nearly in my way, & to have had it snatched from me with the prospect of your residence somewhere far off, is a privation I feel as a very great one—*[12]

Darwin's life in London was not circumscribed by science, however, and through his brother Erasmus Alvey Darwin he met Harriet Martineau, Thomas and Jane Carlyle, and other literary figures. On 21 June 1838, in recognition of his accomplishments, Darwin was elected to the brilliantly populated Athenaeum Club (in the same class as Charles Dickens) and in its library read extensively while keeping Notebooks M and N.[13]

Darwin's chief, and related, occupations during his London years were establishing himself in its scientific world and bringing to completion work from the voyage, which task included publication of his own observations and arranging for the disposition of his collections. On the first count, Darwin chose to make his way in London primarily as a geologist. Early in the voyage he had decided to write 'a book on the geology of the various countries visited', and in keeping with that goal, even before the end of the voyage, had arranged for himself to be put up for election as a Fellow of the Geological Society of London.[14] Immediately upon arriving

[8] See Darwin's letters to his cousin William Darwin Fox in *Correspondence* 1:316, 432, 460.
[9] *Correspondence* 2:443.
[10] *Correspondence* 2:119.
[11] *Correspondence* 1:542; 2:430 (but see p. 8 for a possible 3 March date), 10–11, 435 n. 5, 432, 435.
[12] *Correspondence* 2:299. On Lyell's relationship with Darwin in his London years see L. G. Wilson 1972, chap. 13; also see Sydney Smith 1960:396–98.
[13] *Correspondence* 2:94 n. 2.
[14] Barlow 1958:81; *Correspondence* 1:499, 517.

home he began attending meetings of the Society and very soon presented papers before it and served as one of its officers. Darwin also began a new project during this period, undertaking a trip to Scotland at the end of June 1838 in order to formulate an explanation for the origin of the so-called 'parallel roads' of Glen Roy. This work was published in the *Transactions* of the Royal Society of London, of which he became a Fellow in 1839.[15] Geology aside, Darwin also participated in the more general scientific life of London, including attendance at numerous meetings and lectures, possibly including the 1837 Hunterian lectures of Richard Owen at the Royal College of Surgeons.[16]

With regard to publishing results from the voyage Darwin made rapid progress. He recast his diary from the voyage into book form on his return to London in early March 1837 to about 25 June, sending a manuscript to the publisher in early August and receiving proof later that month and into the autumn. As publication was delayed, he added a preface and addenda to the book in October 1838.[17] His *Journal and Remarks* on the voyage was in his hands in May 1839.[18] In the autumn of 1837, as he was finishing proofs for the *Journal*, Darwin 'Commenced geology'.[19] Although much delayed by illness and by his increasing attention to the species question, the geology emerged in three parts: *The Structure and Distribution of Coral Reefs* (1842), *Geological Observations on the Volcanic Islands Visited during the Voyage of H.M.S. Beagle* (1844), and *Geological Observations on South America* (1846).

On the zoological side Darwin's immediate task following the voyage was to arrange for his collections, which had been sent back to Henslow, to be dealt with by competent specialists. He spent much of the winter of 1836–37 in Cambridge unpacking and sorting his specimens. At the same time he had already begun to place some of the specimens with individuals who would be responsible for classifying and describing them. In August 1837 he received a Treasury grant of £1,000 to support publication of their findings.[20] In the autumn he planned for the work that would emerge, noting in his 'Journal' under the year 1837 that he was in 'October–November.— preparing scheme of Zoology of Voyage of Beagle.'[21] With Darwin assisting the specialists at every step, the *Zoology* from the voyage was published in five parts made up of nineteen numbers over a five-year period from 1838 to 1843. The parts were assigned as follows: *Fossil Mammalia*, Richard Owen; *Mammalia*, G. R. Waterhouse; *Birds*, John Gould (with assistance of T. C. Eyton and G. R. Gray); *Fish*, Leonard Jenyns; and *Reptiles*, Thomas Bell.[22] Reports on other portions of Darwin's collections appeared subsequently in various formats.

In the course of placing his collections, Darwin became a respected member of London's botanical and zoological circles. His working ties with Robert Brown and Richard Owen made him a frequent visitor to the British Museum and to the Royal College of Surgeons. Through his collaboration with John Gould and George Robert Waterhouse in work on

[15] Darwin 1839; *Correspondence* 2:433.
[16] Sloan 1986:398–432. On the institutional milieu of zoology in London see Desmond 1985.
[17] *Correspondence* 2:432.
[18] According to *The Publisher's Circular* of 1 June 1839, Darwin's *Journal and Remarks*, vol. 3 of Fitzroy 1839, was published between 15 May and 1 June 1839. The separate publication of Darwin's book, retitled *Journal of Researches*, was first advertised in *The Publisher's Circular* of 15 August 1839.
[19] *Correspondence* 2:431.
[20] *Correspondence* 2:37–39.
[21] *Correspondence* 2:431, 17. The scheme of publication was altered after its inception. Freeman 1977:26 cites an 1837 prospectus for the work stating that 'a description of some of the invertebrate animals procured during the voyage will also be given. At the conclusion of the work Mr. Darwin will incorporate the materials which have been collected, in a general sketch of the Zoology of the southern parts of South America.' As Freeman points out, neither of these intentions was realized. However, it is possible that some of the entries in the Zoology Notes, Edinburgh Notebook were directed towards them.
[22] *Zoology*; on the assistance of Eyton and Gray see Freeman 1977:29–31.

specimens from the *Beagle* voyage, Darwin came to attend the meetings of the Zoological Society, where descriptions of his voyage specimens were read, and to visit the Society's gardens. Over the notebook years, these biological contacts and interests would expand, and Darwin would become recognized as one of London's most accomplished young naturalists, as well as a well-published geologist.[23]

Each one of Darwin's occupations immediately following the voyage proved instrumental in establishing his stature as an author of wide influence. His narrative of his travels and his articles for scientific journals established his credentials as a writer on both popular and specialized topics. His work on the *Zoology* from the *Beagle* voyage gave him experience in organizing a major editorial enterprise. And all of these activities brought him in contact with the personalities and ideas of the men who formed the community of naturalists that Darwin would always regard as the prime audience for his work.

Amidst all these public activities, in the years Darwin would characterize as the most active ones he ever spent, he opened the private notebooks that comprise this edition. In them he took up the question that would unify biology and give definition to his own life: the transmutation of species.

EDITORIAL CONSIDERATIONS

The primary goal of the editors has been to provide an accurate transcription of Darwin's text. While the aim has not been to produce a facsimile in type, the editors have attempted to capture those elements in the text that are important for meaning. The method of transcription adopted is a modified version of that described by Fredson Bowers.[24] Following that method the text is printed together with a set of textual notes containing editorial comments concerning the transcription. These notes are keyed to the text by a quotation from it followed by a lemma. In the textual notes Darwin's words are printed as in the text, the editors' comments in italics. For an illustration of editorial method the reader may compare the two manuscript pages reproduced as a frontispiece to this volume with the published text for the pages.

In transcribing Darwin's text all words and fragments of words are recorded. Deletions are enclosed by single angled brackets ‹ ›. Superimpositions are noted in the textual notes, as in the phrase ' "a" *over* "o" ' meaning that the letter 'a' is written over the letter 'o'. Superimpositions of the same letter (an 'a' over an 'a' for example) are ignored. Insertions in the text are set off by double angled brackets « ». If their placement on the page is unusual, that fact is recorded in the textual notes. Entries in the notebooks are regarded as insertions if they appear to have been written out of sequence as compared to other entries on the page. Insertions would include carated remarks and afterthoughts generally. Annotations are recorded in the text in bold face type.

To decide whether to record an out-of-sequence entry as an insertion or as an annotation is difficult. An insertion is an addition to the text thought to be contemporary with the original entry. An annotation, on the other hand, is a note or comment made after the fact, that is, at some time after composition of the original entry. The editors have attempted to distinguish between these two categories. The most common sign of annotation has been taken to be a

[23] On the interplay between Darwin's public and private lives in science during these years see S. Herbert 1974–77 and Rudwick 1982.
[24] Bowers 1976. Tanselle 1978 has also been used as a guide to editorial standards.

change in writing medium. Where an out-of-sequence entry is written in the same medium as its surrounding passage, the entry will usually be taken to have been an insertion. Only if there are other signs suggesting addition at a later date—as, for example, having been written diagonally across a page over existing script—or some indication from the content of the entry that it must have been written well after the original passage, will the entry be treated as an annotation. An example of an easily identified annotation is the entry in brown ink on RN38, 'V. back of page 1 of New Zealand Geological Notes'. This entry stands out sharply from the other entries on the page, which are in pencil, and was obviously squeezed on to the page after the original entries were made. Of course, not all cases are so clear cut, and where there has been doubts our inclination has been to treat out-of-sequence entries as insertions rather than annotations. No systematic attempt has been made to date annotations. However, certain annotations which appear in grey ink, discussed below, can be assigned a definite range of dates. Also, brown-crayon numbers, indicating subject portfolios, were added to many excised pages in the 1850s. On this see the 'Table of Location of Excised Pages'. At points in the notebooks, and always at the beginning and end, one must be cautious in inferring temporal sequence from physical sequence. Thus, for example, Darwin used the inside of the covers of notebooks as convenient, and the ordering of entries must there be regarded as problematical. In addition, the entire Questions & Experiments Notebook was filled in a more complicated manner than front to back.

Other difficulties in transcription pertain to symbols, abbreviations, spelling (and its attendant problem of handwriting), punctuation, variation in media, and the treatment of excised pages.

With respect to symbols and abbreviations the editors have preserved Darwin's usages in so far as they can be brought gracefully into print. Darwin's brackets are kept with a superscript 'CD' added; his use of the ampersand and plentiful employment of the dash are unaltered; and his abbreviations 'do' for 'ditto' and 'V.' for 'Vide' remain unexpanded. Darwin's own figures and sketches have been traced or, where feasible, reproduced photographically from manuscript. Two idiosyncratic markings have also been kept: first, a backwards question mark that he borrowed from Spanish, and a cross-hatch he used on occasion for emphasis. Darwin's underlinings have been reproduced according to printers' conventions: one line, italics; two lines, small capital letters; three or more lines, large capital letters. Where underlinings are annotations they are recorded in the textual notes ('*underlined*'). The lines Darwin often used to separate entries, his 'rule lines', have been reproduced according to the following convention: a rule line taking up one quarter or less of the notebook page is transcribed as a short line, a rule line taking up between one quarter and three quarters of the notebook page is transcribed as a medium rule line, and a line taking up three quarters or more of the page is transcribed as a long line. The marginal scorings that Darwin used for emphasis have not been reproduced in the text but are recorded in the textual notes, the passage in question being noted as having been '*scored*'. Similarly, where Darwin circled or enclosed a word or phrase for emphasis, the fact has not been reproduced in the text but recorded in the textual notes, the word or phrase being noted as having been '*circled*'. (However, where Darwin circled or half circled notebook page numbers that fact has not been recorded.) The vertical lines that run through much of the text of the notebooks were Darwin's indications to himself that he had made full use of the material. Passages so marked are recorded in the textual notes as having been '*crossed*'. Finally, some of Darwin's markings elude succinct description. In such cases the editors have attempted to indicate Darwin's intention in so far as they could determine it.

In spelling the editors have preserved Darwin's practice where individual letters are clearly formed. Where individual letters are indistinct, as is often the case, the editors have offered the probable reading of the word without comment. Also, the slurring together of letters amounted to a kind of shorthand (for example, 'by' is often written as one curving stroke rather than as two clear letters, or the 'e' in 'the' appears only as a tail), and here too the words have been spelled out in full. Where two readings of a letter are possible, as is, again, often the case, the editors have chosen what seemed from the context the more probable reading. Such ambiguities are particularly common with the vowels—for example, an 'a' being indistinguishable from 'or'. Where ambiguous readings are plausible in context, the editors have chosen one for the text and listed the other as an '*alternate reading*' in the textual notes. Where the editors can do no more than guess at a reading, they have appended a textual note '*uncertain reading*' to the word or passage. Where a section of text is illegible, bracketed ellipses marks ('[. . .]') appear in the text with a notation being made in the textual notes as to how many letters or words are illegible. In orthography Darwin's forms are standard with the exception of the long 's' (appearing as the first 's' of a double 's'), which has been modernized silently. In capitalization Darwin's usage has been preserved where it is clear. Where it is unclear, preference has been given to conventional usage.

In punctuation, as in spelling, the editors have attempted to follow Darwin's practice in so far as they could determine it. As R. C. Stauffer has pointed out, Darwin followed a system similar to that suggested in Lindley Murray's *English Grammar*.[25] In that system, commas, semicolons, colons, and full stops or periods indicate increasingly longer pauses more than they distinguish different constructions. Thus Darwin might use a colon where a semicolon would now be employed. The editors have not altered Darwin's practice in this regard. More problematical for the editors has been the question of distinguishing commas from full stops and both from stray marks and pen rests. Here the editors have attempted to transcribe literally while being alert for signs of Darwin's intentions. The editors have not added punctuation where there is none.

Another issue in transcription is media. Customarily Darwin wrote in pencil or in pen and ink, with an occasional entry in crayon. The pencil he used was of the standard sort, except in the Glen Roy Notebook where he used a special metallic pencil that came with the notebook. The inks he used are usually shades of brown, grading into one another, presumably as a result of different inks being mixed in a common well. The distribution of pages written in ink and pencil is indicated in the introduction to each notebook. The medium of individual entries is recorded in the textual notes only where a change of medium occurs on the page. The colour of the ink Darwin used is noted systematically in only one case, where knowledge of its presence is helpful in dating entries. Darwin used a greyish ink from 29–31 July 1838 to 19 October 1838 (D21–E26 and M56–N18).[26] During this eleven-week period, Darwin also added numerous grey-ink annotations to Notebooks B and C.

With regard to excised pages, in all cases excisions, whole or partial, have been recorded. Where the excised material has been recovered, the location of the recovered material is given in the 'Location of Excised Pages'. Recovered material is noted as follows: the letter 'e' is given next to the page number in the text and, for partly excised pages, the word '*excised*' appears in

[25] *Natural Selection*:20–21.

[26] These dates have been determined as follows. On 29 July 1838, Darwin set out from Shrewsbury to Maer in Staffordshire (*Correspondence* 2:432). His first known use of grey ink occurs in a page of observations on Staffordshire geology written as the continuation of brown ink notes on Shropshire geology (DAR 5:21–22, 'Alluvium/Shropshire/1838/July'). This document is a single folded piece of paper with the brown notes on 5:21 verso and recto followed by the grey notes on 5:22 recto. Darwin was at or in the vicinity of Maer until 31 July. The grey ink in the notebooks begins with Darwin at Maer on D21—after the second line—and on M56. The last dated use of grey ink is 19 October and occurs in E26. See textual note on E IFC for details. N18 is in grey ink up to, but not including, the entry dated 'Octob. 19th'.

the textual notes together with an indication of the placement of the excision. Unrecovered material is indicated by the phrase 'not located'. On some excised pages, one can detect pin holes that once held pins attaching various sheets together. Pin holes are signalled by an asterisk next to the page number of the excised page in the 'Table of Location of Excised Pages'. It should also be noted in connection with excised pages that a fragment of a letter, a word, or a sentence sometimes appears on the stub of an excised page. The editors have checked all stubs for such material and have incorporated it into the text silently.

To aid the reader the texts in this edition are provided with explanatory footnotes. Most footnotes fall into two categories: (1) persons referred to in the text other than as authors, and (2) references to written work, usually books or articles. The first category of notes includes persons referred to in the text who are not being cited as authors and who thus require identification. Thus the 'Aunt Sarah' referred to on M83 is identified in a note as 'Aunt Sarah: Sarah Elizabeth Wedgwood'. In the biographical index she is further identified as the daughter of Sarah and Josiah Wedgwood I and her birth and death dates given. Readers seeking fuller biographical information on the persons listed should consult standard biographical dictionaries and listings, most particularly the *Dictionary of National Biography* and the *Dictionary of Scientific Biography*. For a list of sources of biographical information on figures in nineteenth-century English natural history readers should consult the *Calendar* to the Darwin correspondence.

The second, and largest, category of footnotes pertains to the published literature. Where possible the identical edition of a work to that used by Darwin has been cited. Frequently Darwin's own copy of a work has provided the citation. Where Darwin's marginal comments on the relevant passage in his own copy of the work were particularly apposite, they have been included in the note. However, marginalia can rarely be dated, and the reader should not infer that they are necessarily contemporaneous with the notebook entries.[27] In some notes, cross-references are made to various of Darwin's publications, or to his manuscripts, but no attempt has been made to provide an exhaustive set of cross-references between the manuscripts published here and Darwin's other writings. Also, the notes do not refer to what has become a large secondary literature on Darwin and his contemporaries, since to have done so would have greatly lengthened the volume.[28]

Titles of journal articles are cited in full in the bibliography and journal titles are abbreviated according to the standard specified in the *List of Serial Publications in the British Museum (Natural History) Library* (London, 1968). Book titles have been reduced in length; those seeking full titles and complete publication data may consult standard library catalogues. In addition, a number of works have been given short titles, by which they are cited throughout the volume. Books and articles marked with asterisks in the bibliography are those that are part of Darwin's personal library. Darwin's library is presently divided between the Cambridge University Library and Down House, his home in Kent. While there is no published list of Darwin's reprint collection—an unpublished list by P. J. Vorzimmer (Ph.D. thesis 1963, vol. 2) can be found in the Cambridge University Library—there does exist a published catalogue of his library: H. W. Rutherford, *Catalogue of the Library of Charles Darwin now in the Botany School, Cambridge* (1908).

Following each entry in the biographical index and bibliography, there is an index of mentions listing every location in the manuscripts where that reference appears. Thus following the biographical entry for Charles Lyell is a list of all those places in the notebooks

[27] A catalogue of Darwin's marginalia is presently being prepared for publication by Mario DiGrigorio and Nick Gil.
[28] On the literature see the bibliography produced with the assistance of Malcolm J. Kottler in Kohn 1985.

where he is mentioned in a guise other than that of author. References in the notebooks to specific publications by Lyell are listed under the appropriate bibliographical entry.

To aid the scholar a colour microfilm of the majority of the manuscripts in this volume has been prepared under the supervision of Peter J. Gautrey. Manuscripts included are the Red Notebook, Notebooks A, B, C, D, E, M, N, the Torn Apart Notebook and Questions & Experiments. In order to keep the microfilm to one roll, the other manuscripts in this volume were omitted, the Glen Roy Notebook being too faint for filming in any case. The microfilm, published jointly by Cambridge University Library and the British Museum (Natural History), is available for purchase from the Museum and from Cornell University Press. Also for the scholar, a concordance to the manuscripts in this volume is being prepared under the supervision of Paul H. Barrett and Donald J. Weinshank and will be published by Cornell University Press.

Red Notebook

Transcribed and edited by SANDRA HERBERT

The Red Notebook[1] forms part of the collection of Darwin manuscripts at Down House in Kent, Darwin's home from 1842 to his death, and, since 1929, a museum in his honour. The notebook (164 × 100 mm) is bound in red leather, blind embossed on both sides, and has a metal clasp. The front and back covers bear the initials 'R.N.' on rectangular pieces of white paper, and on the back cover is written 'Range of Sharks', referring to an entry within the notebook. There is also written in large letters across the back of the notebook 'Nothing For any Purpose'. All the inscriptions are in brown ink in Darwin's hand, with the exception of a circled '16' in the hand of Nora Barlow on the inside front cover and the notation '1.2', made by an unknown cataloguer, on the inside back cover.

There are 90 leaves of a green-edged paper, chain lined, feintly ruled, and bearing a 'T WARREN 1830' watermark. Seventy-eight pages were wholly or partly excised from the notebook of which all but eight have been wholly or partly recovered. The notebook divides into two parts: the first to page 113 is written in pencil across the page parallel to the spine; the second is written in both ink and pencil across the page perpendicular to the spine. The only trace of grey ink occurs on the inside front cover.

The first part of the notebook yields a perfect progression of place-names corresponding to points visited by the *Beagle* from late May to the end of September 1836.[2] The second part, from page 113 onwards, is more difficult to date, and, indeed was once thought to have been produced during the voyage, a significant dating given its transmutationist passages.[3] However, scholars now agree that the second portion of the notebook postdates the voyage.[4] While an exact date cannot be set, this second part opens with comments that suggest a January 1837 dating. On 2 January Darwin was in London dining with the geologist Charles Lyell; the next day work was being done at the Hunterian Museum of the Royal College of Surgeons on the fossil bones Darwin had collected in South America.[5] Presumably Darwin assisted in this work, which would explain his query on page 113 regarding Richard Owen, comparative anatomist at the Hunterian Museum: 'Should Mr Owen consider bones washed about much at Coll. of. Surgeon's?' On the same page Darwin noted, 'With discussion of camel urge S. Africa productions.—' The 'camel' represents Owen's initial judgement of a fossil he would later name *Macrauchenia patachonica*.[6] Owen's earliest known written statement on the affinities of the fossil occurs in a letter to Lyell of 23 January 1837, and presumably Darwin knew of Owen's judgement before this time.[7] Thus January 1837, and earlier rather than later in the month, would appear a likely date for the opening of the second part of the notebook.

[1] For Darwin's use of the title 'Red Note Book' see DAR 29.3:9^v.
[2] S. Herbert 1980:6.
[3] Barlow 1945:260–64; Barlow 1963:277 dates RN127–53 to before April 1836.
[4] S. Herbert 1968:78, 1974:246–49, 1980:6–12; Kohn 1980:67 n. 3; Sulloway 1982:370–86.
[5] *Correspondence* 1:532; also Sloan 1986:422 n. 98 citing two diary entries by William Clift at the RCS, the first noting a Darwin communication of 30 December 1836 asking the College not to open his parcels shipped to the College from Cambridge until his return to London and the second, on 3 January 1837, recording the commencement of the washing of Darwin's fossils at the College.
[6] Darwin to Owen [28 December 1837], *Correspondence* 2:66.
[7] Owen to Lyell, 23 January 1837 in L. G. Wilson 1972:436–37. Against the dating offered in Sulloway 1982:355 see *Correspondence* 2:4.

Darwin filled the notebook sometime from late May to mid-June 1837. His reference to petrified wood on page 178 is tied to its examination by Robert Brown reported in a letter of 18 May.[8] On the same page the reference beginning 'Puncture one animal' also occurs in a letter written sometime between late May and mid June.[9] The correlation between notebook references and the *Journal of Researches* also supports an early June dating.[10]

The year in which the Red Notebook was kept was one of transition. The notebook reflects that transition, and itself served partly as an instrument of adjustment, for Darwin used it to assist future publication. Scattered throughout the first part of the notebook are reminders to himself: 'Introduce part of the above in Patagonian paper' (p. 49), 'In Rio paper . . .' (p. 65), 'In my Cleavage paper . . .' (p. 101), and so on. Entries in the Red Notebook were also directed to another publishing project, the work commonly known as the *Journal of Researches*. From about 7 March to 25 June 1837 Darwin converted his diary from the voyage into the draft of his *Journal* and used the Red Notebook as a storehouse for references.[11]

Most entries in the notebook are geological and of these the majority describe specific land formations and rock types. However, nearly as many entries pertain to the elevation and subsidence of the earth's crust, the attendant issue of the form of the earth, and such patterns of disturbance in the earth's crust as were indicated by the occurrence of earthquakes and the presence of volcanoes and mountain chains. Contemplating the prospects for geological theory in understanding the vertical motion of the earth's crust, Darwin speculated that the 'Geology of whole world will turn out simple.—' (p. 72) In addition to crustal motion, Darwin also made notes on such topics as the distribution of metallic veins, the preservation of fossils, erratic blocks, and life at the bottom of the sea.

However compelling the geology, it is Darwin's remarks on the species question that have drawn the greatest attention to the Red Notebook. The entries, on pages 127–133, form Darwin's first thoughts on transmutation, the notion, as he put it on page 130, that 'one species does change into another'. These entries record his adoption of the transmutationist hypothesis in 'about' March 1837, a date he set down later in his 'Journal'.[12] This date coincides with Darwin's receipt of the views of London zoologists regarding key specimens from his collection.[13] Particularly important was John Gould's description of the new species of Galapagos mockingbirds on 28 February 1837 at the Zoological Society of London; Darwin was impressed to learn the mockingbirds were not 'only varieties' as he had once thought, but, in Gould's opinion, good species. Only slightly less important—since Darwin had presumed it to be a new species on his own—was Gould's work on the smaller rhea, described at the

[8] *Correspondence* 2:17–18.

[9] Charles to Caroline Darwin, 19 May–17 June 1837, *Correspondence* 2:19–21. This letter has been redated from earlier treatments (*Calendar*:30 [#360], Sulloway 1982:384–85) on the basis of a newly discovered letter (*Correspondence* 2:23–24). Since the 'Thursday' referred to in the letter is probably 1 or 8 June, the date of the letter would then be a few days prior to either of these dates.

[10] The last third of *JR* was written between 18 May and 25 June (note 11 and *Correspondence* 2:18). Compare *JR*:521–22 with references towards the end of the notebook.

[11] On 6 March 1837 Darwin departed Cambridge, where he had remained until he finished looking over his geological specimens, to reside in London. On 12 March he was already 'hard at work' on the book, and by 26 June he could allow himself a holiday and a trip to Shrewsbury 'as I had finished my journal'. (*Correspondence* 2:430, 11, 29)

[12] *Correspondence* 2:431; on dating RN:127–33 see S. Herbert 1980:10–11 and Sulloway 1982:370–86. The entries beginning on RN127 were probably written in the second half of March, after Darwin had conferred with the London specialists. See also Sydney Smith 1960.

[13] S. Herbert 1974:241–45, 1980:11–12, and Sulloway 1982:356–69. For current identifications of birds named in RN see David Snow's notes in S. Herbert 1980.

14 March 1837 meeting of the Zoological Society; the two rheas figure prominently in the entries on pages 127 and 130.[14]

Darwin's remarks on species in the notebook are directed towards three general questions: geographical distribution, the relation between the spatial and temporal distribution of species, and generation. The central theoretical notion to emerge with respect to geographical distribution is that of the 'representation' of species (p. 130), or what Darwin referred to in his autobiography as 'the manner in which closely allied animals replace one another in proceeding southwards over the [South American] Continent . . .'.[15] From this notion Darwin drew the tentative conclusion that such representative species as the two South American rheas had descended from a common parent (p. 153). Darwin's second, and critical, point on species is the comparison he drew between the distribution of species through space and time. The passage, simplified by omission of one example, reads, 'The same kind of relation that common ostrich bears to (Petisse . . .): extinct Guanaco to recent: in former case position, in latter time' (p. 130). The first relation—the 'common ostrich' being the common rhea, the 'Petisse' the lesser or Darwin's rhea (note RN127–3)—was based on spatial succession, the geographical ranges of the two birds being contiguous. The second relation, between what Darwin termed the 'extinct Guanaco' (*Macrauchenia*) and 'recent' guanaco, involved temporal succession, though the exact nature of the succession is not specified in the text.[16] The common element binding the two relations derives from the fact that both involved the replacement of one species by an allied species. Moreover, in context it is clear that replacement implied transmutation, for immediately upon asserting an analogy between spatial and temporal succession, Darwin referred to species changing. In doing so he also asserted that allied species do not grade into each other but 'inosculate' (p. 130).[17]

A third topic taken up in the notebook with general relevance for the species question was generation or reproduction. In the notebook Darwin dealt briefly with individuation in the zoophytes. The technical nature of zoophyte generation was not Darwin's primary concern; rather he wanted to see where zoophyte generation might fit in the general analogy he was drawing between the generation of species and the generation of individuals. Although the claim is not made explicitly in this notebook, Darwin presumed that the complementary relationship might also hold, that the birth of new species might be understood by analogy to the birth of individuals.[18]

[14] Gould 1837f, Barlow 1963:262, and RN130–7; Gould 1837b, RN127–3. On Darwin's discussions with Gould see Sulloway 1982:362–69.
[15] Barlow 1958:118.
[16] Initially Richard Owen associated the fossil he named *Macrauchenia* with the order Ruminantia; before writing up the *Zoology* he placed it with the Pachydermata. (Presently it is in the order Litopterna.) Owen's association of the fossil with camels was based on the vertebrarterial canal in the neck being like that in camels while his later assignment of the fossil to the pachydermata depended in large part on the structure of the fossil's astragalus or ankle bone (Rachootin 1985). This analysis is borne out by evidence from Owen's research notebooks. Owen's notebook labelled '1836–1837' contains his description of the cervical vertebrae of the fossil while his notebook labelled '1838–1839' contains his description of its astragalus (Owen MSS, Notebooks 12 ['1836–1837'] and 13 ['1838–1839'], British Museum [Natural History] Archives). The dates on the covers of these notebooks are not in Owen's hand and are presumably approximate. Owen probably examined the fossil for the second time in the autumn of 1837, sometime before he named it in December (*Correspondence* 2:66). Darwin still thought of the fossil as primarily camelid in its affinities when he wrote A9 and JR:208–9, which he sent off to the publisher in August 1837 (*Correspondence* 2:33). For a photograph of Darwin's specimen of *Macrauchenia* see S. Herbert 1980.
[17] MacLeay 1819–21; Sydney Smith 1968; Barlow 1967:62, n. 2; note B8–1. Also see A76 for an alternative reading.
[18] On Darwin' invertebrate programme see Sloan 1985.

Once the Red Notebook was filled, Darwin reorganized his method of taking notes. Where the Red Notebook contained entries on all subjects of interest, subsequent notebooks were more restricted in content. On its own, however, the notebook provides a means not only for gauging the extent of Darwin's geological ambitions and for documenting his early belief in transmutation, but also for observing his passage from H.M.S. *Beagle* to the larger world of science.

FRONT COVER

R.N

INSIDE FRONT COVER **up to 1° / July 1835. the excess of harbor = 180**

See Daubisson both Volumes,[1] and Molina 1st Vol[2] & Lyell[3]
Sailed, 27th
‹Friday, gale 29th›
Friday
Thursday 29th gale
Lyell's Geology[4]
The living atoms having definite existence, those that have undergone the greatest number of changes towards perfection (namely mammalia) must have a shorter duration, than the more constant: This view supposes the simplest infusoria same since commencement of world.—[5]

1e–4e [not located]

5e La. billardiere mentions the floating marine confervæ, is very common within E. Indian Archipelago, no minute description, calls it a Fucus. P «Vol I 287»[1]

P 379. Henslow Anglesea, nodules in Clay Slate. major axis 2.½ ft.—singular structure of nodule, constitution «same as» of slate same.—longer axis in line of Cleavage. laminæ fold round them;[2] Quote this. Valparaiso Granitic nodules in Gneiss.

IFC **1°/]** *alternate reading* **'1s of'**.
up] *feint* '118' *appears beneath, questionably Darwin's hand.*
up . . . 180] *pencil, perpendicular to spine.*
See . . . Vol] *pencil, parallel to spine*
& Lyell] *ink, parallel to spine.*
Sailed . . . gale] *pencil, parallel to spine.*
27th] *uncertain reading.*
Lyell's Geology] *ink, circled, written over* 'Sailed, 27th', *parallel to spine.*
The living . . . world.—] *ink, scored at left, perpendicular to spine, crossed ink. Crossing probably grey ink.*
5 constitution . . . them;] *scored left margin.*
La.billardiere . . . 287] *crossed ink.*

IFC–1 Aubuisson de Voisins 1819.
IFC–2 Molina 1788–95, 1.
IFC–3 Charles Lyell.
IFC–4 Lyell 1830–33 or, possibly, a later edition of the same work or Lyell 1838a. The dates preceding this entry probably pertain to the departure of H.M.S. *Beagle* from England. The *Beagle* sailed from England Tuesday 27 December 1831. The ship encountered heavy seas, caused by gales elsewhere, on Thursday 29 December 1831. For Darwin's description of the ship's departure see his letter to his father of 8 February–1 March 1832 in *Correspondence* 1:201 and his own *Diary*: 18–19.
IFC–5 The probable stimulus for this passage was Ehrenberg 1837c, 1:555–76.

5–1 Labillardière 1800, 1:287, 'Je revis le fucus que j'avois auparavant rencontré tout près de la Nouvelle-Guinée; il ressemble à de l'étoupe très-fine coupée par petis morceaux longs d'environ trois centimètres: ce sont des filamens aussi fins que des cheveux. On les voyoit souvent réunis en faisceaux, et si nombreux qu'ils ternissoient l'eau de la rade.'
5–2 Henslow 1821–22:379, 'The major axis of some of the larger nodules is two feet and a half, and the minor one foot and a half; and the conical structure extends to the depth of three or four inches. The direction of the longer axis is placed parallel to the schistose laminae, which pass round the nodules.' Darwin scored this passage.

6e Epidote seems commonly to occur where rocks have undergone action of heat. it is so found in Anglesea, amongst the varying & dubious granites.—Wide limits of this mineral in Australia. Fitton's appendix[1]

Would Slate. & unstratified rocks show any difference in facility of conducting Electricity? Would minute particles have a tendency to change their position?

7e Carbonate of Lime disseminated through the great Plas Newydd dike.—Mem tres Montes. ((Henslow Anglesea))[1]

---

great variety in nature of a dike.—Mem. at Chonos & Concepcion. P. 417[2]

Veins of quartz exceedingly rare Mem C. [Cape] Turn P. 434 & 419[3]

As Limestone passes into schist scales of chlorites—Mem. Maldonado P 375[4]

Much Chlorite in some of the dikes.—P 432.[5] as in Andes.

8e In Dampier's voyage there is a mine of metereology with respect to the discussion of winds & storms:[1]—«in Volney's travels also»[2]

Dampier's last voyage to New Holland P 127.—Caught a shark 11 ft long.[3] "Its maw was like a leathern sack, very thick & so tough that a sharp knife could not cut it: in which we found the Head & Boans of a Hippotomus; the hairy lips of which were still sound and not petrified, and the

6 Would . . . position] *crossed pencil.*
8 *page crossed.*

6–1 Fitton in King 1827, 2:585, 'The Epidote of Port Warrender and Careening Bay, affords an additional proof of the general distribution of that mineral; which though perhaps it may not constitute large masses, seems to be of more frequent occurrence as a component of rocks than has hitherto been supposed.'

7–1 Henslow 1821–22:403, 'Carbonate of lime is very generally disseminated through every part [of the Plas-Newydd dike].'

7–2 Henslow 1821–22:417, 'The most interesting phenomena exhibited by this dyke, are the various changes which it assumes in its mineral character.'

7–3 Henslow 1821–22:434, 'Through this dyke there run several veins of quartz, which also abound in the surrounding rock, a fact which I do not recollect witnessing in any other dyke in Anglesea.' Also p. 419: 'At its [the dyke's] Northern termination the trap has been removed by the continued action of the sea, and its original walls, composed of quartz rock, form a small bay about eighty feet wide.' In his copy Darwin commented on p. 434 'At C. Turn Quartz broad vein traversed Slate & greenstone' and on p. 419 'just what occurs in the Magdalen Channel at C: Turn T del Fuego'.

7–4 Henslow 1821–22:375, 'As the limestone passes into the schist [at Gwalchmai], it assumes a fissile character, and scales of chlorite are dispersed over the natural fractures.' Next to this passage in his copy Darwin reminded himself to 'Mem: Maldonado Limestone . . .'.

7–5 Henslow 1821–22:432, 'The whole [mass of trap] assumes a greenish tinge, but the colouring substance does not appear to be of a very crystalline nature, and is probably chlorite.' In his copy of the work Darwin scored this passage, adding a comment—now partly cropped—that ends 'of Chili'.

8–1 Dampier 1698–1703, 2, pt 3.

8–2 Volney 1787b, 1, chap. 20 the section entitled 'Des vents', and chap. 21 entitled 'Considérations sur les phénomènes des vents, des nuages, des pluies, des brouillards et du tonnerre'.

8–3 Dampier 1698–1703, 3:125. Exact edition unknown.

9 jaw was also firm, out of which we pluckt a great many teeth, 2 of them, 8 inches long, & as big as a mans thumb, the rest not above half so long; The maw was full of jelly which stank extreamly."—This shark was caught in Shark's Bay. Lat 25°.[1] The nearest of the E. Indian Islands. namely Java is 1000 miles distant! Where are Hippotami found in that Archipelago? Such have never been observed in Australia

10 Dampier also repeatedly talks about the immense quantities of Cuttle fish bones floating on the surface of the ocean, before arriving at the Abrolhos shoals.‖—[1]

N.B. The view of the Volcanos of the chain of the Cordilleras as arising from «the expulsion of fluid nucleus through» faults or fissures, produced by the elevations of those mountains on the continent of S. America is inadmissible «may have happened from incipient elevation.» The volcanos originated

11 in the bottom of the ocean. & the present Volcanos have been said to be merely accidental apertures still open.—The fault like appearance «arising from the manner of horizontal upheaval» of the shore of the Pacifick is 60 miles distant from the grand ancient volcanic axis «of the Andes».—«Has this fault determined side of volcanic activity.» That axis was produced, from a fissure in a deep & therefore weak part of the ocean's bottom.

12 With respect to Sharks distributing fossil remains: Sharks followed Capt. Henry's vessel from the Friendly Isles. to Sydney; know by having been seen & from the contents of its maw, amongst which were things pitched over board early in the passage!!—[1]

M. Labillardiere in Bay of Legrand, (SW part). describes a Small granite Is$^{d.}$ capped by Calcareous rock;[2] following

13e Curvature of hill; states could discover no shells: nothing said about K. Georges Sound

---

9 *page crossed.*
10 *page crossed.*
11 *page crossed.*
12 M. Labillardiere . . . following] *crossed pencil.*
13 . . .] *3 to 5 letters illeg.*
Curvature . . . Sound] *crossed.*
The . . . flowing] *crossed.*
bottom half unlocated.

9–1 Dampier 1698–1703, 3:125–26.
10–1 Dampier 1698–1703, 3:114.
12–1 Samuel P. Henry: personal communication. Darwin met Capt. Henry and his father, a missionary, at Tahiti. See FitzRoy 1839, 2:524, 546, 615; and Williams 1837:471.

12–2 Labillardière 1800, 1:394, 'L'îlot sur lequel nous étions est composé d'un beau granit, où le quartz, le feld-spath et le mica dominent; . . .' and, p. 395, 'La partie occidentale de cet îlot offre, dans un des points les plus élevés un plateau de pierre calcaire. . . .'

The idea of the water at Cauquenes. coming from the [. . .] Cordillera & flowing

---

14e The gradual shoaling of the water to more than 100 fathoms. proves the existence of some moving ‹point› power ? Submarine currents

---

15e Find instances; The whole coast of New Holland shoals much: Dampier remarks on great flats on the NW coast:—[1]

---

8 leagues, from Sydney 90 fathoms La Peyrouse.[2]

South of Mocha; 19 miles. 65 Fathoms

Vide facts in Beechey. on NW coast of America[3]

---

off Cape of Good Hope 70 fathoms 20 miles from the shore? Beagle

---

Coast of Brazil? where not rivers **in my Coral paper**[4]

16e		leagues		Fathoms
	Parallel of St Catherine [27° 30′ S.][1]	18	—	70
	Paranagua [25° 42′ S.]	12	—	40
	St Sebastian [23° 52′ S.]	12		50
	Joatingua SE [23° 22′ S.]	5		35
	R. de Janeiro SE [23° 58′ S.]	18		77
	C. Frio [23° S.]	7		60

Soundings about same as last to N. of C. Frio Except at Abrolhos. [18° S.]

Bahia [12° 57′ S.]	8	200 (bar with dot)

14 *bottom rule line cross hatched.*
The . . . currents] *crossed.*
top half unlocated.
15 *first rule line cross hatched.*
in my Coral paper] *added ink.*
16 **18–20**] *added above ink.*
60–80] '60' *cancelled ink, cancellation* '80' *presumably intended;*

15–1 Dampier 1698–1703, 3:151, 'The Land hereabouts was much like that part of *New Holland* that I formerly described. . . . 'tis low, but seemingly barricado'd with a long Chain of Sand-hills to the Sea, that let's nothing be seen of what is farther within Land.'
15–2 La Pérouse 1798–99, 2:179, 'From Norfolk Island, till we got sight of Botany Bay, we sounded every evening with a line of two hundred fathoms, but we found no bottom till we were within eight leagues of the coast, when we had ninety fathoms of water.'
15–3 Beechey 1832. See note RN45–1
15–4 Eventually published as Darwin 1833–38a.
16–1 The sequence of points on this list runs from south to north along the Brazilian coastline. A bar with a dot over a number indicates that no bottom was found at that depth.

Morro S. Paulo [13° 22′ S.]	9	$\dot{\overline{120}}$
Garcia de Avila [lighthouse] [12° 35′ S.]	9	124
Itapicuru [R.] [11° 46′ S.]	9	200
R. Real [11° 31′ S.] & [R.] Sergipe [11° 10′ S.]	20	190
R. San Francisco [10° 32′ S.]	10	50
Whole coast to Olinda [8° S.]	9–10 =	30–40

at twice or **18–20** ‹60›—80 $\dot{\overline{120}}$ parallel of Olinda

---

Shoaler N. of Olinda.—a little WNW of C. Rock. [5° 29′ S.]
still shoaler, coast composed of sand dunes. 15 — 15

Does not seem to consider this a very shoal coast.[2]

Beyond the 10 or 12 leagues sea deepens suddenly. coast of Brazil generally.—

17 Mrs Power at Port Louis talked of the *extraordinary* freshness of the streams of Lava in Ascencion known to be inactive 300 years?[1]

No Volcanic Earthquakes or Hot Springs in T. del Fuego=The Wager's Earthquake the most Southern one I have heard of[2]

---

18 In a preface, it might be well to urge, geologists to compare whole history of Europe, with America; I might add I have drawn all my illustrations from America, purposely to show what facts can be supported from that part of the globe: & when we see conclusions substantiated over S. America & Europe. we may believe them applicable to the world.—

19e My general opinion from the examination of soundings, from about 80 fathoms & upwards. that life is exceedingly rare, at the bottom of the sea.—«certainly data insufficient, yet good» «(I suspect fragments of shells will generally be found to be old & dead)» «(I have not kept a record)» In looking over the lists of organic remains in De la Beche,[1] for the older

17 M^rs . . . years?] *crossed pencil.*
No . . . of] *crossed ink.*
18 *page crossed.*
19 19] *ink over pencil.*

16–2 Probably Robert FitzRoy.
17–1 Mrs Power: personal communication.
17–2 The shipwrecked crew of the H.M.S. *Wager* identified their position as 47° 00′ S., 81°40′ W. Capt. FitzRoy recalculated the probable position of the ship as 47° 39′ 30″ S., 75° 06′ 30″ W. See Bulkeley and Cummins 1743:48; FitzRoy 1839, appendix to vol. 2:78; and *JR*:287.
19–1 De la Beche 1831, sec. 5–10.

formations I must believe they «the limestones» have been formed in shallow water: so have the Conglomerates: Yet this view is directly opposed to common opinion

20e The Tertiary formation South of the Maypo at one period of elevation must in its configuration have resembled Chiloe

---

In De La Beche, article "Erratic blocks" not sufficient distinction is given to angular & rounded.—[1]

Fox Philosoph. Transactions on metallic veins. 1830 P. 399.—[2] Carne. Geolog. Trans: Cornwall «Vol II»[3]

---

21 It is a fact worth noticing that cryst of glassy felspar in Phonolite arrange themselves in determinate planes ∴ such action can take place in melted rocks

---

The frequent coincidence of line of veins & cleavage is importants; veins appearing a galvanic phenomenon, so probably will the Cleavage be

---

There is a resemblance at *Hobart town* between the older strata & the bottom of sea near T. del Fuego.—

22 Is there account of Baron Roussin's voyage.—[1]

In Europe proofs of many oscillations of level, which in the nature of strata & Organic remains does not appear to have taken place in the Cordillera of S. America.

Study Geolog: Map of Europe

Conybeare. Introduct XII P. silicified bones not common in Britain. Mem Concepcion Says Echinites. Encrinites. Asteriæ, usually petrified into

23 a peculiar cream-coloured Limestone:[1] the strange substitution of matter in

20 In . . . II] *crossed.*
21 &] '&' *over* 'of'.
page crossed.
22 *page crossed.*
23 a . . . rocks] *crossed.*
P XV . . . steps] *crossed.*

20–1 De la Beche 1831, sec. 3, 'Erratic Blocks and Gravel'. In his treatment of the subject De la Beche did not discuss the shapes of individual pieces of gravel.
20–2 Fox 1830:399–414.
20–3 Carne 1822.

22–1 Baron Albin-Reine Roussin did not write a general account of the hydrographical expedition he led in 1819–1820 to South America; but see Roussin 1826.
23–1 Conybeare and W. Phillips 1822:xii, '. . . one instance of a bone penetrated by silex has occurred to the

shells, like Concretions & laminæ, show what movements take place in semiconsolidated rocks

---

P xv. mentions in what formations Conglomerates are found.—[2]

The above oscillations remarkable because the formations are now seen in regular descending steps

24 Mem.; rapidity of germination in young corals.—vide L. Jackson's paper. Philosoph Transact:[1] at R. de Janeiro. Coquimbo. Balanidæ. at Concepcion.

Humb: Pers. N. vii P. 56[2] Serpentine form: of Cuba for comparison (?) with St Pauls

---

25e [not located]

26e [not located]

27 The frequency of shells in the Calc. Sandstone Concret, is connected with frequency of shells in flints in Chalk

---

New Providence more hilly than others of the Bahama consists of rock & sand mixed with sea shells—about 500 Is$^{d.}$ & great banks. effect of Elevation. United service Journal[1]

28 In the Iron sand formation ‹would› wood converted into siliceous pyritous & coaly matter. Mem: Chiloe

---

In the endless cycle of revolutions. by actions of rivers currents. & sea

24 *page crossed.*
27 *page crossed.*

author, on the beach at Reculver. The calcareous substance of shells, echinites, encrinites, corals, &c. in its slightest change seems only to have lost its colouring matter and gelatine; next they become impregnated with the mineral matrix in which they lie, especially if that matrix be calcareous; hence they become much more compact; often at the same time their original calcareous matter undergoes a change of internal structure, assuming a crystalline form, and in some cases, viz. asteriæ, encrinites, and echinites, a calcareous spar of very peculiar character results, of an opaque cream colour:. . .'
23–2 Conybeare and W. Phillips 1822:xv, 'These consolidated gravel beds are called conglomerates, breccias, or puddingstones; we find them among the transition rocks, in the old red sandstone, in the millstone-grit and coal-grits, in the lower members of the new red sandstone, in the sand strata beneath the chalk, and in the gravel beds associated with the plastic clay, and interposed between the chalk and great London clay.'
24–1 Despite the faulty citation the reference is certainly to Lister 1834.
24–2 Humboldt and Bonpland 1819–29, 7:56, 'Farther south towards Regla and Guanabacoa [to the east of Havana], the syenite disappears, and the whole soil is covered with serpentine, rising in hills from 30 to 40 toises high, and running from east to west.' Darwin's copy of Humboldt's *Personal Narrative* is inscribed, 'J. S. Henslow to his friend C. Darwin on his departure from England upon a voyage round the World. 21 Sept 1831.' It consists of vols. 1–2, 3rd ed.; vol. 3, 2nd ed.; vols. 4, 5, 6, 7, 1st ed. (1819–1829).
27–1 'Proteus' 1834:215, '[New Providence] is more hilly than most of the islands, the surface being composed of rock and sand intermixed with sea shells.' See also pp. 216 and 226 for mention of the banks.

beaches. All mineral masses must have a tendency. to mingle; The sea would separate quartzose sand from the finer matter resulting from degradation of Feldspar & other minerals containing Alumen.—This matter

29 accumulating in deep seas forms slates: How is the Lime separated; is it washed from the solid rock by the actions of Springs or more probably by some unknown Volcanic process? How does it come that all Lime is not accumulated in the Tropical oceans detained by Organic powers. We know

30 the waters of the ocean all are mingled. These reflections might be introduced either in note in Coral Paper or hypothetical origin of some sandstones, as in Australia.—Have Limestones all been *dissolved.* if so sea would separate them from indissoluble rocks? Has Chalk

31 ever been dissolved?

Singularity of fresh water at Iquiqui. not from rain, because alluvium saline; Mem: on coast of Northern Chili as springs become rarer, so does the rain, therefore *such* «rain» is cause, hence at least no water is absorbed into the earth

---

‹I did not see one dike in the whole Galapagos Arch; because no sections›

---

same cause as no colour

32 Sir J. Herschels idea of escape of Heat prevented by sedimentary rocks, & hence Volcanic action, contradicted by Cordillera, where that action *commenced before* any great accumulation of such matter.—[1]

---

28 In the endless . . . matter] *crossed.*
29 *page crossed.*
30 *page crossed.*
31 colour] *uncertain reading.*
Singularity . . . colour] *crossed.*
32 *page crossed.*

32–1 John F. W. Herschel: presumably personal communication. Darwin met Herschel sometime between 8–15 June 1836 during the *Beagle*'s call at the Cape of Good Hope. Months before, Herschel had described his new notion of the cause of volcanic action in a letter to Charles Lyell dated 20 February 1836. Probably he repeated the same explanation to Darwin in June. Herschel's letter to Lyell has been published (Cannon 1961). From it see, for example, Herschel's summary comment to Lyell on p. 310, 'I don't know whether I have made clear to you my notions about the effects of the removal of matter from . . . above to below the sea.—1st it produces mechanical subversion of the *equilibrium of pressure.*—2dly it also, & by a different process (as above explained at large) produces a subversion of the equilibrium of temperature. The last is the most important. It *must be an excessively slow process.* & it will depend 1st on the depth of matter deposited.—2^{d} on the quantity of water retained by it under the great squeeze it has got—3dly on the tenacity of the incumbent mass—whether the influx of caloric from below—which MUST TAKE PLACE acting on that water, shall either heave up the whole mass, as *a new continent*—or shall crack *it* & escape as a submarine volcano—or shall be suppressed until the mere weight of the continually accumulating mass breaks its lateral supports at or near the coast lines & opens there a chain of volcanoes.' This passage from Herschel's letter was read at the 17 May 1837 meeting of the Geological Society of London (Herschel 1833–38:550) and also appeared in Babbage 1838:235.

Dr A. Smith says. that Boulders do not occur in the South African plains.—[2] Sydney no

---

33e I believe the secondary? formations of Brazil, all originate from the decomposition of Granitic rocks Mem. Chanticleers voyage at ‹[. . .] Maranh› Pernambuco.[1]

EARTHQUAKE AT SEA.—Extract from the log-book of the *James Cruikshank*, Captain John Young, on her voyage from Demerara to London:—"Feb. 12, 1835. At 10h. 15m. a severe shock of earthquake shook the ship in a most violent manner. Although it lasted about a minute, there was no uncommon ripple on the water. It was quite calm at the time. Latitude 8 deg. 47 min. N: longitude 61 deg. 22 min. W. mid. calm and clear.[2] Caermarthen Journal

34e I look at the cessation northwards of the Coal in Chili as clearly bearing a relation to present position of ‹Coal› Forests. These thick beds of Lignite stratified with substances so like the Coal measures in England (Excepting Conglomerates?) «& absence of limestone?» have been collected on the open

33 . . .] *1 letter illeg.*
EARTHQUAKE AT . . . clear] *passage is a newspaper clipping pasted on the page.*
I . . . Pernambuco.] *crossed.*
EARTHQUAKE AT . . . Journal] *crossed.*
34 ‹[. . .]›] *2 letters illeg.*

32–2 Andrew Smith: personal communication. Darwin's *Diary*:409 records for 8–15 June 1836 while he was at the Cape of Good Hope, 'During these days I became acquainted with several very pleasant people. With Dr A. Smith who has lately returned from his most interesting expedition to beyond the Tropic, I took some long geological rambles.'

33–1 The meaning of this entry is obscure. The H.M.S. *Chanticleer* did not stop at Pernambuco [Recife] during its 1828–31 voyage, nor was Pernambuco on the *Beagle*'s itinerary in June of 1836, when this entry was presumably made. In the narrative from the *Chanticleer*'s voyage, however, there are passages which describe decomposing granitic rock at Rio de Janeiro, and refer to what seem to be related formations at Para [Belém] and Maranham [São Luís]. Given Darwin's apparent uncertainty in this entry about location, as indicated by his two cancellations, it may have been these passages which he had in mind. See W. H. B. Webster 1834, 1:52–53, 'The country about Rio in a geological point of view has large claims to attention. Granite and gneiss are the prevailing formation. . . . The rocks in some parts are decomposed into sand and petunse; the sand having been carried down into the plains, while the petunse remains, and forms extensive beds of porcelain clay admirably adapted for the use of the potter. The lower parts of the granite hills were found chiefly in this condition; the granite having crumbled into micaceous sand and greasy unctuous clay.' Also 2:367, 'The geology of Para will detain us a very little while; as there is very little variety or novelty. Precisely the same materials are found here as at Maranham, so that it would be impossible to distinguish them. It is a rare and unusual circumstance to find such a striking coincidence, in two different places. The soil upon which the city stands is of clay and sand. The beds of clay are very extensive, and frequently thirty or forty feet deep. There is scarcely any rock, and that only in particular and isolated masses; it is a coarse dark iron sand-stone, with numerous particles of quartz in it. . . . This dark iron sand-stone, with fragments of white quartz, is observable at Maranham, and is the predominant formation at St. Paul's, a little to the southward of Rio.'

33–2 The clipping, entitled 'Earthquake at Sea', is from the *Carmarthen Journal*, 3 April 1835. The story was reprinted verbatim from *The Times* (London), 28 March 1835, p. 5, with the unfortunate error of a lost digit in the quotation of the ship's latitude. The ship's coordinates as given in *The Times* were 18° 47′ N., 61° 22′ W., which would place the ship in the Atlantic Ocean to the north-east of the Leeward Islands, rather than, as in the incorrectly printed version, in Venezuela. See Anonymous 1835c, 1835d.

coast. Perhaps as at Concepcion. favoured by basin formed by outlying rocks; (such as between Mocha & main land). ‹[. . .]› At Carelmapu.—Within Chiloe:—

35e On open coast, near where Challenger was lost:[1] I know no reason for supposing these matters are not now collecting, in the bottom of an open & not deep sea.—(Character of coast regular & ‹not very› rather deep soundings, 60–100 fathoms 2 & 3 miles from shore. V. Chart) Every winter torrents must bring much vegetable matter from thickly wooded mountains, probably chiefly leaves.—This position agrees with character of. .

«in Basins from rivers. & natural position»

36e position at N.S. Wales & Van Diemen's land.—

Whole coast S. of Concepcion where there are Tertiary strata there is Coal— ? No shells in all cases. «.Mytilus.—»

«at Guacho» «on N. Chile? Washington.—»[1]
Mem: Micaceous formation of Chonos. interesting from great quantity of altered Carbonaceous shales

---

Examine chart of Patagonian coast to see proportional *cliff* & low or sloping land

---

What are the "palatal Tritores" found in the coraliferous mountain Limestone

37e are they allied to the jaws of the Cocos fish

Rio Shells argument for rise

In Cordillera, the dikes do not generally appear to have fallen into lines of faults

---

36 Whole . . . cases] *double scored ink.*
No] *Boxed.*
What . . . Limestone] *crossed.*
37 are . . . fish] *crossed.*
Rio . . . rise] *crossed.*
In . . . faults] *scored, crossed.*
I. . . raised] *crossed.*
No . . . upheaval] *crossed ink.*

35–1 The H.M.S. *Challenger* ran aground on the Chilean shore at Punta Morguilla [Point Molguilla] (37° 46′ S., 73° 40′ W.) 19 May 1835. See FitzRoy 1839, 2:451–56. Capt. FitzRoy led the party which rescued the crew.

36–1 In this series of place names the locations of Guacho and Washington are uncertain. There is presently a Quebrado del Guacho, a small stream, at 33° 58′ S., 71° 09′ W. in Chile, and a Cerro Guacho, a mountain, nearby. 'Washington' may refer to the Canal Washington at 55° 40′ S., 67° 33′ W. in Tierra del Fuego.

I do not think so many faults in Cordillera, as in English Coal field—because lowered & raised—so on—but gradually & simply raised

---

No Faults in Patagonia[,] enormous extent; if lowered again & covered no sign of upheaval

38e To Cleavage add other instances in old world of symetrical structure. East India Archipelago. «Aleutian Arch.—» V. Fitton. Australia:[1] cases in Europe.—

Auvergne. very little Pumice, though Trachyte. same fact in Galapagos. Daubeny P 24[2]

V. back of page 1 of New Zealand Geological Notes.[3]

at St. Helena. This structure was very clear at base of great lava cliffs[4] [Fig. 1]

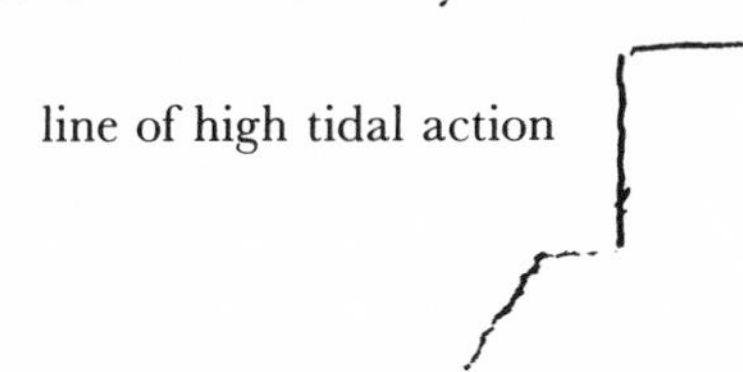

NB. patches of modern Conglomerates [Fig. 2]

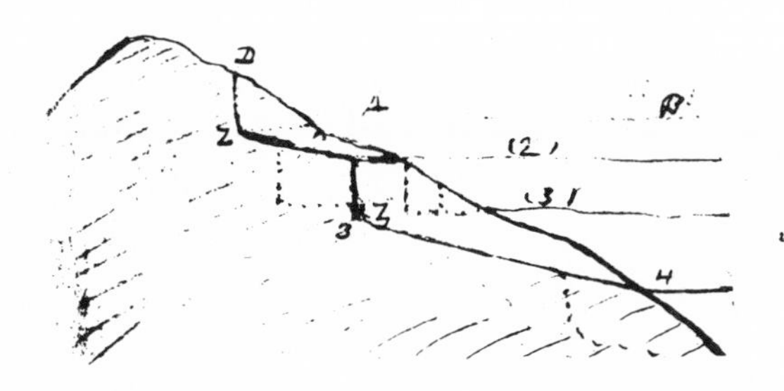

38 **V. back . . . Notes.]** *added ink.*
To . . . P 24] *crossed.*

38–1 Fitton in King 1827, 2:604, 'The tendency of all this evidence is somewhat in favour of a general parallelism in the range of the strata,—and perhaps of the existence of primary ranges of mountains on the east of Australia in general, from the coast about Cape Weymouth to the shore between Spencer's Gulf and Cape Howe.' And on p. 605, 'If . . . future researches should confirm the indications above mentioned, a new case will be supplied in support of the principle long since advanced by Mr. Michell which appears (whatever theory be formed to explain it,) to be established by geological observation in so many other parts of the world,—that the outcrop of the inclined beds, throughout the stratified portion of the globe, is every where parallel to the longer ridges of mountains,—towards which, also, the elevation of the strata is directed.'

38–2 Daubeny 1826:24, 'It [a formation at the hill of Mouton] should be noticed, as one of the few localities in Auvergne where pumice is to be found, which seems the more remarkable, as this substance is a common product of that class of volcanos, which consists of trachyte.'

38–3 The back of page 1 of Darwin's geological notes on New Zealand is DAR 37.2:802. The page contains a sketch of the silhouette of an island in the Bay of Islands, New Zealand. Darwin noted that at high water the island had the figure of a hill and at low water the figure of a hill surrounded by a level ledge of naked rock. He associated the formation of the ledge with the action of the tides. This page in Darwin's geological notes also contains a cross-reference to 'R.N.' page 38.

38–4 See *GSA*: 25–26, for the published version of this description of the origin of the cliffs at St Helena.

39e The action of sea A.B. will be to eat in the land in line of highest tidal action. this will at length be checked by increased vertical ‹height› thickness (DZ) of mass to be removed & from the resistance offered to the greater lateral extension of the waves. by the part beneath the band of greatest action not having been worn away.—If the level of the sea was to sink by very slow & gradual movements to line (2). The part (o) which was before beneath band. of greatest action. would now by degrees be exposed to it, & the result would [be] a uniform slope to base of cliff (Z). to which point the waves would not reach. If now the ocean should *suddenly*

40e fall, (3) the case would be as at first. & according to the greater or less time of rest. so would the size of the triangular mass removed vary.—The gradual rising continuing. a another sloping platform would be made, & so on.—This is grounded on the belief of constant rising with successive periods of greater activity & rest.—Such changes could be shown (as represented), along line of coast.—[Fig. 2] Mem San. Lorenzo; Valley of Copiapò & parts of coast of Chile.—

Must first explain «top of» tidal band of action.

---

41 This case differs. *I think.* from Patagonian steps, because the deposition & accumulation is brought into play

---

As in Ocean & Air; there are «likewise» differences of temperature «at equal distances from centre of rotation» & a ‹circulation owing› rotation in fluid matter of globe. must there not be a circulation «however slow & weak.»; «(cause of not accumulation of Coral limestone in intertropical)» hence varieties of substances ejected from same point. & changes. «(changes in variation?)» as in Cordillera.—

From poles to Equator current downwards & to West.—From Equator to poles. nearer the surface & to the Eastward.—If matter proceeds from *great depth.* from axis to surface must gain a Westerly current:—If great changes of climate have happened. hurricane in bowels of earth cause:—‹exp› does not explain cleavage lines./ possibly general symetry of world.—

42 I feel no doubt. respecting the brecciated white stone of Chiloe, after having examined the changes of pumice at Ascension

---

41 ‹exp›] *uncertain reading.*
As in . . . world.—] *crossed.*

42 Have shells] *feint word possibly follows* 'shells'.
Calcareous] 'areous' *uncertain reading.*
??] *first* '?' *possibly* '!'.
I . . . Ascension] *crossed.*
In . . . instead] *crossed.*

In Calc: sandstone at Ascension, each particles coated by pellucid envelope of Lime.—form resembles the husks at Coquimbo: in that case, may not central and rather differently constituted lime have been removed?—As shell out of its cast which, although not very intelligble is a familiar case: If refiltered with other matter how very curious a structure: Have shells ever *casts* alone in Calcareous. rocks??—if so case precisely analogous: fragments instead

43e Peak of Teneriffe. also Cotopaxi has a ‹[. . .]› cylinder placed on the rim of conical crater: at Teneriffe Wall of Porph. Lava with base of Pitchstone; Mem Galapagos. chiefly red glassy scoriæ.—could walk round base:—not universal: could not climb up many parts, in James Isd.—Mem St Helena—All Trachytic.—**Daubeny**[1] P. 171. Vol I. Humboldt[2]

There is long discussion on Pumice «& Obsidian:» in the I Vol. Humb:[3]

There is rather good abstract of Humboldt. S. American Geolog. in Daubeny. P. 349[4]

Admirable little table showing long PERIODS of great violence volcanic. from Humboldt: Comparison P 361. Daubeny[5]

43 ‹[. . .]›] *3 letters uncertain, middle letter* 'i'.
Daubeny] *added ink.*
page crossed.
chiefly . . . P 361. Daubeny] *excised.*

43-1 Humboldt and Bonpland 1819–29, 1:171, 'The Peak of Teneriffe, and Cotopaxi, on the contrary, are of very different construction. At their summit a circular wall surrounds the crater; which wall, at a distance, has the appearance of a small cylinder placed on a truncated cone.' Also, with respect to the peak of Teneriffe, on p. 176, 'The wall of compact lava which forms the enclosure of the Caldera, is snow white at it's surface. . . . When we break these lavas, which might be taken at some distance for calcareous stone, we find in them a blackish brown nucleus. Porphyry with basis of pitch stone is whitened externally by the slow action of the vapors of sulphurous acid gas.'

43-2 Reference uncertain, possibly to Daubeny's representation of Humboldt's 'unpublished' views. See Daubeny 1826:345–351.

43-3 Humboldt and Bonpland 1819–29, 1:219–32.

43-4 Daubeny 1826:349. Not easily summarized, see note 43-2

43-5 Daubeny 1826:361, 'Humboldt gives us the following series of phænomena, which presented themselves on the American Hemisphere between the years 1796 and 97, as well as between 1811 and 1812.

'1796.—September 27.	Eruption in the West India Islands; volcano of Guadaloupe in activity.
. . . . . . November . . .	The volcano of Pasto begins to emit smoke.
. . . . . . December 14 .	Destruction of Cumana by earthquake.
1797.—February 4. . .	Destruction of Riobamba by earthquake.
1811.—January 30. . .	Appearance of Sabrina Island in the Azores. It increases particularly on the 15th of June.
. . . . . . May	Beginning of the earthquakes in the Island of St. Vincent, which lasted till May, 1812.
. . . . . . December 16 .	Beginning of the commotions in the valley of the Mississippi and Ohio, which lasted till 1813.
. . . . . . December . . .	Earthquake at Carracas.
1812.—March 26. . . .	Destruction of Caraccas; earthquakes which continued till 1813.
. . . . . . April 30.	Eruption of the volcano in St. Vincents'; and the same day subterranean noises at Caraccas, and on the banks of the Apure.'

44e Von Buch is very strong about Trachyte being the most inferior rocks[1]—The stream at Portillo Pass example of do? ‹Poor›

Daubeny good account of ejected granitic fragments. P. 386[2]

Mem. Lyell's fact about sulphuric vapours in East Indian Volcanos[3]

Gypsum
Andes

45e Mem. Beechey. account of regular *change in soundings*. on approaching the coast of NW. America P. 209–13 P & 444 «(Yanky Edit)»[1]

---

44 ‹Poor› *uncertain reading.*
Mem. Lyell's . . . Indian Volcanos] *added ink with short rule line beneath.*
Von . . . Andes] *excised.*
45 At] 'A' *over* 'a'.
‹. . .›] *1 or 2 letters illeg.*
turn over] *indicates passage continues on next page.*
My results . . . over] *scored ink.*
‹I think› . . . over] *crossed.*

44–1 Daubeny 1826:382–83, 'With regard to the mineralogical characters of lava, I shall appeal to the authority of [Leopold] Von Buch. . . . Almost all lavas he conceives to be a modification of trachyte, consisting essentially of felspar united with titaniferous iron, to which they owe their colour and their power of attracting iron. . . . This felspar is derived immediately from trachyte, that being the rock which directly surrounds the focus of the volcanic action; for if we examine the strata that successively present themselves on the sides of a crater, we are sure to find that the lowest in the series is trachyte, from which is derived by fusion the obsidian, as is the case at Teneriffe.'
44–2 Daubeny 1826:386, '. . . in the collection of Dr. Thomson, now in the Museum of Edinburgh, there is said to be a fragment of lava enclosing a real granite, which is composed of reddish felspar with a pearly lustre like adularia, of quartz, mica, hornblende, and lazulite. I have likewise seen among the specimens from the Ponza Islands, . . . a piece of granite, or perhaps rather of a syenitic rock, . . . found in the midst of the trachyte from this locality. But the most interesting fact perhaps of this description, is . . . the presence of a mass of granite containing tin-stone, enveloped in the midst of a stream of lava from Mount Ætna. . . . It may be remarked, that these specimens of granitic rocks have, in general, a degree of brittleness, which accords very well with the notion of their exposure to fire.'
44–3 Lyell 1830–33, 1:318 refers to Java, 'where there are thirty-eight large volcanic mountains, many of which continually discharge smoke and sulphureous vapours.'
45–1 Beechey 1832:209, 'In latitude 60° 47′ N. we noticed a change in the colour of the water, and on sounding found fifty-four fathoms, soft blue clay. From that time until we took our final departure from this sea the bottom was always within reach of our common lines. The water shoaled so gradually that at midnight on the 16th, after having run a hundred and fifty miles, we had thirty-one fathoms.' P. 211, 'We soon lost sight of every distant object, and directed our course along the land [St Lawrence Island], trying the depth of water occasionally. The bottom was tolerably even; but we decreased the soundings to nine fathoms, about four miles off the western point, and changed the ground from fine sand, to stones and shingle. When we had passed the wedged-shaped cliff at the north-western point of the island, the soundings again deepened, and changed to sand, as at first.' P. 212–13, 'In our passage from the St. Lawrence Island to this situation, the depth of the sea increased a little, until to the northward of King's Island, after which it began to decrease; but in the vicinity of the Diomede Islands, where the strait became narrowed, it again deepened, and continued between twenty-five and twenty-seven fathoms. The bottom, until close to the Diomedes, was composed of fine sand, but near them it changed to course stones and gravel, as at St. Lawrence Island. . . .' P. 213, 'Near the Asiatic coast we had a sandy bottom, but, in crossing over the [Beering's] strait, it changed to mud, until well over on the American side, where we passed a tongue of sand and stones in twelve fathoms which, in all probability, was the extremity of a shoal, on which the ship was nearly lost the succeeding year. After crossing it, the water deepened, and the bottom again changed to mud, and we had ten and a half fathoms within two and a half miles of the coast.' P.

‹I think› At Ascension, the laminæ ‹. . .› changes in rocks. connected with & alternating with obsidian must *clearly* be *chemical* differences. & not those of rapid cooling &c &c

---

My results go to believe that much of all old strata of England. formed near surface: Mem Patagonian pebbles beds, most unfavourable to preservation of bones &c &c—Yet ‹silicified› turn over

46e silicified wood. *Cordilleras, Chiloe.* &c seems the organic structure most easily preserved.—

Mr Conybeare introduct to Geolog—"Between the height of same beds, deposited in different basins; little or no relation appears to ‹exist› be made out, but in those belonging to the same district there seems. I think, little ground for skepticism, as to the general truth of the proposition."—[1] If such can happen in troubled England; the more minute equalities

47 of elevation, may well be preserved at Patagonia. The English fact is astonishing *consult* book itself. P. xx: same fact is indeed shewn ? by the parallel bands of formations on any Geolog Map: Quoted from Daubeny P 402:[1] likewise, mean height of tertiary. being less than secondary:—consider arguments for *oscillation* of level independent of mineralogical nature & dependent: & then how wonderful level ««of same beds»» should have been kept; it shows that throughout all England, whole surface oscillated equably.—

48 These facts become easy if we look at the action as a deep & extensive movement of viscid nucleus, which in any one country would produce equable effects.—««though so immense to short breathed traveller»» Mountains, which in size are grains of sand, in this view sink into their proper insignificance; as fractures, consequent on grand rise, & angular displacement, consequent of injection of fluid rock.—
Try on globe. with slip paper a gradually curved enlargement

46 *page crossed pencil.*
47 ?] *added ink, circled*
page crossed.
48 *page crossed.*

444, 'In this parallel [61° 58′ N] the nearest point of land bearing N. 74° W. true, thirteen miles, the depth of water was 26 fathoms; and it increased gradually as we receded from the coast. . . . We made the land [St Lawrence Island] about the same place we had done the preceding year, stood along it to the northward, and passed its N.W. extreme, at two miles and a half distance, in 15 fathoms water, over a bottom of stones and shells, which soon changed again to sand and mud. . . . On the after-noon of the 2d we . . . anchored off Point Rodney . . . in seven fathoms, three miles from the land. . . .'

46–1 The quotation is from Daubeny 1826:402 which summarizes the argument presented in Conybeare and Phillips 1822:xx.

47–1 See note RN46–1.

49 see its increased length. which will represent the dilatation, which dilated cracks must be filled up by dikes & mountain chains.—

Introduce part of the above in Patagonian paper; & part in grand discussion[1]

Consult. reconsult Geolog. Map of Europe

50 Consult charts for distribution of pebbles.—Plains. off coast of Patagonia.— British channel &c &c.

There is a Hill. near Copiapò which is asserted to make a noise,—My impression. is not very distinct, from some of the lower orders; it was connected with movement of sand.—it is called "Bramidor"(?).—it was a strange story; I believe it was necessary to ascend the hill,—but my recollection is imperfect & was recalled by note in

51 Daubeny. P. 438., of similar fact near the Red Sea.—which occurred in a sandy place.—(the sound was long & prolonged).[1] NB, Is it generally known. the acute chirping sound produced in walking over the sand: I am nearly sure, it is necessary to ascend the hill.—

The absence of Second form, except near submarine Volc: in harmony with the prevailing movement being one of elevation alone.—In England much subsidence: hence difference; action on land different

52 Volney, P 351. Vol I. woody bushes, «gazelles» hares, grasshoppers & *Rats*. characteristic of the deserts of Syria ‹chara› ditto for Patagonia, especially rocky parts of central Patagonia[1]

Does Andes in Chili. separate geographical ranges of plants. V. Lyell. Chap XI Vol II.[2]

---

49 *page crossed.*
50 Consult . . . &c &c.] *crossed.*
There . . . in] *crossed.*
51 Daubeny . . . hill.—] *crossed.*
The . . . different] *crossed.*
52 Volney . . . II] *crossed.*
Urge . . . Kelp.—] *crossed.*

49–1 While he did not publish a paper on the subject, Darwin did eventually treat the subject of elevation in Patagonia in *GSA*, chap. 1.

51–1 See Daubeny 1826:438 for the following note, 'Cet endroit [near the Red Sea] recouvert de sable, environné de rochers bas en forme d'amphitheatre, offre une pente rapide vers la mer dont il est eloignè d'un demi mille, et peur avoir trois cent pieds de hauteur sur quatre-vingts de largeur. On lui a donné la nom de Cloche, parcequ'il rend des sons, non comme faisait autrefois la statue de Memnon, au lever du soleil, mais à toute heure du jour et de la nuit et dans toutes les saisons. La premiere fois qu'y alla M. Gray, il entendit au bout d'un quart d'heure un son doux et continu sous ses pieds, son, qui en augmentant ressembla à celui d'une clocha qu'on frappe, el qui devient si fort en cinq minutes, qu'il fit detacher du sable, et effraya les chamaux jusqu'â les mettre en fureur.' See also *JR*:441.

52–1 Volney 1787b, 1:351 with reference to the deserts of

Urge the entire absence of any rock situated beneath low water in the Southern ocean not being buoyed with Kelp.—

53e With respect to degradation of rocks—It may be a question. whether organic remains protect a rock, or that the rock not weathering allows such

54e ———

Compare the elevated estuary of the Plata. to the Bay of Bengal. dimensions?

55e Strong currents off the Galapagos.—strata must be accumulating which like the secondary strata of England, «besides ordinary marine remains» may contains ‹shells few corals Tortoise› «remains of Amphibia, exclusively.» & Turtle bones. & the bones of ‹two graniniverous› a herbivorous lizard.—from the action of torrents. «*marine*» Tortoise & other species of large lizard.—There would probably be no other organic remains.—

56e On Pampas looked in vain for a pebble of any sort; not one was found.—Miers saw then near?[1]

Mem. La Condamaine on the Amazons.[2] Consult

Insist on the frequency of dikes in Granitic countries, enumerate cases.—M. Video exception, but even there, hills of Basalt & other Volcanic rocks. Bahia, Rio de Jan: B. Oriental? level surface not disturbed.—Whole West coast. Chonos to Copiapo.—Sydney. K.G. Sound. C. of Good Hope.—**Carnatic**

53 or] *uncertain reading.*
of rocks . . . such] *located, crossed; remainder page not located.*
54 Compare . . . dimensions?] *located, crossed; remainder page not located.*
55 from] 'from' *over* 'the'.
Strong . . . remains.—] *crossed, remainder page blank, excised.*
56 On Pampas . . . near?] *excised.*
Mem . . . Consult] *crossed.*
Insist . . . Hope.—] *crossed.*
Carnatic] *added ink.*

Syria, 'Presque toujours également nue, la terre n'offre que des plantes ligneuses clair-semées, et des buissons épars, dont la solitude n'est que rarement troublée par des gazelles, des lièvres, des sauterelles et des rats.'

52–2 Lyell 1830–33, 2, chap. 11 bears the following summary heading, 'Theory of the successive extinction of species consistent with their limited geographical distribution—The discordance in the opinions of botanists respecting the centres from which plants have been diffused may arise from changes in physical geography subsequent to the origin of living species—Whether there are grounds for inferring that the loss from time to time of certain animals and plants is compensated by the introduction of new species?—Whether any evidence of such new creations could be expected within the historical era, even if they had been as frequent as cases of extinction?—The question whether the existing species have been created in succession can only be decided by reference to geological monuments.'

56–1 Miers 1826, 1:77, 'About two miles to the eastward of Barranquitos [32° 35′ S., 64° 20′ W.] I picked out of the sand a small fragment of quartz, about half the size of a hazel nut. This was the first pebble or stone of any sort I had seen since I left Buenos Ayres.'

56–2 La Condamine 1747:24, 'Below *Borja*, even for four or five hundred leagues, a stone, even a single flint, is as great a rarity as a diamond would be. The savages of those countries don't know what a stone is, and have not even any notion of it. It is diversion enough to see some of them, when they come to *Borja*, and first meet with stones, express their admiration of them by signs, and be eager to pick them up; loading themselves therewith, as with a valuable merchandize; and soon after despise and throw them away, when they perceive them to be so common.' See *JR*:289.

57 **It has been common practice of geologist.**[1]
Lyell considers (P 84 Vol III.) whole of Etna series of coatings;[2] hence it will be necessary to state all arguments for believing that there must be a central core of melted rock—I think the strongest is the consideration of the state at a grand eruption when whole summit of mountain is blown off; & again when in great crater. different little craters are all burning, surely there must be «somewhere» below a field of fluid rock.—In the discussion it will be better not to refer to Lyell. but merely to

58 state these reasons, & saying that they refer to CENTRAL nucleus & that envelopes no doubt existed. These higher portions probably formed Isl[ds] from which proceeded pebbles & on which trees grew.—? Are not the dikes in upper strata. quite different from the Porphyries: certainly *appearance* leads me to believe mere fissures filled up.—the appearance will here be the strongest argument:—? Consider causes for subaqueous crater being of diff: form subaerial one?—In former not so much; or no rapilli;[1] & from action of water probably not so much aluminated.

---

59 As argument in favor of lines of anticlinal violence crossing lines of crater, ‹arg› state that all the great Volcanoes. have been elevated considerably. which shows an afflux of inferior melted rocks to those parts.

Are not the dikes generally vertical? if so posterior to elevations? & not sources of lava streams.—**Urge not tilted strata.—**

---

57 **It has . . . of geologist**] *added ink.*
page crossed.
58 These . . . grew.—?] *scoring added ink, left margin; circled '?' added ink.*
bottom rule line marked through.
page crossed.
59 **Urge not tilted strata.—**] *added ink.*
Cornwall] 'C' over 'c'.
page crossed.

57–1 Despite the fragmentary nature of the entry, there exists a reference in Darwin's notes from the voyage, again by way of addition made in light brown ink, which identifies the use of 'Carnatic' in this context. See DAR 33:115[v] for citation of the following reference: Allardyce 1836:332–33, 'It has been remarked that granite in America is found at a much lower level than in Europe: this is also the case throughout the south of India, by granite—meaning always granitic rocks; for a regularly crystallized compound of quartz, felspar and mica, is not to be expected. The Carnatic, and several other similar tracts, occurring along both coasts, are, as granitic plains, surprisingly level: the slight tertiary diluvium with which they are covered, cannot be considered as a principal cause of this uniformity, for the rock itself is everywhere found near the surface: every appearance here indicates the granitic formation has at one time been a great deal more flat than it is generally understood to have been.'

57–2 Lyell 1830–33, 3:84, 'It is clear, from what we before said of the gradual manner in which the principal cone [of Etna] increases, partly by streams of lava and showers of volcanic ashes ejected from the summit, partly by the throwing up of minor hills and the issuing of lava-currents on the flanks of the mountain, that the whole cone must consist of a series of cones enveloping others, the regularity of each being only interrupted by the interference of the lateral volcanos.'

58–1 'Rapilli' was equivalent in meaning to 'lapilli'. See, for example, the use of 'rapilli' in Daubeny 1826:251.

It will be well to urge the case of St Helena, where dikes certainly have not been points of eruption.

Nobody supposes that all the dikes in Cornwall or in the coal measures have been conduits to volcanoes.—

60 Talking of the cricket valley «the most remarkable feature in the structure of Ascension»[1] give as an example the great subsidence at the famous eruption of Rialeja, & the more true analogy from the Galapagos—

Mr Lyell. P. 111 & 113. «seems to» considers that successive terraces mark as many distinct elevations;[2] hence it would appear he has not fully considered the subject.—

S. America in the form of the land decidedly

61 bears the stamp of recent elevation. which is different from what Mr Lyell supposes.[1]

Lyell P 116 Vol III, says that in N. Pliocene formation of Limestone, *casts* of shells, as in some older formations:[2] Mem the envelopes at Coquimbo. the analogy is now perfect

---

‹The grand propulsion of fluid rock, which elevates a continent›

We are more abound to take analogy of movements of W coast in explaining plains because such are found in perfection on that side.—

60 Talking . . . Galapagos—] *crossed.*
Mr . . . subject.—] *crossed.*
S. America . . . decidedly] *crossed.*
61 bears . . . supposes.] *crossed.*
Lyell P 116 . . . perfect] *crossed.*
The grand . . . side.—] *crossed.*

60–1 An oval depression towards the eastern end of Ascension Island was described by the resident English marines as the cricket ground because 'the bottom is smooth and perfectly horizontal'. See DAR 38.2:941v.
60–2 Lyell 1830–33, 3:111 begins the section entitled, 'Seacliffs—proofs of successive elevation'. Lyell's point is stated most succinctly on page 113 where he cites the testimony of another author writing on the alterations produced by the sea on calcareous rocks on the shores of Greece 'that there are four or five distinct ranges of ancient sea cliffs, one above the other, at various elevations in the Morea, which attest as many *successive* elevations of the country.'
61–1 In this passage Darwin would seem to be addressing Lyell's argument (1830–33, 3:114) that, '. . . a country that has been raised at a very remote period to a considerable height above the level of the sea, may present nearly the same external configuration as one that has been more recently uplifted to the same height.'
61–2 Lyell 1830–33, 3:116, '. . . we have seen [for the newer Pliocene] that a stratified mass of solid limestone, attaining sometimes a thicknesss of eight hundred feet and upwards, has been gradually deposited at the bottom of the sea, the imbedded fossil shells and corallines being almost all of recent species. Yet these fossils are frequently in the state of mere casts, so that in appearance they correspond very closely to organic remains found in limestones of very ancient date.'

62 Add from M. Lesson. character of Flora to New Zealand, which agrees with St Helena in being unique, yet no quadrupeds.—[1]

Is the white matter beneath pebbles. the degraded matter of such pebbles extending to seaward, the alternating with such matter at St Julians looks like such?—destructive to animal life.—*Patagonia*

63 In the Chonos Isl^ds we must imagine bituminous shales have been metamorphised, as in Brazil feruginous sandy ones have undergone the same process.—

Neither lakes or Avalanches (Glaciers very rare) to cause floods in valleys, which must aid in preserving the terraces ‹. . .› **Molina's Case**[1]

At Vesuvius. Vol III P. 124. Lyell. dikes have a parting of pitchstone; which is described as very rare[2] Mem. St Helena; probably more abundant in this case from intersecting a mass probably cold & not warm as sides of a crater as Vesuvius.—

64 There may have been oscillations in the upheaval of Andes.—but as long as all below water no evidence—The depth of shells (which being packed. in

62 Add . . . quadrupeds.—] *crossed.*
Is . . . *Patagonia*] *crossed.*
63 ‹. . .› **Molina's Case'**] *added ink,* ‹. . .› *illeg 3 letters*
In . . . process.—] *crossed.*
Neither . . . **Case**] *crossed.*
At . . . as Vesuvius.—] *crossed.*
64 There . . . trees] *crossed.*
Grand . . . came] *crossed.*

62–1 Lesson and Garnot 1826, 1, pt 1:14, '. . . mais il est à remarquer que cette île vaste et composée de deux terres séparées par un détroit, quoique rapprochée de la Nouvelle-Hollande et par la même latitude, en diffère si complétement, qu'elles ne se ressemblent nullement dans leurs productions végétales. Toutefois la Nouvelle-Zélande, si riche en genres particuliers à son sol et peu connus, en a cependant d'indiens, tels que des piper, des olea, et une fougère réniforme qui existe, à ce qu'on assure, à l'île Maurice.' Also p. 22, 'Il est à remarquer qu'on ne connaît aucun quadrupède comme véritablement indigène de la Nouvelle-Zélande, excepté le rat, si abondamment répandu sur les îles de l'Océanie, comme sur presque l'univers entier.'

63–1 Molina 1788, 1:30, 'La erupcion mas famosa de que tenemos noticia, fue la del volcan del monte de *Peteroa*, que el dia tres de Diciembre del ãno 1762 se abrió una nueva boca ó *cratéra*, hendiendo en dos partes un monte contiguo por espacio de muchas millas. El estrepito fue tan horrible, que se sintió en una gran parte del Reyno, pero no causó vibracion alguna sensible. Las cenizas y las lavas rellenaron todos los valles inmediatos, y aumentaron por dos dias las aguas del rio *Tingiririca*; y precipitandose un pedazo de monte sobre el gran rio *Lontué*, suspendió su corriente por espacio de diez dias, y estancadas las aguas, despues de haber formado una dilatada laguna que exîste en el dia, se abrieron por ultimo con violencia un nuevo camino, é inundaron todos aquellos campos.' Darwin noted this passage in his own copy of the work with the remark, 'P 30—Piteron Earthquake caused lake & deluge—state of valleys'.

63–2 Lyell 1830–33, 3:124, 'Towards the centre [of the dikes at Somma, the ancient cone of Vesuvius]. . .the rock is coarser grained, the component elements being in a far more crystalline state, while at the edge the lava is sometimes vitreous and always finer grained. A thin parting band, approaching in its character to pitchstone, occasionally intervenes on the contact of the vertical dike and intersected beds. M. Necker mentions one of these at the place called Primo Monte, in the Atrio del Cavallo; I saw three or four others in different parts of the great escarpment.'

beds) lived there, makes it very doubtful whether they could have lived in so deep a sea.—Perhaps agrees with formation of pebbles & vertical trees

---

Grand Seco at B. Ayres; mention about the deer approaching the wells.—the effect of Salt water of the Salado.—Mem. in Owens Africa it is mentioned that the Elephant came

65e towns driving by the want of water.—[1] I believe in all flat countries. years of drought are common.—Mr Lyell has mentioned the drifting of carcases putrid.[2]

In Rio paper. when discussing probable rise of land: Mention M. Gay's fact about shells:[3] Hibernation of fresh water Shells. multitudes.—

The question of shell's concretions, living only in that spot & being cause of concretion; or being only preserved in that part. having lived over whole bottom is important; because in this latter case. we cannot judge whether such fossils. lived in groups or not.

66e Ferruginous veins of this figure (A) in sandstone: evidently depend on a concretionary contraction: the fact is in alliance with those balls at Chiloe, full of sand.—the ‹scale› «quantity of iron» being there in excess.—If veins (A), are secretionary, so are all those plates in Australia. New Red Sandstone. at Bahia in modern sandstone. a circle, (|), had in its middle a short ‹fissure› «vein» terminated each way, which little vein was like the rest of these thin veins which project outwards.—

67e In Patagonia. the blending of pebbles & the appearance of travelling may be owing to successive transportal from prevailing swell, (as Shingle travels on

65 towns] *preceeding line at top of page cut out.*
towns . . . putrid.] *crossed.*
In . . . multitudes.—] *crossed.*
The . . . not.] *crossed.*
67 ‹me›] *alternate reading* '‹ma›'.
When . . . distance?—] *crossed.*

65–1 W. F. Owen 1833, 2:274–75, '[at Benguela] . . . the elephants were likewise common, but at present are scarce. A number of these animals had some time since entered the town in a body, to possess themselves of the wells, not being able to procure any water in the country. The inhabitants mustered, when a desperate conflict ensued, which terminated in the ultimate discomfiture of the invaders, but not until they had killed one man and wounded several others.'
65–2 Lyell 1830–33, 2:189, 'Thousands of carcasses of terrestrial animals are floated down every century into the sea, and, together with forests of drift-timber, are imbedded in subaqueous deposits, where their elements are imprisoned in solid strata. . . .' Also p. 247, '. . . we see the putrid carcasses of dogs and cats, even in rivers, floating with considerable weights attached to them. . . .'
65–3 Gay 1833:371, 'Ces contrées [Rio de Janeiro, Monte Video, Buenos Aires] m'offrirent aussi une assez belle collection d'insectes et plusieurs coquilles fluviatiles et marines, telles que des Mytilus, des Solens, des Ampullaires, etc., qui offraient ce phénomène digne de remarque, de vivre pêle-mêle dans les eaux simplement saumâtres.' See *JR*:24.

the Chesil bank. V. De la Beche).[1] Ask Capt. F.: R:[2] how the swell, generally & during gales would tend to travel on a ‹me› central line of Patagonia. «NB. Mr Lyell P. 211 Vol III. talks of line of cliff marking a pause»[3]

When mentioning pumice of Bahia Blanca, mention black scoriaceous rocks of R Chupat. & fall of Ashes of Falkner, ? how far is the distance?—[4]

68e Fossil bones black as if from peat.—yet cetaceous bones so likewise «of miocene period».—Mem Bahia blanca P. 204 Vol III. Lyell[1]

Owing to «open» faults in mountains: to elevated strata in eocene lakes of France, & unequal action of Earthquakes. «on Chili & delta of Indus», my belief in submarine tilting alone, must be modified. «Moreover, the Volcanos from sea there burst out, after rise from sea: ‹As did› as did those aerial Volcanos in Germany»

In the Valle del Yeso it is probable that point of Porphyry has been upheaved in a dry form

---

It is clear the forces have acted with far more regularity

69 in S. America: in France we have *freshwater lakes* unequally elevated, which movements if present in the Andes, would have destroyed regularity of slope of valleys.—All my observations of period «& manner» of elevation Volcanic

68 modified.] *original rule line beneath.*
as did . . . Germany] *marked connection with* 'It is clear . . .'
Owing . . . regularity] *crossed.*
69 in S. America . . . country.] *crossed.*
Read . . . latter).] *crossed.*

67–1 De la Beche 1831:73, 'The Chesil Bank, connecting the Isle of Portland with the main land, is about sixteen miles long, and . . . the pebbles increase in size from west to east. . . . The sea separates the Chesil Bank from the land for about half its length, so that, for about eight miles, it forms a shingle ridge in the sea. The effects of the waves, however, on either side are very unequal; on the western side the propelling and piling influence is considerable, while on the eastern, or that part between the bank and the main land, it is of trifling importance.'
67–2 Robert FitzRoy.
67–3 Lyell 1830–33, 3:210–11, 'The situation of this cliff [at Dax, France], is interesting, as marking one of the pauses which intervened between the successive movements of elevation whereby the marine tertiary strata of this country were upheaved to their present height, a pause which allowed time for the sea to advance and strip off the upper beds a,b, from the denuded clay c.'
67–4 Falkner 1774:51, 'Being in the Vuulcan, below Cape St. Anthony, I was witness to a vast cloud of ashes being carried by the winds, and darkening the whole sky. It spread over great part of the jurisdiction of Buenos-Ayres, passed the River of Plata, and scattered it's contents on both sides of the river, in so much that the grass was covered with ashes. This was caused by the eruption of a volcano near Mendoza; the winds carrying the light ashes to the incredible distance of three hundred leagues or more.'
68–1 Lyell 1830–33, 3:204, 'Some of these bones [in certain strata in the basin of the Loire] have precisely the same black colour as those found in the peaty shell-marl of Scotland; and we might imagine them to have been dyed black in *Miocene peat* which was swept down into the sea during the waste of cliffs, did we not find the remains of cetacea in the same strata, bones, for example, of the lamantine, morse, sea-calf, and dolphin, having precisely the same colour.'

action, must be more exclusively confined to that country.

———

Read description of channels or grooves in rocks at Costorphine hills. to compare with Galapagos.—Chiloe. M. Hermoso. & Coral reefs (imperfect in latter).

70 ‹At› Lyell. Vol I. P. 316. Earthquake of 1812 affected valley of Missisippi & New Madrid & Caraccas.—[1] Is this mentioned by Humboldt in his account of extensive areas.—[2]

———

P. 322 In any archipelago. & neigbouring Volcanos. eruption from «more than» one orifice ‹. . .› does not occur at same time: this is contrasted to contemporaneous action over larger spaces of the globes & "*periods*" of increased activity.—[3] such as that of 1835.—

———

State the three «or 4» fields of Earthquakes in Chili:—

71 Chiloe. Concepcion. Valparaiso (Copiapò & Guasco). yet whole territory vibrates from any one shock—

———

In S. America—continuity of space in formations & durability of similar causes go together. add. ‹"›" ‹from› "in the *same* line" to "from the epoch of Ammonite to the present day.

———

70 ‹. . .›] *illeg 3 or 4 letters.*
‹At› . . . areas.—] *crossed.*
P. 322 . . . Chili:—] *crossed.*
71 In . . . day.] *crossed.*
at . . . in the] *crossed.*

70–1 Lyell 1830–33, 1:316, 'We have before mentioned the violent earthquakes which, in 1812, convulsed the valley of the Mississippi at New Madrid, for the space of three hundred miles in length. As this happened exactly at the same time as the great earthquake of Caraccas, it is probably that these two points are parts of one continuous volcanic region. . . .'

70–2 Humboldt and Bonpland 1819–29, 4:11–12, 'The extraordinary commotions felt almost continually during two years on the borders of the Missisippi and the Ohio, and which coincided in 1812 with those of the valley of Caraccas, were preceded at Louisiana by a year almost exempt from thunder storms.'

70–3 Lyell 1830–33, 1:321–22, 'Syria and Palestine abound in volcanic appearances, and very extensive areas have been shaken, at different periods, with great destruction of cities and loss of lives. It has been remarked . . . that from the commencement of the thirteenth to the latter half of the seventeenth century, there was an almost entire cessation of earthquakes in Syria and Judea; and, during this interval of quiescence, the Archipelago, together with part of the adjacent coast of Lesser Asia, as also Southern Italy and Sicily, suffered extraordinary convulsions; while volcanic eruptions in those parts were unusually frequent. A more extended comparison . . . seems to confirm the opinion, that a violent crisis of commotion never visits both at the same time. It is impossible for us to declare, as yet, whether this phenomenon is constant in this, or general in other regions, because we can rarely trace back a connected series of events farther than a few centuries; but it is well known that, where numerous vents are clustered together within a small area, as in the many archipelagos for instance, two of them are never in violent eruption at once.'

at Mauritius. (consult Bory[1] «dip of strata on East») cannot believe in a great explosion, nor would sea remove more internally than externally—I did not see any number of dikes in the

72 cliffs.—wide valleys.—central peak small; yet great body of lavas have flowed from centre—

Pisolitic balls occur in the Ashes which fill up theatre of Pompeei (?).—Such have been seen to form in atmosphere.—Mem. Ascencion. concretions & Galapagos.—

«Humboldts. fragmens.[1]»
Read geology of N. America. India.—remembering S. Africa. Australia.. Oceanic Isles. Geology of whole world will turn out simple.—

73 Fortunate for this science. that Europe was its birth place.—Some general reflections might be introduced on great size of ocean; especially Pacifick: insignificant islets—general movements of the earth;—Scarcity of Organic remains.—Unequal distribution of Volcanic action, Australia S. Africa—on one side. S. America on the other: The extreme frequency of soft materials being consolidated; one inclines to belief all strata of Europe formed near coast. Humboldts quotation of instability of ground at present. day.—applied by me geologically to vertical movements.[1]

74 In Cord: after seeing small Bombs. without a vesicle. we may consider appearances of eruption at bottom.—solution under high pressure of gazes. especially the most abundant. Sulp. Hyd: Carb: A. Mur: A.=(& this effect of water thus holding matter in solution must be great: & in the fact of bombs in tufa there is proof of such gaz) steam condensed.—Perhaps these mighty

72 cliffs . . . Galapagos.—] *crossed.*
Humboldts . . . simple.—] *crossed.*
73 *page crossed.*
74 Mur: A.] 'A' *over* 'a'.
this] 'is' *over* 'ese'.
metallic veins . . . other phenomena.] *added ink.*
page crossed.

71–1 Bory de Saint-Vincent 1804, 1, chap. 6 describes the physical geography of Mauritius but does not answer Darwin's question directly. While at Mauritius Darwin was unable to inspect the entire island himself and sought information from other sources. See *VI*:28–31 and pp. 118–120 of this notebook.

72–1 Humboldt 1831.

73–1 The exact quotation is uncertain, but the following sentence suggests Humboldt's views (1831, 1:5–6) 'La *volcanicité*, c'est-à-dire, l'influence qu'exerce l'intérieur d'une planète sur son enveloppe extérieure dans les différens stades de son refroidissement, à cause de l'inégalité d'agrégation (de fluidité et de solidité), dans laquelle se trouvent les matières qui la composent, cette action du dedans en dehors (si je puis m'exprimer ainsi) est aujourd'hui très affaiblie, restreinte à un petit nombre de points, intermittente, moins souvent déplacée, très simplifiée dans ses effets chimiques, ne produisant des roches qu'autour de petites ouvertures circulaires ou sur des crevasses longitudinales de peu d'étendue, ne manifestant sa puissance, à de grandes distances, qui dynamiquement en ébranlant la croûte de notre planète dans des directions linéaires, ou dans des étendues (cercles d'oscillations simultanées) qui restent les mêmes pendant un grand nombre de siècles.'

changes might go on. & not a bubbles on the surface bespeak the changes.—
metallic veins solution of silex & many other phenomena

75 I do not believe that the extraordinary fissures of the ground at Calabria were present at the Concepcion earthquake.—expatiate on difficulty of evidence about eruptions of Volcanos. (where there are no country newspapers)—At the Calabrian earthquake things pitched off the ground. «Ulloa states that Volcanos!! were in eruption at time of great Lima earthquake[1]»

In the Chili earthquakes if rise was more ‹than› inland than on coast it would be invariably discovered; this may be mentioned with general slope of the country; (perhaps generally over whole world)

76 Yet eruptions ‹both› at sea (as wells as in the Cordillera), they may be considered as accidents (if ‹[. . .]› part of a regular system can be called accidental; the proportional force of crust of globe & injecting matter on the great rise).—

The great rains which attend severe Earthquakes «1822 ? 1835?» alone, (& the general belief in N. Chili, where rains are so infrequent; so as to exclaim, «as I have heard» how lucky! when they hear of a place having a pretty severe shock). are much more curious

77 & perplexing. than those that attend Eruptions: Mr P. Scopes explanation of low Barometer?[1]

In a subsiding area. we may believe the fluid matter instead of afflux (always slightly oscillating as that of a spring) moves away.—Will geology ever succeed in showing a direct relation of a part of globe rising, when another falls.—When discussing connection of Pacifick & S. America.—

75 Ulloa . . . earthquake] *over original medium rule line.*
page crossed.
76 ‹[. . .]›] *2 or 3 letters illeg.*
Yet . . . rise.—] *crossed.*
The . . . curious] *crossed.*
77 & perplexing . . . Barometer?] *crossed.*
In . . . S. America.—] *crossed.*

75–1 Juan and Ulloa 1806, 2:84, 'According to an account sent to Lima after this accident, a volcano in Lucanas burst forth the same night and ejected such quantities of water, that the whole country was overflowed; and in the mountain near Patas, called Conversiones de Caxamarquilla, three other volcanoes burst, discharging frightful torrents of water. . . .'

77–1 Scrope 1825, chap. 2, sec. 41–42 including the statement on p. 60, 'It is obvious how the powerful ascending draught of air which constitutes a hurricane, and which acts so strongly in depressing the barometer, will have an equal effect in setting loose the imprisoned winds of the earth.' Also see *JR*:431.

78 Volcanos must be considered as chemical retorts.—neglecting the first production of Trachyte. look at Sulphur. salt. lime, are spread over «whole» surface; how comes it they do not flow out together? How are they eliminated.—«Sulphur last.—» Metallic veins likewise must separate ingredients if we look to a constant revolution.—Are we to consider that the dikes which so commonly (state facts) traverse granites, are granitic materials simply altered by circumstances; & not in chemical nature, or has a subterranean fluid mass itself changed.—No.—

79 Yet the fluid granitic mass under ‹[. . .]› less pressure might have its «proportional» particles altered.—

With respect to Volcanic theory. I want to ground, that the first phenomem. is an inward afflux of melted matter.—Volcanos perhaps may be admittance of water, through the rent strata: «Mr Lyell considers that Plutonic rocks are generated as often as Volcanic. I consider latter as accidental on the afflux of the former.—[1]

Ascension. Vegetation? Rats & Mices. At St Helena there is a native mouse

---

80 Did wave first retreat at Juan Fernandez: the first great movement was one of rise (any smaller prior ones might have been owing to absolute movement of ground). Michell (Philos: Transacts) «seems to» considers that fall first movement (as in Peru 1746).—[1] At great Lisbon Earthquake Loch Lomond water oscillated between 2 & 3 ft. (as in Chili lake). Therefore motion of sea ought to be considered as a plain movement communicated to it as well as by the vertical as lateral movement.—At first one would think movement. owing to water keeping its level whilst land rose up & down.—But from above reasons, do not think so

78 ingredients] 'g' *over* 'd'.
page crossed.
79 ‹[. . .]›] *2 letters illeg.*
Mr Lyell . . . former.—] *written half above, half below entry on Ascension.*
With . . . former.—] *crossed.*
80 Did . . . Lake).] *crossed.*

79–1 Lyell 1830–33, 3:364, 'If . . . we conceive it probable that plutonic rocks have originated in the nether parts of the earth's crust, as often as the volcanic have been generated at the surface, we may imagine that no small quantity of the former class has been forming in the recent epoch, since we suppose that about 2000 volcanic eruptions may occur in the course of every century, either above the waters of the sea or beneath them.'

80–1 Michell 1760:617, 'The great earthquake that destroyed Lima and Callao in 1746, seems also to have come from the sea; for several of the ports upon the coast were overwhelmed by a great wave, which did not arrive till four or five minutes after the earthquake began, and which was preceded by a retreat of the waters, as well as that at Lisbon.' Darwin's own copy of this article was a reprint which had been repaginated by the printer; this quotation appears on p. 54 of his copy.

81 also elevating Earthquake of Valparaiso. (1822) no great wave on record.—«also neighbouring sea must partake in absolute movement» Moreover wave «with same general character» reaches far beyond coast, which has been raised.—It must be considered as an oscillation, from violence. Is it not same as swell travelling across Pacifick.—excepting in number of waves & in wind, instead of sea's bottom being in motion what difference? In watching heavy swell, sea retreats & then breaks: i e to form a wave in ocean. is not this [Fig. 3] form present, i e a part below «mean» level before the higher part.—
Does the

82 sea fall on banks as a Bore wave rushes up? (NB. Earthquake wave is an oscillation, body of water manifestly does not travel up.—) If these view are right the coincidental retreat at Portugal & Madeira (Lyell. vol I. P. 471) is explained.[1] also the similar fact at Concepcion? Read the various accounts & see if *fall* is not the first very evident movement.—The swelling first on beach I cannot understand, without (c^s ‹[. . .]› raised above a^s).—

---

83 In great Calabrian wave did not sea break first? I can imagine from local form of coast (as seen in swell) the undertow & overfall must vary proportionally Partial shrinking after elevation in perfect conformity with ‹Mr Lyell's› idea of an injected mass of fluid rock[1]
In Patagonia plains. long periods of rest & vice versâ more likely to be coincidental than single elevations along whole line of coast

81 *page crossed.*
82 ‹[. . .]›] *2 letters illeg.*
page crossed.
83 more likely] 'more' *over* 'most'.
In . . . proportionally.] *crossed.*
Partial . . . coast] *crossed.*

82–1 Lyell 1830–33, 1:471–72, 'Sometimes the rising of the coast must give rise to the retreat of the sea, and the subsequent wave may be occasioned by the subsiding of the shore to its former level; but this will not always account for the phenomena. During the Lisbon earthquake, for example, the retreat preceded the wave not only on the coast of Portugal, but also at the island of Madeira and several other places.'

83–1 Lyell did discuss 'partial shrinking after elevation', but, as Darwin's cancellation indicates, did not relate it to the existence of an underlying injected mass of fluid rock. See Lyell 1830–33, 1:477, 'It is to be expected, on mechanical principles, that the constant subtraction of matter from the interior will cause vacuities, so that the surface undermined will fall in during convulsions which shake the earth's crust even to great depths, and the sinking down will be occasioned partly by the hollows left when portions of the solid crust are heaved up, and partly when they are undermined by the subtraction of lava and the ingredients of decomposed rocks.' In his own copy of this work Darwin commented: 'if there are hollows left what forces up the lava' and then crossed out his remark. A few pages previously (p. 468) he had challenged Lyell's association of the occurrence of submarine earthquakes with the percolation of sea water to underlying masses of incandescent lava with the remark, 'We may more easily imagine the fluid stone injected (as occurs in every mountain chain) amongst damp strata.' He also questioned whether water could percolate through strata already under great pressure. In short, it would seem that Darwin realized he was describing his own idea rather than Lyell's in the course of writing this entry.

84 Darby mentions beds of marine shells on banks of Red River Louisiana. V. Lyell. Vol I. P. 191[1]
State at St Helena. pebbles entirely coated with Tosca. which implies motion in the «loose» bed of pebbles. (On a sea beach under a cascade, one can understand pebbles thus coated.—The motion is most wonderful, from chemical attraction, as a blade of grass penetrating by action of Organic power a lump of hard clay.—

85 In the History of S America we cannot dive into the causes of the losses of the «species of» Mastodons. which ranged from Equatorial plains to S. Patagonia. To the Megatherium.—To the Horse. = One might fancy that it was so arranged from the forseight of the works of man
Feeling surprise at Mastodon inhabiting plains of Patagonia is removed by reflecting on the nature of the country in which the Rhinoceros lives in S. Africa: the same caution is applicable to the Siberia case

86 We must not think alluvial plains «always» most favourable; In what part of the globe are there such vast numbers of wild animals. both species & individuals as in the half desert country of S. Africa. It would be well to quote Burchell. V. where the Rhinoceros was killed.—[1]

In Patagonia, are all beds same age? is white substance triturated Porphyritic rock. s (mem white tufas with purple Claystones of P. Desire). = Where talking of such substances being worn into channels.

87e mention submarine channels. such as that in front of St^s^. of Magellan

---

In Chiloe curvilinear strata subsidence.—The sudden increased dip is not

84 (On a] 'On' *over* 'in'.
understand] *mark following, possibly an* 'a'.
Darby . . . 191] *crossed.*
State . . . clay.—] *crossed.*
85 works] *alternate reading* 'work'.
page crossed.
86 *page crossed.*
87 *page crossed.*
In Cordill: . . . dikes.] *excised.*

84–1 Lyell 1830–33, 1:191, 'Darby mentions beds of marine shells on the banks of Red River, which seem to indicate that Lower Louisiana is of recent formation: its elevation, perhaps, above the sea, may have been due to the same series of earthquakes which continues to agitate equatorial America.' The work referred to is Darby 1816.
86–1 Burchell 1822–24, 2:71–79 describes the killing of two rhinoceroses south of the Hyena Mountains (30° 10′ S., 24° 0′ E.). In his own copy of the work Darwin scored the passage on p. 78 where Burchell described his sensation of the heat on a day of the hunt, 'Although so chilling at sunrise, the weather had, by noon, changed to the opposite extreme. Exposed in the middle of a dry plain, where not a tree to afford shade was to be seen, I scarcely could endure the rays of the sun, which poured down, as it were, a shower of fire upon us.' See also *JR*:101–102.

parallel case to Isle of White. but rather to one out of a series of faults. [Fig. 4]

In Cordill: should basal lavas be called Volcanic or Plutonic
The cellular state of all the Porphyry specimens, must be well examined

---

At M. Video «facts of Passages marked by do.» discuss quartz veins, there contemp—yet similar ones in Clay. Slates contemporaneous others subsequent. as in dikes.

88e In Granite great crystals arranged on sides. V. Lyell P 355 Vol III. constitution of veins, is there said granite in close contact varies in nature,[1]— Does not granite at C. Tres Montes become more siliceous in close contact?— «Cordillera???» Porphyry at Valparaiso; Epidote—

Must we look at regular greenstone cones at S. T. del Fuego as nucleus of a Volcano or as an injected mass.—From conical form I incline to ‹latter› former; & thus occurring in groups.—As these greenstone rocks are seen to graduate into granites

89 the ‹conta› passage from lava to Granite is much more perfect. than in believing mere agency of dikes: & indeed when do these dikes lead to a conical mass. will this conical mass be granite? Why not more probably greenstone? What probable origin can be given to the numerous hills of greenstone?—

Daubeny. P 95. Glassy & Stony Pearlstones alternate together in contorted layers:[1] Mem: Phillips Mineralogy some such fact stated to exist in Peru.[2]— Ascension

88 In . . . Epidote—] *crossed, excised.*
89 *page crossed.*

88–1 Lyell 1830–33, 3:355, 'The main body of the granite here [in Cornwall] is of a porphyritic appearance with large crystals of felspar; but in the veins it is fine-grained and without these large crystals. . . . The vein-granite of Cornwall very generally assumes a finer grain, and frequently undergoes a change in mineral composition, as is very commonly observed in other countries. Thus, according to Professor Sedgwick, the main body of the Cornish granite is an aggregate of mica, quartz, and felspar; but the veins are sometimes without mica, being a granular aggregate of quartz and felspar.'

89–1 Daubeny 1826:94–95, 'Trachytic porphyry also appears to pass by imperceptible gradations into the next species, pearlstone, which is characterized by the vitreous aspect generally belonging to its component parts. . . . In its simplest form, this rock presents an assemblage of globules, varying from the size of a nut to that of a grain of sand, which have usually a pearly lustre, and scaly aspect. . . . In some varieties the globules are destitute of lustre, and exhibit at the same time sundry alterations in their size, structure, and mode of aggregation, till at length they entirely disappear, and the whole mass puts on a stony appearance, which retains none of the characters of pearlstone. . . . Various alternations occur between the glassy and stony varieties of the pearlstone, sometimes so frequent as to give a veined or ribboned appearance to the rock, at others curiously contorted as though they had been disturbed in the act of cooling.'

89–2 W. Phillips 1823 contains no reference to pearlstone in Peru, but on p. 112 there is the statement that, 'At Tokay in Hungary, [pearlstone] is found enclosing round masses of black vitreous obsidian, and is intermixed with the debris of granite, gneiss, and porphyry, and alternating in beds with the latter.'

90 At Ischia there is a pumiceous conglomerate with small & large fragments, nature of which is doubtful. P. 180.[1] I think my Ascension case *very* doubtful.—

In Iceland Bladders of Lava are described, & many minute craters as at Galapagos. ‹|› Sir George Mackenzie must be worth reading[2]

Some earthquakes of Sumatra no connection with a neighbouring Volcano of Priamang.—Marsden Sumatra.[3]

M. De. Jonnes seems to

91 think that Volcanic eruptions form foundations for Coral reefs.—[1] does he mean in contradistinction to sand??

B. Roussin states that generally in North part of Brazil. ‹gravel becomes› sand less & gravel more common. the shoaler the water & nearer the Banks[2]

Is there not a sudden deepening on E. coast of Africa. as at Brazil

92 [blank]

93e What is nature of strip of Mountain Limestone in N. Wales. was it reef.—I remember many Corals?? Breccia—Stratification?[1]

---

90 *page crossed.*
91 *page crossed.*
93 What . . . Stratification?] *crossed.*
Anomalous . . . branching] *crossed.*

90–1 Daubeny 1826:180, '[The island of Ischia] is composed for the most part of a rock which seems to consist of very finely comminuted pumice, reagglutinated so as to form a tuff. . . . Although the pumiceous conglomerate, as I shall venture to call this rock, is seen in every part of the island, yet at Monte Vico . . . we observe intermixed with it huge blocks of trachyte. . . .'
90–2 Daubeny 1826:221, 'In many places [in Iceland], [Sir G. Mackenzie] says, an extensive stratum of volcanic matter has been heaved up into large bubbles or blisters, varying from a few feet to forty or fifty in diameter.' The original reference is to Mackenzie 1811:389–90.
90–3 As quoted in Daubeny 1826:313, 'In Sumatra, Marsden has described four [volcanos] as existing, but the following are all the particulars known concerning them: Lava has been seen to flow from a considerable volcano near *Priamang*, but the only volcano this observer had an opportunity of visiting, opened on the side of a mountain about 20 miles inland of Bencoolen, one fourth way from the top, so far as he could judge. . . . He never observed any connexion between the state of the mountain and the earthquake, but it was stated to him, that a few years before his arrival it was remarked to send forth flame during an earthquake, which it does not usually do.' The original reference is to Marsden 1811:29–30.
91–1 Daubeny 1826:334, 'The process, by which these islands, according to Moreau de Jonnes, are in many instances formed, is sufficiently curious; first a submarine eruption raises from the bottom of the sea masses of volcanic products, which, as they do not rise above the surface of the water, but form a shoal a short way below its surface, serve as a foundation on which the Madreporites and other marine animals can commence their superstructure.' The original reference is to Humboldt and Bonpland 1819–29, 4:42–43; also Cortès and Moreau de Jonnès 1810:130–31.
91–2 Roussin 1826:47 states that on approaching the banks of Cape S. Roque, '. . . nous croyons avoir observé que le sable est d'autant plus rare et les graviers d'autant plus communs, que les sondes sont plus petites et plus voisines des bancs.'
93–1 Darwin's memory of the geology of Wales stemmed

Anomalous action of ocean.—at Ascension. (where occassionally most tremendous surf & loose sandy beach) deposits «calcareous» encrustations; At Bahia ferruginous.—At Pernambuco (great swell & turbid water) organic bodies protect like peat reef of sandstone.—Corals, & Corallina survive, in the *most* violent *surfs*: in both latter cases become petrified, & increase.—In Southern regions *every* rock is buoyed by Kelp, now Kelp sends forth branching

94e roots which must protect surface; On «hard» exposed rocks near Bahia, whole surface to where highest spray (there pale green confervæ) coated with living beings; In smooth seas (& even turbulent as at St Helena) I have mentioned point of greatest action; I now having seen Pernambuco believe much is owing to protection of Organic productions. = Yet everywhere on coast (Il Defonsos «Kelp») rocks show signs of degradation; (soft substances worn into bare cliffs evident); the action is anomalous; It is wonderful to see Coral reef—or confervæ in the breakers or in waterfall: Excepting by removal of large fragments by mere force of waves: & action on upper tidal band, I do not

95e see how to account for oceans power.—excepting when pebbles are brought into play; most manifest example of degradation I ever saw on beach near Callao.—From Sir. H Davy experiment on the copper bottom. we see a trifling circumstance determines whether an animal will adhere to a certain part.[1] Apropos to question does animal adhere to rock because it does not decompose. or vice versâ. Clay slates unfavourable to attachment of many bodies

96e [blank]

97e Beechey.—changes in bottom in NW coast of America. from shingle to sand &c &c. ‹Vol II› P. 209. 211. 213. 444 «Yanky edition»[1]

Shores of Pacifick, as compared to whole E. America. ‹East› Africa. Australia. profoundly deep: a great fault or rather many faults.—

---

97 Beechey . . . faults.—] *pencil.*
Necessary . . . matter.—] *ink.*

from his tour there in August 1831 in the company of Adam Sedgwick. Darwin's notes from the trip refer to deposits of limestone and to fossil madrepores. See DAR 5(ser. 2):5–14[+15–16?], fols 5–14 published in Barrett 1974.

95–1 Davy 1824:151–58. After describing his experiments Davy concluded on p. 158, 'small quantities of zinc, or which is much cheaper, of malleable, or cast iron, placed in contact with the copper sheeting of ships, which is all in electrical connection, will entirely prevent its corrosion. And as negative electricity cannot be supposed favourable to animal or vegetable life; and as it occasions the deposition of magnesia, a substance exceedingly noxious to land vegetables, upon the copper surface; and as it must assist in preserving its polish, there is considerable ground for hoping that the same application will keep the bottoms of ships clean, a circumstance of great importance both in trade and naval war.'

97–1 See note RN45–1.

Necessary form; as long as coast line fixed.—[Fig. 5]

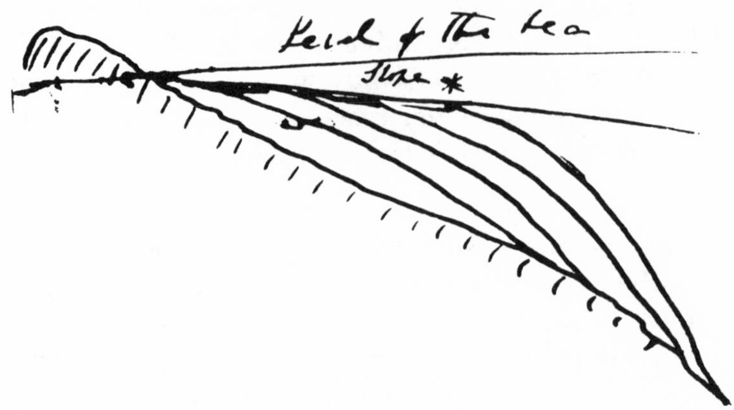

*Slope necessary for seaward transportal of drift matter.—

98e Give various cases. [Fig. 6]

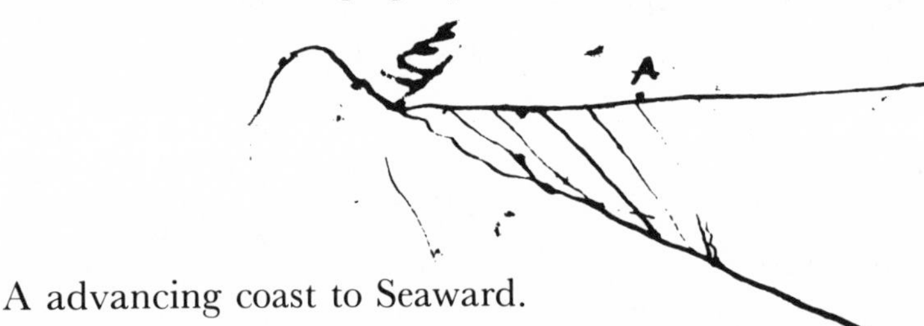

A advancing coast to Seaward.

Retreating case in excess as first case.

When discussing Falkland soundings introduce this discussion.—Brazil bank: (& I believe SE coast of Madagascar. where a $\dot{\overline{40}}$ line ‹shows› runs at equal distance?) 1st cases.—[1]

99e The terraces in Valleys of Chili may be with much truth compared to the step = formed streams of lava at St Jago. C. de Verds

---

Quartz pebbles in the Cordilleras look as if some peaks elevated.—

Greywacke. as a general fact absent in T. del Fuego, excepting in Port Famine

---

Mr Sorrell says that numerous icebergs are commonly stranded on shores of Georgia «Lat° (   )», he has rocks on surface. applicable to Patagonia.[1]

98 ‹shows›] *uncertain reading.*
page crossed pencil.
99 The . . . Verds] *crossed.*
Quartz . . . Famine] *crossed.*
Mr Sorrell . . . Patagonia.] *crossed.*

98–1 A bar and a dot over a number indicates that no bottom was found at that depth.

99–1 Thomas Sorrell: personal communication. See FitzRoy 1839, 2:21. Also see *JR*:282.

100e During a period of subsidence the shinglle of Patagonia would become more or less interstratified with sediment.—**& escarpment worn away like english escarpment**

The great conglomerate of the Amazons & Orinoco mentioned by Humboldt under name of Rothe-todte-liegende is perhaps same with that of Pernambuco?[1]

Quote Miers about shells at Quillota[2]

Lyell, states that contact of Granite & sedimentary rocks, in Alps becomes metalliferous. Vol III Latter Part[3]

101 Are there Earthquakes in the Radack & Ralix Isl^{ds}?

In my Cleavage paper Dr Fittons Australia case must be quoted at length.[1]

The Lines of Mountain appear to me to be effect of expansions acting at great depths (mem: profound earthquakes), which would cause parallel lines, but the rectangular intersections are singular—

M. Lesson considers the Sandstone & Granite districts to be separated by profound valley[.] Sydney.—[2]

100 & escarpment . . . escarpment] *added ink.*
page crossed.
101 Are . . . singular—] *crossed.*
M. Lesson . . . Sydney.—] *crossed.*

100–1 Humboldt and Bonpland 1819–29, 4:384, 'We discover between Calabozo, Uritucu, and the *Mesa de Pavones*, wherever men have made excavations of some feet deep, the geological constitution of the Llanos. A formation of red sandstone [*Rothes todtes liegende*] (or ancient conglomerate) covers an extent of several thousand square leagues. We shall find it again hereafter in the vast plains of the Amazon, on the eastern boundary of the province of Jaën de Bracamoros. This prodigious extension of red sandstone, in the low grounds that stretch along the East of the Andes, is one of the most striking phenomena, with which the study of rocks in the equinoctial regions furnished me.'

100–2 Miers 1826, 1:394–95, 'All around Quintero [near Quillota] . . . the fishermen had employed themselves digging shells for lime-making from a stratum four or five feet thick, in the recesses of the rocks, at the height of fifteen feet above the usual level of the sea, it being evident that at no very distant period this spot must have been buried in the sea, and uplifted probably by convulsions similar to the one now described.' Also p. 458, 'The recent shelly deposites mixed with loam [at Quintero] I have traced to places three leagues from the coast, at a height of 500 feet above the level of the sea. . . .' See *GSA*:35.

100–3 Lyell 1830–33, 3:371, '[According to M. Elie de Beaumont] . . . near Champoleon [in the Alps] . . . [it] is also an important circumstance, that near the point of contact both the granite and the secondary rocks become metalliferous, and contain nests and small veins of blende, galena, iron, and copper pyrites.'

101–1 Fitton 1827 as quoted in note RN38–1. Darwin did not publish a paper on cleavage but see his 36 page discussion of the subject in DAR 41. In these notes, possibly the 'paper' to which he referred, he mentioned Fitton's views twice.

101–2 Lesson and Garnot 1826, 1, pt 1:5, 'Toutes les côtes de la Nouvelle-Galles du Sud [New South Wales] sont, en effet, entièrement composées d'un grès houiller à molécules peu adhérentes; et ce que nous appelons le premier plan des montagnes Bleues est également composé de ce grès, qui cesse entièrement au mont York. Là, une vallée profonde isole ce premier plan du second, qui est composé en entier the granite.'

102 Lesson Zoologie

Grand tertiary formation of Payta[1]: N. part of New Zeeland entirely volcanic!! New Zeeland rich in particular genera of plants[2]: All St. Catherine & coast Granite[3]: P. 199; Falkland account of cleavage differs wonderfully from mine: phyllade covered by quartzose sandstones: refers to broken hill described by Pernetty: account of streams of stones agrees with mine.—[4]At Conception,[5] cleavage E & W! at Payta[6]. *talcose slates*, do at latter place. sandy. sandstone with gypsum, covered by *limestone* with recent shells 200 ft, how exact agreement with Coquimbo;

103e [not located]
104e [not located]
105e Is^{ld} near coast of America not reached. Juan. Galapagos. Cocos—
Ulloas voyage

North of Callao, the country, to the distance of 3 or 4 leagues «from the coast» may be concluded to have been covered by the sea—judge from the pebbles such as those on the beach—"This is particularly observable in a bay about five leagues North of Callao, called Marques, where in all appearances not many years since, the sea covered above half a league of what is now Terra Firma & the extent of a league & a half a long the coast. ‹"› The rocks in the most inland part of this bay are perforated & smoothed like those washed by the waves, a

102 *page crossed.*
105 Ulloas . . . waves, a] *crossed.*

102–1 Lesson and Garnot 1826, 1, pt 1:260–61, 'Le lambeau de sol tertiaire [at Payta] se compose de couches ou bancs alternatifs, dont voici l'énumération, en commençant par la formation de phyllade qui le supporte. 1° *Roches talqueuses phylladiformes*, terrain primordial. 2° *Argiles plastiques.*—Sable argileux, schisteux, traversé par des veines entrecroisées de gypse fibreux . . . 3° *Calcaire grossier* . . .'

102–2 On volcanic formations in the north part of New Zealand, Lesson and Garnot 1828, 1, pt 2:410, 'De nombreux volcans, dont les traces des éruptions sont récentes, existent sur plusieurs points de ces îles [off the north shore of the North Island] . . . Aussi trouve-t-on communément des pierres ponces. . . .' With respect to richness of plant genera in New Zealand see note RN62–1.

102–3 Lesson and Garnot 1826, 1, pt 1:189, 'Le granite forme entièrement la croûte minérale de l'île de Sainte-Catherine et du continent voisin. . . .'

102–4 On the Falkland Islands see Lesson and Garnot 1826, 1, pt 1:198–99, 'Les couches se composent de feuillets fendillés dans tous les sens, dont la direction, au lieu d'être horizontale, est presque verticale, et forme particulièrement sur le pourtour de la baie un angle de 45 degrés: ceux de la grande terre se dirigent à l'Est, et ceux des îlots aux pingoins à l'Ouest. . . . Cette phyllade supporte un grès schisteux. . . .' Also on p. 200 reference is made to the discussion by 'Pernetty' of a 'montagne des Ruines' which looked man-made, and on p. 201, to what Darwin later quoting directly from Pernety called a 'stream of stones' and what Lesson referred to as 'blocs énormes du même grès, entassés pêle-mêle . . .' See *JR*:255 and Pernety 1769, 2:526.

102–5 On the region around Concepcion see Lesson and Garnot 1826, 1, pt 1:231, 'La couche la plus inférieure est formée par une sorte de phyllade noire, compacte et terne; celle qui est moyenne se compose d'un mica-schiste à feuillets très-brillants, dont la direction est de l'Ouest à l'Est.' The presence of talcose slates at Concepcion is mentioned on p. 232.

102–6 Lesson and Garnot 1826, 1, pt 1. Rock cleavage is described as running from east to west on p. 260. On p. 262 Lesson uses the figure 200 feet in describing the change in sea level which would have caused such configurations of strata as seen at Payta.

106e sufficient proof, that the sea formed these large cavities", &c &c &c Vol II. Chapt VIII. p. 97[1]

---

at Potosi the veins run from North ‹inclining› to South. inclining a little to the West: the veins which follow this direction are thought by the ‹oldest› «most intelligent» miners to be the richest Vol II 147[2]

Shells at Concepecion 50 toises above the sea. = talks of them being packed clean. & without earth.—Moreover that such do not occur on the beaches. Perhaps these facts attest a ‹more› decided elevation of sea's bottom. beds of shells. 2–3 toises thick.—Vol II. p. 252[3]

---

107 Urge cliff form of land, in St Helena. Ascension. Azores. («sandstone first gives» half demolished craters).—worn into mud & dust.—connection with age, & agreement with number of craters. No cliffs at Ascension (or modern streams of St Jgo) yet no historical records of eruptions how immense the time!! How well agrees with number of Craters!—At S. Cruz. there is no occasion to wonder what has become of the Basalt. Gone into fine sediment Look at St Helena!!—

108 There are some arguments which strike the mind with force.—the exact yearly rise of the great rivers prove better than any meterological table the precise periods over immense areas. (& the counterbalancing variations) of rain. = The Bulk of sediment «daily» yearly brought down by every torrent proves the decay atmospheric of the most solid rocks.—The grand cliffs of a thousand feet in height, of the solid lavas.—proportionally high to age. (we do not wonder to see tertiary plains consumed) Where slope «plainly» indicates former boundary. (as in other unworn

106 sufficient . . . 97] *crossed.*
at Potosi . . . 147] *circled.*
Shells . . . 252] *crossed.*
107 *page crossed.*
108 *page crossed.*
109 [. . .]] *4 or 5 words illeg, possibly including* 'has', 'out' *or* 'cut', *and* 'limits'.
page crossed.

106–1 Juan and Ulloa 1806, 2:97.
106–2 Juan and Ulloa 1806, 2:147, 'These are the principal mines of Potosi, but there are several smaller crossing the mountain on all sides. The situation of the former of these mines is on the north side of the mountain, their direction being to the south, a little inclining to the west; and it is the opinion of the most intelligent miners in this country, that those which run in these directions are the richest.'
106–3 Juan and Ulloa 1806, 2:252, 'The country round the bay, particularly that between Talcaguana and Conception . . . is noted for . . . a stratum of shells of different kinds, two or three toises in thickness, and in some places even more, without any intermixture of earth, one large shell being joined together by smaller, and which also fill the cavities of the larger. . . . Quarries of the same kind of shells, are found on the tops of mountains in this country, fifty toises above the level of the sea.' Also, p. 254 'All these species of shellfish are found at the bottom of the sea in four, six, ten and twelve fathom water. They are caught by drags; and . . . no shells, either the same, or that have any resemblance to them, are seen either on the shores continually washed by the sea, or on those tracks which have been overflowed by an extraordinary tide.'

109 islands) we take in at once the stupendous mass which has been corroded.— If man could raise such a bulwark to the ocean, who would ever suppose that its age was limited? Who could suppose such trifling means could efface & obliterate so grand a work?—In valleys one is not sure whether fissures may not have helped it, or diluvial waves. but when we see an entire island so encircled, the one slow cause is apparent. «I confess I never see such islands whose inclination natural [. . .] deepest astonishment.» Perhaps scarcely a pebble might remain to tell of these losses.—

110 Cause of chimney. to crater. as at Galapagos. St. Helena.—

[Fig. 7] effect of heat on inner wall, hence resists degradation longer than outer parts.—

The common occurrence of a breccia of primitive rocks between that formation and the secondary (stated in Playfair to be the case p. 51).[1] presupposes an elevated country of granite, not ‹so› great«er» for all Europe, than from the Plata to Caraccas, which is all of granite:

111 In discussing circulation of fluid nucleus,—the similarity of Volcanic products «over whole world» argument, as well as separating causes by water.—Or rather begin & explain how water separates.—(intertropics at present fix lime). ‹Also Volcanos separate.› Volcanos blend all substances together; & products being similar over whole world, general circulation. But Volcanic action separates some sulphur (perhaps lime) salt. & metallic ores.—which mingling & separating is well adapted to

112 use of mankind.—‹Hutton show›[1] Earthquakes part of necessary process of terrestrial renovation & so is Volcano a useful chemical instrument.—Yet

110 ‹so›] *uncertain reading.*
than from] 'than' *over word, probably* 'as'.
Cause . . . parts.—] *crossed.*
The . . . granite:] *crossed.*
111 volcanic products] 's' *over* 'es'.
page crossed.
112 *page crossed.*

110–1 Playfair 1802:51–52, 'Indeed, the interposition of a breccia between the primary and secondary strata, in which the fragments, whether round or angular, are always of the primary rock, is a fact so general, and the quantity of this breccia is often so great, that it leads to a conclusion more paradoxical than any of the preceding, but from which, nevertheless, it seems very difficult to with-hold assent. Round gravel, when in great abundance, agreeably to a remark already made, must necessarily be considered as a production peculiar to the beds of rivers, or the shores of continents, and as hardly ever formed at great depths under the surface of the sea. It should seem, then, that the primary schistus, after attaining its erect position, had been raised up to the surface, where this gravel was formed; and from thence had been let down again to the depths of the ocean, where the secondary strata were deposited on it. Such alternate elevations and depressions of the bottom of the sea, however extraordinary they may seem, will appear to make a part of the system of the mineral kingdom, from other phenomena hereafter to be described.'

112–1 The principle expressed in this passage, that the destruction of the earth's surface is required for its renovation, is consistent with the general content of the

neglecting these final causes.—What more awful scourges to mankind than the Volcano & Earthquake.—Earthquakes act as ploughs [,] Volcanos as Marl-pits:

113e Consider well age of Bones. = slowness of elevation proved at St Julian. = do not these bones differ as much nearly as the Eocene. = Should Mr Owen consider bones washed about much at Coll. of. Surgeon's?[1] I really should think probably that B. Blanca & M. Hermoso contemp:.—Inculcate well that Horse at least has not perished because too cold:—With discussion of camel urge S. Africa productions.—

114e I think in Patagonia white beds having proceeded from gravel proved.—curious similarity of rocks of very diff. ages. at Port Desire on plain. & interstratified.—

Urge fact of Boulders not in lower strata. only in upper. in accordance in Europe with ice theory.—

Capt Ross found in Possession Bay in 73° 39 N. living worms in the mud which he drew up from 1,000 f[athoms], & the temp of which was below *freezing point*!!![1]

115 Remember idea of frozen bottom **or beach** of sea to explain preserved animals.—Mem: stream of water in the country.—

Sir J. Herschel. says. precip. of Sulph. B. all the infinitesimal cryst. arrange

113 *page crossed pencil.*
114 Urge . . . point!!!] *crossed pencil.*
115 **or beach**] *added ink.*
At the first] *uncertain reading.*
Remember . . . country.—] *pencil.*
Sir . . . this.—] *ink.*
State . . . it] *pencil.*
page crossed pencil.

work of James Hutton. However, as Darwin's cancellation would seem to indicate, the application of the principle failed in this instance, for Hutton, speaking providentially, had chosen rather to characterize volcanos as instruments designed 'to prevent the unnecessary elevation of land, and the fatal effects of earthquakes', and his interpreter John Playfair, while not quoting Hutton's words, did not challenge his conclusion. See Hutton 1795, 1:146, and Playfair 1802:116–19.

113–1 Richard Owen, then at the Royal College of Surgeons. Darwin commented on the single fossil quadruped he had found at the port of San Julián that the 'skeleton probably was at first perfect, but the sea having washed away part of the cliff, has removed many of the bones,—the remaining ones, however, still occupying their proper relative position to each other.' (*Fossil Mammalia*: 10) Owen named the fossil quadruped *Macrauchenia*, on which see note RN129–1.

114–1 John Ross 1819:178, 'Soundings were obtained correctly in one thousand fathoms [at Possession Bay], consisting of soft mud, in which there were worms. . . . The temperature of the water on the surface was 34 ½° [F.], and at eighty fathoms 32 °; . . . at two hundred and fifty fathoms [measurement taken aboard another ship], . . . 29 ½° [F.].' In Appendix No. III, p. lxxxv this information is summarized and the coordinates of Possession Bay given as 73° 39′ N., 77° 08′ W.

themselves in planes. «Mem silky lustre»[1] ask Erasmus. whether electricity would affect this.—[2]

State the circumstances of appearance at Concepcion [.] no sign of elevation. Effects of great waves to obliterate all land marks.—At the first it

116 would though be easy to see on beach successive lines of sea weed—

Histoire Naturelle des Indes
Acosta. p. 125. of French «?» Edition states that the same earthquake has run from Chili to Quito a distance of more than 500 leagues. A little time after a bad earthquake in Chili; Arequipa in 82 was overthrown, & 86. Lima. next year Quito. considers these earthquakes travel in order.—[1]

117 If we look at Elevations as constantly going on we shall see a cause for Volcanos part of same phenomena lasting so long.—

The great movements (not mere patches as in Italy proved by Coral hypoth. agree with great continents).

118 Voyage aux terres Australs Vol. I. p. 54. M. Bailly says."en effet toutes les montagnes de cette île se developpent autour d'elle comme une ceinture d'immenses remparts; toutes affectent une pente plus ou moins inclinée vers le rivage de la mer, tandis, au contraire, que vers le centre de' l île, elles presentent une coupe abrupte et souvent tailée a pic. Toutes ces montagnes

116 would . . . weed.—] *pencil.*
Histoire . . . order.—] *ink.*
page crossed pencil.
117 *page crossed pencil.*
118 *page crossed pencil.*

115–1 John Herschel wrote to Charles Lyell on 20 February 1836 as follows (Cannon 1961:310), 'Cleavages of Rocks.—If Rocks have been heated to a point admitting a commencement of crystallization, ie to the point where particles can begin to move inter se—or at least on their own axes—some general cause must determine the position these particles will rest in on cooling—probably position will have some relation to the direction in which the heat escapes.—Now when all—or a majority of particles of the same nature have a general tendency to one position that must of course determine a cleavage plane.—Did you never notice how the infinitesimal crystals of fresh precipitated sulphate of Baryta [barium sulphate] & some other such bodies—arrange themselves alike in the fluid in which they float so as, when stirred all to glance with one light & give the appearance of *silky filaments*. Ask Faraday to shew you this phenomenon if you have not seen it—it is very pretty. What occurs in our expt, on a minute scale may occur in nature on a great one, as in granites, gneisses, mica slates &c—some sorts of soap in which insoluble margarates exist shew it beautifully [added: when mixed with water].' Lyell incorporated Herschel's observation into his next edition of the *Principles*. See Lyell 1837, 4:358. Possibly Lyell showed Darwin Herschel's letter, or discussed its contents with him, sometime in late 1836 or early 1837, or possibly Herschel had discussed this point with Darwin at the Cape in June 1836. See note RN32–1.

115–2 Erasmus Alvey Darwin.

116–1 Acosta 1600:125 refers to 'des tremblemens de terre qui ont couru depuis Chillé, jusques à Quitto, qui sont plus de cinq cens lieues . . .' Acosta continued, 'En la coste de Chillé (il ne me souvient quelle année) fut un tremblement de terre si terrible. . . . A peu de temps delà, qui fut l'an, de quatre vingts deux, vint le tremblement d'Arequipa, qui abbatit & ruina presque toute cette ville là. Du depuis en l'an quatre vingts six . . . aduint un autre tremblement en la cité des Roys [Lima]. . . .' And on p. 125^{v}, 'En apres l'an enfuyuant, il y eut encor un autre tremblement de terre au Royaume & cité de Quitto, &

119 sont formées de couches paralleles et inclinées du centre d'ile, vers la mer; ces couches ont entre elles une correspondance exacte, et lorsquelles se trouvent interrompues par quelque vallées ou par quelque scissures profondes, on les voit se reproduire a des hauteurs communes sur le revers de chacune des montagnes qui forment les vallées ou les scissures.—M. B. thinks these parts incontestably formed the parts of one whole

120 burning mountain, & that the central part fell in.—Says posterior craters in centre:—[1] Bailly talks of much granite on all East side of Van Diemen Land.[2]

All the Calcareous rocks which harden by themselves cannot be pure. for if so Chalk would

121 harden.—Climate.!? or small Proportion of Alum: matter.—all pale cream colour.—
The Brecciated structure of *all* the Pitchstone (which I have seen). is a kind of concretionary structure, for the interlineal spaces are of diff cont: & even in one case contained lime.—All bear close analogy to Obsidian, & all show chemical action as well as effects of cooling

[misnumbering, no page 122]

123 In Igneous rocks.—which have the cryst of glassy F. fractured. have been melted with little pressure. & perhaps cooled suddenly.—

As the rude symmetry of the globe shows powers have acted from great depths, so changes, acting in those lines. must now proceed from great depths.—important.—

119 *page crossed pencil.*
120 burning . . . centre.—] *ink.*
Bailly . . . would] *pencil.*
page crossed pencil.
121 harden . . . colour.—] *pencil.*
The . . . cooling] *ink.*
page crossed pencil.
123 *page crossed pencil.*

semble que tous ces notables tremblemens de terre en ceste coste, ayent succedé les uns aux autres par ordre. . . .'

120–1 Joseph-Charles Bailly, mineralogist to the expedition, as quoted in Péron 1807–16, 1:54–55. Following the passage quoted, the text continues (p. 55), 'De ces observations, il résulte bien incontestablement que toutes ont la même origine, qu'elles datent toutes de la même époque; que réunies jadis, elles n'ont pu être séparées depuis, que par quelque révolution violente et subite. Quelle peut avoir été cette dernière révolution? . . . Tous les fait se réunissent pour prouver que l'île toute entière ne formoit jadis qu'une énorme montagne brûlante; qu'épuisée, pour ainsi-dire, par ses éruptions, elle s'affaissa sur elle-même, engloutit dans ses abîmes la plus grande partie de sa propre masse, et que de cette voûte immense, il ne resta debout que les fondemens, dont les débris entr'ouverts sur différens points, forment les montagnes actuelles de l'île. Quelques pitons de forme conique, qui s'élèvant vers le centre du pays, notamment le Piton du centre, portent les caractères d'une origine postérieure à l'éboulement du cratère. . . .' See also *VI*:29–31.

120–2 Bailly (note RN120–1) as quoted in Péron 1807–16, 1:295, 'De hautes montagnes granitiques . . . dont les sommités étoient presque entièrement nues, forment toute la côte orientale de cette partie de la terre de Diémen. . . .' See also p. 304.

124 Decemb 10. 1802. Earthquake at Demerara. The earthquakes "seem to arise from some efforts in the land to lift itself higher & to grow upwards; for the land is constantly pushing the sea (which of course must retain same level) to a greater distance".—Afterwards speaks of this phenomena in connection with "the shooting upwards" of the ‹ground› land in the W Indies.—p. 200. Bollingbroke voyage to the Demerary[1]

125 Earthquakes at St Helena. 1756. June 1780, Sept. 21st. 1817.—p 371. Webster Antarctic veg:—[1]

Study Ulloa to see if Indian habitation above regions of vegetation.—«I can find nothing.»[2] Mem Carolines quotation from Temple[3]

Urge the *mineralogical* difference of formations of S. America & Europe.—If great chain of Volc. had been in action during secondary period how diff. would the rocks have been. The red Sandstone of Andes fusible?

126 no. mad dogs. Azores. although kept in numbers. p. 124. Webster[1]

Consult W. Parish.[2] & Azara[3] about dry season[.] 1791. seen commonly bad

124 *page crossed pencil.*
125 Earthquakes . . . veg:—] *ink.*
Study . . . Temple] *pencil.*
Urge . . . fusible?] *ink.*
page crossed pencil.
126 *page crossed.*

124–1 Bolingbroke 1807:200, contains the passage Darwin quotes and pp. 200–201 the additional comment, 'This constant shooting upwards of the land, which is so sensible in the West Indies, has been little heeded by European mineralogists.'

125–1 W. H. B. Webster 1834, 1:371, 'Instances of earthquakes occuring in the island [St Helena] are on record. One took place in 1756, and in June 1780. On the 21st September 1817, one occurred, which it is said was particularly noticed by Napoleon . . .' The reference to antarctic vegetation pertains to Webster's discussion of the natural history of Cape Horn, Staten Island, and Deception Island in 2:290–306.

125–2 Darwin apparently searched Juan and Ulloa 1806 for evidence connecting Indian habitation and climatic change, and could 'find nothing'. He was more successful in his reading of Ulloa 1792. He later quoted from that work (p. 302) to the effect that Indians of one arid region in the Andes had lost the art of making durable bricks from mud. This suggested to Darwin that the local climate had once been wetter, which fitted his notion that the South American continent had undergone elevation in geologically recent times. See *JR*:409–11.

125–3 Temple 1830, 2:10, 'In the course of this day's journey were to be seen, in well-chosen spots, many Indian villages and detached dwellings, for the most part in ruins. Up even to the very tops of the mountains, that line the valleys through which I have passed, I observed many ancient ruins, attesting a former population where now all is desolate.' See also pp. 4 and 5, and *JR*:412. From his comment it would appear that this reference was given to Charles by his sister, Caroline Sarah Darwin.

126–1 J. W. Webster 1821:124, 'There is scarcely a man on the island, who has not a dog, and many have half a dozen. It is a remarkable fact that, although these animals are so numerous, no instance of hydrophobia was ever known among them.' See *JR*, p. 436.

126–2 See *JR*:156, 'Sir Woodbine Parish informed me of another and very curious source of dispute [in the province of Buenos Ayres]; the ground being so long dry, such quantities of dust were blown about, that in this open country the landmarks become obliterated, and people could not tell the limits of their estates.'

126–3 Azara 1809, 1:374, 'On voit un exemple aussi étonnant de cette fougue dans les années sèches, où l'eau est extrêmement rare au sud de Buenos-Ayres. En effet, ils partent comme fous, tous tant qu'ils sont, pour aller chercher quelque mare ou quelque lac: ils s'enfoncent dans la vase, et les premiers arrivés sont foulés et écrasés

over whole world. «(Was it so in Sydney, consult history? Phillips.»[4]

1826.27.28. grt. drought at Sydney. which caused Capt. Sturt expedition.—[5]

⁋Another one in 1816 (?).—

127 Mr Owen's curious fact about Crust Bra in Brine[1] Springs. (Henslow)[2]

Speculate on neutral ground of 2. ostriches; bigger one encroaches on smaller.—[3] change not progressif‹e›: produced at one blow. if one species altered: ‹altered› Mem: my idea of Volc: islands. elevated. then peculiar plants created. if for such mere points; then any mountain, one is falsely less surprised at new creation for large.—Australia's = if for volc. isl^d. then for any spot of land. = Yet new creation affected by Halo of neighbouring continent: ∓ as if any

128 creation «taking place» over certain area must have peculiar character:

Contrast low limit of Palms, evergreen trees, arborescent grasses, parasitic

127 neighbouring] *alternate reading* 'neighboring'.
Mr Owen's . . . Brine] *pencil.*
Springs . . . any] *ink.*
Speculate . . . any] *crossed pencil.*
128 grasses, parasitic] *alternate reading* 'grasss, parasite'.
Contrast . . . Zorilla:] *crossed pencil.*

par ceux qui les suivent. Il m'est arrivé plus d'une fois de trouver plus de mille cadavres de chevaux sauvages morts de cette façon.' See *JR*:156.

126–4 Hunter 1793:507, 508, 525, and 535 refer to the drought around Sydney in the first half of the year 1791. 'Phillip's Voyage' refers to Phillip 1789.

126–5 Sturt 1833, 1:1, 'The year 1826 was remarkable for the commencement of one of those fearful droughts to which we have reason to believe the climate at New South Wales is periodically subject. It continued during the two following years with unabated severity.' And p. 2, 'But, however severe for the colony the seasons had proved . . . it was borne in mind at this critical moment, that the wet and swampy state of the interior had alone prevented Mr. Oxley from penetrating further into it, in 1818. . . . As I had early taken a great interest in the geography of New South Wales, the Governor was pleased to appoint me to the command of this expedition.' See also *JR*:157.

127–1 From this entry it would appear that it was probably Richard Owen who referred Darwin to the article by Rackett quoted in *JR*:77, 'In the Linnean [Society of London] Transactions, [1815], vol. xi, p. 205, a minute crustaceous animal is described, under the name of *Cancer salinus*. It is said to occur in countless numbers in the brine pans at Lymington; but only in those in which the fluid has attained, from evaporation, considerable strength; namely about a quarter of a pound of salt to a pint of water. This cancer is said, also, to inhabit the salt lakes of Siberia. Well may we affirm, that every part of the world is habitable!' The phrase 'Crust Bra' is probably an abbreviation for Crustacea Branchiopoda.

127–2 John Stevens Henslow did not publish on the subject of springs, but he may have been the source for two references which Darwin quoted on the subject in the *JR*:78. Both works cited discuss plant life at the location of the springs, a subject which would have interested Henslow. The references were to J. E. Alexander 1830: 18–20, and Pallas 1802–3, 1:129–34.

127–3 The two ostriches are the greater or common rhea. *Rhea americana*, found from north-eastern Brazil to the Rio Negro in central Argentina, and the lesser rhea or Darwin's rhea, *Pterocnemia pennata*, found in the Patagonian lowlands, where Darwin collected portions of a specimen, and in the high Andes of Peru, Bolivia, northern Chile, and northwestern Argentina. The ornithologist John Gould described it and named it *Rhea Darwinii* at a meeting of the Zoological Society of London on 14 March 1837, since when it has been known as Darwin's rhea. However, the species had already been named *Rhea pennata* by Alcide Dessalines d'Orbigny. (Anonymous 1837b; d'Orbigny 1835–47, 2:67, 194, 212, 303; on the dating of d'Orbigny see Sherborn and Woodward 1901.) See also Gould 1837b:35–36; *Birds*:120–25; and *JR*:108–10.

plants, Cacti: & with limits of no vegetation at S. Shetland = [1]

Great contrast of two sides of Cordillera, where climate similar.—I do not know botanically = but picturesquely = Both N & S. great contrast. from nature of climate. =

Perpetual snow.—subterranean lakes, near Volcanoes. lakes of brine all inhabited:

Go steadily through all the limits of birds & animals in S. America. Zorilla:[2]

129 wide limits of Waders: Ascension. Keeling: at sea so commonly seen. at long distances; generally first arrives:—

New Zealand rats offering in the history of rats, in the antipodes a parallel case.—

Should urge that extinct Llama owed its death not to change of circumstances; reversed argument. knowing it to be a desert.—[1] Tempted to believe animals created for a definite time:—not extinguished by change of circumstances:

130 The same kind of relation that common ostrich bears to (Petisse.[1] & diff kinds of Fourmillier)[2]: extinct Guanaco[3] to recent: in former case position, in

129 New Zealand . . . circumstances:] *crossed pencil.*
130 *page crossed pencil.*

128–1 One probable source for this passage was W. H. B. Webster 1834, 2:281–302. In the course of describing the 'exuberant fertility of Staten Island and Cape Horn' (p. 297), Webster mentioned evergreens and a 'parasitic shrub' (p. 292). See also note RN125–1.
128–2 *Zorrilla* is the Spanish word for skunk. The notes on species ranges of South American forms which Darwin suggested making in this entry are presumably those found in DAR 29.1. The 'Birds' list is numbered fol. 41; the 'Animals' list appears between fols. 46–47. The *zorrilla* appears on the list for animals.
129–1 The 'extinct Llama' is the animal that Richard Owen would name *Macrauchenia* (long-necked) in December 1837 (*Correspondence* 2:66). Darwin collected the fossil specimens in January 1834 at the port of San Julían, having 'no idea at the time, to what kind of animal these remains belonged'. *JR*:208. Owen's earliest known comment on the specimens occurs in a letter to Charles Lyell dated 23 January 1837. For the citation from Owen see L. G. Wilson 1972:437; also see *JR*:208. Sometime after his initial assessment of the fossil as a 'Gigantic Llama' Owen restudied the fossil and came to a more complex conclusion regarding its affinities, one emphasizing its non-camelid over its camelid (i.e. llama-like) features. See notes A9–1 and B231–1, and *Fossil Mammalia*: 10–11, 35–56, and plates VI–XV; for current views see Simpson 1980.
130–1 The 'Petisse' is the lesser rhea, or Darwin's rhea. (See note 127–3.) Darwin customarily referred to the two rheas in his field notes as 'Avestruz' and Avestruz Petise' from the Spanish *avestruz* (ostrich) and *avestruz petiso* (small ostrich). See also B153.
130–2 *Fourmilier*, ('antbird') so named for its falsely reported habit of living chiefly on ants (*fourmis*). Since Darwin does not seem to have used the term *Fourmilier* elsewhere in his notes, it is doubtful that it was the antbird, or at least primarily the antbirds, which he had in mind when he made this entry. More likely he was thinking of those birds which he described in his Ornithological Notes as *Myothera*, a term which was given as the equivalent of *Fourmilier* in the systematic work he had with him aboard ship (Bory de Saint-Vincent 1822–31,

latter time. (or changes consequent on lapse) being the relation.—As in first cases distinct species inosculate, so must we believe ancient ones: «.˙.» not *gradual* change or degeneration. from circumstances: if one species does change into another it must be per saltum—or species may perish. = This ‹inosculation› «representation» of species important, each its own limit & represented.—Chiloe creeper:[4] Furnarius.[5] ‹Caracara›[6] Calandria:[7] inosculation alone shows not gradation;—

131 An argument for the Crust of globe being thin, may be drawn. from. Cordillera. rocks.—When beneath water.—together with hypothetical case of Brazil.—

---

132 Propagation. whether ordinary. hermaphrodite. or by cutting an animal in two. (gemmiparous. by nature or accident). we see an individual divided either at one moment or through lapse of ages.—Therefore we are not so much surprised at seeing Zoophite producing distinct animals. still partly united. & eggs which become quite separate.—Considering all individuals of all species. as «each» one individual «divided» by different methods, associated life only adds one other method where the division is not perfect.—

133 Dogs. Cats. Horses. Cattle. Goat. Asses. have all run wild & bred. no doubt with perfect success.—showing non Creation does not bear upon solely adaptation of animals.—extinction in same manner may not depend.—There [...] more wonder in extinction of species than of individual.—

[...] birds he [...] by John [...] the group, [...] and 351– [...] number, all [...] 50–60; and [...] of Darwin's [...] meaning to [...]

130–4 [...] *spinicauda*, the Thorn-tailed Rayadito. [...] the distinctive sub-species *A. spinicauda fulva*, being buff-coloured instead of mainly white below. For further information on Darwin's specimens see Barlow 1963:250; *JR*:301; and *Birds*:81.

130–5 *Furnarius*, the ovenbird, the genus which gives its name to the family Furnariidae. Found from southern Mexico to Patagonia the family shows the greatest measure of diversity in the southern part of its range. Darwin collected a number of species belonging to the family. For further information see Barlow 1963:214, 217–18; *JR*:112–113, 353, and 477; *Birds*:64–65; and Herbert 1980:116.

130–6 Caracaras are large carrion-feeding birds belonging to the family Falconidae. They are very common in parts of South America, and Darwin collected a number of specimens. In his 'Ornithological Notes' Darwin also referred to the Galapagos hawk as a caracara (p. 238), though John Gould later corrected him. For more on these birds see Barlow 1963:233–39; *JR*:63–69, 256, 461; Gould 1837a:9–11; *Birds*:9–31; and S. Herbert 1980:116.

130–7 Calandria, the Chalk-browed Mockingbird which Darwin collected at Maldonado (specimen 1213). In this entry Darwin probably also had in mind other mockingbirds he collected in South America and the Galápagos Islands. For further discussion of the mockingbirds see *Birds*:60–64; Barlow 1963; *JR*:62–63, 461; Gould 1837f:27; and Herbert 1980:116–17.

134e Mr Birchell says Elephant lives on very wretched cou[n]tries thinly covered by vegetation.[1] Rhinoceros quite in deserts.—Much struck with number of animal[s] at Cape of Good Hope

---

Says at Santos «M Birchels» at foot of range some miles from shore. rock of oysters quite above reach of tides.—thinks them same as recent species.—[2]

135e May I not generalize the fact glaciers most abundant in interior channels. there no outer coast.—important effect.—? Capt. FitzRoy.—[1]

Limited Volcanic action & limited earthquakes & great but local elevations of the land in Europe—

136e Urge difference of plutonic rocks & Volcanic metalliferous—

Urge enormous quantity of matter from CREVICE of Andes—therefore flowed towards it. a mass on each side 3000 ft thick & 150 broad. neglecting Cordillera itself now remaining—

137e Lyell « ‹p 419› p 428» states that Von Buch has urged that Java volcanos differ from all others in quantity of Sulph. acid emitted:[1] mem: Grand *gypseous* formation of Cordillera

---

In describing structure of Cordillera it must be said, that lines of elevation have connected ‹lines› «points» of eruption[.] give instance of Etna Stromboli & Vesuvius

134 quite] *alternate reading* 'quits'.
Birchels] *alternate reading* 'Birchel'.
Mr Birchels . . . Hope] *crossed pencil.*
quite in . . . species.—] *crossed pencil.*
135 May . . . FitzRoy.—] *ink.*
Limited . . . Europe—] *pencil.*
page crossed pencil.
136 *page crossed.*
137 Lyell . . . Cordillera] *crossed.*
In . . . Vesuvius] *crossed.*

134–1 William J. Burchell: personal communication as indicated in *JR*:101. Burchell 1822–24, 2:207 is quoted on the subject of the large size of South African animals compared to animals from other continents in *JR*:101.
134–2 William J. Burchell: personal communication. See *GSA*:3. 'Mr. Burchell informs me, that he collected at Santos (lat. 24° S.) oyster-shells, apparently recent, some miles from the shore, and quite above the tidal action.'
135–1 Robert FitzRoy: personal communication. See also *JR*:266-67, 'I have heard Captain FitzRoy remark, that on entering any of these channels [at Tierra del Fuego] from the outer coast, it is always necessary to look out directly for anchorage; for further inland the depth soon becomes extremely great.'
137–1 Charles Lyell: personal communication. The reference is to von Buch 1836:428, 'Ces émanations sulfureuses paraissent donner aux volcans de Java un caractère tout particulier qui n'appartient certainement pas avec le même degré d'intensité et de fréquence à la plupart des autres volcans de la surface du globe.' See *GSA*:239 for Darwin's use of the citation.

138 Investigate with greater care. vegetation & climate of Tristan D. Acuneha. Kerguelen Land. Prince Edwards Is^ld^. Marion & Crozet. L. Auckland. Macqueries.—Sandwich Is^d^—

Specimens of rocks were brought home in Capt. Forster expedition from ‹Deception Isl^d.^› South Shetland Cape Possession. Syenite ? Andite?—[1]

139 Degrading of inland bays. like St. Julian & Port Desire applicable to Craters of Elevation.—The longer diameter of Deception Is^l^ is six Geographical miles and width 2 & ½ miles[1]

S. Shetland. Lat. 62° 55′. ‹only› one lichen. only production. a body which had long been buried, ‹see› from rotten state of coffin «buried in a mound» long consigned to the earth. yet body had scarcely undergone any decomposition: countenance so well preserved. that it was thought not to have belonged to an Englishman.—On 8th of March cove began to freeze. correspond to September[2]

140e ?Did I make any observations on springs at S. Cruz.???—

Form of land shows subsidence in T. del Fuego, and connection of quadrupeds.—although recent elevation, there may have been great subsidence previously. Mem. pebbles of Porphyry.—Falklands.—off East Coast.—Capt. Cook found soundings. (end of 2^d^ voyage outside coast of T. del Fuego. off. Christmas sound.—[1]

---

138 in] 'in' *over* 'by'.
page crossed pencil.
139 ‹see›] *uncertain reading.*
Degrading . . . miles] *crossed pencil.*
S. Shetland . . . September] *crossed.*
140 ?Did . . . sound.—] *ink.*
«(Think . . . 24»] *pencil.*
page crossed pencil.

138–1 See Kendal 1832:64, 'Possession Cape is situated in 63° 46′ S., and 61° 45′ W. We procured specimens of its rock. . . .' Also p. 63 where the land is described as being composed 'of a collection of needle-like pinnacles of sienite.' Henry Foster commanded the *Chanticleer* from 1828–1831, Darwin's misspelling of his name deriving from an identical misspelling in the title of the article cited.
139–1 Darwin's estimate of the dimensions of Deception Island is taken from the map facing p. 64 of Kendal 1832.
139–2 See Kendal 1832:65, 'There was nothing in the shape of vegetation except a small kind of lichen, whose efforts are almost ineffectual to maintain its existence amongst the scanty soil afforded by the penguins' dung.' P. 66, 'Having observed a mound on the hill immediately above this cove, and thinking that something of interest might be deposited there, I opened it; and found a rude coffin, the rotten state of which bespoke its having been long consigned to the earth, but the body had undergone scarcely any decomposition. The legs were doubled up, and it was dressed in the jacket and cap of a sailor, but neither they nor the countenance were similar to those of an Englishman.' Also p. 66, 'We took the hint of the freezing over of the cove, and effected our retreat. . . . We quitted it on the 8th of March. . . .' See *JR*:613.
140–1 Cook 1777, 2. There is, facing p. 177, a full page map of Christmas Sound with numerous soundings included. On p. 200 Cook commented of the entire south-western coast of Tierra del Fuego, 'For to judge of the whole by the parts we have sounded, it is more than probable that there are soundings all along the coast, and for several leagues out to sea. Upon the whole, this is, by no means, the dangerous coast it has been represented.'

«(Think some 60 fathoms, none thicker than thumb» Sea weed said at Kerguelen Is$^{d.}$ to grow on shoals like Fucus giganteus! 24 fathoms deep 24»

141e under 50. Kerguelen Land, = the way it stands gales = very strong. Stones as bigger than a man's head.—

Kerguelen 40 by 20 leagues. dimensions:[1]

Bynoe informs me that in Obstruction Sound, in the narrow parts which break through the N & South lines the tides form eddies with its extreme force.[2] Yet, no outlet at head. Important in forming transverse valleys

---

Ice

142e Sir W. Parish says they have Earthquakes in Cordoba. one of which dried up ‹all› a lake in neigbourhood of town[1]

Mr Murchison insisted strongly. that taking up a piece of Falkland Sandstone. he could not distinguish from stone Caradoc from lower of third Silurian division—Together with same general character of fossils deception complete.=[2]

---

141 under . . . head] *pencil.*
Kerguelen 40 . . . valleys] *ink.*
Ice] *pencil.*
under . . . dimensions] *crossed pencil.*
142 *page crossed pencil.*

141–1 Cook 1784, 1:78–79 records that at Kerguelen Land, 'A prodigious quantity of seaweed grows all over it, which seemed to be the same sort of weed that Mr. Banks distinguished by the name of *fucus giganteus*. Some of this weed is of a most enormous length, though the stem is not much thicker than a man's thumb. I have mentioned, that on some of the shoals upon which it grows, we did not strike ground with a line of twenty-four fathoms. The depth of water, therefore, must have been greater. And as this weed does not grow in a perpendicular direction, but makes a very acute angle with the bottom, and much of it afterwards spreads many fathoms on the surface of the sea, I am well warranted to say, that some of it grows to the length of sixty fathoms and upward.' See *JR*:303–304. Darwin's notebook entry the expression '24' would seem to be a variant of '24'. See note RN16–1.

141–2 Benjamin Bynoe: personal communication.

142–1 Woodbine Parish personal communication. Later published in Parish 1838:242, 'It is related that for many years after its foundation, the inhabitants [of Córdoba] were subjected to much inconvenience from the occasional overflowings of a lake in the neigbouring hills, until an earthquake swallowed up its waters, and drained it apparently forever.'

142–2 Roderick Impey Murchison: personal communication. See Murchison 1839a, chap. 18:216–22 on 'Lower Silurian Rocks.—3rd Formation of "Caradoc Sandstone".' Also p. 583, 'the same forms of crustaceans, mollusks and corals, are said to be found in rocks of the same age, not only in England, Norway, Russia, and various parts of Europe, but also in Southern Africa, and even at the Falkland Islands, the very antipodes of Britain. This fact accords, indeed, with what has been ascertained concerning the wide range of animal remains in deposits equivalent to our oolite and lias; for in the Himalaya Mountains, at Fernando Po, in the region north of the Cape of Good Hope, and in the Run of Cutch and other parts of Hindostan, fossils have been discovered, which, as far as the English naturalists who have seen them can determine, are undistinguishable from certain oolite and lias fossils of Europe.' To this remark Murchison added in a footnote: 'The fossils from the Falkland Islands were discovered by Mr. C. Darwin, and they appear to me to belong to the Lower Silurian Rocks.' See also *JR*:253.

Silliman Journal. year 1835 excellent account of N. American geology. Conybeare[3]

143e Lava in Cordillera & on Eastern plains «by Antuco». Athenæum April 1836 (p 302)[1]

Coleccion de obras. 2 Vols fol: Buenos Ayres 1836:[2] W. Parish?? «by Pedro de Angelis.»[3]

This work is reviewed in present Edinburgh March 1835[4]

Sir W. Parish says. that beds of shells are found on whole coast from P. Indio to Quilmes. & at least seven miles inland.[5]

144e The Cordoba earthquake a very remarkable phenomenon. showing line of disturbance inside Cordillera: It is not therefore so wonderful that

---

volcanic rocks at M. Video «Volcano in Pampas»

Pasto Earthquake. Happened on January 20th. 1834

Mr Sowerby. younger. says that Falkland fossils decidedly belong to old Silurian system.[1]
Apply degradation of landlocked harbors to Craters of elevation.—

143 Lava . . .(p 302)] *crossed pencil.*
Coleccion . . . 1835] *crossed pencil.*
144 Volcano in Pampas] *boxed.*
Apply . . . elevation.—] *circled.*
page crossed pencil.

142–3 In a report to the British Association (Conybeare 1833:396) William Daniel Conybeare expressed a high opinion of *Silliman's Journal* as a source for North American geology. This journal, formally entitled the *American Journal of Science and the Arts*, contained the following articles on North American geology for 1835: (vol. 27) Ducatel and J. H. Alexander, pp. 1–38; Chapin, pp. 104–12; Rogers, pp. 326–35; 'Notice of the Transactions of the Geological Society of Pennsylvania, Part I', pp. 347–55; Shepard, pp. 363–70; (vol. 28) Ball, pp. 1–16; Conrad, pp. 104–11, 280–82; Gebhard, pp. 172–77; S. G. Morton, pp. 276–78; and Totten, pp.347–53.

143–1 From review of Angelis 1836–37 in *Athenæum* 1837:302, 'In many places the large stones which covered the ground were to be cleared away; but the chief obstacles were the cracked streams of lava to be crossed in the Andes, and the numerous banks of rough scoriæ or ashes occurring in the plains as well as the mountains.' Darwin misdated his reference to this review by a year.

143–2 Angelis 1836–37, 6 vols. Darwin's reference was to the first two volumes which were published in 1836.

143–3 Angelis 1836–37. Woodbine Parish would have been a likely owner, and thus a possible lender, of Angelis's work.

143–4 [Cooley] Review of Angelis 1836–37 in *Edinburgh Review* 65, 1837:87–109. The 'March 1835' notation in this entry is puzzling since the date is rather far removed from either the date of publication of Angelis's work or the date of the 'present Edinburgh'.

143–5 Woodbine Parish: personal communication. The distance between Quilmes and Punta Indio is approximately 70 miles (112.63 km). The two points are found along the coastline south of Buenos Aires. See also Parish 1838:168 and *GSA*:2–3.

144–1 James de Carle Sowerby: personal communication. See *JR*:253, 'Mr. Murchison, who has had the kindness to look at my specimens [of fossil shells from the Falkland Islands], says that they have a close general resemblance to those belonging to the lower division of his Silurian system; and Mr. James Sowerby is of [the] opinion that some of the species are identical.' See also Morris and Sharpe 1846.

145e Lyell suggested to me that no metals in Polynesian Islds—.[1] Volcanic plenty in S.America!! Metamorphic

---

146 Volcanos only *burst* out where strata in act of dislocation (NB. dislocation connected with fluidity of rock ∴ «in earliest stage» when covered up beneath ocean).—The first dislocations & eruptions can only happen during first movements, and therefore beneath ocean, for subsequently there is a coating of solidifying igneous rocks which would be too thick to be penetrated by the repeted trifeling injections.—Old vents would keep open long after emersion, but improbably so long, that to be surrounded by continent.—change of volcanic focus.—

147 ‹it is certain, if strata can be›

Problem dislocate strata without ejection of the fluid propelling mass.

If one inch can be raised then all can, for fresh layers of igneous rock replace strata. & it is nothing odd to find them injected by veins & mass[es] [Fig. 8]

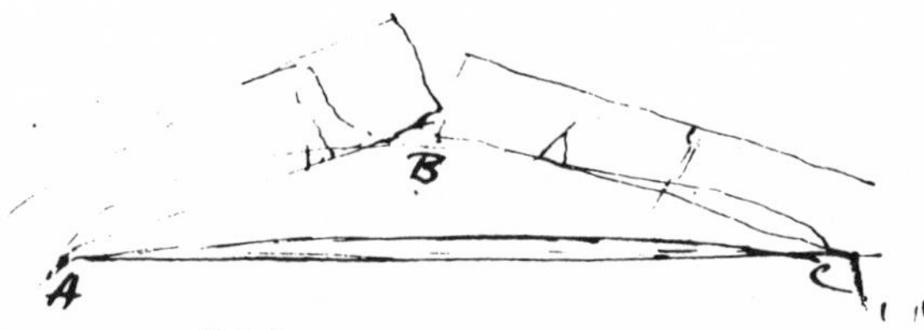

(A.B.C, now grown solid.)

148 Red Sea near Kosir, land appears elevated. Geograph. Journal p 202 Vol IV[1]

When recollecting Gulf of California. Beagle Channel.—One need never be afraid of speculating on the sea

---

145 *page crossed pencil.*
146 *page crossed pencil.*
147 mass[es]] *reads* 'massse'.
[Fig. 8]] *figure and* '(A, B, C . . . solid.)' *appears on unnumbered page facing page 147.*
page crossed pencil.
148 Red . . . IV] *ink.*
When . . . sea] *pencil.*
page crossed pencil.

145–1 Personal communication.
148–1 J. Bird 1834:192–206. The passage from which Darwin inferred that the land in question had been elevated is the following (p. 202), 'The town of old Kosír is situated on the north side of an inlet of the sea, which formerly extended westward into the land about a mile, but is now crossed by a bar of sand, that prevents the ingress of the water into the former channel. . . .' The sea appears to have gradually retired from the land, and left a considerable beach between its present limits and the base of the mountains westward.'

149 The 24 ft. elevation at Concepcion. from impossibility of such change having taken place unrecorded must be *insensible.*

Quantity of matter from Cordillera. HORIZONTAL movement of fluid matter not (for instance) expansion of solid matter by Heat

150 Consider profoundly the *sandstone* of the Portillo line.—connected with ‹*gneiss*›.—(*Mica Slate*) [Fig. 9]

((3) like Bell of Quillota.) (A) in this strata may be older than (B).–– Most important view Urge curious fact felspar melted gneiss/// QUARTZ!!! Analogous to Von Buch. Basalt where Basalt. trachyte where trachyte.[1]

151 There must have been as much conglomerate on West of *Peuquenes* as on East.

Where gone to.?—

There must have been some conglomerate East of Portillo

Where gone to? *Intermediate* space *protected.*—

Oh the vast power of the ocean!

152 Make a *grand* analogy between Wealden & Bolivia

Transportal of conglomerate between two ranges mysterious!—

Mem. SUBSIDENCE Uspallata of which no trace except by trees

---

149 Quantity . . . Heat] *scored right margin.*
page crossed pencil.

150 [Fig. 9]] *labels read from left to right* 'gneiss', 'Porphory', 'gneiss', Porph' *with a* '(3)' *underneath* 'Porph'.
Most . . . view] *circled.*
Urge . . . QUARTZ!!!] *circled, marked connection with* 'Analogous . . . trachyte.'
page crossed pencil.

151 *page crossed pencil.*

152 *page crossed pencil.*

150–1 Citing Leopold von Buch, among others, Darwin wrote (*VI*:121), 'The separation of the ingredients of a mass of lava would, perhaps, sometimes take place within the body of a volcanic mountain, if lofty and of great dimensions, instead of within the underground focus; in which case, trachytic streams might be poured forth, almost contemporaneously, or at short recurrent intervals, from its summit, and basaltic streams from its base: this seems to have taken place at Teneriffe.' To this last point Darwin added a footnote: 'Consult von Buch's well-known and admirable *Description Physique* of this island [Teneriffe], which might serve as a model of descriptive geology.' See Buch 1836:153–228.

153 The structure of ice in columns. show that granite when weathering into balls. must exhibit orbicular structure.—When we recollect connection of columnar & orbicular in basalt.—

When we see Avestruz two species. certainly different. not insensible change.—[1] Yet one is urged to look to common parent? why should two of the most closely allied species occur in same country? In botany instances diametrically opposite have been instanced: it is

154 Let it not be overlooked that except by trees, I could not see trace of Subsidence at Uspallata.—

¶If crust very thick would there be undulation? would it not be mere vibration? but walls & feeling shows undulation ∴ crust thin.—Concepcion earthquake

155 Draw close Analogy Lake of Cordill: of Copiápò & Desaguadero.—*three* ridges in Copiapo, as well as in latter.—

According to Mr Brown,[1] a person (whom I met at S.W.P.) the Cordillera extend to near Salta. & not far from Tucama[n]. & at Chuquisaca. half across the continent.—He states plains of Mendoza smooth. Sir W.P. states that in Helm's travels accounts of travelled boulders. from the Cordovise range.[2] Signor Rozales tells me at seven oclock Novem ‹5th› Concepcion most violently shaken, by earthquake. but no serious injury.—[3]

153 The . . . basalt] *crossed pencil.*
When . . . it is] *crossed pencil.*
154 *page crossed pencil.*
155 *page crossed pencil.*

153–1 The 'Avestruz' was the local name for the rhea. See notes RN127–3 and RN130–1.

155–1 'Mr Brown' was obviously a guest with Darwin at the house of Sir Woodbine Parish (S.W.P.). He may have been William Brown (1777–1857), an admiral in the navy of Buenos Aires, a native of Ireland, and the only Brown mentioned in Shuttleworth 1910. See also Mulhall 1878:166 for information which places Brown in Ireland in 1836 and therefore plausibly in London in 1837.

155–2 Woodbine Parish referring to Helms 1807: personal communication. Parish considered Tucumán to lie in the upper parts of the Sierra de Córdoba, a low range of pampean mountains (Parish 1838:254). On the subject of the travelled boulders, see Helms 1807:45, 'It in a particular manner excited my astonishment here, to find the highest snow-capt mountains within nine miles from Potosi, covered with a pretty thick stratum of granitic stones, rounded by the action of water. How could these masses of granite be deposited here, as there is a continual descent to Tucuman, where the granitic ridge ends, and from Tucuman to Potosi it consists of simple argillaceous shistus? Have they been rolled hither by a general deluge, or some later partial revolution of nature?' Darwin quoted from this passage in the *JR* (p. 290), and added, 'He [Helms] supposes they [the boulders] must have come from Tucuman, which is several hundred miles distant: yet at p. 55 he says, at Iocalla (a few leagues only from Potosi), "a mass of granite many miles in length, rises in huge weatherbeaten rocks:" the whole account is to me quite unintelligible.'

155–3 'Signor Rozales' would also seem to have been a guest with Darwin at the house of Sir Woodbine Parish. Possibly he was Francisco Javier Rosales (d. 1875), Chilean chargé d'affaires to Paris from 1836–1853. Another member of the family probably in Europe at the time was Vicente Pérez Rosales (1807–1886).

156 ‹Analysis of Atacama. Iron in Edinburgh. Phisoph. Transactions. = Mem: Olivine. Volcanic product.=›[1]

———

‹Did Peruvian Indians use arrows or Araucanians?—›
If wood now preserved over world Dicotyledones far preponderant, if so coniferous must formerly have been most abundant tree—

———

Metamorphic action: ‹most› coming so near surface most important

157e There is map of Cordillera by Humboldt in Geolog. Society[1]

———

Sir Woodbine Parish informs me that town near Tucuman and Salta. towards the Vermejo was utterly overthrown by earthquake with great destruction of human life.—[2]Temple mentions some earthquake at Cordova.— There the Cordova earthquake

158e in which lake was absorbed.—[1]Earthquakes felt. different case from shore of Pacific.—Isabelle's volcano, many amygdaloids.—[2] Boussingualt «(Lyell)» cracks mountains falling in.—[3] Earthquakes at Quito. tranquillity «at Mendoza» exception.—«formerly perhaps otherwise» Mendoza never overthrown,—no mountains

156 If . . . important] *crossed pencil.*
157 *page crossed pencil.*
158 *page crossed pencil.*

156–1 Edward Turner as quoted in Allan 1831:226, 'Externally it [the specimen] has all the characters of meteoric iron. The metal in the specimen is tough, of a whiter colour than common iron, and is covered on most parts with a thin film of the oxide of iron. The interstices contain olivine.' The proportions of iron, nickel and cobalt in the specimen are given on p. 228. See also Parish 1838:257–63.

157–1 In a catalogue of its library the Geological Society listed an 1831 contour map of the Cordillera of the Andes done by Humboldt. (Geological Society of London 1846:128) This would appear to be the same map as Plate 5 in Humboldt 1814–34.

157–2 Personal communication: the information was not repeated in Parish 1838.

158–1 Temple 1830, 1:116, '[January] 19th [1826], when about to rise with the sun, as was our custom, we suddenly felt ourselves shaken in our beds . . .' Temple's description of his route (p. 109) places him in the province of Santiago del Estero, just over the border of the province of Córdoba, at the time when the earthquake occurred. For the account of an earthquake in Córdoba causing the disappearance of a lake see note RN142–1.

158–2 Isabelle 1835:454–55, 'Au nord-est du *passo*, à distance de quatre à cinq lieues, est une montagne boisée, appelée *Serra do Butucarahy* . . . Je suis porté à croire que cette montagne est volcanique, parce que les *moradores* du lieu m'ont assuré avoir entendu des détonations très fortes dans son intérieur; ils prétendent encore qu'il y a un lac à la cime, dont les eaux, en filtrant ou en débordant, produisent des éboulemens qui mettent à nu la roche qu'elle semble avoir pour noyau; aussi la partie supérieure est—elle devenue inaccessible à cause de sa dénudation. Après les grandes pluies d'orage, et pendant les gelées, l'eau se trouvant dans les fissures du rocher en détache des fragmens qui tombent avec fracas; sa grande hauteur, ou plutôt son isolement attire le tonnerre, ce qui fait que cette montagne est souvent foudroyée.' The hill described is probably Coxilha which lies to the northeast of the Rio Botucaraí [30° 0′ S., 52° 46′ W.] in Brazil. It is not an active volcano, nor are there any in the area.

158–3 Lyell 1835b, 2:96, 'M. Boussingault declares his belief, that if a full register had been kept of all the convulsions experienced here and in other populous districts of the Andes, it would be found that the trembling of the earth had been incessant. The frequency of the movement, he thinks, is not due to volcanic explosions, but to the continual falling in of masses of rock which have been fractured and upheaved in a solid form at a comparatively recent epoch.' See also Boussingault 1834–35:54–56.

159 Mackenzie has talked of lava flowing up Hill; (what does he mean?)[1] Consult D[r] Holland about bubbles.—[2]

No Volcanic action on coast line of Old Greenland, close to W of Jan Meyen Is[ld].—M[r] Barrow[3] thinks N & S. line connects western isles of Scotland & Iceland.—**Bosh** nor on Norway, or Spitzbergen.—Spitzbergen animals (?). ≠

160 The Hollowness of ‹sep› Chiloe concretions somewhat analogous to septa.—would particle attracted towards space tend to form ring. [Fig. 10]

motion from within and without

H. Kingdom N. Spa. Vol III p. 113
"Nature exhibited to the Mexicans enormous masses of Iron and Nickel, & these masses which are scattered over the surface of the ground are fibrous. malleable & of so great tenacity, that it is with difficulty that a few fragments can be separated from them with steel instruments."[1]

159 **Bosh**] *added pencil in left margin.*
page crossed pencil.
160 *page crossed pencil.*

159–1 Barrow, Jr. 1835:224. 'This supposition [of lava blistering, see note 100] would appear to afford a better solution of the difficult problem of accounting for those blocks of lava that are perched on high ridges, than that given by Sir George Mackenzie, who imagines this lava to have flowed from the lower ground, and calls it the "ascending lava". He says—"It is caused by the formation of a crust on the coating of the surface, and a case or tube being thus produced, the lava runs in the same manner as water in a pipe".' The quotation is from Mackenzie 1811:108.

159–2 Henry Holland. See Barrow, Jr. 1835:223, 'Dr. Holland, in his account of the Mineralogy of Iceland, seems to countenance the opinion of these masses having been thrown up on the very spot they occupy, observing there was one formation of lava which had every appearance of not having flowed. Speaking of these masses of lava, he says:—"It was heaved up into large bubbles or blisters, some of which were round, and from a few feet to forty or fifty in diameter; others were long, some straight, and some waved. A great many of these bubbles had burst open, and displayed caverns of considerable depth".' However, this description, which Barrow attributed to Holland, is rather to be found in Mackenzie 1811:390, chapter entitled 'Mineralogy'. See also *VI*:95–96, and 103.

159–3 Barrow, Jr. 1835:276–77, 'Here, then, we have the plain and undeniable evidence of subterranean or sub-marine fire, exerting its influence under the sea, almost in a direct line, to the extent of 16½ degrees of latitude, or more than 1100 statute miles. If we are to suppose that one and the same efficient cause has been exerted in heaving up this extended line of igneous formations, from Fairhead to Jan Meyen, we may form some vague notion how deep-seated the fiery focus must be to impart its force, perhaps through numerous apertures, in a line of so great an extent, and nearly in the same direction. It may probably be considered the more remarkable, that no indication whatever is found of volcanic fire on the coast-line of Old Greenland, close to the westward of the last-mentioned island, and also to Iceland, nor on that of Norway on the opposite side, nor on that of Spitzbergen; on these places all is granite, porphyry, gneiss, mica-slate, clay-slate, lime, marble, and sandstone.'

160–1 Humboldt 1811, 3:113.

161 In R. Brown (Collect: «of F. W.»)[1] where the stalactiform masses have layers been accumulated, round knobs, or pushed where soft, or redissolved soft.—/ is there any flexure ‹fr› in the fragmentary jasper.—do undulations (as Hutton says)[2] always come from without.—

"True native iron that to which we cannot attribute a meteoric origin & which is constantly found mixed with lead & copper is infinitely rare in all parts of the globe". p. 113[3]

162 How utterly incomprehensible that if meteoric stones simply pitched from mo«o»n, that the metals should be those which have magnetic properties.

Study well products of Solfataras[.] some general laws. association of lead & silver. Sulp. of Barytes: Fluoric. Barytes:—

163e Humboldt. New Spain. Vol III. p. 130[1]
Metals in Mexico rarely in secondary alway in primitive & transition; the latter rarely appear in central Cordillera. particularly between 18° & 22° N. = formations of amph: porphyry. greenstone[,] amygdaloid. basalt & other trap cover it to great thickness. = Coast of Acapulco granitic rock.—in parts of table granits & gneiss with gold veins visible:—"Porphyries of Mexico may be considered for most parts as rock eminently rich in mines of gold & silver." «p. 131»[2]

164e The above porphyries characterized by no quartz & amphibole frequently only vitreous felspar: = gold veins in a phonolitic porphyry. = several parts of N. Spain great analogy to Hungary. = Veins of Zimapan offer zeolite. stilbite. grammalite. pyenite. native sulphur.. fluor spar. bayte. asbestos

161 redissolved] *uncertain reading.*
"True . . . p. 113] *continued from entry on previous page.*
In . . . without.—] *crossed.*
"True . . . p. 113] *crossed pencil.*
162 *page crossed pencil.*

161–1 Robert Brown who supervised the botanical collections of the British Museum, assembled a valuable collection of fossil woods ('F.W.') which he bequeathed to the Museum.
161–2 Darwin was referring here to the opinion of James Hutton respecting the formation of fossil wood. In Hutton's view 'undulations' in silicified fossil wood would be traced to the action of exterior heat and pressure. See Playfair 1802:24–25.
161–3 Humboldt 1811, 3:113.
163–1 Humboldt 1811, 3:129–30, 'The Mexican veins are to be found for the most part in *primitive* and *transition* rocks . . . and rarely in the rocks of *secondary* formation . . . In the old continent *granite, gneiss* and *micaceous slate* (*glimmer-schiefer*) constitute the crest of high chains of mountains. But these rocks seldom appear outwardly on the ridge of the Cordilleras of America, particularly in the central part contained between the 18° and 22° of north latitude. Beds of amphibolic porphyry, greenstone, amygdaloid, basalt and other trap formations of an enormous thickness cover the granite and conceal it from the geologist. The coast of Acapulco is formed of granite rock. . . . Farther to the east in the province of Oaxaca the granite and gneiss are visible in table lands of considerable extent traversed by veins of gold.'
163–2 Humboldt 1811, 3:131.

garnets.—carb & chrom. of lead. orpiment. chrysop[r]ase. opal:—[1]

Veins in Limestone & Grauwacke: Silver appears far more abundant in the upper limestone, which H. calls by several secondary names[2]

165e «Study Hoffmans account of steam acting on trachytes. also Azores. We here have case of such vapours washing a rock[1]» Veins concretionary; concretions ‹dt› determined by fissures as in septaria. (& Chiloe case, at least corelation)—Galapagos vein. vein of secretion.—metallic veins follow mountain chain. there after NW ‹W›.—

«same chemical laws as in concretions perhaps makes intersections richest—Humboldt has urged phenomena in veins, chemical affinities like in composed rock.[2] granites syenite» «strangling &c of veins can only be accounted for by concretionary action, conjoined with other» «(state simplest case. concretions of clay iron stone; iron pyrite in a fossil» Insist strongly on the grand fact of Volcanic & non Volcanic. Then Solfataras. «Mem: Micaceous iron ore.»

N.B. To show how metals may be transported by complicated chemical law & steam of salts, quite curious case of oxided Iron by Mitterschlich. Vol. II Journal of Nat. & Geograph Siciences?—[3]

165 same . . . syenite] *over preceeding rule line.*
Mem: . . . ore.] *circled.*

164–1 Humboldt 1811, 3:131–32, 'They [the Mexican porphyries] are all characterized by the constant presence of amphibole and the absence of quartz . . . We frequently discover only *vitreous felspar* in the porphyries of Spanish America. The rock which is intersected by the rich gold vein of Villalpando near Guanaxuato is a porphyry of which the basis is somewhat a kin to *klingstein* (*phonolite*), and in which amphibole is extremely rare. Several of these parts of New Spain bear a great analogy to the problematical rocks of Hungary . . . These veins of Zimapan offer to oryctognostic collections a great variety of interesting minerals such as the fibrous zeolith, the stilbite, the grammalite, the pyenite, native sulphur, spar fluor, baryte suberiform asbestos, green grenats, carbonate and chromate of lead, orpiment, chrysoprase, and a new species of opal . . .'

164–2 Humboldt 1811, 3:133–34, 'The alpine lime-stone and the jura lime-stone (*jurakalkstein*) contain the celebrated silver mines of Tasco and Teuilotepec in the intendancy of Mexico; and it is in these calcareous rocks that the numerous veins which in this country have been very early wrought, display the greatest wealth. . . . The result of this general view of the metalliferous depositories (erzführende lagerstätte) is that the cordilleras of Mexico contain veins in a great variety of rocks, and that those rocks which at present furnish almost the whole silver annually exported from Vera Cruz, are the *primitive slate*, the *grauwakke*, and the *alpine lime-stone*, intersected by the *principal veins* of Guanaxuato, Zacatecas and Catorce.'

165–1 Hoffmann 1838, the section 'Dämpfe verändern die vulkanischen Gesteine':480–81.

165–2 Humboldt 1811, 3:128, 'How can he [the naturalist] draw general results from the observation of a multitude of small phenomena [regarding metalliferous deposits], modified by causes of a purely local nature, and appearing to be the effects of an action of chemical affinities, circumscribed to a very narrow space?'

165–3 Mitscherlich 1830:302, 'As to the oxide of iron, its history will be best understood by an experiment or two. If a mixture of salt, oxide of iron, and silica, be heated to redness in a tube, and water in vapour be passed over it, much muriatic acid is formed, but very little chloride of iron, and crystallized oxide of iron will be found in the mass: but if muriatic acid be brought in contact with ignited oxide of iron, water and chloride of iron are formed, and sublime; if the chloride of iron come in contact with more water, muriatic acid is first developed, then chloride of iron, and a residue of crystallized oxide or iron remains. The formation of chloride of iron by the action of muriatic acid upon oxide of iron appears, therefore, to depend upon the proportion of water present. M. Mitscherlich applies these experiments and principles

166e H. says in Potosi the silver is contained in a primitive slate, covered by a clayey porphyry, containing grenats. In Peru. on other hand, mine of Gualgayoc or Chota & Pasco in "alpine limestone" = "The wealth of the veins in most part totally independent of the nature of the beds they intersect". = In the Guatemala part. (& Chiloe do) no veins discovered. Humboldt suggests covered up by volcanic rocks.[1]

167e //St Helena has been slightly broken up, & has there not been vein «of iron» discovered?—

Klaproth analysed silver ores from Peru consisted of native silver & brown oxide of Iron in Mexico. sulphuretted silver, arsenical grey copper, and antimony, horn silver, black silver & red silver, do not name native silver because not very abundant.—ǂ muriated silver. which is so rare in Europe. common there accompanied by molybdated lead & «argentiferous lead»; sulfated Barytes very «un»common in Mexico. Fluor spar only in certain mines.[1]

166 166] *ink over pencil*
beds they intersect] 'they' *over* 'that', 'intersect' *over* 'intersects'.
167 argentiferous lead] *circled.*

in explanation of the manner in which volcanic crystallized oxide of iron is formed—all the conditions necessary, according to the above view, being present in those cases, where heretofore it had been supposed the oxide of iron, as such, had been actually sublimed.'

166–1 Humboldt 1811, 3:134, 'Thus it is in a *primitive slate* (*ur-thon schiefer*) on which a clayey porphyry containing grenats reposes, that the wealth of *Potosi* in the kingdom of Buenos-Ayres is contained. On the other hand, in Peru the mines of Gualgayoc or Chota and that of Yauricocha or Pasco which together yield annually double the quantity of all the German mines, are found in an *apline lime-stone*. . . . The wealth of the veins is for the most part totally independent of the nature of the beds which they intersect.' And pp. 142–43, 'The province of Quito, and the Eastern part of the kingdom of New Granada . . . the Isthmus of Panama, and the mountains of Guatimala, contain for a length of 600 leagues, vast extents of ground in which no vein has hitherto been wrought with any degree of success. It would not, however, be accurate to advance that these countries which have in a degree, been convulsed with volcanos are entirely destitute of gold and silver ore. Numerous metalliferous depositories may be concealed by the superposition of strata . . .'

167–1 Humboldt 1811, 3:152–53, 'In Peru, the greatest part of the silver extracted from the bowels of the earth is furnished by the *pacos*, a sort of ores of an earthy appearance, which M. [Martin Heinrich] Klaproth was so good as to analyse at my request, and which consist of a mixture of almost imperceptible parcels of native silver, with the brown oxyde of iron. In Mexico on the other hand, the greatest quantity of silver annually brought into circulation, is derived from those *ores* which the Saxon miner calls by the name of *dürre erze* especially from *sulfuretted silver*, (or vitrous *glaserz*) from *arsenical greycopper* (*fahlerz*) and *antimony*, (*grau* or *schwarzgiltigerz*) from *muriated silver*, (*hornerz*) from *prismatic black silver*, (*spödglaserz*), and from red silver (*rothgiltigez*). We do not name native silver among these ores, because it is not found in sufficient abundance to admit of any very considerable part of the total produce of the mines of New Spain being attributed to it.' Also p. 154, 'The muriated silver which is so seldom found in the veins of Europe, is very abundant in the mines of Catorce, Fresnillo, and the Cerro San Pedro, near the town of San Luis Potosi. . . . In the veins of Catorce, the muriated silver is accompanied with molybated lead, (*gelb-blei-erz*) and phosphated lead (*grünblei-erz*).' And p. 155, 'The true mine of *white silver* (weissgiltigerz) is very rare in Mexico. Its variety *greyish white*, very rich in lead, is to be found however in the intendancy of Sonora, in the veins of Cosala, where it is accompanied with argentiferous *galena*, red silver, brown blende, quartz and sulfated barytes. This last substance . . . is very uncommon among the *gangues* of Mexico . . . Spar-fluor has been only found hitherto in the veins of Lomo del Toro, near Zimapan, at Bolaños and Guadalcazar, near Catorce.'

168e «Vol. III» "In general it is observed both in Mexico & Peru, that those oxidated masses of iron. which contain silver are peculiar to that part of the veins, nearest to the surface of the earth."—p. 156.[1] Mines of Batopilas in New Biscay, "Nature, exhibits the same minerals ‹as› there, that are found in the veins of Kongsberg in Norway.—namely dendritic silver intersecting carbonate of lime—[2] native silver in Mexico

169e is always accompanied by Sulp. silver sometimes by selenite.—[1] in New Spain, contrary to Europe. *argentiferous* lead not abundant. = [2] considerable quantity of silver procured from martial pyrites; great blocks of pure silver not common in ‹S.› America: In all climates distribution of silver «in veins» very unequal sometimes disseminated ‹[. . .]› sometimes concentrated: wonderful quantity of pure silver in S. America.[3]

170e Geology of Guanuaxuato.—Clay slate. passing into talcose & chloritic slate. with beds of syenite & ‹sep› serpentine dipping to SW at 45° to 50°—covered by *conformable* greenstone porphyrys & phonolites do. amphibole quartz & mica very rare.—[1] ancient freestone & breccia is the same with

169 ‹[. . .]› *4 or 5 letters illeg.*
170 ‹sep›] *uncertain reading.*

168–1 Humboldt 1811, 3:156.
168–2 Humboldt 1811, 3:157, '*Native Silver* . . . has been found in considerable masses, sometimes weighing more than 200 killogrammes in the seams of Batopilas in New Biscay. . . . Nature exhibits the same minerals there, that are found in the vein of Kongsberg in Norway. Those of Batopilas contain filiform dendritic and silver, which intersects with that of carbonated lime.'
169–1 Humboldt 1811, 3:157–58, 'Native silver is constantly accompanied by *glaserz* [sulfuretted silver] in the seams of Mexico, as well as in those of the mountains of Europe . . . From time to time small branches, or cylindrical filaments of native silver, are also discovered in the celebrated vein of Guanaxuato; but these masses have never been so considerable as those which were formerly drawn from the mine *del Encino* near Pachuca and Tasco, where native silver is sometimes contained in folia of selenite.'
169–2 Humboldt 1811, 3:158, 'A great part of the silver annually produced in Europe, is derived from the *argentiferous sulfuretted lead* (*silberhaltiger bleiglanz*) which is sometimes found in the veins which intersect *primitive and transition mountains*, and sometimes on particular *beds* (erzflöze) in rocks of *secondary formation*. In the kingdom of New Spain, the greatest part of the veins contain very little argentiferous galena; but there are very few mines in which lead ore is a particular object of their operations.'
169–3 Humboldt 1811, 3:159, 'A very considerable quantity of silver is produced from the smelting of the martial pyrites (*gemeine schwefelkiese*) of which New Spain sometimes exhibits varieties richer than the *glaserz* itself . . . It is a very common prejudice in Europe, that great masses of native silver are extremely common in Mexico and Peru . . .' Also pp. 160–61, 'It appears that at the formation of veins in every climate, the distribution of silver has been very unequal; sometimes concentrated in one point, and at other times disseminated in the *gangue*, and allied with other metals.' And p. 162, 'Although the New Continent, however, has not hitherto exhibited native silver in such considerable blocks as the Old, this metal is found more abundantly in a state of perfect purity in Peru and Mexico, than in any other quarter of the globe.'
170–1 Humboldt 1811, 3:176, 'What is the position of the rock which crosses the veins of Guanaxuato? . . . The most ancient rock known in the district of Guanaxuato, is the *clay slate* (*thon schiefer*). . . . It is of an ash-grey or greyish-black frequently intersected by an infinity of small quartz veins, which frequently pass into talk-state [sic] (*talk schiefer*) and into *schistous chlorite*.' Also, pp. 177–78, 'On digging the great pit (*trio general*) of Valenciana, they discovered banks of *syenite* of *Hornblend slate* (*Hornblend schiefer*) and true serpentine, altering with one another, and forming *subordinate beds*, in the *clay slate*. . . . These strata [of clay slate] are very regularly *directed* h. 8 to 9 of the miner's compass; they are inclined from 45 to 50 degrees to the south west. . . . Two very different formations repose on the *clay slate*: the one of porphyry . . . and the other, of old *freestone* in the ravins, and table lands of small elevation.' And pp. 179–80, 'This porphyry . . . is generally of a greenish colour. . . . The most recent [beds] . . . contain vitreous felspar, inchased in a mass, which

that on surface of plains of Amazon, no relation—there is more modern breccia, chiefly owing to destruction of porphyries. whereas other to ancient rock.—this N° 2. superimposed on N° 1. even No. 2. might be mistaken for Porphyry

171e above ancient freestone, limestone & ‹many› ‹‹other secondary›› rocks.[1]

Vein traverses both Clay slate, Porphyry North 52 W, & is nearly the same with that of the veta grande of Zacatecas, & veins of Tasco & Moran—of Guanaxuato to SW. with respect to latter doubts whether bed or vein (very like that of Spital of Schemnitz in Hungary.) Humboldt says fragments from roof & penetrating overlying beds tells the secret.—[2] p. 189. "The small ravins into which the valley of Marfil is divided, appear to have a decided influence on the richness of the veta madre of

170e [misnumbered page]

172e D[r] D. remarks. bad conductor of Heat do of Electricity[1]
Does not iron, combined with nickel & cobalt (meteoric) resist oxidation?—
Mem Sir W. P. stone[2] It is clear to me, there are laws of solution & deposition under great pressure. (? heat!) unknown to us.
ǂ M. Chladni.—on meteoric Mexican stone. Journal des Mines 1809. No. 151. p. 79.[3]

171 veta madre of] *passage continued on page 175.*
172 D[r].D . . . Electricity] *pencil.*
Does . . . p. 79.] *ink, crossed pencil.*
(?heat!)] *uncertain reading.*
Page number not in Darwin's handwriting.

sometimes passes into the petrosilex jadien, and sometimes into the pholonite [sic] or *klingstein* of Werner. . . . All the porphyries of the district of Guanaxuato possess this in common, that amphibole is almost as rare in them as quartz and mica.'

171-1 On freestone and breccia at Guanaxuato see Humboldt 1811 3:180–83 including the passage on pp. 182–83, 'The most experienced mineralogist, after examining the position of the *lozero* [agglomeration] of Guanaxuato, would be tempted to take it at first view, for a porphyry with clayey base, or for a porphyritic brescia (*trümmer-porphyr*). . . . These formations of old *freestone* of Guanaxuato, serve as bases to other secondary beds, which in their *position*, that is to say in *the order of their superposition*, exhibit the greatest analogy with the secondary rocks of central Europe. In the plains of Temascatio . . . there is a compact limestone . . .'

171-2 Humboldt 1811, 3:185, 'The vein (*veta madre*) [of Guanaxuato] traverses both clay slate and porphyry. In both of these rocks, very considerable wealth has been found. Its mean direction is . . . [N. 52° W.] and is nearly the same with that of the *veta grande* of Zacatecas, and of the veins of Tasco and Moran, which are all western veins (spathgänge). The inclination of the vein of Guanaxuato, is 45 or 48 degrees to the south west.' Also pp. 186–87, 'The *veta madre* of Guanaxuato, bears a good deal of resemblance to the celebrated vein of *Spital* of Schemnitz, in Hungary. The European miners who have had occasion to examine both these *depositories* of minerals, have been in doubt whether to consider them as true veins, or as *metalliferous beds* (*erzlager*). . . . If the *veta madre* was really a *bed*, we should not find *angular fragments* of its *roof* contained in its *mass*, as we generally observe on points where the *roof* is a *slate* charged with *carbone*, and the wall a talc slate. In a vein, the *roof* and the *wall* are deemed anterior to the formation of the *crevice*, and to the minerals which have successfully filled it; but a *bed* has undoubtedly pre-existed to the *strata* of the rock which compose its *roof*. [Hence] we may discover in a bed fragments of the *wall*, but never pieces detached from the *roof*.'

172-1 Erasmus Darwin 1791 (1790), 1:16, 'The air, like all other bad conductors of electricity, is known to be a bad conductor of heat. . . .' See also p. 10 on the subject of shooting stars and fireballs and pp. 1–5 of 'Additional Notes' following p. 214 for a discussion of meteors.

172-2 See Parish 1834:53–54.

172-3 Chladni 1809:79–80.

[misnumbering, no page 173]

174e Under name of Sagitta Triptera D'Orbigny has figured animal with setæ like my undescribed[.] p. 140. Flèche of Quoy et Gaimard.—D'Orbigny has described it with care to 3 species. I think I have much additional information[1]

175e Guanaxuato, which has yielded the most metal, where the direction of ravins, and the slope of the mountains (flaqueza del cerro) have been parallel to the direction & inclination of the vein".—[1]

---

at Zacatecas the veta grande has same direction as Guanax.—the other E & W.—veins richest not in ravins or along gentle slopes. but on the most elevated summits, where mountains most torn.—(?anticlinal line?).—[2]

Mines of Catorce «(Principal veins)» 25° to 30° to NE. vein of Moran 84° NE. of Real del Monte 85° to S. // Tasco 40° to NW (*afterwards said to be «all with some exception» directed NW & SE*).[3]

176e «Vol III» Mexican Cordillera "immense variety of Porphyries which are destitute of quartz, & wh abound both in hornblend & vitreous felspar".—p. 215[1]

Same metal in Tasco vein in Mica Slate & overlying Limestone[2]

---

174 *page crossed pencil.*
175 Guanaxuato . . . vein".—] *passage continued from page 171.*

174-1 A. D. d'Orbigny 1835–47, 5, pt 3:140–44 and plate 10. According to a typewritten list compiled in 1933 by Charles Davies Sherborn of the British Museum (Natural History), the section which includes pp. 140–44 was published in 1835 and plate 10 in 1834. The three species described by d'Orbigny were *Sagitta triptera, Sagitta exaptera*, and *Sagitta diptera*. In this entry Darwin was noting the similarity of one of his unidentified specimens to *Sagitta triptera*. The genus *Sagitta* or 'Flèche' had been established by Quoy and Gaimard 1827:232–33. Presumably Darwin's 'additional information' on the genus appeared in Darwin 1844a. See also ZEd5.
175-1 Humboldt 1811, 3:189.
175-2 Humboldt 1811, 3:205, 'The *veta grande*, or principal vein [at Zacatecas], has the same direction as the *veta madre* of Guanaxuato; the others are generally in a direction from east to west.' And p. 207, 'This wealth is displayed . . . not in the ravins, and where the veins run along the gentle slope of the mountains, but most frequently on the most elevated summits, on points where the surface appears to have been tumultuously torn. . . .'
175-3 Humboldt 1811, 3:210, 'The greatest number of these veins [at Catorce] are *western* (*spathgänge*); and their inclination is from 25° to 30° towards the north east.' P. 223, '. . . the vein of Moran . . . inclined 84° to the north east . . .' P. 226, 'The oldest rock which appears at the surface in this district of mines [at Tasco], is the primitive slate. . . . Its direction is hor. 3–4; and its inclination 40° to the north-west. . . .' Also p. 227, 'The district of mines of Tasco . . . contains a great number of veins . . . all directed from the north-west to the south-east, hor. 7–9.'
176-1 Humboldt 1811, 3:215, 'What relation exists between these last beds [of porphyry], which several distinguished mineralogists consider as volcanic productions, and the porphyries of Pachuca, Real del Monte, and Moran, in which nature has deposited enormous masses of sulfuretted silver and argentiferous pyrites? This problem which is one of the most difficult in geology, will only be resolved when a great number of zealous and intelligent travellers, shall have gone over the Mexican Cordilleras, and carefully studied the immense variety of porphyries which are destitute of quartz, and which abound both in hornblend and vitreous felspar.'
176-2 Humboldt 1811, 3:227, 'These veins [in the mining districts of Tasco and the Real de Tehuilotepec], like those of Catorce, traverse both the limestone and the micaceous slate which serves for its base; and they exhibit the same metals in both rocks.'

Balls of Silver ore occur in do veins.[3] At Huantajaia. Humboldt says, mur of Silv.[,] Sulph. of do.[,]galena[,]quartz, Carb. of Lime. accompany.—Ulloa has said silver in the highest & gold in the lowest. Humboldt states that some of the richest gold mines on ridge of Cordillera near Pataz, also at Gualgayoc. where many petrified shells[4]

177e Bougainville says P 291.—

The Fuegians treat the "chefs d'œuvre de l[']industrie humaine, comme ils traitent les loix de la nature & ses phenomenes."—[1]

Ulloa's Voyage, Shell fish purple die, marevellous statements on, Vol I, P. 168. on coast of Guayaquil, same as Galapagos.[2]

---

no Hydrophobia at Quito. P 281. do do[3]

Australia, C. of Good Hope.—Azores Is^{ds} «nor at St Helena.—»

Humboldt. New Spain Vol. IV. «p. 58» At Acapulco earthquakes are recognized as coming from three directions. from W. NW & S.—last to Seaward[4]

177 Bougainville . . . Is^{ds}] *pencil.*
nor . . . Seaward] *ink.*
page crossed pencil.

176–3 Humboldt 1811, 3:230, 'This formation [of veins, one of four types existing at Tasco and Tehuilotepec] which is the richest of all, displays the remarkable phenomenon, that the minerals the most abundant in silver, form spheroidal balls, from ten to twelve centimetres in diameter. . . .'

176–4 Humboldt 1811, 3:347–48, 'The mines of Huantajaya, surrounded with beds of rock salt are particularly celebrated on account of the great masses of native silver which they contain in a decomposed gangue; and they furnish annually between 70 and 80 thousand marcs of silver. The muriate of conchoidal silver, sulphuretted silver, galena with small grains, quartz, carbonate of lime, accompany the native silver.' Also pp. 348–49 '[Antonio de] Ulloa after travelling over a great part of the Andes, affirms that silver is peculiar to the high table lands of the Cordilleras, called *Punas* or *Paramos*, and that gold on the other hand abounds in the lowest, and consequently warmest regions; but this learned traveller appears to have forgot that in Peru the richest provinces in gold are the *partidos* of Pataz and Huailas, which are on the ridge of the Cordilleras. . . . It [gold] has also been extracted from the right bank of the Rio de Micuipampa, between the Cerro de San Jose, and the plain called by the natives, *Choropampa* or *plain of shells*, on account of an enormous quantity of ostracites, cardium and other petrifications of sea shells contained in the formation of alpine limestone of Gualgayoc.'

177–1 Bougainville 1772, 1:291, 'Ces hommes bruts [the Fuegians] traitoient les chefs-d-œuvre de l'industrie humaine, comme ils traitent les loix de la nature & ses phénomènes.' See *JR*:242.

177–2 Juan and Ulloa 1806, 1:168, 'On the coast [at Guayaquil] . . . is found that exquisite purple, so highly esteemed among the ancients; but the fish from which it was taken, having been either unknown or forgotten, many moderns have imagined the species to be extinct. This colour, however, is found in a species of shell-fish growing on rocks washed by the sea. They are something larger than a nut, and are replete with a juice, probably the blood, which, when expressed, is the true purple; . . . Stuffs died with this purple are also highly valued.' See also ZEd 5.

177–3 Juan and Ulloa 1806, 1:281, 'As the pestilence, whose ravages among the human species in Europe, and other parts, are so dreadful, is unknown both at Quito and throughout all America, so is also the madness in dogs.'

177–4 Humboldt 1811, 4:58, 'It is observed at Acapulco that the shakes take three different directions, sometimes coming from the west by the isthmus [which separates Acapulco from the Bay de la Langosta de la Abra de San Nicolas] . . . sometimes from the north west as if they were from the volcano de Colima, and sometimes coming from

178 partaking of the character of a Araucarian tribe, with point affin of yew & intermediate[1]
Puncture one animal with recent dead body of other. & see if same effects, as with man

---

Does Indian rubber & black lead unite chemically like grease & mercury

---

179 [blank]

180 NB. P. 73. General reflections on the geology of the world
P. 14. } gradual shoaling of coasts
91 }
93 action of sea on coast.

---

27. Bahama Isd[1]

---

181 De Lucs travels[1]
Beauforts Karamania[2]

Capt. Ross.[3] & Scoresby[4] *deep soundings*

---

178 partaking . . . intermediate] *pencil, crossed pencil.*
Puncture . . . mercury] *ink, crossed pencil, written upside down on page.*
180 *page crossed.*
181 De Lucs travels] *crossed.*
Capt. Ross . . . Isd] *crossed.*
Mawes . . . Brazil.—] *crossed.*
Did . . . travels?] *crossed.*
Bellinghausen . . . 1816] *crossed.*

the south. The earthquakes which are felt in the direction of the south are attributed to submarine volcanoes; for they see here that the sea becomes suddenly agitated in a most alarming manner in calm and serene weather when not a breath of wind is blowing.'

178–1 Of the petrified trees he found on the Uspallata range Darwin wrote (*JR*:406), 'Mr. Robert Brown has been kind enough to examine the wood: he says it is coniferous, and that it partakes of the character of the Araucarian tribe (to which the common South Chilian pine belongs), but with some curious points of affinity with the yew.' Also see *GSA*:202 for repetition of the same information. From Darwin's correspondence it is clear that Brown described the specimens of silicified wood sometime during the period from the end of March to mid-May 1837. On 28 March Darwin wrote to J. S. Henslow telling of Brown's general interests in specimens from the *Beagle* voyage; on 10 April Darwin wrote to the English naturalist Leonard Jenyns, 'Tell Henslow, I think my silicified wood has unflintified Mr. Brown's heart'; and on 18 May Darwin wrote to Henslow with Brown's identification of the specimens. For the letters see *Correspondence* 2:13–19.

180–1 Pages refer to the Red Notebook.

181–1 Deluc 1810–11.

181–2 Beaufort 1817.

181–3 John Ross 1819, appendix No. 3, 'Table of Soundings obtained in Davis' Strait and Baffin's Bay'.

181–4 Scoresby, Jr. 1820, 1:184–94, 'Temperature, Depth, and Pressure of the Greenland Sea, with a Description of an Apparatus for bringing up Water from great Depths, and an Account of Experiments made with it'.

Gilbert Farquhar Mathison travels Brazil. Peru. Sandwich Isd[5]

Mawes travels down the Brazil.—[6]

Did Melaspena publish his travels?[7]

Bellinghausen in 1819[8]
Kotzebue 1816[9]

INSIDE BACK COVER

Constant log always additive to convert French Toise into English ft. 0.8058372
French metre into English ft. 0.5159929

	Toises	Pieds		
Myriametre =	5130.,	4.	5 inch	
Kilometre	513.,	0.	5	
Hectometre	51.	1.	10	
Metre		3.	0.	11 lig[nes]
Decimetre			3.	8
Centimetre				4.4

C. Darwin

BACK COVER

R.N.

Range of Sharks
Nothing For any Purpose

IBC 51.1.10] *over* '51.0.10'.
Constant . . . 4.4] *pencil.*
C.Darwin] *ink, upside down from other entries on page.*

BC **Range of Sharks**] *added.*
Nothing For any Purpose] *added.*

181–5 Mathison 1825.
181–6 Mawe 1825.
181–7 Imprisoned upon his return in 1794 from a 5-year circumnavigation of the globe, Alessandro Malaspina did not publish a narrative of the voyage. After his death a narrative of the voyage was published (Malaspina 1885).
181–8 Bellinsgauzen 1831. For an English translation see Debenham 1945.
181–9 Kotzebue 1821.

Notebook A

Transcribed and edited by SANDRA HERBERT

Notebook A (DAR 127, 165 × 100 mm) is bound in tan leather with the border blind embossed: the clasp is missing. The front of the notebook has a label of cream-coloured paper with an 'A.' written in ink, 'Geology' in pencil, and 'Note on *Woolwich*' in ink. Across the cover is the phrase 'Nothing on any Subject' in ink, and the notation by Francis Darwin, 'Geol most O/— FD'. The back cover also has a cream-coloured label on which 'A.' is written in ink. The paper is green edged, feintly ruled, and chain lined. Page 114 bears part of a watermark 'T WA . . .' and page 121 '18 . . .'. There are 90 leaves, or 180 pages, which Darwin numbered consecutively in brown ink. Darwin wholly or partly excised 78 pages, of which all but a portion of one have been recovered. Entries in the notebook are in ink and pencil, grey ink entries running from mid-page 103 to page 127, with an entry on the top of page 131 also in grey ink. A circled '49' in an unknown hand occurs on the inside front cover.

The notebook was begun in mid-1837 and finished towards the end of 1839. The first datable entry, on page 15, is to the August 1837 issue of *L'Institut*. However, Notebook A was probably begun earlier, possibly in June 1837, for the notebook appears to have succeeded the Red Notebook, which it physically resembled and which — like the opening pages of Notebook A — mixed entries on species and geology. It appears to have preceded Notebook B, opened in July, which is on a more specialized topic. Page 41 of the notebook refers to papers published in November and December 1837. Darwin began dating pages of the notebook in August 1838 (see pages 112, 115, 116). A closing date is difficult to determine. Page 140 is dated 25 February [1839], page 141 contains notes to a work probably read about 1 June 1839 (compare A141 to C269 and E166), and page 146 refers to a letter dated July 1839 that had been in Darwin's possession for some time. The important entries on species, A180, were almost certainly made out of sequence and towards the beginning of the period in which the notebook was kept. Annotations in the notebook are of varying dates, one entry on page 33 bearing a date as late as 1877.

As its cover states, Notebook A is devoted to geology. On a first reading one is struck with what is *not* in the notebook. There is little on coral reefs, the geological subject most closely associated with Darwin's name, and while there are some references to fossils, neither paleontology nor the geological record are the leading themes in the notebook. Darwin had presented the outline of his theory on coral reef formation to the Geological Society of London on 31 May 1837, shortly before he opened Notebook A. According to his 'Journal', he did not return to the subject until 5 October 1838, so that for much of the period covered by the notebook he was not actively engaged with the subject. On fossils Darwin was an interested but not obsessed reader. Convinced by Charles Lyell of the imperfection of the geological record — 'each formation being merely a page, torn out of a history' (D60), he did not expect to be able to reconstruct the full history of life from the fossil record and so did not train his interests narrowly on its literature. After initially mixing entries in Notebook A, he eventually came to place paleontological entries in the transmutation notebook series rather than in A.

What one does find in Notebook A is a collection of entries of great diversity and richness. Outside remarks on purely local phenomena, approximately thirty topics are treated in Notebook A. The twelve most frequently discussed listed in order of declining prominence are: the internal structure of the earth including questions of the thickness and heat-conducting capacity of the earth's crust; the elevation and subsidence of the earth's crust; processes involved in rock formation; drift and erratic boulders including the topic of the 'parallel roads' of Glen Roy; geology in relation to species including the question of transmutation; volcanos and their action; dikes; metamorphism; mountains and their formation; rivers and valleys; cleavage; and salt deposits. The notebook contains a series of short reading notes and speculative remarks punctuated by several passages of greater length. The latter include pages 22–4 on meteoric stones, pages 49–53 on the arrangement of particles in rocks, pages 72–5 on the widening of vallies, pages 90–1 on the conduction of heat in the earth's crust, pages 113–14 on the figure of the earth, and pages 121–4 and 137–40 again on the conduction of heat in the earth's crust. These passages show Darwin engaged with the current literature in the physical sciences and, unlike Lyell, ready to speculate on the original state of the earth (A121) and on such astronomical causes as might bring 'planets to an end' (A24). Such topics bore on the age of the earth, though that question was not raised directly. Darwin also declared himself in favour of a thin crust model of the earth's structure (A133).

Notebook A bears a different relationship to Darwin's published work than do the other alphabetically lettered notebooks. Whereas Darwin was writing 'for the drawer' in the other notebooks, awaiting the appropriate audience, in Notebook A he was developing themes freely discussed among geologists. Indeed Darwin himself was a major contributor to contemporary geological discussion: five papers presented to the Geological Society of London from 1837–1839; work from October 1837–June 1838 on what was to become his 1844 book on volcanic islands; a paper on Glen Roy researched and written from the end of June to September 1838; work from October 1838 into 1839 on what was to become the 1842 book on coral reefs; and service from 16 February 1838 until 19 February 1841 as one of the two Secretaries of the Geological Society.[1] Hence Notebook A represented but a small fraction of Darwin's activity in geology during the years it was kept. It served as a counterpoint to his other reading and writing, at times a locus for developing themes that had no other home and at times as a storehouse for references. The latter use is obviously the case in his references to Scotland and to Glen Roy (pp. 110, 115, 133), but references stemming from his other geological writing appear at the appropriate points in the notebook as well. The utility of the notebook for Darwin's publications is indicated by his note to himself on the inside of the front cover of the notebook: 'Feb 24th 1839 As far as p 140—abstracted as far as concerns "Geolog Observat on Volcanic islands & Coral Formation'.[2]

Darwin's later use of the notebook is evident from the 'Location of Excised Pages' and most of the excised pages from Notebook A were placed into three folders: 'Gravel, Valleys Denudation &c &c'; 'Scraps to end of Pampas Chapter'; and 'Scraps Cleavage'. Many of the entries thus excised found their way into later geological writing, particularly the *Geological Observations on South America* (1846). Equally interesting are the passages from the notebook that were not incorporated into later writings. Had Darwin pursued the study of geology in later life with the same zeal that he had earlier, more of the questions raised in Notebook A would have been addressed in print. As it was, after 1846, further researches on a number of geological topics were put aside in favour of other pursuits.[3]

[1] For the sequence of Darwin's activities see de Beer 1959:7–9 expanded in *Correspondence* 2:430–34; for Darwin's papers see CP 1.

[2] *Correspondence* 2:70 n. 3, 'Early in 1838, Smith, Elder & Co., the publishers, advertised a single octavo volume for that year, entitled *Geological observations on volcanic islands and coral formations.*' On 21 January 1838 Darwin was already contemplating dividing that work into two (*Correspondence* 2:69). Eventually, much delayed, the geology from the *Beagle* voyage appeared in three parts: *CR* (1842), *VI* (1844), and *GSA* (1846).

[3] For an introduction to the themes of Darwin's geology see Rudwick 1974–75 and Herbert 1985; on nineteenth century debates concerning the interior of the earth see Brush 1979.

FRONT COVER

A.

Geology

Note on Woolwich

Nothing on any Subject

INSIDE FRONT COVER **As far as p. 33. distributed to several subjects.**
Feb 24th 1839 { **As far as p 140— abstracted as far as concerns "Geolog Observat on Volcanic islands & Coral Formation**

Lyell's Salband p. 86[1]

Shells near Woollich p. 112[2]

1e Speculate on the extension of Patagonia seaward, at mouth of S. Cruz. from ascertained inclination. of plains: Lias in Shropshire. or some other wonderful outlyer.—

Linn: Transact. Vol. 8. p. 288. Salt deposited on windows of houses. & trees all injured on Eastern side, far inland.— even 70 miles from salt water.[1]

2e Mr. Arrowsmith[1] tells me, that Himalayas penetrated like Bolivian Chain. Volcanic islands. from number of craters very ancient. which agrees. with peculiar character of Vegetation.—

FC **A.**] *ink.*
Geology] *added pencil.*
Note on Woolwich] *boxed, added ink*; **'Woolwich'** *underlined.*
Nothing on any Subject] *added ink, underlined.*
'Geol most O/– FD' *added by Francis Darwin, pencil.*

IFC **As far as p. 33 . . . Formation**] *pencil, crossed, perpendicular to spine..*
Lyell's . . . 86] *pencil, parallel to spine.*
Shells . . . p. 112] *ink, parallel to spine.*

1 plains:] ':' *over* '.—'.
Linn: . . . water.] *crossed pencil.*

2 *page crossed pencil.*

IFC–1 See 86–1.

IFC–2 See A112.

1–1 Salisbury 1807:286, 'On the 14th of *January*, 1803, I observed an east window of my house, which had been cleaned a few days before, covered on the outside with an apparent hoar frost. When the servant who was sent to remove it, came and told me it was *salt*, I was astonished.' Pp. 287–88 describes the injuries caused by the spray to the eastern sides of trees. On p. 288 the author also notes, 'The farthest place to the westward that I visited the following summer was Bulstrode, where many of the trees had suffered severely: from thence to the mouth of the Thames, the last place where this wind could have taken up any salt spray, cannot be a less distance than seventy miles.'

2–1 John Arrowsmith.

3 So accustomed to utter confusion in Europe, that the simplicity of Ventana's «Quartz.» unmixed is very pleasing; owing to the movements being of one order.—[1]

There should not be surprise at Horse being found in America, when Mammoth & narrow toothed Mastodon.—

4 argue against the prejudice of not believing recent elevation, yet sea shells at tops of mountains we ought to sympathize with. old doubters of what are fossil shells.— accustomed to such terms "fixed as the land, stable as the water"—

5 It may be worth noticing edentates & camels in deserts & *rodentia*

In Plata Mastodon Toxodon[1]

Is the general saline tendency of America connected with its elevation. vapour from below—

6 *Malte Brun* «Salt Lakes» *Siberia*[1] must be read as well as *Pallas*[2] before Geology is written

---

Cuvier. Europe possessed a great edentata.—[3]

---

7 How much is temperature of world regulated by atmospheric currents?— chiefly clearly by sun's position= If equatorial streams of warm pole; in name of Heaven why are tops of Equatorial mountains so cold.—

---

3 There should . . . Mastodon.—] *crossed ink.* *page crossed pencil.*
4 *page crossed pencil.*
5 *page crossed pencil.*
6 *page crossed pencil.*
7 *page crossed pencil.*

3–1 *GSA*:147. 'The Sierra Ventana . . . consists, up to its summit, of quartz, . . .'
5–1 The Toxodon was first described by Richard Owen at a meeting of the Geological Society of London on 19 April 1837 (R. Owen 1833–38). The specimen had been collected by Darwin.
6–1 Malte Brun 1822–33, 2:393–94 and 399–400 describe salt lakes in Siberia.
6–2 *GSA*:73, 'The salt lakes of Siberia appear . . . to occur in very similar depressions to those of Patagonia.' The reference is to Pallas 1802–3, 1:283–84, 'Those saline lakes, from their adjacency to the Caspian Sea, and their natural formation, appear to have originally been gulphs, which are indebted for the salt they contain, to the sea-water confined in them, and diminished by evaporation. They seem to have lost their former communication with the sea, either by having been filled up with mire, or by the gradual retreat of its waters. Most of these lakes are found in very extensive cavities of the steppe, . . .'
6–3 G. Cuvier 1821–24, 5 (1):193–95, the section entitled, 'Sur une phalange onguéale fossile qui annonce a elle seule un Édenté inconnu, probablement du genre des Pangolins, et de taille gigantesque.' The bone was found (p. 193) 'pres de *Eppelsheim*, canton d'Alzey, dans la partie de l'ancien Palatinat . . .'. The animal in question is no longer regarded as an edentate but as a perissodactyl, *Chalicotherium goldfussi* of Miocene age.

Siberia no plants to it, lately raised above level of the Sea. Lyells Encyclopædia[1]— *Lately elevated*

8 When Siberia went up. Arctic land went down.— Probably more Arctic land would be required to produce climate resembling S. America in Europæan latitudes.—

9 Will it be supposed that the armadilloes have eaten out the Megatherium.—[1] The Guanaco the Camel.?[2]

10 Make note about N. American bone not probably in salt marshes Efflorescence nothing — Study account.— Alluvial plains of Mississippi — **No**
Vol. I. p 212. Cuvier Oss Foss[1]

11 Wide range of Mammalia really very important. harmonizes well with Lyells idea of intertropical land.— Siberia rises. therefore to the South sinks.— —Meditteranean continent corresponding to Europæan risings. Pacific great land.—[1]

12 Will use argument of proof of *slow* corrosion of valley of ‹Patagonia.› S Cruz — from terrace like structure—

8 *page crossed pencil.*
9 *page crossed.*
10 **No**] *circled, added pencil.*
p 212] *last digit uncertain.*
page crossed pencil.
11 *page crossed pencil.*
12 *page crossed pencil.*

7–1 Presumably Darwin was citing an article in an encyclopedia owned by Charles Lyell. Possibly it was the entry on Siberia in *Encyclopædia Londinensis* (Anonymous 1810–29, 23:179), 'Siberia has hitherto been found to possess scarcely any genera of plants; and even all the species of any considerable importance, are those trees which are common to it with the north of Europe.' This exact sentence also appears in Rees 1819, 32: n.p. If either of these entries were the relevant ones, the observation on elevation would have been Darwin's own. Also see Lyell 1837, 1: chap. 7.
9–1 See B54, 69–70. It should also be noted that in mid-1837 the *Megatherium* had not yet been ruled out as a possible bearer of tessellated bony armour like that of the armadillo (*Correspondence* 1:276, 280, 301, 312, 331; Wilson 1972:437; *JR*:181–82; then see Owen 1838–42: 112–13).
9–2 See RN129–1.
10–1 As the emphatic 'No' added to this passage would seem to indicate, the notes on this page do not adequately reflect Cuvier's text, which has to do with mastodon remains in North America (G. Cuvier 1821–24, 1:205–24). In particular Darwin's first entry contradicts Cuvier's explicit claim that mastodon bones are to be found in salt marshes. Thus Cuvier (1821–24, 1:215), 'D'après tout ce que nous racontent les observateurs, les dépôts d'os de mastodontes, ainsi que d'autres espèces fossiles qui les accompagnent d'ordinaire, sont plus généralement dans des endroits marécageux, où il sourd de l'eau salée, qui attire les animaux sauvages, et surtout les différentes espèces de cerfs, et qui, par cette raison, on été désignés par le nom anglois de *Lick*.' Darwin's next phrase probably refers to the fact that Cuvier did not mention anything in regard to saline efflorescence in his discussion of marshes. For an indication of Darwin's interest in the phenomenon of efflorescence see *JR*:91–92. With respect to the plains of the Mississippi see Cuvier's reference (p. 217) to the '*Great-bone-lick*' found 'près de la rivière des *Grands Osages* qui se jette dans le *Missouri*, peu au-dessus de son confluent avec le *Mississipi* [sic].'
11–1 For Lyell's views on the distribution of continents and oceans see Lyell 1837:1, chaps. 6–8 or, in an earlier version, Lyell 1830–33:1, chaps. 6–8.

13e Intersection of veins prove, that there are at least several attempts at elevation

---

From the lost & turned abaut position of strata, prooff thickness not very great; where piece turned over axis or hinge no doubt fluid.—

---

analogy as continental elevations slow. so would line of mountain chain be

14e Mr ‹Lyell›[1] «Waterhouse» has frequently heard that Herons bring eels alive to their nests; & then they may picked up beneath the trees— —[2]

Are any Fish seed-eaters. This important in transport of Fish Let a Hawk fly at Heron.—

15e Ceratophytes common in Northern seas p. 312. Chamisso in Kotzebue.[1]

Study Humboldt. Fragmens Asiatiques account of *American* Volcanic action.—[2]

Fragments of slate converted into crystals of Hornblende p. 248. L.Institut 1837.—[3]

16e Helms remark on common salt being found on low hills East of Cordillera very important[1] V. Malte brun[2] — Main character of Andes Metamorphic action — Mem: red sand of Europe no fossil shells — ? action of Heat

13 *page crossed pencil.*
14 Mr . . . trees — —] *ink.*
Are . . . Heron.—] *pencil.*
fly] *uncertain reading.*
15 Ceratophytes . . . action.—] *crossed pencil.*
Fragments . . . 1837.—] *crossed pencil.*
16 *page crossed pencil.*

14–1 Charles Lyell.
14–2 George Robert Waterhouse: personal communication.
15–1 Darwin's inference derived from Chamisso 1821, 3:311, in the chapter entitled 'Kamtschatka, the Aleutian Islands, and Beering's Straits', 'The place of the southern *Lithophytes* is occuped by the *Ceratophytes*; and the north coast of Umnack, in particular, produces several very distinguished species.'
15–2 Humboldt 1831:144–62, the section entitled 'Sur une nouvelle éruption volcanique dans les Andes de Cundinamarca' records eruptions at Tolima (Nevado del Tolima) in 1595 and 1826. In addition there are scattered references elsewhere in the volume (*e.g.* pp. 111–13) comparing South American and Asian topography and volcanic activity. See also ZEd6.
15–3 Fournet 1837:248, 'Dans un quatrième cas (celui du pont de Gassie sur la route de Chessy, à l'Etrat), des fragmens de schiste argileux gris, qui se sont trouvés en contact avec les porphyres quarzifères, après avoir éprouvé-diverses altérations, se sont convertis définitivement en beaux cristaux d'amphibole vert foncé.' Darwin double scored this passage in his copy.
16–1 Helms 1807:12–13, 'Cordova [Argentina] . . . is very pleasantly situated near a wood at the foot of a branch of the Andes. . . . at Remanso, 60 miles from Cordova, [the mountains] again branch out so far from one another, that from that place to Tucunum [sic] the traveller passes through a saline plain 210 miles in length, . . . The whole ground is covered with a white incrustation of salt, ...' Also see p. 16 for mention of a salt encrusted river bank near Tala, on the way to Salta.
16–2 Malte Brun 1822–33, 5. See p. 454 for reference to salt marshes near San Juan, Argentina, and pp. 454–55 for reference to fossil salt in the Argentinian province of Tucuman.

17 bubbles volatilized at bottom, condensed before rising?— Mem. granite heated.— Metamorphic action in red sandstone.— Certainly Volcanic— CD[Might not bottom of ocean boil; yet heat never reach surface.—

18 Journal de Physique, et D Histoire Naturelle, C«o»urrejolles. 11th Observ.— Les grands tremblemens de terre sont presque toujours precedes et suivis, queque temps avant et apres, par de petites secousses."— Tom 54. p. 106[1] do— p. 110. Mountains on west side of Domingo formed of coral limestone, with interstices yet emptty.— In all the mountains of Saint Marc et des Gonaïves, it is difficult

19 to find stone not thus composed on the NE part more like marble requires polish to see structure,— «He» Thought of erecting machine to see if water fell.—[1]

‹Keys off extreme point of Flori[da]›

20 Excellent paper on Erratic blocks in Alps. Memoires de la Soc. «de Geneva» Vol 3' P. II.—[1]

Bed, of elevated shells on the Senegal. L Institut p. 192.— (1837. Peninsula of Cape Verd. volcanic.— Isle of Gory. rocks encrusted with serpula— Isle of Cayenne. Syenite & diorite, covered with iron clay common to Guyana said to extend to Cordillera[2]

17 *page crossed pencil.*
18 Journal . . . secousses."—] *crossed pencil.*
Tom 54 . . . difficult] *crossed pencil.*
19 *page crossed pencil.*
20 *page crossed pencil.*

18–1 Courrejolles 1802:106, as quoted with minor spelling variation.
19–1 Courrejolles 1802:110–11, 'Les pierres calcaires que l'on trouve sur les montagnes de la partie de l'ouest de Saint-Domingue, laissent appercevoir les pertuis encore vides des polypes marins qui les ont formées. Il y a de ces fossiles dans toutes les montagnes de Saint-Marc et des Gonaïves en si grande quantité, que quand on les rompt à coups de masse, il est plus difficile d'en trouver qui ne portent l'empreinte d'un madrepore, que de celles où elle est effacée. . . . On observe dans la partie du nord-est de Saint-Domingue, que les pierres calcaires des montagnes y sont très-dures, et ne laissent plus appercevoir aucune trace de leur origine; il faut les polir pour y distinguer, comme dans les autres marbres de même espèce, les coraux, les polipiers des madrepores, et les autres marques distinctives des pierres calcaires.' Footnote 1 on p. 111 describes Courrejolles' incompleted attempt to measure the rise and fall of sea level using a machine of his own design.
20–1 Deluc 1826.
20–2 Robert 1837:192. 'En remontant le fleuve du Sénégal, à 5 ou 6 lieues de Saint-Louis, on trouve un vaste terrain formé par une espèce d'Huître. . . . Autant que les sables qui recouvrent toute la presqu'île du cap Vert permettent de voir, M. Robert pense qu'elle est entièrement volcanique, et que les deux mamelles qui forment les deux points les plus élevés du cap, appartiennent aux restes d'un ancien cratère. Ile de Gorée . . . Cette roche, qu'au premier aspect on pourrait prendre, dit M. Robert, pour une coulée, est cependant, bien qu'à 200 pieds environ au-dessus du niveau de la mer, d'origine évidemment aqueuse; elle est incrustée de Serpules. La syénite et la diorite paraissent constituer la base de l'île de Cayenne. . . . le second [diorite], qui offre plus d'étendue. recouvre les pentes des collines et remplit le fond des vallées, paraît presque entièrement formé d'un fer limoneux, roche à ravets des colons, qui, suivant eux, règne aussi dans toute la Guyane, et même jusqu'au pied des Andes.'

21 I see Brewster speculates from believing meteorolite but old Planet, that inside our globe melted magnetic metals.[1] ∴ earthy crust compared to those of falling stones.— ⸮ does this bear upon the sorting of matter. in making trachyte come out before.—

22 What must be the effect of all the meteoric stone which must have fallen on the globe since the Cambrian system In Ures dictionary between 1768 & 1818. that is fifty years— 90 ‹showers of› stones are recorded as falling; many of these were not single, but are described as many, (one even 3000) This ninety includes all actually counted.— The weight «or size» is given of 25 stones.— The total weight

23 recorded is 473. pounds (taking about average when several are given), this will give nearly 19 pounds average for each stone. that fell, that was weighed,; ‹but› carrying on this ratio I can count 90 stones which have fallen in the 50 years. ∴ 90 × 19 = 1710 ÷ 50 = 34 pounds each year.— but instead of 90 stones in many cases there were flights of stones of large numbers (& how few cases recorded if we say «100» ‹5›0 lbs a year too little.— How comes it

24 none in fossil state? suppose «100» ‹5›0£
× 50,000 × ‹50 = 250 0000› × 100
= 50, 0,0,000
= 2500 = tons in fifty thousand years][CD1]
If world increased a tenth; would the perturbation be serious? if so other cause besides thin vapour bringing planets to an end?

25e Fragmentary granite showing schistose structure (& veins appearing): mem.

21 but] *uncertain reading.*
page crossed pencil.
22 this] 'i' *over* 't'.
page crossed pencil.
23 ‹5›0] '‹50›' *intended.*
page crossed pencil.
24 ‹5›0] '‹50›' *intended.*
page crossed pencil.
25 *page crossed pencil.*

21–1 Where exactly Darwin encountered David Brewster's speculations is uncertain. However, the section of his enclyclopedia article on magnetism entitled 'On the nature and causes of the earth's magnetism' contains the following passage that relates, albeit indirectly, to the subject of Darwin's comment (Brewster 1837:280; 1842, 13:754):, 'According to our views, terrestrial magnetism resides wholly in the earth's atmosphere, which contains throughout its whole extent ferruginous and other metallic matter, and sulphureous exhalations, all of which are carried up from the earth by evaporation, by ejection from volcanoes, and by the returning strokes of electricity from the earth to the air. The actual existence of such materials in the atmosphere, particularly sulphureous and ferruginous matter, is proved by the observations of Fusinieri, and by the existence of meteoric stones and other solid substances which fall on the earth.'

24–1 See Ure 1823:587–88 for his chronological list of meteoric stones from which Darwin did his estimates. In his own copy of the work Darwin noted in the margin the weights of stones for cases where Ure had provided that information.

Henslows Anglesea solution of silex also shewn.[1] No 3^{d} of Ed. N. Phil. J. p 194.[2]

Fact of dust blown far out to sea valuable; because transportal of Minute seeds—

26e L. Institut. p. 209. May. 1837 Paper by Humboldt on Quito Volcanoes & another on Mexican Trachyte ‹roc› lava called Andesite.[1]

Red Coral in the Mediterranean 700 feet deep in some of. the twopenny periodical said so. «Campbell the Poet»[2]

27e Accra. Coast of Africa. Clay Slate & Quartz. strike SSW & NNE dip 30°–80° Ed. N. Phil Journ. p. 410. 1828[1]

Ed. N. P. J. p. 105. Oct. 1828. gneiss in India (falls of Garsipa) dip 30°. ‹strike› «direction‹?›»ESE—
CD [In the Darwar. transition Hills & strata SE. direction of transitions clay slate &c nearly vertical[2]

26 Coral] 'a' *under or over* an 'l'
Red . . . Poet] *fancifully scored margin, crossed pencil.*
27 *page crossed pencil.*

25–1 Henslow 1821–22:408 describes crystals, taken from a bed of shale, which chemical analysis showed to be 49 per cent silex.
25–2 Turnbull 1827:194, 'This consolidated debris [composed of fragments of the original granite found in the country around Darwar] is almost every where intersected by small veins of quartz, or of quartz and felspar mixed. Nor have these veins originated from subsequent eruption; for they intersect one another in all directions, and often terminate in two ends, in a small portion of rock. Moreover, this rock often displays, in a slight degree, a schistose structure, especially when acted on by the weather. . . . These facts appear to prove, that a new arrangement of particles may take place in solid bodies, giving rise to crystallization, and to different kinds of structure in rocks.'
26–1 Humboldt 1837. The author's comments on volcanos in the mountains around Quito, on South American trachyte, and on andesite 'au volcan mexicain de Tolucca' (p. 136) are treated together in the *Institut* article. Darwin scored the relevant remarks in his copy.
26–2 Presumably Thomas Campbell provided Darwin with the following reference to an article from the 4 June 1836 issue of *Chambers' Edinburgh Journal*, a low-priced mass circulation weekly selling for one and one half pence (Anonymous 1836:151), 'The coral fisheries form a very considerable trade in several parts of the Mediterranean. . . . The greatest portion [of coral] is procured from a depth of from sixty to a hundred and twenty-five feet; but some fisheries are carried on to the depth of nine hundred feet. . . . The vermilion coloured coral, being the rarest, is the most expensive. The common red, however, brings a high price also, when the quality is good.'
27–1 T. Park 1828:410, 'The mountains bounding the sides of this long valley [of Accra] . . . appear composed of quartz-rock and clay-slate alternating with each other, and disposed in strata ranging SSW. and NNE., the dip from 30° to 80°. . . .'
27–2 Christie 1828–29:105, 'All these varieties [of rock], with the gneiss and granite, pass insensibly into each other. They are distinctly stratified; have a dip of about 30°; and their direction is nearly ESE. They form the sides of the chasm, over which the river is precipitated at the Falls of Garsipa. . . . This is the only place in India where I have met with primitive gneiss; but it is not improbable that it occurs in many other parts of the country.' Also, p. 108, '*Transition Rocks.* These rocks occupy a very large part of the Darwar and Canara districts, and of the territory of Goa. . . . The principal rocks of this series are clay-slate, chlorite-slate, talc-slate, limestone, greywacke, gneiss, and quartz rock. The strata appear to have a general direction of north-west and south-east. They are generally highly inclined, and, in many, instances, quite vertical.'

28e Linear earthquake 500 by 90.— in Syria Geolog. Proc. p. 541. year 1837[1]

In Upper Assam. Geolog Proc p. 566 1837.— Tertiary ‹bea› formation twenty species same as Paris. 1500 ft high[2]

29 Mr Bird in paper to Brit. Assoc: has shewn how electrical currents tend to deposit metals, if in solution. My view of metamorphic in contradistinct to Volcanic will explain their solution. Athenæum M. 516 1837[1]

30 High up the Essequibo, granite & quartz, after passing sandstone Vol II. p. 69.— Geograp Journal[1]

Earthquake at Melville Isld New Holland Augus 1d to 3d & 19 1827 Geograp Journ[2]

31 There are some ideas about order of injected rock being determined by

28 1500] '15' *over a stroke.*
Linear . . . 1837] *crossed.*
29 1837] *boxed.*
page crossed pencil.
30 High . . . p. 69.— Geograp Journal] *ink.*
Earthquake . . . Journ] *pencil.*
page crossed pencil.
31 *page crossed pencil.*

28–1 Moore 1833–38:540–41, 'The dispatch contains a list of thirty-nine villages which had been totally destroyed, and six partially; . . . it had been ascertained that the earthquake [on 1 January 1837, Syria] was felt on a line of five hundred miles in length by ninety in breadth.'

28-2 McCleland 1833–38:566–67, 'At the top of the first stage [of ascent of mountain in Upper Assam], or at about fifteen hundred feet above the sea level, the author discovered a well-defined marine beach, containing shells and other marine exuviæ about two feet deep, and reposing upon sandstone and covered with soil. The shells consist of Pectens, Cardia, Ostreæ, Terebratulæ, and Melaniæ, mineralized by a fine yellow sandy matter, and united together by a brown indurated clay. . . . These shells were compared with a collection of about one hundred and fifty species from the Bay of Bengal and the estuaries of the great rivers, but not one was found to correspond; nor with those found by the late Dr. Gerrard in the secondary strata on the north of the Himalaya; but a small collection of about one hundred species from the Paris basin were at once recognised by the author as familiar objects, from his acquaintance with those from Kossia and Cherraponji: these consisted of about an equal number of species, and on being submitted to systematic comparison about twenty species were found to be identical in the two collections.'

29–1 G. Bird 1837:670, 'From [Bird's experiment] it appeared. that the *mere passage of an electric current*, independent of the presence of poles, was sufficient to effect metallic reductions, supporting, in a satisfactory manner, the experiments of Dr. Faraday on this subject. The metallic crystals thus obtained were very hard and brilliant, resembling in a striking manner those produced in the vast theatre of nature. . . .'

30–1 J. E. Alexander 1832:68–69, 'After passing up the Essequibo, they got into the Mazaroony river, which makes a considerable sweep to the north-west, and then returns, so as to form a large peninsula, inclosing lofty mountains and considerable creeks; across the narrow isthmus is a journey of only three days, so that the sweep may easily be avoided. The travellers passed several creeks, and saw on the left mountain-ranges of white quartz several thousand feet in height. A magnificent waterfall, seen at a great distance, fell over the face of a rock apparently eleven hundred feet high; white sandstone rocks were succeeded by felspar on the river's banks; then granite and quartz formed the highest ridges.'

30–2 Campbell 1834:151, '. . . [I] may also remark that we experienced successive shocks of an earthquake on the 1st, 2nd, 3rd, and 19th of August, 1827; . . .'

fusibility in. L Institut p 247. 1837.— The most infusible first injected.— Basalt: last because it could reach the surface. before being cooled.—[1]

32 Berzelius. L'Institut. [1837 p. 297]CD thinks Olivine a preexisting mineral.— Mem. Galapagos ∴. Basalt deepest??[1]

Marcel Serres L'Institut. 1837. p 331 Considers that Mercury & Sulpuret of Iron has been sublimed into the tertiary limestones of Vendarques. Mem sublimation of sulphur to form salts of America.—[2]

33e The number of minute turbos in red earth with volutas. prove regular mud bank at Bahia Blanca.‹fl› Flustra identical. recent & bone bed.—
November 8th 1877 (Memoranda so far distributed to various subjects)

34e Dr. A. Smith informs me that in the year a Rhinoceros was found ‹emb› in the mud, of the Salt river.— in reference to fossil guanaco of P. St.Julian.—[1]

Mr Scrope seems to consider that elevation & eruptions are antagonist forces. but they are parts of one force, one *locally* relieving the other.—[2]

32 *page crossed pencil.*
33 **November . . . subjects)**] *added pencil.*
page crossed pencil.
34 *page crossed pencil.*

31–1 Fournet 1837:247, 'Ainsi . . . les roches qui ont paru les premières en soulevant et disloquant le sol, sont les granites communs; roches qui, étant les plus infusibles à raison de la silice, élément négatif qu'elles contiennent en excès, avaient besoin pour être fondues d'être plus immédiatement appliquées sur la source de la chaleur. Viennent ensuite, et dans l'ordre proportionnel de diminution de la silice et de l'augmentation de la fusibilité, les granites porphyroïdes ou à grands cristaux de felds-path. et les porphyres, quarzifères, puis les eurites micacés. roche presque entièrement feldspathique, et enfin la minette, roche pyroxénique et amphibolique très-fusible. et pouvant traverser, sans être solidifiée, toutes les roches précédentes.' Darwin scored this passage.

32–1 Hoff 1837:297. In the course of describing J. J. Berzelius' view that meteorites originate as the products of lunar volcanos, Hoff wrote, 'M. Berzélius ne pense pas que l'olivine soit un produit volcanique à cause de son état refractaire: mais il la regarde comme un minéral préexistant, qui a seulement été enveloppé dans la lave liquide.' Darwin scored this passage in his copy.

32–2 M. Serres 1837:331, 'Or, de pareilles sublimations de fer sulfuré n'ont pui s'opérer que par suite de la chaleur centrale ou de la température propre du globe. Du reste, elles font concevoir facilement comment l'on découvre dans le sol tertiaire des environs de Montpellier [including Vendargues] les goutelettes de mercure natif que l'on y observe en si grand nombre.'

34–1 Andrew Smith: personal communication. There is a 'Salt River' ('Soutrivier') at 33° 03′ S. 23° 29′ E. in the Great Karroo, though other streams in the Cape region in South Africa also bear the name.

34–2 The thought is expressed variously in Scrope's book on volcanos but see the representative passage on p. 192 (Scrope 1825), 'But since we have no reason to doubt the influence of subterranean expansion in the phenomena of volcanos, earthquakes, and elevations of the superficial strata, to have taken place for ages past, indeed to have been co-eval with the existence of our planet, under its actual laws, if it be true that the development of the one class of these phenomena, viz. volcanic eruptions, proportionately obviates that of the other, or the absolute elevation, *en masse*, of extensive superficial portions of the earth's crust; and therefore in the same locality, and at the same period, the one class of effects must always have varied inversely with the other; we should expect to find proofs of the operation of this law in the visible traces left by these phenomena on the globe, and, consequently, that, wherever volcanos have existed in the greatest number, and in the most prolonged and energetic activity, there will have happened the least absolute elevation, *en masse*, of the superficial strata; and, vice versâ, that wherever the most remarkable elevations of the solid crust of the globe have taken place, there should seem to have been little or no absolute escape of subterranean caloric through the spiracles of habitual volcanos.'

35 Is the felspar glassy in greenstone dikes which rise through granite.— a most important question with respect to my theory of changes. of granites into Trachytes.—

Mention Osorno in lake. few Volcanos now in lakes.—

---

36 Mr Murchison. M.S. Chapter on drift.— Beyond region of great boulders, pebbles of granite clearly effect of remodelling same manner. as bits of Patagonian boulders might be transported.—[1]

On grooved rocks. **Specimen of rock from Costorphine at Geolog. Soc:**[2]

Colonel Imrie	Transact Wern. Soc. Vol. 2. p. 35[3]
Sir J Hall	Trans. Phils Royal Ed. Vol 7[4]
Dr Buckland	Reliquiæ Diluvianæ p. 201. & seq[5]
Murc	Trans Geolog Soc Vol 2. p 257[6]

35 *page crossed pencil.*
36 **specimen of . . . Geolog. Soc:**] *added pencil.*
page crossed pencil.

36–1 The manuscript would be that which became chapters 38 and 39, entitled 'The Northern Drift', in *The Silurian System* (Murchison 1839a). The manuscript does not seem to have survived. In the published text (p. 532) Murchison referred to a locality south of Worcester above the Severn where there were 'rounded pebbles of several varieties of granite, similar to those traced from the north through Shropshire and Staffordshire. The granite pebbles, however, are much the least abundant and never equal the size of the fragments of rocks derived from the neighbourhood, some of which have the dimensions of a man's head, though the mass of the accumulation is simply what would be termed coarse gravel. Here, therefore, the smaller and finer portions of the northern detritus are commingled with various rocks of the country through which the prevailing current has passed. . . . there can be no doubt, that when this tract was submarine, the lower part of the country, extending south of Gloucester, was also under the sea.' On p. 533 Murchison noted that Darwin 'has investigated the heaps of gravel and shingle around his native town (Shrewsbury), and he assures me, that they are in no way to be distinguished from many shore deposits of the southern hemisphere.' The list of references on the subject of grooved rocks was taken from a portion of Murchison's text that now appears as a footnote on p. 537. Darwin copied the incorrect page citation ('257' rather than '357') in the last entry from Murchison. See Imrie 1818, Hall 1815, Buckland 1823, and Murchison 1829.

36–2 From Corstorphine at Geological Society of London.

36–3 Imrie 1818:35–36, 'I have here mentioned above, that the disappearance of the trap in some of the glens and narrow vales [in the Campsie Hills], seems to have been produced by the effects of the attrition of heavy bodies set in motion by a great force of water in rapid movement. . . . In some of the glens and narrow vales, where the trap had not entirely disappeared, I perceived upon its surface strong indications and marks of attrition. In some places the surface of the trap was smooth, and had evidently received a considerable degree of polish; and this polish is almost always seen marked by long lineal scratches. In other places, there appeared narrow grooves, apparently formed by the rapid movement of large masses of rock having been swept along its surface; and I remarked, that these striæ or scratches, were very generally, in a direction from west to east, excepting where inequalities of the surface, and sudden turns in vales had partially influenced the course of the current.' See *JR*:622.

36–4 James Hall (1815) invoked 'diluvian inundations' (p. 207), caused by the sudden elevation of land, to account for the presence of granite boulders on Mount Jura and for various 'diluvian facts in the neighbourhood of Edinburgh' (p. 169). In this latter category Hall listed (p. 209), '1. The distribution of loose matters in tails and ridges, on this side of the Island, and in knolls on the West. 2. The grooves and scoopings, and obtuse-angled ridges, occurring on the surface of rocks of every description; and, lastly, the scratches and other minute features of abrasion, which are found to accompany the large features, where the rock has been protected from injury.' For Darwin's reaction to Hall's non-gradualist but Huttonian hypothesis see *JR*:621–25. Buckland 1823: 203–5 had also referred to Hall's paper, as did Darwin's St. Helena Model Notebook.

36–5 Buckland 1823:202–5, 'Proofs of Diluvial Action in Scotland' describes and quotes from the work of Imrie (1818) and Hall (1815). See *JR*:622.

36–6 Murchison 1829:357, '. . . these hills [of the Brora

37e The Pota: labiata certainly is found with the Mactra. at Buenos Ayres

---

at the Zoolog: Soc: Terebratula from Hudson's Bay. 2. species[1]

Vol VI. Geograph. Journ. Analysis of Pœnig Voyage Valparaiso[2]

38e Dr. Gillies in MS. letter in Sir. W. Parish Possession. talks of ‹hill› «cerro» of Diamante near stream of same name. with imperfect crater ‹—› near summit,— much pumice —. appears to be outside of the Cordillera— Near the Planchon talks of very much of Gypsum.—[1]

39e The officers of the Bonite. French discovery ship, found clear proofs of shells & waterworn rocks «at Cobija.» At Iquique of elevation to amount of 30 ft.—[1]

37 species] *line following extended in pencil.*
The . . . species] *crossed pencil.*
Vol . . . Valparaiso] *crossed pencil.*
38 Planchon] **n** *added pencil,* 'Planchon' *underlined pencil.*
page crossed pencil.
39 At] 'A' *over* 'a'.
‹Ceylon›] 'l' *crossed like a* 't'.
talks of quantities . . . at Iquique] *crossed pencil.*
The . . . at Iquique.] *crossed pencil.*
Band . . . Heckla—] *crossed pencil.*

district] probably owe their present form to denudation: which supposition is now confirmed by the exposure on their surface of innumerable parallel small furrows and irregular scratches, both deep and shallow,—such, in short, as can scarcely have been produced by any other operation than the rush of rock-fragments transported by some powerful current . . . . the furrows and scratches preserve an uniform direction from N.W. to S.E.; thus indicating the great force of a current, unaltered in its course by any inequalities of the surface over which it rolled.' See *JR*:622.

37–1 At the Zoological Society of London.

37–2 Poeppig 1836:381–85.

38–1 John Gillies's letter, which Darwin recorded as having seen in manuscript, eventually found its way into Woodbine Parish's own book. Gillies wrote (Parish 1838:323–24), 'After reaching the river Diamante, the southern boundary of the province of Mendoza, I crossed that river and ascended the Cerro del Diamante, and at every step found ample evidence of its volcanic origin: the ascent was covered with masses of lava, and near the summit with loose pumice. The upper part of the mountain consists of a ridge elevated a little at each of the extremities into a rounded form, on the north side of which, a little below the summit, is a plateau about 400 yards in diameter, which undoubtedly has been formerly the crater of a volcano. The whole mountain appears to rest on an immense bed of pumice-stone. On the steep banks of the Diamante opposite to it such strata are laid open on both sides:—at one place on the south bank I traced one great mass of pumice-rock, 100 feet long and 145 wide, the whole forming distinct basaltic pillars.' Also p. 325, 'We proceeded from thence [the banks of the Atuel river] towards the Planchon, along a succession of valleys rich in pasturage, but very bare of shrubbery: in several places we saw immense masses of gypsum, . . .'

39–1 The French naval ship *Bonite* circumnavigated the globe from 6 February 1836 to 6 November 1837 under the command of Auguste-Nicolas Vaillant. The source for Darwin's reference would seem to have been a letter written by Yves Eugène Chevalier, an officer on the *Bonite*, from which extracts were read at the 20 November 1837 meeting of the Académie des Sciences in Paris. The relevant passage from the letter reads as follows (Chevalier 1837a:721), 'Après avoir quitté Valparaiso, nous avons relâché successivement à Cobija, Callao, Payta et l'île de Puna, sur la côte d'Amérique. Dans la première de ces relâches, l'examen de la nature des rivages, du terrain qui le borde m'a fourni des preuves que je regarde comme positives de l'exhaussement du sol sur ce point: à 30 pieds environ au-dessus du niveau actuel des eaux, et sur des amas de coquilles de même nature que celles qui vivent sur les lieux, sont des roches qui semblent battues et découpées par les vagues et recouvertes encore du *Guano*, qui, partout ailleurs, ne s'observe que sur les rochers du rivage.' No mention in the extract was made of Iquique which lies to the north of Cobija along the South American coastline. Chevalier's report was also picked up

Mr Bollaert (at Roy. Institut) talks of quantities of shells at Iquique.[2]

‹Ceylon›.

Band of Volcanic action in Iceland parellel to Greenland: Mem. ? Greenland subsiding.) Von Buch Canary Isd. p. 351..[3] NB. Mackenzie talks of gravel on basalt of Heckla—[4]

40e All the Azores Isld. Von Buch p 359 stretched out NE & SW.—[1] Von Buch. Can. Ile p. 406. List of Volcanos Salomon Isld,— New Britain— &c &c[2] In Ascension for centuries afterwards it might be percieved on which side craters were low — ? applicable to Auvergne???

41e The fact of Galapagos Isld. steep side to windward in allusion to St. Helena discussion.

Mr Brayley says he can give me facts respecting lime ‹n› being heated without parting with Carb. Acid.—[1]

Mr Malcolmson in Paper on India gives reason for knowing that Mur. Soda. and Carb of lime decompose each other.—[2]

40 stretched] 'tch' *written as* 'ch' *with* 'h' *crossed.*
page crossed pencil.
41 *page crossed pencil.*
The fact . . . discussion.] *excised.*
Mr Brayley . . . other.—] *excised.*

in *L'Institut* but here the port cited was Callao. (Chevalier 1837b:405) Neither Chevalier's later report on the geology of the voyage (Chevalier 1844) nor the relevant portion of the narrative of the voyage (La Salle 1851) referred to Iquique.

39–2 William Bollaert: personal communication. On a related point see *GSA*:233.

39–3 Buch 1836:351, 'Cette bande volcanique [in Iceland] est dirigée parallèlement á la côte du Groënland qui se trouve vis-à-vis. Cette disposition montre donc encore dans ce cas que les volcans sont généralement en relation avec les continents ou les chaînes de montagnes qui les forment.'

39–4 The exact reference is uncertain. In writing on Iceland, Mackenzie noted (1811:365) that the lava of Mount Hekla cannot be distinguished from some varieties of basalt and (pp. 391–2), 'The circumstance of alluvial sand covering [cavernous lava] to a considerable thickness in many places, particularly near Mount Hekla, and the appearance of a tract of gravel upon it in the Guldbringè Syssel [near Hekla], seem to be sufficient evidence of the sea having once been above it.' Darwin probably took the spelling 'Heckla' from Buch 1836:350.

40–1 Buch 1836:358–59, 'L'île de Pico est allongée du sud-est au nord-ouest, et il en est de même de toutes les autre îles, Saint-Georges, Saint-Michaël, Terceira, et ce qu'il y a de plus remarquable, c'est que toutes ces îles jusqu'à Flores et Corvo sont situées l'une derrière l'autre, exactement dans cette même direction.' Darwin wrote 'NE & SW' rather than 'SE & NW', presumably inadvertently. See also VI:127.

40–2 Buch 1836:406–7 lists, '4° *Sesarga*, au-dessous de îles Salomon, auprès de Guadalcanar. . . . [existence of volcano not established] 5° Volcan de la *Nouvelle-Bretagne*, à l'entrée du canal Saint-George, sur le côte occidental de ce canal. . . . 6° Volcan situé sur la côte orientale de la *Nouvelle-Bretagne*, à peu de distance du cap Gloster.'

41–1 Edward Brayley: personal communication.

41–2 Malcolmson 1833–38:2, 583–84, 'In accounting for the production of the natron [a hydrous sodium carbonate], [Malcolmson] adopts the theory of Berthollet for the formation of that salt in the lakes of Egypt, viz., a mutual decomposition of the muriate of soda and carbonate of lime, when in a pasty state; but as the natron of Fezzan and the Lonar lake contains half an equivalent more of carbonic acid than can be furnished by carbonate of lime, he proposes a modification of that theory, and suggests that the carbonic acid by which the lime is held

42e on Direction of mountains in Brazil L.'Institut No° 221[1]

Lamellar dikes like Mica Slate Von. Buch. Canary Is^{d}. p 170.—[2] Mem. Cordillera

---

Can Greenstone dikes. be residue of quartzose vein in higher parts? & felspathic veins?—

43e Mr Poulett Scrope. talks of Trachyte, "superficially coated by a thin pellicle of a blackish colour like a dull & poor varnish, which I conceive to be analogous to the black glazing observed by Humboldt on the granitic rocks of the Orinoco".— ‹but› on one of the Ponza isles. but no minute description is given.— Vol II. 2^{d} Series. p. 221.—[1]

44e Mr Bollaert tells me, that the upper strata alone at Guantajaya contains salt.[1] **see Geolog. Proceedings**

---

42 felspathic] 'h' *over an* 'i'.
Lamellar . . . Cordillera] *crossed pencil.*
Can . . . veins?.—] *crossed pencil.*
on . . . 221] *excised.*
Lamellar . . . veins?.—] *excised.*
43 *page crossed pencil.*
44 **see Geolog. Proceedings**] *added pencil.*
Proceedings] *line following extended in pencil.*
M^{r} . . . salt.] *crossed pencil.*
Lake . . . Phozgonea] *crossed pencil.*

in solution in the mud, furnishes the acid, and perhaps indicates the existence of an unstable sesquicarbonate of that substance.' Malcolmson's paper was read at the 15 November and 16 December 1837 meetings of the Geological Society of London.

42–1 Clémencon 1837:567, 'Ces couches . . . s'étendent en longueur du nord au midi: c'est non seulement la direction des couches du district, mais aussi celle de la chaîne des montagnes qui, dans le Brésil, sont de la même nature. Cet accord dans la direction des couches et dans celle des chaînes de montagnes, avait été déjà signalé dans les Pyrénées, c'est aussi la direction des hautes Cordilières de l'Amérique méridionale. Il servirait, s'il en était besoin, à confirmer la justesse des observations de M. de Humbold[sic], qui a fait observer que la direction des couches d'un terrain est déterminée par celle des chaînes de montagnes plus élevées, quoique très-distantes.' Darwin scored this passage in his copy.

42–2 Buch 1836:170, 'A peu de distance d'Angostura [at Tenerife], les couches sont traversées par un filon d'une grande puissance. . . . Il se divise en tables peu épaisses, et la masse de ce filon est composée de lamelles de feldspath tellement minces, qu'on ne peut presque point observer de cassure que dans la direction de ce clivage. Cette circonstance donne à toute la roche un aspect feuilleté, et l'éclat perlé du feldspath lui donne beaucoup d'analogie avec une roche de micaschiste blanc. C'est ce qui fait que plusieurs fois cette roche a été considérée comme une roche micacée.' Darwin scored the second and third sentences of this passage in his copy.

43–1 Scrope 1829:221 as quoted except that the sentence ends 'granitic rocks of the cataracts of the Orinoco, and apparently produced by long exposure to sun and moisture.' The land referred to is the insular rock Scoglio della Botte belonging to the Ponza or Pontine Islands. Scrope referred to the rock composing the island as graystone, rather than trachyte, though he described graystone as being 'closely related to trachyte' (p. 214) and left open the question as to whether graystone should be considered as belonging to the family of trachytes (p. 224).

44–1 William Bollaert: personal communication. Bollaert's observations were communicated by Darwin to the Geological Society of London on 31 January 1838. See Bollaert 1833–38:598–99, 'The following section of the principal shaft will illustrate the nature of the Panizo deposit [of the mine]. 1. Caliche. This bed contains

Lake let out by steps in Central France not very conclusive proofs, but certainly probable. Bulletin de la Soc. Geolog: 1833–34. p. 35.— Ancient Lake Lemagne in Auvergne

---

Proofs from Phryganea[2]

45e NB. Sedgwick talks of LAMINATED structure (∴ separation of ingredients) as uniting with cretionary.— it may ‹of› come of use in discussion on Cleavage &c
Geolog Transacts. Vol III. p 1. p. 86. et p 95.—[1]

46e It is easy to prove. (pyrites, agates, calcareous balls) that concretions are connected with a crystalline process.— now cleavage as suggested by Sir J. Hershel is all crystals obeying one law of crystallization.[1] therefore concretions in this case laminar. hence the thick wedges of feldspar in gneiss.—

47e Veins in septaria. a kind of concretionary process (analogous to layers of quartz & feldspar) within other concretion.—

---

state last page thus. point of attempted crystallization, & therefore as a consequence aggregated (I assume the same force which draws together two particles of Carb. of Lime, tends to crystallize them as seen in stalactite).— some force crystallizes minerals in layer. therefore aggregates them in layer.—

48e So that layer of feldspar in gneiss is identical with *layer* of flint on calc.: sandstone. (& as I believe most strata) (Hence endless passages from gneiss to granite): Why not horizontal? Why have particles in such cases moved more

45 *page crossed pencil.*
48 depth] *uncertain reading.*

near the surface a large quantity of common salt, and occasionally a few small papas [nodules of ore] are found in it. . . .' Also see *GSA*:233–35.

44–2 Lecoq 1833–34:34–35, 'Outre les données que Crouelle présente, pour trouver le niveau des eaux de l'ancien lac, on en trouve de nouvelles dans le dépôt des arkoses immédiatement superposées au granite, et formant, presqu'à la même hauteur, des couches inclinées sans avoir été soulevées, et indiquant aussi toute la limite *ouest* des eaux du lac; mais ce sont surtout les calcaires à phryganes qui présentent sous ce rapport les faits les plus curieux. On les rencontre dans un grand nombre de localités, et à des hauteurs très différentes qui indiqueraient que le Léman d'Auvergne aurait éprouvé des changemens notables dans la hauteur de ses eaux.

M.C. Prévost demande si l'on peut conclure de la présence des phryganes a des hauteurs variées, des abaissemens et des élévations successifs dans la surface du lac, . . .

M. Lecoq répond que l'on peut admettre des abaissemens successifs, mais que l'on ne peut affirmer qu'il y ait eu des cours d'eau assez forts et assez prolongés pour élever la surface de l'eau de manière à ce qu'elle soit indiquée par une formation quelconque.'

45–1 Sedgwick 1835:86–87. Also see the section from pp. 94–98 on the relation of laminae and concretions.

46–1 John Herschel's comments on cleavage in relation to the process of crystallisation were contained in his letter of 20 February 1836 to Charles Lyell. For citation of the relevant passage from the letter see note RN32–1. Portions of the letter, though not those on cleavage, had been read at the 17 May 1837 meeting of the Geological Society of London. See Herschel 1833–38:548–50. By that time Darwin was familiar with the content of the letter.

laterally than vertically, in concretions more vertically than laterally.
— ‹In Area of this›
If surface covered with oil should shrink. film parallel to longer axis.

But if great depth

49e NB. Prof ‹Henslow›[1] Sedgwicks lamination parallel to stratification evidently small scale of concretionary action[2]

---

all fluid at once, the films vertical.

Ascertain law of attraction of particles of same nature: then get mathematician to when two particles ‹would› are aggregated, would they not attract strong. a third.— & this would make layers.— **(Gravity can have no effect, on particles of equal weight.—)** ? cleavage not vertical ∵ combined with gravity.—

50e hence changes in dip of no sort of consequence.— Therefore <s of inclination «varies with chemical attraction &c.» becomes measure of force. ‹∴. where little inclination, little force & varying direction.—› Therefore in PILE of mud from Trapiches. inclined

51e layer!!!.— The separation in the Ponza case of Scrope parallel to walls of dykes—[1] Mem. laminated dikes in Cordillera.!!!—

In stratum OP. let force drag particls to line AB, & likewise gravity MN. Then every particle would tend to meet at ‹B. but if particls attract each other in some increasing ratio in proportion to proximity would they not unite in B.K.›

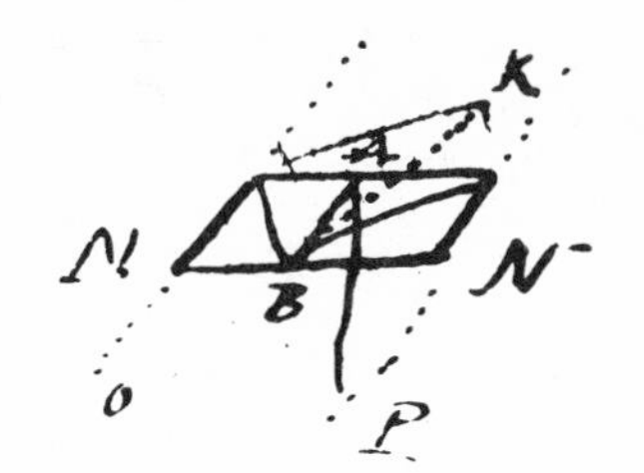

49 all fluid at once.] *follows last line on page 48.*
(**Gravity can . . . equal weight.—**)] *added pencil.*

49–1 John Stevens Henslow.
49–2 See note 45–1 and Sedgwick 1835:94. 'These various modifications of *small concretionary structure* derive a great interest from the consideration, that in them are exhibited, on a minute scale, the same peculiarities of aggregation, which, on a great scale, form the most extraordinary features of the deposit I am describing.'
51–1 Scrope 1829:216–17, 'Here also, as in Ponza, there exists every indication of the zones [of prismatic trachyte of different shades of white, yellow, blue and brown] having been drawn out in their direction while in a half-liquid state, after a partial separation had taken place of the pure feldspathose part from the mixed siliceous base; . . . The direction of the zones . . . seems usually to coincide with that of the bed or dyke itself. . . .'

52e on the diagonal of BK.— ‖ This is not applicable. it does not explain CLEAVAGE of rock— nor the Falkland case, nor. the arrangement of particles of granite in Henslow's Grit[1], yet it is worth consideration. especially effect of gravity, versus some fault explaining vary dip & inclination.—

53e which last is strong character.— A discussion on concretions and cleavage conjoined very good.— It is the Key to the story.— consider stalactites.— agate rings, crystallization transverse.— or rather radiating to central point. can cleavage be radiation from some grand centre.—

A Stalactite of Gypsum, is the best case of cleavage.—

54e Phillips (113) «Lardner Encyclop.—» absolutely considers gneiss an aqueo deposit resulting from disintegrated granite!!![1] Look at gneiss of Rio

---

Concretions in Pumice bed at Ascension instance of hollow concretions & concretion filled with unconsolidated matter—

55 Phillips Lardner p. 197. refers to salt as being produced by local heat,[1] **Ask Capt. Beaufort, whether, water flashing into steam, would Babbage .— Webster**[2]

Phillips insists of analogy between Australia & Oolitic period.— comparison rather loose.— perhaps worth[3] Says from Lardner's (p. 213) form of escarpment relation kept to sea coast ∴ curious exception in Wealden.—[4]

53 centre.—] *short rule line added pencil.*
A Stalactite . . . of cleavage.—] *added pencil.*
55 **Ask Capt . . . Webster**] *added pencil.*
Phillips insists . . . worth] *crossed ink.*
page crossed pencil.

52–1 Henslow 1821–22 does not refer to granite in grit but see plate 16, fig. 8 illustrating the 'Arrangement of particles in the stratified grit at Bodorgan' and also p. 395. In his own copy of this work Darwin has triple scored the bottom paragraph on p. 395 and placed quotation marks around the following sentence, 'The particles of the quartzose fragments appear likewise to have undergone a partial re-arrangement; for several contiguous fragments possess a common cleavage'.

54–1 J. Phillips 1837–39, 1:113–14, 'Now it is impossible to doubt that clay slates and grauwacke slates have been deposited in water: it is equally certain that the gneiss and other felspathic or quartzose rocks, which are associated with it, and occasionally with clay slate, are also of aqueous production; and the composition of gneiss, &c., completes the evidence wanted to prove that the primary strata analogous to sandstones and clays were formed from the waste of granitic rocks.'

55–1 J. Phillips 1837–39, 1:197, 'It is very important to remark that the salt lies always in small narrow patches; therefore, most evidently it was not produced by a *general* extrication from the marine water, and most probably is to be referred to *local heat*, or some other cause at great depths, or else to evaporation from a limited area, filled at intervals by the sea.'

55–2 The three men whose opinions Darwin reminded himself to seek on the subject of water flashing into steam were Francis Beaufort, Charles Babbage, and Thomas Webster. All were known for their knowledge of the physical sciences. Beaufort, the hydrographer, was expert in meteorology; Babbage, the mathematician, was known for his attempts to mechanise computation; and Webster, Secretary of the Institution of Civil Engineers (1837–39), had only recently published on steam (T. Webster 1836). Like Darwin all three men were active in London scientific circles during the late 1830s. It should also be noted that Thomas Webster the authority on steam was a different man from Thomas Webster the geologist, though both were Fellows of the Geological Society of London.

55–3 J. Phillips 1837–39:1. See the section entitled '*Organic Remains*,' pp. 207–11, which includes the statement on p. 208 that '. . . it is interesting to know that the

56 Would crystals arrange themselves in that direction, in which most substance lies ‹.—›?

Phillips. Lardner's p. 270–4, good discussion showing present form of land in Northern England influence dispersion of Boulders.—[1]
See Rogers for Southern limits

57 of Boulders in N. America[1]

---

do/p. 280. the gravel beds in England different from Boulder beds—[2]

What is Osteopora platycephalus (Harlan) found on the Delaware. is it Edentate? Phillips p 289.—[3]

---

58 Alludes to big bones in interior at Falkland Is^{d}.— Peron does as if well attested.—[1]

There is no difference between dike & mountain axis. except in relative ‹strata› size with superincumbent strata. where they have yielded conical axis of mountain.—

---

56 270] 'o' *over* '4'.
page crossed pencil.
57 What is . . . p 289.—] *crossed ink.*
page crossed pencil.
58 axis of] 'f' *over* 'n'.
page crossed pencil.

earliest mammalia, of which we have yet any trace, were of the marsupial divison, now almost characteristic of Australia, the country where yet remain the trigonia, cerithium, isocardia, zamia, tree fern, and other forms of life so analogous to those of the oolitic periods.'

55–4 J. Phillips 1837–39, 1:213, 'The minute flexures, irregularities, and breaks in the ranges of these [lias and oolitic] formations, can only be understood by consulting a good geological map; but the preceding notices will suffice to show how remarkable is the effect, in the geology of England, of their parallel courses from sea to sea—from Yorkshire to Dorsetshire. In this respect their ranges are of great importance, offering to the inquiring mind a proof of the long succession of quiet processes by which the bed of the sea was gradually filled with a regular series of varying deposits. . . . The Wealden formation, in this, as in all else, contrasts very strongly with the truly marine deposits. It makes no part of this parallel series, but lies principally in Kent and Sussex, . . .'

56–1 J. Phillips 1837–39, 1:270–74, including p. 274, 'The most prevalent direction in which the blocks have been transported in the British isles, is from north to south; but . . . the natural configuration of the ground appears to have had considerable influence in determining many minor currents.'

57–1 Rogers 1835b as cited in J. Phillips 1837–39, 1:277. In Darwin's own copy of Phillips the reference to Rogers is scored. See also *JR*:614.

57–2 J. Phillips 1837–39, 1:280, 'Many parts of England are almost totally free from the accumulation of proper diluvium,— as the Yorkshire coal field, the Wealden denudation, large tracts in North Wales, the vicinity of Bath, &c. But these districts contain abundance of local gravel deposits, which sometimes appear to be quité as ancient as the diluvium, and may justly be styled "Ancient Alluvium;" . . .'

57–3 J. Phillips 1837–39, 1:289 lists the fossil as being found in a diluvial formation in Delaware.

58–1 In discussing the formation of coral mountains Peron 1804:476 referred in passing to 'd'ossemens énormes qu'on observe aux Malouines [Falklands], bien avant dans l intérieur, . . .'

59 only when dikes reach near the surface. that strata yield.—

In Undulation in open ocean. as pebbles would be lifted up & down. on coast itself, undertow would draw it outwards.— form of breaker affected some way out to sea.— ? effects on bottom a thing floating some way from coast is driven on to it.— rollers at Tristan d'.Acunha.— **silting up. channels on coast of England—**

60 Any one. who has studied rocks in detail as amygdaloid. calcareous rocks of Ascension, each particle coated. &c will be aware how little common Gravity has to do with arrangement of particles in rock. This applies to cleavage & concretions.—
Septaria in concretion arranged in planes, case of separation.— the branching cracks— only bear relations to VEINS in primitive rocks—

61 Are substances soluble under great pressure? equally with little pressure? An important question! If water yields substances from impact, «it» would look like it.

Are greenstone dike in Granite residual matter of upper quartzose ones & felspar.??

Are the great crystalls, & the layers first of felspar & then quartz &c, owing to separation having taken place most gradually, first the more fusible substance, & then the next being sucked out.

62 In Cleavage discussion, state broadly indication of new law acting in certain directions predominantly, connection with magnetism &c counteracting gravity.—
As volcanic eruptions are accompanied by horizontal elevations, so are injection of mountain chains. accompanied by do.— Give this after supposition

59 **Silting up . . . England—**] *added pencil.*
page crossed pencil.
60 VEINS] *double underlined pencil added.*
page crossed pencil.
61 question! If] *alternate readings* 'question ? If' or 'question: If'.
Are . . . it.] *ink.*
Are greenstone . . . sucked out.] *pencil.*
page crossed pencil.
62 In Cleavage . . . gravity.—] *pencil.*
As volcanic . . . supposition] *ink, brace in left margin.*
page crossed pencil.

63 p. 461 «of Proceedings» List of collections in Geological Society.[1]

Pumice at South Shetland. Geological Society—

Dikes have not been the moving agents, because not wedge-formed.— Hence fill up fissures— If dikes effect of horizontal elevation excepting fissures from above unite with those from below. would always thin out above which explains a difficulty.—

64 All De la Beche's reasoning of mountains being formed by crust being too large & pitching against each other, is, I suspect much weakened by ‹vi› considering how close the dislocations occur & therefore that the crust might be considered a level.—[1]

---

65 Dikes being last action. (effect of horizontal movement) hence generally intersect metallic dikes: It is an important view being subsequent to

---

dislocation of strata.

A capital discussion might be made between dikes & «axis of» mountain-chain in proportion to weight of super [. . .] mass.—

Absence of Caverns, in Plutonic rocks argument against great bodies of vapour. according to Hopkins theory.—[1] general presence of dikes. argues in favour of pressure of liquid rock. **Andes discussion—**

63 p. 461 . . . in Geological Society.] *crossed ink.*
page crossed pencil.
64 ‹vi›] 'i' *uncertain.*
page crossed pencil.
65 super [. . .]] *half a word illeg, plausibly* 'superincumbent'.
Andes discussion—] *added pencil, left margin.*
page crossed pencil.

63–1 At the annual meeting of the Geological Society of London on 17 February 1837 the committee appointed to report on the state of the society's museums and libraries presented a 'Summary of the Foreign Collection' listing the number of drawers of specimens held for various localities. See Anonymous 1833–38: 461–62.

64–1 De la Beche 1834:118–32, including the statement on p. 121, 'If we suppose with M. Élie de Beaumont that the state of our globe is such that, in a given time, the temperature of the interior is lowered by a much greater quantity than that on its surface, the solid crust would break up to accommodate itself to the internal mass; . . . We should also be led to anticipate . . . that under favourable circumstances, broken and tilted masses would be thrust up into ridges or mountain chains.' On the inside back cover of his copy of the book Darwin wrote, 'Every mountain chain may be considered as the ruin of an earthquake, aided or obliterated by time! It is vain to bring first & other causes to bear they are comparati[vely] insignificant.'

65–1 Hopkins 1836:7, 'It is easy . . . to conceive such a[n] [elevatory] force to act as above supposed, if we assume the existence of a cavity beneath the elevated mass, . . . Any vapour or matter in a state of fluidity from heat, forced into this cavity, or expanded there, will produce the elevatory force which I assume to have acted.'

66 Albite certainly contains 6 per cent more silica than common felspar therefore on axis of Cordillera, in Andite – containing 80 per cent of Albite

$$\frac{80}{100} \times \frac{6}{100} = 480$$

In Falkland islands. & generally where rock metamorphic & thickness of ‹strata› not great, one can conceive anticlinal lines near. (lateral pressure would always produce it) but where great thickness **is** affected, **they would be** far off

67e In Discussion on dikes argue impossibility on fissure going right through superincumbent mass (varying hardness,— takes time to trace) from few dikes which have given rise to eruptions.— We must suppose everywhere—, in granitic areas &c &c

volcanos

fissure

dike.—

thus dikes terminated

68e Solubility of fluids varies with temperature ? with pressure?

Salt on surface of plains due to whole moisture being lost by evaporation therefore capillary attraction would bring water with salt to surface

---

Lyell remarked to me that Kylow(?) was astonished with him that ‹th› gneiss, mica-slate of whole kingdom of Norway was contorted yet no mountain chain case parallel to Banda Orientel. ask Lyell for sentence.—[1]

69e Origin of Breccia, introduce in Cordillera discussion, deep sea, fragments fall off cliffs. but then how spread abroad?—

---

66 x] *uncertain reading of symbol.*
is] *added pencil.*
, they would be] *added pencil.*
Albite certainly . . . 480] *pencil, crossed.*
In Falkland . . . far off] *ink, crossed pencil.*

67 everywhere—,] ‘,’ *over* ‘—’.
fissure . . . terminated] *added pencil.*
page crossed pencil.

68 Solubility . . . to surface] *crossed pencil.*

69 bay] *alternate reading* ‘being’.
Origin . . . abroad?—] *crossed pencil.*
There is . . . turbulence.] *crossed pencil.*

68–1 In July 1837 Charles Lyell geologized in Norway in the company of Baltazar Mathias Keilhau. (Lyell 1838b) Presumably Darwin intended to ask Lyell for a reference. On Banda Oriental see *GSA*:165.

There is thus wide difference between erosive power of river & sea.; the former as its channel becomes wider looses its cutting power. (as does it when the inclination becomes less & ∴ tends to finite power) whereas sea. on coast, as long as exposed to waves of sea, cutting power increased with width. for besides more surface exposed. bay more open to turbulence.

70e Bull. Soc. Geolog «1837» p. 320. paper on shrinking of Clay. applicable to Cleavage. C. Prevost.—[1]

In Cordillera. a rush of water will account for filling up of valleys— subsequent opening a medial gorge by slow erosion. but we have evidence in distribution of blocks, that there has been no tumultuous rush.— besides general improbability. stratification, If chain of lake. ‹a›

71e the alluvium would form a succession of flights of steps; if one lake then we must suppose barrier in the very part, where barrier least probable.— The sea harmonizes well with character of mouth of valleys &c; Pampas.— If blocks above their parent rocks. would be prove of subsidence.— removal downwards by successive torrent spread out. by sea— beach action **— no one will dispute. sea. once came to Mendoza—· Will they introduce other causes to explain «alluvi» in valleys**

72e Lowe in his paper says land shells found with calcareous matter & concretions on *coast* of Madeira.?[1] How came it if this powder results from «decomposed sea» shells, that land shells should be preserved in it— some error? **(because more recent)** ———— Coquimbo on. other hand?—

The widening a valley depends on serpentine course.— the latter (it is generally said) is consequence of ‹rapid› slow course, &

73e with slow course small erosive power. therefore tendency of running water to deepen not to widen valley.— Why is serpentine course result of little inclination??— — It is simply as the inclination is little the force required to move ‹it› «stream» aside is not great.— Is there more degradation at first

70 a rush] 'r' *emended in pencil.*
page crossed pencil.
71 &c;] *alternate reading* 'on;'.
no one . . . in valleys] *added pencil, bottom page and right margin.*
72 **(because more recent)**] *added pencil.*
Lowe . . . hand?—] *crossed pencil.*
73 *page crossed pencil.*

70-1 C. Prévost 1836–37.
72-1 Lowe 1833b describes land molluscs in Madeira, those on the coast being the species numbered 13, 19, 27, 31–33, 36, 46, and 48. For number 46, *Helix polymorpha*, one variety (*irrasa*) is listed as found in tufa, another (*arenicola*, sub-variety 2) as found in sandy chalk. See also the footnote in Lowe 1833b:64.

angle owing to momentum. which the water has obtained.— If inclination be great where arrow stands the force immediately deflected from (B) which would not have been case. if inclination small.—

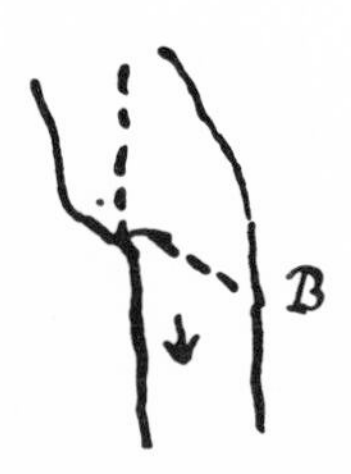

74e The power of widening channel depends on power of deflection with stream retaining its force, now it will be evident that deflected stream cannot retain its force if inclination be great. There could «not» be great deflection in a "rapid".— is a familiar illustration.— Therefore stream has no tendency to widen course until inclination is become comparatively small, & when that is case force is lessened. therefore rivers very ineffectual in widening valley.— it is essentially a deepening agent

75e  Therefore when we have

valleys of this structure. as the inclination in all probability would be greater when flowing over (B) than when at (C) its tendency would ‹cut› be to cut a narrower channel instead of wider.— This applies to all vallies (except mere talus «over cliffs edge» of which limit cannot be great over) with very gently sloping sides This argument is partly taken from Delabechs Theoretical Researches[1].—

76e Athenaeum. 1838— p. 137. Three inosculating rivers in Southern America ? effect of subsidence—[1]

‹Is there same.› Institute. 1838 p. 40 or Phil Mag. Dec 1837. p. 520 Mr Fox on increase of temperature at great depths.[2]

All Earthquake unaccompanied by Volcanos must be sought after proofs of sinking.— No Sweden!! swelling of rock from Heat.

76 *page crossed pencil.*

75–1 In this passage Darwin drew on De la Beche 1834:184–205. In his own copy of the book Darwin wrote atop p. 198, 'as long as stream rapid forms gorge straight (why?) then zazgy [*uncertain reading*], widen it, but could not produce sloping side/?' Alongside Fig. 37 (p. 198), Darwin also referred to p. 200 on which he double scored the passage, 'Undoubtedly the flatter the surface the more irregular the windings of a river, all other things being equal.'

76–1 Humboldt 1838b:137, 'The great rivers of Guyana; the Orinoco, with its tributaries, the Paragua and Caroni; and the Rios Negro and Branco flowing into the Amazons, form so many sytems really distinct; yet the basins of those systems are still so imperfectly developed that in some places it is hard to draw the line of separation between them. In the rainy seasons the high plains between them, or round their sources, are inundated, and great lakes or canals are formed, which unite for a time streams flowing in opposite directions.'

76–2 Fox 1837:520–23 and Fox 1838:39–40. The article, with its tables, reports on Fox's observations on the increase of temperature in mines as one descends from 100 to 300 feet beneath the surface of the earth.

77 Specific gravities of many artificial limestones produced by Sir J. Hal. End of pages. p. 157. Vol VI Edinburgh. Phil: Transacts.—[1]

Does the isothermal subterranean line moves upward from effects of Elevation if not crust much thinner beneath ocean than above it

---

no because heat proceeds from great body of mass.—

78 The last speculation becomes important with respect to thickness of crust broken up.— — My view of Volcanos &c &c

---

This view will bear much reflection on method of cooling — Very difficult subject. PP—

I think from dislocation taking place chiefly beneath water & volcanos. crust must be thinner «under water» but cause most difficult (better conductor)

79 Fitz Roy's Case of S. Maria & Tubul applicable to Andes & Patagonia—[1]

On Lyells idea of whole centre of earth same heat, then change in form of fluid

77 Specific . . . Transacts.—] *crossed pencil.*
Does . . . mass.—] *crossed.*

78 This view . . . cooling—] *marked connection with sentence* 'The last speculation . . . broken up.— '.
Very difficult subject. PP—] *marked connection* ('PP—') *with* 'On Lyells . . . would be thicker.—' *on next page.*
page crossed pencil.

79 On Lyells . . . thicker.— PP] *marked connection* ('PP' *in left margin*) *with* 'Very difficult subject.' *on previous page.*
page crossed pencil.

77-1 Hall 1812:156–57, begins the section where Hall showed how his fusing of limestone under pressure in the laboratory supported the geological theory of James Hutton. Page 180 contains a table of specific gravities for artificial limestones and marbles produced in Hall's experiments; the general discussion of procedures and results is on pp. 177–83.

79-1 FitzRoy 1836:327–8, Darwin applied the information in a paper delivered to the Geological Society of London on 7 March 1838 (Darwin 1833–38b:659), '. . . he [Darwin] also stated his belief, that the earthquake of Concepcion marked one step in the elevation of a mountain chain; and he adduced, in support of this opinion, the fact observed by Capt. FitzRoy, that the island of Santa Maria, situated 35 miles to the south-west of that city, was elevated to three times the height of the upraised coast near Concepcion; or at the southern extremity of the island, eight feet, in the middle, nine feet; and at the northern extremity, upwards of ten feet; and that at Tubal [Tubul], to the south east of Santa Maria, the land was raised six feet; . . .'

centre would lift with it isothermal line, but if heat from centre, then crust of solid earth would be thicker.—[2]
PP

Andes mark the line between sinking & rising areas.—

80 In Earthquake if Subsidence we should not expect volcanos.— not so much horizontal oscillation. or so many *shocks* directly after great shock —

It appears to me unphilosophical to think calcareous springs near coral reefs.— Where vegetation luxuriant it might be almost as well said probably much Carbonic Acid gaz here.—

81e [top portion page excised, not located]

Bull:. Soc: Geolog. Tome IX 1837–8. p. 24. rocks of Chimborazo., & Pichincha. Melaphyre.[1] = **Andesite— Albite & amphibole=**

Cook found Granite at Christmas Sound Vol XIV. (My Edition) p 500. Well described[2]

82e [top portion page excised, not located]
— do— [. . .]

---

80 *page crossed pencil.*
81 Well described] *circled.*
= **Andesite . . . & amphibole** =] *added pencil.*
Bull: . . . Melaphyre.] *crossed pencil, excised.*
Cook . . . described] *excised.*
82 [. . .]] *2 strokes illeg.*
Subaqueous . . . same] *crossed pencil.*
Subaqueous . . . way.—] *excised, cut through line* 'Sea . . . way.—'
Sea . . . same] *excised, cut through line* 'Sea . . . way.—'
83 The preservation . . . lessened.—] '?' *and bracket left margin.*
page crossed pencil.

79–2 Lyell 1837, 2:317, 'If, then, the heat of the earth's centre amount to 450,000° F., as M. Cordier deems highly probable, . . . it is clear that the upper parts of the fluid mass could not long have a temperature only just sufficient to melt rocks. There must be a continual tendency towards a uniform heat; and until this were accomplished, by the interchange of portions of fluid of different densities, the surface could not begin to consolidate.'

81–1 Humboldt 1838a:24, 'Les découvertes géognostiques récentes, ont montré que ces masses intercalées qui tantôt ont la forme de dômes arrondis, tantôt celle de vastes cratères, n'ont point la même composition minéralogique dans tous les pays. Ainsi, aux Canaries et dans les Sept-Montagnes, la roche est un véritable trachyte feldspathique; à l'Etna et à Stromboli, au Chimborazo et au Pichincha, on a un mélaphyre approchant du basalte; dans les volcans du Chili, au Puracé, au Tolucca, c'est l'*andésite* (roche composée d'amphibole et d'albite), qui joue ce rôle; enfin la Somma, qui forme les parois du cratère de soulèvement du Vésuve, est composée de *leucitophyre* (mélange d'amphigène et de pyroxène augite).'

81–2 A note by Georg Forster in an edition of James Cook's second voyage (Kerr 1811–17, 14:500) reads, 'The rock [forming an island in the vicinity of Christmas Sound] . . . is a coarse granite, composed of feld-spath, quartz, and black mica or glimmer. This rock is in most places entirely naked, without the smallest vegetable particle; . . .' Darwin's copy has not been located.

Subaqueous. removal, shown by the number of bones lying at the bottom of sea. off coast of England.—

Sea must always on actual beach act same way.— a little further from beach action probably modified by form of waves & currents.— but this must be continued. no currents & elevation have same

83e effect, a tendency direct (or oblique) outwards may be granted. independent of currents.— *mud* going out can actually be seen.—
? The preservation of dikes & ledges of first-rate importance in showing not subaqueous removal—??? the difficulty of such preservation certainly is lessened.—
Coral flats. argument for Heaping up.— *very good*

---

this will show effects.— analogous to broad flat sand beach.— De la Beches argument of low coast gaining & high loosing answered by this[1]

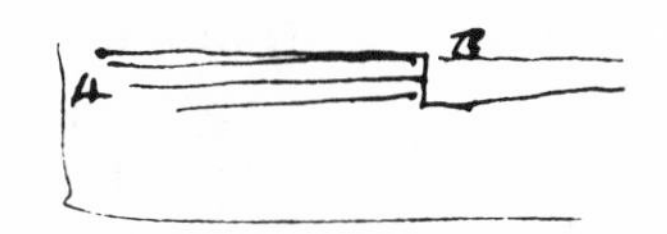

— No one can doubt. A–B once formed low coast.—

84e Annales des Mines. a translation of paper by rose on Greenstone, diorite, &c most important.:— must be studied.—[1]

Scientific Memoirs Edited by Taylor Ehrenbergh on flints in chalk must be studied— though I do not think good p. 411[2] When discussing concretions

---

Carbonate soda. formed by Ca. of L. & Mur. of Soda mixed.— Turner's Chemistry p. 206[3]

84 Annales . . . studied.—] *crossed pencil.*
Scientific . . . concretions] *crossed pencil.*
Carbonate . . . 206] *crossed pencil.*
85 elevation &] '&' *over* '—'.
page crossed pencil.

83–1 De la Beche 1834:58, 'When land rises above the level of the sea, the action of breakers tends constantly to destroy and remove it: when land is low and only rises to the same level, the same action tends constantly to throw detrital matter upon it.' In his copy of the work Darwin wrote 'No' in the margin against this passage and reminded himself to 'Study Mr Palmers Papers in Royal Transactions' (Palmer 1834).
84–1 Rose 1835.
84–2 Ehrenberg 1837b:411, 'It was natural for me now to test again the flint of the chalk, which I had before often examined: . . . The black flint, which broken into small pieces is transparent, showed no evident traces of an inclosure of microscopic organic bodies, but such are easily perceptible in the whitish and yellowish opake pieces. . . . The chalk-like envelope, and white covering of the flint does not effervesce with acids, and is therefore not chalk, but silica . . . ; it does not appear to originate in decomposition, but is like the meally covering of a lump of dough; that is to say, it is that layer of siliceous meal (of evident organisms) which at the formation of the flint has only been touched by the dissolving or metamorphosing matter, but not completely penetrated by it. . . .'
84–3 E. Turner 1837:206, 'But if carbonate of lime and sea-salt are mixed in the solid state, and a certain degree of moisture is present, carbonate of soda and chloride of calcium are slowly generated;. . . The efflorescence of carbonate of soda . . . which in some countries is found on the soil, appears to have originated in this manner.' Mur[iate] of soda is an old term for common salt.

85 Both Beck & Deshayes saw fossil shells from West Indies & declare them to be recent species— Lyell—[1]

Some internal changes are in process. connected with variation of compass & these may cause «or be effect of» elevation & subsidence. examine these «lines»

---

86 Description of rocks in Lyells'. Capital Norway case.— The fragment. consisted of hornblende (?) & felspar, (some crystals being red) «with» cleavage, veins of pyrites, few curious fissures; base in part. block not crystallized[1]
Salband like basalt. full of circular cryst of glassy felspar different from either fragment or dike, blackish grey base. crystals from fragment disseminated on that side of salband. gradually becoming finer grained & more compact on that side— separation DISTINCT from dike junction mechanical: DIKE base reddish feldspathes with grenish. black specks of hornblende, large irregular cryst of reddish felspar. & scales. of mica.— large cryst of Hornblende *blending* into base— Salband might have oozed out of cleavage plates: the crystals

87e must have recrystallized, as such do not occur in either dike or fragment. junction certainly most distinct on dike side.— oozed from one of the true rocks, most probably from the gneiss beds in the mica slate.—[1]

Geograph. Journal. Vol IV (p 321) Mr Hillhouse describes central granitic ridge of Guayana as NW / SE.[2] Vol VI. p. 247. Mr. Schomburgk NW. numerous boulders of GRANITE"[3]

86 feldspathes] 'feldspathes' *over* felsdspathes'.
'feldspathes' *alternate reading* 'feldspathic'.
87 Geograph. Journal. Vol IV . . . p. 316 &328] *crossed pencil.*
VI. p. 365 . . . Journal] *crossed pencil, excised.*

85–1 Charles Lyell: presumably personal communication. Gérard Paul Deshayes and Henrick Beck aided Lyell in his work on Tertiary strata by judging the dates of fossil shells from various localities. See also Lyell 1837, 4:23, 'Of thirty species [of fossil shells from the West Indies] examined by M. Deshayes from this rock, twenty-eight were decidedly recent.'
86–1 Charles Lyell: presumably personal communication since the details Darwin cites differ slightly from those offered in Lyell 1838a:173–74. The fragment in question would, however, seem to be that labelled 'b' in Fig. 94 (p. 173). In his copy of this work Darwin queried whether Lyell had faithfully represented the cleavage in the fragment labelled 'b', and also the relative position of fragment 'b' and the clear salband.
87–1 Compare this paragraph with Lyell 1838a:172–74 for discussion of a salband, or selvage, between a dike and its surrounding country rock.
87–2 Hilhouse 1834b:321, 'Extending south-east and north-west [in Guyana], is a central granitic ridge, unbroken except by the river Massarony, which circumscribes one of its western forks; . . .'
87–3 Schomburgk 1836:246–47, 'Our journey to the southward across the savannahs, on the eastern bank of the river, was to commence next morning (the 19th December): . . . The chain of mountains [Sierra Conocon] is here a short distance from the house. One of the Indian boys brought me a beautiful piece of crystallized quartz, with laminæ of mica. On my return from the Corona, I examined the mountains, and found the crystals were partly embedded in gypsum. Direction of strata northwest, and the place surrounded by numerous boulders of

"direction of strata on the Berbice N. 35°. E. dip to NW to 80° faults with red wacke contorted evidently dike. V. VII. p. 316 & 328[4]

VI. p. 365. Meyen on Chile must be studied Analysis of Voyage: many observations on heights of valleys in Chile Geograph. Journal[5]

88e Vol. VII p. 216.— Guava trees, introduced about twenty years since (1835) from Norfolk Is[d] into[1]

Geograph Journal Vol VII p. 279. Carcases of birds drifting out to sea[2]

---

88 **Geograph Journal**] *added pencil left margin.*
Vol. VII p. 216 . . . into] *crossed ink.*
Vol. VII p. 279 . . . of facts] *crossed pencil.*
See page 101 . . . rock— —] *excised, crossed pencil.*

granite.' Darwin scored the sentence following the ellipsis points in his copy.

87–4 Schomburgk 1837:316, '. . . these [rocks] on the Berbice are more of trappean origin: the direction of their strata is N. 35° E.: they dip to the west by north, and the strata have evidently been disturbed since their deposition: various examples of cross currents are evident, and the beds are sometimes contorted and cut off by faults, which are filled with a species of wacke of a red colour: the angle of the regular beds amounts to upwards of 80°. . . . Where the current, during inundations, has excavated channels in the soil, I observed numerous boulders of about four feet in diameter, decidedly of the same formation, but much more covered with the black coating before mentioned, and exhibiting ripple marks.' Page 328 again refers to boulders: 'We observed some granitic boulders in the river. The latitude observed at noon was 3° 58′N.; . . .' Darwin scored the first sentence of this passage in his copy.

87–5 Meyen 1836. On the heights of valleys see p. 367, 'From hence Dr. Meyen undertook an excursion into the Andes, along the Rio Tinguiririca, towards the pass of Las Damas. The plain rises suddenly fifty or sixty feet, and continues at that elevation on a level to the very foot of the range. The mountains rise here with great steepness, forming in some places almost perpendicular walls of sienite, rising upwards of 1000 feet.' Also, pp. 367–68, 'About five leagues from Tollo, the narrow glen, through which the river runs, widens to a pretty valley, which is covered with the fruit-trees of Europe, its elevation being so high that heavy snow-falls are frequent; and the snow remains for a considerable time on the ground. At the junction of the Rio del Yeso with the Rio Maipù, Dr. Meyen quitted the road conducting to Mendoza, and entered the mountain-passes. About two miles farther up, the Rio Maipù is joined by the Rio del Valian, which, as well as the Rio del Yeso, comes down from the north-east; but the Rio del Valian is much larger, nearly as wide as the Rio Maipù at their confluence. Here Dr. Meyen estimates the elevation of the valley at 4500 or 5000 feet above the sea. The whole course of the Rio del Valian . . . lies in a very narrow glen, . . . The sides of the mountains are mostly bare; where the valley is rather wider, are excellent pasture grounds for cattle and goats, at an elevation of about 9000 feet.' Darwin scored the first two sentences of this passage and underscored 'suddenly fifty or sixty'.

88–1 F. D. Bennett 1837:216, 'The aspect of the lowlands of Tahiti has latterly undergone a considerable change, from the extent to which the guava shrub flourishes on the soil. Scarce twenty years have elapsed since this fruit tree was introduced from Norfolk Island, and it now claims all the moist and fertile land of Tahiti, in spite of every attempt to check its increase.' Darwin scored this last sentence.

88–2 F. D. Bennett 1837:229, 'In 2° 53′S., long. 174° 55′ E., observed a remarkable line of froth on the sea, some yards in width and of great extent, and accompanied by a mass of dead birds, fish, shells, drift wood, &c., which seemed to indicate the limits of a current, and in fact we found that after entering it we lost the strong N.W. current that had hitherto accompanied us.' Darwin double scored this passage.

do p. 358. changed soundings in Mouth of S. Cruz in connection with Fitz Roys fact of elevated block of stone.— & Caldcleughs collection of facts[3]

See page 101. in Note Book (C) for some speculats on conducting powers of rock— —

89 Geograph Journal Vol IV p. 36. on subsidence of the land in Guiana, worthy of consideration.[1]

When discussing nucleuse's of old volcanos within Cordillera— allude to Lyell's view of not discovering dike one end granite & other trap.—[2] It is in the mountain masses we must look for that.— how few isolated volcanos there are. where one alone has been formed — Look at the now active volcanos & see what high they are

90 «See Athenæum. 1838. p 274. probably will be published in the Geograph. Journal.—»
A meeting of the Geograph Soc, April 9 1838. Letter from M. Erhman stating that the mean temp at Yakous in Siberia being –8 Reaumur.— there ought to

89 *page crossed pencil.*
90 *page crossed pencil.*

88–3 Angelis 1837b:358, 'Off the river of Santa Cruz [in 1745] they [the Jesuit Fathers Quiroga and Cardiel] were nearly lost, which leads them to remark upon the great alteration which must have taken place in the depth of water in that river since it was first discovered, and they quote authorities to show, that in former times large ships could safely enter it, whereas when they were there it was blocked up by dangerous sand-banks, upon which they narrowly escaped shipwreck.' Darwin scored this passage. The FitzRoy reference is uncertain, but see FitzRoy 1837:121–22; *JR*:216–18, 224, and 284; and Darwin 1842:415–17. Caldcleugh's facts are probably those in the section 'Evidences of vertical movements' in Caldcleugh 1833–38:445–46, but also see Darwin 1842:427.

89–1 Hilhouse 1834a:36, 'In the centre of George Town, Demerara, Major Staples . . . succeeded in penetrating the depth of the alluvium; and on arriving at the micaceous substratum, which is the indication of the primary formation, a clear spring of water burst out, . . . It was well known, that at ten or twelve feet below the surface, an irregular stratum of fallen trees, of the kind called courida, common on the coast, existed in a semi-carbonized state; but Major Staples discovered, at *fifty* feet below the surface, another immense fallen forest of the same kind of wood, twelve feet thick; the superstratum being blue alluvium, and the substratum reddish ochre, diminishing in shades to yellow, light straw, and again merging into slate-coloured clay. The remainder, to a depth of one hundred and twenty feet from the surface is argil, the lower part of which is of that smooth soapy surface which indicates the purest Wedgwood clay, and would no doubt be of great use in the potteries. It is evident from this, that some few ages ago this continent was habitable fifty feet below the present surface; that it was then covered with an immense forest of couridas, which was destroyed by conflagration, as appears by the ochrous substratum. The sea must, at that time, have been confined to the blue water, where there is now eight or nine fathoms; and whatever may have been the comparative levels then between the Pacific and Atlantic, the level of the water on this side of the isthmus of Darien is now fifty feet higher than it was once—whether before or after Columbus's time is uncertain.'

89–2 Lyell 1838a:217, 'It has . . . been objected, that if the granitic and volcanic rocks were simply different parts of one great series, we ought to find in mountain chains volcanic dikes passing upwards into lava, and downwards into granite. But we may answer, that our vertical sections are usually of small extent; and if we find in certain places a transition from trap to porous lava, and in others a passage from granite to trap, it is as much as could be expected of this evidence.' In his own copy of this work Darwin has scored this passage in the margin and added the remark 'poor'.

be 32° Fah. at a greater depth than 400. & the limit being 400 ft. shows that the strata have very unusual conducting power of heat from

---

centre.—[1] But is this not wrong? we know mean of surface formerly much higher, «so» that we must look at the upper four hundred feet of strata having conducted away the heat of surface. & if conducting powers had been better

91 then 32° would have been found lower.— We have no right to consider the conducting powers either better or worse & the depth of 32°. being little we may confidently infer that time has not been allowed for lower beds to cool down. & then in 50000 years the depth will be greater than ‹5000.› 400.— These facts of SLOW but successive transmission of temperature clearly prove possibility of metamorphic theory

---

92 On the idea of statical equilibrium, the height of lava (habitually) becomes measure of force in that part.— Important as explaining want of levelness

---

Major Mitchell showed me a river ‹near› W. of Port Philip. which had bar at mouth excavated in solid rock.— 4 & 5 fathoms deep. perfectly still water. Major Mitchell inferred subsidence; Mem my remarks on coast of Australia.—[1]

Great NW. dip in SE part of Australia.— Probably a case of rivers turning round & penetrating

91 *page crossed pencil.*
92 *page crossed pencil.*

90–1 Erman 1838b:274. 'I enclose the observations taken three times a day for the year 1827, wherein it results that the mean temperature of the atmosphere at Yakuzk is −5°9′ of Réaumur, which agrees very well with the temperature which I had found near the surface of the ground. . . . The *data* which we hitherto possessed on the increase of the internal heat of the globe . . . indicated from 90 to 100 French feet for an increase of 1° of Réaumur. I did not therefore expect to find the ground thawed at Yakuzk until at a depth of from 500 to 600 French feet . . . and if the actual fact of a thaw at the depth of 400 feet has surprised me, it is only because it has occurred *too soon*; and that it thereby indicates for the strata that compose the ground at Yakuzk a more rapid faculty of conducting heat than is possessed by the strata hitherto examined in Europe.' See notes A117–2, 135–3.

92–1 Thomas Livingstone Mitchell: personal communication. In 1838 Mitchell was on leave in England. Upon publication of his book at the end of August 1838, the information appeared in Mitchell 2:366. 'That changes have taken place in the relative level of land and sea, is evident from the channel of the Glenelg, which is worn in the rock to a depth of five fathoms below the sea level. The sea must have either risen or the earth must have subsided since that channel was worn by any current of water, for it is now as still as a canal, . . .' For Darwin's remarks on Australia see DAR 38.1:812–36. Darwin's St. Helena Model Notebook also records his conversations with Mitchell and includes the entry 'Depth of rivers near mouths'.

93 their own range in Australian Alps.—

Taylors Scientific Memoir, Part IV. p. 403
Ehrenberg on ferrugineous Gallionella[1]
Examine Iron stone of C. of Good Hope & Australia/ and mud of salt-lakes of Rio Negro—**Mr Bowerbank**—[2]

---

94 D^r^. A. Smith's curious specimens of «transversely fibrous» quartz. & iron stone alternating. bear on subject of cleavage[1]

Clay slate. a distinct formation deep «& therefore extensive» water ∴ not formed in modern formation & not ever in Secondary in Europe. gneiss— metamorphosed clay slate.— — shale in shall sea. Lyell confounds these introduce discussion—[2]

I see Lyell talks of *different* composition using difference in metamorphic action which I give at C. of Good Hope.—[3]

95 A bare hill of greenstone, if we know origin of greenstone tells subsidence as plainly as Temple of Serapis.[1] (now we have banished diluvial waves). & likewise ‹tells,› «offers a presumption» it has been excessively slow because beach line chief cause of denudation, but does not tell period.—

I cannot help suspecting that clay-slates have been more frequently metamorphosed than other deposits.— NB. because lowest. first accumulated in bed of ocean

93 **—Mr Bowerbank—**] *added pencil.*
page crossed pencil.
94 transversely] *uncertain reading.*
shall] 'shell' *alternate reading.*
Clay . . . Hope.—] *crossed pencil.*
95 *page crossed pencil.*

93–1 Ehrenberg 1837a:402, 'I had been inclined . . . to assign a great influence in the origin of Raseneisen (bog-iron-ore) to an infusorium discovered by me in 1834, and of which I have . . . given an engraving . . . under the name of *Gaillonella ferruginea*. . . .' Ehrenberg continues, on pp. 402–3, to discuss the association of iron ore with *Gaillonella ferruginea*.
93–2 James Scott Bowerbank.
94–1 The specimens referred to here were presumably collected by Andrew Smith during his sixteen year residence in Africa at the Cape, possibly during the 1834–36 expedition he led from the Cape to central Africa from which he returned with 799 geological specimens (Andrew Smith 1836b:412). See also note RN32–2.
94–2 See Lyell 1838a:223 and 521–23 for discussions of clay slate, including observations that the rock 'resembles an indurated clay or shale' (p. 223); that it can be 'variable in composition . . . and be said to belong almost equally to the sedimentary and metamorphic order of rocks' (p. 522); and that had it been subject to more intense plutonic action it might have been transformed into 'more perfectly crystalline rocks, such as are usually associated with gneiss' (p. 523).
94–3 See *GSA*:165, VI:148–50; and DAR 38.2:904–9.
95–1 A monument of antiquity celebrated by Lyell as showing signs of having undergone elevation and subsidence since its construction. See Lyell 1830–33,1:frontispiece, 449–59.

96 With the exception of sandstone rare to have any *horizontal* non cleaving beds. metamorphosed.

The chemical action which gives polarity to atoms in slates that cleave, & which unite the homogenious crystals., must aid in adding effects to common heat.—

Where there are cliffs there ought to be creeks & mouths of rivers ought to be deep.—

97e Henslow has deposited specimens from Anglesea in Geolog. Soc. if numbered compare them with my rocks. when writing on Falkland Islds[1]

---

p. 94. Von Buch's Travels account of Norway chain being broken through like that near— Obstruction Sound in S. America[2]

The very general absence of *fragments* «& pebbles» in mica slate & gneiss, can only (see «supra» p 94) be accounted for by great molecular attraction of every atom in rock[3]

98e On a coast, the shallower the water, the greater power of oscillations & currents.— if matter was «successively» given of every degree of fineness. then most regular slope—

if not course enough flat top. ended by abrupt slope

each stratum would thin out, both inland & seaward: if matter too coarse, then that form.— All this depending not on absolute ‹force› «size of» of ‹currents› «fragments» but relative to currents. Small lakes have power of levelling their shores

96 *page crossed pencil.*
97 *page crossed pencil.*
98 that form.—] *second* 't' *over* 'n'.
On . . . form.—] *crossed pencil.*

97–1 On Henslow's donation see Anonymous 1824:436. Darwin apparently did make the comparison for in a later paper he noted (1846:274), 'It is singular in how many points the old quartz-rock of Anglesea, as described by Professor Henslow . . . , agrees with that of the Falkland Islands.' The Geological Society's British material is now deposited in the Geological Museum, Exhibition Road, London.
97–2 Buch 1813:94, 'Lessöe is the only valley in all Norway which descends from the east side to the Western Ocean without our being under the necessity of previously ascending high mountains. This is a singular phenomenon. The chains of mountains which run through the whole length of Norway are here intersected by a great valley, and completely separated from one another.'
97–3 Buch 1813:94–95 describes a section of rock in the valley of Lessöe where mica is found embedded in gneiss. A note added by Robert Jameson reads, 'The conglomerated appearance of the gneiss and mica-slate mentioned in the text is probably an original formation, not an instance of gneiss and mica-slate containing fragments.'

99e where currents very weak??— too great an abundance of matter would have same effect as too coarse.

Read Kylau on Granite Edinburgh Philosophical Journal[1]

Rapport on D'Orbigny's Voyage. good section of Rio Negro beds.— — refers to species non decrite de petites corbules analogue living in mouth of Plate. p. 26.

Geology of Arica[2]

100e ‹Schit› Schmidtmeyer travels into Chile p 29. gold is not sought for in Chile in beds of river, but in shelving «successive» banks ‹above› 30 ft or so above bed of river. formed of rounded pebbles— it is clear gold occurs in submarine alluvium, or sublittoral formations.[1]
p. 150. at Portezuelo, extremity of mountains of Cordova project on plain, like ‹re› a reef on a sea beach— «p. 151» first discovered «very small» bits of red granite between 40 & 50 from Portezuelo.[2]

99 where . . . coarse] *crossed pencil.*
page crossed pencil.
100 p. 150 . . . from Portezuelo] *crossed pencil.*
‹Schit› . . . formations.] *crossed pencil.*

99–1 Keilhau 1838.
99–2 Cordier 1834:109 lists tertiary beds of the Pampas reported by d'Orbigny without mentioning the Rio Negro specifically and refers to banks of shells in the vicinity of Buenos Aires 'qui sont composés d'une espèce non décrite de petites corbules, dont l'analogue est vivante à l'embouchure du fleuve de la Plata; . . .' The geology of Arica is discussed on p. 112. From the page number he cited Darwin seems to have read a repaginated offprint of the original report. Cordier's geological report was reprinted verbatim in d'Orbigny 1835–47, 3.
100–1 Schmidtmeyer 1824:28–29, 'In Chile, from Santiago to Copiapo, there are some vallies which are very remarkable, and form a singular feature in that country. These extend from the Andes to the sea, are from one to two miles wide, and although now only watered by very small rivers, shew traces of having been once filled up by them, to a height of thirty or fifty feet: each bank is lined with a high extensive mass of rounded stones mixed with earth, . . . It is in them, and often at a height of twenty or thirty feet from the present bed of rivers, that search is made for gold, and that the washers or *lavadores* are looking for it. To do so in the lowest channel, and by the sides of the streams now flowing, will not reward their labour, unless it be in some mountainous spots known to contain gold, or in some breaks and hollows, where a heavy winter shower may have disturbed the soil or the fragments of rocks.' The comment about gold occuring in alluvium, or sublittoral formations is Darwin's. See *GSA*:235.
100–2 Schmidtmeyer 1824:150, 'After travelling a hundred and twelve miles from Punta de Agua, we reached Portezuelo, a very small hamlet lying at the most southern extremity of the mountains of Cordova. . . . Although this southern projection into the pampas does not rise above the height of a hill, yet we hailed it as a mountain scenery, and as the foreground of those immense masses which we were approaching, but which were not visible yet.' The phrase 'like a reef on a sea beach' was Darwin's own and may have been inspired by the accompanying plate. With reference to 'bits of red granite' see p. 151, 'I could not discover any stones whatever, until within forty or fifty miles of Portezuelo, when a few very small fragments of the red granite, which chiefly forms the mountains of Cordova, begin to mix with a still very considerable, and I believe alluvial, depth of soil.' See *GSA*:79.

101e Bull: Soc. Geolog. 1837. December. p. 91. a classification of Europæan strata according to composition thinks sand with vegetable remains formed near coast, limestone deep water. will bear on formations. during elevation & depression. C. Prevost.—[1]

My views of insensible oscillations of level will alone explain the immense amount of change which must have taken place, otherwise the world would daily be scene of ruin

102e in late Natical Magazine (before June 1838) that 70. F were obtained 100 miles E of Staten land. bringing up pebbles 2 inches long?—[1]

L'Institut. 1838 p. 151. Formations of Payta extend close to Guayaquil.—modern shells of Cobija doubtful.[2]

Examine well shores of lakes. to see effects of degradation, «no» tides, water always falling or at least not rising are there cliffs. Sir L. Dick says (.p 52) fringe

101 *page crossed pencil.*
formed . . . C. Prevost.—] *excised.*
102 in late . . . long?—] *crossed pencil.*
modern . . . doubtful] *crossed pencil.*
L'Institut . . . doubtful] *crossed pencil.*
Examine . . . fringe] *crossed pencil.*
L'Institut . . . doubtful] *excised.*

101–1 Prévost 1838:91–92, 'Si de ce point de vue élevé on embrasse d'une manière générale l'innombrable série des couches qui composent les terrains secondaires du centre de l'Europe, ne voit-on pas se dessiner deux grands groupes dont les caractères particuliers ne peuvent pas être attribués a l'époque, mais au mode de leur formation. D'une part, les argiles et sables à lignite tertiaires se lient aux schistes et grès houillers par: . . . [list follows] D'une autre part, on peut descendre sans interruption du calcaire grossier parisien jusqu'aux marbres dit de transition, par: . . . [list follows] Les embranchements de chacun de ces deux groupes opposés l'un à l'autre et placés parallèlement, se remplacent réellement sur certains points, tandis que sur d'autres ils se confondent, se pénètrent, s'enchevêtrent et alternent; ils représentent pour chacune des époques de la grande période secondaire les effets synchroniques de causes distinctes agissant simultanément. L'argile et le sable, l'abondance des végétaux terrestres, . . . : ce sont des *formations fluvio-marines.* Le second groupe se fait remarquer par la prédominance de la matière calcaire provenant du brisement et de la trituration des corps marins . . . : ce sont des *formations marines pélagiennes.*'

102–1 Hammond (December 1837:822), under heading 'Bank of Seventy Fathoms Off Staten Island'. 'We tacked, had a cast of the lead, and found bottom in seventy fathoms, . . . We got a sight for the chronometers, which . . . placed us in lat. 54° 35′ S., and long. 61° 06′ W. After running W. by N. true eighteen miles, at noon we got another cast in eighty fathoms, which brought up a stone two inches long . . . which must have been the edge of the bank. . . .'

102–2 Cordier 1838:151, 'A Guayaquil, M. Chevalier a eu la preuve que le remarquable terrain calcaire de Payta se retrouvait à plus de 75 lieues vers le nord, aux environs de la pointe Sainte-Hélène, car on tire de cette dernière localité des filtres en grès coquilliers absolument semblabies à ceux qu'on exploite à Payta pour le même usage. Il faut vraisemblablement rapporter au même terrain les roches de grès quartzeux polygénique, d'argile et de marne, contenant quelquefois des rognons de silex, qui ont été recueillis soit à Guayaquil, soit à l'île de Puna, qui est à l'entrée du golfe.' On pp. 150–51 Cordier also stated that Chevalier was unable to report on the age of marine shells, and hence the time of elevation, at Cobija, owing to his specimens having been lost.

103e of sublittoral deposit always equal width —subject of fine paper this would make.—[1]

L'Institut. (1838) p. 216 M. Gay on the Geology of Chile.— **P p217. Pentlands Fossils & Meyens —‹Jura &› Chalk**[2]

When we consider parallelism of dikes (Hopkins) & that every dike. which has not formed volcanos. or become scoriform. has thinned upwards & is now cut off by denudation it gives one grand idea of amount denudation.—[3]

This may be added to any place where dikes described—
Cordillera. St Helena &c &c.—
in Cordillera, it is at once evident only small proportion of dikes have reached the surface

104e Arguments against Herschel's view[1] of cause of continental elevations (1) the alternation of linear bands of movement in Indian & Pacific Oceans.— (2d—)

103 **P p 217. . . . Chalk**] *added pencil.*
This . . . St Helena &c &c.—] *figure added right margin, pencil.*
in Cordillera . . . the surface] *added, brown ink.*
of sublittoral . . . make.—] *brown ink, crossed pencil.*
L'Institut . . . &c &c] *grey ink.*
L'Institut . . . Chalk] *excised, crossed pencil.*
When . . . **surface**] *crossed pencil.*

104 analogous, that] 'that' *over* 'as'.
(& does not . . . that elevation] *crossed.*
page crossed pencil.
in Indian . . . start with] *excised.*

103–1 Dick 1823:52, 'We see that the natural shelves now existing around the borders of our Highland lakes, do not appear to have been very much, if at all increased, beyond the breadth of those remaining from the lakes which we suppose to have been emptied at so very remote a period. The depth, therefore, of the indentation of a shelf, does not form any criterion whereby we may judge of the length of time expended in its formation.'

103–2 Gay 1838:217, 'M. Léopold de Buch, dans la revue de tous les volcans connus qu'il a jointe a l'édition française de son ouvrage sur les îles Canaries, dit, page 471, que M. Meyen, en montant sur le volcan de Maypo, voisin de Valparaiso, y a rencontré des couches immenses, presque verticales, de pierre calcaire, qui contiennent une quantité prodigieuse de pétrifications, et qui s'élèvent au-delà de la limite des neiges perpétuelles. M. de Buch a examiné ces pétrifications, et il paraît résulter de leur nature que ces couches présentent à la fois des rapports avec le calcaire du Jura et la craie. La même analogie se déduit, dit M. de Buch, des pétrifications que M. Pentland a rapportées du pont de l'Inca, au pied du passage de Mendoza.' The above remarks were made by Élie de Beaumont following his reading of a letter from Gay.

103–3 Hopkins 1835:8 in his introduction to the phenomena of physical geology, 'The dykes are usually found in nearly vertical planes, and, when they occur in the vicinity of each other, with a general tendency to parallelism.'

104–1 In Herschel's view continental elevations occur as the result of a cyclical process involving both the sea wearing down the land, depositing the worn-off particles on the ocean floor, and the expansionist power of the earth's interior heat. See the passage from Herschel quoted in the Red Notebook (note 32–1) which continues as follows (Cannon 1961:310–11): 'Thus the circuit is kept up— The primum mobile is the degrading power of the sea & rains . . . above and the inexhaustible supply of heat from the enormous reservoir below always escaping at the surface unless when repressed by an addition of

does not explain first formation of continents, if globe be considered as condensed vapour.— inequlities are required to start with (& does not Hersche theory imply tendency to equilibrium.) 3^{d}. there are mountains in the moon, which though not very analogous (see Edinburgh. Phil. Journal ‹]^{CD}›,[2] no great chains like Andes or Himalayas, but great circular mountains, yet so analogous, that as we see mountains formed (& mountains are effect of continental elevations) we may conclude that elevation

105 is independent of spreading out matter by action of the sea.— as no sea exists there.—[1] But Sir John considers an irregular figure to be that of equilibrium,— What causes that of tendency to irregularity,—. Why does Sir John assume it to be constant.—[2]

It is to be profoundly considered, metamorphic rocks at surface. & great heigth on mountains.— consist of rocks with fossils,, therefore formed near surface. whether they can have been plunged so many miles deep into the bowels of the earth, as would be required by thermometrical scale.— (for the temp must be immense to convert rock into gneiss &c

106 judging from what we see when trap in dike & approach other rocks. & trap at least as hot as lava— of which temperature is partly known— ‹[. . .]› moreover gradation from gneiss to granite shows that the metamorphic rocks have just floated over the absolutely fluid pool.— (this is shown by the softness & curvature of quartz rock?) also by my phenomena of earthquakes.— by the narrowness which

105 thermometrical] *initial* 't' *over* 's' *as in* 'scale'.
106 & approach] *alternate reading* 'a-pproach'.
‹[. . .]›] *illeg 3 letters, possibly* 'but' *or* 'bef'.

fresh clothing at any particular part.— . . . Every continent deposited has a propensity to rise again & the destructive principle is continually counterbalanced by a reorganizing principle from beneath.—'
104–2 Beer and Mädler 1838:40–41, '. . . sometimes [lunar mountains] form a regular circular zone round an enclosed space, which space is on all sides connected with the exterior by lateral valleys. These circles of the mountains (Bergkränze) lead us to very remarkable forms, . . . We allude to the Lunar *Craters*.'
105–1 Beer and Mädler 1838:45, 'Nearly all the forms of the earth's mountains are changed by the action of water and atmospherical variations, whilst these modifying agencies are probably wholly wanting in the moon.'
105–2 Herschel 1833:118–22 including p. 121, 'Land, in this view of the subject, loses its attribute of fixity. As a mass it might hold together in opposition to forces which the water freely obeys; but in its state of successive or simultaneous degradation, when disseminated through the water, in the state of sand or mud, it is subject to all the impulses of that fluid. In the lapse of time, then, the protuberant land in both cases would be destroyed, and spread over the bottom of the ocean, filling up the lower parts, and tending continually to remodel the surface of the solid nucleus, in correspondence with the *form of equilibrium* in both cases. . . . In that of an earth in rotation, the polar protuberances would gradually be cut down and disappear, being transferred to the equator . . . till the earth would assume by degrees the form we observe it to have—that of a flattened or *oblate* ellipsoid.'

107 the anticlinal lines are apart— the curvatures of the strata. ? the enormous faults & facility with which the earth is cracking by vertical planes into small pieces— mem coal-field.— the structure of Andes. where we believe we can trace the outlines of what were fluid undulations— the equal movements of Glen Roy road. (? metamorphic action at the bottom of the sea?) All this profoundly considered. study Hopkins.[1] theory of dikes may throw

108 some light.— thin dikes not cooling if they had travelled some hundred miles through nearly cold rock.— in volcano the pool is not deep. —Hot springs &c &c — then if so, thermometer show it cannt be ordinary heat, then there is something superadded, that which give cleavage to rocks.—, but lava shows the rocks really hot. & therefore I doubt the thermometer.

109e Is not common salt more soluble in ‹hot› cold than hot water with «— especially if very hot under high pressure.—» respect to formation of salt.?.—??? **Footsteps in New Red Sandstone. look as if a surface deposit.—**

The case of the shingle in the great Chilian valleys must be profoundly considered. if elevation near coast more than at interior effect would be such as present. to spread sheet of matter over surface.— if elevation then went on at greater rate, not only river would carry further its own matter. but would cut wide gorge. leaving cliffs, on each side, such as now exist.— caution about action of rivers.— **Excess of matter brought down**

110e Mention absolute elevation of Patagonian blocks (1200 ft??). Scotland at least 2200. Jura 4000 feet.—

The veins of segregation in Greenstone of Salisbury Craigs well worthy of attention— rear Glen Roy Notebook— & scraps on Salsisbury Craigs. Kept amongst ‹old› papers read before societies.—[1]

---

107 planes] 'n' *over* 'i'.
109 **Footsteps in . . . deposit**] *added and rule line extended, brown ink.*
Excess . . . down *added left margin, brown ink.*
Is . . . deposit.—] *crossed pencil.*
110 The veins . . . societies.—] *crossed pencil.*

107–1 Hopkins 1835 and 1836.
110–1 For discussion of the veins of segregation see GR2–4 and 'Salisbury Craigs/June 1838/Geological Notes/' (DAR 5:33–38[2nd ser.]). At present DAR 5 does not contain Darwin's papers read before societies. The Salisbury Crags (Craigs) is a high cliff overlooking the palace of Holyrood in Edinburgh.

111e Sir. J Hall Vol VI. p 173. (Ed. Transact) has seen clay stiff enough ‹to form› for potters to use. in which great Knife formed crystals of ice were formed— (like my gypsum case) shows power of segregation.— & has heated angular fragments of rock, which retained their angles sharp— yet with character completely altered, & a crystalline structure superinduced[1]

Lyell on Sweden p. 5. «& 7.» violet strata from decomposed muscles. .[2] **Smith of Jordanhill has seen same thing—**[3]

112e Consider profoundly How came it. that Glen Roy district could have been elevated without fissure & unequal.— where were cracks?—? How came there ever to be cracks

---

11th August. 1838
Near Woolwich there are plains & valleys just like Patagonia, & many shells in parts on surface, but I saw none embedded this point would be worth examining. to support. shells on surface of Patagonia, yet none in shingle beds.

---

113 Lyell on Sweden. p. 12. proofs of small rise at Stockholm.— analogous to my Valparaiso case.[1]

---

111 **Smith of . . . same thing—**] *added brown ink.*
page crossed pencil.
112 Consider . . . cracks] *crossed pencil.*
11th August . . . beds.] *double scored left margin, pencil.*
113 *page crossed pencil.*

111–1 Hall 1812:173–74, 'The mechanical power exerted by some substances, in the act of assuming a crystalline form, is well known. I have seen a set of large and broad crystals of ice, like the blade of a knife, formed in a mass of clay, of such stiffness, that it had just been used to make cups for chemical purposes. In many of my former experiments, I found that a fragment of glass made from whinstone or lava, when placed in a muffle heated to the melting point of silver, assumed a crystalline arrangement, and underwent a complete change of character. During this change, it became soft, so as to yield to the touch of an iron rod; yet retained such stiffness, that, lying untouched in the muffle, it preserved its shape entirely; the sharp angles of its fracture not being in the least blunted.' Darwin's St. Helena Model Notebook also contains this reference.
111–2 Lyell 1835a:5, 'Portions of the *Mytilus edulis* also occur [near Stockholm]; and there has evidently been a great accumulation of this shell in the stratum, but it is almost entirely decomposed, and is only recognized by the violet colour which it has imparted to the whole mass.' P. 7, 'Here [at Blåbacken, or "blue hills" near Stockholm] the violet colour of the decomposed *Mytilus edulis* is so remarkable as to have given a name to the hill.' Also *GSA*:13.
111–3 J. Smith 1838:393, 'I have often met with beds of shells imbedded in marly clay, which had received a violet colour from the decomposition of the common mussel (*Mytilus edulis*), exactly as described by [Lyell].'
113–1 Lyell 1835a:12–13 describes signs of recent elevation at Fiskartorp near Stockholm that suggest the rise in land (p. 13) 'in each century must have been very slight, although it may undoubtedly have amounted to ten inches in a hundred years, . . .'

Consider profoundly all consequences of EXTREME FLUIDITY of earth.— study different forms of earth as shown by arc.— read Herschels astronomy with oscillations of level.—[2]

will point  be the one which generally yields.—
Will this not explain *littoral* mountains & volcanos.— Why on *one* coast?

---

114 How can Herschel consider figure of earth statical.—[1] if platform of mexico owes its elevation to equilibrium.— it cannot be equilibrium of fluid, but of solid. because if of fluid, the waters of the ocean would obey that Law. & lie over the platform:— On my view the degrading action must prevent internal fluid arriving at equilibrium so soon from; crust being cut of— if part of «cold» crust under ocean, became thicker, then when fluid moved [. . .]

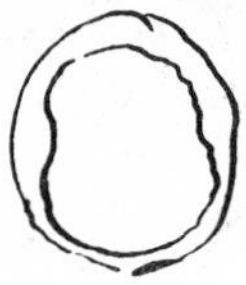

115e August 25. I saw metamorphic conglomerates on shore of Loch Lochy very like those of Andes

---

Speculate under head of Beagle Channel. on origin of mud with stones scattered irregularly.— (Mem near Gregory Bay). Shropshire case where lamination appeared.— Lyells Denmark.—[1]

---

114 [. . .]] *4 or 5 words illeg.*
page crossed pencil.
115 August 25'.] '5' *possibly* '3'.
August 25' . . . Denmark.—] *crossed pencil.*
Lyells . . . Von Buch] *crossed pencil.*
August 25' . . . Gregory] *excised.*
L'Institut . . . Von Buch] *excised.*

113–2 Darwin is instructing himself to read Herschel with his own 'views of insensible oscillations of level' [A 101] in mind. See Herschel 1833: chap. 3.
114–1 For Herschel's views on the figure of the earth see Herschel 1833: chap. 3 and Cannon 1961. More than Darwin, Herschel integrated the oceans and the land in considering the balance of forces determining the figure of the earth. See the extract quoted in note 105–2.
115–1 Lyell 1838a:323, 'The only other mode of transport [of stones] which suggests itself is sea-weed. Dr. Beck informs me, that in the Lym-Fiord, in Jutland, the *Fucus vesiculosus*, sometimes grows to the height of ten feet, and the branches rising from a single root, form a cluster several feet in diameter. When the bladders are distended, the plant becomes so buoyant as to float up loose stones several inches in diameter, and these are often thrown by the waves high up on the beach.'

L'Institut (1838) p. 268. Paper by Humboldt on Bogota. Cordillera,— nothing.— salt & coal near Bogota; p 270.—[2] SPLENDID PAPER on fossil shells of S. America. Von Buch[3]

116e Lyell. (under head of Delta) describes near Alps great beds of rivers which must be like the Chilian ones.—[1]

———

Septemb. 2d.— Sulphur like carbon must go round of dissemination & separation in volcanos.— if so why not metals.

———

The theory of veins will, I suspect be greatly aided by considering space formed— great vacuum— by dike.— Mem. however. veins of segregation in Salisbury Craigs

117e Letter from M Angelis. B. Ayres. 3d. May. states remains found in many part.— great Dasypus near *Canelones* — large quadruped bigger than ox.— at Buenos Ayres 20½ quadras from river; 20 varas from surface in tosca.— remnant of Megetherium in interior..—[1]

‹The theory of [. . .] .› ‹The›
Geographical Journal Vol VIII. (1838) p 212. Facts from Erman about great

116 *page crossed pencil.*
Lyell . . . Chilian ones.—] *excised.*
so why . . . Craigs] *excised.*
117 [. . .]] *illeg stroke.*
Letter . . . ox. —] *crossed pencil.*
at Buenos . . . interior.—] *crossed pencil.*
Geographical . . . period] *crossed pencil.*

115–2 Humboldt 1838c:270 remarks, 'Ce grès [near Bogota] est recouvert par du gypse renfermant du soufre, une argile salée et du sel en roche, et dans d'autres points par du schiste argileux et des couches de houille.'
115–3 Buch 1838.
116–1 Lyell 1837, 1:338 under the heading of 'Deltas in Lakes' describes an alluvial tract at the point where the Rhone River enters Lake Geneva and, under the same heading, on pp. 341–2 suggests that 'The Alpine rivers of Vallais are prevented at present from contributing their sedimentary contingent to the lower delta of the Rhone in the Mediterranean, because they are intercepted by the Leman Lake; but when this is filled, they will transport as much . . . to the sea, as they now pour into that lake. They will then flow through a long, flat, alluvial plain, between Villeneuve and Geneva, from two to eight miles in breadth, which will present no superficial marks of the existence of more than one thousand feet of recent sediment below. Many hundred alluvial tracts of equal, and some of much greater area, may be seen if we follow up the Rhone from its termination in the Mediterranean, or explore the valleys of many of its principal tributaries.'
117–1 *GSA*:106, 'Signor [Pedro de] Angelis, in a letter which I have seen, refers to some great remains found in Buenos Ayres, at a depth of twenty varas from the surface. Seven leagues north of this city the same author found the skeletons of *Mylodon robustus* and *Glyptodon ornatus*. From this neighbourhood he has lately sent to the British Museum the following fossils:— Remains of three or four individuals of Megatherium; of three species of Glyptodon; of three individuals of the *Mastodon andium*; of Macrauchenia; of a second species of Toxodon, different from *T. platensis*; and lastly, of the Machairodus, a wonderful large carnivorous animal.'

depths of frozen soil.[2] p. 211 Consider proved that Siberia must have been in same condition for long period[3]

118e Subsidence in Demarara p. 131 (B.) Wrong Entrance.

Book C. p. 101. On Frozen Soil of Siberia (with refer to Metamor) wrong entrance

---

Athenæum. 1838. p. 652. Dr. Daubeny on mountain Chains in N. America[1]

Erasmus suggested to me that Herschel's theory offers no explanation of intermittent action of elevatory force—[2]

119e Erasmus says he has seen in making brass a piece of copper not melted absorb, zinc thrugout its thickness.—[1] this most curious with respect to epigmous action.— if the zinc were mixed with 90 percent of lead. it would be still more curious to know whether it would be absorbed.— if so exactly parallel to limestone & volcanic rock containing magnesia

120e Lyell. Elements p.119 on such strata[1] 

---

do p. 171. argument against lateral injection. from probability of fissures being prolonged to surface.[2] see p. 181 on do subject[3]

118 *page crossed pencil.*
119 *page crossed pencil.*
120 Lyell . . . subject] *crossed pencil.*
do p. 473 . . . Cruz] *crossed pencil.*

117–2 Erman 1838a:212, 'I see by a report recently published of one of your meetings that some members doubt the reality of the fact that the soil in some parts of Siberia does not thaw till a depth of 400 feet from the surface is reached. Permit me to draw your attention to the observations I have made on this subject, recorded in the Second Volume of my Journey round the World, p. 248, *et seq.* The well at Yakutsk . . . existed when I was in that town; it had then a depth of 50 feet, and in plunging my thermometers into the clods of earth which were dug up before me, and guarding them carefully from the influence of atmospheric temperature, they constantly marked— 6° of Réaumur.'
117–3 Baer 1838a:211, 'The immense thickness of the layer of ground ice (which at Yakutsk is not less than 382 feet) proves that Siberia must have been in the same physical condition for a long period of years . . .'
118–1 Daubeny 1838:652.
118–2 Erasmus Alvey Darwin: personal communication with respect to John Herschel's theory of elevation. See note 104–1; Cannon 1961; and Herschel 1833: chap. 3.
119–1 Erasmus Alvey Darwin: personal communication.
120–1 Lyell 1838a:119–20, 'I have already stated that a geologist must be on his guard, in a region of disturbed strata, against inferring repeated alternations of rocks, when, in fact, the same strata, once continuous, have been bent round so as to recur in the same section, and with the same dip. . . . If, for example, the dark line A H . . . represent the surface of a country on which the strata *a b c* frequently crop out, an observer, who is proceeding from H to A, might at first imagine that at every step he was approaching new strata, whereas the repetition of the same beds has been caused by vertical faults, or downthrows.'
120–2 Lyell 1838a:170–71, 'As fissures sometimes send off branches, or divide into two or more fissures of equal size, so also we find trap dikes bifurcating and ramifying, and sometimes they are so tortuous as to be called veins, though this is more common in granite than in trap.' The top of p. 171 shows a sketch of trap veins with one reaching to the surface.
120–3 Lyell 1838a:181–82, 'Masses of trap are not

do p. 447 & 449. «& 450». On Vertical trees. Uspallata.—[4]
do p. 473. on great Iceland stream. the 90 miles includes opposite *directions*.[5]
Mem. S. Cruz.

121 Assuming from Sir. W. Herschel's views earth originally fluid, then cooling process must go from surface towards the interior,— who knows how far that may have penetrated,— lower down the temperature may be kept up far higher from circulation of heated fluid
or gases under pressure.—[1]
Lyells view of transmission of heat by gases— does not apply it to thickness of crust.—[2]

if crust were metal then thinner if better conductor, then still thinner →

121 *figure reads* 'metal | if fluid'.
→] *arrow indicates page to be turned.*
page crossed.

infrequently met with intercalated between strata, and maintaining their parallelism to the planes of stratification throughout large areas. They must in some places have forced their way laterally between the divisions of the strata, a direction in which there would be the least resistance to an advancing fluid, if no vertical rents communicated with the surface, and a powerful hydrostatic pressure was caused by gases propelling the lava upwards.' Of this passage Darwin wrote in his copy of the book, 'These cases appear to me most wonderful'. He also underlined Lyell's 'if no' and added 'if stone very fluid so as to communicate pressure'.

120–4 Lyell 1838a. On p. 447 Darwin wrote in his copy of this work, 'At Uspallata all vertical' with reference to a discussion of the positioning of fossil trees in coal strata. Describing a fossil tree found in a quarry Lyell wrote, p. 449, 'The tree could not have been hollow when imbedded, for the interior still preserved the woody texture in a perfect state, the petrifying matter being, for the most part, calcareous. It is also clear, that the lapidifying matter was not introduced laterally from the strata through which the fossil passes, as most of these were not calcareous.' Darwin scored this passage, underlining the word 'clear', and added '? Uspallata'. On p. 450 Darwin wrote 'Mem. Bartram.— See scrap of Paper pasted at end of Book A'. There is presently no scrap pasted in Notebook A; the question at issue, however, seems to have been the resistance to decomposition of tree trunks submerged in water. See Bartram 1791:476–77 on petrified trees.

120–5 Lyell 1838a:472–3, 'As sediment of homogeneous composition, when discharged from the mouth of a large river, is often deposited simultaneously over a wide space, so a particular kind of lava flowing from a crater, during one eruption, may spread over an extensive area, as in Iceland in 1783, when the melted matter, pouring from Skaptar Jokul, flowed in streams in opposite directions, and caused a continuous mass, the extreme points of which were 90 miles distant from each other.' Darwin scored the last sentence.

121–1 The well known views of William Herschel as described, for example, in Lyell 1837, 2:308–9, 'It has long been a favourite conjecture, that the whole of our planet was originally in a state of igneous fusion, and that the central parts still retain a great portion of their primitive heat. Some have imagined, with the late Sir W. Herschel, that the elementary matter of the earth may have been first in a gaseous state. . . . Without dwelling on such speculations, . . . we may consider how far the spheroidal form of the earth affords sufficient ground for presuming that its primitive condition was one of universal fluidity.'

121–2 See Lyell 1838a:247–50 for observations on how heated gases might transmit heat through rocks, the observations being (p. 250) 'calculated to meet some of the objections which have been urged against the metamorphic theory on the ground of the small power of rocks to conduct heat; . . .' Lyell did discuss the action of vapours in connection with the thickness of the earth's crust but perhaps not in a manner Darwin found convincing; see p. 250, 'The extent, therefore, of the earth's crust, which the vapours have permeated and are now permeating, may be thousands of fathoms in thickness, and their heating and modifying influence may be spread throughout the whole of this solid mass.'

122 The Problem is, you have temperature known at surface,— you have temperature known far below surface, say 1000— «III but an equilibrium is supposed to have been attained.» how much matter separates them, this is ascertained by conducting powers— we judge from the surface, & say 60 ft to degree.— but this may be very wrong,— The fact of a dumplin being bad conductor is
against my views— if we had rod
thus & judged by increments
at, how wrong, would
our judgement be—
Does *condensed* metal, conduct heat
better than plain?— Mem

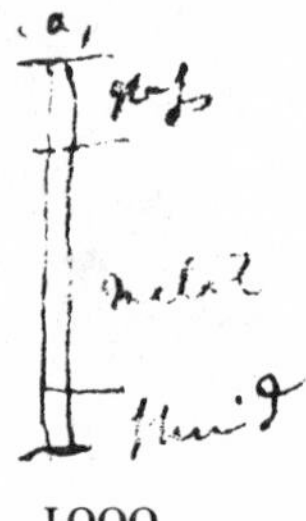

1000

123 how easily water percolates rocks,— when pressure increased or under surface, would not the fluid matter be driven upwards & so conduct heat?—

How comes it in volcanos that have gone on for thousands of years, that surface does not become hot?— this looks as if bad conductor—

III But equilibrium is not attained, & if cold water did not percolate surface, would become hotter.— hence temperature ought to increase rapidly beneath level of sea.— deep seated springs «spring requires connected column.—» of cold water show, that water does percolate, & springs beneath sea— →

124 According to this latter view the rod is reversed, upper part metal «conveying heat in one direction only, like water below 39°» & lower part glass.— then the high temperature would be much nearer the surface. especially at bottom of great ocean, where the circulations from surface can take place.— ∦ the depth of frozen soil is against this view.— however it is said in some of the papers that there are *springs* even in siberia.—
? from water thawed at + in
isothermal curve.—

+

122 III] *indicates connection to similarly marked passage next page.*
The fact . . . rod] *brace in left margin.*
page crossed.

123 III] *indicates connection to similarly marked passage previous page.*
→] *arrow indicate page to be turned.*
page crossed.

124 the high temperature . . . curve.—] *marked with triangular figures in ink left margin to indicate connection with* '—But Siberia . . . less.—' *on next page.*
Read . . . Springs.] *added ink at bottom of page.*
page crossed.

Read Daubeny on Thermal Springs.[1]

125 East-clinal. West clinal. S.-clinal. N-clinal & anticlinal «synclinal—» line.— ‹ditto of synclinal› simply *clinal* lines. dipping so & so or may be used East-clinal lines & c &

.—But Siberia was once thawed. & hence. (when climate hotter) was cooled to greater depth.— Now the ‹inf› subterranean isothermal line must be creeping «pushing» up to «the» line of ice.— Hence further N. when soil frozen for greater length of time *depth* of ice ought to be less.—

126 Memoir of the Irish Academy Vol 8. p. 118 water no—. oil will freeze if cooled in a closed globule of glass. (oil may be cooled to 0°!)—[1] shows effects of pressure in change of form as the result of heat.— will it bear on central fluidity.— do p. 137. Lord Tullamore found Sulph of Soda in peat ashes in Ireland[2]

127e *dikes* in *mountains*. «(not on continents)» prove elevation.— *great* mountain chains. may be effects of subsidence

---

Elie de Beaum. Memoires of French Geolog. Cantal Vol *III* 1? p. 246. on formation of cones beneath sea.— with reference to old submarine orifices in Cordillera[1]

125 .—But Siberia . . . less] *marked with triangular figure left margin to indicate connection with* 'the high temperature . . . curve,—' *on previous page.*
page crossed pencil.
126 p. 137.] '3' *over* '1'.
page crossed pencil.
127 of] 'f' *over or under* 'n'.
dikes . . . subsidence] *crossed pencil.*
Elie . . . Cordillera] *crossed pencil.*

124–1 Daubeny published extensively on thermal springs. See, for instance, Daubeny 1838 (the reading of which may have inspired Darwin's note), Daubeny 1836, and Daubeny 1832.

126–1 Templeton 1802:118, 'The following experiments may throw some light upon the cause of plants remaining unfrozen, when the surrounding water is frozen. Water enclosed in sealed glass globules remains unfrozen, 'till the thermometer descends to twenty-four; unsealed ones freeze and burst immediately on being cooled down to freezing water. Oil enclosed in the same kind of globules continued unexpanded, and consequently the globules unbroken, when placed in a mixture of snow and sal ammoniac, and cooled below O.'

126–2 Bury 1802. Expecting to produce potash from the burning of turf of a peat bog, Bury (Lord Tullamore) instead produced (p. 137) '*sulphat* of *soda* with little or no intermixture.' He found it (pp. 137–38) 'not a little singular, that *marine alkali* combined with *sulphuric acid* should be found in such abundance in turf ashes procured at a great distance from the sea; . . .'

127–1 The reference, which bears an oblique rather than a direct relation to Darwin's question, is to Dufrenoy and Élie de Beaumont 1830–38, 3:245–46, 'Le fait que les terrains de trachytes et de phonolithes présentent partout des dômes accumulés les uns sur les autres; le fait que dans la grande majorité des cas les terrains basaltiques et trappéens ne présentent que des dykes et des nappes horizontales, sans la plus légère trace d'un cône d'éruption comparable soit au Vésuve, soit à l'Etna, soit même aux Puys de Thueys et du Tartaret, n'indiquent-ils pas évidemment que ces roches ont eu des modes d'éruption particuliers? Si chaque district, basaltique, trachytique ou (*continued overleaf*)

128e Geograph. Journal vol II. p 89. at Madras. surrounded by salt water. purest fresh water must be sought for below the sea mark.—[1]

If mountain chains are matter piled up. over crevice from effect of general elevation,— when subsidence takes place.— Mountain will first fall— the problem will be falling of an arch weighted in its centre.—

129 Will not abrasion of land on one side. produce subsidence of water on other. from tendency to regain statical equilibrium

---

This will be only a modifying cause. land

protuberant water to counterbalance

How strongly the Glen Roy case shows that the figure of the world has just that form which forces dilemma.

130 Transactions of the Maryland Academy (at Athenæum.) I. Part. I Vol.— some notices on modern Tertiary strata on coast of do—[1]

I *believe*?? coast of North America., like the Mexican Gulf. is fouled by bars of sand & shallow lagoon.— when describing Coast of. Brazil. Maldonado enter into this case.—

128 Geograph . . . mark.—] *crossed pencil.*
If . . . centre.—] *crossed pencil.*
129 *page crossed pencil.*
130 fouled] *alternate reading* 'fronted'.
page crossed pencil.

(*127–1 continued*)
trappéen, avait présenté dans l'origine un cône d'éruption, comment ces cônes auraient-ils presque toujours disparu? Ne serait-il-pas naturel de présumer que la différence qui a existé entre les éruptions basaltiques, trachytiques et trappéennes et celles des volcans actuels, était de nature à y rendre moins nécessaire et plus difficile la production d'une montagne conique? Or, quelle est la circonstance qui, dans les volcans actuels, entraîne comme conséquence nécessaire, la production d'un cône d'éruption: n'est-ce pas le rôle prépondérant qu'y jouent les dégagemens de substances gazeuses? Les substances gazeuses ont certainement joué un rôle dans les éruptions basaltiques et trappéennes. . . .' Darwin scored the third and fourth sentences of this passage in his copy.

128–1 W. F. Owen 1832:89, 'At Madras, which is surrounded by salt water, the purest fresh water must be sought in wells dug below the sea-mark; elsewhere, to whatever depths the wells are dug, the water is brackish; . . .' Darwin scored this passage in his copy.

130–1 Ducatel 1837:53, 'On the Eastern Shore [of Maryland]: the secondary formation extends to the Chester river. . . . The tertiary deposites lie south of the Chester river, and do not extend further than the Choptank, inclining in the same direction; so that in the upper portions they are found several feet above tide, whereas, in the lower parts, as in the necks of Talbot county, they appear but little above the water-line. Beyond the Choptank, in Dorchester county, they have been reached at the depth of forty-five feet.'

131 Ed. New. Phil. Journal Vol XXI. p. 213. Beyond the limits of Alps size of boulders *sorted*: ditto Murchisons case[1].— ? does it bear on Patagonia?

«Facts about subsided forests.— Many repeated oscillations»

Hitchcock Report on Massacuhssets. p. 133 The most wonderful case of great block of rock moved by gale— When writing on Valleys. «Tertiary strata of S America» read parts of this work, though it is but poor.[2]

132 Athenæum. 1838 p. 791 — Most curious account of great *subsidence* «20 miles long 1 in with.» which must have been from an axis, «20 ft at least in depth» near mouth of Columbia river— Read Mr Parker's Book.—[1]

M. Bichoffs Papers, in Edinburgh New Phil. Journ 1838. several case given of hot heads &c heat beneath the sea.—[2] CD[did not Beechy have some such

131 Ed. . . . Patagonia?] *grey ink.*
Report] 'R' *over* 'r'.
Facts . . . poor] *brown ink, crossed pencil.*
132 *page crossed pencil.*

131–1 Since Charpentier 1836:213 does not refer to the sorting of boulders beyond the limit of the Alps, that statement was presumably Darwin's inference based on such statements by Charpentier as the following (p. 215), 'Wherever stones are deposited by glaciers, they are collected together and heaped up without order, and without any separation according to size and weight.' In contrast, Charpentier argued that water-borne stones are found deposited according to size, with the largest being deposited most closely to the point of origin and the smallest the farthest. On this point see Murchison 1839a:513–15 and Murchison 1833–38, 2:231. Exactly what case of Murchison's Darwin had in mind, and whether it pertained to this point, is uncertain. On Patagonia see *GSA*:19–26 and the table on p. 16 showing 'that the pebbles at the bottom of the sea *quickly* and *regularly* decrease in size with the increasing depth and distance from the shore, whereas in the gravel on the sloping plains, no such decrease in size was perceptible.' Darwin traced the latter fact to the recent and gradual elevation of Patagonia from the sea and took the nature of the superficial deposits as an indication that glaciers had not been the agent of deposition.

131–2 Hitchcock 1835:133–34 quoting Benjamin Haskell on the subject of boulders moved by the sea at Cape Ann, Massachusetts, 'But there is one [boulder] far more interesting than all the rest; both on account of its greater bulk, and comparative regularity of shape, which renders the former easy to be estimated, and thus affords the means of ascertaining the maximum force of the Ocean in its anger. . . . The weight of this bowlder [sic] has been calculated with care . . . and found to be rising of 28 tons.'

132–1 Parker 1838:791–92. 'At the La Dalles [The Dalles on the Columbia River] . . . commences a wood country, which becomes more and more dense as we descend. . . . Noticed a remarkable phenomenon— trees standing in their natural position in the river, in many places where the water is twenty feet deep, or much more, and rising to high, or freshet water mark, which is fifteen feet above the low water. . . . As I approached the Cascades, instead of finding an embankment formed from volcanic eruptions, the shores above the falls were low, and the velocity of the water began to accelerate two-thirds of a mile above the main rapid. On a full examination, it is plainly evident that here has been an uncommon subsidence of a tract of land, more than twenty miles in length, and more than a mile in width. The trees standing in the water are found mostly towards and near the north shore, and yet, from the depth of the river and its sluggish movement, I should conclude the subsidence affected the whole bed. That the trees are not wholly decayed down to low water mark, proves that the subsidence is, comparatively, of recent date; and their undisturbed natural position proves that it took place in a tranquil manner, not by any tremendous convulsion of nature.'

132–2 Bischoff 1836–38 [vol. 23], chap. XI, p. 379, argues that 'The decrease of temperature in the waters of (*continued overleaf*)

133e case]CD[1] what would be the chance in sounding over a continent to fall across a hot.—spring.— Hot water would not lie. at bottom.— Surely we here have proofs of hot bottom.— Study Bishoofs Paper.—[2]

Weelsted told me of some large fresh Water springs off coast of Persia[3]

In Glen Roy paper I show crust yield easily.[4] & if easily must be thin: ‹beside mere fracture›

134e A Elevation as in Patagonia
B subsidence; ‹as in›
be cautious. mud banks

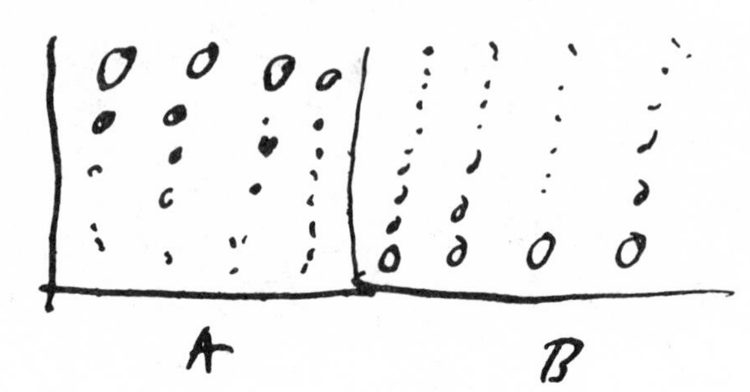

& sand. dunes.— in these littoral deposits there probably would be marked line of separation

---

A Paper by Parrott Mem. Acad. Peters. Scienc Math. Phy— Nat. t. I, 1831. sur le temp du globe on Volcanos &c worth reading.[1]

133 off] *triple underl pencil.*
beside] *uncertain reading.*
page crossed pencil.
134 A Paper . . . reading.] *crossed pencil.*

(*132–2 continued*)
the sea and of lakes, is not contradictory to the hypothesis of an increase of temperature towards the centre of the earth. On the contrary, we can only explain the temperature of sea and of lakes by admitting an increase of temperature towards the centre of the earth.' With respect to isolated points of heat in the sea, Bischoff wrote, pp. 382–83, 'But single spots are to be found in the sea, where the temperature is greater in the depths than at the surface. Hot springs and volcanic action may cause considerable partial elevations of temperature in the sea, as, for example, seems to be the case in the *gulf stream* on the coast of *America*, where, on hauling up the lead from a depth of 80 or 100 fathoms, it is so hot as scarcely to allow of its being handled. Similar spots seem to exist near the *Kurile Islands*, in *Basse's Straits*, and in the *Atlantic Ocean* where Horner in clear weather observed, at about nine miles' distance from the vessel, a cloud of vapour, which, during a quarter of an hour, continued alternately appearing and disappearing from the surface of the sea, and could neither be the smoke of powder nor that of a vessel on fire. Horner considers that this phenomenon may perhaps have been caused by a volcanic eruption.'

133–1 Beechey did not comment in print on hot heads beneath the sea, but there is information on the temperature of the Atlantic and Pacific Oceans at various depths beneath their surfaces in Beechey 1831, 2:731–32. Since Beechey and Darwin were in personal contact after the *Beagle* voyage (*JR*:21), Beechey may also have communicated to Darwin other information on heat beneath the sea such as the data collected in 1818 and later published in Beechey 1843:339–40 showing that there was 'a small increase of temperature with an increase of depth' beneath the ocean between the latitudes of 79° 45′ N. and 80° 27′ N.

133–2 Bischoff 1836–38.

133–3 James Raymond Wellsted: personal communication.

133–4 Darwin 1839:78, 'But first I must remark that the crust of the earth seems to yield easily to the forces which have acted on it from below; . . . If indeed the crust did not yield readily, partial elevations could not be so gradual as they are known to be, but they would assume the character of explosions.'

134–1 Parrot 1831.

135e L'Institut. 1838 p. 360. on orbicular trap thought to be bombs submarine[1]

L'Institut 1838 p. 400. Observations on Mountains of the Moon. by Dr. Nichol— adduces the case to show Sir. J. Herschel's theory wrong.—[2]

Geograph. Journal Vol. 8. p. 402.— ground ice— subterranean isothermal line[3]

136e Athenæum. 1839. p. 52. On Frozen soil of Siberia.— facts of water flowing from beneath frozen crust in America Richardson.—[1]

From strata being not only vertical, but turned over in many parts of the world.— argument strong in favour of thin crust theory.—

135 L'Institut. 1838 p. 360 . . . submarine] *excised, crossed pencil.*
L'Institut 1838 p. 400 . . . line] *crossed pencil.*
136 *page crossed pencil.*
Athenæum . . . from] *excised.*

135–1 Mallet 1838:359–60, 'M. Mallet lit un mémoire sur une structure nouvelle observée dans certaines roches de trapp du comté de Galway. . . . La masse générale de ce trapp possède une structure nodulaire. . . . Cette formation nodulaire est essentiellement différente de toutes celles décrites jusqu'ici, puisque dans le granite orbiculaire de la Corse et du midi de la France, et la pierre d'ognon des chaussées, etc., les nodules et la gangue sont des matériaux différents. La structure en question paraît avoir été produite par l'éjaculation du trapp sous forme fluide au sein de la mer, et dont les jets ou coulées, en se refroidissant dans leur passage, sont retombés aussitôt dans la masse encore fluide, où ils se sont trouvés enveloppés et chauffés à la température de la masse à laquelle ils ont adhéré sans perdre leur forme extérieure.'
135–2 Nicol 1838:400, 'IV. La lune nous permet d'éliminer un certain nombre de circonstances qui ne sont pas essentielles, dans l'énoncé définitif du problème de la cause des soulèvements. Par example, la théorie récemment publiée par sir John Herschel, relative à la dépendance mutuelle entre les convulsions et élévations, etc., les érosions et la stratification, ne peut plus, suivant l'auteur, se soutenir. Considérant que la température de la terre augmente avec la profondeur, au moins jusqu'aux points où il nous a encore été permis d'atteindre, il est indubitable que les érosions et la stratification amèneront des changements de température dans les roches inférieures, et parconséquent des dilatations, des fractures et des mouvements étendus quoique lents; jusque là la théorie assigne une cause vraie à l'origine de certains changements qui ont lieu à la surface de la terre, mais il y a bien loin de là à la théorie définitive et complète de la cause des soulèvements. La lune ne contient aucun agent d'érosion ou de stratification.'
135–3 Baer 1838b:402–3, 'Professor Erman, it appears, is not satisfied with the expression 'ground ice' (bodeneis, Germ.) which I have proposed. I did so because it seemed to me to embrace all the requisite modifications, and is very concise. . . . Perpetual ground ice is, then, that which is found in the arctic regions, in that layer of earth which is immediately below that which is thawed by the summer heat, reaching, as this does, to the depth where the temperature of the earth is at freezing-point. It seems to me very important for physical geography to ascertain the thickness of perpetually frozen ground in countries of which the mean temperature is considerably under the freezing-point.'
136–1 Richardson 1839:52, 'Travellers into the arctic regions of Asia and America have mentioned that the sub-soil of certain districts is permanently frozen, and Gmelin long ago declared that, in Siberia, the thickness of the frozen earth was upwards of 100 feet; . . .' And p. 53, 'The Mackenzie [River] itself, is mostly supplied from districts having a mean heat inferior to the freezing point; its more southerly branches being comparatively small. Now, the waters which sustain the perennial course of many of the tributaries of the Mackenzie, the Great Bear Lake River, for instance, must rise from beneath the frozen stratum.'

137 What a curious investigation it would be to compare, the time of the earthquake of Chile, with that of the passage of the moon.— **Ask Hopkins.—**[1]

M. Parrot, Mem. Acad. Imp. des Sciences. (Sc Math. Phys. et Naturelles. Tom I. p 501.[2]— shows first that data wholly insufficient to *calculate* rate of increase of heats in earth's crust.— yet heat does increase,— but in Ocean

138 does not. (see resumè p. 536)— «NB. I cannot understand the argument, that cold ‹oceans› «lakes» bottom. if not colder than mean of place, shows earth not with central heat.—» «(does M. Parrot suppose there is no volcanicity beneath lakes)?» Suppose ocean represents proper ‹state› temperature of earth. at the freezing point.— accounts for increase on earth by volcanic action.— ‹Why› now as we know volcanic action prevails more beneath the sea, ‹than› «&» on coast lines, than on continents. it ought, (according to M. Parrots argument against central heat to warm the ocean).— and M. Parrot does conjecture that in Scoresby's case volcanicity has warmed it. Is not cold of ocean accounted for, by the circulation

139 being greater, than the transmission from ocean's bottom.— (according to M..Parrots own hypothesis some such explanation appears to me necessary) as M. Parrots shows from variation in strata earth a *very* bad conductor.— shows p. 516 that subterranean springs give result less to be trusted than any others— may not the cold «bottom of» ocean. (with fresh sediment added to bottom) be caused, by absence of circulating water.— & therefore that

137 **Ask Hopkins.—**] *added pencil.*
M. Parrot, . . . Ocean] *crossed pencil.*
138 «NB. I . . . heat.—»] *top margin.*
«(does . . . lakes)?»] *interlined* 'does not . . . at the freezing'.
temperature] *first* 't' *over* 'of'.
action.— . . . warmed it.] *crossed pencil.*
139 as] *possibly* '—'.
page crossed pencil.

137–1 William Hopkins.
137–2 Parrot 1831:501–62. Parrot challenged the presumption of the earth's central heat, chiefly on the grounds that the oceans did not show an increase in temperature proportional to depth. Pp. 536–37 summarize his argument. Parrot's own view favoured attributing evidence of underground heat to volcanic action. P.535 refers to the fact that '. . . M. Scoresby a trouvé une petite augmentation de température avec la profondeur dans la mer entre le Grönland et le Spitzberg, les terrains de ces deux masses de terre-ferme et vraisemblablement le fond de la mer entre deux, étant entièrement volcanique.' Pp.514–16 offers evidence that springs, as compared to certain other underground phenomena, have the highest variability of temperature as measured against depth. The closing paragraph of Parrot's paper reads as follows (p. 562): 'Enfin, le Géologiste doit saisir le problème entier à sa source, à laquelle personne ne paraît avoir songé avant moi. Tous les Géologistes sont aujourd'hui (1825) d'accord sur le principe que nos continens et le fond des mers étaient primitivement dissous dans un océan qui couvrait tout le noyau du globe et se sont formés par voie de précipitation chimique. Mais aucun d'eux n'a recherché lesquelles des matières des roches se trouvaient en dissolution, et quels ont été les réagens qui les ont précipités, ni où ces réagens se trouvaient. Et cependant c'est de l'action successive des réagens, tranquilles ou troublés par les forces mécaniques, que dépend la succession des formations, la suite des différentes espèces de roches, inexplicable d'ailleurs. Ce défaut primitif des systèmes géologiques a fait que le cahos est resté cahos.'

temperature of earth beneath ‹of Sahara de› a dry desert, would be very high.—

140 M. Parrot ends his paper like a fool.—

Feb 25'
All facts show how slowly heat travels; & therefore the abysses where fluid rock has been ejected must remain fluid for an enormous period: now when we see how many points have been penetrated by volcanic & trappean rocks, within say the Tertiary period. one is led, to look at globe as resting on film of molten rock.—

141e Voyages of Adventure & Beagle[1]
vol 1. p. **2 &** 3. Porphyry at St. Elena.

---

p. 6. few «living» shells. on coast of do

---

140 the abysses] 't' *over* 'a'.
Feb 25' . . . rock.—] *crossed pencil.*
141 **2 &**] *added pencil.*
Vol. I . . . St. Elena.] *rule line beneath continued in pencil.*
p. 6. . . .] '6' *over* '8'.
Admiralty . . . dip. much] *added with circled* '×' *and a* '√', *left margin, all pencil.*
p. 375.] *circled* '×' *left margin with* '|| **FitzRoy refers to** ||' *circled,* '**& Rocks**' *added over original entry, all pencil.*
vol. I. . . . Virgin] *crossed pencil.*
p. 136. . . . subsidence] *crossed pencil.*
Rocks of . . .Coast] *crossed pencil.*
Vol. II . . . Falklands] *crossed pencil.*

141-1 FitzRoy 1839. *The Narrative of the surveying voyages of His Majesty's Ships Adventure and Beagle*, of which Darwin's book appeared as vol. 3, was published between 15 May and 1 June 1839 according to *The Publishers' Circular* (vol. 2, no. 41), a trade publication. Presumably Darwin's notes were written near the date of publication as were the notes in E166 and C269. From vol. 1 by Phillip Parker King Darwin noted the following passages. P. 3 describes the countryside around and the coast to the north and south of Port Santa Elena as consisting of a 'fine-grained porphyritic clay slate'. P. 6, 'Among the sea-shells, the most abundant [at Port Santa Elena] was the *Patella deaurata*, Lamk; this, with three other species of Patella, one *Chiton*, three species of *Mytilus*, three of *Murex*, one of *Crepidula*, and a *Venus*, were all that we collected.' Pp. 7–8 describes cliffs near Cape Virgins that were composed (p. 8) 'of soft clay' rather than of chalk, as charts had indicated. P. 59, 'I have before observed that the strata of the slate rocks, in the Strait, dip to the S.E.; and I found that they dip similarly all the way to the bottom of this inlet, which I named Admiralty Sound.' P. 136 describes the greenstone and granite in the vicinity of Mount Maxwell in Tierra del Fuego. P. 204, 'With respect to the geological features . . . all the islands on which I landed, and, I believe, all the others, are composed of green-stone of various characters.' [western Tierra del Fuego] P. 210 contains a chart listing two sets of baronometrical determinations of heights of several points on the road between Valparaiso and Santiago. P. 328, 'The land, near the mouth of the river [San Tadeo], is studded with dead trees (a species of pine, about twenty feet high), which appear to have been killed by the sea overflowing the banks; ['Or by an earthquake wave,' added in a footnote by FitzRoy] as it does at high-water for several miles.' P. [374–]375 as described. P. 385 of a rock found in the vicinity of the Magill Islands, 'It is similar to that of Fury Island and Mount Skyring, apparently metallic, with a sulphureous smell, when struck or broken.' [Note: Geological Society, Coll. No. 197] From vol. 2 by FitzRoy Darwin took note of the following passage (p. 277), 'Some very large bones were seen a long way from the seashore, and some hundred feet above the level of high water, near St. Salvador Bay. How they got there had often puzzled Mr. Vernet, and Brisbane also, who had examined them with attention; Brisbane told me they were whale's bones.'

p 8.— soft Clay beds near C. Virgin
p. 59. dip of Clay slate in T del Fuego
Admiralty Sound. SE dip. much
p. 136. Rocks on Western Coast
p. 204 do. do
p. 210. Height on road from Valparaiso to Santiago
p. 328. dead trees on Isthmus of Pen. Tres Montes.— as by subsidence
|| Fitz Roy refers to || & Rocks
p. 375. on the soundings on outer coast of T. del. Fuego.—
p 385 Rocks of S. Western Coast

——— ———

Vol II p. 277. on whale bones in Falklands

142e Some of the Tosca nodules at Bahia Blanca Mr. Malcolmson says are like Kankaer[1]

———————

———

South Part of Luconia— Phillipines there is volcano on isld in large lake.—

———————

Berghaus Chart of do[2]

143e Journal of Asiatic Society
Vol I.
p. 145. on salt mines of Punjab[1]
p. 149. on the ‹salt mines› «saline deposits» of India[2]
p. 503. On Indian Saline Deposits.[3]
Vol II. p. 23.[4] p. 77 do[5]
Vol III p. 36. do[6]
——— p. 188 do[7]
Vol 5. p. 798 do[8]
Vol 7. p. ‹52› 363. do[9]

144e Journal of Asiatic Soc
Vol V. p. p 96. apparently good geological paper. by Malcolmson— worth

142 South . . . do] *crossed pencil.*
143 *page crossed pencil.*
144 Journal . . . Soc] *short pencilled line underneath.*
Vol V . . . reading—] *crossed pencil.*
Burnetts . . . abundance] *crossed pencil.*

142-1 J. G. Malcolmson: personal communication.
142-2 Berghaus 1832–43 included on his map of the Philippines, dated 1832, an enlarged map of the area around Volcan de Taal [Mt Taal], a volcano in Lake Taal, southern Luzon. Luzon was also known as Luconia.
143-1 Burnes 1832.
143-2 Everest 1832.
143-3 Spry 1832.
143-4 Stevenson (presumably Stephenson) 1833.
143-5 Malcolmson 1833.
143-6 Stephenson 1834b.
143-7 Stephenson 1834a.
143-8 Stephenson 1836.
143-9 Gubbins 1838.

reading—[1]

Burnetts. vol 4. p. 193
in Lat 26° S. Wafer looking for Copiapo. found inland a great many sea shells some miles from coast— quote passage to show abundance[2]

---

145 Bengal Journal. Vol 4. 1835. p. 437. Tours by Benza Neilgherries— Much inform. on. decomposition of granite—.[1]

Bengal. J. vol 7. p. 522. Mountain c near Caubul. parallel ranges. with here & there little branches at ⊥ from each side intercepting plain & dividing it—[2] Hopkins fissure at ⊥.—[3]

146 G. J. Malcolmson has described formation of shore of Coromandel. just same as. at Bahia Blanca— letter in drawer with important letters—[1]

---

145 *page crossed pencil.*
146 *page crossed pencil.*

144–1 Malcolmson 1836.
144–2 Burney 1803–17, 4:193, 'In latitude 26° S, wanting fresh water, they made search for the River *Copiapo*. They landed and ascended the hills in hopes of discovering it. According to Wafer's computation they went eight miles within the coast, ascending mountain beyond mountain till they were a full mile in perpendicular height above the level of the sea. They found the ground there covered with sand and sea-shells, "which," says Wafer, "I the more wondered at, because there were no shell-fish, nor could I ever find any shells, on any part of the sea-coast hereabouts, though I have looked for them in many places,"' Darwin cited this passage in *GSA*:46, where he also mistakenly identified the author of the passage as Burnett rather than Burney.
145–1 Benza 1835:437 continues the description, begun on p. 435, of 'Specimens from the Northern Circars'. For discussion of the decomposition of the granite of the group of hills called the Neilgherriess see pp. 419–21, which includes the following passage (p. 419), 'If observations and facts were wanting to prove that this thick mass of lithomargic earth is owing to the decomposed granitic rock of these hills, the following is conclusive. The original undecomposed rock is . . . traversed occasionally by thick veins of quartz. These veins resisting decomposition . . . are seen *in a continuous course, penetrating from the hard crystalline undecomposed nucleus of the rock into the lithomargic earth, and into the concentric layers of the already decomposed rock*. Therefore, it is impossible to avoid the conclusion, that the red earth and the rock were, at one time, *one mass, traversed by the quartz vein*. . . .'
145–2 Lord 1838:522, 'When the two mountain ranges [near Kábul] have for some time preserved their parallel east and west courses the northern is observed to deflect or send off a branch towards the south, while a corresponding deflexion or ramification of the southern chain comes to meet it, and the plain which otherwise would have been one continued expanse from east to west is thus cut into a number of valleys, the longitudinal axis of which however, is still in general to be found in the same direction.'
145–3 Hopkins 1835:36, 'It is evident, however, that in whatever manner a system of parallel fissures may be produced, that, after their formation, the only tension of the mass between them must be in a direction parallel to them. Consequently, *should any other system be subsequently formed, it must necessarily be in a direction perpendicular to that of the first system*.' Also pp. 34–36, 50, 55–57 and Hopkins 1836:31–32, 'After one system of fissures is formed, there is no difficulty whatever in conceiving the formation of a second system perpendicular to the former.'
146–1 In a letter dated 24 July 1839 (*Correspondence* 2:208–9), J. G. Malcolmson wrote to Darwin describing 'the appearances presented along the Coromandel coast where it is gaining on the sea'. The 'series of actions' described included '1st. a bank of sand [being] thrown up, and increased by the wind & consolidated by creeping grasses—behind this there is a backwater or swamp often salt from the sea water *springing up* from the sand at high tide, when the inside of the sand hills is below the level of the sea—' The area is described as likely burial ground for land animals. Malcolmson also described 'a long extent of flat covered with fresh water in the rains & full of fresh water shells—running parallel to the coast for many miles.' Darwin scored this section and added 'Bahia Blanca'.

147e When I come to treat of the age of the Pampas Deposit, I may properly remark on the superiority of Lyell's classification to that of Phillips as given p. 13. Vol II. Lardner's— Treatise[1]

Phillips in Lardner Vol II p. 73.: some remarks on veins:[2]

Phillips in Ladner Vol. II p. 80— some remarks on dikes: applicable to Cordillera[3]

Phillips in Lardner Vol II. p. 81. «&83» Some remarks on thinness of crust as implied by meeting with granite every-where.[4]

Phillips in Ladner Vol II p. 125. Good discussion on mineral veins p. 125 to 129 & p. 135—160 & 162[5]

147 When . . . Treatise] *crossed pencil.*
Phillips in Lardner Vol II p. 73 . . . & 162] *crossed pencil.*
Phillips in Lardner Vol II. p. 81 . . . every-where] *crossed pencil.*

147-1 J. Phillips 1837–39, 2:13, 'We have therefore the following general classification of the results arrived at in studying fossil mammalia:

Modern period –	Pachydermata almost lost, . . .
Diluvial era –	Pachydermata abound, . . .
Tertiary period –	Pachydermata of *extinct* and *living* genera abound; . . .
Supercretaceous era –	Pachydermata of *extinct* genera first appear, . . .
Secondary period –	Marsupial quadrupeds occur in one place (Stonesfield).'

Phillips then proceeded (pp. 13–14) to compare his classificatory scheme to that of Charles Lyell. In his copy of Phillips, Darwin commented (p. 13) that he believed Phillips' scheme inapplicable to South American fossils and (on the end pages) that Phillips' scheme was 'wretched'. It would appear that Darwin first annotated his copy of Phillips, vol. 2, and then transferred his remarks, in reduced form, to this page of Notebook A.

147-2 J. Phillips 1837–39, 2:72–81 is devoted to the subject of veins. Darwin annotated pages 72–73 heavily and suggested that plutonic rocks, being heated more intensely, would shrink more than volcanic rocks. He asks himself whether most substances, except water, shrink upon solidifying, answers that they do and then queries 'iron shrinks??—'.

147-3 J. Phillips 1837–39, 2:80–81, '. . . *dykes* . . . are very rarely granitic. If this seem a paradox, its solution may lead to important results. Could we behold enormous masses of porphyry, or basalt, below vast breadths of stratified sediments, as granite is commonly seen, there would probably be found porphyritic or basaltic veins passing from them into the cracks of the strata. If this is never the case, does it not show the peculiar mineral character of granite, and its peculiar effects on the adjoining rocks, to be the fruit of the local circumstances of its deep 'plutonic' origin?' In this passage Darwin underlined 'porphyritic' and 'never the case' and added in the margin 'Andes'.

147-4 J. Phillips 1837–39, 2:81. In a paragraph headed 'Amorphous Masses under all the Strata', Phillips called attention to the 'vastness of the masses' of granite from which veins of granite arise. He suggested that these granitic masses deserve 'the title of an universal formation'. Darwin scored the paragraph, adding in the margin 'Argue granite near surface of all ages in all parts of world hence thinness of crust.' On the bottom of p. 83 Darwin noted 'thin crust theory' with reference to a passage including such statements as, 'The induration of the strata is an effect quite distinct from their deposition, and appears to require the supposition of long continued application of heat.'

147-5 J. Phillips 1837–39, 2. Chap. 8, pp. 120–64, is devoted to the subject of mineral veins. Pp. 124–28 bear the sub-heading 'Occurrence of Mineral Veins near Centres of Igneous Action'. On p. 129 in a section bearing the heading 'Relations of Veins to the Substance and Structure of the neigbouring Rocks', occurs the passage '. . . and we believe these differences of character may be distinctly referred to the natural structure of the rocks, and the movements to which they have been subjected.' Darwin scored this passsage and added a 'no' in the margin. In the same section on p. 135 Darwin scored the paragraph beginning, 'The same truth of the dependence of the contents of mineral veins upon the containing rocks is put in a strong light by Von Dechen, in his translation of De la Beche's Geological Manual.' On p. 160 Darwin scored the passage introduced by the sentence, 'Lastly, the

148e [blank]

149 Ed. New. Phil J. 1838. p. 72. on metallic vapours condensed from furnaces[1]

---

do/p. 84 on the effects of veins of slag in iron furnaces affecting to some distance & blending with sandstone «said to be» analogous to granite infiltering some of its constituents into chert.[2]

150 [blank]

151 Ed: New. Phil J. 1838. p. 132. «& 134» Bischoff. On the effects of meteoric waters on the temperature of the interior & p. 142 / p. 155. the increase of temperature beneath the sea, is probably much more rapid than beneath continents[1]

---

151 meteoric] *second* 'e' *over* 'o'.
page crossed pencil.

vague suggestion of electrical agency, in depositing the materials of mineral veins, has been reduced to a regular system by Mr. Fox, . . .' On p. 162, Darwin scored a passage that begins, 'Sublimation and re-crystallisation of metallic matters . . . are common phenomena; and the passage of veins downwards to heated regions is too probable to render it doubtful that such operations have sometimes contributed to fill the fissures of rocks.'

149–1 Hausmann 1838, pp. 71–72, 'Particular notice is due to the appearance of graphite in the cavities of a slag of a blast-furnace at the same time with iron, which in a botryoidal form, and partly with an oxidized surface, forms the basis on which the graphite is deposited. Here we perceive manifestly that the graphite vapour was condensed after the liquid iron had been deposited in that form, on the smooth surface of the rigid slag. There is a circumstance worthy of note attending it,—iron and graphite are found always together, and only on the upper part of the hardened slag: hence we must conclude that the iron too gained access to the slags in the form of vapour, but became liquid before it passed into a solid state; whereas, graphite was immediately crystallized. If this opinion be correct, it follows necessarily that iron requires a higher temperature for its sublimation than carbon.'

149–2 Hausmann 1838:83–84, 'More worthy of note is the phenomenon which I remarked on pieces of the forepart of the building of an extinguished blast-furnace . . . in the Harz. Veins of a silicate of the black oxide of iron, resembling slags of the refining process, cross in different directions the sandstone changed by the heat. The veins which ramify towards the sides are of different sizes . . . and have evidently ensued from the penetration of fused masses into the fissures of the sandstone. . . . The transformed sandstone is partly distinctly separated from the mass of the vein and partly amalgamated with it. These relations display a striking analogy with the phenomena, which accompany granite in several places, where, as takes place in the Harz, it is in contact with the greywacke. The chert which surrounds the granite, and which is often just as intimately associated with it as with the greywacke, appears to have been formed by the penetration of the mass of granite into the adjoining mountain rock.'

151–1 Bischoff 1836–38 [vol. 24]:132–34, on the question (p. 132) 'What influence do the meteoric waters exert on the internal temperature of the earth?' (Meteoric water is that in or derived from the atmosphere.) The opening paragraph on this subject reads (p. 132), 'In Chap. VII. it has been shewn that the temperature of the meteoric waters has no great influence on that of the crust of the earth through which they flow. But small as this influence is, it will yet be found to be very various in different parts of the earth, according as the strata be more or less impervious to water, or as the waters sink to a greater or less depth. But such variations must naturally disturb any supposed fixed law in the increase of temperature towards the centre of the earth.' Pages 142–64 of the article are devoted to the question (p. 142), 'Can the increase of Temperature towards the interior of the Earth follow the same law in all parts of the earth, departing from points all situated in the same level?' On p. 155, arguing from 'general considerations'—that is, mathematical induction—rather than from 'direct observations' Bischoff concluded that, 'The increase of temperature beneath lakes must, therefore, follow a more rapid progression . . . than under the solid earth. The same will be the case under seas and glaciers: under the sea, where it has a great depth, even in high latitudes, provided only that the temperature of the bottom of the sea do not reach that degree which it would have, were it completely filled up with the solid matter of the earth, and under glaciers

In Berlin Transactions (1832. or 3?) there is an account of Sellow Geolog. Observat. in Southern Brazil.[2]

---

152 [blank]

153e «p. 4. (Lyells Book)»
Observaciones sobre El Clima del Lima par Dr. H. Unanùe says he believes the sea has formerly stood three hundred feet above its present level, & in many parts has extended a league inshore both N & S of Lima.— judges from «beds of» sand & gravel & shells.[1]

---

p. 47. do has table of every earthquake, during two years.— will serve for comparison with the moon at some future time[2]

154e [blank]

155 Sir. J. Halls Paper on the consolidation of strata— he heated sand red hot & brine was boiling on the top—[1]

156–175 [blank]

176 Would rotting wood by yielding Carbonic Acid unite with «piece of cabbage» alklali & precipitate silica / or charcoal charged with carbonic acid

---

153 believes the] 't' *over* 'a'.
p. 47 . . . time] *crossed pencil.*
page crossed pencil.
155 *page crossed pencil.*

also, only so long as the original mean temperature of the soil beneath them is not reduced to 32°.'

151–2 On the collections of Friedrich Sellow see Weiss 1827, also d'Alton 1833.

153–1 Unanue 1815:4, 'Es constante que en nuestra costa han ido las aguas en diminucion. Las conchas que se hallan al S. y. N. [of Lima] esparcidas sobre sus colinas, y la composicion de éstas, de arena y despojos marítimos, son monumentos que con otras muchas señales acreditan, que no han pasado muchas centurias despues del tiempo en que nuestros mares se internaban de dos á tres leguas, subiendo á mas de cien varas de altura sobre los cerros de granito, en que terminan las ramas descendentes de la cordillera.' Darwin was apparently using Charles Lyell's copy of this book.

153–2 In Unanúe 1815 following p. 47 there are two meteorological tables for Lima, one for the year 1799, the other for the year 1800. Both tables list earthquakes recorded during the year.

155–1 To account for the consolidation of sandstone from loose sand Hall 1826 adduced and experimentally demonstrated the consolidating action of heat when operating in the presence of salt (p. 325) 'whether in a dry state mixed along with loose materials, or driven in fumes through them, or applied in the state of brine. . . .' On being challenged in his theoretical conclusions respecting the influence of heat at the bottom of the sea, where the influence of heat would be counteracted by the neighbourhood of the cool water, Hall replied (p. 326), 'In answer to this difficulty, I must beg leave to remark, that, in all my experiments above alluded to, the sand . . . was seen to become red-hot during the process of consolidation, while the superincumbent brine remained boiling above; and it was even found easy, by supplying cool brine in sufficient quantity, to maintain the temperature of the fluid permanently such, that the hand could be plunged into it at top, without injury, the sandstone below remaining all the while at a full red heat.'

177–179 [blank]

180 Many interesting experiments might be tried by comparing Zoophite to plants.— grafting length of life &c &c

---

Will any *in*organic substance cause such monstrous growth as oak galls or rose ‹buds› galls.— is it not effect of superadded vital influence?—

INSIDE BACK COVER See End of Note Book. called R.N.—

Massac[h]usset would be well worth visiting really good account of ice.— C. Darwin

BACK COVER

A.

IBC Massac[h]usset] *second part* 'M' *over an* 'a'.
See . . . ice.—] *perpendicular to spine.*
C. Darwin] *parallel to spine.*

Glen Roy Notebook

Introduction by SYDNEY SMITH & PAUL H. BARRETT

Transcribed and edited by SYDNEY SMITH, PETER GAUTREY & PAUL H. BARRETT

This is a tiny notebook (DAR 130, 116 × 73 mm) in which Darwin kept field notes while on a trip through Scotland from Edinburgh to the Glen Roy River near Fort William. There are 63 leaves including the front and back fly leaves of this De la Rue & Co.'s Improved Metallic Memorandum Book. A label on the inside cover remarks, 'The point of the Pencil [a metal stylus now lost from the side pocket] not being liable to break, and the writing being permanent, they will be found of great advantage to Commercial Gentlemen, Short-hand Writers, &c.' The stylus by pressing laminae in the patent paper together leaves a black mark when fresh. There remains today only a very faint impression and reading of the text and sketches is made easier by holding the book obliquely under a bright light and transcribing the faint shadow of the persisting impression of the stylus. There are also some entries in ink.

The binding is of black leatherette with a blind stamped pattern. Both front and back covers bear small cream paper labels with 'Glen Roy' written by Darwin in pencil. The sheet with pages 64/65 had been excised, filed in DAR42, folio 87, but is now restored. All pages except 3, 4, 9, top third of 10, top half of 11, 64e, 119, 120 and 122 are crossed in stylus pencil. Page numbers have been added in pencil by Cambridge University Library staff.

Most of the entries in the notebook were made across each opening parallel to the spine, suggesting that Darwin held the book length way in his left hand while writing. Thus it was possible to write observations in the open whatever the weather.

Two maps and a conversion table are provided to help collate the original Gaelic names with today's Ordnance Survey. One can easily understand Darwin's trouble in the spelling of place names because he would have had, in many instances, to rely on his own interpretation of the local dialect.

The first two pages of the notebook begin with breeding in animals followed by 9 pages of sketches and notes on Salisbury Craigs overlooking Edinburgh. Scattered notes on instincts and behaviour of dogs and sheep are on pages 11, 12, 25, 26, 31 and 125. Glen Roy begins in the middle of page 29. There are 24 sketches of topography and explanatory geological maps in the notebook. These are reproduced here as line drawings because the original sketches are too faint to reproduce photographically. Darwin's labels written within the sketches are transcribed and printed in the figures as close as possible to their original positions.

Darwin spent 8 days at Glen Roy from 28 June to 5 July 1838. He noted in his Journal, '1838 June 23rd. Started in Steam boat to Edinburgh (one day Salisbury Craigs). Spent eight good days in Glen Roy.' He returned by sea through Greenock and Liverpool, slept at Overton (the home of his eldest sister Marianne Parker), and reached Shrewsbury 13 July. Back in London he wrote in his Journal 'August 1st. [1838] London. Began paper on Glen Roy & finished it, 6th September;' he adds 'one of the most difficult & instructive tasks I was ever employed on', *Correspondence* 2 (Appendix II):432. Darwin was elected a Fellow of the Royal Society 24 January 1839 in succession to his grandfather, Erasmus, and his father, Robert Waring, and was married to Emma at Maer on 29 January. His Glen Roy paper was read at the Royal Society on 7 and 28 February 1839.

The field observations on Glen Roy were written for the most part while walking and climbing in the vicinity. It is surprising that he was able to write such a long and detailed article as was published in the *Philosophical Transactions of the Royal Society* pt. 1, 1839, pp. 39–81 (Darwin 1839), from the few notes he recorded in this notebook.

The parallel roads are conspicuous lines along the upper slopes of the valley. In Chile Darwin had studied similar raised shore lines which seemed clearly marine in origin. He attributed the changes in level of the land to the uplifting of sections of the earth's crust. His views about Glen Roy did not persist unchallenged for long; the Glen Roy shore levels were soon rather convincingly shown to be relics of lake shore levels trapped behind ice dams. See Agassiz 1840, 1842, Milne-Home 1849, Jamieson 1863, Barrett 1973 and Rudwick 1974–75.

In his autobiography written nearly 40 years later Darwin admitted 'This [Glen Roy paper] was a great failure, and I am ashamed of it.' (Barlow 1958:84). He called it a gigantic blunder. Nevertheless the explanation of the origin of the Glen Roy parallel roads remained controversial well into the nineteenth century. For additional readings see Chambers 1848, Thomson 1848 and Rogers 1861.

Conversion Legend

The words Darwin used in the notebook are given in italic, the modern Scottish equivalent in roman.

Ballivard Ballieward
Bohunthine Bohuntine
Ben Erin Beinn Iaruinn
Bright [Glen (River) Buck?]
Cairn taw leer peek Càrn Leac
Moel Derry Meall Doire
Collarig Caol Lairig
Corry Coire
Forrest Forres
Glen Bought [Glen Buck?]
Glen Bright [Glen Buck?]
Glen Collarig Glen Caol Lairig
Glen Fintec Glen Fintaig
Glen Glouy Glen Gloy
Glen Guoy [Glen Gloy?]
Glen Tarf Glen Tarff
Glengarry Glen Garry
Glenoe Glencoe
Glen Turrit River Turret
Grant town Grantown
Habercalder Aberchalder
Inverorum Inveroran
King's House Kinghouse
Letter Finlay Leiter Finlay
Loch Docart Loch Dochart
Loch Tring Loch Treig

GLEN ROY MAP

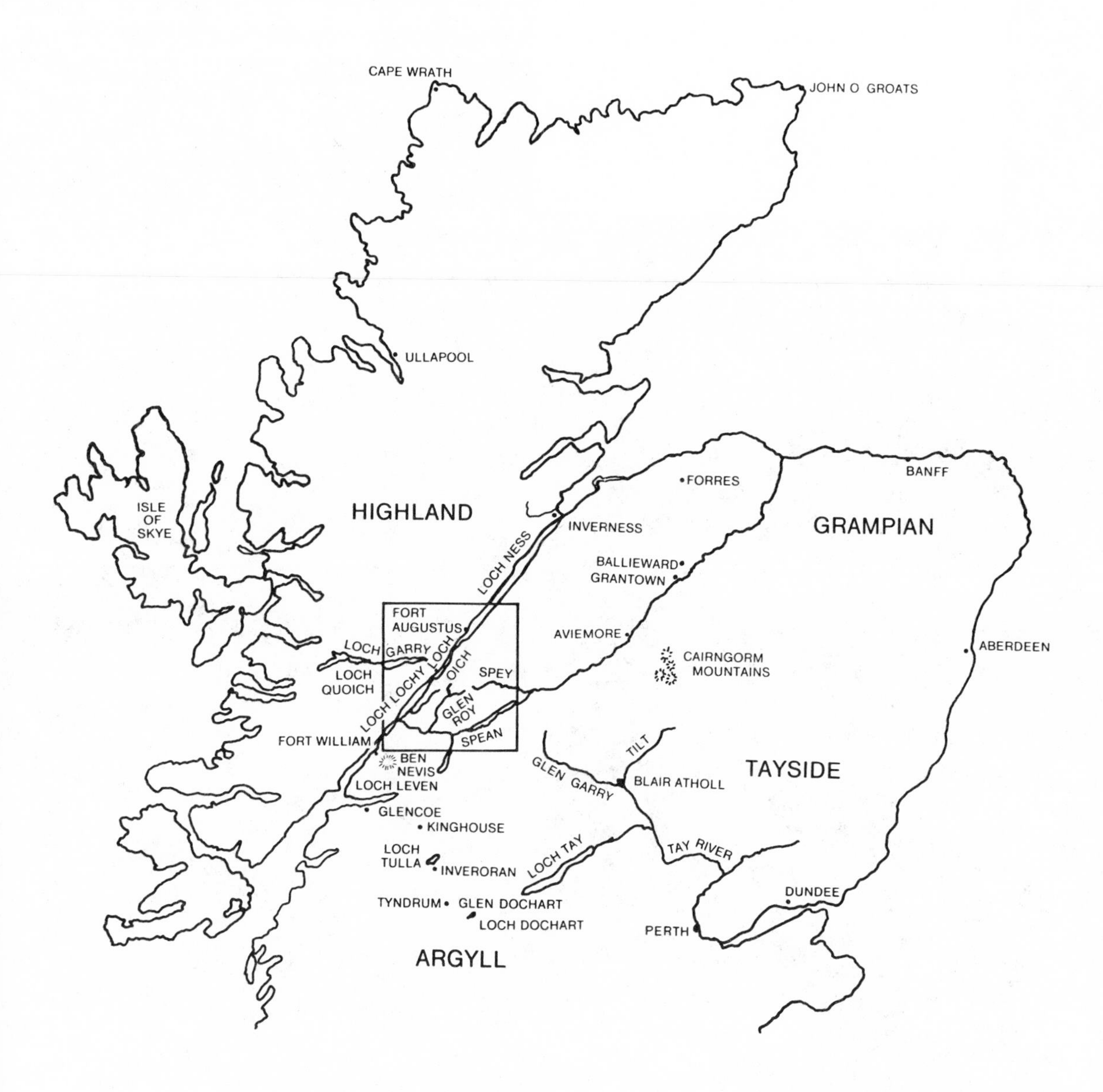

CAPE WRATH
JOHN O GROATS
ULLAPOOL
ISLE OF SKYE
HIGHLAND
INVERNESS
FORRES
BANFF
GRAMPIAN
LOCH NESS
BALLIEWARD
GRANTOWN
FORT AUGUSTUS
AVIEMORE
ABERDEEN
CAIRNGORM MOUNTAINS
LOCH GARRY
LOCH QUOICH
LOCH LOCHY
LOCH OICH
SPEY
GLEN ROY
SPEAN
FORT WILLIAM
BEN NEVIS
LOCH LEVEN
TILT
GLEN GARRY
BLAIR ATHOLL
TAYSIDE
GLENCOE
KINGHOUSE
LOCH TULLA
INVERORAN
LOCH TAY
TAY RIVER
TYNDRUM
GLEN DOCHART
LOCH DOCHART
DUNDEE
PERTH
ARGYLL

OY MAP

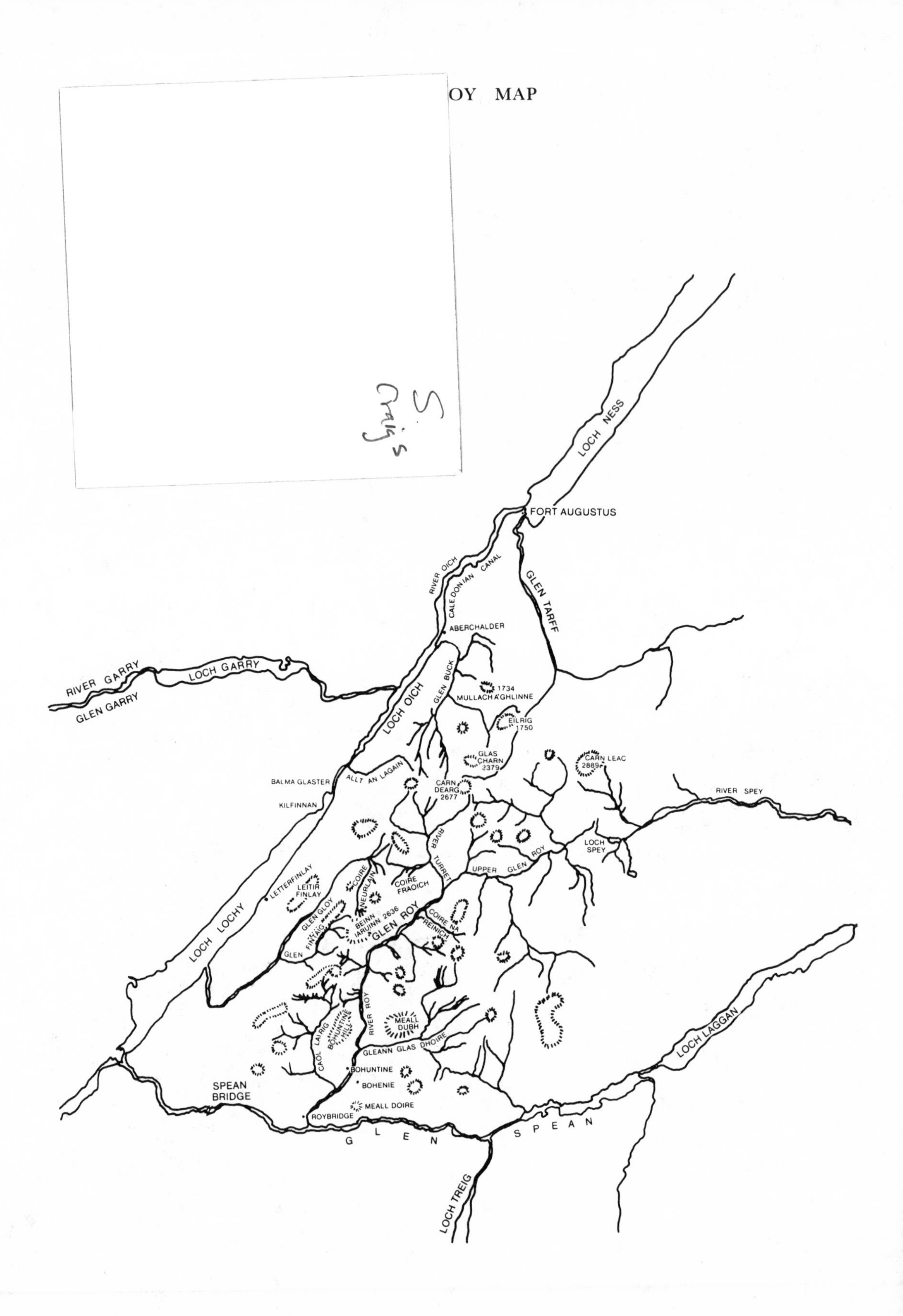
LOCH NESS
FORT AUGUSTUS
RIVER OICH
CALEDONIAN CANAL
GLEN TARFF
ABERCHALDER
RIVER GARRY
LOCH GARRY
GLEN GARRY
LOCH OICH
GLEN BUCK
1734
MULLACH A'GHLINNE
EILRIG 1750
GLAS CHARN 2379
CARN LEAC 2889
RIVER SPEY
BALMA GLASTER
ALLT AN LAGAIN
CARN DEARG 2677
KILFINNAN
RIVER TURRET
LOCH SPEY
UPPER GLEN ROY
LETTERFINLAY
LEITIR FINLAY
COIRE NEURLAIN
COIRE FRAOICH
GLEN GLOY
BEINN IARUINN 2636
GLEN ROY
COIRE NA REINICH
GLEN FINTAIG
LOCH LOCHY
CAOL LAIRIG
BOHUNTINE HILL
RIVER ROY
MEALL DUBH
GLEANN GLAS DHOIRE
BOHUNTINE
BOHENIE
SPEAN BRIDGE
MEALL DOIRE
ROYBRIDGE
GLEN SPEAN
LOCH LAGGAN
LOCH TREIG

FRONT COVER

Glen Roy

1 *Generally received opinion* that male impresses offspring more indelibly than female p 367 Quarterly Journal of Agricl Dec 1837[1] Yet instances given against it—

Mere fact of many races of Animals

2 in Britain shows that either races soon made or crosses difficult

---

Salisbury Craigs

The Highland shepherds dogs coloured like Magellanic fox.. an instance of Provincial breeds.

3–4

[3] Veins of Segregation in Salisbury Craigs

[4] *Salisbury Craigs*

V. Specimens—

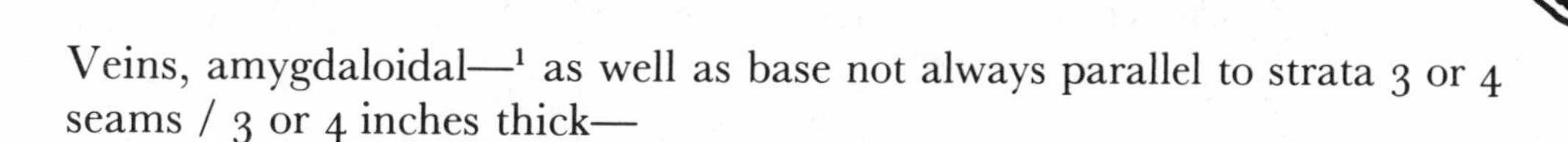

Veins, amygdaloidal—[1] as well as base not always parallel to strata 3 or 4 seams / 3 or 4 inches thick—

5

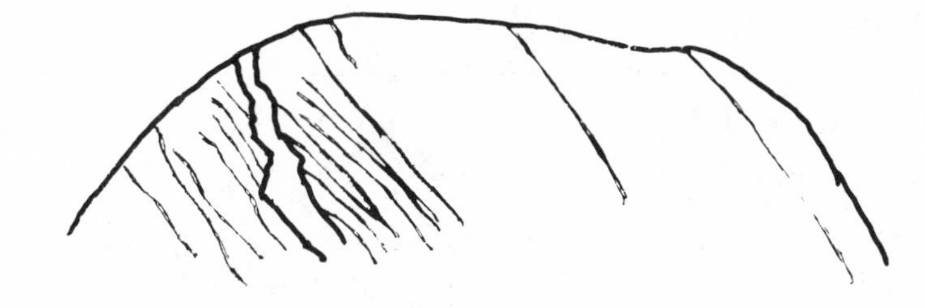

6 35° is I believe about greatest dip of sandstone in upper part «of Salisbury Craigs» 25° perhaps most common—

2 *Lower right portion of p. 2 and upper right portion of p. 3 torn off. No text seems to be missing.* **The Highland . . . breeds.**] *ink*, **'breeds'** *underlined.*

3 *The right half of the sketch is on p. 3; the left half is on p. 4.*

1–1 Anonymous 1838:367, 'According to the generally received opinion, that the male imprints his characters more indelibly than the females on the progeny, there may be a risk of breeding from too large a horse . . . but . . . it is frequently seen that small stallions and bulls produce large stock.'

4–1 Writing in his autobiography at the age of 67 Darwin remembered a dike such as described here and sketched on p. 5, 'Professor Jameson, in a field lecture at Salisbury Craigs, [discoursed] on a trap-dyke, with amygdaloidal margins and the strata indurated on each side, with volcanic rocks all around us and [said] that it was a fissure filled with sediment from above . . .' Barlow 1958:53.

7 Will not curved form of hill be explained by my idea— highest part must project

8 [Blank]

9

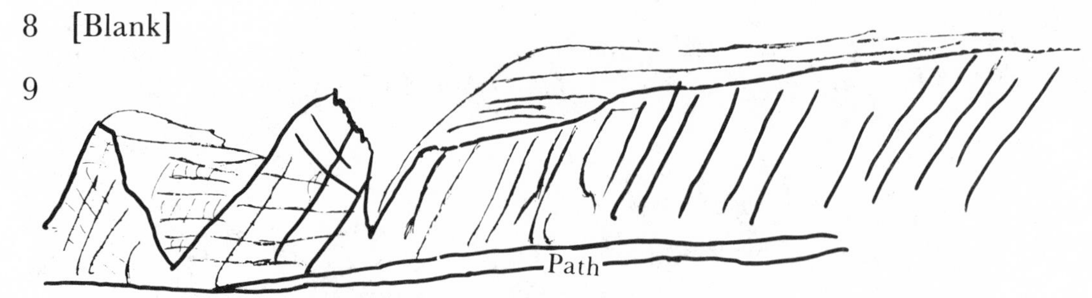

East End near Holyrood Palace

10 In same way at top the trap could be traced
Grey in front on wall perhaps wall oblique

The hill has been well— denuded.— «of hard metamorph» path only covering

---

11 Great Slip, 10 years since three hundred feet in vertical height— enormous mass thunder storm, many ‹hundred› thousand tuns.

Black faced sheep, sometimes mottled with white black legs & tail like species in colouring

12 Strike an analogy between pleasures of association, & passions, such as love— dislike & ‹f› passion of hatred

---

To fulfil an instinct a pleasure; mem. Shepherd dogs

13 The Patches of Conglomerate on S. Ventana,[1] excellent instance, how accidental is the preservation in situ of even imperishable pebbles / I am nearly certain there were none on surface of any hill

---

Thursday
On side of Hill South of upper end of Loch Dochart buttresses of Alluvium or rather mass of well rounded pebbles in

14 yellowish argillaceous or sandy soil—
These Buttresses formed vestige of irregular terrace perhaps near 300 ft above Loch.— From this point could be followed up to neighbourhood of Tyndrum where a large sort of ‹plain› space is

9–10 *Sketch extends across p. 9 and 1/3 across p. 10.*

13–1 *JR*:79, 'On the flanks of the mountains [of Sierra Ventana in Argentina] . . . there were a few small patches of conglomerate and breccia, firmly cemented by ferruginous matter to the abrupt and battered face of the quartz,— traces being thus exhibited of ancient sea-action.'

15 thickly studded with ridges & flat topped hill/ do alluvium.
NB In one part pure sand in current cleavage— in other irregular horizontal strata I suppose these upper patches if prolonged would

16 intersect alley above the 300 ft Alluvium ‹abo› by Loch Dochart—
Rivers could not have deposited it. Barrier of lake very lofty, & no trace of it; to the

17 Sea more probable
I did not look carefully for Marine remains—
Some of the hills almost appeared as if they belonged to double series

---

Whole very obscure but it is certain

18 there must once have been very considerable mass of waterworn pebbles in Alluvium which without lake or sea could not be placed in present position

19 Thursday Evening ½ past 8
Tyndrum
29.‹625› «636» Temp. 62

Friday morning ¼ past
seven o'clock
29.642 Temp 55 Air 50°?

Friday. Inverorum
about 20 ft above Loch Tulla
29.804 Temp 62° Air 60°

20 Below Loch Tulla whole wide valley scattered with few very small & irregular hills of alluvium— nothing very striking yet possibly sea more probably than river—

No exact terraces but appearances, as if valley had been filled with sloping bed of rubbish

21 Friday
Highest part of road between Inverorum & King's House 28.935/82° A
Temp of Air 65°?

Glenoe, 6 ft above high water mark
30.380 . 68°
65°?

For comparison with all the measure before

22 There some of the half rounded gravel nearly as high as highest measurement but nature I am quite doubtful of as I am of all the Alluvium.

At Mouth of Caledonian Cänal opposite Loch Leven

23 two terraces perhaps upper one 100 ft & other one 40— ∦ traces of them all along ‹Glencoe›.— towards Fort William
yet in Glencoe in *parts* no trace of them—

24 Mem Coast of Chile— ⸮ is not Mica Slate too hard & uneven to be impressed

---

Case of Birch Wood by Inverorum being determined by sheep & not deer

---

25 When Black faced sheep are crossed with English my informant said the lambs were nearly like each other «& half between parents» (& not like dogs), but they thought the breed liable to vary— I asked this

---

question in many ways & received same answer

Thought lambs most like MOTHER!—

26 the cross not so hardy but more easily fatten,

This man confirmed the account of the «YOUNG» Shepherd dogs

Saturday.

Before coming to Bridge of Spean, hills of «sea», gravel, current cleavage, & pretty well rounded stones, mixed with some quite irregular

27 very like rubbish at head of Loch Dochart ‹Nea› Above Spean Bridge many flat terraces one above much inclined towards river all these composed— where side ravine entered terraces formed successive bays but plains sloped

28 centre-wards which would not have happened if the side-streamlet had cut them out— In all cases «I urge» deposition marine— because if not chain of lake & if so there would be barrier— recollect

29 the case of loch ‹in› ‹below› «**by**» pass of Glencoe—
the erosion may often be due to rivers—

By Roy Bridge, a tongue of flat land, with terraces of each side of the two valleys

23 '∦' *over* '&'.
29 the case . . . Glencoe—] 'below' *crossed ink*; 'by' *added in ink over* 'below'.

30 corresponding as in Andes, composed of sand & perfectly rounded stones— lake required to deposit this

---

Remember however the great Chilian valley Acongua, must there have deposited much— On other hand remember modelling power of sea N of Valparaiso

31 are those animals subject to much variation which have lately acquired their peculiarities?

The slope of A & B regular & even towards

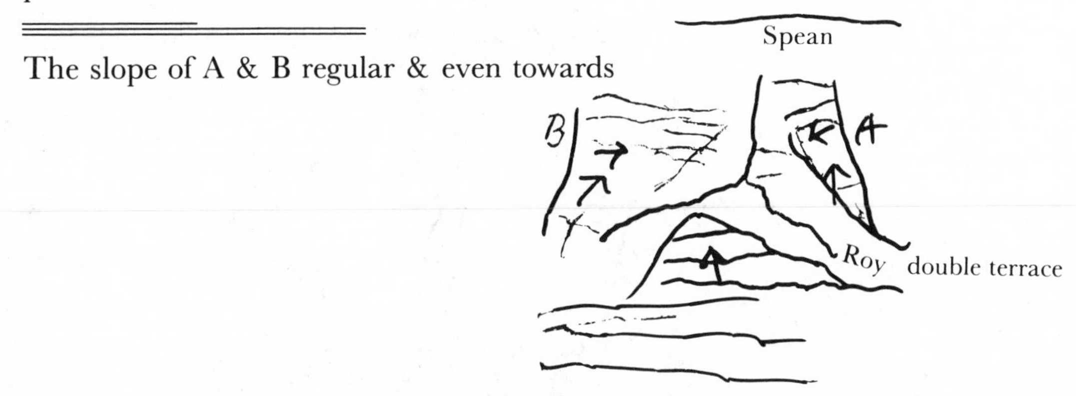

32 river «& to West of Spean» difficult to explain on ‹formation› deposition in lake

---

On the summit «& on Spean side» of Meal—
Derry there were perfectly rounded «base» pebbles of quartz & other rocks not apparently in situ ‹& in› hill being gneiss ‹& also› also near summit on Hill on side of Inn BOULDER of granite above 4th Shelf a little lower down the hillock with beach & channel precisely as with Isl[d]—

33–34

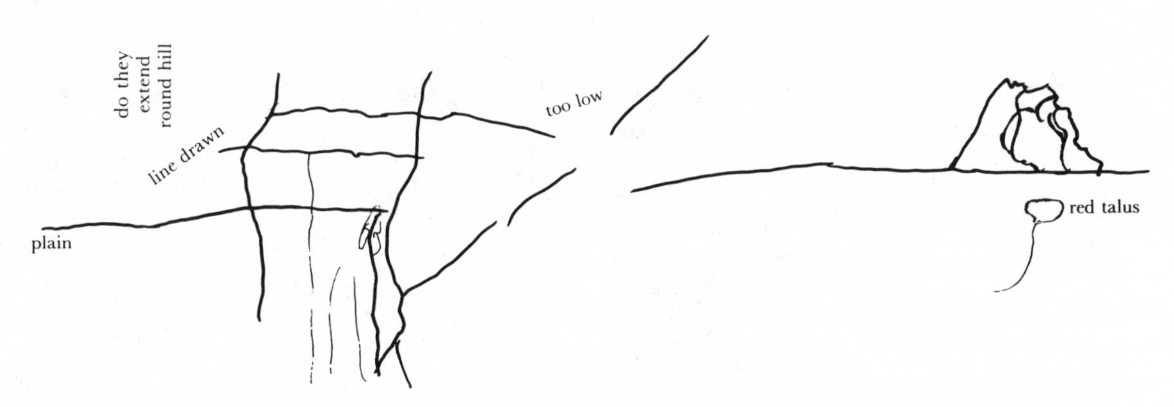

32 Hill on side of Inn] 'side' *uncertain reading; thumb-sized area partially erased between* 'on Hill' *and* 'above'; *three very small ink blobs over* 'side'.

33–34 *3 to 4 words written along diagonal line in middle are illegible.*

35 line on N. side of Spean most clear & upper line[1] running up great bight just as Dick shows

---

NB. Lake gradually draining off would form plains such as those near Bridge Roy (& other cases) but then if gradually drained, where is barrier

36

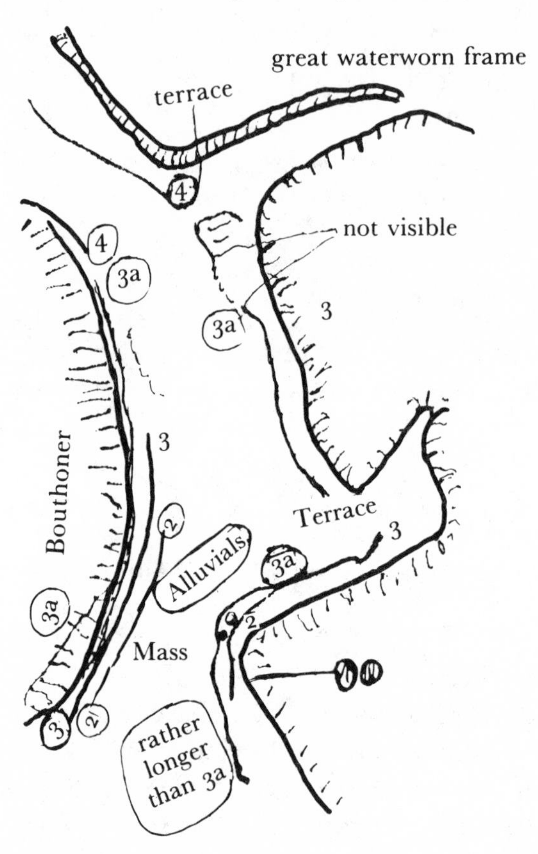

37 *Sunday*
In Glen Collarig, when water up to shelf very shallow channel 50 ft wide & river get formed in centre

---

In Glen Collarig, on side of Hill of Bohunthine upper road (2) extends as far nearly as house, the 3^d below them

38 opposite to where side ravine enters On opposite side of valley both extend below the Houses
The Hills in this neighbourhood appear very round-topped with much drainage & far more earthy than what is usual—
Lines die away where slope

38 *Very small ink blob lower right side.*

35–1 Macculloch 1817:316, 'To avoid any bias which the original and not sufficiently descriptive term, *Parallel Roads*, might preserve in the reader's mind, I have substituted that of *Lines*, a term less exceptionable and sufficiently expressive of their appearance.'

39 less., best developed on *steep earthy* slope, two circumstances rarely united.— die away also, without any cause, must be tides. &c.

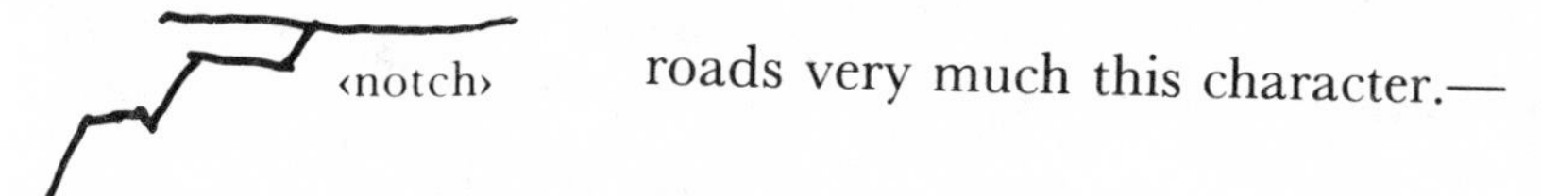

roads very much this character.—

40 The boulders (one of Gneiss remarkably water worn) are often times of rock not in immediate neighbourhood, (as granite or gneiss of Moel Derry) on low hill between Inn & Bouhunthine the *summit* «doubtless worn into coincidence» has beach or band of pebbles on line of 4th shelf.—

41 Even on Lauder Dicks Hypothesis[1] impossible to explain absence of lines in certain parts.—
At the Pass of Glen Collarig two little lines of Hill (judging from external form alluvium) descend from shelf 3^{d} & almost meet,

42 but are separated by flat bottomed strait. connecting flat on one side with irregular gravel plain of other, which must have been waterworn after 3^{d} lake.—

———

4th shelf runs up some way on great sloping plain of

43 alluvium (much corroded by rivers) & not to head of plain.— but below houses where rivulet enters two great projecting butresses, upper slope of which corresponds to shelf the truncation & the

44 upper shores may correspond with some line subsequent to shelf

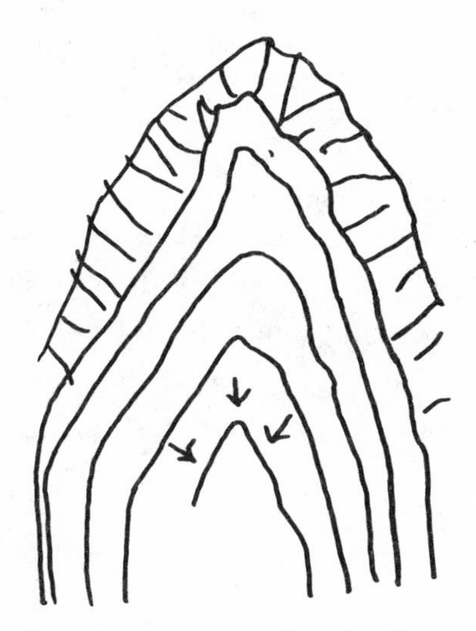

42 waterworn after 3^{d}] '3^{d}' *over* 2^{d}.
44 upper shores may] 'may' *over* 'must'.

41–1 Dick 1823:14, 'That theory seems to me infinitely the most probable, which attributes the formation of the shelves to the action of the waters of a lake.'

45 In Glen Collarig, by Dicks theory lake burst in most improbable part & not in Pass, where shallowest[1]

In Glen Collarig good case of shelves entering «on» one side ravine.

46 Are the lip, or necks of land on level with shelves *effect* of corrosion & not *cause*.

Monday

---

a rapid descent of a terrace except at very head of valley indicates new terrace

47 Ballivard 2 miles North of Grant town to Forrest road comminuted shells

---

Important contingency if elevation from Axis, then rivers might deposit, & afterwards

48 with greater cut through, *not* applicable to Glen Roy

Lake, must have remained very long at 4th shelf from size of buttresses, to upper edge of which they cut near Loch Tring—

49 Tuesday

Bridge of Roy
Level of «bed of» River 30.221/65°/ Temp of air 65°?

There are two terraces on the East side of river & bed of river about 40 ft beneath general plain.

50 30.127 A 72°
Air 65°? at level of upper terrace

The buttresses of Alluvium rise nearly up to Glen Collarig up within 200 ft of level of 4th shelf= argument against river— composition &—

51 stratification argument detritus— 

---

50 A 72°] 'A' *over* '6'.

45–1 Dick 1823. Dick's theory included the propositions that the shelves were formed by lakes confined in the valleys by earthen dams. Catastrophic convulsions in the earth's crust burst the dams and the rush of the flood-waters washed away the barriers.

where buttresses on 4th shelf: others «lines not so level because of upper edge of cliff» Others below it—argument for lake «or sea» at successive levels—

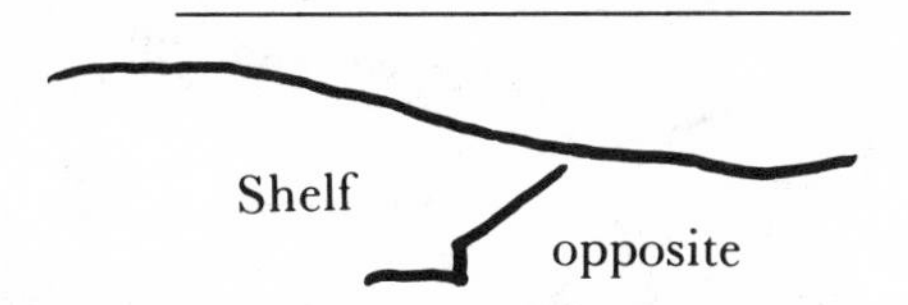

52 Glen collarig at bend & here most accumulations

At gentler bends roads disappear

The normal condition of 4th shelf, some way below House

53 of Glen Roy, seems to be

4th [shelf]
river

which higher up on is corroded

4th [shelf]

Could earthquake cause collection of sediment?

54 Where ravines enter side by, opposite entrance into Glen Fintec a kind of landing place is formed

---

55 Ben Erin summit
27.813. 65
55°?

Boulder of Granite
28.362
68°
60°

Granite— «band» 4 × 3 × 2 «feet» & 2 deep

56 Another rather smaller block 30 ft «above» & other 50 ft lower

---

& other smaller ones «these boulders are decaying.» neighbouring rock gneiss & [. . .] sandstone actually resting on them on summit of hill rounded, site N N W of Ben Erin

56 [. . .]] *5 or 6 lettered word illeg, possibly* 'likely'.

57 Shelf of Glen Guoy
flat peat plain
divortium aquarium—[1]
tidal channel—
12 ft
obscure
obscure

NB In Glen Collarig tidal channel, sides ‹alm› 15 ft above bank or terrace, from terrace of 2^d shelf

58 Level of shelf of Glen Guoy form comparison with granite block «& Ben Erin»
29.287. 72°
Air 65 ?
70 ?

59

Where a buttress projects from side of hill if line suppose continued across to side removed all well & good, but

60 how came river to do this vast quantity when during repose of lake it did but little more

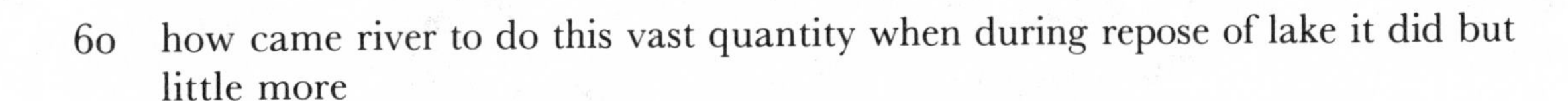

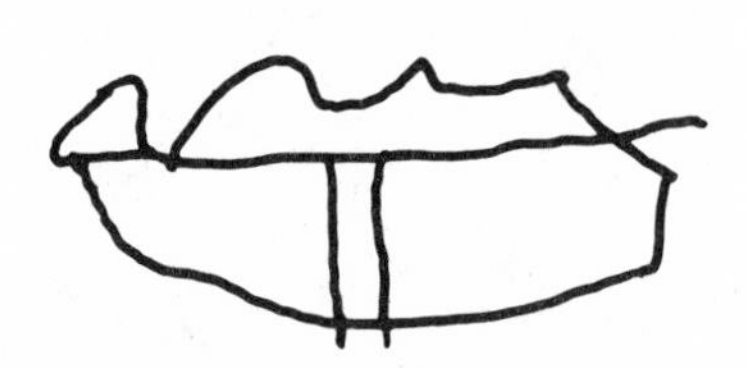

now that it has got to the rock of cols if—. why should it deposits

61 River terraces often descend by flights

the terraces if the largest has hollowed out most

57–1 *Divortium aquarum*: the diversion of water; the separation of water flow at the top of a mountain divide from one side to the opposite drainage. See *GSA*:188, 'Hence, both in composition and in stratification, the structure of the mountains on this western side of the *divortium aquarum*, is far more simple than in the corresponding part of the Peuquenes section.'

62 Wednesday
Shelf 3^d dies away *almost* imperceptibly on Glen Turrit side

---

2nd shelf *very broad* «& cut out, produced» from same «cause» as «great» spit ‹is› or plain ‹now› formed on shelf 4th

---

2^d & 3^d can be traced some way up, but most faintly on East side of Glen Turrit, where

63 I believe they end in *upwards* inclined plains, as in Corry.[1] & as «as I believe in side ravine above houses of Roy»

Maccullochs supernumerary *shelf*[2] I doubt, much about «50 or 60 ft» «no doubt, a mound of Alluvium nearly parallel—»

Inclination of river must constantly alter with falling sea & so corrode

64 plain into terrace as regressed

What ‹alter› a balance there must be in power of rivers either bringing more «detritus» than they corrode or vice versa

---

Same inclination when serpentine might remove, what above straight line «only» cut deep gorge

65 on sea hypothesis, if gullies not *now* formed «(Mac, hypoth,)»[1] the level during any oscillation must have been so carefully preserved as to have thrown water in same «drainage» lines

---

66 Mound of Gneiss though wonderful— ‹that they are preserved› how much more so, these lines & even water-scooped rock «only decay from fragment falling» of no particular hardness no wonder that all «three» lines «should be» EQUALLY preserved

---

2^d or *upper* one more perfect

62 Wednesday] *boxed.*
64 What ‹alter› . . . vice versa] *double score right margin; page excised now restored.*

63–1 Corry: a semi-circular hollowed out valley on the side of a mountain.
63–2 Macculloch 1817:373–74, '. . . the view [viz., the lake theory] explains those irregular appearances of *lines* unconnected with the principal ones . . . sometimes called supernumerary . . . these are the remains of . . . deltas or alluvia . . .'

65–1 Macculloch 1817:382–84. Macculloch discusses the relative ages of the gullies down the sides of the mountains and the lake beds of the former lakes. Where the 'roads' curve along within the gullies, the gullies must therefore be older than the lakes; where the gullies cut through the 'roads', the lakes must have existed prior to the gullies.

67 in this ‹part› «glen» than 3^{d}.

3(a) less perfect than upper & lower but quite as perfect as those lines in Glen Collarig, & some «other parts»

Boulders of same granite, all on these *three* shelves

68 soil is ‹the› usually slaty

Point of *rounded* not scooped rock on ‹bend» of $3^{(a)}$

Cannot ‹see› «make out» composition of shelves: generally angular except near head of valley

---

fragments which had fallen before lake drained could be told from «some of» those since fallen.

69 «on the 3 shelves» Solid rock is much notched

---

on Macculloch's supposition;—[1] the *old* ravine, where water entered are not proportionately large to those now formed in same

70 spot by present torrents

Macculloch wrong in saying no transported materials ‹into›[1] on upper shelves granite & some other rocks

---

at head of shelf 3^{d} almost all granite pebbles

71 Level of plain of 4th shelf at *head* of Lower Glenroy
29.581 A 82
75°?
From this point plain appears like one uniform slope slightly bending up each main valley.— & that river

72 alone had modified it— perhaps however sea also,—
Barometer on shelf
3^{d}. 29.455
A 83°
∴ plain of 4′ shelf slope, above «line of 4th» shelf

This shelf at head where ‹granite &› «veined» gneiss ‹unite› «occurs» abundantly with perfectly *rounded pebbles* of granite & forming «sloping» buttresses

69–1 Macculloch 1817:382–84.
70–1 Macculloch 1817:330, 'But the terraces themselves at the top of the glen vary in composition, and though often composed of the same sharp fragments that overspread the general declivity, they occasionally also exhibit various rolled and transported matters.'

73 Yet certainly shelf 4th ‹near› only usually contains many pebbles, but I believe this is chiefly caused by its being lower,— [no pebbles in parts of Beagle Channel when mica slate, only sand blow away][CD]

---

where lines appear

74 to cross stony parts; appearance chiefly cause by fall of angular masses from above on soft shelf—

29.330 A 84° compare this with last measurement of shelf of 3^{d}:— granite block a yard across.

75 On side of «that» hill, in front of which shelf 3^{d} form beach of *granite* pebbles, & around which shelf 2^{d} «almost» forms it into island— whole hill composed of remarkable gneiss with red granite veins & quartz, &

76 garnets.—
Boulders as before certainly must
have ‹come› «been drifted» here:
on very summit no granite—
(in valley «there are» granite) «boulders»

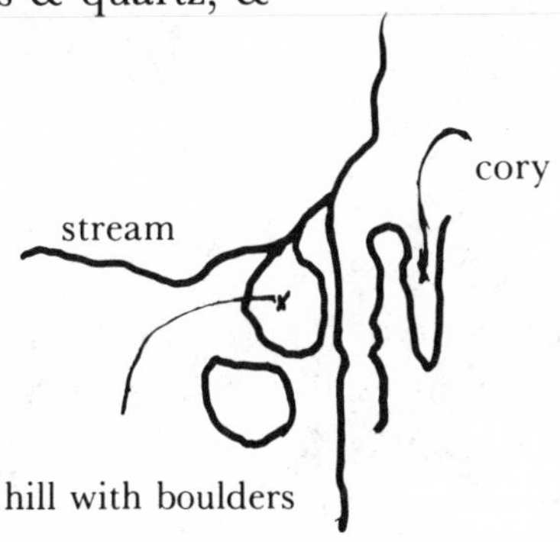

cory
stream
hill with boulders
river

77 Right Hand Cascade has ‹cut› «where two branches unite in upper Glen Roy» very little
back from line 2^{d}; little action since «that shelf» formed

---

Upper terrace near Loch Spey
‹29.35161›
29.360?
A 79° 75°?

78 A little below Divortium on slope towards Loch Spey
29.297
A 79.½

29.316
divortium aquarum «about 12 ft higher than last station»
29.316
true terrace «2^{d}» near divortium aquarum is a lip with it— Dick right— Mac[1] mistook terrace also right—

78–1 Macculloch 1817. The passage of Macculloch to which Darwin refers has not been identified.

79 Granite such as boulder on ‹thes› Divortium aquarum

Peaty Mass of this point very nearly like head of Glen Guoy
nor is horizontal line apparently continuation of upper terrace — on right hand

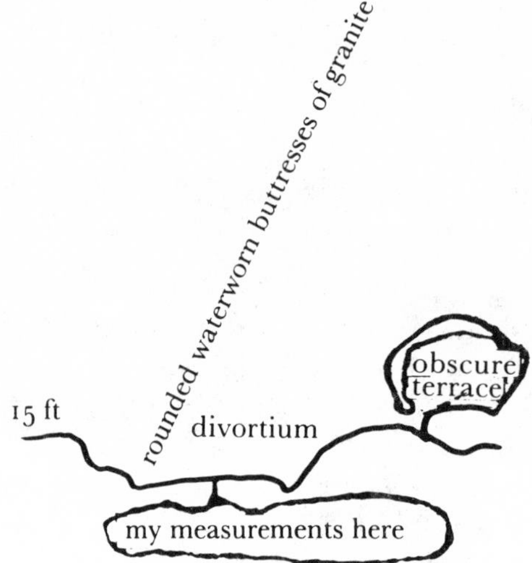

80 side of Loch Spey Forms terrace about 60 feet above Loch trace of this terrace «on ‹will› Granite ridge or a modified Granite ridge» at head of Glen Roy on same side

---

where two rivers unite in *Upper* Glen Roy great

81 plain about 60 ft beneath shelf peat on pebbles tidal plain as sea gradually retired, hard to explain on river doctrine

82 ‹Little Hill with granite blocks almost encircled›

‹fre› Gneiss cut smooth on sides *of hill where Boulder lies.*

---

buttresses «occur» high up on Shelf 2d «in Upper Glen Roy»

83 In this upper part «about junction of Upper & from Glenroy» near the upper shelfs ground strewed with pebbles

---

Shelf 3d runs up with buttresses on each side «very little way» in *Upper* Glen Roy at pass

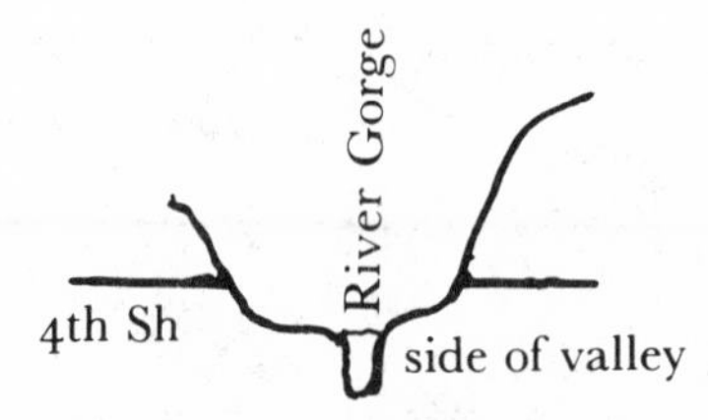

82 «in upper Glen Roy»] *circled.*

84 Granite blocks on *this* side (*return*) between 2^{d} & 3^{d} shelf
Mountain ‹Mica› «composed of» Gneiss

Block on 2^{d} shelf & below it some way; several large ones (one 6 ft across) on top of spit between river & dry Corry

---

85 **Scarcely conceivable. if Hill between Corry so much cut Granite could have remained, no peat supply.—**
Consider profoundly Boulder hypothesis[1]

Thursday, from Glen Turrit to Fort Augustus

Barom on upper (rather above)? shelf
29.290
A. 69°
Air 68°?

---

86 Barom 29.008
A. 75°
Air 70°?
This station a little way down slope of obscure terraces (& conical hills on same) of «semi» waterworn & some partly well worn pebbles— «which river could not have deposited» the slope is continued some hundred

87 feet lower & begins about 60 higher—
There are however fringes of alluvium (?) still higher

Slope of valley much more gentle than in Glen Roy, & partly shut in

---

No Granite blocks in higher parts??

Bought Glen name of Glen by which we

84 river & dry] 'dry' *over* 'C'
85 **Scarcely . . . supply.—**] *ink.*

85–1 See *JR*:288–90, 614–15 for discussions of erratic boulders.

88 descended, it is to the west of Glen Tarf

What I called Alluvium shows the ascending fringes

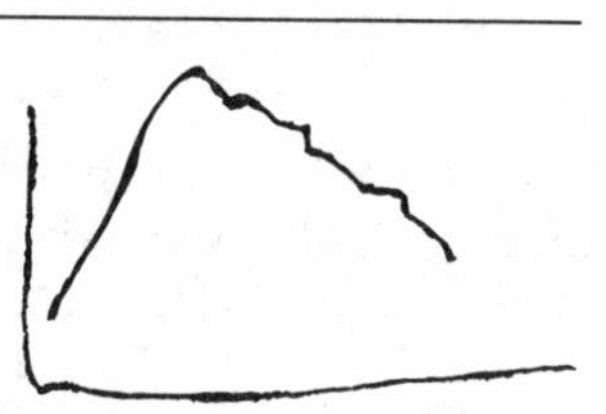

which makes me think it submarine, 400 or more feet above station! There is long straight isthmus connecting

89 E & W connecting Glen Bought & Glen Tarf a perfect old Loch, making ‹several› two divortiums aquarum, viz two branches of River Bought & between one of these & Glen Tarf

90 Hill «Cairn ‹taw› leer peak» Barom 28.700 . A.75°
75°?
Boulder, much covered by turf 2^{ft}. 8- long of syenite with pinkish felspar;— whole hill dark grey fine grained. Much contorted gneiss «narrow sharp ridge with peak» I walked all round

91 hill. Boulder about 20 ft. below summit

‹Isthmus›

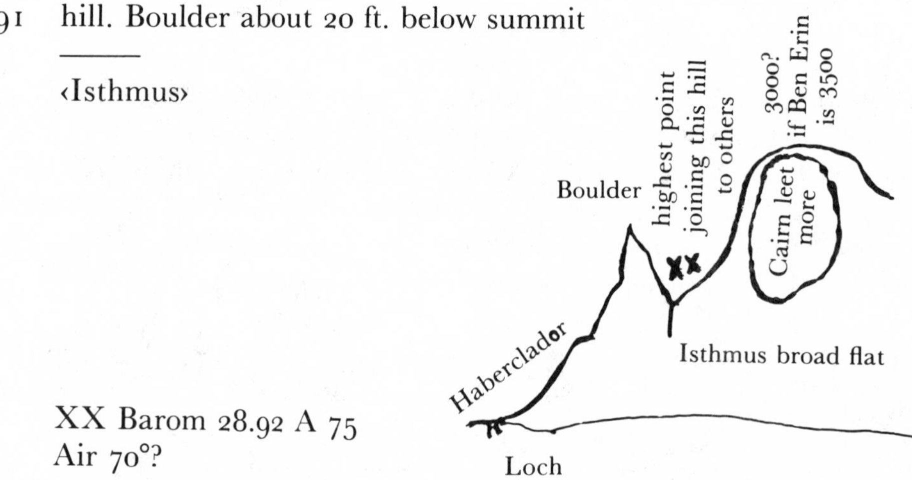

XX Barom 28.92 A 75
Air 70°?

92 Isthmus broad flat peat mass— (general character in these mountains & not ridges) between arm of Glen Bright flowing into E. end of L. Oich, & waters flowing into west end with obscure terraces on one side Barom 29.200 A 80 70°?

93 for about ¾ of mile on ‹one› S side of «this» Isthmus (which runs E & W) broad terrace «of pebbles ? & Alluvium» which appear perfectly level, ‹on op› dies away on gradual slope—
: on N side.. dies away on rocky place, but narrow shelves just like road of Glen Roy— appears to lip with moss

91 Air 70°?] *circled.*

94 On this terrace «station perhaps 6 ft too low» (to test last on Peat-Mass
Divortium aquarum)
Barom. 29.200
A.77°
Air 70°?

Barom. 066 lower than last. but A 77 station was ‹a few› «about 3» feet
‹lower›

---

too high

---

about a quarter of a mile further on, where three [. . .] abutted

95 Having crossed the mouth, (deep) of above valley this road level with Peat moss most distinct then lost by slope, then concealed by fragments, then clear. this bit to eye certainly appears level with road, & with piece of excised rock lost at point of valley chiefly from rockiness

96 When on other side

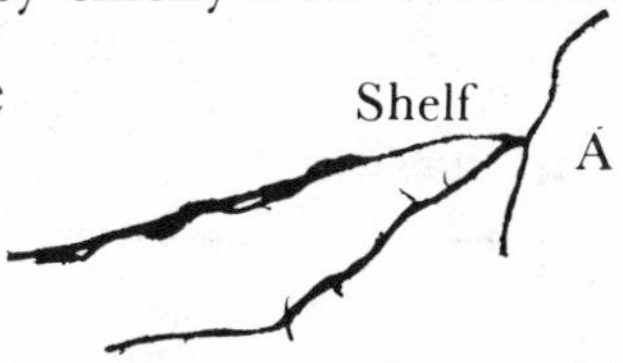

Shelf
A

Shelf A at head of Gentle mossy slope, which from a distance hid it, could be followed for at least 2 miles on dead level «by eye» to moss—

---

on this terrace Barom. 29.264 A 82
75°?

97 This last measurement turns out too low, (NB .260 would have been more correct) there were several obscure but not far continuous flights above it— (NB the buttress or pass at Isthmus appears above level of shelf certainly) I took another measurement on short buttress but not continuous & it

98 was 29.200
minus .008
.192

Loch Lochy 4 ft above water
Barom: 30.372 A 76°
75°?

The River ‹the› of which the source is a lip with the new shelf flows into canal between L. Lochy & Oich.

94 where three [. . .]] *five letter word illeg.*
97 This last . . . too low] *brace right.*
98 '——' *and* '**192**'] *ink.*

99 is a brook on the Lochy side of it— the terraces of which, last measurements belong are so complicated, that nothing can be made out of them— but it may be said that a mound stretches along, parallel to

100 Shelf on opposite side & dies away on the steep & rocky gully of last stream

---

Friday Loch Lochy near Letter Finlay Barom 30.267, A 68
Air 65°?

‹.194 372 about
267 28.75
.105 I reached›
29.090

101 Preservation of form of land *very* much due to Peat & Heather When it did not grow at first— relics destroyed.—

---

the Brook ‹about› Head of which is so interesting. enters by old tower called Glengarry

102 (Nead Roy told me) it is impossible to see my new shelf, from road:

Loch Ness
30.140 . A 66°
30.095
.0458 or 6
difference between bedrock & Loch Ness
‹30.100›
‹Donald Macphee›

103 Saturday Morning
29.958 A 64°,
air 60 «Evening do»

The extreme right arm of River Tarf ‹it› Has a very long, flat divatium aquarum with, left of Bright.— like bed of lake with trace of terraces on each side

---

High up the Tarf

104 (a Granite (boulder), sloping buttresses, an[d] one alternate curved layer of fine sand & small angular— rounded pebbles— dip sideward, & inwards— deposited when water stood at higher

105 Loch Keeper tells me, that Loch Lochy is 8 ft below Loch Oich wh is 92 ft above sea— Loch Ness 40 ft above

106 do. When cutting bank where Locks now are (32 ft rise) they found alternating layers of coarse & fine & many *Sea shells*. My informant saw them himself—

107 Sand with tide ripple Near Fort Augustus hill & fringe as if it has been filled up «at» 30 ft. higher with pebbles now worn away—

The above shells must have been

108 about 60 ft above sea— soon decayed on exposure

Mr H. C. Watson Geographical distribution of British Plants[1]

---

109 Shropshire Quartz what substance is collected in little spots

---

Speculate on «under head of» Beagle Channel.[1] Forchammers[2] (Lyells Denmark) Shrewsbury rubbish.— Speculate on origin pebbles brought by different cause: from mud.—[3] [4]

110–118 [blank]

107 Sand . . . worn away—] *scored left margin.*
108 Mr . . . Plants] *double scored left margin.*
109 **Speculate** . . . **mud.**—] *ink.*

108–1 Watson 1835. This work paid particular attention to Scottish highland plants.
109–1 *JR*:286, 'When we consider the vast dimensions and number of these glaciers, the effect produced on the land must be very great. Every one has heard of the mass of rubbish propelled by the glaciers of Switzerland . . .' P. 300, '. . . in the two large islands cut off by the Beagle channel from the rest of Tierra del Fuego, one has cliffs composed of matter that may be called stratified alluvium, which front similar ones on the opposite side of the channel,— — while the other is exclusively bordered by the older rocks . . .'
109–2 Forchhammer 1828:67–68. The author here summarizes a confusing series of beds of chalk, clay, limestone, sand, gravel, loam and boulders some of which contain fossils. He points out various hypotheses by which the geological history of the region may be reconstructed.
109–3 Lyell 1840a:245, 'Had the deposit been only a few feet thick, and all the boulders of moderate dimensions, it might have been argued, that a violent current of water, or diluvial wave, had thrown together materials of all sizes in one promiscuous mass, and left them as devoid of arrangement as a quantity of rubbish shot from a cart. For my own part, I am unable to suggest any conjecture to account for the phaenomena, except that of islands of drift ice, loaded with earth, gravel, and blocks . . . then melting.'
109–4 Lyell 1837, 4:87–88. Here Lyell discusses the deposition of clay and sand into fissures of chalk. He says p. 88, 'Organic remains are rare in Denmark, except those derived from older strata, and hence the age of the formation is on the whole very doubtful; but it has been supposed by Dr. Forchhammer and Dr. Beck to have been in progress throughout more than one tertiary period.'

119–120

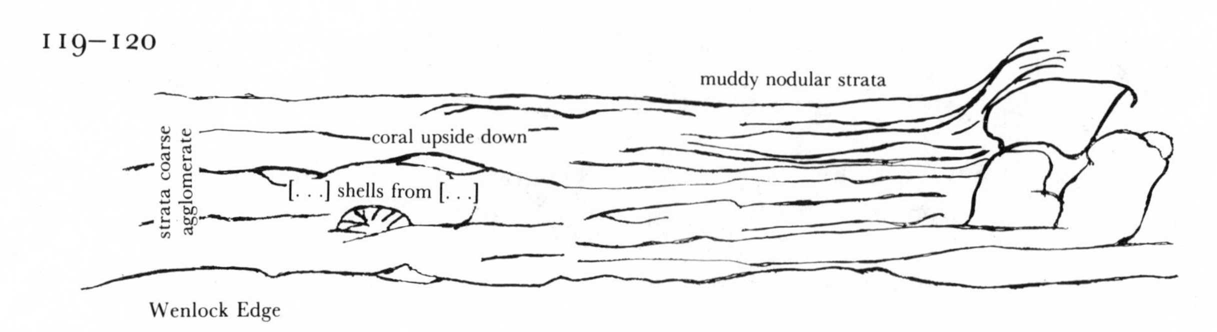

121 [blank]

122 L. Lochy 12 ft
——— 96 L. Oich
12

84

123 29.958
− 1.17
28.788
+28.8
30.372
29.200
1.172
Loch Oich 92
each Loch 8 ft.

119 *Figure extends across pages 119–120.*
[. . .] shells from [. . .]] *illeg words before and after* 'shells from'.
121 *blank, except for short line extending from sketch on page 122.*
122 '**12**' *and* '**84**'] *ink.*

124 The Metamorphic conglomerates near Loch Lochy would be well worth examining— Inverness & waters of the Tarf—

Kilfinnan Tower

---

where stream enters at head of which hill is round

125 & not merely thoughts laying dormant—

Man from Glen Turret said he learnt to know the lambs because most like *Mother* in face— asked stated this generally the case

---

126 Wednesday 12/ & 3/
Why is the Tetrao scoticus & Tetrao— not an American form

The union of two instincts crossing most remarkable ever obseved? Shows that ‹nervous› brain makes thought

BC Glen Roy

124–125 *bottom 1/4 of page torn off.*

Notebook B

Transcribed and edited by DAVID KOHN

Notebook B (DAR 121, 170 × 97 mm) is the first of Darwin's notebooks exclusively 'on "Transmutation of Species".—'.[1] It is bound in brown leather with a metal clasp and has cream-coloured paper labels with 'B' written in ink on each cover. The notebook comprises 280 pages with feint green rules and numbered consecutively by Darwin. He made his notes in brown ink with later annotations on 46 pages in grey; the reader may assume boldface signifies grey ink unless another medium is given in the textual notes. With few exceptions, B1–64 were written parallel to the spine and B65–280 were written perpendicular to the spine. Of 27 leaves excised by Darwin, 23 have been located.

Darwin did not initially date Notebooks B or C and it was only after commencing Notebook D that he took dating seriously. He then went back to his first two transmutation notebooks and assigned approximate dates, in grey ink, for their beginning and end on the inside front covers. His opening date for Notebook B 'about July 1837' proves to be a rough but reasonable estimate—provided we do not take 'about' to exclude mid-June. His closing date 'beginning of February' is early by about one month. Dates for the body of the notebook prove scarce and have to be deduced from Darwin's citations.

Scholars have accepted Darwin's *post hoc* judgement that 'This book was commenced about July. 1837', perhaps without appreciating that this was only an estimate. Since the note is in grey ink, it could only have been made between 29 July and 16 October 1838. Darwin probably estimated the beginning of Notebook B from the first citation of a dated publication. This occurs in B30: 'July 1837. Eyton of Hybrids propagating freely'. The reference is to Eyton 1837a, which appeared in the July number of the *Magazine of Natural History*. Unfortunately, this leaves the Zoonomical essays[2] at the beginning of the notebook undated. While the *latest* opening date is established as after the end of July 1837, the *earliest* plausible opening date is after the estimated completion date of the Red Notebook in late May to mid June.

Darwin put his closing date on the inside front cover of Notebook B as 'probably ended in beginning of February' since 'p. 235 was written in January 183[8]'. This retrospective dating he based on the incorrect assumption that Martens 1838 appeared in January, rather than February. Indeed he ignored his deletion of 'Jan' on B235. He also overlooked his earlier note on B198 'Henslow says. (Feb 1838) that few months since . . .' The firmest date does appear in B235 and comes from the publication of Martens 1838 in the 24 February issue of *Athenaeum*. So a closing date for Notebook B of early March is plausible.

To date the body of the notebook we have to rely principally on citations. After B30, the next unambiguous date is to Gould 1837d (B138), which appeared in the September number of *Magazine of Natural History*. However, since Darwin says Gould's paper is in the 'Septemb or Octob' number the earliest date for B138 is October 1837. On the following page, Darwin refers to information from the breeder Mr Wynne, whom Darwin probably saw during his stay in Shrewsbury 25 September to 21 October.[3]

[1] De Beer 1959:7; *Correspondence* 2:431.
[2] Hodge 1982.
[3] *Correspondence* 2:63–64.

A final group of dates occurs between B198 and B250. Darwin indirectly dated B198 with his note 'Henslow says. (Feb 1838)' and on B199 he cites Weissenborn 1838c from the January number of *Magazine of Natural History*. B235 refers to the Martens article of 24 February, while Falconer and Cautley 1838, which appeared in the January 1838 *Institut*, is cited in B250. Pages B265–271 are blank, so that B272 to the inside back cover cannot be counted as part of the chronological sequence of the notebook.

If the rich species passages of the Red Notebook mark Darwin's first statements as a transformist, with Notebook B he launches into the systematic search for an explanatory theory that will support his new convictions. Indeed the notebook begins with a full-blown theory derived from reading Erasmus Darwin's *Zoonomia*. In this Zoonomical essay of about 36 pages, Darwin formulates the fundamental questions and establishes the answers not only for Notebook B but for his whole enterprise.[4] Darwin's questions: how do species produce adaptations to a changing world, how do new species form, and how does the hierarchy of relationships in the natural system of classification form? form the basis of evolutionary theory; adaptation, speciation, and phylogeny. Darwin's first question, how do species adapt to changing circumstances, reflects a synthesis of problems Darwin recognized in William Paley's natural theology and Charles Lyell's uniformitarian geology. Hence, the key question of the transmutation notebooks is formulated: how to explain adaptation over time by natural means.

Darwin explains adaptation by natural, secondary laws drawn from the phenomena of reproduction as described in *Zoonomia*. There are 'Two kinds of generation'. The ordinary kind, sexual reproduction, produces new individuals that vary, whereas in the asexual kind, new individuals are constant. Darwin concludes 'living beings become permanently changed or subject to variety, according «to» circumstance . . . we see generation here seems a means to vary or adaptation' (B3). Thus problem of adaptation is treated as a problem in finding the evolutionary meaning of reproduction. As Darwin begins to think about reproduction, he asks the question, what is 'the final cause of life' (B5)? His answer assumes that variation automatically adapts the organism, but he does not specify how. Thus his explanation leaves untouched the argument that adaptation is perfect. Initially Darwin's appeal to reproduction is hardly less explanatory than Lyell's invoking the 'introduction' of new species, but as he fills Notebook B and the subsequent transmutation notebooks, Darwin persists in exploring the question of adaptation through the laws of reproduction.[5]

The second question Darwin asks at the outset of Notebook B is how are new species formed? Aboard the *Beagle* Darwin was struck by two phenomena of distribution: representative species on the South American continent and species endemic to oceanic islands. How, Darwin wonders, can species be made in a 'whole country'? Or as he puts it 'with this tendency to vary by generation, why are species constant over whole country'? The question and Darwin's answer connect his view of adaptation by sexual generation with his view of species formation: 'beautiful law of intermarriages ‹separating› partaking of characters of both parents, & these infinite in Number' (B5).

Darwin's concern to show that species can be made without isolation persists and indeed strongly influences his formation of the principle of divergence in the 1850s.[6] However, in

[4] See Kohn 1980, Hodge 1982. Also Herbert 1968, Limoges 1970, Gruber 1974. For a review of the literature see La Vergata 1985.
[5] See Hodge and Kohn 1985, Sloan 1985.
[6] Sulloway 1979, Kohn 1985.

Notebook B isolation dominates Darwin's answer to how new species are produced. The importance of isolation and especially island endemism to Darwin's firm acceptance of that theory is clear in B7.

Darwin's final question early in Notebook B concerns the relationship between transmutation and systematics. How will his theory account for the similarities within and the gaps between natural groups? Darwin's answer is accompanied on B26 and B36 by striking diagrams of the 'coral of life' and the 'tree of life'. He writes on his irregularly branched tree 'Case must be that one generation then should be as many living as now' and he comments 'Thus genera would be formed.—bearing relation to ancient types.—with several extinct forms, . . .' (B36–37). Thus descent from 'ancient types' is to explain the similarities and extinction the gaps. Comparative anatomists, such as Cuvier, Owen and von Baer contributed to the branching view of nature but these thinkers shared an outlook premised on the special creation of species.[7]

Darwin translated their view into a transformist vision: types become ancestors and the tree becomes an emblem for structured change that produces hierarchical relationships.

The ideas Darwin first sketched early in Notebook B, and pursued throughout the notebook, outlined the structure and substance of Darwin's principal contributions to life science. It was a bold sketch for a powerful theory, which as Darwin predicted, would 'give zest' to natural history and encompass 'whole metaphysics' (B228) in the laws of life.

[7] Ospovat 1981.

FRONT COVER

B

INSIDE FRONT COVER

C. Darwin

All useful Pages cut out Dec. 7th. /1856/
(& again looked through April 21 1873)
p. 26 30 41 46 50 54 56 67 69 76 79 91 93 107 Ireland 113 117

This Book was commenced about July. 1837

---

p. 235. was written in January 183[8] : probably ended in beginning of February

1

Zoonomia[1]

Two kinds of generation[2]

---

the coeval kind, all individuals absolutely similar;[3] for instance fruit trees, probably polypi, gemmiparous propagation.[4] bisection of Planariæ. &c &c.— The ordinary kind ‹the› which is a longer process, the new individual passing throug several stages (?typical, ‹of the› or shortened repetition of what the original molecule has done).[5]— This

2 appears highest office in organization[1] (especially in lower animals, where

IFC **All** . . . **/1856/**] *added brown ink.*
(& again . . . **1873)**] *added brown ink.*
p. 26 . . . **117**] *added pencil, crossed pencil.*

Pages 1–60, 62–64, 272, and inside back cover written parallel to spine.

1–1 See Erasmus Darwin 1794–96, vol. 1, sect. 39 'Generation', pp. 478–533.
1–2 Erasmus Darwin 1794–96, 1:487 '. . . buds and bulbs, [are] attended with a very curious circumstance; and that is, that they exactly resemble their parents, as is observable in grafting fruit-trees, and in propagating flower-roots; whereas the seminal offspring of plants being supplied with nutriment by the mother, is liable to perpetual variation.' Passage triple scored.
1–3 See Erasmus Darwin 1794–96, 1:519 '. . . it was observed above, that vegetable buds and bulbs, which are produced without a mother, are always exact resemblances of their parent; as appears in grafting fruit-trees and in the flower-buds of the dioiceous plants, which are always of the same sex on the same tree; hence those hermaphrodite insects, if they could have produced young without a mother, would not have been capable of that change or improvement, which is seen in all other animals and in those vegetables, which are procreated by one male embryon received and nourished by the female. And it is hence probable, that if vegetables could only have been produced by buds and bulbs, and not by sexual generation, that there would not at this time have existed one thousandth part of their present number of species; which have probably been originally mule-productions; nor could any kind of improvement or change have happened to them, except by the difference of soil or climate.'
1–4 Erasmus Darwin 1794–96, 1:488.
1–5 *Cf.* Erasmus Darwin 1794–96. 1:505 '. . . would it be too bold to imagine . . . that all warm-blooded animals have arisen from one living filament, which THE GREAT FIRST CAUSE endued with animality, with the power of acquiring new parts, . . . possessing the faculty of continuing to improve by its own inherent activity, and of delivering down those improvements by generation to its posterity, world without end!' Passage scored.
2–1 Erasmus Darwin 1794–96, 1:514. 'The formation of the organs of sexual generation, in contradistinction to

mind, & therefore relations to other life has not come into play)—See Zoonomia arguements, fails in hybrids where every thing else is perfect;[2] mothes apparently only born to breed.—[3] annuals rendered perennial. &c &c.—

Yet Eunuchs nor «cut» Stallions nor nuns are longer lived

Why is life short, Why such high object generation.—[4] We *know* world subject to cycle of change, temperature & all circumstances which

3 influence living beings.—

We see ‹living beings›. the young of living beings, become permanently changed or subject to variety, according «to» circumstance,— seeds of plants sown in rich soil, many kinds, are produced, though new individuals produced by buds are constant, hence we see generation here seems a means to vary. or adaptation.— Again we ‹believe› «know» in course of generations even mind & instinct becomes influenced.—

4 child of savage not civilized man.—birds rendered wild ‹through› generations, acquire ideas ditto. V. Zoonomia.—[1]

There may be unknown difficulty with *full grown* individual «with fixed organization» thus being modified,— therefore generation to adapt & alter the race to *changing* world.—
On other hand, generation destroys the effect of accidental injuries, ‹on› which if animals lived for ever would be endless

5 (that is with our present system of body & universe

---

therefore final cause of life

With this tendency to vary by generation, why are. species are constant over

that by lateral buds, in vegetables, and in some animals, as the polypus, the taenia, and the volvox, seems the chef d'oeuvre, the master-piece of nature; as appears from many flying insects, as in moths and butterflies, who seem to undergo a general change of their forms solely for the purpose of sexual reproduction, and in all the other animals this organ is not complete till the maturity of the creature.' Passage scored. See Darwin's abstract of MacLeay 1819–21 (DAR 71:128), which bears on reproduction as the 'highest office': 'P. 1. quotes from Kirby (Monog. Apium. Angliae [1802] p. 39 Vol I) that in animal & Veg. Kingdom 'reproduction of the species «compose, the» essence of its being' ˁwhether not to perfect the species? this remark supported by observation that life decays when reproductive organs fail.—is it not vice versâ??' See D70.

2–2 Erasmus Darwin 1794–96, 1:514, 'Whence it happens that, in the copulation of animals of different species, the parts necessary to life are frequently completely formed; but those for the purpose of generation are defective, as requiring a nicer organization; . . .' See D19 where Erasmus Darwin's position is called 'false'.

2–3 Erasmus Darwin 1794–96, 1: 494, 514.

2–4 See *JR*:262; and CD's note 'Shortness of Life' in his copy of Erasmus Darwin 1794–96, 1:IBC; and 1803, Canto 2:43.

4–1 Erasmus Darwin 1794–96, vol. 1, sect. 16 'Of instinct', pp. 135–184. See p. 158 on birds rendered wild after contact with humans.

whole country; beautiful law of intermarriages ‹separating› partaking of characters of both parents, & these infinite in Number

---

In Man it has been

6 said, there is instinct for opposites to like each other

ÆEgyptian cats & dogs ibis same as formerly but separate a pair & place them on fresh isl^{d}. it is very doubtful whether they would remain constant;[1] is it not said that marrying in *deteriorates* a race, that is alters it from some end which is good for Man.—

---

7 Let a pair be introduced & increase slowly, from many enemies. so as often to intermarry who will dare say what result

---

According to this view animals, on separate islands, ought to become different if kept long enough.— «apart, with slightly differen circumstances.—» Now Galapagos Tortoises, Mocking birds; Falkland Fox— Chiloe, fox,— Inglish & Irish Hare.—[1]

8 As we thus believe species vary, ‹in› changing climate we ought to find representative species; this we do in South America closely approaching.— but as they inosculate,[1] we must suppose the change is effected at once, — something like a variety produced—
—[every grade in that case surely is not

6–1 See Lyell 1837, 2:395–98, Discusses the apparent identity of mummified and modern Egyptian animals as a refutation of 'the advocates of the theory of transmutation [who] trust much to the slow and insensible changes which time may work, [and] . . . are accustomed to lament the absence of accurate descriptions, . . . handed down from . . . history, such as might have afforded data for comparing the condition of species, at two periods considerably remote.' (p. 395). Darwin commented: 'Not time to form varieties in America & Australia— «I think this fact coupled with Egyptian shows change suddenly produced» Appeal to any breeder, whether if none imported, some breed would no[t] be cultivated[.] Yet «these» animals in certain countries have [*over* 'are'] been changed, but yet fresh ones now imported do not change—Oxen do not get long horns now in S. Africa' (p. 395). On Lyell's 'In answer to the arguements drawn from the Egyptian mummies, Lamarck said that they were identical with their living descendants in the same country, because the climate and physical geography of the banks of the Nile have remained unaltered for the last thirty centuries. But why, it may be asked, have other individuals of these species retained the same characters in so many different quarters of the globe, where the climate and many other conditions are so varied?' (p. 398), Darwin wrote: 'Grt difficult. Have they?. what is date of Cat of Persia Dog of Australia. sheep of Cape of Good Hope.'

7–1 W. Thompson 1838d reviews discussions of the specific status of the Irish hare by a number of Darwin's acquaintances. According to Thompson, Yarrell (1833) offers no opinion, Jenyns (1835) considers it a variety bordering on a species, Bell (1837) considers it a species, and Eyton (1838b) sees osteological evidence of specific difference. See D61.

8–1 See RN130 'This ‹inosculation› representation of species important, each its own limit & represented.' The term is strongly associated with the quinarian theory of systematics (see MacLeay 1819–21, 396n; Swainson 1835), where an osculant group stands between and radically recombines the characters of two typical circles and inosculant groups are those members of the circle that touch the osculant group. Darwin who was long familiar with the terms (see *Correspondence* 1:280, letter to J. S. Henslow, October 1832), conflates them so that inosculant means radically reorganized.

9 produced?—
‹Granting› Species according to Lamarck disappear as collection made perfect.[1]— truer even than in Lamarck's time. Gray's remark, best known species. (as some common land shells) Most difficult to separate[2]

Every character continues to vanish, bones instinct &c &c &c

10

---

non fertility of hybridity &c &c

‹assuming all› if species (‹a›) «(1)». ‹fr› may be derived from form (2). &c.— ‹(› Then (remembering Lyells arguments of transportal)[1] ‹continents› island near continents might have some species same as nearest land, which were late arrivals

11 others old ones, (of which none of same kind had in interval arrived) might have grown altered
Hence the type would be of the continent though species all different.
In cases as Galapagos & Juan Fernandez. When continet of Pacific existed might have been Monsoons.. when they ceased importation ceased &

12 changes commenced.— or intermediate land existed.— or they may represent some large country long separated.—
On this idea of propagation of species we can see why a form peculiar to continents;[1] all bred in from one parent why. Myothera several

12 separated.—] '—' *over* 'on'.

9–1 Lamarck 1830, 1:57–58 '. . . nous voyons presque tous les vides se remplir.' Darwin commented 'Lower animals where many species'. Passage scored.

9–2 J. E. Gray: probably personal communication. The following citations point to Gray's attitude about the difficulties naturalists face in distinguishing species. Gray 1834:786, 'Land shells are much influenced, as regards their size, by the temperature, altitude, and abundance of food, of the country in which they are found . . . there is so much difference in size between individuals of *Bulimus rosaceus* found on the coast and on the mountains of Chili, that the latter have been described as a distinct species under the name of *Bulimus Chilensis*. There would be no difficulty in multiplying examples of the same kind.' Passage scored. Gray 1835:302, '[Patella and Lottia] . . . are referrible to two very different orders of *Mollusca*, while the shells are so perfectly alike, that after a long-continued study of numerous species of each genus, I cannot find any character by which they can be distinguished with any degree of certainty.' Passage scored.

10–1 Lyell 1837, 3:1–76, chaps 5–7 give detailed arguments on agencies of dispersal. See particularly 3:31 'Quadrupeds found on islands situated near the continents generally form a part of the stock of animals belonging to the adjacent mainland.' Passage scored.

12–1 By 'form' Darwin means generic type, and in his use, the ancestral form of the genus. See Henslow 1837:289 'De Candolle also supposes a definite number of species or typical forms to have been originally created, . . .'

13 species in S. America why, 2 of ostriches in. S. America— This is answer to Decandoelle. **(his argument applies only to hybridity.—**[1] genera being usually peculiar to same country, different genera different countries.[2]

---

14 Propagation explains why modern animals same type as extinct which is law almost proved.— We can see «why» structure is common in certain countries when we can hardly believe necessary, but if it was necessary to one forefather, the result, would be as it is.— Hence Antelopes at C. of Good Hope—

15 Marsupials. at Australia—

Will this apply to whole organic kingdom, when our planet first cooled.—
Countries longest separated greatest differences— if separated from immens ages possibly two distinct type, but each having its representatives— «as in Australia»
This presupposes time when no Mammalia existed; Australian; Mamm were produced from propagation from different set, as the rest of the world.—

13–1 Candolle 1820:412, 'Parmi les phénomènes généraux que présente l'habitation des plantes, il en est un qui me paroît plus inexplicable encore que tous les autres: c'est qu'il est certains genres, certaines familles, dont toutes les espèces croissent dans un seul pays (je les appellerai, par analogie avec le langage médical, genres endémiques) et d'autres dont les espèces sont réparties sur le monde entier (je les appellerai, par un motif analogue, genres sporadiques).' Note that Darwin underlined 'genres endémiques' and 'genres sporadiques'. Darwin's need to 'answer' Candolle was set by the latter's anti-transformist interpretation of geographical distribution. See Candolle 1820:417, 'Toute la théorie de la géographie botanique repose sur l'idée que l'on se fait de l'origine des êtres organisés et de la permanence des espèces.' See Lyell 1837, 2:432–33 as a complementary source for Darwin's understanding of Candolle. 'The most decisive arguments, perhaps, amongst many others, against the probability of the derivation of permanent species from cross-breeds, are to be drawn from the fact alluded to by Candolle, of species having a close affinity to each other occurring in distinct botanical provinces, or countries inhabited by groups of distinct species of indigenous plants: for in this case naturalists who are not prepared to go the whole length of the transmutationists, are under the necessity of admitting that, in some cases, species which approach very near to each other in their characters, were so created from their origin; an admission fatal to the idea of its being a general law of nature, that a few original types only should be formed, and that all intermediate races should spring from the intermixture of those stocks.'

13–2 See Candolle 1820:418, 'Je comprends très-bien, quoique je ne partage pas complétement cette opinion, je comprends et j'admets, dans quelques cas, que, dans un pays où se trouvent rapprochées plusieurs espèces des mêmes genres, il peut se former des espèces hybrides, et je sens qu'on peut expliquer par là le grand nombre d'espèces de certains genres qu'on trouve dans certaines régions; mais je ne conçois pas comment on pourroit soutenir la même explication pour des espèces qui vivent naturellement à de grandes distances. Si les trois mélèzes connus dans le monde vivoient dans les mêmes lieux, je pourrois croire que l'un d'eux est le produit du croisement des deux autres; mais je ne saurois admettre que, par exemple, l'espèce de Sibérie ait été produite par le croisement de celles d'Europe et d'Amérique.' See also Henslow 1837:289 'De Candolle also supposes a definite number of species or typical forms to have been originally created, but he does not imagine any decidedly new form or type to have ever originated from them. He considers that certain hybrids can reproduce their kind, but that in such cases there exists a constant tendency in the offspring to return again into one or other of the original types from which they sprang. Thus we should never have any strictly new type introduced, or any form which differed very materially from what was already in existence, but only a multitude of minute shades of difference, in varieties which were all intermediate between the original species. In this way he proposes to account for the endless varieties of some of our long cultivated fruits, as apples, pears, &c.' Darwin scored 'Thus we . . . pears &c.' Note, Henslow 1837 is based on the botany lectures he delivered at Cambridge, which Darwin attended.

16 This view supposes that in course of ages. & therefore changes. every animal has tendency to change.—

---

this difficult to prove cats &c from Egypt no answer because time short & no great change has happened[1]

I look at two ostriches as strong argument of possibility of such

17 change,— as we see them in space, so might they in time

As I have before said *isolate* species ‹& give even less change› especially with some change probably ‹change› vary quicker

---

Unknown causes of change. Volcanic isl^{d}.— Electricity

---

18 Each species changes. does it progress.

Man gains ideas.

---

the simplest cannot help..— becoming more complicated,; & if we look to first origin there must be progress.

---

if we suppose monads are «constantly» formed ?would they not be pretty similar over whole world under

19 similar climates & as far as world has been uniform, at former epoch— How is this Ehrenberg?[1] every successive animal is branching upwards different types of organization improving as Owen[2] says simplest coming in & most perfect «& others» occasionally dying out;—for instance secondary terebratula may

20 have propagated recent terebratula, but Megatherium nothing.

We may look at Megatheria, armadillos & sloths as all offsprings of some still older type

←————→

some of the branches dying out.— with this tendency to change, (& to multiplications when isolated, requires deaths of species to keep numbers

16–1 Compare Lamarck 1830, 1:70, 'Je ne refuse pas de croire à la conformité de ressemblance de ces animaux avec les individus des mêmes espèces qui vivent aujourd'hui. . . . Il seroit assurément bien singulier que cela fût autrement; car la position de l'Egypte et son climat sont encore, à très-peu près, ce qu'ils étoient à cette époque.' Darwin annotation: '& not isolated pair.'

19–1 Ehrenberg 1837c:564, '1. All infusoria are organized . . . animals. 2. The infusoria . . . can be separated scientifically according to their structure; and permit no identification of their forms with greater animals, however similar they often may appear.'

19–2 Richard Owen.

21 **of forms** equable:— **but is there any reason for supposing number of forms equable: this being due to subdivisions & amount of differences, so forms would be about equally numerous.**

---

changes not result of will of animal,[1] but law of adaptation as much as acid & alkali[2]

---

organized beings represent a tree. *irregularly branched* some branches far more branched.— Hence Genera.— «as many terminal buds dying, as new ones generated»

22 There is nothing stranger in death of species, than individuals

---

If we suppose monad definite existence,[1] as we may suppose is the case. their creation being dependent on definite laws, then those which have changed most. «owing to the accident of positions» must in each state of existence have shortest

23 life.; Hence shortness of life of Mammalia.—[1]

Would there not be a triple branching in the tree of life owing to three elements air, land & water, & the endeavour of each ‹one› typical class to extend his domain into the other domains. & subdivision ‹six› three more, double arrangement.—

24 if each Main stem of the tree is adapted for these three elements, there will be certainly points of affinity in each branch

---

A species as soon as once formed by separation or change in part of country. repugnance to intermarriage ‹increases it—› settles it

21 Genera] 'G' *over* 'g'.

21–1 See B216 'Lamaks "willing" doctrine absurd' and note 216–1, which cites a Darwin annotation on Lamarck 1830, 1:268 that echoes B21.

21–2 Echos Erasmus Darwin 1794–96, 1:499, 'All animals . . . have a similar cause of their organization . . . as essential . . . as chemical affinities are to certain combinations of inanimate matter.'

22–1 RN inside front cover: '«The living atoms having definite existence, those that have undergone the greatest number of changes towards perfection (namely mammalia) must have a shorter duration, than the more constant: This view supposes the simplest infusoria same since commencement of world.—»' Thus the meaning of 'definite existence' is definite duration. In this regard, *cf.* Ehrenberg 1837c:560 '[infusoria are not] undetermined elementary forms of life.'

23–1 See *JR*:97. Discussing the embedding of his extinct fossil mammalia together with recent shells, Darwin concludes, 'We here have a strong confirmation of the remarkable law so often insisted on by Mr. Lyell, namely, that the "longevity of the species in the mammalia, is upon the whole inferior to that of the testacea."' Lyell 1837, 4:40 draws the following inference from the shorter duration of the mammalian life span: 'Their more limited duration depends, in all probability, on physiological laws, which render warm-blooded quadrupeds less capable, in general, of accommodating themselves to a great variety of circumstances, and consequently, of surviving the vicissitudes to which the earth's surface is exposed in a great lapse of ages.'

25 ———

ʕWe need not think that fish & penguins really pass into each other.—

———

The tree of life should perhaps be called the coral of life, base of branches dead; so that passages cannot be seen.— this again offers

26 contradiction to constant succession of germs in progress.— «no only makes it excessively complicated.»

Is it thus fish can be traced right down to simple organization.—

———

birds— not.

27 We may fancy, according to shortness of life of species that in perfection, the bottom of branches deaden.— *so that* in Mammalia «birds» it would only appear like circles;— & insects amongst articulata.— but in lower classes, perhaps a more linear arrangement.—

28 ʕHow is it that there come aberant species in each genus «(with well characterized parts belonging to each)» approaching another.

———

‹Petrels have divided themselves into many species, so have the awks, there is particular circumstances, to which›
is it an index of the point whence, two favourable points of organization commenced branching.—

29e As all the species of some genera have died; have they all one determinate life dependent on genus,. that genus upon another, whole class would die out, therefore not.—

———

Monad has not definite existence.—

———

There does appear some connection shortness of existence, «in» perfect‹ion›, «species from many» ‹therefore› changes and base of branches being dead from which they bifurcated.—

27 *so that*] *underlining crossed.*
28 Petrels . . . to which] *crossed.*
29 «in» . . . changes] 'perfection, therefore changes' *altered to* 'in perfect species from many changes'.
bottom half of page excised: 'not.— . . . bifurcated.—'.
page crossed pencil.

30e Type of Eocene with respect to Miocene of Europe?

Loudon. Journal. of Nat History.—

July. 1837. Eyton of Hybrids propagating freely[1]

In Isl[d] neighbouring continent where some species have passed over, & where other species have ‹come› «"air" of that place. Will it be said, those have been there created there;—»

31

Are not all our «British» Shrews diff: species from the continent look over. Bell, & L. Jenyns.[1]

Falkland rabbit may perhaps be instance of domesticated animals having effected; a change whi[ch] the Fr. naturalists[2] thought was species

---

‹Ascensi› Study Lesson

Voyage of Coquille.—[3]

---

32 D[r]. Smith says he is certain that when White Men & Hottentots or Negros cross at C. of Good. Hope the children cannot be made intermediate, the first children partake more of the mother, the later ones of the father; **is not this**

30 *top half of page excised:* 'Type of . . . freely'.
Type of . . . Europe?] *crossed pencil.*
Loudon . . . freely] '17' *added brown crayon.*
'1c' *added in pencil (not by Darwin) indicates original location of excised portion in DAR 205.*
In Isl[d] . . . there;—] *crossed pencil.*
32 «–:Humboldt. New Spain:»] *smaller hand, presume passage added.*

30–1 Eyton 1837a:359, '. . . the theory of Hunter will be partially disproved; . . . . it will be found that animals of different species, nearly allied in their organisation, will breed together when in a state of domestication, confined together, or occasionally in a wild state, under peculiar circumstances . . .'

31–1 T. Bell 1837:109–10 '. . . notwithstanding the doubts . . . of many Zoologists . . . in which I had . . . participated. . . . there can be no doubt . . . that the Continental and British animals may be identical.' See L. Jenyns 1835:17n '[evidence], induces me to suspect that the *S. Araneus* of the continental authors may be distinct from ours.'; also L. Jenyns 1838a [dated Feb. 1837], where L. Jenyns, relying on dentition based on Duvernoy (1835), takes issue with Bell 1837 to distinguish British and continental shrews. See L. Jenyns 1838b, 1838a [1839] and 1839b for shifts in his opinion.

31–2 The French naturalists concerned are Lesson and Garnot (Duperrey's *La Coquille* voyage), although Darwin erroneously identifies them in B54 as 'Q. & G.', that is Quoy and Gaimard (Freycinet's *Uranie* and *Physicienne* voyage, Dumont d'Urville's *Astrolabe* voyage). Lesson and Garnot discuss the 'Falkland rabbit' in three publications: Garnot 1826:42, Lesson and Garnot 1826–30, 1 [1826]: 168–70, and Lesson 1827:295. Their assessments of the rabbit shift over time. Garnot 1826:42 considers the rabbit only in the context of the Falklands and gives it species rank ('nous croyons pouvoir considérer comme une espèce nouvelle bien distincte'); but he asks 'Serait-ce cependant une variété?' Lesson and Garnot 1826–30, 1[1826]:168–70 and Lesson 1827:295 give the rabbit species rank (*L. magellanicus*, Lesson and Garnot) but rather than treating it as an endemic species think it was seen by Magellan in 1520 in the straits that bear his name.

31–3 Lesson and Garnot 1826–30.

owing to each copulation producing its effect; as when bitches puppies are less purely bred owing to ‹first› having once borne Mongrels he has thus seen the black blood come out from the grandfather, (when the mother was nearly quite white) in the two first children[1]

How is this in West Indies «—:Humboldt. New Spain:—»[2]

33 Dr. Smith always urges the distinct locality or Metropolis of every species:[1] believes in repugnance in crossing of species in wild state.—

No doubt «C.D.» wild men do not cross readily, distinctness of tribes in T. del Fuego. the existence of whiter tribes in centre of S. America shows this.— ‹If› Is there a tendency in plants hybrids to go back?— If so Men & plants together would establish law. ≠as above stated:— no one can doubt that lesser trifling differences are blended «by»

34 by intermarriages, then the black & white is so far gone, that the species (for species they certainly are according to all common language) will keep to their type: in animals so far removed, with instinct in lieu of reason, there would probably be repugnance & art required to make marriage.— as Dr Smith remarked Man & **wild** animals in this respect are differently circumstanced.—

35

?Is this shortness of life of *species* in certain orders[1] connected with gaps in the *series of connection*? «if stating from same epoch *certainly*»
The absolute end of certain form from considering. S. America (*independent of external causes*) does appear very probable:— Mem: Horse, **L**lama. &c &c—

If we ‹suppose› «grant» similarity of animals in one country owing to springing from one branch, & the monucle has definite life, then all die at one. period, which is not case, ∴ MONUCULE NOT DEFINITE LIFE

34 they] *before unused carat.*
35 «if stating . . . *certainly*»] *in margin.*
Llama] **'L'** *added in grey ink over original* 'L'.

32–1 Andrew Smith: personal communication. For Andrew Smith's views on Hottentots see Smith 1831, which however does not discuss the point in question.
32–2 Humboldt 1811, 1:183–86.
33–1 See Andrew Smith 1836a, Appendix 1, pp. 39–40, 'During the journey [from Cape Town to central Africa], we traversed or visited three distinct Zoological provinces, each supplying certain animal forms, which, if not restricted to itself, certainly occurred in that relative proportion, which warranted its being regarded as their favourite, if not their prescribed resort. . . . Thus, most of the species we met with, appeared to have each a natural or chosen domicile, . . .' (p. 39). 'The *Falco Chiquera* may also have its African metropolis in the same direction.' (p. 40). See also Smith 1838–49, Mammalia [1849], *Bubalus caama*, 3d p. Note the term metropolis is used in Swainson 1835.
35–1 See B23 on shortness of life in mammals.

36 I think

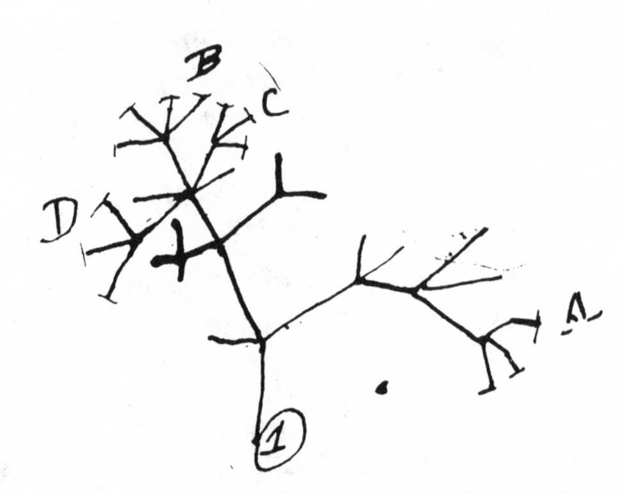

Case must be that one generation then should be as many living as now
To do this & to have many species in same genus (as is). REQUIRES extinction.

Thus between A. & B. immens gap of relation. C & B. the finest gradation, B & D rather greater distinction
Thus genera would be formed.— bearing relation

37 to ancient types.— with several extinct forms, for if each species «an ancient (1)» is capable of making, 13 recent forms.— Twelve of the contemporarys must have left no offspring at all, so as to keep number of species constant.— With respect to extinction we can easily see that variety of ostrich, Petise may not be well adapted, & thus perish out,[1] or on other hand like Orpheus. being favourable

38 many might be produced.—[1] This requires principle that the permanent varieties produced by ‹inter› confined breeding & changing circumstances are continued & produce according to the adaptation of such circumstances & therefore that death of species is a consequence (contrary to what would appear from America)

39 of non adaptation of circumstances.—

Vide two. pages back.

Diagram

---

36 *Two notes added in balloon to Figure. Note 1:* 'Case must be that one generation then should be as many living as now' *brown ink, apparently same as figure. Note 2:* **'To do this & to have many species in same genus (as is). REQUIRES extinction.'** *grey ink.*

37–1 Darwin's *Rhea*; see RN127.
38–1 Mocking bird; see discussion of *Orpheus modulator* and *O. patagonica* in *JR*: 62–3. See RN130.

The largeness of present genera renders it probable that ‹the› «many» contemporary, would have left scarcely any type of their existence in the present world.— or we may suppose only each species in each generation only breeds; *like* individuals in a country not rapidly increasing.—

40 If we thus go very far back to look to the source of the Mammalian type of organization; it is extremely improbable that any of ‹his relatives shall likewise› the successors of his relatives shall now exist,— In same manner, if we take «a man from.» any large family of 12 brothers & sisters «in a state which does not increase»

41 it will be chances against any one «of them» having progeny living ten thousand years hence; because at present day many are relatives, so that by tracing back. the ‹descen› fathers would be reduced to small percentage.— & ‹in› therefore the chances are excessively great against, any two of the 12. having progeny. after that distant period.—

42 Hence if this is true, that the *greater the groups* the *greater the gaps* (or *solutions* of *continuous structure*) «between them.».— for instance there would be great gap between birds & mammalia, Still greater between

43 Vertebrate and Articulata. still greater between animals & Plants

---

But yet besides affinities from three elements, from the «infinite» variations, & all coming from one stock & obeying one law, they may approach,— some birds may approach animals, & some of the vertebrata invertebrates—

Such on few on each side will yet present some anomaly & ‹the g› bearing

44 stamp of ‹some› great main type, & the gradation will be sudden—

Heaven know whether this agrees with Nature: *Cuidado*

The above speculations are applicable to non progressive development

---

which certainly is the case at least during subsequent ages.—

---

45 The Creator has made tribes of animals adapted preeminently for each element, but it seems law that such tribes, as far as compatible with such structure are in minor degrees adapted for. other elements.

---

every part would probably be not complete, if birds were fitted solely for air & fishes for water.

If my idea of origin of

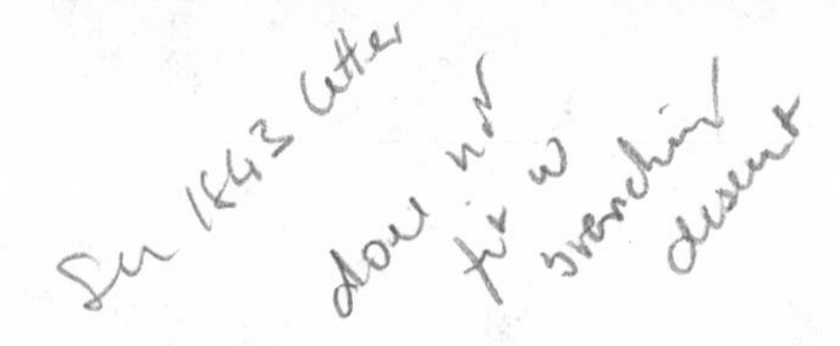

46 Quinarian system is true,[1] it will not occur in plants which are in far larger proportion terrestrial,— if in any in the Cryptogamic flora **but not atmospheric type. Hence probably only four, is not this Fries rule—[2] What subject has Mr Newman the (7) Man studied[3]**

The condition of every animal is partly due to direct adaptation & partly to heredetary taint;. hence the

47 resemblances & differences for instance of finches of Europe & America.

&c &c &c

The new system of Natural History will be to describe limits of form. (& where possible the number «of steps» known). ‹for instance among the Carabidæ.— instance in birds›
Examine «good» collection of insects with this in view.—

---

48 Geogr. Journal Vol VI. P II. p 89.— Lieut. Wellsted obtained many sheep from Arabian coast. "These were of two kinds one ‹with› white with a black face, & similar to those brought from Abyssinia, & others dark brown, with long clotted hair resembling that of goats."[1]

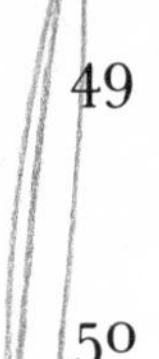

49 Progressive development gives final cause for enormous periods anterior to Man.. difficult for man to be unprejudiced about self, but considering power, extending range, reason & futurity. it does as yet appear clim

50

In Mr Gould Australian work some most curious cases.[1] of close but certainly distinct species between Australia & Van Diemen's land. & Austral & New Zealand

---

Mr Gould says in sub-genera, they undoubtedly come from same countries.— In mundine genera,

46 The condition . . . hence the] *scored pencil.*
49 *bottom third of 49 and top third of 50 excised and not located.*

46–1 Refers to the three element theory of B45. Darwin sought to base a transmutationist refutation of the quinarian system of classification on the idea that the three fundamental environments (earth, air, water) provide the opportunity for branching species formation. Darwin's transformist explanation was, as Darwin was quite aware, one of several contemporary revisions and critques of MacLeay's original quinarism (1819–21). For example, Swainson's (1835) derivation of MacLeay's groups of five from the interaction of three categories of type is alluded to in C73 and Waterhouse's (1837–40) attempt to grapple with the quinarian system is noted in B57.
46–2 See E. M. Fries 1825 and Lindley 1826.
46–3 See Newman 1832.
48–1 Wellsted 1836:89.
50–1 Gould [1837–38]. See C239.

51e the nearest species often come very remote quarters. (NB. if Plata Partridge «or Orpheus» was introduced into Chili. in present states. ‹they› it might continue & thus two species be created) & live in same country.

How is propagation of wolf & Dog.[1] (because being believed same species) if they do not breed readily. point in view.— ʕwhether highly domesticated animals like races of man.—

52e M. Flourens. .Journal des Savants.— April 1837. p. 243[1] it is said as well known fact that "serin avec le chardonneret, avec la linotte, avec le verdier" &, ‹fr› silver gold & common pheasants & fowls..—
"On sait que le "métis" du loup et du chien, que celui de la chevre et du belier, cessent d être feconds. dès les premières generations" go back to type of either animal when crossed with it.—

53 ʕWhether extinction of great S. American quadrupeds. part of some great system acting over whole world,[1] the period of great quadrupeds declining as great reptiles must have once declined.[2]—

Read his theory of the Earth attentively[3]
Cuvier objects to ‹tran› propagation of species, by saying, why not have some intermediate forms been discovered. between palæotherium, megalonyx

51 the nearest species often] *on stub.*
)&] '&' *over* ')'.
page crossed pencil.
52 '17' *added to page brown crayon.*

51–1 Hunter 1792, 'Observations tending to show that the Wolf, Jackal and Dog are all the same species.' Originally published in Hunter 1876. Although included in Hunter 1837, that work did not appear until December 1837 (see B161). In the libraries available to Darwin at Shrewsbury, Maer, and London he had access to both Hunter 1786 and 1792.
52–1 Flourens 1835–37 (1837): 243.
53–1 Note the catastrophist context of 'some great system'. See Flourens 1835–37 (1837):237–50. This fourth, and last part of Flourens' review of the fourth edition of G. Cuvier's *Ossemens fossiles* (1834–36), is devoted to 'la question de la *continuité* ['constance'] et de la *discontinuité* des espèces' (p. 237).
53–2 Flourens 1835–37 (1837):246, 'A ne compter donc que les âges ou successions d'animaux terrestres, trois âges, trois populations distinctes ont précédé la population actuelle: la première est la population des *reptiles gigantesques*; la seconde est celle des *palaeothérium*; la troisième, celle des *mammouths*, des *mastodontes*, des *mégathérium*; et entre chacun de ces âges, entre chacune de ces populations, la mer est venue recouvrir la terre et y a laissé des traces manifestes de son séjour.' See also p. 249 'pour concevoir la transformation d'une espèce en une autre, on est forcé d'admettre des modifications lentes et graduées, et par conséquent des événements, des causes, qui aient agi graduellement aussi. Or de telles causes n'ont point existé. Les catastrophes qui sont venues couper les espèces ont été subites, instantanées. On en a une preuve aussi démonstrative que singulière dans ces grandes quadrupèdes du Nord saisis par la glace, et conservés jusqu'à nos jours . . . Lors même qu'on irait jusqu'à accorder que les espèces anciennes ont pu, en se modifiant, se transformer en celles qui existent aujourd'hui, cela n'avancerait donc rien; car elles n'auraient pas eu le temps de se livrer à leur variations.' See Flourens 1835–37 (1835):178, 'Quant à *l'espèce fossile*, ou *mammouth* des Russes, ... ses os, répandus dans presque tous les pays du monde, se montrent toujours dans les mêmes couches que ceux des *mastodontes*, des *rhinocéros*, des *hippopotames*. Toutes ces espèces sont donc du même âge . . . sont également perdues.'
53–3 See G. Cuvier 1827:280–85, which concerns the fossil faunal associates of the Mammoth. Darwin's annotations enumerate eleven different forms. See *JR*:149, which quotes G. Cuvier 1827:285.

mastodon, & the species now living.[4]— Now according

54 to my view. in S. America parent of all armadilloes might be brother to Megatherium.— uncle now dead.

Bulletin Geologique April 1837. p. 216 Deshayes on change in shells from salt & F. Water—[1] on what is species.[2] *very good* **Has not Macculloch written on same changes in Fish**[3]

Mem. Rabbit of Falklands described by Q. & G. as new Species.[4] Cuvier examined it.

53–4 Flourens 1835–37 (1837):248–49, 'Pourquoi ne découvre-t-on pas, comme le dit M. Cuvier, entre le *palaeothérium*, le *mégalonyx*, le *mastodonte*, et les espèces d'aujourd'hui, quelques formes intermédiaires? Concluons donc, avec lui, que les espèces d'autrefois étaient aussi constantes que les nôtres, ou . . . ce qui caractérise les espèces de chaque âge du globe, prises en elles-mêmes, c'est leur *continuité*, et les espèces d'un âge comparées à celles d'un autre, leur *discontinuité*.' For Darwin's 'answer' to Cuvier see B88–89.

54–1 Deshayes 1836–37:216, 'On doit penser que les coquilles marines sont moins susceptibles de modifications que celles d'eau douce, les conditions extérieures dans lesquelles elles vivent étant moins sujettes à varier; cependant, certaines espèces de Cardium qui vivent aujourd'hui dans la mer, prises à divers âges ou sous diverses latitudes, offrent dans la disposition de leur charnière des différences remarquables. Dans les espèces tertiaires de la Crimée, les modifications dans les caractères des dents de la charnière sont encore beaucoup plus nombreuses. On y observe une multitude de combinaisons dans . . . [les] dents . . ., variations que M. Deshayes attribue à un changement de milieu; il pense que d'abord marines, ces coquilles auront continué de vivre dans les eaux douces qui ont remplacé les eaux salées; et en effet, le même dépôt renferme plusieurs genres essentiellement d'eau douce . . .'

54–2 Deshayes 1836–37:216–17, 'Passant ensuite à la détermination de l'espèce en général, M. Deshayes regarde la science comme étant encore peu avancée sous ce rapport, ce qui lui semble tenir à ce que l'on n'a pas suivi l'espèce sur un assez grand nombre de points dans l'étude des variations qu'elle peut subir; or, l'étude comparative des fossiles, peut fournir plusieurs de ces points intermédiaires qui nous manquent dans la nature actuelle. . . . lorsqu'on en réunit deux ou trois cents individus d'une même espèce, [des Cérites,] présentent aussi des modifications presqu'à l'infini, surtout si on y fait entrer la considération de l'âge. . . . D'un individu à un autre . . . [les caractères] peuvent changer . . . on arrive par une série de passages . . . Si l'on prenait alors les deux extrêmes de cetter série, ou seulement quelques points intermédiaires, il est certain qu'on serait porté à en faire autant d'espèces distinctes. Mais . . . [il y a] des caractères beaucoup plus constants, et ceux par conséquent sur lesquels doit être établie l'espèce.

'Aujourd'hui . . . on fait trop d'espèces . . . l'espèce pourrait être définie, une série de variétées renfermées dans un cercle, et qui, commençant à un point donné, se terminerait à ce même point après en avoir parcouru la circonférence. Les modifications ou variétés sont déterminées par l'âge, la température, la profondeur des eaux, la nature du fond, etc. Si, comme on le dit quelqufois, on voit des espèces passer de l'une à l'autre, ce n'est que parce que la série est incomplète, et que plusieurs chaînons manquent.' See the discussion of species in B9.

54–3 Macculloch 1824.

54–4 Darwin is citing Lesson and Garnot 1826–30, [1830]: 168–70 or that is part of what he is doing. 'Mais cependant, après un examen attentif, et forts surtout de l'opinion du baron Cuvier, nous ne balançons pas à la regarder comme une espèce distincte, dont la souche provient indubitablement de la Patagonie.' (p. 169); the continuation of this passage discusses Magellan. See also Garnot 1826:42 'Le quadrupède le plus multiplié aux Malouines est le Lapin . . . Nous en avons rapporté un de couleur brun roux violacée, parsemée de poils blancs, à oreilles brunes, que nous croyons pouvoir considérer comme une espèce nouvelle bien distincte, et que nous proposons de nommer lapin magellanique, (*Lepus magellanicus*, Garnot). Serait-ce cependant une variété?' Darwin's 'Q' in 'Q. & G.' is quite distinct; apparently he confused Lesson and Garnot with Quoy and Gaimard. See B31 where he recalls them as the 'Fr. naturalists.' The matter is not discussed in Quoy and Gaimard 1824 or 1830–35. For more on the Falkland rabbit see B31, C23, and C29.

55e There certainly appears attempt in each dominant structure to accomodate itself to as many situations as possible.— Why should we have in open country a ground «do. ‹w› parrot.—» woodpecker— a desert. Kingfisher.— mountain tringas.— Upland goose.— water chionis water rat with land structures; **:→law of chance would cause this to have happened in all. but less in water birds—** carrion eagles.— This is but carrying on. attempt at adaptation of each element.—

May this not be explained on principle, of animal having come to island. where it could live— but there were causes to induce great change. like the Buzzard which has changed into Cara cara at the Galapagos.

56e Fernando Noronha Ophyressa bilineata (Gray) new ‹liza› species, belonging to true. American genus

---

Waterhouse says he is certain, that in insects, each family, however many there may be, represent every other;[1] for instance in Heteromera, you have representatives (which at first would be mistaken for) Carabidæ, Crysomela, Scarabadæ, & longicornes.— Again taking a subdivision of Heteromera

57 same. thing occurs with regards to other tribes in that same family.—

(NB I see Waterhouse thinks Quinary only three elements)[1]

---

55 to accomodate . . . possible] *marginal X added pencil.*
:→**law chance . . . birds—**] *in margin.*
May this . . . Galapagos] *double scored grey ink.*
page crossed pencil.
page folded.
56 '19' *added to page, brown crayon.*

56–1 Waterhouse 1837–40:188–89, 'Some time since, about January, 1835, I had collected together a number of specimens of insects to illustrate certain views relating to analogies observable amongst them. . . .

'This collection consisted chiefly of Coleopterous insects, and among them I had most of the more curious *forms* observed in the section *Heteromera*,—my object being to show that the species thus selected were analogous representations of other groups of beetles; that is to say, that they departed from their own group in certain characters of form, colour, &c., and that in these respects they appeared to have borrowed (if we may use such a term) the characters of other groups of the same order, to which they bear such a resemblance that they might at first sight be mistaken for species belonging to those groups; and we often observe that the markings vary according to the habits of the individuals.'

57–1 Waterhouse 1837–40:189–90 '. . . at first when I perceived these resemblances I was inclined to believe that there existed a positive affinity between certain species of one group and those of several other groups; . . . I however found that I never could trace a positive linking of one group to more than two others,—that which preceded it and that which followed. I therefore felt compelled to give up my theory, which I afterwards had approached to one already made known—I mean the '*net-work theory*,' as I have heard it termed. I perceived that these supposed affinities were in fact analogies. My next step was to make notes of these various analogies as I went through each group, and in so doing I found, as I thought, that each group preserved analogous representations to all other groups which are of equal value, and of the same greater section. For instance, I found analogies in one section of the *Coleoptera* to almost every other section of equal value, and I perceived that in the *order Coleoptera* there were analogous representations to almost nearly all the other *orders* of insects; . . . [My views] may go a great way to prove or disprove an exceedingly ingenious and favourite theory—I mean the circular and quinary system; for it may happen that in the formation of

How far Does Waterhouse's representatives agrees with breeding . . in irregular trees. & extinction of forms.?? It is in simplest case saying every species in genus resembles each other,[2] (at least in one point, in truth in all excepting specific character); and in passing from species to genera, each retains

58 some one character of all its family; but why so? I can see no reason for these. analogies; CD[from the principle of atavism, where real structure obliged to be altered, I can conceive colouring retained; therefore probably in some ‹heteromera› colouring of crysomela may be going back to common ancestor of Crysom. & Heterom, but I cannot understand the universality of such law.—

59 It would be curious to know in plants, (or animals) whether, ‹in› *races* have tendency to keep to ‹each› either parent, (this is what French call (*atavism*) Probably this is first step in dislike to union, offspring not well intermediate

---

Lyell Vol III. p. 379.[1] Mammalian type of organization same from one period to another, preeminently Pachydermata, less so in Miocen & so on.—

---

60 As I have traced the ‹type› great quadrupeds to Siberia;[1] we must look to type of organization.—
extinct species of that country parents of American.— Now Genera of those two countries ought to be similar

---

61 ?Law: existence *de*finite without change, superinduced, or new species; therefore animals would perish, if there was nothing in country to superinduce a change?[1]

this theory analogies may in some instances have been mistaken for affinities.' Waterhouse's 'three elements' may refer to the following: 'there appears to me three circumstances, each of which may give an appearance of correctness to the theory of the circular arrangement of animals, and yet that idea may still be erroneous.' . . . [1.] 'a group may be so arranged that the last species may be an analogous representation of the first, . . . [2.] certain species [may be] removed from their natural affinities and wrongly placed, but so disposed that they possess an affinity to the first; . . . [3.] a certain series of species follow in succession according to their affinities, . . . in the middle . . . there may be a species which bears an analogous representation to the group which commences the series; if this species, together with a few others. . . , be removed from their natural situation, and placed at the end of the line, and the case of analogy be called an affinity, the natural way to arrange them would appear to be in a circle . . .'

57–2 Waterhouse 1837–40:190, 'I found, as I thought, that each group preserved analogous representations to all other groups which are of equal value, and of the same greater section.'

59–1 Lyell 1837, 3:379, '*Mammiferous remains of successive Tertiary Eras.*— But although a thirtieth part of the Eocene testacea have been identified with species now living, none of the associated mammiferous remains belong to species which now exist, either in Europe or elsewhere. Some of these equalled the horse, and others the rhinoceros in size; . . . More than forty of these Eocene mammifers are referable to a particular division of the order Pachydermata, which has now only four living representatives on the globe. . . In the Miocene mammalia we find a few of the generic forms most frequent in the Eocene strata associated with some of those now existing, and in the Pliocene we find an intermixture of extinct and recent species of quadrupeds.'

60–1 See *JR*:149–54.

61–1 See *JR*:210, 'It is impossible to reflect without the deepest astonishment, on the changed state of this con-

62 Seeing animal die out in S. America; with no change, agrees with belief. that Siberian animals lived in cold countries & therefore not killed by ‹Siberia a more› cold countries.—[1]

Seeing how horse & Elephant reached S. America.—[2] explains how Zebras reached South Africa—

It is a wonderful fact Horse, Elephant & Mastodon dying out

63 about same time in such different quarters.—[1]

Will Mr Lyell say that some circumstance killed it over a tract from Spain to S. America.—[2] Never

---

They die; without they change;[3] like Golden Pippens, it is a *generation* of *species* like generation of *individuals*.—

64 Why does individual die, to perpetuate certain peculiarities, (therefore adaptation), & to obliterate accidental varieties, & to accomodate itself to change, (for of course change even in varieties is accomodation). Now this argument applies to species.—
If individual cannot procreate, he has no issue, so with species.—

---

65 I should expect that Bear & Foxes &c same in N. America & Asia, but many species closely allied but different, because country separated since time of extinct quadrupeds:— same argument applies to England.— Mem. Shew Mice.— ——

---

63 Never] *in rectangle*.

tinent. Formerly it must have swarmed with great monsters . . . Since their loss, no very great physical changes can have taken place in the nature of the country.' See also RN129. 'Should urge that extinct Llama owed its death not to change of circumstances; reversed argument. knowing it to be a desert.— Tempted to believe animals created for a definite time:— not extinguished by change of circumstances:' See also B22, B29 and B35 on 'definite existence', that is, delimited duration of monads.

62–1 See *JR*:103 '. . . we must grant, as far as *quantity alone* of vegetation is concerned, that the great quadrupeds of the later tertiary epochs might, in most parts of Northern Europe and Asia, have lived on the spots where their remains are now found.'

62–2 See B223, *JR*:151.

63–1 See RN85.

63–2 See Lyell 1837, 3:134–35, *Effects of a General Alteration in Climate on the Distribution of Species*. 'If . . . cold seasons were to become frequent, in consequence of a gradual and general refrigeration [migrations would occur] . . . But although some species might thus be preserved, every great change of climate must be fatal to many which can find no place of retreat when their original habitations become unfit for them.' See also Lyell 1837, 1:144–57 on 'Siberian Mammoths'. Lyell does not discuss South American fauna.

63–3 Compare Darwin's annotation 'Will the theory do, form acquired ∴ change extermination.' on Lyell 1837 3:119 (at end of the concluding remarks to chap. 9).

66 Animals common to South and North America.—

?are there any?

67 Rhinoceros peculiar to Java, & another to Sumatra —Mem Parrots peculiar according to Swainson[1] to certain islets in East Indian archipelago

Dr Smith considers probable two Northern species replace ‹No› Southern kinds—

(1) Gnu reeaches Orange river & says so far will I go and no further:[2]—

68 Prof. Henslow[1] says. that when race once established so difficult to root out.— For instance ever so many seeds of white flower. all would come up white, though planted in same soil with blue.

Now this is same bearing with Dr. Smith's fact of races of men

69e tendency to keep to one line[1]

Dr Smith says very. close species generally frequet slightly different localities, so that they become useful to know what is species.—[2]

66 ?are there any?] *in rectangle.*
67 (1) Gnu . . . further:—] *scored pencil.*
69 tendency . . . what is species.—] *crossed pencil.*
Dr Smith . . . what is species.—] *double scored pencil.*
In proof . . . armadilloes «&»] '7' *added brown crayon.*

67–1 Swainson 1835:22, 'The gallinaceous genera are few. Their wide dispersion is decidedly against the theory, that all birds, with heavy bodies and short wings, are more limited in their geographic range than other terrestrial tribes. This argument has been ingeniously made use of, to account for the very restricted limits nature has imposed upon the greater number of Indian parrots; many species, as it is stated, being confined to particular islands. We must not, however, expect to find a reason for every thing: in the present instance, the above conclusion is particularly erroneous. . . . the wings of nearly the whole of the parrot family are peculiarly adapted for strong and vigorous flight.' Passage scored.

67–2 Andrew Smith 1837:11, Item 8., 'Antelope Taurina. The Brindled or Black-tailed Gnu and the second species of the genus which has been found in South Africa. The Nu Gariep or Black River, appear to form the limit of its Southern range; and though herds often feed almost upon the very banks of that stream, yet not an individual has been known to cross, a circumstance the more remarkable, as the common species (Catoblepas Gnu) regularly passes it by, for the northern districts of the colony.' See also Smith 1838–49, Mammalia [1849], *Catoblepas taurina*, 4th p. 'Both species of *Gnu* inhabit, during a certain period of the year, the extensive grassy plains which exist some considerable distance to the northwards of the *Vaal* River; and at another period a portion of each, at least advances to the southward to feed upon the vegetation which occurs in that direction after the fall of the summer rains. Both species advance simultaneously as far as the southern branches of the Orange River, but on reaching those, the species here figured [*C. taurina*] ceases to advance and the common species (*Catoblepas Gnu*) passes by itself into the Colony.'

68–1 John Stevens Henslow: personal communication.

69–1 Personal communication. Note B32–1.

69–2 See note B33–1.

In proof that structure is not simple adaptation,[3] armadilloes «&»

70e & Megatherium. each with same kind of coat.— If we could tell, I do not doubt even colour hereditary in time as in space. (Mem: Galapagos).

Little wings of Apteryx
Dacelo & Kingfisher same colours

---

71 Strong odour of negroes, a point of real repugnance.—

Waterhouse says there is no TRUE *connection* between great groups.—

---

72 Speculate on land being grouped towards centres near Equator in former periods & then splitting off.—

If *species* generate «other *species*», their race is not utterly cut off:— like golden pippen. if produced by seed go on.— otherwise all die.═
The fossil horse, generated in S. Africa Zebra.— & continued.— perished in America

73 All animals ‹are› of same species are bound together just like buds of plants, which die at one time, though produced either sooner or later.—
Prove animal like plants:— trace gradation between associated & non associated animals.— & the story will be complete.—

---

74 It is absurd to talk of one animal being higher than another.— *We* consider those, where the {cerebral structure / intellectual faculties} most developed, as highest.—
A bee doubtless would when the instincts were.—

70 If we . . . (Mem: Galapagos).] *crossed pencil.*
«Little wings of Apteryx»] *smaller hand, presume passage added; crossed pencil.*
Dacelo & Kingfisher same colours] *added pencil.*
'11' *added to page brown crayon.*
72 in former] 'in' *over* '&'.
is not] 's' *over* 'n'.
73 Prove . . . complete.—] *circled.*

69–3 As one source of this notion, see Andrew Smith 1836a, Appendix 1, p. 41, 'The range of species, generally speaking, appeared to vary considerably as to extent, and in no case was it possible to discover any cause or causes, depending upon external circumstances, which could enable us to account in a satisfactory manner for such a diversity.' See also Lyell 1837, 3:100. 'In other words, the possibility of the existence of a certain species in a given place, or of its thriving more or less therein, is determined not merely by temperature, humidity, soil, elevation, and other circumstances of the like kind; but also by the existence or non-existence, the abundance or scarcity, of a particular assemblage of other plants and animals in the same region.'

75e relation of type in two countries direct relation to facilities of communication

---

Have *races* of Plants. ever been crossed really, if there is any difficulty in such marriages or offspring show tendency to go back— there is an end to species.—

76e «Brown Appendix»
A most remarkable observation of Mr Brown, about peculiarities of Flora. on East & West. ends of New Holland. diminishing towards centre (p. 586)[1]— Parallel 33°–35°, source of forms. reduce towards Northern[2] Eastern end & die away, & partake of Indian character[3]

77 There appears in Australia great abundance of species of few genera or families.[1]—
(long separated.—

Proteaceæ & other forms (?) being common to Southern hemisphere, does not look as if S. africa peopled from N. Africa[2]

78 An originality is given (& power of adaptation) is given by *true* generation, throughe means of every step of progressive increase of organization being imitated in the womb, which has been passed through to form that species.— ‹Man is derived from Monad, each fresh—›

79 Mr Don remarked to me, that he though species became obscurer as knowledge increased, but genera stronger,[1]— Mr Waterhouse says no real ‹separ› passage between good genera[2]— How remarkable spines, like on a

75 *page crossed pencil.*
76 A most . . . (p. 586)—] *scored pencil,* '20' *added brown crayon.*
79 Mr Waterhouse . . . Echidna—] *scored pencil.*

76–1 R. Brown 1814, 2:586, 'I have formerly remarked that nearly half the Australian species of plants, at present known, have been collected in a parallel included between 33° and 35° S. latitude; and it appears, from the preceding observations on the several natural orders, that a much greater proportion of the pecularities of the Australian Flora exist in this, which I have therefore called the *principal parallel*; and that many of them are even nearly confined to it. But these peculiarities exist chiefly at its western and eastern extremities, and are remarkably diminished in that intermediate part which is comprehended between 133° and 138° E. long.'

76–2 R. Brown 1814, 2:586, 'From the principal parallel most of the characteristic tribes diminish in number of species as well as of individuals, not, however, equally in both directions, but in a much greater degree towards the equator.'

76–3 R. Brown 1814, 2:586, 'Within the tropic, at least on the East coast, the departure from the Australian character is much more remarkable, and an assimilation nearer to that of India than of any other country takes place.'

77–1 R. Brown 1814, 2:587, 'The plants of Terra Australis at present known, amounting to 4200, are referable, as has been already stated, to 120 natural orders; but fully half the number of species belong to eleven orders.'

77–2 R. Brown 1814, 2:587, '*Proteaceae* and *Restiaceae*, which are nearly confined to the south appear to be most abundant in the principal parallel of New Holland, are also very numerous at the Cape of Good Hope:'

79–1 The remark could have been made by David Don or by his brother George Don.

79–2 See Waterhouse 1837–40, from which it is implicit that apparent passages are instances of 'representative analogies', not affinities.

porcupine on Echidna—

Good to study Regne Animal for *Geography*.—[3]

80 The motion of the earth must be excessive up & down.— Elephants in Ceylon— East Indian archipelago.— West Indies = Opossum & Agouti same as on continent— **3 Paradupasi in common to Van Diemen's Land & Australia** From the consideration of these archipelagos ups & downs in full conformity with Europæan formations— **England & Europe Ireland common animals**— +++ for instance tertiary deposits between East Indian islets—

(+++. Ireland longer separated., Hare of two countries different.— Ireland & Isle of Man possessed Elk not England. Did Ireland possesse Mastodons?? Negative facts tell for little)

81 Geographic distribution of Mammalia more valuable than any other, because less easily transported— Mem plants on Coral islets.— Next to animals land birds.— & life shorter or change greater—

In the East Indian Archipelago it would be interesting to trace limits of large animals—[1]

82 Owls. transport mice alive?

Species formed by subsidence. Java & Sumatra. Rhinoceros.[1] Elevate & join keep distinct. two species made
elevation & subsidence continually forming species.—

Man & wife being constant together for life, is in accordance with
The Male animal affecting all the Progeny of female insures often mixing of individuals.

83 **Here we have avitism the ordinary event. & succession the extraordinary**

South Africa. proof of subsidence. & recent elevation: Pray ask Dr. Smith.— «to state that most clearly».[1]

---

80 **Did**] *over* '**Does**'.
82 **Man & . . . accordance with**] *crossed grey ink.*
all] *underlined.*
83 **Here we . . . the extraordinary**] *in margin.*

79–3 G. Cuvier 1829–30.
81–1 See Lyell 1837, 3:31, 'The Indian archipelago presents peculiar phenomena in regard to its indigenous mammalia, which, in their generic character, recede, in some respects, from that of the animals of the Indian continent, and approximate to the African.'
82–1 See Lyell 1837, 3:31, 'Sumatra [contains] . . . a rhinoceros resembling the African more than the Indian species, but specifically distinguishable from both.'
83–1 Andrew Smith: personal communication.

Fox tells me, that beyond all doubt seeds of Ribston Pippin, produce Ribstone Pippins, & Golden Pippen, goldens—[2] hence— *sub-varieties* & hence possibility of reproducing any variety, although many of the seeds will go back.— Get instances of a *variety* of fruit tree or plant run wild in foreign country.—

84 When one sees nipple on man's breast. one does not say some use, but ‹no.› sex not having been determined.— so with useless wings under elytra of beetles.— born from beetles with wings .& modified.— if simple creation, surely would have been born without them.=

85 In some of the lower orders a perfect gradation can be found from forms marking good genera— by steps so insensible, that each is not more change than we know *varieties* can produce.— Therefore all genera MAY have had intermediate steps.—

Quote in detail some good instances.

But it is other question, whether there

86 have existed *all* those intermediate steps especially in those classes where species not numerous. (NB in those classes with few species greatest jumps strongest marked genera? Reptiles?)
For instance there never may have been grade between pig & tapir, yet from some

87 common progenitor,— Now if the intermediate ranks had produced infinite species probably the series would have been more perfect. because in each there is possibility of such organization.
[Spines in Echidna & Hedgehog]CD—
As we have one Marsupial animal in Stonefied slate,[1] the father of all

88 Mammalia in ages long gone past. . & still more so «known» with fishes & reptiles.—

In mere eocine rocks. we can only expect some steps.— I may ask whether the series is not more perfect by the discovery of fossil Mammalia than before

89 & that is all that can be expected—
This answers Cuvier[1]—

Perhaps the father of Mammalia as Heterodox as ornithorhyncus. If this last animal bred— might not new classes be brought into play.—

85 In some . . . *varieties*] *double scored.*

83–2 W. D. Fox: personal communication. See D70.
87–1 Stonesfield. See Broderip 1827:408–12; note Lyell's interpretation of this discovery: 'it seems fatal to the theory of progressive development' (1837, 1:239).
89–1 B53.

90 The father being climatized, climatizes the child. ?— whether every animal produces in course of ages ten thousand varieties, (influenced itself perhaps by circumstances) & those alone preserved which are well adapted, This would account for each tribe ‹being› «acting as» in vacuum to each other

91 p. 306.—. Chamisso on Kamschatka quadrupeds
Kotezebues first Voyage[1]

Copied into list
Entomological Magazine paper on Geographical range[2]

Richardson— Fauna Borealis.[3]

It is important the possibility of some isl^d^ not having large quadrupeds[4]—

92 Humboldt has written on the geography of plants. Essai sur la Geographie des Plants. 1 Vol in 4°.[1]—

I have abstracted Mr Swainson's trash.— at beginning of Volume

---

on Geographical distribution of animals[2]

93 **Brown** Geograph Journal. Vol I p. 174. says from Swan river long South coast all the remarkable Australia genera, collected together[1]

Man has no *heredetray prejudices* «or instinc» to conquer or breed together:—

91 p. 306 . . . first Voyage] *scored pencil.*
Copied into list] *crossed grey ink.*
Geographical] 'G' *over* 'D'.
92 I have . . . animals] *double scored and rule crossed grey ink.*
93 **Brown**] *added pencil.*
Geograph . . . together] *scored pencil.*

91–1 Chamisso 1821, 3:306.
91–2 Delta 1835.
91–3 Richardson 1829–37.
91–4 See Lyell 1837, 3:31–32, e.g., 'Quadrupeds found on islands situated near the continents generally form a part of the stock of animals belonging to the adjacent mainland; "but small islands remote from continents are in general altogether destitute of land quadrupeds" [quoted from Prichard 1826, 1:75].' See note B92–2; and *JR*:249–50. 'The only quadruped native to the island, is a large wolf-like fox, which is common to both East and West Falkland. I have no doubt it is a peculiar species, and confined to this archipelago; . . . As far as I am aware, there is no other instance in any part of the world, of so small a mass of broken land, distant from a continent, possessing so large a quadruped peculiar to itself.'
92–1 Humboldt and Bonpland 1805.
92–2 Swainson 1835. Darwin's abstract of the geography section of this work is on the inside front cover and fly leaf of his copy. However, it begins with the following comment relevant to the classification section: 'There is a great deal of nonsense talked about perfection of groups &c. as far as I can discover; some families have mingled characters & varied habits, others confined characters & peculiar structures.—' The remainder of the abstract consists of brief descriptive entries. One note bears directly on Darwin's comment in B91: 'p. 17 No «large» animals in Madagascar'. On p. 17 he underlined Swainson's 'absence of large animals in Madagascar' and added 'New Zealand Caledonia *New Guinea* Contrasted with, Sumatra &c &c & England.'
93–1 R. Brown 1832:18.

Man has no limits to desire, in proportion instinct more. reason less. so will aversion be

94 L. Institut «1837. No 246» a section of fossil "singe", it cannot be made to approach the Colobes which in south Africa, appear to represent the semnopitheque of India.[1]— Tooth of ‹Spi› of Sapajou—

NB Sapajou is S. American form. therefore it is like case of great ‹rodent› edentate [**has been doubted ?**][CD]

95 & opossum found in Europe now confined to southern hemisphere.— **If these facts were established it would go to show a centrum for Mammalia.**[1]— I really think a very strong case might be made out of world before zoological divisions.— Mem. species doubtful when known only by bones.—
Mem: Silurian fossils: ?How are. South American shells?

96 ?Do not plants, which have male & female organs together, yet receive influence from other plants.— Does not Lyell give some argument about varieties being difficult to keep on account of pollen from other plants because this may be applied to show all plants do receive intermixture.[1]— But how with «hermaphrodite» shells.!!!?

97 We have not the slightest right to say there never was common progenitor to Mammalia & fish. when there now exist such strange forms as ornithorhyncus

---

95 **centrum**] *underlined.*
Mem. . . . bones.—] *brace in margin.*

94–1 Blainville 1837:205–6 reports his examination of fossil bones sent by Edouard Lartet from the French tertiary deposit of Sansan; the principal results are as Darwin records. Blainville's interpretation is 'malgré que nous soyons obligés de ne pas admettre le fait extraordinaire de fossiles d'animaux aussi rigoureusement limités dans leurs circonscriptions géographiques que les Singes, les Sapajous, les Makis trouvés à Lefort, en France, dans les mêmes lieux, dans les mêmes circonstances géologiques, la découverte d'ossemens fossiles ayant indubitablement appartenu à un Singe . . . et à une espèce qui a plus de rapports avec les Gibbons, qui sont limités aux portions les plus reculées de l'Asie, qu'a toute autre actuellement vivante, n'en reste pas moins l'une des plus heureuses et des plus inattendues découvertes qui ait été faite en palaeontologie dans ces derniers temps.' There are several marginal scorings on Darwin's copy. See also E. Geoffroy Saint-Hilaire 1837b:242, which comments on 'la haute portée que ce fait [le singe fossile de Sansan] doit avoir . . . dans la science de la philosophie naturelle.' Darwin's copy is scored; B133.

95–1 The grey-ink note shows the subject of interest in B94–95: the possibility of establishing a centre of origin for the Mammalia. See B87–88 where the Stonesfield marsupial fossil is called the 'father of all Mammalia.'

96–1 Lyell 1837, 2:401–2, 'If . . . we suppose in a state of nature the seed of the wild Brassica oleracea to have been wafted from the sea-side to some spot enriched by the dung of animals, and to have there become a cauliflower, it would soon diffuse its seed to some comparatively sterile soils around, and the offspring would relapse to the likeness of the parent stock. . . . But . . . imagine the soil, in the spot first occupied, to be constantly manured by herds of wild animals, . . . still the variety could not be maintained; because . . . each of these races is prone to fecundate others, and gardeners are compelled to exert the utmost diligence to prevent cross-breeds. The intermixtures of the pollen of varieties growing in the poorer soil around would soon destroy the peculiar characters of the race which occupied the highly manured tract; for if these accidents so continually happen, in spite of our care, among the culinary varieties, it is easy to see how soon this cause might obliterate every marked singularity in a wild state.' Passage scored.

The type of organization constant in the shells.—

---

98 The question if creative power acted at Galapagos it so acted that bi[r]ds with plumage «&» tone of voice partly American North & South.— (**& geographical ‹distri› division are arbitrary & not permanent. this might be made very strong. if we believe the Creator creates by any laws. which, I think is shown by the very facts of the Zoological character of these islands** so permanent a breath cannot reside in space before island existed.— Such an influence Must exist in such spots. We know birds do arrive & seeds.—

99 The same remarks applicable to fossil animals same type, armadillo like covering created.— passage for vertebrae in neck same cause, such beautiful adaptations yet other animals live so well.— This view of propagation gives. ‹ro› hiding place for many unintelligible structures. it might have been of use in progenitor— or it may be of use.— like Mammæ on mens' breasts.—

100 How does it come wandering birds. such sandpipers. not new at Galapagos.— did the creative force know that «these» species could, arrive— did it only create those kinds not so likely to wander. Did it create two species closely allied to Mus. coronata, but not coronata.[1]— We know that domestic animals vary in countries, without any assignable reason.—

101 Astronomers might formerly have said that God ordered, each planet to move in its particular destiny.— In same manner God orders each animal created with certain form in certain country, but how much more simple, & sublime power let attraction act according to certain laws such are inevitable consequen let animal be created, then by the fixed laws of generation, such will be their successors.—

102 let the powers of transportal be such & so will be the form of one country to another.— let geological changes go at such a rate, so will be the numbers & distribution of the species!!

103 It may be argued representative species chiefly found where barriers «& what are barriers but» interruption of communication. or when country changes. Will it said that Volcanic soil of Galapagos under equator that external conditions would produce species so close as Patagonian ‹Chat› & Galapagos orpheus.= Put this strong so many thousand miles distant.—

104 Absolute knowledge that species die &. others replace them— two hypotheses fresh creations is mere assumption, it explains nothing further, points gained if any facts are connected.—

---

99 ‹ro›] *began* 'room' ?

100–1 *Muscicapa coronata.*

No doubt in birds, mundine genera, **Bats .Foxes. Mus** are birds that are apt to wander & of easy transportal.— Waders &

105 Waterfowl.— scrutinize genera, & draw up tables— Instinct may confine certain birds which have wide powers of flight; but are there any genera, mundine, which cannot transport easily.— **it would have been wonderful if the two Rheas had existed in different Continents** In plants I believe not. .—

106 It is a very great puzzle why Marsupials & Edentata should only have left off springs ‹ne› in or near South Hemisphere.

Were they produced in several places & died off in some? Why did not fossil horse breed in S. America— **it will not do to say period unfavourable to large quadrupeds—horse not large—**

107e Ed. New. Philosop J. No 3. p. 207 "It is not generally known that Ireland possesses varieties of the furze, broom, & yew very different from any found in great Britain, British varieties are also found in Ireland[1]—

108e There must be progressive development; for instance none—?, of the ‹vebtetrata› «vertebrates» could exist without plants & insects had been created; but on other hand creation of small animals must have gone on since from parasitical nature of insects & worms.— In abstract we may say that vegetables & mass of insects could live without animals

109 but not vice versâ. (could plants live without carbonic acid gaz.)— Yet unquestionably animals most dependent on vegetables. of the two great Kingdoms.—

110 Principes de Zool: Philosop:— I deduce from extreme difficulty of hypothesis of connecting Mollusca & vertebrata, that there must be very great gaps.— yet some analogy.[1]

The existence of plants, & their passage to animals appears greatest argument against theory of analogies.

---

106 **large**] *underlined.*
109 Yet] *altered from* 'Yes' *by* 't' *over* 's'.
110 de] *over* 'of'.
theory] 'the' *over 2–3 illeg letters.*

107–1 Anonymous 1827.
110–1 E. Geoffroy Saint-Hilaire 1830 records the debates in the Académie Royale des Sciences between Georges Cuvier and Etienne Geoffroy Saint-Hilaire. Darwin's comment reflects an overall assessment of the debate, which was sparked by a report of evidence that the cephalopods formed a linking group between the molluscs and the vertebrates. Cuvier insisted on great gaps, and Geoffroy's doctrine of unity is epitomised by the principle of analogy.

111 States **there** is but one animal: one set of organ:— the others «animals» *created* with endless differences:—[1] does not say propagated, but must have concluded so= Evidently «or hints» considers generation as a short process, by which man «one animal» passes from worm to man; «highest» as typical of ‹sa›. changes, which can be traced in *same* organ in *different* animals in scale.— In monsters «also» *organs*

112 of lower animals appear.[1]— yet nothing about propagation= I see nothing like grandfather of Mammalia & birds &c ╫
p. 32. reference to M Edwards. law of crustacea[2]— with respect to mouth those beautiful passages from one to other organ.— Cuvier on opposite side; 1s Vol of Fish[3]
p. 59.]CD Cuvier has said each animal made for itself does not agree with old & modern types being constant.[4] Cuvier's theory of *Conditions* of existence is thought to account resemblances[5] & ∴ quinary system, or three elements
p 66]CD

112 66] *second* '6' *over* '4'.

111–1 E. Geoffroy Saint-Hilaire 1830: e.g. 218–19, 'Ainsi le *principe de connexions* et celui du *balancement des organes*, expliqués l'un par l'autre, conduisent M. Geoffroy à cette conclusion: que les animaux sont tous créés *sur le même plan*; qu'il y a, pour le règne animal, *unité de composition organique*, et cette conclusion est le corollaire le plus général de la *théorie des analogues*.

'Telle est la doctrine philosophique de M. Geoffroy Saint-Hilaire. Elle semble, comme il le dit lui-même, être la confirmation du principe de Leibnitz qui définissait l'univers: *l'unité dans la variété*.' Exclamation mark in margin, 'créés' underlined.

112–1 E. Geoffroy Saint-Hilaire 1830:216–17, 'Les *monstres*, qu'on a si long-temps regardés comme d'étranges caprices de la nature, ne sont que des êtres dont le développement régulier a été arrêté dans certaines parties; et, chose admirable, il n'arrive jamais à un organe de perdre, dans un individu, les caractères normaux de l'espèce à laquelle il appartient, sans que cette déformation n'imprime à cet organe les caractères normaux d'une espèce inférieure. Il en est de même pour le développement naturel des corps animés. Ainsi, l'homme, considéré dans son état d'embryon, dans le sein de sa mère, passe successivement par tous les degrés d'évolution des espèces animales inférieures: son organisation, dans ses phases successives, se rapproche de l'organisation du ver, du poisson, de l'oiseau.' Darwin underlined 'développement naturel', scored the passage, and marked it with a large Q (for quote).

112–2 E. Geoffroy Saint-Hilaire 1830:32–33 quotes Milne-Edwards 1833, 'On connaît . . . deux groupes principaux de crustacés, les crustacés à vie errante qui ont la bouche armée d'organes masticateurs forts et tranchans, et les crustacés qui vivent en parasites, dont la bouche est destinée à livrer passage aux liquides . . . La conclusion du mémoire est que *la composition organique décrite est toujours restée analogique. Les mêmes élémens constituans sont retrouvées dans l'un et l'autre cas; c'est une tendance remarquable à l'uniformité de composition*.'

112–3 G. Cuvier in E. Geoffroy Saint-Hilaire 1830:56–57 cites G. Cuvier and Valenciennes 1828–49. Scored.

112–4 E. Geoffroy Saint-Hilaire 1830:59 paraphrases G. Cuvier 1825, 'Ces vues d'*unité* sont renouvelées d'une vieille erreur née au sein du panthéisme, étant principalement enfantée par une idée de causalité, par la supposition inadmissible que *tous les êtres sont créés en vue les uns des autres*; cependant chaque être est fait pour soi, a en soi ce qui le concerne.' Darwin scored the passage and noted 'ancient & modern types'.

112–5 E. Geoffroy Saint-Hilaire 1830: 65–66, quotes Cuvier 'En un mot, si par unité de composition . . . on entend *ressemblance analogie*, . . . ce n'est qu'un principe subordonné à un autre bien plus élevé et bien plus fécond, à celui des conditions d'existence, de la convenance des parties, de leur coordination pour le rôle que l'animal doit jouer dans la nature; voilà le vrai principe philosophique d'où découlent les possibilités de certaines ressemblances et l'impossibilité de certaines autres; . . .' Darwin marked this passage extensively: he put a 'Q' (for quote) in the margin, scored the passage, underlined 'bien plus élevé et bien plus fécond'. He also wrote 'I demur to this alone' (p. 65), 'All this will follow from relation', interlined before 'voilà', and 'The unity of course due to inheritance.' (p. 66).

113 With unknown limits, every tribe appears fitted for as many situations as possible. for instance take birds animals, reptiles fish— **Conditions will not explain status** (Perhaps consideration of range of capabilities past & present might tell something)

---

p.111 G. St Hilaire
Insects & Molluscs allowed to be wide hiatus: states in one the sanguineous system, in other nervous developed.[1] (Owen's idea[2]) states these class approach on the confines? Balanidæ?— ——
I cannot understand whether. G. H. thinks developent in quite straight line, or branching[3]

114 S. H
What does the expression mean used by Cuvier, that all animals (though some may be) have not been created on the same plan.[1] [Second resumé well worth studying[2]] CD says grand idea god giving laws & & then leaving all to follow consequences.[3]— I cannot make out his ideas about propagation His work. Philosophie Anatomique. 2d Vol about monsters worth reading[4]

---

115 NB well to insist upon ‹different animals› large Mammalia not being found on all isld, (if act of fresh creation why not produced on New Zealand; if generated ‹No› «an» answer ‹could› «can» be given.—

It is a point of great interest to prove animals not adopted to each country.— Provision for transportal **otherwise** not so numerous: quote from Lyell:[1]

113 With . . . tell something)] *brace in each margin, original ink.*
Conditions . . . **status**] *added pencil.*
G. St Hilaire] *in box.*
114 2d Vol . . . reading] *circled, original ink.*
115 ‹No›) 'N' *over* 'p'.

113–1 E. Geoffroy Saint-Hilaire 1830:111, 'Le système sanguin est en excès et au contraire le système nerveux est frappé d'atrophie chez les mollusques; c'est l'inverse chez les insectes. Cela explique le large hiatus que l'on a remarqué entre ces familles'. Passage scored.
113–2 Richard Owen.
113–3 Continuation of quote in note B113–1 'spécialement à l'égard des êtres du milieu de chaque série, et aussi les rapports si nombreux qu'elles montrent à leur confins'.
114–1 Cuvier in E. Geoffroy Saint-Hilaire 1830, e.g., p. 56 'je me suis bien gardé de dire que cette organisation, . . . de celle des animaux vertébrés, fût composée de même, ni arrangée sur le même plan; au contraire, j'ai toujours soutenu que le plan, qui jusqu'à un certain point est commun aux vertébrés, ne se continue pas chez les mollusques.'
114–2 E. Geoffroy Saint-Hilaire 1830:204–22.
114–3 E. Geoffroy Saint-Hilaire 1830:219, 'La puissance créatrice, par des combinaisons aussi simples a produit l'ordre actuel de l'univers, quand elle eut attribué à chaque chose sa qualité propre et son degré d'action, et qu'elle eut réglé que tant d'élémens, ainsi sortis de ses mains, seraient éternellement abandonnés au jeu, ou mieux, à toutes les conséquences de leurs attractions réciproques.' Passage scored.
114–4 E. Geoffroy Saint-Hilaire 1818–22. On monsters see B112–1.
115–1 Lyell *Principles* 1837, 3:27–76, chaps 6 and 7 on geographical distribution and dispersal of animals are replete both with examples of transportal and with Darwin marginalia.

116 assuming truth of quadrupeds being created on small spots of land, of the same type with the great continents we get a means of knowing movements.—

How can we understand excepting by propagation that out of the thousand of new insects all belong to same

117 types already established. why out of the thousands of forms, should they all be classified.— Propagation explains this.——

Ancient Flora thought to more uniform than existing. Ed. N. Philos J. p. ‹191› «p 191» No. 5. Ap. 1827[1]

118 F. Cuvier says. "But we could only produce domestic individuals & not races, without the occurence of one of the most general laws of life, the transmission of a fortuitous modification, into a durable form, of a fugitive want into a fundamental propenity, of an accidental habit into an instinct." Ed. N. Phi. J. p 297, No 8 Jan–Ap. 1828.[1]—

I take higher grounds & say life is short for this object & others, viz not too much change

119 In Number 6′.? of E.d. N. Philos. Journ.— Paper by Crawford on Mission to Ava.[1] account of HAIRY man. **because ancestors hairy** with one hairy child, and of *albino* **DISEASE** being banished, & given to Portuguese. priest.—In first settling a country.— people very apt to be split up into many isolated races. ⸮are there any instance of peculiar people banished by rest?—

∴ most monstrous form has tendency to propagate, as well as diseases.

120 In intermarriages; smallest differences blended, rather stronger tendency to imitate one of the parents; repugnance «generally» to marriage ‹if offspring not fertile› ‹,but producing› «before domestication, afterwards none or little with ‹fertile offspring›» fertile offspring; marriage never probably excepting from «strict» domestication offspring not fertile. or at least most rarely & perhaps never female.—no offspring: physical impossibility to marriage.—

116 on] *over* 'f'.
119 account of . . . by rest?—] *brace left margin, original ink.*

117–1 Sternberg 1827:191, 'Ferns, scaly trees, and their attendant *calamites*, are to be found wherever black bituminous coal of the older formation is discovered. The species, however, are frequently different, and they therefore follow, in their climatic and geographical distribution, the same laws observed by the plants of the present world, and according to the relations of a higher and more uniform temperature, which must be supposed to have existed at some former period.'
118–1 F. Cuvier 1827–28, 4:297, In copying this quotation, after 'laws of life' Darwin dropped 'transmission of the organic or intellectual modifications by generation.'
119–1 Crawford 1827:368.

121 ⸮whether those genera which unite very different structure as. *petrel* & *alk*. do not show the possibility of common branching off.?

Accra, Coast of Africa. Clay slate. strike. SSW. & NNE. dip 30°–80° (?).— Ed. Phi.l.. N. J. p. 410 ‹Nov› 1828[1]

---

122 It is daily happening; that naturalist describe animals as species, for instance Australian dog: ‹yet when that› or Falkland rabbit.— There is only two ways of proving to them it is not; one when they can proved descendant, which of course most rare, or when placed together they will breed.— But what a character is this?

123e Race permanent, because every trifle heredetary, without some cause of change,, yet such causes are most obscure. without doubt:— Vide cattle:

The grand fact is to establish whether in crossing very opposite races, whether you would expect equal fertility— ditto in Plants.‹==›

124e It will be well to refer to Chamisso Vol III p. 155. about quantities of seeds in sea;[1] also Holman: ‹at› Keeling these are most important facts.—[2] As soon as island large enough for land birds, seeds picked from the beach by the birds; most seeds germinating.——

125e It would be curious experiment to know whether soaking seeds in salt water &c has any tendency to form varieties?

Ed. N. Phil. J. Morse found in Virginia p. 325. July 1828. Animal now confined to extreme North.—[1] ‡.do p. 326. 2 Fossil species of ox in N. America; as well as 2 recent[2]

121 Accra . . .1828] *crossed.*
123 *page crossed pencil.*
124 It will . . . important facts.—] *double scored pencil.*
'18' *added to page brown crayon.*
125 *page crossed brown crayon.*

121–1 T. Park 1828:410.
124–1 Chamisso 1821, 3:155–56, 'The sea brings to these islands the seeds and fruits of many trees, most of which have not yet grown there. The greater part of these seeds appear to have not yet lost the capability of growing, and we have frequently [planted them]. We have collected them . . . The greatest part . . . belong to the arborescent or creeping siliquose plants . . . We observed that the seeds which, being cast by the tide over the reef, reach the inner or lee side of an island, find more protection, a better soil, and circumstances more favourable for their growth than those which the surf throws upon the outside.'
124–2 Holman 1834 4:378–79 '. . . seeds and plants from Sumatra and Java have been driven up by the surf on the windward side of the islands . . . [lists several plants]. Large masses of . . . wood, have also been found, besides immense trees. . . All the hardy seeds, such as creepers, retain their germinating power, but the softer kinds . . . are destroyed in the passage.'
125–1 Morse *sic* Cropper. Mitchill, Smith, and Cooper 1828:325 reports discovery by Mr Cropper of a fossil 'Walrus or Sea-Horse' in Virginia.
125–2 Dekay 1828:326.

126e See Geolog. Proc. p. 569. 1837. Account of wonderful fossils of India.[1]— & p. 545 «*great monkey*»[2]

Mr Johnston says Mag of Zooly & Bot. p 65 Vol II talking of annelidæ.— ‹“›The fact is an additional illustration of that axiom in Natural History that all aberrant & osculant groups are not only few in species, but every two or three these form genera[3]— **this is «from unfavourable conditions» there are many gaps. & those forms which «nevertheless» have produced species, have produced fe[w]**

127e–128e [not located]

129 The relation of Analogy of Maclay[1] &c. appears to me the same, as the irregularities in the degradation of structure of Lamarck, which he says depends on external influences.—[2] For instance he says wings of bat, are from external influence.—[3] . . . . . .

126 See Geolog. Proc . . . p. 545] *crossed brown crayon.*
Mr Johnston . . . form genera] ‘11’ *added brown crayon.*
Mr Johnston . . . form genera] *marginal X added pencil.*
produced fe[w]] *trails onto excision stubs of 126 and 127.*
129 &c.] ‘.’ *on 131.*

126–1 Cautley and H. Falconer 1833–38:568–69 reports discovery of a quadrumanous fossil compared to *Semnopithecus entellus* among the Sewalik Hills fossils; also found were a variety of other vertebrate fossils, including *Anoplotherium Sivalense* and *Sivatherium giganteum.*
126–2 Cautley 1833–38:544.
126–3 G. Johnston 1838a:65. Passage scored in ink and pencil note in margin: ‘Hereditary | descendan[t] many kil[led] Analogic[al] X’; crossed in pencil. At bottom of the page: ‘much extinction’. See C185–86, where G. Johnston 1838a is also reviewed.
129–1 See Darwin’s abstract of MacLeay 1819–21 (DAR 71:128–38). Vol. 1: ‘p. 363 “Relations in analogy consist in a correspondence between certain insulated parts of the organization of two animals which differ in their general structure. These relations, however, seem to have been confounded by Larmarck, and indeed all Zoologists, with those upon which orders, sections, families . . . immediately depend. . . . the test of a relation of affinity is its forming part of a transition continued from one structure to another by nearly equal intervals, and that the test of a relation of analogy is barely an evident similarity in some one or two remarkable points of formation, which at first sight give a character to the animal and distinguish it from its affinities.” ’ (DAR 71:133) ‘p. 364 & 365 Cimicidae & «some of the» Orthoptera so many points of similarity that Linnaeus united them into one order “. . . more careful examination assures us there is little similarity either in their organs of manducation or in their internal structure”.— “This relation, therefore, which exists manifestly (& to me C.D manifestly more than ordinary case of analogy) between a cimex & a Gryllus is one of analogy & not of affinity”.— Now all this is grounded on the assumption that one animal «or group» can only have relation to two others, & not with more than two.— it is quite gratuitous.—’ (DAR 71:134)
129–2 Lamarck 1830, 1:134–36, ‘La progression dans la composition de l’organisation subit, ça et là dans la série générale des animaux, des anomalies opérées par l’influence des circonstances d’habitation, et par celle des habitudes contractées.’ (1:134–35). Passage scored. ‘Or, comme nous prenons la série générale des animaux en sens inverse de l’ordre même qu’a suivi la nature, en les faisant successivement exister, cette gradation se change alors, pour nous, en une *dégradation* frappante qui règne d’une extrémité à l’autre de la chaîne animal, sauf les interruptions qui résultent des objets qui restent à découvrir, et celles qui proviennent des anomalies produites par les circonstances extrêmes d’habitation.’ (1:136). Darwin’s comment relevant to ‘degradation’ on 1:140, ‘There appears to me to be some confusion in these ideas of degradation. What makes perfection, except «that» towards the end wanted. Look at [three words illegible] | Scale (of many kinds) of *complication* exists.’
129–3 Lamarck 1830, 147–48, ‘En effet, le sternum des *oiseaux* donnant attache à des muscles pectoraux que des mouvemens énergiques, presque continuellement exercés, ont rendu très-épais et très-forts, est devenu extrêmement large, et cariné dans le milieu. Mais ceci tient aux habitudes de ces animaux, et non à la dégradation générale

130 Hence name of analogy, the structures in the two animals bearing relations to a third body., or common end of structure

---

A Race of domestic animals made from influences in one country is permanent in another.—[1] **Good argument for species not being so closely adapted.**

131 Near the Caspian «Province of Ghilan» wooded district cattle with humps as in India. Geograph J. Vol III. P. I. p. 17[1] (Lat 37° about)

Vol IV P. I. Geograp. Journal. Voyage up the Massaroony by W. Hillhouse.— Demerara. In note. Demerara. 10 «12» feet beneath surface forest trees fallen «kind well known, carbonized»—; clay fifty feet, then forest 120 ft Micaceous rocks. subsidence appears indicated.—— p. 36.—[2]

132 Geograp. Journ. Vol IV. P II. p. 160. Melville Is^{d}.— "The buffaloes, introduced from Timor, herded separate from the English cattle, nor could we get them to associate together"[1]

133 There is long rigmarole article by S Hilaire on wonder of finding Monkey in France.— of genus peculiar to East Indian isles.—[1] Compares it to fossil Didelphis (S. American genus) in plaster of Paris.[2]— Now this is exception to *law of type*. like horse in S. America. or like living Edentata in Africa &c &c.— Now if suppose world more perfectly continental. we might have

134 wanderers. (as Peccari in N. America) then if it is doomed that only one species of family has offspring the *chance* is that these wanderers would not, but where original forms most numerous there would be wanderers.— Some however might have offspring, & then

130 country is] 'is' *emmended pencil.*
Good argument . . . closely adapted.] *added pencil.*
131 Lat . . . about] *in rectangle.*
Vol IV . . .—p. 36.—] *crossed.*
10] '0' *over* '5'.
133 S Hilaire] 'S' *over* 'G'.

que nous examinons. Cela est si vrai, que le mammifère qu'on nomme *chauve-souris*, a aussi le sternum cariné.' Passage scored from 'Mais ceci . . . cariné.' Darwin noted at bottom of 1:148–51 'The «economy» of world could have gone on without Bats or ostriches.— It can only be following out some great principle[.] It is clear birds made preeminently for air, yet if no birds, Mammalia would have taken place. There limit to this Adaptation. Fish could hardly have lived out of water. Though Crabs.— spider under water.—' This note shows that B129 has to be considered in the same frame of reference as B45–46 'my idea of origin of Quinarian system.'

130–1 See note B131–1.

131–1 Montieth 1833 [1834]:17, 'The people [of Ghilan] . . . more resemble Indians, and the cattle are small, have also the hump peculiar to that country.'

131–2 Hilhouse 1834a:36n, 'It is evident . . . ages ago this continent was habitable fifty feet below the present surface.'

132–1 Campbell 1834:160.

133–1 See B94 on the fossil 'singe' of Sansan. E. Geoffroy Saint-Hilaire 1837b:243; extracted from 1837c '. . . ce n'est point un Singe généralement parlant que M. Lartet a découvert dans notre Europe, mais précisément l'analogue de l'une de ces formes qu'on ne rencontre que dans cette région décidément à part, ou la spécialité d'essences animales propres aux Indes Orientales est soumises à l'influence d'un milieu ambiant d'une sorte determinée.'

133–2 E. Geoffroy Saint-Hilaire 1837b:242–43.

135 «V. L. Institut p 245. 1837» we should have anomalies. as Cape Anteater,.[1]— This supposes world divided into Zoological provinces— united— & now divided again— Weakest part of theory death of species without apparent physical cause:— Mem: Mastodon all over S. America. Hilaire does not seem(?) to consider the monkey as a wanderer, but as produced by climate?—[2]

136 M. Baer (thinks) the Aurock, was found in Germany & thinks even now in central & Eastern Asia beyond the Ganges & perhaps even in India p. 261. L. Institut. 1837[1]

Mem. Sir F. Darwin cross breed boars were wilder than parents. which is same as Indian Cattle. ∴ tameness not heredetary?, having been gained in short time.[2]

137 Milvulus forficetus[1] ‹has a wide range› is a tyrant flycatcher doing the service of a swallow

---

I think we may conclude from Australia & S. America. that only some mundine cause has destroyed animals over the whole world— For instance gradual reduction of tempereture from geographical or central heat.— But then shells—

138 Mr Yarrell[1] says that old *races* when mingled with newer, hybrid variety partakes chiefly of the former
Eyton's paper on Hybrids Loudon's Magazine.[2]
Gould on Motacilla,. «species peculiar to Continent & England» Loudon Mag: Septemb or Octob 1837[3]

136 Mem. Sir F. Darwin . . . time] *marginal* 'X' *added in pencil.*

135–1 E. Geoffroy Saint-Hilaire 1837b, *sic* p. 243.
135–2 E. Geoffroy Saint-Hilaire 1837b:243, 'Or . . . n'allez pas . . . conclure . . . par construire une route géographique . . . afin que les espèces, Sarigue et Gibbon, aujourd'hui vivantes en leur contrée respective, aient fourni des voyageurs vers un point voisin de leur antipode et soient venus ainsi déposer en France les débris . . .: non, il n'en est point ainsi. . . . Je m'en tiens pour dénégation, . . . aux principes et aux données philosphiques de mes mémoires sur les milieux ambians . . .' See also, E. Geoffroy Saint-Hilaire 1837a, *inter alia.*
136–1 Baer 1837a:261. Passage scored.
136–2 Sir Francis Sacheveral Darwin, Darwin's half uncle, bred wild boar at Syndcope, Derbyshire. This we know from Darwin's abstract of Prichard 1836–47 DAR 71:139–42, which comments on Prichard 1836–47, 1:112 'Particular species have, in general, one limit in regard to . . . the circumstances connected with the reproduction of their kind, such as the number of their progeny'; Darwin commented as follows 'p. 112 Reginald Darwin says the Wild Boar in his fathers Park produces only two offspring.' See *Correspondence* 1:207, letter from Susan Darwin, 12 February [–3 March] 1832. 'Sir Francis & Lady Darwin will I conclude leave their mountainous abode [Syndcope] & come to the [Breadsall] Priory now, which the latter must prefer to the society of Eagles and Wild Boars.—' Reginald Darwin (son of Sir Francis, ergo Charles' half first cousin) may be the source of the information in B136.
137–1 *Milvulus forficatus.*
138–1 William Yarrell.
138–2 Eyton 1837a.
138–3 Gould 1837d:459–61, p. 460 scored.

139 Westwood has written paper on affinity and analogy in Linnæan Transactions[1]

Mr Wynne[2] distinctly says that the mixture between Chinese & English Breed. decidedly exceedingly prolific & hybrid about half way.
Eyton says Hybrid about half aways & result the same[3]

Indian cattle & common produced very fine

140 Hybrid offspring, much larger than the dam, from those imported by Ld. Powis

---

Hybrid dogs offspring seldom intermediate between parents.— How easily does Wolf & Dog cross?
Mr Yarrel thinks oldest variety impresses the offspring most forcibly— Esquimaux dog & Pointer «Game-fowls have courage independently of individual force»

141 Mr Wynne has crossed Ducks & Widgeon & offspring either amongst themselves or with parent birds.—[1] «W. Fox. knew of case of male widgeon, winged & turned on pool, first season bred readily with common ducks.—»

Kirby all through Bridgewater errs greatly in thinking every animal born to consume this or that thing.[2]— There is some much higher generalization in view.

In Marsupial division «do» we not see a splitting in orders, Carnivora, rodents &c, JUST COMMENCING.

139 Indian . . . very fine] *marginal 'X' added in pencil.*
140 «Game-fowls . . . force»] *smaller hand, in rectangle at bottom of page; presume passage added.*
141 «W. Fox . . . ducks.—»] *smaller hand, presume passage added.*
In Marsupial . . . COMMENCING.] *triple scored brown ink; marginal 'X' added in pencil.*

139–1 No paper quite fits this description; however, a possibility is Westwood 1837:285. 'But this lateral prolongation of the head into ocular peduncles is not confined to insects, strictly so called, but is found in a few instances in other classes and orders; and as these instances involve in some degree the doctrine that every affinity is connected with, and must be tested by, a corresponding analogy, I shall detail them, without, however, offering any opinion upon the doctrine itself.'
139–2 Mr Wynne remains unidentified notwithstanding several archival traces of his connection with Darwin. Darwin's comment here could reflect an answer to one of the 'Questions for Mr Wynne' (Gruber and Barrett 1974:423, *Correspondence* 2:71): 'Cross of Chinese pigs, are they intermediate in form, as in dogs.— does Mr Wynne believe in dogs'. See also B141.
139–3 Eyton 1837a:358. Reports crossing Chinese and English geese; the goslings were '. . . precisely (as far as being intermediate in external character . . .) similar to their parents . . .'
141–1 Possible answer to 'Case of heterogeneous offspring, in fowls, pidgeons, rabbits.—' ('Questions for Mr Wynne', Gruber and Barrett 1974:423, *Correspondence* 2:71).
141–2 Kirby 1835, e.g. 1:141, '. . . another function . . . devolved upon animals with respect to the vegetable kingdom . . . [is] to keep the members of it within due limits, and to hinder them from encroaching too much upon each other. All organised beings have a natural tendency to increase and multiply; and while there is space this tendency is beneficial; but when plants or animals exceed certain limits, they stand in each other's way, and prevent all further growth or healthy progress. The herbivorous animals . . . countercheck . . . this tendency, and keep the vegetable tribes from encroaching too much upon each other.'

142 Kirby says (not definite information) West of Rocky Mountains Asiatic types discoverable.—[1]

Bridgewater Treatise p 85.[2] Parasite of Negroes different from European.— Horse & Ox have different parasite in different climates.—

Hun^bt^. **Humboldt?** Vol V. P II. p 565. Consult— Says types most subject to vary where intermixture precluded.—[3]

143 **Kirby Bridgewater Treatise**

There are some good accounts of passages of legs into mouth-pieces of Crustacea. Vol II. p 75[1]

a Fish which emigrates over lands is a siluris, p. 123[2]

A climbing fish. p. 122

A Terrestrial annelidous animal p. 347. Vol I.—[3] compare with my planariæ Leaches out of water

144 Does the odd Petrel of F. del F. take form of awk, because there is no awk in Southern hemisphere. does this rule apply?

---

145 A Treatise on Form of Animals by Mr Cline

"The character of both parents are observed in their offspring, but that of the male more frequently predominates," p. 20. do "If hornless ram be put to horned ewe almost all the lambs will be hornless.—" **does this apply to where same animal breeds often with same female**

p. 28 "It ‹is› wrong to enlarge a Native breed of animals. for in ‹the› proportion to their increase of size they become worse in form, less hardy, & more liable to disease"[1]

142 **Humboldt?**] *added pencil in rectangle.*
145 A Treatise . . . Cline] *marginal* 'X' *added in pencil.*

142–1 Kirby 1835 1:52.
142–2 Kirby 1835 1:85–86, 'Little stress will be laid on the parasite of the negroes, being specifically distinct from that which infests the whites, when we reflect that the horse and the ox have different insect parasites and assailants in different climates.'
142–3 Humboldt and Bonpland 1819–29, 5:565, 'We were assured, that the whole [Guaica Indian] tribe were of this extreme littleness; but we must not forget, that what is called a tribe constitutes, properly speaking, but one family. The exclusion of all foreign mixture contributes to perpetuate varieties, or the aberrations from a common standard. The Indians of the lowest stature next to the Guaicas are the Guainares and the Poignares. It is singular, that all these nations are found close to the Caribbees, who are remarkably tall. They all inhabit the same climate, and subsist on the same aliment. They are varieties in the race, which no doubt existed previously to the settlement of these tribes, (tall and short, fair and dark brown) in the same country.' Darwin scored 'We were . . . remarkably tall.'
143–1 Kirby 1835 2:75.
143–2 Kirby 1835 2:123.
143–3 Kirby 1835 1:347. See also Darwin 1844b.
145–1 Cline 1805. Darwin owned Cline 1829, a later edition of this often reproduced pamphlet.

146 If population of place be constant «say 2000» and at present day, every ten living souls on average are related to the (200^{dth} year) degree. Then 200 years ago, there were 200 people living who now have successors.— Then the chance of 200 people ‹might be› being related within 200 years backward might be calculated & this number elimanated say 150 people four hundred years since were progenitors of present people, and so on backwards to one progenitor, who might have continued breeding from eternity «backwards.—»

147 If population was increasing between each lustrum,[1] the number related at the first start must be greater, & this number would vary at each lustrum, & the calculation of chance of the relationship of the progenitors would have different formula for each lustrum.—

We may conclude that there will be a period though long distant, when of the present men (of all races) not more than a few will have successors. at present day. in looking at two fine families one with

148 successors «for» centuries, the other will become extinct.— Who can analyze causes, dislike to marriage, heredetary disease, effects of contagions & accidents: yet some causes are evident, as for instance one man killing another.— So is it with *varying* races of man: these races may be overlooked mere variations consequent on climate &c— the whole races act towards each other, and are acted on, just like the two fine families «no doubt a different set of causes must act in the two case,» May this not be extended to all animals first consider species of cats.— ‹& other tribes›.— &c &c **Exclude mothers & then try this as simile**

149 In a decreasing population at any one moment fewer closely related; ∴ (few species of genera) ultimately few genera (for otherwise the relationship would converge sooner) & lastly perhaps some one single one.—
Will not this account for the odd genera with few species which stand between great groups, which we are bound to consider the increasing ones.—
NB As Illustrations are there many anomalous lizards living; or of the tribe fish extinct. or of Pachydermata, or of coniferous trees; or in certain shell cephalopoda.— «Read Buckland[1]»

150 L'..Institut «1837.» p 319 — Brongniart.— no dicotyledenous plants & few Monocot in Coal formation?[1]
p. 320 ‹Think› States Cryptogam. Flora formerly common to New Holland?![2]

146 *page crossed.*
148 «no doubt . . . case,»] *circled.*

147–1 Census.
149–1 Buckland 1836. Darwin's abstract deals almost exclusively with fossils.
150–1 Brongniart 1837:319, 'Or, la première de ces classes [Dicotylédones] . . . manque complétement dans notre flore primitive, et à peine si l'on y trouve quelques indices des Monocotylédones.' This and associated passage triple scored.
150–2 Brongniart 1837:320 refers to the vascular cryptogam flora 'Lycopodes, Fougères et Prêles' in the 'couches de houille.': '. . . cette grande végétation primitive a dû couvrir pendant longtemps de ses épaisses forêts toutes les

p. 320. Says Coniferous structure intermediate between vascular or cryptogam. (original Flora) and Dicotyledenous,[3] which «nearly» first appear «(p 321)» at Tertiary epock[4]
p. 330. Fossil Infusoria found of unknown forms, a circumstance undiscovered by Ehrenbergh.—[5]

151e ‹Marcel Serres p. 331. L'Institut— considers that mercu[1]› Geo. Joun. p. 325. Vol. IV. Ducks on rivers in Guiana. build top of trees carry duckling to the water in their beaks, & the young one ‹inland› directly *by instinct.* can dive & conceal themselves in the grass.—[2]

Beatson St. Helena says no trees succeed so well at St. Helena. as. Pineaster & Mimosa called Botany Bay Willow[3]

V. Dr Royle introductory remarks to Himalaya Mountains—[4]

152e Bory St. Vincent Vol. III. p. 164. Lîle de la Reunion presente elle seule plus d'especes polymorphes que toute la terre ferme de l ancien monde".—[1]

Considers forms in recent volcanic islets not well fixed.—[2]

Peron thinks Van Diemen's land long separated from Hobart Town— {from difference of races of men and animals}[3]

151 'I' *pencil top right (not added by Darwin).*
152 toute la] 'a' *over* 'e'.
page crossed blue crayon.

parties du globe qui s'élevaient au-dessus du niveau des mers; car elle se présente avec les mêmes caractères en Europe et en Amériques, et l'Asie équatoriale, ainsi que la Nouvelle-Hollande, sembleraient même avoir participé alors à cette uniformité générale de structure des végétaux.' Darwin scored and put a question mark on this passage.
150–3 Brongniart 1837:320, 'L'existence de ces deux familles [conifers and cycads], pendant cette période, est d'autant plus importante à signaler qu'intimement liées entre elles par leur organisation, elles forment le chaînon intermédiaire entre les Cryptogames vasculaires qui composaient presque seules la végétation primitive de la période houillère, et les Phanérogames dicotylédones proprement dites, qui forment la majorité du règne végétal pendant la période tertiaire.' Passage scored.
150–4 Brongniart 1837:321. 'Cette classe de Dicotylédones, dont on pouvait à peine citer quelques indices dans les derniers temps de la période secondaire, se présente tout-à-coup durant la période tertiaire, d'une manière prépondérante.' Passage scored.
150–5 Ehrenberg 1837d:330, '. . . [Ehrenberg] a pu distinguer [in Agassiz's material from Oran] 9 autres formes organiques dont plusieurs constituent des genres nouveaux et sans correspondant parmi les types génériques actuels, phénomène qui ne s'etait pas encore présenté parmi les Infusoires fossiles.' Passage scored.
151–1 M. Serres 1837:331 begins 'Dans une précédente communication sur le mercure . . .'
151–2 Hilhouse 1834b:325.
151–3 Beatson 1816:32.
151–4 Royle 1835, Darwin's copy is unopened.
152–1 Bory de Saint-Vincent 1804, 3:64.
152–2 Bory de Saint-Vincent 1804 3:162–63, '. . . la forme des végétaux des îles volcaniques semblait à peine déterminée.' See C268.
152–3 Péron 1807–16, 2:163–65. Since Hobart is the principal town of Tasmania (Van Diemen's Land), presumably Darwin meant New Holland not Hobart Town.

153e See R. N. p. 130 Speculations on range of allied species. **p. 127. p. 132**[1]

There is no more wonder in extinction of individuals than of species

---

Paris Tertiary Shells in India!? A p. 28[2]

Dr Beck. & Lyell. most curious law of species few in Arctic.[3] in proportion to genera. agrees with late production of those regions, & consequently

154e not many yet multiplied: NB How does this bear with law referred to by Richardson in Report about each genus having its parent type in hotter parts of world[1]

Is monkey. peculiar to C. de Verd's.—? **NO Macleay Name given in Congo Expedition**[2]

We need not expect to find ‹species›, varieties, intermediate between every species.— Who can find trace or history of species between

153 *page crossed brown crayon.*
154 not many . . . world] *crossed brown crayon.*
Is monkey . . . Verd's] '19' *added brown crayon.*
We need . . . species between] *crossed brown crayon.*

153–1 Red Notebook. The pages cited form the core of Darwin's theoretical speculations on species in the Red Notebook. The addition of page numbers in grey ink indicates that Darwin's autumn 1838 review of his notebooks included the species passages of the Red Notebook.

153–2 A28. McCleland 1837:567 '. . . a small collection of about one hundred species from the Paris basin were at once recognised by the author as familiar objects, from his acquaintance with those familiar objects, with those from Kossia and Cherraponji.' As Darwin's '!?' indicate, he considered this a significant claim. See Lyell 1830–33, 3:58, 'Thus, for example, if strata should be discovered in India or South America, containing the same small proportion of recent shells as are found in the Paris basin, *they* also might be termed Eocene, . . .'

153–3 Charles Lyell and H. H. Beck: probable personal communication.

154–1 Richardson 1837:223–24, 'Zoology, as Cuvier has remarked, is now and must continue to be for many years, a science of observation only, and not of calculation; and no general principles hitherto established will enable us to say what are the aboriginal inhabitants of any quarter of the world. It seemed therefore hopeless to attempt to elicit the laws of the distribution of animal life from results yielded by a fauna so very imperfectly investigated as that of North America; consequently in the preceding report, the ranges of the species have been generally stated, as recorded by observers, and without reference to the opinions which have been heretofore advanced by theoretical writers. Buffon hazarded the remark that none of the animals of the Old World exist in the New, except the few which are capable of propagating in the high northern latitudes. Temminck adduces circumstances which favour a modern opinion almost directly opposed to Buffon's; namely, that all the genera which people the earth (a small number belonging to the polar regions only excepted) are to be found in the equatorial zone, or at least within the tropics; and that the genera are spread abroad by means of analogues or species possessing exactly similar generic characters, which range in the same parallels of latitude, through all the degrees of longitude, and that not withstanding the barrier which a wide ocean may be supposed to interpose. The comprehensiveness of this law will evidently be modified by the number of generic divisions admitted by naturalists, and it will be scarcely tenable if the geographical groups of species be raised to generic rank as has been of late frequently done.' Darwin scored 'Buffon hazarded the remark . . . possessing exactly similar generic characters'.

154–2 Tuckey 1818:36. 'The only species here [Porto Praya, Cape Verde Islands] is the green monkey (*Cercopithecus sabaeus*).' Presumably either the original or clarifying information was given to Darwin by W. S. MacLeay.

155 Indian cow with hump & Common;— between Esquimaux & European dog? Yet man has had no interest in perpetuating these particular varieties.

If species made by isolation; then their distribution (after physical changes) would be in rays— from certain sports.— **Agrees with old Linnæan doctrine & Lyells. to certain extent**[1]

---

156 Von Buch,.— Canary Isles: French Edit. Flora of Isl^{ds} very poor «(p. 145)» 25. plants.[1] 36 St Helena, without ferns.—[2] analogous to nearest continent: poorness in exact proportion to distance (?).[3] & similarity of type (?).— [«Mem:» Juan Fernandez]^{CD}. From study of Flora of islands; "ou bien encore on pourrait au plus en conclure quels sont les genres qui, sous ce climat, se divisent le plus aisément en espèces distinctes et permanentes." p. 145.[4] In Humboldt great work

157 de distribut. Plantarum. relation of genera to species in France is 1 : 5.7 in Laponia 1 : 2,3:[1] Mem. Lyell on shells.—[2]

	Genera[3]
In North Africa.	1 : 4,2
Iles Canaries	1 : 1,46
St. Helena	1 : 1,15

Calculate my Keeling Case: Juan Fernandez Galapagos =Radack Isl^{ds}.= ∴ Islands & Artic are in same relation. We find

155–1 Lyell 1837, 3, chap. 8, *Theories respecting the original introduction of species*. Lyell begins by discussing 'Theory of Linnaeus' (3:77–78); his reference is to De terra habitabili incremento. In Lyell's version, Linnaeus '. . . imagined the habitable world to have been for a certain time limited to one small tract, the only portion of the earth's surface that was as yet laid bare by the subsidence of the primaeval ocean. In this fertile spot he supposed the originals of all the species of plants which exist on this globe to have been congregated, together with the first ancestors of all animals and of the human race.' (3:77–78). In 3:81–82 Lyell discusses '*Supposed centres, or foci, of creation*': 'Now this congregating, in a small space, of many peculiar species, would give an appearance of *centres* or *foci* of creation, as they have been termed, as if there were favourite points where the creative energy has been in greater action than in others, and where the numbers of peculiar organic beings have consequently become more considerable.' Darwin scored p. 81 and at the bottom wrote 'All this agrees perfectly with my theory'.

156–1 Buch 1836:144, 'Les conditions extraordinaires dans lesquelles cette station se trouve placée [Canary Islands, 5,000–10,000 ft], sont cause que, des 23 espèces [total number of plants found], 19 sont entièrement propres à ces îles, et jusqu'à présent n'ont été trouvées nulle autre part.'

156–2 Buch 1836:144–45 '. . . leur nombre à Sainte-Hélène, d'après le catalogue de Roxburgh, ne monte pas à plus de 36 espèces.' In note: 'Beatson [1816] (Tracts on St.-Helena, p. 295 et suiv.), De Candolle [1820:395] et Schouw [1823] (Géogr. des Plantes, 494) disent 61: mais ils n'en ont pas déduit les fougères.'

156–3 Buch 1836:144 '. . . ce résultat caractérise la nature des îles, où le nombre des plantes diminue en proportion de l'éloignement des continents.'

156–4 Buch 1836:145.

157–1 Humboldt 1817:39–40, cited in Buch 1836:147n.

157–2 See Lyell 1837, 3:55–61, '*Geographical Distribution and Migrations of Testacea*'.

157–3 Buch 1836:147. The ratio is genera to species, hence Darwin's 'Genera' added carefully over the '1' column.

158 species few in proportion to difficulty of transport. For instance the temperate parts of Teneriffe, the proportion of genera 1 : 1. I can understand in «one» small island species would not be manufactured, ‹but why they should be manu› Does it not present analogy to what takes place from time? Von Buch distinctly states that permanent varieties. become species. p.147 «p. 150»— not being crossed with others.—[1] Compares it to languages[2] But how do plants cross?— = admirable discussion

159e Von Buch says from Humboldt, in Laponia. genera to species 1. 2,3—[1] From Mackenzie Iceland there 144 genera & 365 species of plants not cryptogamic but 1 . 2,53.—[2]

In know varieties. there is analogy to species & genera.— for instance three kinds of greyhound.— In plants. do the seeds of marked varieties produce no difference. if they do.—there probably will be this relation also. ||**Yes «Fox»**||[3]

160e The creative power seems to be checked when islands are near continent: compare Siicily & Galapagos**!!**—

Some of the animals peculiar. to Mauritius are not found at Bourbond Zoolog. Proceedings∤.‹p› 1832.p.111[1]

---

161 Mr Owen[1] suggested to me, that the ‹cas› production «of monsters» (which, Hunter says owe their origin to very early stage[2]) & which, follow certain laws according to species. present an analogy to production of species.—

---

159 *page folded.*
160 Some of . . . p. 111] *marginal* 'X' *added in pencil.*
Some of . . . p. 111] '19' *added brown crayon.*

158–1 Buch 1836:147–50, 'Sur les continents, les individus d'un genre se dispersent fort au loin, et, par la diversité des stations, de la nourriture, du sol, forment des variétés qui, à cette distance, n'étant pas croisées par d'autres variétés et ramenées par là au type primitif, deviennent à la fin des espèces constantes et particulières' (147–48). 'Presque chaque vallée peut en montrer une nouvelle [joubarbe] espèce . . .' (150).
158–2 Buch 1836:148.
159–1 Humboldt 1817:39–40, cited in Buch 1836:147. See note B156–3.
159–2 W. J. Hooker 1811 (in Mackenzie 1811) lists genera and species of Icelandic plants. Darwin may have made the tabulation himself.
159–3 W. D. Fox: personal communication.
160–1 Desjardins 1832:111, 'Among the novelties which have occupied that Society during the season of 1830–31 have been some observations by M. J. Desjardins on the Zoology of the Mauritius as compared with that of the Isle of Bourbon, from which has resulted the curious fact, that notwithstanding that these islands are situated in such close proximity to each other, are of the same formation, and present a most remarkable analogy in their soil, their animals are not universally the same, some species being met with in the one which never occur in the other.' Passage scored.
161–1 Richard Owen: personal communication based on Owen's edition of Hunter 1837, which appeared 1–15 December 1837 (*Publishers Circular* 1, 1838). This passage must have been written either slightly before or after that publication.
161–2 Richard Owen 1837b:xxv–xxvi, 'With respect to the cause or origin of monsters, Hunter referred it to a condition of the original germ, or, as he expresses it, "each part of each species seems to have its monstrous form originally impressed upon it . . . I should imagine,"

Animals have no notions of beauty, therefore instinctive feelings against other «for sexual ends» species, whereas Man has such instincts very little.

---

162 in Zoolog. Proceedings. Jan 1837, «by Eyton» Account of three, kinds of pigs. difference in skeletons:[1] VERY GOOD.

Apteryx. a good instance probably of rudimentary bones.[2]— As Waterhouse remarked Mere length of bill does not ‹indicate affinity with snipes›

‹————————›

indicate affinity, because similar habits produce similar structure.[3]— Mem. Ornitho Rhyncus

---

Would not *relationship* express, a real affinity & affinity— whales & fish.—

163 Progeny of Manks cats without tails: some long & some short: therefore like dogs.—

Ogleby says,[1] Wolves at Hudson bay breed with dogs.— the bitches never being killed by them, whilst they eat up the dogs.—

L.' Institut. Curious paper by M. Serres on Molluscous animals representing fœtuses of Vertebrata, &c 1837 p. 370[2] **Owen says Nonsense**

164 The distribut of big Animals in East Indian Archipelago—,[1] very good in connection with Von Buch Volcanic chart[2] & my idea of double line of intersection.—

Dr. Horsfield[3]

162 in Zoolog . . . VERY GOOD] *marginal* 'X' *added in pencil.*

he writes, "that monsters were formed monsters from their very first formation, for this reason, that all supernumerary parts are joined to their similar parts, as a head to a head, &c. &c." '

162–1 Eyton 1837b:23, the three kinds of pig are English, African, and Chinese; 'I was surprised to find that a very great difference existed in the number of vertebrae from that given in the *Leçons d'Anatomie Comparée* . . . From what has been stated the result appears to me to be that either the above three *Pigs* must be considered as distinct species, and which . . . would do away with the theory of Hunter, that the young of two distinct species are not fruitful, or we cannot consider osteological character a criterion of species.' Darwin wrote 'Quoted' at the beginning of this article; he also recalculated Eyton's table of the number of vertebrae to show more clearly the extent of variation in the dorsal, lumbar, and sacral vertebrae of the pig varieties.

162–2 See Keir 1837:24, which appears on the page after Eyton's article (note B162–1) in *Proc. Zool. Soc. Lond.*

162–3 G. R. Waterhouse: probably personal communication.

163–1 William Ogilby: probably personal communication at the Zoological Society.

163–2 A. Serres 1837:371, 'Ces animaux sont donc des embryons permanens des animaux Vertébrés, et leur composition, de même que leur nature, de même que leur développement, sont des déductions rigoureuses, ou des corollaires de la loi *centripète* des developpemens organiques.' Passage scored and '?' added.

164–1 See B82, 91–92.

164–2 Buch 1836, Atlas, Pl XII, 'Volcans des Molucques et des Isles de la Sonde' shows chains of volcanic islands.

164–3 Thomas Horsfield is cited here as the source of information in this passage or as an authority to verify it.

At India House, collection of Birds from Java.— at Leyden series from several islands.— Bear peculiar to Sumatra & not found on Java— Monkey peculiar to. latter not to former—

165e Mr Martens of Zoolog Soc[1] told me an Australian dog he had, used to burrow like fox.— a sort of internal bark. would remain for long time together in tub of water with only nose projecting.— would pull the garden bell, & then run into Kennel to watch who would come to the door— would constantly do this, so was obliged to be removed.— In L.' Institut. 1837: p. 404. account of instinct of dogs.— agreement & reason[2]

166e Some animals common to Mauritius & Madagascar.[1] ?

Proceedings of Zoolog. Soc June 1837 p. 53. an Irish Rat.— different from English.[2]

Waterhouse has information respecting the Water Rat.—[3]

167 ⁋Consult Dr Smith History of S. African Cattle[1]

Phillips Geology «p 81» in Lardens Encyclop. Proportions between fossils & recent shells between herbivorous & zoophagous Mollusca according to periods.—[2] NB. Was Europe desert (like S. Africa) after Coal Period.—

168 ⁋In those divisions of molluscs. where species now least in number (as cephalopods,) in[1] last tertiary epochs most genera dead? —Examine into this «in Phillips».—[2] According to this, formerly there would have been many genera

---

165 'I' *pencil top right (not added by Darwin).*
166 Proceedings . . . English.—] *marginal* 'X' *added in pencil.*

165–1 William Charles Linnaeus Martin.
165–2 Anonymous 1837a:404. Anecdotes about dogs' behavior, including 'agreement' of a street pack to ambush another dog and a case where 'la réflexion semble jointe à la mémoire'.
166–1 See Desjardins 1832:111. [*Platalea* sp.] was . . . stated by mistake to be a native of the Mauritius; its true *habitat* as pointed out by M. Desjardins, is Madagascar . . . Of another bird, which is common in Madagascar, . . . a single specimen has been shot in the Mauritius. See B160. The grey ink '?' indicates Darwin wanted authoritative resolution of the question.
166–2 W. Thompson 1837a:52–53 describes *Mus Hibernicus*.
166–3 See B55.
167–1 See *Variation* 1:88, 'Sir Andrew Smith several years ago remarked to me that the cattle possessed by the different tribes of Caffres, though living near each other under the same latitude and in the same country, yet differed, and he expressed much surprise at the fact.'
167–2 J. Phillips 1837–39, 1:81 gives tables showing such proportions.
168–1 Passage is clearer if 'in' is read as 'by'.
168–2 J. Phillips 1837–39, 1:82–83, 'The analogy of the tertiary to the actual system of organic nature is very apparent in these numerical proportions, and the distinctness of both from the older types in the lower strata is one of the most remarkable and important generalisations in geology.

'Nearly all the fossil mollusca, even in the tertiary system, belong to extinct species, a large proportion to extinct genera, particularly among the cephalopoda, brachiopoda, and mesomyona.' (p. 82). 'Most of the fossil cephalopoda belong to extinct genera:' (p. 83). Table III, (p. 83) shows that almost all the fossil cephalopod genera were extinct before the tertiary.

of monotrematous animals.— p. 82 «There are many tables in Phillips of numerous genera in fossil & recent state, well worth consideration—»

169 Tabulate Mammalia on this principle.

‹Varieties› ‹«Races»› Man in *savage* state may be called, ‹species› **species.** in *domesticated* ‹species› **races.**— If all men were dead then monkeys make men.— Men makes angels—

170 Those species which have long remained are those ?Lyell?, which have wide range and therefore cross & keep similar.[1] But this is difficulty; This immutability of some species.

In Phillips. p. 90. it seems the most organized fishes lived far back, fish approaching to reptiles at Silurian age—[2]

171 How long back have insects been known?

As Gould remarked to me, the "beauty of species is their exactness,' but do not known varieties do the same, May you not breed, ten thousand grey hounds & will they not be greyhounds?—
Yarrell's remark about old varieties affecting the cross most well worthy of observation.—

172 I think it is certain strata could not now accumulate without seal-bones & cetaceans.— both found in every sea, from Equatorial to extreme poles.— Oh. Wealden.— Wealden.—[1]

---

169 ‹«Races»›] *deletion in grey ink.*
called, ‹species›] ‹species› *deletion in grey ink.*
domesticated ‹species›]. ‹species› *deletion in grey ink.*

170–1 Lyell 1837, 3:142–44, *Successive Extinction of Species consistent with their limited Geographical Distribution*, 'If the views which I have taken are just, there will be no difficulty in explaining why the habitations of so many species are now restrained within exceedingly narrow limits. Every local revolution, such as those contemplated in the preceding chapter, tends to circumscribe the range of some species, while it enlarges that of others; and if we are led to infer that new species originate in one spot only, each must require time to diffuse itself over a wide area. It will follow, therefore, from the adoption of this hypothesis, that the recent origin of some species, and the high antiquity of others, are equally consistent with the general fact of their limited distribution, some being local, because they have not existed long enough to admit of their wide dissemination; others, because circumstances in the animate or inanimate world have occurred to restrict the range which they may once have obtained.'

170–2 J. Phillips 1837–39, 1:90–91. Placoid and ganoid fish occur from the Silurian to the present, but '. . . the true solution [to the function of the heterocercal tail] may be found in the analogy which placoid fishes in general, and certain ganoid fishes, present to the class of reptiles' (p. 91).

172–1 Wealden is a Cretaceous formation with terrestrial plants and animals but no mammals. The extraordinary thing about Wealden was the absence of mammalian fossils. Lyell 1837, 4:310–11, 'It is certainly a startling proposition to suppose, that a continent covered with vegetation, which had its forests of palms and tree-ferns, and its plants allied to Dracaena and Cycas, which was inhabited by large saurians, and by birds, was, nevertheless, entirely devoid of land quadrupeds. . . . But . . . we can hardly refuse to admit that the highest order of quadrupeds was very feebly represented in those ages, when the small Didelphys of Stonesfield was entombed.'

Do the N. American. Tertiary deposits present analogies to shells of living seas.?—

173e Roxburgh. list of plants in *Beetsons St.Helena.* —[1] Galapagos—Juan Fernandez Falkland Isl^ds^— Kerguelen land.—
Phillips. Lardner Encyclop. insists on analogy between Australia and «fossils of» Oolitic Series[2]
does not appear to me very strong
What is Osteopora platycephalus. (Harlan) found on Delaware
is it Edentate? Phillips. Lardner p. 289[3]

174e It is certain, that North American fossils bear the closest relation to those now living in the sea.— See Rogers report to Brit Assoc ‹to› on N. American Zoology—[1]

---

175 A breed of Blood Hounds from Aston Hall close to Birmingham, and supposed to be descended from a breed known to be there since the time of Charles.,— and now in the possession of Mr Howard Galton,[1] have one of the vertebra, about 2/3 from base of tail, enlarged two

176 very considerably, so that any person would say the tail was broken— This came so often «that» it was difficult to obtain a litter without this defect, Very curious case = W. D. Fox.[1]

---

173 *Beetsons St. Helena*] *underlined brown crayon,* '20' *added brown crayon.*
Phillips . . . Series] *brace left and right margin.*
What] *part of* 'W' *on stub.*
page crossed brown crayon.
174 *page crossed brown crayon.*

(p. 310). Darwin commented in the margin, 'very strong & very honest' and at the bottom of the page, 'As long as Didelphis | Monkey, no progressive'; 'Main strong fact on opposite side | you lean'. Lyell goes on 'So far, then, as our present inquiries enable us to judge, there are strong indications that, during the periods of the Wealden, the Oolite, and Lias, there was a large development of the reptilia, at the expense, as it were, both of the cetaceous and terrestrial mammalia.' (pp. 310–11). Darwin wrote at top of p. 311 'I think it is an argument for prevalence of centrum and classes at former times, the prevalence of orders now.— as pachyderms in Tertiary— Deer now'.

173–1 Roxburgh 1816.

173–2 J. Phillips 1837–39, 1:208 '. . . but it is interesting to know that the earliest mammalia, of which we have yet any trace, were of the marsupial division, now almost characteristic of Australia, the country where yet remain the trigonia, cerithium, isocardia, zamia, tree fern, and other forms of life so analogous to those of the oolitic periods.'

173–3 J. Phillips 1837–39, 1:289 listed in 'General Table of Vertebral Remains in Post-Tertiary Accumulations'.

174–1 Rogers 1835b:33, refers to fossil shells of the 'Newer Pleiocene' on the Atlantic coast.

175–1 John Howard Galton.

176–1 W. D. Fox: personal communication. Darwin visited Fox on the Isle of Wight in late November 1837 ('Journal'; Letter to Fox, [11 December 1837], *Correspondence* 2:63–64). Presumably the Bloodhound case was discussed during that visit. However, the account in B175–76 cannot have been written before December 1837 since at B165 Darwin read the December issue of *L'Institut.* Probably this passage and the Fox information in B182–84 was written from memory—perhaps stimulated by the case of the tailless Manx cat (B163). This passage is related to a letter from Fox, (*Correspondence* 2:111–13, probably written after 14 November 1838). In the letter Fox verifies information he has given Darwin during their November 1837 meeting (recorded in B175–76 and B182) and in the July 1838 meeting (on geese, recorded in D9). With respect to Galton's Bloodhounds,

When dogs are bred into each other, the females loose desire, and it is required to give the canthairides

177e and milk—Fox tells me that it is generally said.= How came first species to go on.— **There never were any constant species**
Both males & females. lose desire.

Native dog not found in V. Diemen's land J. de Physique. Tom 59. p 467. Peron[1]

178e G. St. Hilaire has written "opuscule" entitled "Paleontographie" developing his ideas on passage of forms.— Deshayes states Lamarck priority refers to introduction to Animaux Sans Vertèbres as latest authority.—[1]

The case of the tailess cat of Isle of Man mentioned in Loudons (analogue of Blood hound—[2]

177 and milk . . . desire.] *crossed brown crayon.*
Native . . . Peron] '19' *added brown crayon.*
178 *page crossed brown crayon.*

Fox wrote: 'First about H: Galtons Dogs— I wrote to him & have the following answers. "I have found from breeding in & in that there is considerable difficulty in keeping up the breed. Many of the females have never exhibited any sexual appetite & those which do so at all, very rarely.

The Knot in the tail appeared by accident in one of the finest Dog puppies I had, so fine that I kept it notwithstanding this imperfection & all his descendants had it until at last I got a cross with one of Lord Aylesfords Bloodhounds, since which time it has disappeared. The knot was always in the same part of the tail. Another consequence of breeding in & in is that the animals become prematurely old." I give you this in his own words— They agree pretty much with what I told you, do they not?' Across this passage Darwin wrote in brown crayon '(Breeding In & In)', accordingly the information was used in *Variation* 2:121.

177–1 Péron 1804:466. Given as zoological evidence of 'cette distinction sinon primitive, du moins prodigieusement ancienne de la Nouvelle-Hollande d'avec la terre de Diémen.'

178–1 Deshayes 1833–34a:99–100, 'M. Deshayes, répondant à ce que M. Geoffroy Saint-Hilaire a exposé dans la séance précédente en présentant son opuscule intitulé: *Paléontographie*, réclame en faveur de notre célèbre Lamarck la priorité de cette idée, que les animaux sont modifiés dans leur organisation par les circonstances ambiantes. Cette thèse, dit-il, a été développée par Lamarck, non seulement dans sa philosophie zoologique, en 1809, mais encore dans sa belle introduction à l'histoire des animaux sans vertèbres, 1815. M. Deshayes fait observer qu'il n'est pas juste de citer, comme l'a fait M. Geoffroy Saint-Hilaire, l'Hydrogéologie de Lamarck, ouvrage antérieur aux deux précédens, et dans lequel cette idée n'est exposée que très accessoirement; enfin, M. Deshayes termine en affirmant de la manière la plus positive, et en citant les pages 129 et 130 de l'introduction précitée, que jamais Lamarck n'a partagé les idées systématiques de Telliamed, reproduites par Bonnet et Roedig, comme paraît le croire M. Geoffroy Saint-Hilaire.' The Geoffroy Saint-Hilaire 'opuscule' has not been traced, but Deshayes evidently referred to E. Geoffroy Saint-Hilaire 1833–34: 89–90, 'M. Geoffroy Saint-Hilaire, à l'occasion de la découverte par lui signalée en Auvergne, de plusieurs nouvelles espèces de mammifères fossiles, entre dans quelques détails théoriques sur la manière de concevoir l'organisation de l'ensemble des animaux. Il considère la production successive des différentes organisations comme pouvant expliquer l'apparition des êtres de l'ancien monde. Cette apparition aurait été lente, graduelle, sans secousses, et surtout produite par des changemens dans les milieux qu'ont habités les êtres organisés, changemens faibles si l'on compare les êtres de périodes voisines, plus forts si l'on rapproche les êtres de périodes éloignées.'

178–2 The basic analogy is between the tailless Manx cat (first mentioned at B163) and Fox's case of Bloodhounds with deformed tails (B175–76). Further, the same effect (hereditary taillessness) produced by inbreeding results from the analogous processes accidental geographic isolation and domestication (see B6–7). There are multiple sources for the Manx cat in Loudon (*Mag. Nat.*

179 Bull. Soc. Geolog. 1834. p. 217. Java Fossils 10 out of twenty have ANALOGUES **uses this word «for» similar** in the Indian sea.— Deshayes.—[1]

Mr McClay[2] is inclined to think that offspring of Negro & white will return to native stock[3] (the cross often whiter‹)› than white parent) the mulattos themselves explain it by intermarriages with people. either a little nearer black

180 or white as it may happen.— Dr Smith[1] says he is sure of the case at Cape.— McClay argues from it Black & White species.—[2] For, says he Seeds of hybrid lillies &c &c &, (V Herbert on hybrids[3]) thus act.— Now the point will be to find whether know varieties in plants do so.— As in cacti &c &c.— as in dogs investigate *c*ase of pidgeons. fowls. rabbits

181 cats &c &c.— When black & white men cross some offspring black others white which is more closely allied to case of cross of dogs.— See paper in Philosoph. Transaction on a quagga & mare crossing by Ld. Moreton,[1] where mare was influenced in this cross to after births, like aphides.— Case of boy with fœtus developed in breast.— looking as if many ova— impreg

182 nated at once.— Dr. Smith considers the Caffers (like Englishmen) men of many countenances, as hybrid once. Is not this contradiction to his view of races not mingling?—[1]

In Foxes[2] case of Blood Hounds, a little mingling would probably have been good, namely such as blood hounds from other parts of England.[3]

183 Mr Bell of Oxford St.[1] had a very fine blood hound bitch which would never

179 Bull . . . Deshayes.—] *marginal* 'X' *added pencil.*
ANALOGUES] *triple underlined grey ink.*
uses this word «for» similar] *added pencil.*
182 men] 'e' *over* 'a'.

Hist.); the most relevant is W. B. Clarke 1834:140 '. . . a vessel from Prussia . . . was wrecked . . . and . . . two or three cats without tails made their escape . . . if it be the truth, the original breed is not of Manx extraction, but must be sought out in the north of Europe.' Passage scored. See also Mancuniensis 1832:717.
179–1 Deshayes 1833–34b:217. Refers to fossils in a 'terrain tertiaire très moderne'. See note B174–1 for a similar report.
179–2 William Sharp MacLeay.
179–3 See B32 and B69.
180–1 Andrew Smith.
180–2 W. S. MacLeay: probably personal communication.
180–3 See Herbert 1820, 1822, 1837.
181–1 G. Morton 1821; several passages scored and underlined. Lord Morton's mare was an example of what was later called telegony (Weismann 1893: xii, 383). See D41 for continued connections between Morton's telegony and parthenogensis in aphids.
182–1 Personal communication.
182–2 W. D. Fox, see B176.
182–3 See notes B176–1 and D5–15.
183–1 J. Bell: personal communication. See Thomas Bell 1837:209, 'The race of [blood-hounds] has been gradually diminishing and is now very rarely to be met with in its purity. Amongst the very few instances of its present existence, I may mention a fine breed in the possession of Mr. J. Bell, of Oxford Street, who retains them in great purity.'

take the dog. But at last a rough-haired shepherd dog lined her & produced a very large litter.— never afterwards went in heat.— This is good instance of same fact in Mr Galtons case.—[2] It explains the loss **& expense, (must probably have occurred to every one)** of rare breeds of dogs, from owners great care of them. Fox says when two dogs of opposite breeds are crossed, sometimes offspring quite intermediate

184 sometimes take strongly after either parent. about ‹half & half tim› as often one way as other.— He has known case of good pointer & rough water spaniel «produce litter like both parents» & Mr Bell has half bloodhound & greyhound.—
Where two dogs have lined bitch directly one after the other, puppies differ, & like both parents.— Fox told me of case of mare covered by blood horse & Carthorse two folds

185e–186e [not located]

187e Mr Don[1] gave me instances of one species of Australian genus being found in Sumatra; again another of other Genus in Sandwich islands— A genus with species in Van Diemen's land and Tierra del Fuego.— Araucaria, species. Brazil Chile, Norfolk Isl.— Isle of Pines—

Australia.— A ‹South› American «form of» Lathyrus has one species in Europe Madagascar has several American forms—
The above facts evidently show that Mr D.[2] wonders

188e at these species being wanderers.—

Iceland no species to itself, a remark common to all northern islds.— This is interesting, because Iceland, must have been all ice in time of ice transported.— This gives room to fine speculation.— Are there many *Northern genera* peculiar to itself—

189e on hybrids between grouse & pheasant— Magazine. Zoology & Botany Vol 1 p. 450[1]

---

185 *one illeg letter on stub.*
187 *page crossed pencil.*
189 on hybrids . . . p. 450] '17' *added brown crayon.*
on hybrids . . . p. 450] *crossed pencil.*
There is . . . to plants] *crossed brown crayon.*

183–2 See note B176–1.
187–1 David Don published on floristic botany and wrote at least one paper on Australian plants, Don 1833.
187–2 David Don.

189–1 W. Thompson 1837b:450, 'In four instances only am I aware of similar hybrids being recorded.' Passage scored, 'four' underlined. One of these instances was a report by Darwin's friend T. C. Eyton (1835).

There is in nature a «real» repulsion «amounting to impossibility, holds good in plants» between all different forms; therefore when from being put on island. & fresh species made. parents do not cross— we see it even in men); the possibility of Caffers & Hottentots coexisting. proves this— but when Man makes variety these are vitiated.— This barely applies to plants

190e Female pig apt to produce monsters in Isle of France— — Madagascar oxen with hump.— p 173. Voyage par un Officier du Roi[1]

Mem. Capt. Owen's story of cats on West coast of Africa.— changing hair—[2]

The Edinburgh. Journal of Natural History— Preface appeared good with facts about changes when animals transported.)[3]

191 Mr Herbert's papers are in the Horticultural Transactions and a distinct work on Hybridity under title of Amaryllidæ & Narcissus.[1] Mr Donn[2] considers Mr H. rather wild

---

Mr Donn remarks to me. that give him a species from Ireland, England, Scotland & other localities & each one will have a peculiar «constant» aspect. That is varieties, though of trifling order are formed by nature.

192 Carmichael. Tristan D'Acunha, a list of its Flora. is given[1] Mr Don remarked to me. that some good African & some good S. American forms. (& daresays some of these ‹African forms› forms would have some peculiarity.—[2] Now when we hear that the whole island is volcanic surmounted by water & studded with others.—[3] we see a beginning to isl^d. Graham isl^d.— we know many seeds, might be transported some blown—floating trees

190 *page crossed brown crayon.*
191 Mr Donn considers . . . wild] *in balloon.*

190–1 Saint-Pierre 1773:172–73, 'Dans les quadrupèdes domestiques; il y a . . . des boefs dont la race vient de Madagascar. Ils portent une grosse loupe sur leur cou. La femelle de cet animal est sujette dans cette isle à produire des monstres.'
190–2 W. F. W. Owen 1833 2:180–81. 'A cat . . . was landed at Mombas . . . [after] eight weeks, it had undergone a complete metamorphosis, having parted with its sandy-coloured fur, and gained . . . beautiful short white hair.'
190–3 Anonymous 1835a, 1:5. 'We are astonished [at there] being domestic plants, which follow man in his improvement and change of soil, or wanderers seeking to inhabit distant regions . . .'
191–1 Herbert 1820, 1822, 1837.
191–2 David Don.
192–1 Carmichael 1818:502–13.
192–2 Carmichael 1818:498, 'The Flora of Tristan da Cunha . . . with the exception of the cryptogamous . . . plants, . . . offers nothing that is possessed of any peculiar interest.'
192–3 Carmichael 1818:484, 491, 'The whole island is apparently a solid mass of rock in the form of a truncated cone, rising abruptly from the sea . . . This mass is surmounted by a dome upwards of five thousand feet high, on the summit of which is the crater of an old extinguished volcano.

'The island is of a circular form . . . In various places the sea beats home against the salient angles of the mountain, rendering it impossible to walk round the island.' (p. 484). 'In viewing the general structure of the island, and comparing its diminutive size with the great number of spiracles crowning its summit, . . . [there is] little doubt that the whole of it is of igneous origin.' (p. 491).

193 Thrushes & bunting & coots— «(Turdus Guyanensis?) (*Emberiza Brasiliensis*?) (Fulica Chloropus)—»[1] might bring in stomach— &c &c. (Mem discover what kinds of seeds. these plants) [Mem Fact stated by Mr Don in island, Teneriffe, St. Helena. J. Fernandez. Galapagos. Many trees Compositæ, because seeds first arrived «Ferns ditto.—» & hence formed trees[2]][CD] & would creator ‹on volcanic island.› *make* plants ‹grow closely› When this volcanic point appeared in the great ocean, have made

194 plants of American & African form, merely because intermediate position.— We cannot consider it as adaptation because volcanic isl[d]. whilst ‹neig› Africa, sandstone, & granite, (that is genera near Cape) see if there are any species same as T. del Fuego & C. of Good Hope show *possibility* of transport. If some cannot be explained more philosophical to state we do not know how transported.—

195 (Glaciers might have acted at Tristan d'Acunha— Carmichael Linn. Transacts. Vol XII.—[1]

The Alpine plants of the Alps. must be ‹Alpine› «new formations» because snow formerly descended lower, therefore species of lower genera altered. or northern plants
«No» [CD][Mem. the antarctic flora must formerly have been separated by short space from mountains low down, therefore plants common take an example from T. del Fuego.—

196 Ellis (?) says Tahitian kings. would hardly produce from Incestuous intercourse.[1] a parallel fact to Blood-Hounds.[2]

Before Attract of Gravity discovered. it might have been said it was as gret a difficulty to account for movement of all, by one law. as to account for each separate one, so to say that all Mammalia, were born from one stock, & since distributed by such means as we can recognize, may be thought to explain nothing.— it being as easy to produce «for the creator» two quadrupeds at S. America Jaguar & Tiger

196 Ellis . . . Blood-Hounds.] *marginal* 'X' *added in pencil.*

193–1 Carmichael 1818:496, 'The only land birds on the island are a species of thrush (*Turdus Guianensis*?), a bunting (*Emberiza Brasiliensis*?), and the common moorhen (*Fulica Chloropus*).'
193–2 See Don 1830 on the compositae of Peru, Mexico, and Chile.
195–1 Carmichael 1818.
196–1 Ellis 1831. See 2:345, 'The hui arii, or highest class, included the king or reigning chieftain in each island, the members of his family, and all who were related to them. . . .
'Whenever a matrimonial connexion took place between any one of the hui arii with an individual of an inferior order, unless a variety of ceremonies was performed at the temple, by which the inferiority was supposed to be removed, and the parties made equal in dignity, all the offspring of such an union was invariably destroyed . . .'
196–2 See B175–76 and related passages.

197e & Europe, as to produce same one. ‖Although in plants, you cannot say that instinct perverted, yet organization «especially» connected with generation certainly is.= The dislike of two species to each other is evidently an instinct— & this prevents breeding. now domestication depends on perversion of instinct (in plants domestication on perversion of structures especially reproductive organs) & therefore the one distinction of species would fail. But this applies only to coition & not production. But who can say, whether offspring does not depend on mind or instinct of parent. Mem Lord Moreton's Mare.[1] The fact of plants going back

198e hybrid plants; analogous to Men. & dogs. Now if we take structure as criterion of species Hogs different species, dogs not, but if we take character of offspring. Hogs not different. some dogs different.—

Henslow says. (Feb 1838[1]) that few months since in Annales des Sciences, paper on Botany of Tahiti[2]

In Charlesworth Magazine Jan: 1830. most curious paper on heredetary fear (like rooks with guns) of the

199e Bustards in Germany.—[1]

Athenæum. No. 537. Feb. 1838. p. 107.[2] Mr Blyth states that all «genera of» birds in «N.» America & Europe, which have not their representative species in each other, are migratory species from warmer countries. When will this paper be published it will be curious.— Some general statements about mundine & confined genera.—

200e Lyell has remarked about no confined species in Scicily.[1]

---

197 'I' *pencil top right (not added by Darwin).*
page crossed pencil.
198 hybrid plants . . . Tahiti] *crossed pencil.*
In Charlesworth . . . of the] *double scored pencil.*
199 Athenæum . . . genera.—] '19' *added brown crayon.*
200 Lyell . . . Scicily] *crossed brown crayon.*
Islds X] *added brown crayon.*

197–1 G. Morton 1821. This classic report of telegony (Darwin's 'after births' B181) is framed as a critique of 'the old doctrine of impressions produced by the imagination; . . . I can hardly suppose that the imagination could pass by the white tufts on the quagga's mane . . .' (p. 22).
198–1 Refers to the date of this passage.
198–2 Guillemin 1836–37, the final installment appeared in the last issue of 1837 (n.d.).
199–1 *Sic* 1838. Weissenborn 1838c:51, 'I have often admired the sagacity, which enables this large and heavy creature [the bustard] to exist and thrive amid so many dangers, in a thickly peopled country; but in adverting more particularly to the means through which it effects this purpose, we shall find that every generation learns instinctively from the former, what objects the experience of the latter has taught them to shun.' Passage scored.
199–2 Blyth 1838a:107.
200–1 Lyell 1837, 3:444, 'The newly emerged surface [of Sicily], therefore, must, during this modern zoological epoch, have been inhabited for the first time by the terrestrial plants and animals which now abound in Sicily. . . . The plants of the flora of Sicily are common,

Jan: 1838 L'.Institut. Bats, in Eocene beds, very like present species., p 8.[2]
ʕAre mundine forms, longest persistent??

---

do.— The most perfect Plants Composites.— !!«good» *those which have undergone most metamorphosis*[3] **Islds X**
Is this applicable to insects &c &c?— (p. 23

---

do.— On animal – Confervæ— p. 23[4]

201e p. 267. Dela Beche. Geolog. Researches. facts of salt-water shells living in absolutely fresh water.—[1] origin of Fresh-water genera?

The absence of lime in Plutonic & Volcanic rocks. most remarkable.—ʕ[2] Have the changes been so slow., that all have existed for ages as metamorphic; & therefore according to Lyells doctrine removed??[3]

202e Is the prevalence of Coniferous Woods before Dicotyledenous a fact analogous to reptiles before Mammalia
Think about Miocene fossils some species being recent agreeing with Senegal.

201 '19' *added to page brown crayon.*
The absence . . . removed??] *crossed grey ink.*
202 Is the . . . Mammalia] *brace left and right margin.*

almost without exception, to Italy or Africa, or some of the countries surrounding the Mediterranean'. See E105.
200–2 Blainville 1838a:8, 'Ces chauve-souris étaient très probablement contemporaines des Anoplotherium, des Palaeotherium, puisque leurs ossements se trouvent dans les mêmes conditions géologiques.'
200–3 See B170.
200–4 Meyen 1838a:23. This paper summarizes E. M. Fries 1836, who gives seven indices of 'perfection' in plant families, of which the first is 'Plus est grand le nombre des degrés de métamorphoses par lesquelles une plante doit passer avant que son fruit se développe, plus elle est parfaite.' Fries, who criticized A. P. de Candolle's ranking of Ranunculaceae as the most perfect family, considered 'les Composées . . . les plus complètement développées.' Meyen 1838a also summarizes Mohl 1836, who shows that in *Conferva glomerata* '. . . les cellules s'accroissent par la formation de cloisons . . .', which is taken to support Confervae as plants rather than animals.
201–1 De la Beche 1834:267, 'Should . . . a *Voluta* or an *Arca* be detected among the organic contents of a rock, it would at once be considered as marine, because both genera are generally known to us as marine . . . there is, however, a chance for error . . . for *Voluta magnifica* is known to live high up in the brackish waters near Port Jackson in Australia, and an *Arca* inhabits the freshwater of the Jumna, near Hamirpúr, 1000 miles from the sea.' Passage scored.
201–2 See De la Beche 1834:297–300, 'If the theory of central heat be [well] founded . . . there must have been a time when the mineral crust first became solid, . . . [Later a time] must come when the heated rock would . . . allow . . . water to remain liquid upon it . . . Such a condition of things would necessarily be one in which . . . [no life can] exist. Hence no organic exuviae could become entombed in any rocks which could result from this state of the earth's surface. It would also be one highly unfavourable for the production of carbonate of lime, since the carbonic acid would be driven out of the water, and, consequently, no carbonate of lime could exist in solution.' (pp. 297–98). 'We should expect the inferior rocks also to correspond in their general chemical characters with . . . igneous products . . . [of] different geological epochs . . . Now this resemblance does exist, more particularly among the older igneous products, precisely where we should expect to find it.' (p. 300).
201–3 Lyell 1837 3:302, 'The constant transfer, therefore, of carbonate of lime from the inferior parts of the earth's crust to its surface, must cause at all periods and throughout an indefinite succession of geological epochs, a preponderance of calcareous matter in the newer, as contrasted with the older formations.'

whilst Crag ‹agrees with› according to Beck has none recent, yet genera same.—[1]
Speculate on multiplication of species by travelling of climates & the backward & forward introduction of species.—

203 When species cross & «hybrid» breed, their offspring show tendency to return to one parent, this is only character., & yet we find this same tendency (only less strongly marked) between what are called varieties.—
NB. one mother bringing forth young having very different characters is attempt at returning to parent stock.— I think we may look at it so— **?? It holds good even with trifling differences of expression — one child like father another like mother**

204 Has Lowe written any other papers besides one in Latin one ‹of› «on» Madeira—[1] any general observations— difference of species between land shells of Porto Santo & Madeira— I believe very curious—

My idea of propagation almost infers, what we call improvement, —All Mammalia from one stock, & now that one stock cannot be supposed to be «most» perfect (according to our ideas‹)› of perfection); but intermediate

205 in character, the same reasoning will allow of decrease in character. (which perhaps is) Case with fish— as some of the most perfect kinds the shark. lived in remotest epochs.—[1] ? lizards of secondary period in same predicament. It is another question, whether whole scale of Zoology may not be perfecting by change of Mammalia for Reptiles, which can only be adaptation to changing world:— I cannot for a

206 moment doubt, but what cetaceæ & Phocæ now replace Saurians of Secondary epoch: it is impossible to suppose such an accumulation at present day & not include Mammalian remains.— The Father of all insects gives same argument as father of Mammalia; but have improvement in system of articulation. ? whether type of each order may not be supposed that form, which has wandered least from ancestral form. If so are present typical

207 species most near in form to ancient; in shells alone can this comparison be instituted.

People often talk of the wonderful event of intellectual Man appearing..— the

204 Latin] 'L' *over* 'l'.

202–1 Lyell 1837, 4:72, 'Dr. Beck, of Copenhagen, well known for his profound knowledge of recent shells, has lately seen 260 species of crag shells in Mr. Charlesworth's cabinet in London, and informs me, that although a large proportion of the species approach very near to others which now live in our northern seas, he regards them as almost all of distinct species and not recent. I attribute this discordance of opinion between the Danish naturalist and M. Deshayes, chiefly to the different estimate which they have formed of the amount of variation necessary to constitute a distinct species.'
204–1 See Lowe 1833b.
205–1 See B170.

appearance of insects with other senses is more wonderful. its mind more different probably & introduction of Man. Nothing compared to the first thinking being. although hard to draw line.— —

208 not so great as between perfect insect & former hard to tell whether articulate or intestinal, or even a mite.— a bee «compared with cheese mite» with its wonderful intincts ‹might well say how› The difference is that there is wide gap between Man & next animals in mind, more than in structures.—

If the skeleton of a Negro— had been found what would Anatomists have said.— ⁽where is Pentland's account of

209e Bolivian human species?—[1]

Small «new» animal mentioned from Fernando Po Zoolog. Proceedings October (?) 1837[2] **Contrast New Zealand with Tasmania**

The reason why there is not perfect *gradation* of change in species,, as physical changes are *gradual*, is this if after isolation (seed blown into desert) or separation by mountain chains &c the species have not been *much* altered they will cross (perhaps more fertility & so make that sudden step. species or not.

210e A plant submits to more individual change, (as some animals do more than others, & cut off limbs & new ones are formed) but yet propagates varieties according to same law with animals??

Why are species not formed. during ascent of mountain or approach of desert?— because the crossing of species less altered prevents the complete adaptation which would ensue

211 A. B. C. D. — (A) crossing with (B) (& B having crossed with (C) prevents offspring of A. becoming a good species well adapted to locality A. but it is instead a stunted & diseased form a plant, adapted to A. B. C. D.— Destroy plants B. C. D. & A will soon form good species!

The increased fertility of slightly different species & intermediate character of offsprings accounts for *uniformity* of species & we Must confess. that we canot tell, what is the amount

209 **Contrast** . . . **Tasmania**] *added pencil.*
Small . . . 1837] *marginal* 'X' *added in pencil.*
Small . . . 1837] '19' *added brown crayon.*
The reason . . . or not.] *crossed brown crayon.*

209–1 Pentland 1835:623–24, 'The author . . . stated . . . there existed [known from tombs at Titicaca] at a comparatively recent period a race of men very different from any of those now inhabiting our globe, characterized principally by the anomalous form of the cranium, in which two thirds of the entire weight of the cerebral mass is placed behind the occipital foramen, and in which the bones of the face are very much elongated.'
209–2 W. C. L. Martin 1837.

212 of difference, which improves. & checks it.— It does not bear any precise relation to structures
Mem. Eyton's Hogs & dogs.—[1]

The passage in last page explains that between Species from «moderately» distant countries. there is no test but generation, «(but experience according to each group)» whether good species, & hence the importance Naturalists attach to Geographical range of species.—

213 Definition of Species: one that remains «at large» with constant characters, together with other ‹animals› beings of very near structure.— Hence species may be good ones & differ scarcely in any external character:— For instance two wrens forced to haunt two islands one with one kind of herbage & one with other, might change organization of stomach & hence remain distinct.

214 Where country changes rapidly, we should expect most species.—

The difference intellect of Man & animals not so great as between living thing without thought (plants) & living thing with thoughts (animal).

«∴ my theory very distinct from Lamarcks[1]»
Without *two* species will generate common kind, which is not probable; then monkeys will never produce man. but

215 both monkeys & man may produce other species., man already has produced marked varieties & may someday produce something else., but not probable owing to mixture of races.— When all mixed & physical changes (? intellectual being acquired alters case) other species or angels. produced «&»

216 Has the Creator since the Cambrian formations gone on creating animals with same general structure.— miserable limited view.—

212–1 See Eyton 18[obscured] & animals.' This may bear on B214 'The
214–1 See marginali[obscured] . . . (animal).' The third comment of Darwin's notes summarize his [obscured] relates to pp. 261–62, 'Mais des mélanges animals & plants cha[obscured]s entre des individus qui n'ont pas les mêmes plants & animals— [obscured]rités de forme, font disparoître toutes les par- want of habit cause [obscured]s acquises par des circonstances particulières. De Explains how crossi[obscured]ut assurer que si des distances d'habitation ne Man— & supposes [obscured]nt pas les hommes, les mélanges pour la généra- would not be many [obscured]ient disparoître les caractères généraux qui dis- view.' The first of th[obscured] les différentes nations.' Passage scored and on les végétaux, où il[obscured]er Darwin wrote 'p 261 on effects of intermarriage séquent, point d'h[obscured]enting multiplication of species'. Darwin's mar- changemens de cir[obscured] on Lamarck 1830 show several other distinctions grandes différence[obscured]n their theories.
parties;' Darwin c[obscured]

With respect to how species are. Lamaks "willing" doctrine absurd.[1] (as equally are arguments against it— namely how did otter live before being made otter— why to be sure there were a thousand intermediate

217 forms.— Opponent will say. show them me, I will answer yes, if you will show me every step between bull Dog & Greyhound). I should say the changes were effects of external causes, of which we are as ignorant. as why millet seed turns a Bullfinch black,[1] or **iodine on glands of throat,** (or colour of plumage altered during passage of birds (where is this statement I remember. L. Jenyns. talking of it) or how to make Indian Cow with bump & pigs foot with cloven hoof)

218 Ask Entomologists whether they know of any case of *introduced* plant, which any insects hav[e] become attached to.— that insect «not» being called ‹Phitophagous› omniphitophagous.

But it will be said there are latent insects.— as crows against man with gun. & Bustards &c &c!!![1]

An American & African form of plant being found in Tristan D'Acunha. may be said to deceive man. as likely as fossils in old rocks for same purpose.!![2]

219 Can the wishing of the Parent produce any character on offspring?[1] Does the mind produce any change in offspring? if so adaptations of species *by generation* explained?

NB. Look over Bell on Quadrupeds for some facts.—[2] about dogs &c &c NB. Animals very remote. ass & Horse. produce offspring exactly intermediate.— Reference to Pig & Dogs.

My theory will make me deny the creation of any new quadruped since days of Didelphis in Stonefield ∴ all lands united (Falkland Fox. ice). . Mauritius what a difficulty— where elevation Subsidence New is only hope.— New Zealand «compare to Van Diemen's land.» glorious fact. *of absence of* quadrupeds East India Archipelago very good on opposite tendency.—

220 Study Ellis & Williams.[1] zoology of South Sea isl^{ds}. any animals?—[2] I believe none.— Canary isl^{ds}.? Madeira? «Tristan d'Acunha?» «Iceland?—»

216–1 See Darwin's comments on Lamarck 1830, 1:268, at end of chap. 7. 'It is absurd the way, he assumes the want of habit cause annihilation of organ & vice versâ.—' See also E150.

217–1 See Anonymous n.d. (*The British Aviary*):43 '. . . they [bullfinches] are particularly fond of hemp-seed, but it should only be given in small quantities, as it tends to destroy the beauty of their plumage.' Passage scored.

218–1 See B199.

218–2 See B192–95.

219–1 See B197.

219–2 See Darwin's abstract of T. Bell 1837.

220–1 Ellis 1831 and Williams 1837 (see E10).

220–2 That is, any large mammals.

The Connection between Mauritius & Madagascar very good.— Fernando Po & Coast of Africa. equally good.— Small isl^d off New Guinea same fact see Coquille's Voyage.—[3] Galapagos mouse (?) **brought by canoes** Ceylon & India.— Van Diemen's land Australia. England & Europe.— It will be well worth while to study profoundly the origin & history of every terrestrial Mammalia.— Especially moderately large ones.—

221 Is the flora of ‹S. America› Tierra del Fuego like that of North Europe, many genera & few species.

The number of genera on islands & on Arctic shores evidently due to «the» chance of some one of the different orders being able to survive or chance having transported them to new station.— When the new island splits & grows larger species are formed of those genera.— **& hence by same chance few representative species. this must happen. & thus acquired will explain representative system** Of this we see example in English & Irish Hare.— Galapagos.— shrews, & when big continent many species belonging to its own genera

222 Therefore if in small tract we have many species, we may insure mass continental or many large islands.— Hence this must have been condition of Paris basin land.—
(How is this with Fernando Po.). with plants of St. Helena & Tristan D'Acunha, resolves itself into question of proportion of species to genus
If on one isl^d several species of same genus, subsided land.— Mauritius? ‹In plants where do most species occur.)› Although the Horse has perished from S. America, the jaguar has been left & Fox, & bear.— If I had not discovered

223 channel of communication[1] by which great Edentate[2] might have roamed to Europe & Pachydermata from Europe to America., How strange would presence of Jaguar been in S. America.—
‹East.› «W» Coast of Africa & ‹West.› «E» of America, ought to present great contrast in forms.— India; intermediate, see how that is.—
?are shell-boring Molluscs, like Carnivorous Mammalia in their wide range & in their duration of species. (?are carnivorous Mamm: in Paris basin «allied to present») **more like present carnivora than Pachydermata**

222 (**How is . . . Po**)] *underlined.*

220–3 Lesson and Garnot 1826–30.
223–1 See *JR*:151, 'The separation, therefore, of the Asiatic and American zoological provinces appears formerly to have been less perfect than at present. . . . On the opposite shores . . . of the narrow strait which divides these two great continents . . . the remains of both animals [elephant and ox] occur abundantly . . . With these facts, we may safely look at this quarter, as the line of communication . . . by which the elephant, the ox, and the horse, entered America, and peopled its wide extent. ‡ I do not here mean to fix on the more northern parts of the old world as the parent country of these two animals. I only want to point out the channel of communication . . .'
223–2 *Fossil Mammalia* referred the majority of Darwin's fossil mammalia to the Edentata: *Glossotherium*, *Mylodon*, *Scelidotherium*, *Megalonyx*, and *Megatherium*.

224 If my theory true, we get (1) a *horizontal* history of earth «within recent times.». & many curious points of speculation; for having ascertained means of transport, we should then know whether former lands intervened.— (2[d]) By character of any «two» ancient fauna, we may form some idea of ‹origin under› connection of those two countries **Hence India, Mexico & Europe. one gret sea (Coral reefs ∴ shallow water at Melville Is[d].** (3[d]) We know that structure of every organ in A. B. C. three species «of one genus» can pass into each other «by steps we see»: but this cannot be predicated of ‹genus.› structures in two genera.— ‹we then cease to know the steps.› although D E F. follow close to

225 A. B. C. we cannot be sure that structure (C) could pass into (D).— We may foretell species. limits of good species being known.— It explains the blending of two genera— It explains typical structure.— Every species is due to adaptation + heredetary structure. **latter far chief element ∴ little service habits in classification. or rather the fact that they are not far the most serviceable.** We may speculate of durability of succession from what we have seen. in old world, & on amount changes which may happen— || It leads you to believe the world older than *geologists* think. it agrees with excessive inequality of numbers of species in divisions. look at Articulata!!!—!

226 It leads to Nature of physical change between one group of animals & a successive one.—
It leads to knowledge what kinds of structure may pass into each other: now on this view no one need look for intermediate structures ‹between› say in brain. between lowest Mammal & Reptile. (or between extremities of any great divisions) thus a knowledge of possible changes is discovered, for speculating on future.

227 !.fish never become a man.— Does not require fresh creation!— If continent had sprung up round Galapagos on Pacific side. the Oolite order of things might have easily been formed.—
With belief of ‹change.› transmutation & geographical grouping we are led to endeavour to discover *causes* of change.— the manner of adaptation (wish of parents??) instinct & structure becomes full of speculation & line of observation.—
View of generation. being condensation, test of highest organization intelligible.—may look to first germ—

228 —led to comprehend true affinities. My theory would give zest to

recent & Fossil Comparative Anatomy, & it would lead to study of instincts, heredetary. & mind heredetary, whole metaphysics.— it would lead to closest examination of hybridity «to what circumstances favour crossing & what prevents it—» & generation, causes of change «in order» to know what we

225 **that** *and* **not**] *underlined.*

have come from & to what we tend.— this & «direct» examination of direct passages of ‹species› structure in species, might lead to laws of change, which would then be main object of study, to guide our ‹past› speculations

229 with respect to past & future. **The Grand Question, which every naturalist ought to have before him, when dissecting a whale, or classifyng a mite, a fungus, or an infusorian. is "What are the laws of life".—**

Where we have near genera far back, as well as at present time, we might expect confusion of species.— Important. For instance take Voluta & Conus (??) which now run together, were not both genera formerly abundant.

Seed of Ribston Pippin tree ‹go› producing crab. is the offspring of a male & female animal of one variety going back ?whether this going back may not be owing to cross from other trees.????

230 Do the seeds of Ribston Pippin & Golden Pippin &c produce real crabs, & in each case similar or mere mongrels.—

It really would be worth trying to isolate some plants, under glass bells & see what offspring would come from them. Ask Henslow for some plant, whose seeds go back again, not a monstrous plant, but any marked. variety.— Strawberry produced by seed«s»??—[1]

Universality of generation strongly shown by hybridity of ferns.— hybridity showing connexion of two plants.

231 Animals— whom we have made our slaves we do not like to consider our equals.— «Do not slave holders wish to make the black man other kind?[1]» Animals with affections, imitation, fear ‹of death›. pain. sorrow for the dead.— respect

---

We have no more reason to expect the father of man kind. than Macrauchenia[2] yet he may be found:— We must not compare «chances of embedment in» man in present state. with what he is as former species. His arts would not then have taken him over whole world.—

232 «—the soul by consent of all is superadded, animals not got it, not look forward» if we choose to let conjecture run wild then ‹our› animals our fellow brethren in pain, disease death & suffering «& famine»; our slaves in the most laborious work, our companion in our amusements. they may partake, from

230–1 This question is in the spirit of those recorded in *Questions & Experiments*, e.g. QE11 '(31) Ask Henslow for list of annuals to place in hot-house to see effect on generative organs of gret Heat'. See D15.

231–1 Cf. *JR*:28, 'This man had been trained to a degradation lower than the slavery of the most helpless animal.' See B34.

231–2 See RN129–130, where the fossil 'llama', subsequently named *Macrauchenia* by Richard Owen, plays an integral role in Darwin's early transformist theory. See *Fossil Mammalia*: 35–46.

our origin in ‹there› one common ancestor we may be all netted together.—

Hermaphrodite animals couple: argument for true molluscs coupling.—

233e *Dr. Smith's Information*[1]

Long Horned (very) aboriginal at Cape: crossed with English Bull. offspring very like common English.—
Hottentots say great tailed sheep aboriginal at Cape & a thinner-tailed kind further inland.—

NB. There is division of snakes. with hinder teeth perforated for poison channels, but not having them, instance of useless structure—

Smith thinks several species of Rhinoceros range from Abyssinia to extreme South coast. Elephant he believes is mentioned by old writers on extreme Northern Coast.

234e Hippopotamus do.— Giraffe do.—

Range of East Indian Rhinoceros (?)— Some paper in Institute[1] on range of Bos in India.— Range of Zebra?—

The Crocodile & Tortise former inhabitants of Mauritius Freycinet Voyage,[2] agrees.[3] with *several mammalia* being peculiar[4] (?)

If. Henslow discusses possibility of seeds of Keeling standing transport.—[5]
‹tr but› Get him to discuss those mention[ed] by Lesson & Chamisso.—[6]

---

233 & a thinner] '&' *on stub.*
page crossed brown crayon.
234 Hippopotamus . . . zebra?—] *crossed brown crayon.*
The Crocodile . . . peculiar] *brace left margin.*
The Crocodile . . . peculiar] '19' *added brown crayon.*
several mammalia] *underlined grey ink.*
‹tr but›] *uncertain reading.*
If. Henslow . . . Chamisso.—] *crossed brown crayon.*

233–1 Andrew Smith: personal communication, follows up on B167.
234–1 Reference not traced.
234–2 Reference not in Quoy and Gaimard 1824 (*Uranie* voyage, of which Freycinet was captain) and Péron 1807–16 (continued by Freycinet). See Saint-Pierre 1773:90, 'On trouvoit autrefois sur le rivage beaucoup de tortues de mer, aujourd'hui on y en voit rarement.'
234–3 Agrees with Duméril and Bibron 1834–54, 2[1835]:509, 'Cette espèce est la seule Emyde Africaine que l'on connaisse encore; on la trouve à l'île Bourbon, et elle doit y être rare, puisque'il ne s'en est jamais trouvé d'exemplaire dans les envois zoologiques qui ont été adressés de ces deux pays à notre établissement.'
234–4 See Saint-Pierre 1773, 1:87–88 suggests the 'singe' is unique, p. 90 implies there is a unique bat.
234–5 Henslow 1838:337–38.
234–6 See Chamisso 1821 3:155–56, 'The sea brings to these islands [Radack] the seeds and fruits of many trees, most of which have not yet grown there.' (p. 155).

235 Geograph Journal Vol V. P. I. p. 67. Dr. Coulter on decrease of population in California cessation of female offspring:[1] applicable to any animal—

Athenæum. ‹Jan› «p. 154»— 1838. Hybrid Ferns[2]

It may be argued against theory of changes that if so in approaching desert country or ascending mountain you ought to have a gradation of species, now this notoriously is

236 not the case, you have stunted species,[1] but not such as would make species (except perhaps in some plants & then a chain of steps is found in same mountain).—
How is this explained by law of small differences producing more fertile offspring.— 1st. All variation of animal is either effect or adaptation, ∴ animal best fitted to that country when change has taken place, Nature

237e–283e [not located]

239 Any change suddenly acquired is with difficulty permanently transmitted.— ‹It will admit› a plant will admit of a certain quantity of change at once. but afterwards will not alter. This need not apply to very slow changes. without crossing.— Now a gradual change can only be traced geologically (& then monuments imperfect) or horizontally & then cross breeding prevents perfect change.—

240 It is scarcely possible to get evidence of two races of plants run wild.— (for we know that such can take place without impregnating each other), for if they are different then, they will be called species & mere producing fertile hybrids will not destroy that evidence, as so many plants produce hybrids, or else whole fabric will be overturned.— Hence extreme difficulty, argument in circle.— Falkland Isd case good one of animals not soon being subjected to change in Americas. perhaps merely gone back previous

241 to fresh change

Get a good many examples of animals «or plants» very close (take Europæan birds. Mr Goulds' case of Willow wren[1]) & others varying in wild state to

235–1 Coulter 1835:67. Referring to the Indian population, 'their decrease is greatly hastened by the failure of female offspring,—or the much greater number of deaths amongst the females in early youth than among the males, . . .' Passage scored in pencil and red pencil; 'failure of female offspring' underlined in red pencil.
235–2 Martens 1838.
236–1 See Royle 1835:363, 'In approaching the base of the Himalayas a close jungle is everywhere met with . . . the jungle becomes short and scrubby, in ascending the mountains . . .'

241–1 Gould 1832–37, vol. 2, Genus *Sylvia*, Pl. 131–33. Pl. 131: 'Willow Wren. Sylvia trochilus, *Gmel* . . . We here present, on a single Plate, three little birds, which are nearly allied to each other in habits, manners, and plumage, and which form the British portion of a genus to which the generic title *Sylvia* is truly applicable. . . .
'The Willow Wren is by far the most abundant in England: it is also dispersed throughout the greater portion of Europe, from Sweden to Italy. At the same time its localities are less strictly confined, inhabiting not only groves, woods, and willow plantations, but gardens,

show that we do not know what amount of difference prevents breeding;—or as others would exjudge it amount of varying in wild state.—

When breaking up «the primeval.» continent. Indian Rhinoceros. Java & Sumatra ones all different.— Join Sumatra & Java ‹to› together, by elevations now in Progress, & you will have two.

242 Tapir existing in East Indian Seas. Marsupials animals all show greater connexion in quadrupeds, ‹bu› *plants do not follow* by any means.— Ostriches.— Hippotamus only African.—
American & African forms mingle in India & East Indian isl[ds]— Monkeys different not travellers??

Royles case of Himalayan, plants ⁽migratory birds, he told me some story of crane from Holland!!! in stomach —or in feathers[1]—**seeds.—**

---

243 Two inhabitants of the Tropics, (whether one fossil or not) are related by real relationship. as well as effect of similar temperature.— now those of temperate regions & tropics are only related by one connection.— viz descent.— Hence far greater discordance in latter— Have change in form.— This probably explains crag & miocene.— The descendants left in cooling climate might change twice over, whereas those which migrated a little to the southward would merely be specifically different if so.— Now this is difficult to explain by creation— or we must suppose a multitude of small creations.—

244 Will Dromedaries & Camels breed?—

As man has not had time to form good species, so cannot the domesticated animals with him.!—

Modern origin shown by only one species, far more than by non-embedment of remains— ⁽agrees with non-blending of languages?—
Till man acquired reason, he would be limited animal in range— & hence probability of starting from one point.—

244 As man . . . him.!—] *marginal* 'X' *added in pencil.*

hedge-rows, and commons covered with bushes. Chiff-chaff. Sylvia hippolais. . . . The Chiff-chaff so nearly resembles the former species as to be frequently confounded with it . . .

'A little variation frequently occurs in the size of each of these birds. . . . Its habitat is extended to the greater portion of Europe, from Sweden to France. Wood wren. Sylvia sibilatrix, *Bechst*. . . . Secluded woods and groves are its general places of resort, . . . It inhabits the same countries as the two preceding species.' Pl. 132: 'Yellow willow wren. Sylvia icterina. *Vieill.* . . . In form and colouring this species of Willow Wren so nearly resembles the British members of this interesting group, that it requires an intimate knowledge of its habits, manners, and song to distinguish it from them with any degree of certainty.'

242–1 John Forbes Royle.

245 In the crag we see the process of change of those forms, which have succeeded in becoming habituated to colder climate whilst others died out, or moved towards equator.— «or some species might then have been wanderers.—» There ought to be fewer species in proportion to genera than in present seas, «All» The ‹one› species which survives any change may undergo indefinite change., (marking in their history an eocene miocene & pliocene epoch), whilst others may die out or move South ward. . . . species must be compared

246 to neighboring sea.— For change of species does not measure time but physical changes (we assume like weather on long average tolerably› uniform).— Comparing fossils with whole world. would be like in a Meteorologic table «in comparison of temperature of two countries» finding a very hot day, «in one», oh we will take a day from the equator to add to the mean of the other.—

247 If the the world had cooled by secular refrigeration in chief part instead of change from insular to extreme climate, ‹more northern› Iceland would have possessed a most peculiar Flora «& north of Europe»— As Europæan forms have travelled towards Equator, ‹th› so would the plants from extreme north, which according to all analogy would have been very unlike Southern Europæan ones.— "a variation played on secular refrigeration".—

‹The Phenomena of the S. Hemisphere look as if heat gained›

248 Experimentise on land shells in salt water & lizards do.— Ask Eyton[1] to procure me some

---

Get Hope[2] to give me an account of parasitic animals of beasts varying in different climates

---

Those will not object to my theory, those the philosophers who soar above the pride of the savage, they perceive the superiority of man over animals, without such resorts

249e Mr Waterhouse[1] has most curious facts about the distribution of Lemurs in Madagascar, on neighbouring islets & a sub-genus in Southern Africa

---

In same manner. Cuscus, (a *sub* genus of Phalangista New Holland form) is found in many island Celebes «Waggiou» &c &c. (See Lyell. Vol III p. 30[2])

245 There ought . . . present seas] '?' *added in margin and scored, grey ink.*
246 Comparing . . . the other.—] *scored and two* '?' *in margin.*
249 Mr Waterhouse . . . Africa] *marginal* 'X' *added in pencil.*
Mr Waterhouse . . . Africa] '19' *added brown crayon.*

248–1 Thomas Campbell Eyton.
248–2 Frederick William Hope.
249–1 George Robert Waterhouse.

249–2 Lyell 1837, 3:30, 'Thus the Phalangista vulpina inhabits both Sumatra and New Holland; the P. ursina is found in the island of Celebes; P. chrysorrhos, in the

different species in different isl^d. (as far East as New Ireland. see Coquilles Voyage[3]), Waterhous remark Australia Fauna so far. Indian all the rest. Timor according to Mountain chain ought to be Australian.?—
Mr Gould has been struck with similar extension of form in birds.— |

250e Waterhouse thinks two main divisions of cats. Tortoise shell—& grey-banded. ?species?— thinks offspring of cats sometimes heterogenous.— Australian dog jumped into tub leaving only nose above it— pulled bell.—[1]—
It was most curious to observe, that all the species of mice in S. America. which were hard to distinguish came from closely neighbouring localities.—[2]

Institute 1838. p 38. account of fossils of Sewalick «India» *Monkeys of old World.* Crocodiles. *Anoplotherium.*—[3]

251 ‹M. Jerrod› & Dumeril great work on Reptiles. M. J. says some reptiles same from Maurice & Madagascar & C. of Good Hope.— His book Probably worth studying.—[1]

Wingless birds S. Continents— Ostriches. Dodo. Apteryx Penguin— Logger-headed Duck— Large proportion of Water & small of land or few quadrupeds—

Study Productions. of great Fresh water lakes of North America

252 If Parasite different, whilst man & his domesticated quadrupeds are not so. greater facilities of change in the articulate than ‹M› Vertebrate. But how does this agree with longevity of species in Molluscs!!!

When we talk of higher orders, we should always say, intellectually higher.— But who with the face of the earth covered with the most beautiful savannahs & forests dare to say that intellectuality is only aim in this world

250 Waterhouse . . . pulled bell.— —] *crossed brown crayon.*
It was most . . . localities.—] *marginal* 'X' *added in pencil.*
It was most . . . localities.—] *double scored brown crayon.*
Institute . . . Anoplotherium] *crossed brown crayon.*
251 ‹M. Jerrod›] *deletion in grey ink.*

Moluccas; P. maculata, and P. cavifrons, in Banda and Amboyna.' Passage scored; Darwin added 'Ask Lyell for authority'. Lyell cites Temminck 1827–41 as his authority.
249-3 Lesson and Garnot 1826–30, 1 [1826]:150–60; see pp. 158–59 for *Cuscus* of New Ireland.
250-1 See Darwin's note on Sebright 1836:11. 'Mem. Mr Martens story «of Australian dog» of sitting in a tub, pulling Bell.—' and B165.
250-2 Waterhouse 1837, *Mammalia* describe Darwin's South American mice. Darwin added his specimen numbers to his copy of Waterhouse 1837.
250-3 H. Falconer and Cautley 1838:38, 'Indépendamment de l'intérêt attaché à la première découverte dans l'état fossile d'animaux qui offrent dans leur organisation autant de points de rapprochement avec l'Homme que les Quadrumanes, ce fait est encore plus digne de remarque pour les espèces Sewalik à cause des autres fossiles avec lesquels ces débris sont associés.' These include *Anoplotherium* and crocodiles.
251-1 Duméril and Bibron 1834–54, 3 [1836]:278, 'Parmi ces cinq dernières espèces [Geckotiens, Platydactyles] Africaines, une a pour patrie commune le cap de Bonne-Espérance, Madagascar et Maurice . . .' See also p. 279, 'Le seul Hémidactyle qui soit particulier à l'Afrique est originaire de l'Ile-de-France . . .'

253e–254e [not located]

255e T. Carlyle,[1] saw with his own eyes. new gate. opening towards pig.—latch on other side.— Pigs put legs over, & then with snout lift up latch & back.—

Frogs attempted to be introduced I isle of France p. 170.[2] «Fish introduced[3]» Hump backed race of cows from Madagascar— p 173.[4] Vol I. Voyage à France— Par un Officier du Roi.—

Mackenzie Travel. p. 280. says cattle in Iceland. «"are very» like ‹those of Ice Highlands of› the largest ‹sort› of our highland Sort, except in one respect, that those of Iceland. are seldom seen with horns"[5]—— p. 341. Black Fox sometimes introduced by ice[6]

256e ‹no› «only few» pigs.—[1] birds mentioned. but few.—[2] CD **[There was notice in report of British Association of 1838. (Newcastle) about somebody who had made great collection of birds of Iceland. —M. Gaimard, however, will settle this.—[3]**

Waterhouse says he is certain there are local varieties «of colour & size, but not for[m]» of animals.— He says Stephens say he can at once tell by general colouring a group of Nebria complanata from. Devonshire, from another from Swansea.—[4] Again Waterhouse finds certain varieties of a Harpalus. common at South end, but ‹rare› «absent from» near London. = Dr. Smith, he says, is deeply

257e–260e [not located]

261 of ‹all› genera, «in all classes» are not a few only cosmopolites, & in genera peculiar to any one country do not species generally affect different stations;— this would be strong argument for propagation of species—
Again is there not similarity even in quite distinct countries in same hemisphere. more than in other.

255 T. Carlyle . . . & back.—] *crossed brown crayon.*
Frogs attempted . . . Roi.—] '19' *added brown crayon.*
Mackenzie . . . by ice] *crossed brown crayon.*
256 Devonshire] 'D' *over* 'S'.
Again Waterhouse . . . deeply] *marginal* 'X' *added in pencil.*
page crossed brown crayon.
261 peculiar] 'pe' *over* 'in'.

255–1 Thomas Carlyle.
255–2 Saint-Pierre 1773, Lettre XV, *Animaux apportés à l'Isle de France*, p. 170, 'On a essayé, mais sans succès, d'y transporter des grenouilles . . .'
255–3 Saint-Pierre 1773:169, 'On a fait venir ici jusqu'à des poissons étrangers.'
255–4 Saint-Pierre 1773:172–73, 'Dans les quadrupèdes domestiques, il y a . . . des boefs dont la race vient de Madagascar. Ils portent une grosse loupe sur le cou.' See B190.
255–5 Mackenzie 1811:279–80.
255–6 Mackenzie 1811:341.
256–1 Mackenzie 1811:342.
256–2 Mackenzie 1811:345–48.
256–3 Hancock 1838:613; Gaimard 1838–51.
256–4 James Francis Stephens.

262 Are there any cases, where «domesticated» animals separated. & long interbred ‹p› having great tendency to vary? Is not man thus circumstanced; varieties of dogs in different countries a case in point.—

———

All cases like Irish & English Hare bear upon this.—

263 Why do Van Diemens land people require so many, imported animals?— At what. part of tree of life, can orders like birds & animals separate &c &c Work out Quinary system according to three elements

264 How is Fauna of Van Diemens Land & Australia

265–271 [blank]

272 Falconer's remarks on influence of climates, situations &c[1] on 242. Hook[2]
Smellie Philos of Zoolog.[3] 842
White regular gradat in Man[4] **Poor trash Lyell** 1024
Flemings Philosophy of Zoolog[5]
Royle on Himalaya Plants—[6]

273 Would it not be possible to work through all genera, & see how many confined to certain countries—

———

so on with families.—

———

Ask Royle[1] about Indian Cattle with humps.—

274 ?To be solved if horses sent to India. & long bred in & no new ones introduced would not change be superinduced— why is every one so anxious to cross. animals from different quarters to prevent them taking peculiar character— Indian Bull?—

275 Do species of any genus. as American or Indian genus inhabit different kind of localities.— if so change

———

The GRAND QUESTION Are there races of plants run wild or nearly so, which ‹breed› do not intermix,—any cultivated plants produced by seed.— Lychnis.— Flax.—

263 Why] 'W' *over* 'D'.
272 White . . . 1024] *crossed pencil.*
Poor trash Lyell] *added pencil.*
page crossed grey ink.

272–1 W. Falconer 1781.
272–2 '242. Hook' refers to Hooker; presumably the library of W. J. and, or, J. D. Hooker (if the latter this note must have been written after 1842). See Darwin's Reading Notebook (DAR 119:1). 'Brown at end of Flinders & at the end of Congo voyage (Hooker 923).'
272–3 Smellie 1790–99.
272–4 C. White 1799b.
272–5 Fleming 1822.
272–6 Royle 1835.
273–1 John Forbes Royle.

276 Read Swainson[1]

277 [blank]

278 In production of varieties is it not per saltum.—
Isl[d] bordering continents same type collect cases.— African isl[d].— «How is Juan Fernandez— Humming Birds»

---

types of former dogs. character of Miocene Mammalia—of Europe

---

279 Mem. M[r] Bell's case of *Sub Himalayan* land emys, decidedly an Indian form of Tortoise.—[1]

On other hand. fresh water tortoise from Germany. (where M[r] Murchison fox was found[2]), decidedly next species to some South American kinds.—

---

280 Are the closest allied species always from distant countries, as Decandœlle says,[1] **no he only says sometimes** we might expect disseminated species to vary a little, but such should not be general circumstance.—In.insects «in England» surely it is not— intermediate genera we might expect.—

INSIDE BACK COVER

Lindley Introduct[1]

Dict. Science. Naturelle
Geographie Botanique
De Candoelle.[2] Geol. Soc

Horæ Entomolgicæ[3]
Linn: Soc.

Geoffry-St. Hilaire
Philosophy of Zoology[4]
Waterhouse

BACK COVER

B

276 Read Swainson] *crossed grey ink.*
277 'photo p 228' *added ink (not by Darwin): for photograph in Francis Darwin 1887, 2: facing p. 5.*
IBC *page in pencil, crossed pencil.*

276–1 Swainson 1835.
279–1 T. Bell 1834.
279–2 Murchison 1835.
280–1 Candolle 1820:402, Candolle stresses the opposite point. 'Il ne seroit peut-être pas difficile de trouver deux points dans les Etats-Unis et l'Europe, ou dans l'Amérique et l'Afrique équinoxiale, qui présentent toutes les mêmes circonstances . . . cependant, presque tous, peut-être tous les végétaux seroient différens . . .' Passage scored.
IBC–1 Lindley 1830.
IBC–2 Candolle 1820; at the Geological Society Library.
IBC–3 MacLeay 1819–21; at the Linnean Society Library.
IBC–4 E. Geoffroy Saint-Hilaire 1830; G. R. Waterhouse's copy.

Notebook C

Transcribed and edited by DAVID KOHN

Notebook C (DAR 122, 164 × 99 mm), second in the transmutation series, is bound in maroon leather with a blind-embossed border and a metal clasp. The front cover has faded to a brown. Cream-coloured paper labels with C written in ink are affixed to each cover. The notebook comprises 276 green-edged pages numbered consecutively by Darwin. He made his notes in brown ink with later annotations on 70 pages in grey; the reader may assume boldface signifies grey ink unless another medium is given in the textual notes. Of 55 leaves excised by Darwin, 42 have been located.

On the inside front cover of Notebook C, Darwin noted 'written between («beginning of» February & July 1838)'. Here, as in Notebook B, the grey-ink dating was added between 29 July and 16 October 1838. Darwin's estimated beginning date uses the same words he used for the closing of Notebook B and is equally questionable. Given that B235 was written after 24 February, it is unlikely that Notebook C was opened before mid March. C15 and C36 include references to the 15 February *Institut* (Ogilby 1838a, Ehrenberg 1838a) and the 10 February *Athenaeum* (Blyth 1838a). It should be noted that C13–29 are given over to Darwin's detailed, and no doubt time consuming, reading of Quoy and Gaimard 1830–35 and Lesson and Garnot 1826–30. At C55 Darwin cites Whewell's Presidential Address to the Geological Society, which was read on 16 February and appeared in No. 55 of the Society's proceedings. One may suppose that C55 was not written before April and this supposition is supported by the reference in C67 to Bonaparte 1838a, which the *Publishers' Circular* lists as appearing between its 2 April and 16 April issues. In C95 and C96 we find a cluster of April references to the *Annals of Natural History* (C95: Berkeley 1838, Gunn and Gray 1838; C96: Wiegmann 1838, Hodgson 1838, and Hill 1838). These pages may date from early May since C100 seems to reflect Darwin's visit of 10–13 May to Henslow in Cambridge.[1] On 23 June, Darwin departed London for Edinburgh and the Glen Roy tour.[2] Thus C100 to about C256 were concentrated into this five week, mid-May to mid-June, period. The Journal records that in the beginning of June Darwin again did 'some little Species theory, & lost very much time being unwell.—'[3] C223 refers to Royle's 1 June lecture at the Royal Institution and at C224 he has read Turpin 1838 in the 2 June issue of *Athenaeum*. On the next page he refers to Bonaparte 1838b in the June number of *Annals of Natural History*, while C241 refers to Bongard 1838 in the June number of *Annals of Natural History*. Both C224 and C249 refer to St Jago, which correlates with Darwin's work on the geology of this island (21 May to beginning of June).[4] C257 refers to a copy of Home 1774 at Maer and to Blainville 1838c, which appeared in the 19 July *Institut*. Hence this page was either written out of sequence or is the effective end of Notebook C. Unless Darwin consulted *Institut* at Shrewsbury or Maer, this page was probably written after Darwin's return to London on 1 August and thus overlaps with the beginning of Notebook D, which was opened in Shrewsbury. C258–264 are blank. C265–276 were filled with Darwin's lists of 'Books examined: with ref: to Species' and books 'To be read'. For a

[1] *Correspondence* 2:431.
[2] *Correspondence* 2:432.
[3] *Correspondence* 2:431.
[4] *Correspondence* 2:431.

chronological reconstruction of these reading lists see note C276–1. The conversation with John Bachman recorded in C251–256 may reflect the June meeting of the Zoological Society. Thus, Darwin probably completed the chronological sequence of entries in Notebook C shortly before he embarked for Edinburgh on 23 June 1838.

Notebook C is a remarkable testament to Darwin's determination and versatility as a scientific theorist. By far the longest notebook in the transmutation series, it is also the richest in sustained exploration of the literature as the reading lists at the end of the notebook attest (C276–265). In Notebook C, Darwin carries forward the coherent framework of ideas established in Notebook B. These lead him to empirical laws and clarifying definitions in a variety of fields; the hereditary transmission of form, the distribution of local and wide-ranging species, the distinction between systematic affinity and analogy, and the relation between habit (behaviour) and structure. With respect to heredity, the major themes are reversion, the presence or absence of intermediacy in the offspring of crosses, and the notion that hereditary characters become fixed with time. The latter view underlies Darwin's interest in Yarrell's law, which becomes his leading hereditary generalization. According to Darwin's transformist reworking of the ideas of William Yarrell, older varieties should predominate in crosses with newer ones, likewise natural species should predominate over domesticated varieties. Yarrell's law suits Darwin's transformism because it guarantees the stability of hereditary change. Darwin's exploration of heredity is paralleled by a growing exposure to such writers on the laws of animal breeding as Sebright and Wilkinson (C133–134). For the time being this exposure does not undermine his conclusion that 'picking varieties [is an] unnatural circumstance' (C120). With respect to distribution, we find Darwin intensively studying the results of voyages of exploration, notably the French reports such as Lesson and Garnot 1826–30 (C16–28). This study yields a rich harvest of examples of geographic isolation, and it also leads Darwin to tentative generalizations on whether local species and genera or wide-ranging ones are the principal vehicles of transformist change (C59–60). In the field of systematics, Darwin's careful analysis of the quinarian school of MacLeay and Swainson, a school which Darwin roundly rejects (C170), nevertheless leads him to formulate a transformist version of their distinction between affinity and analogy (C61). Taxonomic affinity reflects descent. Analogy is the taxonomic mark of evolutionary adaptations to similar environmental circumstances. With respect to behavioural adaptations, which become a major concern of Notebook C, Darwin's thinking is shaped by his view that behaviour, including instinct and thought, are hereditary. He formulates the idea that adaptive changes in habit precede changes in structure as a guiding theme (C63). This Lamarckian law is an important extension of the adaptive mechanism developed in Notebook B and becomes the dominant expression of Darwin's search for a transformist explanation in the latter part of Notebook C. The related questions of behaviour and materialism become so important that upon completing Notebook C, Darwin establishes Notebooks M and N as a separate series of 'metaphysical enquiries'.

It should be noted that the four major fields of interest summarized above are often discussed simultaneously. The laws of heredity, distribution, systematics and behaviour are treated as related dimensions of a common problem. And the problem that dominates Notebook C is the relationship between adaptation and heredity, between evolutionary change and evolutionary continuity. In the process, the framework of Notebook B is expanded in range and its explanatory mechanism is transformed in direction.

FRONT COVER

C

INSIDE FRONT COVER **N[a]me of two pigeons,— «with specks» which cross & keep colour on wing**[1]
Effects of colour on parent, white room[2]
How are varieties produced, by picking offspring?
Instances of old Breeds taking greatest effect
Account of the [. . .] the world.—

Charles Darwin

written between («beginning of» February & July 1838)

All good References selected Dec. 13 1856

Also looked through April 23. 1873

Books

About amount of difference: where hybrids produced have any close species ever yet failed. About trades affecting form of man.

Could you get racehorse from Carthorse by picking without change of habits[3]

1 Mr Yarrell «Give it as his theory» tells me. he has no doubt that oldest variety, takes greatest effect on offspring. Thus presuming those varieties to be oldest which have long been known in any country; he states that Esquimaux dog when crossed with pointer produces offspring much nearer Esquimaux than Pointer.— He has no doubt that same thing would happen with Australian dog & any of our common varieties. He has no doubt that

2 Chesnut, for many generations back, were crossed with Bay mare, only bay a few generations, that offspring would be chesnut.— On this principle I may add, that fact of half cross with parents, going back to either parent is lucidly explained.— Mr Yarrell states that if any odd pidgeon crossed with common pidgeon, offspring most like latter, because oldest variety.— —He says of two varieties of

FC C] *upside down.*

IFC **N[a]me . . . world.—]** *added pencil.*
N[a]me . . . wing] *crossed pencil.*
white room] *crossed or underlined pencil.*
Instances . . . world.—] *crossed pencil and part erased.*
Account of [. . .] the world.—] *circa five words illeg.*
written . . . July 1838)] *added grey ink over erased* **'Account of [. . .] the world.—'**.
All good . . . 1856] *added pencil.*
Also . . . 1873] *added brown ink.*
Books . . . man.] *added pencil, crossed pencil.*
Could . . . habits] *added pencil, crossed pencil.*

1 Give] 'Gives' *alternative reading.*
presuming] 'pr' *over* 'Es'.

2 only bay] 'b' *over* 'a'.
— — He] He *over* '—'.

IFC–1 Probably William Yarrell: see C2 and C71.
IFC–2 Probably William Yarrell: see C68–69 and D104.
IFC–3 See E118.

3 pidgeon, although having skulls so different, that they would be called genera., yet retains markings of wings like the wild rock pidgeon.— fact analogous to Owen's Phil: remark of Apteryx having feathers.—[1] It is possible, time being an element in the transmission of form, may explain mule & pig being half way. Yet dogs sometimes like father, sometimes like mother. The fact of

4 great monstrosities being produced, & handed down, with ease, is analogous to what occurs in plants.— All these facts clearly point out two kinds of varieties.— One approaching to nature of Monster, heredetary. other adaptation.— Mr Yarrell says, that after breeding in pidgeons with very much care that, it requires the greatest difficulty to rear them, eggs hatched under other birds & brought up by hand.— These facts all account for

5e–12e [not located]

13e Falkner Patagonia no description of wild animals, nor in Dobrizhoffer Abipones.—[1]

Voyage. de L'Astrolabe Zoologie. p. 60. Vol I. Cynocephalus. niger. comes from the Moluccas «Matchian» & Celebes.—[2] ‖ Amboina; Viverra Zibetha. ‖—[3] All the Moluccas, Waggious New Guinea. New Ireland, have phalangista, which differ in «form & head &» colour from those of New Holland.—[4] The New Holland species are not found in the Archipelago— Former statements to such effects false[5]
In New Guinea. a Kangaroo d'Aroe (Didelphis Brunii) which as yet had only been found in isle of Aroe & Solor),

14e «Vol I» likewise new species of Parameles, which joined to Casoars, perroquets, establishes its «zoolog» alliance with New Holland.[1] The Barbaroussa, (when young very like the Siam race with long nozzle & few hairs) inhabits Celebes & few of the larger islands.——[2] Antelope in Celebes,

4 clearly . . . heredetary.] *scored and '?' in margin, grey ink.*
13 Falkner . . . Abipones.—] *crossed pencil.*
Voyage . . . tail like Woodpecker.— [14]
Cynnocephalus] 'y' *over* 'n'. '19' *added brown crayon.*

3–1 This passage illustrates Yarrell's hereditary law operating as a theoretical justification of systematic affinity based on descent. See R. Owen 1838, prepublication discussion of which may have sparked Owen's 'Phil: remark'. The paper was read to the Zoological Society 10 April, 12 June, and 14 August 1838 and published as R. Owen 1841a. In 1841a:258, Owen referred to the feathers of *Apteryx*, 'It is clothed with a plumage having the characteristic looseness of that of the terrestrial birds deprived of the power of flight.'

13–1 Although the statement is very largely true for Falkner 1774, it does not strictly hold for Dobrizhoffer 1822, which has several chapters of at least general description of wild animals.

13–2 Quoy and Gaimard 1830–35, 1 [1830]:60.

13–3 Quoy and Gaimard 1830–35, 1 [1830]:61.

13–4 Quoy and Gaimard 1830–35, 1 [1830]:61–62.

13–5 Quoy and Gaimard 1830–35, 1 [1830]:62.

14–1 Quoy and Gaimard 1830–35, 1 [1830]:62, 'joined to' is Darwin's rough translation of 'joint aux', that is 'together with'.

14–2 Quoy and Gaimard 1830–35, 1 [1830]:64.

Bourou new species of Axis.—[3] «Cervus moluccensis is different from that of the Marianna islands & at Amboina»[4] I fancy there is marked wild breed of oxen at Java.—[5] .p. 140, calls it Bos. leucoprymnus. does not say whether wild or not[6] p. 156.— Parroket with stiff tail like woodpecker.—[7]

15 Birds of Australia. Many in common ʕspecies? with New Guinea.—[1] Many ‹genera›. kinds common to New Guinea & rest of isle in E. Indi: Arch:

———

In New Zealand. a sturnus of American form— a Synallaxis. ʕAmerican?).[2] p. 159. & 160 «162» list of some birds of Tongatabou. & New Ireland.—[3]
Gould will hereafter know about birds of N. Zealand

———

L'Institut. 1838. A Dipus. & other rongeur in Australia.— p 67 ʕAmerican forms?[4]

———

All Infusoria. not extinct species. good
Resumé do/p. .62[5] **???**

16 Age of Deinotherium. p. 23.. Bull: Soc. Geolog. 1837–8. Tom: IX.—[1]

———

M. D'.Urville on the Distrib of Ferns in South Sea (Indio Polynes: ‹)› vegetation far East) Ann: des Sciences. Semptemb. 1825[2]

———

Get Henslow[3] to read over the pages from about 8 to 20 of Zoologie of Coquille's Voyage to see if Lessons' remarks on the Floras can be trusted[4]

15 of American] 'of' *over* 'in'.
page crossed pencil.
16 1837] '7' *over* '6'.
page crossed pencil.

14–3 Quoy and Gaimard 1830–35, 1 [1830]:64, refers to the island of Bourou.
14–4 Quoy and Gaimard 1830–35, 1 [1830]:133–34.
14–5 Quoy and Gaimard 1830–35, 1 [1830]:64, 'Nous avons pris à Java un Boeuf . . . On suppose que cet animal est le produit de l'accouplement d'un Boeuf sauvage du pays avec un individu provenant d'Europe.'
14–6 Quoy and Gaimard 1830–35, 1 [1830]:140–42.
14–7 Quoy and Gaimard 1830–35, 1 [1830]:156.
15–1 Quoy and Gaimard 1830–35, 1 [1830]:157.
15–2 Quoy and Gaimard 1830–35, 1 [1830]:158.
15–3 Quoy and Gaimard 1830–35, 1 [1830]:159–62.
15–4 Ogilby [February] 1838a:67, reported the discovery by T. L. Mitchell of a *Dipus* ('une vraie Gerboise') in central Australia, a group known previously from Asia and Africa. Darwin questions whether there are also American gerbils.
15–5 Ehrenberg [February] 1838a:62. Darwin means all *fossil* infusoria are not extinct species. Ehrenberg gives a different emphasis, 'Il résulte de l'ensemble des faits, que maintenant on ne peut plus prétendre avec certitude, ni même avec vraisemblance, que toutes les Infusoires fossiles sont des espèces encore vivantes actuellement . . .'
16–1 Bronn 1837–38 [1838].
16–2 Dumont d'Urville 1825.
16–3 Darwin would have viewed Henslow a competent judge given his work on the *Beagle* Keeling plants (Henslow [July] 1838).
16–4 Lesson and Garnot 1826–30, 1 [1826]:12–19. The *Table des Matières* meticulously accounts for the authorship of each section of the work. Thus Darwin would have known that Lesson wrote most of this work, including almost all of chapter 1. 'Considérations générales sur les îles du Grand Océan et sur les variétés de l'espèce humaine qui les habitent.'

17e The changes in species must be very slow, owing to physical change, slow & offspring not picked.— as men do. when making varieties.—
Voyage of Coquille. Zoolog. p 19.. Tapir, «des» couroucous et rupicole vert instances of American forms in East. Ind: Archipelago.[1] ǂRaffles. Horsfield. Diard. Duvaucel. Leschenault Kuhl. Van-Hasselt, Reinwardt «Forrest» authors on E. I«ndian». A«rch.»ǂ[2]
Borneo & Sumatra both seem to have elephant & has orangs, ǂ[3] Tapir common to Sumatra & Malaccaǂ Borneo & Malacca «& Cochin China» are said to have orang-utang & Pongo in commonǂ—[4] Galiopithecus common to Moluccas & Pelew Is^ds^.—[5] p. 22. New Calidonia— New Ireland & Britain[6] same kind of dog, with those of New S. Wales. ‹V.› p. 123[7]

18e Crocodile at New Guinea. All the isles of Oceania have the Scincus with golden streaks— the lacerta vittata extends ‹to› from Amboina to New Ireland[1] p. 23

Voyage of Coquille Lesson

No (p. 24) batrachian in isles of Great ocean says in conformity with Bory's Views.—[2] ‹Says› D'Orbigny is said to have brought a tortoise & toad from S. America & identical with those from S. Africa.[3] «M. Bibron doubts fact.—»[4] **My toad is same species**

17 The changes . . . varieties.—] *circled.*
Archipelago . . .Galiopithecus common] *marginal brace with* 'p 21' *inserted in margin*; 'p 2' of 'p 21' on stub.
New Ireland & Britain . . . ‹V.› p. 123] 'New Ireland ‹V.› p. 123' *circled.*
page crossed pencil.

18 Crocodile . . . New Ireland] *crossed pencil*
have . . .Ireland] *marginal brace with 'p 23' inserted in margin.*
Voyage of Coquille Lesson] *added pencil.*
No (p. 24). . . fact.—] '19' *added brown crayon.*
Bibron] 'i' *over* 'ri'.
My toad is same species] *added grey ink.*

17–1 Lesson and Garnot 1826–30, 1 [1826]:19.
17–2 The authors listed are cited in Lesson and Garnot 1826–30, 1[1826]:19. See Raffles 1830, Horsfield 1824, Blume 1827–28, Reinwardt 1824, and Forrest 1780.
17–3 Lesson and Garnot 1826–30, 1 [1826]:20, 'seem to have' is Darwin's translation of 'paraissent renfermer'.
17–4 Lesson and Garnot 1826–30, 1 [1826]:21, 'are said to' corresponds to Lesson's qualification 'à ce qu'on assure'.
17–5 Lesson and Garnot 1826–30, 1 [1826]:22.
17–6 New Britain.
17–7 Lesson and Garnot 1826–30, 1 [1826]:22. Darwin circled 'New Ireland ‹V› p. 123', referring to 1[1826]:123, which also mentions the New Ireland dog.
18–1 Lesson and Garnot 1826–30, 1 [1826]:23.
18–2 Lesson and Garnot 1826–30, 1 [1826]:24. Bory de Saint-Vincent 1804.
18–3 Thomas Bell: personal communication. Bell described Darwin's reptiles and amphibia in *Reptiles*. Bell says of *Pyxicephalus Americanus* Bibron, 'This curious species has, I believe, only once before been found. A single specimen exists in the French Museum, which was brought from Buenos Ayres by Mons. d'Orbigny, and which formed the subject of Mons. Bibron's description . . . Of the three species of this remarkable genus at present known, two are inhabitants of Africa, from whence they were brought by Delalande.' (*Reptiles*:40–41). Darwin scored 'Of the three . . . Delalande'.
18–4 Gabriel Bibron: probably personal communication through Thomas Bell. Bibron visited London and was in close professional contact with Bell. Duméril and Bibron 1834–1854 describes a tortoise, *Emys Dorbigni* (2 [1835]:272–75) and the toad *Pyxicephalus americanus* (8 [1841]:446–47) as both sent by M. D'Orbigny from Monte Video.

19 **Coquille Voyage**

p. 25 Mais il n'y a pas jusqu'aux îles Macquarie et Campbell (52° S) qui n'aient egalement leur espèces; et certainment on eût été bien éloigné, il y a peu d'années, d'admettre que ces oiseaux eussent leurs représentants dans de si hautes latitudes".,— ‹translate?)[1]
All Australian forms have representatives (& instances given) in East Ind. Arch:—[2] Birds of New Zealand absolutely different.— —*Philedon circinnatus* not found in Australia only New Zealand—[3] Norfolk. Isd. & New Caledonia

20 peculiar species of cassicans: (‹cassicans Australian form?[1] p. 27. many fish of Taiti found at ‹New› Isle of France:[2]
xx instance of wide range, where means of wide range[3] work this out—
L. Jenyns, about my fish[4]
New Zealand & New Holland fish very similar.—[5]
NB. Lesson method of generalizing without tables or references highly unphilosophical[6]
xx Says same remark with regard to shells.—[7] But he says shells towards extremities of the continents peculiar to the different points.—[8]

21 Consult Voyage aux terres Australes Chap XXXIX tom IV p. 273 2d Edit[1]

---

19 **Coquille Voyage**] *added pencil.*
20 xx instance . . . my fish] *circled.*
NB. Lesson . . . unphilosophical] *brace in both margins.*
21 Consult Voyage . . . Plants.—»] *crossed grey ink.*
taken] 'en' *over* 'ing'.
Coati roux common] *crossed.*
very stiff] 'v' *over* 'w'.

19–1 Lesson and Garnot 1826–30, 1 [1826]:25.
19–2 Lesson and Garnot 1826–30, 1 [1826]:25–26.
19–3 Lesson and Garnot 1826–30, 1 [1826]:26.
20–1 Lesson and Garnot 1826–30, 1 [1826]:26. Lesson does not mention an Australian cassicans.
20–2 Lesson and Garnot 1826–30, 1 [1826]:27.
20–3 Lesson and Garnot 1826–30, 1 [1826]:27–28. Darwin is responding to Lesson's comments on distribution of fish: '. . . l'ensemble de l'icthyologie du Grand-Océan, des mers d'Asie et des Indes, se compose presque entièrement d'espèces analogues. C'est ainsi que nous avons retrouvé à l'île de France un grand nombre des poissons de Taiti, et que nous avons pu très-souvent les suivre d'archipel en archipel. On doit donc conclure que les espèces sont identiques, depuis les Marquises jusqu'à Madagascar . . . Entre les tropiques, les récifs de coraux, . . . sont habités par des poissons . . . dont l'éclat est vraiment fantastique . . . Mais plus on s'engage dans les canaux étroits et sans cesse réchauffés par le soleil équatorial, qui séparent en tout sens les îles innombrables de la Polynésie, plus le nombre des poissons augmente; et là seulement on observe certains genres ou certaines espèces qui n'existent sur aucun autre point.'
20–4 Darwin's reminder to discuss the subject with Leonard Jenyns, who prepared *Fish.*
20–5 Lesson and Garnot 1826–30, 1 [1826]:27.
20–6 Lesson and Garnot 1826–30. Lesson provides abundant distribution information, which Darwin was eager to extract in order to test generalizations such as 'wide range, where means of range'. However, Lesson's descriptive 'esquisse à grands traits' stymied and frustrated Darwin.
20–7 Lesson and Garnot 1826–30, 1 [1826]:28, 'La partie intertropicale de l'Océanie est très-pauvre en testacés. Plus on se rapproche des îles de la Polynésie, plus le nombre des espèces s'accroît d'une manière rapide.'
20–8 Lesson and Garnot 1826–30, 1[1826]:29.
21–1 Péron 1824, cited in Lesson and Garnot 1826–30, 1 [1826]:29.

Consult Latreille. Geographie des insectes, in 8°. p. 181 «who says insects Indian, like. Plants.—»[2]

It would be very important to show wide range of fish & shells in tropical sea, it. would demonstrate.; not distance, makes species but barrier.— —it would make strong contrast with southern regions.— «it would now represent. what ‹actually is› «has» taken place with quadrupeds»

---

p. 118. wild pigs of Falklands, generally "red of brick" hair, very stiff,[3]

---

p. 120— Coati roux common. near Concepcion. some *tatous*!!![4]

---

p. 120.— Most of the dogs of Payta— belong to the hairless kind, «said» to come originally from Africa[5]

22 p. 122. Mus decumanus, at Caroline Is^ld^, & a Roussette[1]
p. 136. Isle of France.— the Tenrecs from Madagascar. Monkey from Java.— Hairs, & deer.— Procured two makis alive from there.—[2] Mem Waterhouse knows of some species which escaped there.—

---

p. 139. Vespertilio bonar«i»ensis (from Buenos Ayres) replaces ‹Vesp.› holds same relation with equator—that Vesp. lasiurus does in North. Hemisphere.—[3]

23e p. 158 Cuscus albus. New Ireland[1]
——— maculatus— Waigiou[2]

Speaking of Lepus Magellanicus says; ‹after› "après un examen attentif, et forts surtout de l'opinion du baron Cuvier, nous ne balançons pas a la regarder comme une espèce distincte![3]

---

22 replaces] *crossed grey ink.*
23 Speaking . . . distincte!] *crossed pencil.*
wild pigs . . . teats] *double scored brown crayon.*
Coquille Voyage] *added pencil.*
'18' *added to page brown crayon.*

21–2 Latreille 1819, cited in Lesson and Garnot 1826–30, 1 [1826]:30.
21–3 Lesson and Garnot 1826–30, 1 [1826]:118.
21–4 Lesson and Garnot 1826–30, 1 [1826]:120.
21–5 Lesson and Garnot 1826–30, 1 [1826]:120.
22–1 Lesson and Garnot 1826–30, 1 [1826]:122.
22–2 Lesson and Garnot 1826–30, 1 [1826]:136.
22–3 Lesson and Garnot 1826–30, 1 [1826]:139. That is, they live at an equal distance from the equator in the two hemispheres.
23–1 Lesson and Garnot 1826–30, 1 [1826]:158.
23–2 Lesson and Garnot 1826–30, 1 [1826]:152.
23–3 Lesson and Garnot 1826–30, 1 [1826]:169. Contrary to Lesson, Darwin rejected *Lepus Magellanicus* as a distinct species. He held this opinion consistently, from the time of the *Beagle's* visit to East Falkland Island in 1834 through the publication of *Zoology*. See *'Beagle' Zoological Diary* (DAR 31.1: MS p. 236). 'The black rabbit

p.171. Sus papuensis «partly domesticated[4]» like in general appearance to Siamese kind.— but considered good species from dental characters, wild pigs.[5] said by Forrest to swim from one isl^d to another—╫[6]

«It is a good species, with different number of teats»[7]

Coquille Voyage

24e Durville has written Flora of Falkland Isl^ds,[1] where is it?

All the Society isles have the same productions p 293— is very strong about this Lesson insists much.—[2]

The (p. 296) Columba Kurukuru found in all Malasia— & oceania, offers many varieties in each place to puzzle naturalists.—[3]

---

p. 372. Bourous. the Barbyrousa; a Cervus near Marianus new,[4] & some rats & mice.[5] In Amboina only Cuscus & Barbyroussa[6]

25e «NB»
[isl^ds. Springing up more likely to ‹M› have different species than those sinking, because arrival of any one plant might make conditions in any one isl^d different]^CD.—[1]

25 isl^ds. . . account for this.—] *crossed pencil.*
p. 414 . . . bark] *circled.*
Coquille Voyage] *added pencil.*
Says no . . . *Mauritius*] *double scored brown crayon.*
in Mauritius] *underlined pencil.*
'19' *added to page brown crayon.*

of these islands has been described by M. ‹Lesson› «Rang» as a distinct species, the Lepus Magellanicus. — I cannot think so: my reasons are. — The Gauchos, who are most excellent practical Naturalists, say they are not different: & that they ‹breed› & the grey breed together: that the black are *never found* in *distinct situations* from others: they have seen piebald ones: there other varieties such as white &c but not common (. . .)' [material in parentheses quoted in note C29–1]. Darwin is referring to Lesson 1827, a small manual, similar in format and from the same series as Rang 1829, with which work Darwin confuses it. See note C29–1 for further discussion, and versions in *JR*:248–250, and *Mammalia*:92; also B31 and C29 'Rabbits . . . 70 years'.

23–4 Lesson and Garnot 1826–30, 1 [1826]:174.

23–5 Lesson and Garnot 1826–30, 1 [1826]:171.

23–6 Forrest 1780:97, cited in Lesson and Garnot 1826–30, 1 [1826]:174.

23–7 Lesson and Garnot 1826–30, 1 [1826]:175.

24–1 Dumont d'Urville 1826, cited in Lesson and Garnot 1826–30, 1 [1826]:119.

24–2 Lesson and Garnot 1826–30, 1 [1826]:293, 'Mais nous devons dire cependant que toutes les îles de la Société, séparées les unes des autres par de courtes distances, habitées par la même famille humaine, soumises aux mêmes influences, ont, d'une manière exclusive, les mêmes productions.'

24–3 Lesson and Garnot 1826–30, 1 [1826]:296–97, '*columba kurkuru* . . . offre partout des nuances variées qui ont déjà cent fois torturé les naturalistes systématiques, aux définitions précises desquels elle semble vouloir échapper.'

24–4 That is, a new *Cervus* allied to *Cervus Marianus* (Lesson and Garnot 1826–30, 1 [1826]:372).

24–5 Lesson and Garnot 1826–30, 1 [1826]:371–72.

24–6 Lesson and Garnot 1826–30, 1 [1826]:377. However, Lesson says 'Le babi-russa n'y vit point'.

25–1 Note the rudimentary connection of this passage to the principle of divergence, which Darwin sketched in Notebook E shortly after formulating natural selection, but only fully developed from 1854–58 (DAR 205). See *Natural Selection* (chap. 6) and *Origin* (chap. 4) for the matured expression of these ideas.

p. 414. dogs of New Zealand of large size, resemble, chien-loup.—long, black & white, ears short & straight— do not bark[2]
p. 433. birds & bats have certainly travelled from East Indies, isl^{d}, as far as Oualan.—[3] Wide space of sea, to East of America. would account for this.—

Coquille Voyage

Says no reptiles. p 460[4] & very doubtful whether any birds Except. Dodo!!—[5]
in Mauritius

26e Lesson &c p. 620. Centropus (Coucal) of Java & Phillippines, has variety at Madagascar, Calcutta & Sumatra,.[1] but I do not see how it is known that they are varieties & not species.—[2]

Vol :694. King-fisher of Europe, (Alcedo ispida) from Mollucca s. scarcely differs at all from those of Europe, but beak rather sharper., & rather longer in proportion, colour slightly different.[3] Who can say whether species & varieties[4]

---

p. 708. Columba Oceanica (Less) inhabits Caroline

27e ‹«NB. The»› isl^{d}. (.perhaps Phillippines & perhaps, Friendly Isles «& Hebrides») is very closely allied to C. muscadivora., which lives in the Eastern Moluccas, New Guinea.—[1](Case of replacement)—

Coquille Voyage

The caswary, inhabits Ceram, Bourou & especially New Guinea[2] (replaces, Emeu) in North of New Holland.—

New Guinea scarcely differs more from, ‹Van Diemen's land.› Australia more than Van Diemen's land.—

Vol II p. 8 no snakes on isles of central Pacific, yet there appears to be one at Botouma[3] from account of natives, & probably on Oualan.—[4]«Mitchill says

26 *page crossed pencil.*
27 isl^{d} . . . Van Diemen's land.—] *crossed pencil.*
Coquille Voyage] *added pencil.*
'19' *added to page brown crayon.*

25–2 Lesson and Garnot 1826–30, 1 [1826]:414–15.
25–3 Lesson and Garnot 1826–30, 1 [1826]:433.
25–4 Lesson and Garnot 1826–30, 1 [1826]:460.
25–5 Lesson and Garnot 1826–30, 1 [1826]:459.
26–1 Lesson and Garnot 1826–30, 1 [1826]:619–20.
26–2 Lesson's distinguishing characters are, '*Variéteé de Sumatra*. De moitié plus petite; le corps en-dessus brun-noir terne; ailes d'un roux sale. *Var. de Madagascar*. Même teinte dans la livrée, mais taille encore plus petite. *Var. de Calcutta*. Parties inférieures brun-sale.' (Lesson and Garnot 1826–30, 1 [1826]:620).
26–3 Lesson and Garnot 1826–30, 1 [1826]:694.
26–4 As with the Centropus of Madagascar, Calcutta, and Sumatra, Lesson considers the Mollucan kingfisher a variety.
27–1 Lesson and Garnot 1826–30, 1 [1826]:708–9.
27–2 Lesson and Garnot 1826–30, 1 [1826]:718.
27–3 Rotouma.
27–4 Lesson and Garnot 1826–30, 2 [1830]:8–9.

snakes on Friendly isles. p 50. LX. Journal of Silliman[5]» «Study Silliman.—»

28e Vol II. p. 10. it seems that Crocodile was washed on shore at one of the Pelew Isl^ds.— killed a woman.[1] Chamisso p. 189 Tome III: Kotzebue.—[2] p 22. a Gecko on St Helena.—[3] ‹in 1813, a venemous snake was[4]› one Gecko on Isle of France[5] Scincus multilineatus (p 45) Moluccas & New S. Wales[6]

Scincus Cyanurus «p 8 &». p 49 on all the Moluccas «New Guinea, New Ireland» & «even» Java. & very common on Otaheite— according «stated in note to p 21» to Quoy & Gaimard in Sandwich isl^d. & according to Chamisso in Radack isl^d.—[7]

---

p. 69. Sharks very generally distributed:[8] Mem of great geological age[9]— Gastrobranchus «only» 2 species one in Northern Hemisphere 2^d in southern[10] —p. 71 Chimera— Antarctica «also Tæniatole austral» «caught» Chile, Van Diemen's land & Cape of Good Hope[11] **V. p. 44 of this Note Book**

29 Rabbits introduced in 64, of very many colours, like the cattle, ‹"›which I say "are as variously coloured as a herd in England"— Black & Grey varieties of rabbits thus handed down for nearly 70 year.[1]

---

28 Vol II . . . St Helena.—] *scored brown crayon.*
«stated . . . p 21»] *carated to two locations: after* 'according to Quoy' *and after* 'according to Chamisso'.
p. 69 . . . austral] *crossed pencil.*
v. . . . Note Book] *added brown ink.*
29 Rabbits . . . 70 year.] *crossed pencil.*
Galapagos . . . ages.—] *crossed pencil.*

27–5 S. L. Mitchill 1826:50, 'I offer . . . [a] description of a two-headed serpent, I received from one of the Fejee Islands, a few years ago.'
28–1 Lesson and Garnot 1826–30, 2 [1830]:10.
28–2 Lesson and Garnot 1826–30, 2 [1830]:10, cites Chamisso 1821 (Kotzebue 1821, 3:189), 'M. de Chamisso dit que le carolin *Kadu* crut le ['Une grande espèce de lézard'] reconnaître dans la figure du *lacerta monitor*.'
28–3 Lesson and Garnot 1826–30, 2 [1830]:22.
28–4 Lesson and Garnot 1826–30, 2[1830]:22.
28–5 Lesson and Garnot 1826–30, 2[1830]:22.
28–6 Lesson and Garnot 1826–30, 2 [1830]:45–46.
28–7 From Lesson's mentions of the widely distributed skink *Scincus cyanurus*, Darwin assembled a tabulation, whose absence from Lesson he lamented in C20. Lesson and Garnot 1826–30, 2 [1830]:8, 21–22, 49, citing Quoy and Gaimard 1830–35, and Chamisso 1821, 3:157.
28–8 Lesson and Garnot 1826–30, 2 [1830]:69. On the range of sharks see also the back cover of Red Notebook.
28–9 On the great age of sharks and the difficulty posed by their apparent immutability, see notes B170–2 and B205–1.
28–10 Lesson and Garnot 1826–30, 2 [1830]:69.
28–11 Lesson and Garnot 1826–30, 2 [1830]:71.
29–1 'Rabbits . . . cattle' is based on Lesson and Garnot 1826–30, 1 [1826]:168–169, the black rabbit was brought to the Falklands in 1764, along with horses and cattle, by French settlers. Darwin's 'are as variously . . . England' quotes the *'Beagle' Zoological Diary* (DAR 31.1:MS p. 236). See B31, where Darwin says that 'Falkland rabbit . . . instance of domesticated animals having effected a change', a position Lesson (Lesson and Garnot 1826–30,) entertained and rejected. In C23 Darwin quotes Lesson's rejection, without comment, then carries on abstracting Lesson. In C29 he comes back to Lesson obviously with his *Zoological Diary* in hand, and his note 'Black & Gray varieties of rabbits there handed down for nearly 70 years' (calculated from 1764, Lesson's date for introduction of the rabbit, to the *Beagle's* visit in 1834) answers a question raised in the *Zoological Diary* '(it would be curious to see how long varieties have remained, if the time of introduction was certain; the same idea applies to the cattle & horses which are of as varying colour as a herd in England)'.

Galapagos Mouse not the same section, with house mice.[2] It is wonderful how it could have been transported. ‽What section does the New Zealand Rat belong to

---

There is this great advantage in studying. Geograph. range of quadrupeds.: that either created in each point, or migrated from those quarter, where we know quadrupeds have existed for ages.—

30 ∴ The most hypoth: part of my theory, that «two» varieties of many ages standing, will not readily breed together: The argument must thus be taken, as «in» wild state (where instinct not interfered with, or generative organs affected as with plants) no animals VERY different will breed together, so when we grant «(which can be shown probable,)» varieties may be made in wild state, there will be presumption that they would not. breed together.— We see even in domesticated varieties a tendency to go back to oldest race, which evidently is tending to same end, as the law of hybridity, namely the

31e–32e [not located]

33 animals unite, all the change that has been accumulated cannot be transmitted;.— hence the tendency to revert to parent forms, & greater fertility of hybrid & parent stock, than between two hybrids.—

As we see external influences first affect external [for]m, so will the internal parts be of longest [consta]nt & therefore most permanent

---

Owen [re]markable laws of Brain & manner of generation «& primary divisions of insects»[1]

---

2. Relation, of external conditions, & of succession: the ‹first› latter is most intimately connected with important structures. ‹which are less obviously affected by external circumstances› these therefore will be chiefly heredetary.—

34 If varieties «produced by slow causes, without picking» become more & more impressed in blood with time, then generation will «only» produce an

30 «two» varieties] 'ies' over 'y'.
33 [for]m] *tear.*
[re]markable] *tear.*
34 of[fspri]ng] *tear.*

29–2 The 'sections' are the subgenera of *Mus* proposed in Waterhouse 1837, which, however did not describe *Mus galapagoensis*. Darwin's note reflects a communication from Waterhouse on the progress of *Mammalia*, in which Waterhouse, having completed an extensive study of Rodentia (Waterhouse 1839b), no longer used his 1837 subgenera. Instead he proposed a separate genus for American mice *Hesperomys*. (*Mammalia*:75).

33–1 R. Owen 1835a:5, 'Organs of generation. Changes effected in the nervous and other systems during the metamorphoses of insects.'

offspring capable of producing ‹two› such as itself.— therefore two different varieties will produce hybrids

---

but not varieties. which are not deeply impressed on blood., will cross & produce fertile offspring

---

in the first case it will either produce no of[fspri]ng or such as not capable of producing again

35 The Varieties of Cardoon are cases ‹sp› like those of Primrose & Cowslip run wild,

---

The two species of Clenomys. case of replacing species. Dr. Smith[1] will give me some capital information

---

?Carnivora of New & Old word. do not form two sections is this not connected with wide range of animals. Follow this out, where species of same *genera* in two words. have not species, generally wide range? Mice.—[2]

---

36 Waterhouse's remarkable fact of no forms peculiar to *word* **to special districts????** north of 30°.—[1], may be connected with, Mr Blyth's statement of birds of Europe & America, which are of different forms being migratory;[2] also with Temminks fact of forms being within Tropics.—[3] Europæan birds at Japan. connected with Europæan forms on Himalaya??— This is very remarkable, when we consider number of quadrupeds in Eocene period. Have the Edentata & Marsupial forms been chiefly preserved,— where shut up by themselves without other animals? but they were not shut up!!

36 *word*] *underlined pencil.*
to special districts ????] *added pencil.*

35–1 Andrew Smith.
35–2 Waterhouse 1839a:75, 'These considerations have induced me to separate the South American mice from those of the Old World.'
36–1 The remark remains obscure. If it refers to mice, then note C35–2 is relevant and Darwin may mean no *mice* peculiar to *New* World north of 30° S latitude: 'The South American *Muridae*, which form the chief part of Mr. Darwin's collection, were none of them procured further north than latitude 30°, with the exception of those from the Galopagos Archipelago.' (*Mammalia*:77). If the remark refers to Rodentia, then Waterhouse 1839c:173 is relevant: 'The first thing that strikes the attention . . . is, that the great mass of South American Rodents belong to a different section from those of the northern portions of the globe . . .' and Waterhouse's 'tabular view of the distribution of the *Rodentia*' (*ibid.*:172) suggests that Darwin may mean no *rodents* peculiar to New World north of 30° N latitude. Also see C116.
36–2 Blyth 1838a:107, 'Mr. Blyth . . . took a rapid survey of the modern theories of zoological provinces . . . and called particular attention to the following fact, which he was not aware had been previously announced — viz. that those North American birds which have no generic representative in Europe, and those European genera which have no species proper to America, are, almost without exception, migratory, belonging to types of forms characteristic of those regions where they pass the winter.
36–3 Temminck 1813–15.

37 Extreme southern points of S. Hemisphere fully characterized, of each continent. Try amongst Europæan quadrupeds if Africa destroyed would not then some forms be peculiar to it, so on, & so on.—
Whatever destroyed great ‹quadrupeds›; Pachyderm in S. America destroyed great Edentata or American form.—
Is the Australian Dipus an American form?
The climate having grown more extreme both in, N & S. America, is only common cause I can conceive of destruction of Great Animals in Europe & America

38 Some portion of the world «Africa» being left more equable. **yet America preeminently equable.** might have allowed fresh species to have been formed & spread to other Africa & East India Arch.— but where these great animals had not spread then such tribes as Marsupial & Edentata increased most. Certainly Africa approaches Nearest to what is supposed to have been condition of former whole world. America might have been string of islands.—

39e ?Europe has many species but not genera distinct from rest of world???

Lyells Principles, must be abstracted & answered

Much might be argued what is *not* cause of destruction of large quadrupeds.— **common to two types of animals**

What reptiles coexisted with Palæotherium in Paris quarries & at Binstead. Mem. recent Crocodiles with Palæotherium in India—: connection with Latitudes!?

40e Zoological Journal.——
Vol I. p. 81. Capromys, West Indian isl^{d}.[1]
p. 120. «ref.» Philosop. Transacts 1823. Read June 5th) important paper by Dillwyn, on replacement of Cephalopods & Trachilidous Molluscs. by each other in secondary & Tertiary periods.—[2]
p. 125. ref. to Phil Transacts. (read November 20th) Paper by Jenner, on

39 **common . . . animals**] *added brown ink and circled.*
page crossed pencil.
40 Zoological . . . periods.—] *crossed pencil.*
full] ‘u’ *over* ‘i’.
p. 125 . . . nest] ‘18’ *added brown crayon.*

40–1 Desmarest 1825.
40–2 Dillwyn 1824, which is a review of Dillwyn 1823, see p. 397; Darwin scored ‘From the occurrence in such great numbers of the carnivorous Trachelipodes in the formation above the chalk, it therefore appears, that the vast and sudden decrease of one predaceous tribe has been provided for by the new creation of many genera, and a myriad of species possessed of similar appetencies, and yet formed for obtaining their prey by habits entirely different from those of the Cephalopodes.’

birds seen far at sea, migrations of species, geese killed in Newfoundland, with crops full of maize.[3] (get limits of latter from ‹Tarton› Barton.—[4] swifts return after years to nest[5]

Vol II. p. 49.[6] on the localities of certain parrots habitations India & Africa.— NB. Any monograph like Gould on Trogons worth studying.—[7]

41e «*do*» **Zoolog Journal Vol 2.**
p 221. Horsfield on two bears *very* close species, inhabiting Borneo & Sumatra. differ only in form of white mark on breast: p. 234.— good case.—[1]
p. 526. (ref) To Temminck Monograph. Mammal; «4to» good facts about distribution of Cats[2]
Vol III. ‹p.› p 233, stated that the "Asseel Gazāl. (Bos Gazœus) does not mix with the Gobbah or village Gazāl.—[3] ʕis latter same species domesticated, strangely contradictory to Azaras fact of conduct of wild & tame horses.—[4]
p. 246— Gmnura—[5] new genus of Mam: found in Sumatra
p. 452 Append to Denham Clapperton &c[6] on Mammalia no doubt will all be included in Smiths work[7]

42e «*do*»
Vol. IV p 273. Macleay on Capromys.[1] 4 species probably in Cuba (p 271 Viedo says American dogs silent.[2] Mem contrary assertion of Molina[3])

41 **Zoolog . . . Vol. 2.**] *added pencil.*
p 221 . . . good case.—] '19' *added brown crayon.*
p. 526 . . . Smith's work] *crossed pencil.*
42 *page crossed pencil.*

40–3 Jenner 1824:12-13. Darwin marginalia: p. 13 scored, 'no corn of this kind is cultivated within a vast distance of that island' underlined. Jenner 1824 was reported in Jenner 1825.
40–4 Barton 1827 does not give a northern limit for maize; however, this essay is much concerned with the geographic limits of plants. See Barton 1827:86, 'Although, as I have mentioned, our European corn does not succeed within the tropics, several other kinds of grain are plentifully cultivated there. One of these is *Maize*, or Indian Corn, which I have already referred to as raised in the south of Europe.'
40–5 Jenner 1824:16, 'At a farm-house in this neighbourhood I procured several swifts, and by taking off two claws from the foot of twelve, I fixed upon them an indelible mark. The year following . . . I found several of the marked birds . . . at the expiration of seven years . . . a bird . . . proved to be one of those marked for the experiment.' Passage scored.
40–6 Vigors 1826a:49–56.
40–7 Gould 1838a.
41–1 Horsfield 1826:234, distinguishes between *Helarctos Malayanus* from Sumatra and *H. euryspilus* from Borneo, which differ in two characters (p. 234).
41–2 Temminck 1826.
41–3 Hardwicke 1828:232, 'The [*Bos Gayaeus*] is considered an untameable animal, extremely fierce, and not to be taken alive. It rarely quits the mountainous tract of the S. E. frontier, and never mixes with the *Gobbah*, or *Village Gayal* of the Plains.'
41–4 Azara 1801, 2:298, 'Les Chevaux sauvages . . . qu'on voit quelquefois de ces troupes, composées de dix mille individus . . . apercoivent à peine des Chevaux domestiques, qu'ils courent au galop, et, passant au milieu d'eux ou près d'eux, ils les appellent, les caressent avec des hennissemens graves et prolongés qui expriment l'affection; ils parviennent ainsi à les séduire;' Passage scored, 'de dix mille individus' underlined.
41–5 Gymnura.
41–6 Denham, Clapperton, and Oudney 1828.
41–7 Andrew Smith 1838–49.
42–1 MacLeay 1829a:273.
42–2 Fernandez de Oviedo y Valdes 1547, MacLeay 1829a:271.

(p. 277.). probably another in Jamaica & perhaps one extant at Leeward Isles.[4]
p. 388 Reference to Rüppel. travels (what language?)[5] Hyena «venatica» ‹of› Cape found in Desert of Korto & Steppes of Kordofan
p. 401. Admirable letter from Macleay to Bicheno much excellent detail & fine, views about Species— MUST BE STUDIED: genera founded in nature[6]

43e–44e [not located]

45 The systematic naturalists get clear indication of circumstances in Geography to help in distinguishing empirically what is species.— The Collector is directed to study localities of isl^{d}.—«immense importance of local faunas foundation of all our knowledge especially great continents»
Give Specimen of arrangement,

Rhinoceros
3 ‹5› Species Cape town good species
Indian species so distinct that all analogy
from each other I do not know how different { Sumatra / Java ———————— do from Indian

increase of knowledge would probably tell more certainly

———

Get Closer species— FOX IS & Mice of America «good case on account of varieties in N America» «some doubt, from want of knowledge of time— Analogy from three first will give one *almost* certain guide ∵ because time required too separate isld very long»

45 *page crossed pencil.*

42–3 Molina 1788–95, 1:302, 'Es verdad que estos perros ladran como los originarios de Europa: mas no por esto deben ser reputados por extrangeros, mediante a que la opinion de ser mudos los perros americanos . . .' Passsage scored, 'ladran' [bark] underlined. Darwin commented in the margin: 'Do the early voyages say anything about dogs in T. del fuego'. His copy is inscribed 'Charles Darwin Valparaiso 1834'.
42–4 MacLeay 1829a:277.
42–5 Rüppell 1829:388.
42–6 MacLeay 1829b is highly critical of Bicheno 1827, whom he accuses of currying favour with the 'Linneans'. The 'fine, views about species' refers to pp. 404–406, e.g. 'when . . . species are ascertained "*to run into one another*", [Zoologists] are accustomed to doubt the fact of their being distinct species; we call them *varieties*, and search for some general characteristic which will include and insulate the whole of these varieties, and then call that the specific character' (p. 405); and 'By the bye, on the subject of species you [Bicheno] settle the question, by deciding that "in cases of difficulty the assumed law ought to be brought to the test of experiment, or the species should be rejected." Now . . . the only "*assumed law*" I can perceive mentioned is as follows: "A species shall be that distinct form originally so created, and producing, by certain laws of generation, others like itself," and unfortunately you have forgotten to inform us how we are to ascertain by experiment, "a distinct form originally so created." This, however, is clearly the essential characteristic laid down in law, since a Negro "produces by certain laws of generation others like himself," and yet is not very generally accounted to be a distinct species. But I ought to recollect, that in spite of Mr. Wilberforce, you have your doubts on this particular point; that in fact it still remains with you "the most difficult problem of all." ' (pp. 405–406). MacLeay refers here to William Wilberforce's abolitionism.

46 America & India deer.= Africa not.— Africa Camels?? Africa *Bears*??— *Plantigrade Carnivora*??— «compare rodents of two countries» & «Monkeys.»

Fact of Elephant same species in Borneo. Sumatra. India Ceylon— perhaps shows great persistency of character. Hence Elephas primigenious over so wide a range, & Mastodon angustidens.—

Ogleby has facts to show that Australian dog introduced by savages into Australia.— What are they?[1]

Colonel Montagu probably contains some facts about close species of Birds.[2]

47e Zoolog. Transact. Vol I p. 165.— ‹a› "an account of the MANELESS lion of Guzerat by Capt. W. Shee. considered merely variety.— red[1] form of skull very slightly different.—[2]

Q[3]

---

46 *page crossed pencil.*
47 Zoolog. Transact . . . different.—] '3' *added brown crayon.*
Q] *added pencil and circled.*
Zoolog. T . . . structure??—] *crossed pencil.*

46–1 Possibly based on the report in the February 1838 *Institut* (Ogilby 1838a:67, cited in C15), '. . . le Chien indigène que M. Ogilby regarde comme une importation . . . opinion qu'il cherche à démontrer au moyen de divers arguments.' For the facts Darwin sought see Ogilby 1841:121–22, 'I think, that there are strong grounds for believing that the *Dingo*, or native dog, the . . . solitary exception . . . [to] this position [that all non-marsupial Australian mammals are rodents], is not an aboriginal inhabitant of the continent, but a subsequent importation, in all probability contemporary with the primitive settlement of the natives. Many circumstances might be advanced in support of this opinion; the simple fact of this anomaly is itself a strong corroboration of it; and his absence from the contiguous islands of Tasmania and New Zealand, inhabited by races of human beings differing in language and origin from the natives of Continental Australia, appears almost to demonstrate his introduction from the north, where he is found in New Guinea, in Timor, in many of the smaller groups scattered throughout the Pacific Ocean, and in all the great islands of the Indian Archipelago. The extirpation of the *Thylacinus Harrisii* and *Dasyurus Ursinus* from the continental portion of Australia, is a strong corroboration of this supposition. It is contrary to all the principles of Zoological Philosophy, and to what we already know of the laws which regulate the geographical distribution of animals, to suppose that these species, two of the largest Mammals in that part of the world, should have been originally confined to so small an island as Tasmania, to the exclusion of the neighbouring continent. The more probable theory is, that they were extirpated from the latter locality by the introduction of some more powerful adversary: this could have been no other than the native dog, to whose attacks these two species were more peculiarly exposed, from being the slowest, most cowardly, and least protected animals in the country.'

46–2 See Montagu 1804 and 1831.

47–1 Smee 1835:168, 'Both the *African* and *Guzerat* Lion are subject to considerable variation in intensity of colouring. In both the colour is fulvous; but in some individuals this is much paler than in others, and in the darker specimens there occurs a tinge of red.'

47–2 Smee 1835:170, 'The variation in the form of the *cranium*, perhaps not sufficiently made out, would probably, even were it certain and constant, be scarcely adequate to establish a specific distinction: and the other differences to which I have adverted are all, as it will have been observed, differences in degree alone.'

47–3 *Natural Selection*: 116, esp. n. 2, 'Look to the King of

(*continued overleaf*)

Zoolog. T. V. I. p 389. Owen remarks on Entozoa, the organs of generation, afford the least certain indication of the perfection of the Species—![4] How does this agree with grand fact of Marsupial, low Cerebral structure??—

48e «do»
p. 390. All classes of *Acrite* exhibit lowest stages of animal organization, "& are analogous to the earliest conditions of the higher classes, during which the changes of the ovum or embryo succeed each other with the greatest rapidity"— so we find species each class successively present modifications, typical of succeeding classes & likewise those much higher in scale.[1]
So Owen actually believes in this view!!!
p. 392.— except generation & digestion in Acrite Kingdom

49e all organs blended together, & same organ «where eliminated is» often repeated, as mouths in Polypi, **Surely not correct view of Flustra or Ascidia** spicule in sponge. stomachs in infusoria, generation in each joint of tænia worm.— formative energies easily expended & no one system developed ‹is› not surprising to find many forms in Acrita,— typical of other,[1] (surely rather parents[2]). (NB These views must lead to spontaneous generation??) This whole Paper Must be studied.—

48 *page crossed brown crayon.*
49 **Surely . . . Ascidia**] *added grey ink.*
infusoria] *before unused carat.*
page crossed pencil.

(*47–3 continued*)
the beasts, as popularly called, how naturalists have doubted whether «or not» the Maneless Lion of Persia is a distinct species'; n. 2, 'Capt. Smee in Zoolog. Transacts. Vol III. [actually Vol. I, error caused by Darwin's hasty reading of "Vol 1" in C47] concludes that «maneless» lion of Guzerat is only a variety; . . .'

47–4 R. Owen 1835d:389, 'With respect to generation, the organs of which function afford in their varieties the least certain indications of the relative perfection of the species, . . .'.

48–1 R. Owen 1835d:390, 'But as all the classes of the *Acrite* division exhibit the lowest stages of animal organization, and are analogous to the earliest conditions of the higher classes, during which the changes of the *ovum* or embryo succeed each other with the greatest rapidity, so we find that the species in each class successively present modifications of their peculiar types, which come into close approximation, not with the *Acrite* classes immediately succeeding them, but with some one or other of the classes of higher groups in the animal kingdom, of the typical form of which the *Acrite* classes represent, as it were, the germs.'

49–1 R. Owen 1835d:392, 'We thus perceive in the *Acrite* subkingdom that, with the exception of the generative and digestive organs, all the other systems are more or less blended together, and the corporeal *parenchyma* seems to possess many functions in common. Where a distinct organ is eliminated, it is often repeated almost indefinitely in the same individual. In the *Polypi* we frequently find the nutritious canals supplied with a thousand mouths and the *Polygastrica* derive their name from an analogous multiplication of the digestive organ itself. Among the *Sterelmintha* the generative system becomes the subject of this repetition, each point of the *Tæniæ* being the seat of a separate ovary . . . Again, the calcareous and siliceous *Sponges*, . . . are limited to the repetition of the same *spiculum*. . . .
'The formative energies being thus expended on a few simple operations, and not concentrated on the perfect development of any single system, it is not surprising that we should find in the *Acrita* the greatest diversity of external figure; all the leading types of animal organization seem to have their origin in this division; and it has been well observed, that "Nature, so far from forgetting order, has, at the commencement of her work in these imperfect animals, given us, as it were, a sketch of the different forms which she intended afterwards to adopt for the whole animal kingdom." [MacLeay 1819–21, 1:223].'

49–2 For 'parents' read 'ancestors'.

50e ‹Three[1] p. 7. Am.› D'orbigny. Birds of prey, are distributed in S. America like other forms, but those inhabiting 3ᵈ zone of ‹latit› height & 3ᵈ of latitude more commonly are the same species, instead of analogues.—[2] ‹as› in other classes this evidently relates to greater range of such forms.—
p. 56: **Ornithological Part of Voyage of ???** A Urubu, (with one leg) attended the distribution of food, at the Mission of Mojos (even 20 leagues apart from each other.— this bird was well known for its impudence.[3] «This excellent case of memory without association..»

51 Instinct goes before structure (habits of ducklings and chickens) **Young water ouzels** hence aversion to generation, before great difficulty in propagation.—

Feathers on, Apterix because we may suppose longest part of structure.— shape of wings have altered many times, but all have had feathers.— if wing totally obliterated. This may account for permanence in many trifling marks.— such as the bands on pidgeons back.— According to this description of class is description

52 of ancestor of all birds, & so for birds. We thus obtain an abstract idea of a bird— An animal with skeleton of such general forms.—

The hybridity of ferns bears on my doctrine of cross-generation.

The infertility of crosse & cross, is method of nature to prevent the picking of monstrosities as Man does.— One is tempted to exclaim

53 that nature conscious of the principle of incessant change in her offspring. ,has invented all kinds of plan to insure stability; but isolate your species her plan is frustrated or rather a new principle is brought to bear.

---

50 D'orbigny . . . such forms.—] '19' *added brown crayon.*
Ornithological . . . ???] *added pencil.*
p. 56 . . . association . .] '23' *added brown crayon.*
51 Instinct . . . propagation.—] *crossed pencil.*
Feathers . . . class is description] *crossed pencil.*
53 is frustrated] 'is' *over* 'are'.
If man . . . homogenious.—] *pencil.*
There must . . . other animals.—] *scored grey ink.*
?] *added grey ink.*

50–1 Refers to Orbigny's three zones of latitude and elevation (1835–47).
50–2 Orbigny 1835–47, 4 (pt 3, Oiseaux) [1835–44]:6–7, 'On peut conclure de tout ce qui précède, que les oiseaux de proie suivent toujours la même loi de distribution géographique que les autres séries d'oiseaux; c'est même parmi eux que, dans la 3.ᵉ zone en élévation, ou dans la 3ᵉ zone en latitude qui lui correspond, nous retrouvons, le plus souvent, la même espèce, au lieu des espèces seulement analogues que nous présentent quelques genres des ordres suivans.'
50–3 Orbigny 1835–47, 4 (pt 3, Oiseaux) [1835–44]:36. Darwin's question marks suggest he had difficulty in locating the passage, because he had noted the wrong page number. The bold face 3 used in French publications for page numbers looks like a 5.

If man created as now. languages. would surely have been more homogenious.—

There must be some sophism. in Lyells statement that some species vary more,, than what makes species in other animals.—[1]?

54 Forster on South Sea, will probably contain descriptions of domesticated animals in those regions[1]

Species so far are not natural, that they are *either* A. B. C. D E., *or* A C D E H.

Very striking to see M. Bibron looking over reptiles he often had difficulty in distinguishing which were species,[2] (theory admirably) yet a glance would tell from which country,— I often disputed for a moment,— Galapagos, S. American— — genus.— The circumstance of having

55 two sexes is the check to distribution of birds & animals

Mr Strickland & Hamilton— found tertiary formation amongst Græcian isles, ?See if type continued?— See to Boblaye & Virlet.—[1]

Whewell thinks (p 642) **anniversary** Speech. **Feb 1838** thinks gradation between Man & animal, small point in tracing history of Man.— granted.—but if all

55 two sexes . . . Virlet.—] *crossed pencil.*

53–1 Lyell 1837, 2:388–92, *Some mere varieties possibly more distinct than certain individuals of different species.*— Evidently Darwin regarded the following as Lyell's sophism. 'We must suppose that when the Author of Nature creates an animal or plant, all the possible circumstances in which its descendants are destined to live are foreseen, and that an organization is conferred upon it which will enable the species to perpetuate itself, and survive under all the varying circumstances to which it must be inevitably exposed. Now the range of variation of circumstances will differ essentially in almost every case.' (p. 389). 'Now, there is good reason to believe that species in general are by no means susceptible of existing under a diversity of circumstances, which may give rise to such a disparity in size, and, consequently, there will be a multitude of distinct species, of which no two adult individuals can ever depart so widely from a certain standard of dimensions as the mere varieties of certain other species —the dog, for instance.' (p. 391, passage heavily scored).

54–1 Forster 1778. This work has several references to domestic animals; Darwin's marginalia suggest the instances he took particular note of. 'The two domestic quadrupeds are the hog and the dog. The Society-Isles alone are fortunate enough to possess them both: New Zeeland and the low islands must be content with dogs alone; the Marquesas, Friendly-Isles, and New-Hebrides have only hogs; and Easter-Island and New Caledonia are destitute of both. The hogs are of that breed which we call the Chinese . . .' (p. 188, Darwin commented: 'implies same var.'). 'The dogs of the South Sea isles are of a singular race: they most resemble the common cur, . . . They are chiefly fed with fruit at the Society Isles; but in the low isles and New Zeeland, where they are the only domestic animals, they live upon fish.' (p. 189, passage scored, Darwin again commented: 'implies same var.'). Darwin also prepared an 'abstract' on the inside back cover: '187 Besides «two» domestic Mammals only *Bat* in Western isla[d]; & Black Rat in Society, Friendly & New Hebrides p 188 in Tanna 2 species of Bats p 188 Hogs of same breed in the several islands 193 Natives of Society & Friendly isl[ds] catch & tame Pigeons & Parrots.'

54–2 Gabriel Bibron. According to Bell, Darwin's ophidians were placed in Bibron's hands (*Reptiles*:vi).

55–1 Hamilton and Strickland 1837 [1838]; Boblaye and Virlet 1833. Both works are mentioned in Whewell's Geological Society Presidential Address (1833–38:640).

other animals have been so formed, then man may be a miracle, but induction leads to other view.—[2]

56 Till we know uses of organ's clearly, we cannot guess causes of change.— hump on back of cow !!— &c &c

---

D'orbigny (p 108) says having observed B. Tricolor[1] in Patagonia. then in Chile & lastly 12,000 ft above sea in Bolivia; he examined— all species & found "beaucoup des mêmes oiseaux. que nous avions déjà observés en Patagonie. ou au moins des espèces tres-analogues,— quand ce nétaient

57 pas tout à fait les mêmes."[1] This good case. of replacement under peculiar conditions— of «nearly» same kind country distant. ‹Study›

The circumstance of ground woodpeckers.— birds that cannot fly &c &c. seem clearly to indicate those very changes which at first it might be doubted were possible,— it has been asked how did the otter live before it had its web-feet— all Nature answers to the possibility.—

My views will explain no Mammalia in secondary-epocks, & developement of lizards.—

58 As we have birds impressions in Red Sandstone. great lizards in do.— ‹Wood› ‹Dicot wood› Coniferous wood in Coal Measure.— highest fish in Old Red Sandstone.— Nautili in——. it is useless to speculate «not only» about beginning of animal life.: generally, but even about great division, our ‹only› question is not, how there come to be fishes & quadrupeds, but how there come to be, many genera of fish &c &c at present day.—

59 It is ASSUMPTION to say generation produces young ones capable of producing young ones like itself, but ⸮whether great assumption? not solely producing like itself, not applicable to monsters:— Are monstrosity heredetary??.? Does not atavism relate to this law.?—

55–2 Whewell 1833–38:641–42, 'Not only had no human bones been found in genuine strata, but as it had been generally held, no traces of those creatures which most nearly imitate the human form. This rule now no longer holds good; for during the past year the bones of monkeys have been discovered both at Sansan, in France, in the Sewalik Hills in the north of Hindostan, and more recently under the city of Calcutta. . . . I do not know if there are any persons who lament, or any who exult, that this discovery tends to obliterate the boundary between the present condition of the earth, tenanted by man, and the former stages through which it has passed. For my own part I can see no such tendency. I have no belief that geology will ever be able to point to the commencement of the present order of things, as a problem which she can solve, if she is allowed to make the attempt. The gradation in form between man and other animals, a gradation which we all recognise, and which, therefore, need not startle us because it is presented under a new aspect, is but a slight and, as appears to me, unimportant feature, in looking at the great subject of man's origin.' On the Sansan fossils see B94 and B133; on the Sewalik fossils B250; and on the relationship between monkey and man B169: 'If all men were dead then monkeys make men.— Men make angels—'.

56–1 *Buteo tricolor.*

57–1 Orbigny 1835–47, 4 (pt 3, *Oiseaux*) [1835–44]:108.

Local varieties formed with extreme slowness, even where isolation, from general circumstances effecting the area equably.—

60 Animals having wide range, by preventing adaptation owing to crossing, with unseasoned people. would cause destruction.— simile Man living in hot countries, if continually crossed with people from cold, children would not become adapted to climate.—

Descent. or true relationship, tends to keep to species to one form, (but is modified), the relationship of Analogy is a divellent[1] power & tends to make forms remote antagonist powers.— Every animal in cold country has some analogy in not gaudy colours so all changes may be considered in this light.—XX

61 Zoolog. Journal— Parrots in Macquarrie isl^{d}. vol III p 430 alluded to by Capt. King[1]

---

do. p. 434. Table of birds from Cuba Vigors.—[2] nothing of much interest

---

XX. Hence relation of analogy may chiefly be looked for in the aberrant groups.— It is having walking fly catcher, woodpecker &c & which causes the confusion in this system of nature— Whether species may not be made by a little more vigour being given to the chance offspring who have any slight peculiarity of structure. «hence seals take victorious seals, hence deer victorious deer, hence males armed & pugnacious (all order; cocks all warlike)» «thiis wars against in any class:, those points which are different from each other, & resemble some other class analogy»

The resemblances relationship, the dissenblances analogy,— **See Abercrombie p. 172. for definition of Analogy.**[3]

62 All the discussion ‹after› about affinity & how one order first becomes developed & then another— (according as parent types are present) must follow after there is proof of the non creation of animals.—[1] then argumen

61 XX] *corelates with passage marked* 'XX' *on C60.*
(all order . . . warlike)] *parentheses interpolated by Ed. to represent MS circle.*

60–1 In using the term divellent, which means drawing asunder or decomposing, Darwin is relating descent and chemical affinity.
61–1 King 1828–29:430. See C99.
61–2 Vigors 1828.
61–3 Abercrombie 1838:172, 'Relations of resemblance and analogy, arising out of a comparison of the qualities of various individual substances or events. These admit of various degrees. When there is a close agreement between two events or classes of events, it constitutes resemblance: when there are points of difference, it is analogy. . . . On the relations of resemblance, also, depend the arts of arrangement and classification; and the use of those general terms by which we learn to express a great number of individual objects by a single term, derived from certain characters in which they agree, such as solids, fluids, quadrupeds, &c.' Darwin double scored 'When . . . it is analogy.' The passage goes on to discuss classification of individuals into species.
62–1 See C55 for Darwin's reaction to miraculous creation as inferred in Whewell 1833–38.

May be.— subterranean lakes, hot spring &c &c inhabited therefore mud wood be inhabited, then how is this effected by— for instance, fish being excessively abundant

63 & tempting the Jaguar to use its feet much in swimming, & every developement giving greater vigour to the parent so tending to produce effect on offspring— but WHOLE race of that species must take to that particular habit.— All structures either direct effect of habit, or heredetary «&combined» effect of habit.— perhaps in process of change.— Are any men born with any peculiarity. or any race of plants—Lamark's willing absurd, ∵ not applicable to plant,[1]

64 Epidemics of South Sea, wonderful case of extermination of species— Epidemic amongst trees. Plane trees all died certain year.

Extreme difficulty of TRACING change of species to species «although we see it affected» tempts one to bring one back to distinct creations.— it is only be recollecting that the ground woodpecker &c.—, fresh water animals of great lakes are American form, that one is brought to admit the possibility

65 (any great change in species is reduced by atavism)

Even a deformity may be looked at as the best attempt of nature under certain very unfavourable conditions.— as an adaptation, but adaptation during earliest existence; if whole life then real adaptation

---

The case of heredetary disease, is on the same principle that, cut a sheeps tail off plenty of times & you will have no tail (example probably not true).— or again healthy parents have healthy children

---

the other case is ‹adaptation› «change» during life of parent, & therefore being always necessary may be called adaptation

66 With respect to my theory of generation, fact of armless parent not having armless child, shows than there is reference to more than offspring (like atavism) & shows my «view of» generation right?—

If puppy born with thick coat monstrosity, if brought into cold country, «& there acquired» then adaptation.—

---

65 healthy parents] *first* 'h' *over* 'p'.

63–1 See Lamarck 1830, 1:223, 'Dans les végétaux, où il n'y a point d'actions, et, par conséquent, point d'*habitudes* proprement dites, de grands changemens de circonstances n'en amènent pas moins de grandes différences dans les développemens de leurs parties; en sorte que ces différences font naître et développer certaines d'entre elles, tandis qu'elles atténuent et font disparoître plusieurs autres.' Scored. Darwin, '. . . not same theory to plants & animals.'

67 No Carrion Vultures in Australia!! Wilsons Ornithology, Vol III. p. 226[1]

Wilson's Ornithology, D'orbigny, Spix,[2] &c might compare birds of N. America & South— Any how temperate regions— crows in N. America‖ **Study Bonapartes list**[3]

In the Zoological Journal I read a curious account to show that *very* many birds of different kind have been know to assist in feeding young cuckoo;[4] as if there was storge,[5] which could not be resisted, when hearing crys of hunger of little bird, in same way Wilson

68 (p. 5). describes many kinds of bird uniting together in pursuit of Blue-Jay, when ‹one› birds hears ‹dis› crys of distress of other parents.—[1] **Shows community of language**

Desert country. is as effectual as a cold one. in checking beautiful colours of species— Mem. St. Jago—solitary Halcyon bird of passage.—[2] M. Coronata of Latham, wrong,—[3] Mr Yarrell says— that some birds or animals are placed in white rooms to give tinge to offspring.—[4] Darkness effect on human offspring.—

69 white, snow.— the fine green of vegetation,— ?account for colour of bird in district. which they frequent!!?—

Wilson's American ornithology a mine of valuable facts,. regarding habits range & all kinds of information— instinct[1]

---

68 ‹one› birds hears ‹dis›] *altered from* 'one bird hears dis'.
69 Wilson's . . . of all species] *crossed grey ink.*

67–1 A. Wilson 1832, 3:226n, 'The vultures are comparatively a limited race, and exist in every quarter of the world, New Holland excepted; *I have said "New Holland excepted," because we have yet no well authenticated instance of any thing approaching this form from that very interesting country. The New Holland vulture of Latham rests, to a certain extent, on dubious authority, and cannot now be referred to. I have no doubt that some representing group will be ultimately discovered, which may perhaps elucidate the principal forms wanting to the *Raptores*, and I know that Mr Swainson possesses a New Holland bird, whose station he has been unable to decide whether it will enter here, or range with the gallinaceous birds.'

67–2 A. Wilson 1832; Orbigny 1835–47, 4 (pt 3, *Oiseaux*) [1835–44]; Spix 1824. See Note C67–3.

67–3 Bonaparte 1838a. Darwin noted on fly leaf: 'Go through this list with D'Orbigny & self & see what birds common to N & S America & Europe.—' Bonaparte's list shows which birds are common to Europe and North America and which are unique to each continent.

67–4 Blackwall 1829. Note p. 298 quotes from *Zoonomia*, octavo edn, section on instinct.

67–5 Natural affection; usually that of parents for their offspring. See C96.

68–1 A. Wilson 1832, 1:5, '. . . the blue jay . . . sneaks through the woods . . . plundering . . . and spreading alarm and sorrow around him. The cries of the distressed parents soon bring together a number of interested spectators, (for birds in such circumstances seem truly to sympathize with each other,) . . .'

68–2 *Birds*: 41–42, *Halcyon erythrorhynca*, 'As Mr. Gould informs me it is an African species; it is probably only a winter visitant to this archipelago. . . . This Halcyon was the only brilliantly coloured bird which I saw on the island of St. Jago.'

68–3 See Latham 1821–28, vol. 6.

68–4 See C–IFC and D104.

69–1 A. Wilson 1832.

Swainson's remarks in Fauna Borealis must be studied.[2] There is capital table of extent of all species[3]

Accumulate instances of one family sending out structures into many genera.— like Synallaxis[4] or Marsupial animals of N. America

70 Hence it is universally allowed that the discrimination of species is empirical. **show this by instances**

Once grant my theory, & the examination of species from distant countries., may give thread to conduct to laws of change of organization!

The little turtle, without its parent running to the water, is a good instance of connate[1] instinct, better than child sucking or even ducklings & fowls—

When talking of races of Man.— black men, black bull finches, from linseed—[2] not solely effects of climate on some «antecedent races, perhaps not on now existing»

71e Mr Gould says wherever any mark like red patch on wings of Furnarius, Synallaxis &c &c.[1] sure to unite the birds into group.— it is same as Yarrell's remark about rock Pidgeons.—[2] & the latter most important in obviating a great apparent difficulty— preservation of colouring, when form has changed.

Can be said that animals no notion of beauty, when do**es** prefer most powerful buck

72e Owen talking of Plesiossaurus **Plesiosaurus.**[1] alludes to some structure in head, which he says (evidently is an *exception*) can only be explained by direct adaptation to animals wants & not as change in typical structure?!!

---

70 give thread] 'g' *over* 'b'.
«antecedent . . . existing»] *inserted by circle.*
71 do**es**] '**es**' *emmended grey ink.*
Mr Gould . . . has changed.] '5' *added and crossed in brown crayon,* '11' *added brown crayon.*
72 Plesiossaurus] *crossed grey ink.*
evidently] *first* 'e' *over* 'an'.
in Comment/] *circled.*
page crossed pencil.

69–2 Swainson and Richardson 1831.
69–3 Swainson and Richardson 1831. Introduction by Richardson, Table of species . . ., pp. xxii–xxxiii.
69–4 *Birds*: 76–82, which describes ten species of *Synallaxis* and their habits.
70–1 Congenital, simultaneous with birth.
70–2 Blyth 1835:44 'The influence of particular *sorts* of food may be exemplified by the well-known property of madder (Rubia tinctorum), which colours the secretions, and tinges even the bones of the animals which feed on it of a blood-red colour; and, as another familiar instance, may be cited the fact, equally well known, of bullfinches, and one or two other small birds, becoming wholly black when fed entirely on hempseed.'
71–1 Gould described Darwin's specimens of *Synallaxis* and *Furnarius* (*Birds*).
71–2 See C-IFC and C2.
72–1 *Fossil Mammalia.*

Whewell «in Comment/ few will dispute—» says civilization heredetary;[2] ie instincts of wisdom virtue? «like senses of savages» (How come its some countries patriotic?)— but more especially the powers of reasoning &c &c.—

73 Study the wars of organic being.— the fact of guavas having overrun— Tahiti. thistle. Pampas.[1] show how nicely things adapted—.— These «aberrant[2]» varieties will be formed in any kingdom of nature, where scheme not filled up, (most false to say no passages; nature is full off them.— wading birds partially webbed. &c &c.—)— & in round of chances every family will have some aberrant groups.— but as for number five in each group absurd.— the mere fact of division of lesser & more power (2.typical 3.subtypical[3])

74 where power arbitrary.[1] leaves door open for Quinarians to deceive himself.—

72–2 Whewell 1833: 254–68, bk 3, chap. 1, *The Creator of the Physical World is the Governor of the Moral World* (p. 254). 'These are all parts of the constitution of the earth. But these would all remain mere idle possibilities, if the nature of man had not a corresponding direction. If man had not a social and economical tendency, a disposition to congregate and cooperate, to distribute possessions and offices among the members of the community, to make and obey and enforce laws, the earth would in vain be ready to respond to the care of the husbandman. . . . Must we not suppose that He who created the soil also inspired man with those social desires and feelings which produce cities and states, laws and institutions, arts and civilization; and that thus the apparently inert mass of earth is a part of the same scheme as those faculties and powers with which man's moral and intellectual progress is most connected?' (p. 260–61).

73–1 *JR*:138, 'Near the Guardia we find the southern limit of two European plants, now become excessively common. The fennel . . . covers the ditch banks in the neighbourhood of Buenos Ayres . . . But the cardoon (*Cynara cardunculus*) has a far wider range: it occurs in these latitudes on both sides of the Cordillera, across the continent. In [Chile alone] . . . (probably several hundred) square miles are covered by one mass of these prickly plants . . . I doubt whether any case is on record, of an invasion on so grand a scale of one plant over the aborigines.' On the guava in Tahiti, see C94.

73–2 Darwin uses aberrant here and C82, to mean adapted to a peculiar specialized behaviour (habit) or function. See Swainson 1835, pt 3, On the First Principles of Natural Classification, chaps 1–2, pp. 224–266, 'Every group, whatever may be its rank or value, . . . contains, according to our theory, *three* other *primary* groups, whose affinities are also circular. One of these is called the *typical*, the other the *sub-typical*, and the third the aberrant group.' (p. 226); '. . . but there is a marked difference between the types of a typical circle, and the types of an aberrant one. In the first we find a combination of properties concentrated, as it were, in certain individuals, without any one of these preponderating, in a remarkable degree, over the others; whereas in the second it is quite the reverse: in these last, one faculty is developed in the highest degree, as if to compensate for the total absence, or very slight development, of others.' (p. 242). Swainson then uses the woodpecker to exemplify an aberrant circle: 'Here, instead of finding a combination of diversified characters, . . . the whole structure becomes adapted for one particular purpose—that of climbing trees, and extracting from them the allotted food. The energies of nature are concentrated, as it were, to the production of that form most adapted for one especial purpose.' (pp. 243–44). At the end of this section Darwin commented 'It would appear that some circles unite many characters & varied adaptations ‖ others more confined'. Note that in his abstract on the inside front cover he wrote 'some families have mingled characters & varied habits, others confined character & peculiar structures.—' See also B92.

73–3 This numbering derives from Swainson; see Swainson and Richardson 1831:xlix and Swainson 1835: 231, *inter alia*. However, Swainson's numbering is 1. typical, 2. subtypical, 3. aberrant.

74–1 See Swainson 1835:245–49, where subtypical groups are defined as 'always those which are most powerfully armed' (p. 245). By 'power arbitrary' Darwin means power is an arbitrary characteristic for distinguishing major systematic categories and the mere fact of basing one's theory on such a division makes Swainson's theory absurd.

Give the case of Apterix— split, depress & elevate & enlarge New Zealand; a division of nature of Apterix, many genera & species—

The believing that monkey would breed (if mankind destroyed) some intellectual being though not MAN.— is as difficult to understand as Lyells doctrine of slow movements &c &c.

75 this multiplication of little means & bringing the mind to grapple with great effect produced, is a most laborious, & painful effort of the mind (although this may appear an absurd saying) & will never be conquered by anyone (if has any kind of prejudices) ‹without› who just takes up & lay down the subject without long meditation— His best chance is to have profoundly over the enormous difficulty of reproductions of species & certainty of destruction; then he will choose & firmly believe in his new faith of the lesser of the difficulties

76 Once grant that «species» one genus may pass into each other.— grant that one instinct to be acquired (if the medullary point in ovum. has such organization as to ‹per› force in one man the developement of a brain capable of producing more glowing imagining or more profound reasoning than other— if this be granted!!) & whole fabric totters & falls.— look abroad, study gradation. study unity of type— Study geographical distribution

77 study relation of fossil with recent. the fabric falls! But Man— — wonderful Man. "divino ore versus cœlum attentus[1]" is an exception.— He is Mammalian.— his ‹has› origin has not been indefinite— he is not a deity, his end «under present form» will come, (or how dredfully we are deceived) then he is no exception.— he possesses some of the same general instincts, ‹as› & ‹moral› feelings as animals.— they on other hand can reason— but Man has reasoning powers in excess. instead of

78 definite instincts.— this is a replacements in mental machinery— so analogous to what we see in bodily. that ‹I› it does not stagger me.— What circumstances may have been necessary to have made man! Seclusion want &c & perhaps a train of animals of hundred generations of species to produce contingents proper.— Present monkeys might not,— but probably would.— the world

75 means] *the* 'e' *though typical of Darwin's writing is perfectly superimposed on the last stroke of the* 'm', *thus he wrote* 'mans'.
choose] *faintly crossed or underlined in pencil, possibly by a later reader.*

77 He is Mammalian] *carat but nothing inserted after* 'is'.
his ‹has›] *altered from* 'he has'.
dredfully] 'r' *over* 'e[*one letter illegible*]'.

77–1 With divine face, turned towards heaven.

79 now being fit, for such an animal.—man, (rude, uncivilized man) might not have lived when certain other animals were alive, which have perished.—

Let man visit Ourang-outang in domestication, hear expressive whine, see its intelligence when spoken; as if it understood every word said— see its affection.— to those it knew.— see its passion & rage, sulkiness, & very actions of despair; «let him look at savage, roasting his parent, naked, artless, not improving yet improvable» & then let him dare to boast of his proud preeminence.— «not understanding language of Fuegian, puts on par with Monkeys»

80 Gould[1] seems to think that many species when close come from different localities, as my Furnarii.— some genus of yellow & brown-breasted bird in Australia &c &c— but of course they might be blended, if archipelago turned into continent &c &.—

There is beautiful gradations of forms in Australia leading on one side into shrikes & at the other into into Crows. yet all forming, according to Gould, good genus

81 Gould seems to doubt how far structure & habits go together. This must be profoundly considered.— Structure may be obliterating, whilst habits are changing— or structure may be obtaining, whilst habits slightly preceed them— From this view habits must form most important element in considering to which tribe,— structure without corresponding habits clearly showing true affinity, for instance

82 tail of ground woodpecker— — but tail of some ducks aberrant[1] from— habits.—

Gould I see quite recognizes habits in making out classification of birds[2]

Birds vary much (more than shells) owing to variety of station inhabited by them—

Timor. Australian forms amongst birds

83 Java. not so much—

Peculiarities of structure. as six fingered people are sometimes heredetary,—

80–1 Mentions of John Gould in C80–82 and C88 reflect conversations with Gould and Darwin's observations of Gould's scientific practice.
82–1 See discussion of aberrant in note C73–2.
82–2 Barlow 1963:216 (DAR 29.2:8v). Darwin wrote: 'It appears to me a curious circumstance as showing the fine shades of difference in habit that when I first saw the second species «inhabiting the plains» near the Rio I thought it was different from the Maldonado species. Having procured a specimen, they were so similar, that I changed my opinion, but now Mr Gould, ‹who was not aware of these facts› pronounces them to be distinct kinds, in conformity with the trifling difference of habits, of which however he was not aware.—' This post-voyage comment is on the verso of *Beagle* notes on specimen 1213 (Calandria). See also Gould 1837.

yet these not adaptations «they are counteracted by nature by crossing with other varieties»— but ‹accidental› changes after birth do not effect progeny— many dogs in England must have been lopped off & sheeps tails cut yet there is no record of any effect.— New Hollanders have gone on boring their noses. &c & This congenital changes show that grandson is determined, when child is.—

84 shows that generation implies more than mere child, but that child should produce like children. ǂLyell has story from.— Beck about six fingered children heredetaryǂ

With respect to question what is adaptation.— Ermine, ptarmigan hare becoming white in winter of Arctic countries few will say it is direct effect, ‹of› according to Physical laws, as sulphuric acid disorganizes

85 wood, but adaptation.— albino however is monster. yet albino may so far be considered an adaptation as best attempt of nature, colouring matter being absent.— again dwarf plant in alpine district & dwarf plants from seed, one adaptation, other monster.—

The only way of judging whether structure is owing to habits, or heredetary is to see, whether a large family has it, & one member of that

86 family, having it with very different habits— Thus bill & nostril of Puffinuria I think we may clearly attribute to heredetary origin & not adaptation. to its habits.— Few will dispute that it is possible to have structure without habits— after seeing beetle with wings beneath soldered wing-cases— Yet these wings may be of some use,— Nature

87 is never extravagant though clearly not of the use to which wings are generally applied.— Therefore argument not destroyed even if these shrivelled wings could be shown to be of some use.

If we only had Puffinuria Garrottii & no other species— as we have only ornithorhyncus, then we should never know how much structure was connected with habits, & how much heredetary. The circumstance of aberrant groups being small it is truism. for it not so not aberrant.—

88 Tenioptera rufiventris, is instance of bird belonging to family with peculiar coloured plumage, where colours have changed in accordance to habits.—,[1] one is tempted to suppose from beholding the ground.— why do beetles &

88–1 See Darwin's habit notes in *Birds*:55 (*Xolmis variegata*, called *Taenioptera variegata* on pl. 11), 'This bird feeds in small flocks, often mingled with the icteri, plovers, and other birds on the ground. . . . I may observe that its plumage (in accordance with these habits) is different from that of the rest of the genus. . . . At Bahia Blanca I saw these birds catching on the wing large stercovorous [dung-feeding] Coleoptera; in this respect it follows the habits, although in most others it differs from those of the rest of its tribe.'

birds & ‹f› become dull coloured in sterile countries.—

Gould insist much upon knowing to what type a bird belongs.— I conceive without knowing from which country many birds come it would be impossible to classify them.— I would

89e–90e [not located]

91e *Musalman's* of the Peninsula, are, generally speaking, a much fairer race than the Hindu's, in the same tracts;. & that in their appearance & manners they are as opposite as day & night: yet we know how remote the periods at which both left the lands of their forefathers;— the first to escape the doctrines of Muhanmad, the last to extend, their dominion, armed alike with the Koran and the sword[1]" ‖quote Whewells Bridgewater treatise, (p..26). about plants from Cape of Good Hope continuing for some time to flower at their own periods.—[2]

92e Arcana of Science & Art. 1831. p 160. account of Bulbous root from Mummy: after 2000 years, germinating.—!![1] **Henslow doubts?**

GEOGRAPHICAL JOURNAL. Vol V. p 201 Wellsted. Memoir on isl[d] of Socotra.[2] Cattle
generally marked like those of the Alderney breed, but size not larger than those of Black cattle. Not have hump like those of *India* & *Arabia*
p. 202— sheep have not the enormous tails, which disfigure those of Arabia & Egypt.— CIVETS CATS only wild animals on isl[d].— Niether Hyenas; jackals monkeys— common to either coast not found here **not even Antelopes, though common on coast of Arabia**

93e not even antelopes though common on islets off Arabian Coast.— ‖Vol VI. p. 89.— Lieut Wellsted "on coast of Arabia between Ras Mohamned &

91 *page crossed pencil.*
92 Arcana . . . germinating.—!!] *crossed pencil.*
Henslow doubts?] *added grey ink.*
CIVETS CATS] *double underlined pencil.*
not found here] 'not even Antelopes, though common on cost of Arabia' *added pencil from C93 when latter page excised.*
GEOGRAPHICAL . . . not found here] '19' *added brown crayon.*
not even . . . Arabia] *added pencil.*
93 "on coast . . . of goats"] *crossed pencil.*
Geograp . . . out to sea [94]—] '20' *added brown crayon, applies to lower portion of C93 and top portion of C94.*

91–1 See Prudhoe 1835 on the Penninsula of Sennar. Although the passage quoted does not appear in this work, Darwin was evidently going through past volumes of the *J. R. Geog. Soc.* Prudhoe 1835 is based on extracts from Prudhoe's private memoranda, which were communicated by Sir John Barrow. Darwin knew Barrow and may have seen the complete memoranda.
91–2 Whewell 1833:26, 'Plants from the Cape of Good Hope, and from Australia, countries whose summer is simultaneous with our winter, exhibit their flowers in the coldest part of the year, as the heaths.'
92–1 Houlton 1831.
92–2 Wellsted 1835:201–3, relevant passages scored or underlined: cattle (p. 201), sheep (p. 202), civets, hyenas, jackals, monkeys, and antelopes (p. 203).

Jeddah".[1] sheep numerous "of two kinds one white with «a» black face, & similar to those brought from Abyssinia; the others dark brown with long clotted hair resembling that of goats"

Geograp. Journ. Vol VII. p. 216. Mr Bennett Voyage round world, 20 years have scarcely elapsed since the Guava introduced from Norfolk Isld— "& it now claims all the moist & fertile land of Tahiti, in spite

94e of every attempt to check its increase. The ‹bush› woodlands for miles in extent are composed solely of this shrub".—[1] p. 229. carcases of birds drifting out to sea—[2]
Vol VII. p. 325— Wild dogs of Guayana always hunt in packs «30 or 40 together» colour reddish ‹brown›. ears long.— like bull terrier— Indian secured one, as they always like to cross their breed[3] p. 333— alludes to the Macúsie breed no description given—[4] **Ch. 2. dogs**

95e L'Institut. 1838. p. 67. Australian rodents[1]

Abstract of Infusoria. ‹p› do p. 62. Ehrenberg[2]

Annals of Nat. Hist. **precursor of magazine???** p. 75. roe of Asterias in stomach. of Sammon remain after rest of animal digested.—[3] **Important**
do p. 98. on a quaternary arrangement of Cryptogamic plants.—[4]**Owing to Plants not being adapted to Air!!**

---

94 '18' *added brown crayon, applies lower portion of C94.*
Ch. 2. Dogs] *added brown crayon.*
95 1838] *second* '8' *over* '7'.
L'Institut . . . Ehrenberg] *crossed pencil.*
Annals . . . digested.—] '19' *added brown crayon.*
precursor of magazine???] *added grey ink.*
Important] *added pencil, boxed.*
do p. 98 . . . Australia] *crossed pencil.*
Owing . . . Air!!] *added grey ink.*
(**Waterhouse . . . differently**)] *added grey ink.*

93–1 Wellsted 1836:89, passage scored.
94–1 F. D. Bennett 1837:216, relevant passage scored. See C73.
94–2 F. D. Bennett 1837:229, 'In 2° 53′ S., long. 174° 55′ E., observed a remarkable line of froth on the sea, some yards in width and of great extent, and accompanied by a mass of dead birds, fish, shells, drift wood, &c.'
94–3 Schomburgk 1837:325, relevant passage scored.
94–4 Schomburgk 1837:333, 'I regretted the dog very much, as he had shown much attachment to me, and being of the Macusie breed, with all the marks of that variety, I bought him, with the intention of sending him, on my return, to the Zoological Gardens.' Passage scored.
95–1 Ogilby 1838a.
95–2 Ehrenberg 1838a:62.
95–3 Stark 1838:75, '. . . when these fishes [salmon] prey upon animals in roe, such as the *Asterias*, the ova often remain in the stomach and intestinal canal after the other portions of the food are wholly digested.' Passage scored.
95–4 Berkeley 1838:98–99, 'A tabular view of the affinities of Fungi', with five circles (p. 98). 'The facts stated above confirm in a striking degree the theory that a quaternary arrangement prevails in Cryptogamic plants.' (p. 99). Passage scored, 'quaternary arrangement' underlined.

p. 101—&c. valuable paper on quadrupeds of Van Diemen's land, which appear diff. from Australia[5] (**Waterhouse ‹disputes this)› says differently)**

96e do. do. on the genus Procyon.— by Wiegman[1]
Classified catalogue of animals of Nepal read before Linnæan Soc. Feb. 1838.—[2]

Annals of Natural History. «vol I» p. 159 curious account of Tit mouse feeding young of redstart & actually driving away parent birds.—[3]showing how blind a storge

---

From what I see of S. American birds. genera blend into each other in very same district.— ‹The same› ‹Mem› Tennioptera[4] &Tyrannula (NB work out how many forms Tyrannula can we worked out into. Milvulus

97e–98e [not located]

99 element geographical distribution is.— ʕPelagic forms similar—birds??—

We must always bear in mind proofs of most equable climate both in S. & N. Hemisphere just anterior to present. ʕcause of destruction of «great» animals?—

Show independency of shells to external features of *land* by seeing how many species common to Patagonia **desert** & Tierra del Fuego. **& forest** «Parrots in Macquarrie Is^{d}.—[1]» very good. Study D'Orbigny.[2] & range on West Coast «Guayaquil & Peru»

100 Henslow in talking of so many families on Keeling seemed to consider it owing to one of each, being fitted for transport[1] ʕmay it not be explained by mere chance?— or it like each great class of animals having its aquatic, aerial &c type?— This of consequence, because applicable to N. Hemisphere

96 *MS page number* '46' *in error.*
page crossed pencil.
100 type] 'y' *over* 'r'.
page crossed pencil.

95–5 Gunn and Gray 1838.
96–1 Wiegmann 1838.
96–2 Hodgson 1838.
96–3 Hill 1838:158–59, 'My attention was at first attracted by the violence with which I frequently saw the titmouse drive away both the parent redstarts when approaching their own nest with food for their young . . . I was at last astonished to find that the object of the titmouse was actually to feed the young redstarts together with its own. . . . In this way did the titmouse indiscriminately feed the young in both nests . . .' Darwin scored 'In this way . . . both nests'. See C67.
96–4 Taenioptera.
99–1 King 1828–29:430, 'French writers remark on the singularity of *Parrots* being found in high latitudes. . . Here however is fact against theory. . . . *Parrots*, as you are well aware, are brought from Macquarrie Island, which is in latitude 54 3/4 South, while the spot where I procured *P. smaragdinus* is in 53 1/2 only.' See C61.
99–2 Orbigny 1835–47, 4 (pt 3, *Oiseaux*) [1835–44].
100–1 Conversation with Henslow during Darwin's visit to Cambridge, 10–13 May 1838 ('Journal'). Henslow 1838 may have been drafted during this visit.

(NB.— Examine Abrolhos Flora with this view[2]) Tristan D'Acunha, St Helena &c &c. Juan Fernandez

101e A communication to Geograph. Soc. in February or March 1838 on soil in Siberia being frozen to 400 ft in depth, (& Erman's surprise that it is not 700)[1] is applicable to metamorphs theory suppose when rhinocerose lived. mean temp 60° mean, then temp at depth of four hundred feet would be 60° + 6°??., therefore 34° degrees of change have travelled that thickness in that period & no ways assisted by fluid currents which, may take place in Metamorphic action.—

102e Geograph Journal. vol I. p. 17 &c excellent sketch of plants of New Holland,[1] supplementary to Appendix to Flinders Voyage by Brown.—[2] great space seems to act per se as barrier—[3] Mem. Tartary & China.—, both coasts of New Holland.—«Compare birds of Australia with plants, with this object in view»

The intimate relation of Life with laws of Chemical combination, & the universality of latter render— spontaneous generation not improbable.—[4]

---

103 After reading "Carus on the Kingdoms of Nature, their life & affinity" **in Scientific Memoirs** I can see that perfection may be talked of with respect to life generally.—[1] where ‹"›unity constantly develops multiplicity‹"› [(his definition "constant manifestation of unity through multiplicity" this

101 *page crossed grey ink.*
102 Georgraph . . . Holland.—] '20' *added brown crayon.*
The intimate . . . improbable.—] *crossed pencil.*

100–2 See letter to J. S. Henslow, 23 July–15 August 1832, *Correspondence* 1:251, 'My collection from the Abrolhos is interesting as I suspect it nearly contains the whole flowering Vegetation, . . .'
101–1 Baer 1838a, Erman 1838a:213, 'The *data* which we hitherto possessed on the increase of the internal heat of the globe, and which have been collected together by Mr. Delabeche, in his excellent treatise on geognosy, indicated from 90 to 100 French feet for an increase of 1° of Réaumur. I did not therefore expect to find the ground thawed at Yakutzk until at a depth of from 500 to 600 French feet . . ., and if the actual fact of a thaw at the depth of 400 feet has surprised me, it is only because it has occurred *too soon*; and that it thereby indicates for the strata that compose the ground at Yakutzk a more rapid faculty for conducting heat than is possessed by the strata hitherto examined in Europe.' See also Erman 1838b.
102–1 R. Brown 1832.
102–2 R. Brown 1814.
102–3 R. Brown 1832:18, 'In comparing the Flora of the district of Swan River with more distant regions of the same continent, it may be remarked, that probably not more than four or five species are common to this part of the west coast, and to the same parallel of the east coast of New Holland; and that even the existence of some of these species at Swan River is not altogether certain.'
102–4 See Carus 1837:232–34, which discusses spontaneous generation and contains the reference to Treviranus that Darwin notes in C104.
103–1 In addition to Carus 1837, cited below, also note Darwin's abstract of MacLeay 1819–21 (DAR 71:136, written almost contemporaneously), 'P. 405. quotes G. St. Hilaire to show that every animal has some part of structure ‹p› in greatest perfection. NB. 1 in old ideas of creation this language of talking of perfection is nonsense, but it is sense when my theory comes into play.— from abortive to absolutely perfect organs there must be series.— N.B. 2. The assumption that only one pair of ‹anim› every species was created is it founded on Genesis or observation—if latter it bears on my theory.'

unity,—[2] this distinctness of laws from rest of]CD universe «which Carus considers big animal[3]» becomes more developed in higher animals than in vegetables.[4]

p. 243 radiate animals ‹tu› plants turned inside out. have position of organs of generation!!!.[5] Mem. Agaziz.(1 No Annals of Nat. Hist) *spiral* structure in Echinodermata.—[6]

104 Agassiz says Infusoria «are» insecta.—[1]

G. R. Treviranus, Biologie referred to,. as compilation of action of organic nature on inorganic[2]

It is very remarkable as shown by Carus how intermediate plants are between animal life & "*inorganic life*".— animals only live on matter already organized.—[3]

This paper might be worth consulting,[4] if any Metaphysical speculations are entered in upon life. **Namely Carus.—**[5]

103–2 Carus 1837:225, 'As it follows from the foregoing observations that life is not a single isolated reality, we shall be obliged to define it generally as the constant manifestation of an ideal unity through a real multiplicity, that is, the manifestation of an internal principle or law through outward forms. . . . Thus we find in fact the idea of life, that is, the constant manifestation of unity through multiplicity, exhibited by universal nature; and are therefore bound to consider nature collectively as one vast and infinite life, in which, through the extinction of any one of its various modifications, or the merging of a single external form of life in the universal life, is possible, an absolute and proper death is inconceivable.' Passage scored and 'the constant manifestation of unity through multiplicity' underlined. Darwin also commented at the bottom of p. 227: 'The whole universe a life, the planet, a crystal, a life— ie. his definition, but What commonly called life, a unity producing a different class of complexity than other unities.— Good idea. to show life only laws. like universe.'

103–3 Carus 1837:226, 'Now it is clear that the idea of life and that of an organism are essentially the same; . . . Universal nature is consequently to be considered as the highest, the most complete, the original organism; . . .' Darwin underlined 'Universal nature is', 'highest', 'most complete, the original organism'.

103–4 Carus 1837:242, 'As the plant may be considered a crystal continually developing itself in a constant change of its matter, in like manner the living animal body so nearly represents a plant which has reached a higher unity and faculty of self-determination, that although the animal still remains a part of a higher unity . . . yet this hold . . . is even less in degree than that which we observe in the plant as compared with the unorganized body. For this very reason, the animal presents, among natural bodies, the most perfect idea of an organism; . . .'

103–5 Carus 1837:243–44, 'Here we observe how very much in this metamorphosis the plant has assumed the type of the animal body, such as we observe it among the lower classes of animals. In this way we may now see why in the Medusa, the Sea-star, the Echinus, and other inferior kinds of animals, the aperture of the mouth is turned downwards, and the alimentary duct upwards . . . and we are thereby enabled to account for the usual place assigned by nature to the organs of generation.'

103–6 Agassiz 1838:41–42, discusses development in *Echinus*. 'The young *Echini* have a small number of plates in each of their vertical series; . . . the new plates are developed in spiral lines . . . It appears to me well worthy of remark, that in these animals, holding so low a rank among organized beings, we should find the succession of the solid parts composing their integument so strikingly analogous to the arrangement of the leaves around the stems of plants;—'

104–1 Agassiz 1838:30, '. . . a great part, if not the whole, of the Infusoria should be restored to the section of the articulated animals.'

104–2 See Carus 1837:232–33, which discusses the 'succession of the changes which take place in the formation of the Infusoria' (p. 232) and quotes Treviranus 1802–22 at length.

104–3 This is a recurrent theme in Carus 1837:234–45.

104–4 On the inside back cover of Taylor's Scientific Memoirs for 1837, Darwin wrote 'Nothing October /56/'.

104–5 See Carus 1837.

105e How remarkable that Turdus Magellanicus. in the. S. Hemisphere. (replaced to the North by other species.—) should build a nest lined with mud, in forest where not a tree in which it builds, a berry on which it feeds. or insects it devours is same species. yet that it should so strictly ‹f› agree in habits with the Turdus *Musicus* «not found in N. America» whose Southern range is?[1] ‹One› The black & white thrush of Azara builds its nest in ‹same country› something same manner, much mud.—[2] These facts show, habits heredetary whilst species have changed

106e *Argumentum ad absurdum*. The creative American halo has extended to Juan Fernandez in birds. but ⸮whether to same island in plants?— What is this halo.—[1] Continents are not stationary «unerring proofs not always continents».— it is a plastic virtue.—[2] it is expression for ignorance

---

Two grand classes of varieties.; one where offspring picked, one where not.— the latter made by man & Nature; but cannot be counteracted by Man.— effect of external contingencies & long bred in— Mem, ‹an› a statement in Mr Wynne's book,[3] about not altering breed of animals in certain countries.—

107e Fraser remarked to me at Zoological Society,, that you never find two «similar» groups of birds in two countries, without intermediate ones occurring in intermediate country— ie. mundine groups.—

Waterhouse tells me in insects there are many plenty of instances of insects of one tribe taking on structure (probably accompanied by habits) of other,[1]

106 *page crossed pencil.*
107 Fraser . . . mundine groups.—] *crossed pencil.*
mundine] 'n' *over* 'm'.
Waterhouse . . . will tell this.—] '11' *added brown crayon on C108.*

105–1 Barlow 1963:250, '2125 «cop» Turdus: male: this bird is found whole west coast. (la Plata? [*added in margin*] Falkland Isd & in Chili to its Northern limits, is very common—feeds chiefly on seeds & berrys. Is said to have its nest smoothly lined with mud. I presume like our thrush.' See also Darwin's habit notes in *Birds*: 59 (*Turdus Falklandicus*), 'I obtained specimens from the Rio Negro, Falkland Islands, Tierra del Fuego and Chiloe: I believe I saw the same species in the valleys of Northern Chile; I was informed that the thrush there lines its nest with mud, in which respect it follows the habits of species of the northern hemisphere. In the Falkland Islands it chiefly inhabits the more rocky and dryer hills. It haunts also the neighbourhood of the settlement, and very frequently may be seen within old sheds. In this respect, and generally in its habits, it resembles the English thrush (*Turdus musicus*): its cry, however, is different.'
105–2 Azara 1809, 3:209–11, describes nests of Azara's 'grive rousse et noirâtre' and 'grive blanche et noirâtre', which are synonyms for *Turdus rufiventer* in *Birds*: 59.
106–1 See RN127, 'Mem: my idea of Volc: islands. elevated. then peculiar plants created. . . . Yet new creation affected by Halo of neighbouring continent.'
106–2 See Lyell 1830–33, 1:23, '. . . the absurdity of having recourse to a certain 'plastic force', which it was said had power to produce stones into organic forms.'
106–3 Probably a book owned by Mr. Wynne, which Darwin may have examined during his September 1838 visit to Shropshire.
107–1 See B56–58, C108, and Waterhouse 1837–40:189, 'This collection [Darwin's 'new species of exotic insects'] consisted chiefly of Coleopterous insects, and among them I had most of the curious *forms* observed in the section *Heteromera*,—my object being to show that the species thus selected were analogous representations of other groups of beetles; that is to say, that they departed from their own group in certain characters of form, colour, &c.,
(*continued overleaf*)

thus in Chalcididous insect, which I brought from Australia,[2] *probably* live in flowers & has Elytra. formed from developement of some other part of body.—[3] there are hemipterous insects, having spiny legs & running quick & generl appearance of blattae

108e other Hemiptera stikingly resemble Coleoptera.— Donacia.— some orthopterous insects & some third, have got thighs with same peculiar structure & habits of clinging to rushes similar.—[1] The question which I more immediately want are there Heteromera, which have habits & part structure like Curculionidæ.— Are there any Crysomelidæ, with similar habits. But the Horæ Entomologicæ will tell this.—[2]

What peculiar conditions the Staphylinidæ on St. Pauls Rocks must be placed under

109e Gould[1] says most subgenera confined to continent, though we have seen species «of subgenera» scattered over it.—

We have abundant instances of remarkable structure which as far so species is concerned superabundant,— the tail in cock peacock,. widowbird.— Birds of Paradise. Trogons.— the one feather in wing the curious feathers in tail of Edolius.—

Remarkable how small detail in structure prevails amongst the same species & subgenera in families.— thus the banded tarsi is common to all the Laniadæ & Muscicapidæ of new World, but not found in Old-World—.

109 Gould . . . over it.—] *crossed pencil.*
We have . . . Old— World—] '19' *added brown crayon.*

(107–1 continued)
and that in these respects they appeared to have borrowed (if we may use such a term) the characters of other groups of the same order, to which they bear such a resemblance that they might at first sight be mistaken for species belonging to those groups; and we often observe that the markings vary according to the habits of the individuals.' Note that this paper was read 5 December 1836.

107–2 Darwin's Chalcidites are described in Walker 1838. Close to the time that Darwin wrote C107, he evidently wrote to Walker about differences between European *Chalcidites* and those in his collection. (See *Correspondence* 2:92–93, From Francis Walker, 6 July [1838]; Walker's letter of 18 December 1843, *Correspondence* 2:418–19 gives an indication of Darwin's 1838 interest in *Chalcidites*.)

107–3 Waterhouse 1837–40:191, 'The last is perhaps one of the most remarkable instances [of analogous representation]. This is one of the *Chalcididae*, in which the thorax is produced posteriorly into two processes, like the elytra of a Coleopterous insect (and they appear to answer the same purpose); and so strong is the case of analogy, that when viewed only from above, the insect might be mistaken for a species of the genus *Mordella*.

108–1 Waterhouse 1837–40:191, where Darwin's *Belus testaceus (Curculionidae)* is said to represent *Lixus*; *Leptosomus acuminatus (Curculionidae)* represents the *Brentidae*; *Allelidea Ctenostomoides* represents *Ctenostoma (Cicindelidae)*.

108–2 MacLeay 1819–21, chap. 8, General reflections on the Synthetical Method. Although the particular representation of habits and structure that Darwin sought are not discussed, MacLeay mentions the following coleopteran genera: *Curculio, Donacia, Chrysomela*, and *Heteromera*. MacLeay divided the coleoptera into five groups (1819–21:439); Darwin's 'question' discusses members of three contiguous MacLeayan groups: 'many *Heteromera* of Latreille' are in *Thysanuriformes, Chrysomela* is in *Anopluriformes* and *Curculio* is in *Vermiformes*.

109–1 References to John Gould in C109–114 reflect conversations at the British Museum (see C115).

110e ++ If in any «well developed» family (Gould says) there is any marked colouring of plumage (as «black & white» bars on wings of trogons are lengthened rump feathers.—; ‹then› & one species has small band & others large, then he says from long experience, you may be almost sure, that there exist intermediate species,— This is remarkable & would lead one to suppose, that species in same group generally contemporary

---

«++.» «This would lead one to expect that fossil forms would generally fill up genera & not species, which is not true, with shells?.?.?» **It looks as if animals perished by errors.—**

111e It is most wonderful how in every family of bird,, even the most ‹m› strongly marked, there is a preeminently aerial,— formed for flight & great *movement* in the air, & likewise rasorial species, & likewise perching[1] (Gould), but the latter is obscure because nearly all are so.— Thus in Hawks, there is a swallow, both in structure & habits (it cannot be doubted that if swallow perished) hawks & Milvulus &c would instantly fill up their place.— Humming bird there is strongly marked variety,:.in the Tyrannidæ.— Milvulus— Even flying woodpeckers., with powerful wings, but

112e tail stiff.— Swallow & goatsuckers likewise exaggerated.— There is one most remarkable connection between these aerial representatives of the different families.— that sexes ‹are› «have» same plumage.— ‹no› this is applicable to swallow-hawk, «this not the case in swallow??? which is most wonderful of all. ˁwhether in most aerial of swallows.» Milvulus, & still more wonderfully to the Humming bird, which is one instance of its whole family where female is not dull.— I must observe that this preeminent structure is not always applicable to same habits, though swallow hawk[1], ,milvulus,,[2] may catch insects on the wing & pratencole[3] (ˁconnecte[d]

113e with Chionis[1]), yet the Tropic bird,[2] has very different habits, though

110 contemporary] *first* 'o' *over* 'r'.
If . . . contemporary] '11 & Back' *added brown crayon, refers to* '19' *on C109; alternative reading* '& Book' *referring to Natural Selection.*
It . . . errors.—] *added grey ink.*
111 variety] 'y' *over* 'ies'.
:] *added pencil.*
111–115 *formerly pinnned together.*
112 swallows.] 's.' *on stub.*
«this not . . . swallows.»] *carat added in pencil.*
,milvulus.,] *commas added in pencil.*

111–1 See B57.
112–1 *Falco furcatus*, swallow-tailed hawk (Alexander Wilson 1832, 2:275).
112–2 *Milvulus forficatus* Swainson, *Muscicapa forficata*, swallow-tailed flycatcher (Alexander Wilson 1832, 3:275).
112–3 Swallow-plovers (Swainson 1836–37, 2:175).
113–1 See Swainson 1836–37, 2:350n, 'the *Columbidae* [pigeons] are the most aberrant of the rasorial order, so is *Chionis* the most aberrant of the *Columbidae*.' See B55.
113–2 (*See overleaf*)

preeminently belonging to this type, ?the Humming bird? the woodpeckers Gould says, he believes does. but also on fruit.—
The Rasorial type is wonderfully shown in long legged cuckoos with claw like lark, (one in Australia is called swamp pheasant) Goatsucker—, parrots with claw like lark (NB The La jeune veuve **parrot** though so much on the ground has not this structure., instance of habits going before structure).—

114e even one kingfisher— Gould has seen with long tarsi.— «Ground woodpecker» Secretary bird.— & Millisuga. Kingii very rasorial for type.— Now here I must observe these characters vary in degree in last instance hardly at all developed. Not confined to one species, but generally small genus. ?are there not many ground parrots? are there not many ground woodpeckers?—
In each division Gould thinks he can trace structure for insects & structure for vegetation.—

115e In conversation in Museum— I could not discover any other clear relations besides aerial, & terrestial,— How is it in water birds, there are walking forms in water birds,— but no web forms in ‹water› land birds,,— Grups of very different value have their represetatives;, the rasorial may be observed even in Lessonia[1] &c &

In relations of affinity all organs change together, in analogy certain parts perfect of typical structures certain ‹imper› parts changed

116e Have ‹not›. S. Africa, Australia, & S. America very few forms in common.,— but each several with Europe & northern Asia & Northern America.— may we not look to these Northern regions as the receptacles of the wanderers out of the rest of the world?— Will this not agree with Waterhouse & ‹birds› Mammalia.—[1] We have clear indication

117e–118e [not located]

119 alone, but on all the general arguments—

---

113 Gould] 'G' *over* 'g'.
shown] 'n' *emmended pencil.*
116 *page crossed pencil.*
119 *page crossed pencil.*

113–2 See Swainson 1836–37, 2:194, *Phaeton*: Ord. Natatores (The Swimmers), Fam. Alcadae (Auks), Subfam. Laridae (Gulls). 'The same regions, as their name implies, are inhabited by the tropic-birds (*Phaeton*), whose flight, . . . is frequently as high as that of the frigate-bird.'
115–1 See Swainson 1836–37,2:48, Lessonia, Sw. Subfam. Motacillinae, Wagtails. Fam. Sylvianae, Warblers). Ord. Insessores, Perching birds, Tribe Dentirostres. See p. 41 'Hitherto none of the wagtails had been detected in America; but the recent researches in Chili have brought to light a most singular bird, long ago figured by Buffon, but which had been lost to our modern collections, until very lately. This is our *Lessonia erythronotus*, which at present stands as a solitary example of the fisserostral type of the *Motacillinae*.'
116–1 Probably personal communication. Also see C36 and Waterhouse 1839c.

Lamarck was the Hutton of Geology.[1] he had few clear facts, but so bold in many such profound judgment, that he forseeing. consequence., was endowed with what may be called the prophetic spirit in science—. the highest endowment of lofty genius

---

Using geograph distribution of animals, ‹as› I use (new step in induction) as keystone of ancient geography species tell of Physical relations in time «& forms» distribution tells of horizontal barriers—

120 Mr Yarrell.— says my view of varieties is exactly what I state.— &c picking varieties. unnatural circumstance

---

Ld Orfords had breed of greyhounds fleestest in England lost courage— **(Bull-dogs are used because they have no scent!) Mr Wynne)** at end of chase would not run up hill— he took thorough bred ‹greyhound›. bull-dog. & crossed & recrossed, till there was a dash of blood, with whole form of grey hound.— picking out finest of each litter & crossing them with finest greyhounds.—[1] Sir. J. Sebright first got point on hackles on Bantams by crossing with common *Polish cock* «is not that old variety» & then recrossing off spring. till size diminished, but feathers continued by picking chickens of each brood.— These bantam feathers

121 at last got ducky, then took white Chinese Bantam crossed & got some yellow & others yellowish & white varieties by picking the yellow one & crossing with duck bantams procured old variety.—[1] The pidgeons which have such different skulls, but same marks on wings are Blue Pouters & small Bald Heads Mr Yarrell will mention in his work[2]

I am sorry to find Mr Yarrell's evidence about old varieties is reduced to scarcely anything.— almost all imagination— He says he recollects all half Bred **cattle** of L'Darnleys were most like parent Brahmin bulls— Mr. Y. is inclined to think that the male communicates the *external* resemblances, than the female.

120 Mr Yarrell . . . circumstance] *crossed pencil.*
121 The pidgeons . . . his work] *marginal brace.*

119–1 If one considers Hutton the forerunner of Lyellian uniformitarianism, then Lamarck would be the Hutton of evolutionary zoology. This would make Darwin the Lyell of zoology. This comment may reflect Darwin's continued reading of MacLeay 1819–21:326–29, where Lamarck is both criticised and praised (pp. 326–29) in passages that are cited in C157. See also Lamarck 1802 for his actualist geology.

120–1 *Variation* 1:41, 'Lord Orford, as is well known, crossed his famous greyhounds, which failed in courage, with a bulldog—this breed chosen from being deficient in the power of scent; . . .'

121–1 *Variation* 2:96, 'The Sebright bantam, which breeds as true as any other kind of fowl, was formed about sixty years ago by a complicated cross.'

121–2 Yarrell 1843.

122 The expression hybrid & fertile Hybrids, may be used to varieties, as well as species—

---

as formation of species ‹of› gradual, so may we suppose., that something intermediate, between no offspring & ordinary offspring:— this gradation is infertile offspring. **without organs of generation?!**

By profound study of local varieties laws of change— whether beak (as it appears to me). colour of plumage & laws, which might probably be reduced

---

123 What the Frenchman.[1] did for SPECIES— between England & France. I will do with *forms.*—

Mention persecution of early Astronomers.— then add chief good of individual scientific men is to push their science a few years in advance only of their age. (differently from literary men.—) must remember that if they *believe* & do not openly avow their belief. they do as much to retard, as those, whose opinion they believe have endeavoured to advance cause of truth

124 It is of the utmost importance to show that habits sometimes go before structures.— the only argument can be, a bird practising imperfectly some habit, which the whole rest of other family practise with a peculiar structure, then Milvulus forficatus **Tyrannus Sulphureus** if compelled solely to fish. structure would alter.—

It is a difficulty how a different number of vertebræ are produced, where, (& in all such structure) there cannot be gradation. See what Eytons young pigs— if vertebræ much lengthened &c there may be tendency to divide, which often enough

125 repeated would cause an unequal number of vertebræ—[1]

⁽Where two very close species inhabit same country are not habits different, (Mem: Gould's Willow Wren[2]) but where close species inhabit different countries habits similar ⁽law?— probable.— if habits & structure similar would have blended together

---

122 **without . . . generation?!**] *added pencil.*
123 Frenchman] 'Frenchmen' *alternative reading.*
species] *underlined grey ink.*
124 Milvulus forficatus] *crossed grey ink.*
125 **Mem Mr Herberts . . . fertility**] *pencil* '**habits**', *underlined.*

123–1 Possibly Lamarck.
125–1 See Eyton 1837b.
125–2 Gould 1837. See note B241–1.

Mem Mr Herberts law; habits determining fertility[3]

---

126 Scheme for abolishing specific names & giving subgenera. true value.— as in
Opetiorhyncus. fulginosus.
(a) Falklands
(b). F. del Fuego differ from
(C) Chiloe
(E) Chile..
rupestris — good species
it is reverting to old plan, but reason now assigned for doing so
There should be mark to every species. only known by analogy
genera of course distinct. analogy from every country & class tells us that
?O. Modulator & O. Patagonicus. till neutral ground ascertained, call them varieties. but two ostriches good species because interlock

127 analogy to be guide. in islands. species.— each describer giving his test namely differ as much as those (naming them) which are found together.— If two species come over to this country, without range, or habits ascertained— put them as (a). (b) until data be given.— This will aid in preventing the chaos.— will point out what to observe.— will aid us in physiology, tell traveller what to observe.— if he knows he has done least part.— that he will not have brought home new species. until, he can show range & habits—

Take instances of most disputed shells, such as Cyrena

128 This is reform which probably will be slow. but must take place— such a classification would answer every purpose, & would present many ideas of causes of change.— The mark of analogy would be empirical because as soon

126 There . . . tells us that] *in braces.*
128 solely by] 'by' *over* 'of'.

125–3 Herbert 1837:335–80, 'On crosses and hybrid intermixtures in vegetables.' Herbert's law is the relation between 'constitution' and fertility, put forth several times in this chapter. Darwin replaces the term 'constitution' with his term 'habit', which here comes close to meaning adaptive situation. The following statement and example of 'Herbert's law' show the grounds for Darwin's formulation: 'It was my opinion that fertility depended much upon circumstances of climate, soil, and situation, and that there did not exist any decided line of absolute sterility in hybrid vegetables, though from reasons, which I did not pretend to be able to develope, but undoubtedly depending upon certain affinities either of structure or constitution, there was a greater disposition to fertility in some than in others. Subsequent experiments have confirmed this view to such a degree as to make it almost certain that the fertility of the hybrid or mixed offspring depends more upon the constitutional than the closer botanical affinities of the parents.' There are many marginal notes on Herbert 1837, no doubt from many dates. One layer, however, corresponds to the period of Notebook C. At the bottom of p. 343 Darwin wrote: 'Habits &c determining sterility (, & hence probably intermarriage in some degree) is very important, as solving case of Willow Wrens & explaining «*great*» importance «generally solving question» of habits in determining what is species — a fact tacitly admitted by all naturalists.—' Note the striking parallel reference to willow wrens in this comment and C125. In a very similar hand Darwin remarked on the use of 'constitutional' in the Herbert passage quoted above: '+ is this not error: does not constitutional difference, infer some difference in *innermost* organization, or rather on *whole* organization'.

as two species were placed in different subgenera, then it would be useless, but the formation of subgenera is empirical, & is judged solely by comparison with other genera in other families.— it will however be much ‹shu› surer, when false species banished by this test.—

129 Excepting where an Andrew Smith, «Richardson» a Vaillant[1], a D'orbigny has travelled this will be most difficult.

Sub-genera so far may be eliminated. where every species of a section is confined to one continent & every species to another then those sections & subgenera; are analogical, because we do not know, whether nearest species of each might not breed:— Genus must be a *true cleft*— putting out of case the Analogys.— If genus does not Mean this it means nothing.—There should be some term used, when there is series.

130 Could I not give Catalogue of Mammalia arranged according to my own methods.

Dasyurus being found fossil in Australia, & only one tree species (Mitchell's authority[1]) in Australia, & several in Van Diemen's land is most important as showing former connection of two continents & death of form in one. The caves are at a height of more than 1000 ft. & many hundred miles

131 from the sea, associated with teeth of seals and dugong, therefore immense age since breccia accumulated— surely ask Owen to see whether species same, excessive improbability. Mem in Clifts list a rat said to have been found!![1] rodents old inhabitants most important!! like Dipus of present day??! Major Mitchell does not think that dog was found in Van Diemen's Land.— **V. 1^s. Number of Geographical Journal to discover whether dog found at Swan river.**[2]

132 The change in England from Rhinoceros Elephants &c in the most modern period, compared to Faunas of these countries, greter than Toxodon, Macrauchenia, &c compared to America.— the wonder is that the Europæan forms, were able to escape to some more fitting country,— if Toxodon had been found in Africa, the wonder have been same for S. America & Europe.— The difficulty is how came it animals not preserved, in central S. America & yet Africa & India???— **& Indian Isl^ds.—**

129 Sub . . . breed:—] *scored and '?' in margin.*
130 Dasyurus . . . hundred miles] *crossed pencil.*
131 **V. 1st. . . . Swan river.**] *added pencil.*
132 **& Indian Isl^ds.—**] *added pencil.*

129-1 John Richardson. François le Vaillant.
130-1 T. L. Mitchell: personal communication. See Mitchell 1838.
131-1 Probably personal communication from T. L. Mitchell. See Clift 1831; however, rats are not mentioned in Clift's list.
131-2 See Nind 1832:28–29, who mentions dogs domesticated by natives.

133 Sir J. Sebright— pamphlet— most important, showing effects of peculiarities being long in blood.—++[1] thinks difficulty in crossing race.—[2] bad effects of incestuous intercourse..—[3] excellent observations of sickly offspring being cut off— so that not propagated by nature.—[4] Whole art of making varieties may be inferred from ‹this› facts stated.—[5]

++. Fully supported by Mr Wilkinson.[6] =Milking heredetary, developemen of important organ (.see marks on pages).—[7] Crosses of diff: breeds succeed,

133 back] 'b' *over* 'p'.

133–1 Sebright 1809:7, 'Regard should not only be paid to the qualities apparent in animals, selected for breeding, but to those which have prevailed in the race from which they are descended, as they will always shew themselves, sooner or later, in the progeny: it is for this reason that we should not breed from an animal, however excellent, unless we can ascertain it to be what is called *well bred*; that is, descended from a race of ancestors, who have, through several generations, possessed, in a high degree, the properties which it is our object to obtain.' Passage quadruple scored in pencil.
133–2 Sebright 1809:17–18, 'Although I believe the occasional intermixture of different families to be necessary, I do not, by any means, approve of mixing two distinct breeds, with the view of uniting the valuable properties of both: this experiment has been frequently tried by others, as well as by myself, but has, I believe never succeeded. The first cross frequently produces a tolerable animal, but it is a breed that cannot be continued.' Darwin commented: 'Disputed satisfactorily «or nearly so» by Wilkinson | Differs from Mr Wynne' (ink). Passage scored ink.
133–3 Sebright 1809:8, in particular, 'If a breed cannot be improved, or even continued in the degree of perfection at which it has already arrived, but by breeding from individuals, so selected as to correct each other's defects, and by a judicious combination of their different properties, (a position, I believe, that will not be denied), it follows that animals must degenerate, by being long bred from the same family, without the intermixture of any other blood, or from being what is technically called, *bred in-and-in*.' (p. 8). Passage scored in ink. Darwin commented: 'does not take into account loss of desire'.
133–4 Sebright 1809:15–16, 'Many causes combine to prevent animals, in a state of nature, from degenerating; they are perpetually intermixing, and therefore do not feel the bad effects of breeding *in-and-in*: the perfections of some correct the imperfections of others, and they go on without any material alteration, except what arises from the effects of food and climate. The greatest number of females will, of course, fall to the share of the most vigorous males; and the strongest individuals of both sexes, by driving away the weakest, will enjoy the best food, and the most favourable situations, for themselves and for their offspring.
'A severe winter, or a scarcity of food, by destroying the weak and the unhealthy, has had all the good effects of the most skilful selection. In cold and barren countries no animals can live to the age of maturity, but those who have strong constitutions; the weak and the unhealthy do not live to propagate their infirmities, as is too often the case with our domestic animals. To this I attribute the peculiar hardiness of the horses, cattle, and sheep, bred in mountainous countries, more than to their having been inured to the severity of climate; . . .' Passage scored: 'Many causes . . . and the unhealthy do' [ink], 'not live . . . climate;' [pencil]. See D135, where Darwin's 'Even a *few* years plenty, makes population in Men increase, & an *ordinary* crop. causes a dearth then in Spring, . . .' may echo Sebright's 'A severe winter, or a scarcity of food'.
133–5 Although Darwin's comment applies to Sebright 1809 as a whole, he specifically remarked 'This shows the whole principle of making varieties' on 1809:10, 'Mr. Bakewell had certainly the merit of destroying the absurd prejudice which formerly prevailed against breeding from animals, between whom there was any degree of relationship; had this opinion been universally acted upon, no one could have been said to be possessed of a particular breed, good or bad; for the produce of one year would have been dissimilar to that of another, and we should have availed ourselves but little of an animal of superior merit, that we might have had the good fortune to possess.' Darwin also noted: 'In plants. man prevents mixture, varies conditions & destroys, the unfavourable kind—could he do the last *effectively* & keep on same exact conditions for *many* generations he would make species, which would be infertile with other species—' (1809:14–15 in pencil).
133–6 Wilkinson 1820:5, 'The longer also these perfections have been continued, the more stability will they have acquired, and the more will they partake of nature'.
133–7 Wilkinson 1820:24, 'Such a procedure in the formation of a breed, clearly adds very considerably to the expense in the first instance; but the advantages afterwards derived are more than a sufficient compensation, as the property of milking is inherited as readily as that of peculiarity of shape.' Passage quadruple scored, ink.

yet seems to grant, that difficult & other go back to either parent.—[8]

134 Shows instinct (Sir J. Sebright admirable essay) heredetary[1]
Young wild ducks.— lose as well as gain instincts.[2] Wild & tame rabbits good instance—[3] instincts of many kinds in dogs, is clearly applicable to formation of instincts in wild animals, many species in one genus—[4] external circumstances in both cases effect. it.— Sir J.. Sebright excellent authority because written on dog-Breaking.—[5] applies it to national character.—[6]

135 N.B. If two species were excessively old, they would not make hybrids, whereas two newer ones, even if more different might do so.— ‹whi› is this true?? My views, which would even lead to anticipate mules is very important for L.**yell** said to me, the fact of existence of mules appeared to him most strange.— This even might be said.— My theory thus explains a grand apparent anomaly in nature.‹t›——

Many animals not breeding at all in domestication. throws great difficulty in way of ascertaining about hybrids.— **& is a very remarkable fact. show influence of mind**

136 It is not difficult to see that it is less repugnant to nature to produce one offspring unlike itself, than to produce that capable of producing itself alike.— in one case it changes one, in other it changes thousands in futurity.— This is right way of viewing it.—
Variety when long in blood, gets stronger & stronger, so that though by great effort ‹pr› one unlike can be produced, yet to produce whole generation unlike would go against the tendency.— it tries to go back to grandfather, but if too unlike its own parent this impossible— (Hence we might expect even if two mules bred or two certain varieties, they would go back to grandfather, which is

135 **yell**] *added pencil.*
& is . . . of mind] *added pencil.*

133–8 Wilkinson 1820:29–30.
134–1 Sebright 1836.
134–2 Sebright 1836:7–8, 'If the eggs of the wild duck be hatched by a hen, although the young ones will, from the first, show signs of wildness, that are never seen in the domestic breed, they may, if pinioned, be made apparently tame, but if suddenly alarmed, or if removed to a strange place, their natural wildness will show itself.'
134–3 Sebright 1836:10, 'The domestic rabbit is, perhaps more easily tamed than any other animal, excepting the dog; . . . The wild rabbit, on the contrary, is by far the most *untameable* animal that I know . . . I have, when a boy, taken the young ones from the nest, and endeavoured to tame them, but never could succeed.' Passage scored ink, also pencil quote marks on 'The wild . . . succeed.' At the bottom of the page Darwin commented in ink 'Mem. Mr Wynne's Malay fowls. such different habits . . ‹go to› «‹leave their›» ‹roost—› hunts the fields late in the evening.'
134–4 Sebright 1836:11–16. Many passages marked by Darwin.
134–5 Sebright 1836:11–12 on Sebright's experience in taming a wild Australian puppy. See B250.
134–6 Sebright 1836:16, 'How far these observations may apply to the human race I do not pretend to say; I cannot, however, but think that part of what is called national character may, in some degree, be influenced by what I have endeavoured to prove, namely, that acquired habits become hereditary.'

137 true) & infertility is consequence.—

The simple expression of such a naturalist "splitting up his species **& genera** very finely" show how arbitrary & optional operation it is.— show how finely the series is graeduated.—

D[r] Beck doubt if local varieties should be remembered, therefore do not consider it as proved that they are varieties, (though that would be best).—

Argue the case theoretically if animals did change excessively slowly. whether geologists would not find fossils such as they are—

138 My theory explains that *family* likeness, which as in absolute **human** family is undescribable, yet holds *good*, so does it in real classification

---

The relation of all cock birds in Gallinaceous having tendency to lon[g] or peculiar tails strange

---

?? / Genus only natural from death or slow propagation of forms.— just same way as ‹we› «all men» not all equally related to each other

I cannot help thinking good analogy might be traced between relationship of all men now living & the classification of animals— talking of men as related in the third & fourth degree.—

139 a species must be compared to family entirely separated from any degree; the tailors— «in each branch would be analogous to each other— &c &c.— V. p.140»

I should think meaning of circular arrangement was only so far true as avoided linear arrangement **the central twigs dying, affinities would be ‹circular›, in broken circles.—** which in each group is quite fatal.— ǂRelations of analogy[1]

---

being those last obtained.— less firmly fixed & therefore most subject to change.— may account for certain organs not being fixed, ‹whi› in some genera, which are most fixed in others.

---

137 **& genera**] *added grey ink over added pencil.*
Argue the case . . . as they are] *added pencil.*
whether] *crossed grey ink.*
139 «in each . . . V. p. 140»] *inserted above rule line.*
the central . . . circles.—] *added pencil.*

139–1 See Darwin's abstract of MacLeay 1819–21:363–64 (DAR71:133).

140 In analogy it is not the relation to bear to each other, but to some external contingency.— affinity is the sum of all the relations, analogy is the close relationship in some one.—

---

imagine the men to have greater power of change yet, as external conditions over whole world. similar— & constitution of man originally similar limits of change, would be same— Yet each family might have its own character,— we here suppose these changes of ‹use› adaptation greater than those heredetary ones.— which would elapse; during time such changes

141e had elapsed.— let these families take ‹dogs› «domestic animals» with them. they might be supposed to change. & make genera of bird. analogous. animals would be possessed by the different races of man, yet altogether different.— To make this case perfect, we must suppose men instead of mere colour & trifling form & head &c to become greatly changed. in structure & even to certain degree in habits, yet we might have these analogies.— We must two races of such men living in same country but separated, now

142e if one of these races had become eminently aquatic—; «NB, aquatic, i.e relation to elements & not minding particular trades.—» then the second race would not obtain a cast of washing men— but might hire the preexisting race,[1] thus the analogy would not in all cases be produced, but would depend upon exclusion.—

The same characters which are analogical in a genus with respect to rest of its family as in ground cuckoos, is affinity with respect to species of each other, because we suppose all descended from same.— but if two original species, each became ground then the relation of all the ground cuckoos. would not be affinity, but the truth would never be discovered

143 When one reads in Ehrenbergs Paper on Infusoria on the the enormous production— millions in few days—[1] one doubt that one animal can really produce so great an effect.— the spirit of life must be every where ambient. & ‹in› merely determined to such points by the vital laws.—

---

so that all character originally may «must» have had the character of analogical.—

---

141 *page crossed pencil.*
143 eagle] *crossed grey ink.*

142–1 Note Darwin's use of the division of labour.
143–1 Ehrenberg 1834:207, 'Dans l'espace de peu de jours, il peut naître plusieurs million d'individus, soit au moyen d'oeufs, soit par division. Une observation directe démontre qu'en mettant en expérience un Rotateur, on peut obtenir au dixième jour un million d'êtres, quatre millions le onzième, et seize millions le seizième jour. La progression est plus rapide encore chez les Infusoires polygastriques. Le premier million est obtenu, en effet, dès le septième jour.'

Gould says it is only in large groups. where you have representations.—

The aerial type in each family is relation to elements & not habits as shown by frigate Bird & flying eagle.— **Hawk**

144 Gould seemed to think, that widow bird. replaced Birds of Paradise— if such fantastic «sexual» ornaments. have so intimate a relation to two continents as to be ‹replaced› called into existence in two continents our ignorance is indeed profound & such it appears.—

Is there not some statement about diversity of forms in aberrant circles.— explained by such not having been long in blood?—

145 My theory agrees with unequal distances between species. some fine & some wide. which is strange if creator had so created them.—

People will argue & fortify their minds with such sentences as "oh turn a Buccinum into a Tiger."— but perhaps I feel the impossibility of this more than anyone.— no turn the Zebra into the Quagga.— «let them be wild in same country with their own instinct & (even though ‹fertile› when compelled to breed hybrids produced—)» & then all that I want is granted.—
For. at Galapagos. make ten species of Orpheus— one of which has very short legs & long tail «short much curved beak.—», other very long beak, with short., let these only have progeny with species & there will be two genera.— let short billed one, be exaggerated, & all rest destroyed far remote genera. will be produced.

146 As we know from Ehrenbergh, there are fossil (see Scientific Memoirs & L'Institut.) that there are Tertiary fossil Infusoria, of same forms with recent,[1] we have nothing to do with CREATION.—[2]

‹On› The end of formation of species & genera, is probably to add to quantum of life possible with certain preexisting laws.— .If only one kind of plant not so many.—

146 **laws]** *emmended grey ink.*

146–1 Ehrenberg 1837a, 1837b, and 1837d.
146–2 Connected with the question of Creation is that of spontaneous generation. See Ehrenberg 1834 (cited in C143), p. 207, 'Il n'est pas nécessaire d'admettre une génération spontanée pour expliquer cette immense formation d'êtres, ces nouvelles observations ayant démontré que cette hypothèse dont on peut se passer d'ailleurs, est loin de reposer sur des observations rigoureuses.'

147e The *quantity of life* on planet at different periods, depends,— on relations of desert, open ocean, &c this probably on long average, equal quantity, 2d on relations of heat & cold. therefore probably fewer now than formerly.— ***The number of forms depends*** on the external relations (a fixed quantity) & on subdivision of stations & diversity—.— this perhaps on long average equal.—[1]

148e The Cocos do Mar on the Mahé island,, one the higher parts & only on those, & the islets separated at high water.— not other islands, nor any any other part of world.— no other *plants peculiar to these* isld.— ?**Brown** can «*not*» bear the least salt water.— Nuts prodigiously heavy (where trees of such Nature far apart. Must have travelled by ‹dead› «each» trees dying & mountain torrents.— but to crawl up an hill, then by deaths?!)— looks like subsidence.— on the islets

149 Mr Blyth remark that a resemblanc between some form in birds is visible, when young, but not when old.—[1] thus speckled form of young blackbird. good remark if general.—

Where any structure is general in all species in group we may suppose it is oldest, & therefore lest subject to variation.— + **‹being› good for generic divisions [+] ought genus to be founded on such characters, as do not vary in the species of it.— «where does such occur?»** now some such «characters» rule are used by Naturalists in their test of value of character— **Macleys rule is converse, ‹when› value of character depends on non-variation, & not on extension ?these go together?**
Therefore value of organs vary in different group. & Not known in single ones—. viz. Macleay letter to Fleming p. 32 "where it (mode of generation) varies according to the species, it is manifestly of less importance, as affording natural characters than among those groups, where it remains less subject to Variation"[2]

150 Dr. A. Smith. knows lots of instances of replacement of one species by another. supply place in each others' economy

---

147 *Quantity of life*] *underlined grey ink.*
The] 'T' *in grey ink over* 't', *passage underlined grey ink page crossed pencil.*
148 Brown?] *added pencil.*
'20' *added brown crayon.*
149 **Macleays . . . together?**] *added in margin, grey ink.*

147–1 See B36 and E85.
149–1 Blyth 1836b:507, 'The blackbird has, when young, a spotted breast; and, in fact, the characters of its nestling plumage alone forbid its alienation from the spotted thrushes. Where, indeed, can we trace the line of separation between Mérula and Philomèla even?'
149–2 MacLeay 1830a:32. Darwin's pagination refers to the proof copy. In 1830b the passage is on 8:205. The 'letter' criticizing Fleming 1822 was addressed to N. A. Vigors.

Dr. S. showed that savages are not born with any capacity for observation of tracks &c &c

---

Dr. S. has some remarkable crochets about instincts. whenever instinct is mentioned some definition must be given

---

It would not be difficult to arrange children of same parents in a circle,— & «hermaphrodites» father « ‹mother› » & grandfather ‹Might› Must be introduced & made young.—╫ father must be left out of case, that difference occurring.—

151 It will be necessary to show hybridity **from few forms, parents of all species** not possible in some detail,— the relations to islands close species, «on these isld» &c will probably upset it—

The space which one branch of the tree if live occupied after its decay, will be occupied by the vigorous shoots from each branch **No: because decay in that species ‹shows› is effects of unfavourable conditions.** (hence rise & depression of importance, in each group & connection of even distant ones) the characters will be first those of analogy, but will grow into affinity. **—but whether ever arrive at true affinity doubtful**

152 A species is only fixed thing with reference to other living being.— one species May have passed through a thousand changes, keeping distinct from other & if a first & last individual were put together, they would not according to all analogy breed together.—

The bottom of the tree of life is utterly rotten & obliterated in the course of ages.—

As species is ‹certain› real thing with regard to contemporaries— fertility must settle it.—

153 Changes in structure being «necessarily» excessively slow, they become firmly embedded in the constitution, which other marked difference in the varieties «made by» of Nature & Man.— The constitution being heredetary & fixed, certain physical changes at last become unfit, the animal cannot change quick enough & perishes.—
Lyell has show such Physical changes will be unequally rapid, with respect to their effects

---

The ÆEgyptian animals domesticated «??», & therefore Most especially under care of Man. & external circumstances not variable.—

151 **because**] *final* ‘e’ *over* ‘d’.
152 **As species . . . settle it.—**] ‘**species**’ *underlined.*

154 Animals have voice, so has man. Not *saltus*. but *hiatus* animals expression of countenance. «[s]hare of sickness,— death, unequal life,— stimulated by same passions— brought into the world same way» they may convey much thus, Man has expression.— animals signals. (rabbit stamping ground) Man signals.— animals understand the language, they know the crys of pain, as well as we.—

It is our arrogance, to raise on the same shelf— to (look at common ancestor, (scarcely[)] conceivable in savages) Has not the white Man, who has debased his Nature «& violates every best instinctive feeling» by making slave of his fellow black, often wished to consider him as other animal— it is the way of mankind. & I believe those who soar above Such prejudices, yet have

155 justly excalted nature of man. like to think his origin godlike, at least every nation has. done so as. yet.—

We now know what is the natural arrangement, it is the classification of ‹arrangement› relationship; latter word meaning descent.—

A tree is taken by Fleming as *emblem* of *dichotomous* arrangement which is false[1]

There is same difficulty in arranging animals in paper as drying plant, all brought in one plane

---

156 Fleming Quarterly review says nat: fam: of Willows contains many Linnæan genera.—[1] Now are the characters which unite these of older standing than constant number of stamens.— in order, or in next family?

In considering fossil animals, what relation in *classification* in books, ought they to hold,—

Birds having web-feet, where we see scarcely any traces of passage a difficulty, but after all a slight one

---

It will be necessary from manner Fleming treats subject to put in alternative of Man created by distinct miracle.[2]

154 (scarcely[)]] *close parenthesis implicit in stroke of* 'ly'.
him] 'h' *over* 'it'.

155–1 Fleming 1829:311, 'This method of division in natural history has been termed 'Dichotomous,' because every group, as it is arranged or analyzed, is cut in two. The class, in this system, may be compared to the trunk of a tree, the subordinate orders and sections to the branches, and the species to the buds or leaves on the sprays.' See MacLeay 1830b:439, 'Man in this system may be compared to the trunk of a tree, Dominies and D.D.s to the branches, and John Fleming to the bud or leaf on the spray.'

156–1 Fleming 1829:316.

156–2 Fleming 1829.

157 Macleay letter to Dr. Fleming. Philosophical Magazine & Annals. 1830 (?).[1]

"if she has put man on the throne (*of reason*), she has also placed a series of animals on the steps that lead up to it" p. 20 ╫[2] +++ hiatus & saltus not syn.— Linn: Transact Vol XIV.—╫. p. 24.[3] Lamarck bears to Cuvier that relation of theoretical astronomer to plain observer╫[4] +++. between Mammalia & fishes, one penguin, one tortoise shows hiatus— but not saltus—[5] when Linnæus put whale between cow & hawk a frolicsome saltus. «p. 19»╫[6]

158 Macleay seems to limit Lamarck definition of relations to settling the relative importance of the organs *in same state.* in different animals, & the value of those organs, when changed in different animals.—[1] + whether variations in eye of vertebrate afford better character, than variations in eye of mollusca. [+] **These questions may be all disputable, but the one end of classification to express relationship. & by so doing discover the laws of change in organizaton. But the classfication must chiefly rest on these same organs,— habits, range. &c &c—**

158 **These . . . &c &c—**] '**chiefly**' *underlined.*

157–1 The MacLeay (1830a, 1830b) critique of Fleming is a letter addressed to N. A. Vigors. Although Darwin cites MacLeay 1830b, his subsequent page references are to the more rebarbative MacLeay 1830a.

157–2 MacLeay 1830a:20 (1830b, 8:57).

157–3 MacLeay 1830a:23 (1830b, 8:136–37) cites MacLeay 1825:54, 'Some years ago, in a paper in the fourteenth volume of the Linnean Transactions [Macleay 1825] —which if Dr. Fleming had read, marked, and inwardly digested, he would not thus have exposed himself— I stated it as an undoubted fact, that *hiatus* or chasms are every where in nature presenting themselves to the view; and I think I have now satisfactorily explained how the more numerous they are they produce the more continuity. "But this truth by no means contradicts the Linnaean maxim, that no *saltus* exists in nature, although such has been esteemed its effect by certain naturalists, who have been in the habit of taking the words *hiatus* and *saltus* as synonymous terms." ' This quote is continued in note C157–5.

157–4 MacLeay 1830a:24 (1830b:137). After dismissing Fleming's 'unfeeling sneer' on Lamarck, MacLeay cites *Horae Entomologicae* (1819–21:328–29), where he rejects Lamarck's transformism, but states 'we may be permitted to assign due praise to the labours of Lamarck, as being those of the first zoologist France has produced; . . . whose merits in natural history bear much the same relation to those of M. Cuvier, that the world has been commonly accustomed to institute between the calculations of the theoretical and the observations of the practical astronomer.'

157–5 MacLeay 1830a:29 (1830b, 8:202), 'Thus if the only animal existing between Mammalia and Fishes were one Penguin, it would still be in the path of passage. But if a Tortoise existed in addition, the chain would be more complete, and if one Frog existed also, the chain would scarcely escape notice.'

157–6 MacLeay 1830a:23 (1830b, 8:137) cites MacLeay 1825, 'Thus the series of the *Systema Naturae* and of the *Regne Animal* is not natural where the *Cetacea* intervene between quadrupeds and birds, but is perfectly consonant with nature where the tortoises are made to follow these last. In the first case there is a *saltus* or leap from quadrupeds to birds over a group totally dissimilar to the latter; there is, in short, an unnatural interruption of the law of continuity . . . In the other case there is only an *hiatus* or chasm, which the discoveries of a future day may fully occupy.'

158–1 MacLeay 1830a:24 (1830b, 8:137), 'Dr. Fleming, by the way, seems to hint that I borrowed the distinction of affinity and analogy from Lamarck. But this only proves that he reads as he reasons. Lamarck says, "On distingue les rapports en ceux qui appartiennent à differens êtres comparés, et en ceux qui ne se rapportent qu'à des parties comparées entre des êtres différens." Now the first of those kinds of *rapports* may be either relations of affinity or of analogy, for both affinity and analogy present resemblances between different objects compared with one another; and the second kind of *rapports* I shall speedily explain as having no connexion with relations of analogy.'

Macleay rests his whole groundwork of analogy on its concurrence in parallel parts of his series,[2] ie, cannot be discovered till circles completed

159e Major Mitchell, does not know whether breeds of oxen have deteriorated, or altered, but it is certain that rams & bulls from England fetch very ‹go› large price. as is evident to be worth introducing, instead of breeding from original Durham breed.— Native dogs & English cross readily— think about half way in appearance.— bark about half way «in tone»— the native dogs howl most dismally, very rarely bark.— are almost useless not the least notion of hunting, or keeping watch.[1] how completely «nature & instinct» modified.—

The partial migrations of birds in same country. may explain greater migrations, if American & intersected wider & wider
at Rio Plata. birds which had originally crossed would continue to cross, means of knowing directions: mysterious.

160e Were the woodcocks,. which came Madeira & ‹seized› ceased their migrations lost??[1] I conceive a bird Migrating from Falkland Is^{d} regularly to main land, proof of. land having been formerly nearer.—

«Selby» Magazine of Zoology & Botany No XI p. 390. a slight change in enclosing a common seems in part of— to have almost banished the Grasshopper Warbler—[2] —Yellow Wagtail never seen in one district, though

159 *page crossed pencil.*
160 Were the . . . formerly nearer:—] *crossed pencil.*
Magazine . . . might follow.—] '7' *and* '23' *added brown crayon.*

158–2 MacLeay 1830a:25–26 (1830b, 8:139), cites MacLeay 1825:51, 'The *theoretical* difference between affinity and analogy may be thus explained. Suppose the existence of two parallel series of animals, the corresponding points of which agree in some one or two remarkable particulars of structure. Suppose also that the general conformation of the animals in each series passes so gradually from one species to the other, as to render any interruption of this transition almost imperceptible. We shall thus have two very different relations, which must have required an infinite degree of design before they could have been made exactly to harmonize with each other.

'When therefore two such parallel series can be shown in nature to have each their general change of form *gradual*, or, in other words, their relations of affinity uninterrupted by anything known; when moreover the corresponding points in these two series agree in some one or two remarkable circumstances, they afford relations of analogy, and there is every probability of our arrangement being correct. It is quite inconceivable that the utmost human ingenuity could make these two kinds of relation tally with each other, had they not been so designed at the Creation.'

159–1 T. L. Mitchell: personal communication.

160–1 Heineken 1835:77, 'Forty years ago they [woodcocks, *Scolopax rusticola*] were unknown here [Madeiras]. One was then accidentally met with in the South, and afterwards abundance in the North of the island, where they were for many years plentiful, and since that time have never disappeared.' See also *Natural Selection*: 491, 'The migratory instinct in birds is occasionally lost; as in the case of the Woodcock . . . In Madeira the first arrival of the Woodcock is known & it is not there migratory; . . .'

160–2 Selby 1837–38, 2:390–91, 'Of the genus Salicaria, the sedge warbler (Sal. phragmitis) is still frequent upon the margins of the brooks and moist bushy situations; but the grasshopper warbler (Sal. locustella,) which, during the early growth of many of the plantations, then abounding in whin, broom, and other undergrowth, might be heard in various directions, pouring forth its sibilous note, now that they have attained a considerable growth, is rarely heard, and then only in the brushwood adjoining the moor and other open ground.' Scored, ink and pencil.

common on another,[3] (golden creted wren so rare in some countries— nightingale do.— all shows how nicely adapted species to localities╫— p. 390,. young ‹grebes› «ring ouzels» dive instant touch the water.—[4] capital instances of typical land bird, having habits of a Grebe, structures might follow.—

161e examine structure of this bird & get account of habits

My definition ‹in wild› of species. has nothing to do with hydridity,, is simply, an instinctive impulse to keep separate, which no doubt be overcome, but until it is the animals are distinct species

---

If any one is staggered at feathers & scales. passing into each other let him look at wings & orbits of penguin & then he will cease to doubt **:Scales into Teeth in Bering Pike (Waterhouse)**

162e Magazine of Zooly & Bot— Vol II p. Dr Johnston[1] ‹on› Entomostraca Daphnia, produce young, capable of producing young many times[2] & lay two sorts of eggs— one remaining through winter.[3] «It would be curious to know whether a variety could be transmitted more easily in those born without coitus, than with.»
╫Might be given as a hopeless difficulty, except as distinct creation.— Generation may be viewed as condensor, «+++ must (on my theory) =supported by fœtal lower developed forms.=» (NB waterhouse says of *affinity* of many insects may be told by their larvæ) but the acts of condensing must alter method of generation— Heaven knows how.— This reaction takes

161 examine . . . distinct species] *crossed pencil.*
'11' *added brown crayon at bottom of page.*
:Scales . . . (Waterhouse)] *added pencil. This note written around, hence after,* '11' *added in brown crayon.*
162 *page crossed pencil.*

160–3 Selby 1837–38, 2:390, 'The yellow wagtail is never seen in this district, though far from uncommon upon the dry hilly grounds a few miles to the westward.' Passage scored, ink.
160–4 Selby 1837–38, 2:390, 'The young [dipper, or water-ouzel], even before they leave the nest of their own accord, if disturbed and made to quit it, dive instinctively the moment they touch the water, . . .' Passage double scored, pencil.
162–1 Darwin has confused G. Johnston's paper 'The Natural History of the British Zoophytes' (1838b) with Baird's paper 'The Natural History of the British Entomostraca' (1838), which appeared in the following number of *Mag. Zool. Bot.*
162–2 Baird 1838, 2:405–6, 'The males are very few in number compared with the females, and are only met with at certain seasons. From this circumstance Schoeffer and others have considered them as Hermaphrodites; and Sulzer (as quoted by Straus), though he oppugns this, gives a more singular opinion still, being of opinion that a copulation might take place with the young before they see the light of day! These authors had never seen the males, nor ever witnessed the act of copulation. Muller and others, however, detected the male, and witnessed the act; and it is ascertained that one single copulation is sufficient not only to fecundate the mother for her life, but all her female descendants for several successive generations.' Passage scored. See Straus-Durckheim 1819–20.
162–3 Baird 1838, 2:410, 'Straus . . . has found it [the epihippium of *Daphnia*] to be a substance containing two eggs, destined, he says, for the future generations of the insect in the spring, these eggs resisting the cold of the winter, which proves fatal to the perfect insect.' Passage scored.

place in every organ|| «Hence «method» of generation is very good «general» character in those animals, where much change has been added. «as it speaks to amount of change only & not kind» insects— & vertebrata & plants. At first classification on generation might appear an analogy

163 NB Pyrrho-alauda (bird of St Jago) of brown colour; lives on ground, colour of habitation.[1] Must have some effect.— Maldonado as good forests for beautiful birds.—

═══

heredetary ambling horses, (if not looked at as instinctive) then must be owing to heredetary power of Muscles.—[2] then we SEE structure gained by habit

═══

Talent &c in man not heredetary, because crossed with women with pretty faces

═══

When horse goes a round, the minute gets into the road at right anglles, how pleased it is, just like man. emotions very similar.—

Geology. Transact. Vol V. Birds bones— in strata of Tilgate forest[3]

164 Seeing common gull in garden at Zoology Soc. it's pale ash grey back, like a black bird washed, Whilst tips of primaries black, by examining series I cannot doubt laws of change, Will be known.—

It appeared to me that half between fowls & pheasants, is most like pheasant., I think so because very 3/4 bred.— (hence hybrids in this case have bred). White & common pheasants. have crossed.— ‹/I saw› .—[1]

---

165 The attachment of dogs to man. not altogether explained by F. Cuvier, «— .Mem. Hensleighs objection.—[1]» it is more, he cuts the matter short by saying man cannot be companion but master.—[2] heredetary tameness as well

163 *page crossed pencil.*
164 White] 'W' *emmended grey ink.*

163–1 See *Birds*:87. *Pyrrhalauda negriceps*.
163–2 See *Natural Selection*: 485 and E127.
163–3 Mantell 1837.
164–1 Fuller 1836:84, 'Several specimens of hybrids, from the preserved collection in the Museum of the Society, were placed on the table for exhibition and comparison. These had been bred between the *Pheasant* and *common Fowl*, the *common Pheasant* and the *silver Pheasant*, and the *common Pheasant* with the *gold Pheasant*.' Passage marked with 'Q', for quote, ink.

165–1 Hensleigh Wedgwood: personal communication.
165–2 Frédéric Cuvier 1827–28, 3:315, 'The absolute submission which we require of animals, and the sort of tyranny with which we govern them, have led to the idea that they obey us as absolute slaves, that the superiority which we have over them is sufficient to constrain them to renounce their natural love of independence, to bend them to our pleasure, to satisfy such of our wants as their organisation, their intellect, or their instinct permit us to employ them for.'

as wildness— cf Sir J. Sebright.— love. of man gained & heredetary.[3] «problem solved»

---

habits become important element in classification, because structure has tendency to follow it, or it may be heredetary & strictly point out affinities. conducct of Gould,[4] remark of D' orbigny point out importance of habits in classification.—[5]

166 Thought (or desires more properly) being heredetary‹)›.— it is difficult to imagine it anything but structure of brain heredetary,. analogy points out to this.— love of the deity effect of organization. oh you Materialist!—[1] Read Barclay on organization!![2]
Avitism in mental structure or disposition. & avitism in corporeal structure are facts full of meaning.—
Why is thought. being a secretion of brain,[3] more wonderful than gravity a property of matter? It is our arrogance, it our admiration of ourselves.—

167 The idea of fœtus being of one «both» sex«es». is strongly supported by wonderful fact of bees changing the sex by feeding.— **no it is developing an hybrid female** it is a wonderful relation going through all Nature.— Makes hermaphroditisms. one step in ‹scale›. Series— in plants we have a step between ‹mon› monœecious & diœcious plants in animals it may be difficult to imagine how sexes were separated.— **in plants we have some flowers monœcious & others diœcious. some flowers hermaphrodites & others not???** ╫
The death of some forms & succession of others, (which is almost proved. Elephant has left no descendant in Europe‹)› Toxodon in S. America) is absolutely necessary to explain genera & classes. if extinct forms were all fathers of present, then there would be

168 perfect series or gradation.— It is easy to see if South America grew very much hotter, then Brazilian species would migrate south ward being ready made.— & so destroy individuals, wheras in Falkland Is^{d} they would change & make new species.— alpine species being destroyed at Falkland Is^{ds}++.— Mem Lyell hypothesis of change in Scicily.—[1] Splendid Harmony these

165–3 Sebright 1836:11, 'Of all the propensities of the brute creation, the well known attachment of the dog to man is, perhaps, the most remarkable, arising, I conceive, from his having been for so many years his constant companion, and the object of his care. That this propensity is not instinctive is proved, by its not having existed even in the slightest degree in my Australian.' Passage scored, ink and pencil. See C134, B165.
165–4 See note C80–1.
165–5 A. D. d'Orbigny: personal communication.
166–1 See Fleming 1822, 1:xiii, which refers to 'Dr Barclay's Treatise on Life and Organization. It should be perused with care by every student of anatomy and Natural History, as an effectual preservative against the doctrines of Materialism, and deserves a place as well in the library of the Divine as in that of the Physiologist?' See also M57, 'To avoid stating how far, I believe, in Materialism, say only that emotions, instincts degrees of talent, which are hereditary are so because brain of child resembles parent stock.—'
166–2 Barclay 1822. Darwin's copy has no marginalia; however, his youthful inscription may date from his Edinburgh days. Only certain sections are cut, including: chap. 3, sect. 2 (opinions of Darwin) to sect. 12 (opinions of Maupertuis).
166–3 See GR126 '. . . brain makes thought'.
168–1 See B200, note B200–1 and Lyell 1837, 3:444.

views— did Lamarck connect extermination of some forms with his views.—[2]
as genera are large probably only few of extinct

169 forms have generated species. & of 100 extinct species the greater number probably have no descendants on earth.—
+++. Even at Falklands some probably would stand change better than others.—

The more complicated the animal the more subject to variation. therefore sexes or two animals:—
«**When**» sexes ‹being in one normal— when so the› **are united (which probably is first stage) the** tendency to change cannot be great, otherwise it would be unlimited.

We absolutely know that tendency is greater in Mammalia, than in shells ? univalves or bivalves.—

170 Anyman No VI. Magazine of Zoology & Botany p. 566 wants to see absurdity of Quinary arrangement let him look at abstract of Swainson on Classification "Let anyone even with a very superficial knowledge «like myself» of ‹classi› real affinities. ie structure of the whole animal let him read Mr Swainson's on the Classification of animals. & observe the character of the *demonstrations* offered of the singular views there offered, & he must be a zealous man in the cause if his faith is not staggered ‹"›=[1] I confess. no dissertation against these views, could possibly have had ‹to so convin› brought so much conviction to my mind.—

171 Reflect much over my view of particular instinct being memory transmitted without consciousness «a most possible thing. see men walking in sleep».— an action becomes habitual is probably first stage, & an habitual action implies want of consciousness & will & therefore may be called instinctive.— But why do some actions become heredetary & instinctive & not others.— We even see they must be done often «to be habitual» or of great importance to cause long memory.— structure is only gained slowly.— therefore it can only be those actions, which *Many* successive generations are impelled to do in same way— The improvement of reason implies diversity & therefore would banish individual, but general ones might yet be transmitted.=

170 faith is] 's' *over* 'n'.
have had] 'd' *over* 've'.

168–2 See Darwin's comment on Lamarck 1830, 1:78, 'Lamarck argues, species of shells, not killed by man, no apparent cause of death; but causes of change are present therefore fossil same species with modern.'— Also on bottom of 1830, 1:80, 'thinks every fossil species direct father of existing analogues & no extinction except through man!— [Hence cause of innumerable error in Lamarck]CD.'

170–1 Swainson 1837. Darwin is not quoting a passage, rather he is composing one.

172 Memory springing up after long intervals of forgetfulness.— after sleep, «strong» analogies with memory in offspring.— Some association in such cases recall the idea. «or simple structure in brain people in fevers recollecting things utterly forgotten» —it is scarcely more wonderful, that it should be remembered in next generation. [NB what are those Marvellous cases, when you feel sure you have heard conversation before. is strong association recalling up image which had been past— so great an anomaly in structure of brain not probable) put note. Sir W. Scott has written about it]CD If we saw a child do some action— which its father. had done habitually we should exclaim it was instinct.— Even if savage takes. & was given a Great coat & this he put on & we afterwards could understand «(language better instance)» he had done this without reflection or consciousness of reasoning to tell back from front. &c or use of button holes it would

173 be instinctive.— My view of instinct explains its loss ? if it explains its acquirement.— Analogy. a bird can swim without being web footed yet with much practice & led on by circumstanc it becomes web footed, now Man by effort of Memory can remember how to swim after having once learnt, & if that was a regular contingency the brain would become webfooted & there would be no act of memory.—

[There is no corelation between individual objects as Ichneumon & caterpillar, though our ignorance, may make us think so, but only between laws—]CD.

174 Many diseases in common between man & animals— Hydrophobia &c cowpox, proof of common origin of Man.— different contagious diseases, where habits of people nearly similar. Curious instance of difference in races of men.—

Wax of Ear, bitter perhaps to prevent insects lodging there. Now these exquisite adaptations can hardly be accounted for by My method of breeding there must be some corelation. but. the whole Mechanism is so beautiful.— The corelations are not, however, perfect, else one animal would not cause misery to other.— else smell of Man would be disagreeable to Musquitoes

175 We never may be able to trace the steps by which the organization of the eye, passed from simpler stage to more perfect. preserving its relations.— the wonderful power of adaptation given to organization.— This really perhaps greatest difficulty to whole theory.—

There is breed of tailless cats near Bath. Lonsdale[1]

175–1 William Lonsdale: personal communications C175–177.

do. says Sheep could not live for some time at New York «instance of the fine relations of adaptation of animals & the country they inhabit.—» & the first one that bred one was diseased in its loins & all were so afterwards, (forgets authority).—

176 Lonsdale is ready to admit, permanent small alterations in wild animals,[1] & thinks Lyell has overlooked argument, that domesticated animals change a little with external influence— & if those changes permanent so would the change in animal be permanent.— It will be easy to prove persistent Varieties in wild animals— but how to show species— I fear argument must rest upon analogy & absence of varieties in a wild state— it may be said argument will explain very close Species in isl^{ds}. near continent, Must we resort to quite different origin when species rather further.— once grant good species as

177 carrion crow & rook formed by descent. or two of the willow wrens[1] &c &c. & analogy will necessarily explain the rest,—

Lonsdale says he has seen in old Book last Bear **in England** killed in year 1000.[2] reference to succession of types ʕdifferent species;— Horse— &c

---

‹Lonsdale says. that first shee›
State broadly scarcely any novelty in my theory, only slight differences. «the opinion of many people in conversation.» the whole object of the Work is its proof,[3] «its limiting, the allowing at same time true species.» & its adaption to classification & affinities, its extension.—

Travels, p. 306. account of trees ceasing to grow far N. becomes ered, & lose **(mere sickness)?** fertility[1] ʕbecause offspring too

---

v' *over* 'a'.
'm' *over* 'w'.
se Species] 'S' *over* 's'.
over 'ts'.
' *over* 'c'.
lose] *emmended grey ink.*
(**mere sickness**)?] *added grey ink.*
fertility] *emmended grey ink.*
No] *added brown ink.*
analogy of man] *added grey ink.*

176–1 This degree of heterodoxy may have contributed to Lonsdale's being considered as a possible editor of the *1844 Essay* (Letter to Emma Darwin, 5 July 1844, *Calendar*: 761, *Correspondence* 3).
177–1 See B241 and C125.
177–2 See Thomas Bell 1837:122, 'the extirpation of the Bear, *Ursus arctos*, of the existence of which mention is made in Scottish history as late as in the year 1073.' Noted in Darwin's abstract of Bell (DAR 71:117).
177–3 Echoed in *Origin*: 459, 'this whole volume is one long argument'.
178–1 See Buch 1813:303 quotes Wahlenberg, 'On approaching the Lapland Alps (Fjall) we first arrive at the line where the spruce-fir, (*pinus abies*,) ceases to grow. This tree had previously assumed an unusual appearance; that of a tall slender pole, covered from the ground with

Memoire by Charles D' orbigny on Plastic Clay of Paris contains many genera of Pachydermata &c & other Mammals.—[2] «otter; civet cat, rodents.» (Pachyderm in Portland stone of Alps!!!? **No**) p. 15 (Lyell's Pamphlet)

Is man more hairy than woman. because ancestors so, or has he assumed that character — — female & young seem most like mean characters the others assumed—+++ ||Daines Barrington says cock birds attract females by song[3]. do they by beauty, **analogy of man** if so war not

179e–182e [not located]

183e Erasmus says he has seen old Stallion tempted to cover old mare by being shown, young one.—[1]

Many African monkeys in Fernando Po— no new *forms* only species!!

No salamanders (D' orbigny Rapport. p. 11[2]) in S. America so highly developed in North.—[3]

Icthiology of S. America. more peculiar than its ornithology X p. 12 do.[4] excepting salmons

---

L' Institut. Sorex from Mauritius. p. 112.[5] & paper on genus[6]

---

183 Erasmus . . . one.—] *crossed pencil.*
X] *added brown ink along with a line connecting to* 'D'orbigny Rapport'.
L'Institut . . . germs] '19' *added brown crayon.*

short, drooping, dark branches . . . The *rubus arcticus* had already, before we arrived at this point, ceased to bring its fruit to maturity. . . . Higher up are hardly any bears to be met with, and the berries of *vaccinium myrtillus* . . . do not ripen well. . . . A little before the birch ceases we miss the *sorbus aucuparia*, which for some time had not presented us with any fruit; the *rubus arcticus*, already likewise barren, *erica vulgaris*, &c.' (pp. 306–7). References to plants not bearing fruit are underlined by Darwin. See E152.

178–2 Charles d'Orbigny 1836a and 1836b.

178–3 Barrington 1774:263.

183–1 Erasmus Alvey Darwin: personal communication.

183–2 Four reports on the results of Orbigny's voyage to South America were read at the Paris Academy of Sciences on 21 April 1834. I. Geoffroy Saint-Hilaire and Blainville 1834 is the zoological report. The geological report (Cordier 1834) is discussed in A99. From the page numbers he cites, Darwin read an offprint of the original report or a repaginated offprint of the reports published in *Nouv. Ann. Mus. Hist. Nat. Paris.*

183–3 Isidore Geoffroy Saint-Hilaire and Blainville 1834:94, 'La classe des Amphibiens est encore moins riche que la précédente dans les collections de M. d'Orbigny. Il n'a rencontré qu'un énorme Crapaud d'un pied de long, quelques espèces nouvelles de Grenouilles et de Reinettes, mais aucun Salamandre terrestre ou aquatique, point de Sirènes ou autres genres voisins si répandus dans la Nord-Amérique.'

183–4 I. Geoffroy Saint-Hilare and Blainville 1834:95, 'Ainsi, sauf les saumons qui son assez communs dans les rivières de l'Amérique méridionale, on peut dire que la physionomie ictyologique de ce pays est plus particulière que celle des oiseaux.'

183–5 Duvernoy 1838:112.

183–6 Duvernoy 1838:111, 'M. Duvernoy lit la suite de ses fragments d'histoire naturelle sur les Musaraignes, qui comprend des additions à son premier mémoire (Duvernoy 1835) et une révision des espèces.'

184e Magazine of Zool. & Bot.— Vol I. p. 450. 4 instances of hybrids between pheasants & Black fowl.—[1] use as argument possibly some few hybrids in nature.—

«p. 473» Webb &. Berthelot. must be studied on Canary islands— **Endeavour to find out whether African forms. (anyhow not Australian) on Peaks. Did Creator make all new yet forms like neighbouring Continent. This fact speaks volumes.** 2 Chapters.[2] translated by Hooker.—[3] my theory explains this. but no other will.— **St. Helena (& flora of Galapagos?) same condition. Keeling Is[d] «shows where proper dampness seeds can arrive quick enough»**
Vegetation of peak— altogether original. owing to being oldest. & having undergone changes «no near lofty country»?? p. 475[4]
NB. This bears on fossils of Europe., those species which can migrate remaining constant in form, others altered much.— **these others will be plants & land animals. & land shells.— all in short** Extreme North = = to peak of Teyde in relation to surrounding countries & present tropical countries.

185e p. 564. an abstract of M[r] Swainsons views. which if **abstract** true are *wonderfully* absurd.—[1]
p. 565 ‹breed› Scotch wild Cattle. breed freely with the tame[2]
Vol II. Magaz**ine of Zoology**
p. 56. Peregrine Falcon holds *birds* for some time alive ⁅therefore other

184 Webb . . . p. 475] *double scored grey ink.*
Webb . . . islands] '20' *added brown crayon.*
This . . . volumes] *on stub.*
Vegetation of peak . . . p.475] *brace in left margin.*
185 p. 564 . . . tame] *crossed pencil.*
Scotch] emmended grey ink.
Magaz**ine of Zoology**] '**ine**' *added grey ink and* '**of Zoology**' *added pencil.*
p. 56 . . . hunger] *crossed pencil.*
divided] *first* 'd' *over* 's'.
p. 65 . . . group aberrant] '11' *added brown crayon.*

184–1 W. Thompson 1837b:450, 'In four instances only am I aware of similar hybrids being recorded.'
184–2 Barker-Webb and Berthelot 1837.
184–3 William Jackson Hooker.
184–4 Barker-Webb and Berthelot 1837:475 '. . . above this elevation, when the traveller reaches the peak itself . . . the vegetation of these wild regions is found to be altogether original.' Darwin double scored this passage and commented, 'Can this be owing to summit being oldest part of isl[d]?' See C170.
185–1 Swainson 1837:564. Darwin 'abstracted' Swainson. That is he put a quotation mark along with a score, and a question mark on the following passage, 'The Solipedes pass into the Ruminants by the camel; the tapirs, which show an intermediate form between the ruminants and the armadilloes, connect the *Anopletheres* with the *Edentates*; the Sloths, which belonging to the latter, lead directly to the fossil genus *Megatherium*, in the circle of the *Pachydermes*, and the return to the Solipedes is supposed to be effected by means of the hippopotamus, though we confess the affinity seems distant, and that some intermediate forms are wanting to bring these divisions into juxtaposition.'
185–2 Swainson 1837:565, 'In his [Swainson's] observations upon the *Bos Scoticus* or *wild ox*, existing in some of our parks we do not agree, as we consider them merely varieties of the *Bos taurus*, and very nearly akin to the kyloe, nor do we think them naturally more savage and untameable than the latter would show themselves if placed in a similar situation, and treated in the same manner. We know, besides, that they breed freely with the common ox, and that the progeny produced from the cross is also productive, and it is a well ascertained fact, that, if taken young, they become as tame as any of the common breed.' Passage double scored.

species mice & only kills them when urged by hunger.—[3]
p. 65. Aberrant groups few in numbers & vary much in character, divided into many small genera[4] ∵circumstances not favourable to many species., same circumstances., which by causing death, makes the group aberrant

186e When species rare we infer extermination, when group few in number of kind, extermination.— New forms *made* through probably an infinite number of forms.— therefore an isolated form probably a remnant.— Pachydermata. & Horses few forms. & they are remnants.— Cephalopoda ditto.—

Mag of Zoolog. & Bot. Vol. II p. 125
Allusion to abortive spiracles in Hemiptera[1]

---

do. p. 160. soft plumage of night jar. like owls.[2] analogy in habits adaptation to nocturnal habits— to cats &c.— must be acquired by my theory— else my theory not applicable

187e–188e [not located]

189 p. 428. Ouzel sometimes builds nest without doom.[1] **Vol 2. Mag of Z. & B.**
p. 431. Missel thrush lately increased in numbers over whole of England &

186 When species . . . ditto.—] *crossed pencil.*
Mag . . . applicable] *crossed pencil.*
189 **Vol . . . & B.**] *added pencil.*
Gould says . . . question.—] *crossed pencil.*

185–3 William Thompson 1838a:56.
185–4 George Johnston 1838a:65. 'Of the four genera which Audouin and Milne-Edwards include in this family [Ariciadae], we have two native species of one only; but it is remarkable that our other species, which as yet are limited to the number of three, constitute two new genera in it very distinct from any hitherto characterized. The fact is an additional illustration of an axiom in natural history,—that all aberrant and osculant groups are not only comparatively few in species, but at the same time these species are so dissimilar among themselves that each, or every two or three of them, will be found to have characters which are properly generical.' Darwin scored this passage in ink and commented, in pencil, 'Hereditary | descendan[ts] many kil[led] | analogica[l] | O X'. Darwin's note is crossed in pencil. At the bottom of the page he wrote 'much extinction'. Bits of Darwin's marginal note have had to be interpolated because the writing was cut off when he had the volume bound. Hence, marginalia in *Magazine of Zoology and Botany* 2 (1838) are likely to be contemporaneous with the Notebook entries. See Darwin's abstract, pinned at the end of the volume, '[p.] 65 «Johnston on» aberrant genera very distinct & *few* species'.
186–1 Westwood 1838:125, 'The genus Nepa offers a still more remarkable modification in the structure of its respiratory organs and mode of respiration. On examining an insect of this genus, the spiracles appear at first sight to be in the ordinary position and of the ordinary form; but we learn from M. Dufour's admirable Recherches Anatomiques sur les Hemipteres, that these spiracles have no orifice and are quite useless, the only spiracles being two, which are placed at the base of the anal setae.' Passage scored. At the bottom of the page he noted '‹Owen talks›', and in a separate note, 'Expatiate that abortive organs show, that such an insect has descended from an air breather'.
186–2 W. B. Clarke 1838:160, 'The plumage is peculiarly soft, which enables it . . . to pass rapidly through the air without the vibrations of its wings being heard, and . . . during its diurnal rest . . . the colours of the plumage much resembling the tints of the bodies amidst which it secretes itself.' Passage scored. Darwin commented at the bottom of the page, 'Good case of adaptation like owls'.
189–1 W. Thompson 1838b:429, 'The first nest of this bird I remarked was placed in a hole in the clayey bank of a pond, where, owing to the shelter afforded, there was no occasion for the display of its domed architecture, and this was consequently dispensed with.' Passage scored.

Ireland.—[2] curious in so wild a bird.—

Annals of Natural History Vol. I. p. 185 case of tit lark placing withered grass over nest, when often looked at.—[3] this most puzzling whether instinct, or reason??

Gould says he believes that he has seen half fox & dog. & that it was most like fox.— He felt sure the half breed of Australian dogs, would be «most» like Australian.— Curious this +ready answer, without any leading question.—[4] [+] **This might be mentioned in note.—**

190 try to trace from simplest reasoning in lower animals many times produced, a general tendency produced. Such as man getting habitually into passion, becomes habitually passionate.— the Key to the affections might perhaps thus be found— a person who is habitually kind to children,—‹get› «increases» general instinctive feeling.—

191 There is great difficulty in Making an alpine species from one in lower country during gradual elevation of isl^d^.— We must imagine

«———————————»

a considerable range of one species «on — mountain side» of which the central parts become occupied by a third best adapted kind.— lower species would then revert to pristine form (which must have been altered by crossing) with alpine form) lower species afterwards would probably often be destroyed.— or regrafted with fresh arrivals..— &c &c **—Climate altering as island increases.— upper parts attracting all the moisture.—**

192 Henslow thinks if leaf of plant varies, ‹whole cross› «all organs» vary in plant.[1]
The variation in character of leaf of plants is remarkable what is analogous to it in animals?—

Babington says in most plants, even those on Guernsey & on West coast of Ireland, are absolutely (& who better authority) similar with those over whole of country.— some species are larger &c in different countries.[2]
These facts show how very permanent plants

191 «on— mountain side»] *and preceding rule added in same watery brown ink used in C192–C197. C192–197 written watery brown ink.*

189–2 W. Thompson 1838b:431. Relevant passage scored. For Darwin's attention to population fluctuation see C160.
189–3 W. Thompson 1838c:185.
189–4 John Gould: conversation.
192–1 C192–197 are written in a distinct ink, which may reflect conversations with Henslow and Babington. Henslow 1830b:537, 'Though we cannot say the following law is certain in botany, yet it seems to me very likely to be true, viz. "That if a change takes place in one of the organs of a plant, a simultaneous change may be expected in some or all of the other organs considered to be modifications of the same organs."' Quote from W. J. Hooker. Passage scored.
192–2 C. C. Babington: personal communication. See Babington 1838.

193 are. & this conclusion must be arrived at, when one sees a plant like Paris quadrifolium growing in one wood far from any other plants of same species

———

Channel Isl^{ds} (& probably Isle of Man) no plants peculiar to themselves. this remarkable compare it with Canary Is^{lds}. Galapagos.— Iceland has same uniformity

194 Primrose & Cowslip, quite wild, but they affect different localities,— latter on banks & in damp parts.— both propagated by seeds.—[1] There are two Dandelions, which just lately have been shewn to be same.— one grows in marsh & other dry; yet if T. *palustris* be sown in dry station it will for some generations come up so.— there are not Many intermediate shades in these cases,— but absolute species formed.
The Anagallis perhaps, offers another case of permanent varieties in wild state[2]— The two.

195 former produced by difference of ‹locality›. station &

‹═══›

varieties chiefly produced by cultivating parent in *rich* soils & their seeds produce ‹offspring› variety.

═══

wild carrot. made into biennial domesticated kind with large root by sowing it at wrong time of year & manuring it.—

═══

Epigonous. Perigonous &c— very important in classification. here we have generative organs. first character.—

═══

In diœcious plants many of the Female flowers unimpregnated   Babington[1]

196 We see gradation to mans mind in Vertebrate Kindgdom in more instincts in rodents than in other animals & again in Mans mind, in different races, being unequally developed.— ʕis not Elephant intellectually developed amongst Pachydermata. like Man amongst Monkeys— or dogs in Carnivora.—

196 ʕ . . . Carnivora.—] *brace in left margin.*

194–1 Henslow 1830a:408, 'Mr. Herbert remarks:— "I raised, from the natural seed of one umbel of a highly manured red cowslip, a primrose, a cowslip, oxlips of the usual and other colours, . . ." . . . I confess that I had myself given very little credit to this experiment of Mr. Herbert's, until it was recalled to my mind by a circumstance which I noticed in April, 1827, a few miles from Cambridge, at a place called Westhoe. I there found in great plenty a peculiar variety of Primula, which I scarcely knew whether to call the oxlip or the cowslip . . .'

194–2 Henslow 1830b:537, 'Dr. Hooker . . . still keeps the Anagallis caerulea distinct from the A. arvénsis; . . . From these seeds I have raised a dozen plants, nine of which have blue flowers, and three have red. Hence it should seem that in future Anagallis caerulea must be considered as a variety of A. arvénsis.' Passage scored and marked with 'Q' for quote.

195–1 C. C. Babington was either the source of the above information or Darwin is noting him as a potential botanical informant.

Man in his arrogance thinks himself a great work. worthy the interposition of a deity, more humble & I believe true to

197 consider him created from animals.—

Insects shamming death,[1] most difficult case to iimagine how art acquired.— They reason however on this to a degree.
Mem Spider only dropping where ground thick.— shamming death it is but being motionless.
How is instinctive dread **it is exceedingly doubtful whether animals have any fear of death, only of pain.** of death acquired?.

The S. American dung beetles will each become the fathers of many species.— a few eggs transported to the St^r of Magellan.— Change of habits in Van Diemen's land.

198 Study Mr Blyth's papers on Instinct.— His distinction between reason & instinct very just,[1] but these faculties being viewed as replacing each other it is hiatus & not saltus.—
The greater individuality of mind in man, is analogous to greater individuality of bodies of some animals over those of others.— the mind of different animals less divided.— But as man has heredetary tendencies. his mind is still **only** a divided body. ǂp 3. language seems to supply instincts,— & those powers which allow of, acquirement of language. heredetary & acquirable.—[2] therefore mans mind not so different from that of brutes[3]

197 **The S. . . . land.**] *pencil.*
198 only] *emmended grey ink.*

197–1 See Blyth 1837b, which discusses 'counterfeiting of death', but not in insects.
198–1 Blyth 1837a:2, 'The distinction is, that, whereas the human race is compelled to derive the whole of its information through the medium of its senses, the brute is, on the contrary, supplied with an innate knowledge of whatever properties belong to all the natural objects around, which can in anywise affect its own interests or welfare;' and *passim*. Passage triple scored. Darwin noted at the bottom of the page, '⁋ ⁋Child fears the dark—before reason has told it—' and in the margin, 'Connexion is less obvious between man. yet real'.
198–2 Blyth 1837a:3, 'The human infant, too applies instinctively to the breast, like the young of all other mammalians; but unlike those, it has to attain all its after-knowledge through the medium of its external senses. It looks to its nurses, and those about it, for information; and these are capable of so communicating their attainments, as very materially to assist the infant learner in its acquisition of knowledge. It is preposterous to assert the contrary, as has been done; or to pretend that it rests on the choice of the infant whether or not it will learn. Practically, it cannot help doing so; and it is equally monstruous to deny that human beings can so communicate the results of their experience, that, with what in addition is ever accumulating, each generation must necessarily rise in knowledge above the last.' Passage double scored.
198–3 Blyth 1837a:2–3. Darwin marked the following passage with four exclamation marks, 'Witness a thrush that has captured a wasp, first squeezing out the venom from its abdomen, before it will swallow it. Or see a spider trying to cut it clean away. Can aught analogous be traced in the actions of inexperienced man? Whence, then, the acquired knowledge on which these animals could reason to act thus?' (p. 2). He drew a line to connect this passage to the following one on p. 3, 'The human infant, too, applies instinctively to the breast, like the young of all other mammalians; . . .'

p. Hard to say what is instinct in animals. « & what reason» in precisely same way not possible to say what habitual in men & what reasonable[4]— Same action may be either in same individual

199 p. 7. is not squirrel hoarding, & killing grains. acquirable through hoarding from short time.— My theory must encounter all these difficulties.— Knowing that animals have some reason, & actions habitual. it surely is not worthy interposition of deity to teach squirrel to kill ears of corn[1]

═══

according to my views, habits give structure, . . habits precedes structure, . . habitual instincts precede structure.—duckling runs to water. before it is conscious of web. feet.—
p. 7. Mr Blyths arguments against squirrel using reason in hiding its food is applicable to any habitual action.[2] even which Man performs.— child striking a post in passion.—

═══

Habit instinct gained during life.— do Elephants easily acquire habits is this the Key to their mental powers.?

———

p. 8. mistakes of instinct are external contingencies, where the habit is not applicable.[3]

200 The degree of development of all animals of same class being about equal.— organs of generation about equally complicated.—

An Entomologist going into a country & collecting thousands & tens of thousands New insects, perhaps scarcely one new family & no new orders,—

198–4 Blyth 1837a, probably pp. 3–7, 'I wish not to defend the untenable doctrine, that the higher groups of animals do not individually profit by experience; nor to deny to them the capability of observation and reflection, whereby to modify, to a considerable extent, their instinctive conduct: neither do I assert that the human race is totally devoid of intuition, when I see the infant take naturally to the breast; when I perceive the force of the maternal attachment, and the ardour of the several passions: which latter, however, are, of course, but incentives to conduct common to both man and animals. In only the human species are the actions resulting from them unguided by intuitive knowledge. All I contend for is, that the ruling principle of human actions is essentially distinct from that which mainly actuates the brute creation, whence the general influence of the two is diverse in kind; and I mistake if I cannot establish the position.' (pp. 3–4). Last sentence scored.

199–1 Blyth 1837a:7–8, 'Proceed we, then, to examine into the presumed sagacity of those provident creatures, as the ant and harvest mouse, that habitually lay up a store for future need, and even provide against all possible injury from germination, by carefully nibbling out the corcule from each grain. [Hence, CD's reference to 'killing'.] . . . I have a tame squirrel, which, though regularly fed all its life from day to day, nevertheless displays the intuitive habit of its race, in always hoarding the superfluities of its food. . . .

Thus it plainly appears, that the instinct of each animal is adapted to its proper sphere; for the mode of life it was destined to pursue, and for that only.'

199–2 Blyth 1837a:7, 'I have repeatedly seen the same animal act . . . upon a smooth mahogoney table . . . deposit its nut, give it a few quick pats down, and finally thus leave it wholly unconcealed.'

199–3 Blyth 1837a:8, 'a tame marsh tit . . . used habitually to drop the remainder of the almond . . . he had been picking, into the water-glass attached to the cage, although he never could thence reobtain it, and though his water was thus daily rendered turbid.'

Wonderful, partly explained on my theory, = other wise mere fact creator chooses so to create.—

It is very remarkable, with so much death, as has gone on,, No greater gaps.— external conditions, to be sure, have remained somewhat similar.—!!!

---

201 My theory drives me to say that there can be no animal at present time having an intermediate affinity between two classes.— there may be some descendant of some intermediate link.— the only connection between two such classes will be those of analogy, which when sufficiently Multipliied become affinity yet often retaining a family likeness, & this I believe the case. = any animal really connecting the fish & Mammalia, must be sprung from some source anterior to giving off of these two families, but we see analogies between fish.— Birds same remarks.

202 Characters of analogy.— last acquired,— or aberrant, therefore more easily modified[1] == this is not easily told, for any small family. having analogous characters, might be multiplied.— we must argue reversely. WHERE CHARACTER VARIABLE it is (one of analogy or) LATELY ACQUIRED. In pigs number of vertebræ. subject to variation. therefore lately acquired.—
I fear «great evil» from vast opposition in opinion on all subjects of classification, I must work out hypothesis, & compare it with resuts. if I acted otherwise, my premises ‹in di› would be disputed.—

---

according to Principles of last page. ‹an› osculant groups between two circles «of equal value» must be so from characters of analogy.— see my notes on p. 37. of Macleay.[2] wonderfully accordant. with fact there stated, only in most discordant groups.

203 The formation of genera may sometimes be due to accident as submersion of land containing all of intermediate Father-species, & not, therefore, solely owing to such interm: father-species, being little adapted to some physical change.— If Patagonia became fertile all intermediate species living there would be destroyed, & N & S. existing species becomes father of genera—

202 *where*] *altered from* 'whether'.

202–1 Darwin is synthesizing 'Yarrell's law' with Swainson's interpretation of aberrant. See note C3–1 for the use of Yarrell's law in the context of systematics.

202–2 MacLeay 1819–21:37, quoted here from Darwin's abstract (DAR 71:128), 'These *genera osculantia*,— such as *Linodendron, Lathrus, Platycerus*, and as I suspect, also Aesalus, have in preference to all others have a special right to be termed natural, and appear in general to possess a remarkable character, which is the fewness of species of which they are composed.*(a)'. Double scored and a finger points to '(a)', which refers to note on verso, '*(a) It ought to be observed that this peculiarity is not so remarkable in the genera which connect the two circles of *Petalocera* with each other, and therefore it may perhaps belong solely to those singular insects which serve to connect the more discordant groups.'

whatever the cause is. ‹the› any osculant species. which survived would be few in number.—

Parallel of Japan, near Himalaya, & Europæan forms on that isl^{d}.—

204 The *races* of men differ chiefly in ‹size› colour, form of head «& features» (hence intellect?) & what kinds of intellect) quantity & kind of hair forms of legs— hence the father of man kind probably possessed a structure in these points for ‹t› a less time than other. points. **female genital organs «in some monkeys clitoris wonderfully produced».— make abstract on this subject from Lawrence. Blumenbach & Prichard**[1] — Now we might expect that animal halfway between man & monkey, would have differed in hair colour +++ form of head «& features»;. but likewise in length of extremities, **how are races in This respect** upper & lower, which I do not know whether it ‹would have› differs in present races, & form of feet.= (Negro **or father of negro** probably was first black at base of nails & over white of eyes,—

+++, Will he say creation is at end, seeing that Tertiary geology has obeyed rules of modern causes. & considering over the viccissitudes of present animals.— He-will be bold. I will venture to say unphilosophical.

205e L' Institut 1838. p. 128. Extraordinary genus. Mesites bird from Madagascar uniting pidgeons & gallinaceous birds & parrots.—[1] legs of pidgeons perfect.— &c &c.— do p. 136. Ichthyosaurus in the Chalk[2]

Those who say «philosphically to a certain extent,— nothing but experience. will, tell us. when group is true,» there are no genera. if Mammalia are adduced. say oh look to your fossils, now if extinction had gone, without creation this would have been fair, but to place all that ever lived ‹on› into one list is unfair, [moreover what will become of the future creations, if the list is now perfect.—]^{CD}. the creator so creates animals, «, it will be said» that although at any one there. are gaps. yet ‹what› altogether «he» has created a perfect chain. ‹Icthyo› .+++. supra & next page

204 ‹would have› differs] *altered from* 'differed' *when* 'would have' *deleted.*
205 L'Institut . . . chalk] '19' *added brown crayon.*
Those who . . . next page] *crossed pencil.*

204–1 Blumenbach 1792, W. Lawrence 1822, Prichard 1836–47. (See abstract of latter, DAR 71:139–42.)

205–1 I. Geoffroy Saint-Hilaire 1838:128, 'M. Isidore Geoffroy a donné à l'espèce type de ce genre remarquable le nom de Mésite variée, *Mesites variegata*. . . . M. Isidore Geoffroy signale comme très digne de remarque que la coloration si caractéristique de la tête chez la Mésite variée offre la plus grande analogie avec la coloration de la même région chez ces Hélicornes ou Grébifoulques dont la Mésite se rapproche tant aussi par les formes de son bec et les dispositions de ses narines.'

205–2 Montbrun and Courcier 1838:136.

206e It is a fact pregnant with SOMETHING.? that intermediate. species have generally perfect organs. **do changes of habit affect particular organs.—** of two adjoining families & not *all* organs blending away.—

+++ Hopeless work to systematist, who believed that all his divisions merely marked his own ignorance.— the collector who plodding at making a series, which would render our knowledge a chaos: who will *doubt this if series now existed from Man to Monads*— though physiology would profit. if the series were believed to past into each other—

Different classes Keep to their types. with different degrees of closeness. — look how close birds! look at Mammals: how wide.— therefore birds younger???? or «have» not «been» exposed to so many contingences???

207 A Question of immense difficulty is, whether Apterix descends from same parent with other birds,, or branched off anteriorly think what principles are there to guide in this opinion?— EXCELLENT PRINCIPLE OF ABORTION ISOLATION of range «‹far more prob›» tends to alteration views.— ostriches do— but then there may have existed series between apterix & other birds.— will having many trifling characters, in common with other birds reveal the secret.—
Now all the different forms of Synallaxis. trifling characters as red band on wing show to be from one parent.— same forms of beak &c without these trifles. it would not then be told whether not descended from long way back.— aberrant forms produced where many species

208 ‹osculant› but, where much death, may be inferred much time elapsed & therefore descended from branch high up.— Such probabilities only guides.—
Yet trifles are produced by circumstances. *Spines* on Echidna.— when it can be traced through series then probably heredetary & not produced by circumstances
In ostrich which is not isolated, we must suppose the changes from typical structure have either been more rapid than in all other birds, or that it sprung from a branch high up.— this argument not applicable to apterix.— but source of error for if

206 It is . . . blending away.—] '11' *added brown crayon.*
& not *all*] 't' *over* 'ne' *of* 'none'.
+++ Hopeless. . . contingencies???] *crossed pencil.*
207 trifling characters] *order altered from* 'characters trifling'.
208 yet trifles . . . circumstances.] *brace, left margin.*

209e some of the ostriches were to die, then they would *appear* isolated.

In my birds from S. Hemisphere there are some godwits which are close to Europæan species, and the sexes of which vary in colour of plumage in same remarkable manner as Europæan species = singular coincidence if distinct creation.— ie.— a mere statement nothing is explained.— this is fact analogous to mocking thrushes of Galapagos having tone of voice like S. American.—

Have not Ruffs & Reeves a remarkably varying plumage for wild birds—

210e At Zoolog Gardens there is half Jackal & Scotch Terrier.— certainly more like Jackall in *gait*, size, fur.; manner in which *ears droop like dog* **older character** & manner of wagging tail.—[1] **habitual movements connected with mind**

There is no progression in the developement in instincts in the ‹classes› «orders» of insects, so is there none of reason in order of ‹ver› Mammals.— Mem Elephants & dog.—

There is one living spirit, prevalent over this word, (subject to certain contingencies of organic matter & chiefly heat), which assumes a multitude of forms «each having acting principle» according to subordinate laws.—[2] There is one thinking «‹& Creat› sensible» principle (intimately allied to one kind of

211 organic matter.— brain. & which ‹prin› thinking principle. seems to be given or assumed according to a more extended relations of the individuals, whereby choice with memory. *or reason*? is necessary.—) which is modified into endless forms, bearing a close relation in degree & kind to the endless forms of the living beings.— We see thus Unity in thinking and acting principle in the various shades of ‹dif› separation between those individuals thus endowed, & the community of mind, even in the tendency to delicate emotions between races, & recurrent habits in animals.—
—Animal Magnetism—principles of irritations *sleep walking*. fits, laught &c &c Man & Man may have some relation together, as well as Man & child, polypus & polypus, bud & bud, polypus & germ, plant & seed.—[1]

209 *page crossed pencil.*
210 At Zoolog . . . wagging tail.—] '‹5› 17' *added brown crayon.*
droop like dog] *underlined pencil.*
true character] *added pencil.*
habitual . . . **mind**] *added grey ink.*
There is no . . . kind of] *crossed pencil.*

210–1 See *Natural Selection*: 456 and *Variation*, 2:67, 'With animals, the jackal is prepotent over the dog, as is stated by Flourens who made many crosses between these animals; and this was likewise the case with a hybrid which, I once saw between a jackal and terrier.'
210–2 See notes on Carus 1837 in C103–4.
211–1 See abstract of MacLeay 1819–21:487–88 (DAR 71:138v).

212 instincts in young animals, well developed, just like, habits easily gained in child hood.— Young salmons. first a species which lived in estuaries its taste. taught it to go to ‹sea› salter water (& its necessities teach it taste, but that is much more general argument) & therefore down the stream followed ebb tide, therefore got into habit of going down stream which would last were the stream 1000 miles long.—

---

a monkey. (Baboon) at Z. Gardens upon being beaten behaved very differently from a dog.— more like man. continued long in a passion & looked out for him to come again very differently from dog. perhaps being in passion chief difference

213e Major Mitchell[1] is not aware that Australian dogs ever hunt in company — marked difference with dogs of La Plata & Guyana‖ people will say. not species.—

---

organs of generation a captial character. (Owen) not for first & grandest divisions. but for ones of very high order. not for vertebrata, but mammalia & reptiles &c[2]

Timor is connected with Australia «map to King's Australia[3]» by a bank of soundings of which «there appears to be one line, in which» greatest depth ‹appears to be› is not more than 60F. & «in» the whole area,. 120 is greatest **(about 200 miles distant)**.— directly beyond produced line of Timor 215. What productions Sandal. Wood Is^{d}.? ought to agree with Java??

214e Terrestrial Planariæ assuming bright colours., good instance of colours dependent on localities.— —

Hamilton will give an account in his Travels in Asia Minor of the domestic animals— At Angora «centre of Asia Minor» are the fine-haired goats.[1] which it is said cannot be transported from their country.— the long-haired cats are supposed to come from there.— All the sheep are thick-tailed
The dogs called Persian «greyhounds» are Kurdish & come also from Asia Minor.— tail like setters. long ears— colours vary, but form constant.—

213 Major . . . not species.—] '1' *added brown crayon.*
organs . . . reptiles] *crossed pencil.*
Timor . . . Java??] '18' *added brown crayon.*
214 *page crossed pencil.*

213–1 T. L. Mitchell: personal communication.
213–2 Owen 1835d:390, 'Thus the generative system fails to afford a character applicable to the whole of the Acrita of Mr. MacLeay; . . . In general it may be observed, that it is only with respect to the nervous system that we can attribute a community of structure to a primary division of the animal kingdom.'
213–3 King 1827. Chart in vol. 1 indicates no bottom found at 215 fathoms west of Timor.
214–1 William John Hamilton: personal communication. See Hamilton 1842, 1:415.

215e The females of some moths, like glowworm ‹are› have «These abortive organs in some Males animals, Mammæ in Men, capable of giving milk)» rudimentary wings. so nature can produce in sex, what she does in species of Apterix[1]

‹————————————›

This is important for if these abortive wings in the female are allowed to the fully organized wings of the male rendered abortive in the womb.— if these apparently useless organs do indicate such origin, then we are bound to consider abortive organs of same tendency in species, this is capital & novel argument.—

═══

Are there any abortive organs in neuter bee, **(There is paper by Yarrell «in Zoolog Transactions[2]» & Hunter on this subject[3]**) because if so as she can be converted «into female», it will be splendid argument old female, turning into cock, abortive spurs. growing.—

216e Are there any abortive organs produced in domesticated animals, in plants I presume there are.? get examples.— for instance where a tendril passes into a mere stump.—
Shall abortive organs «of very same kind» in these cases, have plain meaning & none in other case!

Savigny has shown same fundamental organs even in Haustellata & mandibulata.—!![1] **—Argument, when general argument is extended from species to genera & classes.**

---

p. 479. fragment of tusk «& Molar tooth» of Hippotamus from Madagascar!!!!!! Proceedings of Geol. Soc Vol I[2]

---

215 Are there . . . spurs. growing.—] *marginal* 'x' *and* '10' *added brown crayon.*
old] 'o' *on stub.*
216 '19' *added brown crayon.*
It is capable . . . some perished] *crossed pencil.*
page heavily crossed pencil.

215–1 See Hunter 1837 ('An account of an extraordinary Pheasant') and Yarrell 1827, which describe changes in secondary sexual characters produced by the obliteration of sexual organs in fowl.
215–2 *Sic* Philosophical Transactions.
215–3 Hunter 1780, cited in Yarrell 1827:268, 'A Paper on this subject, by Mr. John Hunter, published in the 70th Volume of the Philosophical Transactions, and afterwards reprinted in his Animal Economy, details the appearance of several female birds having the feathers of the male, in which account he is led to observe, "that this change of character takes place at an *advanced age* of the animal's life, and does not grow up with it from the beginning."' Passage scored and 'grow up with it' underlined.
216–1 See Savigny 1816. Given the mentions of W. S. MacLeay in C202 and C218, see MacLeay 1819–21:423–24 as a possible source of this reference to Savigny, 'If further observations should prove Savigny to have erred in his analysis of the mouth of these last animals [*Strepsiptera*], they must then no doubt take a situation near to the *Diptera*. But at present, all who confide with me in the consummate accuracy of this gentleman's microscopical dissections can only consider the resemblances which the *Strepsiptera* bear to certain *Haustellata* as so many relations of analogy.'
216–2 Telfair 1826–33.

It is capable of demonstration that all animals have never at any one time formed chain, since if cretceous period assumed, then some perished before, Carboniferous some perished

217 before, then there always have been gaps, & there now must be, ∴ extinction of species bears relation to existence of genera &c &c

---

Two savages, two species.— «discussion ustil, unless it were fixed what a species means» civilized Man, May exclaim with Christian «we are all» Brothers in spirit— all children of one father.— yet differences carried a long way.

—Case of Habit I kept my tea in right hand side for— some month, & then when that was finished kept it in— ‹right› left, but I always for a week took of cover of right side, though my hand ‹vibrate› would sometimes vibrate— «seeing no tea brought back memory» old habit of putting tea in pot made me go to tea chest almost unconsciously.— **why do absent «Dr. Black. tea & sugar» people. reverse habits**

218 Insects & birds are the only two tribes fitted for water, air, & land, (Macleay has this remark)[1]
Mem. number 5 here most evident!!? examine into this case

D. Jeffrey (life of Mackintosh Vol II. p. 495)— in fact, in all reasonings, of which human nature is the object, there is really no natural starting place, because there is nothing more elementary than that complex nature itself with which our speculations must end as well as begin"[2] &c &c then centre is every where & then circumference no where— as long as this is so— !!Metaphysics!!!

219 Mrs Somerville, Connection of Physical Sciences p 276[1] May be worth glancing at, as she has no original ideas., it will show state of knowledge «Negroes existed since time of earliest Egyptian drawings & Old Testament»

Domesticated animals having *same* idiosyncrasy, cause of fertility.— varieties not produced as by nature. if so. the habits which would. have formed them, would have arisen under different climates &c. **Do I mean that ideosyncracy of wild animals is generally different, because their difference. arise a good deal from**

219 «Negros . . . old Testament»] *written in box, presume added, possibly grey ink.*

218–1 MacLeay 1819–21:265, 'It is unnecessary here to describe creatures so well known as birds . . . Some of them, as the *Palmipedes*, are also the most gifted with the power of locomotion, since they command three elements, and can make way equally well on land, in the air, or in water. With the exception of some insects, we are acquainted with no other instance of this in Nature.' See note B46–1.

218–2 Mackintosh 1835, 2:495.

219–1 Somerville 1834:276.

climate & habits, & therefore less fertile. according to Mr Herbert's views.—[2]

Argue ‹argue› case of abortive organs to mules in their genitals & even to a limb not used

---

The only cause of similarity in individuals «we know of:», is relationship, children of one parent, races of animals— argue «opening» case. «thus»

220 Educate all classes— avoid the contamination of ‹cl› castes. improve the women. (double influence) & mankind must improve—

The areas of subsidence marked out by animals of same ‹spe› genera. is not equal to areas of elevation: Marked out by existence of elevated «extinct?» genera of shells.— duration in two classes however different.—

221e Male glow-worm knowing female good case of instinct. bees turning neuter into Queen, more wonderful case

---

Dwights' Travels in America, speaks of short legged sheep. heredetary proceeding from an accident. New England farmer,—[1] useful could not leap fences:—[2] Dr Lang **on Polynesian nations** (quoted) p. 4.—[3]

---

do. p. 186. quotes Burkhardt to show black colour of certain Arabs.— **NB** avoid quoting these hackneyed cases

---

222e Mr Ed Blyth does not believe in circular or linear arrangement.—[1] Thinks passages very rare., in anatomical structure.— «the passages between— owls & hawks only external» intermediate groups often have full structure «of one

221 *page crossed pencil.*
222 '11' *added to page brown crayon.*
If I . . . acquired] *crossed pencil.*

219–2 Herbert 1837:343. 'In further confirmation of the fact that the sterility depends on constitutional discrepancy, or difference of what medical men call idiosyncrasy, may be adduced the curious plant figured in the Botanical Magazine under the name of Crinum submersum . . .' See note C125–3.

221–1 Dwight 1821–22, 3:122, '. . . an ewe, belonging to one of the farmers, had twins, which he observed to differ in their structure from any other sheep in this part of the country, particularly the forelegs, which were much shorter . . . they were . . . unable to climb the stone walls, with which this region abounds . . . The progeny had all the characteristics of the parents, and, although they have since multiplied to many thousands, have exhibited no material variation.'

221–2 Dwight 1821–22:185.

221–3 Dwight 1821–22:4 cites Lang 1834a.

222–1 Blyth 1836a:401, 'All organised matter is, of course, intrinsically allied in its nature, as contradistinguished from that which is not organised; . . . Next, we have a grand primary distribution of all organic matter into the animal and vegetable kingdoms; a division too obvious to be for a moment called in question, and universally allowed; admitted even, inconsistently enough, by those who hold that every natural assemblage of species, great or small, forms part of some quinary circle.'

class» & full of second—this class of facts «analogous to petrel-grebe. external» appears to be a puzzle against my theory,—[2]

If I be asked by what power the creator has added thought to ‹an› so many animals of different types. I will confess my profound ignorance.— but seeing such passions acquired

223 & heredetary & such definite thoughts, I will never allow that because there is a chasm between Man (— ‹a› «&» chasm«s» ‹necessary to account› «consequence of» for the scheme of nature) and animals that man has different origin.

«Royal Institution»
Dr Royle[1] seems to think Botanical Provinces will turn out not nearly so confined as now thought.— N. American, Europæan & Chinese Genera & some species on Himalaya.— some English beetles, birds & a fox most close

224 The most curious case is Saxifrage, almost ‹same› «closely allied» species Himalayas, 13,000 & Melville Is^{d}.—

West Africa & India some plants same.

—America.— See Brown Congo Expedition: 400 Australian plants found in other parts of world[1]

Athenæum June 3^{d} 1838. quotes M. Turpins assertion that globules of milk produce a plant capable of growing!! & propagating itself.[2]

In *Tropical* countries (as St Jago Cape de Verds) the shells in equal periods with Europe would probably have changed much less.— Here is an

225e element of extreme difficulty in mundine geological chronology

---

Annals of Natural History «Vol I??» p. 318. some remarks on Bonaparte's

225 element . . . chronology] *crossed pencil.*
Annals . . . consult] '19' *added brown crayon.*
p. 326 . . . Van Diemen's land diff.—] *crossed pencil.*

222–2 Blyth 1836b:509–10, 'Very lately the American scaup (pochard) was found, on comparison, to be distinct from that of Europe, although the difference almost wholly consists in the obliquity of its wing spectrum; a character which, however, proved to be fixed and constant. Had there not been this diversity, the two species would have been, of course, equally distinct: yet how should we have discriminated them apart.'

223–1 J. F. Royle lectured at the Royal Institution, 1 June 1838, on 'The vegetation of the Himalayan chain in connection with climate' (Royal Institution Archives). See letter [24 May 1838] to Royle, requesting the date of the lecture (*Correspondence* 2:89).

224–1 R. Brown 1818 (Tuckey 1818).

224–2 Turpin 1838:396.

list of birds in Europe & N. America, on closely allied species.[1] «replacing each other». good to consult

---

p. 326 wild ass extending over 90° of Long. & Col. Sykes alludes to some other case of 180° & great diff of Lat^{d}.[2]

---

p 355 Echidna of Van Diemen's land & Australia different[3] Temminck Fauna Japonica (?!) 82 mammalia
293 Phalangista of Australia & Van Diemen's land diff.—[4]

226e Habits can only be used «in classification» as indication of structure (including brain & other organs difficult to analyse) will not this separate facts about abortive organs &c

---

The doctrine of monsters is preeminently worthy of study on the idea of those parts being most easily mostrified, which last produced —insane men in civilized countries— this is well worthy of investigation.—

227e Institut 1838. p. 174. Apercu very good on insectiferous ‹insects› **quadrupeds** geographical range very good.— Blainville[1]

Ovington's Voyage to Surat, floating isl^{d}. off coast of Africa .p 69. with tall grass.[2] & p 72 hairy sheep—[3]

Edinburgh. Transact. Vol IX p. 107. an Ascaris inhabits the eyes of horses in India in which it may be seen swimming about.[4]

A Smith[5] is firmly believed in representation. certain birds in many families,

226 *page crossed pencil.*
227 Institut . . . Blainville] '19' *added brown crayon.*
Ovington's . . . sheep] '18' *added brown crayon* '8' *over* '9'.
Edinburgh . . . about.] *crossed pencil.*
A Smith . . . value] *crossed pencil.*

225–1 Bonaparte 1838b:319. 'Several of the above-named birds we have examined, and at the time we thought very minutely, and considered identical: a comparison of many specimens from each country might induce a change of opinion; and we have now to regret that the distinctions between all those so closely allied had not been shortly given, by which we should have been at once enabled to judge of the propriety or impropriety of the Prince's separations.' Passage scored.
225–2 Sykes 1838b:326, '. . . with respect to the extent of geographical distributions, I have elsewhere proved that it is no bar to the identity of species inhabiting mean temperatures varying 40° of Fahr., and separated by half the earth in longitude.'
225–3 Gray 1838:335 cites Home 1802. Passage scored.
225–4 Temminck 1838 cites Temminck in Siebold 1833–50. Passage scored.
227–1 Blainville 1838b:174. The article is on 'Mammifères Insectivores', hence the grey-ink correction.
227–2 Ovington 1696:69, 'A floating Island washed from the Shoar, sailed by our Ship, . . .' Darwin noted: 'See p. 64. probably 4 leagues from shore', passage scored.
227–3 Ovington 1696:72, 'Instead of the soft Wool which Cloatheth Sheep, a harsh kind of hair, not unlike that which grows upon Dogs, is the usual excressence; . . .' Passage scored.
227–4 Kennedy 1823.
227–5 Andrew Smith: personal communications C227–28.

«+*very often in number* 5» will have long tail.— in raptorial birds, & tigers & sharks, being spotted, & *colours* of little value

228e Dr Smith if black & white Man crosses; children heterogenous, he feels sure of this, first offspring most like ‹parents› Mother.— like dogs Smith knew chinese hairless dog & common spaniel crossed.— 3 puppies PERFECTLY like chinese & 3 perfectly like spaniel even when grown up.— Are mules homogenious owing to no attempt to keep up offspring, are not half lion & tigers ditto. (see Griffith[1]) & half Muscovy ducks, «black cock & pheasant see Jardines Journal.»— consult on this point— pigs always go against this, without «number of vertebræ» new acquisition, we must

229e–230e [not located]

231 Henry Thompson tells me best way to improve cattle is to cross between a good bull & ‹be› the provincial breed, & that first offspring thus produced are better, than those bred in & in.— which looks as if qualities were not permanent, in the new cross.— In the Bantam clubs, they used to fix on the kind wanted, colouring of each feather weight & size & they would produce number agreeing almost to the point in question.— «—merely picking opposite qualities, with no other means whatever.—»

232 Individual Men «& animals» could only exist by habit.— therefore same principle transferable., not wonderful

---

According to my view ‹ins› beccause actions are constant they are instincts, & not ∵ instincts, constant.

¿ whether mutilations non-heredetary & variations produced in short time in some extent counterpart, mutilation being «variation» produced in shortest possible time.

Mr Willis[1] Long eared little dogs, I am told, go to heat, take dog. but do not become impregnated

233 & puppies delicate— they cross sister & brother of same litter, those. of different litters or of father & child are thought long breeding in.

Must not trust him
Hope[1] says that genus of parasite to genus of animals different «+++ p 234», different species to different,— inguinal louse African & Europæan.

228 *page crossed pencil.*
233 **Must not trust him**] *added margin, grey ink.*
Read Entomological Transactions] *crossed grey ink.*

228–1 Griffith 1827.
232–1 Personal communication.
233–1 F. W. Hope: personal communication.

different.— thorax & head differ Africa Australia
Parasites die, when brought over on tropical animals which accounts for the species changing which accounts for the species changing « ∴ because mammalia can subsist where parasites cannot»

Read Entomological Transactions

234 Why if louse created should not new genus have been made, & only species, good argument for origin of man one.—

Is the *extinction* & *change of species* two very different considerations, with respect to law of mammals shorter duration, than molluscs, argue case both in Europe & S. America. «very difficult case»

Does this law of duration apply to utter extinction or rapidity of specific change.? ‹One› the first would be called. generic & other specific extinction—

235 In the Entomostraca (Magazine of Zoology & Botany) where several generations are produced in succession (13?) without impregnation,[1] therefore sexual passion must arise after long interval very good case.— habit is awakened by association (case of Elephant, which had run wild in India. in Heber?[2]) is analogous to dormant instinct.— (How wonderful a case bees developing sex of neuters) species may have had their infancies, as well as men— when habits much more firmly impressed.— we see in the Entomostraca. The sexual curiosity of the orang outang of (in June 1838 when young male was added good instance of instinct showing itself, not from instruction

236 Even the action of the viscera. under sympathetic nerve may be instinct or habits. ˁare sympathetic nerves & nervous system of insects analogous.?— («Even plants have *habitual* actions.—» «this very important in considering how children come to suck or other actions in fœtus of Mammalia, or chick eat»)

Generation becomes necessary, when organs of parent are concentrated in different parts, & scission cannot effect the process.— but why two sexes **scission in all cases probably gemmation (.Ehrenberg)** —not necessary to generation (lateral with no relation to time) as in buds.— I can scarcely

234 *page crossed pencil.*

235–1 Baird 1838:406. '. . . [Straus] thinks it probable it might extend in some species to the fifteenth generation.' Scored. See Straus-Durckheim 1819–20, 5:390; and C162.
235–2 Heber 1828.

doubt final cause is the adaptation of species to circumstances by principles, which I have given[1]

237e ∴ Those animals, which only propagate by scission can not alter much.??

M[r] Brown showed me Bauer's drawings of a curious plant where a tube consisting of pistils & stamens united into long organ, moved on being touched, so as to protect itself, one segment of the corolla being (probably) small to allow it to lie on one side.— but in other species, this segment is converted into hood which possesses power of movement. & not the organ itself[1]

238e How except by direct adaptation has such a change been effected.— the consciousness of the plant that this part must be protected however it may be effected.—

Prodromus Floræ Norfolkicæ. 1833 Steph. Endlicker[1] (He will give sketch of botany of islands of south seas says so in preface.— Mr Brown says character of Flora, N. Zealand & N. Caledonia with a dash of New Holland. **As in N. Zealand— Some species of Australian Genera** Some species

239e same (Palm & Phormium tenax) as in New Zealand & Australia, some SPECIES of *Australian* GENERA ∴ good case. rather large flora. (150?)

M[r] Brown did not observe scarcely any Australian character in Timor plants, yet it seems there may be Eucalyptus!— **(Hostile fact)**

Be cautious about Goulds case of birds of Van Diemens land & Australia.—[1] The wombat **(Brown)** is found in Is[d] of Bass' Straits

237 . . . much??] *crossed brown ink.*
page crossed pencil.
238 How . . . effected.—] *crossed pencil.*
Prodromus . . . Some species] '20' *added brown crayon.*
As in . . . Australian Genera] *added pencil from C239 when the latter was excised.*
239 same . . .(150?)] *crossed pencil.*
M[r] Brown . . . Eucalyptus!—] '20' *added brown crayon.*
(Hostile fact)] *added pencil.*
Be cautious . . . Straits] *crossed pencil.*
(**Brown**)] *added grey ink.*

236–1 See B1–5, 'Why is life short, why such high object generation.— . . . hence, we see generation here seems a means to vary or adaptation.—'; and D135, 'The final cause of all this wedging, must be to sort out proper structure & adapt it to change.—'
237–1 Robert Brown: personal communications C237–39. R. Brown 1833. See also *Synaphea dilatata* in Brown 1814, 2:606–7.
238–1 Endlicher 1833.
239–1 Reflects a warning from Brown about Gould's propensity to splitting. See B50.

240e The common Mush room & other cryptogamic plants same in Australia & Europe.—[1] if creation be absolute thing, the creation must take place on[ly] when creator sees. the means of transport fail.— otherwise no relation between means of Transport & creation exists..— pooh. May have been Created at many spots & since disseminated

---

See. Habits of Malay fowls p 5. (note) in some papers on instincts[2]

241e L.' Institut «1838» p. 184 Botany of *Bonin*. "grande *analogie* avec le flore du Japon", some Europæan & Sandwich species & some of Japan.[1] I do not understand any new ones.—

Menoir will be published St. Petersburgh Academy Imperial— Paper read in 1837. semestre

---

I suspect some valuable analogies might be drawn between habitual actions of plants «when exciting cause is absent» & memory of animals.— (surely in plants

242e movements effects of irritability, though means injection of fluid different from contraction of fibre)— it is most remarkable habitual actions in plants, it allows of any degree in lowest animals —habitual action, in intestines subject to sympathetic nerves—

The vividness of first ‹thoughts› «memory» in children or rather their memory. very remarkably— scenes in themselves accidental— My first thought of sea side—

243 Study Bell on Expression[1] & the Zoonomia,[2] for if the former shows that a man grinning is to exposes his canine teeth. no doubt a habit gained by formerly being a baboon with great canine teeth.— **(This may be made capital argument if man does move muscles for uncovering canines) —. Blend this argument with his having canine teeth at all.—** This way of viewing the subject important.— Laughing modified barking., smiling modified laughing. Barking to tell other animals in associated kinds of good news. discovery of prey.— arising no doubt from want of assistance.— crying is a puzzler— Under this point of view. expression «of all animals» becomes very curious.— a dog snarling in play.—

240 *page crossed pencil.*
241 L.'Institut . . . semestre] '20' *added brown crayon.*
I suspect . . . in plants] *crossed pencil.*
242 **The vividness . . . sea side.—**] *added pencil.*
page crossed pencil.

240–1 See R. Brown 1814:594 in 'A list of plants natives both of Terra Australis and of Europe'.
240–2 French 1825:5.
241–1 Bongard 1838 reports on plants from Litke's voyage.
243–1 C. Bell 1806.
243–2 Erasmus Darwin 1794–96.

244 Hensleigh says the love of the deity & thought of him «or eternity», only difference between the mind of man & animals.—[1] yet how faint in a Fuegian or Australian! why not gradation.— no greater difficulty for Deity to choose, when perfect enough for future state, that when good enough for Heaven or bad enough for Hell.— «+glimpses bursting on mind & giving rise to the wildest imagination & superstitions.— +York's Minster story of storm of snow after his brothers murder.—» **good anecdote**

—.Sowerby.—.[2] Geographical range of shells **like Cryptogamic plants.** of Marine kinds, there are some restricted genera, but then they appears always very small ones as Trigonia in Australia or Concholepas in America,— yet many countries have far more species than other countries «++ p. 246»

245 as Cyclostoma in Phillippines & Anphidesma in S. America— yet there are a few Cyclostomes & a few Anphidesmas.— this is remarkable.— **Fish & drift sea weed— may transport ova of shells.— Conchifera. hermaphrodites— eggs in groups.. Have Diœcious plants more restricted ranges than other plants.—**
Many «Some» genera confined to hot countries & many to cold.— Hence latitude is more important element than longitude.— But in land & F W shells there is more confinement. thus the Naiads (study De Ferrussac[1]) are confined to ‹S.› America.— Mr Sowerby says

246 there are some shells common to West coast of Africa & E. S. America.— get instances.— very good anomaly in range

---

++ What circumstances have led to formation of new species some few have been scattered over whole world

---

Many shells at present day same (or according to Sowerby fine species) on coasts of N. America & England.— but the fossils are not like, except in very few cases,

247 those of Tertiary Europæan fossils— «(so much the more remarkable ∵ carboniferous ones similar?)» Now this is very remarkable.— **(connect these facts with identity of land animals.— these however come from Siberia** it cannot be said American fossils more resemble those of America than of Europe, because the recent ones are so close.
Was there continent between N. America & Europe?— ‡.*Norton* has written on fossils of N. America.— [1]

245 **eggs in groups**] *underlined*..

244–1 Hensleigh Wedgwood.
244–2 James de Carle Sowerby: personal communications. See J. Sowerby 1820–25.
245–1 See Ferrusac 1819–51.
247–1 Possibly Joseph Granville Norwood.

248 At the end of "*White's Selbourne.*" many references very good.[1] also "*Rays Wisdom of God.*"[2] Often refer to these.— Also some few facts at end of "*The* British Aviary" or Bird Keepers Companion[3]
Study Appendix (& only appendix) of Congo Expedition[4, 5]

249e NB. I met an old man—, who told me that the mules between canary birds & goldfinches differed considerably in their colour & appearance

---

Every now & then a short-tailed *cat.*⁅cut? has its offspring short tails /one born at Maer.

Tuckeys voyage— p. 36 "Cercopithecus sabœus" said to be monkey of St Jago C. de Verd; same as on coast of Africa.—[1] **Macleay tells me same thing**

---

p. 55. 40 leagues from land several patches of reed & trees[2]
p. 259 120 ft in length, some branches of Justicia still growing,) passed us.[3]

250e do. p. 243 (, Professor Smith's Journal) on the heights of St Jago found a Euphorbia so near Piscatoria as scarcely to be distinguished from it.— & several old acquaintances.[1] which grow on the lower region of the Canary islands—
p. 250 admirable table of plants of St. Jago showing many common to Canary. isl^d^., Europe, & St Jago upper region, & some to Cape.—[2] some proper well-worth studying, with respect to forms.—

251e Study Appendix to Tuckey's Expedition[1]

Journal of the Academy of Natural Sciences of Philadelphia Vol VII. Part II /. 1837 account of the various hares «some since discovered» of N. America,[2] & of the shrews.—[3] Dr Bachman[4] told me. that near Charlestown ?three species, near New York. (600 miles N.?) replaced by three other species.— Says all the hares West of ‹all› Rocky Mountains have peculiar character in extreme length of ears & length of limbs, so that he first thougt only one species. & all hares on East side have other

249 NB . . . Maer] *crossed pencil.*
Tuckeys . . . Africa.—] '19' *added brown crayon.*
250 '20' *added to page brown crayon.*
251 '19' *added to page brown crayon.*

248–1 G. White 1825. The book is extensively annotated.
248–2 Ray. 1692.
248–3 Anonymous, n. d.
248–4 R. Brown 1818.
248–5 Tuckey 1818.
249–1 Tuckey 1818:36. See B154.
249–2 Tuckey 1818:55.
249–3 Tuckey 1818:259.
250–1 Tuckey 1818:243–44.
250–2 Tuckey 1818:249–52.
251–1 R. Brown 1818 (Tuckey 1818).
251–2 Bachman 1837a.
251–3 Bachman 1837b.
251–4 Bachman: personal communications, C251–56.

252e peculiar appearances— Now this is precisely the case with the mice of S. America, with respect.. to the Cordillera,—— Bachman has seen webbed Shrews. case of adaptation.— (case of Squirrel from extreme north turning white like Hares??—)
I never saw more beautiful adaptation for snow— like snow shoes. than feet & hind legs of these white hares, fitted for regions of snow.—

———

253e **Acclimatisation.—**
Bachman tells me in Audubon there is most curious history of first appearance of the S. American Pipra Flycatcher, which is now becoming common—[1] likewise of the *Hirundo fulva* (added by Audubon in Appendix)[2] showing WHAT CHANGES are taking place & how birds are extending their ranges. «even migratory birds, lik swallows» —

═══

°/ migrations of birds he mentioned many most curious case— — the birds seem to follow narrow bands, certain kinds as gallinules taking the low country near coast & others the mountains, & these

254e appearing to remain about a fortnight, «See Silliman's Journal 1837. Paper by Bachman.[1]» that is succession of birds.— «in» some species «a Tanagra» Males come first & the females in flocks. «as in English Nightingales» — other birds (& this seems common «kind» migration of America) migrate singly flying few miles every day «generally by night» — other birds which is strictly diurnal, migrates singly by night.— others in flock, these birds seem clearly directed by kind of country; «kinds of migration quite different in species of same genus.» The Muscicapa solitaria stay about a fortnight in one particular part of country, like White of Selbournes Rock Ouzels[2].— ..If the line or bands of country (These facts show the Normal condition of Migration)

255 gradually separated the birds might yet remember which way to fly.— There is a kind of Wren (Bebyk??) which seems common in Rocky Mountains & on one lofty isolated spot on the Alleghanies to which it migrats every year,; and probably a «chance» wanderer like the first pair of Pipra flycatcher.—

———

252 peculiar . . . webbed] *scored brown crayon.*
253 **Acclimatisation**] *added pencil, marked with brown crayon finger, underlined.*
'9' *added to page brown crayon.*
254 migrate . . . day] *scored pencil.*
'23' *added to page brown crayon.*
255 *page crossed pencil.*

253–1 See Audubon 1831–49, 2 (1834): 392–96.
253–2 See *Natural Selection*: 287, 572–73, where C253 is published together with related ms notes.
254–1 Bachman 1836.
254–2 G. White 1825 has many references to migration of ring-ouzels, but none to rock-ouzel.

Bachman says he thinks the Mocking thrush beats all English birds in song.— one of their thrushes exceeds our blackbird, but our blackbird exceeds their other thrushes— yet they have one with very sweet notes.—

256 Their soft-billed birds are inferior to ours, & our lark ranks very high.— Upon the whole thinks ‹many› more birds sing in England than in America, but the few of N. America are quite as beautiful.[1] The thrushes of N. America. singing so well. & the mocking thrush being so very beautiful gret contrast with South America.—

257e In Home's History of Man at Maer, it is said the Samoyed women (? north end of the Oural mountains) have black nipples to their breasts.—[1]

L' Institut, 1838, p. 230 says the Macrotherium of Europe is between the Anteater of C of Good Hope & those of S. America,—[2] Are not some of the Australian fossils intermediate between those of Van Diemen's land & Australia proper.—
Irish Elk case of fossil geographical range.—

258e–264 [blank]

276 Books examined: with ref: to Species[1]

Most of those which have references at end; is so said to have

---

257 In Home's . . . breasts.—] *crossed pencil.*
L'Institut . . . geographical range.—] '21' *added brown crayon.*
276–265 *all reading-list pages crossed brown ink.*
Many entries with marginal brace.
276 Whole of Geographical] *over pencil* 'Geograp Journal Vol V. P II'.

256–1 See ZEd 19.
257–1 H. Home 1774:12, 'M. Buffon, from the rule, that animals which can procreate together, and whose progeny can also procreate, are of one species, concludes, that all men are of one race or species; and endeavours to support that favourite opinion by ascribing to the climate, to food, or to other accidental causes, all the varieties that are found among men. But is he seriously of opinion, that any operation of climate, or of other accidental cause, can account for the copper colour and smooth chin universal among the Americans, the prominence of the *pudenda* universal among Hottentot women, or the black nipple no less universal among female Samoides?'
257–2 Blainville 1838c:230.
276–1 Darwin maintained two reading lists at the end of Notebook C: 'Books examined with ref: to Species' (C276–269); and 'To be read' (C268–265). To preserve continuity the pages are presented here in the order in which each list was written. The first list begins with references from the Red Notebook and Notebook B and both lists were continued as Darwin's record of reading as he filled Notebooks C, D, E, M, N and Notebook A. From internal references, variations in Darwin's hand and ink, and from the dates he began to supply from October 1838, it is possible to reconstruct a rough chronology of the entries.
Books examined with ref: to Species
The neat and uniform writing of the first page (C276) gives the strong impression of having been written at one sitting—except for two grey-ink additions. If so, a clue to the earliest possible date is the entry 'Zoological Transactions. ‹done› up to parts published. March 1838'. It seems safe to infer that Darwin began the reading list by April 1838, nine months after first commencing his *Notebooks on the Transmutation of Species* and two months after beginning Notebook C. The entries on C275 are in a variety of hands and inks. The first half of the page is in brown ink, up to 'The British', while the remainder of the page and the first two entries on C269 are in grey ink. Thus the brown-ink half of C275 was written between April and 1 August 1838. The grey section of C275–270,
(*continued overleaf*)

Mackenzie's Iceland[2]
Molinas Chile[3]
Falkners Patagonia[4]
Azaras Voyage[5] & Quadrupeds of Paraguay[6]
Dobrizhoffer. Abipom‹e›nes.[7]
Edinburgh New. Phil Journal. about 13 numbers have been read
Voyage a l'isle de France[8]
Voyage de l'Astrolabe Partie Zoologique[9]
Pernety. voyage a l isle Malouines[10]
Zoological Journal 5 Vols
Voyage de la Coquille[11]
Zoological Transactions. ‹done› up to parts published March 1838
Whole of Geographical Journal
{Asiatic Journal to end of 1837. re«a»d— contains very little
{Macleay's letter to Dr. Fleming.[12] & Review of latter in Quarterly[13]

275 Sir J. Sebright's Pamphlets[1]
Wilkinson on Cattle[2] **not abstracted**
Scientific Memoirs. published by Taylor[3]
Magazine of. Zoology & Botany & Continuation
«Annals of Natural History»
Skimmed Von Buch Travels[4]

275 J] *over* 'S'.
Voyage Congo . . . :Plants] *boxed.*

(*'Books examined' continued*)
which accounts for the majority of entries, was written between 1 August and 12 October. The section is divisible in two parts. The first section Darwin did not date ('Aviary . . . Julie'). This was written 1 August to 2 October 1838. The second, dated, section at the bottom of C270 and very top of C269, runs from 2–12 October 1838. Darwin's dating began again on 10 January 1839. Thus C269 'Lutke's . . . Brown travels' was written 12 October to 31 December 1838. From there to the end of the list (the remainder of C269) Darwin dated from 10 January through 1 June 1839.

To be read:
This list begins on C268 in grey ink, which suggests it was commenced at least four months (April–July 1838) after 'Books examined'. However, two excised sheets corresponding to C274–271 have not been located. Since there seems to be no information lost in the 'Books examined' list, it is possible that Darwin began both lists in April 1838, but did not leave enough space for 'Books examined'. On this premise, he would have excised C274–271 and recopied the start of the 'To be read' list on C268. The grey ink continues through the middle of C266 and the list is completed by a series of brown-ink entries through C265. The *inside back cover* is not a continuation of the 'To be read' list as it includes a mixture of references to works read and to be read, as well as substantive notes.

The reading lists in Notebook C were later copied into Darwin's Reading Notebook, with one significant addition, 'N.B. These books have been read since I thought of my transmutation theory' (DAR 119:1).

276–2 Mackenzie 1811.
276–3 Molina 1788–95.
276–4 Falkner 1774.
276–5 Azara 1809.
276–6 Azara 1801, 1838.
276–7 Dobrizhoffer 1822.
276–8 Sainte-Pierre 1773.
276–9 Quoy and Gaimard 1830–35.
276–10 Pernéty 1769.
276–11 Lesson and Garnot 1826–30.
276–12 MacLeay 1830a (1830b). Here, as in C157, Darwin means MacLeay's letter to Vigors *about* Fleming.
276–13 Darwin means review *by* 'latter': that is Fleming 1829, which is a review by Fleming of Bicheno 1827.
275–1 Sebright 1809.
275–2 Wilkinson 1820.
275–3 Taylor 1837 (Ehrenberg 1837a, 1837b; Carus 1837).
275–4 Buch 1813.

Whites Natural History of Selbourne[5] **References at end**
D[r]. Lang Australia[6] «trash» skimmed
Macleay's Horæ Entomologica[7]
Ray's Wisdom of God[8] **references at end—**
The British **Aviary.[9]——do——**
&
Lisle's Husbandry[10]
Tuckeys voyage reread Appendix[11]
Ovington Voyage to Surinam.[12]
{Voyage Congo expedition: Zaire[13] except Brown's Appendix[14] & excellent table of Canary isl[d]: Plants[15]
Home's History of Man[16]
{Transactions of the Entomological Society Vol I. & 1[s] No of Vol II. (read remainder) when out

274e–271e [not located]

270 ‹**Rays Wisdom of**›[1]

———

Lisiansky's Voyage round world. 1803–6.[2] Nothing
Lyells Elements of Geology[3]

———

Gibbons life on himself[4]
Hume's do, with correspond. with Rousseau[5]

———

Miss M«artineaus» How to observe[6]

———

Mayo Philosophy of Art of Living[7]
Several of Water Savage Landons Imaginary Conversations—[8] very poor
Sir T. Browne's Religio Medici[9]

———

270 **on Etna. Almost reread**] *crossed brown ink.*
previous volume.] *crossed brown ink.*
269 **skimmed well**] *underlined.*
Discourses] 'D' *over* 'on'.

275–5 G. White 1825.
275–6 Lang 1834b.
275–7 MacLeay 1819–21.
275–8 Ray 1692.
275–9 Anonymous, n. d.
275–10 Lisle 1757.
275–11 Tuckey 1818, R. Brown 1818.
275–12 Ovington 1696.
275–13 Tuckey 1818.
275–14 R. Brown 1818.
275–15 Christen Smith 1818:249–52, 'Dispositio geographica plantarum quas legi in insula Sti. Jacobi . . . circa portum Prayae . . . et montibus Pico St. Antonio . . .'
275–16 Home 1774.
270–1 Ray 1692.
270–2 Lisiansky 1814.
270–3 Lyell 1838a.
270–4 Gibbon 1830.
270–5 Ritchie 1807.
270–6 Martineau 1838.
270–7 H. Mayo 1837.
270–8 Landor 1824–29.
270–9 T. Browne 1835–36.

Lyell Book III[10] There are many marginal notes

‹Rengger[11] &› Mitchell's Australia[12]

Walter Scotts life[13] 1 & 2^d & 3rd Volumes

Abercrombie on the Intellectual Powers.[14]

{ **Hunters Animal Œconomy. edited by Owen.[15] read several papers— all, that bear on any of my subjects**
{ **Elie De Beaumonts. 10 Vol. of Memoirs on Geology of France.=[16] on Etna. Almost reread the previous volume.[17] & C. Prevost on L'Ile Julie[18]**
Waterton's Essays on Natural History.[19] Octob 2^d
Transactions of Royal Irish Academy. ——do
Lavater's Physiognomy[20] —— Octob 3^d
Malthus on Population[21]
W. Earls'. Eastern Seas.[22] .Octob12^th.—

269 **Sir G. Staunton's Embassy to China.[1] Oct. 12^t**
Kotzebue's two voyages, skimmed well.[2] do
Lutke's Voyage.[3] carefully read.—
Reynolds Discourses[4]
Lessing's Laocoon[5]
Whewells— inductive History.—[6] References at end of each Vol
Herschels. introduction to Natural Philosophy[7]
R. W. Darwin's Botany.—[8] References at end
Mayo Pathology of the Human. Mind[9]
Evelyns Sylva.[10] skimmed, stupid
Brownes travel in Africa;[11] «well skimmed.»

270–10 Lyell 1837 (vols 2 & 3).
270–11 Rengger 1830.
270–12 T. L. Mitchell 1838.
270–13 Lockhart 1837–38.
270–14 Abercrombie 1838.
270–15 Hunter 1837. This edition of Hunter's *Animal Oeconomy*, edited by Richard Owen, appeared both as vol. 4 of *Hunter's Works* (Hunter 1835–37, John Palmer general editor) and separately (Hunter 1837). Since neither of these identical versions are in Darwin's library, it has not been possible to distinguish which one he used. Hunter 1837 is used here as the standard reference.
270–16 Dufrénoy and Elie de Beaumont 1830–38, vol. 4 (1838).
270–17 Elie de Beaumont 1838.
270–18 Prévost 1835.
270–19 Waterton 1838.
270–20 Lavater 1804, 1820.
270–21 Malthus 1826.
270–22 Earl 1837.
269–1 Staunton 1797.
269–2 Kotzebue 1821, 1830.
269–3 Litke 1835–36.
269–4 Reynolds 1831.
269–5 Lessing 1836.
269–6 Whewell 1837.
269–7 Herschel 1831.
269–8 R. W. Darwin 1810.
269–9 Thomas Mayo 1839.
269–10 Evelyn 1664.
269–11 W. G. Browne 1799.

1839

Jan 10t.— All life of W. Scott.—,[12] except the V Volume.—
—— 19t. Mungo Park— travels[13]
Feb 12. Sir. H. Davy Consolations in Travels[14]
—— Observations on morals by Eugenius[15]
Feb 14th. Bo«s»well's life of Johnson.[16] 4. Vols
25th Phillips. Geology.[17] Larder 2d vol.—
March 16. Gardner's Music of Nature[18]
—— Herbert on Hybrid Mixtures:[19] Marginal notes.
—— 20th. Carlyle's French Revolution[20] 3? vols. oct:
—— 26th Blumenbach's Essay on Generation.[21] Englis Transla
—— The Revd. A. Wells. Lecture on instinct[22]
—— Cline on the Breeding of Animals[23]
—— Spallanzani's Essays on Animal Reproduction[24]
—— Treatise on Domestic pidgeons[25]
30th Lives of Hayd & Mozart[26]
«Apri 25th Lockarts life of Napoleon.[27]»
April 5d Dr. Edwards of ‹ter› influence of Physical causes[28]: well skimmed
Bartrams Travels in N America[29]
May 18th Stanley familiar History of Birds[30]
—— Mackintoshs' Ethical Philospohy[31]
—— Bell's Bridgewater Treatise[32]
—— Wilkinsons Egyptian[33] remains skimmed
—— Pliny Nat. Hist of World[34] do
—— Lamarck. II Vol. Philo. Zoology[35] «references at end of each Chapter»
Crabbes Life[36]
June 1s. King & FitzRoy[37]

269–12 Lockhart 1837–38.
269–13 M. Park 1800.
269–14 Davy 1830.
269–15 Herries 1838.
269–16 Boswell 1799.
269–17 J. Phillips 1837–39, vol. 2 (1839).
269–18 Gardiner 1832.
269–19 Herbert 1837.
269–20 Carlyle 1837.
269–21 Blumenbach 1792.
269–22 Wells 1834.
269–23 Cline 1829.
269–24 Spallanzani 1769.
269–25 J. Moore 1765.
269–26 Haydn 1817.
269–27 Lockhart 1833.
269–28 Edwards 1832.
269–29 Bartram 1791.
269–30 Stanley 1835.
269–31 Mackintosh 1837.
269–32 C. Bell 1833.
269–33 Wilkinson 1837.
269–34 Plinius 1601.
269–35 Lamarck 1830.
269–36 Crabbe 1834.
269–37 FitzRoy 1839.

268 **To be read**

Humbold. New Spain—[1] Much about *Castes* &c
Richardson's Faun. Borealis[2]
Entomological Magazine (paper on Geograp. range[3]
Study Buffon on Varieties of Domesticated[4] **animals see if law's cannot be made out**
{ **Find out from Statistical Society— where M. Quetelet**[5] **has published his laws about sexes relative to age of Marriages**
Brown at end of Flinders & at end of the Congo Voyage[6]
Decandoelle. Philosophie. or Geographical distrib.[7] **«in Dict. Sciences. Nat. in Geolog Soc.»**
F.. Cuvier on instinct[8]
L. Jenyns paper in Annals of Nat. History[9]
Prichard.—[10] **Lawrence**[11]
Bory St. Vincent. Vol III p. 164. on unfixed forms.[12]
Dr. Royle on Himalayan. types.[13]
Smellie. Philosophy of Zoology[14]
Flemming. ditto[15]
Falconers remarks on the influence of climate[16]
White's regular gradation in Man.[17]
Lindlys introduction to the Natural system[18]
Bevan on Honey Bee[19]
{ **Dutrochet Memoires sur les Vegetaux et animaux.—**[20] **on sleep & movements of Plant** / 1£:4s
Voyage aux terres australes.[21] **Chapt. XXXIX. tom IV. p 273**
Latreille Geographie des insectes 8vo p. 181.—[22]
See (p. 17) for references to authors about E. Indian Islands.[23] **consult Dr Horsfield**[24]

268 Humbold . . . *castes* &c] *added brown ink.*
& at end of the Congo Voyage] *added brown ink.*
Dutrochet . . . Plant] '**sleep**' *underlined.*

268–1 Humboldt 1811.
268–2 Richardson 1829–37.
268–3 *Entomological Magazine*, published 1832–38 (vols 1–5).
268–4 Buffon 1762.
268–5 Quetelet 1835a.
268–6 R. Brown 1814 (Flinders 1814), 1818 (Tuckey 1818).
268–7 Candolle 1820, 1821.
268–8 Frédéric Cuvier 1822.
268–9 Jenyns 1838b.
268–10 Prichard 1836–47.
268–11 William Lawrence 1822.
268–12 Bory de Saint Vincent 1804.
268–13 Royle 1835.
268–14 Smellie 1790–99.
268–15 Fleming 1822.
268–16 William Falconer 1781.
268–17 Charles White 1799b.
268–18 Lindley 1830.
268–19 Bevan 1827.
268–20 Dutrochet 1832.
268–21 Péron 1807–16.
268–22 Latreille 1819.
268–23 Raffles 1830, Blume 1827–28. Reinwardt 1824, Forrest 1780.
268–24 Horsfield 1824.

267 **Silliman's Journal[1]**
Rengger on Mammalia of Paraguay.[2] account of wild cattle &
Montagu on birds[3] (facts about close species)
Wilson's American Ornithology[4]
Read Aristotle to see whether any my views very ancient?
{Study with profound care, abortive organs produced in domesticated plants; where function has ceased to be used as tendril into stump}
Library of useful knowledge. Horse, Cow, Sheep.—[5]
Verey. Philosophie d'Histoire Naturelle[6]
{Marcel de Serres Cavernes d'Ossements[7] 3^d^. Edit. Octav. (good to trace Europèan forms compared with African}
‹A› Annals
{Histoire Generalle et Particuliere des Anomalies de l'organization des Hommes. & les animaux.— by Isid. Geoffroy. S^t^. Hilaires.[8] 1832. contains all his fathers views.— Quoted by Owen.—[9]}
{Hunter has written quarto. work on Physiology[10] besides the papers collected by Owen.[11] (at Shrewsbury)}
{Yarrells Paper on change of plumage in Hen Pheasants ‹Zoological› Philosop Transactions. 1827[12]}
Paxton on the culture of Dahlias[13]
M^rs^. Gore on Roses might be worth consult.[14]
Paper on Consciousness in Brutes in Blackwood. June 1838[15]
H. C. Watson on Geograph. Distrib: of British Plants.[16]
Humes Essay on H. Understanding (some time)[17]
Du Stewart works. & lives of Reid, Smith & giving abstract of their views[18]
Mackintosh Ethical Philos:[19]

266 **{Prostitution of Paris.[1] with respect to licentiousness, destroying children. —it is not effect, as Lyell suggested, of organ being worn out as. otherwise old whores would not have children}**

267 **Study . . . plants**) '**abortive**' *underlined.*
266 Lessings Laoccaon . . . lik it.] *crossed brown ink.*
Brit. Assoc.—] 'A' *over* 'g'.
D^r^. Mayo . . . Mind.—] *crossed brown ink.*

267–1 Benjamin Silliman's *American Journal of Science.*
267–2 Rengger 1830.
267–3 Montagu 1804, 1831.
267–4 A. Wilson 1832.
267–5 Youatt 1831, 1834, 1837.
267–6 Virey 1835.
267–7 M. Serres 1838a.
267–8 I. Geoffroy Saint-Hilaire 1832–36.
267–9 R. Owen 1837:xxv, in Hunter 1837.
267–10 Hunter 1786 or 1792.
267–11 Hunter 1837.
267–12 Yarrell 1827.
267–13 Paxton 1838.
267–14 Gore 1838.
267–15 Ferrier 1838a.
267–16 Watson 1835.
267–17 Hume 1750.
267–18 Stewart 1829. That is, Dugald Stewart's works, of which vol. 7 contains his life of Adam Smith and his life of Thomas Reid.
267–19 Mackintosh 1837. Darwin's copy, which bears many marginal notes, is inscribed 'Charles Darwin 1840'.
266–1 Parent-Duchâtelet 1836. Charles Lyell: personal communication.

Turners embassy to Thibet, perhaps worth reading quoted by Malthus.—[2]
Heberdens Observat. on increase & decrease of different diseases. 4to. 1801.— quoted by do.—[3]
There appears to be good art.. on ‹Etn› Entozoa by Owen[4] in Encyclop. of Anat. & Physiology.—
Dampier.[5] probably worth reading
Lessings. Laoccaon.—[6] (translated in 1837) on limits of painting & poetry.— Erasmus thincks I should lik it.[7]
The Sportsman's Repository. 4to.[8] contains much on dogs.—
Reports of Brit. Assoc.—[9] some important Papers.
Dr. Mayo. Pathology of Human. Mind.—[10]
Audubons. Ornithological Biography.[11] 4. Volumes well worth reading
Bevans work on Bees,[12] new Edit 1838
Harlaam. Physical & Medical Researches. on Horse in. N. America.—[13] Owen has it.—[14]
Ld. Brougham. Dissertations on subject of Science connected with Natural Theology.—[15] on instinct & animal, intelligence.— *very good.*
Endlicher has published in first volume of Annales of Vienna. sketch of south Sea. Botany[16]
R. Brown. has curious *coloured* maps. by Copenhagen Botanist of range of plants[17]

265 Books quoted by *Herbert.*[1] p. 338
Schiede[2] in 1825. & Lasch[3]. Linn. in 1829 has given list of Spontaneous Hybrids. where?
Sweet. Hortus Britannicus.[4] has remarks on acclimatizing of Plants.
Herbert. p. 348. gives reference to Kolkreuter's Papers[5]
Wiegman has published German Pamphlet on crossing Oats, &c[6]

266–2 Malthus 1826, 1:201–5 discusses S. Turner 1800.
266–3 Malthus 1826, 2:302–6 discusses Heberden 1801.
266–4 R. Owen 1836–39.
266–5 Dampier 1698–1703.
266–6 Lessing 1836.
266–7 E. A. Darwin: personal communication.
266–8 J. Lawrence 1820.
266–9 The British Association for the Advancement of Science.
266–10 T. Mayo 1839.
266–11 Audubon 1831–49; 5 vols, vol. 4 (1838).
266–12 Bevan 1827.
266–13 Harlan 1835.
266–14 Richard Owen.
266–15 Brougham 1839.
266–16 Endlicher 1836.
266–17 Possibly J. F. Schouw.
265–1 Herbert 1837.
265–2 Schiede 1825.
265–3 Lasch 1829.
265–4 Sweet 1830.
265–5 Herbert 1837:348, 'Kolreuter's experiments are detailed in the transactions of the Petersburg Acad. in 1777, and the five or six following years.' Darwin commented, 'must be read' on Herbert's note giving rough publication details for Kolreuter's hybridization experiments in *Lychnis, Cucubalis, Digitalis, Lobelia, Lucia, Verbascum, Datura,* Malvaceous plants, and Flaxes.
265–6 Wiegmann 1828.

«Horticultural Transacts.—»
{Mr Coxe "Views of the Cultivation of Fruit trees in. N. America" in Lib. of Hort. Soc[7]

Mr Neil. has written good article on Horticulture in Edinburgh. Encyclop.—[8]
{The ‹Edin› British & Foreign Medical Review No XIV. April 1839.— Review on "Walker on intermarriage" price 14s.[9]

Marh. 20t. 1839. Philosophy of Blushing[10] lately advertised. /6s

---

Mrs Necker on Education[11] preeminently worthy of studying in Metaphysical point of view

---

{Henslow[12] has list of plants of Mauritius with locality in which each one is found. very good to see whether peculiar plants— in high points

INSIDE BACK COVER Read Volney's travels in Syria[1]
Vol I. p. 71. account of Europæan plants transported.—
Crawford. Eastern Archipelago.[2] probably some account
Raffles. Sir. S do. do—[3]
Buffon Suites[4]
Cline on the improvement of domesticated animals[5]
Fries de plantarum prœsentum crypt. transitu et analogia commentatia[6]
Library of Useful Knowledge on Horse & Cow & Sheep[7]
Clarke's Travels.—[8] Temminck Hist. Nat. des Pigeons et des Gallinaces.[9]

IBC Read Volney's . . . **Waterhouse has it)**] *crossed brown ink.*
Cline . . . domesticated animals] *crossed brown ink.*
Library . . . & sheep] *crossed brown ink.*
Wowett . . has it)] *crossed brown ink.*
shells from . . . same time] *added pencil*; '**living**' *underlined.*
Study . . . **Gemmæ**] *crossed brown ink.*

265–7 Coxe 1817.
265–8 Neill 1830.
265–9 Walker 1838, reviewed in Walker 1839.
265–10 Burgess 1839. Listed in books received for review in *Brit. For. Med. Rev.* 7 (1839):594. Walker 1839 appeared in the same number.
265–11 Saussure 1839–43 (Necker, née Saussure).
265–12 Henslow: personal communication.
IBC–1 Volney 1787a, 1:71, '. . . but it is worthy of observation, that this soil appears extremely unfavourable to all exotics. foreign plants degenerate there rapidly: . . .'
IBC–2 Crawfurd 1820.
IBC–3 Raffles 1830.
IBC–4 *Suites à Buffon* is not a work, but a large-scale publishing enterprise comprising twenty five, often multi-volumed, works on natural history. The series was published in Paris from 1834 to 1890 under the aegis of Nicholas Edme Roret. Darwin cites two of the ten works published or under way in the series by 1839: Duméril and Bibron 1834–58 and Milne-Edwards 1834–40. His note may well be a memorandum to keep an eye out for works published as *Suites à Buffon.*
IBC–5 Cline 1829.
IBC–6 E. Fries 1825.
IBC–7 Youatt 1831 (Horse), 1834 (Cattle), 1837 (Sheep).
IBC–8 E. D. Clarke 1816–24.
IBC–9 Temminck 1813–15.

Silliman's Journal. during 1837. paper by Bachman on migration of birds[10]
Temminck has written "Coup d'Œil sur la Faune des iles de la sonde et de L'empire du Japon[11]
Wowett on Cattle—[12] (Waterhouse has it[13])

---

shells from Barrier isl^{d} many relations with a living Matica

---

& many shells of Genera Corlula Cham. Cardium. Porcellus Turbo. Cerithium Jardin du Roi

---

Java fossils at same time

Study Botanical work on Buds & Gemmæ.

BACK COVER

C

IBC–10 Bachman 1836.
IBC–11 Temminck 1833.
IBC–12 Youatt 1834.
IBC–13 G. R. Waterhouse.

Notebook D

Transcribed and edited by DAVID KOHN

Notebook D (DAR 123, 175 × 97 mm), third in the transmutation series, is bound in red leather. It is similar to Notebook M, with blind-embossed borders, a metal clasp, and cream-coloured paper labels with clipped corners on each cover. D is written on the labels in ink. The notebook comprises 180 green-edged pages numbered consecutively by Darwin. Pages D1 to the first entry of D21 are written in brown ink.[1] The remainder of Notebook D is written in grey ink. Of 43 leaves excised by Darwin, 27 have been located.

On D1 Darwin made two separate dating notes: 'July 15th' and 'Finished. October 2d'. Here, as in Notebooks B and C, the grey-ink dating was added between 29 July and 16 October 1838. Though Darwin's opening date is an estimate, it must be very close to correct. The first date in the notebook is 23 July on D4. Darwin probably took the 'July 15th' date from M1, which is so dated. Likewise, Darwin may have drawn on N1 with its clear 'October 2d' for the closing date of Notebook D. The last dated entry is 'Sept. 29th' on D136 and the first date in Notebook E is 'Octob. 4th' on E3.

Notebook D presents a marked contrast to Notebooks B and C with respect to Darwin's attention to dating. With his arrival in Shrewsbury on 13 July 1838[2] a new awareness of time is evident in the 25 direct dates he entered in the notebook. Thanks to these entries, the chronology of D1–136 (15 July–29 September) is straightforward. For this period, the abundant datable citations confirm Darwin's direct dates; in addition we learn from references to the *Proceedings of the Zoological Society* that D27–35 correlates with the Society's 14 August 1838 meeting.

However, after D136, dating is difficult. Along with Darwin's differentiated sense of time comes a complex use of writing space. Thus the last quarter of the notebook involved extensive backfilling, and the internal chronology is complicated by the following factors: 1, gaps produced by unlocated excised pages D141–146 and D149–150; 2, dated and undated interpolated essays dealing with generation on D152–159 and D174–179; and 3, citations to the October 1838 number of *Annals of Natural History* on D151, 167 and 169.

Notebook D was filled in three months, by far the most accelerated pace of the transmutation series. The dominant theme of the notebook is reproduction. It includes a detailed abstract of Hunter's *Animal Economy* edited by Richard Owen[3] (D112–116 and D154–161) and several attempts to bring the complex laws of generation under a unified view. In so far as the notebooks on transmutation are a record of the conceptual growth of Darwin's theory, Notebook D is the climactic document of the series. While Darwin constructed the basic framework of his theory in Notebook B and deepened and extended that theory in Notebook C, by the end of Notebook D he formulated a new answer to the fundamental question of his theoretical enterprise: what is the origin of adaptation? In evocative language, Darwin expresses a metaphorical yet decisive grasp of the adaptive role of competition that attends 'the warring of species as inference from' the Malthusian law of population: 'One may say there is a force like a hundred thousand wedges trying force ‹into› every kind of adapted

[1] There are grey-ink annotations, indicated in bold face, on D1, D9, and D17–19.
[2] *Correspondence* 2:432.
[3] Hunter 1837.

structure into the gaps ‹of› in the oeconomy of Nature, or rather forming gaps by thrusting out weaker ones.' (D134–135).[4] The full articulation of the theory of natural selection from this first formulation takes place over the subsequent months and is to be found in Notebook E, but the image of natural selection as a wedging force persists through the *Origin*.[5] By his reliance on the classical political economy of Malthus as a resource Darwin made competition in man the model for his understanding of nature.[6] Moreover, Darwin conceives of his wedging force in teleological terms. 'The final cause of all this wedging, must be to sort out proper structure & adapt it to change.—' (D135). However, from the beginning of Notebook B Darwin sought a strictly naturalistic solution to the problem of adaptation. While he formulates his new theory in Notebook E, he examines at the same time its metaphysical significance in his abstract of Macculloch to conclude that final causes are 'barren virgins'.[7] Darwin's 'theory by which to work'[8] is rooted in a teleology that relies on utilitarianism rather than on providence. Just as the problem of adaptation, inherited from Paley and Lyell, has theological and political underpinnings, so does Darwin's solution to the problem: natural selection.

[4] See Limoges 1970, Herbert 1971, Gruber and Barrett 1974, and Kohn 1980 for interpretations of the Malthus passages.
[5] See Ospovat 1981, Hodge and Kohn 1985, and *Origin*: 67. See also Stauffer 1975:631–32.
[6] See Young 1985.
[7] Macculloch Abstract DAR 71:58r.
[8] Barlow 1958:120.

FRONT COVER

D

INSIDE FRONT COVER

Charles Darwin
36 Great Marlborough St

Did Eytons ‹intermediate›. «hybrids, when» interbred. show any tendency to return to either parent.?
Is the first cross, which makes hybrids. productive like geese?—[1]
Are the number of kittens between Lion & Tiger at litter as numerous as in common lion?[2]
Are the number of nipples in domesticated very fertile animals increased?—
Where offspring, heterogenous, in plants are the number of seeds greater.?—
Mem. for Eyton.— Sir. R. Heron's case of breed of pigs with solid feet.—[3]

1838

[In this Book some curious notes on Monkeys recognising Sexes of animals:]CD
[All Selected Dec. 14— 1856]CD
Towards close I first thought of selection owing to struggle

IFC Is the] 'Is' *over* 'On'.
Are the] 'A' *over* 'H'.
[In this . . . of animals:]CD] *added pencil.*
[All Selected Dec. 14— 1856]CD] *added pencil.*
Towards . . . struggle] *added pencil.*

IFC–1 Refers to Eyton's report of 'a hybrid male and female, derived from the Chinese and common goose' (1837a:357). See B30, B139, 'Eyton says Hybrid about half aways' and E169.

IFC–2 See C228 and D8.

IFC–3 Refers to an extract of a letter from Sir Robert Heron to William Yarrell: 'There is a breed of Pigs, size «of the» Chinese, with feet undivided—internally—: Like a horses «foot» some had them, and on crossing them, had some with two whole «feet» and two divided feet.' (*Correspondence* 2:141). The later uses of this passage occur in the context of characters that do not blend. We can trace these uses because the passage was scored in 1850's brown crayon and annotated '(*Dorkings feet.* P. Chron.)' which refers to *The Poultry Chronicle* (F. K. P. 1855). Heron's report figures in *Natural Selection* and in *Variation*, always together with Dorkings' feet. In *Natural Selection*, chap. 9, Hybrids, the context is set by Isidore Geoffroy Saint-Hilaire, who 'has stated that hybrids from between two species generally present fixed & constant characters partly those of the father & partly those of the mother: on the contrary that mongrels are either intermediate like hybrids, or resemble entirely one of their parents. (p. 453) . . . One case, however, seems to occur frequently with mongrels, almost in accordance with Is. Geoffroy's remark, and which as far as I am aware has not been noticed in hybrid animals from between species in a state of nature; . . . namely either the perfect transmission or entire absence of some marked character of one of the parents in the mongrel . . . such cases as the mongrel offspring from the Dorking & other fowls, having five toes on one foot & four on the other—the cross from the ‹solid› whole-hoofed & common pig, which with Sir R. Heron had two feet whole & two normally divided—are probably due to this same difficulty of fusion in certain characters.' (p. 456). In *Variation* 2:93, the reference occurs in the section *On certain Characters not blending*: 'When Dorking fowls with five toes are crossed with other breeds, the chickens often have five toes on one foot and four on the other. Some crossed pigs raised by Sir R. Heron between the solid-hoofed and common pig had not all four feet in an intermediate condition, but two feet were furnished with properly divided, and two with united hoofs.' Heron's letter was received 17 December 1838, which indicates that the ink portion of the Inside Front Cover (Did Eyton . . . solid feet.) dates from December 1838 or later, that is at least two and one half months *after* the 'completion' of the notebook.

1 **July 15th. 1838 Finished. October 2d**
As a proof. what ‹trifling› «unknown» causes act upon people. My Father mention, than for ten years he never saw one case of malignant erysipelas spreading over the head, not caused by a wound, when suddenly during one time he had three patients at very distant quarters of the county, who had had no sort of communication, were seized with it, & for ten years afterwards, he then did not see other cases.—
He thinks apoplexy affects people all over England at same periods

2 When he began practice, he remember during a year or two he saw many cases of virulent cancer in women, & since that time it has been rare disease.— but now (July 1838) he has seen more case in a month, than in several previous years, two having consulted him on one day.—

3 Mark at Shrewsbury thinks the half bred Alderney Cows take more after Alderney that the Durham,, with which they have been crossed—is Alderney oldest breed— He believes all pretty much alike.—

My Father
Water-in the hair a century since used to be called Worm Fever, as used much more latley diseased Mesenteric glands.— My Father has seen case of pleurisy, broken limb «in children» & other such disorders accompanied with some fever, be attended by the transmission of large number of worms

4 the child not having passed them before.—
Hence disordered intestines are not healthy to worms, (like parasites of Tropical countries cannot endure this climate— .) —

July 23d. Eyton, a stone blind horse, seemed to perceive turn on road where No houses to Eaton Mascott,[1] where he had been accustomed to turn down.— — applicable to birds migrations & Australian Savages.—

5e W. D. Fox[1] has a cat. which he bought in Portsmouth, said to come from coast of Guinea, — *ears* bare. skin black & wrinkled— fur short. (tail cut off in

1 were seized] 'w' *over* '&'.
4 seemed] 's' *over* 'pe'.
No] 'n' *over* 'h'.
5 W. D. Fox . . . of the Common cat] '1' *added brown crayon, crossed faded brown ink.*
Ch IX Mongrels Hybridism] *added brown crayon.*
half] *underlined grey ink.*

4–1 Thomas Campbell Eyton personal communication. Eaton Mascott: 5 miles SSE. of Shrewsbury in the Atcham district.

5–1 D5–15 are based on conversations with William Darwin Fox, as is evident from his comment, 'no leading question was put' (D8). No direct evidence of a meeting survives; however, by July 1838 Fox was probably established as the Rector of Delamere and either man may well have made the *c.* 50 mile trip between Delamere and Shrewsbury during the period 23–29 July 1838. (See note B176–1).

progeny peculiar) limbs very long, eyes very large, very fierce to dogs «otherwise habits not different; tone of voice. perhaps rather different».— crossed with ‹un›common cat, exact variety unknown., three kittens, alike each other, partaking ‹more› «very closely» of form of mother: more than of the Common cat.— **Ch IX Mongrels Hybridisim**
Fox has *half* Persian cat. which bred with unknown common house cat.— had four Kittens. two appeared

6e «so» very like common cat, that they were killed, & other two very closely resembled in form of tail, fur &c to the half bred Persian.— Here then we have clear case of heterogenous offspring from one impregnation ?is this one impregnation, or two impregnations one giving half character & other more of English, but the effect is the same.—
Fox thinks that when a *wild* animal is crossed with

7e tame, offspring always take most after wild.— i.e that ‹alw› «no» domesticated ones have been so long as wild one under present form.— Fox has seen several cases of foxes & dogs crossed, offspring always more resembled foxes than dogs (Mem Jackall in Zoolog Garden)

———

He has seen in a show half Wolf & «half Esquimaux» dog which ‹likewise more resembled the wolf than dog.—› appeared to be intermediate between two parents.— this is very interesting as Esquimaux dog approaches to species.
Again he has seen several crosses between Esquimaux dog & common dogs & Fox thinks they decidedly take

8e much most after Esquimaux.— this agrees perfectly with Yarrell & no leading question was put.—
Fox thinks half Lion & Tigers are exactly intermediate in character & Kittens alike each other.—[1]

———

Even in children of parents ‹some› one sometimes resembles one parent & one another & are not exactly intermediate.—

———

9 Where two dogs line the same bitch & «perfect» spaniels & setters are produced. one would argue the whole effect of race was determined by male:[1]

6 *page crossed faded brown ink.*
7 «Esquimaux»] 'E' *over* 'e'.
appeared to] 'to' *over* 'be'.
'16' *added to page brown crayon.*
8 Fox . . . intermediate.—] *crossed pencil.*
9 is Lor[d]] 'i' *over* 'o'.
they] *crossed grey ink.*
page crossed pencil.

8–1 See IFC.
9–1 See Darwin's implicit critique of male determination in his comments on Erasmus Darwin in D19.

& How completely is Lord Moreton's case opposed to this fact & views.—[2]

Fox says[3] a cousin «one of Mr Strutt» of his used to breed to Common & Muscovy Ducks.— English. ‹Common› «China» & Canada Geese, & that they **this first cross** were *equally* fertile with pure bred animals.— Mem. number of Mules.—
«He recollects one hatch of hybrid geese very fine.—»
How is it

10 with plants? This indicates a remarkable law, that first cross ‹not se› plentiful, second absolutely sterile.—

My case of Stallion, according to Erasmus preferring young mare to old, explained by Stallions, (according to Fox) being guided entirely by their smell.—

Fox says he knew «a» carter well, who placed his stallion as second horse between ‹whe› shaft mare

11e & another leader mare,— this stallion though eager to all other mare had been entirely broken from their mares, (though horsing every month) & worked in the same cart in loose chains, by being at first beaten from her, & always accustomed to her.— case parallel to brothers & sisters in Mankind.—

The case of all blue eyed cats (Fox has seen repeated cases) being deaf curious case of corelation of imperfect structure.—

12e Fox says in «Lord» Exeter's Park «or in the Duke of Marlborough» there is a breed of white-tailed squirrels, which form a marked *wild* variety. doubtful whether all are white.
Fox says the Half Muscovy

Fox says a settler near Swan river, lost his ‹on› two cows entirely, changed his residence a great many miles.— yet one day ‹th› a cow walked in, then disappeared, & three days afterwards came again, bringing with her the other & younger cow.—

13 Fox says when common & China goose are crossed the neck is not intermediate in its peculiar long neck, but much nearer to common goose.—

---

11 *page crossed pencil.*
12 Fox says in «Lord» . . . white.] '3' *added brown crayon.*
Fox says a settler . . . younger cow.—] *crossed pencil.*

9–2 G. Morton 1821.
9–3 See *Correspondence* 2:111–12, from W. D. Fox, c. November 1838.

What has long been in blood, will remain in blood.—
—converse, what has not been, will not remain,— yet offspring must be somewhat like parents,— therefore offspring will tend to go back, or have none— the argument does not apply to first parents, because they are not new breed.— the first hybrids may be

14 compared to animal with amputated limb.—
Heredetary ‹thr› Six fingered people, ‹Hill› «Lord Berwick» family with defective palates. heredetary & therefore exceptions. to above law.— Study what these monsters are:— are they «abortive» twins.— ǂ The fertility of first cross, as stated by Fox, is very important, as showing above facts as first cross being new species, ǂ —

Are not dreadful monsters, abortive, just like *mules*.

15 Fox's half bred Persians «cat» favour the Persian side.—

Theory of abortive hybrids.— If mules did breed, the offspring would «as in all other animals» be like either parent, or intermediate within certain small limits (within which limits they might return to either parent), then according law, that in proportion as things are long in blood so will they remain, a mule «being new species» will have no tendency to have offspring like parent, but as they must like or there will be none, therefore a mule can have no offspring.= but as «badly» deformed people & as mutilations «(produced very quickly)» sometimes have similar offsprings, so will the worst mules (as real mule) have offspring,— slight deformities «as supernumerary fingers» (that is slight alterations of primitive stocks «relative to changes which every species undergoes») & hybrids between very near *species* (that is slight alterations of primitive stock) are heredetary: «Hybrids of» Varieties is different because not long in blood.= The case of union of perfect animals is

16 distinct case,— gradation from physical impossibility to (*perhaps increased*). fertility.— (but many animals are fertile, when offspring infertile,— two considerations are here combined). In last page, we have seen mules could have no offspring, & this being case, owing to the corelations of system, the organs of generation would necessarily fail.—

In last page. I should have said, "an animal ‹acquires ‹th› any new› is ‹only› able to transmit «only» those peculiarities, to its offspring, which have been *gained slowly*, now all the mules

15 the] 'e' *over* 'this'.

17 have their whole ‹body› form of body gained in one generation, so it is impossible to transmit them, & as offspring must be like parent, therefore mule has no offspring & therefore no generative organ.—

Same Prop. better enunciated.— "An animal **Either parent** cannot transmit to its offspring any ‹peculiarity› change from the form which it inherits from its parents «stock» without it be small & slowly obtained
NB. The longer a thing is in the blood, the more persistent.— «any amount of change» shorter time less [s]o.— the result of this is that animal would endeavour to return to parent stock. but if both parents are alike, offspring must be like

18 Hence mutilations not heredetary,, but size of particular Muscles— **When two animals cross. each sends his own likeness, & the union makes hybrid, in fact the parents beget child like themselves.** expression of countenances, organic diseases, mental disposition, stature, are slowly obtained & hereditary; ‹but if› if the change be congenital (that is most slowly obtained with respect to that individual) it is more easily inherited.— «but if change be in blood long, it becomes part of animal &» by a succession of ‹such changes› generations, these small changes become multiplied, & great change be effected, but

19 in a mule these conditions are not fullfilled.— «[My grandfather's theory of Mules not hereditary, because generation — highest point of organization[1]] ^{CD}» false.—[2]

The creator would thus contradict his own law.
So far is there any appearance of animals being created. it is probable if created at once. ‹w^{d}› according to ordinary laws, the character of offspring would vary, or rather they would not have offspring—

On the idea of generation being a ‹slip› «bud» from parent.[3] if whole parent not entirely embued with the change, a bud could not be taken, without it either went back, or not being perfect would perish.—

19–1 Erasmus Darwin 1794–96, 1:514, 'The formation of the organs of sexual generation . . . seems the chef d'oeuvre, the masterpiece of nature; . . . Whence it happens that, in the copulation of animals of different species, the parts necessary to life are frequently completely formed; but those for the purpose of generation are defective, as requiring a nicer organization; or more exact coincidence of the [maternal] particles of nutriment to the . . . appetencies of the original [paternal] living filament.' Note that Erasmus Darwin did not accept biparental formation of the embryo and, at least by implication, biparental inheritance: 'the embryon is secreted or produced by the male, and not by the conjunction of fluids from both male and female . . .' (1:485).

19–2 Darwin is rejecting the position in B2: 'See Zoonomia arguements, [generation] fails in hybrids where everything else [all other systems] is perfect'.

19–3 See Erasmus Darwin 1794–96, 1:487, 'This paternal offspring of vegetables, I mean their buds and bulbs, is attended with a very curious circumstance; and that is, that they exactly resemble their parents, as is observable in grafting fruit-trees, and in propagating flower-roots; whereas the seminal offspring of plants, being supplied with nutriment by the mother, is liable to perpetual variation.'

20 The Varieties of the domesticated animals must be most complicated, because they are partly local & then the local ones are taken to fresh country & breed confined. to certain best individuals.— scarcely any breed but what some individuals are picked out.— in a really natural breed, not one is picked out, & few even of local varieties approaches quite to wild local variety.— our Europæan varieties must be very unnatural— Italian Greyhound is probably the effect of ‹sev› local variety many times changed

21 together with some training in the earlier branches «as in common greyhound» & much intermarriage.—

In my speculations.[1] Must not go back to first stock of all animals, but merely to classes where types exist[2] for if so. it will be necessary to show how the first eye is formed.—[3] how one nerve becomes sensitive to light.— (Mem whole plant may be considered as one large eye— have they smell, do plants emit odour solely for others parts of creation) & another nerve to finest vibration of sound.— which is impossible.—[4]

22 Mr Spence remarks that the Fringilla domestica of North Europe is replaced by the F. cisalpina in Italy, which is so like that difference would not be discovered by an unscientific observer.—[1]

‹Transactions of the Entomological Soc›

A capital passage might be made from comparison of Man, with expression ‹of a› of Monkey, «when offended» who loves, who fears, who is curious &c &c &c who imitates.— who will say there is distinct Creation required if he believes «hyæna & squirrel» seal & mouse, elephant, come from one stock.—

23 *Theory of Geograph. Distrib: of ‹ani› organic beings.—*
Animals «of same classes» differ in different countries in exact proportion to the time they have been separated; together with physical differences of

21 others] 's' *over* 'r'.

21–1 The following passage, Darwin's first in grey ink, could have been written at Maer during the first three days of his courtship of Emma Wedgwood (29–31 July 1838); see *Correspondence* 2:94–95 for Darwin's letter of 7 August, to Emma, looking back on their cozy chat (goose) by the Maer Hall library fire.

21–2 See *1844 Essay*: 128, 'But if the eye from its most complicated form can be shown to graduate into an exceedingly simple state, it is clear (for in this work we have nothing to do with the first origin of organs in their simplest forms) that it may *possibly* have been acquired by gradual selection of slight, but in each case useful deviations.' In the margin of the MS fair copy (DAR 113:89), Emma Darwin wrote 'a great assumption E. D.' Perhaps in response to his wife's criticism, Darwin pencilled in the following after 'graduate': 'through the animal Kingdom & that each eye is not only most useful, but *perf*ect for its possessor'.

21–3 Darwin's strategy was in part rooted in his uniformitarianism. See Lyell 1837, 1:89, '[Hutton's *Theory of the Earth* (1795)] was the first in which geology was declared to be in no way concerned about "questions as to the origin of things;" '.

21–4 See Abercrombie's critique of materialism, 1838: 26–27.

22–1 Spence 1836:6.

country: the time of separation depends on facility of transport in the species itself, & in the local circumstances of the two countries in times present & past.

The effect of physical conditions of country is not perhaps so great, as separation on ‹be› inter-breeding, for otherwise we could not understand the vast number

24 of domesticated races.—

Athenæum. p. 505. some (very poor account) of plants of Nova Zenbla — in review of Baers work[1]

Edinburgh. Royal. Transact.— p. 297. Vol 9. Dr. Ferguson seems most clear that the ideosyncracy of the Negro (& partly Mulatto) prevents his taking any form of Malaria— adaptation & species-like,— — Says Negro— thick skinned[2]

My hairdresser (Willis) says ‹black› «that strength of» hair goes with colour. black being strongest.—

25 V. p. 63. Note Book M′. for case of change in food in insects entered by mistake[1]

Surely the fossil Mamalogy of Britain & Europe is African. & the only difference is by the *extinction* of certain forms from Northern part & not by fresh creation of new forms.— what is range of Hyæna? Hippotamus.? Indio-African, or pure Africa?— ||Fossil Elephant of Africa Most important under this view, & Hippotamus of Madagascar: because. contemporaries.

In introduction to Eytons Anatidæ.— recurs to idea of only animals from distant countries breeding!.[2] «Mem 3 species of grouse»! Has not Goldfinch &

24–1 Ferrier 1838b:506, note, 'The struggle between the cryptogamic vegetation and the plants of more developed form strikes the observer forcibly when he turns his attention to the borders of the plains. The former, deriving more subsistence from the soil, advances unrelentingly and threatens to exterminate every other kind.' Ferrier is reviewing Baer 1837c, 1837d and 1837e.

24–2 Ferguson 1823:297–98, note 'On the negro skin': 'The adaptation of the Negro to live in the unwholesome localities of the Torrid Zone, that prove so fatal to Europeans, is most happy and singular. From peculiarity of idiosyncrasy, he appears to be proof against endemic fevers; for to him marsh miasmata are in fact no poison, and hence his incalculable value as a soldier, for field service, in the West Indies. . . . One of the most obvious peculiarities of the Negro, compared with the European, is the texture of his skin, which is thick, oily and rank to a great degree.' (p. 297).

25–1 See M63 'do. p. 157. Westwood remarks . . . *sylvestris*' and note M63–1.

25–2 Eyton 1838a:1. 'The generally received . . . [definition of species] is that of John Hunter, viz. that hybrids between true species will not be productive; and this, we are inclined to believe, is partially correct, but not entirely so, as some birds, in a state of domestication, have bred together, and their offspring been productive,

Greenfinch bred, & surely wild Duck & «pintail» Widgeon!— Divides animals «world into Zoological Provinces» according to varieties of Man.?[3] «In Australia. plants E & W very different.— Man not so, but N. & S. New Zealand & New +++ Caledonia. two races of Men, but not plants» will it hold good.— Thinks Temmink doubtful when he says No genera.—[4] thinks

26 there are some small divisions.—[1] does not seem to think any improbability to animals being distributed after *flood* (!) according to affinities!.[2] confounds, *like*

26 **No]** *added pencil.*

although differing most materially in external form.

'It may be advanced, however, and with truth, that those animals upon which this experiment has been tried have invariably been brought from countries far apart, and that consequently in a wild state the experiment has never been tried; no fact, that we are aware of, can be brought forwards in answer to this objection. Should it prove true, that animals inhabiting different countries, and with slightly different forms and colouring, are of the same species, . . . it can only be accounted for in this mode, namely, that at the universal distribution of animals after the Deluge, those of the same species, and derived from the same parents, going to different localities, have in a succession of ages been influenced by various local circumstances, as climate, the plentitude, the want, or nature of food, which causes have changed their form, colouring, and, in many instances, their habits.' Note that by 'recurs to', Darwin is referring to the similar point made in Eyton 1837a:359, 'All true hybrids that have been productive have been produced from species brought from remote countries, and in (or partially) a state of domestication.' (Passage scored.) Eyton goes on to explain this circumstance, as follows: 'May we not, therefore, suppose that it is a provision of Providence, to enable man to improve the breeds of those animals almost necessary to his existence; and that it is almost a necessary provision, as it is universally found that breeding in, as it is called, . . . tends to diminish and dwindle the race which has been subjected to it?' Darwin subsequently tested Eyton's proposition in his 'Questions for Mr Herbert' (*Correspondence* 2:181), *c.* 1 April 1839), Question 7: 'It has been said (apparently with little foundation) that amongst birds, species originally coming from distant parts of the world, are more likely to breed together, than those from nearer countries.— Has M^r^. Herbert observed anything of this kind in plants?' William Herbert replied: 'I think species from remote parts are only thus far more likely to breed together than those from neighbouring localities, that in the latter case there is greater probability that they have been already approximated & have not bred together. I do not think the distance of their natural location can facilitate the disposition to interbreed.' (*Correspondence* 2:183.)

25–3 Eyton 1838a:2, 'Mr. Swainson divides the earth into five Zoological provinces, corresponding with the continents of Europe, Asia, Africa, America, and Australia; these are well known divisions of the world, and as such are convenient, but we must consider that those countries occupied by the different races of mankind . . . form much more natural Zoological provinces than those mentioned by Mr. Swainson.'

25–4 Eyton 1838a:4, 'Many parts throughout the group appear to be in favour of Mons. Temminck's opinion [1817], that there are no such divisions as Genera in nature, the transition from one extreme of form to another being so gradual that it is difficult to say where to draw the line of division.

'Upon a minute examination, however, as far at least as we are at present acquainted with the species, there is always found some break as it were between the forms constituting contiguous Genera or Subgenera, and some tangible distinction between them, although in many particulars they appear closely to approach. It is probable, however, that many new forms will yet be discovered; therefore, in the present state of science, it is impossible to say whether this opinion of Mons. Temminck's will eventually prove true or not.'

26–1 Eyton 1838a:4. See note D25–4, 'there is always . . . approach.'

26–2 Eyton 1838a:1. See note D25–2, 'Should it prove . . . their habits.'

Whewell[3] affinity with analogy—[4] Good table at end of distrib: of ‹birds›. Anatidæ.—[5] Consult this book again.—

Mine is a bold theory. which attempts to explain, or asserts to be explicable every instinct in animals.

Heard at Zoolog Soc their Pintail & Common Ducks, breed one with another— & hybrids fertile inter se—**No** directly against Eyton's rule. ?Are the hybrids similar inter se.—

27e–28e [not located]

29e the «4» Struthionidæ, **Mr Blyth remarked that greater difference in** than in many large orders of birds.[1] The Emu & Cassowary closest.— Ostrich & Rhea

29 **Mr Blyth . . . difference in**] *added pencil, presumably from 28 when the latter was excised.* '11' *added to page brown crayon.*

26–3 Whewell 1837, 3:353–55, 'It will appear, . . . that those steps in systematic zoology which are due to the light thrown upon the subject by physiology . . . have been, . . . led to and produced by the general progress of such knowledge. We can hardly expect that the classificatory sciences can undergo any material improvement which is not of this kind. Very recently, however, some authors have attempted to introduce into these sciences certain principles which do not, at first sight, appear as a continuation and extension of the previous researches of comparative anatomists. I speak, in particular, of the doctrines of a *circular progression* in the series of affinity; of a *quinary division* of such circular groups; and of a relation of *analogy* between the members of such groups, entirely distinct from the relation of *affinity*.' (p. 353). 'But the doctrine of a relation of analogy distinct from affinity, in the manner which has recently been taught, seems to be obviously at variance with that gradual approximation of the classificatory to the physiological sciences, which has appeared to us to be the general tendency of real knowledge. It seems difficult to understand how a reference to such relations as those which are offered as examples of analogy[14] can be otherwise than a retrograde step in science.

[14] For example, the goatsucker has an *affinity* with the swallow; but it has an *analogy* with the bat, because both fly at the same hour of the day, and feed in the same manner. Swainson [1835], *Geography and Classification of Animals*, p. 129.' (pp. 354–55).

26–4 Eyton 1838a:1, 'Much has been said and written on analogy and affinity, and the connection by one or the other of them between the groups and species of the animal kingdom. We have not, however, been able to distinguish between them in any other manner than that the former is generally applied when the groups or species between which a connexion is supposed to exist are far removed from each other, and the latter when nearly related, we shall use the terms indiscriminately, as convenient.'

26–5 Eyton 1838a, Appendix.

29–1 Part of the missing D27–28 and D29–35 (through the passage on Owen) are based on Darwin's attendance at the 14 August 1838 meeting of the Zoological Society, and permit reconstruction of the evening's formal and informal discussions. In the first lines of D29 ('the . . . closer'), Darwin made note of a comment by Edward Blyth on the paper Owen read that evening on the *Apteryx* (Owen 1838). This is followed by a question and answer exchange between Blyth and Owen (D29–30). Next Darwin noted mention of 'Animals from Hobart Town' (D30), which *Proc. Zool. Soc.* 6 (1838): 105 records, as follows: Mr. Waterhouse then directed the attention of the Meeting to an interesting series of skins of Marsupial animals, brought from Van Diemen's Land by George Everett, Esq., and presented by that gentleman to the Society; the collection includes a specimen of the *Thylacinus*, two species of Kangaroo, and two of the genus *Parameles*, besides others of more common occurrence.' (*Proc. Zool. Soc. Lond.* 6 (1838): 105). Next Darwin noted a 'New species of Moschus, characterized by Ogleby' (D30), which is recorded in *Proc. Zool. Soc. Lond.* 6 (1838): 105, as follows: 'Mr. Ogilby [1838b] pointed out the characters of a new species of Muntjac Deer, which lately died at the Gardens. . . . A female specimen which accompanied that here described, is still living and has lately produced a fawn, which is interesting from exhibiting the spotted character common to the generality of the young in this extensive group.' D31–34 records a conversation with John Bachman, who read a paper at the meeting (Bachman 1838). The conversation was interspersed with at least one remark by Blyth (D33). Then in D34, the record of conversation moves to William Yarrell, who was in the

closest.— (& two Rheas still closer).— Mr Blyth asked whether structure of pelvis & was not adaptive structure,[2] like little wings of Auks which does not make that bird a Penguin.— (i.e. whether relation in one point or many) Owen answered that all characters might be considered as adaptive & that he did not see where the line could be drawn— thus the most remarkable character in Apteryx, small respiratory system; even much smaller

30e than in other Struthios. was adaptation to little Movement.— nocturnal crawling bird.— Wings reduced to rudiment.— clavicle scapula &c strongly developed to aid in breathing.—[1]

Animals from Hobart Town mentioned, it seems most of species from there now found in Australia

---

New species of Moschus, characterized by Ogleby, who observed that the young of this animal, which is so anomalous among true deer, yet is spotted like so many deer.— very curious like some facts of Mr Blyth on birds.—

31e Dr. Bachman tells me line of Rocky Mountains separate almost all Mammals of N. America & many birds.—[1] which however are most closely represented.— Thus the red breasted thrush is represented by one not differing except by black line,— A Bunting by one only differing by some permanent white streaks.— &c &c

---

Dr. Bachman has crossed cock Guinea Fowl with Pea ‹cock› Hen.— offspring female, yet so infertile never even in seven years produced even an egg.—

32e a most curious bird, did not seem to know itself,. at last associated with the ducks.— most *strange voice* often in the night, like peacock.— tail as long as Pea hen.— about intermediate.— (In Zoolog Gardens there is hybrid of Penguin duck a variety of Muscovy) with *goose*!!)

31 Dr. Bachman tells . . . &c &c] '19' *added brown crayon.*
Dr. Bachman has . . . egg.—] '17' *added brown crayon.*
32 the ducks] 't' *over* 'd'.

Chair that evening, with interspersed comments from William Ogilby. Finally in D35, Darwin records a more theoretical comment by Richard Owen.

29–2 See R. Owen 1838:107, 'The *iliac* bones in size and shape present the character of the struthious birds. The *pubic* element is a slender bony style connected by ligament to the end of the *ischium*, but attached by bone only at its acetabular extremity. A short pointed process extends from the anterior margin of the origin of the *pubis*. The *acetabulum* is produced anteriorly into an obtuse ridge.'

30–1 See R. Owen 1838:71–72. This section of Owen's paper on *Apteryx*, read 12 June, was devoted to the flightless bird's respiratory system: 'Mr. Owen remarks, that the system of respiration in birds is so obviously framed with especial reference to the faculty of aerial progression, and the peculiarities in the former exhibit so marked a physiological relation to the latter, that in the *Apteryx*, where the wings are reduced to the lowest known rudimentary condition, the examination of the accompanying modifications in the respiratory apparatus presented a most interesting subject for inquiry.'

31–1 John Bachman.

Dr. Bachman regularly breeds «in Carolina» for his table Muscovy & common ducks— they are produced in full equal Numbers with pure bred (just like common mules) & lay many eggs but never produce inter se or with

33e parent species.— The hybrids do not vary (ie the hens all alike & Cocks all alike) More than parent species— Mr Blyth remarked only near species or varieties produce heterogenous offsprings.— «are not the hybrid pheasants & grouse different.—» (if so chinese pigs & common must be considered as distant species?? or is time the varying element). Then do those SPECIES which breed most freely. & produce somewhat fertile offspring produce heterogenous offspring.

It appears certain that hybrid Muscovy & Common duck have been shot *wild* (escaped from Carolina?) off New York. therefore instincts not imperfect.— **Are Pheasant & Grouse homogeneous?**

34e I observe Bachman calls these *Hybrids new* species.

Yarrell says the bird fanciers say the throw of any two species crossed is uncertain—

Yarrell remarks he has somewhere met conjecture that all salt-water fish were once salt water (as they almost must have been on elevation of continents) but Ogleby well answers that nearly all F. W. Fish are Abdominals. ∴that order first converted— is it an old order Geologically?

---

35 Owen says relation of Osteology of birds to Reptiles shown in osteology of young Ostrich.[1]

16th. D Israeli (Cur of Literat. Vol II p 11) accidentally says "—is distinctly marked as whole dynasties have been featured by the Austrian lip & the Bourbon nose".[2] if this be not imagination.— then old peculiarity overbears the crossing with females not thus characterized.—

---

36 16th Aug.— What a magnificent view one can take of the world Astronomical ‹& unknown› causes, modified by unknown ones. cause changes in geography & changes of climate superadded to change of climate

33 which breed] 'b' *over* 'a'.
Are Pheasant & Grouse homogeneous?] *added pencil in margin.*
35 *page crossed pencil.*

35–1 See R. Owen 1841a:289, 'The close resemblance of the Bird to the Reptile in this skeleton is well exemplified in the young *Ostrich*, in which even when half-grown the costal appendages of the cervical region of the vertebral column continue separate and moveable, as in the *Crocodile*.'

35–2 Disraeli 1835, 2:11.

from physical causes.— these superinduce changes of form in the organic world, as adaptation. & these changing affect each other, & their bodies, by certain laws of harmony keep perfect in these themselves.— instincts alter, reason is formed, & the world peopled «with Myriads of distinct forms» from a period short of eternity to the present time, to the future— How far grander than idea from cramped

37 imagination that God created. (warring against those very laws he established in all ‹nature› organic nature) the Rhinoceros of Java & Sumatra, that since the time of the Silurian, he has made a long succession of vile Molluscous animals— How beneath the dignity of him[1], who «is supposed to have» said let there be light & there was light.— «bad taste {whom it has been declared "he said let there be light & there was light".— »

August 17th Two regions may be Zoolo-geographically divided either by developement of new forms in one., or apparently so. by the extinction of prominent ones in ‹latte› one: The latter will take place when Conditions are unfavourable to numbers of animals. as in changing from ‹hot›. Warm to

38 cold, damp to dry.— Thus Tierra del Fuego has ‹not› «only» one «Guanaco» of the characteristic forms of S. America.

With respect to future destinies of mankind, some of species or varieties are becoming extinct. others though the negro of Africa is not loosing ground. Yet, as the tribes of the interior are pushing into each other from slave trade, & colonization of S. Africa, so must the tribes become blended & prevent that strong separation which

39 otherwise would have taken place. otherwise in 10,000 years Negro probably a distinct species— We know how long a Mammal may go on as one species from Egyptian Mummies & from the existing animals found fossil when Europe must have worn a quite different figure

---

19th. With respect to the Deluge it may be worth adding in note than amongst the Mammalia of Europe the shells of do— shells of. N. America.— shells of S. America.— there is no appearance of sudden termination of existence.— nor is there in the Tertiary ‹older› geological epochs.—[1]

37–1 See Erasmus Darwin 1803:54 note, 'Perhaps all the productions of nature are in their progress to greater perfection! an idea countenanced by modern discoveries and . . . consonant to the dignity of the Creator of all things.' Passage scored.

39–1 See *JR*:212n, 'The *Elephas primigenus* is thus circumstanced, having been found in Yorkshire (associated with recent shells: Lyell, vol. i., chap. vi.), in Siberia, and in the warm regions of lat. 31°, in North America. The remains of the Mastodon occur in Paraguay (and I believe in Brazil, in lat. 12°), as well as in the temperate plains south of the Plata.'

40 There are some admirable tables on Geograph distribution of reptiles in Suites de Buffon.—[1]

Vigors has given list in Linnæan Transactions of birds of Java[2]

Caterpillars not being fertile is same as children not being so.— consider this with reference to "new species & hybrid doctrine"— I have read there are exceptions to this in some larvæ of insects— (?glowworms) breeding— ‹beet› imago state fertile at once.—[3] Consider this with reference to those insects, which have fertile offspring. Entomostraca & Aphides.[4]

41 The extreme difference of sexes. is probably arrived at in case of insects as glowworm

---

The case of one impregnation sufficing to several births analogous to superfœtation,[1] & to successive fertile offspring in Entomostraca & Aphides

---

Developement of sexes in Caterpillar. very valuable facts— they are eating fœtuses, as young of Marsup. is sucking fœtus.—[2]

August 23d The Rev R. Jones gave an admirable harrier from Ireland to Brighton Park—*first rate* bitch— tried to breed from her, but

42 her offspring came out one big & one small. Now Jones, before this happened from her looks thougt she was halfbred Beagle Staghound. «++» ∴the grandchildren went back to either paret, & breed not fixed. though she resembled a harrier & her husband was pure Harrier.— «The peculiarities of our breeds must have been acquired, & hence this is then case of avitism.++»

---

42 «The peculiarities . . . avitism»] in margin with marked connection to 'Staghound', *presume passage added.*

40–1 Duméril and Bibron 1834–54. See 2(1835):28, 196; 3(1836):47, 280; 4(1837):59–60.
40–2 See Horsfield 1822 on birds of Java. Darwin's reference to Vigors may be a slip for Vigors and Horsfield 1827, which is also in the *Transactions of the Linnean Society*, but is on the birds of Australia.
40–3 See Kirby and Spence 1818–26, 2:410–11, 'If you take one of these glow-worms home with you for examination, you will find that in shape it somewhat resembles a caterpillar, only that it is much more depressed; and you will observe that the light proceeds from a pale-coloured patch that terminates the underside of the abdomen. It is not, however, the larva of an insect, but the perfect female of a winged beetle, from which it is altogether so different, that nothing but actual observation could have inferred the fact of their being the sexes of the same insect.'
40–4 For related discussion of entomostraca (*Daphnia*), see C162. For related discussion of aphids, see B181. See also Kirby and Spence 1818–26, 4:161 as representative of treating *Aphides* and *Daphnia* as analogous cases.
41–1 See B181, 'Ld Moreton . . . mare was influenced . . . to after births, like aphides'. (Morton 1821.)
41–2 Kirby and Spence 1818–26, 3:58 attributes a similar analogy to Virey (1816–19, 20:247–76 article on Metamorphosis): 'In them [vertebrate animals], he observes, a state analogous to the *larva* state begins at the exclusion of the foetus from the womb; . . . the digestive system now preponderates, and the great enjoyment is *eating*.'

Three gentlemen of party all thought with pigs &c, that hybrids were uncertain.

Mr Drinkwater thought that a ‹pure blooded› «"first blood"» animal must have gone on for many years, before deserves ‹name› «to be so called»,— the short horned cattle have gone on for 50 or 70? years— now «well fixed» breed,: Jones says Sussex cattle

43 were all white headed, but this was bred out & now all are pure red, yet calf every now & then born with white head (,or «short-horned with» black lip) & then calf «in both cases» is killed.

Notes from *Glen Roy* Note Book.—[1]
Why is not Tetrao Scoticus. an american form (if so)?.—[2]

A Sphepherd of Glen Turret. said he learnt to know lambs, because in their faces they were most like their mothers believes this resemblance general. ?depends upon mother bein[g] oldest breed?.— —[3]

Quarterly Journal of Agriculture p. 367. Dec. 1837. *Generally*— received

44 opinion that male impresses offspring more than female, yet instances given on opposite side,—[1] «The theory of males impressing most is in harmony with their wars & rivalry.—»

The very many breeds of animals in Britain shows, with the aid of *seclusion* in breeding. how easy races or varieties are made.—[2]

The Highland Shepherd dogs, coloured like Magellanic Fox.— peculiar hair & appearance— good case of Provincial Breed—[3] Highland Sheep jet black legs, & face & tail, just like species.— high active breedin[g][4]

43–1 Immediately after Darwin's field trip to Glen Roy (23–29 June 1838) he began Notebooks D and M in Shrewsbury. On the field trip, Darwin 'tested' two important hereditary questions against the experience of shepherds he encountered on his way: are the progeny of a cross 'heterogeneous' or are they uniformly like one parent, and are the characters of males dominant over those of females. The shepherds' responses were transferred from the Glen Roy Notebook and the these questions are pursued in early Notebook D. In most cases Darwin rewrote the entries from the Glen Roy Notebook without conceptual change. However, it is clear from notes D44–2 and D44–4 that Darwin added an interpretative gloss on selective breeding when he 'copied' those two entries.

43–2 See GR126. 'Why is the Tetrao scoticus & Tetrao—not an American form'.

43–3 See GR125.

44–1 See GR1. Anonymous 1838:367, 'According to the generally received opinion, that the male imprints his characters more indelibly than the females on the progeny, there may be a risk of breeding from too large a horse for the usual purposes of the farm. . .'

44–2 See GR1–2, 'Mere fact of many races of Animals in Britain shows that either races soon made or crosses difficult'. Note Darwin's emphasis on '*seclusion*' in rewriting GR2.

44–3 See GR2.

44–4 See GR11, 'Black faced sheep, sometimes mottled with white black legs & tail like species in colouring'. Note Darwin's emphasis on 'high active breeding' in rewriting GR11.

45e–46e [not located]

47e half breed liable to vary. I asked this in many ways, but received same answer.— Thought lambs were more like father than Mother.— The cross not so hardy as Black faced, but more tendency to fatten— This man confirmed my account of the Shepherd dogs.—[1]

Aug. 24th. Was struck with pink shade on plumage of the Pelican.— Mem pink spots on Albatross, on some Gulls. Flamingo— (Spoonbill Wader. Ibis)— laws of plumage might possibly be made out.—

48e August 25th Athenæum (1838) p. 611. Ld. Tankerville account of wild cattle of Chillingham,— habits peculiar,—*young* one 203 days old butted violently. & fell.— *gore to death the old & wounded,*— **see Annals. vol. 2. 1839.—**[1] are bad breeders & subject to the rush as all animals which breed, in & in are— colour white, uniform.—crafty, go in file, hide their young., bold.—[2] a Mr W: Hall remarked that it was against all rules their preserving character & breeding in & in—[3] Nonsense[4] a flock of more than 100.—[5] **Agrees, «nearly» with. the account. given by Boethius of ancient Caledonian Cattle.**[6] **Ch 3.**[7]

Instinct

47 *page crossed pencil.*
48 August 25th . . . fell.—] *crossed ink.*
gore to . . . wounded] *underlined brown crayon.*
see Annals. vol. 2. 1839.—] *added brown ink.*
a Mr. W: Hall . . . 100.—] *crossed faded brown ink.*
a Mr. W: Hall . . . 100.—] *crossed pencil.*
Agrees . . . Caledonian Cattle.] *added faded brown ink.*
Ch 3] *added brown crayon, underlined.*
Instinct] *added to page brown crayon.*
pin mark.

47–1 See GR25. Note the parallel concern with character transmission in early Notebook D. Also note that Darwin did not copy the speculative comment in GR31, 'are those animals subject to much variation which have lately acquired their peculiarities??'
48–1 Hindmarsh 1839a.
48–2 Tankerville and Hindmarsh 1838.
48–3 Tankerville and Hindmarsh 1838:612, 'Mr. Webb Hall thought this an important paper, although opposed in its results to the received opinions of cattle-breeders. Here was a race breeding in and in, yet retaining all its beauty, strength and vigour. This was opposed to all known facts.— Mr. Hall's remarks excited considerable interest; and it appeared, that in the present instance great care had been taken to prevent the deterioration of the breed, which had undoubtedly taken place in other herds.'
48–4 Notwithstanding Darwin's strong comment, Mr. Hall's problem persisted as a difficulty. See *Natural Selection*, chap. 3, 'On the possibility of all organic beings occasionally crossing', p. 37. 'On the other hand some competent judges have doubted the ill effects of interbreeding. . . . Again the case of the half-wild cattle in Chillingham which have gone on interbreeding for the last 400 or 500 years seems a strong case; but Lord Tankerville, the owner, expressly states that 'they are bad breeders' (Hindmarsh 1839a).
48–5 Tankerville and Hindmarsh 1838:612, 'There are about 80 in the herd, comprising 25 bulls, 40 cows, and 15 steers, of various ages.'
48–6 Boece 1526.
48–7 Darwin maintained a long standing interest in the 'wild' Chillingham cattle. Accordingly D48 forms part of a complex archival record. This page was filed with Darwin's material for *Natural Selection*, chap. 5, 'The Struggle for Existence' (DAR 46.1), where it appears to have been pinned to the following note on Falkland cattle, which is based on information provided by Darwin's

49 L'Institut. p. 249. (1838). Eggs discovered to Tænia.— hard so as to resist external influence.—[1]

27th. August. There must be some law, that whatever organization an animal has, it tends to multiply & IMPROVE on it.— Articulate animals must articulate. ‹i› in vertebrates tendency to improve in intellect,— if generation is condensation of changes. then animals must tend to improve.— yet fish same as, or lower than in old days: «for a very old variety will be harder to vary, & therefore more apt to be extinguished.— ???»

Mayo (.Philosop of Living) quote Whewell as profound. because he says length of days adapted to duration of sleep of man.!!![2] whole universe so adapted!!! & not man to Planets.— instance of arrogance!!

50 August— 29th.— Macleay in A. Smith's Zoolog.— of Africa.—[1]
p. 4. sticks to genus or group of any kind not being perfect till circular.[2]
p. 5 Most clearly shows that genus expresses as now used almost any

49 27] '7' *over* '6'.
«for a . . . extinguished.—???»] *altered hand, presume passage added.*

former *Beagle* shipmate B. J. Sulivan: 'Ap. 1850. Sulivan says in the Spring of 1849, numbers of the *wild* Cattle «Falkland Isd» died being so weak & getting into bogs— «winter unusually severe» & from disease with hair falling off— Is reported same deaths occurred 12 or 14 years before. It is a general opinion that the white cattle are the finest.— Wild Horses stood the winter well.—'

49–1 Dujardin 1838:249, 'M. Dujardin conclut de là que les oeufs de Toenia protégés par une coque très résistante peuvent résister aux causes extérieures de destruction et attendre dans les lieux où ils ont été disséminés un instant favorable pour se développer, et que, parconséquent, pour expliquer l'apparition de ces Entozoaires dans les animaux, il n'est pas nécessaire de recourir à l'hypothèse de la génération spontanée, comme l'a fait Rudolphi, . . .'

49–2 H. Mayo 1838: 146–48, 'Mr. Whewell observes [1833:39], in reference to sleep, 'Man in all nations and ages has taken his principal rest once in twenty-four hours, and the regularity of this practice seems most suitable to his health. . .

'*That* sleep . . . so curiously adjusted to the length of diurnal revolution, Mr. Whewell has shown to be an additional proof of the existence of God, is felt by every one to prove His benevolence. . .'

50–1 MacLeay 1838. Darwin's copy was annotated in 1838, as is evident from the close fit between the marginalia and D50–53. One should not be misled by the title page: *Illustrations of the Zoology of South Africa*, Andrew Smith, *Invertebratae*, 1849, which was no doubt added when Darwin's subscription parts were bound in volumes. In another copy, at Cambridge University Library, the Smith 1849 title page is followed by a separate title page with *Illustrations of the Annulosa of South America*, W. S. MacLeay, 1838. It is not clear that MacLeay's work was separately published. Notwithstanding the 1838 title page, it does not seem to have been separately advertised. *Publishers' Circular* 1 (1838) lists no separate work for MacLeay; however, 'Part 3 Insects' is listed as 'just published' in the 1 October 1838 issue.

50–2 MacLeay 1838:4, 'Every one knows that sometimes sub-genera, and at other times even sub-sections of genera, are in the most unphilosophical manner published as genera. Too often we find every thing a genus which some gnathoclast, with Scapula in hand, has thought proper, in his good pleasure, to call so. Some persons again there are, who on a first inspection can oracularly decide that this groupe is a sub-genus, and that another groupe is of "full generic value." To such clearness of vision I can lay no claim; yet I cannot help thinking that there is a mode of discovering the true subordination of these several kinds of groupes—nay, I am sure this discovery will ever be the result of calm patience, of keeping before our view a great number of the species of any family, and finally of following up that *aphorism* of a distinguished botanist, which says, "Omnis sectio naturalis circulum, per se clausum, exhibet."' Darwin scored the lines with the Latin aphorism.

group.—[3] ǂall groups *natural* (p 6) as expressing *natural* affinitiesǂ[4] Macleays plan of arrangement depends on the organs judged to be of importance in inverse ratio to their variability.— (Now cæteris paribus these will be the oldest)[5] ǂ"The most important characters break down in certain species & become worthless— Mammalia Edentataǂ[6] We do (p 6) say such is group. because it has such characters of importance, "but we say such happens to be the character, of no matter of what importance, which prevails throughout the group & serves to insulate ‹them› it".— i.e what characters

51 chance to be heredetary whether important or not,).[1] p. 7. "The Natural arrangement of animals themselves is the question in point." Now what is *natural arrangement*,— affinities, what is that, amount of resemblance,— how can we estimate this amount, when ‹value› no scale of value of difference is or can be settled,—[2] I believe *affinity* may be taken literally,, though how far we can ever discover the real relationship is doubtful.— not till much knowledge

51 ascend] 'as' *over* 'des'.

50–3 MacLeay 1838:5, 'Nay, has the word *genus* any signification which is universally deemed definite? I fear in all such cases of assertion, there is a latent disposition of the human mind to erect an arbitrary standard, founded on the supposed value of some point of structure. Thus one person says that the genera of *Mammalia* ought to be established on the differences in their system of dentition; and yet there are some genera of *Mammalia* where almost every species varies in the number and form of its teeth; so that to adopt the rule, we must consider every species of such genera to be a genus itself. Another person will tell us, like Linneus, that there are as many genera, as aggregations of different species present similar constructions of some arbitrarily selected organs, such as those of fructification in phaenogamous plants, or the teeth in *Mammalia*. In this sense it is evident that a genus [Darwin underlined 'genus'] may be made to signify any groupe whatever, as its extent will depend on the nature of the structure selected.' Passage scored.

50–4 MacLeay 1838:6, 'But here some one may observe that all groupes are arbitrary and artificial, since after all they must depend on the selection and good pleasure of man. To this I answer that affinities are natural; and if all these affinities are expressed by any mode of grouping, it follows also that the groupes must be natural; although certainly, these last must in some degree have depended on our selection.' Darwin scored 'To this I answer . . . must be natural'.

50–5 MacLeay 1838:6, 'My plan, as is well known, has ever been not to estimate the value of any arrangement by the value in animal economy of the structure upon which this arrangement is founded, but to make the importance of every organ or structure for purposes of arrangement, rise in inverse proportion to its degree of variation.' Passage scored.

50–6 MacLeay 1838:6, 'Indeed, it is obvious in every part of natural history, that the most important characters break down in certain species, and become at times perfectly worthless. Comparatively constant as is the structure of the teeth in the genera of *Mammalia* generally, we find in some groupes, such as the *Edentata*, or the genus *Rhinoceros*, that the dentition varies extensively in almost every species.' Darwin double scored 'Indeed . . . perfectly worthless.'

51–1 MacLeay 1838:6–7, 'We truly make use of a process of *tatonnement*. We do not argue that such must be the groupe, because such and such are, in our opinion, good and distinct characters; but we say, such happens to be the character, of no matter what importance, which prevails throughout the groupe, and which serves in some degree to insulate it from other groupes. But it is evident that we must previously have arrived at the knowledge of the groupe; and this is effected by a close watching of the variation of affinity, and by considering the groupe to be complete only when the series of natural objects returns into itself.' Passage scored.

51–2 MacLeay 1838:7, 'If even we were right in any such comparative estimate of the importance of organs in general economy, we ought not to forget that the true question under consideration is, the natural arrangement of the animals themselves; and that this is to be attained only by the expression of every affinity, and every analogy that can be detected.' Darwin underlined 'the natural arrangement' and commented: 'It may be asked what is meant by Natural arrangement— first step vague.— if it is said affinities of animal.— what does affinities mean?' At the bottom of the page he noted: 'Most resemblances— evidently disputed,—sum of difference[.] I conceive object is real relationship'.

is elicited.— It will rest upon the discovery what characters VARY most easily:— those which do not vary being foundation for chief divisions.≠ p. 7. «In» Some ‹of› cases the circular arrangement from fewness of forms— Cannot be discovered «un»till ‹in› «we ascend to» subgenera & families, ‹even in Cetionidæ› «in the Cetoniadæ»,—[3] when will ornithorhynchus come in circle?!!!

52 p. 8— Anomalous structures, as in Hippotamus, solely owing to number of lost links.╫[1] if all species know they would be innumerable╫ does not know any difference between *permanent variety & species*!! (given in note.)—[2] Macleay ‹met› uses term *genus* when it is so many steps from a head, as subkingdom.— — evidently artificial, as interlopement of Marsupials will change all.— & so on no one will settle *number* of primary divisions.— Complains (p. 53) of M. Edwards, thinking any group good, though not circular, if characters can be established— clearly so.—[3] NB. This paper worth referring to again.— According to my theory, every species in any subgenus will be. descended from one stock, & that stock with other subgenera

53e will come from. common stock.— all genera, common stock.— so that value can only be judged of in each «separate» line of descent.— ‹& here limits of varieties being constant. it would be exceedingl wrong to call,, one group

53 «separate»] 's' *on stub.*
NB. . . . no way.?—] *double scored, most of scoring on stub.*
page crossed pencil.

51–3 MacLeay 1838:7, 'Now, if we start from the principle that when a few species *first* agree in some particular character, they combine into a series that will return into itself, we shall probably imagine every such series, so forming a circle in practice, to be in theory the first natural assemblage of species. Yet this will be an incorrect mode of viewing the matter; for owing to the rarity of its species, the first known circular grouping of the species of *Cryptodinus*, for instance, is into sub-genera; whereas the first known circular grouping of the species of *Cetoniidae* is into certain sub-sections.' Darwin scored and added a '!' in the margin on 'Yet this . . . sub-sections.'

52–1 MacLeay 1838:8, 'Thus, when the naturalist talks of any anomalous structure, I understand merely that so many links, that is, so many groupes, of the great plan of creation are wanting, as would connect this singular being with some other and better known form. If I say that the *Hippopotamus* forms a stirps by itself, I only mean that it is the sole species of its stirps known; and that, speaking theoretically, four families are wanting, or rather twenty-four genera to connect it well with the other tribes of Pachyderms. It is of no consequence whether the families and genera supposed according to this theory to be wanting, have disappeared, or whether they have never been created. I merely suppose them to be wanting, in order that I may obtain something like a just notion of the relation which the *Hippopotamus* bears to the other Pachyderms.' Darwin scored: 'Thus, . . . its stirps known'.

52–2 MacLeay 1838:8, note, 'Some persons have imagined that I only assign five species to the lowest groupe in nature; but the above theory evidently proceeds on the assumption that if we knew *all* the species of the creation, their number would be infinite, or in other words, that they would pass into each other by infinitely small differences. This actually takes place sometimes in nature; and as yet I do not know any good distinction between a species and what is called "a permanent variety." ' Passage scored.

52–3 MacLeay 1838:53, 'And indeed this very arrangement [of *Crustacea*] of Edwards is not natural, since he unfortunately conceives that every groupe he can invent, provided he can furnish it with a character, must be therefore a good one. As, on the contrary, the true definition of a complete natural groupe is, that it must be a series returning into itself, many of the groupes of Milne Edwards, when weighed by this scale, will be found wanting.' Passage scored. See Milne-Edwards 1834–40.

genus & other subgenus,,—› Propagation, best rule for genera, & so mount upwards.— «judged by analogy»—
Consider all this

NB. How can local species as at Galapagos., be distinguished from temporal species as in two formations? by no way.?—

54e "Natura nihil agit frustra", as Sir Thomas Browne says "is the only indisputable axiom in Philosophy‹"› Religio Medici. Vol II. Sir T Browne's Works p. 20

There are no grotesques in nature; not anything framed to fill up empty cantons, & unnecessary spaces" p 23. "for Nature is the art of God"[1]

Septemb 1,. It has been argued Man first civilized. ‹note› add this in note. ?mere conjecture?— Australians.— Americans. &c **After Decandolles idea**

55e Septemb. 1st. Macleay & Broderip were talking of some Crustacean, like Trilobite.[1] (Polirus??) female blind & of quite different form from male with eyes!— (are not these differences in sex confined to annulosa?) Remarked that young of Cirrhipedes can move & see, parent fixed,— young of sponges move.—*young of Cochineal* insects move about & see, *parent* «(2)», *female* «(1)» *fixed* & blind: — Macleay observed all these *facts* prove that *perfection* of organs have nothing to do with *perfection* of individual, though such relation seems common, but that perfection consists in being able to reproduce

56e Here there is some error— Observed, nature does nothing in vain, therfore organ fitted to animals place in creation.— thus senses, especially sight connected with locomotion.— «Mem. Dr. Blackwell (Abercrmbies) comparison of sight to threads.—[1]» Hence the *Pecten*, which move imperfectly

54 There are . . . God"—] '10' *added brown crayon.*
After Decandolles idea] *added pencil.*
Septemb . . . Americans] *crossed pencil.*
55 *young of Cochineal*] *underlined brown crayon.*
parent, female fixed] *underlined brown crayon.*
'11' *added to page brown crayon.*
56 Here] 'H' *over* 'th'.
Spondylus] 'lus' *on stub.*
page crossed pencil.

54–1 Browne 1835–36, 2:20–23, *Religio medici* [1642], '*Natura nihil agit frustra*, is the only indisputable axiom in philosophy. There are no grotesques in nature; not any thing framed to fill up empty cantons, and unnecessary spaces.' (2:20). 'Now, nature is not at variance with art, nor art with nature; they being both the servants of his providence. Art is the perfection of nature. Were the world now as it was the sixth day, there were yet a chaos. Nature hath made one world, and art another. In brief, all things are artificial; for nature is the art of God.' (p. 23).

55–1 W. S. MacLeay and W. J. Broderip: personal communications, possibly at the Zoological Society.

56–1 *Sic* Blacklock, see Abercrombie 1838:288–89, 'Smellie mentions of Dr. Blacklock, who lost his sight at

has eye-point, but Broderip added it has been stated that stationary Spondylus has eye-points— Macleay then answered, because nature leaves vestiges of what she does— does not move per saltum— yet does nothing in vain!!

57 Fœtus of man undergoes metamorphosis., heart altered & umbilical cord,— Broderip alluded to Hunter's views on this subject.— Monstrosities, kind of determined by *age* of fœtus.—[1]

---

the age of a few months, that, in his dreams, he had a distinct impression of a sense which he did not possess when awake. He described his impression by saying, that when awake there were three ways by which he could distinguish persons,—namely by hearing them speak, by feeling the head and shoulders, and by attending to the sound and manner of their breathing. In his dreams, however, he had a vivid impression of objects, in a manner distinct from any of these modes. He imagined that he was united to them, by a kind of distant contact, which was effected by threads or strings passing from their bodies to his own.' Passage scored. The following annotation at the bottom p. 289 may be relevant: 'have a distinct recollection of solving some geological puzzle in my sleep.— what it was I forget, which I am surprised at, for I have so *clear* an *indistinct* notion.'

57–1 Hunter 1837:45, 'each species seems to have its monstrous form originally impressed upon it'. This is the first of a series of references to correlated variation, a class of phenomena that Darwin soon grouped as 'Hunter's Law' (See D67 and D112). Through the remainder of Notebook D, the opening paragraphs of Hunter's 'An account of an extraordinary pheasant' from the *Animal Economy* (Hunter 1837:44–45) play a comparable role to key passages of Erasmus Darwin's *Zoonomia* in Notebook B. That is, they form a rich source of direct and indirect reference as Darwin seeks to understand the laws of generation. Hence, they are cited here in extenso. 'Every deviation from that original form and structure which gives the distinguishing character to the productions of Nature, may not improperly be called monstrous. According to this acceptation of the term, the variety of monsters will be almost infinite[a]; and, as far as my knowledge has extended, there is not a species of animal, nay, there is not a single part of an animal body, which is not subject to an extraordinary formation.

Neither does this appear to be a matter of mere chance; for it may be observed that every species has a disposition to deviate from Nature in a manner peculiar to itself[a]. It is likewise worthy of remark, that each species of animals is disposed to have nearly the same sort of defects, and to have certain supernumerary parts of the same kind: yet every part is not alike disposed to take on a great variety of forms; but each part of each species seems to have its monstrous form originally impressed upon it[b].' Richard Owen's notes follow:

'[a] [p. 44] Mr. Hunter attempted, notwithstanding, to reduce this variety of monsters to definite groups, and left the following outline of a classification of monsters, in an explanatory introduction to the extensive series of those objects in his collection:

"1. Monsters from preternatural situation of parts.
"2. ———— addition of parts.
"3. ———— deficiency of parts.
"4. ———— combined addition and deficiency of parts, as in hermaphroditical malformation." . . .

'[a] [p. 45] The value of the principle here enunciated will be appreciated, when it is stated that it is the basis of the latest and most elaborate work on the subject of monsters. It is claimed for Geoffroy St. Hilaire as the most important of his deductions in Teratology, and the chief point in which his system differs from, and is superior to, those of his predecessor. "C'est de principes précisément inverses que mon père a pris sur point de départ; et c'est aussi, comme cela devait être à des résultats inverses qu'il est parvenu. Etablissant, par un grand nombre de recherches, que les monstres sont, comme les êtres dit normaux, soumis à des règles constantes, il est conduit à admettre que la méthode de classification que les naturalistes emploient pour les seconds, peut être appliquée avec succès aux premiers." [Isidore Geoffroy Saint-Hilaire 1832–36, 1:99]

'[b] [p. 45] In this principle Mr. Hunter is opposed to Geoffroy St. Hilaire, who attributed the production (*l'ordonnée*) of monstrosities to the operation of exterior or mechanical causes at some period of foetal development. Defective formation in parts of a foetus has indeed been produced by destroying a portion of the respiratory surface of an egg during incubation; but this result by no means affords adequate grounds for assigning as the sole cause of every malformation accidental adhesions between the foetus and its coverings. Mr. Hunter also made experiments with reference to monstrosities . . . It is evident, however, from the expression in the concluding paragraph of the text, that he regarded the cause of congenital malformation as existing in the primordial germ.'

As Larvæ may be more perfect (as we use the word) than parent, so may species retrograde, but these facts are rare.—

2d Sept
Those animals which have many ABORTIVE organs, might be expected to have larvæ more perfect— this is applicable to young of Cochineal??

58 Is there some law in nature an animal may acquire organs, but lose them with more difficulty, «contradicted by abortive organs, but number of species with abortive organ of any *kind* few.— » hence become EXTINCT, & hence the IMPROVEMENTS of every type of organization. such law would explain every thing.— PURE HYPOTHESIS be careful.—

Argument for circularity of groups. When ‹species of› a group of species is made. father probably will be dead— hence there is no central radiating point, all united . (links in circle must be granted unequal, because fossil) Now what is group without centre but circle, two or three

59 lines deep— with respect to Macleay's theory of analogies—[1] ‹be› when it is considered the tree of life must be erect not pressed on paper, to study the corresponding points.—

The present geographical distribution of animals countenances the belief of their extreme antiquity (ie much intervening physical change).— distribution especially of Mammalia

---

As every organ is modified by use, every *abortive* organ must have been once changed.— what is abortive? when it does not perform that function which *experience* shows us it was for.— Most important law.— Penguins wing perhaps not abortive???. Apterix certainly.—

60 Lyell's excellent view of geology, of each formation being merely a page torn out of a history, & the geologist being obliged to fill up the gaps.—[1] is possibly

59 As every . . . certainly.—] *double scored both margins. page crossed pencil.*

59–1 See D50–52, also notes B129–1, C158–2 and MacLeay 1830a.

60–1 The immediate reference appears to be Lyell 1838a:272, 'So, of a series of sedimentary formations, they are like volumes of history, in which each writer has recorded the annals of his own times, and then laid down the book, with the last written page uppermost, upon the volume in which the events of the era immediately preceding were commemorated. In this manner a lofty pile of chronicles is at length accumulated; and they are so arranged as to indicate, by their position alone, the order in which the events recorded in them have occurred.' This work was advertised as 'just published' in August 1838 (*Publishers' Circular*). See *Origin*, p. 310–11, 'For my part, following out Lyell's metaphor, I look at the natural geological record, as a history of the world imperfectly kept, and written in a changing dialect; of this history we possess the last volume alone, relating only to two or three countries. Of this volume, only here and there a short chapter has been preserved; and of each page, only here and there a few lines. Each word of the slowly-changing language, in which the history is supposed to be written, being more or less different in the interrupted succession of chapters, may represent the apparently abruptly changed forms of life, entombed in our consecutive, but widely separated, formations.' Darwin's joint lingusitic-historiographic metaphor appears to be a combination of Lyellian metaphors (history of nature as book: from the

the same with the ‹Zoologist› «philosopher», who has trace the structure of animals & plants.— he get merely a few pages.—

Hence (p. 59) looking at animal, if there be many others somewhat allied whether «like» parent stock, or not. Now wings for flight— therefore ostrich not.

The peculiar ‹Malacca› «Malacca» bears, ‹are› belong to same section with with those of India—

61e Waterhouse knows three species of Paradoxurus common to Van Diemen's land & Australia[1]

---

well developed ‹tits› «Mammæ» in male ourang-outang. other point of resemblance with man.—

September 3d Magazine of Natural History. 1838
vol II p. 402. Mr Gould on Australian birds— *all Eagles.* of Australia characterized by wedge tails.—[2] many of the hawks ‹to› are analogues to ‹Bustard› Europæan birds. also «do» p. 403. & 404[3]
vol II. do (p. ‹69› «71»). alludes to Eyton's discovery of different number of vertebræ in Irish & English Hare.—[4] good case these hares compared to North American hares. Many species, separated by Mountains. & & &c.—

61 Waterhouse . . . man.—] *crossed pencil.*
September 3d . . . Mountains. & & &c.—] '19' *added brown crayon.*

Elements, Lyell 1838a; and geological processes as language: *Principles* 1830–33, 1:461–62; 1837, 2:352).

61–1 G. R. Waterhouse: personal communication.

61–2 Gould 1838b:402, 'Of the genus *Aquila* only one species has as yet been discovered, viz., the *Aquila fucosa* of Cuvier, which doubtless represents in Australia the Golden Eagle of Europe, from which it may be readily distinguished by its more slender contour, and by its lengthened and wedge-shaped tail.'

61–3 Gould 1838b:403, 'Of the genus *Falco*, the *Peregrinus* is replaced by a species most nearly allied to, and hitherto considered identical with that bird: the experienced eye of the ornithologist will, however, readily distinguish an Australian specimen when placed among others from various parts of the globe, so that there will be but little impropriety in assigning to it a separate specific name.' Passage scored ('Of . . . distinguish'). 'The Hobby, so familiar as a European bird, is represented by the Falcon, for which I now propose the specific name of *rufiventer*, as I believe it to be undescribed.' Passage scored. 'The *Cerchnis cenchroides* (*Falco cenchroides* of Messrs. Vigors and Horsfield,) exhibits a beautiful analogy with the Common Kestril of our island, but although nearly allied possesses several important and permanent differences.' Passage scored.

61–4 W. Thompson 1839a:71 [No. 7, Sept. 1838], 'On looking to their [*Lepus Hibernicus* and *L. timidus*] osteology, some slight differences are observable in the head; the comparatively more horizontal direction of the lumbar vertebra in the Irish hare is conspicuous, and likewise the relative shortness of its tail, which, as first recorded by Mr. Eyton [1838b], contains three vertebrae less than that of the English species, thirteen only being possessed by the former, and sixteen by the latter animal.' Passage scored. This paper is a notice of W. Thompson 1838d, which Darwin had read; see note B7–2 and E184.

62e do. p. 69. A Dr Macdonald believes the Quaternary arrangement & not the Quinary.—[1] any one may believe anything in such rigmaroles about analogies & number

---

L'Institut p. 275. (1838) M. Blainville has written paper to show Stonesfield Didelphis not Didelphis[2] «Answered satisfactorily by. Valenciennes.[3]»

The change from caterpillar to butterfly— is not more wonderful than the body of a man undergoing a constant round,—each particle is placed in place of last by the ordering of the nerves, but in different parts according to age of individuals— (see Mammæ of Women) in different parts when age

63e changes caterpillar to Butterfly.—

When two Varieties of dogs cross, Erasmus says[1] it look lik[e]

64e Institut. 1837. p. 351. Paradoxurus Phillippensis. Philippines[1]

---

65 Man have varies the range— Argue the case of Probability. has Creator made rat for Ascension.— The Galapagos mouse probably transported like the New Zealand one— It should be observed with what facility mice attach themselves to man.

Sept 7th. — I was struck looking at the Indian cattle with Bump. together with Bison,[1] at some resemblance as if the "*variation* in one, was analogous to *specific character* of *other species* in *genus*."— Is there any law of this. Do any varieties of sheep «evidently artificial» approach

62 *page crossed pencil.*
63 *top 2cm extant.*
crossed pencil.
64 *top 2cm extant.*
'19' *added brown crayon.*
65 *page crossed pencil.*

62–1 Macdonald 1839:69 [No. 7, Sept. 1838], 'The author had scarcely time to do justice to himself or subject, and we have still less in our limited space. He stated he thought zoologists attended too little to anatomy, those especially who gave themselves to tracing analogies throughout the scale of animated nature. He avowed himself an advocate for the quaternary not the quinquennary grouping of the series.' The author goes on to propose anatomical analogies between fishes and insects. Passage scored *in extenso*.

62–2 Blainville 1838d:275, 'Les deux seuls fragments fossiles de Stonefield attribués au genre *Didelphis* de la classe des Mammifères n'ont aucun des caractères des animaux de ce genre, et ne doivent certainement pas y être rangés.'

62–3 Valenciennes 1838a:297, 'M. Valenciennes lit les observations sur les mâchoires fossiles découvertes dans les couches de Stonefield, et attribuées par M. Cuvier aux Mammifères. M. de Blainville . . . n'avait eu sous les yeux d'autres pièces que des dessins. M. Valenciennes a été plus heureux: M. Buckland lui a fait voir . . . deux mâchoires d'animaux fossiles des couches de Stonefield. La comparaison qu'il a faite . . . lui a permis d'établir avec certitude que ces débris appartiennent réellement à des animaux de la classe des Mammifères, et qu'ils doivent former un nouveau genre appartenant à l'ordre des Marsupiaux.'

63–1 Erasmus Alvey Darwin: personal communication.

64–1 Jourdan 1837:351.

65–1 At the Zoological Society Gardens.

66 in character to goats.— or dogs to foxes. (yes Australian dog) or donkeys to Zebras.— «Mr Herberts variety of horse, dun-coloured with stripe approaches to ass.[1]» or fowls to the several aboriginal species «or ducks» (here argue if it be said domestic fowls are descended from several stock, then species are fertile‖; as long as opponents will «are» not «able to» tie themselves down, they can find loopholes) "It is well worthy of examination whether variations are produced only in those character which are seen to ‹vary among› be different in species of same genus." Law of monstrosity not prospective, but retrospective as showing

67 what organs are little fixed— (‹also› Hunters law of monstrosity with regard to age of fœtus. distinct consideration)[1] Now in different SPECIES of genus Sus. do vertebræ vary? «See Cuvier Ossemens Fossiles[2]»

Although no new fact be elicited by these speculation even if partly true they are of the greatest service, towards the end of science. namely prediction.— till facts are grouped. & called. there can be no prediction.— The only advantage of discovering laws is to foretell what will happen & to see bearing of scattered facts..—

68 What takes place in the formation of a bud— the very same must take place in copulation— (Man & woman separate parts of same plant‹s›)— now in some Polypi we see young bud changing into ovules.—

Captain Grants. Himalaya. shells (see Paper in Geolog Transacts) same appearance with Secondary Species distinct— but close.—[1] Mem. Von Buch on Cordillera fossils same remark.[2] ⸮was there formerly one great sea, & two Polar Continents Marsupial. Edentata.— Pachydermata &c &c—

69 It is important with respect to extinction of species, the capability of only small amount of change at any one time

---

67 will happen & to] 't' over 's'.

66–1 Herbert 1837:339–40, 'I feel satisified that the fox and the dog are of one origin, and suspect the wolf and jackall to be of the same; nor could I ever contemplate the black line down the back of a dun pony without entertaining a suspicion that the horse, unknown in a wild state except where it has escaped from domesticity, may be a magnificent improvement of the wild ass in the very earliest age of the world: . . .' Darwin underlined 'black line down the back of a dun pony' and noted 'instance of my law of variations agreeing other species of genus'.

67–1 Hunter 1837:45, see note D57–1.

67–2 See G. Cuvier 1821–24, 2:124. Cuvier surveys the osteology of the extant 'cochon'; on p. 124 he describes the vertebrae, but gives no mention of variation. He devotes two pages to 'Des os fossiles de cochons', and barely mentions vertebrae.

68–1 Grant 1840:297 [read 22 February 1837], 'In its mineralogical character and general appearance, this formation [upper secondary] greatly resembles the English lias; but its fossils have been found, after a careful examination by Mr. James Sowerby, to assimilate very closely to those of the oolitic beds; and a very few belonging to the green sand.'

68–2 Buch 1838:270, 'La majeure partie consiste en une espèce du genre *Pecten*'.

Seeing what Von Buch (Humboldt). G. St. Hilaire, & Lamarck have written I pretend to no originality of idea—[1] (though I arrived at them quite independently & have used them since) the line of proof & reducing fact to law only merit if merit there be in following work.—

The history of Medicine, the extraordinary effects of different Medicines on organs, leads one to suspect any amount of change from eating different kinds of food: grazing animals who eat every species new.—

70 Sept. 8′. A Golden Pippen or Ribston do producing occasionally (as Fox says)[1] same fruit trees is analogous to some hybrids breedings— there is tendency to reproduce in each case, but something prevents the completion.— ǂSay my Grandfathers expression of generat. being highest end of organization *good expression* but does not include so many facts as mineǂ[2]

71 The facts about half breed animals being wilder than parents is very curious as pointing out difference between acquired & heredetary tameness.—[1]

In comparing my theory with any other. it should be observed not what comparative difficulties (as long as not overwhelming) What comparative solutions & linking of facts—[2]

Savages over whole world. (Major ‹1› Mitchell p. 244. vol I) spit & throw dust[3]

‹according to my theory of generation (p. 175) if›[4]

72 8th Sept
Yarrell told me he had just heard of Black game & Ptarmigan having crossed in wild state— & the English & Some African dove.—[1]

The extinction of the S. American quadrupeds is difficulty on any theory— without God is supposed to create & destroy without rule— But what does he

72 S. America &] 'S' *over* 'A'.

69–1 See Buch 1836, referred to at B156. Presumably Darwin is referring to Humboldt's broad contributions to natural history, e.g. Humboldt and Bonpland 1819–29. However, by placing Humboldt in parentheses after von Buch, he may have been referring to Humboldt 1817, which is cited in Buch 1836. See E. Geoffroy Saint-Hilaire 1830 and Lamarck 1830.

70–1 W. D. Fox: personal communication. See B83, 'Fox tells me, that beyond all doubt seeds of Ribston Pippin, produce Ribstone Pippins, & Golden Pippen, goldens—'

70–2 Erasmus Darwin 1794–96, 1:514. See B1–2, 'This appears highest office in organization (especially in lower animals, where mind, & therefore relation to other life has not come into play)— See Zoonomia arguements . . .' Also see notes D19–1 and D19–2.

71–1 See B136 and C165.

71–2 See Herschel 1831 and Whewell 1837 on 'consilience of inductions' as a test of a scientific theory's validity.

71–3 T. L. Mitchell 1838, 1:243–44.

71–4 This passage indicates that Darwin's summary 'Proved facts relating to Generation' (D176–179, 174–175) was written before 8 Sept. 1838. See discussion of chronology in the introduction to Notebook D.

72–1 William Yarrell: personal communication.

in this world without rule? The destruction of the great Mammals over whole world shows there is rule.— S. America & Australia appear to have suffered most with respect to extinction of larger forms.—

From observing way the Marsupials of Australia have branched out into orders one is strongly tempted to believe, one or two were landed

73e as at present in New Ireland & continent since grown.— This will explain. S. American case & Didelphis being Mundine form., & the less developement of Marsupials in S. America. from presence of Edentata— Edentata & Marsupials have been almost destroyed wherever other animals existed.—

Athenæum 1838. p. 654. Reason given for supposing Tetrao **media or** Rakkelhan is hybrid (produced commonly in Nature. both in Sweden & anciently in Britain) between *hen* Caperailkie & cock Black-cock.—[1] (Curious the readiness with which this genus becomes crossed. ⸮is red game an hybrid?—

74e When I show that island would have no plants were it not for seeds being floated about.— I must state that. the ‹p› mechanism by which seeds are adapted for long transportation, *seems* «?» to imply knowledge of whole world— if so doubtless «part of» system of great harmony.[1]

73 as at . . . existed.—] *crossed pencil.*
Athenæum . . . an hybrid?—] '17' *added brown crayon.*
media or] *added pencil,* 'm' *on stub.*
74 *page crossed pencil.*

73–1 Charlton 1838:654, 'Dr. Charlton exhibited a specimen of *Tetrao Rakkelhan*, of Temminck, and read a short notice, to prove that this bird, though described as a distinct species, by so great an authority as Temminck, was in fact nothing but a hybrid, between the hen capercailzie and blackcock. This he acknowledged to be an old theory, but it was also, held by the greatest living naturalist of Sweden, Prof. Nillson [*sic* Nilsson], of Lund. Dr. Charlton brought forward in favour of his opinion the fact, that the female of this bird has never yet been described or discovered, and yet that every year *males* were transmitted to England. Mr. Tunstall, on the authority of some old Scottish gentleman, had stated, that the *hybrid* bird as well as the capercailzie was formerly met with in Scotland. This Dr. Charlton considered as an argument in favour of its being a hybrid, for were it otherwise, it would in all probability, being a much smaller bird, have survived the extirpation of the capercailzie.' See *Natural Selection*: 434 (cross between *Tetrao urogallus* Capercailzie and Lygurus [*sic*] tetrix Black Grouse) and p. 436, n. 16.

74–1 Compare Darwin's grey-ink addition to C184, which must have been written within days of D74: 'Webb &. Berthelot. must be studied on Canary islands— **Endeavour to find out whether African forms. (anyhow not Australian) on Peaks. Did Creator make all new yet forms like neighbouring Continent. This fact speaks volumes.** . . . my theory explains this. but no other will.— **St. Helena, (& flora of Galapagos?) same condition. Keeling Is[d] «shows where proper dampness seeds can arrive quick enough»**'. Reading D74 and the grey-ink additions to C184 together both the ironic tone of 'system of great harmony' and the familiar theme of the absurdity of creationism become evident.

The peculiar character of St. Helena.—[2] contrast with otaheite in relation «See Gaudichauds Volume on the Botany of the Pacific.—[3]» to nearest continent.— With respect to ancient geography of Atlantic Tristan D'Acunha ditto, Juan Fernandez do

---

75 Mitchell. Australia Vol I. p. 306 "The crows were amazingly bold, *always accompanying us from camp to camp*; it was absolutely necessary to watch our meat, while in kettles on the fire, & on one occasion, not withstanding our vigilance a piece of pork 3 lb was taken from a boiling pot, & carried off by one of these birds"[1] Case of bird of different family. having very same habits in some respects as this Caracara.—

76 Sept. 9th. It is worthy of observation that in insects where one of the sexes is little developed, it is always female which approaches in character to the larva, or less developed state.— the female & young of all birds resemble each other in plumage «(that is where the female differs from the male?)».— children & women = "women recognized inferior intellectually"= Opposed to these facts are effects of castration on males & of age or castration on females.—

77e–84e [not located]

85e hen freely.— here we have beautiful proof of the breeding in & in (like «courage in dogs[1]» EFFEMINATE men),— if carried much further, if by the process this were possible, the organs doubtless would shrivel up.— **Yet odd they should have so much sexual character as they have** This character of not having *sexual plumage* is very common by hybrids, that are infertile.— thus the common. pheasant & fowl when crossed never even lay eggs. & the men[2] cannot «hardly» tell any sex by appearance.— The *silver & common pheasant* crossed, has a cock (infertile) ‹with› the breast of

86e which is like common pheasant & back like silver.— But the hen hybrid of this bird, has long tail figure, & some degree of whiteness like a Male.— Thus castration, hybridity, & breeding in & in tend to produce same effects.—

85 hen freely . . . sex by] *crossed pencil.*
Yet odd . . . they have] *added pencil.*
sexual plumage] *underlined pencil.*
silver & common pheasant] *underlined pencil.*
'17' *added to page brown crayon.*
86 *page crossed pencil.*

74–2 See B157 and B222 (also B173 and B193).
74–3 Gaudichaud 1826–30.
75–1 T. L. Mitchell 1838, 1:306.
85–1 See C120, 'Ld Orfords had breed of grey hounds fleetest in England lost courage.'
85–2 Workers in the Zoological Society Gardens. From this point (and perhaps parts of the missing D77–85) through D94, Darwin records his observations and reflections at the Zoological Society Gardens. These observations, which bear principally on generation, are paralleled by those on behaviour in M137–141.

CD[May it be said, that breeding in & in tends to produce unhealthiness,— «or» to perpetuate some organic difference.— it may be so, but this assumption as long as animals are healthy

87e which is often the case, & why should organic affections always influence the sexual organs alone.—

═══

It is singular pheasant & fowl being so totally infertile whereas animals further apart have bred inter se.—

‹———›

These hybrids are very *wild* & take ‹very little› in disposition after their «pheasant» parents.—[1] (There are some 3/4 birds «of», which I think there must be some mistake in their origin)

———

Saw cross between Penguin Duck «from Bombay» & Canada Goose.— Former strange mishaped bird— looks very artificial breed— but Mr Miller says[2] that breeds larger numbers, & rears an

88e unusual number out of any one nest. even more than common duck— Male Penguin was crossed with hen Canadian offspring, I should say in every respect most like Penguin duck.— *which is strange anomaly in Yarrells law.*— it probably is explained by the vigour of their propagating powers. (as if they were a good species, or local variety, & not effect of breeding in & in like our pidgeons)

———

The male of every animal certainly seems chiefly to impress the young most with its form & disposition

———

89e Saw three young ducks, like each other,— (& not very like either either wild or Pintail duck) from which they were descended— they descend from 1/2 pintail ‹into› «by» duck, into pintail.— Of these there were four, two like each other & two dark-coloured & different.—— the former were the parents of the three little ones.—[1]

———

87 which . . . alone.—] *crossed pencil.*
wild] *underlined pencil.*
It is . . . their origin)] '17' *added brown crayon.*
Saw cross . . . rears an] *double scored brown crayon.*
88 Male Penguin . . . law.—] *double scored brown crayon.*
The male . . . & disposition] *crossed pencil.*
89 *page excised and cut in half at* 'Keeper said'.
Saw three . . . little ones.—] '17' *added brown crayon.*
Keeper said . . . is concerned] '16' *added brown crayon.*

87–1 See *Natural Selection*: 457, 'The Pheasant preponderates over the fowl in those hybrids which I have seen.' See also *Variation* 2:67–68.

87–2 Alexander Miller.

89–1 See *Natural Selection*: 433, n. 1.

Keeper said in ‹two› crosses ‹‹twice made›› between terrier & hairless dogs of Africa,— some puppies hairless. some in patches, & some hairy— the former preponderated ‹which seems owing ‹‹determined›› by the sex›

Individual instances trouble Yarrels law. chief trust must be in general knowledge of breeders, where their interest is concerned.

90e Same man had crossed Jackal & dog— (offspring did not go to heat. but parts swelled, though no fluid came from them.— showing how gradually every ‹thing› ‹‹change›› is effected)— the one in garden is from ‹bitch dog do› father dog. & hence general appearance of face & tail somewhat like dog— though it has full share of Jackall shape of body.— disposition wild, & fearful. though not so much as in Jackall.— In case where Jackall was father resemblance much nearer to Jackall.—

91 This Keeper has seen when sickly tigers have first come over, insects somewhat like ‹‹between›› lice & fleas. sticking on them, but never in an animal, that had long been in confinement— is this effect of climate, or state in which they are kept?—

Is there any mistake about Yarrell's law, is it *local* (not *artificial* variation) which impresses offspring most.— ‹‹*& not time*›› thinking of the Penguin duck & Herberts law of ideosyncrasy

92 I have hitherto thought that a small difference ‹of any kind›, if very firmly fixed from long time, made no difference what its kind was.— but if it were *opposed* to the difference in other sex, it would be much more difficult to propagate— ‹now› ‹‹as›› if one bird had very bright red breast & other very bright blue, it might be harder ‹to tr› for

93 *both* parents to transmit there peculiarities; that if ‹one had a› both had mottled breasts, ‹when› of a sort that would allow the offspring to have some different kind of mottle, each feather partaking of character of other.— ‹so› the *most* aquatic & most terrestrial species, might be harder to cross than two less opposed in habits, though externally similar.— this however is a sophism for

94 their brain or stomach would be different.— Or if one species left its type in having very long legs, & another in having very long tail, & other in having very *short* tail.— I can readily see that two first might cross easier than two last.

90 *page excised and cut in half, crossed pencil.*
91 Is there . . . ideosyncracy] *brace right margin, and arrow to 92.*

95e Sept. 11. **N**
Mr. Blyth, at Zoolog. Meeting stated, that Green-finch, all linnets red-pole, goldfinch, hawfinch— in nursling plumage resembled that of Cross-Beak—[1]
In lark if I understand right, all species have same character which is mottled, & not like any existing species—
[In two herons, ‹both› *plumage* of both (nursling) quite similar.— one species retained this character in adult stage, other alters entirely[2]]CD
In common sparrow young & female similar plumage.— in tree sparrow, (if I understand rightly) young cock &

96e hen, all nearly similar.— in blackbird group young like some of the species— (?do these facts indicate that the change is effected through the male??)—
Yarrell observed that female of some water birds, (as Phalarope) assume for breeding a more brilliant plumage than male.—[1] «My case of Caracara. N. Zelandiæ.—[2]»
Mr Blyth stated «that there are» two ducks, which have pretty close repesentative species in England & N. America.— the teal which some authors

97e–98e [not located]

95 '12' *added to page brown crayon.*
N] *added blue crayon.*
96 Yarrell observed . . . some authors] *crossed pencil.*
Mr Blyth . . . some authors] *crossed pencil.*

95–1 Blyth 1838b:115, 'Mr. Blyth made some remarks on the plumage and progressive changes of the Crossbills, stating that, contrary to what has generally been asserted, neither the red nor saffron-tinted garb is indicative of any particular age. . . .

'He also exhibited a Linnet killed during the height of the breeding season, when the crown and breast of that species are ordinarily bright crimson, in which those parts were of the same hue as in many Crossbills; and observed that the same variations were noticeable in the genera *Corythraix* and *Erythrospiza*.' The remainder of Blyth's information in D95–96 is not reported in the 11 September *Proceedings* and, thus, reflects informal discussion at the meeting. Note that the lark was also discussed in the preceding paper (Yarrell's comments on Sykes 1838c).

95–2 Yarrell 1830–31:27.

96–1 See Yarrell 1843, 3:46, 'The females of this species [the Grey Phalarope] appear to assume more perfect colours in the breeding-season, and to retain them longer than the males.'

96–2 Caracara N. Zelandiæ refers to *Milvago leucurus* (*Birds*: 15–18) among whose synonyms are *Polyborus Novæ Zelandiæ*: 'The plumage in the two sexes of this species differs in a manner unusual in the family to which it belongs. The description given in all systematic works is applicable, as I ascertained by dissection, only to the old females; namely, back and breast black, with the feathers of the neck having a white central mark following the shaft, . . . MALE of smaller size than female: dark brown; with tail, pointed feathers of shoulders and base of primaries, pale rusty brown. On the breast, that part of each feather which is nearly white in the female, is pale brown: . . . As may be inferred from this description, the female is a much more beautiful bird than the male, and all the tints, both of the dark and pale colours, are much more strongly pronounced. From this circumstance, it was long before I would believe that the sexes were as here described. But the Spaniards . . . constantly assured me that the small birds with gray legs were the males . . .' (p. 16). See note D114–2.

99 September 13th The passion of the doe to the victorious stag. who rubs the skin off horns to fight— is analogous to the love of woman (as Mitchell remarks seen in savages) to brave men.—[1] Effect of castration horns drop off., replaced by hairy ones. which never «dry up &» peel off their skin (not being wanted for war) & hence never fall off.‖ Curious the rapidity of the change in 5 or 6 weeks after castration, fresh horns begin to grow.—

Mr Yarrell says the «male» Axis of India, breeds at times when horns not perfect— (is not this so in S. America with C. Campestris‹)› refer to my notes)[2] & Mr Yarrell supposes this a consequence of the female breeding all the year round. ask Colonel Sykes.—

100 Even our domesticated cattle have *tendency* to breed at particular times.

Mr Yarrell has old book 1765? Treatise on Domestic Pidgeon,[1] in which it appears that all the ‹bird› varieties «,now know» were then ‹pr› existing.— he has also some very fine recent drawing «of prize pidgeons» in 1834— now this would be most curious to show that in sixty years— (how many generations) the strangest peculiarities have been kept perfect— also to trace the laws of change in this time.— the impossibility of discovering their origin.— I see only some «but very strange *races*» of them have the forked black mark of the Rock Pidgeon,—[2] several have a group

100 *page crossed pencil.*

99–1 T. L. Mitchell 1838, 1:304, '. . . the possession of [the women] . . . appears to be associated with all their ideas of fighting; while, on the other hand, the gins [women] have it in their power on such occasions to evince that universal characteristic of the fair, a partiality for the brave. Thus it is, that after a battle, they do not always follow the fugitives from the field, but not infrequently go over, as a matter of course, to the victors, even with young children on their backs; and thus it was, probably, after we had made the lower tribes sensible of our superiority, that the three gins followed our party, beseeching us to take them with us.' See note D113–2.

99–2 *Cervus campestris.* Darwin's *Beagle* zoology notes (DAR 30.2: MS p. 196) and his comments in *Mammalia* on this species speaks indirectly to the question: 'This specimen was killed at Maldonado, in the middle of June; another specimen was killed at Bahia Blanca, . . . in the month of October, with the hairy skin on the horns: there were others, however, whose horns were free from skin. At this time of the year, many of the does had just kidded. I was informed, by the Spaniards, that this deer sheds its horns every year.' (*Mammalia*:30.)

100–1 Moore 1765. See St. Helena Model Notebook (Down House), 'Mr Yarrell 1763 [*over* 1780], has book history of Pidgeon. Treatise on Domestic Pidgeons—very curious in comparison for time,— Mr [Yarrell] has comparison' (p. 47). D100 shows Darwin's early and excited interest in pigeons as a model animal for the study of transmutation—an interest that comes to fruition in the *Origin* and *Variation*. QE4 records a further early step in this direction: 'Keep. Tumbling pigeons. cross them with other breed.—' Darwin repays this promissory note with experimental studies on pigeon variation and embryology in the 1850s, whilst writing *Natural Selection.*

100–2 See St. Helena Model Notebook (Down House), 'Has rock Pidgeon ‹pouter's› specks on shoulder, pouters have specks. Have any new varieties of Pidgeons been established? There must be laws of variation chance would never «produce» «feathers or make breed»' (p. 48). See also *Variation* 1:183, 'We now come to the best known rock-pigeon, the *Columba livia*, which is often designated in Europe pre-eminently as the Rock-pigeon, and which naturalists believe to be the parent of all the domesticated breeds. This bird agrees in every essential character with the breeds which have been only slightly modified. It differs from all other species in being of a slate-blue colour, with two black bars on the wings, and with the croup (or loins) white.'

101e of white speckles on elbow joint— in Bewick[1] drawing the the rock Pidgeon has not: now how many wild pidgeons have *spangles* on this part: this will be well worth working out.— Study Temminks work on Pidgeons—,[2] & see whether feathered legs.— «Carruncles on beak & in Muscovy duck» crested feather, pouters, fan tails are found in any *colours* of plumage &c &c «Pouting pidgeon exaggeration of cooing.—» & compare them with all the varieties.—[3] Habits of rock pidgeon. (I suspect *Pennant* has described them[4])— [Study horns of wild cattle.— plumage of fowls— *long* ears of rabbits. & long fur.— feathers on legs of Ptarmigan & in Bantam.—]^{CD} ^{CD}[In the Pidgeons, trace the *washing* out of the forked band, like in plumage of ducks.—

102e Mr Yarrell says[1] in very close species, of birds, habits when well watched always very different.— the two redpoles can hardly be told apart, so that after differences were pointed out Selby confounded them,[2] yet can readily be told by incubation & other peculiarities.— (Mem.— Goulds Willow Wren.—[3]) (Goulds story of Water-Wagtails mistake both species scattered over Europe)— The habits of some «same» North American & Europæan birds «slight» different— Barn Owl ‹the› in the former place breeds in ‹flags› «thick vegetation» in swamps— (owing to barns, perhaps, not being left open to them,— ⫲. In singing birds, part instinctive & part acquired,— thus Yarrel has Lark & Nightingale which both sing their own songs, though imperfectly.— Male birds always *second* their songs, *the* ++

103e Cervus Campestris spotted white when a fawn compare with *fallow*? deer. & Moschus &c & — like young blackbirds

---

Dr Bachman[1] told me that 1/2 Muscovy & common duck were often caught wild off coast of America.— showing hybrids can fare for themselves.||

---

101 *page crossed pencil.*
102 the ++] '++' *on stub.*
'7' *added to page brown crayon.*
103 Cervus . . . themselves||] *crossed pencil.*
++ first year . . . killed itself.—] *passage continues from 102* 'In singing . . . songs, the ++'.
Q] *added pencil in circle.*
'12' *added brown crayon.*
Sir. J. Sebright . . . by crossing—] *crossed pencil.*

101–1 Bewick 1794–1804, 1:267. 'The Wild Pigeon'.
101–2 Temminck 1813–15.
101–3 Temminck 1813–15, 1:43. See Darwin's abstract (DAR 71:6; probable date 1840s): 'Columba australis Lath. legs partially feathered: other species with carruncles at base of bill & others naked skin round eyes— [NB domestic vars. analogous these species]^{CD}'.
101–4 Pennant 1773:28, 'Swift and distant flight, walking pace. Plaintive note, or *cooing*, peculiar to the order. The male inflates or swells up its breast in courtship. Female, lays but two eggs at a time. Male and female sit alternately; and feed their young, ejecting the meat out of their stomachs into the mouths of the nestlings. Granivorous, seminivorous. The nest simple, in trees, or holes of rocks, or walls.'
102–1 D102–103 is based on conversation with William Yarrell, interspersed with information from John Gould.
102–2 P. J. Selby.
102–3 Recurs to theme of B213, C125, C177 and C241.
103–1 John Bachman: personal communication.

++ first year.— The bird fanciers match their birds to see which will sing *longest*, & they in evident rivalry sing against each other, till it has been known one has killed itself.—[2] **Q**

Sir. J. Sebright— has almost lost his Owl-Pidgeons from infertility,—[3] Yarrell says in such case they exchange birds with some other fancier, thus *getting fresh blood*, without *fresh feather*, & consequent trouble in obliterating the fresh feather, by crossing—

104e It seems from Lib. of Useful. Knowledge that sheep originally. black.[1] & Yarrell thinks the occasional production of black lambs is owing to old ‹story› return.— The Revd R. Jones told me precisely same story about some Southern «see p. 43 supra» breed of cattle with white heads; which years afterwards occasionally went back—[2] (Effect of imagination on mother. white peeled rods mentioned in old Testament placed before sheep—[3] it has been thought that silver Pheasants about a house made other pheasants have white feathers).—[4]

It certainly appears in domesticated animals, that the amount of variation is soon reached— as in pidgeons no new races.—[5]

105e In Scandinavia besides the Rakhekna, before mentioned[1] between Capercailzie & Black Cock.— The latter has crossed with the Ptarmigan. *subalpina* in wild state.— Neilson has given figure of it.—[2] In England no

104 *page crossed pencil.*
105 In Scandinavia . . . of it.—] '17' *added brown crayon.*
In England . . . hardly be] *crossed pencil.*

103–2 See Stanley, 1835 1:72, 'The bird-fanciers in London, who are in the habit of increasing the singing powers of birds to the utmost, by training them by high feeding, hot temperature of the rooms in which they are kept, and forced moutling, will often match one favourite Goldfinch against another. They are put in small cages, with wooden backs, and placed near to, but so they cannot see, each other: they will then raise their shrill voices, and continue their vocal contest till one frequently drops off its perch, perfectly exhausted, and dies on the spot.'
103–3 See C120 and D85.
104–1 Youatt 1837:17–18, 'This is the first intelligence which the Scriptures afford of the kind of sheep in these early times, or at least of those of which these flocks were composed: they were of one uniform colour, brown or dingy black, and the exceptions were accidental and of few in number. From the experiment or policy of Jacob, sheep of a new colour arose: . . . and the better appearance of the fleece . . . would lead to a selection from those that had the most white about them, until at length the fleece was purely white.' Darwin scored 'and the better . . . purely white.'
104–2 See D43.
104–3 *Genesis* 30:37–39, 'and Jacob took him rods of green poplar and of the hazel and chestnut tree; and pilled white strakes in them, and made the white appear which *was* in the rods. And he set the rods which he had pilled before the flocks in the gutters in the watering troughs when the flocks came to drink, that they should conceive when they came to drink. And the flocks conceived before the rods, and brought forth cattle ringstraked, speckled, and spotted.'
104–4 See C68: 'Mr Yarrell says, that some birds or animals are placed in white roooms to give tinge to offspring.'
104–5 See D100, but also note Lyell's emphasis on the limits of variation under domestication (Lyell 1837, 2:407–20). 'The alteration of the habits of species has reached a point beyond which no ulterior modification is possible, however indefinite the lapse of ages during which the new circumstances operate. (2:408).' Darwin commented: 'assumption'.
105–1 Refers to mention of *Tetrao Rakelhan* in D73.
105–2 Nilsson 1817–21, 1:300–5 discusses *Tetrao tetrix*,

doubt the cross between Pheasant & Black game is owing to their rarity., a single female in wood with Pheasants would sure to be trod,[3] & in many parts of Scandinavia these birds are very far from common.— Under this predicament, probably, alone would species cross in wild state.— Is English red Grouse. a cross between Black Game. &, the *subalpina* of Sweden, (which in summer dress somewhat resembles Red Grouse) it may be so— but very improbably, for it can hardly be

106e thought that the cross would have adapted it to changing circumstances.— More probably during known changes climate became unfit for. subalpina, or some Northern species, & being restricted species has been Made.—

In the hybrid grouse between Black Cock & Ptarmigan (probably *subalpina*.) former has *blue* breast, latter reddish, hybrid purple— be careful, See to hybrids between Pheasant & Black Cock, & other hybrids—

The fact of Egyptian animals not having changed is good—[1] I scarcely hesitate to say that if there had been considerable change, it would have been greater puzzle, than none, for the «e»normous time

107 which it must have taken to separate. Van Diemen's land from Australia &c &c

---

Sept. 14th. When Macleay says their is no difference between ‹t› "permanent varieties" & species.[1] he overlooks— ‹restric› those ‹restr› restricted in their range by men & by art.— the former only giving average of effects of country, (& no monstrosity, or adaptation to unhealthy state of womb).—[2]

One can perceive that Natural varieties or species., all the structure of which is adaptation to habits (& habit second nature) may be more in constitutional.,— more conformable to the structure which has been adapted to former changes. than a mere monstrosity propagated by art.

106 *page crossed pencil.*

Black grouse, Blackcock with varieties and hybrids including Rackelhane. No figures of ptarmigan are given. Nilsson 1835, 2:42–60 describes Capercaillie, *Tetrao Urogallus*; the Black-Grouse (Black-cock), *Tetrao Tetrix*; Ptarmigan, *Lagopus alpina*, and hybrids the Rackelhane, *Tetrao hybridus Urogalloides*, and *Tetrao hybridus Lagopoides*. See *Natural Selection*: 436, nn. 18–19.

105–3 Copulated with.
106–1 See B6, B16, and D39.
107–1 See D52.
107–2 Darwin is distinguishing between free-ranging domestic animals, which live in semi-natural conditions (e.g. Highland sheep and Chillingham cattle) and those which are entirely dependent on man (e.g. fancy pigeons).

108 Yarrell told me of a cat & of a dog, born without front legs—— the former of which had kittens with imperfect ones.— now Sir J. Sebright.[1] thought if he had had a pair he could have produced from these.— this instance of monstrous variety. which could not have been persistent in nature.—

According to my view, the domesticated animals would cease being fertile inter se., or at least show repugnance to breeding if instincts unchanged, & if their characteristic qualities were all deeply imbued in them from long permanence, so that all their peculiarities must be transmitted if their

109e–110e [not located]

111 ‹The› every case common to many good species; & therefore to genera (& the uncles & aunts) & therefore does not tell against transmutation of species— Will it against genera.— How long will the wretched inhabitants of NW. Australia, go on blinking their eyes. without extermination, & change of structure.— When will the musquitoes of S. America take an effect.— would perfect impunity from muskitoes bite influence propagation of species.—

Case of Association very disagreeable hearing maed servant cleaning door outside, as often as she touched handle, though really fully aware she was not coming in, could not help being perfectly distracted
«Referred to ‹other› Book M.[1]»

112 Is there any *law of variation.* — «(as Hunter supposes with *Monsters*)[1]» if armless cat can propagate,[2] ie with the chance of two being born at same time, & make breed, one would doubt any law.— Yet seeing the feathers along one toe of the Pouter one thinks there is a law.,—[3] that there must have been a tendency for feathers to grow there

---

«That Mutilations will not alter form may be inferred from Australians knocking out teeth.—[4] the account of the people on the NW. Coast blinking to keep out flies might be used»

The wild ass has no *cross.* how comes it that the tame donkey has. CD[old Buffon should be read on Mare[5]

---

111 Case of . . . distracted] *crossed.*
«Referred to ‹other› Book M.»] *presume added.*

108–1 William Yarrell and Sir J. S. Sebright: personal communications.
111–1 Darwin cross references this case in M142.
112–1 For 'Hunter's Law' of correlated variation in monsters see Hunter 1837:44–45, which is quoted in note D57–1. See also note D113–2 for the textual relation of that quote to Hunter's definition of primary and secondary sexual characters.
112–2 Refers to the case in D108.
112–3 That is, a broader law of correlated variation than Hunter's, which emphasises monsters.
112–4 See T. L. Mitchell 1838, 2:339, 'But still more remarkable is the practice of striking out one of the front teeth at the age of puberty . . .'
112–5 Buffon 1762.

My view, why hybrids are infertile. supposes that when fœtus is forming the ovum within it, is forming «& this must be so, else avitism could hardly ever occur.—».— & if that cannot be formed, genetal organs by that co-relation of parts, will not be produced.—

113e Sept. 17th. Saw mule. apparently fathered by a donkey. with all four legs ringed with brown.— animal like large, heavily made cream coloured ass.— stripe on back also.— legs reminded me strongly of Zebra.— **Mem. Quagga & Ld Moreton Mare[1] ringed**

Owen says that Bell in Encyclop of Anat & Phys. describes, a high-flying bat, which has the power of inflating its body like balloon— by air cells connected with cheek pouches.—[2]

Hunter's Animal Œconomy p. 45 "One of the most general marks is the superior strength ‹of› «of make in» the males; & another circumstance, perhaps, equally so, is this strength being directed to one part more than another, which part is that most immediately employed in fighting"[3] instances thighs of cock & Neck of Bull.— is most common in vegetable feeders. **because males always armed in carnivora.** *Where* females, are peacable—[4] (Mem Lucanus & Copris &c)[5].— In birds singing

113 **Mem. Quagga . . . Mare ringed]** *added pencil.*
Saw mule . . . Zebra.—] *crossed pencil,* '5' *added brown crayon and crossed pencil.*
Owen says . . . pouches.—] *crossed blue crayon.*
Hunter's . . . birds singing] *crossed pencil,* '12' *added brown crayon, double scored blue crayon.*
because males . . . in carnivora.] *added pencil,* **'armed'** *underlined.*

113–1 Morton 1821.
113–2 Richard Owen: personal communication. T. Bell 1836, 1:599, 'In the genus Nycteris a curious faculty is observed, namely, the power of inflating the subcutaneous tissue with air. . . . By this curious mechanism the bat has the power of so completely blowing up the spaces under the skin, as to give the idea, as Geoffroy observes, "of a little balloon furnished with wings, a head, and feet".'
113–3 Hunter 1837:45–46 (An account of any extraordinary pheasant), 'One of the most general marks is, the superior strength of make in the male; and another circumstance perhaps equally so, is this strength being directed to one part more than another, which part is that most immediately employed in fighting. This difference in external form is more particularly remarkable in the animals whose females are of a peaceable nature, as are the greatest number of those which feed on vegetables, and the marks to discriminate the sexes are in them very numerous. The males of almost every class of animals are probably disposed to fight, being, as I have observed, stronger than the females; and in many of these there are parts destined solely for that purpose, as the spurs in the cock, and the horns in the bull; and on that account the strength of the bull lies principally in his neck; that of the cock in his limbs.' This passage immediately follows Hunter's famous definition of primary and secondary sexual characters (1837:44–45): 'It is well known that many orders of animals have the two parts designed for the purpose of generation different in individuals of the same species, by which they are distinguished into male and female; but this is not the only mark of distinction, in the greatest part the male being distinguished from the female by various other marks. The varieties which are found in the parts of generation themselves I shall call the first or principal marks, being originally formed in them, and belonging equally to both sexes; all others depending upon these I shall call secondary, as not taking place till the first are becoming of use, and being principally, although not entirely, in the male.' The latter passage immediately follows the statement of 'Hunter's Law' of correlated variation, which is given in note D57–1.
113–4 Hunter 1837:45–46, quoted above.
113–5 See *Natural Selection* p. 131, 'In the Stag-Beetle, & indeed generally in the Lucanidæ, the mandibles in the males are enormously developed & are eminently variable . . .'

114e of cocks settle point.— (do the females then fight for male) & are merely most attracted). — singing best sign of most vigorous males.— «(NB. most strange cocks & hens. being either alike or very different in recently altered genera. Guinea Fowl & Peaccocks.!!» other birds display beauty of plumage.— (The females (as Owen observes) in Raptorial birds largest.— p. 47.[1] (‹“› is evidently the male which recedes from the species all females being most like offspring, **Q** (how is this with those females which put on (like some waders) the bright plumage.—[2] «thinks» Hence specific character most perfect in ‹male› «hermaphrodite».—[3]

---

114 p. 47 . . . like offspring] *double scored brown crayon.*
Q] *added pencil in circle.*
Peacock & spurs] 'P' *over* 'e'.
'11' *added to page brown crayon.*
Chapt I. Also Latent Character] *added to page in mixture of brown crayon and pencil:* **'Chapt I. Also'** *and underlining of* **'Latent'** *in pencil,* **'ent'** *in pencil over* **'ent'** *in brown crayon.*

114–1 Hunter 1837:46. Hunter's text: 'In carnivorous animals, whose prey is often of a kind which requires strength to kill, we do not find such a difference in the form of the male and female, very little being discernible in the dog and bitch, in the he or she cat, or in the cock and hen of the eagle[a].' Owen's note [a]: 'The difference in the size of the two sexes is sufficiently marked in most of the Raptorial birds; but it is the female which has the advantage in this respect.'

114–2 Hunter 1837:47–48, 'In some species of animals that have the secondary properties we have mentioned, there is a deviation from the general rule, by the perfect female, with respect to the parts of generation, assuming more or less the secondary character of the male.

'This change does not appear to arise from any action produced at the first formation of the animal, and in this respect is similar to what takes place in the male; neither does it grow up with the animal as it does to a certain degree in the male, but seems to be one of those changes which happen at a particular period, similar to many common and natural phenomena; like to what is observed of the horns of the stag, which differ at different ages; or to the mane of the lion, which does not grow till after his fifth year, &c.'

114–3 Hunter 1837:46–47, 'To bring the foregoing observations into one point of view, I here beg leave to remark, that in animals just born, or very young, there are no peculiarities to distinguish one sex from the other, exclusive of what relates to the organs of generation, which can only be in those who have external parts; and that towards the age of maturity the discriminating changes before mentioned begin to appear; the male then losing that resemblance he had to the female in various secondary properties; but that in all animals which are not of any distinct sex, called hermaphrodites, there is no such alteration taking place in their form when they arrive at that age. It is evidently the male which at this time in such respects recedes from the female, every female being at the age of maturity more like the young of the same species than the male is observed to be; and if the male is deprived of his testes when young, he retains more of the original youthful form, and therefore more resembles the female.

'From hence it might be supposed that the female character contains more truly the specific properties of the animal than the male; but the character of every animal is that which is marked by the properties common to both sexes, which are found in a natural hermaphrodite, as in a snail, or in animals of neither sex, as the castrated male or spayed female.

'But where the sexes are separate, and the animals have two characters, the one cannot more than the other be called the true, as the real distinguishing marks of each particular species . . . are those common to both sexes, and which are likewise in the unnatural hermaphrodite. That these properties give the distinct character of such animals is evident, for the castrated male and the spayed female have both the same common properties; and when I treated of the free-martin, which is a monstrous hermaphrodite, I observed that it was more like the ox than the cow or bull; so that the marks characteristic of the species which are found in the animal of a double sex are imitated by depriving the individual of certain sexual parts, in consequence of which it retains only the true properties of the species.

'They are curious facts in the natural history of animals, that by depriving either sex of the true parts of generation, they shall seem to approach each other in appearances, and acquire a resemblance to the unnatural hermaphrodite.'

(Fishes have no secondary characters.—[4])p. 49. (wonderful case of Pea hen. taking *feathers* of Peacock & spurs—[5] *no final cause here.*—[6] & therefore different from Hunter I should say females recede in organization from specific character.—[7]

Chapt I. Also Latent Character

115 **Hunter Animal Economy**
p. 482 (Same book) Owen says "the necessity of combining observation of the living habits of animals, with anatomical & Zoological research, in order to establish entirely their place in nature, as well as fully to understand their œconomy, is now universally admitted." "—[1] p. 483. Owen thinks from climate of Australia, & from Ornithorhyncus & Hydromys not being Marsupial. («but» *also mice*) & these being water animals ‹that this› structure ‹connected with animals being compelled to travers› "May have reference to the Great distances which the Mammalia of N. S. Wales are generally compelled to traverse in order to quench their thirst"—[2] But *New*

115 Hunter Animal Economy] *added pencil.*
Zoological] 'Zoo' *over* 'physio'.

114–4 Hunter 1837:47 note, 'in fishes there is no great difference'. See D169 and T25–26, where Darwin notes a sexual difference in *Syngnathus.*

114–5 See Hunter 1837:48–49. In the descriptive or evidentiary part of his essay, Hunter analyses a series of cases where post-reproductive female pheasants acquire male secondary sexual characters. The case Darwin calls 'wonderful' is the final one in the series and becomes the epitome for Hunter's theoretical summation, 'Lady Tynte had a favourite pied pea-hen which had produced chickens eight several times; having moulted when about eleven years old, the lady and family were astonished by her displaying the feathers peculiar to the other sex, and appeared like a pied peacock. In this process the tail, which became like that of the cock, first made its appearance after moulting; and in the following year, having moulted again, produced similar feathers. In the third year she did the same, and, in addition, had spurs resembling those of a cock. She never bred after this change in her plumage, and died in the following winter during the hard frost in the year 1775–6.' (p. 49). See D154 and Yarrell 1827.

114–6 Since the female is past her reproductive period there is no fulfillment of the 'final cause' of reproduction, which Darwin here regards to be the perpetuation of the species. See also E147 on the absence of final cause: 'no one can be shocked at absence of final cause mammæ in man & wings under united elytra.'

114–7 Hunter 1837:49, 'The female, at a much later time of life, when the powers of propagation cease, loses many of her peculiar properties, and may be said, except from mere structure of parts, to be of no sex, even receding from the original character of the animal, and approaching, in appearance, towards the male, or perhaps more properly towards the hermaphrodite.'

115–1 Hunter 1837:482 (Descriptions of some animals from New South Wales). Hunter's text: 'The subjects themselves may be valuable, and may partly explain their connection with those related to them, so as in some measure to establish their place in nature[a], but they cannot do it entirely; they only give us the form and construction, but leave us in other respects to conjecture, many of them requiring further observation relative to their oeconomy.' Owen's note [a]: 'It is interesting to meet with these indications of the spirit in which Hunter prosecuted his zoological researches. To ascertain the affinities of the animals whose structure he explored, or, in other words, to establish a natural system of classification, was not less the aim of Hunter than the determination of the functions of the different organs in the animal frame; and the truth of the remark of the necessity of combining observation of the living habits of animals, with anatomical and zoological research, in order to establish entirely their place in nature, as well as to fully understand their oeconomy, is now universally admitted.'

115–2 Hunter 1837:483 note, 'Many have been the conjectures respecting the final intention of the premature birth of the marsupial animal, and the various singular modifications of structure necessitated by, and adapted to that circumstance. Since it obtains in quadrupeds of almost every variety of form, and with various modes of locomotion and diversity of diet, it must result from some more general law than individual proportions or habits of the parent. It is associated with a marked inferiority of

Guinea.— !! S. America.— Such difficulties will always occur if animals are thought to have been created.—[3] it might as well be attempted to be shown from peculiarities of climate cause of N. Zealand not having any Mammalia.— Type of geographical organization. no more can be said. . . .

116 In paper on bees in same work. (it is said that some kind lay ‹pu› up honey even for single rainy day—[1] & from case of wasps, is supposed cells properly are made for larvæ.—[2] CD[(p. 451.)— Wasps breed many females, but almost all die.— bees breed but few, because they are kept in security.—[3] Hunter doubts about production of Queens.—[4] Neuters are bred first, «then males—»[5] how has this been arranged— Neuters are true females, but with parts little developed.—

Sept. 19th ‹Are› There is no scale, according to importance of divisions in arrangement, of the perfection of

117 their separation.— thus Vertebrate blend with Annelidæ by some fish.— But birds quite distinct.—

Collect cases of difficulty of growing plants in all parts of world, thus tea tree in Brazil must have degenerated. as must spices &c &c

---

The line of argument «often» pursued throughout my theory is to establish a point as a probability by induction, & to apply it as hypothesis to other points. & see whether it will solve them.—

cerebral organization; . . . Long-continued droughts and a scarcity of freshwater streams are amongst the most striking features of the climate and territory of Australia; and when we reflect that the principal exceptions to the marsupial organization, viz. the Ornithorhynchus and the Hydromys, or water-rat of the colonists, habitually inhabit the freshwater ponds, the peculiarities of the re-production above described may have reference to the great distances which the mammalia of New South Wales are generally compelled to traverse in order to quench their thirst.'

115–3 Note that Darwin disparages Owen's functional teleology while maintaining his higher order teleology.

116–1 Hunter 1837:432 (Observations on bees), 'When they [a swarm] have fixed upon their future habitation, they immediately begin to make their combs, for they have the materials within themselves. I have reason to believe that they fill their crops with honey when they come away; probably from the stock in the hive.'

116–2 Hunter 1837:437, 'The comb seems at first to be formed for propagation, and the reception of honey to be only a secondary use; for if the bees lose their queen they make no combs; and the wasp, hornet, &c. make combs, although they collect no honey; and the humble bee collects honey, and deposits it in cells she never made.'

116–3 Hunter 1837:451, 'This circumstance, that so few queens are bred, must arise from the natural security the queen is in from the mode of their society; for although there is but one queen in a wasp's, hornet's, and humble bee's nest or hive, yet these breed a great number of queens; the wasp and hornet some hundreds; but not living in society during the winter, they are subject to great destruction, so that probably not one in a hundred lives to breed in the summer.'

116–4 Hunter 1837:451–52, 'Mr. Riem asserts, he has seen the copulation between the male and the female, but does not say at what season. I should doubt this; but Mr. Schirach supposes the queen impregnated without copulation. I know not whether he means by this that she is not impregnated at all, and supposes, like Mr. Debraw, that the eggs are impregnated after they are laid, by a set of small drones, who pass over the cells, and thrust their tails down into the cell, so as to besmear the egg. . . . It is probable that the copulation is like that of most other insects. . . . The circumstances relative to impregnating the queen not being known, great room has been given for conjecture, . . .'

116–5 Hunter 1837:452, 'The males, I believe, are later in being bred than the labouring bee.'

118 It is less wonderful that childs nervous system should build up its body, like its parent, than that it should be provided with many contingencies how to act— so with the mind. the simplest transmission is direct instinct. & afterwards enlarged powers to meet with contingency.—

Sept. 23rd. Saw in Loddiges Garden. 1279 varieties of roses!!! proof of capability of variation.—[1] Saw his collection of «Humming» birds, saw several fully developed tails, & one with beak turned up like Avocette. here is what

119e–126e [not located]

127 that it shall beget young different in colour, form, & so altered in disposition, as to be more easily trained up to the (required) offices" &c &c Owen illustrates case of Dingo (he alludes to the dholes or wild dogs of India) in Zoolog. Garden having coloured offspring.—[1] but surely in all these cases an unseen change is produced in parents—colour is a doubtful subject, but what other instances are there of *such* changes, *not* acquired by parent, being handed down?

128 Are not Loddiges 1279 roses kept in same soil. same atmosphere?—[1] may they produced not be transplanted?, & yet year after year, successive roses & bud are produced, like parent stock, or if different dieteriorating very slowly.— I presume most of these roses, without circumstances very unfavourable, will ‹deteriorated› continue of same variety as long as life lasts, yet they cannot transmit through seeds these characters though transmitting them with such facility to bud.— this must be owing to their unity in one stem.—

118 It is . . . contingency.—] *pencil.*
121 *brown crayon score on stub.*
123 *brown crayon score on stub.*
127 *page crossed pencil.*

118–1 See Loddiges 1823.
127–1 Hunter 1837:330 (Observations tending to show that the wolf, jackal, and dog, are all of the same species). Hunter's text: 'As animals are known to produce young which are different from themselves in colour, form, and dispositions, arising from what may be called the unnatural mode of life, it shows this curious power of accommodation in the animal oeconomy, that although education can produce no change in the colour, form, or disposition of the animal, yet it is capable of producing a principle which becomes so natural to the animal that it shall beget young different in colour[a] and form, and so altered in disposition as to be more easily trained up to the offices in which they have been usually employed, and having these dispositions suitable to such change of form.' Owen's note [a]: 'This has recently been exemplified in the produce of a male and female Dingo, or wild dog of Australia, brought forth at the Zoological Gardens, and under circumstances which precluded the possibility of connection between the female and any other dog than the male with which she was kept confined. Two, out of the litter of five puppies brought forth, had the uniform red-brown colour of the parents, the rest were more or less pied, brown and white.'
128–1 See Loddiges 1823.

129 a bud may be transplanted & carry all these peculiarities not so a seed.— Bud probably is like cutting off tail of Planaria, the whole grown to that part.— **claw[1] added to crab, tail to lizard,— healing of wound.— reproductive faculty** + in the separated part every element of the living body is present— in generation something is added from one part of the body «(or of other ‹like› «similar» body)» to another part of body.— [in plants does not whole individual change into generative organs?]CD it is of no consequence if it does= ‹I do not doubt, the› Do plants loose any qualities by being buds— , more than if whole branch transplanted? +**.simplest forms of budding. Why does Gecko produce always different tail?**

130 An Individual bud may be thus produced from the growth of one part, (not strictly new individual), or he may produced by having undergone, the endless changes, which its parents have.— — Not this is effected by short method in *generation*.—

———

Ehrenberg considers artificial division of animals, as gemmation, I consider gemmation as artificial division.—[1] On this view each particle of animal must have structure of whole comprehended in itself.— it must have the knowledge how to grow, & therefore to repair wounds— but this has nothing to do with generation.— **Why crab can produce claw. but man not arm. hard to say—**

131 **if it were possible to support the arm of Man, when cut off , it would produce another man.— That the embryo the thousandth of inch should produce a Newton is often thought wonderful. it is part of same class of facts that the skin grows over a wound.—**
Does likeness of twin bear on this subject?

———

A mans arm would produce arm if *supported*., ‹so› & in making «true» bud some such process is effected.— a *child* might be so born. but it would be very different from true generation.— there is no caterpillar state; the vast difference of two kinds of generation shown by their happening in same plant.—

129 **claw added . . . faculty +**] *added pencil*, **'reproductive faculty'** *underlined*.
another] 'an' *over* 'the'.
+ **.simplest . . . different tail?**] *added pencil*.
130 **Why Crab . . . hard to say—**] *added pencil*.
131 **if it were . . . twin bear on this subject?**] *added pencil*, **'support'**, **'thousandth'** *underlined*.
twin] *emmended grey ink*.

129–1 Read the pencil additions to D129–131 as a unit. The fact that 'twin' in D131 was emmended in grey ink proves that this pencil unit is contemporaneous with the ink text.

130–1 Ehrenberg 1838b:653, '*Paramecium aurelia* was not one of the examples best adapted for the examination of the gastric vesicles, which he believed to be digestive cavities capable of great dilatation; and moreover he had witnessed the expulsion of a kind of exuvium apparently thrown off from their interior: he believed the process of multiplication by division to be merely the developement of a gemma or bud, a view which would much simplify our ideas concerning that singular mode of propagation.'

132 The Marsupial structure shows that they became Mammalia, through a different series of changes from the placentates, Having Hair. like true Mammalia, no more wonderful. than Echidna. & Hedgehog having spines.—

Does not male Pidgeon (yes) surely) secrete *milk*? from stomach. analogous to other males feeding young, & to abortive ‹organs› «mammæ» in male Mammalia:— ʕis not this argument, for Mammalia recent creation.—

---

why. what tendency can there be for abortive organ ever disappearing??—

Have Marsupiata abortive Mammæ?.—
My view would make every individual a spontaneous generation: what is animalcular semen— but this— — the ‹nerve› living nerve nursed in Mould.—

133e Lyells Elements. p. 290. Dr. Beck on numerical proportion in shells in Arctic Ocean.[1] p. 350 Grallæ in Wealden. oldest birds.[2] p. 411— Decapod Crust in Muschelkalk, & 5 genera of reptiles.—[3] ‹M› p. 417. Magnesian Limestones

133 Lyells . . . p. 411] '19' *added brown crayon.*
411] '11' *over* '41'.
Decapod . . . organized] *crossed pencil.*

133–1 Lyell 1838a:289–90, 'Now, it has been suggested by Dr. Beck, in order to form such an estimate of the comparative resemblance of the faunas of different eras, we may follow the same plan as would enable us to appreciate the amount of agreement or discrepancy between the faunas now existing in two distinct geographical regions. It is well known that, although nearly all the species of mollusca inhabiting the temperate zones on each side of the equator are distinct, yet the whole assemblage of species in one of these zones bears a striking analogy to that in the other, and differs in a corresponding manner from the tropical and arctic faunas. By what language can the zoologist express such points of agreement or disagreement, where the species are admitted to be distinct?

'In such cases it is necessary to mark the relative abundance in the two regions compared of certain families, genera and sections of genera; the entire absence of some of these, the comparative strength of others, this strength being sometimes represented by the numbers of species, sometimes by the great abundance and size of the individuals of certain species. It is, moreover, important to estimate the total number of species inhabiting a given area; and also the average proportion of species to genera, as this differs materially according to climate. Thus if we adopt comprehensive genera like those of Lamarck, we shall find, according to Dr. Beck, that, upon an average, there are in arctic latitudes nearly as many genera as species; in the temperate regions, about three or four species to a genus; in the tropical, five or six species to a genus.' Darwin scored 'proportion of species to genera . . . six species to a genus.' This may bear directly on D134 'We ought . . . changes in number of species, from small changes in nature of locality.' That is, changes in number of species per genus with different locality—as Lyell expresses it above. See also E59.

133–2 Lyell 1838a:350–51, 'The bones of birds of the order Grallae or waders have been discovered by Mr. Mantell in the Wealden, and appear to be the oldest well-authenticated examples of fossils of this class hitherto found in Great Britain. But no portion of the skeleton of a mammiferous quadruped has yet been met with.' Passage scored: 'The bones . . . Britain'. See B172 'Oh. Wealden.— Wealden.—'.

133–3 Lyell 1838a:410–11, 'There are also some encrinites in the Muschelkalk, and some teeth of cartilaginous fish, a few decapod crustacea, and no less than five genera of large extinct reptiles, all peculiar to the Muschelkalk, as Phytosaurus, Dracosaurus, and others.' Darwin underlined 'decapod' and scored 'a few . . . reptiles'.

& Zechstein oldest rock in which reptiles have been found.[4] p. 426 Sauroid fish in Coal, true fish, & not intermediate between fish & reptiles— yet osteology closely resembles reptiles.—[5] p. 432 some plants in coal supposed to be intermediate between Coniferous trees & Lycopodiums.—[6] p. 437. Many. existing genera of shells in the mountain limestone (how different from plants!) But the Cephalopoda depart more widely from living forms.—[7] p 458 Upper Silurian, fishes oldest formation highly organized.—[8]

134e do. p. 461.— Lower Silurian— several existing genera. Nautilus turbo. buccinum. turritella. terebratula, orbiculas, with many extinct forms & Trilobites[1]

Sept 25th. In considering infertility of hybrids inter se, the first cross generally brothers & sisters, & therefore somewhat unfavourable—[2]

28th. «I do not doubt, every one till he thinks deeply has assumed that increase of animals exactly proportiona[l] to the number that can live.—» We ought to be far from wondering of changes in number of species, from small

134 proportiona[l]] *part of* 'l' *on stub.* *page crossed pencil.*

133–4 Lyell 1838a:417, 'The remains of at least two saurian animals of new genera, Palaeosaurus and Thecodontosaurus have been lately discovered in the dolomitic conglomerate near Bristol. They are allied to the Iguana and Monitor, and are the most ancient examples of fossil reptiles yet found in Great Britain. The Zechstein of Germany is also the oldest rock on the continent in which Saurian remains have been found.' Passage scored.

133–5 Lyell 1838a:425, 'No bones of mammalia or reptiles have as yet been discovered in strata of the carboniferous group. The fish are numerous, and for the most part very remote in their organization from those now living, as they belong chiefly to the Sauroid family of Agassiz; as Megalichthys, Holoptychus, and others, which were often of great size, and all predaceous. Their osteology, says M. Agassiz [1833–43, bk 4:62, bk 5:88], reminds us in many respects of the skeletons of saurian reptiles, both by the close sutures of the bones of the skull, their large conical teeth striated longitudinally (see Fig. 248.), the articulations of the spinous processes with the vertebrae, and other characters. Yet they do not form a family intermediate between fish and reptiles, but are true *fish*.' Passage scored: 'Yet they . . . *fish*.'

133–6 Lyell 1838a:431–32, 'Another class of fossils, very common in the coal-shales, have been named Lepidodendra. Some of these are of small size, and approach very near in form to the modern *Lycopodiums*, or club-mosses, while others of much larger dimensions are supposed to have been intermediate between these and coniferous plants.' Passage scored: '*Lycopodiums* . . . plants'. Darwin underlined 'intermediate between these and coniferous'.

133–7 Lyell 1838a:437, 'There are also many univalve and bivalve shells of existing genera in the Mountain Limestone, such as *Turritella, Buccinum, Patella, Isocardia, Nucula*, and *Pecten*. But the Cephalopoda depart, in general, more widely from living forms, some being generically distinct from all those found in strata newer than the Coal.' Passage scored. Darwin underlined 'existing genera', 'Cephalopoda', and 'more widely'.

133–8 Lyell 1838a:458, '*Ludlow formation*.— This member of the upper Silurian group . . . is of great thicknes. . . The most remarkable fossils are the scales, icthyodorulites, jaws, teeth, and coprolites of fish, of the upper Ludlow rock. As they are the oldest remains of vertebrated animals yet known to geologists, it is worthy of notice that they belong to fish of a high or very perfect organization.' Passage scored: 'The most . . . organization'.

134–1 Lyell 1838a:461, 'There are also several genera of mollusca in this deposit, and it is an interesting fact, that with many extinct forms of testacea peculiar to the lower Silurian rocks, such as orthoceras, pentamerus, spirifer, and productus, others are associated belonging to genera still existing, as nautilus, turbo, buccinum, turritella, and orbicula.' Passage scored.

134–2 Thus hybrids would suffer from the combined disadvantages of inbreeding and hybridization. Compare with D10 and D14.

changes in nature of locality.[3] Even the energetic language of ‹Malthus› «Decandoelle[4]» does not convey the warring of the species as inference from Malthus.— «increase of brutes, must be prevented soley by positive checks, excepting that famine may stop desire.—[5]» in Nature production does not increase, whilst no checks prevail, but the positive check of famine & consequently death..[6]

135e population in increase at geometrical ratio in FAR SHORTER time than 25 years— yet until the one sentence of Malthus no one clearly perceived the great check amongst men.—[1] «Even a *few* years plenty, makes population in Men increase,[2] & an *ordinary* crop. causes a dearth then in Spring, like food used for other purposes as wheat for making brandy.—» take Europe on an average, every species must have same number killed, year with year, by hawks. by. cold &c— .. even one species of hawk decreasing in number must effect instantaneously all the rest.— One may say there is a force like a hundred thousand wedges trying force ‹into› every kind of adapted structure into the gaps ‹of› in the œconomy of Nature, or rather forming gaps by thrusting out weaker ones. «The final cause of all this wedgings, must be to

135 *page crossed pencil.*

134–3 See note D133–1.
134–4 See Lyell 1837, 3:87–88, '"All the plants of a given country," says De Candolle, in his usual spirited style, "are at war one with another".'
134–5 See note D134–6. See also Darwin's previous interest in inbreeding stopping desire, e.g. B176, D177.
134–6 Malthus 1826, 1:12–13, 'These checks to population, which are constantly operating with more or less force in every society, and keep down the number to the level of the means of subsistence, may be classed under two general heads— the preventive, and the positive checks.

'The preventive check, as far as it is voluntary, is peculiar to man, and arises from that distinctive superiority in his reasoning faculties, which enables him to calculate distant consequences. The checks to the indefinite increase of plants and irrational animals are all either positive, or, if preventive, involuntary.'
135–1 Malthus 1826, 1:5, 'In the northern states of America, where the means of subsistence have been more ample, the manners of the people more pure, and the checks to early marriages fewer, than in any of the modern states of Europe, the population has been found to double itself, for above a century and half successively, in less than twenty-five years. Yet, even during these periods, in some of the towns, the deaths exceeded the births, a circumstance which clearly proves that, in those parts of the country which supplied this deficiency, the increase must have been much more rapid than the general average.' The single sentence that caught Darwin's attention may have been: 'It may safely be pronounced, therefore, that population, when unchecked, goes on doubling itself every twenty-five years, or increases in a geometrical ratio.' (1:6). Or: 'A thousand millions are just as easily doubled every twenty-five years by the power of population as a thousand. But the food to support the increase from the greater number will by no means be obtained with the same facility. Man is necessarily confined in room.' (1:7).
135–2 See Malthus 1826, 1:18–19, which discusses the dynamics of population oscillations, '. . . ultimately the means of subsistence may become in the same proportion to the population, as at the period from which we set out. The situation of the labourer being then again tolerably comfortable, the restraints to population are in some degree loosened; and, after a short period, the same retrograde and progressive movements, with respect to happiness, are repeated.'

sort out proper structure & adapt it to change.—[3] to do that, for form, which Malthus shows, is the final effect, (by means however of volition) of this populousness, on the energy of Man[4]»

136e D.'Orbigny. Comtes Rendus p. 569. 1838 says the cross between the Guaranis & Spaniards are almost White from first generation., that with Quichuas the American character is more tenacious. & does not disappear for Many generations[1]

———

Sept. 29th Dr. Andrew. Smith «Remarks on extraordinary *curiosity* of Monkeys».[2] The Baboon of which anecdotes have been told is Cyanocephalus Porcarius.— this Monkey did not like a great coat made for it at first, but in two or three days learn its comfort & though could not put it *on*, yet threw it over

137 it, & made it meet in front.— Dr Smith every baboon & monkey, big & little that ever he saw knew women.— he has repeatedly seen them try to pull up petticoats., & if woman not afraid clasp them round waist & look in their faces & Mak the st. st noise.— The Cercopithecus *chinensis*: (or bonnet faced monkey he has seen do this.— These Monkeys had no curiosity to pull up trousers of men. Evidently knew ‹men› women, thinks perhaps by smell.— but monkeys examine sexes of every

138 Has repeatedly seen one he kept pull up feathers of tail of Hen; which lived with it.— also of ‹a› dog«s». *but did not seem to evince more lewdness for bitch than dog*: Monkey thus examine each other sexes,— «by taking up tail» Mem: Ourang Jenny with Tommy.— Good evidence of knowledge of Woman— ǂǂ

136 D'. Orbigny . . . generations] '16' *added brown crayon.*
does not] 'd' *over* 'n'.
Sept 29th . . . threw it over] *crossed pencil.*

135–3 This passage is Darwin's first formulation of natural selection, particularly as the origin of adaptation. It is unambiguously crystallized by his reading of Malthus on 28 September 1838. Darwin completed reading Malthus 1826 on 2–3 October. (See E3, 'Journal', and 'Books examined with ref. to Species.—' C 270: 'October 3d Malthus on Population'). The theory continued to take form in Notebook E.

135–4 See Malthus 1826, 1:94–95. 'The combined causes soon produce their natural and invariable effect, an extended population. A more frequent and rapid change of place then becomes necessary. A wider and more extensive territory is successively occupied. A broader desolation extends all around them. Want pinches the less fortunate members of society: and at length the impossibility of supporting such a number together becomes too evident to be resisted. Young scions are then pushed out from the parent stock, and instructed to explore fresh regions, . . . Restless from present distress, flushed with the hope of fairer prospects, animated with the spirit of hardy enterprise, these daring adventurers are likely to become formidable adversaries to all who oppose them. The inhabitants of countries long settled, engaged in the peaceful occupations of trade and agriculture, would not often be able to resist the energy of men acting under such powerful motives of exertion. And the frequent contests with tribes in the same circumstances with themselves, would be so many struggles for existence . . .'

136–1 Orbigny 1838:569, 'Il existe une inégalité étonnante entre le mélange des Espagnols avec telle ou telle race américaine. Avec les Guaranis, les Métis sont de belle taille, presque blancs; leurs traits sont beaux dès la première génération, tandis qu'avec les Quichuas, les traits américains sont plus tenaces et ne disparaissent qu'après plusieurs générations.'

136–2 Andrew Smith: personal communication.

The noise st st. which the C. Sphynx makes is also made by the C. porcarious., together with a grunting noise, the former signifies recognition with pleasure, as when food is offered, as much as to

139 say give me— the other when Dr. Smith more distant.— But he thinks other monkeys make st.— noise‖ In case of woman instinctive desire may be said to be more definite than with bitch, for some feeling must urge them to these actions. «These facts may, be turned to ridicule, or may be thought disgusting, but to philosophic naturalist pregnant with interest»

Hyæna. thinks, when pleased cocks his ears., when frighten depresses them.—

England was united to Continent, when elephants lived. & when present animals— lived.— we know the great time, necessary to form channel & (& Basses St) yet no change in English species— time no element in *making* change, only in *fixing* it: only circumstances. a contingency of time.[1]

140 When we multiply the effects of ‹earthquakes›, elevating forces in raising continents, & forming mountain-chains, when we estimate the matter removed by the waves of the sea, on beaches— we really, measure the rapidity of change of form, & instincts in the animal kingdom.— It is the unit of our calendar.— epochs & creations, reduce themselves to the revolutions of our system in the Heaverns.—

Is not *puma*, same colour as *Lion*. because inhabitant of *plain* & Jaguar of woods &c like ground birds

---

141e–146e [not located]

147e :Hence, also structure not really fitted for water, only habits & instincts—[1] The young of the ‹p› Kingfisher (.p. 169) has the colour on its back bright

141 'E' *on stub.*
147 :Hence . . . & instincts—] *crossed pencil.*
Sexual Selection] *added pencil, underlined.*
If masculine . . . Female alike] *added pencil.*
Good Ch 6] *added brown crayon, circled in pencil.*
Keep] *added pencil.*
Is it . . . habits] *crossed pencil.*

139–1 See D174, 'How completely *circumstances* «alone» make changes or species!!'.

147–1 See Waterton 1838:2166 (Notes on the habits of the kingfisher), 'Modern ornithologists have thought fit to remove the kingfisher from the land birds, and assign it a place amongst the water-fowl. To me the change appears a bad one; and I could wish to see it brought back again to the original situation in which our ancestors had placed it; for there seems to be nothing in its external formation which can warrant this arbitrary transposition. The plumage of the kingfisher is precisely that of the land bird, and, of course, some parts of the skin are bare of feathers; while the whole body is deprived of that thick coat of down so remarkable in those birds which are classed under the denomination of water-fowl. Its feet are not webbed; its breast-bone is formed like that of land birds; and its legs are ill calculated to enable it to walk into the water. Thus we see that it can neither swim with the duck, nor dive with the merganser, nor wade with the heron. Its act of immersion in the water is quite momentary, and

blue.—[2] «thus young of» Many of the pies assume the metallic tints, such as Magpie, Jay, & perhaps all the rollers—[3] «He says» whenever metallic brilliancy is present in Young birds, one may be sure cock & hen will be alike—[4] I presume converse is not true for he says Hen & cock Starling alike, yet young ones *brown*.—

Sexual Selection

If masculine character. added to species,. we can see why young & Female alike

Good Ch 6 Keep

═══

Is it Male that assumes change, & is the offspring brought back to earlier type by Mother?— do these differences indicate, species changing forms, ‹& loosing do› if so domestic animals ought to show them.—[5] Anyhow not connected with habits

148e According as child is like parent, so is *species* old: Hence ‹young› Kingfisher & pies, have long had their present plumage.— How is this in Pidgeons & fowls.— ???

———

Wate[r]ton «p. 197» put 12 wild duck's eggs under common ducks, the young crossed amongst themselves, & I presume with common ducks. so often, that it was impossible to say which was origin of any identical bird— for they were of all colours.— they were "half wild-half tame, they came to the windows to be fed, but still they have a wariness about them *quite remarkable*".[1] instance of old Species transmitting so much longer its Mental peculiarities. **Wildness Reversion Q**[2]

149e–150e [not located]

148 Mental peculiarities.] *full point on stub.*
Wate[r]ton . . . Mental peculiarities.] '23' *added brown crayon.*
Wate[r]ton . . . Mental peculiarities.] *crossed pencil.*
Wildness Reversion] *added pencil.*
Q] *added pencil.*
them quite . . . Mental peculiarities.] *double scored brown crayon.*

bears no similarity to the immersion of those water-fowl which can pursue their prey under the surface, and persevere for a certain length of time, till they lay hold of it. Still the mode of taking its food is similar to that of the gulls, which first see the fish, and then plunge into the deep to obtain it; but this bird differs from the gull in every other habit.'

147–2 Waterton 1838:169, 'There is not much difference in appearance betwixt the adult male and female kingfisher; and their young have the fine azure feathers on the back before they leave the nest.'

147–3 Waterton 1838:169, 'This early metallic brilliancy of plumage seems only to be found in birds of the pie tribe. It obtains in the magpie, the jay, and, most probably, in all the rollers.' See also D160 on metallic tints.

147–4 Waterton 1838:169, 'Wherever it [metallic brilliancy] is observed in the young birds, we may be certain that the adult male and female will be nearly alike in colour.'

147–5 Darwin is explicitly using domestication as a model for testing ideas on species formation. This is the pattern he followed through the rest of his career.

148–1 Waterton 1838:197–98.

148–2 Quoted in *Natural Selection*: 486 with other cases of reversion to 'wild' instinct in hybrids.

151e The present age is the one for large Cetacea, as the past for other Mammalia. & still further back reptiles & Cephalopoda:
Old Jones remarked to me,[1] that one of the children of Sir J. H. was so very like Sir W. whilst Sir J. himself is not like—[2] now this is clear case of avitism. but then ? was «not» the expression of ‹father› Sir W. itself received from his father so that case ceases to be true avitism[3]

Annals of. Natural. History. p. 135. Natural History of the Caspian. Fresh Water Fish!! ?adapted to salt water?— peculiar species, crabs & molluscs few.—[4] ?are not some same— what is the alliance with the Black sea.— it would be ocean, what is land to continent— Original Paper, worth studying. Archiv. fur. Naturgeschichte.

152e September 11' *Generation*
Mr Yarrell says it is well known that in breeding very pure South Down that the ewe must never be put to any other breed else all the lambs will deteriorate.—[1] Lord Moreton's Case.—

When cows have twins, ‹one› though capable of producing both ‹male› pair of male & female.— if there be one female, she will be free Marten. ‹Owen.› See Hunter's Owen—[2]
In the Athenæum Numbers 406, 407, 409, Quetelet papers are given, & I think facts there mentioned about proportion of sexes, at birth & causes.[3]

151 Cephalopoda:] ':' *emmended pencil.*
The present . . . avitism] *crossed pencil.*
Old Jones . . . avitism] *pencil.*
Annals . . . Naturgeschichte.] '19' *added brown crayon.*
152 **In the Athenæum . . . causes.**] *added brown ink.*
page crossed pencil.

151–1 Thomas Jones: personal communication. Perhaps the remark was made at the Athenaeum Club, where Jones had been a member since 1830 (Waugh 1888). See *Correspondence* 2:97, letter to Charles Lyell, 9 August [1838]. 'I met old Jones this evening at the Athenaeum. . .'
151–2 Possibly John Frederick William Herschel and his father William. As a founder of the Astronomical Society, Jones would have known the Herschels well.
151–3 See D180, 'Ask my father to look out for instances of Avitism'.
151–4 Eichwald 1839 (No. 8, October 1838):135, 'Most of the fish found in the Caspian are fresh-water fish; there are however several peculiar species from genera which hitherto have been observed in salt water only. . . The sea is very poor in Crustacea. . . It is also exceedingly poor in Mollusca compared with the Black Sea. . .'
152–1 William Yarrell: personal communication.
152–2 See Hunter 1837:38 (Account of the free-martin), 'It is a fact known, and I believe almost universally understood, that when a cow brings forth two calves, and one of them a bull-calf and the other to appearance a cow, that the cow-calf is unfit for propagation, but the bull-calf grows up into a very proper bull. Such a calf-cow is called in this country a FREE-MARTIN. . .' From the context, this reference to Owen's edition of Hunter's *Animal Oeconomy* may be based on a personal communication from Yarrell.
152–3 Quetelet 1835b. Since this reference to the *Athenaeum* review of Quetelet 1835a is in brown ink, it was written after 16 October 1838, very likely during the early Notebook E period, post E26. It was not until after the 28 September 1838 reading of Malthus (D134–35) that Darwin's new interest in populations led him to pursue Quetelet's work.

153 If an animal breeds young her growth is immediately checked— the *vis formativa* goes entirely to the offspring— this is clearly the converse of annual being rendered biennial—[1] the hardness of life in female Moth &c

---

Mr Y.[2] says that Macleay considers the house bug, as a ‹female which have› larvæ which have bred before the vis formativa had completed them— (but this argument is VERY WEAK without knowing whether if kept they would have wings.—).— Seep p. 84.[3] Hens «like»— Cocks from effect of breeding in & in.— Mr Yarrell does not know of any case of old Male. becoming like female, though many

154 of old female becoming like cocks.—[1] It is very singular. so many Gallinaceous birds have cock & hen plumage so different, yet the Cassowary & Guinea Fowl cannot be distinguished.—

A capon will sit upon eggs, as well as, & often better than a female.— this is full of interest; for it shows latent *instincts* even in *brain* of male.— Every animal surely is hermaphrodite—[2] (as is seen in ‹fe› plumage of hybrid birds)

155 After animal has copulated., though no offspring, Milk sometimes comes in Mammæ & even when bitch is in heat.— Yarrell believes Gestation is always some multiple of seven— if woman does not menstruate in the month, she will in 5 weeks.—

A Bull is never taken from his own field to bull a cow.— — a dog if led in string will not.— some of the tigers.— cat, though caterwhalling. & put into female, when muzzled, he is disabled.— so Elephant in confinement, & so *imagination* in Man, has strange effect.—

155 A Bull . . . strange effect.—] *brace left margin.*
to bull] 'b' *over* 'c'.
page crossed pencil.

153–1 See B2, D165, D176, and E184.
153–2 William Yarrell: personal communication.
153–3 Presumably D84.
154–1 See C215 and D114. Yarrell 1827, carrying on from Hunter's 'An account of an extraordinary pheasant', claims that the phenomenon is not restricted to *old* females. The Owen edition refers to Yarrell's paper (Hunter 1837:48).
154–2 See Hunter 1837:46–47 (An account of an extraordinary pheasant), 'To bring the foregoing observations into one point of view, I here beg leave to remark, that in animals just born, or very young, there are no peculiarities to distinguish one sex from the other, exclusive of what relates to the organs of generation, which can only be in those who have external parts; and that towards the age of maturity the discriminating changes before mentioned begin to appear; the male then losing that resemblance he had to the female in various secondary properties; but that in all animals which are not of any distinct sex, called hermaphrodites, there is no such alteration taking place in their form when they arrive at that age.'

156 *Directly* a Capon is cut, it increases in size *prodigiously*—
Animal Œconomy by, Hunter. (edited by Owen) p. 34.— Owen classifies Hermaphrodites.[1]
Cryptandrous. (only female organs visible). Oyster. cystic Entozoa. Echinoderms. Acalephes. Polyps. Sponges
Heautandrous, male organs formed to fecundate female (as in plants)[2]
Cirrhipeds rotifers, trematode & cestoid Entozoa
Allotriandrous.[3] ‹or M› Mollusca, with pectinibranchiate order—[4] the Annelida.
All others, ‹animals,› «are Diœecous as» Cephalopods, Pectinibranchiate molluscs.— insects. spider crabs.— (all these however do not require coition every generation)[5]— Epizoa & the nematoid Entozoa—
Therefore highness in scale has no «constant» relation to separtion of sexes, as may be

157 seen in Monoœcious & Diœcious plants.— NB. in Heautandrous animals ‹are› is there gradation of structure leading to supposition, that the Cryptandrous are really, Heautandrous.— How is fecundation effected in latter; are ‹it› «organs» open to water? Would not Ferns according to this doctrine be considered as really cryptandrous, & they have hybrids— this is most important support to my views— Seeing sexes separate in some of the lowest tribes, leads one to suppose still more that they must in effect be so in all.— 2 NB. In Pectinibr Mollusca.— «or Cephalopoda» are there abortive traces of other «sexual» organs; for if so, separtion of sexes very simple.—[1] as in plants even in same genus some diœcious & some monoœcious— (& cultivation might make one set of organs barren in one plant & not in other),

157 are ‹it› «organs»] 'are' *over* 'is'.

156–1 Hunter 1837:34–35 (Account of the free-martin). Owen's classification is given in a note on the following passage by Hunter: 'The natural hermaphrodite belongs to the inferior and more simple genera of animals, of which there is a much greater number than of the more perfect; and as animals become more complicated, have more parts, and each part is more confined to its particular use, separation of the two necessary powers for generation seems also to take place.' Owen's note begins: 'The animals in which the organs of the two sexes are naturally combined in the same individual are confined to the invertebrate division, and are most common in the molluscous and radiate classes. If the term hermaphrodite may be applied to those species which propagate without the concourse of the sexes, but in which no distinct male organ can be detected, as well as to those in which both male and female organs are present in the same body, then there may be distinguished three kinds of hermaphroditism.' Darwin's abstract of the rest of the note runs from 'Cryptandrous' to 'nematoid Entozoa'.

156–2 The information in parentheses is not in Owen's note.

156–3 Darwin leaves out Owen's definition (Hunter 1837:35): 'the male organs are so disposed as not to fecundate the ova of the same body, but where the concourse of two individuals is required, notwithstanding the co-existence in each of the organs of the two sexes.'

156–4 Owen (Hunter 1837:35) has 'with the exception of the pectibranchiate order'.

156–5 The information in parentheses is not in Owen's note.

157–1 See the complemental males case developed in *Living Cirripedia* 2:584–86.

Hunter ‹asks› p. 36 is thought by Owen to ask. whether a Heautandrous animal is ‹evidently› actually split in two— keeping sexes separate. Owen say such view worthy of a Lamarckian.—[2] Mine is much simpler.—

158 Hunter shows almost all animals subject to Hermaphroditism,—[1] those organs which perform nearly same function in both sexes.— are never double, only modified.[2] those which perform very different, are both present in every shade of perfection[3]—How came it nipples ‹are› «though» abortive, are so plain in Man, & yet no trace of abortive womb, or ovarium.— or testicles in female.—[4] the ‹add› presence of both testes & ovaria in Hermaphrodite,— but not of pœnis & clitoris,[5] shows to my mind.— , that both are present in every animals, but unequally developed.— surely analogy of molluscs.[6] & neuter bee would shew this—[7] (Do any male animals give suck)—But this not distinctly stated by Hunter.— Do testes, & ovaria when

157–2 Hunter 1837:36. Hunter's note: '*Quere: Is there ever, in the genera of animals that are natural hermaphrodites, a separation of the two parts forming distinct sexes? If there is, it may account for the distinction of sexes ever having happened*[c].' Owen's note [c]: 'The separation of the two sexual organs from one another in the same body occurs in many of that class of natural hermaphrodites which we have termed 'allotriandrous'; and there are many examples in the Hunterian collection showing the fact. What, therefore, Mr. Hunter seems here to refer to is a spontaneous fission of the body in the interval separating the two sexual parts, so that one portion of the body shall contain the male and the other the female organs. Some annellides, as the Naïs. exhibit the phenomenon of spontaneous fission, but the separation never occurs so as to divide the two sexual organs from one another, and appropriate one to each division; and were even such an occurence to be supposed ever to take place, the application of the fact to explain the occurrence of the distinct sexes in the naturally dioecious classes seems more worthy of a speculatist of the Lamarckian school than of a sober observer of Nature.'

158–1 Hunter 1837:35–36, 'The unnatural hermaphrodite, I believe, now and then occurs in every tribe of animals having distinct sexes, but is more common in some than in others; and is to be met with, in all its gradations, from the distinct sex to the most exact combination of male and female organs.'

158–2 Hunter 1837:36, 'There is one part common to both the male and female organs of generation in all animals which have the sexes distinct: in the one sex it is called the penis, in the other the clitoris; its specific use in both is to continue, by its sensibility, the action excited in coition till the paroxysm alters the sensation. . . .

'Though the unnatural hermaphrodite be a mixture of both sexes, and may possess the parts peculiar to each in perfection, yet it cannot possess in perfection that part which is common to both.'

158–3 Hunter 1837:35. Hunter's text: 'The unnatural hermaphrodite[b], I believe, now and then occurs in every tribe of animals having distinct sexes, but is more common in some than in others;*' Hunter's note *: '*Quere: Is there ever, in the genera of animals that are natural hermaphrodites, a separation of the two parts forming distinct sexes? If there is, it may account for the distinction of sexes ever having happened.*' Owen's note [b]: 'The unnatural hermaphrodites may be divided into those in which the parts peculiar to the two sexes are blended together in different proportions, and the whole body participates of a neutral character, tending towards the male and female as the respective organs predominate, and into those in which the male and female organs occupy respectively separate halves of the body, and impress on each lateral moiety the characteristics of the sex.' Darwin (see note D174–6) substitutes 'abortive hermaphrodite' for 'unnatural hermphrodite'.

158–4 See Hunter 1837:36–37, 'Although it may not be necessary, to constitute an hermaphrodite . . . so readily as the scrotum.' Darwin's comment reflects his reading, but not abstracting, this passage.

158–5 Hunter 1837:36–37, 'Although it may not be necessary, to constitute an hermaphrodite, that the parts peculiar to the one sex should be blended with those of the other, in the same way that the penis is with the clitoris, yet this sometimes takes place in parts whose use in the distinct sexes is somewhat similar, the testicle and ovarium sometimes forming one body, without the properties of either.'

158–6 Refers to D157 above, 'In Pectinibr Mollusca.— «or Cephalopoda» are there abortive traces of other «sexual» organs; for if so, separation of sexes very simple.—'

158–7 See Hunter 1837:453–54 (Observations on Bees), 'The queen and the working bees are so much alike that the latter would seem to be females on a different scale: . . .'

159e they first appear occupy their *proper* positions,— this would be argument for developement of either.—[1] (Mammæ or sheath of Horses pœnis reduced to extreme degree of abortion).— Insecta.— hermaphrodite, being not only dimidiate, but quarter-grown seems to show whole body imbued with possibility of becoming either sex.—[2] ꟷ In my theory I must allude to separtion of sexes as very great difficulty, then give speculation to show that it is not overwhelming.—[3]

Seeing in Gardens of Hybrids between Common & Silver Pheasant, one like cock & other like Hen.— one doubts whether they are not Hermaphrodites, like J. Hunters. Free Marten[4]

N.B. the common mule must often have been dissected[5]

160e Zoolog. Garden. Sept 16." Hybrid between Silver & Common Pheasant. Male bird, said to be infertile.— spurs rather smaller than in ‹ma› silver male— Head like silver except in not having tuft,— back like do.— but the black lines on each feather instead of coming to point [sketch] are more rounded. # & much broader., & ‹more ro› «three, I believe instead of» two lines.— «faintly edged with reddish brown» black marks on tail much ‹blacker› «broader.— » Breast red like Common pheasant.— lower part of breast, each feather is fine metallic green. ‹from› with tip & part of shaft metallic green.— This green doubtless is effect of Metallic hue of silver pheasant.[1] yet why green? & not purple?— legs pale coloured.— In the back feathers, we have character very different from either parent bird—

159 they first . . . overwhelming.—] *crossed pencil.*
Seeing . . . Marten] '17' *added brown crayon.*

159–1 Hunter 1837:36–37, 'Although it may not be necessary, to constitute an hermaphrodite, that the parts peculiar to one sex should be blended with those of the other, . . . yet this sometimes takes place in parts whose use in the distinct sexes is somewhat similar, the testicle and ovarium sometimes forming one body, without the properties of either. This compounded [abortive] part in those animals that have the testicle and ovarium differently situated is generally found in the place allotted for the ovarium; but in such animals as have the testicle and ovarium in the same situation, as the bird tribe, the compound of the two, when it occurs, will also be found in that common situation.'

159–2 Hunter 1837:35, continuation of Owen's note [b] cited above in note D158–3, 'This latter and very singular kind of hermaphroditism has hitherto been found only in insects and crustaceans.' Owen cites reports of 'dimidiate hermaphrodites' by Alexander MacLeay. He also cites Westwood's report of both dimidiate and 'quartered hermaphrodites': 'a specimen of the stag-beetle (*Lucanus Cervus*), in which the left jaw and right elytrum are masculine, and the right jaw and left elytrum feminine.'

159–3 In his published work, Darwin was frequently to adopt this argumentative strategy. See *Origin*, chap. 6, (Difficulties on the theory). Darwin's theory of separation of sexes is developed in *Natural Selection*, chap. 3 (which was not included in the *Origin*) and later in *Crossing*.

159–4 This analogy to abortive hermaphrodites is reflected in Darwin's 'Theory of abortive hybrids' (D15).

159–5 Darwin is becoming interested in conducting anatomical dissections and hybridisation experiments of his own to test his views on generation, including those on the separation of sexes. See D165 ('splitting animal experiment' and the contemporaneous list of queries and experiments in D180 and DIBC, and possibly the E15 strawberry experiment).

160–1 See note D147–2 on metallic tints.

161 Hunters Animal Œconomy. (by Owen) p. 44.
Classification of Monsters. (1) From præternatural situation of parts (2) addition of parts, (3) deficiency of parts (4) combined addition & deficiency of parts, as in Hermaphrodites,[1] (shows my doctrine of Hermaphrodite differs from Hunter)[2]— Hunter (p. 45) observes "every species has a disposition to deviate from Nature in a manner peculiar to itself"[3] ‹Is this so› Each part ‹not› of each species not similarly subject— ǂDivides sexual marks into primary & secondary, the latter only being developed, when the first ‹are› become of useǂ ‹Great characteristic of male greater strength, (p 45) & that strength›[4]
In speaking of generation alway put female first[5]

———

Will not even a fruit tree or rose degenerate during its life so that successive buds do differ— any variety is not handed down. but is handed down for some generations

162 Theory of sexes (woman makes, bud, man puts primordial vivifying principle) one individual secretes two substances, although organs for the double purpose are not distinguished. —yet may be presumed from hybridity of ferns)[1] afterwards they can be seen distinct. (in diœcious plants are there abortive sexual organs?): they then become so related to each other, as never to be able to impregnate themselves (this never happens in plants[2] «only in subordinate manner in the plants which have male & female flower on same stem.—» so that Molluscous hermaphroditism takes place.—[3] thus one organ in each becomes obliterated, & sexes as in Vertebrate tak place.— ∴ Every man & woman is hermaphrodite:— ∴ developed instincts of Capon. & power of assuming male plumage in females.,[4] & female plumage in castrated male.— «Men giving milk—»

161 Will not . . . some generations] *triple scored both margins.*

161–1 Hunter 1837:44. See Owen's note [a] (note D115–1), which quotes Hunter MSS.
161–2 According to Darwin's doctrine the ancestors of both plants and animals were hermaphroditic, and the various grades of separate sexes arose by abortion (deficiency) of parts.
161–3 Hunter 1837:45.
161–4 Hunter 1837:45, 'The varieties which are found in the parts of generation themselves I shall call the first or principal marks, being originally formed in them, and belonging equally to both sexes; all others depending upon these I shall call secondary, as not taking place till the first are becoming of use, and being principally, although not entirely, in the male.
'One of the most general marks is, the superior strength of make in the male; and another circumstance, perhaps equally so, is this strength being directed to one part more than another, which part is that most immediately employed in fighting.' See D113.
161–5 The female is to be put first because she tends to show only the primary sexual marks and is thus most like the species' ancestor.
162–1 See Martens 1838 cited in B235.
162–2 Much of Darwin's later work in floral anatomy was devoted to elucidating the complex mechanisms that prevent self-fertilization in plants. Here he is expressing the common view of the time that monoecious flowers are self-fertilizers. Hence, this passage marks a critical point: he has developed a theory for the separation of sexes, but does not recognise its implications for flowering plants. See Henslow 1837.
162–3 See D156–57. Mollusca are dioeceous.
162–4 See D96 and D114 on old females assuming male plumage.

163 Sept. 25th Young man at Willis «Grt. Marlborough Str, Hair dresser, assures me he has known many cases of bitch going to mongrel, & all subsequent litters having a throw of this mongrel.— I did not ask the question.— His bitch will not take «& if she did take, probably would not be fertile» without she know & LIKES HIM & then is actually obliged to be held.— like she wolf of Hunter.—[1] young take distemper very readily & are subject to fits.— «there is great difference between hybrids & inter se offspring in latter being unhealthy.—» *males* «bred in & in» *never lose passion.*
(Mem: so it was said little cock «yet very odd loosing visible powers» in Zoolog Gardens. & Kings at Otaheite[2]) ‹Think› Last litters are considered the most valuable. because smallest sized dogs.— one litter big & then second small & so.— Says, there is breed of Fowls called everlasting layer— . or Polish breed. (he thinks

164 half pheasant, half fowls.— eggs fertile, but parent bird will never sit on them.—

May be just worth remembering that ovarium of women (Paper in Vol I of Irish Royal Academy) have contained perfect *teeth* & hair—[1] showing fœtus has *gone on growing*— I believe same has happened in boys bodies.—

Lavaters. Essays on Phy. transl by Holcroft Vol I. p. 195. says children resemble parents in their bodies "It is a fact equally well known, that we observe in the temper, *especially of the youngest children*, a striking ‹resemblance› similarity to the temper of the

165 father, or of the mother, or sometimes of both."[1] If L. can be trusted, this is Lord. Moretons law.— "How often do we find in the son, the *character*, constitution, & most of the moral qualities of the father!. In how many daughters does the character of the mother revive! Or the character of the mother in the son, & of the father in the daughters![2] This last remark good. because showing probably not education.—

---

165 Cannot I . . . Pippen trees!] *brace left margin.*

163–1 Hunter 1837:323 (Observations tending to show that the wolf, jackal, and dog, are all of the same species), '. . . Mr. Gough having an idea of obtaining a breed from wild animals, as monkies, leopards, &c., he was desirous to have the wolf lined by some dog; but she would not allow any dog to come near her, probably from being always chained, and not accustomed to be with dogs. She was held, however, while a greyhound dog lined her, and they fastened together exactly like the dog and bitch.'
163–2 See B196. 'Tahitian kings. would hardly produce from Incestuous intercourse.', which contradicts the 'Young man at Willis'.
164–1 See Cleghorn 1787.

165–1 Lavater 1804, 1:195, 'It is likewise a fact universally acknowledged, that new born children, as well as those of riper growth greatly resemble their father or mother, or sometimes both, as well in the formation of the body as in particular features. . . . It is a fact, equally well known, that we observe, in the temper, especially of the youngest children, a striking similarity to the temper of the father, or of the mother, or sometimes both.'
165–2 Lavater 1804, 1:195, 'How often do we find in the son the character, constitution, and most of the moral qualities of the father! In how many a daughter does the character of the mother revive! Or the character of the mother in the son, and of the father in the daughter!'

Cannot I find some animal with definite *life* & split it, & see whether it retains same length of life— like Golden Pippen trees! How is this with buds of plants, does *annual* give buds.— life may be thus prolonged bud being formed & one part dying for great length of time.—

166 There is probably law of nature that any organ. which is not used is absorbed.— this law acting against heredetary tendency causes abortive organs.— the origin of this law is part of the reproductive system.— of that knowledge of the part, of what is good for the whole.—[1] if cut off nerves in snail. (Encyclop of Anat & Phys) can make a head;[2] the other part may surely absorb a useless member.— in fact they do it in disease & injury.— The *sympathy* of part is probably part of same general law, which makes two animals out of one

167 & heals piece of skin.— if the tail knows how to make a head. & head & tail, & the belly both head & tail,—no wonder there should be *sympathy* in human frame.—

«one of» The final cause of sexes to obliterate differences. final cause of this because the great changes of nature are slow. if animals became adapted to every minute change, they would not be fitted to the slow great changes really in progress.—[1]

Annals of Natural History. 1838. p. 123. Ehrenberg. makes gemmation in animals very different from that of plants. (though latter does sometimes occur in animals). latter the division taking place from outside inwards. & in animals from inside to the outside.—[2] is this not owing simply to more importance of internal organs in animals‖. One «invisible» animalcule in four days could form 2. cubic stone. like that of Billin.—[3]

166–1 See the parallel comment in D159 'whole body imbued with the possibility of becoming either sex'.

166–2 Jones 1836–39, 2:402, 'If . . . the head is cut quite off, a new one will succeed: the new head however, does not at first contain all the parts of the old one, but they are gradually developed. . .'

167–1 See note B1–5.

167–2 Ehrenberg 1839 (No. 8, October 1838), 2:123, 'A vegetable cell apparently capable of self division always become one, or contemporaneously many exterior warts (gems) without any change in its interior. An animal which is capable of division first doubles the inner organs, and subsequently decreases exteriorly in size. Self division proceeds from the interior towards the exterior, from the centre to the periphery; gemmation, which also occurs in animals, proceeds from the exterior towards the interior, and forms first a wart, which then gradually becomes organized.' Darwin added this passage and those in notes D167–3 and D169–1 after reading the October issue of *Annals of Natural History*.

167–3 Ehrenberg 1839 (No. 8, October 1838), 2:123, '. . . an imperceptible corpuscle can become in four days 170 billions, or as many single individual animalcules as contained in 2 cubic feet of the stone from the polishing slate of Bilin.' This passage, added after the 28 September reading of Malthus (D134–135), reflects Darwin's new attention to superfecundity.

168 ‹Generation—› V. p. 152

It is very singular the same difference from parental stock having been repeated several times, that it becomes fixed in blood.— Looking at ovum of mother & ovum in offspring, as similar to the several ova in mother. (with only difference of time) is the above law anyways connected with the case of successive copulation impresses offspring more & more with the added «like Lord Moretons case & D^{r}. Andrew Smith,» difference.—[1] If A. B. C. D. E be ‹offspring› «animals»: if «x» male impresses ovum ‹of› in A, «with some peculiarity» that in (B) to ‹a slight› «some» degree, & likewise ovum in (B) ‹an C› that in (C) «in lesser degree» — Then when (C) unites with Male (X) «assume that every peculiarity has a tendency to descend to several generations»

169 If A & B be two animals which have some peculiarity for first time & if their ‹D & E› «all their offspring» inherit the same peculiarity in lesser degree C. & theirs again in lesser degree—now if the ‹tw› second race both have this peculiarity strongly; they transmit with same force as first pair, but to this tendency is added ‹that› the 3^{d} tendency from first pair.— Now if two of third pair of same peculiarity breed they will have same influence as first pair + tendency they inherited from second pair, + the influence they themselves inherit./[1]

Annals of Natural History .p. 96. Vol I. Notice the Syngnathus, or Pipe fish the male of which receives ‹young› «eggs» in belly.—[2] analogous to men having mammæ.—

170 There is an analogy between caterpillars with respect to moths, & monkey & men.— each man passess through its caterpillar state. the monkey represents this state.—

---

168 If A. B. C. D. E . . . Male (X)] *heavily crossed.*
to ‹a slight›] ‘to’ *over* ‘in’.
169 degree C] ‘C’ *over* ‘G’.
Vol I] ‘I’ *over* ‘II’.

168–1 See D152: ‘Mr Yarrell says it is well known that in breeding very pure South Down that the ewe must never be put to any other breed else all the lambs will deteriorate.— Lord Moreton’s Case.—’

169–1 The thrust of Darwin’s argument is to show how the proven facts of generation (superfoetation, male impression, and crossing) cumulatively give the perpetuation of hereditary change required for the formation of new species.

169–2 This passage was added in the considerable space Darwin left when he first filled D168–169. Fries 1839a (No. 8, October 1838), 2:96–97, ‘The discovery of the remarkable peculiarity existing in the sexes, by which the males are not only destined as protectors of the eggs and of the birth, but are also for this purpose endowed with a peculiar organ in which the eggs are deposited, developed, and hatched, and in which the young in their tender state find a sure protection, has obtained for this genus of late a greater attention than would else have probably been the case.’ Passage scored and ‘deposited, developed, and hatched’ underlined. See also E57 and T25–26.

When it is said. that difference between bud & seed, that latter carries with stock of food.— the generalization begins low.— it goes through transformation, nearly independently of its parent & therefore wants independent supply of food.— is real. difference— but this does not apply to potato.—

171 With respect to offspring being determined by imagination of Mother.— We see in a litter every possible variation from being very near mother, & some very near father.— now if one of these staid in the womb, when it came out. it might partake of shade of fathers character.— according to this view more semen to one child. more like father.— stuff.!—[1]

172 How much opposed. the Quagga case appears to that of «2» dog begetting different puppies out of same mother.—

The view that man & «or cock» pheasant &c is abortive hermaphrodite is supported by change which takes place in old age of female assuming plumage of cock, & beards growing on old women = Stags horns & testes curious instances of corelation in structure = Neuter bee having both sexes abortive fact of same tendency. — Mammæ in man. having given milk. — testis & ovaria

---

The following views show that transmission of mutilation impossible.— it should be observed that transmission bears no relation to *utility* of change— hence *harelips* heredetary, *disease*. extinction.

Animals in domestication (even Elephant) not breeding— remarkable

---

Athenæum 1838. p 653. Ehrenberg‹h› thinks multiplication by division ‹only› is developement of gemma.—[1]

173e The manner in which Frogs copulate & fish shows how simply instinctive the feeling of other sex being present is— it also shows that semen. must actually reach the ovum.— [Why in making a bud, which is to pass through all transformations, should there need two organs; whilst in common bud there is no such need.— one would ‹one› suppose that the vital portion ⁽nerves? passed through transformation, & was received into bud matured by

172 both] 'b' *over* 'n'.
173 The manner . . . complicated animals.][CD] *crossed pencil.*
p. 310 . . . Eœconomy] '17' *added brown crayon.*
So with . . . by Willis] *added pencil.*
Q] *added brown ink in circle.*

171–1 See Morton 1821, which Darwin first cited at B181.
172–1 Ehrenberg 1838b:653, '. . . he believed the process of multiplication by division to be merely the developement of a gemma or bud, a view which would much simplify our ideas concerning that singular mode of propagation.'

female;‹][CD]› such view no ways explains Ld. Moretons case: without the nervous matter consists of infinite number of globules: generally sufficient for one birth or rather][CD] ǂIt should be observed that the constant necessity for change. in process by generation applies only the more complicated animals.][CD] p. 310 She wolf took dog. but had such aversion to it, that she was held Hunters Eœconomy[1] **So with inter-breeding as told by Willis Q**

176 Proved facts relating to Generation[1]

One copulation may impregnate one or many offspring.—[2] it affects the subsequent offspring, ‹when› though other male may have copulated.— two animals may unite & each have offspring by same mother.— one animal will fecundate female for several births, & even produce fertile offspring— DESIRE LOST when male & female too closely related:[3] this most important with regard to theory, showing generation connected with whole system, «as if there was, a superabundance of life, like tendency to budding, which wishes to throw itself off.—» as might be inferred from annual plant being prolonged till it has bred.— Offspring like both father & mother, or very close to either.— Male & female as fœtus one sex; & therefore both capable of propagating, but one is rendered abortive

177 as far as parturition is concerned.—[1] Generation being means to propagate & perpetuate differences (of body, mind & constitution)[2] is the end frustrated, when near relations, & therefore those very close are bred into each other.— **This is somehow connected** (This seems case, for by careful observing cattle can be bred in & in.[3])— [The loss of *passion* in hybrids. perhaps connected with this same case (& not merely as I have stated it) it is certainly very remarkable that too much difference should produce same effect as too little.— in (latter case female often takes males but does not produce) tendency to deformity ʕthis does not happen with hybrids?][CD] Plants must stand much breeding in & in[4] (those which have solitary flower) exotics

177 **This is somehow connected**] *added.*
ʕthis does] 't' *over* 'ʕ'.

173–1 Hunter 1837:310. See *Natural Selection*: 427, n. 1.
176–1 D174–175 have been placed after D179 to preserve the writing order of Darwin's attempt to synthesize his views on generation. In the original the pages are numbered sequentially.
176–2 Note the mention of aphids in D175. In B181 Darwin made the analogy between parthenogenesis and telegony explicit: 'See paper . . . by Ld Moreton, where mare was influenced in this cross to after births, like aphides.—' (Morton 1821).
176–3 It is notable that in Notebook D, which spans the major part of Darwin's courtship of his cousin Emma Wedgwood, Darwin pays particular attention to evidence that inbreeding leads to loss of sexual desire.
177–1 See D174 'abortive hermaphrodite'.
177–2 As in D174 Darwin's language recurs to the opening pages of Notebook B.
177–3 See discussion of the Chillingham cattle in D48.
177–4 Darwin is operating under the same assumption as in D162 that hermaphrodite plants self fertilise. His comment that 'Plants must stand much breeding in & in' suggests he recognises that hermaphroditic flowering plants pose a difficulty to his general theory of sexes. This language is transformed into the view that although organisms can stand much inbreeding, there must be occasional outcrossing (see note D175–1 below).

brought from foreign country.[5] (‹annuals› & so must those forms which are produced by budding «only» as cryptogamia & hydras,— (this repugnance to breeding in & in seems connected with more developed *forms*[6]) Study buds— gemmæ— & monocotyledenous, do those which are Monocotyledenous have many flowers same Spath, as they have only one *bud*.—[7]

178 Every individual fœtus would reproduce its kind was it not for the necessity of some change.— ‖ Without some small change in form. ideosyncrasy or dispositions were added or substracted at *each*, or in *several* generations, the process would be similar to *budding*. which is not object of generation.— therefore passions fail.— In fruit trees no doubt there is tendency to propagate the whole difference of parent, tree, but it fails.— therefore «each» seedling of one apple ought to differ from those of other.— The upshot of all this is that effect of Male is to impress some difference: to make the *bud* of the woman, not a bud in every *respect*.—[1] [Is this connected with the physical differences in almost all Male animals?]CD If the male «in the course of some generations» has gained some difference «from what it received» (for it is probable that breeding in & in would not be deletereous if the relative had come from different quarters) then it causes ‹to› a secretion of something someways different from himself, for it should be observed that from

179 Books to read
Buffon Suites de.—[1]
Horse & Cattle[2] Library of Useful Knowledge
Bell's Quadrupeds[3]

---

the effects of breeding in, it is not merely the too close animals, which will not breed, but the female at least (♀male?) looses all appetite.— It is the comparison of each animal with its‹elf› «ancestors», & not its comparison «of difference» with other sex. = The highest bred Blood-hound. would be infertile with highest bred of other ♀ breed.= Therefore it is not really breeding in & in, but « ‹on› » breeding animals that have neither varied from their stock, for to breed (as Sir J. Sebright urges?[4]) one with opposed characters is by impliance to breed two which have each varied from parent stock.— The very theory of generation being the passing through whole series of forms to acquire differences: if none are added object failed, & then

179 Books . . . Quadrupeds] *crossed*.

177–5 Their small initial populations would require close inbreeding.
177–6 See B34.
177–7 Thus they would be all alike and equivalent to solitary flowers.
178–1 See Eras Darwin 1794–96, 1:519 and note B1–2.
179–1 See note C1BC–4.
179–2 Youatt 1831, 1834.
179–3 T. Bell 1837.
179–4 Sebright 1809, see C133.

by that corelation of structure desire fails. Every individual except by incestuous marriage has acquired from father some differences. V. Supra

174e ‹v. infra› p 179, continued from
Is a flower bud produced by union of two common buds??? Amongst buds each one exactly like its parent. ‹but these buds do not procreate› & all alike in one parent or tree, (but not in other trees.—[1] — Why should there be a necessity that there should be something «each time» added to that kind of generation, which typifies the whole course of *change* from simplest form.— (Because by this process it separates those differences which are in harmony with all its previous changes, which mutilations are not[2]). but why should it demand some further change?— Man properly is hermaphrodite (hence monstrosities tend that way[3] «& from frequency of this tendency all mammalia must long have so existed.» with double union.—
At present I can only say the whole object being to acquire differences «indifferently of what kind, either progressive improvement or deteriorate[4]» that object failing, generation fails.— How completely *circumstances* «alone» make changes or species!![5] CD[The view of ‹In› each Man or mammalia being abortive hermaphrodite simplifys case much;[6] & originally ‹her› each hermaphrodite being simple (Are not Coniferous trees generally diœcious oldest forms)

175 why are twin in man more like «each other» than twins «or triplets &c &c in» in litter. Why is there some law about sexes of twins in former case.— (many monster are really twins.)—

It is absolutely necessary that some «but not great» difference (for even brother & sister are somewhat different) should be added to each individual before he can procreate.[1] these changes may be effect of differences of parents, or external circumstances during life.— if the circumstances which induce «which must be external» change are always of one nature species is formed if not.— the changes oscillate backwards & forwards & are individual differences (hence every individual is different). (All this agrees well with my view of those «forms» slightly favoured, getting the upper hand. ‹]CD› &

174 *page crossed pencil.*
175 It is . . . external circumstances] *double scored.*
circumstances which] *carat deleted before* 'which'.

174–1 See D177: 'do those which are moncotyledenous have many flowers same Spath, as they have only one *bud.—*'
174–2 This question was Darwin's point of departure in the opening 'Zoonomical' pages of Notebook B, see especially B2–4.
174–3 Darwin is applying the reasoning of Hunter's law. See notes D57–1, D112–1, and Hunter 1837:44–45.
174–4 Note the independence of variation from utility.
174–5 Note that Darwin emphasises the relativity of adaptation in this passage.
174–6 See D15 and notes D158–3, D159–4.
175–1 This is the first formal statement of what is known in the botanical literature as the Knight-Darwin law (Goebel 1909:421). See notes D162–2 and D177–4. Darwin's views clearly stem from his theory of separate sexes, which derives from his analysis of Owen's notes on Hunter 1837:34–35 in D156–159. It is rooted in B4–5.

forming species)—[2] [Aphides having *fertile* offspring without coition or addition of differences. shows that difference need not be added EACH TIME. but after some time[3]]CD

What kind of plants are Monoœcious or diœcious.— very curious how this was superinduced? (Surely all are really *diœcious*..) only simple form of life are monoœcious.

180 Will ova of fishes & Mollusca «& Frogs» pass through birds stomachs & live?[1]

In Muscovy ducks do young take most after father or Mother according as they are crossed? & How is it with China & Common Geese «how are their instincts?» «Chineses & Common Pigs.—»

Experimentize on crossing of the several species of wild fowl ‹in Z› of India «with our common ones» in Zoolog. Gardens;

‹Buffalo & common cattle— Esquimaux (& Australian) dogs with common dogs—›

Ask my father to look out for instances of Avitism

Examine English weeds in Hot. Houses will they flower

Make Hybrids with moths, where fecundation can be made artificially.—

---

180 wild fowl] *carat before* 'fowl'.
Experimentize . . . Gardens;] *crossed.*
Ask my . . . avitism] *pencil.*

175–2 The selectionist tone is striking, which may date Darwin's essay 'Proved facts relating to Generation' (D176–179, D174–175) to after the D134–135 reading of Malthus. This would be consistent with viewing the independence of variation and adaptive ability noted in D174 as first fruits of Darwin's new theory of natural selection. If so the passage could have been written as late as 16 October 1838. The passages referred to in notes D177–2 and D174–2 are so strongly reminiscent of early Notebook B that they may reflect a rereading of his 'Zoonomia arguements'. In which case, 'my view of those «forms» slightly favoured, getting the upper hand . . . & forming species' in D175 has to be compared with 'variety of ostrich, Petise may not be well adapted, & thus perish out, or on other hand like Orpheus. being favourable many might be produced.—' of B37–38.

175–3 See D178 'small change . . . added or subtracted at *each*, or in *several* generations'.

180–1 D180 and DIBC comprise a list of queries and ideas for experiments and observations for Darwin and others to pursue. This list was begun, and from the uniform hand may have been completed, before Darwin reached D179. Lack of space forced him to go back to a blank page before D176 to continue 'Proved facts relating to Generation'. He chose D174 and D175, which may have been the only pages left in *Notebook D*. See *Questions & Experiments* and Darwin's breeding questionaires where this method is extensively developed: *Correspondence* 2:446–49, Appendix v (printed questionnaire); 2:187–89 (Ford replies); 2:190–92 (Tollet replies). See also 2:70–71 (Wynne questions); 2:179–81, 201–2 (Herbert questions); 2:182–84, 202–3 (Herbert replies).

INSIDE BACK COVER

Are hybrids pintail & common ducks. similar inter se? Zoolog. Gardens

---

Are the hybrids of those species. which cross & are fertile heterogenous?

When bird fanciers say the throw of two varieties is uncertain do they mean they cannot tell first result., or that «hybrid» breed is uncertain

---

Is there any peculiarity or variation common to any zoophyte «born in succession» which is not transmitted by generation??

Is it «chiefly» in high. bred dogs ie. (bred in & in) that one copulation with other dogs renders subsequent progeny faulty. Does male fail in passion.—

Disposition of half bred Cattle at Cinbermere? How is Jackall & dog at Z. Gardens

BACK COVER

D

IBC Are . . . Gardens] *crossed.*
When bird . . . uncertain] *crossed.*
How is Jackall . . . Gardens] *crossed.*

Notebook E

Transcribed and edited by DAVID KOHN

Notebook E (DAR 124, 170 × 99 mm), fourth in the transmutation series, is bound in rust-brown leather. It is similar to Notebook N, with plain covers, a broken metal clasp, and cream-coloured labels with clipped corners on each cover. E is written on the labels in ink. The notebook comprises 184 pages numbered consecutively by Darwin in grey ink. Pages E1 to E24 are written in grey ink, the remainder in brown ink. Of 39 leaves excised by Darwin, 28 have been located.

Darwin assigned no opening date for Notebook E unless he did so on the unlocated excised pages E1–2. However, E3 continues the references to Malthus 1826 begun on 28 September 1838 in D134 and E4 is dated 'Octob. 4th.' So the 'October 2d. . 1838' date on N1 provides a good nominal date for the beginning of Notebook E. The last dated page is 'June 26th' on E174. On the inside front cover, in brown ink that is indistinguishable from the last pages of the notebook, Darwin wrote 'Finished July 10th 1839.—' Hence this may well be Darwin's direct date for the completion of Notebook E. As with Notebook D, Darwin's datable citations confirm his direct dates.

Notebook E opens with the continuation of Darwin's reading notes on Malthus (E3). Darwin's commitment to 'my Malthusian views' (E136) is affirmed and refined in isolated passages throughout the notebook. In E58 he summarizes his theory in three principles:

'(1) Grandchildren. like. grandfathers
(2) Tendency to small change. .
(3) Great fertility in proportion to support of parents'

But this summary is not expanded into an essay. Rather Darwin explores some of the implications of his new mechanism and its compatibility with his prior assumptions. Four leading issues emerge: the relationship between variation and adaptation, the rate and pace of transformist change, the separation of sexes, and the analogy between selection in nature and under domestication.

In the course of Notebook E Darwin gauges the strength of the Malthusian force. At first he stresses its power: 'No structure will last. without it is adaptation to *whole* life . . . it will decrease & be driven outwards in the grand crush of population' (E9). Later he emphasizes its subtlety: 'there is a contest. & a grain of sand turns the balance' (E115). The fulcrum for selective wedging is small hereditary variation: 'if a seed were produced with infinitesimal advantage it would have better chance of being propagated' (E137). Adaptation is relative: 'chance & unfavourable conditions to parent may be become favourable to offspring' (E26). The sources of variation are complex but independent of adaptive need: 'Whether the . . . Nisus formativus . . . succeeds in altering . . . form of body, or whether it merely has tendency . . . to do so, effects are equally handed to offspring' (E127). Adaptation is contingent and the potential pool of variability is large: 'Every structure is capable of innumerable variations, as long as each shall be *perfectly* adapted to circumstances of times' (E57).

In Notebook B Darwin defines gradual geological change as the context in which adaptive transmutation takes place. In Notebook C he stresses the balance between adaptation and heredity. In Notebook E both concepts are reaffirmed: 'It is curious that geology. by giving proper ideas of these subjects. should be *absolutely* necessary to arrive at right conclusion about species' (E5). However, there is an important clarification of emphasis. Small geological changes accumulating gradually may not immediately alter a species' circumstances and its adaptation. But these effects multiply and in Lyell's terms 'give rise, during a very brief period, to important revolutions.'[1] Selection operates so that organisms are adapted to local circumstances; hence species may remain stable over long periods of time. Having recognized the Malthusian force as a principle of change, he also stresses its contribution to stability, indeed 'if the change could be shown to be more rapid, I should say there was some link in our train of geological reasoning, extremely faulty' (E5).

Much attention is devoted to the problem of the separation of sexes. His conclusion that the 'formation of sexes is rigidly necessary' (E49) has roots in early Notebook B and derives from his work in Notebook D on the broader subject of generation, but it is crystallized in Notebook E as a major deduction from his new theory, with its stress on populations: 'it was absolutely necessary that Physical changes should act not on individuals, but on masses of individuals . . . this could only be effected by sexes' (E50). As always with Darwin's secondary laws, the ramifications are wide ranging. The necessity of separate sexes determines the formation of social instincts and this Darwin hopes 'to show is the foundation of all that is most beautiful in the moral sentiments' (E49).

In Notebook E Darwin clarifies the analogy between the formation of species in nature and under domestication.[2] In Notebooks C and D he recognizes that domestic varieties produced by man's art are the analogues of species produced by nature, but he vigorously rejected the idea that they are formed by analogous means: 'One can perceive that Natural varieties or species, . . . more conformable to the structure which has been adapted to former changes, than a mere monstrosity propagated by art' (D107). They are analogous effects *not* produced by analogous causes. In Notebook E he reverses his position: 'It is a beautiful part of my theory, that «domesticated» races. of ‹a› organics are made by precisely same means as species—but latter far more perfectly & infinitely slower.' (E71). Here he constructs an analogy that flows, as did the development of his thinking, from nature to art and his former objection is subsumed as a qualification. Finally, in E118 he inverts the analogy and sketches the strategy of exposition in the *Origin*: 'greyhound. & poutter Pidgeons «race-horse» . . . produced by crossing & keeping breed pure.— «& so in plants *effectually* the offspring are picked & not allowed to cross.—» Has nature any process analogous . . . Then give my theory.— excellently true theory'.

Like the notebooks that preceded it, Notebook E is open ended, concluding with questions for further speculation and opening new lines of empirical observation. The new 'excellently true theory' is consolidated but does not culminate in a full exposition. That would only come with the *1842 Sketch* and the *1844 Essay*.

[1] Lyell 1837, 3:124.
[2] Limoges 1970, Herbert 1971, Kohn 1980. For an alternative interpretation see Ruse 1975 *J. Hist. Ideas* 36:339.

FRONT COVER

E

INSIDE FRONT COVER **Finished July 10th 1839.—**
Selected. Dec 15 1856

1e–2e [not located]

3 Epidemics— seem intimately related to famines., yet very inexplicable.—[1] do p. 529. "It accords with the most *liberal*! spirit of philosophy to believe that no stone can fall, or plant rise, without the immediate agency of the deity. But we know from *experience*! that these operations of what we call nature, have been conducted *almost*! invariably according to fixed laws: And since the world began, the causes of population & depopulation have been probably as constant as any of the laws of nature with which we are acquainted."—[2] this applies to one species— I would apply it not only to population & depopulation, but extermination & production of new forms.— their number & corelations

4 Octob. 4th. It cannot be objected to my theory, that the amount of change within historical times has been small—[1] because change in forms is ‹al› solely adaptation of whole of one race to some change of circumstances; now we know how slowly & insensibly such changes are in progress.— we feel interest in discovering a change of level of a few feet during last two thousand

IFC **Finished . . . 1839.—**] *added ink.*
Selected . . . 1856] *added pencil.*
VARIATIONS IN MEDIA
1–19 grey ink.
20–23 up to 'God.—' *brown ink.*
Remainder of 23–24 grey ink.
25 brown ink.
26 grey ink.
Remainder of notebook brown ink.
Exceptions: pencil on portions of 145–152, 155–156, and 183 as indicated in textual notes.
Pages numbered throughout in grey ink.
4 geological] 'ical' *over* 'y'.

3–1 See Malthus 1826, 1:512, 'In contemplating the plagues and sickly seasons which occur in these tables after a period of rapid increase, it is impossible not to be impressed with the idea, that the number of inhabitants had in these instances exceeded the food and the accommodations necessary to preserve them in health. The mass of the people would, upon this supposition, be obliged to live worse, and a greater number of them would be crowded together in one house; and these natural causes would evidently contribute to produce sickness, even though the country, absolutely considered, might not be crowded and populous.'

3–2 Malthus 1826, 1:529. Instead of 'deity' Malthus wrote 'divine power'. The underlinings and exclamation marks are Darwin's. The passage begins: 'In New Jersey the proportion of births to deaths, on an average of 7 years, ending with 1743, was 300 to 100. In France and England the average proportion cannot be reckoned at more than 120 to 100. Great and astonishing as this difference is, we ought not to be so wonder-struck at it, as to attribute it to the miraculous interposition of Heaven. The causes of it are not remote, latent and mysterious, but near us, round about us, and open to the investigation of every inquiring mind.'

4–1 See Darwin's prior discussions of the Egyptian mummified animals: B16, C219, D39, and D106.

years in Italy,[2] but what «changes» would such a change produce in climate vegetation &c.— It is the circumstance of small physical changes & oscillations, not affecting organic forms, that the whole value of the geological chronology depends, that most sublime discovery of the genius of man

5e Those who have studied history of the world most closely, & know the amounts of change now in progress, will be the last to object to this theory on the score of small change.— on the contrary islands separated with some animals, &c.— «if the change could be shown to be more rapid, I should say there was some link in our train of geological reasoning, extremely faulty»

———

The difficulty of multiplying effects & to ‹ponder› conceive the result with that clearness of conviction, absolutely necessary[1] as the «basal» foundation stone of further inductive reasoning is immense.

———

It is curious that geology. by giving proper ideas of these subjects. should be *absolutely* necessary to arrive at right conclusion about species

---

Changes of level &c are easily recorded, but changes of species *«are» not* as «without every animal preserved.».— the latter pages in the history are perfect,

6e we obtain a glimpse only of the changes which the government is subject to.— further back we obtain here & there in order a scattered page; we find ‹great› sensible change in the institutions. & we suppose not only revolutions, but entire obliterations & fresh laws created., & yet with ‹gov› symmetry «& regular laws» that baffles idea of revolution.—[1]
My very theory requires each form to have lasted for its time: but we ought in

5 «if the change . . . faulty»] *inserted above rule line & trails along margin, hence written after succeeding passage.*
«are» not] *'a' on stub.*
page crossed pencil.
6 we obtain . . . idea of revolution.—] *crossed pencil.*
My very theory . . . objection *to my theory*:] *double scored.*
Look at whole Glacial period?] *added pencil,* '22' *added brown crayon.*

4–2 See *inter alia* Lyell 1837, 3:124, '. . . if we regard each of the causes separately, which we know to be at present the most instrumental in remodelling the state of the surface, we shall find that we must expect each to be in action for thousands of years, without producing any extensive alterations in the habitable surface, and then to give rise, during a very brief period, to important revolutions.'

'*Illustration derived from subsidences.*— I shall illustrate this principle by a few of the most remarkable examples which present themselves. In the course of the last century, as we have seen, a considerable number of instances are recorded of the solid surface, whether covered by water or not, having been permanently sunk or upraised by subterranean movements. Most of these convulsions are only accompanied by temporary fluctuations in the state of limited districts, and a continued repetition of these events for thousands of years might not produce any decided change in the state of many of those great zoological or botanical provinces of which I have sketched the boundaries.'

5–1 Note the parallel trope D134, 'I do not doubt every one till he thinks deeply has assumed . . .' and E114.

6–1 See D60 for Darwin's earlier working of Lyell's metaphor.

same bed if very thick to find some change in upper & lower layers.— good objection *to my theory*: a modern bed at present might be very thick & yet have same fossils.[2] does not Lonsdale[3] know some case of change in vertical series: **Look at whole Glacial period?**[4]

7e–8e [not located]

9e Study introduction to Cuviers Regne Animal[1]

No structure will last. without it is adaptation to *whole* life of animal, & not if it be solely to womb, as in monster. or solely to childhood, or solely to manhood,— it will decrease & be driven outwards in the grand crush of population.—[2]

Octob 10th. Saw. two **undoubtedly** rabbits in poulterer shops., of same colour as a Hare, but paler & buffer.— with long ears— **& longer hind legs???** so that I was almost doubtful which it was.— do hind legs increase in any rabbits[3]

10e One may strongly suspect, that breeding in & in, produces bad effects solely, because of similarity, because in every country, where only pair has been introduced, & have freely bred, they have not lost power of producing.[1]

Williams. Narrative of Miss. Enterprise, p. 497. Vampire bat abound in the Navigators & at Manguia, but are unknown Eastward of the Navigators. Snakes occur there,, but are unknown in Hervey or Society isles.[2]

11e Hope says positively he has seen. a Calosoma. (very like American form) in Stonesfield slate., & a Melolonittha—[1] In marl from «Lake» Constance species of Europæan genera=.— Hope has ideas about generic characters.

9 10th] '0' *over* '1'.
undoubtedly] *added brown ink.*
& longer hind legs???] *added brown ink.*
page crossed pencil.
10 One may . . . producing.] *crossed pencil.*
Williams . . . Society isles.] '19' *added brown crayon.*
11 *page crossed pencil.*

6–2 See *Origin*:280, 'Why then is not every geological formation and every stratum full of such intermediate links? Geology assuredly does not reveal any such finely graduated organic chain; and this, perhaps, is the most obvious and gravest objection which can be urged against my theory. The explanation lies, as I believe, in the extreme imperfection of the geological record.'
6–3 William Lonsdale.
6–4 See *Origin*:294, 'And in the distant future, a geologist examining these beds [formed during the glacial period] might be tempted to conclude that the average duration of life of the embedded fossils had been less than that of the glacial period, instead of having been really far greater, that is extending from before the glacial epoch to the present day.'
9–1 Cuvier 1829–30, 1:1–51.
9–2 See Macculloch Abstract 205.5:28, 'Suppose six puppies are born . . . & in the Malthusian rush for life, only two of these live to breed . . .'
9–3 Passage continues at the top of E12.
10–1 See Notebook D for extensive discussion of the ill effects of inbreeding.
10–2 Williams 1837:497. This brown ink passage filled in a blank space at the bottom of the page.
11–1 *Melolontha.*

dominant. predominant &c having relation to geographical distribution—[2] Thus Hattica is great genus.— because found in all quarters: his ideas not clear. In Australia from approach to Asiatic [. . .]t in part near Timor, & to Europæan in Van Diemens land, where there is close species of elater.— Where this collection is particularly rich. «as in Lucanidæ» ‹no› less difficulty in establishing good groups.—

12e ears varying so much,— kind of fur— (do tips of ears take any colour?)— length of tail varies, & character of fur—[1] I am sure a very good case, might be made out of variation analogous to specific variations.—

Kerr's Collect of Voyages Vol 8 «p. 46» Capt Davis in 1598 found cattle in Table Bay with Humps on their backs & big tailed sheep[2]

---

do Vol 10. p. 373, «& 374» Spaniards says *no Tortoises* in other «places»[3] **besides Galapagos**

13e do. p. 376. Isle Tres Marias off Mexico with **small** Hares & *raccoons*.— «S. American form.—» off province of Guadalaxura—[1]

October 11th.— Uncle John—[2] says Decandoelle, distributed seeds of Dahlia all over Europe same year.—[3] he sowed them for four generation before they broke.—[4], showing effects of cultivation gradually adding up. & four more generations before they began to double.—[5] at present time Uncle J. does not suppose one aboriginal variety.— for they are all made by fertilizing

12 ears varying . . . variations.—] *crossed pencil.*
Kerr's . . . other «places»] '19' *added brown crayon.*
Spaniard says] 'Sp' *over* 'No'.
no Tortoises] *underlined pencil.*
besides Galapagos] *added pencil.*
12 and 13 formerly pinned together for notes on Kerr.
13 **small**] *added pencil.*
October 11th.— . . . fertilizing] *crossed pencil.*

11–2 F. W. Hope: personal communication. Hope's ideas may have been adumbrated in his paper at the eighth meeting of the British Association (Hope 1839 outlines his talk).
12–1 Continutation of E9 observations on rabbits.
12–2 Kerr 1811–24, 8:46.
12–3 Kerr 1811–24, 10:373. See also *JR* 1845:385, 'Wood and Rogers also, in 1708, say that it is the opinion of the Spaniards, that it is found nowhere else in this quarter of the world.'
13–1 Kerr 1811–24, 10:376.
13–2 Much of the information conveyed by John Wedgwood in E13–E16 is to be found in the early volumes of *Trans. Hort. Soc. Lond.* Wedgwood himself cultivated Dahlias (Wedgwood 1812:113–15).
13–3 See A. P. de Candolle 1810, cited in Sabine 1820:225.
13–4 Sabine 1820:225, 'it seems as if some period of actual cultivation were required, before the fixed qualities of the native plant gave way & began to sport into those changes, which now so much delight us.' (Quoted in Darwin's abstract. DAR 74:62.)
13–5 See E142, 'NB time *is element* in change, see *Dahlias*'. Thus Darwin's comment continues the theme of E4 that 'change within historical times has been small'.

14e one plant with another— Uncle John says he has no doubt bees fertilize enormous number of plants— it is scarcely possible to purchase seeds of any cabbage, where a great many will not return to all sorts of varieties, which he attributes to crossing.— Cape Broccolli can hardly be reared without greatest care be taken to prevent fertilization from turnips or other stocks. Says if any variety of apple be sown, all

15 sorts come up from it. lately saw a nonpareil sowed by Mr Tollet so produce.—[1] thinks it probable that great part of those varieties may be due to impregnation from other apple trees.— now seeds of crab produce crab, so that some effect from apple trees is produced.—
Thinks probably experiment was never tried of separating apple tree entirely from all others— so my experiment of strawberry not so absurd.—[2]
Thinks— that such variety as red cabbage

16 produced from passage from *many varieties*, & probably would take long before all the stain would be got out of it.—[1] Now this is curiously different from primrose suddenly produce *cowslip*, one is tempted to think here some anomaly— I can fancy cowslip producing primrose return to old stock, but not primrose producing cowslip[2]

17 Uncle J. says common belief. that female plant impresses main features on offspring. & male the lesser peculiarities.—[1] brilliancy of inflorescence

---

Gardeners. by chance ‹often› sometimes graft pears on apples. they will live but not flourish— a medlar may be Grafted on pear. Mountain-ash & white Thorn!

---

14 *page crossed pencil.*
15 Thinks . . . absurd.—] *marginal brace.*
page crossed pencil.
17 *page crossed pencil.*

15–1 George Tollett, a family friend of the Wedgwoods, became a source of breeding information. See QE 19 and *Correspondence*, 2:190–92.
15–2 See Barton 1827:67–68, 'In the Hautboy Strawberry these two sorts of blossoms are produced by distinct plants; and of course the stamen bearing plants yield no fruit; . . .' (Darwin had read Barton by C40.) Thus, the experiment may have involved isolating female plants of the dioecious strawberry. Alternatively, the strawberry experiment may have involved cutting the runner between two strawberry plants and either isolating the plants from crossing or growing them in different environments to induce variation. See D165 for what may have been an analogous experiment on splitting 'some animal with definite life'; also D1BC where there is the hint of an idea for an experiment on associated life. See also QE5: 'Sow seeds & place cuttings or bulbs in several different soils & temperatures & see what the effect will be.— will seedlings vary much more than cuttings &c'. See B230.
16–1 See QE5: 'Place pollen of Red Cabbage «mixed with own pollen» on flowers of other cabbages & see whether there will result hybrids—'.
16–2 See Darwin's discussion of the specific identity of the Primulas in *Natural Selection*:128–33. The cowslip-primrose problem was set for Darwin by J. S. Henslow (1830a). See also C35, C194, E113, E141, and QE1v. Darwin's persistent interest in *Primula* leads also to his paper on heterostyly (*CP* 2:45–63).
17–1 Note the numerous instances in Notebooks B–D where Darwin recorded the opposite view, namely male dominance.

Species not being observed to change «is very great difficulty» in thick strata, can only be explained, by such strata being merely leaf, if one river did pour sediment in one spot, for ‹*whole*› many epochs— such changes would be observed.—[2]

18 G. W. Earl's Eastern seas. p. 206— shot a monkey, ceased their cries. "many of them descending to examine their defunct companion".—[1]
p. 229. Borneo.— only animals he heard of pigs, small bears or badgers, deer, apes, baboons, monkeys & an animal probably a tapir[2]
p. 233. dogs in Borneo— ‹brought probably by Chinese›, "the breed being of the latter being the same as the fox-like animals. which are met with near Canton" "Here, as in all Malay countries, I noticed a peculiarity

19e in the cats «p 10[1]» the joints near the tip of the tail are generally crooked, as if they had been broken".[2] are born so **in all Malay Countries W. Earl Eastern Seas. p 233**[3]

Octob 12th Kotzebue's second «1st» Voyage. Vol II p. 344. account of insects of St. Peter & St. Pauls in Lat′ 53° yet fauna like that 60° & 70° of Europe.— Many Europæan insects— list given,= some peculiar[4]

‹M› p. 359. At Manilla a small Cercopithecus., & skins of galiopithecus.—[5]

Malte Brun. Vol ‹I› II p.,133: at Samar SE of Luçon, many monkeys, buffaloes &c &c—[6] Malte Brun. would be worth skimming over with regard to this archipelago

20e Octob. 13th.— Kotzebues first Voyage. Vol II. p 367. "The Fauna of the Sunda islands presents us, for the most part, with the same families and genera, that are natives of S. Asia, but many of the *species* are peculiar to them"[1] do— p. 368. "Several kinds of animals have spread from the N. end of Borneo to the adjacent island— In Sooloo we find the elephant— in Magindanao several kinds of the large monkeys.— Fewer ‹of th› «Mammalia» have passed to Paragua & in Luçon the most northern **of the group the number is limited**[2]

19 **in all . . . Seas. p 233.**] *added pencil.*
in the cats . . . **p 233**] *crossed pencil.*
Octobr 12th . . . archipelago] '19' *added brown crayon.*
20 Kotzebues . . . to them"] *crossed pencil.*
of the . . . is limited] *added pencil from 21 when latter excised.*

17–2 See D60, E5–E6 and E127 for additional reworkings of the Lyellian metaphor of the history of the earth as a book. Also see *Origin*, chap. 9 (On the imperfection of the geological record).
18–1 Earl 1837:206.
18–2 Earl 1837:229.
19–1 Refers to E10 on bad effects of inbreeding.
19–2 Earl 1837:233.
19–3 Earl 1837:233.
19–4 Escholtz 1830, 2:344–45 in Kotzebue 1830.
19–5 Escholtz 1830, 2:359 in Kotzebue 1830.
19–6 See Malte-Brun 1822–33, 3:490.
20–1 Kotzebue 1821, 2:367.
20–2 Kotzebue 1821, 2:368.

21e of the group, the number is very limited.— **Kotzebue's Second Voyage**[1] do Vol III p. 77. Many foreign plants have been cultivated in Guahon. (Mariannes), "for example the prickly *Limonia trifoliata*, which cannot now be checked".—[2]

Mar[s]den p. 94 (1st Edit) of Sumatra has given account of Buffalo of the East which differs from that of S. Europe[3]

---

p. 189. The gaut, kind of crocodile, sometimes wanders from Pellew to Eap.— There is another great lizard. Kaluz. which is found at Pellew & Eap, but not at Feis (near island)[4]

22e do p. 190. The inhabitants of Summagi, a territory in the small isld of Eap in the Carolines, are remarkably short.— & Deformations are particularly common.— without arm, ‹skin› hands thumb,— one leg, hare lip &c &c.[1]

---

in Vol II p 363 account of Flora of pacific,[2] given in my coral paper[3]

Oct 14th Macleay says, that ‹every› «any» character even colour is *good.* (ie invariable) in some classes—[4] it is because every part is under change, now one part now another[5]

23 Macleay says it is nonsense to say take a tooth of any animal (as Toxodon) & say its relations.—[1] if we know its congeners then we can.— now on my theory this «certainly» can be accounted for, on any other it is the will of God.—

Octob. 16th. A very strong passage might be made— why seeing great variation in external form of varieties, do we suppose bones will not change in *number.* (even *species* do not this). because it has been so pronounced ex cathedrâ. let us look at facts. considering few domestic animals few. that have not ‹which› not, cows hornless, (horses not)

24 If they give up infertility in largest sense. ‹es› «as» test of species.— they must deny species which is absurd.— their only escape is that rule applies to

21 of the . . . limited.—] *crossed pencil when page excised.*
Kotzebue's Second Voyage] *added pencil.*
Mar[s]den . . . S. Europe] *marginal brace, part on stub.*
'19' *added in brown crayon to bottom part of page.*
22 Oct 14th . . . another] *crossed pencil.*
23 so pronounced] 's' *over* 'pr'; *page crossed pencil.*

21–1 Kotzebue 1830, but Darwin's quotes are from the first voyage (Kotzebue 1821).
21–2 Kotzebue 1821, 3:77.
21–3 Marsden 1783:94–97.
21–4 Kotzebue 1821, 3:189.
22–1 Kotzebue 1821, 3:190.
22–2 Kotzebue 1821, 2:363.
22–3 Darwin 1833–38a.
22–4 William Sharpe MacLeay: personal communication.
22–5 Darwin seems to assume considerable variability in nature. Compare *1842 Sketch*:6 (DAR 6:19), 'But if every part of a plant or animal was to vary infinitesmally . . . Who, seeing how plants vary in garden, what blind foolish man has done in a few years, . . .'
23–1 Cuvier was famous for claiming just that (1821–24, 1:60–63). See E55 and note E72–2.

wild animals only. from which plain inference might be drawn that whole infertility «of hybrids receive no explanation» was consequent on mind or instinct, now this is directly incorrect

---

The case of my mice is good, because it is an involuntary variation made by man, common to every individual & therefore effect of climate.—[1]

25e Octob 19th. When reading. L'Institut: .1838 p. 329— Milne Edwards, description of curious mechanism of respiration, or rather ventilation peculiar to ‹the class› «some orders» of crustacea, one is tempted to think that it must have been invented all at once.—[1] but naturalists if they had series perfect, would expect this structure would become obscure & therefore it might thus have arisen, & M. Edwards p. 330 distinctly states that the flipper is a mere simple modification of an organ present in whole class.[2]

26e Case of Mexican greyhounds.— young being habituated.[1] instance such as Hunter,[2] or some one mentions of influence on parent affecting offspring.—⫲ & as *adaptation*,— however mysterious such is case⫲. therefore chance & unfavourable conditions to parent may be become favourable to offspring:⫲[3] Australian dogs have mottled coloured puppies case of this.— tendency in «manner of» life to be mottled + heredetary *tendency* determines the puppies to be so.—

25 '11' *added brown crayon.*
26 tendency . . . to be so.—] *scored.*
page crossed pencil.

24–1 See Waterhouse 1837, *Mammalia*:31–68 and *1844 Essay*: 100.

25–1 Milne-Edwards 1838d:330, 'Si maintenant, dit l'auteur, on compare au jeu de l'appareil respiratoire des autres animaux le mécanisme que je viens de décrire, on verra qu'il diffère essentiellement de tout ce qui est connu jusqu'ici. Chez les Crustacés Décapodes, cet appareil ne répresente plus une pompe . . . simplement foulante comme chez certains Reptiles, mais un instrument d'hydraulique particulier à parois immobiles dans lequel un système de palettes vient battre le fluide de façon à en rejeter sans cesse une certaine quantité au dehors, et, parconséquent, à déterminer dans la cavité, située derrière lui, un courant rapide qui s'alimente par d'autres orifices dans le milieu ambiant. Ce mécanisme curieux rappelle d'une manière frappante celui de certains appareils de ventilation dont les ingénieurs se servent pour renouveler l'air vicié dans l'intérieur des mines ou des égouts.'

25–2 Milne-Edwards 1838d:330, 'Il est aussi digne de remarque que l'instrument affecté à cet usage insolite n'est pas un organe nouveau introduit *ad-hoc* dans la structure des Crustacés à branchies inférieures, mais un appendice qui existe dans tous les animaux de cette catégorie, et qui est seulement en partie détourné de sa destination ordinaire et légèrement modifié dans sa conformation pour devenir apte à remplir ses fonctions nouvelles.'

26–1 See Lyell 1837, 2:409–11, *Acquired instincts of some animals become hereditary.*—, 'Some of our countrymen, engaged of late in conducting one of the principal mining associations in Mexico . . . carried out with them some English greyhounds of the best breed to hunt the hares which abound in that country. . . . It was found that the greyhounds could not support the fatigues of a long chase in this attenuated atmosphere, and before they could come up with their prey, they lay down gasping for breath; but these same animals have produced whelps which have grown up, and are not in the least degree incommoded by the want of density in the air, but run down the hares with as much ease as the fleetest of their race in this country.' (2:410–11).

26–2 John Hunter.

26–3 Darwin is applying his new and developing principle of natural selection. Note the very clear statement of the adaptive relativity of characters. See D175.

27e–30e [not located]

31 Did *man* spread over world as early as Elephants &c &.— if in next 20 years none of his remain found in the Americas probably did not.—[1]

Octob. 25th. I observed in Windsor Park.— the «Fallow» Deer. which were of a nearly uniform ‹dusky› blackish brown.— yet retained a trace of horizontal mark on flank.; & tail. & kind of semilunar [sketch] mark on each side darker,, so that whole colour is changed, these best marked characters are partly retained, therefore colours vary in same Manner as they would vary, if in wild state; thus mark on ear of cats, colour can be brownish

32 do Saw what was said to be hybrid between silver & gold fish—

Octob. 26th. If. hereafter. M. angustidens[1] be found to be inhabitant of S. America & as it is ‹falle› embedded with almost recent shells.— shows that progression of change in Mollusca is somewhat similar in two hemispheres.— It might be worth investigate whether. Megatherium & Mastodon are coembedded in N. America. see my Journal for references[2]

In such cases as at Galapagos. where different islets have different forms it is either effects of having been long separated, or having never

33e–34e [not located]

35e ARGUMENT REAL of antiquity of reasonable cosmopolite man.—[1]

L'Institut. 1838. p. 338.«V[ide]» *Important account* of cross of sheep & Moufflon

31 Did *man* . . . did not.—] *crossed pencil.*
32 *page crossed pencil.*
35 ARGUMENT . . . man.—] *crossed pencil.*
L'Institut . . . breed.—] '17' *added brown crayon.*
p. do . . . paper.] '19' *added brown crayon.*
page cut in two to distribute passages marked '17' & '19' to appropriate portfolios.
L'Institut 1838. p. 338] *added pencil.*
A most . . . stated.—] *crossed pencil.*

31–1 See E35: '*argument real* of antiquity of reasonable cosmopolite man—'.

32–1 *Mastodon angustidens.*

32–2 *JR*:152, 'The occurrence of the fossil horse and of *Mastodon angustidens* in South America, is a much more remarkable circumstance than that of the animals mentioned above in the northern half of the continent; for if we divide America, not by the Isthmus of Panama, but by the southern part of Mexico, in lat 20°, where the great table-land presents an obstacle to the migration of species . . . we shall then have two zoological provinces strongly contrasted with each other.'

35–1 See Lyell 1837, 3:213–17, *Caverns in the South of France.*—, 'In the controversy which has arisen on this subject MM. Marcel de Serres [1823], De Christol [1829], Tournal [1833], and others, have contended, that the phenomena of this and other caverns in the south of France prove that the fossil rhinoceros, hyaena, bear, and several other lost species, were once contemporaneous . . . with man; while M. Desnoyers [1831–32] has supported the opposite opinion.' (p. 215).

of Corsica.[2] ‹would not›, sadly against Yarrell's law.— not so much against my modification of it—[3] Goat & Moufflon will not breed.—[4]

---

p. do.— Fish of Teneriffe. St. Helena & Ascension most species like & *identical* with S. America. & many very *close*:[5] see full paper.[6] **L'Institut 1838. p. 338**

A most grave source of doubt. in distinguishing ‹after› which parent impresses offspring most is whether mother has had any offspring before.—— now this is never stated.—

36e Regarding the similarity of offspring to Parents same laws appear to hold good. with regard to marriage of individuals, & varieties of same species & to different species— sometimes like one parent & sometimes other & sometimes ½ way.

Ed. New-Phil. Transact. Rabies, common to men, dogs, horses cows, pigs & sheep.— diseases common to men & animals cow pox.— case in Spain of pustular disease following handling sheep— all cases: d° p. 354— The most vicious dog. will not attack any animal except, dog when absent

37 from its master.—[1] dogs when strayed hang their tails.—[2]

November 1st..— Addenda to Journal. I show erratic blocks transported far S. in Northern. Hemisphere.— likewise far North in Southern.—[3] Great animals. of same two great orders destroyed about same time in North & South. America.— Whole wor[l]d, formerly possessed a climate compared to

36 one] 'o' *over* 'p'.
page crossed pencil.
page cut in two to preserve passages in 35.
37 *page crossed pencil.*

35–2 Flourens and Serres 1838:338, 'M. Durieu . . . a fait venir des Moufflons de Corse, et il a donné à l'une des femelles, à l'époque du rut, un bélier de Mérinos. Ces deux animaux . . . se sont accouplés et ont donné un métis femelle, mais bien plus semblable au père qu'à la mère, Il n'était plus recouvert en effet de poils jars roussâtres comme ceux qui caractérisent le Moufflon, mais bien de laine blanchâtre parsemée seulement et par intervalles de poils jars. Ce métis femelle a été ensuite accouplé avec un bélier Moufflon de race pure, et le produit obtenu ressemblait cette fois bien davantage au Moufflon. . . . Ce nouveau métis a été accouplé avec une femelle de Mérinos, et il en est résulté, cette fois, un individu du même sexe, qui a retenu tous les caractères de sa mère: . . .'

35–3 In Darwin's modification of Yarrell's law male determination can overpower age predominance.

35–4 Flourens and Serres 1838:338, '. . . toutes les tentatives faites pour accoupler des Boucs privés de leur liberté avec des femelles de Moufflons ont été infructueuses. Il semble donc résulter de ces faits, . . . qu l'on ne peut pas toujours triompher de la répugnance que les espèces différentes éprouvent pour s'accoupler mutuellement; et puisque le Moufflon et le Mouton se sont réunis d'eux-mêmes, c'est que très probablement l'un et l'autre appartiennent à une seule et même espèce.'

35–5 Valenciennes 1838b:338.

35–6 Valenciennes 1838d.

37–1 Wagner 1838:354, 'No other dog [than a rabid one], even the most vicious, when accidentally separated from its master, will attack any save animals of its own kind, unless in self-defence.' Passage scored.

37–2 Wagner 1838:355, 'On the other hand, it is a well known fact that some dogs, in a state of health, constantly foam at the mouth, and that all look dejected and hang their tails when they have wandered astray.' Passage scored.

37–3 *JR*:614, addendum to p. 289.

S. America at present days,, which S. America now does to North. America & Europe.— S. America favourable to Tropical productions.[4]

38 The world formerly much more so. yet climate of same order as that of S. America.— (Explained by profound views of Lyell[1]) Now «Equatorial» America from the «low» limits of blocks both North & South, has probably undergone a greater change, than any part, (except Europe. in which all Tropical forms have been obliterated) of the world. from the ‹Tropical› Equable kind of

39 climate to the extreme.— Therefore species, which were fitted for such a preeminently equable climate. might not have been able to have survived a change, (& become transmuted), although other parallel species in other continents might have survived this mundine change..— Therefore I argue from this that Africa «& East Indian Archipelago»— formerly were not so «very» EQUABLE. or so *tropical*, & therefore present state of world is not so different, with

40 ‹d› regard to their productions.— Hence it is, from the ancient preeminently equable & temperate climate, ‹that› of America, that the Mammalia of S. America are as diferent from the existing orders, as the Eocene of *Paris*! (Great Edentata at that period) Analyse this,— consider state of

41e *vegetation*, & conchology,— shells of Africa ought most to resemble fossil ones of Europe, Consider probable form of land,— — S America, an *island*, «connects with Asia» between two polar lands,—; Africa not so *equatorial*..—

The fact of No. Mam: Placent: insectivore being in S. America & Australia. reason, why: Marsupiata, when first introduced live & multiplied, specifically & individually.—

42e I see clearly from F. R. it will be highly necessary to show that if species fall, genera must.[1] **Lesson I remember says Mariana Deer very close to a Molucca species.—**[2]

L'Institut 1837. p. 253, on animals of Antilles.—[3] (see Macleay in Zoolog.

39 *page crossed pencil.*
41 *page crossed pencil.*
42 I see . . . genera must.] *crossed pencil.*
Lesson . . . Molucca species.—] *added pencil.*
L'Institut . . . S. American forms.] '19' *added brown crayon.*
low] 'l' *over* 'ex'.
The climate . . . Americas.—] *crossed pencil.*

37–4 *JR*:615–17, addendum to p. 294. The theme of a former equable climate is repeated in E37–40, and E42. See also C99.
38–1 See Lyell 1837.
42–1 Personal communication with Capt. Robert FitzRoy, who is referred to in M43 as 'Capt. F. R.'.
42–2 Darwin is probably referring to Lesson and Garnot 1826–30. However see C14, which is based on Quoy and Gaimard 1830–35, where Darwin seems to make the opposite point, 'Cervus moluccensis is different from that of the Marianne islands & at Amboina'.
42–3 (*See overleaf*)

Journal. for those of Cuba.—[4] It is important to understand well the relation of passage from N. to S. American forms.

The climate of N. America, must have been *equable* & *low*— more so than any other part of the World.— Europe perhaps less so, that either Americas.—

43 If species change, we see external conditions have great effect on them, & therefore extermination becomes part of same law.—

When we know what a great effect. light has in colouring plants,— who can say. what ‹light› «colours». acting. by a most delicate organ, on the whole system may. produce— ?[1]

When a species becomes rarer, as it progresses towards extermination. some other species must increase in number where then is the gap, for the new one to enter?—[2]

44 The wonderful species of Galapagos, must be owing to these islands, having been purely result of *elevation*,— «all» modern & wholly volcanic— Azores might be prophecied to have this character.— worth going there for.— «Gales of wind would blend species»

Buckland. Reliquiæ Diluvianæ. p. 222. Bones of Horse.
Bear & Deer at 16000 ft. with Snow on Himmalaya—[1] Humboldt bones

45 at 7800 in Andes— parallel & curious facts.—[1] The Himmalaya. case, bears on the vast changes even in that quarter of the world—— Mem. elevation & subsidence of East Indian Archipelago. now rising.

On a particular part of coast of Somersetshire the Cockles are all apt to be diseased., & some of them symmetrically.— it is easy to get 50 of same kind of monstrosities.— G. B. Sowerby.—[2]

---

46 Looking over Lamark surprised to see how many Tropical genera come from New Holland, ˹Sydney?[1]

42–3 Gervais 1837:253, '. . . le nombre de [Mammifères] qui sont propres aux Antilles . . . est plus considérable qu'on ne le penserait d'abord.' Darwin scored the first two paragraphs of this paper. At the bottom of the page he wrote: 'Cuvier says genus Capromys confined to Cuba!, West Indian Genus.'

42–4 MacLeay 1829a:273–78.

43–1 See QE5: **'Raise seedlings surrounded by various bright colours, any effect? and silk caterpillars'**.

43–2 See D135: 'One may say there is a force . . .'

44–1 Buckland 1823:222, 'But in central Asia the bones of horses and deer have been found at an elevation of 16,000 feet above the sea, in the Hymalaya mountains.'

45–1 Buckland 1823:222, 'Mr. Humboldt has also found the tooth of the fossil elephant, resembling that of the northern hemisphere, at Hue-huetoca, on the plain of Mexico; . . .'

45–2 George Brettingham Sowerby (Elder): personal communication.

46–1 See Lamarck 1822–24.

The dog being so much more intellectual than fox, wolf &c &c— is precisely analogous case to man, exceeding monkeys;—

47 Having proved mens & brutes bodies on one type: almost superfluous to consider minds.— as difference between mind of a dog & a porpoise was not thougt overwhelming.— yet I will not shirk difficulty— I have felt some difficulty in conceiving how inhabitant of Tierra del Fuego is to be converted into civilized man.— ask the missionaries about Australians yet slow progress has done so.—[1] Show a savage a dog, & ask him, how wolf was so changed.

48 When discussing extinction of animals in Europe. :the forms themselves have been basis of argument of change.— now take greater area of water & snow line descent.

My theory gives great final cause «I do not wish to say only cause, but one great final cause,— nothing probably exists for one cause» of *sexes* «in separate» «*animals*»: for otherwise, there would be as many species, as individuals,[1] & though we may not trace out all the ill effects. — we see it is not the order in this perfect ‹uni› world, either

49 at the present, or many anterior epochs.— but we can see if all species, there would not be social animals. «this is stated too strongly. for there would be innumerable species. .& hence few only social there could not be one body of animals, living with certainty on other» hence not social instincts, which as I hope to show is «probably» the foundation of all that is most beautiful in the moral sentiments of the animated beings.— &c—[1]

If man is *one* great object, for which the world was brought into present state.— «whether he was or not. He is present a social animal» a fact few will dispute, [although, that it was the sole object, I will dispute, when I hear from the geologist the history, & from the Astronomer that the moon probably is uninhabited]CD & if my theory be true then the formation of sexes rigidly necessary.—

47 I have] 'I' *over* 'a'.

47–1 See *JR*:235, note. 'I believe, in this extreme part of South America, man exists in a lower state of improvement than in any other part of the world. . . . But the Australian, in the simplicity of the arts of life, comes nearest the Fuegian.'

48–1 See B4–5: 'generation destroys the effect of accidental injuries, ‹on› which if animals lived for ever would be endless (that is with our present system of body & universe — therefore final cause of life'. Also see D157 and D167: 'the final cause of sexes to obliterate differences. final cause of this because the great changes of nature are slow.' This question is the subject of chap. 3, 'Possibility of all organic beings crossing', in *Natural Selection* and a key theme of *Crossing*, where he says, 'The most important conclusion at which I have arrived is that the mere act of crossing by itself does no good. The good depends on the individuals which are crossed differing slightly in constitution. . .'

49–1 See discussions of the moral sense in Notebooks M, N, Old & Useless Notes and *Descent*, chap. 3.

50 Without sexual crossing, there would be endless changes, & hence no feature would be deeply impressed on it, & hence there could not be *improvement*. «& hence not «be» higher animals» — it was absolutely necessary that Physical changes should act not on individuals, but on masses of individuals.— so that the changes should be slow & bear relation to the whole changes of country,[1] & not to the local

51 changes. = this could only be effected by sexes:

All the above should follow after discussion of crossing of ‹species› individuals. with respect to representative species., when going N.orth & South

---

Thinking of effects of my theory, laws probably will be discovered. of co relation of parts, from the laws of variation of one part affecting another.—[1] (I from looking at all facts as inducing towards law of transmutation, cannot see the deductions which are possible.)— Ascertainment of

52 closest species (& naming them) with relation to habits, ranges. & external conditions of country, most important & will be done to all countries,—[1] but naming mere «single specimens in» skins worse than useless.— yet there is no cure «I may say all this, having myself aided in such sins» (do not add name, without reference to description), except describers having some high theoretical interest,— "the great end must be the law & causes of change".— A philosopher, would as soon turn tailor, as, mere describer of species, from its garments, without some end.— *Respect* good describers like Richardson.—[2]

53 The relations of numbers of species to genera &c &c can never be told, without species being described.—[1] but then permanent varieties in same country, must be distinguished, from permanent varieties not in same country.—
The traces of changes in forms of organs, will care little for species, except so far as wanting names to refer to, to those forms. where the termination of change occurs.— those discovering the *formal* laws of the corelation of parts in individuals, will care little, ‹in› whether the individual be species or variety, but to discover *physical* laws of such corelations, & changes of

52 would as soon . . . Richardson.—] *scored & '?' added.*

50–1 See B5: 'With this tendency to vary by generation, why are. species are constant over whole country: beautiful law of intermarriages ‹separating› partaking of characters of both parents, & these infinite in Number'. See also *1842 Sketch*:66.

51–1 See D112 *inter alia* on correlated variation.

52–1 See Darwin's 'Scheme for abolishing specific names & giving subgenera. true value.—' in C126.

52–2 See Richardson 1829–37.

53–1 That is, 'described' by the standards given in E52. Indeed Darwin did not undertake to work out the 'relations of numbers of species' until the early 1850s in his researches leading to the principle of divergence. See *Natural Selection*:134–71. See also D133.

54 individual organs, must know whether the individuals «forms» are permanet, all steps in the series, their relation to the external world, & every possible contingent circumstance.— ǂthe laws of variation of races, may be important in understanding laws of specific changeǂ.— When the laws of change are known.— — then primary forms may be speculated on, & laws of life,— the end of Natural History, will be approximated to.—

Treating of the formal laws of corelation of parts & organs it may serve perfectly to

55e specify types, & limits of variation., & hence indicate gaps.—[1] by this means the laws probably would be. generalized, & afterwards by the examination of the special cases, under which the individual steps in the series have been fixed, to study the physical causes. «All Cuviers generalization. of teeth to kind of extremities come under this head[2]»

27th November
When summing up argument against my theory, doubtless, the presence of animals in ‹own› «the present» orders (not so in S. America, however) is very remarkable & none discovered before them in any part of World.— Wealden to boot.—[3]

56e When one sees in Coralline powers of multiplication of individuals, & yet another means for individuals (Mem: transportation will be answered) one looks to analogy for cause in plants. where *innumberable* individuals can be produced. & yet sexual apparatus.—

My account of Circus cinereus of the Falklands Isl^d^. is interesting as showing some change in habits before form.—[1] **I have already given various examples**[2]

55 «All Cuviers . . . head»] *inserted in a box.*
27th November] '2' *on stub.*
page crossed pencil.
56 When one . . . apparatus.—] *crossed pencil.*
I have already given various examples] *added pencil.*
various] *alternative reading* **'enough'**.
'5' over '3' *&* '23' *added to page brown crayon.*

55–1 See *Origin*: 143–50 on the correlation of growth.
55–2 Cuvier 1821–24, 1:60–63.
55–3 See B172.
56–1 *Birds*:30–1, 'My specimens were obtained at the Falkland Islands, and at Concepçion in Chile. M. D'Orbigny states that it is a wild bird; but at the Falkland Islands it was, for one of its order, very tame. The same author gives a curious account of its habits: in a different manner from other raptorial birds, when it has killed its prey, it does not fly to a neighbouring tree, but devours it on the spot. It roosts on the ground, either on the top of a sand hillock, or by the bank of a stream: it sometimes walks, instead of hopping, and when doing so, it has some resemblance in general habit to the *Milvago chimango*. It preys on small quadrupeds, molluscous animals, and even insects; and I find in my notes, that I saw one in the Falkland Islands, feeding on the carrion of a dead cow. Although in these respects this *Circus* manifests some relation in its habits with the *Polyborinæ*, yet it has the elegant and soaring flight, peculiar to its family; and in form it does not depart from the typical structure.'
56–2 May refer to *Natural Selection*:347–49 where Darwin gives many examples of change in habits without change in structure in birds.

57 The Pipe-fish is instance of part of the hermaphrodite structure being retained in the male.—[1] ‹like› «far» more than marsupial bones,. & even more than Mammae, which have given milk.— is secretion from Pidgeons stomach true milk.— ‖ ‹Species. are innumerable variations›. Every structure is capable of innumerable variations,[2] as long as each shall be *perfectly* adapted to circumstances of times.[3] & from persistency «owing to their slow formation» these variations tend to accumulate. «on any structure.»[4]

58 L'Institut. 1838. p. 384. List of fossil Mamm: from Poland. &c.—[1]

Three principles, will account for all[2]
(1) Grandchildren. like. grandfathers[3]

57 *page crossed pencil.*
58 especially] *circled.*

57–1 See discussion of *Syngnathus* in Fries 1839a, 2:96–97, D169 and T25–26.

57–2 See E22: 'every part is under change, now one part now another'.

57–3 See Macculloch Abstract, 205.5:58^{v}, 'I look at every adaptation, as the surviving one of ten thousand trials.— each step being perfect to the then existing conditions.—«or nearly so . . . although having hereditary superfluities Man could exist without Mammæ.»'

57–4 See E4, E9, and E17. Darwin's concern with the concept of adaptation here and in E71 is reflected in his notes on two works he was reading at the time: Müller's *Physiology* (1838–42) and Whewell's *History of the Inductive Sciences* (1837). See note E69–1. For example, where Müller writes, 'This law of organic conformation,— adaptation to an end,— regulates the form, not only of entire organs, but also of the simplest elementary tissues.' (Müller 1838–42, 1 [June 1838]:20), Darwin altered— in brown ink—'adaptation to an end' to'«(as the effect of) adaptation to « ‹(varying)› circumstances»'. Similarly Darwin 'rewrites' Whewell on adaptation: ' "I take care", says Geoffroy, "not to ascribe to God any intention." And when Cuvier speaks of the combination of organs in such order that they may be in consistence with the part which the animal *has to play* in nature; his rival rejoins, "I know nothing of animals which *have to play* a part in nature." ' (Whewell 1837, 3:461–62), Darwin inserts after Cuvier's '*has to play*'— 'thus qualified is correct: Owing to external contingencies & numbers of other allied species & not *owing* to mandate of God'. Darwin goes further than rejecting providential adaptation. On Whewell's 'The use of every organ has been discovered by starting from the assumption that it must have *some* use.' (3:468), Darwin substituted 'relation' for 'use' and commented: 'Shrivelled wings of those non-flying Coleoptera?! In every science, one may trust that every fact has some relation to whole world.—' (top of page) and 'In every animal, final cause,, or adaptation is applicable to far greatest proportion of structure for otherwise it would be pressed' (bottom of page). Whewell goes on to quote Kant: ' "It is well known that the anatomisers of plants and animals, in order to investigate their structure, . . . assume, as indispensably necessary, this maxim, that in such a creature nothing is *in vain*, and proceed upon it in the same way in which in general natural philosophy we proceed upon the principle that *nothing happens by chance*. In fact, they can as little free themselves from this teleological principle as from the general physical one; . . .' (3:470). Darwin scored the passage and commented: 'All this reasoning is vitiated; when we look at animals on my view.' Whewell concludes: 'It appears to me, therefore, that whether we judge from the arguments, the result, the practice of physiologists, their speculative opinions, or those of the philosophers of a wider field, we are led to the same conviction, that in the organized world we may and must adopt the belief, that organization exists for its purpose, and that the apprehension of the purpose may guide us in seeing the meaning of the organization.' (3:470–71). Darwin commented: 'When a man inherits a harelip, or a diseased liver is this adaptation.— «as much as Bullfin to linseed» doubtless it is in one sense, but not in that, in which these philosophers mean.'

58–1 Eichwald 1838b.

58–2 The three principles are: (1) heredity (reversion), (2) variation, (3) Malthusian 'population pressure' (Darwin's wedging force). See also note E150–2.

58–3 See B32, C136, and D42 for the basis of this principle. See also 'Questions for Mr Herbert' (*Correspondence* 2:179–82), '(8th)— In Mr Herbert's great experience, has he ever known a character appear for the *first* time in a plant, and afterwards transmitted to the second generation, or grandchild and not to the child? There being, however, always this great source of doubt to guard against, that the same cause, which produced the character in the grandfather, may again act in the grandchild.'

(2) Tendency to small change..[4] «especially with physical change»
(3) Great fertility in proportion to support of parents[5]

---

59 December 2[d]
Lyell tells me Beck considers the characteristics of the Tropical Forms in shells. are numerous species, numerous individuals, & ‹*individuals*› «*species*» of *large size*,—[1] consider this (Cetaceæ) with reference to my theory

---

Babbage 2[d] Edit, p. 226.— Herschel calls the appearance of new species. the mystery of mysteries. & has grand passage upon problem.! Hurrah.— "intermediate causes"[2]

60 The Sexual system of the Cirrhipedes is the more remarkable from their alliance to Articulata, which are all truly bisexual.[1]

Buckland's Reliqu: Diluv. says *Africa* only place, where, Elephant, Rhinoceros, Hippot, Hæna &c are found together.—[2] Read this Work—

---

59 *page crossed pencil.*

58–4 This principle is emphasised throughout much of early Notebook E and is a frequent theme in late Notebook D.
58–5 See the Malthus passages D134–D135.
59–1 H. H. Beck. See note D133–1 on numerical proportions.
59–2 J. F. W. Herschel letter of 20 February 1836 to Charles Lyell, in Babbage 1838:225–27, 'I am perfectly ashamed not to have long since acknowledged your present of the new edition of your Geology, a work which I now read for the third time, and every time with increased interest, as it appears to me one of those productions which work a complete revolution in their subject, by altering entirely the point of view in which it must thenceforward be contemplated. You have succeeded, too, in adding dignity to a subject already grand, by exposing to view the immense extent and complication of the problems it offers for solution, and by unveiling a dim glimpse of a region of speculation connected with it, where it seems impossible to venture without experiencing some degree of that mysterious awe which the sybil appeals to, in the bosom of Aeneas, on entering the confines of the shades—or what the Maid of Avenel suggests to Halbert Glendinning,

'He that on such quest would go, must know nor fear nor failing;
'To coward soul or faithless heart the search were unavailing.

'Of course I allude to that mystery of mysteries, the replacement of extinct species by others. Many will doubtless think your speculations too bold, but it is well to face the difficulty at once. For my own part, I cannot but think it an inadequate conception of the Creator, to assume it as granted that his combinations are exhausted upon any one of the theatres of their former exercise, though in this, as in all his other works, we are led, by all analogy, to suppose that he operates through a series of intermediate causes, and that in consequence the origination of fresh species, could it ever come under our cognizance, would be found to be a natural in contradistinction to a miraculous process— although we perceive no indications of any process actually in progress which is likely to issue in such a result.'
Darwin had met Herschel in Cape Town in June 1836. The 'mystery of mysteries' passage first appeared in print in the first edition of Babbage's treatise, which was published in 1837. However, E59 is the first evidence that Darwin read, or knew of, the passage. However, RN32 shows that Darwin and Herschel discussed at least the geological content of the letter, which was published in Herschel 1833–38. In the opening passage of the *Origin* Darwin alludes to Herschel's letter. *Origin*:1, 'When on Board H.M.S. 'Beagle', as naturalist, I was much struck with certain facts . . . These facts seemed to me to throw some light on the origin of species—that mystery of mysteries, as it has been called by one of our greatest philosophers.'
60–1 See D156, Darwin's notes on Owen's classification of hermaphrodites (Hunter 1837:34–5). 'Heautandrous, male organs formed to fecundate female (as in plants) Cirrhipeds. . .'
60–2 Buckland 1823:170.

Decb. 4th.—
Why has the organization of fishes & Mollusca (& plants???) been so little progressive «!Agassiz makes it wonderfully *changed*, since Cretaceous period,[3] whether progressive I know not.» (& insects.— Stonesfield????). Have Mammalia??[4] My theory certainly requires progression, otherwise

61e–62e [not located]

63 Are the feet of water-dogs at all more webbed than those of other dogs.— if nature had had the picking she would make ‹them› such a variety far more easily than man,—[1] though *man's practiced* judgment. even without time can do much.—[2] (yet one cross, & the permanence of his breed is destroyed)— —

———

When two races of men meet, they act precisely like two species of animals.— they fight, eat each other, bring diseases to each other &c, but then comes the more deadly struggle,, namely which have

64 the best fitted organization, or instincts (ie intellect in man) to gain the day.— In man chiefly intellect, in animals chiefly organization: though Cont of Africa & West Indies shows organization in Black Race there gives them preponderance.[1] intellect in Australia to the white.— The peculiar skulls of the men on the plains of Bolivia— strictly fossil «& in Van Diemen's land»— they have been exterminated on *principles*. strictly applicable to the

65 universe.— The range of man is not unlike that of animals transported by floating ice.— I agree with Mr Lyell., man is not an *intruder*.—[1] : the geological history of man is as perfect as the Elephant, if some genus. holding same relation as Mastodon to Man. were to be discovered.

———

63 *page crossed pencil.*

60–3 Agassiz 1833–43, 1:xxvii, 'Les espèces de la craie appartiennent pour plus des deux tiers à des genres qui ont entièrement disparu.' See also Macculloch abstract, note 167v–2. We know from E69 that Darwin was reading Whewell's *History of the inductive sciences* during this period. Hence, a source for the Agassiz reference may be Whewell 1837, 3:374, 'Thus only the two first orders, the *Placoidians* and *Ganoidians*, existed before the commencement of the cretaceous formation: the third and fourth orders, the *Ctenoidians* and *Cycloidians*, which contain three fourths of the eight thousand known species of living Fishes, appear for the first time in the cretaceous formation: and other geological relations of these orders, no less remarkable, have been ascertained by M. Agassiz.' Passage scored.
60–4 See Broderip 1827.
63–1 In Darwin's personification of nature, there is an implicit analogy from natural to artificial selection, which is pursued more directly in E71. See the *1842 Sketch*:45, where nature becomes 'a being infinitely more sagacious than man'. For an example of the tradition of personification see the introduction to Cuvier 1829–30.
63–2 See *1842 Sketch*:45, 'Who, seeing . . . what blind foolish man has done in a few years, will deny [what] an all-seeing being in thousands of years could effect. . .'
64–1 May refer to immunity from 'miasmata'. See D24.
65–1 See Lyell 1837, 1:248–58, *Introduction of man, to what extent a change in the system.*—, 'We have no reason to suppose, that when man first became master of a small part of the globe, a greater change took place in its physical condition than is now experienced when districts, never before inhabited, become successively occupied by new settlers.' (p. 250). Darwin agreed that man is not an 'intruder', but did not agree with two other Lyellian positions: denial of any genuine human antiquity and denial of the transformist implications of man's recent origin.

Man acts on. & is acted on by «the» organic and inorganic agents of this earth. like every other animal.—

66 Would anyone raise an argument against, my theory, should no fossil «very distinct species» of the Ornithorhyncus be found.;— yet until man became cosmopolite, he would probably be confined in locality like Ornithorhyncus,: since being cosmopolite, we do find his remains.— Lima.— caves.—[1]

There being no fossils, the only way, that I can see to discover whether the parent of man was quadruped or bimanous,, is to see, what

67 parts of structure abortive.— Remember my fathers remark about the Bladder.—[1]

The numbers of fatal diseases in mankind, «the more valuable domesticated animals» no doubt is owing to the rearing up of every heredetary tendency towards fatal diseases, & such constitutions only being cleared off by fatal diseases.—

---

68 The Value of a group does not depend on the number of the species.: therefore Man & monkeys have equal chance that progenitor was bimanous, or quadrumanous.— What a chance it, has been, (with what attendant organization, Hand & throat) that has made a man.— CD [any monkey probably might, with

69 such chances be made intellectual, but almost certainly not made into man.— It is one thing to prove that a thing has been so, & another to show how it came to be so.— I speak only of the former proposition.— as in «races of» Dogs, so in species, & in Man

---

December 16th. The end of each volume of Whewells Inductive History. Contains many most valuable references[1]

70 See if any law can be made out, that varieties are generally additive, & not abortive: with reference to the non-necessity of the «so called» progressive tendency law.—[1]

---

69 December 16th . . . references] *both margins double scored.*

66–1 See *JR*:451–52, 'On the island of San Lorenzo . . . I was much interested by finding embedded, together with pieces of sea-weed in the mass of shells, in the eighty-five foot bed, a bit of *cotton-thread*, plaited rush, and the head of a stalk of Indian corn. This fact, coupled with another, which will be mentioned, proves I think the amount of eighty-five feet elevation since man inhabited this part of Peru'. (Cited in Lyell 1838a:295–96.)

67–1 R. W. Darwin: personal communication.

69–1 Whewell 1837. The references at the end of each volume are Darwin's lists of pages that have marginalia. Darwin's syntax indicates that he completed Whewell before he wrote E69—how much before is uncertain. It is assumed that E57 reflects his reading of Whewell.

70–1 See Whewell 1837, 3:576–80, *Sect.4.—Hypothesis of Progressive Tendencies*. Darwin's marginalia: 1, 'been already intimated . . . apparently imply most clearly' (all p. 577) scored; 2, 'which tendency . . . are successively developed'

(*continued overleaf*)

In animals analogy leads one to suppose that seminal fluid fluid, (& not dry as in plants) therefore, great difficulty in crossing [*& this most important obstacle to my theory*] CD without the hermaphrodites mutually couple,— now how is it— in Planaria, they couple— CD [lowest terrestrial animals.— in shells?—

71 *insects*?.— all!??!?— Worms? [Barnacles, aquatic., ‹yet› Crustacean, & true hermaphrodites] CD «It may be said that true hermaphroditism is a consequence of non-locomotion— (contradicted by Plants). & as there are no fixed. land animals. so there are true hermaphrodites.— I suspect this rather effect of liquid semen: therefore animal life commenced in the Water!»

It is a beautiful part of my theory, that «domesticated» races. of ‹a› organics. are made by percisely same means as species— but latter far more perfectly & infinitely slower.— No domesticated animal is perfectly adapted to external conditions.— (hence great variation in each birth) from man arbitrarily destroying certain forms & not others.—[1] Term *variety* may be used to *gradation* of changes

72 which gradation shows it to be the effect of a gradation in difference in external conditions.—— as in plant up a mountain—[1] In *races* the differences depend upon inheritance & in *species* are only ancient & perfectly adapted races

---

L'Institut 1838. p. 394. Rhinoceros «tichorhinus» in Paris basin.— its relation to African Species ‹good observations.— ›, larger than any living[2]

73e–74e [not located]

75 A Greyhound might be made «almost» without any relation to running hares.— **as in Italian Greyhound** not so species every part of newly acquired structure is fully practised & perfected
Hence difference between *races* & *variety*? «Man picks the Male, instead of allowing strength to get the day»

---

75 **as in Italian Greyhound**] *added pencil.*

(*70–1 continued*)
(p. 578) scored with question mark in margin and 'constantly' underlined; 3, 'the additional assumptions' underlined and scored (p. 579); 4, Darwin commented at the bottom of p. 579 'These are not assumptions, but consequences of my theory, & not all are necessary'; 5, citation of E. Geoffroy Saint-Hilaire (p. 580) is scored.
71–1 Note that the flow of Darwin's analogy is from natural to artificial selection, as it is in E63.
72–1 See B235.
72–2 Valenciennes 1838c:394, 'Cette découverte, dit M. Valenciennes, vient confirmer la prévision de Cuvier qui regardait le Rhinocéros bicorne fossile comme un animal plus gros et plus trapu que le Rhinocéros d'espèces actuellement vivantes sur le globe. Enfin, en comparant l'humérus récemment découvert à celui des environs d'Abbeville et à ceux des deux Rhinocéros vivants cités plus haut, on acquiert une nouvelle preuve de la justesse des lois auxquelles Cuvier était arrivé, que l'on peut, par l'étude d'un seul os, déterminer le genre ou l'espèce de l'animal auquel il a appartenu, que l'on peut reconnaître les affinités des espèces entre elles. Dans le cas actuel cet humérus montre que le Rhinocéros à narines cloisonnées, et qui portait deux cornes sur le devant de la tête, a plus d'analogie et plus de ressemblance avec l'espèce du bicorne d'Afrique qu'avec les autres Rhinocéros unicornes ou bicornes de l'Inde.'

The fertility of Indian & Common Oxen, which one must think deserve the name of species, may be owing to the little fixity of organization, in the two races,. owing to the domestication of both.— Now in the ass— there is little tendency to vary. & hence offspring are hybrids,.—

76 Mr G. B. Sowerby[1] ‹tel› showed me many land shells of the common species: from one locality, all left whorled.— He kept two to see if they would breed,

---

It is difficult to think of ˁˁ*Plato & Socrates*, when discussing the Immortality of the Soul as the linear descendant of ‹Mammferus› «Mammiferous» ‹vert› animal, which would find its place in the Systema Naturæ.—

77 **Mr. Knight makes this analogy between grafting & sexual union—**[1]
Looking at simple generation as being the action of two organs in one body.— or in two bodies, ‹th› we can as well understand the necessity of a relation between the fluids of the two as in the grafting of trees.—]CD CD[The similarity of child to parent appears to follow same law in two of «the» *same* ‹species›, variety, as in *two varieties*, & this we might expect, as the difference between man & woman is «indeed» (independent of sexual differences) a *variety*. The offspring of true

78 ‹pare› hermaphrodite, would of course be like either, that is both parents, for they are one.— The laws, therefore, of likenesses of fathers to children of mankind, no doubt are applicable to likenesses, when species & races are crossed.— Now these laws are, that child may be either like father or mother, independently of its sex, or half way between, or someway different from either: & or like progenitors.— in some families all the children like mother & in some like father «What is cause of this.— »

79 (Lord Moretons law holds with *different species*, & individuals of *same species*.—[1]). some races of men. D'Orbigny. affect the common progeny more than others.—[2] does this more refer to length of time that the resemblance is permanent, or the similarity at first births.— it is the latter only that one refers to in speaking of resemblances of children to their parents.—

Lord Moreton's law cannot hold with fishes, «& there are mule fishes» & reptiles & those which ‹lay› «have» their eggs, ‹inter›, impregnated externally; nor can it be a *necessary* concomitant, with moths, which can be impregnated externally—

77 **Mr. Knight . . . sexual union—]** *added parallel to spine.*

76–1 C. B. Sowerby (Elder): personal communication.
77–1 Knight 1818a:160, 'few if only varieties of fruit can with strict propriety be called permanent when propagated by buds or grafts.' Darwin commented: 'So that propagation by seed merely exposed the plant to external influences more completely, than it is exposed during life.— No doubt these changes «during life» might be propagated.' (DAR 74:58). This contrasts with Darwin's view in B1–5 *et seq.*
79–1 See B181 and Morton 1821.
79–2 A. D. d'Orbigny 1838:569. See note D136–1 for text.

80 My view of every animal being Hermaphrodite— probably will recieve illustration from domestication of Monoœcious plants, & abortion of others.— ? in hemi-hermaphrodite insects is it not easier to understand ?*perfect*?? developement of one sex on one side, than the addition of other organ, in which case the hermaphroditism would not be perfect as in Ox. the amount of double sexual developement is spread over

81e–82e [not located]

83 it utterly untold,— what is added to the composition of the atom, to make it alive, & how the laws of generation were impressed on it.—

Seeing that ‹Man› «all vertebrates. [Müller's Physiolog. p. 24.][CD]»[1] can be traced to a germ, endowed with the vital principle,[2] which gives rise to the sexual organs, different in each species,— & knowing from analogy, that all these very animals are descended from some one single stock,—one is led to suspect that the birth of the species & individuals in their present forms, are closely related— By birth the

84 the succesive modifications of structure being added to the germ, at a time, (as even in childhood) when the organization is pliable,[1] such modifications, become as much fixed, as if added to old individuals,, during thousands of centuries,— each of us, then ‹is as old, as the oldest animal›, have passed through as many changes, as has any species.—

85e Decemb. 21th.— L'Institut 1838. p. 412.[1] M. Eichwald has published Fauna

81 'any' *on stub.*
'ne' *on stub.*
85 Decemb. 21th] '1' *over* '0'.
page crossed pencil.

83–1 Müller 1838–42, vol. 1 [June 1838]. It is perhaps not coincidental that shortly after Darwin read Whewell 1837 we find him reading Müller. Where Whewell writes: 'We have, I think, been led, by our survey of the history of Botany, to this point;—that a Natural Method directs us to the *study of Physiology*, as the *only* means by which we can reach the object.' (3:342, Darwin's underlining), Darwin first responds in defense: 'Systematic Naturalists are the heralds of Nature' (3:342 top), then he pours out a long series of 'physiological' notes and queries.

83–2 Müller 1838–42, 1 [June 1838]:24, '. . . as Baer first discovered, liver, salivary glands, and pancreas are in the further progress of the vegetative process really developed from that which appears to be merely the rudiment of the intestinal canal. It can no longer be doubted that the germ is not the miniature of the future being with all its organs, as Bonnet and Haller believed, but is merely "*potentially*," this being, with the specific vital force of which it is enbued, and which it becomes "actually" by development, and by the production of the organs essential to the active state of the "actual" being. For the germ itself is formed merely of amorphous matter, and a high magnifying power is not necessary to distinguish the first rudiments of the separate organs, which from their first appearance are distinct and pretty large, but simple; so that the later complicated state of a particular organ can be seen to arise by transformation from its simple rudiment.' Passage scored in pencil. Darwin commented in brown ink 'very good' (in the margin) and 'Now in a *bud*, we must suppose there is one particle of *old* organized structure.— a filament of «old» nerve.' (at top of page). See also Whewell 1837, 3:419.

84–1 See B4: 'There may be unknown difficulty with *full grown* individual «with fixed organization» thus being modified'.

85–1 Eichwald 1838c:412, 'Un travail de M. Eichwald récemment publié sur la faune de la mer Caspienne a donné à ce savant l'occasion de combattre l'opinion que la mer Caspienne aurait été primitivement unie à la mer Noire. Il se fonde dans cette conclusion sur la différence

of Caspian.—[2] fishes fresh water kinds. (yet living in the salt?.)— very few animals of any kind— Fauna, must be very curious.—

With respect to the non-development of Mollusca, which I have sometimes speculated might be owing to absolute quantity of vitality «in the World»,— *the production of vitality*, as argued by Müller from propagation of infinite numbers of individuals from one, is adverse.—[3]

86e Decemb. 25th.— Lyell says the elevated shells in Bayfields district are much more like those of Scandinavia, than of the N. American species—**Glacial period** Dr. Beck says the shells in Scandinavia from height of 200 & 300 ft are identically same as those of present seas.—[1] now in this country we have better means of judging of *slowness* of physical changes, than in any other, & yet 200–300 ft elevation & no change & even no loss of species.—

87e It must never be overlooked that the chronology of geology rests upon amount of physical change ‹affecting whole bodies of species›, & only secondarily,, by assumption well grounded, on time;— therefore the mere loss of species, which may be the work of a few years as with the Lamantin of Steller[1] tells much less, (though ‹the› it also the effect of change) than a slow gradation in form, «which must be effect of slow change + & Therefore precludes effects of catastrophes, which must serve to confound our chronology» «CONSIDER ALL THIS» Extinction & transmutation, two foundations, hitherto confounded,. of geology.—

86 **Glacial period**] *added pencil.*
87 '22' *added brown crayon.*

qui résulte de la comparaison des faunes des deux mers. Le plus grand nombre des Poissons de la Caspienne sont des Poissons d'eau douce. Cette mer est de la plus grande pauvreté en animaux marins, surtout quand on la compare à la mer Noire. Et cependant, dit M. Eichwald, si les deux mers avaient été autrefois en communication, on ne devrait trouver dans l'une aucune espèce qui ne fût également dans l'autre.'

85–2 Eichwald 1834–38.

85–3 Müller 1838–42, 1(June 1838):39–40, *The organic force also is increased during the organisation of new matter* (Darwin underlined 'organic force also is' in this section heading), 'Now, by the growth and propagation of organised bodies, the organic force seems to be multiplied; for from one being many others are produced, and from these in their turn many more; while, on the other hand, with the death of organised bodies the organic force also seems to perish. But the organic force is not merely transmitted, as it were, from one individual to another,— on the contrary, a plant, after producing yearly the germs of very many productive individuals, may still remain capable of the same production,—the source of the increase of the organic or vital force seems therefore also to lie in the organisation of new matter; and, this being admitted, it must be allowed that plants, while they form new organic matter from organic substances under the influence of light and caloric, are also endowed with the power of increasing the organic force from unknown external sources, while animals also in their turn would generate the organic force from their nutriment under the influence of the vital stimuli, and distribute it to the germs during propagation.'

86–1 Lyell: personal communication.

87–1 *Hydrodamalis gigas* (Sirenia), Steller's sea cow, formerly existing in the Bering Straits. See R. Owen 1839a.

88e L'Institut 1838. p. 414; M. Guyon thinks Monsters more common in Africa than in Europe especially with Europæans settled there[1]

L'Institut do. p. 419, «long» account of Hyænodon, a fossil dog— leading towards Hyæna.—[2] see Comte Rendu.—[3] I suspect good case of fossil filling up blank.— CD[not between existing series of species of dogs & Hyæna.— but a common point, whence both may have descended.—

89 Jan. 6th
The rudiment of a *tail*, shows man was originally *quadru‹manous›«ped.—* ».
Hairy.— could move his ears
The head being six metamorphosed vertebræ, the parent of all vertebrate animals.—[1] must have been like some molluscous «bisexual» animal with a vertebra only & no head— !!
Handwriting is determined by most complicated circumstances, as shown by difficulty in forging, yet handwriting said to be heredetary. shows well what minute details of structure heredetary'—

90 Athenæum .1839. p. ‹8›36.— A crustaceous animal is mentioned which inhabits the Pinna of Rio Janeiro, (like some Mediterranean species).—[1] might these fertilise other shells, as insects do flowers.— Mem. Spallanzani's experiments showing how little of the spermatic fluid fertilized spawn of frogs.—[2]

---

88 L'Institut 1838 . . . settled there] *crossed pencil.*
L'Institut do. p. 419 . . . descended.—] '21' *added brown crayon.*
89 The head being . . . no head—!!] *marginal brace.*
page crossed pencil.

88–1 Guyon 1838:414, 'M. Guyon conclut, d'après ce qu'il a entendu dire, qu'en Afrique les monstruosités sont plus communes qu'en Europe, et que les Européenees qui viennent habiter l'Afrique y sont plus sujettes que les Arabes.'

88–2 Laizer and Parieu 1838a.

88–3 Laizer and Parieu 1838b.

89–1 See Pictet 1839:320 (cited at E92), 'The principle of the head being composed of vertebrae, that remarkable application of the law of homology, had also struck Goethe before the time when first it was submitted to the examination of anatomists; . . .' Passage scored. See also Whewell 1837, 3:447–48, 'Göethe declares that, at an early period of these speculations, he was convinced that the bony head of beasts is to be derived from six vertebrae. . . . and Meckel, in his Comparative Anatomy, in 1811, also resolved the skull into vertebrae. But Spix, in his elaborate work *Cephalogenesis*, in 1815, reduced the vertebrae of the head to three. . . . Geoffroy Saint-Hilaire put forth a similar doctrine in his "Philosophie Anatomique" in 1818. . . We cannot fail to recognise here the attempt to apply to the skeleton of animals the principle which leads botanists to consider all the parts of a flower as transformations of the same organs [Goethe's aphorism, 'Alles ist Blatt']. . . .

'By these and similar researches, it is held by the best physiologists that the skull of all vertebrate animals is pretty well reduced to a uniform structure, and the laws of its variations nearly determined.' Darwin scored: 'By these and . . . nearly determined.'

90–1 Mittre 1839:36, 'A surgeon of the French navy, M. Mittre, just arrived at Brest, among several new and interesting objects of natural history, has brought a new Maclura, which he found at Rio Janeiro in the *Pinna nobilis*. The existence of this crustacean in the seas of the New World is a curious fact in the geography of zoology, for since the time of Aristotle, it has only been found in the Mediterranean.'

90–2 Spallanzani 1769. . . See QE1^{v}: 'Really good subject for experiment. «to»— repeat Spallanzani. Raise only single Plants & only allow ‹few› one flower.'

Annals of Natural History. (p 225. 1838.) account of metamorphosis in the young of Syngnathus.=[3] curious as showing generality of law. even in fish: ‖.do. p. 236— on Hybridity in ferns.—[4] ‖.do p. 250— «speaking of» the terrestrial mollusca of Morocco «Mr Forbes says the Fauna»— (near Oran) approach in character to Canary Isld.— ie Canary Isld approaches *more* to neighbouring coast of Africa, than to other parts of that

91e continent.[1] in like manner as Madagascar does to other side of Africa.— (& Juan Fernandez to Chile??) Falklands to southern portion.— ‖.do p. 269. **Annals of Nat. Hist 1838** on «a» freshwater fish peculiar to Ireland.[2] ‖do p. 283. on the dark ears of the wild Chillingham Cattle,[3] with reference to Mr Bell's statement of the tame ones.—[4] ‖an instance of a trifling peculiarity not to be eradicated.— ‖.do. p. 305.— Mr Owen says ‹tha› «in abstract» in his paper on the Dugong, "The generative organs being those which are most remotely related to the habits & food of an animal, I have always regarded as affording very clear indications of its true affinities. We are least likely in the modifications of these organs to mistake a merely *adaptive* to an *essential* character—"[5] How little *clear* meaning has this compared to what it might have.—

91 **Annals of Nat. Hist 1838**] *added pencil.*
peculiar . . . tame] *double scored brown crayon.*
Mr Owen says . . . it might have.—] '11' *added brown crayon.*

90–3 B. F. Fries 1839b:225, 'Metamorphosis observed in *Syngnathus lumbriciformis*, by Prof. B. Fries. This interesting paper, which will find its place in one of our following numbers contains a most curious fact hitherto unobserved in the class of fish; namely, that the young of this beautiful species at their development from the egg have the entire tail covered with a fin-like membrane and possess pectoral fins. These at subsequent unknown period are thrown off in a way similar to that of the larvae of frogs rejecting their tails.' See also D169 and T25–26.

90–4 Martens 1839b:236, 'M. Martens observed in the Botanical Garden of Louvain, a fern which he regarded as a hybrid between *Gymnogramma calomelanos* and *G. chrysophylla*, to which Bory de St. Vincent proposes to apply the name of *G. Martensii*. At the same time the latter gentleman observes that this hybrid formation appears to occur quite commonly in nature, for he had received several well-preserved specimens of this plant through L'Herminier from Guadaloupe, where it grows in nature between the two above-mentioned *Gymnogrammae*. He also enumerates several other ferns which might be considered as hybrids, which are only grounded on supposition: to these however Dr. Meyen rather inclines to assent.' Passage scored.

91–1 Forbes 1839:250, '. . . but a great part of these are not correctly speaking from Algiers, but from Oran (near Morocco), where the Fauna of Barbary assumes a different aspect, approximating to that of the Canaries on the one hand, and to that of Spain on the other.' Passage scored.

91–2 W. Thompson 1839b:269n, 'Since my account of the pollan appeared, I have been favoured by Dr. Parnell with a specimen of the *Coregonus* of Loch Lomond . . . and by Sir Wm. Jardine with one of the Ullswater species; both of which are distinct from the *Cor. Pollan*, this having not as yet been found in any of the lakes of Great Britain.' Passage scored.

91–3 Hindmarsh 1839a:283, 'It is true that in the colour of the ears there is a trifling difference, but this appears to be an occasional variety in the species; for Bewick states that about 40 years ago some of those at Chillingham had black ears, that the keeper destroyed them, and that since that period this variation has not recurred.' Passage scored. See D48.

91–4 T. Bell 1837:423.

91–5 R. Owen 1839a:305. Passage scored.

92e What is the difference between an *essential* character & an *adaptive*. one.— are not the essential ones eminently adaptive.— does it not mean *lately* adapted or transformed. & hence not indicative of true affinity.— — — Owen says Dugong connected with Pachydermata.—[1] ‹it was a Pachyderm. which was the origin of the aquatic Mammifers› p. 306, the Dugongs cannot be united with true Cetacea or whales.— but are aquatic Pachyderms. & Walrus— aquatic seal.—[2] (Consult this passage, when considering origin of northern Cetaceæ).— — ‖.do. p. 318 M. Pictet of writings of Goethe.— who maintains, that[3] «Alludes to difference between fossil & recent Bull;[4] like fossil & recent shells of the ‹new› raised beaches»

93e–94e [not located]

95 The enormous *number of* animals in the world depends, of their varied structure & complexity.—[1] hence as the forms became complicated, they opened *fresh*, means of adding to their complexity.— but yet there is no «NECESSARY» tendency in the simple animals to become complicated[2] although all perhaps will have done so from the new relations caused by the

92 Pachyderm . . . maintains, that] *crossed pencil.*

92–1 R. Owen 1839a:305. 'The affinity of the Dugong to the *Pachydermata* is thus again illustrated by the great number of the ribs.' Passage scored.

92–2 R. Owen 1839a:306–7, 'The junction of the Dugongs and Manatees with the true Whales cannot therefore be admitted in a distribution of animals according to their organization. With much superficial resemblance they have little rẹal or organic resemblance to the Walrus, which exhibits an extreme modification of the amphibious carnivorous type. I conclude, therefore, that the Dugong and its congeners must either form a group apart, or be joined, as in the classification of M. De Blainville, with the Pachyderms, with which the herbivorous *Cetacea* have the nearest affinities, and to which they seem to have been more immediately linked by the now lost genus *Deinotherium*.' Darwin underlined 'extreme modification of' in ink, scored 'The junction . . . modification of the' in pencil, and commented: 'With respect to Creator's forming The Mamentes [Manatees] tropical' in pencil. See E87.

92–3 Presumably this passage was continued in the missing sheet E93–94. See Pictet 1839:318–19, 'The type being once created, Goethe applies himself to its comparison with individual forms, and, in this analysis, sets out from the principle that diversity has no other origin than this; that, in the development, one part becomes predominant at the expense of some other, and *vice versa*. He admits with respect hereto the influence of surrounding media and of exterior causes generally, by the force of which the nutritive matter is directed in superabundance and under certain forms to particular parts, so as to produce there a hypertrophy, always followed by an atrophy in some other part of the same being, because the nutritive matter is diverted from it to the gain of the former. He supposes that a certain formative or *plastic* force is given to every being, and that if it be directed to one point the consequence must necessarily be inverse modifications with regard to the others.* "The general total," says he, "in the budget of nature is fixed; but she is free to dispose of particular sums by any appropriation that may please her. In order to spend on one side, she is forced to œconomize on the other, and nature can therefore never run in debt nor become bankrupt." It is easy here to recognise the principle put forth by M. Geoffroy Saint Hilaire under the name of the *balance of organs*.' Passage scored and marked with a large 'Q'; at the bottom of p. 318 Darwin commented 'important'. The text of the footnote (*), not given here, is also heavily scored.

92–4 Pictet 1839:321, 'We may moreover notice among the special labours of Goethe, his observations on the researches of Dr. Jaegger upon the subject of the fossil bulls found in the neighbourhood of Stuttgard. Goethe seeks to prove in this article, that the differences which exist between fossil and recent bulls may be looked upon as the result of the perfecting of the species during the centuries which separate the two periods. His argument affords interest; but it seems to us that the poet plays almost as leading a part in it as the naturalist.'

95–1 See note E92–3.

95–2 See E70.

advancing complexity of others.— It may be said, why should there not be at any time as many species tending to dis-developement (some probably always have done so, as the simplest fish &), my answer is because, if we begin with the simplest forms & suppose them to have changed, these very changes ‹len› tend to

96 give rise to others.— Why then has there been a retrograde movement in Cephalopoda & fish & reptiles.?— supposing such to be the case, it proves the law of developement in partial classes[1] is far from true.— I doubt not if the simplest animals could be destroyed, the more highly organized ones. would soon be disorganized to fill their places.—
The Geologico-geographico changes must tend sometimes to augment & sometimes to simplify structures:= Without enormous complexity, it is impossible to cover *whole* surface of world with life.— for otherwise a frost if killing the vegetable in one quarter of the world would kill all of the one herbivorous. & its one carnivorous devourer.;— it is

97 quite clear that a large part of the complexity of structure is adaptation. though perhaps difference between jaguar & tiger may not be so.—[1] Considering the Kingdom of nature as it now is, it would not be possible to simplify the organization of the different beings, (all fishes to the state of the Ammocœtus) Crustacea to—— ? &c) without reducing the number of living beings— but there is the strongest possible to increase them, hence the degree of developement is either stationary or more probably increases.—

98 Jan 29th. Uncle John[1] says he feels sure, that the reason people send for their seeds to London is that people in the southern Counties have whole fields, some for cauliflower &c.— Uncle John believes one single turnip in a garden is sufficent to spoil a bed of Cauliflower.— (How curious it would be to make enquiries of some of these great seed-growers—).—[2]

Feb. 24th. Monoceros, which Sowerby says, is an American form.— has several species in my

99 fossils—[1] CD[If cases of one variety in upper part of bed— & another in lower is very rare, the conclusion will be that our greatest formations ‹are› have been deposited in a period (say 10,000 years) which is sufficient only to have most slightly modified organic forms.— we know not rate of deposition has been equal even in one bed, much less in alternating strata of sand & limestone &c &c.—

96–1 Compare Pictet 1839:322, 'Cuvier, . . . denied to the unity of organic composition the right of being erected into a general law. He acknowledged it within certain limits, but would not admit of any other analogies than those which were rigorously demonstrated, . . .'

97–1 See Darwin's marginalia on Whewell 1837, 3:468 quoted in note E57–4. See also E144.

98–1 John Wedgwood: personal communication.

98–2 See QE8. Henslow query (9). 'In the nurseries, when «seeds of» the varieties of cabbages, peas, beans, as raised, do the seedsmen select at all from the plants? If not, I am surprised ‹plan› such plants do not degenerate,—'

99–1 See G. B. Sowerby (Elder) 1839b:161.

100 L'Institut 1838.— p. 290— admirable paper on geographical distribution of Crustaceæ.— (I forget whether I have already referred to it.—[1] also on spermatic animalcules in Musci frondosi, et hepatici,— in Chara, in Marchantia & Hypnum—[2]

«Prof:» Don would have known the Composites of Galapagos were South American.— several cases of species peculiar to separate islets.—[3]

March 5th. Lyell says[4] «fossil» shells from North America, Scotland, Uddevalla. Many species same. & Northern forms— & American ones & Europæan— agree very much

101 closer, than the present ones., which according to Beck[1] are different.— Subsidence of Greenland— case of splitting of two regions— — are there any cases of union of two regions in modern times.— this would depend on negative evidence of fossil remains, & therefore not to be trusted.— — Lyell tells me, on authority of Beck, that Hooded crow & Carrion crow. have in Europe different ranges— latter not going

102 North of the Elbe.— yet they meet in one wood in Anhault. & there every year produce hybrids— now this is independent good case, but very odd since these crows are mixed in England— for I presume Carrion Crow is found in Edinburgh.— Why does Fleming consider them varieties[1] & what says Jenyns to it?—[2] — In argument of origin of Wolf, difference of mind is most relied on,

103e but Bell has some account of wolf in Zoolog. Gardens, which brought its puppies to be fondled.—[1] and we see in the Australian dog an instance of a half reclaimed animal.— The dogs, which have run wild have, have done so in hot countries.— CD[Camel does not vary «one ought not to be able to hybridise the Camel» like ass «same way some plants vary more than others» & horse in lesser degree,— how different to dog!— (Hybrids of

103 *page crossed pencil.*

100–1 Milne-Edwards 1838c.
100–2 J. Meyen 1838b.
100–3 David Don: personal communication. Evidently Darwin or Henslow showed his Galapagos specimens to Don. These were eventually described by J. D. Hooker (1847a, 1847b). See *JR* 1845, chap. 17 in which Darwin uses Hooker's information extensively. See also E104 for a parallel case in Labiatae. (Bentham 1832–36).
100–4 Lyell: personal communication.
101–1 H. H. Beck: informant of Lyell.
102–1 Fleming 1828:87. '113. *C. corone.* Carrion Crow.—Plumage black. Throat feathers small, narrow, adpressed, the barbs loose at the margins. Tail slightly rounded. . . . Is this species different from the Hooded Crow? 114. *C. Cornix.* Hooded Crow.—Head, throat, wings, and tail black, with blue and green reflections. Neck and the rest of the body smoke-grey. Tail rounded.'
102–2 Jenyns 1835:146 treats *C. corone* and *C. cornix* as separate species and says the carrion crow is 'Common throughout the kingdom. . .' and the hooded crow is 'Resident all the year in the western and northern parts of Scotland, and in some parts of Ireland, but only a winter visitant in England. . .'
103–1 T. Bell 1837:199.

Calceolaria.)—[2] «:CD[Does the Power of, «easily» making tolerably fertile hybrids, bears relation to capability of variation?? my theory says so.—»[3]

104e March 6th. Mr Bentham says in Sandwich Islds. he believes, there are, many cases of genera peculiar to the group having species peculiar to the separate islands—[1] In his work on the Labiatæ, some of these species are described.—[2] Capital case,— for Sandwich Isld are very similar to Galapagos—[3] study Flora— what general forms.— are the Labiata nearest to American, or Indian groups?— = Believes some Mediterranean, but chiefly *mountainous* «this is very important. (Sicily exception)— see if this can be generalized.—» islds., have peculiar

105 forms.— on the southern flanks of Alps.— many peculiar plants on single mountains, though these are connected with other mountains laterally.—

Owen. Fossil Mammalia. p. 55. talks of *Tapirus* American form. found in Eocene beds of Paris[1]

Lyell has remarked species never reappear when once extinct—[2]

103–2 The mention of *Calceolaria* hybrids connects this and 'Does the power . . . say so.—' to Darwin's reading of Herbert 1837 and to his 'Questions for Mr Herbert' (*Correspondence* 2:179–82); see note E103–3.

103–3 Herbert 1837:363, 'One of the most interesting genera, on which the process of intermixture has been successfully attempted, is that of Calceolaria, because it embraces plants of a decidedly shrubby and tender habit, and others which are completely stemless, and capable of retiring to rest under ground in the temperature of a British winter, and colours very dissimlar, the yellow and the brownish purple; and because most of the numerous species which have been imported appear to intermix with the greatest readiness, producing an endless variety of forms.' See also Herbert 1837:372, 'For instance, there seems little prospect of being able to answer why the hybridizing process is so easy in some genera and so difficult in others, if equally facile of access, unless it shall be found to arise from greater or less constitutional conformity.' Passage triple scored. See 'Questions for Mr Herbert' (*Correspondence* 2:180). '(4th)— In Mr Herberts work (p. 372) it is said some genera are much more difficult to hybridise than others. As a general fact are not such genera, slow in sporting, or yielding many varieties?— Is the converse true: that is, is there any case of tolerably allied species, which have given rise to many varieties, which will not hybridise each other, and generally produce fertile offspring?— In short is there any relation between a facility in varying (a capability which seems very different in different animals) and a facility in giving hybrids, and *especially fertile hybrids*.'

104–1 George Bentham: personal communication. See Bentham 1832–36:xlii, 'The *Sandwich region* . . . although included in my table in the New World, as being nearest to the American continent, are yet at a considerable distance from it, and have a very peculiar Flora. They possess 20 Labiatae, of which the *Plectranthus parviflorus* alone is found in the Oceanic and North Australian regions; the other 19 are endemic, belonging to two genera, *Phyllostegia* and *Stenogyne*, both of them peculiar to these islands.'

104–2 See Bentham 1832–36:650–55 for descriptions of *Phyllostegia* and *Stenogyne*; some are reported as 'Hab. in ins. Sandwich', but others give specific islands.

104–3 See E100.

105–1 *Fossil Mammalia*:55, 'It is well known how unlooked-for and unlikely was the announcement of the existence of an extinct quadruped entombed in the Paris Basin, whose closest affinities were to a genus, (*Tapirus*), at that time, regarded as exclusively South American.'

105–2 Lyell 1838a:275, 'It appears, that from the remotest periods there has been ever a coming in of new organic forms, and an extinction of those which pre-existed on the earth; some species having endured for a longer, others for a shorter time; but none having ever reappeared after once dying out.'

Lyell's argument about ‹Tertiary› Isl[d] «*neighbouring*» **formed in the Tertiary «epoch» like Sicily** not having species,[3] if true important on my view.—

106 March 9th—
Is there any relation between the fact that different species produce abundantly infertile hybrids, & the fact that old varieties do not so much affect first race, as it does indelibly the many subsequent ones.

My views, «V ‹see› p. 103»[1] would lead me to think that a ‹do› variety of one species would cross easier with 2[d] species, than two perfect species; but facts of grouse, & pheasant, & hooded crow goes against this. & *wild* hybrid plants.

107 If many wild animals were crossed, there would probably be perfect series,[1] from physical impossibility to unite to perfect prolifickness.— (‹a series might be obtained›: but the intervention of domesticated ie new varieties destroys the appearance of this series & makes one think that one large body of varieties are fertile & make mongrel, & other great series quite otherwise & make on[ly] *true* hybrids.— but this is false, [give instance of series from wild animals & plants][CD].—

108 M[r] Marsh[1] has some nephews, who are *astonishingly* like to some distant cousins, the nearest blood being a *great great*-grandfather.— — Little Miss Hibbert case of Hindoism coming out more than in mother or indeed grandmother; what is M[r] S.S.[2] parentage?—

Wonderful as is the possession of voice by Man. we should remember, that even birds can imitate the sounds surprisingly well—

---

104 '20' *added brown crayon.*
105 «*neighbouring*»] *underlined pencil.*
formed in . . . like Sicily] *added pencil.*

105–3 Lyell 1837, 3:445, 'We are brought, therefore, to admit the curious result, that the flora and fauna of the Val di Noto, and some other mountainous regions of Sicily, are of higher antiquity than the country itself, having not only flourished before the lands were raised from the deep, but even before they were deposited beneath the waters. Such conclusions throw a new light on the adaptation of the attributes and migratory habits of animals and plants, to the changes which are unceasingly in progress in the inanimate world. It is clear that the duration of species is so great, that they are destined to outlive many important revolutions in the physical geography of the earth; and hence those innumerable contrivances for enabling the subjects of the animal and vegetable creation to extend their range, the inhabitants of the land being often carried across the ocean, and the aquatic tribes over great continental spaces. It is obviously expedient that the terrestrial and fluviatile species should not only be fitted for the rivers, valleys, plains, and mountains which exist at the era of their creation, but for others that are destined to be formed before the species shall become extinct.' Darwin scored the passage and wrote: 'Capital!'

106–1 See E103: 'Does the Power of, «easily» making tolerably fertile hybrids, bears relation to capability of variation?? my theory says so.—'

107–1 Compare Herbert 1837:358, '. . . this deterioration of the descendants may perhaps be in part attributable to the fertility of the mule being less vigorous and perfect than that of the original parents, when there exists some constitutional difference between them, . . .' Darwin double scored the passage and commented: 'good step in series of infertility'.

108–1 Possibly Arthur Cuthbert Marsh.

108–2 Possibly Sydney Smith.

In early stages of transmutations, the relations of animals & plants to each other would rapidly increase, & hence number of forms. once formed. would remain stationary, hence all present types are ancient.[3]

109 According to my views of ‹Diœcious p› *all* plants, being occasionally diœcious; & really diœcious plants being effect of abortion of one sex.— Linnæan class Diœcia & Monoœcia. ought to be preeminently artificial.—

Would not subsidence of Greenland render climate less extreme. (& so account for descent of snow line *there* «& there & there only: as stated by Capt. Graah»)[1] & break up. N. American Conchology from Europæan., & the climate being now less extreme, than before arctic forms would retreat: effect on snow of arctic climate in far north regions? Arctic forms have travelled S.

110 From the analogy of the animal kingdom I should suppose, that the pollen of crab, would POSSIBLY «No, for pollen of any kind would fertilize it» fertilize an apple somewhat more readily, «than other apples» but probably would more indelibly stain offspring—[1] it would not *reach* one apple sooner than ‹other:› «that of another» apple. only effect produced would be different.— same way one variety of ‹animal› dog does not prefer other. but produces greater effect on offspring—

111 M^r^. Herbert says «p 347. Amyyralidæ» Plants do not become acclimatised by crossing, or by *accidental production of seedling with hardier constitution.*—[1] Now Sir.J. Banks. says Zizania in 16 generations did become, acclimatized. & says

109 Conchology] *after deleted carat.*
of arctic . . . travelled S.] *scored.*

108–3 See E95–97.
109–1 Graah 1837, results discussed in Lyell 1837, 2:302, relevant passage scored.
110–1 See 'Questions for Mr Herbert' (*Correspondence* 2:179–80). Query 1, which is devoted primarily to Yarrell's law: 'Does M^r^. Herbert consider a character, produced by crossing, or more especially if arising from simple variation, which has passed through some generations, very much more likely to continue in the offspring than one which has appeared for the first time. If one flower of a plant were impregnated by the pollen of another species, retaining its normal character, and a second flower, by the pollen of the same species but after it had undergone considerable variation, would the seedlings in the first case be impressed more persistently, (that is in a greater number of successive generations) with the character of the plant from which the pollen had been derived *in its wild or old* than in its *new or cultivated state?*'
111–1 Herbert 1837:347, 'In constitution the mixed offspring appears to partake of the habits of both parents; that is to say, it will be less hardy than the one of its parents which bears the greatest exposure, and not so delicate as the other; . . . We now possess a further cross by the impregnation of Altaclarae by arborum, which will probably come so near the father in its colour, that if, as expected, it should be able to endure our winters, we shall have nearly attained the result, which would be otherwise most likely impracticable, of acclimating the magnificent Nepal plant; for it does not appear that in reality any plant becomes acclimated under our observation, except by crossing with a hardier variety, or by the accidental alteration of constitution in some particular seedling;' Darwin underlined 'In constitution . . . both parents'. On 'We now . . . endure our winters' (and the previous sentence) he commented: 'How exactly similar to giving dash of courage to greyhounds by Bull-dogs blood' (see C120). He also underlined 'accidental . . . seedling' and commented: 'This is like *sudden* appearance of cowslip from primrose it is analogous to Australian dogs, producing piebald young.'

Laurels have not been so.[2] (which is case adduced by Herbert)[3] because not reared by seedlings.— Now my principle does not apply to any plant reared artificially, & only very partially to the Zizanias in in Sir. J's ponds— my principle being the destruction of all the less hardy ones. & the ‹*accidental*›

112 preservation of *accidental* hardy seedlings: (which are confessed to by Herbert)[1] to sift out the weaker ones: there ought to be no weeding or encouragement, but a vigorous battle between strong & weak

---

March 11th. Yarrell's law must be partly true, as enuntiated by him to me,[2] for otherwise breeders who only care for first generations,, as in horses, would not care so much about breed.— what can «however» be more striking, about indelibleness, than the

113 number of good race-horses, which *Eclipse*? has begotten ‹?› «Walker attributes this to effect of male sex on locomotive system[1]»

I am bound to insist honestly that the *sudden*,[2] change from Primrose to Cowslip is great difficulty. «I should doubt if wild species ever formed like short-tailed cat or dog has been without recurrent tendency in external

111–2 Banks 1812.

111–3 Herbert 1837:347. The passage quoted in note E111–1 continues: 'nor that any period of time does in fact work an alteration in the constitution of an individual plant, so as to make it endure a climate which it was originally unable to bear; and, although we are told that laurels were at first kept in hothouses in this country, it was not that they were less capable of supporting our seasons than at present, but that the cultivators had not made full trial of their powers of endurance.' Passage scored and Darwin commented: 'against my theory.— change, however, is sudden— & not *many generations*.— From foregoing facts, about constitution we should expect such changes to be slow. & otherwise any attempt to change aquatic to dry plant—'

112–1 See note E111–1: 'accidental alteration of constitution in some particular seedling' (Herbert 1837:347).

112–2 Darwin is distinguishing between Yarrell's law and Darwin's modification of it. See E35.

113–1 A. Walker 1838:150–151, 'I. Law of Selection where both parents are of the same variety. 1. Organs communicated by one parent — the Anterior Series. . . . 2. Organs communicated by the other parent — the Posterior Series. The other parent communicates the posterior part of the head, the cerebral situated within the skull immediately above its junction with the back of the neck, and the whole of the locomotive system (the bones, ligaments and muscles or fleshy parts).' A four page abstract of Walker is tipped into Darwin's copy: 'I reject Mr Walkers theory of one parent giving (see p. 150) one series of organs & ‹another› the other a different set.— because (1) the propagation of plants as we see in their crossing &c & is closely similar to that [of] animals, now, in plants we cannot ‹tra› separate the organs into any two analogous divisions— ‹even if› In plants, according to Mr K. either father or mother can give "excitability" (& I daresay other similar case could be gathered) now excitability or constitutional peculiarities would scarcely be given in one system of organs.— 2^{d}. The kind of argument in favour of it, are such as Phrenologists advance (one series affecting muscles and another their supports, ie only in the face).— does not hold good in my experience, or rather a double answer might be given.'

113–2 See Herbert 1837:348, '. . . but upon the whole an intermediate appearance may be generally expected, but with a great disposition to sport, especially in the seminal produce of the fertile crosses, as in plants which are apt to break into cultivated varieties.' Passage triple scored and Darwin commented: 'Now does this tendency to sport in *hybrids* decrease after several generations of same appearance same fact as in varieties of animals where crossed.—' This comment is the textual basis for Query 2 in 'Questions for Mr Herbert' (*Correspondence* 2:180): '(2^{d}) As an inference from the above questions, it may be asked does the *tendency* to sport with respect to any particular character shall have been permanet through several successive generations?' See E16 and E141.

conditions» sudden loosing of horns.— I do not believe this Nature's plan.—

Whether we can or not trace history of first appearance of varieties of domesticated animals, yet as we know how many plants have been produced (look at the Dahlias.[3] we may infer it in animals.)— Azara gives account of production of hornless cattle— ?& others?—[4]

114 March 12th— It is difficult to believe in the dreadful «but quiet» war of organic beings. going on the peaceful woods. & smiling fields.—[1] we must recollect the multitudes of plants introduced into our gardens (opportunities of escape for foreign birds & insects) which are propagated with very little care.— & which might spread themselves, as well as our wild plants, we see how full nature. how firmly each holds its place.— When we hear from authors (Ramond. Hort. Transact Vol I. p. 17 Append)[2] that in the Pyrenees, that the

115e Rhododendron ferrugineum. begins at 1600 metres precisely & stops at 2600. & yet know that plant can be cultivated with ease near London.— what makes the line, as trees in Beagle Channel.—[1] it is not elements.— we cannot believe in such a line., it is other plants.— a broad border of Killed trees would form fringe.— but there is a contest. & a grain of sand turns the balance.—

Hort. Transact Vol I.
M. Ramond. p. 19. do says lofty Alpine plants of & Pyrennees[2] agree with those of Norway. Lapland & Greenland, but not

116e with those Kamtschatka, Siberia, or even of polar regions of N. America.— if true curious on my view— because these points were last connected with those northern regions—

---

do p. 21[1] says. many plants skirt each side of the great N & S. valleys, which penetrate Pyrenees but are found no where else not even in branch valleys— M. Ramond offers no explanation.—

115 **Hort. Transact Vol I.**] *added pencil.*
'20' *added brown crayon.*

113–3 See E13. Note also Darwin's abstract of Salisbury 1812:90, 'There are three species of Dahlias.— a vast number of varieties appear to have come from D. sambucifolia.—' (DAR 74:56).

113–4 Azara 1801, 2:372, 'On voit également par-là que les individus singuliers que la nature produit quelquefois par accident se perpétuent commes les autres; mais il faut dire que cela n'arrive pas toujours; car j'ai vu dans quelques Taureaux, nés sans cornes que lorsqu'ils sont adultes, ils commencent à avoir des cornes qui ne sont ni grandes ni droites, mais petites, tombantes et attachés à la peau uniquement; de manière qu'elles remuent lorsque l'animal marche, comme si elles avoient été arrachées? Passage scored pencil and marked 'Q' in ink.

114–1 See *Origin*:62, 'We behold the face of nature bright with gladness, we often see super abundance of food; we do not see, or we forget, that the birds which are idly singing round us mostly live on insects or seeds and are thus constantly destroying life;'.

114–2 Ramond de Carbonnières 1815:15.

115–1 See Darwin's Plant Notes (cited in Porter 1981:13).

115–2 Ramond de Carbonnières 1815:19.

116–1 Ramond de Carbonnières 1815:21.

117 Poet Cowper, describes his tame Hares, attacking a sick one like Chillingham bulls are described.—[1] His three have had VERY different dispositions: this is important as showing small variations in offspring of wild animals.— *grateful* & intelligent.—[2]

The theory that all animals have sprung from few stocks. does not bear, the least on ancient generic forms.— the animals in Eocene period could not have been direct parents of any of ours,— even if extinction is denied.— it will not account for all species. even if it will for all.—

118 Varieties are made in two ways— local varieties, when whole mass of species are subjected to some influence, & this would take place from changing country: but greyhound. & poutter Pidgeons «race-horse». have not been thus produced, but by training, & crossing & keeping breed pure.— «& so in plants *effectually* the offspring are picked & not allowed to cross.— » Has nature any process analogous— — if so she can produce great ends— But how.— — «— .Make the difficulty apparent by cross-questioning.— » even if placed on Isl^d— if &c &c.— Then give my theory.— excellently true theory

119e Examine list of St. Helena Plants & see whether those which grow in low grounds are those, which are common & nearest being common to other parts of the world—[1]

March 16th. Mr Lonsdale[2] showed me two specimens of an Inoceranus from the Gault of Folkstone, which is exactly intermediate between *I. concentricus* & *I. sulcatus*.— the beak of this one has concentric striæ, all the lower part rayed longitudinally (give woodcut) like *I.* sulcatus.— Both species are

120e found at Folkstone.— it is unnamed this intermediate one.— Mr Lonsdale evidently inclines to think it *Hybrid*.!!!
Ask Woodward[1]

Mr Lonsdale says *Trigonia costata* & *elongata* thougt considerably different, in proportional dimensions, must ‹almost› be considered merely varieties. &

117 The theory . . . for all.—] *crossed pencil.*
119 *page crossed pencil.*
120 found . . . *Hybrid*.!!!] *crossed pencil.*
Ask Woodward] *added pencil.*
'21' *added brown crayon.*
120–121 formerly pinned together.

117–1 Cowper 1784:413. '[Puss] was ill three days, during which time I nursed him, kept him apart from his fellows that they might not molest him (for, like many other wild animals, they persecute one of their own species that is sick) . . .'
117–2 Cowper 1784:413.
119–1 Beatson 1816: 295–326, Appendix (plant seen by Roxburgh in 1813–14).
119–2 William Lonsdale: personal communication.
120–1 See S. P. Woodward 1851–56. See also de Beer 1967:172, n.41, 'Samuel Pickworth Woodward (1821–65) was 18 years old in 1839 which shows that this pencil note was added at a date later than that at which the Notebook was written.'

even Mr Sowerby[2] is coming to this conclusion, from specimens in grades, now L. says the T. costatus

121e is in England found in the Inferior Oolite, & the T. elongata in the uper formations Portland Stone &c &c.—if? so «it is» good case:— in Sowerby Min. Conch. it is however, said they have been found *together* on coast of France.—[1] L. doubts.[2]— Lonsdale thinks Ammonites would afford instance of such facts.— Ask Phillips.—[3]

122e The more I think, the more convinced I am, that *extinction* plays greater part then *transmutation*.— Do species *migrate* & *die* out.?— **In the place where any species is most common, we need not look for change, because its number show it is perfectly adapted; it where few stray ones. are, that change may be anticipated, & this would look like fresh Creation. the gardener separates a plant he wishes to vary— domesticated animals tend to vary.**

March 20th. Phillips in Lecture in Royal Institution says shells become less in number. (? species, or individuals) the deeper one goes—[1] surely is this true?— most strange.—

123e Does not spermatic animalcule in Mosses,[1] render my view of the crossing of mosses & all others by action of wind difficult.—

Cline on the breeding of Animals, p. 8. size of fœtus in proportion to male parent p. 8.[2] his whole doctrine of the advantage of crossing consists in the idea of the male being smaller, & the female larger than average size:[3] (surely this is very limited view, though perhaps a true element) «give examples, pigs, with small chinese boars &c &c &c»

---

122 **In the . . . tend to vary.**] *added with heavy scoring.*
would look like] *on stub.*
page crossed pencil.
123 Does not . . . difficult.—] *crossed pencil.*
p. 10] 'p' *on stub.*
offspring . . . said.—] '17' *added brown crayon.*

120–2 J. de C. Sowerby: personal communication.
121–1 J. de C. Sowerby 1812–46, 5:40.
121–2 William Lonsdale or Charles Lyell.
121–3 John Phillips.
122–1 John Phillips.
123–1 See Meyen 1838b cited in E100.
123–2 Cline 1829:8, 'The size of the foetus is generally in proportion to that of the male parent; and, therefore, when the female parent is disproportionately small, the quantity of nourishment is deficient, and her offspring has all the disproportions of a starveling.'
123–3 Cline 1829:8, 'The proper method of improving the form of animals, consists in selecting a well-formed female, proportionally larger than the male. The improvement depends on this principle; that the power of the female to supply her offspring with nourishment, is in proportion to her size, and to the power of nourishing herself from the excellence of her constitution.'

p. 10 offspring take more after father than mother; illustrated by the crossing of hornless sheep with horned.—[4] compare this with what highland shepherds said.—[5]

124e p. 12. Attempts to improve the native ‹breed› animals of any country must be made with great caution; owing to its adaptation to the surrounding circumstances[1]

According to my theory no *land* animal with *fluid* seeds can be true hermaphrodite

---

Man probably assumes the hairy character of his forefathers only when advanced in age, & therefore the children do not, (& in *hairless* kittens we see same fact) go back, & this is argument against Blyth's

125e doctrine of young birds retrogressing—[1]

Uncovering the canine teeth, or sneering, has no more relation to our present wants. or structure, than the muscles of the ears to our hearing powers

---

E.[2] frowns prodigiously when drinking very cold water «frowns connected with pain, as well as intense thought.—»

No one but a practised geologist can really comprehend how old the world is, as the measurements refer not to revolutions of the sun & our lives,, but to period necessary to form heaps of pebbles &c &c: the succession of organisms tell nothing about length of time, only order of succession.—

126e Splendid Pamplet. (published in Philosop. Journal ‹Mar› April 1st 1839) by Sedgwick & Murchison; which is a beautiful instance of *forms*, intercalated

124 p. 12 . . . circumstances] *crossed pencil.*
According . . . hermaphrodite] *crossed pencil.*
125 Uncovering . . . intense thought.—] *marginal brace, (part on stub).*
the ears] 't' *over* 'e'.
page crossed pencil.
126 '22' *added brown crayon.*

123–4 Cline 1829:10. See *Natural Selection*:455, 'If we confine our view to the races of one species, or perhaps even to the species of the same group, some such rules may hold good; for instance it seems . . . in sheep that the ram gives the character of the fleece and horns'.
123–5 See D43 extracts from Glen Roy Notebook.
124–1 Cline 1829:12, 'Attempts to improve the native animals of a country, by any plan of crossing, should be made with the greatest caution; for, by a mistaken practice, extensively pursued, irreparable mischief may be done.
'In any country where a particular race of animals has continued for centuries, it may be presumed that their constitution is adapted to the food and climate.
'The pliancy of the animal economy is such, as that an animal will gradually accommodate itself to great vicissitudes in climate, and alterations in food, and, by degrees, undergo great changes in constitution; but these changes can be effected only by degrees, and may often require a great number of successive generations for their accomplishment.'
125–1 Blyth 1836a.
125–2 Emma Darwin or Erasmus Alvey Darwin.

between two great distinct formations.— particular air given. p. 246.— 248[1] & p. 258[2] A beautiful case, showing the gradation from one grand system to another: in each system, the changes from limestone to sandstone &c. show some great change who can say how many centuries elapsed between each of these gaps, far more probably than during the deposition of the beds— The argument must

127 be thus put, shall we give up whole system, of transmut., or believe that time has been much greater, & that systems, are only leaves out of whole *volumes*.—[1]

The fact of tumbling pidgeons; flying high all together & then tumbling, far more wonderful than heredetary *ambling* horses.[2]

---

127 The fact . . . *ambling* horses.] *crossed pencil.*
page crossed pencil.

126–1 Sedgwick and Murchison 1839:246–48, passages scored, 'Without entering, on this occasion, into specific details, we may state that the zoological groups of the Devonian rocks are all of characters intermediate between those which mark the Carboniferous and Silurian epochs. Thus, for example, among the Cephalopoda, *Goniatites* have hitherto been considered as typical of the carboniferous system, while the researches of one of the authors have shown that they never occur in the Silurian system. They do, however, appear in some of the older Devonian rocks; and, just as we should expect, they are associated with analogues of an entirely new type, the *Endosiphonites*.

'Again, there are many large and broad Spirifers in these Devonian rocks, which closely approach to the forms of that genus, so abundant in the carboniferous system. But this genus is feebly developed in the Silurian system, and the few species that do occur are entirely unlike the large typical Spirifers of the carboniferous æra; while the *Orthis*, or real Silurian Spirifer, is rarely if ever seen in Devonshire. The large round spinose *Producti* are among the best-marked fossils of the carboniferous system. Now the closest researches have not hitherto brought to light the existence of one species having this character in the Silurian system; while in Devonshire we find several associated with other species, which are analogous both to the Silurian and carboniferous types. On the other hand, the families and genera which predominate so much more in the Silurian than in any other system, viz. *Trilobites* and *Orthoceratites*, are here just of the intermediate character which ought to be detected in deposits connecting that system with the carboniferous. Some of them approach very closely to upper Silurian species, if indeed there be not some undistinguishable; while others, particularly some of the Trilobites, are of forms entirely different from any species hitherto found, either in the Silurian or Carboniferous systems.' (pp. 247–48, scored).

'In regard to the corals, Mr. Lonsdale informs us, that the few which he can identify with published species (the most abundant and certain being *Favosites polymorpha*, *Porites pyrimorphis*, and *Stromatopora concentrica*), belong to the upper Silurian rocks; while there are several which are new and undescribed. Again, the chain coral (*Catenipora escharoides*), and many of the most remarkable Silurian types, are entirely absent, nor has a single species common to the carboniferous limestone been yet detected among the numerous polypifers of South Devon.' (p. 248). Darwin scored the passage and commented 'analogous to Mammalia changing at diferent rates'.

126–2 Sedgwick and Murchison 1839:258–59, 'He [De la Beche] must, however, (after the recent publication of such large groups of Silurian fossils) before long perceive that the formations of South Devon not merely contain fossils approaching those of the *mountain limestone* (a fact long known), but that their whole suite of fossils is intermediate between those of the Silurian and Carboniferous Systems; a fact which at once defines their true place in the sequence of British rocks.' Passage scored.

127–1 See note E17–2.

127–2 See C163, QE4[a], Experiment (11), 'Keep. Tumbling pigeons. Cross them with other breed.—' See also QE20 and *Natural Selection*:485, n.2.

Whether the body of parent be altered, that is the Nisus formativus. (what does Muller call it)[3] succeeds in altering ‹or› form of body, or whether it merely has tendency (as effects of cultivation on successive generations of plants)[4] to do so, the effects are equally handed to offspring.—

128 Whewell's anniversary address 1839, p. 9.,— talks about fossil Infusoria becoming extinct not so soon as other forms.—[1] p. 36.. speaking about the controversy on Didelphys says. "If we cannot reason from the analogies of the existing to the events of the past world, we have no foundation for our science".—[2] ‹it is only analogy.› but experience has shown we can & that analogy is sure guide & my theory explains why it is sure guide.—

129e **Lychnis**

April 3d.— Henslow tells me following facts: believes that «only» red Lychnis grows in ‹south› Wales & certainly ‹old› only white in Cambridge, in some counties sometimes one & sometimes other.—[1] there is some difference of habit between these varieties, so that they have been thought to be different species. Lychnis dioica, generally dioicous. yet parts only very slightly abortive & bed of female flowers will sometimes produce a few seeds,— — Ruscus aculeatus. a diœcious plant, in which the Male plant sometimes

130e bears female flowers, the organs are most clearly abortive, so that they become so by suppression of one organ. (here language forces on us the change, which ‹to› seems to have taken place.— Almost all Diœcious & Monoœcious plants have rudimentary abortive organs, even more so than

128 can &] '&' *over* 'th'.
page crossed pencil.
129 **Lychnis**] *added across page pencil.*
page crossed pencil.
130 than Polygamia] 'th' *over* 'P'.
'10' *added brown crayon.*

127–3 Müller 1838–42 calls it the '*organic force*, which resides in the whole, and on which the existence of each part depends, [and] has however also the property of generating from organic matter the individual organs necessary to the whole.' (1:22), the '*creative force [that] exists already in the germ, and creates in it the essential parts of the future animal*' (1:23), the 'formative or organising principle, . . . [that in contrast to mind], is a creative power modifying matter, blindly and unconsciously, according to the laws of adaptation.' (1:25), the 'vital force' (1:24 *et seq.*), the 'vis essentialis' (1:48), etc. In Darwin's evolutionary translation this 'organizing' force is capable of varying under the influence of 'external agencies' that Müller calls 'the vital stimuli' (e.g. 'caloric, water, atmospheric air, and nutriment . . . that . . . induce constant changes in the composition of the organised body', 1:29). In animals these include the 'internal vital stimulus' of the nerves through 'their active force, probably an imponderable agent' (1:30). Darwin comments at the bottom of 1:30, 'The vital principle. produces the organs.— as the latter [vital stimuli] vary, so must this vital Principle.—'

127–4 See the report that *Dahlia* required successive generations of cultivation to sport varieties (E13 etc.).

128–1 Whewell 1838–42:9 discusses the work of C. G. Ehrenberg, 'Of about eighty species of fossil infusoria which have been discovered in various strata, almost the half are species which still exist in the waters: and thus these forms of life, so long overlooked as invisible specks of brute matter, have a constancy and durability through the revolutions of the earth's surface which is denied to animals of a more conspicuous size and organization.' Passage scored.

128–2 Whewell 1838–42:36. Passage scored.

129–1 J. S. Henslow: personal communication.

Polygamia: Monoœcia & Diœcia, preeminiently artificial, so that even some species only in genera ‹are› have this structure.—

Some willow trees have been observed to change their sex,— this effect from age, what Mr Knight[1]

131e–132e [not located]

133 the stigma retains its power.—[1]

R. Brown found the ‹poll› masses of pollen of Asclepias placed on Orchis (so very different) that the granules exserted their tubes:[2] now Mr Herbert has shown that stigma swells, when pollen even most remote is put to it.—[3]

April 6th
"Dr. Edwards on the Influence of ‹external› Physical agents". «translated by Dr. Hodgkins» p. 54. The axolotl, siren, & Proteus, affinity to tadpoles.[4] p. 210. Shows. that the action of light is concerned with the developement of form; but that tadpole increased in size

134 now the Proteus anguiformis. he remarks lives in dark caverns of Carniola[1] p. 112. Man. "standing alone in the gift of intellect, he resembles, other mammalia in the effects produced on organization. by *physical agents*."[2] p. 466. Many facts given of high temperature at which fish &c can live.—[3]

130–1 Thomas Andrew Knight.
133–1 For clues as to the meaning of this fragment, see note E133–3 and Darwin's comment on Herbert 1837:352 (bottom of the page): 'does the *multitude* plants preserve them, by allowing very many impregnations, the stigma keeping its power—with respect to wheat'. See also note E143–1 (Darwin's comment on Herbert 1837:371).
133–2 R. Brown 1833:728, 'Pollen masses of *Asclepias purpurascens* being applied to the stigma of *Epipactis palustris*, and immersed in its viscid secretion, the dehiscence, contrary to expectation, not only took place, but even more speedily than usual, that is within twenty-four hours. Some of the grains were also found discharged from the mass unchanged, while others, both discharged and still inclosed, had begun to produce tubes.' See abstract (DAR 74:169–72) and Darwin's questions for Brown and replies (DAR 74:173–74).
133–3 See Herbert 1837:350, 'The whole of my observations has led me to think, that at any period before the decay of the stigma the access of the natural pollen may supersede the influence of the foreign that may have been previously applied, if not from a closely allied species or variety; but that on the other hand no foreign pollen can act upon the germen after the stigma has been fertilized naturally.' Passage double scored.
133–4 Edwards 1832:54, 'There are three remarkable animals, which have a strong affinity to tadpoles, and have been considered as belonging to the family of batrachians, these are the axolotl, the siren, and the proteus.'
134–1 Edwards 1832:206–11, chap. 15 (On the influence of light upon the development of the body), 'We see then that the action of light tends to develope the different parts of the body, in that just proportion which characterizes the type of the species. This type is well characterized, only in the adult. The deviations from it are the more strongly marked the nearer the animal is to the period of its birth. If, therefore, there were any species existing in circumstances unfavourable to their further development, they might possibly long subsist under a type very different from that which nature had designed for them. The *proteus anguiformis* appears to be of this number. The facts above mentioned tend to confirm this opinion. The *proteus anguiformis* lives in the subterraneous waters of Carniola, where the absence of light unites with the low temperature of those lakes, in preventing the development of the peculiar form of the adult.' (p. 210).
134–2 Edwards 1832:112, Darwin's underlining in E134.
134–3 Edwards 1832:465–66, Thomas Hodgkin's notes for p. 56, *On the existence of fish, &c. in water of high temperature.*

Lyell says that naked cuttle fish now bear a very large proportion to other mollusca in cold parts of sea, like Cetaceæ,—[4] although the

135 Cephalopods, seem to have decreased since earliest times—

Apterix has a most perfect Struthio head pulled out. yet feathers retain character?

If separation[1] in horizontal direction is far more efficient in making species, than time (as cause of change) which can hardly be believed, then, uniformity in «geological» formation. intelligible.—

«— No. but the wandering & separation of a few, probably would be most efficient in producing new species; also one being reduced in numbers, but not so much these, because circumstances»

136 ‹April 12th ..› Cestracion, *Port Jackson Shark*— Owen thinks Australia part of Old World[1]

‹If› It «may» be said, that wild animals will vary, according to my Malthusian views, within certain limits, but beyond these not.— argue against this—— analogy will certainly allow variation as much as «the» difference between ‹pi› species,— for instance pidgeons— : then comes question of genera

---

It certainly appears that swallows have decreased in numbers, what cause??[2]

137 Seeing the beautiful seed of a Bull Rush I thought, surely no "fortuitous" growth could have produced these innumerable seeds— yet if a seed were produced with infinitesimal advantage it would have better chance of being propagated & so &c.

The greatest difficulty to my theory, is same type of shells in oldest formations:— The Cambrian formations do not however, extend round world.— Quartz of Falkland.— Old Red Sandstone— Van Diemen's land.— Porphyries of Andes.

138 A familiar History of Birds by the Rev. E. Stanley Vol I. p. 72.— Goldfinches placed near, but not in sight of each other will sing till they drop off their perch.—[1]

134–4 Charles Lyell: personal communication.
135–1 That is, geographical separation.
136–1 R. Owen: personal communication.
136–2 See Stanley 1835, 2:60–61, 'But we fear we are suggesting a needless remedy for an inconvenience not likely to recur; for, within the last few years, particularly since 1809, these pretty social Summer visitors, like our Starlings, have been decreasing in numbers, in the most unaccountable manner, not only in England, but in almost every part of the Continent.'
138–1 See note D103–1.

p. 101— Kingfisher in northern part of England stationary, in southern stays only winter.— Jays & chaffinches sometimes migratory.—[2]
p. 103. Turtles finding their way to the Caymans from Honduras.[3]
good case of migrating.— shows my theory insufficient.—
p. 120 An Eagle is said to have been seen carrying a lamb two miles towards the Morne Mountains, it

139e then dropped it & was found alive.—[1] **Stanleys Familiar History of Birds** several cases on record of stoats being carried (p. 121) & dropped having wounded the bird.[2] p. 124— Mr Willoughby found a dead lamb «& hare» by the side of Eagles nest,[3] which shows power of carrying great weight. p. 125 is said that Eagles bring rabbits & hares to the young ones to exercise them in killing them.— "Sometimes it seems hares, rabbits, rats & not being sufficiently weakened by wounds got off from the young ones while they were amusing themselves with them, and one day a rabbit escaped into a hole, where

140e the old Eagle could not find it..— The parent bird another day brought to her young ones the cub of a fox, which after it had fought well & desperately bitten the young ones, would in all probability have escaped".— if it had not been shot by ‹some› «a» shepherds, who was watching the scene.—[1] «In Shiant Isl^d. it is said, that an Eagle always procured its prey from another island.— »[2]
p. 175., 28 sho[r]t eared owls were counted in a field, where there was great swarm of mice.—[3]

141 May 4th.— The Brussels Sprout returning suddenly to type when brought back to home. (& yet all the varieties of Brassica certainly not becoming Brussels Sprouts) is analogous to Primrose & Cowslip *suddenly* changing into each other,[1] & depends on character of antecedent races.—[2] «& yet in all

139 **Stanleys Familiar History of Birds**] *added pencil.*
'18' *added brown crayon.*
140 was watching] 'was' *over* 'were'.

138–2 Stanley 1835, 1:101–2.
138–3 Stanley 1835, 1:103–4.
139–1 Stanley 1835, 1:120.
139–2 Stanley 1835, 1:120–21.
139–3 Stanley 1835, 1:124–25.
140–1 Stanley 1835, 1:125.
140–2 Stanley 1835, 1:126–27.
140–3 Stanley 1835, 1:175.
141–1 Mons 1820:199–200. See Darwin's abstract (DAR 74:61–62), 'p. 199. "M. J. B. Van Mons on the cultivation of Brussel Sprouts" This plant has been «much» said to ‹sp› degenerate.— in soil of Brussels it remains true & at Louvain, but at Malines (same distance from Brussels as Louvain), where greatest attention paid to growth of vegetables deviates from its character after first sowing. (yet at Brussels grow in any soil, sandy or clayey, fields, or gardens)—true seeds sent to Malines produced the sports in their true form: seeds from these being preserved ‹did not yield› gave plants, "in which the Sprouts did not form «little» cabbages, but were expanded; nor did they shoot again at the axils of the stem". Plants raised from these "only produced lateral shoots with weak pendant leaves, & tops similar to the shoots; so that in 3 generations the entire character of the original was lost".— [NB the influence must be on plant in its life time]^CD Seeds from the degenerated plant were sent back to Louvain & planted, the young retained their
(*141–1 continued & 141–2 overleaf*)

probability the Brussels Sprout was slowly formed.— »

---

if it shall be difficult to show that ‹time› the fixity of characters «from antiquity» prevents their variation, which is not improbable as Mr Herbert does not seem to recognize any difference in crossing between varieties & species,[3] yet the amount of

142 may depend on many circumstances, time of domestication [see Wikinson on dogs of Egypt[1] & Cuvier on Mummies[2]]CD [NB TIME *is element* in change, as in *Dahlias*[3]]CD all much varied breeds both plants & animals have long been subjected to domestication.— the constitution of some may resist the means Man can offer of changes.— as desert «or rock» plant probably would do—

(*141–1 continued*)
character throughout their growth; sowed their seed, the plant regained as much of its character, as it had lost in second generation: resown, & in 3d generation not to be distinguished from the true kind: [this sudden assuming character analogous to primrose & cowslip case.—?]CD.—CD[very curious case. showing effects of external conditions.—'

141–2 For further primrose to cowslip analogues from this period, see Darwin's abstract of Wilbraham 1818:60 (DAR 74:58), 'p. 60. The French call Nectarine a smooth Peach.— certainly same.— the appearance of Nectarines on branch of Peach trees. analogous to production of cowslips from primroses. with no intermediate gradations.—' and see Herbert 1837:347, '. . . it does not appear that in reality any plant becomes acclimated under our observation, except by crossing with a hardier variety, or by the accidental alteration of constitution in some particular seedling;' Darwin underlined 'accidental . . . seedling' and commented: 'This is like *sudden* appearance of cowslip from primrose it is analogous to Australian dogs, producing piebald young'. See also E111–1 for this passage in the context of acclimation.

141–3 See letter of 5 April 1839 from William Herbert (*Correspondence* 2:182–85), 'It stands to reason that the pollen of a plant which is not disposed to produce varieties must be less likely to produce hybrids of uniform appearance, than that of a plant which is not disposed to sport; but I consider the pollen of a permanent garden variety just as likely to produce an uniform effect as that of any easily convertible natural species. Take for instance the hollyhocks of the garden, without question cultivated varieties of one plant, yet *steadily reproducing their respective colours by seed*. I apprehend that the pollen of one variety of hollyhock on another would be neither more nor less decisive than that of one wild species of Calceolaria on another, at least taking the species wh. have similar constitutions & intermingle easily.' This forms part of Herbert's reply to Queries 1 and 2 of 'Questions for Mr Herbert' (*Correspondence* 2:182). Darwin's marginal comment on Herbert 1837:358, probably played a part in the way Darwin framed Queries 1 and 2: 'seems to leave out of question, greater indelibility of some stocks than others'. The importance Darwin attached to Herbert's rejection of 'Yarrell's law' is to be seen in Darwin's notes on his exchange with Herbert (*Correspondence* 2:204, *CD memorandum* following Herbert's letter of *c.* 27 June 1839), 'Mr Herbert considers «*without doubt*» old variety more likely to reproduce itself than new—but he throws doubt on this, as if not necessarily effect of ages, but of the chances.—

'Negatives a relation between facility in sporting and hybridisation.— . . . but believes there would be no difference between a permanent variety and easily convertible species: all facts about the reappearance of character must be inferred from the animal kingdom.

'The non relation of hybridizing power & variation goes far to overturn my views of fixity of character, being dependent on age, & consequently on time after certain period having any further effect on characters.— . . .

The ‹fixity of character› «resistance to hybridisation» being consequence of ‹old› «*age* of» species «(inexplicable by Mr Herbert)» is so simple an explanation, that I can hardly give it up, though ‹these› «some» species varying & yet not hybridising almost overthrows it.—'

142–1 Wilkinson 1837, 3:32–33.

142–2 Cuvier 1804.

142–3 See E13: 'Decandoelle distributed seeds of Dahlia all over Europe same year.— he sowed them for four generation before they broke.— , showing effects of cultivation gradually adding up.' See also Darwin's abstract of Sabine 1820:225 'Mr. Sabine on the genus dahlia (There are two species.) "it seems as if some period of actual cultivation were required, before the fixed qualities of the native plant gave way & began to sport into those changes, which now so much delight us".—' (DAR 74:62).

or be with difficulty be kept alive.— Nevertheless much probably depends on circumstances favouring the reappearance of characters, formerly possessed— ‹that is animals› «or rather the parents» having passed through many changes.—

143 It is very important Mr Herberts fact about the hybrids (mentioned in letter to Henslow) fertilizing each other, better than the pollen of same flower,—[1] as it tends to show my view of ‹i›nfertility of hybrids «with parent species» false, which makes it determined by a facility in returning to old type

---

Mr Herbert showing the extreme facility of crossing, in plants proves how much depends on instincts in animals.— yet the existence of wild close species of plants shows there is tendency to prevent the crossing.—

144 in animals where there is much facility in crossing there comes the impediment of instinct—

---

the possibility of rearing by seeds Holyoaks— (how far is this so) shows either there is not so much crossing as I think,, or that these varieties have become as fixed as species, & prefer their own pollen to that of other variety.—[1]

«Elizabeth & Hensleigh.[2] seemed to think it absurd. that the presence of of the Leopard & Tiger together depended on some nice qualifications each possess., & that tiger springing an inch further would determine his preservation— if killed by some other animal, then that quality which saved him, would be the one encouraged— »

145 Wilkinsons Manners & Customs of the Ancient Egyptians Vol III. p. 33— They had several breeds of dogs.— like greyhound— fox-dog— turnspit & two other kinds[1]

---

144 «Elizabeth . . . encouraged»] *read as comment on 145 second passage.*
145 'It seems . . .' *to end of 152 written in pencil.*

143–1 *Correspondence* 2:202–3, 'The experience of 4 seasons has shewn that it is certain that, if taking two hybrid Hippeastra wh. have (say) each a 4-flowered stem, 3 flowers on each are set with the dust of the plant itself & one on each with the dust of the other plant, that one on each of them will take the lead & ripen abundant seed, & the other 3 either fail or proceed more tardily & produce an inferior capsule of seed.' See also Herbert 1837:371, 'It is only from the superior efficacy of the pollen of another plant, that we can account for the circumstance of some hybrid plants, which breed freely with plants of either parental stock and fecundate them, not producing seed readily when left to themselves; . . .'. Darwin wrote two notes: 'Probably stigma would actually prefer pollen of other plant; as stigma remains open to choice— & as in Mammalia «bred in & in», loose passion (but I do not know whether perhaps other kind)—' and 'My theory explains this: because offspring differ in the two cases, in one going back to parent, & in other remaining constant.'

144–1 See reference to hollyhocks in note E141–3. See QE2, Question 5.

144–2 Sarah Elizabeth Wedgwood and Hensleigh Wedgwood. See E144.

145–1 Wilkinson 1837, 3:33. See E142.

It seems absurd proposition, that every «budding» tree, & every buzzing insect & grazing animal owes its form, to that form being «the one *alone*» out of innumerable other ones, ‹*alone*› «which has been» preserved.— but be it remembered. how little part of the Grand Mystery is this,— the law of growth, that which changes the acorn into the oak.— In short all which «Nutrition, growth & reproduction» is common to all living beings. vide Lamarck Vol II.[2] p. 115. 4 four laws[3]

146 Who can say, how much structure is due to external agency, without final cause. either in present, or past generation.— thus cabbages growing like Nepenthes.—[1] cases of pidgeons with tufts &c &c here there is no final cause yet it must be effect of some condition of external circumstances. results of complicated laws of organization: as we see these strange plumage in pidgeons yet no change of habits, so no ‹cause›

147 corresponding change in Birds of Paradise.— All that we can say in such cases, is that the plumage has not been so injurious to bird as to allow any other kind of animal to usurp its place.— & therefore the degree of injuriousness must have been exceedingly small.— This is far more probable way of explaining, much structure, than attempting anything about habits— no one can be shocked at absence of final cause mammæ in man & wings under united elytra

148 The law of ‹growth› «generation» is only modification, though important one, of growth «Lamark. Vol II. p. 120. observes it commences only, when growth stops».—[1] Spallanzani's facts in connection with buds.—[2] They differ

148 **it is doubtful . . . grafted.—**] *added ink.*
No] *added pencil in box.*

145–2 Lamarck 1830, vol. 2. According to the 1875 MS catalogue of Darwin's library (Darwin Archive, Cambridge University Library), Darwin possessed both volumes of Lamarck 1830. By the time Rutherford 1908 was compiled, vol. 2 was missing. Note that the pagination is identical to the more familiar 1809 edition of *Philosophie Zoologique*.

145–3 Lamarck 1830, 2:115, 'Les facultés communes à tous les corps vivans, c'est-à-dire, celles dont ils sont exclusivement doués, et qui constituent autant de phénomènes qu'eux seuls peuvent produire, sont:

'1°. De se *nourir* à l'aide de matières alimentaires incorporées; de l'assimilation continuelle d'une partie de ces matières qui s'exécute en eux enfin, de la fixation des matières assimilées, laquelle répare d'abord avec surabondance, ensuite plus ou moins complétement, les pertes de substance que font ces corps dans tous les temps de leur vie active;

'2°. De *composer leur corps*, c'est-à-dire, de former eux-mêmes les substances propres qui le constituent, avec des matériaux qui en contiennent seulement les principes, et que les matières alimentaires leur fournissent particulièrement;

'3°. De se développer et de s'accroître jusqu'à un certain terme, particulier à chacun d'eux, sans que leur accroissement résulte de l'apposition à l'extérieur des matières qui se réunissent à leur corps;

'4°. Enfin de se régénérer eux-mêmes, c'est-à-dire, de produire d'autres corps qui leur soient en tout semblables.'

146–1 *Nepenthes* = pitcher plant.

148–1 Lamarck 1830, 2:119–20, 'Ainsi, la *reproduction*, troisième des facultés vitales, tire, de même que l'accroissement, son origine de la nutrition, ou plutôt des matériaux préparés pour la nutrition. Mais cette faculté de reproduction ne commence à jouir de son intensité que lorsque la faculté d'accroissement commence à diminuer: on sait assez combien l'observation confirme cette considération; puisque les organes reproducteurs (les parties sexuelles), dans les végétaux comme dans les animaux, ne commencent à se développer que lorsque l'accroissement de l'individu est sur le point de se terminer.'

148–2 Spallanzani 1769, see E90, QE1[v].

from possibility of concourse of two «individuals» & the action *always* of two organs— instead of one part «as» in producing bud.— Fewer of the *lately acquired* peculiarities are transmitted **it is doubtful whether any are transmitted, for the changes in fruit trees. mentioned by Mr K[3] may be caused by the diversity of stocks, on which they are grafted.— No** than by growth— generation; & more of the effects of conditions on the «propagating» constitution. but not structure of the parents.— Thus would a Crab

149 tree vary if planted in rich soil, I presume not, but its seeds, I presume, probably would— at least the experiment of the carrot seems to show this.—[1] This would be curious law, Certainly Australian Dog is not affected by domestication, yet offspring are,— if Australian Dog, could bud, analogy tells us, ‹be› «offspring» would be similar to ‹f› first form.— The great effect of conditions on offspring, but not on individuals is very curious & important.—

150 The existence of "laws of organization" had better be shown— soil on colour of flowers, Hydrangea[1] — black bullfinches— & all varieties must be presumed to be result of such laws.— The effect of one part being greatly developed on another, must not be overlooked.— it makes fourth cause or law of change.—[2]

———

The weakest part of my theory is, the absolute necessity, that every ‹animal› «organic being» should cross with

151 another.—[1] to escape it «in any case» we must draw such a monstrous conclusion, that every organ is become fixed. & cannot vary.— which all facts show to be absurd.— As there are plants, in northern latitudes, which are generated by buds alone or roots, & never flower, so there may be animals as Coralline, or others. which only generate once in a thousand

149 The great . . . important.—] *double scored pencil in both margins.*
150 ‹animal›] *unusually heavy deletion.*

148–3 Thomas Andrew Knight.
149–1 See QE1v: 'Repeat the French experiment of Carrot'.
150–1 See Darwin's abstract of Hedges 1820:173–77, 'Mr Hedges, experiments on changes of colour in Hydrangea Hortensis.— He found the light brick earth from Hampstead Heath always turned the red flowers of the H. into blue—the next year if put in garden compost they would turn ‹blue› red again— & again returned— No cause could be discovered in nature of soil.— Bog-soil will sometimes effect it.—' (DAR 74:61).
150–2 Presumably a reference to Lamarck's four laws. See note E145–2. Perhaps Darwin intended to write 'fifth law'. Alternatively this was a fourth principle added to the three of E58.
151–1 See draft of letter to William Herbert, *c.* 27 June 1839 (*Correspondence* 2:201–2), '. . . I have been led to believe, that amongst organic being producing seminal offspring there exist no such thing as a true permanent hermaphrodite—ie. that every individual occasionally, though perhaps *very rarely*, after long intervals is fecundated by a other indiv, in short that almost ‹every› plant is occasionally fecundated as in as in dioecious genera—' Herbert's reply may account for Darwin's discouragement in E150: 'The objection to your theory wh occurs to me is this. In some genera the facility of breeding crosses is great; in Hippeastrum there seems even a decided preference for foreign pollen: but if the fecundation by another individual was essential, the cases of wild cross-bred plants would be frequent instead of being rare.' (*Correspondence* 2:202–5, from Herbert, *c.* 27 June 1839).

generations.— any amount of generation may take place by gemmation «My theory will not admit this, now that tulips break by cultivation can a form become permanent?» because its very essence is

152 that little change is produced.— The fact just alluded to of Northern flowers, throws enormous difficulty in the way of Mr Knights. theory[1] «without seeds are freshly transported»— throw over this theory, & the sexual reproduction of species may stop for any number of generations— Gorze in Norway, which never flowers!!— ‹How did it get there? whether›

153 According to the above suggestion my theory would require, that ‹species› «individuals» propagated by gemmation should be absolutely similar; [all the gorze in Norway ought to be thus characterized study Von Buch.][1CD] Now Mr Knights statements about fruit trees. grafted. altering is hostile to this:[2] but on other hand, fruit trees are propagated by means, which wild plants never are, namely on stocks of other varieties & we know that the kind of

152 the way . . . number of] *scored & '?' added pencil.*

152–1 Knight 1799:202–3, 'Many [crossing] experiments, of the same kind, were tried on other plants; but it is sufficient to say, that all tended to evince, that improved varieties of every fruit and esculent plant may be obtained by this process, and that "nature intended that a sexual intercourse should take place between neighbouring plants of the same species." . . . An examination of the structure of the blossoms of many plants, will immediately point out, "that nature has something more in view, than that its own proper males should fecundate each blossom"; . . . But the farina is often so placed, that it can never reach the summit of the pointal, unless by adventitious means; and many trials have convinced me, that it has no action on any other part of it." In promoting this sexual intercourse between neighbouring plants of the same species, "nature appears to me to have an important purpose in view; for, independent of its stimulative power, this intercourse certainly tends to confine within more narrow limits, those variations which accidental richness or poverty of soil usually produces."' Darwin marked the passage by adding quotation marks (").

153–1 See C178 for discussion of Buch 1813:306–7; gorse is not mentioned. Also see QE2: Questions Regarding Plants. (11), 'Is not non-flowering gorze common in Norway No'.

153–2 Knight 1818a:160. 'New varieties of fruit are generally supposed, by gardeners, to be obtainable from seedling plants only; and every part of each seedling tree, which has been detached as a graft or bud, is usually believed to be capable of affording fruit of the same kind, if subsequently grafted upon the same stock, and cultivated in the same manner. This opinion I also formerly entertained; though I was always at a loss to account for the existence of many kinds, which were obviously different, and which yet much more closely resembled each other, than any varieties which I had ever been able to obtain from seeds. But I am now most perfectly satisfied, that many varieties of fruit, which are supposed to be totally distinct, have been propagated from branches of the same original tree; and that few, if any, varieties of fruit can, with strict propriety, be called permanent, when propagated by buds or grafts.

'I have witnessed many instances of the variations above mentioned, but much the most extraordinary of these occurred in my garden, in the last autumn. A tree of the *Yellow Magnum Bonum Plum* . . . had always borne fruit of the usual colour; but, in the last year, one of its branches produced red fruit in every respect perfectly similar to the well known *Red Magnum Bonum Plum*. . . . I am most perfectly confident, that . . . the branches abovementioned had [not] sprung from an inserted bud or graft; and I can not therefore hesitate to decide, that the *Red Magnum Bonum Plum* is a variety only of the *Yellow*.' See also Darwin's abstract (DAR 74:58). 'p. 160 Mr. K. says tree of Yellow Magnum Bonum, 40 years old, had always borne ordinary fruit, produced a branch yielding Red Magnum Bonum.— Also he had a My Duke Cherry. one branch of which (*certainly* not grafted) always produce more oblong fruit & later than the other branches.— The Green-gale. Golden Pippin & Non-pareil have *sported considerably* by being often grafted. "few if any varieties of fruit can with strict propriety be called permanent when propagated by buds or grafts".— p. 160.— CD[so that propagation by seed merely exposes the plant to external influences more completely, than it is exposed during life: .—No doubt these changes «during life» might be propagated.'

stock greatly affects the Graft.— ‖Plants circumstanced as the Gorze must be propagated by its roots: now it is curious Mr K. has observed that to graft from the roots is the best way to get young trees, from worn-out

154 kinds,[1] & quotes from Pliny, that it is bad to graft from top shoots.—[2] If prolongation of life by gemmation ‹can be› being impossible. can be overturned,[3] then the conclusion that the two kinds of generation have some most important difference is forced on us.—[4] My theory only requires that organic beings propagated by gemmation do not now undergo metamorphoses, but to arrive at their present structure they must have ‹done› been propagated by

155 sexual commerce «The fact of Corallina & Halimeda is case in point».—[1] The relation of these «sexual» functions to complexity is evident, yet the inference from *sọme* plants & *some* mollusca being hermaphrodite is, that intercourse every time is of no consequence in that degree of developement.— [It is singular there is no true hermaphrodite in beings with **which have** fluid sperm.—]CD

I utterly deny the right to argue against my theory, because it makes the world far *older* than what Geologists, think: it would be doing, what

156 others but fifty years since to geologists.— & what is older— what relation in duration of a planet to our lives— Being myself a geologist, I have thus argued to myself, till I can honestly reject such false reasoning

157 Bell Bridgewater's Treatise on the Hand.— p. 94.—[1] "The resemblance of the foot the Ostrich to that of the Camel has not escaped Naturalists." Before

155 **which have**] *added pencil.*
'I utterly . . .' *to end of 156 written in pencil.*

154–1 Knight 1818b:253, 'I obtained plants from some detached parts of the extremities of the roots of old ungrafted *Pear* and *Apple*-trees, and as soon as these were large enough to afford grafts, I selected other grafts of similar size from the bearing branches of the same trees, and some of each were inserted in similar stocks, and in several instances two in the same large stock; and wherever inserted, the grafts which had been taken from the bearing branches proved by no means able to contend with their more hardy vigorous rivals.'

154–2 Knight 1818b:252 quotes Virgil's *Georgics* to this effect.

154–3 See Knight 1818b:252, 'The progressive influence of debility and decay upon old varieties of fruit trees is now so generally admitted, that it is wholly unnecessary to advance facts or arguments to prove it: the general law of Nature appears to be, that no living organized being shall exist beyond a limited term of years; and that law must be obeyed. It is nevertheless in the power of man to extend the lives of individual vegetable beings far beyond the period apparently assigned by nature; and parts of the same annual plant may be preserved through many years, perhaps through ages, though it cannot be rendered immortal.' Also see Darwin's abstract (DAR 74:60). 'p. 252. Mr. K. allows even life of annual might be prolonged for vast period. though not for eternity—the roots appear younger than the branches, or more vigorous.' see also E184.

154–4 See B1.

155–1 See E151: 'As there are plants, in northern latitudes, which are generated by buds alone or roots, & never flower, so there may be animals as Corallina, or others. which only generate once in a thousand generations.' See also RN132.

157–1 C. Bell 1833:94.

he alludes to the resemblances of the snout of the mole & Pig in having two additional bones to give strength to it.—[2] p. 139. Doubts altogether the law of balancing of organs.—[3] In the Batracian Order the «32» ribs are wanting. p. 144 in the Icthyosaurus 60 or 70 bones in the paddle, yet all in the arm are perfect.—[4] p. 144.— Alludes to two theories;— that species are the result of circumstances,;— or the will of the Animal.[5]

158 p. 145. Seems to argue, that as the transformations from the egg, or larva. or fœtus to perfect animal are adapted by foreknowledge, so must the mutations of species.!!—[1] p. 203 Chætodon squirting water at fly.— instinct, for how could experience teach distances in air, in which it never touches objects.—[2]

157–2 C. Bell 1833:68–69n, 'The [mole's] snout may vary in its internal structure with new offices. Naturalists say that there is a new "element" in the pig's nose; it has, in fact, two bones which admit of motion, whilst they give more strength. Moles have those bones also, as they plough the earth with their snouts.'

157–3 C. Bell 1833:139, 'In the batrachian order, the ribs are wanting: where then are we to look for them? Shall we follow a system which informs us that when a bone is wanting in the cavity of the ear we are to seek for it in the jaw; and which, yet, shall leave us in the contemplation of this class of animals deficient in thirty-two ribs, without pointing out where they are to be found, or how their elements are built up in other structures? If, on the contrary, we take the principle that parts are formed or withdrawn, with a never failing relation to the function which is to be performed, we see that no sooner are the compages of the chest removed, and the shoulder thus deprived of support, than the bones to which the extremity is fixed are expanded and varied, both in form and articulation, so as to fulfil their main object of giving security and a centre of motion to the arm.'

157–4 C. Bell 1833:144.

157–5 C. Bell 1833:144–45, 'It is, above all, surprising with what perverse ingenuity men seek to obscure the conception of a Divine Author, an intelligent, designing, and benevolent Being—rather clinging to the greatest absurdities, or interposing the cold and inanimate influence of the mere "elements," in a manner to extinguish all feeling of dependence in our minds, and all emotions of gratitude.

'Some will maintain that all the varieties which we see, are the result of a change of circumstances influencing the original animal; or that new organs have been produced by a desire and consequent effort of the animal to stretch and mould itself—that, as the leaves of a plant expand to light, or turn to the sun, or as the roots shoot to the appropriate soil, so do the exterior organs of animals grow and adapt themselves. We shall presently find that an opinion has prevailed that the organization of animals has determined their propensities; but the philosophers, of who we are now speaking, imagine the contrary,—that under the influence of new circumstrances, organs have accommodated themselves, and assumed their particular forms.'

158–1 C. Bell 1833:145–49, 'We do perceive surprising changes in the conformation of animals. Some of them are very familiar to us; but all show a foreknowledge and a prospective plan,—an alteration gradually taking place in preparation for the condition, never consequent upon it.— (pp. 145–46). 'For example, if we examine the larva of a winged insect, we shall see the provisions for its motion over the ground, in that condition, all admirably supplied in the arrangement of its muscles, and the distribution of its nervous system. But if, anticipating its metamorphosis, we dissect the same larva immediately before the change, we shall find a new apparatus in progress towards perfection; the muscles of its many feet are seen decaying; the nerves to each muscle are wasting; a new arrangement of muscles, with new points of attachment, directed to the wings instead of the feet, is now visible; and a new distribution of nerves is distinctly to be traced, accommodated to the parts which are now to be put in motion. Here is no budding and stretching forth under the influence of the surrounding elements; but a change operated on all the economy, and prospective, that is, in reference to a condition which the creature has not yet attained.' (pp. 147–48). See also p. 149: 'Every thing declares the species to have its origin in a distinct creation, not in a gradual variation from some original type; and any other hypothesis than that of a new creation of animals suited to the successive changes in the inorganic matter of the globe—the condition of the water, atmosphere, and temperature—brings with it only an accumulation of difficulties.'

158–2 C. Bell 1833:202–203, 'We have a more curious instance of the precision of the eye and of the adaptation of muscular action in the *choetodon rostratus*. This fish inhabits the Indian rivers, and lives on the smaller aquatic flies. When it observes one alighted on a twig or

far better case than chicken pecking fly.— "whilst the shell stuck to its tail" as mentioned by Sir. J. Banks. p. 212.—[3] p. 282. Allows this instinctive power in chicken, yet says it is evidently acquired by experience in baby[4]

159 Lamarck. Vol II p. 152.— Philosophie Zoologie.

———

says it is not sufficiently proved that any shell fish is really hermaphrodite. & ‹thinks› even oyster may fecundate each other, by the means of the medium in which they live[1]

———

do. "Additions". p. 454.— does really attribute metamorphoses to *habits* of animals & takes series of flying mammifers—[2] says lemur.— volans, has skin between its legs.— — strangely consider existing «long-organized» forms as parent forms of existing highly organized forms—[3] this resulted from the necessity of supposing some inward progressive developing power.—[4]

flying near (for it can shoot them on the wing) it darts a drop of water with so steady an aim as to bring the fly down into the water, where it falls an easy prey. . . . Whether led to admire the wonderful power of instinct in these inferior creatures, or the property acquired by our own eye, we must acknowledge here a compound operation.*' Bell's note *: 'In these instances a difficulty will readily occur to the reader; how does the fish judge of position, since the rays of light are refracted at the surface of the water? Does instinct enable it to do this, or is it by experience?'

158–3 C. Bell 1833:212, 'The late Sir Joseph Banks, in his evening conversations, told us that he had seen, what many perhaps have seen, a chicken catch at a fly whilst the shell stuck to its tail.'

158–4 C. Bell 1833:282, 'This faculty of searching for the object is slowly acquired in the child; and, in truth, the motions of the eye are made perfect, like those of the hand, by slow degrees. . . . It is no contradiction to this, that the faculty of vision is made perfect in the young of some animals from the beginning; no more than that the instinct of the duck, when it runs to the water the moment that the shell is broken, should contradict the fact that the child learns to stand and walk after a thousand repeated efforts.'

159–1 Lamarck 1830, 2:151–52, '. . . il faudra distinguer l'hermaphroditisme parfait qui se suffit à lui-même, de celui qui est imparfait, en ce qu'il ne se suffit pas. En effet, beaucoup de végétaux sont hermaphrodites, en sorte que l'individu qui possède les deux sexes, se suffit à lui-même pour la fécondation: mais dans les animaux en qui les deux sexes existent, il n'est pas encore prouvé, par l'observation, que chaque individu se suffise à lui-même; et l'on sait que quantité de *mollusques*, réellement hermaphrodites, se fécondent néanmoins les un les autres. A la véritée, parmi les mollusques hermaphrodites, ceux qui ont une coquille bivalve, et qui sont fixés comme les *huîtes*, semblent devoir se féconder eux-mêmes: il est cependant possible qu'ils se fécondent mutuellement par la voie du milieu dans lequel ils sont plongés. S'il en est ainsi, il n'y a, dans les animaux, que les hermaphrodites imparfaits; et l'on sait que dans les animaux vertébrés, il n'y a même aucun individu véritablement hermaphrodite. Ainsi, les hermaphrodites parfaits se trouveront uniquement parmi les végétaux.'

159–2 Lamarck 1830, 2:454–55, 'En effet, les écureuils volans (*sciurus volans, aerobates, petarista, sagitta, volucella*), moins anciens que ceux que je vais citer, dans l'habitude d'étendre leurs membres en sautant, pour se former de leur corps une espèce de *parachute*, ne peuvent faire qu'un saut très-prolongé lorsqu'ils se jettent en bas d'un arbre, ou sauter d'un arbre sur un autre qu'à une médiocre distance. Or, par des répétitions fréquentes de pareils sauts dans les individus de ces races, la peau de leurs flancs s'est dilatée de chaque côté en une membrane lâche qui réunit les pattes postérieures à celles de devant, et qui, embrassant un grand volume d'air, les empêche de tomber brusquement. Ces animaux sont encore sans membranes entre les doigts.'

159–3 Lamarck 1830, 2:455, 'Les galéopithèques (*lemur volans*), plus anciens sans doute dans la même habitude que les écureuils volans (*pteromis* Geoffr.), ont la peau des flancs plus ample, plus développée encore, réunissant non-seulement les pattes postérieures aux antérieures, mais en outre les doigts entr'eux, et la queue avec les pieds de Derrière.' This passage is in 'Additions Relatives aux Chapitres VII et VIII de la première partie'.

159–4 See *inter alia* Lamarck 1830, 1:65, 'Que la faculté d'accroissement dans chaque portion du corps organisé étant inhérente aux premiers effets de la vie, elle a donné lieu aux différens modes de multiplication et de régénération des individus; et que par-là les progrès acquis dans la composition de l'organisation et dans la forme et la diversité des parties, ont été conservés;'.

160 My theory leaves quite untouched the question of spontaneous generation.—[1]

Introduction to Bartram's Travels p. XXIII.
‹Some birds› Both sexes of some birds sing equally well. and ‹in› these reciprocally assist in domestic cares, as building nest, sitting on eggs. & feeding & defending their young..— The oriolus (icterus Cat.) is an instance of this, & the female of the icterus minor is a bird of more splendid plumage than the male.—[2]

161 Athenæum May 18. 1839. p. 377.— Statement that the climate is on the decline, as far as vegetation is concerned, in parts of the Northern «French» expedition,—[1] rather the reverse of facts stated by Smith of Jordan Hill.—[2]

---

162 May 27th.—
Henslow[1]
One of the 4 species of Lemna only reproduces itself «in England, as yet observed» by buds— (the other three by buds & seeds «though by the latter very rarely»)) here is a case in answer to Mr Knights doctrine.—[2]
Case like Corallina— «Does it flower anywhere?— **Yes on the continent** is there more variation in its character.?» **No—well characterized.—**

Tulips are cultivated during several years & then they break—— each tulip is the ‹of› product of fresh bud—[3] here then is case of change analogous to change in grafted trees «:so is not effect of different stocks in this case»[4].— & strong case showing analogy of production by gemmation & by seed— which Henslow is inclined to think very close.— «A fruit tree by certain treatment will suddenly send forth quantities of blossoms—»

163 The case of the Lemna, «and the vivaparous grasses, which no doubt are propagated during hundreds of years, without fresh seeds arriving.»— throws a very great difficulty in my theory, here we have a plant remaining constant, without crossing.— & propagation by buds does not insure constancy of form.— is the constancy owing to similarity of conditions— & that no change would affect them in short period & hence no change would effect them, without affecting all the individuals— «— hence there would be real

162 **Yes on the continent**] *added brown ink.*
No—well characterized.—] *added brown ink.*

160–1 See Lamarck 1830, 2:61–90, chap. 6, Des générations directes ou spontanées.
160–2 Bartram 1791:xxiii.
161–1 Anonymous 1839a:377, 'We learned one thing, however, which is not without interest, concerning the climate. It has long been believed that vegetation, in the more northern parts of Lapmark, is constantly on the decline; and large tracts of land are found under the lee of the mountain, formerly covered with fir woods, where now only stumps and rotten roots of fir trees, with a few miserable birch, are to be seen.'
161–2 See J. Smith 1838:393.
162–1 J. S. Henslow: personal communications, E162–165.
162–2 See E151–152.
162–3 See E151.
162–4 See Knight's *Red Magnum Bonum Plum* case in note E153–2.

gradations in species from one region to another.—» — these simple forms perhaps oldest in world & hence most persistent— if form exceedingly difficult to vary.— the run of chances, would prevent it varying.

164 A plant ‹producing› propagating itself by buds is in same predicament, as one, in which structure does not allow of crossing with other individuals, «with facility»— such as cryptogamic plants & true hermaphrodite Mollusca, & probably corals.— these forms then ought to be very persistent,, & then necessity of crossing is much less,— now certainly in the higher animals; changes seem to have been more rapid, & the facility for inter marriage is greater (Hence Diœcious plants highest,— Palms &c &c)— Is there greater resemblance between carboniferous. «& recent» mollusca, than between the corresponding acalepha?— But if Acalepha do not cross there would by my theory gradation of form from one species to other: therefore my theory does require crossing.— The case of Lemna shows dispersion of germs is not end of seminal reproduction.— likewise grasses. &— very heavy seeds.— as Cocos do mer.— Analogy shows some most important end.— **Festuca vivapara F ovina— propagated like oni[on] Poa alpina because vivaparous. Henslow has seen this— (Poa alpina vivaparous sometimes seeds**

All species of Lemna sometimes though very rarely flower [bu]t the one does on the continent— well characterised species Periwinkle wants insects to impregnate allied to Asclepias

---

Turpin cell is individual

165e May 29th.— — Henslow says he has not the slightest doubt that Festuca vivapara is the same species with *F. ovina*, ‹& this› rendered vivaparous by growing on heights.—[1] yet he has seen it propagated in a garden, which is case precisely analogous to the Canada onion mentioned in Hort. Transact.[2] *Aira cæspitosa* becomes vivaparous on mountains & yet can be raised in gardens.— Poa alpina, thougt generally vivaparous sometimes seeds.—

164 A plant . . . important end] *written in brown ink over* '**Festuca vivapara . . . is individual**' *which was written in pencil and erased.*
All . . . Asclepias] *brace left margin.*
165 on heights] 'o' *on stub.*
raised] 'r' *on stub.*
sometimes] *first* 's' *on stub.*
Turpin says] *first* 's' *on stub.*
page crossed pencil.

165–1 J. S. Henslow: personal communication.
165–2 See Darwin's abstract of Strachan 1820:378, 'Mr C. Strachan in his "Account of the different varieties of the Onion". says the Tree or Bulb-bearing Onion was long erroneously supposed to be a species. but Mr Gawler (1812) in Botanical Magazine page 1635 plate 1469 has shewn origin to be the *Allium cepa* ‹growing› «being cultivated» in the gardens of Canada, the climate was too cold for it "to ‹produce› «allow it to» flower & seed «freely» & so becomes vivaporous (bearing bulbs instead of flowers) & retained its habit when brought here".— [good case of origin of new variety, also very curious]CD' (DAR 74:63). See also QE2, Question 3.

╫There are endless curious facts about every part of plant producing buds, so that Turpin says each cell of plant is individual.—[3] Most plants which propagate rapidly by buds, layers &c & &— do not seed freely.—

The periwinkle seldom produces seeds, because it is thought to require insects to impregnate it.— it is allied to Asclepias, where this is always the case according to Brown.—[4]

166e Voyage of Adventure & Beagle
Vol I. p. 306 *Shells*, as well as plants «of Juan Fernandez» differ from American Coast Vol II.— ‹Reference› p. 251. about the drifting of animals on ice[1]

---

p. 643— very curious table of all the castes from Stephenson at Lima[2]

The same numerical relation (both in species and subgenera) between the Crag & Touraine beds, the one with neighbouring & Arctic sea, & the other with neighbouring & Senegal as sea.— is remarkable.—
Again the resemblance between the Superga & Paris, numerically

167e the same with recent & yet almost wholly different, is same, as if Isthmus of Panama.— These two cases highly improbable.— yet I can see no other way of accounting for them.— Think over this— The Superga beds have many shells in common «& are not far distant» with Touraine «which as L. says is strong argument for their contemporaneous»— how is this with the Eocene beds.— see Lyells tables[1]

Bennetts Wandering Vol II. p 155. By inference I imagine that there are Baboons in St Thomas on W. coast of Africa[2]

168e Owen Linn. Soc. April 2^d^. 1839[1]
The Lepidosiren— Amblyrhyncus & Toxodon, ‹all› equally aberrant— the two former connecting classes like Toxodon «In orders»— Fish & reptiles in former case— Reptiles & Birds & Mamm. in ornityhyrhycus— is not this right?—

---

166 Voyage . . . Lima] '19' & '18' *added brown crayon.*
page cut in two: the section beginning 'The same numerical' *was formerly pinned to 167;* '22' *added brown crayon to this section.*
168 *page crossed pencil.*

165–3 Turpin 1827.
165–4 R. Brown 1833:717 cites Sprengel 1793 to this effect.
166–1 FitzRoy 1839, 1:306; 2:251.
166–2 FitzRoy 1839, 1:643.
167–1 Lyell 1830–33, vol. 3, Appendix 1.
167–2 G. Bennett 1834, 2:155 (quoting J. Ogilby 1671), 'This island, which Hanno then found, can be no other but that which we call St. Thomas; and the hairy people which he makes mention of were babeons, or baboons, which Africa, in this place, breeds large, to the amazement of the beholders.'
168–1 See R. Owen 1841b.

June 18th. Eyton[2] tells me, that Yarrell[3] knows of a Gull, which has laid in domestication eggs of two shapes & colour.— Eyton has observed same thing in *Brent Goose*.

169e Eyton says some of the pidgeons in common Dovecot are very like a Himalaya species — *leuconotes*—

Magazine of Nat. History. 1839. p. 106.— Waterhouse refers to fossil remains of the Hamster.— is not this Siberian animal?—[1]

Eyton says that the young of *two* hatches «all alike» between the male Chinese & female common goose took after the common goose[2] thus contradicting (probably) Yarrells law & Walkers of the male giving form— they interbred. & the young kept constant. & all alike[3]

170e Waterhouse says some of the Galapagos Heteromerous insects come very near to Patagonian species—[1]

---

p. 18. of Temmincks. Preliminary discourse to Fauna of Japan— that the «animals of» islands N. of Timor are allied to the «type of genera in» islas de Sonda as well by those which are identical, as those which are different.—[2] now this is same, as Galapagos facts &c &c.— & it shows the causes which give same species to different isl^{d}. is the same as that which gives genera.— ‹it is not transportation› now in case of large

171e–172e [not located]

173e Mr Greenough[1] on his Map of the World, has. written. Mastodon found at Timor.— thinks he has seen specimen at Paris Museum.—

Athenæum: 1839. p. 451. Sheep Merinos from Cape of Good Hope,. has different constitution from those of Europe— for they stand India. better than the latter—[2]

---

169 Eyton says some . . . animal?—] *crossed pencil.*
Eyton says that . . . & all alike] '17' *added brown crayon.*
170 shows] *second* 's' *on stub.*
page crossed pencil.
173 Mr Greenough . . . Museum] '19' *added brown crayon.*
Forrest Voyage] 'y' *over* 'l'.
Forrest . . . parrots.] '19' *added brown crayon.*

168–2 T. C. Eyton.
168–3 William Yarrell.
169–1 Waterhouse 1839b, 3:279 n.1.
169–2 See Eyton 1837a:358, B30 and DIFC
169–3 See Darwin's note on end paper of Walker 1838, 'Experiments.— To cross some very artificial male, with old female—according to Mr Walker, the former ought to preponderate in body—according to my Yarrell's theory, «old age time as described» the father, ought, either in first breed or permanently.—'
170–1 See Waterhouse 1845.
170–2 Temminck 1833:18.
173–1 George Bellas Greenough.
173–2 Anonymous 1839b.

Forrest Voyage p. 323. Sooloo. imported elephants. wild hogs— spotted deer, no loonies, but cocatores & small green parrots.[3]

174e June 26th— Yarrell.:— Black Swan «in domestication & nature» strictly monogamous— geese polygamous (?when wild) but only some birds are so when wild— wild ducks monogamous; tame ones highly polygamous— change of instinct by domestication.—[1]

"Notices of the Indian Archipelago" Published at Singapore in 1837. by Mr. J. H. Moor— — p. 1. Elephant. Rhinoceros Leopard (but not Royal Tiger), &c are found but only in one part the northern peninsula of Borneo.— Ox & hog natives of Borneo[2]

175e **Notices of Indian Arch. Singapore 1837. By J. H. ?** do. p. 189. «190» No full sized horse is found East of ȳ Burramposter & S of Tropic—
By J. H. Moore
after quitting Bengal this fact is noticed in Cassay Ava Pegue— seldom equals 13 hands— those of Lao & Siam inferior to those of Pegu— in Sumatra two breeds both small —Java pony occasionally reaches 13 hands.— Phillipines Pony somewhat resembles that of Celebes is somewhat larger than the Sambawa, Java & Sumatra breeds, (.Hence it appears there are shades of difference in all the isld, like in wild animals).— There are prevailing colours in the different islands.— The horse is only found wild in the plains of Celebes. (but language shows that probably not original there)— shows these isld not fit for horse.[1] Forrest—. (p. 270) says many wild horses, bullocks, & deer South part of Mildanao.—[2]

Q **Horse**

176e do. Appendix. p. 43. «& 45» the *Breed* of elephants ‹of› in little isld of Sooloo.— said to have been imported: shows they will propagate get dimensions—[1]

---

do App. p 73 State of Muar in Malacca.— speaks of Rhinoceros as well as *Tapir*.—[2]

---

174 *page crossed pencil.*
175 **Notices of . . . J.H. ?**] *added pencil.*
By J. H. Moore] *added pencil.*
Q] *added in circle.*
Horse] *added across page brown crayon.*
176 do. Appendix . . . dimensions—] '19' *added brown crayon.*
do App. p 73 . . . my theory] *crossed pencil.*

173–3 Forrest 1780.
174–1 William Yarrell: personal communication.
174–2 Moor 1837:1.
175–1 Moor 1837: 189–90, Notice 'On the different races of the horse in the Malayan archipelago and adjacent countries.', reprinted from Singapore Chronicle August 1825.
175–2 Forrest 1780: 270.
176–1 J. Hunt 1837 in Moor 1837, Appendix: 43–45.
176–2 Newbold 1836 in Moor 1837, Appendix: 73.

‹do do p 75› «Journal of Asiatic Soc.. Vol V. p. 565. in a Paper by Lieut. Newbold.—» A Malayan albino described "To this day the tomb of his grandfather, who was also an albino is held sacred by the credulous natives, & vow made at it. Both his parents were of the usual colour. His sister is an albino like himself said not to be common—[3] probably, I should think grandfather first of race & if so, fact for my theory

177 Cocos Isl[d] & Preparis between Andaman & Pegu. ‹have› abound with monkeys & squirrels.— Horsbrugh E. I. Directory. Vol II. p. 46[1] Carimon Java. (between Borneo & Java) Lat 5°. 50′ S. adjoining it are several small islands. abounding with deer— Horsburgs. Vol II. p. 527.—[2]

‹Scientific Soci› Journal of Asiatic Society Vol I. p. 261. ‹J› Catalogue of Birds of India.[3] — p. 555. Lieut. Hutton[4] counted, the ova of a tick «in India» & found there were 5,283 attached to its body—

178 Journal of the Asiatic Soc. vol.I.p. 335. Catalogue of animals of Nepal by. B. Hodgson.[1] p. 336 In the *most pestiferous* region (mentioned by Heber) «from» which «*all*» man*kind* «(& yet afterwards says native tribes can live there)» flee during 8 months out of 12.— the largest mammifers in the world consistently reside & are bred.[2] "take tame animals into this region between April & October & like man *almost* (*this looks inaccurate* C.D) they will catch the Malaria & die.— On the other hand there are breeds of Men the Thârû & the Dhangar who can live there & do not pine visibly. p. 337.[3] it would appear as if

179e p. 345. The Ceylonese Elephant [. . .]
saul forests by having a smaller, lighter head, carried more elevated & higher forequarters: is said to be of a bolder & more generous temper— **Hodgson**[1]

180e Koloff. voyage through the Moluccas 1825— "No wild animals in Moa.—" Chapt.— V.—[1] : do. Chat XXI. Wild cattle & Hogs

---

on Timor-land— monkeys do not exist. there & it is a singular thing

177 Cocos . . . p. 527.—] '19' *added brown crayon.*
‹Scientific . . . body—] *crossed pencil.*
178 *page crossed pencil.*
179 **Hodgson**] *added pencil.*
179–180 cut: bottom portion extant.
180 '19' *added brown crayon.*

176–3 Newbold 1836 in Moor 1837, Appendix: 75.
177–1 Horsburgh 1836, 2:46.
177–2 Horsburgh 1836, 2:527.
177–3 Warlow 1832.
177–4 Hutton 1832.
178–1 Hodgson 1832:336.
178–2 Hodgson 1832:336.
178–3 Hodgson 1832:337.
179–1 Hodgson 1832:345, 'The elephant is that so well known as the Indian variety, and as such is contra-distinguished from the African variety. But it may be questioned, if there be not two distinct varieties or species in India alone, viz. the Ceylonese, and that of the saul forest.'
180–1 Kolff 1840:79.

181e that throughout the *Moluccas* Archipelago they are only to be found on the isl[d] of Batchian near SE. end of Gilolo.— "—[1]

Forrest Voyages. p. 39— deer but no wild animals in Gilolo.—[2] p. 134: Birds of Paradise were first procured from Gilolo[3]

---

p. 253 In isl[d] of Bunwood (18 miles in circum) there are hogs & *monkeys* ‹at› near shore of Magindanao[4]

Journal of [Asiatic Soc] [. . .] p [. . .]
— most wonderful instinct, how could it have originated— spins thread of cotton.—[5] do p. 583, It APPEARS *probable*,«?» that the Hippopotamus occurs in India. in the Jungles of Borabhum & Dholbum.—[6]

182e Vol do. p. 634, alludes to fact stated by M. Tournal that skulls found near Vienna appoximat to Negro form; those from Rhine to the Caribs.—[1] Vol II

181 *page cut in two portions*: 'that . . . Magindanao' *and* 'Journal . . . Dholbum.—' *180 and top portion of 181 were pinned to preserve* 'Koloff . . . Gilolo.—'.
[. . .] p [. . .]] *part of one MS line lost in cutting page.*
Journal . . . Dholbum.—] *crossed pencil.*

182 Vol do. p. 634 . . . India.—] *crossed pencil.*
Windsor . . . is the form] *crossed pencil.*
[. . .]] *one MS line lost in cutting page.*
Dampier . . . isl[ds]—] '19 ‹**Cocks**›' *added brown crayon.*
Humboldt . . . isl[ds].—] *crossed pencil.*

181–1 Kolff 1840:361.
181–2 Forrest 1780:39.
181–3 Forrest 1780:134.
181–4 Forrest 1780:253.
181–5 Hutton 1833:503 describes nests of the Tailor Bird (*Sylvia Sutoria*, Lath.), 'The first was neatly formed of raw cotton and bits of cotton threads, woven strongly together, thickly lined with horse-hair and supported between two leaves on a twig of the *amaltás* tree (*cassia fistula*).' See T2 and *Natural selection*:473. 'But in all cases in which intelligence comes into play, the animal must to a certain extent know what it wants to do. . . . The Tailor Bird weaves threads of cotton, with which to sew up the edge of a leaf to form its wonderful nest; but it has been seen to pick up and use pieces of artificially made thread, which shows that it before hand knows for what purpose it spins the cotton; though it cannot know that it makes its suspended nest that its eggs may be hatched, and its young reared safe from snakes and other enemies.'
181–6 Tickell 1833:582–83.
182–1 Prinsep 1833a:633–34, 'M. Tournal and other French naturalists, further suppose that several races of men have successively had possession of our continents. The form of the skulls found at Vienna is stated to approach to the African or Negro type. Those discovered in the fluviatile marl of the valley of the Rhine and Danube exhibit a close resemblance to the heads of the Karaibs or those of the ancient inhabitants of Peru and Chili.' This note is added by Prinsep to Captain E. Smith's report of fossil human bones in the bed of the Jumna River. Prinsep's paper, which is an editorialized 'sketch' of Tournal 1833, maintains that caves with fossil bones 'were by no means filled by any brisk transient or universal wave of transport: and there is no ground deducible from them for the separation of organic remains into the two classes of *ante* and *post*-diluvian' and that although 'it became a dogma of the science that man existed not in a fossil state. The recent discovery however of the caverns of Aude, Herault, and Gard exposed a vast magazine of human bones and antique pottery inclosed in the self-same matrix with the hyaena, lion, tiger, stag, and numerous other animals, *all of extinct species*.' Prinsep summarizes the French debate on the existence of human fossils; siding with the supporters of this view, he concludes 'man must also be included among the fossil species, or rather that the sudden transition from one condition of being to another must be disallowed, and that the same gradual alteration of species, already so fully developed by M. Deshayes in his comparison of the fossil shells of the different periods of the tertiary formations, must be extended to animals, and perchance to man himself; . . .'. Prinsep criticizes English geologists

p. 650. Long attested account of fall of fish in India.—[2]
Windsor Earl— Eastern Seas p. 229. Believes the ‹Rhinoceros› «Tapir» is found in Borneo.—[3] «p. 233» There, as well in all Malay countries «the» cats are born with the joints near the tip crooked.—[4] is the form [. . .]

Dampier. Vol I. p. 320. says no wild (carnivora) beasts on Phillipines.[5] Forrest somewhere says same.—[6] do p. 393. ‹"›The *wild*, small fowls at Pulo Condore "crow like ours, but much more small & shrill".—[7]

Humboldt. Vol I. p. 275. says Teneriffe does not countenance the theory of polymorphous plants, abounding in volcanic islds.—[8] **‹Cocks›**

183 The possibility of different varieties being raised by seed is highly odd— as it is not so with the esculent vegetables— how is it with hollyoaks, flaxes &c &c?

Mr Herbert in letter says distinctly, that Hollyoak reproduce each other.[1] & yet I presume seed raised in same garden.— now this good question— single, or half double.— anyhow fertile because they «are» raised by seed.—

Where has Duchesne described Atavism.— ask Dr Holland[2] cases where peculiarity has first appeared.—

"Storia della Riproduzione Vegetale". by Gallesio. Pisa 1816 p. 27. Dr. Holland.[3]

---

183 *page written parallel to spine.*
The possibility . . . flaxes &c &c?] *pencil.*
page crossed pencil.

(Conybeare and De la Beche) for ignoring the new discoveries and urges Indian geologists to search for human remains. See also E35.
182–2 Prinsep 1833b:650.
182–3 Earl 1837:229.
182–4 Earl 1837:233.
182–5 Dampier 1698–1703, 1:320, 'In this island [Mindanao] are also many sorts of Beasts . . . I never saw or heard of any Beasts of Prey here . . .'
182–6 Forrest 1780:39.
182–7 Dampier 1698–1703, 1:393.
182–8 Humboldt and Bonpland 1819–29, 1:275–76, 'A question highly interesting to the history of the progressive display of organization on the Globe has been very warmly discussed in our own times, that of ascertaining whether the polymorphous plants are more common in the volcanic islands. The vegetation of Teneriffe is unfavourable to the hypothesis, that nature in new countries appears less subjected to constant forms. M. Broussonet, who resided so long at the Canaries, asserts, that the variable plants are not more common there than in the south of Europe. Ought it not to be presumed, that the polymorphous species, which are so abundant in the Isle of Bourbon, are owing rather to the nature of the soil, and to the climate, than to the newness of the vegetation.'
183–1 *Correspondence* 2:183. Letter from William Herbert, 5 April 1839, 'Take for instance the hollyhocks of the garden, without question cultivated varieties of one plant, yet *steadily reproducing their respective colors by seed.*'
183–2 Holland 1839a:23. 'A singular variety in this general law is that which Duchesne and others have termed Atavism.' Darwin underlined 'Duchesne'. See QE1 Temporary Question 1. 'Where has Duchesne described Atavism alluded to by Dr. Holland—'.
183–3 Gallesio 1816, cited in Holland 1839a:27. 'Gallesio takes the genus Citrus as his subject, and, from his experiments upon these plants, draws conclusions which apply largely to the production of varieties, monsters, and hybrids in vegetable life. See Darwin's abstract of Gallesio 1816 (DAR 71:95–111). See QE16, (6) for Dr Holland.

184 Are there instances of plants, in becoming double[1] loosing fertility if, sometimes one, sex & sometimes. other, so as to become ‹all› monoœcious.— Are there not wild plants, some partly diœcious?

Mushroom Hybrids?

Any «wild» plants in England, which do not perfect their seed?—

What annuals can be budded «& rendered of great age» as must be inferred from what Mr Knight says. Hort. Transat. V. II p. 252.[2]

Is there any very sleepy Mimosa, nearly allied to the Sensitive Plant.—[3]

---

p. 290. Dr. Edwards in his essay on Spermatic animalcule. has described instrument for galvanize ‹ani› them—[4]

Cross Irish & Common Hare[5]

Decandoelle has chapter on sensitive plants; Physiology[6]

---

184 *page written parallel to spine.*
Are there instances . . . diœcious?] *crossed pencil.*
Mushroom . . . perfect them] *crossed pencil.*
What annuals . . . p. 252.] *crossed pencil.*
Are there instances . . . Sensitive Plant.—] *crossed ink.*
p. 290. Dr. Edwards . . . them] *set off in a box.*
Cross Irish . . . Hare] *crossed ink.*
Decandoelle . . . **Physiology**] *pencil, written upside down.*
Cross Irish . . . **Physiology**] *crossed ink.*

184–1 See QE[5]a, QE6 and QE11.
184–2 Knight 1818b:252. See note E154–3.
184–3 See EIBC–3.
184–4 Darwin found an indirect reference to J. L. Prévost and Dumas 1821 in J. L. Prévost and Dumas 1832, which is an Appendix in Edwards 1832. See Prévost and Dumas 1832:289–90, 'For this purpose, we take a muscle recent and thin, the sterno-pubic muscle of the frog, for example. We place it under the microscope, and submit it to galvanic influence by means of the very simple arrangement, described in our Essay upon Spermatic Animalcula.' See J. L. Prévost and Dumas 1821:103–104 (*sic* 203–204), 'Nous avons fixé sur une glace deux fils de platine dont les extrémités vis-à-vis l'une de l'autre étoient séparées par quelques lignes d'intervalle. Cet appareil a été mis sous le microscope et les fils ont été placés en communication avec deux branches de laiton qui se rendoient dans des capsules pleines de mercure et portées par une table indépendante de l'appui du microscope. L'une d'elles communiquoit à demeure avec l'un des pôles d'une forte pile, l'autre servoit à établir ou rompre le circuit, au moyen de l'immersion ou de l'émersion du fil polaire. On a mis alors une goutte de liqueur spermatique entre les deux fils de platine et le mouvement des animalcules étant bien perçu, l'on a établi le circuit galvanique. Mais soit qu'il ait été continu, soit qu'on ait donné des secousses, on n'a pu voir aucune altération dans le mouvement [des animalcules spermatiques].'
184–5 See B7 and D61.
184–6 See A. P. de Candolle 1832, 2:817–71, (Livre 4, chap. 5, *De la direction des plantes ou des parties des plantes* and chap. 6, *Du mouvement des plantes*).

INSIDE BACK COVER Get Habberley to try experiments. about raising plants. where they cannot «crossed» etc.—[1]

Make Hybrid mosses.— Leighton or some one.[2]

Father— diseases common to men & animals.— :likenesses of children

---

CD[Does any annual give buds, or tubers. Yes— but these are same as trees.—

Shake some sleeping mimosa— do stamina of C. Speciosus. collapse at night. if so irritate them, «as by an insect coming always at same time» see if by so doing can be made sensitive[3]

The function of sleeping someway useful.— it is only the association which is useless.

Granfather's Handwriting, to compare with my own.—

BACK COVER E

IBC :likenesses of children] *pencil.*
Get Habberley . . . trees.—] *crossed pencil.*
Granfather's . . . my own.—] *crossed pencil.*
page crossed ink.

IBC–1 Abberley, R. W. Darwin's gardener conducted several experiments for Charles. See T151, QE11, QE14.
IBC–2 W. A. Leighton.
IBC–3 See QE5 (Remote Experiments— Plants): '(1) Shake a sleeping mimosa, or half bred mimosa between sensitive & sleeping species, & see whether association can be given (2) do the stamina of C. Speciosissimus collapse during sleep & do of Berberis— latter *I think* certainly not)'.

Torn Apart Notebook

Transcribed and edited by SYDNEY SMITH & DAVID KOHN

The existence of this notebook may be inferred from the exiguous sheets that remain. The quality of the paper and the dimensions of the sheets conform with that of the Transmutation Notebooks B to E. When the contents of Box B came to be examined (see Historical Preface p. 5), the C.40 envelope-files (now DAR 205) containing the majority of the excised sheets were found. The sheet numbers missing from each notebook were entered in their place, but it was found there were yet more sheets bearing numbers in Darwin's hand and sharing one peculiarity—they had been *torn* out of a notebook, so that in most cases the stitch marks and corrugations caused by the process of binding were still there. The pages excised from the other notebooks had been cut out with scissors.

The Torn Apart Notebook, so named by the editors,[1] was assembled in 1966 (see Table of Location of Excised Pages, pp. 650–51) and was reported in Sydney Smith's Sandars Lectures at Cambridge University in May 1967. There are 51 pages, including 11 fragments of pages without numbers. In most cases Darwin wrote only on the odd-numbered pages in brown ink except where otherwise indicated.

What survives of the Torn Apart Notebook was written in three sections. The first, T1–89, was written between July 1839 and May 1840 when Darwin was reading in and around the volumes he refers to as the Bengal Journal (*Journal of the Asiatic Society of Bengal*) T1–2. The last entries in Notebook E (E176–181) are from the same series and it was Darwin's custom at this time to read and take notes on all the volumes extant. His reading and speculative notes are in the same spirit as the other Transmutation Notebooks B to E. A few pages relate to answers from animal breeders which were being received at the time, as well as odd pages of integrative speculation which looked forward to the first 'Origin' draft of 1842.

The second and third sections comprise notes drawn from observations on plant breeding, floral anatomy and insect pollination, and the habits of bees. These notes were made at Maer and Shrewsbury during the summers of 1840 and 1841. Here Darwin had time to read in the libraries of his father-in-law and father, and study plant breeding and bee behaviour in the garden. T176–178 were written in July–August 1840 and T91–135 in June–July 1841. No doubt Darwin considered both sorts of notes as 'species work'.[2] As the last of Darwin's notebooks devoted to reading with 'reference to species' it ended in May 1840 on T89, appropriately enough with his reflections on the imperfection of the fossil record.

These activities were carried forward in the summer of 1842 when Darwin again visited Maer and Shrewsbury in June. His observations were recorded on notepaper significantly different from that of the Torn Apart Notebook, but this set of notes, written in pencil during the same month that he composed the *1842 Sketch*, was clearly a continuation of the 1841 theme and is concerned with variation, the habits of bees, floral anatomy and palaeontology. The seven unnumbered sheets (DAR 46.2, 76, 208) appear in this volume under the title of Summer 1842.

[1] See Kohn, Smith and Stauffer 1982.
[2] *Correspondence* 2:434.

1 Bengal Journal Vol 7. p. 658— Falconer on Sub. Him. fossils— Ruminants. & Tortoises gigantic— hyæna— bear & ruminants all of larger size.— *the law of large size established*—[1] «Australia,. S. America—» These strange forms., camels, giraffes. Sivatherium & *Anoplotherium*, with existing, or nearly existing forms of aquatic reptiles most strange, & shows as in shells some forms are long preserved.—[2]

---

vol VI. p. 539. Dr Cantor's account of fossil frog, 40 inches in length—! alludes to ancient gigantic salamanders—[3]
Every order (except whales) have great prototype!!.—

2 **Copied**

Vol II p. 502. «Bengal Journal» The Taylor Bird uses pieces of thread, picked «up-» instead of spinning—[1] better case than English birds, using cotton &c instead of natural substances— useful perversion of instincts—

Beechey's Voyage Vol I. p. 499. «4to. Edit»— Horses in Lao Choo so small, that person with long legs can hardly ride on them.[2]

Mr Miller— in Zoological Gardens.[3] informs me that a hybrid between ass & Zebra, crossed with pony mare & produced a very pretty little animal, showing something of Mule in its ears— ((this is good case as showing gradations,

All extant even numbered pages are blank, except 2, 26, 80, 96, 100, 104, 106, 112, and 178.
2 **Copied**] *added pencil.*
gradations,] *pencil comma added over original comma.*
page crossed ink & pencil.

1–1 Cautley 1838:658, 'When we look at the number of species of Proboscidan Pachydermata which swarmed in the primeval forests; . . . the imagination naturally places before our eyes forms of corresponding magnitude in other genera; . . . Amongst the Ruminants the discovery of the Sivatherium gianteum has most amply tended to prove the truth of this induction, exhibiting a ruminating animal bearing the same proportion to the rest of its genus, as the Mastodon and Elephant do to that of the Pachydermata. Amongst the Carnivora we have the Ursus Sivalensis, an animal far exceeding in dimensions its congener of the present period, or the Ursus Spelaeus and bears of the German caves; with a species of hyaena at least one-third larger than that now existing. The reptiles also have their gigantic representative in an entirely new genus of the tortoise, for which we propose the generic name of Megalochelys, from the enormous proportions of its remains as yet discovered, and the size of its femoral and humeral extremities equalling those of the largest rhinoceros. The question however does not appear to be whether the animals of former periods were larger than those now existing, but whether the genera of larger animals were not more numerous?'

1–2 Cautley 1838:659, '. . . it will be interesting to remark on the co-existence of the Sivatherium, Camel and Giraffe, with Quadrumana, Anoplotheria, Mastodons, and reptiles so closely resembling those of the present rivers, that it is not possible to discover in their osteological pictures, at least, any remarkable deviation from the type which has been left to us.'

1–3 Cantor 1837:540.

2–1 Hutton 1833, See E181. Text used in *Natural Selection*: 473, n. 5; hence T2 is marked '**Copied**'.

2–2 Beechey 1839, 1:499.

2–3 Alexander Miller.

FRAG 1r

Boteler's Narrative Voyage East coast of Africa—[1] Vol II. p. 256— wild cattle at Madagascar—[2] «p. 121» No beasts of Prey.[3]

FRAG 1v

any country should during [. . .] conditions— every spot is occupied & has been occupied [. . .] species, which has undergone all the changes. [im]portant view, **copie[d]**

FRAG 2

———

Gleanings of Sciences. Vol. III p. 83. Paper translated from Meckel. Comp. Anat.—[1] From Buffon cross of he-goat & sheep, it seems male gives form. admitted by Linnæus.— seems to doubt its applicability to common mule & hinnus— in one case bastard of wolf & dog had more form of male, & another of both progenitors— the hinnus, resembles horse in its head ears, tail limbs— in the mules, these parts resemble ass. (& part of body mare)—— this may be, perhaps. squeezed into Mr Walker's law

9 Gleanings of Science Vol III. p 320. Mr Hodgson on Musk Deer— young spotted ‹like in› "prettty much as we see in the young of the wild hog & of several species of deer, which are altogether immaculate when grown up".[1]

13 Saw at Mr Bell's at Hornsey[1] the offspring of a black & white ‹duck of pecu› «drake» with the penguin duck. it took after the *Penguin* in the form of its body & in the manner of walking but not waddling; its colour was darker than the penguin & the bright feathers on its wing resembled the drake.— another of same half breed resembled the plumage of drake still more.— **So Penguin impresses its form both on vars & species**

———

FRAG 1 *page heavily crossed.*
copie[d]] *added pencil, vertical to text.*
9 *bottom portion missing.*
13 *page cut in two:* 'Saw . . . still more.—' *and* 'The ‹male› . . . appearance'.
Penguin] *underlined pencil.*
So Penguin . . . vars & species] *added pencil.*
Bell at Hornsey] *added pencil.*

FRAG 1r–1 Boteler 1835.
FRAG 1r–2 Boteler 1835, 2:256.
FRAG 1r–3 Boteler 1835, 2:121, 'Beasts of prey are unknown in Madagascar, . . .'
FRAG 2–1 Meckel 1831:83, 'From certain experiments, by Buffon, upon the production of bastards between the he-goat and sheep, in which the offspring had greater resemblance to the former, we should be inclined to believe, that the form depends, in a great measure, on the male. This is positively admitted by Linnaeus, with respect to the outward form at least; but so little is at present known upon the subject, that nothing can be laid down with certainty.

'Thus, the *hinnus*, the progeny of a stallion and a she-ass, resembles the latter more in its slender neck, dorsal stripe, as well as in the form of its hind quarters, whilst the mule, bred between a mare and male-ass, receives from the former the strength of hoof, rounded form of body, and the size beauty, and strength of its posterior parts.'
9–1 Hodgson 1831:320.
13–1 Bell at Hornsey not traced. See DAR 205.7:226, 'March 25/55/ Bell tells me that he bred hybrids of China & common goose, & that these wd never copulat inter se, but wd receive & tread both pure parents. They wd. never even dally together.—'

The ‹male› swan-gander with common goose produce full as many eggs as pure bred common.— the half of the cross, as above, take «generally» after the swan-gander.

———

one of these half-bred ganders. crossed with common goose ‹to has› «produce offspring with» so much of the swan-goose in appearance **Bell at Hornsey**

15 (though only ¼ of blood). that it appears about half way between swan-goose & common goose.— the stripe down back pretty plain in in these ‹half› «¾» bred ones—
The brothers & sisters half-breed showed no sexual inclination for each other—[1]

═══════

17

Aug. 20th
The Echnida & Hedgehog **Tenrec** both having spines, is the effect, partly of the same external conditions (ie. *analogical* structure) & partly the laws of organization (ie those laws which prevent infinite variation in every possible way.— the laws which determine the *kinds* of monstrosity, & determine the *kind* of variation «& sporting» in flowers & domestication of animals

19 Aug. 26th.—
When it is said that there is evidence in the organic world of infinite & growing complexity from a few types, it must not be supposed that this refers to time.— Marsupial in Oolite.— insects, of do orders— cheiroptera & cætacea in Eocene— dicot. plants in coal measures.— Shells in Cambrian & Crust show how long since present forms existed, but if it be asked how this complexity from a few types originated, we must go to the first origin of the world.— our present organic beings are the descendants, ‹slightly› «a good deal» modified ‹& Many Forms lost; if› «of this old stock (which from action & reaction grew more complex)» some perhaps rendered more complex & some simplified.—

15 *bottom portion missing.*
bottom portion of 13 and top portion of 15 formerly pinned to preserve 'The ‹male› . . . each other—'.
17 *top portion missing page number inferred from date of 19.*
Tenrec] *added pencil.*
analogical] *underlined pencil.*
19 existed] *first* 'e' *over* 'li'.
'22' *added brown crayon.*

15–1 See DAR 205.7:209 (formerly pinned to T13 and T15), 'His hybrids from common **goose** & Swan or China Goose (L. Jenyns put it in genus Swan) differ much, some intermediate, SOME VERY CLOSELY INDEED after common goose «mother & Swan-gander» & some after Swan-goose— & having no other gander, within some miles there can be no doubts about 2d impregt. The hybrids bred with parents but not inter se **Fox June /46/ Q**'. See *Natural Selection*:439, 'many cases are on record of hybrids from these birds breeding readily with the pure parent species.'

25 Annals of Natural History. ‹no. XII.› **Vol. 2. p. 96**[1] **&** p. 451. 1839—[2]
Translation of P. Fries most curious paper on the Pipe-fish— which he divides into two divisions, one of which are marsupial & the other have young which undergo metamorphosis & are provided with fins, & hence do not require sac.— but the male in these hatch young— are there not some. Marsup. Mammalia, which ‹do› have not sack,— Most curious facts & this paper deserves fresh study & whole order of the fish.— **Embryology**

26 **p. 97. for Man Chapt**[1] **see Yarrell Syngnathus**

37 **Ch 6** I presume, from my theory, as long as any structure can be handed down without being absolutely injurious «(or requiring nutrition)» to a certain amount it will be so handed down«(» . . as mammae of men «callosities on Camels & Horses—».—«)» & therefore probably any structure would rather become accomodated to new circumstances than it would be eliminated, & hence, the application of structure to purpose after purpose would tend to render complex the series.— **Ch 6**

39 **Upland geese would transplant seeds very far.—**

41 Sept 31. The identity ‹of› (or only closeness) of some species— (especially of mammifers) in *old* beds & existing species is valuable because it shows no *innate* power of change & it also shows, what enormous changes of conditions, some species will undergo & yet remain adapted.— it does away with difficulty of rabbits of England remaining same (if so) with those of Spain & such facts— **This unequal duration is exactly same as some species extending much further geographically than others.**

FRAG 3

Athenæum p. 605
Mr. Macgillivray says "‹A Thrush &› Blackbird have been known in ‹their› «its» natural state to mate with a thrush"—[1]

---

25 no. XII] *crossed pencil.*
Vol. 2. p. 96 &] *added pencil.*
Embryology] *added brown crayon.*
26 **p. 97 . . . Syngnathus**] *added pencil as annotation on 25.*
39 *bottom portion missing.*
Upland . . . far.—] *added pencil.*
41 *innate*] *underlined pencil.*
Sept 31 . . . such facts—] '21' *added brown crayon.*
This unequal . . . than others.] *added pencil and* '5' *added in brown crayon to this annotation.*

25–1 B. F. Fries 1839a:96.
25–2 B. F. Fries 1839c:451, relevant passage scored.
26–1 Manuscript chapter; B. F. Fries 1839a:97 refers to a 1785 manuscript cited in Yarrell 1836, 2:327–28.

FRAG 3–1 Macgillivray 1839b:605, 'He has been known, in his natural state, to mate with a thrush; such mis-alliances between different houses being, as all naturalists know, uncommon among birds save when in captivity.'

FRAG 4^{r}

Athenæum 1839. p. 708.— Shrew, found by M. Lartet same as existing species.[1]

FRAG 4^{v}

We see the same object gained by the Mataco-armadillo & the woodlouse— — a good analogy— sea-Crustacea— Tullus.

FRAG 5

Athenæum 1839 p. 772— A curious theoretical French book review on politics in relation to the different races of men, some more intellectual than others—[1] is incidentally said that a mongrel man may lose *all* traces of his parentage in «about» *seven* «7» generations.—[2] so many!!

51 Hensleigh objects to transmut. theory, on the grounds of similarity in condition in Java & Sumatra & dissimilarity of forms—[1] yet how valueless this objection, when one thinks of different kinds of cattle in every part of England. &c &c

---

NB. In botanical geography, there can be no sharp division of partition as between Mammalia in cases such as that of Java & Sumatra

FRAG 4 **We see . . . analogy—**] *added gold ink.*
the woodlouse] *underlined pencil.*
sea-Crustacea— Tulus.] *added pencil, underlined.*
page cut in at least three portions: 'Athenæum . . . species.' *and* 'We see . . . **Tulus**.' *are bottom two portions.*

51 *top*: '‹5› 18' *added brown crayon.*
bottom: 'NB . . . Sumatra'.

FRAG 4^{r}–1 Lartet 1839:708, 'M. Lartet announces that among his recently found fossils he has met with a Desman, or musk Shrew, of the same size as that now living in the Pyrenees. If this opinion should be confirmed it will be the first example of a living species existing also among the mammiferae of the tertiary formations.'

FRAG 5–1 Courtet de l'Isle 1839:771, 'Whether the coloured races are original creations, or degenerated descendants from one original stock—whether their moral and intellectual characteristics are derived from an originally defective organization, or both proceed from an unhappy combination of externals—their condition as it is, is a physical fact possessing its definite bearings on their social position, wherever they come in contact with the white man; and this contains the sum of what is sought in the present inquiry, as far as they are concerned in it. We may therefore set aside M. Courtet de l'Isle's views on such points, who maintains a diversity of original races; and leaving that point undecided, admit the dependence of the intellectual powers of the different varieties of man, such as they now exist, on their respective organizations.'

FRAG 5–2 Courtet de l'Isle 1839:772, 'With respect to the cerebral development of the several coloured varieties, and their other organic differences, we know that the product of a cross breed is a middle term between the conditions of the two parents; and if the descendants of such a cross intermarry only with one of the original stocks, the cross blood will, at the end of about seven generations entirely disappear.'

51–1 See E144 for the reactions of Hensleigh Wedgwood and Sarah Elizabeth Wedgwood to Darwin's selectionist ideas.

55 **Nov 15th**
Waterhouse showed me the component vertebræ of the head of Snake[1] wonderful!! distinct!!— He would not allow such series showed passages— yet in talking, constantly said as the ‹brain› spinal marrow expands, so do the bones ‹are created› expand— instead of saying as brain is created &c &c

---

Bats are a great difficulty not only are no animals known with an intermediate structure, but it is not possible to imagine what *habits* an animal could have had with such structure.— perhaps greatest **Could anyone. have foreseen, sailing, climbing & mud-walking fish?**

57 difficult— yet suggested. (vipers tooth also a difficult), the whole mind is constituted that a difficulty makes greater impression, than the grouping of «many» facts with laws & their explanation will probably reject this theory— (I must answer it by rooting out curious cases of intermediate structure,, & supposing much extinction. give a parallel case)

Waterhouse remarked, that any argument for transmut, from one organ graduating into other is lost, ‹be› (as vertebræ into skull., two bones of tibia into one.—) because if the animals were taken from which these series were drawn they *would not be intermediate, but this is not required..—*

59 Waterhouse says perhaps animals of Fernando Noronha are found *unknown* coast in front of it.—

---

63 Cuvier has grand sentence about the Animaux fossiles— being a mere fragment of the discoveries to come—[1] Owen in his description of my fossils makes same such remark[2] & before the conclusion of his work— Lund makes

55 **Nov 15th—]** *added pencil.*
Could anyone . . . mud-walking fish?—] *added pencil.*
'11' *added pencil.*
do the bones] 'do' *over* 'as'.
57 *would not . . . not required. .—]* *underlined pencil.*
55 and 57 formerly pinned to preserve 'Bats . . . required'.
59 *bottom portion missing.*
63 *bottom portion missing.*

55–1 G. R. Waterhouse: personal communication.
63–1 See Cuvier 1821–24, 1:liii, 'Malgré ces nombreuses et scrupuleuses observations, nous avouons qu'il reste encore beaucoup à faire pour completer un travail tel que celui que nous avons entrepris. Il faudra encore beaucoup de temps, beaucoup de recherches et le concours de circonstances favorables qu'il n'est pas en notre pouvoir de fair naître, pour donner aux détails de ce travail toute l'étendue et l'exactitude qu'on doit y désirer.'
63–2 *Fossil Mammalia*: 14–15, 'But independently of these indications, the abundance and variety of the osseous remains of extinct Mammalia in South America are amply attested by the materials for the following descriptions, collected by one individual [Darwin], whose sphere of observation was limited to a comparatively small part of South America; and the future traveller may fairly hope for similar success, if he bring to the search the same zeal and tact which distinguish the gentleman to whom Oryctological Science is indebted for such novel and valuable accessions.'

his wonderful discoveries=[3] negative facts are valueless= monkeys=

65 Owen has described a greatt Struthonidous Bird from New Zealand—[1] ‹so› not an Apteryx, yet it shows the Apteryx is not «quite» isolated in its present locality— there have been at least other birds, with small wings, & surely the Apteryx is more closely allied to the Struthonidae than any other forms—

FRAG 6r

Lund's Antilope **in Brazil** another point of agreement with. N. America & S., (? is the peculiar. N. American form)— ?**Hunting leopard, how strange, anyone, would have thought isolated species** Mr Blyth, however, believes in the existence of Molina's Pudu—[1] **or goat**

FRAG 6v

There is ibex of Alp Pyrenees &c— (see Blyth's work on Ruminants,—[1] these species must have migrated to these mountains, when the cold was intense just like the alpine plants—

FRAG 6 **in Brazil**] *added pencil.*
?**Hunting . . . isolated species**] *added pencil.*
or goat] *added pencil.*
page cut in at least three portions: 'Lund's . . . Pudu' *and* 'There . . . plants—' *are bottom two portions.*

63–3 See Lund 1840, which is a translation of Lund 1838. Lund's extensive collection of Brasilian mammalian fossils was made in 1837, whilst Owen was describing Darwin's Patagonian collection for *Fossil Mammalia.* Among Lund's discoveries were the bones of an extinct ape: 'I am at length enabled to solve the important question as to the existence of the highest class of mammals in those ancient times to which these fossils belong; a question which has as yet been unanswered, or which most philosophers have thought right to answer in the negative. It is certain this family was then in existence; and the first animal of the class recovered is of gigantic size, a character belonging to the organization of the period. It considerably exceeds the largest individuals of the orang-outang, or Chimpanzee, yet seen; from which also, as well as from the long-armed apes (*Hylobates*), it is generically distinct. As it equally differs from the apes now living here, I would place it for the present in a genus of its own, for which I propose the name *Protopithecus*; with the specific distinction *Prot. brasiliensis*, from the quarter where the first representative of this family saw the light of day.' (p. 315). Darwin scored 'It considerably . . . *Protopithecus*'.

65–1 Richard Owen 1839b.

FRAG 6r–1 See Blyth 1841:255–56, '*O.*——: *Ixalus Probaton*, Ogilby. I stated in my former paper an opinion, to which I am still disposed to adhere, that this animal is no other than a genuine sheep, but specifically distinct from any at present known: . . . I have been favoured, however, by Col. Hamilton Smith with a drawing of an animal observed by himself on the banks of the Rio St. Juan in Venezuela, which appears to accord so nearly with *Ixalus Probaton*, except in the particular of bearing horns similar to those of the Rocky Mountain Goat, that its absolute identity is probable, in which case it would be curious that a species so very nearly allied to the genus *Ovis* should yet differ from it so considerably in the character specified. The South American animal adverted to is the *Aplocerus Mazama* of Col. Smith, and is probably congenerous with the *Pudu* of the Chilian Andes mentioned by Molina (the existence of which would appear to have been lately re-ascertained by M. Gay), and also with the fossil *Antilope Mariquensis* of Dr. Lund: there would indeed appear to be other living species of this type, more or less distinctly indicated by different authors.'

FRAG 6v–1 Blyth 1841. Darwin scored the description of *O. Aries*, the domestic sheep (p. 256) and noted at the end of the article: '14 wild species'.

79 In S. America. it appears from Lund more Mammals, than at present[1] «in Europe we know there has been several successions of Mammals.—» yet only two monkeys, ‹there are now› have been found fossil in S. America, there are now— — species in S. America. — so see what a «mere» vestige, is preserved in this country— same argument to India & Europe— & Africa!,— any negative argument against— monkey-man, valueless.— **May not several generations have been confounded in the caves?**

It is highly important, to bear in mind that enormous periods may elapse, even in situations apparently favourable for the preservation of shells; where land broken, rivers entering.— & yet no shells— now look

80 at Scotland— coasts of Chile, excepting Concepcion— Patagonia— Beds of La Plata. (except close to B. Ayres).— If we may take this as guide, the shells preserved must be as much a casualty as, bones of Mammalia in caves:— :argue first case of bones (New Red Sandstone) & then go on to shells—

81 A profound consideration of method by which races of men have been exterminated (see Pritchards paper) (Ed. Phil. Journ. end of 1839)[1] very important. it seems owing to immigration of other races, so it is with domestic breeds. (though in this case crossing has had somewhat to do with it. mem. dogs «& pigs» in Polynesia; & dogs in S. America «Rengger.[2]» — now it is this very immigration which tends to make the destroyers vary; so that we here see reasons— why no perfect gradation can be expected in any one country.— **in a descending series of strata** **This again shows how much forms depend on other forms**

89 Lyell's Paper, in Taylor's Journ.— Phil. Mag. May. 1840 p. 362.—[1] some Mammals of Norfolk Crag. mentioned— allied Beaver to present forms.— —

How many «tertiary» estuary **& Lacrustine** formations contain fossils,— mammals— a few only — & how many estuary formations are there in old Secondary Series— few—

79 **May not . . . the caves?**] *added pencil.*
monkey-man] *in a rough brace.*
81 A profound . . . any one country.—] '22' *added brown crayon and circled pencil.*
very important . . .Rengger.] *scored brown crayon.*
in a . . . of strata] *added pencil.*
This again . . . other forms] *added pencil.*
89 estuary] 'y' *over* 'ies'.
& Lacustrine] *added pencil.*

79–1 See Lund 1840:374–89, where the discussion is accompanied by tables showing the ratios of past and present mammalian genera and species.

81–1 Prichard 1840:16.

81–2 Rengger 1830:154. Darwin's abstract: 'Although Spaniards introduced many races, now so generally mingled that it is rare to find trace, though that sometimes can be done.'

89–1 Lyell 1840b:361–2, 'Among the mammalian remains found on the beach [at Cromer] and chiefly *in situ* in the blue clay . . . Mr. Owen has recognized the following: . . . 7. lower jaw left ramus of the beaver, a species larger than the living one and apparently distinct.'

91 Maer June/41/, observed 3 plants of Caltha Palustris alone together. one had seed-pods turning brown, whilst both others were in nearly full flower

93 Maer June /41/
Rhododendrum— nectary marked by orange freckles on ⓐ upper petal; bees & flies seen directed to it— The Humbles in crawling out brush over anther & pistil & one I SAW IMPREGNATE by pollen with which ‹bees› «a bee» was dusted over.

Stamens & pistils curve upwards, so that anthers & stigma lie in *fairway to nectary*.— Is not this so in Kidney Bean. How is it generally.—
In Azalea ‹do› «it is so» ‹Though I saw no Bees «several» visiting it›.— In yellow day lily, the Bees visit base of upper petal, though not differently coloured— & stamens bend up a little

94 In a wild purple Geranium, I see Bees visit always base ⓐ of upper petal **from facility of alighting?** which is not differently coloured & to which stamen & pistils have no relation. In Monk's Hood, a bee entering long nectary, would «necessary» cross directly over the bunch of anthers & pistils, but these ‹do› do not bend up— In Lark-spur, if Bees put proboscis *within* nectary «they do» they must disturb all anthers, wh otherwise lie protected by the hairy black lip of lower division of nectary: «wh. itself resembles a Bee, but does not *prevent bees visiting it*.» In Columbine nectaries are placed all round flower as they are in Crown-Imperial Lily & many other flowers—

95 My view of ‹variety acquired› «‹character›» of characters being inherited at corresponing age & sex, opposed by cantering horses having colts which can canter— & DOGS trained to pursuit having PUPPIES with the same powers instinctive & doubtless not confined to sex.— «Is not cantering a congenital peculiarity improved.» Probably every such «new» quality becomes associated with some other, as pointing with smell.= **These qualities have been given to fœtus ‹fr› before sex developed— Double flowers & colours breaking only hereditary characters, wh. come on in**

96 **after life of Plants— also goodness of flavour in fruit— all affected by cultivation during life of individual.**

99 June 1st 1841. Maer Examined the Lemon-thyme.— equally abortive as it was in autumn: filaments united in whole length to corolla—anthers minute,

91 *bottom portion missing.*
93 I SAW IMPREGNATE] *scored brown crayon.*
In Azalea . . . up a little.] *added pencil.*
94 **from facility of alighting?**] *added pencil; page crossed pencil.*
95 **These qualities . . . of individual.**] *added pencil.*
to fœtus . . . flowers &] *scored pencil.*
96 **after life . . . individual.**] *double scored pencil.*
99 **Common Thyme . . . abortive**] *added pencil.*

distinctly doubled, brown, but with no pollen.—
Common Thyme growing close by is equally abortive—and both growing within Kitchen Garden.— As we see in Hybrids that although anther «nor filaments» shrivel, yet stigma does not, so we may feel *somewhat* «but little» less surprised at Henslow's remark that pistil does not become abortive.

100 Examined in microscope—some of the stigmas of

shape of ordinary Labiatæ —the chief part with ordinary divisions, & a few with one lobe again divided «Have dried some».— some with no division in young flowers. The abortive stamen are of useful height.—

103 In Lupine, Bees «frequent» & seem to act, something like on Kidney Bean, they go to nectar at foot of upper petal standing on «I saw Bee go to two species of Lupine,» two wings. & when the Lupine flower is *perfectly ripe* & pollen abundant filaments & stamens all protrude «there is a brush at end of stigma, which forces out from extremity pollen, or pollen comes out with anthers & stigma in slit» — As I think they do in Broom & certainly when over-ripe & half withered— I saw Bees going to clover & once this happened.— And in common Beans it is wonderful ⓐ how the Humbles force down the wings most violently: in Beans the wings seem beautifully to protect sheath

104 ⓐ In all these nectar seems to be at base of upper petal & the curvature of ‹an› pistil, etc lies in gangway= In Lotus corniculatus saw Humble press down wings which ejects pollen from tip of sheath.— «Also in Lathyrus pratensis yellow saw stigma project» In common Pea saw Humble so press down sheath, that stigma covered with pollen was pressed & rubbed along whole breast—
ⓑ pressing either one or both of Pea's wings, stigma & mass of yellow pollen protrudes at sheath.— **At last I saw Bee collecting pollen from ‹sheath› Keel of Lupine— Seen Bees on Potato & several times on Beans**

105 Rough.—green-cabbage «in flower»— *swarmed* with meligethes & small *Staphylinidæ on all their bodies* pollen— on a sulphur Broccoli not many do— pollen not *very* abundant. not *very* small— **Saw one small Bee; saw another on Cabbage—white Butterflies suck nectar:**
«Maer June 41»

100 stigmas. of] *followed by pencil carat.*
103 *page crossed pencil.*
104 ⓐ In all . . . sheath] *crossed pencil.*
Pea . . . sheath] *scored brown crayon.*
At last . . . Beans] *added ink.*
105 *Staphylinida . . . bodies*] *underlined brown crayon.*
Saw one . . . suck nectar:] *added ink.*
winged thrips] *underlined brown crayon.*
take flight] *underlined brown crayon.*

Rhubarb. pollen very minute—not excessively abundant flowers not attractive, very small—stigma rather large & rough— flowers common— many *winged thrips*, covered with pollen— «Thrips» about as large as bit of chopped horse hair with legs & *take flight*— Yet we have crosses— I see Bees almost

106 every flower— Blue-bells— wild-raspberry—leeks— Flowers which thought very unattractive— **Found Rhubarb blossom swarming with small Staphylinidae— Anapsis, Melegethes, Leptuse— Diptera & small Hymenoptera**

111 Saw Humble go from great Scarlet Poppy to Rhododendron— from Larkspur to Lupine **two species of Larkspur** — two varieties of Cistus Speedwell to Rhododendron— ‹Loasa› «Anchusa»— speedwell Iris— Azalea. Rhodendron. **Fraxinella to Anchusa** ‹never› «once» P on Fraxinella ‹Heartease› «small. Humble alighted on base of filaments & reached nectar =again= between them, hence quite below stigma. & so avoided it.» On certain days Humble seem to frequent certain flowers, to day early, the great scarlet Poppy—

112 So that, finally Fraxinella. with respect to nectary is same case as Azalea or Rhododendron

---

xx after several gloomy days. hot one, Bees almost P every minute to Fraxinella «& from ‹flower› plant to plant.»— to my grt surprise— I found all, stamens straightened pollen *profusely* shed; lengthened & *turned up* «more than stamens», so that all were brushed by Bees & especially stigma after bee had brushed over the anthers of long stamens

as stamens grow old «& shed some pollen». they turn upwards & bend over stigma:— but stigma «is» almost roofed by united filaments.— This flower hostile to intermarriage!!xx

106 **Found . . . Hymenoptera**] *added ink.*
111 **two species of Larkspur**] *added ink,* '**species**' *underlined.*
Fraxinella to Anchusa] *added ink.*
page crossed pencil.

119 In Phil Transact. about year 1778. Paper by Camper on Ourang-outang, has examined 7 says one specimen had on *one* foot, a toe-nail & two joints— as it is on one foot probably monstruous & not a second species.—[1]

135 ‹Saw› Maer. June 15./41/. Watched plants of Fraxinella, with seven flower stalks for ten minutes. it was visited by 13 Bees— & each examined very many flowers.= 22^{d}.— /during several succeeding days ‹many› «most numerous» bees visited this same bunch & on this day in five minutes eleven *Humbles* came & each visited many flowers— Saw Bees frequent these flowers till late in evening— On rough calc. 280 flowers— allowing each Bee visits 10 flowers in «minute» each flower will be visited in 28 minutes— say then each flower is visited 30 times a day is considerably under mark, & this has now gone on 14 days. (except some wet ones/ & w^{d} go on longer—

151 Woodfords Marrow fat, Early frame, Groom's Dwarf. planted in *rows* «close to each other» & seeds gathered «all» came up in 1840 true. Shrewsbury.— Abberley—[1]

Early Magazine— &c. double-blossomed «& dwarf-fan Bean» bean, were planted in *rows*, & seeds gathered same year came up true «in 1840»: **All in together blossomed together**

The seeds of these plants will be collected & resown.—

119 *pencil.*
135 ‹Saw› . . . evening] *ink.*
On rough . . . longer—] *pencil; page crossed pencil.*
151 *pencil.*
All in together blossomed together] *added pencil in margin.*

119–1 Camper 1779:146, 'The want of these nails, and of the second phalanx of the large toes, is beyond any doubt a very remarkable character in this animal. Nature, however, seems to be inconstant sometimes; for, upon the great toe of the right foot of the Orang in Dr. Van Hooey's collection, there was a little nail and two phalanges.'

151–1 This page was originally pinned to a letter, of 18 October 1841, from Darwin's father enclosing a report from his gardener, Abberley, on seeds, peas, beans, and Thyme. (See *Correspondence* 2:306, DAR 162.) The following note was pinned to this material:
'Shrewsbury July 1st 1842
'Saw ‹some› «a» rows of Early Frame and «of» Woodfords Green Marrow Peas, whose «respective» fathers & grandfathers had grown close together. & these peas compared with two rows «of the two kinds» raised from bright seed, showed no differences in their present state, (fit for eating). nor did they when first springing above the ground. But the differences between them are so slight, as not to be easily discoverable until pod is ripe when one pea is white & other green.— Last year Abberly fancied that «a few of» the white kind were streaked a very little with green.—
«Beans (?)»
'Also saw rows of Early Long Pod & Early Masgrow whose fathers & grandfathers had grown close to each other & close to Fan Bean & all appeared to come up true.— These two, however, are very little different, but the ‹Earl› Fan Bean is widely different.—
'These Peas & Beans were in flower together—
'Abberly will look at seeds.—
'Most hostile facts to me. Are Humbles driven away by tame Bees ???!—
'The two Peas differ in the earliness of maturity but an intermediate form might escape.—
'I must experimentise myself.
=
'And Mem: one scrap says. "March 1842 years before last Beans & Peas were planted in rows adjoining & seeds gathered & these were planted last year. pall-mall, without sticks & seeds gathered and are now planted this year"— now the result has been described.—'

153 Humble[1] 22 flowers of Egg Tree in one minute

Great Humble 17 flowers of Larkspur on two plants in do

Humble 24 flowers of small Linaria in do

Domestic do 6 *Campanula* (two *species*)— in do—do
3 of do in about ‹3/4› of minute
These latter were pollen gatherers & they seem slow=

176 Maer 1840
My Father formerly planted «Turkey or» Palmated and English, planted within few yards of each other actually produced hybrids— My Father remembered when in the gardens, he knew there was none but English,—the Palmated was introduced about '65 years ago—& soon after mules abounded—so that palmated has now nearly disappeared. ‹& old English› But these *mules* ‹in our garden› show no trace of palmation!!?

177 Bees at
Wild St Johns Wort—Scabies, Cyanoglossum—*Reseda*[1] *wild very many Bees & Humbles*—on Thistles many (curious because a Composite) Asparagus very small flowers & as *much* shut up, frequented by «many» Bees & Humbles— «Humbles & common» On silene, many plants of wh. have abortive stamens= Many Humbles on hedge Linaria=
(Plenty of Humble Bees on Phlox Down, 1854, Sept.)

178 **In Spanish Broom by pulling back Wings, pollen is ejected with violence in shower** On many Papilionaceous; all wh. are in flower «I saw Bees;»— on Monk's Hood, **brushing over stamen** «Egg Tree»—I think *never* on the Galeum saxatile & other common kind—I think not on Phlox though they examine it.—«Little Dusty & Blue» Butterflies at Clover,—Veronica—, Ranunculus in numbers =what insect can get honey out of long, curved nectar of Butterfly Orchis & Listera?

Bryony saw common Bee on:

153 *Campanula*] *scored brown crayon.*
177 *Reseda . . . Humbles*] *underlined red crayon.*
'X' *added brown crayon.*
178 *This page (the verso of 177) is actually numbered '1' in the MS.*
In Spanish . . . in shower] *added ink over erased passage.*
On many . . . Bees;»—] *scored brown crayon.*
Clover . . . Ranunculus] *scored brown crayon.*
page crossed pencil.

153-1 See Darwin 1841.
177-1 See letter to W. A. Leighton of 1–23 July 1841. *Correspondence* 2:294–95.

FRAG 7

Linn. Trans 18. p. 133 Westwood on the Fulgoridæ enumerates the strange forms which the thorax & head displays.—[1] most fantastic & use unknown.— "‹when we find such an endless variety of form in the same› organ "manifestation of divine power"?.— "of their use difficult to conceive any idea"

FRAG 8

Linn. Trans. 18. p. 163. "D. Dod on two new genera of coniferæ".— referring to the 3 main divisions & speaking of their similarity «in structure» he says "indeed it w[d] be difficult to point out a family so completely natural & one whose groups pass so insensibly into each other".[1]

FRAG 9

Phillips (Lardner's E. vol. II p. 18.) capital list of all the fossil Mamm. of Europe—[1]

FRAG 10

Large Lizards in Navigatores. Williams. Narrative of Missionary enterprises

FRAG 11

D[r] Andrew Smith says in the larks from S. Africa he can almost make series from end to end— so that he is almost led to doubt. whether there is such a thing as a species—

FRAG 7–1 Westwood 1841:133, 'In some *Homoptera* the thorax is armed with balls and spines, crescents, sabres, and other mimic instruments of war: in others the same part is transformed into a singularly dilated globe, concealing the rest of the body, or swelled out into an enormous casket which would be far too heavy to bear were it not hollow. In others quite again, the head is produced into an elongated and swollen rostrum of the most singular construction, varying in the different species, which is occasionally armed with spines or the use of saws, and sometimes bent over the back. Of these curious modifications it is difficult to form any idea. We are not indeed to suppose that aught has been made in vain; but when we find such an endless variety of form in the same organ, we must be led to conclude either that the use for which it is bestowed upon the creature is always modified in accordance with the modifications in its structure, or that the production of so many extraordinary variations in organs not having a material influence upon the habits of the animals must be considered as a manifiestation of Divine power; in which point of view the contemplation of such productions is not without use.'

FRAG 8–1 D. Don 1841:163, 'All three [*Abies*, *Cupressus*, and *Taxus*] will be found to correspond remarkably in the structure of their male flowers; and the differences presented by their female inflorescence are more apparent than real, for they consist rather in the degree of reduction of parts than in actual structure. Their organs of nutrition present a remarkable degree of uniformity in their structure, and indeed, it would be difficult to point out a family so completely natural, and one whose groups pass so insensibly into each other.'

FRAG 9–1 J. Phillips 1837–39, 2:18–24.

FRAG 10–1 Williams 1837:498, 'Very large *lizards* are found on the mountains of Savaii and Upolu; and from the description I received, I should conclude that they were guanas.'

Summer 1842

Transcribed and edited by DAVID KOHN

1r Jun 1. 1842[1]
Allen W.[2] sowed some years since gathered the seeds of Papaver bracteatum, & the Papaver oncitate was growing in same garden. & out of 60 seedlings not one came up true.— colour of flower & foliage not «being» like the true *P.* bracteatum; all supposed to have been hybridised == Has tried *several year* to obtain seed, but the pods have (except this one year (1827), always been empty.— See separate note—

1v Elizabeth[1] says several years ago seeds were procured with the P. orientale in garden & all came up hybridised. It is possible to raise them pure for Miss Bent three years since gave her some
She means to try this year. *Little variation* in the 60 one brighter with mere traces of black spot at base, one paler with less riged foliage & no black spot & a third considerably paler, all rest very similar—

2r *June 2. 42* Maer ‹Thursday› Thursday
After watching 14 days. many times every day. many clumps of heartseases, never saw any Bee go to them. Yesterday remarked that many flowers had suddenly withered, & to day saw very odd dusky humble (with pollen) on legs go from clump to clump, & insect proboscis in many flowers, on one of which pollen was routed. wh. was not case, on several flowers I examined some days ago— This Bee flew from yellow to yellow & purple heartease without doubt.— Bee, not large, very dusky & broad never saw such a one before—
Saw Fly 21 «this Heartease withered on Monday.—» alight on upper petals & insert proboscis, under sigma & draw it out over & over again & wipe off pollen. (as a needle becomes

2v covered) so whole sides of flower & stigma dusted.— ‹I think› When It first alights, it cleaned sucker & ‹I think› pollen was scraped off, which appeared like Heartease pollen.— the pollen appeared chaffy, as if sucked?! opens & shuts end of sucker, after having withdrawn it.— Saw 4 more Bees at work— another odd genus— & a small common Humble— & more of same fly
Two more of the flowers withered.—

1r gathered] *ink.*
60] *emended in ink.*
P.] *ink.*
See separate note—] *ink.*
page crossed pencil.
1v Elizabeth . . . this year] *ink.*
2r This Bee . . . without] *underlined red crayon.*
pin mark.

1r–1 At this time, Darwin was based at Maer and Shrewsbury. See 'Journal', 'During my stay at Maer & Shrewsbury 5 years after commencement wrote pencil sketch of my species theory.' *Correspondence* 2:435.
1r–2 John Allen Wedgwood.
1v–1 Sarah Elizabeth Wedgwood.

3^r Sillimans Journal ‹vo› 1842. p. 142— Sus americana & Hippotamus «with Megatherium & Mylodon[1]» in post pliocene strata![2] Mastodon longirostris in miocene, like in Europe— Cuvier never found remains of Sus with Elephants— Lyell says New Red Sandstone

3^v of. N. America is Red Sandstone.[1] & Birds true! Plants in Devonian— How strange no plants in our Devonian— Fish one step lower in America— *How curious all negative laws of America of depth of organisms holding in America as in Britain.*

4 If there has been «as» much subsidence as elevation then all continents of cretaceous periods, together with their littoral deposits are probably buried in the depths of the sea— Maer. June /42/

5^r June /42/— M^r. Bunbury says has heard the Trout from different lakes of N. Wales can be distinguished—[1] & Jackson[2] here (Capel-Curig) says that he can certainly tell Trout from Ogwen, Capel Curig & some other lakes, (different waters) He cannot, however, tell them from L. Groznerat, «on road to Bethgellert» wh

5^v flows by Tremadoc. but can tell them from lake S. of Moel Siabod. wh. flows into Conway by Bettws & there joins streams from Capel-Curig—

6 Mr Bunbury says Miers has described in Linn: Transacts. a Brazilian plant «in Lin. Transacts. it has three stamens» intermediate between Orchis & other plants—[1] & Wallich has described Indian Plant.—[2] June /42/—

7^r June/42/
You can select cattle & sheep for horns & yet no difference in calves—how is this in young pigeons—dogs—cattle? As we see the frame of animals can adapt itself to course of life, «as in trades» there is no reason, why the peculiarities sh^d be born,— may come in corresponding time of life of offspring— No peculiarity in external

7^v structure can be concepcional, as limbs &c &c only appear late in pregnancy, & then may just as well be born a tendency to alter or assume some form late in youth,— only facts can decide— some peculiarities may be early impressed & others later— All poultry with same down-feathers.

5 pinned to note on *Salmonia.*

3^r–1 Harlan 1842:142, 'In this paper Mr. O. [Owen 1839c] has constructed a new genus under the name of "Mylodon," from my description of the *Megalonyx laqueatus*, (vid. Med. and Phys. Researches,) together with an inferior jaw of another species obtained by Dr. Darwin, in South America; the former he names *Mylodon Harlani*, the latter *Mylodon Darwinii.*

3^r–2 Harlan 1842:143.

3^v–1 Harlan 1842:143.

5^r–1 C. J. F. Bunbury: personal communication. Darwin visited Capel Curig 18 June 1842. See 'Journal', 'Carnarvon to Capel Curig altogether ten days, examining glacier action.' *Correspondence* 2:435.

5^r–2 Jackson, not traced.

6–1 Miers 1841.

6–2 See Wallich 1830–32.

Zoology Notes, Edinburgh Notebook

Introduction by SYDNEY SMITH & PAUL H. BARRETT

Transcribed and edited by PAUL H. BARRETT

This notebook (DAR 118, 208 × 136 mm) is known as the 'Edinburgh' notebook for Darwin first opened it while a student at Edinburgh University, dating it 'March 1827' on the flyleaf. He wrote his name out in full, Charles Robert Darwin, above the date. Of several series of entries in the notebook, the one labelled 'Zoology' is transcribed here. The arrangement of the notebook is as follows: at the beginning are 21 pages numbered by Darwin. The first 17, numbered in ink, contain notes on marine life made at Edinburgh. This section, about fishes and invertebrates collected along the North Sea shore, has been transcribed and published in *Collected Papers* 2: 285–291. The flyleaf and pages 1–18 were at one time fastened together with a piece of paper and sealing wax. Pages 18 through 21, numbered in pencil, contain entries made after the *Beagle* voyage[1]. Then follows 168 blank pages. At the back of the notebook are two additional sets of entries, written with the book having been turned over and used from the back. The first set (written on facing pages) is a 6-page list of beetles, along with dates, places, or from whom received; these entries were recorded during Darwin's undergraduate days at Cambridge. Each species is numbered, and there are 56 different names on the six pages. The second set of entries at the back of the notebook was made after the *Beagle* voyage when Darwin, forgetting his beetle list, began a series of notes, 'Zoology', on the first page following the flyleaf but in front of the beetle list. To allow himself to continue with 'Zoology' without being inconvenienced by the 6-page set of intervening beetle names, he fastened the beetle sheets together with a folded piece of paper and sealing wax, and resumed with page 2 on the back of the last page of the beetle list.

Darwin did not number the back pages in this notebook. For index purposes the pages of 'Zoology' have here been numbered consecutively from 1 to 20. The third page in this series is blank and has no designated page number, and the page numbered 3 is therefore a left hand page.

The binding of this notebook is half leather with marbled paper sides. On the front cover is pasted a square of paper, the lower left and upper left corner being torn away, with '*35*' in blue crayon being crossed out by a double scoring in pencil. Also, on this square is '42' in blue crayon and 'IX' in pencil. The 'IX' may be the only extant portion of other writing lost on the torn-away part; a pencil mark is visible at the torn edge indicating writing on the missing section. The numerals '35' and '42' are in Francis Darwin's hand and probably refer to a bookshelving scheme used within the family.

[1] Transcription of pages 18–21:

18 'It would be curious experiment to put box with *boiled* earth on top of house & see how soon any plants would come there'
19 'F. Hope. Carabidæ; F. Hope—Australia; Van Diemen's Land; King George's Land. Waterhouse minute insects from do Westwood Otaheite Babington. Water insects from all parts of the World'
20 [Blank]
21 'I suspect some curious experiments might be made by tying bladder over bottle with different kinds of salts & observing whether vapour does not carry them up? Mem bottles with sallt from Patagonia after having been kept for some time.—'

On the back cover 'Zoology' is written in ink in Darwin' hand. The notebook contains 110 leaves of high quality paper bearing 'J. Green & Sons 1824' and 'I. Annandale 1825' watermarks. The 'Zoology' notes were written in ink with one section on page 19 and one on page 20 in grey ink, and with one entry on page 20 amended in pencil. Darwin excised a small piece of page 1 (DAR 205.2:30), and the lower portions from pages 8/9 and 12/13, both of which were replaced in the notebook October 1980. An additional leaf from the blank section is also excised.

The opening date for 'Zoology' was after the first 1837 issue of the *Magazine of Zoology and Botany*, a work cited in the first entry. The last passage is dated March 1842.

The first entries relate to Darwin's preoccupation with the marine zoology of his Edinburgh studies under Robert Edmond Grant. Jenyn's reference published in 1837 defines the beginning. The second item relates to the debate between Westwood (1835) and J. V. Thompson (1835a,b) on the larval stages of pelagic crustacea—especially floating stages of barnacles. The notes contain extensive zoological data extracted from the literature especially relating to geographical distribution, taxonomy and ecology. As the reading list proceeds, it is obvious the entries summarize an account of what his immediate precursors in South America, Lesson, d'Orbigny, as well as the Russian circumnavigator Kotzebue with Chamisso on board had written.

The value of the list is twofold: it helps justify his inclusion of the words Natural History before Geology in the title of the revised edition of his *Journal of Researches* of 1845; and secondly it shows how early the ideas of much of his later work on barnacles and corals were activated. The precise goal he had in mind is not clear. Perhaps these are notes for his projected book (never published) on the zoology of the southern parts of South America mentioned in an 1837 printed prospectus prepared as part of his application for government funds to allay expenses of publishing his 5 volume *The Zoology of the Voyage of H.M.S. 'Beagle'*.[2]

[2] Freeman 1977:26.

FRONT COVER

Zoology

INSIDE FRONT COVER

December 1856
Skimmed through &
extracted
Zoology

1 Some excellent references in L. Jenyn's introduct to Mag of Zoology and Botany.[1]

Philosoph. Transacts. 3. papers connected with transform of Crust— Westwood[2] & Thompsons— Part II.— 35. Phil Trans[3] [4]

Burrowing & boring marine animals— CXVI. P 111 do[5]

Observations on Planariæ by Johnson CXII.[6] & CXV do[7]

Azara Voyage Vol I p. 196. According to Charpentier de Cossigny. only 10 years ago ‹no› snail was introduced to Mauritius.[8] **18**

2 Azara Voyage Vol. I. p. 279
Thinks the Moruffetes of Chile different from those of La Plata or Paraguay.—[1]
do. p. 365. 3 cats (mbara caya. le negro, et le pajero) l'yaguaré «the zorilla-arskink» le quiyá (Coipu) viscacha.— A. Patagonicus les tatous (.4 pichye, pelud, mulita et mataco.) are all found south of 26° 30′. Lat —[2] — do. p. 207. La punaise was not known amongst Indian. introduced in Paraguay in 1769 introduce in Governor's tran??[3]

3 Azara
Las Vinchuca or Benchuca. "Les individus ailes peuvent avoir ‹quatre› cinq lignes de long et volent. p. 208[1]

---

1 Burrowing . . . do] 'do' *boxed.*
Observations . . . do] 'do' *boxed.*
Azara . . . Mauritius.] '**18**' *brown crayon*; *excised*; located in DAR 205.2(30).

1–1 L. Jenyns 1837.
1–2 Westwood 1835.
1–3 J. V. Thompson 1835a.
1–4 J. V. Thompson 1835b.
1–5 Osler 1826.
1–6 J. R. Johnson 1822.
1–7 J. R. Johnson 1825.
1–8 Azara 1809, 1:196, 'Selong Charpentier de Cossigny, il y a dix ans qu'on ne connaissait pas les limaces à l'Ile-de-France; personne n'y en a porté, et aujourd'hui on y en trouve en abondance.'
2–1 Azara 1809, 1:279, '. . . il y a un animal extrêmement ressemblant au yaguaré, sous le nom de *mouffette du Chili*; et je ne douterais pas que ce ne fût le même animal . . . qu'il est de la famille des martes . . . et qu'il répand à volonté une odeur d'une puanteur incroyable.'
2–2 Azara 1809, 1:365, 'Trois espèces de chats, savoir le mbaracayá, le negro et le pajéro, l'yaguaré, le quiyá, la vizcacha, le lièvre patagon, les tatoús appelés pichý peludo, mulita et mataco, tous animaux du pays que je décris, se trouvent au sud des 26° 30′ de latitude . . .'
2–3 Azara 1809, 1:207, 'La punaise . . . cet insecte y fut introduit dans l'équipage d'un gouverneur.'
3–1 Azara 1809, 1:208–9, 'La vinchuca incommode

Fleas only appear in winter in Paraguay p 207[2]

Slight notice on habit of Iguana. not pass Lat. 28° North p. 239[3]

In ocean between Lat 56° and 57° only inhabitant crust Entomost of the genera— Cyclops p. 134. and p. 115[4]

In white Cape Pidgeon's stomach small shells (patella) sea weed & many pebbles[5]

Mentions stinging Millepora. Quoy. Freycinets Voyage Vol p. 597[6]
Many descriptions about lower animals of Falklands &c &c[7]

4 Bennett on Chinchillidae Zoolog Transacts. worth reading[1]

Cuvier's Memoire 133 1803. on Pennatula showing it to be one animal[2]

In Australia I was assured wild dog copulates ‹.› freely with tame: comes to houses on purpose

---

Mr J. Murray has given paper to Royal Soc on glow worm. luminous property—[3]

Curious arrangement of animals in rays Par un officier du Roi[4]

Rapid growth of Coral— RN. p 24[5]

---

3 In white . . . pebbles] *crossed.*
Cyclops p. 134. and p. 115] 'p. 134.' *boxed*; 'p. 115' *boxed.*
(Many . . . Falklands &c &c)] *partially circled.*
4 Rapid . . . p 24] 'RN. p 24' *partially circled.*

beaucoup ceux qui voyagent de Mendoza à Buenos-Ayres . . . C'est un escarbot ou scarabée, dont le corps est ovale et très-aplati, et qui devient gros comme un grain de raisin, du sang qu'il suce . . . Cet insecte ne sort que de nuit; les individus ailés peuvent avoir cinq lignes de long, et volent; ce qui n'arrive pas aux petits.'

3–2 Azara 1809, 1:207, 'Ce n'est qu'en hiver qu'on voit des puces au Paraguay; d'où l'on doit conclure que la grande chaleur est contraire à cet insecte.'

3–3 Azara 1809, 1:239, 'L'yguana est un lézard qui ne passe pas les 28 degrés vers le nord.'

3–4 Reference not traced.

3–5 Reference not traced.

3–6 Quoy and Gaimard in Freycinet 1824:597, 'Cependant ce qui nous est arrivé à l'égard de ce millépore, prouve évidemment qu'il est recouvert de polypes; car un instant après l'avoir touché, nous ressentîmes une cuisson insupportable suivie de rougeur comme celle que font éprouver les physalies et certain méduses . . . Nous vérifiâmes la nature caustique de ce millépore sur une matelot . . .'

3–7 Quoy and Gaimard in Freycinet 1824:604–71. Flustra, Obelia, and Polypes lithophytes mentioned.

4–1 E. T. Bennett 1835.

4–2 G.L.C.F.D. Cuvier 1803.

4–3 Murray 1828.

4–4 Saint-Pierre 1773

4–5 See RN24.

5 Bougainville Voyage round world no land animal besides Wolf at Falkland ∴ black rabbits not indigenous p 112[1]

M Lesson—Voyage of *Coquille* wide *limits* of Nullipora[2]

Discussion good on Falklands birds[3]

Discussion of Firola,—[4] Salpa[5] Anatifs without shells.! p 442.—[6] Planariae p 451.—[7] many molluscs

---

Under the name of Sagitta Triptera M. D'Orbigny has described my animal with teeth p. 140. Fléche of Quoy et Gaimard[8]

Ulloa shell fish Purple die Marvellous stories Ulloa's Voyage Vol I, p. 168[9]

---

6 Ceratophytes common in Northern sea. Chamisso in Kotzebue p. 312[1]

Leaches on leaves in Sumatra Marsden. p. 311[2]

D'.Orbigny considers Dasypus villosus is true Peludo[3]

Cavia Australis. Dorbigny Vol II, p 24[4]

---

5 my animal . . . Gaimard] *crossed.*

5–1 Bougainville 1772, 1:112, 'On ne voit qu'une seule espèce de quadrupède sur ces îles; elle tient du loup & du renard.'
5–2 Lesson 1830, 2 (pt 2):138, 139. No mention of wide distribution here of Nullipora, but see Bory 1822–31, Vol. 12, 1827:7, 'Nullipores se trouvent dans toutes les mers et à toutes les latitudes; c'est une des productions la plus répandue dans la nature; il en existe également de fossiles.'
5–3 Lesson 1826, 1:196–229.
5–4 Lesson 1830, 2 (pt 1):249–54, Firola described (*Pterotrachea adamaster* Less., and *P. placenta*, Less).
5–5 Lesson 1830, 2 (pt 1):256–78. *Salpa* species described.
5–6 Lesson 1830, 2 (pt 1):442, '*Triton* (*alepas fasciculatus*, Less.) La découverte que les premiers nous avons faite (en septembre 1823) des anatifs sans coquilles, exige que nous tracions l'historique des faits qui les concernent.'
5–7 Lesson 1830, 2 (pt 1):451–54. Planaria mentioned.
5–8 A. D. d'Obigny 1835–47, 5:140, 'Genre Flèche, *Sagitta*. Quoy et Gaimard;' 142, 'Flèche a Trois Nageoires, *Sagitta triptera*, d'Orb. . . . bouche inférieure terminale, renflée par d'énormes mâchoires, munies, chacune, de six à huit dents de chaque côté . . .' See also Darwin 1844a.
5–9 Juan and Ulloa 1806, 1:168. See RN177.

6–1 Chamisso 1821, 3:311–12, 'The place of the southern *Lithophytes* is occupied by the *Ceratophytes*; and the north coast of Umnack, in particular, produces several very distinguished species. The fishermen frequently draw up . . . from the bottom of the sea, large twigs, six feet long, which, from their near resemblance, they consider to be the beard of a gigantic animal, and which appeared to us to be the skeleton of a sea-pen (*Pennatula*).' See A15.
6–2 Marsden 1811:311, 'We were much plagued by a small kind of leech, which dropped on us from the leaves of the trees, and got withinside our clothes. We were, in consequence, on our halting every day, obliged to strip and bathe ourselves, in order to detach them from our bodies, filled with the blood they had sucked from us. They were not above an inch in length, and before they fixed themselves, as thin as a needle, so that they could penetrate our dress in any part.'
6–3 Orbigny. Reference not traced.
6–4 A. D. d'Orbigny 1835–47, 2:24, '. . . vivent en famille un grand nombre de petits coboyes ou cochons d'Inde d'une espèce nouvelle [i.e., Cavia australis] . . .; leur poil est des plus soyeux, et leurs yeux sont bien plus grands que ceux des cochons d'Inde ordinaires . . .'

Proceedings of Zoolog Soc. Important account of habits of Tubularia. p 52. May 1836[5]

---

dimensions of immense Tortoises p. 81[6] & p 113[7] of 1834

On the passeres of S. America. D'.Orbigny. L'.Institut. N°.— 221[8]

Good account of Condor by Humboldt Zoologie Recuiel—[9]

Meyen has written account of Guanaca. In transaction of Bonn Society[10]

7 M. Edwards on Corallines L'Institut 1837. No 212[1]

Observations on the Raptores of S. America translated from D Orbigny no IV Mag. of Zoolog & Botany p. 356[2]

Lesson on Berre. do—[3]

Magazine of Zoolog & Botany. Vol I p. 358. D'.Orbigny ‹considers› states that young birds of prey have longer tails than old ones— in America & sexes not of different size—[4] How does this apply to pale brown Caracara

8 Krauss on Corallinæ from S. Seas written in German.—[1] Stuttgart ranks these bodies amongst Vegetables in Linn. Soc.— M^r^. Donn[2]

Carmichael Linn. Transacts Vol XII. p 496. Birds at Tristan d'Acunha.— (Turdus Guayanensis?)) Emberiza Brasiliensis (?)) Fulica Chloropus.[3]

---

says some of the «species of» smaller petrels, are night birds[4] agree. with ‹pe›

8 says . . . found in] *excised; restored.*

6–5 Harvey 1836:54–55.
6–6 Telfair 1833:81.
6–7 Cary 1834:113.
6–8 A. D. d'Orbigny 1837a:347–48.
6–9 Humboldt and Bonpland 1811, 1 (pt 2):26–45, 'Essai sur L'Histoire Naturelle du Condor ou du *Vultur Gryphus* de Linné.'
6–10 Meyen 1833:552–567ff. Here are descriptions of three varieties of *Auchenia Guanaco* viz., Guanaco, Llama, and Alpaca.
7–1 Milne-Edwards 1837:178–79.
7–2 A. D. d'Orbigny 1837b:356. On this page the author discusses the comparative feeding habits of hens, ducks, finches, Icteri, flycatchers, caracaras, vultures, falcons, eagles, and condors.
7–3 Lesson 1830, 2 (pt 2):95–109, 'Description de la Famille des Beroïdes.' Pp. 104–8, Béroé.
7–4 A. D. d'Orbigny 1837b:358–59, 'We think that we have noticed in the case of certain American birds of prey, a much less striking disproportion between the size of the male and female, than in those of Europe . . . . . . among [birds of prey] the tail of the young is always longer than in the adult, the only example of the kind occurring among birds.'
8–1 Krauss 1837, 1843. See letter of C. F. A. Hartman to Darwin of 23 August 1839 regarding Dr Krauss 'a german traveller in the southern part of Africa . . .' *Correspondence* 2:214–15.
8–2 Donn. Reference not traced.
8–3 Carmichael 1818:496, 'The only land birds on the island are a species of thrush (*Turdus Guianensis*?), a bunting (*Emberiza Brasiliensis*?), and the common moor-hen (*Fulica Chloropus*).'
8–4 Carmichael 1818:497, 'There are in six species of Procellaria, among which are the *P. gigantea*, *cinera*, and *vittata*. The last, and the other three, which are smaller,

nocturnal habits of Crustaceæ

Mr Broderip says that Voluta found in not less than 7 fathoms water.[5] Mem Bahia Blanca. De la Beche theoretical researches[6]

Compare land shells of Galapagos different islds.—

9 Waterhouse remarks that no insectivore in S. America or Australia—[1] very curious.— replaced by didelphidæ

Skunk inhabitant of Patagonia. Mem:— S. Cruz.

Molina Vol. I. p. 244. Baccalao. migratory fish.—[2] See Kings drawings.—[3] for real name

---

Birds of Iceland. Mackenzie. p 345[4] for comparison with Falkland. good also for Journal.— **18**
Admirable engravings in Meyen Zoology on animal of Campanularia[5]

---

10 Alcedo stellata. Meyen p. 92.— great Kingfisher of Tierra del Fuego killed in Chile.[1]

Dobrizhoffer., Vol I p. 310. History of the Abipones— says «the Condor» ‹it› is found in the Tucuman mountains[2]

The fourth Vol. «in Lyell's possession» of Zoolog. of Voyage of Astrolabe must be studied for anatomy. of. corals.— nevertheless the details appear very trifling Also Berre «p. 8» (I think Planariæ) Sagittella, or Fleche «p. 8» my little animal with horns. Madrepores p. 26 Nullipora p. 29—[3]

In Meyen. Voyage round World German a reference to a luminous Sertularia[4]

9 Birds of . . . Journal.—] '**18**' *brown crayon; excised, restored.*

are night birds . . . Attracted by the light [of a fire], they . . . flutter round it . . . till . . . dazzled by the glare, plunge into the flame and perish.'
8–5 Broderip 1834:407.
8–6 De la Beche 1834.
9–1 Waterhouse. Probably personal communication.
9–2 Molina 1788–95, 1:244, 'Es tal la abundancia de bacalao que hay al rededor de las Islas de Juan Fernandez, que alli se verifica lo que se dice del banco de Terranova . . . se acerca en grandes camadas por los meses de Octubre, Noviembre y Diciembre hácia las playas de Valparaiso . . .'
9–3 King [Philip Parker?]. Reference not traced.
9–4 MacKenzie 1811:345–48.
9–5 Meyen 1834, Tabs XXX–XXXIII.
10–1 Meyen 1834:93, and Tab. XIV.
10–2 Dobrizhoffer 1822, 1:310, 'The condòr, a bird of the hawk species, frequently inhabits the very highest summits of the Tucuman and other mountains . . .'
10–3 Dumont d'Urville 1830–35, *Zoologie* 4:8, 'Béroé'; 9, 'Le genre Sagittèle, que nous avons nommé Flèche'; 26, 'Madrépores'. Nullipora does not appear, under that name, on p. 29. See footnote p. 29, 'Les Polyphyses diffèrent des Acétabules . . . a probablement fait que Dawson Turner les avait placès parmi ces végétaux.'
10–4 Meyen 1834:145, '. . . und das von Tilesius mehrmals gesehene Leuchten der Sertularien.'

11 Lesson Zoolog. Coq: *p. 120* Coati Roux. Tatous & perhaps Yagourundi near Concepcion!!!.—[1] (no species mentioned)

---

p. 205. only 9. Terrestrial birds at *Falklands* Is^{d} 8 waders. 22 palmipedes:[2] out of the first 9.:4 raptores. Falco poliosoma
— novozelandiæ
— histrionicus
Vultus aura

Excessively inaccurate talks of nine terrestrial & then 9. passeres![3]

Saw a Chouette a huppe courte[4]
Turdus falklandii

Says the thrush & another species! birds of passage!! sylvia macloviana, 2^{d} like sylvia cisticola.—[5] Embriza melanodera— a linnet not caught.— Troglogdytis Furnarius.— Sturnus Magellanicus.— p. 210. Scolopax very close to ours[6]

Rengger's work of Mammali: of Paraguay must be most important a discussion of geographical distribution of Mammalidae &c &c &c—[7]

---

12 The French ‹Jerrold?› «Bibrons coworker of Dumeril» who is writing with Dumeril says that two species of Tortoises come from Galapagos!!![1]

Azara. Voyage dans l'Amerique Merid. Tatu noir. abundant from Paraguay to 27°, then the Mulita from 41° to 26°[2] CLOSELY allied species, therefore interlock.—

---

11 Excessively . . . passeres!] *circled.*
12 Zoology. Journ . . . of Birds".] '**18**' *brown crayon; excised, restored.*

11–1 Lesson and Garnot 1826, 1 (pt 1):120. 'Nous ne vîmes guère que le coati roux, qu'on dit être commun aux alentours de Penco, quelques tatous et une sorte de chat, peut-être le yaguarundi de D'Azara . . .'
11–2 Lesson and Garnot 1826, 1 (pt 1), 205, '. . . neuf oiseaux terrestres sur ces îles solitaires antarctiques, huit échassiers, et environ vingt-deux palmipèdes.'
11–3 Lesson and Garnot 1826, 1 (pt 1):207, 'Les omnivores passereaux . . . n'ont guère que neuf espèces sur les îles Malouines; encore deux ou trois d'entre elles sont-elles de passage.'
11–4 Lesson and Garnot 1826, 1 (pt 1):206, 'Enfin, nous ne vîmes qu'une seule chouette à huppe courte, dont le plumage ressemblait à celui du moyen duc de France.'
11–5 Lesson and Garnot 1826, 1 (pt 1):207. 'Les environs de l'ancien établissement francais du Port-Louis sont peuplés par deux fauvettes, dont l'une est nouvelle, et sera décrite sous de nom de *sylvia macloviana*, N., tandis que la deuxième est très-voisine de la *sylvia cisticola* de la Sicile . . .'
11–6 Lesson and Garnot 1826, 1 (pt 1):210, in section 'Île de la Soledad, une des îles Malouines' (pp. 196–229), 'Une bécassine, en tout semblable à celle de France, mais un peu plus grosse, variété peut-être du *scolopax longirostris*, est fort multipleé.'
11–7 Rengger 1830:368–83, 'Uber die Vertheilung der Säugethiere in Südamerika.' In his copy of Rengger, Darwin heavily annotated these pages.
12–1 Duméril and Bibron 1835, 2:144, 'Il paraît qu'il y en a deux espèces . . . car celles de Saint-Charles ont une carapace fort allongée et relevée au dessus du cou, . . . au lieu qu'à Saint-James elles sont arrondies et d'une couleur plus foncée ou même d'un noir d'ébène.' The individual mentioned here, and also in B251, as 'Jerrold' or 'Jerrod' remains unidentified.
12–2 Azara 1809, 1:334–51. p. 346, 'Je n'ai jamais vu le tatoú noir au sud de la rivière du Paraná ou des 27 degrés; mais il est très-commun au Paraguay.' Pp. 348–49, 'D'après ce que j'ai observé, le tatoú-mulita ne remonte

Testudo INDICUS not fossil at Isle of France: ‹Jerrold?› Bibron[3]

Zoolog. Journ Vol I. p. 125, owls seen crossing Atlantic.[4] fact taken from Jenner (1825) Phils: Transact.— "on Migrations of Birds".—[5] **18**

---

do. Vol III p. 422. letter from Capt King on birds of St of Magellan. Very inaccurate & Vol IV p. 91.—[6]
Vol IV p. 388. Domestic mouse of Egypt is Mus Cahirimus. of Geof.— reference from Rüppel travels[7]

13 All Owens papers on Intestinal worms must be studied in Vol I, Zoolog: Transact. before writing on Planariæ or Polypi & is especially grand paper. p. 387. "on Classification of such animals.."—[1]

Voyage. Coquille's Voyage p 302 Vol II p. 302. Vaginulus of Lima described[2] "Arion" of Ascension. p. do.—[3]

---

some S. American Reptiles are described[4]

Shells from Tahiti and Chile[5]

The North & S. Range of shells might perhaps be worked out with advantage. with Cumms collections[6] & my own: & Capt. King's

---

p 453— Planariæ velellæ (Less) parasitical on Vellellæ in Atlantic Ocean[7]

---

13 Shells . . . King's] *excised, restored.*

pas au nord des 26 degrés et demi; mais du côté du sud, on le trouve au moins jusque vers les 41 degrés.'

12–3 Duméril and Bibron 1835, 2:568, 'Il est absolument impossible de déterminer si les Tortues trouvées fossiles, appartiennent réellement aux mêmes espèces que celles qui ont été observées vivantes . . . et à l'Ile de France en particulier, des portions de carapace et de plastron qui par leur étendue, unie à leur peu d'épaisseur, à leur forme et à leur légèreté ou defáut de poids, paraissent avoir la plus grande analogie avec les grandes Tortues des Indes . . .'

12–4 Jenner 1825:125, '. . . an Owl, seemingly the common brown owl, flying above the Atlantic wave, with as much agility as if pursuing a mouse in the fields . . .'

12–5 Jenner 1824:13, '. . . in crossing the Atlantic [Jenner], observed an owl (of what species he could not precisely ascertain, but he believes it to be the common brown owl) gliding over the ocean with as much apparent ease as if it had been seeking for a mouse among its native fields.'

12–6 P. P. King 1828, 1829, 1831.

12–7 Rüppell 1829:388, '. . . the common Domestic Mouse of Egypt, the Mus Cahirinus, Geoff. . . .'

13–1 R. Owen, 1835b, 1835c, 1835d.

13–2 Lesson 1830, 2 (pt 1):302–3, 'Vaginule de Lima. *Vaginulus limaryanus*, Less.'

13–3 Lesson 1830, 2 (pt 1), 303, 'Arion de L'Ascension. *Arion Ascensionis*, Less.'

13–4 Lesson 1830, 2 (pt 1):34–65, 'Description de quelques reptiles nouveaux ou peu connus; par R. P. Lesson.'

13–5 Lesson 1830, 2 (pt 1):239–448, in Chapter XI, 'Mollusques, Annélides et Vers.'

13–6 Cuming. See *JR* 1845:390–91 and 490 for Darwin's comments on the indentification and natural history of sea-shells by Mr Cuming.

13–7 Lesson 1830 2 (pt 1):453–54, '*Planaria vélellæ*, Less]. Cette planaire s'attache aux vélelles dont elle dévore la partie charnue . . . dans l'océan Atlantique . . .'

14 Gould agrees with D'Orbigny, that Serpent Eater— or Secretary is S. African representative of Caracaras of Americas.— manner of walking—[1] foot bill crest feathering on legs— habits—[2]

Does the Secretary, make noise & throw head back

M Edwards,—on polypi of Tubulipores L'Institut— 1838 p. 75[3]

A detailed comparison of production without Tropics in Northern & Southern America— valuable & practicable deed

---

Caricaridæ wanderers.—?? in N. America??

Wilson N. American Ornitholog must be studied before writing my general account—[4]

---

15 ⸮ Do not the Penguins replace the ‹Auk› Guillemost of the northern Hemisphere, & the Puffinuria, the Awks.— What structure do the auks bear traces of.— like Puffinuria does of Petrel?—

Study Birds of Europe for other representatives of this class—.[1]

Pyrocephalus & many Tyrannulæ— replaces warblers of Europe—

Study profoundly shells of Bahia Blanca & Southern Hemisphere

16 It is most interesting the way Synallaxis leads into Furnarius. by Patagonian Furnarius.— into Oxyurus, by Maldonado creeper of same plumage.— general red mark on wings of all— Spix has described Philedon. allied to some of my birds— These groups strictly American. Colouring on under side of wings[1]

It would be interesting comparison to find how many of the small finches walk

15 Do not . . . this class—.] *crossed.*

14–1 *Birds*:18, 'On the ground they [*Milvago leucurus*] run with extreme quickness, putting out one leg before the other, and stretching forward their bodies, very much like pheasants . . . It perhaps, indicates an obscure relation ship with the Gallinaceous order— a relation which M. D'Orbigny suggests is still more plainly shown in the Secretary Bird, which he believes represents in Southern Africa, the *Polyborinæ* of America.'

14–2 A. D. d'Orbigny 1835–47, 4:49, '. . . *secrétaire* ou *messager* (*falco serpentarius*, Lin.) . . caractérisé également par la forme de son bec sans dents, de la partie nue du tour des yeux, et même de la huppe, remplacée, chez certains caràcaràs, par des plumes frisées; . . . la nudité de son tarse . . . il est, avec les caràcaràs, le seul oiseau marcheur, et plutôt omnivore que carnassier.'

14–3 Milne-Edwards 1838a.

14–4 A. Wilson 1832.

15–1 Gould 1832–37, 5:396–99, Guillemots; 400–2, Auks; 403–4, Puffins; 446–49, Petrels.

16–1 *Birds*:68, 'In the genus Furnarius, the wing feathers are marked in an analogous manner [viz., with a red band].'

at Maldonado & Patagonia[2] compared with those of England.—

or ground birds— rather indefinite letter
Mem Orpheus—becoming tyrant— flycatcher— shown by habits & plumage so very similar to some of the Fluvicolæ?—[3]

17 The Birds seem to move much further North on West coast of S. America. than on East.— not being replaced by Brazilian Species.—
Mem Turdus Magellanicus.— C, ‹Chingolo› Chimango— Diuca??[1]

See Report ‹by› on D'Orbigny on species of Mephites[2]

4 distinct Camelidæ. do not breed together[3]

Mag: of Zoolog & B. Vol. II. p. 127.[4]

List of submarine insects. Staphylinidæ &c &c. with reference to those of mine from T. del Fuego[5]

p. 141. How comes it salt water so soon putrifies??[6] p. 319. on Hydra— polypi—[7]

18 ‹Rep› do p. 324. Polypi shorter duration than cells.— reproduced.—[1]

Milne Edwards p. 138 on Polypi.— Berenica &c &c L'Institut, 1838[2]

p. 46 Macleay Horæ Entomolog. insects swarm in Lapland & Stizbergen wherever there is extreme heat,[3] the tropical forms extend further north, because during winter they can bear the cold when torpid.—[4] On this principle

16–2 *Birds*:64–65. Here are discussions by Darwin and Gould of the morphology and habits of *Furnarius rufus* (the Casaro) and of *Furnarius cunicularius* (the Casarita) both of which walk.
16–3 *Birds*:50–54. Members of Subfamilies Tityranae and Fluvicolinae are discussed.
17–1 *Birds*:91. Chingolo bunting— resembles in habits and general appearance the English sparrow. See also *JR*:64–69 for a discussion of Chimango.
17–2 Orbigny. Reference not traced.
17–3 Reference not traced.
17–4 Westwood 1838.
17–5 Westwood 1838.
17–6 Baird 1838:141, 'My opportunities for observing these insects [viz., *Cypris strigata*] have been . . . limited . . . from the rapidity with which sea water becomes putrid when kept in a room in a small vessel . . .'
17–7 G. Johnston 1838b.
18–1 G. Johnston 1838b:324, '. . . the life of the individual polypes is even more transitory than their own cells; that like a blossom they bud and blow and fall off or are absorbed, when another sprouts . . . to occupy the very cell of its predecessor . . .' Passage scored in Darwin's copy.
18–2 Milne-Edwards 1838b:139, '. . . des Berenices . . .'
18–3 MacLeay 1819–21:45, 'We see also how insects may swarm in the very coldest climates, such as Lapland and Spitzbergen, where the short summer can boast of extraordinary rises in the thermometer . . .'
18–4 MacLeay 1819–21, 1:44, 'Animals also are subject to the same sort of limitation with plants: that is they have to fear extreme cold rather than extreme heat. Such animals therefore as can avoid this cold— either by passing it in a state of torpidity, or by the habit of burrowing in the earth, or by living in the sea, or by artificial clothing,— will in general be found the most widely dispersed . . .'

tropical forms in N. America extend much further N. in N. America than in Europe—[5] Coleoptera especially require a greater duration of Heat. hence musquitoes & knats abound during short summer far N. where this other order is comparatively rare.—[6] These views clearly explain rarity of insects in T. del Fuego.— Hence it is odd that *Amber* insects of Europe have Tropical Forms[7]

19 See p. 256 of Note Book (C) for comparison of singing powers of birds of N. America & Europe.

Entomolog. Transact. Vol I. p. 130. Col Sykes on balls made by dung beetles, like those from Chiloe.[1]

Amblyrhyncus de marlin James Is^{d}—[2]

Lutke Voyage Vol III p 322 Dr Martens says only one Reptile in Kamtchatka (Salamandra aquatica).[3] Compare with T del Fuego

---

Compare birds of do[4] with ‹T› N American & T. del. Fuego & Iceland

20 Spix & Martius talk of birds singing in the forests of Brazil[1]

H. Wedgwood says in ‹14th› «**13th**.» Vol of Archæologia arrow=heads described in Suffolk as lying under *strata* of gravel & clay about 10 feet in thickness.—[2] (March, 1842)

---

19 See . . . Is^{d}.—] *grey ink.*
20 Spix . . . Brazil] *grey ink.*
«**13th**.»] *pencil;* '14th' *crossed pencil.*

18–5 MacLeay 1819–21, 1:44, 'Tropical plants will therefore thrive better in Thibet and other inland parts of Northern Asia than they would do were we to transport them to places of the same latitude in America.' See also Anonymous 1839c, 27:101, 'The extremes of heat and cold, as has been remarked by Mr Macleay, are much more essential in determining a locality than the mean annual temperature . . . intertropical forms of insects are prolonged much farther north in the New than in the Old Continent, which is the reverse of what botanists have observed in regard to plants.' Darwin wrote in the margin, 'Chiloe interesting under this head.'

18–6 MacLeay 1819–21:46, 'And accordingly we find that insects, such as gnats, mosquitoes, &c., which pass their larva state in water, thus avoiding extreme cold, and whose existence in their perfect state being naturally ephemeral, must therefore suffer little from the shortness of the summer,— are no where more troublesome than in the very coldest climates. Whereas the number of coleopterous insects, which, being naturally longer lived, require a longer continuance of warmth, is sensibly diminished in these dreary countries.'

18–7 Berendt 1839:211, '. . . it would appear that all the species [of Crustacea, Myriopoda, Arachnidæ, and Aptera] found in amber are now extinct, and that but a small number of the genera at present exist. Many new genera have therefore been formed, and also one entirely new family.' In his copy Darwin scored this page.

19–1 Sykes 1836b.

19–2 See Duméril and Bibron 1834–54, 4(1837):197, for reference to *Amblyrhinque de Demarle.*

19–3 Litke 1835–36, 3:322, 'L'on ne connaît au Kamtchatka qu'un seul reptile; c'est une espèce de Triton (ou *Salamandre aquatique*), de couleur noire.'

19–4 Litke 1835–36, 3:275–87. Birds of Kamtchatka discussed.

20–1 Spix and Martius 1824:246–48: rattles and plaintive notes of toucans; screams of orioles; beautiful melody of the thrush; chattering of the manakins; full tones of the nightingale; metallic tones of the uraponga, and cries of the macucs, capueira and goat-suckers.

20–2 Frere 1800:204, 'The strata are as follows: 1. Vegetable earth 1½ feet. 2. Argill 7½ feet. 3. Sand mixed with shells and other marine substances 1 foot. 4. A gravelly soil, in which the flints are found, generally at a rate of five or six in a square yard, 2 feet.'

Questions & Experiments

Introduction by Paul H. Barrett, Sydney Smith & Sandra Herbert

Transcribed and edited by Paul H. Barrett

This account book (DAR 206, 158 × 198 mm) has 40 pages (20 leaves) bound in a limp marbled paper cover. Darwin wrote 'Questions & Experiments' on front and back covers in a large formal hand. A few entries in the notebook are in pencil, but most are in ink of widely varying shades: some entries in ink are supplemented or deleted in pencil. There is nothing to show continuous use in the notebook, ink colours vary at random, and any single page seems to have been written at many times so that it is at present impossible to establish a dated sequence. In this the notebook is unlike the others in this volume, where entries from front to back are on the whole temporally ordered.

Setting an opening and closing date for this notebook is problematical, though assigning general dates of use less so. There are scattered dates in the notebook, some being later additions: March 1842 (page 10a), July 1842 (inside front cover, pages 11^{v}, 15), May 1844 (page 22), and October 1844 (page 22). However, as an enterprise, the notebook would seem to have been begun earlier, for it connects with Darwin's efforts in 1839 to learn, in a systematic manner, the opinions of breeders and agriculturists regarding variation and inheritance. Notable here is Darwin's printed questionnaire entitled 'Questions about the Breeding of Animals' done some time between 1 January and 6 May 1839.[1] The name of Richard Sutton Ford, one of Darwin's respondents to the questionnaire, appears on page 3 of Questions & Experiments. More generally, it is significant that in Questions & Experiments, a notebook of fluid organisation, pages 1–6 follow a systematic format listing questions to ask and experiments to perform with regard to the crossing of plants and the breeding of animals. In addition there is overlap between some entries in Questions & Experiments and more easily datable writing. Darwin's entry regarding atavism on page 1 correlates to E183; page 5 refers to entries on E184 and EIBC; folio 6^{v} relies on a letter from William Herbert of *c.* 27 June 1839; and in Questions & Experiments as at the end of Notebook E Darwin was perusing issues of the *Transactions* of the Royal Horticultural Society. Hence one can assign a probable date of mid-1839 for Darwin's opening of Questions & Experiments. There is no firm date for the closing of the notebook. Most entries were made by 1844, but the notebook continued in use after that date. The first paragraph in the entry under 'Hooker' on page 17 follows from Joseph Dalton Hooker's letter to Darwin of [23] March 1845 while another entry on that page, referring to Isidore Geoffroy Saint-Hilaire, correlates with a letter from Darwin to Hooker of [3 September 1846].[2]

In function Questions & Experiments served as a storehouse for questions on animal and plant breeding that Darwin wished to ask of various authorities and for lists of experiments he wished to perform, or see performed. In the transmutation notebooks some entries similar to these occur, but in this notebook questions and experiments are grouped together into a programme of research. Answers to Darwin's queries appear partly within the text but more often in letters between Darwin and those whose opinion he sought as a consequence of the

[1] The questionnaire is reprinted in *Correspondence* 2:446–49. In his Accounts book, now kept at Down House, under June 1839 Darwin wrote 'Messrs Stewart and Murray for printing Questions 2..5..6' [£2 5s 6d]. For two replies to Darwin's questionnaire see Freeman and Gautrey 1969 and also *Correspondence* 2:187–192.

[2] Compare QE17 to *Correspondence* 3:162–64, 340.

lines of inquiry developed in the notebook. The notebook is thus the public and experimental side of Darwin's species work and complements his more private inquiries in the other notebooks.[3]

In organisation the notebook is comprised of a series of lists, three headed by the names of places—Maer (page 13), Shrewsbury (page 14), and the gardens of the Zoological Society of London (page 20)—where experiments could be carried out, but most headed by the names of persons to whom the questions listed below would be addressed. On page 17 the name changes in mid-page where John Edward Gray is displaced by Hooker. After the Darwin family moved to Down House in September 1842, London friends, hitherto close, were now 12 miles away, and questions were accumulated against the next visit to London or posed in letters. The inside front cover of the notebook holds Darwin's index to the notebook; the names listed there comprise a diverse but impressive company of contemporary British figures in natural history.

Darwin's notes in Questions & Experiments demonstrate his early interest in all factors affecting variation. The range and penetration of these notes also attest compellingly to his ingenuity as an experimenter and constitute a master plan for his treatment of variation in the *Origin*, and later in the massive two volumes on *Variation of Animals and Plants under Domestication* (1868).[4] More immediately, Darwin's use of his Questions & Experiments notebook is evident in the text to his *1842 Sketch* of his species theory, where species mentioned in the notebook also appear in the *Sketch*; and the thrice-mentioned 'July 1842' date in the notebook confirms that he was actively using it about the time he wrote the *Sketch*.

For discussion of Darwin's relation with the world of breeders see Secord 1985; on Darwin's approach to the subjects of generation, variation, and inheritance see Bowler 1974, Geison 1969, and Hodge 1985.[5]

[3] For this point we are indebted to Anne Secord who has also provided us with the correlations between Darwin's correspondence and Questions & Experiments.

[4] See Barrett, Weinshank, and Gottleber 1981 for page locations and frequency of mentions in the *Origin* of words associated with 'variation'.

[5] Bowler, P. J. 1974 *J. Hist. Med.* 29:196; Geison, G. L. 1969 *J. Hist. Med.* 24:375; Hodge, M. J. S. 1985 *In* Kohn 1985; Secord, J. A. 1981 *Isis* 72:163.

FRONT COVER

Questions & Experiments

INSIDE FRONT COVER

	Gowen, Royle, & Horsfield **Sykes**	p. 12
	Maer.	p. 13
Questions &c. July. 1842.—		
	Shrewsbury	p. 14
	Henslow. (2d time)	p. 14.—
Father. And. Smith	Dr. Holland	p. 16
Babington— Gould ————————		**10.(a)**
	J. Gray ————	17
	Yarrell————	18
	Blyth————	19—**Mr. Tollett**
	Zoolog Soc «Gardens» ——— .	. 20 **& Breeders**
Dr. Boott:	R. Brown	p. 21
	Horticulturists	p. 21—**23**
	Eyton	p. 22
	Schomburgk.————	1
	Jordan Smith.	p 1.
	Sowerby Cuming.	—p 1
	Owen	p 17

Hooker p. 17
Mrs. Whitby. Newlands Lymington Hants. Habits of different caterpillar races.[1] —Name of Italian who sold eggs.—[2]

IFC *base text*: 'Royle . . . Eyton'; *remainder added at intervals.*
p. 12] '2' *over* '1'.
p. 13] '3' *over* '2'.
p. 14] '4' *over* '3'.
Dr. Boott:] *pencil.*
Jordan Smith. p 1.] *pencil.*
Hooker p. 17] *scored by pencil finger.*

IFC–1 *Calendar*: Letter 1113. 'To M. A. T. Whitby 2 Sept. [1847] Down Questions Mrs W on difference in flight capacity of male and female silkworm moths . . .' See also *Variation*, 1:302, 303 for further discussion of Mrs Whitby and breeding silkworms.

IFC–2 Probably 'Mr. Antony Tagliabue, 31, Brook Street, Holborn', recommended by Mrs Whitby as a supplier of silkworms' eggs (Whitby 1848, p. 55) or 'Signor Arregoni, 16, Church-street, Soho' from whom Mrs Whitby purchased silkworms in 1844 (Whitby 1848, p. 62).

I

Temporary Question

1 Where has Duchesne described Atavism alluded to by Dr. Holland—[1]
‹Jordan› Smith of Jordan Hill— character of the extinct land-shells of Madeira—[2] analogous or quite distinct from recent ones— I presume some recent not found fossil (perhaps not embedded
? are there any very *common* recent ones not embedded?—

═══

Do the Tame Parrots breed amongst the Indians

═══

Do the Savages select their dogs

═══

Sowerby Entomologist[3]
{Does individual Shell or insect or group vary more in one country or district than in another?
Character of shells of Sandwich group
{Sowerby monstrous Cardium—[4] does it remind him of other species[5]

Hooker says the species of Aquilegia vary much in their *spurs* & Ranunculus in the nectaries.[6] The former best for my experiment on Selection.

I^v

Experiments in crossing &c Plants

1 Repeat the French experiment of Carrot[1]

I *base text: title and question 1, remainder added at intervals.*
‹Jordan› Smith . . . not embedded? —] *pencil, crossed pencil.*
Do the Tame Parrots . . . their dogs] *crossed.*
Sowerby Entomologist] *added in margin opposite first brace.*
Inside front cover, p. 1, and several other pages have pin marks where loose notes were attached; this can be reconstructed at least in part.

I^v *base text: title — question 3; remainder added at intervals.*
3. To apply . . . Transact.—] *crossed pencil.*
4 May we . . . flower] *pencil, written on a slip of paper pasted into notebook.*
(6) flower on] 'on' *altered from* 'in'.
The «above] 'e' *over* 'is' *in* 'The'.
The French . . . negative.—] *change of ink.*

1–1 Holland 1839a:23, 'A singular variety [of heredity] is that which Duchesne and others have termed Atavism; where a bodily peculiarity, deformity or disease, existing in a family, is lost in one generation; reappearing in that which follows.' Duchesne's original citation has not been traced. See E183, which is the source of this question.
1–2 J. Smith 1838–42 [1841]:354, 'The [land] shells have been most carefully examined by the Rev. Mr. Lowe, and one sixth ascertained to belong to species not now living in the island; the Canical sands therefore are assigned by Mr. Smith to the Pleistocene or newest tertiary æra.'
1–3 J. Sowerby 1806, 1838. Includes many pages on insects. Sowerby (Elder) 1812 is also on insects.
1–4 G. B. Sowerby, Jun. 1840. Does not mention monstrous *Cardium*. See however Sowerby, G. B. (Elder) 1839a, for discussion of monstrous *Encrinus*. Darwin may have discussed the monstrous *Cardium* at the 8 Sept. 1840 meeting of the Zoological Society of London, when a paper on *Cardium* by Sowerby, Jun. was presented.
1–5 Sowerby (Elder) 1833. Discusses great variability of species of Mollusca and Conchifera.
1–6 Hooker: reference could not be traced. Probably a personal communication.
1^v–1 See E149.

2 {also try Primrose & Cowslip in rich soil & propagate from their seed
3. To apply pollen of different genus & then some hours afterwards of nearly related plant & see if first pollen produces any effect, as in case of woodpidgeon & Hen. mentioned by Mr Knight. Vol IV Hort. Transact.—[2]
4 May we no suppose, that certain plants, like Aphides produce impregnated young ones; & that it is in these that male organs (not being always useful). fail— Really good subject for experiment.—«to repeat Spallanzani»[3] Raise only single Plants & only allow ‹few› one flower
(5) Dr Fleming. Philosop. of Zoolog. vol 1. p. 427— says *biennial*-wall-flowers & scarlet Lychnis can be propagated by cuttings.—[4] Try.— Important as discovering function of seeds—
(6) To hybridise EVERY flower on melon & see whether fruit affected. Mr. B. seemed to say impregnation ‹caused› of some seeds, caused symmetry in cone— The «above Exper» explains apples on side near other tree being affected.— does one branch of Cabbage being mongrelized affect other branches— **The French Apple tree «with abortive stamens» answers first question in negative.—**

2 Questions Regarding Plants.

1. Uniformity of hybrid & Mongrel offspring
2. How have *late* varieties of Peas &c been obtained?
3.. Whether the viviparous grasses & onion, produce flowers, like the Oxalis from C. of Good Hope mentioned by Mr Herbert in vol IV. Hort. Transact.—[1]

2 *base text: title — question 9; questions, 10 and 11 added in a different ink.*
5. Whether Roses . . . 6 . . . on them?] *crossed pencil.*
7... Bananas] *second* ‘a’ *over* ‘n’.
8. Can any . . . individuals—] *crossed pencil.*
9. **Mal[e]]** *added pencil in box.*
11. non-flowering] *crossed.*

1v–2 Knight 1822. Makes no mention of successive pollination of a flower by pollen of different genera at different times; see however: p. 367, ‘great numbers . . . of species . . . may be made to breed together, with greater or less degrees of facility . . .’; p. 371, ‘. . . a single plant is often the offspring of more than one, and, in some instances, of many male parents.’; pp. 372–73, discussion of a cross of a wood-pigeon and a ‘common hen’ which as reported produced a mule. Knight thought a mistake had been made, as the mule bred freely. Knight, in 1822, believed in creation of species and that hybrids from distinct species could not be fertile (pp. 371–72). But see Knight 1824:293, ‘. . . that real mule plants have in some instances, and under certain circumstances, produced offspring . . .’ See Darwin's abstract (DAR 74:62) of Knight 1822:372, ‘Mentions curious case of Hen mounted by woodpidgeon—which had fertile offspring.—but this one (out of eleven eggs) had no comb and fleshy nostrils and whole profile of head exclusive of point of beak.—looks like as if past impression had been produced.—’
1v–3 Spallanzani 1769. See E90 and E148.
1v–4 Fleming 1822, 1:427, note, ‘Sir James E. Smith, Introd. Bot. p. 138 and 139 seems to consider it as established, that “propagation by seeds is the only true reproduction of plants.” ’ But according to Fleming, ‘The wall-flower and sweet-william plants, whose natural term of life rarely extends beyond two years . . . may be continued for many years, by being propagated by means of cuttings of the slips. Even the annual stem of the Scarlet Lychnis, may be converted into separate plants of many years duration.’
2–1 W. Herbert 1822:33–34, ‘This is the first instance I have known, or heard, of an embryo, either in the vegetable or animal kingdom, drawing its support directly

4.. Are any varieties of Cabbages not attacked in bad years from Caterpillars.
5. Whether Roses impregnate each other. when close planted together: ‹do› **Can** Holyoak **be raised distinct by seed**— Heartease.
6. — Do not species of wild Roses run into each other very much.— Has not some one written on them?
7... Are the wild Bananas of Otaheite seedless;— are all varieties seedless— if so. how have varieties been formed?—
8. Can any annuals be budded. with reference to extension of age of individuals—
9. Do plants in becoming double ever become monoœcious— loosing one sex & not other: which generally fails first?— **Mal[e]**
10. Henslow says semi-doubl flowers are those whose stamens are monstrous, how then are seeds ever raised?
11. Is not non-flowering gorze common in Norway **No**[2]

3 Questions regarding Breeding of Animals

If two half bred animals exactly alike be interbred will offspring be uniform.— Mr Ford[1]

Has M. Sageret WRITTEN on crossing of Cabbages, *quoted* by (as if oral) Decandoelle in V. Vol of Hort. Transacts[2] **& M. Sageret is referred to with doubt by Herbert**[3]

===

Do forest-trees sport much in nursery gardens?

‹are the› is the ground much manured

===

3 *base text: title and first question; remainder added at intervals.*
Do forest-trees] 'No' *added over* 'Do'.
In species . . . **see notes**] *crossed.*
In varieties . . . important.] *pencil, double scored in pencil.*
In crosses . . . versâ] *crossed pencil.*
Good observation . . . comparison with] *crossed pencil.*
Weigh skeleton . . . good] 'good' *in box.*
Weigh skeleton . . . fatness.—] *crossed pencil.*

from the parent, without the intervention and assistance of an intermediate body, such as the cotyledon, the yolk of the egg, or the placenta, to afford it nourishment. I have indeed found one Oxalis from the Cape of Good Hope . . . to be viviparous . . . [and this] amounts . . . only to a habit of premature vegetation in the seed . . .' See also E165.
2–2 See E153.
3–1 R. S. Ford, farmer and neighbour of Wedgwoods at Maer; a respondent to Darwin's questionnaire about the breeding of animals. Ford's answer was dated 6 May 1839. See Freeman & Gautrey 1969:223–24; Freeman 1977:55 and *Correspondence* 2:187–89.
3–2 Candolle 1824:2–3, 'M. Sageret, an enlightened member of the Agricultural Society of Paris, has also sent me the results of his experiments on cross-bred Cabbages . . .' P. 39, 'The cultivated Cabbage, *Brassica oleracea*, according to M. Sageret [is] incapable of receiving fecundation from any but its own species . . .' See also p. 25.
3–3 W. Herbert 1837:353. Questions Sageret's report of seed-pods, some long, some short, produced by a cross of a cabbage and a horse-raddish.

In species of close genus do more than three primary colours occur in relation with species— **answered «by Henslow»**[4] **see notes**
In varieties is there any difference in off spring from A. into B. from B. into A. as takes place in *mules* ass & horse— important.
{In crosses does male offspring take after male parent & vice versâ
= History of Tortoise-shell Cats. as only one sex so coloured =
I have grey-cat «wh was female» with tinge of tortoise-shell «on back.—»
{= Length of intestine in Persian Cat— , in Brazilian «toothless» dog—
I. St. Hilaire says length differs in different cats.—[5]

Good observation— examine semen of Hybrid animal. in comparison with

Weigh skeleton of Tame Duck & Wild Duck, & then weigh their wing bones & see if *relation* is same good, avoids effects of fatness.—

4 *Experiment in crossing animals.*— &c

(1) To cross some artificial male with ‹old› female of old breed & see result.— According to Mr Walker the form of male ought to preponderate;[1] according to Mr Yarrell the latter ought: either in first breed or permanently.—

(2) Cross two half-bred animals. which are exactly alike & see result.—

(3) Cross the Esquimaux dog. with the hairless Brazilian **or Persian** animals of different heredetary constitution, to see whether offspring infertile.—

(4) Does the number of pulse, Respiration, period of gestation differ in different breeds of dogs. Cattle, (Indian & Common) &c: length of life.

4 4] *page number over* '3'.
base text: title — question 2; remainder added at intervals.
(1) form] 'f' *over* 'm'.
(5) Does my Father . . . heredetary.] *crossed pencil.*

3–4 See letter Darwin to Henslow [24 Jan. 1840], 'Do you recollect our discussion about varieties of same plants not having *three* primary — Surely I have seen pale yellow hyacinth, & certainly blue & pink ones.—' *Correspondence* 2:252.

3–5 I. Geoffroy Saint-Hilaire 1841:298, 'Parmi les viscères intérieurs, le canal alimentair s'est notablement allongé chez le chat, en raison du régime en partie végétal, imposé par l'homme à cet animal exclusivement carnassier dans l'état de nature.' In Darwin's copy of Geoffroy, there are double score marks in the margin beside the passage.

4–1 A. Walker 1838:202, 'The second law, namely that of Crossing, operates where *each parent* is of a *different breed*, and when, supposing both to be of equal age and vigour, the *male* gives the *backhead and locomotive organs*, and *the female* the *face and nutritive organs*.' Darwin liberally annotated his copy of *Intermarriage* inserting in the book a four-page outline of his own views of heredity vis-a-vis Walker's. The essay begins 'I reject Mr Walkers theory of one parent giving (see p. 150) one series of organs & ‹another› the other a different set.—'

(5) Does my Father know any case of quick or slow pulse being heredetary.

═══

(6) In the last 1000 years how many generations of man have there been.— on what principles calculated.— in order to guess how many generations in Mammalia. in group effect of crossing.—

═══

(7) Are the Eggs of the Penguin Duck quite similar to those of another Duck. ⁋in *Pidgeon*?— **Mr. Miller said yes with regard to former**

4a (8) Is form of globule of blood in allied species similar.— if not how is it in ‹allied› varieties

(9) Cross largest Malay with Bantam— will egg kill Hen Bantam.— Cross common Fowl with Dorking

(10) Statistics of breeding in Zoolog. Gardens— with respect to conditions of animals & their general healthiness— Fox's, Bears Badgers,— How few wild animals are propagated,, though valuable as show. & curiosities!! What is price of fox. otter. Badger &c &c &c.—

(11) Keep. Tumbling pigeons. cross them with other breed.—

(12) About the blended instincts

---

5 Remote Experiments— Plants

Raise seedlings surrounded by various bright colours, any effect?[1] **and silk caterpillars**[2]

(1) Shake a sleeping mimosa, or half bred mimosa **(a)** between sensitive & sleeping species, & see whether association can be given

(2) do the stamina of C. Speciosissimus collapse during sleep & do of Berberis— **(latter I think certainly not)**

5–[5]a *base text: title — question 2; remainder added at intervals.*
Remote] *crossed.*
Raise seedlings . . . silk caterpillars] *followed by double separation lines.*
(a)] *added as cue to* [5]*a* 'The Leptosiphon . . . see effect—'.
(2) **I think**] *underlined.*
(3) Sow seeds . . . will be.—] 'will seedlings . . . cuttings &c' *crossed pencil.*
(3) bulbs] *altered from* 'bubs'.
(3) seedlings] *altered from* 'seeds'.
(4) Raise] 'i' *over* 'a'.
(6) **This in fact . . . Cabbages**] *added pencil in margin.*
a on ditto] 'on' *over dash.*
such diœcious . . . orifice] *partially erased pencil relating to first question on p. 6.*
12. **Bee**] *underlined.*
done] *added pencil.*

5–1 See also *Correspondence* 2:190 about visual stimuli at time of conception affecting coloration in offspring. This comes up in connection with R. S. Ford's replies to the questionnaire on animal breeding. Freeman and Gautrey 1969.

5–2 See E43.

(3) Sow seeds & place cuttings or bulbs in several different soils & temperatures & see what the effect will be.— will seedlings vary much more than cuttings &c

(4) Raise annuals or common English plants in Hothouse & see what effect on organs of generation

(5) Place pollen of Red Cabbage «mixed with own pollen» on flowers of other cabbages & see whether there will result hybrids—

(6) Dust flowers of one branch of Cabbage with pollen of other, count seeds, & see how great a proportion springs up true.— **This in fact always takes place in natural Hybrids of Cabbages**

(7) Sow ‹daisy› seeds of wild cabbage in VERY rich soil, will plants abort?, does it require successive generations to accustom them to such soil.— Sow weeds in such soil.—

7(a) { Experimentise on Primrose seeds— it really is an important case— cross with cowslip pollen.— as these are wild varieties. Is any intermediate form found wild

[5]a a The Leptosiphon densifolium «an annual» ‹sleep› «closes flower» on all gloomy days.— The «garden» Coronella also sleeps on ditto— Cover them up periodically & see effect— **such diœcious individ—small orifice**

(8) { Carry Bees, powdered with starch & Carmine & experimentise on their returning powers— then carry them in Electrical machine, reversing the poles **test by suspending magnet within** & see which way they fly.—

(9) { **I have noticed leaves covered with Honey-dew dusted with pollen of neighbouring grass**= Spread sheets of Paper. covered with some sticky stuff in flat places & see whether wind, on «dry» *windy day*, «flower garden on gravel walk» will drift many seeds= Necessary to answer Wiessenborns doctrine of Equivocal Generation Charlworth p. 377.[1] Have paper ruled in squares to facilitate investigation.— Capital in middle of ploughed field— on hills.—

10 { Shoot tame duck on pond with Duck-weed— coots— waterhens— examine dog, which has swum— on pools & rivers— every kind of seed must be distributed.— Examine scum of pond for seeds.—

11.{ Soak all kinds of seeds for week in Salt. artificial water.—

12.{ Plant two races of Cabbages near each other— & enclose one twig of each in bell-glass— sow these seeds & see if they will come up true— whilst others are crossed.—

[5]a–1 Weissenborn 1838a:369, 'The ancient doctrine of *generatio spontanea seu æquivoca* [is opposed to *omne vivum ex ova*] ...' P. 373, 'Most ... animals ... spontaneously generated, afterwards propagate their species by ova ...' P. 377, '... the seeds of the *Hypericum* not being furnished with a *pappus* or wings, I do not see how we are to account for [its germination and growth], except by the *generatio æquivoca* ...'

Are Bees guided by smell— or sight.— —. touching Mr Brown theory of insect-like Orchis—[2] **& final cause of beauty of flowers— contrasted by Kirby— with animal reproductive system.—** [3] — cover flower— put artificial flowers— also do with honey— **What is use of Bee Larkspur= =Toad Orchis=**
How many flowers in minute do they visit?? good=!!
Examine pollen of double flowers. compared with single & see whether grains flaccid, as Koelreuter describes[4]

Kill Sparrow after feeding on oats, give body to Hawk & sow pellet. ejected. **done**

Examine pollen of such flowers as do not seed or seed rarely— Magnolias. «Azaleas» & plants grown under unfavourable circumstances, as Hyacinths in glasses &c &c

6

Experiments

Questions concerning Plants

Is the common Fig Diœcious— are its female flowers always barren— if not how does impregnation take place **male & female flower in same receptacle**

(8) Make Duck eat Spawn, eggs of snail, row of fish & kill them in hour or two «My Father made hens cast Holly-seed & they grew»

(9) Place. Snap-Dragon. (I have seen one monstrous) Fox Glove & such like in very rich soil— As they have little tendency to double; what would be effect—

(10) Try in how many generations. daisy. Fever-fuge Groundsil.— gilly flower will break & become double.— **There is a double Crows-foot. or Ranunculus.=**

(11) Try..Nitrate of Soda— Salt. Gypsum. Magnesium Iron Rust Carb. of Ammonia.— Horse Urine &c &c on associated plants. when proportional number appear equal— & see whether proportions will vary, which will show that such proportions not effect of Chance

6 *base text: title and first question.*
Questions concerning Plants] *crossed.*
Is the common Fig . . . **receptacle**] *crossed.*
(8) Spawn] 'S' *altered from* 's'.
(12) **Maer.=**] *added in margin above* '(12)'
(12) **by birds**] *added pencil.*

[5]a–2 R. Brown 1833:740–41, '. . . the remarkable forms of the flowers [viz., a striking resemblance to insects] in this genus [*Ophrys*] are intended to deter not to attract insects . . . [and] the insect forms in Orchideous flowers, resemble those of the insects belonging to the native country of the plants.'

[5]a–3 Kirby 1835, 1:139, 'The most beautiful and admired, and odorous and elevated parts of the plant are its reproductive organs and their appendages, while in the animal they are the very reverse of this.'

[5]a–4 Kölreuter 1761–66, often discusses fertilization of double flowers. In his own copy, Darwin made marginal notations on this subject on the following pages: 1761:46; 1764:110–11; 1766:67, 72–74, 84–89, 119, 122. See especially 1766:71–73 for discussion of 'Der saamenstaub . . . theils aus eingefallenen und leeren Kügelchen.'

Maer.=

(12) Take Bag of soil from centre of woods «especially if date of wood be known» & other odd places & see what plants will spring up which will show, how seeds are transported, or how long they remain dormant. if kinds come up, not found in wood.—
but seeds continually dropping in woods. by birds

[6v] 13. Mr Herbert says Crocuses are very difficult to cross.—[1] are there races— if so plant them together. & raise. seed.— In letter Mr Herbert says do about Œnothera.—[2]

(14) Examine pollen of those genera of which wild hybrids have been formed.

(15). What is History of Viburnum. or snow-ball-tree. what would result from seeds being sown=

See in Cultivated Plants, as Pentstemon, which have abortive parts, whether such vary.—

Do Bees go to Sweet Peas, IMPORTANT, for if so, as these can be raised true, there is no crossing by Bees.—

7 *Henslow.—*

(1) Character of alpine Flora of Tierra del Fuego and Entomology of.— most important, as furthest removed possible point.— ‹**genera in intermediate country**

(2) Any known changes in Flora of countries during last century or two.— where agency of man not known.—

(3) How is Iris impregnated.; which part of stigma?—

(4) As Papil. flowers appear difficult to cross, are there unusually many species in genera of Leguminosæ.— Herbert explains numerous spec. of Cape Heath by facility.[1] ‹Knight takes opposite view.[2] **Gærtner talks of the several great & natural Families, as being difficult to cross.**[3]

7 base text: *title — question 1; remainder question 2–36 (p. 12) added in same hand and ink.*
(1) Character of . . . **country**] *crossed pencil.*
(4) **Gærtner . . . to cross.**] *added pencil in margin.*

[6v]–1 W. Herbert 1822:27; see also Darwin-Herbert 1839 correspondence, Herbert letter 5 Apr. 1839, 'I have failed in all attempts to cross Crocuses . . .' *Correspondence* 2:183.

[6v]–2 W. Herbert to Darwin letter [*c.* 27 June 1839], 'Flowers of Œnothera can scarcely expand from the bud without impregnating the stigmas . . .' *Correspondence* 2:203.

7–1 W. Herbert 1822:27, 'It has been conceived that the African Heaths consist of different genera, which might be distinguished by the shape of their pods: but I have found no difficulty in intermingling species with different shaped pods, which proves that such a division would be erroneous . . .'

7–2 Knight 1822:371–72 (see also QE1v–2). Believed mules to be sterile. Herbert 1822:16, 'It is not even true that all mules amongst animals are entirely sterile.'

7–3 K. F. Gærtner. See numerous publications on plant hybridization.

(5) It is most important to ascertain amount of variation in plants raised by Scions, as Elms. &c &c— I have some reason to suspect Elms.— **& Orchidacæous plants** no other case.—

(6) Will plant accustomed to rich soil, when placed in very poor flower, but not fruit— — Do not orchards become unproductive from poorness of soil.—[4] yet crabs probably would grow there

(7) Where parts of fructification ‹lat› retrograde into leaves— is this ever effect of want of nutrition.— Horned oranges so?[5] **—Yes, my Father lost this character in grt degree from charcoal & good treatment**

8 (8) Do bees frequent Cabbages «& Cowcumber's out of doors.» much— or the minute Orthopt.— important, as we know how readily they cross.—[1]

(9) In the nurseries, when «seed of» the varieties of Cabbages, peas, beans, as raised, do the Seedsmen select at all from the plants? If not, I am surprised ‹plan› such plants do not degenerate,— as the Bees will mingle the infinitesimal varieties which must occur.— ?is «it» not these infinitesimal varieties, which counterbalance each other?

(10) Is number of pollen-grains necessary to impregnate ordinary number of seeds known?— Linnæus has shown that each pistil is connected with separate division of germen ‹?›—[2]

(11) Must pollen grain be whole, to impregnate?— **I presume only stigma impregnable.—**

(12) At Maer Cowcumbers in *frames* are not artificially impregnated. Abberley says Ants— Enquire

(13) Do any of same species of Willows grow in same situation & flower at same time. Has H. seen group of different species growing

[8v] White Mullein good plant to sow & try to get other species

9 ‹near› close to each other.— As they are diœcious, if no hybrids were produced by seed, we might feel sure, that pollen of own kind is much

8–[8v] (9) as raised] 'as' *altered from* 'is'.
White Mullein . . . species] *pencil, only entry on 8v.*
9 ‹near› close . . . of Willows.—] *continued from page 8.*
14 **(a)**] *relates to* '(a)' *on* [9v].
(18) Œnothera] 'n' *over* 'e'.

7–4 Henslow 1838:338. Discusses relation of soil to plant growth: 'From the character of the soil and the condition of the islands we might expect *a priori* to meet with a purely littoral flora, and with none but extensively sporadic species.'

7–5 See letter Darwin to Henslow 3 July [1840], from Shrewsbury, 'I remember in your lecture you said *monsters* were sometimes curious.— We have a largish orange tree, covered with oranges & nearly all therse are annually horned . . . The tree has long been without manure.— *Correspondence* 2:271

8–1 For Abberley's work on pollination of cucumbers by insects, see *Correspondence* 2:306.

8–2 Linnaeus 1775:192–93, '. . . the *stigma* . . . is double, when the fruit consists of two cells, as in the masked and umbelliferous plants; triple, when the seed vessel has three cells, as in the lilies . . . [The flower] is furnished with as many receptacles for the seed as there are *stigmata*.'

more effective than of foreign— Eyton has such a grove of Willows.—[1]
(14) **Bowman female branch**[2] At What distances from males, will female
(a) Willows or Yews **some poplar's** produce.—
(15) Would Yew *fruit* without impregnation.—[3]
(16) Any calculation of number of grains of pollen in any one flower
(17) Catch Bees, Butterflies— Syrphus— Meligethes & see whether they are dusted with pollen— in what *state* (whole or broken) is ball of pollen on Bees thighs
(18) Place pin's heads with Bird lime near male yew tree & see whether they catch pollen— ‹Ne› In Œnothera bush.—
(19) Theory of mock flowers in Hydrangea
(20) As Hop is Diœcious— seedsmen who raise Hop-seed— may know something about proportion of plants necessary &c &c

[9v] (a) Mercurialis— Frog Bit, Valerian— Urtica Dioica Sorrell. Lychnis. Butchers Broom— «also, Vinca,» Examine all these, are they much frequented by Bees or Butterflies or little insect?= or is pollen excessively minute or abundant? do they seed plentifully? Look for isolated females.— Also any plants which are known easily to be crossed & all monoœcious plants.—
Hooker says Rafflesia is diœcious & Pollen must be carried by some insect—[1]

10 (21) Are there many instances of single clumps of plants in counties, as of rare green Cotton Plant— How large «area» clump there? Distinguishable from ‹other› clumps from other parts? Don says Irish, Scotch & English plants generally distinguishable.=[1] What structure of seeds.— **(Paris)**
(22) When Linnæus says so great percentage of seeds have contrivance for transportal,[2] does he include seeds good to eat. (even Nux Vomica is eaten by a Buceros in East Indies— Asiatic Researches)[3]

10 (21) **(Paris)**] *added pencil in margin.*
(25) yellow **white** Butterfly] 'yellow' *crossed pencil and* **'white'** *added pencil.*

9–1 Eyton-upon-the-Wild Moors, the home of Thomas Campbell Eyton, 2½ miles north of Wellington, Shropshire.
9–2 Bowman 1837.
9–3 See *Calendar* letter 750 of 5 May [1844] from [P. de M. G. Egerton] on propagation of Irish and common yews.
[9v]–1 Hooker. References not traced. Probably personal communication.
10–1 David Don: probably personal communication. See Don 1817–20 for a discussion of geographical distribution of plants in Scotland and in various European and Asian countries.
10–2 Linnaeus. A particular reference to this citation could not be traced, but for a discussion of the subject, see Henslow 1837:277–78.
10–3 C. White 1799a:125, 'But what may be probably deemed the most extraordinary circumstance relating to this curious bird [viz., the Hornbill, *Buceros*], is its feeding upon the *Nux vomica*.'

(23) Talk about Thyme. Horned Oranges. Spallanzani Essay—[4] Figs **2 kinds of flower annually.— Periwinkle. (not asclepiadæ. «in» Lindley)**[5]
(24) Do Bees distinguish *species*, they do not varieties.—
(25) Does the yellow **white** Butterfly deposit eggs in all varieties of Cabbage.
(26) Do deer Keepers cross the breed— desirable as in Cattle in Chillingham Park—[6] What Book on varieties &c of deer. Contests of sexes.—

10a Q.30) *March 1842.* ‹Last› Year «before last» beans & peas were planted in rows adjoining & seeds gathered there were planted «last year» pell mell, without sticks & seeds gathered & these are now to be planted *this* year **copied**

Gould.—
Number of species of Birds in New Zealand, plants so few—

Range of mundane genera, «in Birds» in accordance with range of species?—

Are there any fine doubtful species from Van Diemen's Land? or New Zealand? Babington about differences of Irish & British Species & British & distant parts of Europe.—[1]
Gould— go over the Pigeons, Philotis, Dacelo. Alcyone, where there are very close species & see whether they come from isl^{ds}. or different parts or same district.—[2]
About ‹endemic &› wandering species of confined genera

By my theory in volcanic or rising isl^{d}, there ought to be a good many races or doubtful species; how is this at Canarys Arch— it is so at Galapagos.—
Ireland, doubtful species—

10a *all entries added at intervals from March 1842.*
Q.30) *March 1842 . . . this* year] *addition to question 30, page 11; crossed.*
copied] *added in margin.*
or New Zealand?] 'or' *over dash.*
Babington about . . . of Europe.—] *connected by* 'x' *and marginal line to* 'Does any . . . in another region—'.
Gould— go over . . . confined genera] *crossed pencil.*
By my theory . . . doubtful species—] *crossed pencil.*
Does any genus . . . another region—] *crossed pencil.*
Hooker?] *added pencil.*
Gould— go over . . . less in another region—] *crossed pencil.*

10–4 Spallanzani 1769.
10–5 Lindley 1839:137, '*The Periwinkles, Vinca major*, and *minor* . . . are the plants of this order [Apocynaceæ] which inhabit Europe. They are readily known by their opposite leaves, and bifollicular fruit, from all orders except Asclepiadaceæ, and from that order by their separate anthers having powdery pollen.'
10–6 CD received information from W. Yarrell on wild cattle of Chillingham Park. Correspondence 2:134.
10a–1 Babington authored various botanical articles on this subject.
10a–2 Gould 1837a:144; 1837c; [1837–38]; 1839:144; 1840a:114; 1840b:150, 160; 1845:18–19.

Does any genus of Plants. vary & hard to separate specifically in one country & not in other: Rosa is hard in Europe, Walnut in America.— Heaths in Africa; **Hooker?** are these genera less difficult, in other countries, where species are either numerous or even where few are they constant: this very important for it w[d] show that such variation is not a generic or specific character,, but contingent on country.— How is it in Patella or Oysters or Helix. Or does any «one» species of plant, vary in one region of Europe & less in another region—

11 (27) Which sex in Mules generally fails— perhaps indexed by secondary characters— in double flower. do
Henslow Speaking of Thyme doubts about stigma in similar manner ever failing.—[1] **answered by Gærtner**[2]

(28) Can any annual or Biennial be grafted or cuttings taken or tuber— talk about Mr Knights theory with Henslow.—[3] **D[r]. Fleming says yes.**

(29) Are there RACES of Lupine, Stocks Clover, to experimentize on by sowing near each other & see whether cross can be obtained— I name these three plants. because they cannot be crossed, I think, I expect, except by very minute insects.—

(30) Get Abberley to plant SINGLE Peas, Kidney Bean & Bean, intertwined, «without sticks»—[4] in reference to what M[r]. Herbert observe on this subject—[5]

(31) Ask Henslow for list of annuals to place in Hot house to see effect on generative organs of great Heat

(32) Can Henslow ask question of Col. Le. Couteur about Wheat— Change of Soil— crossing— when seeds raised.— His Book.—[6]

11 **answered by Gærtner]** *pencil in margin.*
D[r]. Fleming says yes.] *in margin*
(30) Get Abberley . . . this subject—] *marginal* 'x' *added to show connection with page 10a* 'Q. 30'.

11–1 For letter from J. S. Henslow, 21 November 1840, on this subject see *Correspondence* 2:276.

11–2 K. F. Gærtner. Reference not traced.

11–3 See Knight 1822:369 for a discussion of grafting wild and cultivated varieties of the same species; no mention of grafting annual or biennial plants or tubers. See also Knight 1809:393, '. . . I have observed that seedling plants, when propagated from male and female parents of distinct characters and permanent habits, generally, though with some few exceptions, inherit much more of character of the female, than of the male parent, and the same is applicable, in some respects, to the animal world . . .'; and Knight 1837:369, 'Whenever I have obtained cross-bred animals by propagating from families of dogs of different permanent habits, the hereditary propensities of the offspring have been very irregular, sometimes those of the male, and at other times, those of the female parent being prevalent . . .'

11–4 *Correspondence* 2:306.

11–5 W. Herbert 1837:352, '. . . the closely allied genera Faba, pisum, vicia and ervum cannot be upheld as distinct . . .'; p. 353 '. . . I have seen cultivated in Yorkshire a plant having the growth of a vigorous field pea (Pisum), which produces seeds that no man would hesitate to call beans . . .'

11–6 Le Couteur 1836:3, '. . . winter, or beardless wheat was [according to Romans] best suited to dry uplands, and bearded wheat to low, or moist lands.' Le Couteur served with Henslow on the Wheat Museum of the Royal Agricultural Society and also visited Henslow in Hitcham. *Correspondence* 2:274. See M155.

[11v] 32. Would wheat from ÆEgypt ripen in Scotland?— to show acclimatisation.—
July ‹1842›
When nettle leaf. put into spirits, poison-drop exudes— does not elm. does it «in» melon— «Loasa» Anchusa «Campanula» &c & dead-nettle.— **Lithospernum. Blue Gloss.** it is not possible to see orifice of poison-tube— so put carmine in spirits & then experimentise: for gradation in structure

═══

Compare flowers of wild & tame carrot— Parsley & Fennel. **Verbena**

═══

Compare flower of different Cabbages most carefully to see if variation equal in flower with leaves.— strawberries

═══

How ‹soon› «early» do characters of races of different vegetables & animals come on.— Compare calves.: Compare young. beans. cabbages.—

═══

History of Pheasant-fowl. Hen coloured like cock-pheasant: said not to sit on own eggs

═══

Flowers in short turf. for abortion. or for sterility

═══

{ Land Birds Madeira Migratory— ask Gould about N. Zealand, as Cuculus lucidus is.—[1] Ask Sulivan about Falklands Isds.— Snipe Migratory— probably united by Land to S. America

12 (33) Ornithologum commonly but improperly called Canadense— would it grow in open air in Sweden. Linnæus found 2 flower. which had anthers removed, did not become impregnated.[1]

(34) Any recent information about pollen of Subularia

11v *all entries added at intervals.*
32. wheat] *altered from* 'what'.
Lithospernum. Blue Gloss.] *added pencil.*
Verbena] *added pencil.*
Land Birds Madeira . . . S. America] *triple scored brown crayon.*
12 (35) **no light . . . seedless—**] *added pencil.*
(35) **In Royle's . . . . information**] *added in margin.*
(36) Ask Gray . . . to back] *pencil.*
37 about Chetah . . . Dogs] *pencil square brackets added with* **'does not know'**.

[11v]–1 Gould 1845:18–19. Cuculus of Australia mentioned.

12–1 Linnaeus 1786:40–41, 'The interior petals of the *Ornithogalum*, commonly, but improperly, called *Canadense** [**Albuca major*, Sp. Pl. ed. 2.] cohere so closely together, that they . . . will scarcely permit the pollen of another flower to pass . . . I therefore . . . extracted the antheræ . . . and . . . this single flower proved barren. This experiment was repeated . . . with the same success.'

Royle & Horsfield

(35) Talk about races of Banana & yet seedless— **no light Henslow or Royle, latter says seedless—** Also about Sugar-Cane Edwards says does not seed—[2] «Bruce says does» Royle **In Royle's Productive Resources Book no information**[3] & *Hope*[4] about *Silk worms. Varieties* effects of domestication— said to require Selection

(36) Ask Mr Gowen to ask Mr Herbert,[5] how many generations any hybrid has ‹been› reproduced itself.—
Ask Gray[6] to ask Mr Riley[7] to experimentise on hybridising ferns, tying them back to back

37 Col. Sykes fertility of men & Europæan animals in India?— about Chetah & other tame animals not breeding when tame in India?— **does not know**[8] About Yaks. & other Hybrids— Dogs &c &c[9]

———

38 Does only *male* yak cross with cow: is not reverse possible??

13 *Maer*

(1) Yew Trees near Boat House «ANY male branch.» — ?number of seeds in beginning of November 1841.— Trees above male?

(2) Result of Edwards experiment in Cabbages[1] **given**

(3) _____ in Heartease

13–[13v] *base text: title — question 4; remainder added at intervals.*
(2) Result . . . Cabbages] *crossed in same ink as* **'given'**.
(4) //**Yes**//] *added over dash.*
(5) Examine . . . is open— —] *parentheses added before* '& Menyanthes' *and after* **'No'**.
(5) Examine . . . **No**] *crossed pencil.*
(6) There is apple . . . (7) . . . Dodecatheon ╫.] *crossed pencil.*
(7) Bladder-Nut . . . Dodecatheon ╫] *pencil.*
Castrate apple . . . Also PEAS—] *crossed pencil.*
Maer (1) . . . (7) . . . Also PEAS—] *crossed pencil.*

12–2 Edwards. Not traced.

12–3 Royle 1839,1:118, 'As they [embassies of Lord Amherst and Dr. Abel] approached Canton, groves of orange-trees, of bananas . . . relieved extensive fields of rice.' P. 163, '[In the northern parts of India] are cultivated limes, lemons, and oranges, the jujube, and pomegranate, with bananas . . .' P. 355, 'Mr. B. [i.e. Brown] further adds, that it is not even asserted, that the types of any of those supposed species of American Banana, growing without cultivation, and producing perfect seed, has anywhere been found.' Royle also discusses the taxonomy, geography, and economics of bananas, plantains and the genera Musa of Musaceæ. See also Royle 1840:115–39, 'Silk Culture in India'; pp. 85–94, 'Culture of Sugar in India'. See discussion of bananas 94, 'Culture of Sugar in India'. See J. D. Hooker's letter 12 December 1844 on geographical distribution and R. Brown's claim that bananas are indiginous only to East Asia, not America. *Correspondence* 3:91,93.

12–4 Hope 1836

12–5 In his list of questions about hybridization sent via Henslow to Herbert, Darwin inquired about the influence of various factors, eg., male vs female, wild vs domestic strains on persistence of characters in successive generations of hybrids. See *Correspondence*:2:179–81 and for further correspondence with Herbert pp. 182–84, 201–4.

12–6 J. E. Gray. President of the Botanical Society of London; presided at the 16 Feb. and 16 March, 1839 meetings of the society when papers of Martens 1839a, and Riley 1839, on hybridity of ferns were read.

12–7 Riley 1839. Doubts Martens' 1839 contention that ferns reproduce sexually and hybridize.

12–8 Sykes 1835, 409, '. . . Captain Oakes . . . had a *Colsun* in his possession alive for a considerable time, and was never able to subdue its natural savageness in the slightest degree.'

12–9 Sykes, 1835, 1838a. Mentions members of the dog and cat families of India.

13–1 Edwards. Not traced.

(4) Does the Thyme bear abortive stamens every year & Spring. **& within garden //Yes//**[2]

(5) Examine the Parnassia whose stamens move one after other to flower & Menyanthes whose pollen bursts before flower is open— — **No**

(6) {There is apple with branch in middle of tree with flowers near end of orchard.= **At Shrewsbury one branch of Rhod. flowered later.— effect of accident??**

(7) Which. Rhododendrum seeds??—
Bladder-nut ǂ. Laburnum ǂ. Dodecatheon ǂ .

———

Castrate apple & pear to see if pollen naturally carried, on account of Van Mons[3] views— Also PEAS—

[13v] N.B. I think *very likely* the Peas to cross ought to be placed far *from all other Peas*, from Wiegman[1]

---

14 *Shrewsbury*

(1) Peas.— Beans **seeds alone remain to be compared**— Cabbages.— **kept true** Try experiment (30/p.11)

(2) Yew Berries germinate?— Yew trees sexes—

(3) Get Holyhoaks. races **planted & Linum Perenne.**— Herbert's. fact.=[1]

14 *base text: title — question 4; remainder added at intervals.*
(1) **kept true**] *added pencil.*
(2) Yew Berries . . . sexes—] *crossed.*
(3) Get Holyhoaks . . . fact.=] *four scores pencil.*
(4) Effects . . . Beech.—] *crossed.*
(5) Open . . . oranges.=] *crossed.*
(6) Passion . . . Asclepias—] *crossed.*
(7) History of Potato field =] *crossed.*
(9) Melons . . . hybridised] *crossed.*
(10) unimpregnated] 'un' *over* 'im'.
(11) Abberley . . . Cynoglossum.] *crossed.*
12 Does the . . . by care] *pencil,* **'good'** *added ink.*
13 Arum before . . . old flower=] *crossed.*
(15) Abberley . . . **breed from it**] *crossed pencil.*
(15) **intends . . . it**] *added pencil.*
(15) = **failed to germinate**] *added pencil.*
(Skim . . . Cyclopædia)] *added left margin.*

13–2 *Correspondence* 2:276.
13–3 Mons 1835–36.
[13v]–1 Wiegmann 1828. See also W. Herbert 1837:352, 'In 1823 [Dr. Wiegmann] sowed Pisium sativum (the field pea) and Vicia sativa (the common vetch) together; the seedlings showed a departure from the natural colour, and yielded grey seeds.'
14–1 W. Herbert 1837:366, '. . . in some cases the seminal varieties of plants preserve themselves almost as distinct in their generations as if they were separate species: for instance, the cultivated double holyoaks, of which at least the orange, the yellow, the white, the black, the red, and the pink, may be raised with certainty by seed from plants of the several colours, although planted near together in the garden . . .' See also Herbert to Henslow in answer to Darwin's queries 5 Apr. 1839, 3rd and 4th paragraphs, *Correspondence* 2:182–83.

(4) Effects of Nitrate of Soda under Beech.— **Lychnis dioica answers this question=**
(5) Open more Horned oranges.=
(6) Figs, flower.—Passion Flower. (as it is required to impregnate it artificially.)— Asclepias— Flowers not seeding= Put pot of boiled earth on top of House =**Aristolochia, plant wh require insects to impregnate it**
(7) History of Potato field=
(8) Abortive Thyme seeds **weather wet**—? Linum flavum **put in Spirits** which plant seeds?
(9) Melons fruit itself hybridised
(10) **one had no seeds, & two had plenty of seed & these** Seeds of unimpregnated Cowcumbers will they seed.?—
(11) Abberley has planted seeds of pale green Cynoglossum[2]. **never germinated**
12 Does the horned orange. wh. never has seeds produced **good** pollen? **Yes** «From cultivation lost their horns» is impregnation necessary to fruit—; become well shaped by care
13 Arum before pollen is shed can you find flys dusted with pollen from other flowers? Can flys' escape from old flower=
(14) Has planted seeds of Geranium pyrenaicum. small white-flowered var. with abortive stamens.— show crossing & ?heredetary?

---

(15) Abberley has a hooked Pea.— **intends to breed from it** and large Asparagus: result? =**failed to germinate**

---

16 Will plant some of the Thyme with abortive stamens by Terrace to see, whether stamens will be produced in individual plants

---

17 A dead-nettle in Hot-house. will it seed?—
(Skim through Penny Cyclopædia)

Abberley says that some Bees are smaller & more vicious. Will try to get me some to look at:— Was once offered a hive. of these small Bees— **at Sundorne[1] has large Bees**

July /42/
Mark has six day's puppy of small *true* Bull-Dog— length from nose over head to *root* of tail 28½. inches. From sole of foot to shoulder on line of back, height 17½/.

‹15› *all entries added at intervals.*
—At Sundorne has large Bees] *added pencil.*

14–2 See *Correspondence* 2:294 for CD's discovery of the pale-green Cynoglossum (*C. sylvaticum*) in Shropshire. Letter to W. A. Leighton, [1–23 July 1841].

‹15›–1 Sundorne Castle, 3 miles north-east of Shrewsbury in Atcham district, home of Dryden Robert Corbet. See *Correspondence* 1:188, 620.

The Greyhound. was in length (measured same way) 47½— in heigt 30 inches

---

Examine Keel of Common & Wild Duck— Black Duck & Penguin

15 *Henslow &c*

(36) Has not H. raised races of white & Blue Linum— did parent plants grow near each other.— ? **Cannot remember at all.**

(37) Any cases of plants. which will not produce seed in this country— where cause not apparent— Any where pollen is not produced **or small in quantity** — Any unproductive, where germen does not swell, although there be pollen.— **or FEW. or bad seeds formed; badness may be merely not ripening=**

(38) Have Diœcious plants any secondary, sexual characters.— Stature, position of flowers— Their smell— form of flowers— Nectaries— In Monoœcious «order» flower occupy particular position.—

(39) What does he think of Dr Flemings statement of Sweet Williams & Stocks, being propagated many years, by cuttings.—[1]

(40) Ask Henslow to distribute some of my questions[2] amongst agriculturists. whom he know.— Col. le Couteur on Wheat.—[3]

(41) Have any monoœcious or diœcious plants the Papilionaceous structure of flower— **Ground nuts**

(42) How are Orchidiæ fecundated, as *mass* of pollen is requisite.— **Brown's paper**[4]

[15v] 43. Any flowers of Keeling Diœcious, or Monoœcious, besides the Nettle.[1] at Galapagos— Diœcious.— Carex.— **We may presume Nettle spreads by seeds=**

(44) Zostera. Has he seen it in flower? does he know Botanist who does— What is Ruppia Bennett[2] says in same state. of flower

45. Charlsworth. vol II. p. 670— oats cut down turning into Rye.—[3]

15 *base text: title — question 50; writing becoming progressively smaller on page [15v].*
(36) Has not . . . each other.—?] *crossed.*
(41) **Ground nuts**] *added pencil.*
(42) How are . . . **Brown's paper**] *crossed.*
[15v] (44) What is Ruppia . . . of flower] *crossed.*
(.49) **Gærtner de fruct:**] *added ink,* 'Gærtner' *added in faint pencil in margin.*
(.49) **Geum. Galium Burrh≡**] *added ink, carated in pencil.*
31. Plant . . . ,**Teazle**] '31' *sic, CD misled by misshapen* '5' *in* '50'.
31. ,**Teazle**] *added pencil.*

15–1 Fleming. See note QE1v–5.
15–2 See *Correspondence* 2 (Appendix v):446–49 for a transcription of Darwin's questionaire 'Questions about the Breeding of Animals' which was privately printed and distributed to agriculturalists. See also Freeman and Gautrey 1969 for answers to the questionnaire.
15–3 Le Couteur 1836:12, 14, 15, 27, 55. Wheat crops could be greatly increased if 'we' follow Celsius and '. . . pick out the best ears of corn, and lay up our seed separately by itself.'
15–4 R. Brown 1833:740. Discusses degrees of viscidity of the pollen mass and the flower parts in the orchid *Bonatea*.
[15v]–1 Henslow 1838.
[15v]–2 Probably John Joseph Bennett.
[15v]–3 Weissenborn 1838b.

46). Book describing amount of Horticultural Variation? **Henslow knows only on Citrons**

47. Ficus carica **Henslow presumes females produce**. Polygam. triœcia. (are female flowers ever productive) Smith says many trees in Tropics are of this class.—[4]

(48) .Where «published» list of spontaneous Hybrids— to see whether any Papilionaceous plants,— whether many mono or Diœious plants, & any with peculiarities of structure rendering cross impregnation difficult or reverse

(.49) List of seeds **Gærtner de fruct:**— for woodcut— 1 double hook— — **Geum. Galium Burrh** ≡ single hook; curved spines— simple spines— or seed-cases with similar structure.= good case as showing how simple, but beautiful adaptation might be arrived at.=[5] **Any book with drawing of Seed. Anemone with, tuft— Bull Rush— Dandelion— Sycamore. & seeds with «mere» border— & Humboldts spinning seed.—**[6]

(50) Any cases of wild varieties plants growing together. under same conditions.— like cowslip & primrose, but less strongly marked.—

31. Plant seeds of the Fuller's plants[7] **,Teazle**

16 Dr. Holland ; **My Father. Andrew Smith**

(1) Are cross-births, or other accidents of delivery— inheritable.?— **Bell c^{d} ask Accouchers**[1]
Is any peculiarity in milk teeth inheritable!!! very good
Any peculiarity in the males of a family—
Where one tooth aborts, do you know whether any trace in germ.

16 (1) **Is any . . . good**] *added, top left margin, boxed.*
(1) **Any peculiarity . . . family—**] *added, top left margin, boxed.*
(1) **Where one . . . germ.**] *added, top left margin, boxed.*
(2) through] 'th' *over* 'in'.
(4) Prolifickness . . . father] *crossed.*
(5) metropolis] 'm' *over* 're'.
(5) **Andrew Smith.**] *added in margin.*
(6) What . . . Reproduzione.—] *crossed*; '**D. Holland**' *added in margin.*
(7) than] 't' *over* 'o[r]'.
(10)–(12) *pencil.*
(10) **Paper . . . Memoirs**] *added ink.*
(11) And. Smith . . . in dogs=] *crossed pencil.*

[15v]–4 J. E. Smith 1826:43. *Ficus Carica* is given as an example of Polygamia.
[15v]–5 K. F. Gærtner 1805. See plates 181–225 showing drawings of seeds most of which have wings, spurs, or hooks; see pp. 253–56 for *Index Generum.*
[15v]–6 Humboldt and Bonpland 1819, 4:172–73, 'We found near the *bano* of Mariara the *volador* or gyrocarpus. The winged fruits of this large tree turn like a fly-wheel, when they fall from their stalk.' Mariara is between Valencia and Caracas, Venezuela.
[15v]–7 Fuller's plants: a thistle, *Dipsacus fullonum*, with curved and barbed bracts, used for napping wool; known as Fuller's teasel.
16–1 Thomas Bell was a dental surgeon at Guy's Hospital, London, where, presumably, he would have plenty of opportunity for questioning accoucheurs.

(2) Any more cases of diseases, generally occurring in *man* being transmitted through females, like Hydrocele **Dr. H. thinks asthma in females takes place of gout.**[2]— How are livers **obscure organ. no answer**?—

3 Andrew Smith, about tamed wild animals breeding at the Cape.—[3]
About two vars: of Lion: Annales des Sciences[4]

(4) Prolifickness of female, relation to healthiness? & father
answered

(5) About cross-bred races of men taking after sex. A Smith.
About species of Rhinoceros. becoming rare beyond limits of the metropolis of each—[5] Cause?—
Andrew Smith.

(6) What size book Gallesio storia del Reproduzione.—[6]
D. Holland

(7) Is Hæmorragic tendency, independent of heredetary cases, more common in man than in female—[7]

(8) In Hump-back ever heredetary

(9) Are the works of Berhave (treating of heredetary diseases) translated.[8]

(10) About Daltonism in the MALE Troughtons.— **Paper in Taylors Scientific Memoirs**[9]

(11) And. Smith Savages at Cape any selection of Males in «**cattle**» or in Killing the worst =or in dogs=[10]

(12) Do Hottentots generally resemble each other very closely, more closely than Caffres.=[11]

13 Where are there any medical Statisics, proportion of diseases (heredetary?) in diff. countries in same races

16–2 Holland 1839a, Chapter X:116–144 (especially p. 117, p. 125, p. 125, note, and p. 129), discusses gout in males and females in relation to heredity and other conditions including asthma; in the book he does not say asthma in females takes the place of gout; see however p. 117, '. . . the same state of habit, or predisposition, which in some persons produces the outward attack of gout, does in others, and particularly in females, testify itself solely by disorder of internal parts, and especially of the digestive organs.' Very likely Darwin discussed the issue with Holland.

16–3 Andrew Smith: probably personal communication.

16–4 Reference not traced.

16–5 Darwin had visited Andrew Smith at Cape Town during the voyage of the *Beagle* and also met him in England: see *Correspondence*, 2:176, 311. Smith was Chief Medical Officer at Fort Pitt, Chatham, 1837–45. Christol 1835 and Vrolick 1837 had published in the *Annales des Sciences Naturelles* on rhinoceros; Dubreuil 1837, had published in the same journal on human races. No published references to varieties of lion or to breeding cattle at the Cape were traced. See Andrew Smith 1838–49, *Mammalia*, [no pages numbered] Rhinoceros bicornis, '. . . the time may arrive when the various species [of Rhinoceros] which formerly may have been scattered, each, in a peculiar locality of a large continent, will be huddled together . . .' See also unnumbered pp. '. . . the Keitloa has not . . . been in the habit of generally extending his range higher than about 250° south latitude.' And '. . . the species [Rhinoceros Keitloa] appeared rare when compared to others.'

16–6 Gallesio 1816. Holland 1839a:27 mentions Gallesio's book.

16–7 Holland 1839a:21, '. . . hæmorrhagic diathesis [when hereditary] . . . [is] confined to the lungs . . . [of] the male sex.'

16–8 No mention of hereditary diseases and their treatment could be traced in any of Boerhaave's publications, either originals or translations. Prof. Dr G. A. Lindeboom, Amsterdam, personal communication, biographer of Boerhaave, confirms this.

16–9 Daltonism. See Wartmann 1846:164, '. . . all the male members of the Troughton family are similarly affected [with color blindness].'

16–10 Andrew Smith: personal communication.

16–11 Andrew Smith 1831:124, 'The coolness and indifference with which almost the whole of the Hottentot race regard the approach of death, has often been

17 Mr. Gray **General Questions**

(1) Particulars about Sierra Leone. cow. taking bulls.[1] **is it Domesticated African Animal= Knows nothing**

[‹. . .›]

It is very important to know, whether Gould's observation holds good, that in the mundane genera, the species ‹are› have wide range— How is this in «Plants??»

Are abortive organs as ‹young› teeth, more plain in young Rhinoceros or Whale, than in old?? **Falconer says all in cases.**

Owen. Have talked partially with him Ask him to introduce me to some Human Anatomist.

Has he dissected any animal often, which has abortive bone. **(ask more about the lowest cervical vertebrae process developed into ribs.)** & does its abortion vary, according to Bentham's Remark.[2] Horse or cow.— degree of soldering of tibia & fibula: in Man any abortive bones???

do.

Wing in Apteryx ||| **no** as Os Coccygis— Turbinated bones? **False ribs**

Wings of Apterix: clavicle in—? Combs in combless Poultry— Teeth in fœtal state:

Mr. Horner. On Mr Tremenheres Scottish Colliers,[3] when men & women have long worked, whether children, who have *not* worked have any peculiar configuration.—

---

17 *base text: title — question 1; remainder added at intervals.*
Mr. Gray] *crossed*
(1) Particulars . . . **African Animal=**] *crossed.*
Knows nothing] *added right margin.*
[‹. . .›]] *illeg. word in margin, crossed.*
Owen.] *added left margin.*
Have talked partially with him] *added left margin.*
do.] *added left margin under* **'Owen'**.
as Os Coccygis— Turbinated bones?] *pencil* **'False ribs'** *added ink;* **'Wing in Apteryx |||'** *added left margin ink;* **'no'** *added ink with pointer to previous annotation.*
On Mr Tremenheres . . . configuration.—] *crossed.*
Metaphysics of Morphology . . . of Java?] *crossed pencil.*
Sabine says . . . by wind.] *pencil.*
close] 'c' *over partially formed* 'H'.
Aug. St. Hilaire . . . varieties.] *scored in right margin.*

commented upon . . .' P. 127, 'The disposition to laziness [is] . . . decidedly characteristic of the more regular Hottentots . . .' P. 198, 'That clapping noise occasioned by the various motions of the tongue, [is] truly characteristic of the Hottentot language . . .'

17-1 J. E. Gray 1839, 1846a, 1846b. On mammals of Sierra Leone and on cattle.

17-2 George Bentham on abortion is referred to 8 Dec. 1844 in a Darwin memorandum on a slip of notepaper DAR 100:40.

17-3 Tremenheres. Reference not traced, but see Horner 1836.

Hooker[4]

‹Meta›
Metaphysics of Morphology[5] ‖— Schelgel is he serpent man?[6] about zones separated by non-inhabited spaces: has he published? does he understand English.— Miguel to collect facts for me—[7] what? What does Blume say on alpine Flora of Java?[8]
Has Schow written on double creations & where?[9] How are current & winds in Antarctic ocean: are they from West, like as between Australia & S. America?[10] Sabine says North of Siberia, no sea-current, icebergs travel by wind.
Aug. St. Hilaire Bot. p. 787. position of embryo in close species of Hilianthemum differs greatly—[11] how very interesting to see if any variation in varieties.
G. St. Hilaires law of Balancement[12]

18 *Wm Yarrell*

(1) About non-breeding of animals in confinement, curious.— foxes—[1] English animals. **[Made no import. remark]**[CD]

18 *base text: title — question 5; remainder added at intervals.*
Wm Yarrell] 'Wm' *over* 'Mr'.
(2) Secondary . . . & white] *crossed pencil.*
(3) **Has since . . . characters appeared.=**] *added in margin.*
(3) **believe NO**] **'believe'** *underlined.*
(9) About German . . . Madeira] *pencil*; 'Bhem' *over* 'Brem'.

17–4 Whole entry related to J. D. Hooker — CD letters: CD to H 19 March [1845] see end of letter; H to CD [23] March 1845; CD to H [3 Sept. 1846]. *Correspondence* 3.
17–5 Murray 1845. See letter of J. D. Hooker [23] March 1845 to Darwin in which St. Hilaire 1841, Murray 1845, Schlegel, Blume, Miguel and Schouw are discussed at length. *Correspondence* 3:162–64.
17–6 Schlegel 1834, and 1836.
17–7 Miguel: F.A.W. Miquel.
17–8 Blume 1824a.
17–9 Schouw published on geographical distribution of plants. But see Weissenborn 1838a:369–81; 621–24. See especially p. 370, '. . . in 1823, Professor Schouw, of Copenhagen, in his "Journal Tidsskrift fur Naturwidenskaberne," tried to invalidate the observations . . . favourable to [*generatio æquivoca*].'
17–10 E. Sabine 1840. No reference to an *absence* of sea currents could be traced, but see p. lvi, '. . . currents [moved] to the west at the rate of three-fourths of a mile an hour.' Regarding icebergs, see p. lxx, 'They were at a distance from the shore, surrounded by enormous masses of ice, between which they were driven about by wind and current. . .'
17–11 A. Saint-Hilaire 1841:787, '. . . des dissections faites avec soin ont prouvé que la direction de l'èmbryon était fort différente dans des espèces d'*Helianthemum* très-voisines . . .' In his copy of the book, Darwin double-scored this passage in the margin.
17–12 I. Geoffroy Saint-Hilaire 1832–36, 3:458–59, 'Il n'est point d'anatomiste qui ne reconnaisse immédiatement l'analogie ou plutôt l'identité parfaite de cette généralité tératologique, avec la Loi du balancement des organes, dont mon père et tant d'autres ont fait des applications si multipliées et souvent si heureuses à la zoologie et à l'anatomie comparée.' In his own copy of the book, Darwin scored the margin beside this passage. See also vol. 3:593; vol. 2:344; 1841:165–66; Geoffroy Saint-Hilaire [Étienne] 1830:215; Saint-Hilaire, August de 1841:145, 'Mais plus souvent, comme vous l'avez vu, l'avortement du limbe coïncide avec la dilatation du pétide, et ici vous avec un exemple de cette loi de balancement ou de compensation, qui ne régit pas moins le règne végétal que le règne animal, et qui veut que, quand un organe avorte, l'organe voisin prenne plus de développement.' ['Loi de balancement'—printed in margin of page.] In a letter to J. D. Hooker [3 September 1846] Darwin asks Hooker if he has ever thought of St. Hilaire's 'loi du balancement' as applied to plants, for example in the case of double flowers. *Correspondence* 3:340.
18–1 *Variation* 2:151, 'Many members of the Dog

(2) Secondary male characters.— does male transmit to male more of his features— in negro & white
(3) About the Bantams at Zoolog Soc.— did Sir. J. Sebright select to destroy secondary character **believe NO** or did result appear without his wish[2] **Has since recrossed this breed.— Have secondary male characters appeared.=**
(4) Does he know any seed-raisers
(5) List of qualities in birds & animals for prizes.= Pidgeons. Canary birds— Bantams.—
(6) ‹Mad› Porto Santo Rabbit. Descript. of colour «& length of ears» & skeleton, & skin=[3] **Van. Voorst[4] often writes to Lowe[5]**

---

(7) In breeding. pointers. Bull-Dogs. **Spaniels— Grey-hounds—** is there ever any degeneration?? **HOUNDS. Eyton Mr Wynne, &c Could by selection a different looking animal be formed— not caring whether good or bad.— are any actually rejected??**
(8) Get Sir. R. Heron to give me Pigs foot undivided, & more particulars regarding effects of crossing them with common pigs=[6]
[it is a Lincolnshire Breed]CD— Sir. R. H. supposes is now extinct=
(9) { About. American & Europæan common species, having somewhat of different appearance.— **{will introduce it in work}**
{ Whether ‹Yar› knows whether Shaws hybrids between Trout & Salmon were fertile & whether homogeneous[7]
{ About German ornithologists, Bhem[8] & Glöger[9]
Consul Hunt,[10] birds from Azores or Madeira

Family breed readily when confined. The Dhole is one of the most untamable animals in India, yet a pair kept there by Dr. Falconer produced young. Foxes, on the other hand, rarely breed, and I have never heard of such an occurrence with the European fox; the silver fox of North America (*Canis argentatus*), however, has bred several times in the Zoölogical Gardens.'

18–2 See *Correspondence* 2:331, letter to William Yarrell [5 or 12 Sept. 1842] 'Do not forget when you see Sir J. Sebright to ask him whether the cross with the white Bantam brought back any of the "secondary male characters" to the hen-cock breed.' See also *Variation* 1:252 and 2:96.

18–3 *Variation*, 1:114, 'In colour the Porto Santo rabbit differs considerably from the common rabbit; . . .' See also Lowe 1833a:102.

18–4 John Van Voorst.

18–5 Richard Thomas Lowe. See Lowe 1833a.

18–6 Heron. Mentioned in *Variation* 2:92–93 regarding cloven and solid-hoofed pig-feet, and breeding of rabbits. See also *Correspondence* 2:141 where Yarrell sends information from R. Heron about pigs' feet and rabbits to Darwin.

18–7 No reference has been traced which mentions Shaw's hybrids between trout and salmon. But see Shaw 1836, 1838, and 1840. Shaw points out that parr are in fact the young of salmon; on p. 558 of Shaw 1840, is a comparison of the young of *Salmo salar* [Atlantic salmon], *S. trutta* [Brown trout], and *S. fario* [River trout].

18–8 Bhem. Should be Christian Ludwig Brehm. See letter 17 Oct. [1846] Darwin to L. Jenyns (*Correspondence* 3:354), 'Andrew Smith once declared he would get some hundreds of specimens of larks & sparrows from all parts of Great Britain & see whether with finest measurements he cd detect any proportional variations in beaks or limbs &c. This point interests me from having lately been skimming over the absurdly opposite conclusions of Glöger & Brehm; the one making half-a dozen species out of every common bird & the other turning so many reputed species into one.'

18–9 Constantin Gloger.

18–10 Thomas C. Hunt 1845.

19 Mr. Blyth

(1) Mentions some breeder who raises many English birds— will young wild ones breed as well as,, as those already bred in cages. **Get direction write to—**
(2) Does he believe. Stanley's fact of Hawks distributing live Mamals[1]
(3) Do most Hawks eat stomach. of finches— do they throw up pellets—
(4) About hybrid pheasants treading— any treadèe?—[2]
Difference in lambs of different breeds
Is there any difference in breeds of Cattle & sheep in the sprouting of the horns. at different periods in different breeds—?? or in individual case: subject to disease in youth.—
Mr Tollett— about selection for milking[3]— loss of early habits in Dorsetshire sheep— migration of coots— variation in hounds=
An ugly calf ‹turns› sometimes turns into fine beast. would its offspring have ugly calves. also turning into fine beasts.—

═══

For comparison with hybrids, is offspring of short-horn bull & hereford cow similar to reverse cross.—

═══

Sow cast-up-balls of Hawks or even owls.— How long do seeds remain in stomach of birds— Mem: how many miles they fly in few hours

20 Zoological Soc

(1) Do the animals there, sometimes couple but not conceive **:Bears /Yes/**
(2) Foxes & English animals & birds breed
(3) In cases where Lions have bred, have they been raised from young ones, bred in captivity **—Mr Miller says Wombwalls were**
(4) About fertility of ass-zebra-horse=

19 *base text: title—question 1; remainder added at intervals.*
(1) **Get direction write to—**] *added pencil in margin.*
(2) Does he . . . Mamals] *pencil.*
Difference in lambs of different breeds] *circled.*
Is there any difference . . . in youth.—] *circled.*
An ugly calf . . . fine beasts.—] *scored left and right margins.*
For comparison . . . reverse cross.—] *pencil, double scored in margin.*
Sow cast-up-balls . . . few hours] *scored left and right margins; crossed pencil.*
20 *base text: title — question 3; remainder added at intervals.*
(1) **: Bears /Yes/**] *added pencil.*
(3) **—Mr Miller says Wombwalls were**] *added pencil.*
(5)–(7) *pencil.*
or tailless dogs] 't' *cross bar drawn in pencil.*

19–1 Stanley. Not traced. Possibly Stanley 1835.
19–2 See D105.
19–3 Freeman & Gautrey 1969:223. George Tollet, agriculturalist and neighbour of the Wegwoods at Maer, answered Darwin's 'Questions about the Breeding of Animals' in part, saying, 'From one or other of these crosses [of various breeds already mentioned], (which may all be reckoned *short-horned varieties* the best sort of milking cow may be obtained N.B The first Cross generally gives so much vigour that the produce is apt to be superior to either of the Parent breeds.' See also *Correspondence*, 2:190–92, where Tollet's answers are printed.

(4) About fertility of ass-zebra-horse=
(5) About callosities on Camels-horses. &c &c Rhinoceros=
(6) Cross. Sus Barlyroussa with tame.—
(7) About fertility of Bantams from different countries=
Do the Peacocks cross.= Young Chinese or Penguin Duck in very young state for skeleton==
Does the tumbling of pigeons vary in manner & perfection &c &c &c—if so probably a variety, not specific character=
Cross Rumpless fowls & Dorking fowls,— or tailless dogs & fox, to see whether the characters are then intermediate or «sometimes» all on one side, as in crossing varieties
Amongst varieties cross one with abortive tail or horn, with another & see result, for comparison with natural species, as diœcious plants, when crossed

21.. R. BROWN— will pollen act on any flower before stigmas expanded— in reference to Lobelia & Clarkia—[1] Peas time of impregnation.— **says many flowers are dichogamous** Zostera— Knights notion of pollen & stigma generally not being mature at same time on same plant[2] **—Flora of Australian Mountains.— Is setting of fruit. cross Conception—(|| I could extract nothing from him)**|| Does impregnation ever regularly take place in unopened flower— ‹doubt› disbelieve this in Bauers case of orchidiæ[3] Where does J. Hunter use expression of "male principle of arrangement."—[4] would not male or female

21.. *all entries added at intervals.*
says many flowers are dichogamous] *added pencil in margin.*
Zostera] *circled pencil and carated to* 'Does impregnation . . . orchidiæ'.
Knights notion . . . same plant] *crossed.*
—Flora of Australian Mountains . . . nothing from him)||] *added pencil*; **'Mountains'** *underlined.*
Is the setting of fruit. cross Concepcion—] *crossed pencil.*
R. BROWN . . . same plant] *crossed.*
Where does J. Hunter . . . **answered**] *crossed.*
Does Mormodes . . . **answered=**] *pencil, crossed pencil.*
R. BROWN . . . **answered=**] *crossed.*
Bunbury . . . spur.—] **'spur'** *underlined.*
Alpine Australia Flora=] *added pencil.*
names of Plants . . . Lapland Plants] *pencil,* **'—will get answer='** *added ink.*
Norfolk Is^{d} . . . *Fauna?.*] *pencil, crossed ink.*
Australian Alps . . . there—] *added pencil.*
Lindley . . . pineaple] *crossed.*

21..–1 R. Brown, in Flinders 2, 1814:560 (Appendix 3), 'At the period of bursting of the Antheræ the stigma in Lobelia is almost completely evolved and capable of receiving impregnation from the pollen of the same flower . . .' See also, 'On the contrary in Goodenoviæ the stigma at the same period, is hardly visible, and is certainly not then capable of receiving impregnation from the pollen of its proper flower: it is therefore either impregnated by the antheræ of different flowers, or in some cases, at a more advanced stage by the pollen of its own antheræ, which is received and detained in the Indusium.'

21..–2 Knight. Reference not traced.

21..–3 Bauer 1830–38:xii, '. . . a union is effected between the caudicula of the pollen masses and the stigmatic glands (which precedes the expansion of the flower) . . .' See also Robert Brown 1833:692, '[Bauer proposed in Orchideæ] impregnation to take effect long before the expansion of the flowers . . .'

21..–4 Hunter 1835–37, 4:34–43, 'Account of the free-martin,' p. 34, note, '. . . it is well known that the seed is the female production in [the vegetable], and that the principle of arrangement for action is from the male. The same operation and principles take place in many orders of animals . . .'

"constructive principle" be better. or "constructive action on germ." '=??
answered

{ Does Mormodes (one of the Catasetums) really always hit stigma by
projecting pollen-masses?— = **answered** =
Has Ophrys nectary?= **Bunbury says no «hollow» spur.—**[5] **Ask about Pinks & Solanum impregnation before flower open. (An. des Sci** Where is Boerhaave's paper on impregnation of violets.=[6] Zostera= Are dwarf plants on Wellington Mountain described in Flinders= **Alpine Australia Flora=**[7] Banana's seedless— 20 varieties in mountains of Tahiti.

═══

Dr. Boott— says caricas from every isld differs— do they also differ in different countries— on flora of African Islds—[8]

═══

names of Plants found on mountains of N. America similar to Lapland Plants
—will get answer=

═══

Is pollen of cultivated Orchis & Asclepias &— carnosa?— **good—**

———

Norfolk Isd— geology. volcanic? Applies to my geology & Species theory—
peculiar Fauna?. {**Australian Alps—; are any Europæan forms found there—**

———

Lindley says that only one pineaple

21 *Horticulturists*

(1) Are sterile hybrids healthy: number of generations: about crossing of plants; especially Papilionaceous order

21 *base text: title — question 1; remainder added at intervals.*
(2) History . . not ripen.—] *pencil.*
(2) or does fruit] 'or' *over* 'f'.
(2) **scarcely any**] *added pencil.*
(5) Do the . . . commoner kinds—] *pencil.*
(5) Heartease] 'r' *over* 't'.

21..–5 C. Bunbury: probably personal communication.
21..–6 No mention of impregnation of violets could be traced in any of Boerhaave's publications. Prof. Dr G. A. Lindeboom, biographer of Boerhaave, personal communication, states, 'Neither did Boerhaave write a paper on impregnation of violets.' Boerhaave was an early leader in the theory of sexual reproduction in plants. See Preface by Boerhaave in Marsilli 1725. Darwin's father, Robert W. Darwin, had earned his MD in 1785 at University of Leiden, where Boerhaave had been Professor of Botany and Medicine.
21..–7 Flinders, 1814. No mention of dwarf plants on Mount Wellington elevation 3100 feet, near Hobart, Tasmania, could be traced in Flinders, or in Robert Brown 1814. Darwin climbed Mount Wellington in February, 1836, but did not mention dwarf plants in his *Diary*, 1933:390. On the Blue Mountains, elevation over 3000 feet, near Sydney, Australia, Darwin found 'scrubby' Eucalyptus (*Diary*, 1933:379). Also, J. D. Hooker 1847–60, Part III, Vol. 1(1860) Flora Tasmania:356–57, mentions *Podocarpus alpina*, a small, scraggy bush, or sometimes a tree up to 13 feet high, on Mount Wellington.
21..–8 Boott. Reference not traced; undoubtedly a personal communication. However, see Boott 1845–47, who read a paper in 1842 on Carices of North America.

(2) History of fruit trees far north in Scotland— do they flower— do they live healthily, or does fruit merely not ripen.— The point to attend to is whether good & plenty of pollen is produced. & 2^{d} if so, whether concepcion takes place,— the mere fact of seeds ripening has **scarcely any** no relation to hybrids.—

(3) As peaches *sport* into Nectarines (does reverse happen?) what is effect of crossing peaches & nectarines: same question with regard to Primroses.

(4) Do apples "sport" in fruit, or time of leafing

(5) Do the most cultivated show Heartease produce as large capsules of seed, as the commoner kinds—

Cattle are horned, Suffolk have ‹abortive› «**no**» horns by abortion, but sometimes have dangling ones.— Is there any genus of plants, «in» which some organ is absent by abortion, but appears in abortive state either in the species, or in the individual by chance & under domestication.—

N.B. Benthams remarks, where parts of flower are reduced from normal number, they are apt to vary in number in individuals of same species

---

22 Eyton

(1) Number of eggs— of half-bred geese— inter se, & with parents & of Chinese geese.

(2) Anatomy of muscles of stumps of tailess dogs & cats.—

(3) Hounds— varying—

(4) {About blended instincts of the geese which he crossed; especially if the hybrids were recrossed with either parent.—

May. 44 These Hybrids differ in colour of beak, taking after male & female parent.— Will they grow up in other respects different?—. Important.—

Oct. 44 {Tell J. Anderson's statement of English Horses having fewer vertebræ in tail, than Continental horses.[1]

— {About the leaping of Irish Horses, bred in this country.

{Chinese Dog's Head to send

Cover common Pea (& Sweet Pea) for several generations under net & see if get sterile— Cover that little Ervum[2] in Sand-walk,[3] on which I think I have never seen Bee visit.

22 *base text: title — question 2; remainder added at intervals.*
(3) Hounds— varying—] *pencil.*

22–1 Anderson 1799–1803, 1:69, '. . . in England . . . the horses naturally produced have fewer joints to the tail than those of other countries . . .'

22–2 *Ervum* is a genus of vetch.

22–3 Sand-walk was a gravel path at Down House, leading to the back of the property and through a little wood. Darwin walked there often for reflection and exercise. See F. Darwin 1902:70–71.

23 Experiments in Garden

Sow stones of Standard Apricot grafted on what, & see what comes up.—

[24] [Unnumbered blank]

[24v] Experiment

Cover patch of ground, with different salts & poisons & see in what order plants would reappear after ‹th› being killed

[23v] Experiments not connected with Species Theory

(1) Will an extract of peat do to preserve fungi or animal substances— (Athenæum (40) p. 823 chemical analysis of Peat[1]

(2) Athenæum 1840 p. 777. Decaying wood absorbs oxygen & forms Carbonic Acid. will this bear on Petrifaction?—[2]

INSIDE BACK COVER [blank]

BACK COVER Questions & Experiments

23 Sow . . . up.—] *crossed pencil.*
[24v] Experiment . . . killed] *entered from back of book; upside down with respect to front of notebook.*
[23v] Experiment . . . Petrifaction?—] *crossed pencil; entered from back of book, upside down with respect to front of notebook.*

[23v]-1 J. F. W. Johnston 1840:823, '. . . some varieties of peat . . . were illustrative of a transition from the comparatively fresh and vegetable matter to a substance resembling coal, but which [were] affirmed to be ulmic acid.'

[23v]-2 Darwin meant pp. 773–74. See Liebig 1840:773, 'Woody fibre, in a state of decay, is the substance called *humus*. This body possesses the property to convert oxygen into carbonic acid.' See also p. 774, 'Carbonic acid, water, and ammonia, contain the elements necessary for the support of animals and vegetables. The same substances . . . are the ultimate products of the chemical processes of decay and putrefaction.'

Notebook M

Introduction by SANDRA HERBERT & PAUL H. BARRETT

Transcribed and edited by PAUL H. BARRETT

Notebook M (DAR 125, 168 × 100 mm) is the first of several sets of notes (M, N, OUN) on the general subject of the biological origin of behaviour. It is bound in dark red leather with the border blind embossed; the clasp is now missing. Darwin wrote 'Private' on the inside front cover of M and N, no doubt because the notes contained much about mental qualities, both normal and abnormal, of himself, family and friends.

The front cover bears a cream-coloured label on which 'M' and the word 'Expression' are written in ink; the back cover is identical to the front except 'Expression' is written on the cover rather than on the label. The notebook originally contained 78 leaves, or 156 pages, which Darwin numbered consecutively in brown ink. He later excised thirteen and a half leaves, of which seven and a half have been located. Darwin used both ink and pencil in keeping the notebook. The portion of the notebook definitely kept in grey ink runs from the top of page 56 to the end of the notebook except for page numbers, later annotations, and several entries on page 156 and on the inside back cover. However, beginning two-thirds down page 51 the ink has a greyish tint that may indicate a mixture of inks. Two entries, both in grey ink, record the opening and closing dates of the notebook: 'July 15th 1838' on page 1 and 'Finished. Octob. 2^{d}.' on the inside front cover. The first entry, being in grey ink, was postdated. Notebook D, which resembles Notebook M in appearance, was also begun on or about 15 July and finished on 2 October 1838. The two were probably purchased together.

While Notebook M is like the other alphabetically-lettered notebooks in containing a mixture of Darwin's speculations, inquiries, and reading notes, it differs from them in arrangement, for the first section of the notebook, to the middle of page 62, contains the record of Darwin's speculations and conversations with his father and other family members, and a few reading notes. Entries in this first section of the notebook were written in the two to three week interval from 15 July to the end of July or early August 1838. This dating is consistent with the presence of grey ink from page 56 onwards, for Darwin only began to use this ink at the end of July. At the latest the notes in the first section of the notebook were completed by 7 August, the date recorded on page 62. Darwin was with his father at the family home in Shrewsbury from 13 July, when he arrived on his way back from the Glen Roy geological trip to Scotland, to 29 July when he left for Maer, his Wedgwood cousins' home. On the first of August he was again in London. In his 'Journal' he noted, 'Very idle at Shrewsbury, some notes from my Father. & opened note book, connected with Metaphysical Enquiries'.[1] Notebook M is that metaphysical notebook and, presumably, the primary reservoir of the 'notes from my Father'. Darwin's father's name appears prominently in the notebook to page 55; page 56 refers to his sister Caroline (who had married Josiah Wedgwood III in 1837 and whom he may have seen at Maer); pages 58 and 61 mention his cousin Hensleigh Wedgwood.

The primary subject for Darwin's conversations with his father in July 1838 was human thought and behaviour, particularly as it touched on abnormal states including insanity. Darwin approached his father, Robert Waring Darwin, as an authority on the subject since,

[1] *Correspondence* 2 (Appendix II): 432.

in addition to his skill as a physician, his father possessed a talent for 'reading the characters, and even the thoughts of those whom he saw even for a short time'.[2] Notebook M also contains notes on books in medicine, philosophy, and political economy that were common reading within Darwin's family circle. Medicine was the profession of his father and grandfather and was tied to philosophy within the family: Erasmus Darwin's theory of disease derived in part from the work of David Hartley. Notebook M contains references indicating that the associationist psychology of Hartley, with its accompanying empiricism, remained an oral tradition within the Darwin household. The name of David Hume on Darwin's reading list also reflects a family interest in empiricism. In political economy, there was a similar correspondence between Darwin's reading list and the interests of his extended family. It is interesting to note that Charles's brother Erasmus was then a friend of Harriet Martineau, who was known for her popularization of the writings of Thomas Malthus. When Charles Darwin began to study human behaviour he looked to the Darwins and Wedgwoods for advice; their tastes directed his readings.[3]

Several lines of inquiry begun in Notebooks B and C are carried forward in M.[4] Darwin's comments on the origin of man reveal that the subject held no terror for him; the liberal views of his family in religion helped him accept the consequences for man of transmutationist theory. From Notebook B onwards he was prepared to consider that 'monkeys make men' (B169). In Notebook C he treated the subject with greater frequency and in more detail (C55, 72, 74, 76–79, 196), speculating on the circumstances necessary for the origin of mankind (C78). In Notebook M his interest in human origins—aside from rhetorical flourishes (M84, 123)—was focused primarily on the topic of expression. Darwin had already brought up the topic in Notebook C 'Let man visit Ourang-outang in domestication, hear expressive whine—see its intelligence . . .' (C79), but in Notebook M he developed the subject at greater length. He observed and inquired about expression in a wide array of different animals (M142–43, 152). Darwin believed the similarity of expressions in other animals to man strengthened, even 'proved', the transmutationist case (M84).

Another important line of inquiry connecting Notebooks B, C, D, and M was the search for an explanation for the origin of adaptation in nature. In Notebook B Darwin had discarded the notion of progressive development which included the premise of directed growth. He was thus left to begin again on the subject of adaptation. He started his inquiry with an attribute of organisms which appeared to him quite malleable, viz., behaviour. In adopting this approach he knew he was on dangerous ground, since Jean Baptiste de Lamarck had gone the route before him. Nevertheless, if he avoided the Lamarckian presumption that organisms acted consciously in altering their behaviour, and hence their constitutions, he thought the approach had merit (C63, 163, 173). He concluded that 'according to my views, habits give structure, . . . habits precedes structure, . . . habitual instincts precede structure.—' (C199). Darwin's point of departure in Notebook M was thus slightly altered Lamarckianism. In the notebook he was eager to show that mind affects body, and that human beings act without consciousness, that is from habit or instinct, on all sorts of occasions. For Darwin, understanding unconscious mental activity was a first step towards understanding the origin of adaptation.

[2] Barlow 1958:32
[3] On the intellectual background to Notebooks M and N see Gruber and Barrett 1974; Herbert 1974–77; Manier 1978; and Beer 1985.
[4] Gruber and Barrett 1974; Herbert 1974–77; Richards 1979. Gruber and Barrett 1974 also contains interpretive line by line commentary on Notebooks M and N.

Apart from its scientific content, Notebook M marks a transitional moment in Darwin's intellectual development. The notebook reveals him looking inward and outward, inward to organize and reflect on his experience, and outward to place what he was doing in a larger philosophical and social setting. The inward movement of his thought is signalled by his ordinary but significant act of beginning to date his notes. Before opening Notebooks D and M Darwin did not systematically date his notes, which makes dating some of his early work difficult. While he did not record a reason for his change in practice, he probably did it to make his notes more useful to him in the future. Darwin also opened a new notebook in August that he styled his 'Journal'.[5] In it he listed significant events in his life up to that time, including what has become a well-publicized account of his first acceptance of transmutation theory. Thereafter, until 20 December 1881, he used his 'Journal' to record important events in his private and professional life, and it served as his main guide when writing his full autobiography in 1876. His first autobiography, however, also dates from the D and M notebook period.[6] In August 1838 he wrote a 1700 word account of his life from the date of his earliest memory when he was four up to the year 1820, when he was eleven.

While keeping Notebooks M and D, Darwin began to consider the network of ideas he had assembled from what might be described as a 'metatheoretical' aspect. The most obvious sign of Darwin's new perspective was the alignment of his transmutation view with materialism and determinism. He embraced materialism enthusiastically (M19, 57) and argued, using associationist language, that thought originated in sensation (M61–62). On determinism he examined the traditional position that free will was illusory (M27, 126), and that free will and chance were synonomous (M31). In a second 'metatheoretical' effort Darwin sought to frame his theory within the setting of human knowledge generally. As in Notebooks B and C, he urged the acceptance of secondary causation in accounting for the origin of species (M154), and measured his biological thinking against the criteria for positive knowledge advanced by Auguste Comte (M69–70).

Upon completing Notebook M Darwin replaced it with Notebook N. As indicated in the 'Location of Excised Pages' material from the two notebooks was carried over directly to *The Descent of Man* (1871) and *The Expression of the Emotions in Man and Animals* (1872).

Additional readings on Darwin's treatment of expression in Old & Useless Notes and in M and N notebooks may be found in Swisher 1967, Schweber 1977, Herbert 1977, Manier 1978, Richards 1979, Montgomery 1985, and Browne 1985.

[5] *Correspondence* 2 (Appendix II): 430–35; de Beer 1959.
[6] 'An Autobiographical Fragment' in *Correspondence* 2 (Appendix II): 438–41. The original manuscript in DAR 91: 56–62 is entitled '*Life*. Written August— 1838'.

FRONT COVER

Expression
M

INSIDE FRONT COVER

49

Charles Darwin Esq
36 Grt. Marlborough St[r].—

(p. 64. On ‹insect› Ants getting on Table. Col. Sykes[1])

Private
Finished. Octob. 2[d].

**This Book full of Metaphysics on Morals & Speculations on Expression—
1838**

Selected Dec 16 1856

1 **July 15th 1838**
My father says he thinks bodily complaints «& mental disposition» oftener go with colour, than with form of body.— thus the late Colonel Leigton resembled his father in body, but his mother in bodily & mental disposition.—[1] My father has seen innumerable cases of people taking after their parents, when the latter died so long before, that it is extremely improbably that they should have imitated.— when attending M[r] Dryden Corbet, he could not help thinking, he was prescribing to his father & old M[rs] Harrison, said, although constantly seeing him, she was often struck with this fact.—[2] the resemblance

2 was in odd twiching of muscles, & general manner of holding hands &c &c.— M[r] Dryden Co said he could not remember his father.—

———

My father thinks. people of weak minds, below par in intellect frequently ‹are› have very bad memories for things which happened in early infancy— of this fact M[r] Dryden C. is good instance as he is very deficient, he was nearly 9 years old. when his father died.—

———

The omnipotence of habit is shown about meals, no

FC Expression] *black ink.*
IFC **49**] *circled, added pencil.*
p. 64. . . . Sykes] *added pencil.*
Private **Finished. Octob. 2[d].**] *boxed*; '**Finished. Octob. 2[d].**' *grey ink.*
This book . . . Expression—] *added pencil.*
1838] *grey ink*; *circled pencil.*
Selected . . . 1856] *added pencil.*
1 **July 15th 1838**] *grey ink*; *partially boxed.*

IFC–1 Sykes 1836a. See M 62 for Darwin's discussion of Sykes' article.
1–1 '. . . I am very sorry to hear of poor Col. Leighton's death.' Letter of Darwin to Susan Darwin 23 April 1835. (Barlow 1945:118.)
1–2 Dryden Robert Corbet. See John Bernard Burke 1925:395, and C., A.E. 1915, 2:191.

3e–6e [not located]

7 There is a case of Mr Anson. who told a story of hunting «— habitual fits.—» which my Father thinks is mentioned in the Zoonomia.—[1]

Now if memory «of a tune & words» can thus lie dormant, during a whole life time, quite unconsciously of it, surely memory from one generation to another, also without consciousness, as instincts are, is not so very wonderful.—

‹Now is not epilepsy an *habitual* disease of the muscles.???›

8 Miss Cogan's memory of the tune, might be compared to birds singing, or some instinctive ‹or› sounds.— Miss C. memory cannot be called memory, because she did not remembered, it was an habitual action of thought-secreting organs, brought into play by morbid action.—[1] Old Elspeth's «in Antiquary» power of repeating poetry in her dotage is fact of same sort.[2] Aunt. B. ditto.—[3]

9 ===

Case of Mr Corbet of the ‹Hall› «Park», after paralytic stroke. intellect impaired. ‹after paralytic stroke› : . could converse well on any subject when once started,— could receive a new train through eyesight, though, not through hearing,—[1] Thus when dinner was announced he could not understand it, but the watch was ‹seen› shown him.— «‹Mr Corb› the servant showed him watch & said dinner is ready, what, what.— then showed the watch upon which he exclaimed, why it is dinner time.— »
My father asked him whether he had gardener of name A.B., &c &c. & he maintained he had never heard of such a man & had no gardener.— My F. then asked Mr C. to come to the window & pointed out the Gardener & said, who is tha? Mr C. answered

10 why do you not know, that is A. B my gardener.— Thus was he in every respect, no communication could be held by means of hearing.—

===

8 *page crossed pencil.*

7–1 Erasmus Darwin 1794, 1:437, 'Master A. about nine years old, had been seized at seven every morning for ten days with uncommon fits . . . he began to complain of pain about his navel, or more to the left side, and in a few minutes had exertions of his arms and legs like swimming. He then for half an hour hunted a pack of hounds; as appeared by his hallooing . . .'

8–1 See Darwin's personal copy of Abercrombie 1838: 143, for his marginal notation, 'These cases like Miss Cogans, & serve to show that affections of brain will *recall* facts in ‹those› an individual life after long periods.—' The accompanying passage of Abercrombie 1838:143, reads, 'A case has been related to me of a boy, who, at the age of four, received a fracture of the skull, for which he underwent the operation of trepan. He was at the time in a state of perfect stupor, and, after his recovery, retained no recollection either of the accident or the operation. At the age of fifteen, during the delerium of a fever he gave his mother an account of the operation . . .' Darwin made a pencil score beside the passage.

8–2 Walter Scott 1815, 3:220,'. . . shrill tremulous voice of Elspeth chaunting forth an old ballad in a wild and doleful recitative . . .'

8–3 Aunt Bessey: Elizabeth Wedgwood.

9–1 Erasmus Darwin often used the expression, 'train of ideas'. See, for example, Erasmus Darwin 1794, 1:46.

Mr Corbet, however, in conversation could catch up a new train if *early* association were called up.— My F. asked him, did he know whom Mr Child «of Kinlett» had married.— Answered never heard of such a man.— (My Father explained who he wa & all about him, but still maintained he had never heard of him).—[1] My F. then said you remember

11 Jack Baldwin at school.— Answered To be sure I do.— What became of him.— Answ Had large fortune left. him, took name of Child «of Kinlet» & married Miss A. B.— all the same names as a few minutes before he maintained he had never heard of.— Thus in many things if he began at one end, he knew the whole subject.— if at the other nothing.— He could repeat the alphabet straight, but did not know [Z]^CD when heard isolately.—[1]

12 In old people. (Aunt. B.) when they hear a thing it often does not take any effect at the time, but some time afterwards it calls up pain. or pleasure. & is often recurred to & mentioned as a thing which had just taken place.— as if the idea of time had been disturbed.—

These foregoing cases of «mental» failure very general effect of ‹early› «slight habitual» intemperance.— often accompanied by extreme anger, at not being understood.—

13 My F. says there is perfect gradation between sound people and insane.— that everybody is insane. at some time. Mania is quite distinct, different also from delirium, a peculiar complaint stomach not acted upon by Emetics.— people recognized,— sudden changes of disposition, like people in violent intoxication, often ends in insanity or delirium.— In Mania all idea of decency & affection are lost.— most delicate people do most indelicate actions,— as if «these emotions» acquired.— this may be doubted, whether rather not going against natural instincts.—

My Grand F. thought the feeling of anger, which rises almost

14 involuntarily when a person is *tired* is akin to insanity.— «I know the feeling also of depression, & both these give strength & comfort to the body» I know the feeling, thinking over somebody who has, perhaps, slightly injured me, plotting speeches, yet with a sort of consciousness not just.— From habit the

12 In old people . . . when they hear] 'hear' *over* 'heard'.

10–1 See Anonymous 1854:6, for an account of a Captain Childe (son of W. Lacon Childe) of Kinlet Hall, Salop, who was committed by his family to a lunatic asylum at Hayes Park in about 1842 for having written letters since 1838 to the Queen stating his infatuation for her, and saying that the Queen 'had a marked attachment to him'.

11–1 See also Abercrombie 1838:96–97, '. . . when [Dr. Leyden] wished to recollect a particular point in any thing which he had read, he could do it only by repeating to himself the whole from the commencement till he reached the point which he wished to recall.' Erasmus Darwin 1794, 1:134, expresses a similar view, 'In respect to freewill, it is certain, that we cannot will to think of a new train of ideas without previously thinking of the first link of it . . .'

feeling of anger must be directed against somebody.— Have insane people any misgivings of the injustness of their hatreds, as ‹if› in my case.— It must be so from the curious story of the Birmingham Doctor praising his sister who confined him. & yet disinheriting her.— This

15 «N B. I have read paper somewhere on horse being insane at the sight of anything scarlet.—[1] dogs ideotic.— dotage.—»

Doctor communicated to my grandfather his feeling of consciousness of insanity coming on.— his struggles against it, his knowledge of the untruth of the idea, namely his poverty.— his manner of curing it. by keeping the sum-total of his accounts in his pocket, & studying mathematics.— My Father says after insanity is over people often think no more about it than of a dream.—

---

16 Insanity is produced by moral causes (ideotcy by fear. Chile earth quakes). in people, who, probably otherwise would not have been so.— In Mr Hardinge, was caused by thinking over the misery of an illness at Rome, when by *accidental* ‹was› delay of money, he was «only» NEARLY thrown into a hospital.— My father was nearly drowned at High Ercall, the thoughts of it, for some years after, was far more painful than the thing itself.[1]

17 Asked my F. whether insanity is not distinguished from whims passion &c by coming on suddenly. Ans no.— because often, if not generally, does not really come on suddenly.— Case of Mrs. C.O. who threw herself out of the window to kill herself from jealousy of husband connection with housemaid two years before, to prove she was not insane, answered she had known it at time & had bought arsenic for that purpose.— this found to be true.— Her Husband never suspected during these two years that she had been insane all the time.—

18 There are numberless people insane of particular ideas, «Case of Shrewsbury gentleman, unnatural union with turkey cock,— was *restrained* by remonstrances on him» which are never generally, if at all discovered.— ‹Sup› Sometimes comes on suddenly from ‹I› (in one case ipecacuhan— not acting) in others from drinking cold drink.[1]— then brain affected like getting suddenly into passion.— There seems no distinction between enthusiasm passion & madness.— ira furor brevis est.—[2] My father quite believe my grand F doctrine is true, that the only cure for madness is forgetfulness.—

17 Case of ‹Mrs.›] ‘Mrs.’ *over* ‘Miss’.

15–1 In his personal copy of Abercrombie 1838:159, Section II, ‘Abstraction’, Darwin wrote in the margin, ‘animals have idea of colour.— mad horse (Cline) dread of scarlet. of any kind.—Smells, do.—’

16–1 High Ercall: about 6 miles NE. of Shrewsbury.

18–1 Erasmus Darwin 1794, 1:55, ‘. . . preparations of mercury particularly affect the salivary glands, ipecacuhana affects the sphincter of the anus, cantharides that of the bladder . . .’

18–2 Ferrier 1838a:445, ‘Consciousness is extinguished; and hence the expression of the poet—*Ira brevis furor est*—“Rage is a brief insanity”—is strictly and pathologically true . . .’ See also Harbottle 1897:111, ‘Ira furor brevis est . . ., Rage is a brief insanity.’

19 which does appear a real difference, between oddity & madness.— but then people do not well recollect what they have done in passion.—

═══

People are constantly well aware that they are insane & that their idea is wrong.— (Dr Ashe, the Birmingham Doctor), in this precisely like the passion, ill-humour & depression, which comes on from bodily causes.—

———

It is an argument for materialism. that cold water brings on suddenly in head, a frame of mind, analogous to those feelings. which may be considered as truly spritual.—

———

20 a person twitching when a disagreeable thought occurs, is closely analogous to Epilepsy & convulsion.— affections of the thinking organs ǂ — the action of brain which gives sensation of pain, emits its power on the muscles in the twitching.

═══

Pride & suspicion are qualities, which my F says are almost constantly present in people, likely to become insane.— now this is well worth considering, if pride & suspicion can be well understood.

═══

21 In insanity, the ideas do not go back to childhood, (but appear most capricious) as in delirium after epilepsy, but in the failing from old age, they constantly do.— In Mrs P. ‹. . .› of B. ‹talked of.› thought herself near Drayton & Ternhill,[1] (where she was born) though she never naturally talked of these places.— My F. says, shows that early impressions are most durable.— (but Miss Cogan shows that repetition is not necessary)— the words second childhood full of meaning:— Dreams do not go back to childhood— People, my Father says, do not dream of what they

22 think of *most.* intently.— criminals before execution.— Widows not of their husbands— My father's test of sincerity.—

———

People in old age. exceedingly sharp in some things, though so confused in others.— Mrs P. when in state as above described, (forgetting that her husband was dead) yet instantly perceived when my Father to distract her attention took her «left» hand to pretend to feel her pulse.—
What fails first?— How is this?— Does memory bring in old ideas

21 In insanity . . . in Mrs P. ‹. . .›] *4 or 5 letter word crossed, illeg.*

21–1 Drayton: Market Drayton, 18 miles northeast of Shrewsbury, on the road to Maer.
Ternhill: 3 miles southwest of Market Drayton.

23 ———

‹I have elsewhere remarked do› Dogs take pleasure, when doing. what they consider their duty.— as carrying a basket, bringing back game, or picking up a stone, though only acquired rules by art.— like the law of honour.— they feel pleasure in obeying their instincts naturally.— (generosity in defending a friendly dog).— they feel shame, when doing anything which is wrong.— as eating meat., doing their dirt, running home.— in these cases their actions do not look like

24 fear, but shame.— I cannot remember instances, but I feel sure I have seen a dog doing what he ought not to do, & looking ashamed of himself.— Squib at Maer,[1] used to betray himself by looking ashamed before it was known he had been on the table,— guilty conscience.— Not probable in Squib's case any direct fear.—

My father thinks that selfishness, pride & kind of folly like (Mr George S.) is very heredetary.—

25 My father says on authority of Mr Wynne that bitch's offspring is affected by previous marriages with impure breed.—

A cat had its tail cut off at Shrewsbury & its kittens ‹h› (in number 3) had all short tails; but one a little longer than rest «they all died»:— she had kittens before & afterwards with tails.

My father says, perfect deformity, as an extra number of fingers.— hare lip or imperfect roof to the mouth «stammering in my Father family» (as in Lord Berwick's family) are heredetary.—[1]

---

other deformities are illnesses of the fœtus.— some mothers. have first dead children, then children which were short term, & lastly healthy ones.—

26 Insanity & Epilepsy remain many generations in families.— My fathers does not know whether trains of insanity are heredetary in any one family.—

In Aunt— B. the affections «& N B affections very soon go in Maniacs» seem to have failed even more than the memory.— therefore affections effect of organization which can hardly be doubted, when seeing Nina with her puppy.— The common remark that fat men are goodnatured, & vice versa Walter Scotts remark how odious an illtempered fat man looks, shows same connection between organization &

23 game, or] 'or' *over* '&'.

24–1 Maer Hall, at Maer, home of Josiah Wedgwood II, a few miles south of Stoke-on-Trent, Staffordshire.

25–1 Darwin's grandfather Erasmus '... stammered greatly, and it is surprising that this defect did not spoil his powers of conversation.' Erasmus' eldest son Charles (1758–1778) also stammered. See Krause 1879:40; 80–81.

27 mind.— thinking over these things, one doubts existence of free will every action determined by heredetary constitution, example of others or teaching of others.— (NB man much more affected by other fellow-animals, than any other animal & probably the only one affected by various knowledge which is not heredetary & instinctive) & the others ‹are› learnt. what they teach by the same means & therefore properly no free will.— we may easily fancy there is, as we fancy there is such a thing as chance.— chance governs the descent of a farthing, free will determines our throwing it up.— equall true the two statements.—

28 Catherine remarks that pleasure received from works of imagination very different from the inventive power,—[1] this, though very odd is perhaps true.— mem Erasmus & mine taste for music.— Children like hearing a story told though they remember it so well that they can correct every detail, yet they have not imagination enough to ‹up› recall up the image in their own mind,— this may be worth thinking over.— it. will perhaps show differences between memory & imagination. «Catherine thinks that children like looking at ‹ani› pictures, an early taste, of animals. they know.— pleasure of imitation (common to monkey), & not imagination.—»

29 ———

Thinking over the scenes which I *first recollect*,[1] «at Zoos» they are all things, which are brought to mind, by memory of the scenes, (indeed my American recollections are a collection of pictures).— when one remembers a thing in a book, one remember the part of page.— one is tempted to think all memory consists in a set of sketches. some real— some fancied.— this fact of early memory consisting of things seen, quite agrees with my Fathers case of Mr Corbet of the Hall understanding. (on hearing old association brought up) by sight & not by hearing

30 One is tempted to believe phrenologists are right about habitual exercise of the mind, altering form of head, & thus these qualities become heredetary.— When a man says I will improve my powers of imagination, & does so,— is not this free will,— he improves the faculty according to usual method, but what urges him,— absolute free will, motive may be anything ambition, avarice, &c &c An animal improves because its appetites urges it to certain actions, which are modified by circumstances, & thus the

31 appetites themselves become changed.— appetites urge the man, but indefinitely, he chooses (but what makes him fix!? ‹)›— frame of mind, though perhaps he chooses wrongly,— & what is frame of mind owing

28 Catherine thinks . . . not imagination.—»] *boxed.*
29 Thinking over . . . are all things,] 'things' *over* 'thinking'.
31 Shake ten . . . according to law.] *brace, left margin.*

28–1 Catherine: Emily Catherine Darwin.
29–1 See *MLI*: 1–5, 'An autobiographical fragment' written 1838 which contains a detailed account of Darwin's earliest memories.

to.—‹)›— I verily believe free-will & chance are synonymous.—
Shake ten thousand grains of sand together & one will be uppermost:— so in thoughts, one will rise according to law.

How strange ‹all› «so many» birds singing in England, in Tierra del Fuego not one.— now as we know birds learn from each other «though different species» when in confinement, so may they

32 learn in a state of nature.— Singing of birds, not being instinctive, is heredetary knowledge like that of man, & this agrees with the stated fact, that «birds from» certain districts have the best song. [Migratory birds return to same quarter for many years][CD].—
Beauty is instinctive feeling, & thus cuts the Knot:— Sir J. Reynolds explanation may perhaps account for our acquiring «the *instinct*» our notion of beauty & negroes another;[1] but it does not explain the *feeling* in any one man.—

33 Music & poetry opposite ends of one scale.— former pleases from instinct the ears (rhythm *& pleasant sound per se*) & causes the mind to create short vivid flashes of images & thoughts.— Poetry. the latter thoughts are in same manner vivid & grand. the frame of mind being just kept up by the music of the poetry.— (therefore singing intermediate, who has not had his blood run cold by singing).—
Granny says she never builds castles in the air—[1] Catherine often, but not of an inventive class.—

34 Now that I have a test of hardness of thought, from weakness of my stomach I observe a long castle in the air, is as hard work (abstracting it being done in open air, with exercise &c no organs of sense being required) as the closest train of geological thought.— the capability of such trains of thought makes a discoverer, & therefore (independent of improving powers of invention) such castles in the air are highly advantageous, before real train of inventive

35 thoughts are brought into play & then perhaps the sooner castles in the air are banished the better.— The facility with which a castle in the air is interrupted & utterly forgotten—, so as to feel a severe disappointment «in real train of thought this does not happen. because papers, &c &c round one. one recalls

34 no organs of sense] 'no' *over* 'not'.

32–1 Reynolds 1798, 1:219–20, 'Now this appeal implies a general uniformity and agreement in the minds of men. It would be else an idle and vain endeavour to establish rules of art; it would be pursuing a phantom to attempt to move affections with which we were entirely unacquainted. We have no reason to suspect there is a greater difference between our minds than between our forms; of which, though there are no two alike, yet there is a general similitude that goes through the whole race of mankind; and those who have cultivated their taste can distinguish what is beautiful or deformed, or, in other words, what agrees with or what deviates from the general idea of nature, in one case, as well as in the other.' Thus Darwin's idea about the general sense of beauty which only differs in degree in different races is supported by Reynolds: but neither gives an explanation of how it is acquired.

33–1 Granny: Susan Elizabeth Darwin.

the castle by going to beginning of castle» because train cannot be discovered— is closely analogous to my Fathers positive statement that insanity is only cured by forgetfulness.— & the approach to believing a vivid castle in the air, or dreams real again explains insanity.—

36 Analysis of pleasures of scenery.—

There is absolute pleasure independent of imagination, (as in *hearing* music), this probably arises from (1) harmony of colours, ‹whi› & their *absolute* beauty. (which is as real a cause as in music) from the splendour of light, especially when coloured.— that light is a beautiful object one knows from seeing artificial lights in the night.— from the mere exercise of the

37 organ of sight, which is common to every kind of view— as likewise is novelty of view even old one. every time one looks at it.— these two causes very weak.— (2^{d}) form. some forms seem instinctively beautiful «as round, ovals»;— then there the pleasure of perspective. which cannot be doubted if we look at buildings, even ugly ones.— the pleasure from perspective is derived in a river from seeing how the serpentine lines narrow in the distance.— & even on paper two waving *perfectly parallel* lines are elegant.—

38 Again there is beauty in rhythm & symmetry, of forms— the beauty of some as Norfolk Isd fir shows this, or sea weed, &c &c— this gives beauty to a single tree,— & the leaves of the foreground either owe their beauty to absolute forms or to the repetition of similar forms as in angular leaves,— (this Rhythmical beauty is shown by Humboldt from occurrence in Mexican & Græcian to be single cause)[1] this symmetry & rhythm applies

39 to the view as a whole.—[1] Colour «& light» has very much to do, as may be known by autumn, on *clear* day.— 3^{d} pleasure association *warmth, exercise,* birds singings.—

4th. Pleasure of imagination, which correspond to those ‹he› awakened during music.— connection with poetry, abundance, fertility, rustic life, virtuous

38–1 Humboldt 1811, 1:212–13, 'The remains of the Mexican sculpture, those colossal statues of basaltes and phorphyry, which are covered with Aztec hieroglyphics, and bear some relation to the Egyptian and Hindoo style, ought to be collected together in the edifice of the academy, or rather in one of the courts which belong to it. It would be curious to see these monuments of the first cultivation of our species, the works of a semibarbarous people inhabiting the Mexican Andes, placed beside the beautiful forms produced under the sky of Greece and Italy.' Humboldt 1811, 2:67–69 '... the fantastical hypothesis which M. Witte has [proposed] as to the origins of the monuments of colossal forms in Egypt, Persepolis, and Palmyra [when compared favourably to the pyramids of Mexico] ... What analogies with monuments of the old continent! ... whence did they take the model of these edifices? Were they a Mongrel race? Did they descend from a common stock with the Chinese ...?' See also Humboldt and Bonpland 1819–29, 1:xxxi, '... their monuments of architecture ... [resemble] the arabesques which cover the ruins of Mitla, idols in basalt, ornamented with the calantica of the heads of Isis ...'

39–1 Erasmus Darwin 1794, 1:145, 'A Grecian temple may give us the pleasurable idea of sublimity ...' and '... when any object of vision is presented to us, which by its waving or spiral lines bears any similitude to the form of the female bosom, whether it be found in a landscape with soft gradations of rising and descending surface, or in the form of some antique vases ... we feel a general glow of delight ...'

happiness.— recall scraps of poetry;— former thoughts, & in experienced people— recall pictures & therefore imagining pleasure

40 of imitation come into play.— the train of thoughts vary no doubt in different people., an agriculturist, in whose mind supply of food was evasive & ill defined thought would receive pleasure from thinking of the fertility.— I a geologist have illdefined notion of land covered with ocean, former animals, slow force cracking surface &c truly poetical. (V. Wordsworth about science being sufficiently habitual to become poetical)[1]

41 the botanist might so view plants & trees.— I am sure I remember my pleasure in Kensington Gardens has often been greatly excited by looking at trees at [*i.e.*, as] great compound animals united by wonderful & mysterious manner.—[1]

There is much imagination in every view. if one were admiring one in India. & a tiger stalked across the plains, how ones feelings would be excited, & how the scenery would rise. Deer in Parks ditto.—

42 My Father says there is case on record he believes in Philosoph. Transactions, of ideot 18 years old eating white lead. who was most violently purged «believe worms were passed off.» & vomited, but who when he recovered. was found to be ignorant, but quite sensible & no ways an ideot.— «in this case must have been functional.—» He has some idea of a son of Dr. Prietly who was cured from a fall of ideotcy.—[1]

The story of the Corbets & big noses, quite conjectural, in Blakeways book of Sheriffs.—[2]

---

43 July 22d. 1838
No Deliriums, yet in some inflammatory diseases, where there has been no cloud on the mind, every occurrence for a day or two are absoluteley

40–1 Wordsworth 1802, 1:xxxvii–xxxviii, 'Poetry is the first and last of all knowledge—it is as immortal as the heart of man. If the labours of men of Science should ever create any material revolution, direct or indirect, in our condition, and in the impressions which we habitually receive, the Poet will sleep then no more than at present, but he will be ready to follow the steps of the Man of Science, not only in those general indirect effects, but he will be at his side, carrying sensation into the midst of the objects of the Science itself.'

41–1 Viewing plants in a metaphorical sense as if they were animals was a family tradition. See Erasmus Darwin 1794, 1:102, 'The individuals of the vegetable world may be considered as inferior or less perfect animals; a tree is a congeries of many living buds, and in this respect, resembles the branches of coralline, which are a congeries of a multitude of animals.' See also Litchfield 1915, 2:177, 'At present [Charles] is treating Drosera just like a living creature, and I suppose he hopes to end in proving it to be an animal.' See also Raverat 1960:157–58, for a discussion of the 'Elephant Tree' that stood beside the sand-walk at Down.

42–1 Blakeway 1831:37–38, '. . . an inspection of the ancient monuments at Morton Corbet [ancestral home of the Corbets north of and near Shrewsbury] might induce suspicion that the first Corbet . . . may have taken his name from the same features with the Sioux Indian [viz., Nez Corbean: raven-nose]. This is of course all conjectual . . .' Portraits and statues of Corbets, both male and female, at the Corbet home prove that very large noses were a prominent hereditary feature in this family for many generations.

forgotten.— My father signed a bond, yet when he paid the Attorneys bill, he asked what bond he could have had. yet during whole illness, he had been able to direct about his own health.— his complaint was carbbuncl on ‹Head› Neck.— He has seen other cases of similar nature.— —like FitzRoy in sleep giving directions,— & forgetfulness

44 after bad accidents:— After journey, a fit of = gout, has affected his memory of everything in ‹he a [. . .] Mr B› journey. short time previous,— because, pain prevents repetition of idea.—

Mr Blakeway has mentioned in Antiquities of Shrewsbury something about big noses & name Corbet, perhaps nonsense.— look to it[1]

My father has somewhere heard (Hunter?) that pulse of new born babies of labouring classes are slower than those of gentlefolks.[2] & that

45 peculiarities of form in trades (,as sailor tailor blacksmiths?) are likewise heredetary, & therefore that their children have some little advantage in these trades.—[1]

Delirium seems to rest the sensorium.— — analogous to sleep—; some doctors care it, by stimulus & afterwards patient sinks.—

---

46 When a muscle is moved very often, the motion becomes habitual & involuntary.— when a thought is thought very often it becomes habitual & involuntary,— that is involuntary memory, as in sleep.— a new thought arises?? compounded of the involuntary thoughts.— An intentionally recollection of anything is solely by association, & association is probably a physical effect of brain the «similar remark» thoughts, being functions of same part of brain, or the tendency to habit of producing a train of thought.—

47e–48e [not located]

49 Fox[1] believe cats discover birds nests & watch them till the young are big enough to eat.— There was blackbirds nest, near hot-house at Shrewsbury,

44 [. . .]] *3 to 4 letters illeg.*
45 care] 'a' *over* 'u'.
46 An intentionally] 'An' *over* 'Are'.
49 There was . . . abstinence.—] *crossed*; 'inaccurate' *scrawled over passage.*

44–1 Blakeway 1831.
44–2 Hunter 1837:194, 'This alternate motion of the heart is quicker in some classes of animals than others, in some being extremely quick, in others very slow. In all the more inferior orders of animals I believe it is the slowest . . . The pulse is also found to be quicker in the young than in the old of each species . . .'

45–1 See Erasmus Darwin 1794, 1:356, 'Now as labour strengthens the muscles employed, and increases their bulk, it would seem that a few generations of labour or of indolence may in this respect change the form and temperament of the body.'
49–1 Fox: William Darwin Fox.

which the cat was seen by Hubberley[2] to visit daily to see how the young got on. this nest the cat could
If cats will «ever» eat little birds, this most curious instance of reason & abstinence.—

50 My Father remarks that things of great importance are easily forgotten, (if unconnected with fear &c) because people think that the importance of the event by itself will make it to be remembered. whereas it is the importance.— people very often forget where money is placed.— (How often one forgets where put one key. where all keys are placed)

———

Memory cannot solely be number of times repeated, because some people can remember poetry when once read over.—

51 The extreme pleasure children show in the naughtiness of brothers children shows that sympathy is based as Burke maintains on pleasure in beholding the misfortunes of others.—[1]
In young children, the violent passions they go into, shows how truly an instinctive feeling, ‹may not pa.› In reflecting over an insane feeling of anger which came over me, when listening one evening when *tired* «— how true the heart the scene of anger.—» to the pianoforte, it seemed solely to be feelings of discomfort, especially about heart as of

52 excited action, accompanying violent movement; may not passion be the feeling «consequent on the violent muscular exertion» which accompanies violent attack,— Even the worm when trod upon turneth,, here probably there is no feeling of passion, but muscular exertion consequent on the injury & consequently excited action of heart.— now this is the oldest ‹her› inherited & therefore remains, when the actual movement does not take place.— A start is HABITUAL movement to avoid any danger— Fear, shamming death, or running

53 away. accompanied with want of muscular exertion, palpitation, voiding urine because done by some animals in defence, &c

———

Starting must be habitual «*involuntary*» movement from wish to avoid some danger— but it is instinctive because Nancy[1] tells me very young babies start at anything they hear or see. which frightens. them.— Now every animal moves quickly away from any sudden sound or noise, & therefore brain has been accustomed to send a mandate to the muscles & when the noise comes it

51 scene of anger.—] 'a' *over* 'p'.
pp. 1–51, July 15th 1838 . . . ‹may not pa›] *black ink.*
p 51, In reflecting over . . .] *to end of book, except where otherwise indicated, grey-green ink.*
52 A start] 'A' *over 2 vertical lines, probably part* 'H'.

49–2 Hubberley: probably Abberly.
51–1 E. Burke 1823:55–58, Part I, Section XIV, 'The effects of sympathy in the distresses of others'.
53–1 Nancy: The children's nanny at Shrewsbury.

cannot help doing it.— **Fanny Hensleigh[2] doubts whether young babies start.— ⦀ If children wink. it is instinct**

Fear must be simple instinctive feeling: I have awakened

54 in the night. being slightly unwell & felt so much afraid though my reason was laughing & told me there was nothing, & tried to seize hold of objects to be frightened at.— (again diseases of the heart are accompanied by much involuntary fear) In these cases probably the system is affected, & by *habit* the mind tries to fix upon some object:— When a man, child or colt has once been frightened & started much more apt, this partly owing to heart? readily taking same movements, senses being on the look out, & the conveying means

55 from the senses to the mind being more alive.— How is it. with people nervous from illness., ‹the› it must be an excited action in the involuntary mind which is startled.—

My Father says he should think that in old people, in their dotage, who sing the songs «& tales» of infancy, it is very doubtful whether they could recollect these same things from any effort of will whilst their minds were sound.

---

56 Caroline[1] tells me that Nina[2], when brought from Shrewsbury to Clayton,[3] (though so fond of her & of servant of Richard & of Mary & her bed brought from Shrewsbury) yet for a fortnight continued wretchedly unhappy, constantly whined, would not remain quiet in any room, would not sleep at night even when in bed room— grew very thin, would not go out of house except with Caroline— After fortnight. continued to grow thin & did not seem quite happy. in five weeks was so thin, that she was sent back to Shrewsbury,, then immediately fell into her old ways & became fat! What remarkable affection to a place.— How like strong feelings of Man.—

57 The sensation of fear is accompanied by «troubled» beating of heart, sweat, trembling of muscles, are not these effects of violent running away, & must not ‹this› «running away» have been usual effects of fear.— the state of collapse may be imitation of death, which many animals put on.— The flush which accompanies passion & not sweat is the ‹state› effect of short — but violent action.—[1]

To avoid stating how far, I believe, in Materialism, say only that emotions,

53 **Fanny . . . instincts**] *added grey ink.*

53–2 Fanny Hensleigh. See Wedgwood, Frances [Fanny].
56–1 Caroline: Caroline Sarah Wedgwood.
56–2 Nina: One of the Darwin family dogs.
56–3 Clayton: Caroline Wedgwood's home, near Etruria, Stoke-on-Trent.

57–1 Note similarity of this paragraph to Erasmus Darwin 1794, 1:57, '. . . the whole skin is reddened by shame, and an universal trembling is produced by fear; and every muscle of the body is agitated in angry people by the desire of revenge.'

instincts degrees of talent, which are heredetary are so because brain of child resemble, parent stock.— (& phrenologists state that brain alters)

58 It is known that birds learn to sing & do not acquire it instinctively. may not this be connected with their power of acquiring language.—

Hensleigh. W.[1] says that babies know a frown very early in life, ‹before they› (I think I have seen same thing before they could understand. what frowning means) if so this is precisely analogous or identical, with bird knowing a cat, the first it sees it.— it is frightened without knowing why— the child dislikes the frown without knowing why—[2] a man

59 as in Guy. Mannering. feels, pleasure. in seeing the scenes of his childhood without knowing why—[1] had not conscious of recollecting it— this may be nearest approach to ‹the› such instincts which full grown men can experience—

Instinctive walking of animals. that is the ready movement & co-relation of the proper muscles. may be illustrated by the extreme difficulty of moving muscles in different way from what they have been accustomed to, in certain actions— the difficulty of getting on a horse on the left side (not good example,) because leg is right handed.—

60 In Review (Edinburgh) of Froude's life. that author remarks, that writing down his confessions of sins. did not make him more humble.—[1] it has obscurely occurred to me that Capt. F. R.[2] candour & ready confession of error made him less repentant.— In making too much profession, or rather in only *fully* expressing momentary feelings of gratitude, I had a sort of consciousness I was not right; though I never realized the idea that I was tending to make myself in *act* less grateful.— How comes this tendency in these cases? How did my mind feel it was wrong (& it was not

61e merely morally wrong, but hurting my character I felt it)— this is kind of conscience, is obscure memory of having read or thought of some such remarke as now advanced; for I caught it like a flash.—. strange if judgment

61 *page crossed blue crayon.*

58–1 Hensleigh Wedgwood: (1803–1891), son of Bessy & Josiah Wedgwood II.
58–2 See also Erasmus Darwin 1794, 1:146, '. . . children long before they can speak, or understand the language of their parents, may be frightened by an angry countenance, or soothed by smiles and blandishments.'
59–1 Walter Scott, 1815, 3:26, 'It is even so with me while I gaze upon that ruin; nor can I divest myself of the idea, that these massive towers and that dark gateway . . . is not entirely strange to me. Can it be that they have been familiar to me in infancy, and that I am to seek in their vicinity those friends of whom my childhood has still a tender though faint remembrance, and whom I early exchanged for such severe taskmasters?'
60–1 Stephen 1838:525–35. Here Darwin summarizes various statements in the article, e.g., the author says Froude was hauty, intolerant, had a perverted sagacity, and considered burning his confessions. About himself Froude wrote, pp. 528–29, 'These records . . . are so far from exercises of humility that they lessen the shame of what I record just as professions and good will to other people reconcile us to our neglect of them.'
60–2 F.R.: Robert FitzRoy.

remains, where reason is forgotten. it is conscience, or instinct.

Hensleigh[1] says to say. *Brain* per se thinks is nonsense; yet who will venture to say germ within egg, cannot think— as well as animal born with instinctive knowledge.— but if so, yet this knowledge acquired by senses,— then thinking consists of sensation of images before your eyes, or ears (language mere means

62e of exciting association.)— or of memory of such sensations, & memory is repetition of whatever takes place in brain. when sensation is perceived.= =

Aug. 7th—38.
Transactions of the Entomological Society of London Vol. I. p. 106. Col. Sykes on *Formica indefessa* placed table in cups of water which they waded. or swam across.— they then stretched themselves from wall to table.— table being removed a little further, they ascended about a foot & then leapt across. (Col Sykes compares this with pidgeons finding their way home—[1] there is something wrong in comparing these cases, when agency is unknown, with simple exertion of

63e intellectual faculty) if ants had at once made this leap it would have been instinctive, seeing that time is lost & endeavours made must be experience & intellect.—

===

do. p. 157. Westwood remarks that some imported plants are attacked by insects & snails of this country (thus Dahlias by snails)— ‹The› *Apion radiolum* undergoes transformation in the stem of Hollyhock, although ordinary Habitat is *Malva sylvestris*.[1]

===

do. p. 228 Newport says Dr Darwin mistaken in saying common wasp cuts off wings of flies from intellect. but it does it always instinctively or habitually.—[2]

62 of exciting . . . perceived.=] *crossed blue crayon.*
63 ‹tea›] ‘that’ *over* ‘tea’.

61–1 Hensleigh: Hensleigh Wedgwood.
62–1 Sykes 1836a:105–6, ‘. . . the legs of the table were placed upon low pedestals in little stone pans filled with water. . . . [The ants] boldly pushed over, and succeeded in catching hold of the opposite bank with their fore-legs ere they sunk . . . The edge of the table was about an inch distant from the wall, . . . the largest ants now essayed to pass it, holding on the wall by the hind legs, whilst the front legs were stretched out to touch the edge of the table, and the contact enabled very many to cross. The table was now removed from the wall beyond the maximum stretch of the largest ants . . . an ant upon the wall, about a foot above the level of the sweets . . . fell upon the table. . . . another and another followed. . .’

63–1 Westwood 1836:157, ‘. . . many imported plants afford the most congenial food to our strictly native insects . . . the flowers of the dahlia* (*. . . the hollyhock is attacked by *Apion radiolum*, which undergoes its transformations in the stem, although its ordinary habitat is *Malva sylvestris*) are gnawed, almost as soon as they have burst the calyx . . . but it is to the snails that the greatest portion of the damage must be laid.’
63–2 Newport 1836:228, ‘Dr. Darwin . . . seems to think [the wasp] affords a strong proof of a faculty of reasoning in insects; but I think it will appear that the fact he observed, and upon which he founded his opinion, was only one of those occurrences which form part of the instinctive predaceous habits of the species.’ See Erasmus

good Heavens is it disputed that a wasp has this much intellect. yet habit may make it act wrong, as I have done when taking lid off ‹tea› side of tea chest, when no tea[3]

64e do. p. 233. Mr Lewis describes case of insects «a Perga» of Terebrantia, laying eggs on leaves of Eucalyptus, watching few days till larva excluded, then though not feeding them «nor helping larva from egg» watching them, brooding over them, preserving them from «the» sun & enemies— would not fly away, but bit pencil when touched with it— do not know their own larvæ, but one female may be moved to other larvæ, when two groups near. mother desert one sometimes & go to other, so that two mothers to one group.— (as in birds blind storge— They continue till death, thus acting 4 to 6 weeks. The deserted broods appeared healthy— This remarkable case may be normal. with insects, but habit forgotten in all older species. The earwig & a doubtful one of Acanthosoma grisea described[1]

65e–68e [not located]

69 as first caused by will of Gods. «or God» secondly that these are replaced by metaphysical abstractions, such as plastic virtue, «&c» (Very true, no doubt savage attribute thunder & lightening to Gods anger.— (∴ more poetry in that state of mind: the Chileno says the mountains are as God made them,—[1] next step plastic ‹virtue› natures. accounting for fossils). & lastly the tracing facts to laws. without any attempt to know their nature.— Reviewer considers this profoundly true.— How is it with children.— Now it is not a little remarkable that the fixed laws of nature should be «universally» thought to be the *will* of a superior being; whose natures can only be rudely traced out. When one sees

70 this, one suspects that our will may ‹be› «arise from» as fixed laws of organization.— M. le Comte argues against all contrivance— it is what my views tend to.—[1]

Darwin 1794, 1:183, 'One circumstance I shall relate which fell under my own eye, and showed the power of reason in a wasp, as it is exercised among men. A wasp, on a gravel walk, had caught a fly nearly as large as himself; kneeling on the ground I observed him separate the tail and the head from the body part, to which the wings were attached. He then took the body part in his paws, and rose about two feet from the ground with it; but a gentle breeze wafting the wings of the fly turned him round in the air, and he settled again with his prey upon the gravel. I then distinctly observed him cut off with his mouth, first one of the wings, and then the other, after which he flew away with it unmolested by the wind.'

63–3 See C 217, and Darwin's notes in his copy of Abercrombie 1838:57, 'an action becomes habitual if repeated, without at same time, without much attention at first as taking off cover to tea-chest.' Abercrombie subsequently discusses habit-formation.

64–1 Lewis 1836:232–34. Darwin's synopsis is accurate.

69–1 *JR*:435–36, 'My geological examination of the country generally created a good deal of surprise amongst the Chilenos . . . [who] thought that all such inquiries were useless and impious; and that it was quite sufficient that God had thus made the mountains.' As late as 1861 Darwin again used the same expression in a letter to Lyell: 'It reminds me of a Spaniard whom I told I was trying to make out how the Cordillera was formed; and he answered me that it was useless, for "God made them".' *ML*1:192.

70–1 Darwin agrees with Comte's view that neither divine nor metaphysical contrivances are necessary to explain natural phenomena. See Brewster 1838: 275, 305.

When a man is in a passion he puts himself stiff, & walks hard.— «He cannot avoid sending will of action to muscles, any more than «prevent» heart beat» remember how Pincher[2] does just the same; I noticed this by perceiving myself skipping when wanting not to feel angry— such efforts prevent anger, but observing eyes thus unconsciously discover struggle of feeling.— It is as much effort to walk then lightly as to endeavur to stop heart beating: one ceasing, so does other.—

71 What an animal like taste of, likes smell of, ∴. Hyæna likes smell of that fatty substance it scrapes off its bottom.— it is relic of same thing that makes one dog smell posterior at another.—

Why do bulls & horses, animals of different orders turn up their nostrils when excited by love? Stallion licking udders of mare strictly analogous to men's affect for womens breasts. ∴. Dr Darwin's theory probably wrong, otherwise horses would have idea of beautiful forms.—[1]

---

72 With respect to free will, seeing a puppy playing cannot doubt that they have free will, if so all animals., then an oyster has & a polype (& a plant in some senses, perhaps, though from not having pain or pleasure actions unavoidable & only to be changed by habits). now free will of oyster, one can fancy to be direct effect of organization, by the capacities its senses give it of pain or pleasure, if so free will is to mind, what chance is to matter «(M. Le Compte)»—[1] the free will (if so called) makes change

73 in bodily organization of oyster. so may free will make change in man.— the real argument fixes on heredetary disposition & instincts—.— Put it so.— Probably some error in argument, should be grateful if it were pointed out.— My wish to improve my temper, what does it arise from but organization. that organization may have been affected by circumstances & education, & by choice which at that time organization gave me to will— Verily the faults of the fathers, corporeal & bodily are visited upon the children.—

74 The above views would make a man a predestinarian of a new kind, because he would tend to be an atheist. Man thus believing, ‹yet› would more earnestly pray "deliver us from temptation,' he would be most humble, he would strive ‹to do good› «to improve his organization» for his children's

72 «(M. Le Compte)»] *four marks following ')' resemble quotation and/or footnote marks.*

70–2 Pincher: pet dog.

71–1 Erasmus Darwin 1794, 1:145, 'Our perception of beauty consists in our recognition by the sense of vision of those objects, first, which have before inspired our love by the pleasure, which they have afforded to many of our senses; as to our sense of warmth, of touch, of smell, of taste, hunger and thirst; and secondly, which bear any analogy of form to such objects.' And on p. 253, 'So universally does repetition contribute to our pleasure in the fine arts, that beauty itself has been defined by some writers to consist in a due combination of uniformity and variety.' See also M 37–39.

72–1 Compte: Comte. See Brewster 1838.

sake & for the effect of his example on others.ǂ It may be doubted whether a man intentionally can wag his finger from real caprice. it is chance, which way it will be, but yet it is settled by reason.—ǂ

75 How slow habits are changed may be inferred from expression. "relict of bad habit." as child is cured of sucking his finger by rubbing them with alum, so more slowly does animal leave off ‹t› instinct, when attended with bad effects

---

Martineau. How to observe, p. 21–26. argues «with examples» very justly there is no universal moral sense.— «from difference of action of approved»[1] Yet as, I think, the opposite side has been shown— see Mackintosh.— Must grant, that the conscience varies in different races.—[2] no more wonderful than dogs should have different instincts.— Fact most opposed to this view, where

76 the moral sense seems to have changed suddenly— but are not such «sudden» changes rare,— as when Polynesian mothers ceased to destroy their offspring—ʕ yet perhaps if they had murdered their children, this moral sense, would have been so much, as in other races of mankind..— p. 27. Mart. allows *some* universal feelings of right & wrong «(& therefore in *fact* only *limits* moral sense)» which she seems to think «are» to make others happy & wrong to injure them without temptation.—[1] This probably is natural. consequence of man, like deer &c, being social animal, & this conscience or instinct may be

77 most firmly fixed, but it will not prevent other being engrafted.— No one doubts patriotism & family pride are heredetary., & therefore he has these strong, & does not act up to them, no doubt disobeys & hurts conscience more than other.— A Scotchman will his country or Swis.— it may be answered effects of education, may be opposed undoubted cases of heredetary pride & in single families.

75–1 Martineau 1838:22, 'A person who takes for granted that there is an universal Moral Sense among men, as unchanging as he who bestowed it, cannot reasonably explain how it was that those men were once esteemed the most virtuous who killed the most enemies in battle, while now it is considered far more noble to save life than to destroy it.' And on p. 23, '. . . every man's feelings of right and wrong, instead of being born with him, grow up in him from the influences to which he is subjected.'

75–2 Mackintosh 1837:58, 'There is no tribe so rude as to be without a faint perception of a difference between right and wrong. There is no subject on which men of all ages and nations coincide in so many points as in the general rules of conduct, and in the qualities of the human character which deserve esteem.'

76–1 Martineau 1838:27, 'The traveller having satisfied himself that there are some universal feelings about right and wrong, and that in consequence some parts of human conduct are guided by general rules, must next give his attention to modes of conduct, which seem to him good or bad, prevalent in a nation, or district, or society of smaller limits. His first general principle is, that the law of nature is the only one by which mankind at large can be judged. His second must be, that every prevalent virture or vice is the result of the particular circumstances amidst which the society exists.'

78 Edinburgh. Phil. Transact. p. 365. Case of double consciousness, one only «little» less perfect than other, absolutely two people.[1] Consider this profoundly, may throw light on *consciousness*, explained by Dr Dewar on principle of association.—«fully bears out my fathers doctrine about people forgetting their insanity»[2] there seem other cases somewhat analogous, & which I think will lead to fact of old people singing songs of their childhood. & certainly of Miss Cogan, & fully corroborates the fact of her not ‹remembering which› «repeating song» when she had recollected it in perfect senses.— These things, & drunkedness, show what trains of thought depend on state of turn

79 In drunkedness same disposition recurs, such as —— of Trinity always thinking people were calling him a bastard.— when drunk.— having really been so.— some always sentimental, some quarrelsome as B.^e^ on board Beagle,[1] some merry goodhumoured as self.— «When Miss Cogan has remembered her song, then the song was to her like one which though learnt in infancy, had often been repeated: Now it is remarked that A. Bessy repeated things, which none about her had EVER before heard, so very probably forgotten.» Such facts bear on such characters as Allen W.[2] & Babington, both half ideotic in some respects & with store of accurate & even profound knowledge or other & unusual line— both odd appearance about eyes.— one botanist & great knowledge of Irish Politics, «both bad jokers.—» the other army officer, horticulture & religious sects.— yet Allen. W. remark about his slippers bad for fires, what is wrong in his head. & Babington's silly joking

80 The possibility of the brain having whole train of thoughts, feeling & perception separate, from the ordinary state of mind, is probably analogous to the double individuality implied by habit, when one acts unconsciously with respect to more energetic self, & likewise one forgets. what one performs habitually.— Agrees with insanity, as in Dr Ash's case, when he struggled as it were with a second & unreasonable man.— If one could remember all ones farthers actions, as one does those in second childhood, ‹they› or when drunk they would not be more different, & yet they would make one's father & self one person— & thus eternal punishment explained.

81 These facts showing what a train of though[t] action &c will arise from physical action on the brain, renders much less wondefful the instincts of animals—

---

81 I was interested as was I] 'was I' *over* 'I in'.

78–1 Darwin's cited reference not traced; see however, Ferrier 1838a: 199, 'It is reserved for man to live this *double* life. To exist, and to be *conscious* of existence; to be rational, and to *know* that he is so.'

78–2 Abercrombie 1838: 301–2, discusses Dr Dewar's case of an ignorant servant girl who during paroxysms showed surprising knowledge of geography and astronomy.

79–1 Benjamin Bynoe

79–2 Allen W.: John Allen Wedgwood.

Aug. 12th. 38. At the Athenæum Club. was very much struck with an intense headache «after good days work» which came on from reading «review of» M. Comte Phil. which made me «endeavour to» remember, & to think deeply, & the immediate manner in which my head got well when reading article by Boz.—[1] now in this I was interested as was I in the other, & read so intently as to be unconscious of all around, yet there was no strain on the intellectual powers— the difference is of a man wagging his foot & working with his toe to perform some difficult task.—

82 Aug. 12th. When in National Institution[1] & not feeling much enthusiasm, happened to go close to one & smelt the peculiar smell of Picture. association with much pleasure immediately thrilled across me, bringing up old indistinct ideas of FitzWilliam Musm.[2] I was amused at this after seven years interval.

———

Augt. 15th. As child gains habit «or trick» so much more easily than man, so may animal obtain it far more easily, in proportion to variableness or power of intellect.— Some complicated trades can *hardly* be considered as actions otherwise than habitual.— instances??

83e The possibility of two quite separate trains going on in the mind as in double consciousness may really explain what habit is— In the *habitual* train of thought one idea. calls up other, & the consciousness of double individual is not awakened.— The habitual individual remembers things done in the other habitual state because it will (without direct consciousness?) change its habits.—

Aug. 16th. As instance of heredetary mind. I a Darwin & take after my Father in heraldic principle. & Eras a Wedgwood in many respects & some of Aunt Sarahs. cranks[1], & so is Catherine in some respects—. good instances.— when education same.— My handwriting same as Grandfather.[2]

84e *Aug. 16th* Anger «Rage» in worst form is described by Spenser (Faery Queene. **CD 25** (Descript of Queen) «O» of Hell Cant IV or V.) as *pale* & *trembling*. & not as flushing & with muscles rigid.—[1] How is this? **dealt with p. 241**[2] Origin of man now proved.— Metaphysic must flourish.— He who understands baboon ‹will› would do more towards metaphysics than Locke

———

83 The possibility . . . done in the other] *excised, crossed pencil.*
84 **C.D. 25**] *added pencil, boxed.*
dealt with p. 241] *added pencil, circled.*
Aug. 16th . . . How is this?] *excised.*

81–1 Boz: Charles Dickens.
82–1 Darwin probably meant the National Gallery or possibly the Royal Institution.
82–2 FitzWilliam Museum, University of Cambridge.
83–1 Aunt Sarah: Sarah Elizabeth Wedgwood.
83–2 See E 89 for discussion of inheritance of handwriting.
84–1 Spenser 1589, Booke I, The Fourth Canto, verse 33:27, 'As ashes pale of hew and seeming ded; And on his dagger still his hand he held, Trembling through hasty rage, when choler in him sweld.'
84–2 Reference not traced.

A dog *whines*, & so does man.— dogs laughs for joy, so does dog bark. (not shout) when opening his mouth in romps, ‹so› he smiles.

Many of actions as hiccough & yawn are probably merely coorganic as

85 connexion of mammæ & womb.— We need not feel so much surprise at male animals smelling vaginæ of females.— when it is recollected that smell of ones own pud. not disagree.— Ourang outang at Zoolog Gardens touched pud. of young male & smelt its fingers. Seeing a dog & horse & man yawn, makes me feel how ‹much› all animals ‹are› built on one structure.—

He who doubts about national character let him compare the American whether in the cold regions of the North,— the elevated table land of Peru

86 the hot plains of the Amazons & Brazil— with the negros of Africa, (or again the black man of ‹B› Van Diemens land & the energetic copper coloured natives of New Zealand)— the American in Brazil is under same conditions as Negro on the other side of the Atlantic. Why then is he so different— in organization.—

87 Same cause as colour & shape & ideosyncracy.— Look at the Indian in slavery & look at the Negro— look at them both savage— look at them both semi-civilized—

Perhaps one cause of the intense labour of *original inventive* thought is that none of the ideas are habitual, nor recalled by obvious associations. as by reading a book.— Consider this.—

88 "The fledge-dove knows the prowlers of the air" &c &c &c so is conscience &c &c Coleridge,— Zapoyla p. 117, Galignani Edition[1]

Fine poetry, or a strain of music, when the mind is rendered ductile by grief, or by bodily weakness, melts into tears, with sensations of sorrowful delight, very like best feeling of sympathy.— Mem: Burke's idea of Sympathy.[2] being real pleasure at pain of others, with rational

89 desire to assist them,— otherwise as he remarks sympathy could be barren. & lead people from scenes of distress.— see how a crowd collects at an accident,— children with other children naughty.— Why does person cry for joy?

17th. August Montaigne (Vol. I) has well observed, one does not fear death

89 allowing my mind] 'a' *of* allowing *over* 'm'.

88–1 Coleridge 1829:117, 'The fledge-dove knows the prowlers of the air,/ Fear'd soon as seen, and flutters back to shelter.'

88–2 Edmund Burke 1823:55–58.

from its pain, but one only *fears* that pain, which is connected with death!—[1] How has this instinctive fear arisen?

19th. When I went down to Woollich I was trying to unbend my mind as much as possible (testing success by decreasing headache) & found best plan was allowing my mind to skip from subject

90 to subject as quick as it chose.— although thinking «& talking» for the moments with interest on each.— ∴ my father. is right in saying delirium rest— therefore dreams thus act.— ∴ weak minded people are fickle & full of levity (? do I not confound action & thought here?) The opposite extreme of this desultory thought is following out such an idea, as effect of sea on coves when waters had fallen, as in my Glen Roy paper.—[1] this greatest mental effort, of which I am capable— I suspect from these facts that whole effort consists in keeping one idea before your

91 mind steadily., & not merely thinking intently; for that one does with novel for a length of time.— Then if one endeavur to keep any simple idea as scarlet steady before mind for period, «if the scarlet was before one effort less» one is obliged to repeat the word, & think of qualities as flowers, cloth &c & with all this difficult EXPERIMENTIZE upon this effort.— it looks so analogous to muscle in one position great fatigue.— may explain excessive labour of inventive thought.— Examine frame of mind in following changes during fall of sea.— Is the effort

92 greater if the idea is abstract as love, (or an emotion *not so*) than if simple idea as scarlet?— How can people dwell on pain ? no definite idea. nor is an emotion.— People who can multiply large numbers in their head must have this high faculty, yet not clever people.

Aug. 21st. 38
When a dog in play has his mouth open ready to bark, & lip twisted up, in that peculiar manner they do, even more than in a real snarl, they are enjoying a satirical. laugh.— when snarling real bitter sarcasm.—

93 ‹These› Seeing how ancient these expressions are, it is no wonder that they are so difficult to conceal.— a man «insulted» may forgive his enemy & not wish to strike him, but he will find it far more difficult to to look tranquil.— He may despise a man & say nothing, but without a most distinct will, he will

91 upon this effort.—] 'u' *over* 'a'.
93 man & say . . . satisfied with himself] *double scored pencil.*

89–1 Montaigne 1588, 1:82–83, 'Si c'est une mort courte et violente, nous n'avons pas loisir de la craindre; si elle est autre, je m'apperçois qu'à mesure que je m'engage dans ses avenues et dans la maladie, j'entre naturellement et de moymesme en quelque desdein de la vie. . . . j'ay trouvé que, sain, j'avois eu les maladies beaucoup plus en horreur que lors que je les ay senties. . . . le sault n'est pas si lourd du mal estre au non estre, comme il est d'un estre doux et fleurissant à un estre penible et douloureux.'

90–1 Charles Darwin 1839:73, '. . . the entire hollow . . . must have existed as an indentation or little cove on the line of ancient sea-beach.'

find it hard to keep his lip from stiffening over his canine teeth.— He may feel satisfied with himself, & though dreading to say so, his step will grow erect & stiff like that of turkey.— he may be amused, he need not express it, he may most earnestly wish [not] to do it, but an involuntary laugh will burst forth, this & yawning. (common to other

94 animals) scream of agony, sigh of discomfort & weariness. & meditative tranquility. «whine of children. puppies do so dogs nearly silent, so with men.— How is crying— peculiar not common?—» no bark of anger nor have monkeys & many other animals,— but yet when angry it is hard not to growl out some sound even if it be inarticulate.— the maniac shouts & bellows with passion.— It is not a little remarkable that those sounds which are involuntary, are common to animals.— Curious to trace, which of these actions are habitual, & which now connected physical relations.— CD[like sighing to relieve circulation after stillness.— Now I conceive if organization were changed, I conceive sighing might

95 yet remain just like sneering does.— is yawning habitual from awaking from sleep see how a dog yawns when he awakes. & streching & yawning can be explained from too long rest of muscles.— evidently habitual when transferred, (also how often) to the tale of a wearisome man.— Is frowning, result of straining vision, as savages without hats put up their hands, & as attention would amongst lowest savages clearly be directed chiefly by objects of vision.— Does the contraction & wrinkling of the skin contract iris?— same way as one lifts up eyebrows to see things in dark. & hence is this the cause of expression of surprise— viz seeing something obscurely with the wish to make it out?—

96 Seeing a Baby (like Hensleigh's) smile & frown, who can doubt these are instinctive— child does not sneer. because no young animal has canine teeth.— A dog when he barks puts his lips in peculiar position, & he holds them this way, when opening mouth between interval of barking, now this is smile.‖ With respect to sneering the very essence of an habitual movement is continuing it when useless.— ‹&› therefore it is here continued when the uncovering the canine useless.—

The distinction «as often said» of language in man is very great from

97 all animals— but do not overrate— animals communicate to each other.— Lonsdale's[1] story of Snails, Fox of cows, & many of insects— they likewise must understand each other expressions, sounds, & signal movements.— some say dogs understand expression of man's face.— ‹That› How far they

94 «whine . . . common?—»] *inserted between* 'scream . . . tranquility'.
CD[] '[' *pencil.*

97–1 Lonsdale: William Lonsdale.

communicate not easy to know,— but this capability of understanding language is considerable, thus carthorse & dog.— birds many cries. monkeys communicate much to each other.—

---

98 Waterhouse says far more instincts in *all* of the Hymenoptera;. ‹therefore› than in other orders (study Kirby with this view)[1] therefore there is Instinctual developement in one order, as there is Intellectual in human— probably some genera in different orders more advanced than others just as dog & Elephant most intellectual.— Hymenoptera typical insects. ie have all parts. Waterhouse

99 Study well the greater number of insects in insecta— not connected with transformation because Spiders have many,— great powers of communicating knowledge to each other—

August 23d. Jones[1] said the great calculators, from the confined nature of their associations (it is not so in punning) are people of very limited intellects, & in the same way are chess Players— A man at Cambridge, during his time, almost an absolute fool used to play regularly with D'Arblay of Christ[2] of *great genius*, & yet invariably used to beat

100 him— The son of a Fruiterer in Bond St. was so great a fool that his Father only left him a guinea a week. yet. he was inimitable chess player.— Peacocks[1] remark about mathematicians not being profound reasoners.— all same fact— for, as Jones observed, in playing chess however many places, & contingency a man has keep in mind. all is certain.— there is judgment of probabilities, therefore this judgment gives a man common sense, & the highest intellectual powers of perceiving

101e & classifying distinct resemblances.—

The facts of half instincts. when two varieties are crossed as in Shepherd dogs— **Inherited Habits: Have Effect in Bones** is valuable it shows that new instinct can originate.— strong argument for brain bringing thought, & not merely instinct, a separate thing superadded.— we can thus trace causation of thought.— it is brought within ‹our own› limits of examination.— obeys same laws. as other parts of structure. **C.D.27**

Can an analogy be drawn between «heredetary» associated pleasures &

101 distinct] *crossed pencil*; 'Can't go with this FD' *added by Francis Darwin.*
Inherited . . . in Bones] *added pencil*; 'in' *possibly* 'on'.
C.D. 27 *added pencil.*
Can an . . . fear of death.] *double scored left margin.*

98–1 Kirby 1835.
99–1 Jones: not identified.
99–2 D'Arblay: probably Alexander d'Arblay.
100–1 Peacock: George Peacock.

pains & emotions— such as child sucking, gives pleasure, & always has done therefore sight of own child. (when frame in condition to receive pleasure) gives pleasure, ie. love.— & so pain gives fear of death.

102e Mayo Philosophy of Living. p. 140— Dreams good account of «thinks» are recollected when intense, or when so near waking. that an associated is kept up with waking thought.— Ld Brougham thinks no dreams except at this time.[1] how does he account for dogs & men speaking in their sleep.— Characters of dreams no surprise, at the violation of all ‹rules› relations of time, ‹identity,› place & personal connections— ideas are strung together in manner ‹they› quite different from when awake.— peculiar sensation as flying. (No memory of past events?) or influence on our conduct, the links which when *conscious* connect past, present & future

103 thoughts are broken— Sir J. Franklin when *starved*, all party dreamt of ‹goo› feasts of good food—[1] The mind wills to do this & hears that, but yet scarcely really moves.— the willing therefore is ideal, as all the other perceptions.— The mind thinks with extraordinary rapidity— We may conclude that neither number, vividness, rapidity, novelty of separate ideas cause fatigue to the mind,— it is solely the comparison, with past ideas. which makes consciousness— & which tells one of reality— castle in the air, is more prolonged than dream. never fatiguing,— else it is only our consciousness, & senses tell us it is not real.== dreaming appears clearly rest of the mind, with all other faculties: «Vide page 110, by mistake.»

104 N B. Everything which happens to man who does not produce children. or after he has useless. does not affect race. argument for early education.— fear of death!!! as Montaigne observes. distinct from pain, for one hates pain from this fear— & not death for the pain.— How was this instinct gained.? by conversation— ∴ modified in those races, where it is customary to die—

---

102 *page crossed pencil.*
104 «Hume»] 'Hume' *circled, connected by line with* 'impressions,' *also circled.*

102–1 Herbert Mayo 1838:139, '[According to Lord Brougham] "we only dream during the instant of transition into and out of sleep." . . . For my own part, I am disposed to adopt the opposite opinion, and think that in sleep we always dream . . .'

103–1 Herbert Mayo 1838:141, 'Sir John Franklin remarks, when his party was in the extremity of physical exhaustion and physical suffering,— "Although the sensation of hunger was no longer felt by any of us;— yet we were scarcely able to converse on any other subject than the pleasure of eating." *But their dreams at this period, while they were starving were of plentiful repasts.*' See also Franklin 1823:564–66, 'The dreams which for the most part, but not always accompanied it [i.e., comfort of a few hours of sleep during prolonged starvation], were usually (though not invariably,) of a pleasant character, being very often about the enjoyments of feasting.' In Erasmus Darwin 1794, 1:23, is the following penciled marginal notation made by Charles Darwin (in his personal copy of *Zoonomia*), 'This is strange as hungry men never dream of hunger'. The text passage reads, '. . . in his dreams [a man about fifty years old who had been deaf for nearly thirty years] always imagined that people conversed with him by signs or writing, and never that he heard any one speak to him. From hence it appears, that with the [loss of] perceptions of sounds, he has also lost the ideas of them . . .' See M 21–22, 'People, my Father says, do not dream . . .'

August 24th. As some impressions «Hume» become unconscious.[1] so may some *ideas*.— ie habits, which must require idea to order muscles to do ‹certain› the actions.? is

105 it the ‹becom› impression becoming *very often* unconscious, which makes the *idea* unconscious, if so (think of this). study what impressions become unconscious those which are viewed with little interest, & those which are viewed very often.— former do not give rise to ideas so much. as objects of interest.—

---

do/ I was much struck with observing how the Baboon (‹Macaco› «Cyanocephalus Sphynx Linnaeus») constantly moved the skin of forehead over eyes, at every emotion & ‹look› «turn» of the head.

106 I could not perceive «any» distinct *wrinkle*, but such movements in skin of eyebrow important analogy with man.— I see monkeys *grin* with passion, that is show all the teeth: «& make noise not like pish, but like chit-chit-chit, quickly uncovering their teeth, this the Keeper thinks is from pleasure, & may be compared to laughing» they dance with passion, ie. nervous impulse to action is sent so fast to limbs that they cannot remain still.— I do not doubt this Baboon. knew women.— Another little old American monkey «(Mycelis)» I gave nut, but held it between fingers, the peevish expression was

107 most curious ‹like› «remember» the expostulatory angry look of black spider monkey when touched, also another monkey to dog. I showed nut & then closed my mem. expression of fury, jump to scratch my face. The ourang outang, under same circumstances, threw itself down on its back & kicked & cryed like naughty child.— Do monkeys cry?— «they *whine* like children.—»

Expression, is an heredetary habitual movement consequent on some action, which the progenitor did, when excited or disturbed by the same cause, which «now» excites the expression.—

108 Habitual actions are the reverse of intellectual, there is no comparison of ideas— one follows other as in blindest memory— also low faculty of understanding.

---

105 Cyanocephalus Sphynx Linnaeus] *written in fine pen.*
106 «& make . . . laughing»] *written in fine pen*; 'chit-chit-chit' over otit-otit-otit
107 Expression, is an . . . the expression.—] *double scored left margin.*

104–1 Hume 1817, 2:16, 'By the term *impression*, then, I mean all our more lively perceptions, when we hear, or see, or feel, or love, or hate, or desire, or will. And impressions are distinguished from ideas, which are the less lively perceptions, of which we are conscious, when we reflect on any of those sensations or movements above mentioned.' See also: Hume 1826, 1:15–45, bk 1, pt 1, 'Of ideas, their origin, composition, connexion, abstraction, &c'.

Adam Smith (.D. Stewart life of. p. 27), says ‹sympathy› we can only know what others think by putting ourselves in their situation, & then we feel like them—.[1] hence sympathy very unsatisfactory because does not like Burke explain pleasure.

August 26th. I cannot help. thinking horses admire a wide prospect.— The very superiority of man perhaps depends on the number of sources of pleasure & innate tastes, he

109 partakes, taste for musical sound with birds. & ♀ howling monkeys— smell with many animals— see how a dog likes smell of Partridge—, man's taste for smell of flowers, owing to *parent* being fruit eater.— origin of colours?—

Nothing shows one how little happiness depends on the senses.; than the ‹small› fact that no one, looking back to his life, would say how many good dinners or . . . . he had had, he would say how many happy days he spent in such a place.—

110 Vide page 103, supra (by mistake)

---

have lower animals these vivid thoughts
In same book (p. 143) wonderful case of perfect double consciousness Mayo compares it with Somnambulism.—[1] the young lady almost equally in her senses in either state.— does this throw light on instinct, showing what trains of action may be done unconsciously as far as the ordinary state is concerned.?—

Mr. Mayo told me the case of a lady, (whose name was told me, who told the fact to Mr Mayo himself.[2] she was one day reading a book, with ivory paper cutter, which she valued, & she was suddenly called to go on the lawn to see something, on her return could not find paper cutter, hunted in vain

111 for it— ten years afterwards whilst at a meal, she suddenly like a flash without any assignable cause, remembered she had put it in branch of tree, & apologising to party, went out & found it there!!! Lady in perfect «mental» health.— «Erasmus had almost same thing happen to him about a knife. which he had hid some years before.— was greatly astonished, at the time. &

111 characterized» dream] 'dr' *over* 't'.

108–1 Adam Smith 1795:xxvii, 'Although, when we attend to the situation of another person, and conceive ourselves to be placed in his circumstances, an emotion of the same kind with that with which he feels, naturally arises in our own mind, yet this sympathetic emotion bears but a very small proportion, in point of degree, to what is felt by the person principally concerned.'

110–1 Herbert Mayo 1838:145, 'I believe [double consciousness] to exemplify sudden transitions to and from the state of somnambulism.'

110–2 Darwin very likely discussed the issue with Herbert Mayo at the Athenaeum where both were members.

could trace no chain of association»

Mayo Philos. seems certain that muscular, mental, ‹&› digestive nervous influence replace each other[1]

August 29th. Went to Bed. & built «common» Castle in the air, of being compelled, from some quite imaginary cause to start at once to Shrewsbury., vaguely thought of packing up.— was lying on my back fell to sleep for second & wakened.— had very clear & pretty vivid «& perfectly characterized» dream, in continuation of waking thought— my servant was in the room. with my trunk out & I was engaged in hurriedly giving orders.— Now what was difference between Castle & dream

112 No answer shows our profound ignorance in so simple case.— There was memory, for it related to past idea.— there was a kind of ideal consciousness for moment, implied by «presence» my servant, «box» my own manner of ordering things to be done.— The senses are closed probably by sleep & not vica versa. anyhow I might have been quite still, & not attending to bodily sensation & yet the Castle would not have turned into dream.— It appears to me, that the mind is wholly absorbed with *one* idea (hence *apparent* vividness) & there being no other parallel trains of ideas connected with past circumstances.— as whether I really was going to Shrewsbury, whether I had rung for Covington.[1] whether he had come & opened box, whether I had thought what clothes to take (how often

113 one cannot tell whether one has rung the bell., when one recollects circumstances were such one naturally would so so!) Now all these parallel trains of thought necessary heirs of every action, & always running on in mind, being absent. one could not compare the castle with them, therefore could not *doubt* or *believe*.— When I say trains, it may be instantaneous changes in order ‹to every› calling up ideas of every late impression.— (do the ideas, direct effect of perception by senses fail first, as whether I had pulled the bell??)— It may be deception to say the mind ‹thinks› quicker in sleep, it may do less work & yet do so, from the exertion of keeping up the memory of every late impression. & likewise gaining new ones from senses. & ‹comparing their› «calling up» old ones, to be sure of ones consciousness.—

114 Mayo observe no improbabilities in a dream,[1] effect of doubting nor believing, effect of not reasoning. effect of not having ‹all› other trains of thought, or memory from innumerable late events.— the fatigue of thinking is keeping up these trains,— especially if they are invented as in imagination,

111–1 Herbert Mayo 1838:4–15, discusses four varieties of temperament, viz., sanguine, lymphatic, bilious, and nervous, with mixed or equal amounts also existing. Darwin's statement is a general summary of Mayo's treatment of these 'Temperaments' as influenced by each other and by states of mental disposition.

112–1 Covington: Syms Covington.

114–1 Herbert Mayo 1838:140, 'In dreams, that which most strikes us are their monstrous and capricious combinations, and our want of surprise at their improbability.'

& in rigidly comparing each step as in reasoning— hence delirium & sleep mental rest. though. most vivid & rapid thought.— There may be some «two or three» trains of thought, therefore one may be imperfectly reason — ‹In a› Abercrombie's case of «in Botanical Student» somnabulism, did reason about himself— but not about, facts gained or gaining by senses.—[2] As sleep ‹is› only one idea is awake,

115 when one is awake many necessarily are., when one is deeply reasoning besides these (which must be present, though one is not conscious of them, else one would not stand) a crowd of other trains of thought are in progress— In castle of air the *trouble* «I well recollect» is in making things somewhat probable. in comparing every step, & inventing new means,— therefore works of imagination *hard* work,— Keeping one idea present is, perhaps, hard work— though dreams do that

---

One Reflective Consciousness is curious problem., one does not care for the pains of ones infancy.— one cannot bring it to one self.— nor of a bad dream, when that is not recollected, nor of the Botanical Somnambulist.[1] (if he had been unhappy)— it is because in this

116 state, the consciousness does not go back to former periods so «as» to ‹make› «give» one individuality in this case.— But now in Mayo's «p. 140» case of double consciousness,[1] one would pity suffering in one state almost as much as in the other,— though she when well did not recollect ‹it› «anything».— if one was subject to this disease oneself, one would only feel sympathy. as for for the heard suffering of a dear friend— this gives one strong idea of what individuality is.— Insanity is ‹much› «somewhat» the same as double consciousness, as shown in the tendency to forget the insane idea; & ones expression of

117 double self, though as in Dr Ashe's case, one here was conscious of the two states.—
August 30th.— It is singular when looking at a table one has vague idea something is not there, & then when one begins eating one perceives butter or salt is not there.— the reality does not resemble the picture in one mind, but does not stop to *reason* what there should be & discover loss

---

Definition of happiness the number of pleasant ideas passing through mind in given time.— intensity to degree of ‹happi› pleasure of such thoughts

117 reality does] 'does' *over* 'is'.

114–2 Abercrombie 1838:296–98.
115–1 Herbert Mayo 1838:142–44, quotes Abercrombie 1838 discussion a botanist with somnambulism.
116–1 Herbert Mayo 1838:145–46: case of double consciousness.

118 We give no credit to instinctive feelings.— for man losing his children, any more than to dog losing his puppies— This looks like free will.—

V. last page. A healthy child is «more» entirely happy (contentmt is different it refers to *wishes* for future) than perhaps well «regulated» philosopher— yet the philosopher has a much more intense happiness— so is it ‹with an› when same man is compared to peasant.— To make greatest number of pleasant thoughts, he must have contingency of good food, no pain,— ‹but the› «&» the sensual

119 enjoyment of the minute add to the happiness.— but as they are not recollected whether from frequency, or inherent structure of mind. they make, either in themselves, or if recollected, such part of thoughts innumerable, which past through mind.— These thoughts are most pleasant. when the conscience tells our [mind], good has been done— «& conscience free from offence» — pleasure of intellect affection excited, pleasure of imagination— therefore do these & be happy— & these pleasures are so very great, that every one who has tasted them, will think

120 the sum total of happiness greater. even if mixed with some pain.— than the happiness of a peasant, with whom sensual enjoyments of the minute make large ‹parts› portion of daily ‹happiness› «pleasure». A wise man will try to obtain this happiness. though he sees some «intellectual» good men, from insanity &c unhappy— perhaps not so much as they appear & perhaps partly their fault.— Whether this rule of

121 happiness agrees with that of New Testament is other question.— little is there said of intellectual ‹ple [. . .] hope› cultivation, main source of the intense happiness.— it is again another question, whether this happiness is the object of living.— or whether if we obey literally New Testament future life is almost the sole object—. — I doubt whether the last be right. The two rules come very near each other.→

122 The rules to mortify yourself do not tend to this— though *believing* it to be true, & *then acting* on it, will add to happiness.—

Men having some instincts as revenge «& anger», which experience shows it must for his happiness to check— that is external circumstances are so conditioned as they are effecting a change in his instincts— like what is happening with other animals— is far from odd

123 nor is it odd he should have had them.— with lesser intellect they might be necessary & no doubt were preservative, & are now, like all other structures slowly vanishing— the mind of man is no more perfect, than instincts of

121 ple [. . .]] *probably* 'pleasu'.
123 Our descent] 'Our' *over* 'their'.

animals to all & changing contingencies, or bodies of either.— Our descent, then, is the origin of our evil passions!!— The Devil under form of Baboon is our grandfather!—

124 A man, who perfectly obeys his conscience or instinct, would probably feel but little that of anger or revenge.— they are incompatible & the former, the more pleasant.—
Simple happiness «as of child» is large proportion of pleasant to unpleasant mental sensations in any given time «— compared to what other people experience.—» But then sensation may be *more* or *less* pleasant & unpleasant, in same time,— therefore degrees of happiness— *Entire happiness*. not being so desirable as ‹broken› *intense*

125 happiness even with some pain,— compared to what others experience in same time.— *Pleasure* more usually refers to the sensations ‹it› when excited by impressions, & not mental or ideal ones, ‹which› «& these» must occupy greater proportion of ‹each› «every» man's time.—

Begin discussion— by saying what is Happiness?— When we look back to happy days, are they not those of which all our *recollections* are pleasant.—

126 Browne Religio Medici, p. 21–24. Curious passages showing how easily chance & will of Deity are confounded.—[1] well applicable to free will.

Mayo. Philosop. of Living p. 293. Animals "have notion of property" — their own property. (—regarding food & in birds of place for nest.)— with dogs "have notion of masters property"—[2] is not this rather more friendship.—
Scott's Life. Vol I, p. 127. Talks of difficulty of his own drawing compared to a friend, whose who family can draw— says friend viewed him as Newfoundland dog would Greyhound about dread of water—[3] *innate*

127 Septemb 1— If one performs some actions, which are pleasant, every concomitant circumstance calls up pleasure. or pleasure or pain of association.— now if one has these feelings, without being aware of their association «ie heredetary», does one not call them instinctive emotions?—

Dr Holland remarked that *insanity* like *sleep*[1] does not doubt the reality of the

127 people] *first* 'e' *over an* 'o'
to me vague—] 'm' *over* 'O'.

126–1 T. Browne 1835, 2:24, 'Surely there are in every man's life certain rubs, doublings and wrenches, which pass a while under the effects of chance; but at the last, well examined, prove the mere hand of God.'
126–2 Herbert Mayo 1837:293, 'Honesty is the recognition of the principle of property. It is remarkable that animals have this idea in its simplest form . . .'
126–3 Walter Scott 1837–38, 1:127, 'He [Will Clerk] to whom, as to all his family, art is a familiar attribute, wondered at me as a Newfoundland dog would at a greyhound which showed fear of the water.'
127–1 Holland 1839a:216, 'Yet if it were an object to obtain a description of insanity, which might apply to the greatest number of cases, I believe this would be found in the conditions which most associate it with dreaming; viz, the loss, partial or complete, of power to distinguish

impression on its senses.— insane people believe they *hear* as well see things which have no existence.— He compared spectral illusion & insanity the connexion appears to me vague—[2]
Delirium of every degree of intensity— «in old man, he had just seen mind went on RAMBLING till excited by question.»

128 Sept. 4th. Lyell in his Principles talks of it as wonderful that Elephants understand contracts.—[1] but W. Fox's dog that shut the door evidently did, for it did with far more alacrity ‹than› when something good was shown him, than when merely ordered to do it.—

Plato «Erasmus»[2] says in Phædo that our "*necessary ideas*" arise from the preexistence of the soul, are not derivable from experience.—[3] read monkeys for preexistence ‹"›—

129 The young Ourang in ‹Zoolog» Gardens *pouts*. partly out displeasure (& partly out of I do not know what when it looked at the glass) when pouting protrudes its lips into point— man, though he does not pout. pushes out both lips in contempt ‹&› disgust & defiance.— different from sneer—

How easily. horses associate sounds may be seen by omnibuss Horses starting, when door shut or cad cries out "right." or Drinkwater's horse jumping when word Jump said—

I saw the ourang. take up a stone & pound the earth.

Lockarts life of W. Scott Vol VII p. 35 "as ideas come & the pulse rises, or as they flag & something like a snow-haze. covers my whole imagination."[1]

between unreal images created within the sensorium, and the actual perceptions drawn from the external senses; thereby giving to the former the semblance and influence of realities:— and, secondly, the alteration or suspension of that faculty of mind by which we arrange and combine the perceptions and thoughts successively coming before us.'

127–2 Holland 1839a:217, "[Spectral illusions], while connected on the one side with dreaming, delerium, and insanity, are related on the other, by a series of gradations, with the most natural and healthy functions of the mind."

128–1 Lyell 1837, 2:418, "Some favourite dainty is shown to them, in the hope of acquiring which the work is done; and so perfectly does the nature of the contract appear to be understood that the breach of it, on the part of the master, is often attended with danger."

128–2 Erasmus Alvey Darwin.

128–3 Plato 1977:27 [76, d–e], 'If those realities . . . exist, the Beautiful and the Good and all that kind of reality, and we refer all our sense perceptions to them, and we discover that it existed before and we had knowledge of it, and we compare our perceptions with it, then, just as they exist, so our soul must exist before we are born. If these realities do not exist, then this argument is altogether futile. Is this the position, that there is an equal necessity for those realities to exist, and for our souls to exist before we were born? If the former do not exist, neither do the latter?'

129–1 Walter Scott 1837–38, 7:35–36, 'May 28 [From diary]— Another day of uninterrupted study; two such would finish the work with a murrain. What shall I have to think of when I lie down at night and awake in the morning? What will be my plague and my pastime— my curse and my blessing— as ideas come and the pulse rises, or as they flag and something like a snow-haze covers my whole imagination? I have my Highland Tales— and then— never mind— sufficient for the day is the evil thereof.'

130 Septembe. 3^{d} Why when one thinks of any object, (or having looked at any object‹)› one Shuts ones eyes) is the image not vivid as in sleep— (one can dream of intense scarlet??) is it because one then has no immediate comparison with perceptions, & that on[e] fancies the image more vivid? Surely the image in a dream cannot truly be ‹more› «as» vivid, «a reality» as in Spectral images—

131e Mem Chiloe ‹pi› Sow, who carried from all parts straw to make its nest.[1] Pigs & Elephants, (both Pachyderms) much intellect.—

---

mem: Yarrell's[2] story of wheel horse in drays, scraping against cornice stone to cause friction

---

Athenaeum 1838. p. 652. Dr Daubeny on the direction of mountain chains in N. America[3]
Fear probably is connected with *habitual* stopping of breath to hear any sound.— attitude of attention «So intimately connected is passion with sending force to muscles, that in my grandfather remark, a tired man. involuntarily feels angry, when brain is pumping force to legs & body, & especially, when to whole body, being failed, & not to any particular muscle

132e Sept. 8th. I am tempted to say that those actions which have been found necessary for long generation, (as friendship to fellow animals in social animals) are those which are good & consequently give pleasure, & not as Paleys rule is those that on long run *will* do good.—[1] alter *will* in all cases to *have* & *origin* as well as *rule* will be given.— **Descent of Man Moral Sense**[2]

Mitchell Australia Vol I, p 292 "Dogs learn sooner to take kangaroos than

131 131] *pen.*
Mem Chiloe . . . cause friction] *pencil.*
Athenaeum . . . particular muscle] *pen.*
Athenaeum . . . N. America] *crossed pen.*
page crossed blue crayon.

132 Sept. 8th.] *lines above and below.*
emu, &] *after '&' partial stroke.*
Descent of Man Moral Sense] *added blue crayon.*
‹Paleys . . . given.—] *scored left margin blue pencil.*
Mitchell . . . emu,&] *crossed blue crayon.*

131–1 No mention of pig's nest in *JR* or *Diary*.
131–2 Yarrell: William Yarrell.
131–3 Daubeny 1838:652.
132–1 Paley 1839:36, '. . . the good of mankind is the subject, the will of God the rule, and everlasting happiness the motive and end of all virtue. Yet . . . a man shall perform many an act of virtue, without having either the good of mankind, the will of God, or everlasting happiness in his thought. . . . Man is a bundle of habits. . . . [and thus is] the influence of this great law of animated nature.'
132–2 *Descent*, chap. III, pp. 70–106.

emu, although *young* dogs get sadly torn in conflicts with the former. But it is one thing for a swift dog to overtake an emu, &[3]

133e–134e [not located]

135 notion, are not effects of impressions long repeated, without the powers of the mind being EQUAL to the smallest casuistical doubts.— The history of Metaphysicks shows that such a view cannot be, anyhow, easily overturned.— so ready is change from. our idea of causation, to give a cause (& no one being apparent, one fixes on imaginary beings, many vicarious, like ourselves) that savages (mem York Minster) consider the thunder & lightning the direct will of the God (‹thus› & hence arises the *theological* age of science in every nation according to M. le Comte).—[1] Those savages who thus

136 argue, make the same mistake, more apparent however to us, as does that philosopher who says the innate knowledge of creator ‹is› «has been» implanted in us (‹by› ? individually or in race?) by a separate act of God, & not as a necessary integrant part of his most magnificent laws. of which we profane «degnen» in thinking not capable to ‹do› produce every effect, of every kind which surrounds us. Moreover «it would be difficult to prove that» this innate idea of God in civilized nations has not been improved by culture « ‹was› who feel the most implicit faith that through the goodness of God knowledge has been communicat to us».— & that it does exist in different degrees in races.— whether in Ancient Greeks,

137 with their mystical but sublime views, or the wretched fears & strange superstitions of an Australian savage or one of Tierra del Fuego.—

Mr Miller (superintendent of the Zoological Gardens)[1] remarked that ‹exp› the expression & noises of monkeys go in groups. thus the pig-tailed baboon, shoved out its lip, looking absurdly sulky «as» often as keeper spoke to it,— but he thinks not sulkiness— this expression he believes is common to that group.— this is very important as showing ‹connection› that expression mean SOMETHING.—

138 Hunt (the intelligent Keeper) remarked that he had never seen any of the *American Monkey* show any desire for women— «very curious. as they depart in structure» **The monkeys understand the affinities of man, better than the boasted**

136 «degnen»] *alternate reading* 'deign'.
138 **The monkeys . . . philosopher himself**] *added brown ink.*

132–3 T. L. Mitchell 1838, 1:292, [Darwin's quotation, which is correct, continues] '. . . another thing to kill, or even seize it. Our dogs were only now learning to seize emus, although they had chased and overtaken many.'

135–1 Brewster 1838:280, '. . . each branch of knowledge, passes successively through three different theoretical states— the theological or fictitious state, the metaphysical or abstract state, and the scientific or positive state.'

137–1 Miller: Alexander Miller.

philosopher himself it is chiefly shown in old male.— A very green monkey (from Senegal he thinks Callitrix Sebe??) he has seen place its head downwards to look up womens petticoats— just like Jenny with Tommy ourang.— Very curious.—

Mr Yarrell has seen Jenny, when Keeper was away, take her chair & bang against the door to force it open, when she could not succeed of herself.—

139e ‹The male› «I saw» Jenny untying a very difficult knot— the sailor on board the ship could not puzzle her— with aid of teeth & hands.— **Descent 1838** It was very curious to see her take bread from a visitor, & before eating «everytime», look up to «keeper» see whether, this was permitted & eat it.— good case of association.— «Listened with great attention to Harmonicon. & readily put it. when guided to her own mouth.— seemed to relish the smell of Verbena & Pocket Handerchief & liked the taste of Peppermint.—» Perfect understand voice.— will do anything.— will take & give food to Tommy, or anything of any sort.— I saw Tommy picking his

140e nose with «a» straw.— Jenny will often do a thing, which she had been told not to do.— when she thinks keeper will not see her.— ‹but is› then knows she has done wrong & will hide herself.— I do not know whether fear or shame.— When she thinks she is going to be whipped. will cover herself with straw, or with a blanket.— these cases of commonly using, foreign bodies, for end. most important step in progression.—

141 The male Black Swan is very fierce when female is sitting the Keeper is obliged to go in with a stick, if he drops it, the bird will fly at him— Knowledge.—

Sept. 13th It will be good to give Abercrombie's definition of "reason" & "reasoning,"[1] & take instance of Dray Horse going down hill.— (argue sophism of association. Kenyon,[2] & then go on to show, that if Cart horse argued from this into a theory of friction & gravity. it would be discoverer

139 ‹the Male› . . . & hands.—] *scored left margin blue crayon.*
Descent 1838] *added blue crayon.*
140 going to be whipped . . . in progression.—] *double scored blue crayon.*

141–1 Abercrombie 1838:178–79, 'The process of mind which we call reason or judgment, therefore, seems to be essentially the same, whether it be applied to the investigation of truth or the affairs of common life. In both cases, it consists in comparing and weighing facts, considerations, and motives, and deducing from them conclusions, both as principles of belief, and rules of conduct.' Darwin's marginal notation: 'Perhaps mathematical reasoning does not.— each step then does not require the memory & knowledge of all contingencies,— it is merely to find the step, & then to pursue this deep train.' Abercrombie continued: 'In doing so, a man of sound judgment proceeds with caution, and with due consideration of all the facts which he ought to take into the inquiry.' Darwin's marginal notation: 'requires properly arranged memory'.

141–2 Kenyon 1838:61, 'Due honour to the stout-built Man of Prose!/ Reasoner on facts! who scorns to feel, but knows!/ Yet it be mine, who love not less the true,/ To lead, well-feigning bards! my hours with you;/ And sick, long since, of facts that falsify,/ And reasonings, that logically lie,/ With you live o'er my wisely-credulous youth,/ And in your fictions find life's only truth.'

"reasoning" or "reasoning"— only rather more steps.— dispute about words.—

142 Miss Martineau (How to Observe p. 213) says charity is found everywhere[1] (is it not present with all associated animals?) I doubted it in Fuegians, till I remembered Bynoes story of the women.—[2] The Chillingham cattle (& Porpoises) have not charity— is it in former case instinct to destroy contagious disease.—[3] (Useful to use term instinct, when origin of heredetary habit cannot be traced)

V.D. p. 111, case of Association.[4]

Sept. 16th Zoological Gardens— Endeavoured to classify expressions of monkeys— I could only perceive that the American ones, often put on a peevish expression, but not nearly so often ‹that› hardly ever the expression

143 of passion with open mouths like the old world ones.— Though the[y] move whole skin of head they do not move eyebrows.— (I see some of the old world ones move skin of head & ears,— ∴ *some* men have this power *abortive* muscles) The black Spider Monkey, very different disposition from others, slow cautious, angry cross look, followed by protrusion of lips, in which respect resembles some of the old ones.— — S. American group sneer.—

Sept 21st Was witty in a dream in a confused manner. thought that a person was hung & came to life, & then made many jokes. about not having run away &c having faced death like a hero, & then I had some confused idea of showing scar behind (.instead of front) (having changed hanging into his head cut off) as kind of wit, showing he had

144 honourable wounds.— all this was kind of wit.— I changed I believe from hanging to head cut off. «there was the feeling of banter & joking» because the whole train of Dr Monro experiment about hanging came before me showing impossibility of person recovering from hanging on account of blood.

143 wit showing he had] 'he' *over* 'Ive'.

142–1 Martineau 1838:213, 'When [the traveller] has ascertained the conditions under which the national character is forming . . . he will proceed to observe the facts which indicate progress or the reverse. . . . The most obvious of these facts is the character of charity. Charity is everywhere. The human heart is always tender, always touched by visible suffering, under one form or another. The form which this charity takes is the great question.'

142–2 Darwin seems to be referring to the following which later appeared in *Origin*, p. 36, 'We see the value set on animals even by the barbarians of Tierra del Fuego, by their killing and devouring their old women, in times of dearth, as of less value than their dogs.' But this belief was later denied. *Correspondence* 1:306.

142–3 Hindmarsh 1839a:280, ". . . when any one [of the cattle] happens to be wounded or has become weak and feeble through age or sickness, the rest of the herd set upon it and gore it to death. This characteristic is an additional and strong proof of their native wildness.' See also *Variation* 1868, 1:83–86; 2:19, and Whitehead, 1953:50, "It would seem, however, that today the cattle show more feeling towards other members of the herd."

142–4 See D 111.

but all these idea came one after other, without ever comparing them, I neither doubted them or *believed* them.— Believing consists in the comparison of ideas, connected with judgment.
[What is the Philosophy of Shame & Blushing[1]]CD
«Does Elephant know shame— dog knows triumph.—»
Sept. 23rd. Horses in Omnibus *instantly* start when they hear ready, but if they see anything ahead. which cad cannot see, they do not move muscle.— reason

145e CD[The laughing noise which C. Sphynx made at Z. Gardens may be described as partaking of ‹st.› made by ‹ret› inspiration & quickly retracting tongue from behind upper & little between incisors.— like ‹W[. . .] what› person says "what a pity"—
Lavater's Essays on Physiognomy translated by Holcroft «Vol I» .p. 86 "We ought never to forget— —; that every man is born with a portion of phsiognominical sensation, as certainly as every man who is not deformed. is born with two eyes. ."[1] I think this cannot be disputed anymore in men. than in animals.—

———

In the drawings of Voltaire why is under lip curled over upper with mouth shut. expressing *cool* irony, *not biting*?

146e What is **Emotion** analysis of expression of desire— is there not protrusion of chin, like bulls & horses.— **1838** good instance of useless muscular tricks accompanying emotion.— when horses fighting, they put down ears, when «turning round to kick» kicking they do the same.[1] although it is then quite useless— **Cats kneeding when old, like kittens at the breast** now if horns were to grow on horses, they must yet continue to put down ears, when kicking.— — good case of expression showing real affinity in face of donkey, horse & zebra. when going to kick.— Why does dog put down ears, when pleased.— is it opposite movement to drawing them close on head, when going to fight, in which case expression resembles a

145 ‹W[. . .]] 4–5 *illeg letters.*
page crossed blue crayon.
146 **Emotion** *added blue pencil.*
1838] *added blue crayon.*
Cats . . . breast] *added pencil.*
donkey, horse & zebra.] *underlined blue crayon.*

144–1 See C 265 and Burgess 1839.
145–1 J. C. Lavater 1804, 1:86, 'We ought never to forget that the very purport of outward expression is to teach what passes in the mind, and that to deprive man of this source of knowledge were to reduce him to utter ignorance; that every man is born with a certain portion of physiognomonical sensation, as certainly as that every man, who is not deformed, is born with two eyes; that all men, in their intercourse with each other, form physiognomonical decisions, according as their judgment is more or less clear . . .'
146–1 Erasmus Darwin 1794, 1:152, '. . . the horse, as he fights by striking with his hinder feet, turns his heels to his foe, and bends back his ears, to listen out the place of his adversary, that the threatened blow may not be ineffectual.' In his personal copy beside this statement, Darwin wrote, 'Sir C. Bell says because he looks back'.

147 fox— I can conceive the opposite muscles would act, to when in a passion.—[1] dog tail curled when angry & very stiff. back arched. just contrary. when pleased tail loose & wagging— if as (I believe) Hunter says. neither fox. nor wolf wag their tails,[2] &c. it is very curious, recurrence of pleasure so teaching expression «as constant smiles, cheerful face».— Man when at ease has smooth brow contrary to wrinkled: (a horse when winnowing & pleased pricks his ears?—).— How is expression of anger in species of swans, in parrots &c &c — — peacock & turkey cock in passion.— Cat when pleased, erect its tail & make it very stiff «& back» when savage «no» & ready to dash at prey streched out & flaccid, when furious «with fright» back absurdly arched. & tail stiff.—

148 is shame, jealousy, envy all primitive feelings, no more to be analysed than fear or anger? I should think shame would be more easily analysed than jealousy, because less discoverable in animals than latter.— Yet I think one can remonstrate with a dog, & make him ashamed of himself, in manner quite different from *fear*; there is no inclination to jump away,— it is, ill-defined fear.— Yet one knows oneself it is quite different from that.— like «slight» passion from blood rushing in face, with less action of the heart.—

149 tendency to muscular movement, hence shy people (shame of ridicule) are singularly apt to catch tricks.— so are people in passion my F. rubbing hands.— stamping. grinding teeth.— in shame frowning, & anguish,— shyness not so.— affected laughter.—
A dog who goes home from shooting. runs away. is not afraid the whole way. but ashamed of himself.—
Jealousy probably originally entirely sexual; first try «to» attract female, (or object of attachment) & then failing to drive away rival.—

Fear is open mouthed to *hear*.[1] though in individual case. nothing can be heard.—

150 Shame would never make person tremble, like fear.—

Why does any great mental affection make body tremble. Why much laughter tears.— & shaking body.—

Are those parts of body, as heart, & chest (sobbing) which are most under

149 Jealousy probably . . . away rival.—] *brace left margin.*

147–1 See Erasmus Darwin 1794, 1:430, for a discussion of antagonistic muscles.
147–2 Hunter 1837:323, '[Mr. Gough's wolf-bitch] had all the actions of a dog . . . raising the tail in anger or love, depressing it in fear, and moving it laterally in friendship . . .'

149–1 Erasmus Darwin 1794, 1:153, '. . . when we hear any the smallest sound, that we cannot immediately account for, our fears are alarmed, we suspend our steps, hold every muscle still, open our mouths a little, erect our ears, and listen to gain further information . . .'

great sympathetic nerve. most subject to habit, as being less so will.—

May not moral sense arise from our enlarged capacity ‹acting› «yet being obscurely guided» or strong instinctive sexual, parental & social instincts, giving rise "do unto others as yourself". "love thy neighbour as thyself". Analyse this out.— bearing

151 in mind many new relations from language.— the social instinct more than mere love.— fear for others acting in unison.— active assistance. &c &c. it comes to Miss Martineaus one principle of charity.—[1] ¶ May not idea of God arise from our confused idea of "ought." joined with necessary notion of "causation", in reference to this "ought," as well as the works of the whole world.— Read Mackintosh on Moral sense & emotions.—[2]

The whole argument of expression more than any other point of structure takes its value. from its connexion with mind, (to show hiatus in mind not saltus between man & Brutes) no one can doubt this connexion.— look at faces of people in different trades &c &c &c

152 I observed the Asiatic Leopard. *quarrelling*. mouth wide open, each [lip] drawn back & driving air out of mouth «hairs erect on back» «wide open» with prodigious force.— making growling, guggling noise. Puma did same & & some others— Thus ‹sudden› «forcible prolonged» expulsion of air «dogs snarl much the same way» generic manifestation of great passion.— I do not think they arch their backs— Bengal tiger. when slightly angry. curls tip of tail.— do *two* cats arch their back when *fighting*, & not with dog. when fear might enter?—

I believe common Swan, arch raises neck & depresses chin— strikes with

153e wing arches wings— as does black Swan.— Goose do all species put their necks straight out & hiss.— [Hyæna pisses from fear so does man.— & so dog]:[CD1] Man grins & stamps with passion. can expression be used more correctly than this for C. Sphynx.—

In the wild ass there is a curious drawing out of the side part of nostril, when passion commences.—

‹All› Nearly all will exclaim, your arguments are good but look at the immense difference. between man,— forget the use of language, & judge only

152 «hairs erect on back»] *circled.*

151–1 Martineau 1838:213. See M 142.

151–2 Mackintosh 1837:262, 'The words *Duty* and *Virtue*, and the word *Ought*, which most perfectly denotes *Duty*, but is also connected with *Virtue* . . . become the fit language of the acquired, perhaps, but universally and necessarily acquired, faculty of Conscience.'

153–1 Erasmus Darwin 1794, 1:148, '. . . the passion of fear produces a cold and pale skin, with tremblings, quick respiration, and an evacuation of the bladder and bowels . . .'

by what you see. compare, the Fuegian & Ourang & outang, & dare to say difference so great . . . "Ay Sir there is much in analogy, we never find out."

154 This unwillingness to consider Creator as governing by laws is probably that as long as we consider each object an act of separate creation. we admire it more. because we can compare it to the standard of our own minds. which ceases to be the case when we consider the formation of laws invoking laws. & giving rise at last even to the perception of a final cause.—

155 Read.

Paper on consciousness in Brutes & Animals. in Blackwood's Magazine June. 1838.[1] **Copied**

Mr H. C. Watson[2] on Geographical distribution of British Plants

---

A Volume published by Colonel in army on "Wheat." in Jersey.— very curious facts about early production of foreign seeds.— many varieties.— Rev R. Jones has it.— very curious book.—[3]

Hume's essay on the Human Understanding well worth reading[4] **Copied**

‹Smith› «D. Stewart» lives of Adam Smith Reid,[5] &c worth reading. as giving abstract of Smith's views

156 «Take & pound up inflorescent parts of mosses & see if Hybrid can be made & ferns.—»
Would a sensitive plant if irritated very regularly at one time every day.— naturally close at that time after long period.—

My Father about double consciousness.— & somnambulism.

---

154 *page crossed pencil.*
155 **copied**] *added.*
copied] *added.*
Read . . . British Plants] *crossed.*
A Volume . . . curious book.—] *crossed.*
Hume's . . . Smith's views] *crossed.*
156 «Take & . . . ferns.—»] *grey ink.*
Would a . . . long period.—] *brown ink.*
My Father . . . & of eyes.—] *grey ink.*
Do the Ourang . . . & of eyes.—] *brace left margin.*
Do they pout . . . of bare nails—] 'base' *alternate reading for* 'bare'.

155-1 Ferrier 1838a:784–91.
155-2 Watson 1835.
155-3 Le Couteur 1836:14, '. . . the only chance of having pure sorts, was to raise them from single grains, or single ears.' See also pp. 64–65, chap. XI, 'On the disposition of wheat to sport'.
155-4 Hume 1750.
155-5 Adam Smith 1795.

Do people when inhaling Nitrous oxide, forget what they did when in this state, or remember what they did in former one.

———

about heredetary tricks & gestures, other cases like D. Corbet; «do» ideots form habits readily??

———

Do the Ourang Outang like smells «peppermint» «& music».— Have monkeys lice?— picture.—
Do female monkeys not show signs of impatience when woman present?
Do they pout, or spit, or cry.— ‹fe› Shame, independent of fear: colour of bare nails—, & of eyes.—
Do female monkeys care for men.—

———

Have we any ferns in the hothouse at home

INSIDE BACK COVER Natural History of Babies—
Do babies start, (ie useless sudden movement of muscle) very early in life Do they wink, when anything placed before their eyes, very young, before experience can have taught them to avoid danger
Do they frown, when they first see it?[1]

Charles Darwin
36 Great Marlborough St

Has my Father ever known ‹intemperance› «disease» in grandchild, when father has not had it. but where grandfather was the cause by his intemperance. **‹No.› Cannot say.—**

Private.

BACK COVER Expression
M

IBC Natural . . . it?] *grey ink.*
Do they wink . . . avoid danger] *brace, left margin.*
Charles St] *brown ink, upside down.*
Has . . . intemperance.] *grey ink, brace left margin.*
‹No.› . . . say.—] *added, ink colour uncertain.*
Private.] *brown ink, boxed.*

BC Expression] *black ink.*

IBC–1 See Darwin 1877 for observations and theoretical discussions of early expressions of Darwin's first child, William Erasmus, born 27 December 1839.

Notebook N

Introduction by Sandra Herbert & Paul H. Barrett

Transcribed and edited by Paul H. Barrett

Notebook N (DAR 126), is the second notebook on 'metaphysical enquiries' and expression. The notebook (167 × 98 mm) is bound in red-rust-coloured leather with a metal clasp. On the front and back covers are labels of cream-coloured paper bearing in ink 'N', and 'Expression' written in ink on the covers. There are 92 leaves (184 pages) numbered consecutively, and Darwin excised six leaves of which all but one partly excised leaf have been recovered. Entries in the notebook are in ink, pencil, and blue crayon. The grey ink material runs from the inside front cover to an 'Octob— 19th' entry half-way down page 18. The opening entry on the first page of the notebook is dated 'October 2d. 1838'; the bulk of the notebook was completed by mid-1839 (page 97 is dated 20 July), with later entries being made sporadically. The last dated entry in the body of the text is 28 April 1840 on page 121.

Notebook N is the sequel to Notebook M, as the names signify. In physical appearance Notebook N is nearly identical to E, as Notebook M has its counterpart in D. When Notebooks D and M were filled on 2 October 1838, Notebooks E and N immediately took their place. There are other similarities between Notebooks M and N. Both bear the inscriptions 'Expression' and 'Private'; both reveal details of the personal lives of the author and his family; and both are little excised compared to other alphabetically lettered notebooks. Also, Darwin clearly carried his programme of reading over from one notebook to the next. Thus the reading list on C270 used for Notebook M ends with references to works mentioned at the beginning of Notebook N, as for example, Charles Waterton's 1838 *Essays on Natural History*.

With the exception of a few stray remarks, the entries in the notebooks fall into five categories: (1) the spectrum of human activity ranging from thought and emotion through expression and behaviour; (2) habit, instinct, and heredity (heredity belonging in this group by virtue of Darwin's belief that habits and structural changes due to habits could be inherited); (3) evolutionary origins, including questions regarding the origin of language, reason, conscience, religious belief, taste, evil passions, and chastity in women; (4) continuities between humans and other animals; and (5) epistemology. Within these categories are subordinate questions that receive greater attention in one or the other of the notebooks. Thus the subject of insanity, on which Darwin and his father spoke at length in July 1838, is abundantly represented in Notebook M but not at all in Notebook N. In other areas, however, and notably on the subjects of expression, habit, and instinct, remarks are divided more evenly between the two notebooks.

As a sequel to Notebook M, N shows a gradual decline in attention to metaphysical topics. Both were kept in a period of increasing excitement, M as accumulating data supported transmutation and N as causal agencies of transmutation became more revealed. The high spirits and hard work of the late summer carried over through the month of October and into the late months of 1838 when Darwin's thoughts turned to marriage. After that Darwin's use of Notebook N fell off sharply. In content, the notebook also conveys a sense of closure. There are fewer exclamatory and self-reflective remarks in Notebook N, and more interest in providing definitions and refining terms (see, for example, N57, N76–80, and N87). With the continuity between Notebooks M and N, however, there is one noticeable insight in

Notebook N not previously so succinctly formulated. In an arresting passage on N42–43 Darwin broached the question of the relative influence of 'habitual action' versus 'chance' in determining inheritance.

Notebook N like the other notebooks provides insight into Darwin's processes of thought, his associations, the range of his reading, and his intentions. Of his associations, it is interesting to observe that Emma replaced Robert Waring Darwin as an intimate presence in Notebook N (Darwin married Emma Wedgwood 29 January 1839). She figured as the presumed object of speculation in an entry on page 59 and she herself entered a notation on page 113. Darwin's range of reading is also particularly interesting in Notebook N, including as it does Malthus, whose book he read at this time for its contribution to ideas of the origin of morals, and Hume, whose views on the natural history of religion were sympathetic to his own.

As to Darwin's later use of the notebook, the short title 'Expression' on the covers of the notebook announces the connection between this notebook and his book on the same subject published in 1872. Darwin also noted on the inside of the front cover of the notebook that he 'Selected' from it for his species theory in December 1856 and looked through the notebook once again in May 1873, probably for clues to revisions of *Descent* and *The Expression of the Emotions in Man and Animals.*

FRONT COVER

Expression
N

INSIDE FRONT COVER

What are sexual difference in monkeys.—
Charles Darwin
[Private.][CD]
(Metaphysics & Expression)
Selected «for Species Theory»
Dec. 16 1856

Looked through & all other Books May 1873—

1 October 2[d] . . 1838 Essays on Natural History
Waterton describes. pheasant springing from nest & leaving no tracks.—[1] **My Father says pea-hens do** Wood pidgeons building near houses. yet so shy at all other times.— **Birth Hill shows it is evergreens they seek**[2] Cock Pheasant claps his wings *before*? crowing & only in breeding season & on the ground.— Cock fowl. on the ground, at roost, in all seasons, & *after*? he has done ‹g› crowing.—[3] instances of *expression.*—
Octob. 3[d]. Dog obeying instinct of running hare is stopped by fleas, also by greater temptation as bitch: or dogs

2 defending companion. (mem Cyanocephalus. Sphynx howling when I struck the Keeper) may be tempted to attack him from jealousy. (Pincher & Nina)—[1] or to take away food &c &c— Now if dogs mind were so framed that he constantly compared his impressions, & wished he had done so & so for his interest, & found he disobeyed a wish which was part of his system, & constant, for a wish which was only short & might otherwise have been

IFC What . . . [Private.]] *grey ink.*
(**Metaphysics . . . 1856**] *added pencil.*
Looked . . . 1873—] *added blue crayon.*
1 **My Father says pea-hens do**] *added pencil.*
Birth Hill . . . they seek] *added pencil.*
Cock fowl.] 'f' *over* 't'.

1–1 Waterton 1833:212, '. . . in the wild state . . . the pheasant first covers her eggs, and then takes the wing directly, without running from the nest . . . By this instinctive precaution . . . there is neither scent produced, nor track made. . .'

1–2 Birth Hill: Berth Hill, 8½ miles north-west of Shrewsbury.

1–3 Waterton 1838:292, 'By the way, though the pheasant will unite with our barn-door fowl, and produce a progeny, still there is a wonderful difference in the habits of these two birds. The pheasant crows *before* it shakes or claps its wings; the barn-door fowl, *after*. The pheasant never claps or shakes its wings except in the breeding season, and when it is on the ground; but the barn-door fowl will clap its wings, either on the ground or on the roost, at all times of the year.'

2–1 Pincher and Nina: pet dogs.

relieved, he would be sorry or have a troubled conscience.— Therefore I say grant reason to

3 any animal with social & sexual instinct «& yet with passion» he *must* have conscience— this is capital view.— Dogs conscience would not have been same with mans because original instincts different.—Mem. Bee how different instinct a solitary animal still different.— ǂ
Different nations having different moral sense, if it were proved instead of militating against the existence of such an attribute would be rather favourable to it—!!
Man moreover who *reasons* much on his actions, makes his conscience far more sensitive. ulitmate effects of actions.→

4 till at last he face «instinct of» hunger, «of» death & for the satisfaction of following conscience, obeying habits, & dread of misery of future thinking of injured moral sense.—

Notion of deity effect of reason acting on (‹not social instinct›) but a *causation.* & «perhaps» an instinct of conscience, feeling in his heart those rules, which he wills to give his child.—

Octob 3^{d}. Was told by W of Downing. Coll.[1] that he had seen chicken only hatched few hours placed on table & when fly ran past it. cocked its head, & picked it— Here then, that faculty,

5 whether for position of axe of eyes, state of surface, or other means by which eyes, aided by experience is supposed in man to guide to knowledge, was transmitted *perfectly* to chicken so as to seize small moving object like fly.— young partridge can run even with its shell on back.—

To study Metaphysic, as they have always been studied appears to me to be like puzzling at Astronomy without Mechanics.— Experience shows the problem of the mind cannot be solved by attacking the citadel itself.— the mind is function of body.— we must bring some *stable* foundation to argue from.—

6 Octob. 4th. Seeing some drawings «in Lavater, P. cii Vol III» of excessively cross-half furious faces[1] «which may be described as an exaggerated habitual sneer» the manner in which whole skin or muscles are contracted between eyes & upper lip., is most clearly analogous to a panther I saw in garden

3 Different nations . . . favourable to it—!!] *brace, left margin.*
6 Looking . . . as one] ‘as one’ *written in pencil over* ‘as one’.
the] *added pencil.*
noise,]‘,’ *added pencil.*

4–1 Thomas Worsley, Fellow 1824–36. *Correspondence* 1:659. See also N74.

6–1 See Lavater 1804, 3:297–98, Fig. 1, Plate C11, for an illustration and discussion of Darwin’s comments.

uncovering its teeth to bite.— the *senseless* grin of passion, is like the grin of the Hyæna from fear, no actual intention to bite at moment, but mere symbol of readiness, & therefore done in extreme.—
Looking at ones face ‹&› «whilst» laughing in glass. & then as one ceases, or stops **the** noise , the face clearly passes into smile— laugh long prior to talking, hence one can help speaking, but laughing involuntary.—

7 When one fear any bad news, «though in a letter› why is person painted with mouth open.— why when person is listening is mouth open to hear well «as one will perceive if in night trys to listen to growl of hounds». ‹when› as fear to «man as» animals. comes at distance, mouth is placed open.— Hence becomes instinctive to fear., as ears down to horse.— Horse snuffs «& snorts», the air «& raises its head, & pricks its ears» when afraid, though not every time really wishing to smell its enemy.— Man & dogs show triumph (& pride) same way walk erect & stiff, with head up.— Why does suspicion look obliquely.— who can analyse suspicion— yet who does not recognise look of suspicion, even child will do so.— Contempt look obliquely so does dog. when a little one attacks him

8 Contempt, when there is some anger «& respect to opponent» is showed by same movement as sneering,— it is then more ‹emblem› manner of *hurting* opponent by insulting his pride & is therefore of the snarling order.— But contempt mingled with disgust, when ones opponent is considered as quite insignificant, & when pride makes person extremely self-sufficient,— the corner of lower lip are depressed & *opposite* muscles used to when angry sneering is in progress.‹—› the hypothesis of opposite muscles will want much confirmation.

A grave person close those muscles, which wrinkle

9 when smile.— Hope is the expectant eye. looking to distant object, brightened & moistened by emotion,— why does emotion make tears fall?? Lavater says derision lies in wrinkles about the nose, & arrogance in upper lip.[1] ‹The› Children having peculiar expression is remarkable. the *pouting*, & blubbering— *sulkiness* is same as pouting, ‹but› lesser in degree, no smile, no frown showing thought, no compression of mouth showing action,— sulkiness all negative expression? Expression of affection is accompanied by slight protrusion of lips, as if going to say "my dear," just what smile is to laugh.—

10 I must be very cautious. Remember how Lavater ran away with new Lavaters,—[1] Ye Gods!:— says fleshy lips denote sensuality (p 192 Vol. III

10 with new Lavaters,—] 'Lavaters' *uncertain reading.*

9–1 Lavater 1804, 3:21, 'The nose is the seat of derision; its wrinkles contemn. The upper lip when projecting speaks arrogance, threats, and want of shame: the pouting under lip ostentation and folly.'

10–1 'new Lavaters' is short-hand for the 1820 'Nouvelle édition' vis-à-vis the 1804 first edition.

Octav. Edit)—[2] certainly neither a Minerva or Apollo would have them because not beautiful— is there — anything in these absurd ideas.— do they indicate mind & body retrograding to ancestral type of consciousness &c &c.— Lavater. (Holcroft Translat) Vol III. p.37, quotes from Burke, who says on mimicking expression of emotions, he has felt the passions of a face «& mind» sympathetics with internal organs, as action of heart[3]

Malthus on Pop. p. 32, origin of Chastity in women.—[4] rationally explained.— on the wish to support a wife a ruling motive.— Book IV, Chapt I on passions of mankind, as being really useful to them:[5] this must

11 be studied. before my view of origin of evil passions.—

Man getting sight slowly,, but when in grown years, thinking he instinctively knows distances:, is good instance of obtaining ‹that› «a» faculty in the *form* of a true instinct, which is a *real* instinct in the chicken, just bursting from egg.— Animals have necessary notions. which of them? & *curiosity* «strongly shewn in the numerous artifices to take birds & beasts».— very necessary to explain origin of idea of deity.— Animals do not know they have 'these necessary notions any more than «a» Savage

12 M. Le Comte's[1] idea of theological state of science, grand *idea*: as before having analogy to guide one to conclusion that any one fact was connected with law.— as soon as any enquiry commenced, for instance probably such a thing as thunder. would be placed to the will of God. Zoology itself is now

11 «strongly . . . beasts»] *red-brown ink.*
12 grand *idea*: as] 'as' *over* '—'.

10–2 Lavater 1804, 3:192. 'Very fleshy lips must ever have to contend with sensuality and indolence.'
10–3 Lavater 1804, 3:37–38, 'From Burke, on the Sublime and Beautiful. "Campanella had not only made very accurate observations on human faces, but was very expert in mimicking such as were any way remarkable . . . he was able [thereby] to enter into the dispositions and thoughts of people as effectively as if he had been changed into the very men . . . Our minds and bodies are so closely and intimately connected, that one is incapable of pain or pleasure without the other." '
10–4 Malthus 1826, 1:31, 'Women treated in this brutal manner must necessarily be subject to frequent miscarriages, and it is probable that the abuse of very young girls, mentioned above as common [in New Holland], and the too early union of the sexes in general, would tend to prevent the females from being prolific.' P. 32, 'Women obliged by their habits of living to a constant change of place, and compelled to an unremitting drudgery for their husbands, appear to be absolutely incapable of bringing up two or three children early of the same age.'
10–5 Malthus 1826, 2:256, 'Natural and moral evil seem to be the instruments employed by the Deity in admonishing us to avoid any mode of conduct which is not suited to our being, and will consequently injure our happiness. If we are intemperate in eating and drinking, our health is disordered; if we indulge the transports of anger, we seldom fail to commit acts of which we afterwards repent; if we multiply too fast, we die miserably of poverty and contagious diseases.' Pp. 263–64, 'It may be further remarked . . . that the passion is stronger, and its general effects in producing gentleness, kindness, and suavity of manners, are much more powerful, where obstacles are thrown in the way of very early and universal gratification . . . in European countries, where, though the women are not secluded, yet manners have imposed considerable restraints on this gratification, the passion not only rises in force, but in the universality and beneficial tendency of its effects; and has often the greatest influence in the formation and improvement of the character, where it is the least gratified.' P. 264, '. . . much evil flows from the irregular gratification of it [i.e., the passion between sexes]. . . .'
12–1 See M135–1 and Brewster 1838.

purely theological.—

Origin of cause & effect being a necessary notion is it connected with ‹our› the willing of the

13 simplelst animals, as hydra towards light. being direct effect of some law.— have plants any notion of cause & effect, «they have habitual action. which depends on such confidence» when does such notion commence?— Children understand before they can talk, so do many animals.— analogy probably false, may lead to something.—
October. 8th. Jenny[1] was amusing herself—, by getting out ears of corn with her teeth from the straw, & just like child not knowing what to do with them, came several times & opened my hand, & put them in— like child. Tommy's[2] face, now ill, has *expression* of languor & suffering

14 The Cyanocephalus when fondling the keeper., clasping «& rubbed» his arm. & show signs of affecting something like man.

Has an oyster necessary notion of space— plant though it moves doubtless has not.—

Turkey cock in passion & sends blood to its breast &c &c

All Science is reason acting «systematizing» on principles, which even animals practically know «art precedes science— art is experience & observation.—» in balancing a body & an ass knows one side of triangle shorter than two. V. Whewell. Induct. Sciences— Vol I p. 334[1]

15 Does a negress blush.— I am almost sure Fuegia Basket did. **& Jemmy, when Chico plagued him**— Animals I should think would not have any emotion like blush.— when extreme sensation of heat shows blood is pumped over whole body.— is it connected with surprise.— heart beginning to beat— children inherit it ‹ins› like instinct, preeminently so— who can analyse the sensation, when meeting a stranger. who one may like. dislike, or be indifferent about, yet feel shy.— not if quite stranger.— or less so.—

16 When learning facts for *induction*. one is obliged carefully to separate its memory from all ordinary lines of association.— is totally distinct from

15 **& Jemmy, . . . him**] *added pencil. Alternate reading*, 'Jemny'.
quite stranger.] 'quite' *alternate reading* 'quiet.'
16 two sides . . . ass has it.—] *brace, left side.*

13–1 Jenny: orang-utan at Zoological Gardens.
13–2 Tommy: orang-utan at Zoological Gardens.
14–1 Whewell 1837, 1:334, 'The Epicureans held, as Proclus informs us, that even asses knew that two sides of a triangle are greater than the third. They may be said to have a practical knowledge of this; but they have not, therefore, a science of geometry. And in like manner among men, if we consider the matter strictly, a practical assumption of a principle does not imply a speculative knowledge of it.'

learning *it by heart*. Do not our necessary notions follow as consequences on *habitual* or instinctive assent to propositions, which are the result of our senses, or our *experience*.— Two sides of a triangle shorter than third. is this necessary notion, ass has it.—

17 When one is «*simply*» habituated «in life time» to any line of action, or thought one feels pain, at not performing it, (either if *prevented*, or *overtempted*.— «animals have shyness with strangers»«as in case of temperance, or real virtue, that is action which experience shows will be for general good, or in case of any fantastic custom» «Probably bashfulness is connected with some disturbed habit» [Thus shepherd dog. has pleasure in following its instinct & pain if held.— if tempted not to follow it, by greater temptation, if *memory of its own emotions*. (which must be intimately united with *reason*) it would feel «subsequent» sorrow, whatever the cause had been]CD— «Also» When one is prevented performing *heredetary habit*, (or moral sense, or instinct,) one fẹels pain, & vice versa pleasure in performing it.—

18 As soon as memory improved. direct effect of improving organization, comparison of sensations would first take place, whether to pursue immediate inclination or some future pleasure.— hence judgment, which is part of reason

---

Octob. 19th. Did our language commence with singing— is this origin of our pleasure in music— do monkeys howl in harmony— frogs chirp in do— union of birds voice & taste for singing with Mammalian structure. «— American monkeys utter pleasant plaintive cry—» The taste of recurring sounds in Harmony common to t[he] whole kingdom of nature.

19e If I want some good passages against, opposition of divines to progress of knowledge. see Lyell on Scrope, Quarterly Review. 1827?[1]

In Water Scotts life.. Tom Purdie, (beginning of Vol V) «finally» says "he knew no more what was pretty & what ugly than a cow—"[2] «so it is with all uneducated.—»— Old man at Cambridge observed the ignorant. merêly

17]CD]‘]’ *over* ‘)’.
18 As . . . reason] *grey ink*; Octob. . . . nature] *brown ink*.
19 If I want . . . 1827?] *crossed pencil*.
Descent of Man *added pencil*.

19–1 Lyell 1827:440, note, ‘In short, Mr. Scrope's *elastic vehicle* is a counterpart of Lamark's *nervous* fluid, that "subtle and invisible agent," to which he attributes not only muscular motion, but ideas, sentiment and intelligence. (*Philosophie Zoologique*, Part 3, Chap. 2). If in attempting to trace back the phenomena of heat, as well as those of the vital functions, we ultimately reach a point which eludes the gross apprehension of our senses, why not unreservedly avow our utter inability to solve such problems?’ P. 475, ‘But the discoveries of astronomy were most pre-eminently beneficial, not so much from their practical utility, although in this respect their services were of inestimable value, but because they gave the most violent shock to the prejudices and long-received opinions of men of all conditions . . . [Galileo's] sufferings in the cause of truth did not extinguish the spirit of the Inquisition, but gave a death-blow to its power, and set posterity free, at least from all open and avowed opposition, to enlarge the boundaries of the experimental sciences.’

19–2 Lockhart 1837–38, 5:396, ‘Straightway we had an anecdote of Tom Purdie [Scott's] gamekeeper and *factotum*.

looked at picture as works of imitation.— Hence pleasure in the beautiful. (distinct from sexual beauty) is acquired taste.— Whilst music extremely primitive.— almost like tastes of mouth & smell. **Descent of Man**

20e Understanding languages seem simplits case of Association.— Elephant often given food & word open your mouth said, recognizes that sound as perfectly as a man.— Probably, language commenced in some necessary connexion between things & voice, as roaring for lion &c &c. (in same way alphabet. arose from letters, symbol of word beginning with the sound of letter)— crying yawning laughing being necessary sounds . . . not produced by will ‹by› but by corporeal structure.—

Devotional feelings, probably some distant power of the mind— superstition & charity & prayer, or *eloquent* request.

21e Reason in simplets form probably is single comparison by senses of any two objects— they by VIVID power of conception between one or two absent things.— reason probably mere consequence of vividness & multiplicity of things remembered & the associated pleasure &c accompanying such memory.—

A Melody on flute & Epic poem, opposite ends of series or harmonious prose.—

22e Lutké Voyage in Carolinas Vol II p. 132. offered to take a savage, said his wife would be grieved— "il leva les epaules et dit qu'il valait mieux rester a Farroïlap quelque mal qu'on y fût."—[1]
Expression common to Savage & Frenchman, unaccompanied by dignity— "no mon dieu," with a shrug— "all I can say, I am very sorry so it is"— does not accompany *I will not.* I am sorry I *cannot.*— **Expression** **leave «this» out not in Library no good** **There is a Lutké's Voyage autour du Monde (1826–9) Paris. 1835 Quoted repeatedly by Waitz (In Theil. V) in describing Caroline Archipelago.**[2] **D^n 75 cf., p 268**

23 without, however, very sincere grief— "there is nothing more to be said."— "made no reply, but shrugged his shoulders & went away."— he implies

20 necessary sounds . . .] '. . .' *Darwin's dots.*
page crossed pencil.
21 *page crossed pencil.*
22 **Expression**] *added pencil.*
leave . . . no good] *added pencil.*
slip of paper inserted here: **'There is a Lutké's Voyage . . . Archipelago.'**

Tom has been many years with Sir Walter, and . . . has insensibly picked up some of the taste and feeling of a higher order. "When I came here first," said Tom to the factor's wife, "I was little better than a beast, and knew nae mair than a cow what was pretty and what was ugly." '

22–1 Litke 1835–36, 2:132, 'Il répondit que sa femme en aurait du chagrin. Sur mon offre de la prendre avec lui, il leva les épaules, et dit qu'il valait mieux rester à Farroïlap, quelque mal qu'on y fût.'
22–2 Waitz 1859–72, 5, pt 2: 49–51, 100, Lutké mentioned 7 times.

negation, without violence, without assigning or understanding reason.— surprise with negation.— like shaking something off shoulder— or is it from *inspiration*, which accompanies surprise.— & why does one inspire, when surprise, can one resist blow better with body distended.— intolerable to be poked behind, without ones chest, being distended. touch a person on the ribs & how he gulps in air.—

24 Again a master says I will see you damned first." the man shrugs his shoulders & replies nothing. if he did go to reply. he would throw back his shoulders. he wishes to show, he is determined not to say anything. he presses his lips together & shrugs his shoulders & walks off,— I think shrugging connected with many emotions.— (Explanation of sighing is probably correct, to relieve respiration when immensely immersed— mechanic apt to sigh.— & hence carried on as trick) «Shrugging aroused acting»

25 Octob 25.

Why is modesty, mixed with triumphant feeling so similar to shame after asinine.— both accompanied by depending head., & active vessels of skin.— What diffference is there between Squib[1] after having eaten meat on table, & criminal,— who has stolen. neither, or both may be said to have fear, both both have shame— Animals have not modesty. analyse this.— «*Excellent*— my theory of blushing solves this.—»

26 The similarity of men's reasons: shewn by similarity of the earliest arts.— Mem.— Stokes— arrow heads &c &c[1]

October 27th

Consult the VII discourse by Sir J. Reynolds.—[2] Is our idea of beauty, that which we have been most generally accustomed to:— analogous case to my idea of conscience.— deduction from this would be that a mountaineer ‹takes› born out of country yet would love mountains, & a negro, similarly treated would think

27 negress beautiful,— [male glow worm doubtless admires female. showing. no connection with male figure]CD— As forms change, so must idea of beauty.— [Old Græcians living amongst naked figures, & observing powers common to

24 Again . . . emotions.—] *scored and with phrase* 'Shrugging aroused acting' *in left margin.* immensely immersed] 'immensely' *alternate reading* 'immoveably'.

25–1 Squib: pet dog.
26–1 J. L. Stokes. See references to Stoke's collection of sphaerulites and obsidians, *VI*:62,70.
26–2 Reynolds 1798, 2:235–37, '[A study of Italian Masters] will shew how much their principles are founded on reason, and, at the same time, discover the origin of our ideas of beauty . . . To distinguish beauty, then, implies the having seen many individuals of that species . . . a Naturalist before he chose one as a sample [blade of grass] . . . selects as a Painter does, the most beautiful, that is, the most general form of nature.'

savages???].[CD]— The existence of taste in human mind. is to me clear evidence, of the general ideas of our ancestors being impressed on us.— Surely we have taste naturally all has

28 not been acquired by education. else why do some children acquire it soon. & why do all men. agree ultimately?— We acquire many notions unconsciously, without abstracting them & reasoning on them (as *justice*?? as ancients did high forehead sign of exalted character???) Why may not our heredetary nature thus acquire some general notions, which are taste?

29 Real taste in mouth, according to my theory must be acquired, by certain foods being habitual— & hence become heredetary; on same principle we know many tastes become acquired during life time:— the latter correspond to fashions in ideal taste & the former to true taste.—

Everything that is habitual, if heredetary, is pleasant.— Mental & Bodily

---

Consider case of grazing animals knowing poisonous «herbs:» & man not.— ⸮no vegetable good «for man» to eat poisonous?— How did animals in «Australia» & America manage;— This shows doctrine «of instinct» has been carried too far

30 In all the foregoing cases most difficult to distinguish. between prejudices of youth from ‹here› *habits*. & heredetary habits. & perhaps even latter may be vitiated. or rather altered.

The Reason why New Buildings look ugly is because there is some connection between them, & great masses of rock.— I was much struck with this, when viewing Windsor Castle which rises naturally & hence

31 sublimely from natural rise— I was also much struck in great avenue, resemblance to gloomy aisle of Churche.— these are Mayo's ideas.—[1]

In language. the possibility of poets describing gentle things in gentle language, & vice versa.— almost proves that at earliest times there must have been intimate connection between *sound* & language.—[2] Chinese. simplest language. Much pantomimic gesture?? which would naturally happen.—

29 Everything . . . Bodily] *double scored left and right margins.*

31–1 H. Mayo 1838:264, 'Nature is beyond art. For nature is divine art. Yet human art may select and combine her elements, and reproduce some of her conceptions . . . Architecture is a fine amplification of two ideas in nature: a developement of the thoughts expressed in Fingal's cave, and in the arched and leafy forest. To learn its powerful influence on the imagination, let anyone visit York Cathedral, for an interior;— or, which is not less deeply moving, view in bright moonlight, at some silent hour, the magnificent elevation of St. Paul's.'

31–2 See H. Wedgwood 1866.

32 Reynolds Works. Vol I, p. 226— "The general idea of showing respect is by making yourself less, but the manner, whether by bowing the body, kneeling, prostration «uncovering body» &c &c is matter of custom."—[1] this all applies to *bodily* weakness & inferiority, but now we carry it on to mental inferiority— when we do not expect any bodily harm— case of habitual action.—

33 L'Institut. 1838. p. 340. Mr Carlyle says that negro certainly has less reasoning powers than Europæan.—[1] Ideots. defective brains.—

Erasmus[2] does not liken term instinct to muscular movement.— say instinctive *actions. senses. notions* &c

---

Octob 30th— Dreamt somebody gave me a book in French I read the first page & pronounced each word distinctly. woke instantly but could not gather general sense of this page.— Now ‹awake› «when awake» I could not picture to myself reading French book quickly, & ‹running› «running» over imaginary words: it appears

34 as if the mind had dwealt on each word separately, neglecting time, & general sense, anymore than connected with general tendency of the dream.— It does not hurt the conscience of a Boy to swear, though reason may tell him not, but it does hurt his conscience, if he has been cowardly, or has injured another bad, vindictive.— or lied &c &c

35 Are the facts (about communication of ideas, &c) of expression lawless, whilst they are the only steady & universal means. recognized— no one can say expression was invented to conceal one's thought.—

Macculloch in his Chapter on the Existence of a Deity has an expression the very same as mine about our origin of a notion of a Deity[1]

---

34 It does not ... &c &c] *brace, left margin.*

32–1 Reynolds 1798, 1:226, '... the general principles of urbanity, politeness, or civility, have been the same in all nations; but the mode in which they are dressed, is continually varying. The general idea of shewing respect is by making yourself less; but the manner, whether by bowing the body, kneeling, prostration, pulling off the upper part of our dress, of taking away the lower,* is a matter of custom. (*Put off thy shoes ... for the place whereon thou standest is holy ground. Exodus, iii, 5.)'

33–1 Carlile 1838:340, 'comme exemple de ce mode vicieux d'investigation, M. Carlile cite un mémoire de M. Tiedemann dans lequel cet anatomiste conclut, d'après la mesure de la dimension de la cavité cranienne dans le Nègre et das l'Européen, que tous deux jouissent des mêmes facultés, tandis qu'on sait parfaitement bien, au moins ceux qui ont eu l'occasion d'observer des enfants nègres et européens éléves ensemble dans une même école, que tant qu'on ne met en jeu que les facultés de la perception, ces deux classes d'enfants font les mêmes progrès, mais qu'aussitôt qu'on cherche à développer les facultés de la réflexion et de la comparaison, comme dans l'enseignement des mathématiques ou des autres sciences inductives, l'infériorité de la race nègre devient aussitôt manifeste.'

33–2 Erasmus Alvey Darwin.

35–1 Macculloch 1837, 1:95, 'To proceed a further step, somewhat more rapidly than metaphysics do, the proof of the existence of a Supreme Creator depends therefore on our belief in a cause, or what has been termed causation.' See N4.

36 We can allow «satellites», planets, suns. universe, nay whole systems of universe ‹of man› to be governed by laws,, but the smallest insect, we wish to be created at once by special act, provided with its instincts its place in nature. its range, its— &c &c:—*must be a special act, or result of laws. yet we placidly* believe the Astronomer, when he tells us satellites &c &c «*The Savage admires not a steam engine*[1], *but a piece*» of coloured glass ‹&admires› is lost in astonishment at the artificer.—» Our faculties are more fitted to recognize the wonderful structure of a beetle than a Universe.—

37 November 20th

Saw the youngest child[1] of H.W.[2] constantly. when refusing food, turn his head first to one side & then to other. & hence rotatory movement negation.— he dropped his head when he meant to eat, hence assertion.— but nodding is less strongly marked than negation

---

Marianne.[3] says. that she has constantly observed that very young children. express the greatest surprise at emotions in her countenance— before they can have learnt by experience, that movements of face are more expressive than movements of fingers.— like Kitten with mice.—

38 A person with St Vitus' dance badly, told should have shilling to walk to door without touching table.— cannot avoid it.— curious mixture of voluntary & involuntary movements.—

Person with sore-throat told not swallow spittle. will have involuntary flow & desire to swallow.— tells himself not to turn in bed. will turn in bed.— in case spittle, effect of thought is to make saliva flow, & therefore thinking of subject, even when

39 wishing not to flow— flow it will.—[1]

My father told Miss. C. of the bad conduct of Mrs C. (her brother's wife). & she said nothing but shrugged her shoulders.— analyse this.— Miss C. quite aware & indignant with Mrs C. but had no influence over her.—

---

36 *must be*] *underline crossed.*
37 nodding is] 'is' *over* 'its'.
39 & Spencer] *circled.*

36–1 Davey 1830:211, discusses perplexity of savages in understanding steam engines.
37–1 Must be Ernest Hensleigh Wedgwood, born earlier in 1838.
37–2 H.W.: Hensleigh Wedgwood
37–3 Marianne Parker
39–1 See Erasmus Darwin 1794, 1:419, for similar statements: 'If any one is told not to swallow his saliva for a minute, he soon swallows it contrary to his will, in the common sense of that word. . . . In the same manner if a modest man wishes not to make water, when he is confined with ladies in a coach or an assembly-room; that very act of volition induces the circumstance, which he wishes to avoid, as above explained; insomuch that I once

Hensleigh[2] says. Douglas.[3] «& Spencer»,[4] an old Scotch Poet, has numerous lines. of poetry.— ‹signs› sounds singularly adapted to subject see ‹A› ‖ I think this argument might be used to show language had a beginning, which my theory requires.[5]

40 There probably is some connection between very limited reasoning powers & the fixing of habits,— for instance the Birgos opening a Cocoa nut shell at one end.— Children & old people get into habits.— we probably can hardly form an idea of a mind so limited as Birgos to become absorbed by one end of Cocoa nut.—

41 November 27th.—
Sexual desire makes saliva to flow «yes, *certainly*»— curious association: I have seen Nina licking her chops.— someone has described slovering ‹gum›«teeth»less-jaws. as picture of disgusting lewd old man. ones tendency to kiss, & almost bite, that which one sexually loves is probably connected with flow of saliva, & hence with action of *mouth* & jaws.— Lascivious women. are described as biting: so do stallions always..= No doubt man has great tendency. to exert all senses, when thus stimulated, smell, as Sir. Ch. Bell says, & hearing music. to certain degree sexual.—[1] The association of saliva, is probably due to our distant ancestors having been like *dogs* to bitches.—[2] How comes such an association in man.— it is bare fact, on my theory intelligible

42 An habitual action must some way affect the brain in a manner which can be transmitted.— this is analogous to a blacksmith having children with strong arms.— The other principle of those children. which *chance*? produced with strong arms, outliving the weaker ones, may be applicable to the formation of instincts, independently of habits.— the limits of these two actions either on

43e form or brain very hard to define.— Consider the acquirement of instinct by dogs, would show habit.—

Take the case of Jenner's ‹Hyæna› Jackall.—[1] an animal not destined by

41 desire] 'd' *over* 'p'.
having been like *dogs*] *short line between* 'been' *and* 'like'.
43 *page crossed pencil.*

saw a partial insanity, which might be called voluntary diabetes, which was occasioned by the fear (and consequent aversion) of not being able to make water at all.'
39–2 Hensleigh: Hensleigh Wedgwood.
39–3 Douglas: Gavin Douglas (1474?–1522).
39–4 Spencer: Edmund Spencer (*c.* 1552–99).
39–5 See H. Wedgwood 1866:62, 74–75.
41–1 C. Bell 1824:152, '[Professor Cuvier] conceives that there is a necessary connection . . . between the organs of smell and the sexual propensities.'
41–2 See *Expression*: 119, 'This habit probably originated in the females carefully licking their puppies— the dearest object of their love— for the sake of cleansing them . . . Thus, the habit will have become associated with the emotion of love . . .'
43–1 Hunter 1837, 4:329–30, 'The following account

nature to exist. & carrying «like other hybrids» with ‹the› it the provision for death.— can we deny that brain would be intermediate like rest of body? Can we deny relation of mind & brain. «Do we deny the mind of a greyhound & spaniel. differs from their brains» then can we deny that the grand child dug for mice from some peculiarity of structure of brain.?— is this more wonderful than memory. affected by diseases. &c &c, double consciousness? What other explanation— can we suppose some essence.

44e The facts about crossing races of dogs on their instincts, *most important*, because they obey the same laws, as the crossing of jackall & Fox & wolf & dog.— the only test this is most important: can there be stronger analogy that the tendancy to hybrid greyhound to hunt hares. «& leave the sheep» & jackall to skulk about & hunt mice— **Jenners Jackall** Have we somewhat right to deny identity of instinct.— **Habits import to Bowen**[1]
No one doubts that a cross of bull dogs— increase the courage & staunchness of greyhounds.— bull-dogs being preferred from not having any smell.—

45 27th. November.— Think, whether there is any analogy between grief & pain— certain ideas hurting brain, like a wound hurts body— tears flow from both, as when one burns end of nose with a hot razor.— joy ‹p› a mental pleasure. with pleasure of senses. The shudder of pleasure. from pleasure of music

---

Audubon IV Vol of Ornith. Biog. case of Newfoundland dogs. who will not enter water, till he sees. whether birds badly wounded, or only winged.— fetches two birds out at once.—[1]

46 *Old* People— (Antiquary Vol II. p. 77)[1] remembering things of youth, when new ideas will not enter. is something analogous. to instinct, to the permanence of old heredetary ideas.— being lower faculty than the acquirement of new ideas.—

---

44 **Habit import to Bowen**] *uncertain reading; added pencil, top margin.*
Jenners Jackall] *added pencil.*
identity of instincts.—] *pencilled circle added, viz.*, 'O'.

from Mr. Jenner, of Berkeley, to whom I gave a second remove, viz., three parts dog, is very descriptive of this propensity [i.e., to fall back into original instinctive principles]: "The little jackal-bitch you gave me is grown a fine handsome animal; but she certainly does not possess the understanding of common dogs. She is easily lost when I take her out, and is quite inattentive to a whistle. She is more shy than a dog, and starts frequently when a quick motion is made before her . . . her favourite amusement is hunting the field-mouse, which she catches in a particular manner." '

44–1 Kneeland 1860.

45–1 Audubon 1838,4:8, 'If a duck falls dead, [large dogs of the Newfoundland and water-spaniel mixture] plunge to bring it; but many of them wait to see *how* he falls, and whether he swims . . . and will not make the attempt [of capture of] a bird merely winged. . . . These dogs usually bring one duck at a time out of the water; but a real Newfoundland . . . was seen . . . to swim twenty yards further, and take a second . . .'

46–1 Walter Scott 1829, 6 (vol. 2 of *The Antiquary*):72, '[Old Elspeth] minds naething o' what passes the day— but set her on auld tales, and she can speak like a prent buke [i.e., printed book].'

Walter Scott «(Antiquary)» Vol II p. 126 says seals knit their brows when incensed.—[2]

A Dog may hesitate to jump in to save his masters life,— if he meditated on this, it would be conscience. A man, might not

47 ‹t› do so even to save a friend, or wife.— yet he would ever repent, & wished he had lost his life in doing so.— nor would he regret «having acquired» this sense of right (& Whether wholly instinctive as in the dog, or chiefly habitual as in man), for it added much to the happiness of his life, & the chance, of so dreadful a consequence to each man is small.

Man's intellect is not become superior to that of the Greeks.— (which seems opposed to progressive. developement) on account of dark ages.— «effects of *external* circumstances» Look at Spain now.— man's intellect might well deteriorate. «CD[in my theory there is no absolute tendency to progression, excepting from favourable circumstances!»[1]

---

48 We must believe, that it require a far higher & far more complicated organization to *learn* Greek, that to have it handed down as an instinct.— Instinct is a modification of bodily structure «(connected with locomotion.)» «no, for plants have instincts» «either» to obtain a certain en‹s›«d»: & intellect is a modification of ‹intellect› «instinct»— an unfolding & generalizing of the means by which an instinct is transmitted.—

49 Arguing from man to animals is philosophical. viz. man is not a cause like a deity, as M. Cousin says.[1] because if so ourang outang.— oyster & zoophyte:

47 CD[in my theory . . . circumstances!»] *written across right margin and upside-down across top of page.* circumstances!] '!' *possible colon.*

48 «no, for plants have instincts»] *circled and connected by line to* '«(connected . . .)»'.

49 man is . . . says.] *parentheses marks around sentence, pencil.*
I presume—] 'I' *over* 'a'.

46–2 Walter Scott 1829, 6 (vol. 2 of *The Antiquary*): 127, 'In truth, the seal, finding her retreat intercepted by the light-footed soldier, confronted him manfully, and having sustained a heavy blow without injury, she knitted her brows, as is the fashion of the animal when incensed, and making use at once of her fore paws and her unwieldy strength, wrenched the weapon out of the assailant's hand, overturned him on the sands, and scuttled away into the sea. . .'

47–1 This statement reflects Darwin's early rejection of Progressionism as presented by Erasmus Darwin 1794, 1:505, '. . . all warm-blooded animals have arisen from one living filament, which THE GREAT FIRST CAUSE endued with animality, with the power of acquiring new parts, attended with new propensities, directed by irritations, sensations, volitions, and associations; and thus possessing the faculty of continuing to improve by its own inherent activity, and of delivering down those improvements by generation to its posterity, world without end!' And on p. 509, 'This idea [viz., THE CAUSE OF CAUSES] is analogous to improving excellence observable in every part of the creation; such as the . . . progressive increase of the wisdom and happiness of its inhabitants. . . .' See also *ML*, 1:41, where Darwin in a letter written 11 January 1844, to J. D. Hooker, said, 'Heaven forfend me from Lamarck nonsense of a "tendency to progression, adaptations from the slow willing of animals," etc.'

49–1 Cousin 1828:26, '. . . toute la différence de notre création à celle de Dieu est la différence générale de Dieu à l'homme, la différence de la cause absolue à une cause relative.'

it is (I presume— see p. 188 of Herschel's Treatise) a "travelling instance" a— "frontier instance".—[2] for it can be shown that the life & will of a conferva is not an antagonist quality to life & mind of man.— & we do not suppose an hydatid to be a cause of itself.— [by my theory no animal. as now existing can be cause of itself.]CD & hence there is great probability against free action.— on my view of free will, no one could discover he had not it.—

50 The memory of Plants, must be association,— a certain round of actions take place every day, & closing of the leaves, comes on from want of stimulus, after certain other actions, & hence becomes associated with them.— The establishment of this principle of Association will help my theory of sensitive Plants[1]

51 Habitual actions, (independent of mind) in the intestinal functions &c &c &c.— bears. the same relation to true memory, that the formation of a hinge «in a bivalve shell» does to reason.—[1] an inflamed membrane from local irritation to passion.—

———

Blushing is intimately concerned with thinking of ones appearance,—does the thought drive blood to surface exposed, face of man, face, neck— «upper» bosom in woman: like erection

52 shyness is certainly very much connected with thinking of oneself.— «blushing» is connected with sexual, because each sex thinks more of what another thinks of him, than of any one of his own sex.— Hence, animals. not being such thinking people. do not blush.— sensitive people apt to blush.—— The power of vivid mental affection, on separate organs most curiously shown in the sudden cures of tooth ache before being drawn,—

53 My father «even» believes that the general talking about any disease tends to give it, as in cancer, showing, effect of mind on individual parts of body.==

49–2 Herschel 1831:188, 'Bacon's "travelling instances" are those in which the *nature* or quality under investigation "travels," or varies in degree; and thus . . . afford an indication of a cause by a gradation of intensity in the effect . . . The travelling instances, as well as what Bacon terms "frontier instances," are cases in which we are enabled to trace that general law which seems to pervade all nature— the law, as it is termed, of continuity. . . . "Natura non agit per saltum."'

50–1 Erasmus Darwin 1794, 1:104, also discusses sensitive plants: 'The divisions of the leaves of the sensitive plant have been accustomed to contract at the same time from the absence of light. . . .' In the margin of page 105 (CD's personal copy), in Charles Darwin's handwriting, is, 'does habit imply having ideas?' In the adjacent text passage is the following: 'And it has been already shewn, that these actions cannot be performed simply from irritation, because cold and darkness are negative quantities, and on that account sensation or volition are implied, and in consequence a sensorium or union of their nerves.'

51–1 Lamarck 1830, 2:328–29, 'Il n'y a donc pas plus de merveille dans *l'industrie* prétendue du fourmi-lion (*myrmeleon formica leo*) qui, ayant préparé un cône de sable mobile, attend qu'une proie entraînée dans le fond de cet entonnoir, par l'éboulement du sable, devienne sa victime; qu'il n'y en a dans la manoeuvre de l'huître qui, pour satisfaire à tous ses besoins, ne fait qu'entr'ouvrir et refermer sa coquille. Tant que leur organisation ne sera pas changée, ils feront toujours l'un et l'autre ce qu'on leur voit faire, et ils ne le feront ni par volonté, ni par raisonnement.'

(if you ‹think› «fear» you shall not have e——n, «or wish extraordinarily to have one» you wont. ==)== No surer way to blush, than particularly to wish not to do so.== How directly personal remark will make any one blush.— Is there not some saying about a person even blushing in the dark— «so modest a person.» A person who blushes in the dark is proverbially a most modest person

54 one carries on, by association, the question, "one will anyone, especially a women think of my face,"? to one moral conduct.— either good or bad. either giving a beggar, & expecting admiration or an act of cowardice, or cheating.— one does not blush before utter stranger,— or habitual friends.— but half & half. **Miss F.A.[1] said to Mrs. B.A. how nice it would be if your son would marry Miss. O.B.— Mrs. B.A. blushed. analyse this:—**

55 Let a person have committed any «concealed» action he should not, & let him be thinking over it with sorrow,— let the possibility of this being discovered by anyone, especiall if it be a person. whose opinion he regards, ‹& see how› feel how the blood gushed into his face,— "as ‹she› «the» thought of *his* knowing «it», suddenly came across her, the blood rushed to her face,"— One blush if one thinks that any one suspects one of having done either good or bad action, it always bear some references to thoughts of other person

56 Decemb. 27th.— Fear loose the sphincter muscles, only on the principle **like does an injury of the spine—** that it paralyzes all muscular action — «in man & animals» Blubbering of a child (different in different ones?) **in the most perfect fainting, sphincters are loosed** is a convulsive action to remove disagreeable impression like true convulsion. (Hence pass into convulsions?)— squeeze out tears. replaced & squeezed out again— as power of mind by habit gets more perfect over voluntary muscles, these

57 convulsive actions— (except in weak people & hysterical people inclined to convulsive actions).— But, the lachyrmal gland is «not» under voluntary power, (or only very little so) & hence by association, there pour out tears, & there is slight convulsive wrinkling of some of the muscles «or *twitching*».— But why does joy & OTHER EMOTION make grown up people cry.— What is emotion?

At end of Burke's essay on the sublime & Beautiful there are some notes.[1] &

54 **Miss F. A. . . . this:—**] *added pencil.*
55 one of] 'of' *over* 'ofa'.
other person] 'other to one' *alternate reading.*
56 only on] 'on' *over another short word, perhaps* 'any'.
like . . . Spine—] *added pencil.*
in the . . . loosed] *added pencil.*

54–1 F.A.: Fanny Allen (1781–1875).
57–1 E. Burke 1823. Darwin's notes written on the back flyleaf and on the inside of the back cover of his copy of Burke are on 'ambition, pride, fame, vanity, arrogance, conceit, sense of beauty, instinct, sublimity, triumph, and pleasure.'

likewise on Wordsworth's dissertation on Poetry.—[2]

58 **The expression of shame-facedness for shyness, having been invented, prove of the difference, which my theory believes in.—** From the manner short-sighted people frown, frowning must have some relation to short-sightedness.— do not short sighted people squinny— when they consider profoundly,— this will be curious if it is so.— frown with grief,? bodily pain? frown *shows the* mind is *intent on one object.*—

With respect to my theory of smile. remember children smile before they laugh.— Has frowning anything to do with ancient movement of ears

59 A man shivers, from fear, sublimity, sexual ardour.— a man cries from grief, joy. & sublimity.

January 6th.—
What passes in a man's mind. when he says he loves a person— do not the features pass before him marked, with the habitual expressemotions, which make us love him, or her.— it is blind feeling, something like sexual feelings— love being an emotion does it regard «is it influenced by» other emotions?

When a man keeps perfect. time in walking, to chronometer, is seen to be muscular movement.

60 The *Blushing* of Camelion & Octopus; strong analogy with my view of blushing— in former irritation on a piece of skin cut off made the blush come.— it is an excitement of surface *under the will*? of the animal.(—

Jan 21. 1839. Herchel's Discourse p. 35. On origin of idea of causation; «succession of night & day does not give notion of cause,»[1] do p. 135.— on the importance of a name,[2] with reference to origin of language

---

58 **The expression . . . believes in.—**] *added pencil, top and right margin.*

57–2 William Wordsworth 1802.
60–1 Herschel 1831:35, 'If every thing were equally regular and periodical, and the succession of events liable to no change *depending on our own will*, it may be doubted whether we should ever think of looking for causes. No one regards the night as the cause of day, or the day of night. They are alternate effects of a common cause, which their regular succession alone gives us no sufficient clue for determining.' Darwin double scored the margin beside this passage in his copy, and the underline (italic) is his.

60–2 Herschel 1831:35–36, 'The imposition of a name on any subject of contemplation, be it a material object, a phenomenon of nature, or a group of facts and relations, looked upon in a peculiar point of view, is an epoch in its history of great importance. It not only enables us readily to refer to it in conversation or in writing, without circumlocution, but, what is of more consequence, it gives it a recognized existence in our own minds, as a matter for separate and peculiar consideration . . . and . . . fits it to perform the office of a connecting link between all the subjects to which such information may refer.'

My father says old people first fail in ideas of time, & perhaps of space— in latter respect he thinks

61 he certainly has observed that some people of very weak intellect (As Miss Clive) have only possessed very loose ideas.—[1] Have children loose ideas of time?— Characteristic of one kind of intellect is that when an idea once take hold of the mind, no subsequent ones modify it.— «Weak people say I *know* it because I was always told so in childhood.— hence the belief in the many strange religions.»

Emma W.[2] says that when in playing by memory. she does not think at all, whether she can or can not play the piece, she plays ‹f› better than when she tries is not this precisely the same, as the double-conscious kept playing so well.—[3]

62 L^r. Brougham «Dissert.» on subject of science connected with Nat. Theology.— says animals have abstraction because they understand signs.— very profound.— concludes that difference of intellect between animals & men only in Kind.— probably very important work.—[1]

Feb. 12. 1839. Sir. H. Davy — Consolats: "the recollections of the infant likewise before two years are soon lost; yet many of the habits acquired in that age are retained through life" p. 200.—[2] "The desire of glory, immortal fame, &c so common in the young are symptoms of the infinite & progressive nature of intellect indication of better life p. 207[3]

61 idea once] *alternate reading* 'idea ones'.

61–1 Miss Clive: Probably related to the family of Lord Robert Clive (1725–74), of whom there is a large statue in Shrewsbury. A Miss Clive is mentioned in a letter dated 15 December 1824, written by Mrs Josiah Wedgwood to her sister Fanny Allen: 'I went as chaperon to the Drayton Assembly with Miss Clive, Susan Darwin, Charlotte and Fanny, with Joe and William and Edward Clive, but it was a bad and very thin ball and double the number of ladies to the gentlemen.' Litchfield 1915, 1:163. Lord Clive was born near Market Drayton. See *Correspondence*, 1, for references to many of the Clive families, some of whom were friendly with the Darwins, and who lived at Styche, 2 miles north-west of Market Drayton.

61–2 Emma W.: Emma Wedgwood.

61–3 Playing piano.

62–1 Brougham 1839. 1:196, 'Now connecting the two together [i.e., a particular action with a sign], whatever be the manner in which the sign is made, is Abstraction; but it is more, it is the very kind of Abstraction in which all language has its origin— the connecting the sign with the thing signified; for the sign is purely arbitrary in this case as much as in human language.' In Darwin's personal copy, there is a vertical pencil line in the margin beside the text, and in Darwin's handwriting are the words, 'don't understand'. P. 197, '. . . a rational mind cannot be denied to the animals, however inferior in degree their faculties may be to our own.'

62–2 Davy 1830:199, 'The whole history of intellect is a history of change according to a certain law; and we retain the memory only of those changes which may be useful to us;— the child forgets what happened to it in the womb; the recollections of the infant likewise before two years are soon lost; yet, many of the habits acquired in that age are retained through life.'

62–3 Davy 1830:206, 'The desire of glory, of honour, of immortal fame and of constant knowledge, so usual in young persons of well-constituted minds, cannot I think be other than symptoms of the infinite and progressive nature of intellect— hopes, which as they cannot be gratified here, belong to a frame of mind suited to a nobler state of existence.'

63 March 16th.— Is not that kind of memory. which makes you do a thing properly, even when you cannot remember it. as my father trying to remember the man's Christian name,[1] writing for the surname,, analogous to instinctive memory, & consequently instinctive action.— Sir. J. Sebright. has given the phrase "heredetary habits."[2] very clearly, all I must do is to generalize it, & see whether applicable to all cases.— & analogize it with ordinary *habits that is* my new part of the view.— let the proof of heredetariness in habits. be considered. as grand step if it can be generalize.—

64 The tastes of man, same as in Allied Kingdoms— "*food, sm‹e›ll.* (ourang-outang), *music,* colours we must suppose ‹we› «Pea-hens» admire peacock's tail, as much as we do.— touch apparently. ourang outang very fond of soft, silk-handkerchief— cats & dogs fond of slight tickling sensation.— in savages other tastes few.

———

March 16th. Gardiner's Music of Nature. p. 31. remarks children have no difficulty in expressing their want, pleasure, or pains long before they can speak—

65 or understand— thinks so it must have been in the dawn of civilization— thinks many words, roar, scrape, crack, &c, imitative of the things.—[1] CD[I may put the argument,, that many learned men seem to consider there is good evidence in the structure of language, that it was progressively formed. (—names like sounds)—. Horne Tookes tenses, &c &c —[2] ‹also g› if so & seeing how simple an explanation it offers of radical diversity of tongues.—

66 [Emotions are the heredetary effects on the mind, accompanying certain bodily actions]CD.⸮ but what first caused this bodily action. if the emotion was not first felt?— «without «slight» flush, acceleration of pulse. or rigidity of

66 «He may . . . sensations»] *circled.*

63–1 This incident is written out in greater detail in a note dated 13 January 1838. See OUN31.

63–2 Sebright 1836:15–16, 'No one can suppose that nature has given to these several varieties of the same species such very different instinctive propensities, and that each of these breeds should possess those that are best fitted for the uses to which they are respectively applied. . . It seems more probable that these breeds having been long treated as they now are, and applied to the same uses, should have acquired habits by experience and instruction, which in course of time have become hereditary. . . . I am led to conclude, that by far the greater part of the propensities that are generally supposed to be instinctive, are not implanted in animals by nature, but that they are the result of long experience, acquired and accumulated through many generations, so as in the course of time to assume the character of instinct. . . . How far these observations may apply to the human race I do not pretend to say; I cannot, however, but think that part of what is called national character may, in some degree, be influenced by what I have endeavoured to prove, namely, that acquired habits become hereditary.'

65–1 Gardiner 1832:31, 'Children have no difficulty in expressing their wants, their pleasures, or pains, long before they can speak or understand the meaning of a word.' P. 32, '. . . in all probability the first words that were uttered bore some resemblance to the things described, as the boisterous roar of the sea would call for a boisterous expression.* (*The very word Roar, when forcibly pronounced, carries with it the imitative sound. The same may be said of most of our primitive words, as splash, scrape, crack, crush, and the like.)'

65–2 Tooke 1798:47, 'In English, and in all Languages, there are only two sorts of words which are necessary for the communication of our thoughts . . . 1. Noun, and 2. Verb.'

muscles.— man cannot be said to be *angry*.—» «He may have pain or pleasure these are sensations»

‹Gardner in his work›[1] In the life of Hayd & Mozart. fine music is evidently considered as analogous to glowing conversation of several people.—[2]

Children have an uncommon pleasure in hiding themselves & skulking about in shrubbery. when other people are about: this is analogous to young pigs hiding themselves; & heredetary remains of savages state.—

67 N B. According to my view marrying late, will make average of life longer.— for short-lived constitutions will then be cut off.—

‹Horses› Colts cantering in S. America capital instance of heredetary habit:—[1] there must, however, be a *mental impulse* (though unconscious of it) to move its legs so, as much as in the young salmon to go towards the sea. or down the stream; which it does unconsciously of any end.— N B. There is wide difference, between the means by which an animal performs an instinct, & its impulse to do it.— [the means must be present on any hypothesis whatever][CD] an animal may so far be said to *will* to perform an instinct that it is uncomfortable if it does not do it.—

68 My theory explains how it comes that the heart is the seat of the emotions.— but are not love & hate emotions; what are their characteristics;— they are more truly sensations??. a kind of mental pain & pleasure.—

The Revd. Algernon Wells Lecture on animal instinct. 1834: p. 15. "To act from instinct is to be guided to the performance of a number of prearranged actions, which will bring about a certain result, while the creature performing those actions neither knows nor intends the result they will effect,.—"[1] this not wholly true, for we must grant a bird knows what is about when building its nest; it knows its object **but not result**

69 (first time of building?), but not the means of performing it.— p. 14. There is scarcely a faculty in man not met with in the lower animals.—[1]

68 but are not love] 'are' *over* 'is'.
but not result] *added pencil*; **'result'** *partially boxed*.
69 general aim] 'aim' *uncertain reading*.

66–1 Gardiner 1832.
66–2 Haydn 1817:115, 'A musical composition is a discourse expressed by sounds instead of words.'
67–1 See C163, M101.
68–1 Wells 1834:15. The remainder of the sentence being incompletely quoted by Darwin is, '. . . [they will effect,] nor of course, could plan the arrangement of means with a view to its accomplishment.'
69–1 Wells 1834:13–14, '. . . there is scarcely a faculty of mind or quality of character prevailing among men, but its type or resemblance may be found in some one of the tribes of inferior creatures. Hence, their actions and dispositions have ever furnished the moralist with those striking and instructive fables . . . The industry of the bee and the ant— the cunning of the fox . . .'

hence the general aim of fable, & expression as cunningness of fox, industry of bee &c &c—
p. 15. "instincts act with *unerring* precision".— no[2]
p. 17. Contrast the *invariability* of *instinctive* powers in individuals of the same species with *variability* of *reasoning* power in one species man.—[3] false instinctive pointing varies.—
p. 18. Animals possess strong imitative faculty: pure instinct is not imitative: imitations seems invariably associated with reason:[4] [N B. insects which have never seen their parents offer best cases of instincts].CD all this may be true,, but relation of imitation & reason must be thought of.—[5]

70 p. 19. animals capable of education;[1] (this is again assumed as more allied to reason than instinct.) Mr Wells I can see mentally refers by reason *knowledge gained by reason*: & then these qualities of imitation & education may be used as argument.— for *instinctive knowledge* is not gained by instruction, or imitation.—[2]
p. 20. Animals may be called "creatures of instinct" with some slight dash of reason so mean are called "creatures of reason", more appropriately they would be "*creatures of habit*."— CD[as the bee makes its cells, by means of ordinary senses & muscles, we cannot look at him, as machine to make cell of certain form. (& especially as it adapt its cell to circumstances), it must have impulse to make

71 a cell in certain way, which way its organs are sufficient for hence it must some way be able to measure the cell;[1]

---

70 capable of education;] 'educ' *over* 'idea'.
dash of reason] 'reason' *uncertain reading*.
Animals may be] 'An' *over* 'M'.
dash of reason so mean] *read* 'men' *instead of* 'mean'.

69–2 Wells 1834:15. 'Instinct acts its part with unerring precision, without intelligently knowing what or why it does so . . .' N.B.: Darwin indicates his disagreement with the first part of this sentence by saying, 'no'.
69–3 Wells 1834:16–17, 'Instinct is confined to narrow limits, but within them it never mistakes . . . it is observed, That the processes of reason and contrivance in men are capable of almost endless degrees of imperfection or improvement. . . . But instinct reaches its full perfection at once: and never afterwards receives, or admits of, any improvement. . . . the texture and shape of a bird's-nest, or of the cells and masses of honey-combs, are now what they ever were; and ever will be, without variation of improvement, or degeneracy.'
69–4 Wells 1834:18–19, 'Besides which, many animals possess a strong imitative faculty . . . pure instinct is not an imitative faculty. . . . But imitation seems invariably associated with reason; is one of the most powerful laws by which it acts; and one of the most effectual means of its acquisitions and advancement.'
69–5 Wells 1834:18–19, 'In those processes of instinct which are most difficult and surprising, it is impossible any part of the skill . . . should have been gained by imitation; especially in the case of numerous insect tribes, which ever knew their parents . . .'
70–1 Wells 1834:19, 'Moreover, animals are capable of education: they may be, and often are, taught things that greatly surprise every beholder.'
70–2 Wells 1834:19, 'Now, instinct is neither knowledge gained by instruction, nor a faculty capable of being improved by instruction.'
71–1 Wells 1834:20, '. . . the inferior creatures (inasmuch as they perform by far the greater number of their actions, especially in their wild, native state, by innate, blind instinct) may be properly denominated *creatures of instinct*; although . . . they are not bound down to instinct as their only means of knowledge and action. Just as, on the other hand, man is properly denominated the *creature of reason*

p. 22. instincts & structure always go together:[2] thus woodpecker: but this is not so,, the instincts may vary before the structure does; & hence we get over an apparent anomaly,, for if anyone has taken the Woodpecker as an example fitted for climbing, his arguments partly fall, when a species is found which does not climb[3] CD[.instinct may be divided into migration,— subsidiary to *food* & temperature **molting & breeding** instincts, sexual, social, «subordinate to,» self preservation, (knowledge of enemies). use of muscles, progression.— use of senses.— knowledge of location ducks & turtles running to water,— young crocodile snapping—

72 p. 28. how curious the means of guiding themselves through the air,— waterbirds, the bee to its nest,—[1] cats when carried in confinement,— carrier pidgeons proverbially carried to long distance in dark "it is inspiration."—[2] this is class of so called instincts to which my theory no way applies.— it is the acquirement of a new sense,— bats avoiding strings «in the dark» as well might be called instinct,— migrating to one spot, this is indeed instinct.— Australian man, may be called instinctive: the facts of memory of roads long after once visited by horse & dogs. (even blind horses & dogs) shows it is somewhat analogous

73e to memory.

Shrugging shoulders seems sign of helplessness

E.[1] says she can perceive sigh, commences as soon as painful thought crosses mind, before it can have affected respiration

---

V E. p. 125 Wrong Entry[2]

71 **molting & breeding**] *added pencil.*
73 Entry] *remainder of page excised, not located*

. . . some of his actions are instinctive; performed, especially in infancy. . . .' The expression '*creatures of habit*' does not appear on this page in Well's text.

71–2 Wells 1834:22, 'Distinct notice should be taken of the curiously perfect adaptation of the instincts of animals to their senses and bodily structure; and of both to those scenes or portions of the external world in which it is designed they should dwell. . . .'

71–3 Darwin has reference to the 'Woodpecker of the plains,' *Colaptes campestris*, which he observed on the north bank of the Plata, in Banda Oriental, South America, and which rarely visited trees. See Charles Darwin 1870 (*Collected Papers* 2:161) '. . . I repeatedly saw many specimens living on the open and undulating plains, at the distance of many miles from a tree. I was confirmed in my belief, that these birds do not frequent trees . . .'

72–1 Wells 1834:28–29, 'But to observe a bee, at the distance of a mile or more from its hive, busy among the flowers, without the least anxiety lest it should be lost amidst its mazy flights; and, when loaded, wing its direct way to the hive, without thought, and yet without error, is to us amazing.'

72–2 Wells 1834:29–30, 'No faculty we possess [as the carrier-pigeon] helps us to any analogy by which to enable us to form any notion of such a power. It is intuition— it is inspiration— it is something we do not possess, and cannot conceive of. . . . It is one of those wonders with which the works of God abound . . .'

73–1 Emma Darwin.

73–2 See E125

74e Madagascar Lemur seemed to *like Lavendar Water* «very much» Henslow.

———

N.. Necker has remarks on the means. by which children learn (probably not only experience,[1], but also «by an» instinct‹ing› «which is only present in youth» (Mem. Mr Worsley's story of chicken) to know that which we touch & what [. . .] the same.—(this Hensleigh

75e therefore problem is how we know that thing is same, which touches two parts of our bodies, «or touches one part. very quickly successively.—» [& we know from experiment of crossing fingers, that we only do know that it is one, when applied in peculiar manner.—]^CD
April 3^d. 1839
The Giraffe kicks with front legs & knocks with back of Head, yet never puts down its ear. good to contrast with horses[1], asses, ‹mi› Zebras &c &c.— Here there is kicker but not bite.—

76e Henslow remarks that Chimpanze pouted & whined, when, man went out of room.— all theories of magnetic powe in birds, seeing the sun &c are absolutely useless when applied to birds, which have been carried in hampers. if they have not known the direction in which they STARTED, they cannot return.— Hence I conclude. pidgeon taken little way, *whirled*, & then taken other way— would not find

77 its way back.— ?? this is not instinct, but a faculty, or sense— "We know not how, stonge henge raised, yet not instinct, but if all men placed stones in same position, it would be instinct— instinct is heredetary knowledge of things which might be «possibly» acquired by habit. so bees in building cells, must have some means of measuring cells, which is faculty, they use this faculty instinctively; watchmaker has faculty by his instruments to make toothed wheel. he might by instinct make watch, but he does it by

74 Madagascar . . . Hensleigh] *pencil.*
Hensleigh] *remainder of page excised, not located.*
75 the Giraffe . . . bite] 'C.D.' *added pencil*; 'done p 113' *added ink; crossed pencil.*
April 3^d. . . . bite.—] *excised, crossed.*
76 have been carried . . . find] *excised, crossed blue crayon.*
77 ??] *added pencil.*

74–1 See Saussure 1839,1:4: 'St. Paul tells us that we have two laws within us* (*Romans, vii, 23); and our inward feelings, our experience, our reason, all confirm this declaration. A blind instinct, necessary perhaps to the physical order of things, impels us to seek after pleasure, and thus favours the developement of our faculties. . . .' P. 40: '. . . amongst all these philosophers [astronomers, etc.,] there is not one father who has taken the trouble to note down the progress of his own child.' Perhaps it was due to this suggestion of Mme Necker that Darwin did observe and record the progress of his oldest child from infancy. See 'A biographical sketch of an infant' 1877 (*Collected Papers* 2:191–200). Note also in Abercrombie 1838 (Darwin's copy):163, in Section III, 'Imagination,' Darwin's marginal notation, at pounces & runs after feather, it knows it is not mouse, but does it not use imagination & picture to itself it is.— quote Madam Necker. on playing of children—'

75–1 See M146.

78 reason & experience, or habit.— so bird migrating to certain quarter is instinct, but his knowledge of that quarter,, is faculty, whether by sun, & heavens, or magnetic virtue,— the most probably supposition. with respect to pidgeons, is that they do know from look of Heavens, points of compass, & they do know which way they go; & so return.— «but does not apply to dogs.—» they may do all this instinctively «yes because power varies in breeds,» something of kind oneself knows in walking [one feels inclined to stop at right number of house though one cannot remember it.]CD

79 back, without consciousness & by *habit*, such habit of knowledge of points of compass may be instinctive.

———

it is a test to know how much of the wonder consists in the action being performed or emotion felt in early childhood (before experience or habit) could be formed or afterwards.— child sucking whole wonder instinctive.— carrier pidgeon just as wonderful in old bird as new.— migration, ‹only› «only» more wonderful in young, because can not have been taught, where to go— the act of crossing the

80 ———

sea in dark night & not loosing its direction, equally wonderful in young & old.— These facts point out some essential difference, which clearly ought to be separated— We apply instinct to one part. or another— but (an *instinctus* means *stained in*?). had better refer to to the heredetary part of it,— & *faculty* (faculty «being» always heredetary helps this confusion.—)

81 Hensleigh considers breathing instinctive, certainly heart beating may be considered also such.— heredetary habit, is a part never subject to volition.— like plants going to sleep.— "A bird has the faculty of finding its way, which in certain species is *instinctivedly* «not least by experience» directed to certain quarter"— "An animal has faculty of walking. which in man is learnt by experience is in other is acquired instinctively" So with ‹sight› sight— so a Bee has the faculty of

82 building «regular» cells— [but this faculty ‹may possibly be› «probably is» instinctive, namely the knowledge of size is merely judged by eye, & use of limbs &c, or it result from mere impulse to save wax.]CD which it instinctively exerts in concert with others in building comb— My faculty often will turn out to be instincts, & so in some senses, is sight—CD [The faculties bear so close a relation to the senses, that one feels no more surprise at it & feels no more inclined to ask

83e–86e [not located]

79 could be formed] *written* 'beformed'.
81 Hensleigh . . . to sleep.—] *partially circled.*

87e If dislike, distaste. & disapproval. were not something more than the unfitness of the objects then viewed. to organs adapted to other objects. (as that senna is necessarily disagreeable to organs adapted to like sugar, acid, &c, which may be doubted for possibly even taste of senna. might be acquired. as the Turks have of Rhubarb: again on other hand, it is said people, who like sweet things dislike others.— dogs dislike perfume) I should think, great principle of liking, as simply *heredetary habit.*—

88e ———

A blind man might be born with idea of scarlet, as well as remember it.—

———

Why do children pout & not men— orang-outang & chimpanze. pout.— Former, whines just like a child.

———

Get a Dictionary & make a list of every word, expressing a mental ‹desire› «quality» &c &c

---

89 Mackintosh Ethics
p. 97. on Devotional feeling[1]
p. 103— Abstraction[2]
p. 152. Perception very different from *emotion.*— The former is used with regard to the senses. Reason does not lead to action.—[3]

---

p. 248. Theory of Association. owing to time when entered brain, try contiguity of parts of Brain.— Mackintosh first clearly insisted on assoc of ideas & emotions. rather ideas & bodily actions make the emotions.—[4]

87 dogs dislike perfume)] *parenthesis closure blue crayon.*
88 *page crossed blue crayon.*
89 *text written parallel to spine.*

89–1 Mackintosh 1837:96–97, 'As all devotional feelings have moral qualities for their sole object; as no being can inspire love or reverence otherwise than by those qualities which are naturally amiable or venerable, this doctrine would, if men were consistent, extinguish piety, or in other words, annihilate religion.'

89–2 Mackintosh 1837:102–3, 'The controversy between the Nominalists and Realists, treated by some modern writers as an example of barbarous wrangling, was in truth an anticipation of that modern dispute which still divides metaphysicians, whether the human mind can form general ideas, and whether the words which are supposed to convey such ideas be not general terms, representing only a number of particular perceptions?—questions so far from frivolous, that they deeply concern both the nature of reasoning and the structure of language; on which Hobbes, Berkeley, Hume, Stewart, and Tooke, have followed the Nominalists; and Descartes, Locke, Reid and Kant, have, with various modifications and some inconsistencies, adopted the doctrine of the Realists.'

89–3 Mackintosh 1837:152–53, 'Perception and emotion are states of mind perfectly distinct; and an emotion of pleasure or pain differs much more from a mere perception, than the perceptions of one sense do from those of another . . . Reason, as reason, can never be a motive to action. It is only when we superadd to such a being sensibility, or the capacity of emotion or sentiment (or what in corporeal cases is called sensation), of desire and aversion, that we introduce him into the world of action.'

89–4 Mackintosh 1837:248, 'Both [Berkeley and Hume] agree in referring all the intellectual operations to the *association of ideas*, and in representing that association as reducible to the single law, that ideas which enter the mind at the same time, acquire a tendency to call up each other, which is in direct proportion to the frequency of their having entered together.' Darwin wrote, 'try theory of place in brain' beside this latter statement.

p. 272. Some remarks applicable to my theory of happiness.—[5]

Bell on the Hand

---

p. 191 Says ‹childr› babies have an instinctive fear of falling.—[6] &p. 193. that they perceive the difference on being carried up or downstairs, or dangled up & down— in latter case they struggle their arms.—[7]

90 do. p. 306 "the eyes are rolled upwards during mental agony, & whilst strong emotions of reverence & piety are felt." it appears to me mere consequence of stooping, as sign of humility.—[1]

I suspect very strong argument might be advanced, that animals have reasons, because they have memory.— what use this faculty if not reason.— or does this reasoning apply chiefly to recollection. yet a dog hunting for a bone shows he has recollection.— Lamarck. Phil. Zoolog.— Vol II p. 445. If we compare the judgments & actions of a young animal with an old.— (dog, horse, sow) we perceive great difference.—[2] «(& is not this difference same, but less in degree, as between man & child.—)» what differs— not ‹reason› «instinct», for its character is invariability.— if explained by habits, useful to itself, how gained. reason? or some unnamed faculty—

91 Lamarck. Philosop. Zoolog. «p. 284. Vol. II» — gives explanation & instance of starting identical with mine,—[1]

---

90 do. p. 306 . . . humility.—] *written parallel to spine.*

89–5 Mackintosh 1837:272, '. . . without pre-supposing Desires, the word Pleasure would have no signification; and that the representations by which he [Tucker] was seduced would leave only *one appetite* or *desire* in human nature. He had no adequate and constant conception, that the translation of Desire from the end to the means occasioned the formation of a new passion, which is perfectly distinct from, and altogether independent of, the original desire.' In the margin beside this passage, Darwin wrote, 'with respect to life,' and 'Music?'

89–6 C. Bell 1833:191, '. . . the dread of falling is shewn in the young infant long before it can have had experience of violence of any kind.'

89–7 C. Bell 1833:193, 'The nurse will tell us that the infant lies composed while she carries it in her arms up stairs; but that it is agitated in carrying it down.' P. 194, 'Children, therefore, are cowardly by instinct: they show an apprehension of falling; and we may gradually trace the efforts which they make, under the guidance of this sensibility, to perfect the muscular sense.'

90–1 C. Bell 1833:304, 'The muscles which move the eye-ball are powerfully affected in certain conditions of the mind: independently altogether of the will, the eyes are rolled upwards in mental agony, and whilst strong emotions of reverence and piety are felt. This is a natural sign stamped upon the human countenance, and is as peculiar to man, as any thing which distinguishes him from the brute. The posture of the body follows necessarily, and forms one of those many traits of expression which hold mankind in sympathy.'

90–2 Lamarck 1830, 2:445, 'En effect, si l'on compare les idées et les jugemens de l'animal intelligent, qui est encore jeune et inexpérimenté, aux idées et aux jugemens du même animal, parvenu à l'âge de l'expérience acquise, on verra que la différence qui se trouve entre ces idées et ces jugemens, se montre, dans cet animal, tout aussi clairement que dans l'homme. Une rectification graduelle dans les jugemens, et une clarté croissante dans les idées, remplissent, dans l'un et dans l'autre, l'intervalle qui séparé le temps de leur enfance de celui de leur âge mûr. L'âge de l'expérience et de tous les dévelopemens terminés, se distingué éminemment de celui de l'inexpérience et du peu de développement des facultés, dans cet animal, de même que dans l'homme. De part et d'autre, on reconnoît les mêmes caractères et la même analogie dans les progrès qui peuvent s'acquérir; il n'y a que du plus ou du moins, selon les espèces.'

91–1 Lamarck 1830, 2:284–85, '. . . les *émotions intérieures*

Lamarck. Vol II p. 319.— Habits more prevalent in proportion to intelligence less.—[2] p. 325 «to 29».— Habits becoming heredetary form the instincts of animals.— almost identical with my theory— no facts, & mingled with much hypothesis.—[3] see M.S. notes, where strong argument in favour of brain forming the instincts,— could brain make a tune on the pianoforte, yes if every individual played a little, & something destroyed bad brain.

92 see p. 90.—[1] The relation of reason to organs of locomotion— or that our faculties have been given us to exist, is clearly seen. in the absurdity of a tree having reason: or dog, having high powers without hand or voice.— there is some great puzzle in what Sir. J. M. says of pure reason not leading to action[2] & yet our emotions being only bodily actions associated with ideas.—

A sigh, is an abortive groan.— more power over muscles of voice than respiration.— like sigh before false sneeze.—

d'un animal sensible, consistoient en certains ébranlemens généraux de toutes les portions libres de son fluide nerveux, et que ces ébranlemens n'étoient suivis d'aucune réaction, ce qui est cause qu'ils ne produisent aucune sensation distincte. . . . lorsque ces émotions sont foibles ou médiocres, l'individu peut les dominer et en diriger les mouvemens; mais que lorsqu'elles sont subites et très-grandes, alors il en est maîtrisé lui-même . . . Qui n'a pas remarqué qu'un grand bruit inattendu, nous fait tressaillir, sauter en quelque sorte, et exécuter, selon sa nature, des mouvemens que notre volonté n'avoit pas déterminés?'

91–2 Lamarck 1830, 2:319, 'Qui ne sent alors que le pouvoir des habitudes sur les actions doit être d'autant plus grand, que l'individu que l'on considère est moins doué d'intelligence, et a moins, par conséquent, la faculté de penser, de réfléchir, de combiner ses idées, en un mot, de varier ses actions.'

91–3 Lamarck 1830, 2:325–26, '. . . l'habitude d'exercer tel organe, ou telle partie du corps, pour satisfaire à des besoins qui renaissent souvent, donnoit au fluide subtil qui se déplace, lorsque s'opère la puissance qui fait agir, une si grande facilité à se diriger vers cet organe, où il fut si souvent employé, que cette habitude devenoit en quelque sorte inhérente à la nature de l'individu, qui ne sauroit être libre d'en changer.

'Or, les besoins des animaux qui possèdent un système nerveux, étant, pour, chacun, selon l'organisation de ces corps vivans:

1. De prendre telle sorte de nourriture;
2. De se livrer à la fécondatin sexuelle que sollicitent en eux certaines sensations;
3. De fuir la douleur;
4. De chercher le plaisir ou le bien-être.

'Ils contractent, pour satisfaire à ces besoins, diverses sortes d'habitudes, qui se transforment, en eux, en autant de penchans, auxquels ils ne peuvent résister, et qu'ils ne peuvent changer eux-mémes. De là, l'origine de leurs actions habituelles, et de leurs inclinations particulières, auxquelles on a donné le nom d'*instinct* (*).

(*) 'De même que tous les animaux ne jouissent pas de la faculté d'exécuter des actes de volonté, de même pareillement l'*instinct* n'est pas le propre de tous les animaux qui existent; car ceux qui manquent de système nerveux, manquent aussi de sentiment intérieur, et ne sauroient avoir aucun *instinct* pour leurs actions.'

'Ces animaux imparfaits son entièrement passifs, n'opèrent rien par eux-mêmes, ne ressentent aucun besoin, et la nature, à leur égard, pourvoit à tout, comme elle le fait relativement aux végétaux. Or, comme ils sont irritables dans leurs parties, les moyens que la nature emploie pour les faire subsister, leur font exécuter des mouvemens que nous nommons des actions. [End of footnote.]

'Ce *penchant* des animaux à la conservation des habitudes et au renouvellement des actions qui en proviennent, étant une fois acquis, se propage ensuite dans les individus, par la voie de la reproduction ou de la génération, qui conserve l'organisation et la dispositin des parties dans leur état obtenu; en sorte que ce même *penchant* existe déjà dans les nouveaux individus, avant même qu'ils l'aient exercé.

'C'est ainsi que les mêmes habitudes et le même *instinct* se perpétuent de générations en générations, dans les différentes espèces ou races d'animaux, sans offrir de variation notable, tant qu'il ne survient pas de mutation dans les circonstances essentielles à la manière de vivre.'

92–1 See N90.

92–2 Mackintosh 1837:152–53, see N89, note 3.

93 "A Dissertation on the Influence of the Passion."—[1]
p. 37. The increase of Bilary secretion attends passion[2]
p. 39. The sweat that accompanies fear is the same, as that which attends great weakness.— ‹Diarrhæa› & syncope[3]
p. 42. Sighing from grief. is method of increasing languid circulation—[4] no, for ‹grief› sighing comes on before circulation is affected.
p. 44.— Jealousy. causes spasm in bile duct, & throws bile in circulation[5]
p. 75. Haller says tooth ache, even from carious tooth cured by sight of instrument.—[6]

94 Bennett's Wanderings, Australian Dog does not Bark— quotes Gardner's Music of nature to show barking not natural. (Vol I. p. 234)[1,2]
Vol. II p 153. «do». an account of a monkey in a passion like Jenny.—[3] Dr. Abel has given an account of an Ourang.— see his Travels.—[4]

———

When one sees in Cowper,[5] whole sentences spoken & believed to be audible. one has good ground to call imagination a faculty, a power, quite distinct from self. «or will»

94 to call] 'ca' *pencil under* 'call'; 'C' *pencil preceding* 'ca'.

93–1 W. Falconer 1791.
93–2 W. Falconer 1791:37, 'The increase of the biliary secretion by this passion [viz., anger] is a remarkable . . . circumstance . . . Epileptic fits, the iliac passion, fever, and sudden death, are also numbered among the direful consequences of anger.'
93–3 W. Falconer 1791:39, '[perspiration] . . . is indeed sometimes excited by fear . . . and resembles that which attends syncope, and great weakness. Diarrhœa, jaundice, scirrhus, and gangrene, are said to have been hereby produced.'
93–4 W. Falconer 1791:41–42, 'Grief diminishes the bodily strength in general, and particularly, the force of the heart and circulation; as appears by the frequent sighs and deep respirations which attend it, which seem to be necessary exertions, in order to promote the passage of the blood through the lungs.'
93–5 W. Falconer 1791:44, 'The peculiar effects of jealousy in producing a spasm on the biliary ducts, and throwing the bile into the circulation, are very remarkable, and well attested.'
93–6 W. Falconer 1791:75, 'The sight of the instrument for extracting the tooth, often gives a perfect, though only a temporary relief, and this even though the pain has arisen from a carious tooth.'
94–1 G. Bennett 1834, 1:234, '. . . it is remarked by Mr. Gardiner, in a work entitled the "Music of Nature," "that dogs in a state of nature never bark; they simply whine, howl, and growl: this explosive noise is only found among those which are domesticated." '
94–2 Gardiner 1832:199, 'Dogs in a state of nature never bark; they simply whine, howl, and growl. . . .' See also Erasmus Darwin 1794, 1:154–55, for a discussion of dogs of Juan Fernandes and Guinea, that do not bark.
94–3 G. Bennett 1834, 2:153, 'He [a male Ungka gibbon (*Hylobates syndactyla*) taken on board ship] could not endure disappointment, and, like the human species, was always better pleased when he had his own way; when refused or disappointed at anything, he would display the freaks of temper of a spoiled child; lie on the deck, roll about, throw his arms and legs . . . dash everything aside that might be within his reach . . . he reminded me of that pest to society, a spoiled child . . .' P. 154 [footnote], 'The account of the orang-utan, given by Dr. Abel, in the *Narrative of a Journey in the Interior of China*, accords with the habits of this animal, and the comparison is very interesting.'
94–4 Abel 1818:326, 'If defeated again by my suddenly jerking the rope, he [the orang] would at first seem quite in dispair, relinquish his effort, and rush about the rigging, screaming violently.' P. 328, 'If repeatedly refused an orange when he attempted to take it, he would shriek violently and swing furiously about the ropes; then return and endeavor to obtain it; if again refused, he would roll for some time like an angry child upon the deck, uttering the most piercing screams; and then suddenly starting up, rushing furiously over the side of the ship, and disappear.'
94–5 Cowper 1836–37.

95e–96e [not located]

97 ‹& other cows—›

———

Mr. *Hamilton* on vital laws (in the Athænæum Library) describes effects of emotions— fear giving goose skin— & hair standing on end.—[1]

———

July 20th
Intelligent Keeper . . . Zoolog. Garden told me. he has often watche tame young wolf & it never dropped its ears like dog— wagged its tail «a little» when attending to anything or excited.— so do young dingos, as I saw, wag tail when watching anything— Keeper does not think

98 they drop their ears.— — George the lion is extraordinarily cowardly.— the other one nothing will frighten— hence variation in character in different animals of same species.—

———

99e ———

The general «(as I believe)» contempt at suicide. (even when no relatives left to lament) is owing to the feeling that the instinct of self-preservation is disobeyed— I often have «as a boy» wondered why *all abnormal* sexual actions or even impulses. (where sensations of individual are same as in normal cases) are held in abhorrence it is because instincts to woman is not followed; good case of instinctive

100e conscience.— Why does not man eating cause disgust, because he does not go *against instinctive feeling*, only does not fullfil, like continent man.— a man eating what others by *habit* (not instinct) think not fit, as cannabalism, is held in *abhorrence*.— all this makes analogy of actions with ‹&› against benevolent & parental instincts very clear.— even to the cold or benevelo- continent man

101 Hume has section (IX) on the Reason of animals[1] Essays Vol 2.— «also on origin of religion or polytheism, at p. 424 Vol. II «Sect XV. Dialogue on Natural Religion.»[2] however, he seems to allow it is an instinct.»

———

97 Mr. *Hamilton* . . . on end.—] *pencil.*
99 *upper half of page to* The general . . .' *excised, not located.*
100 *upper half of page to* 'conscience.— Why . . .' *excised, not located.*
101 *pencil.*
«also . . . an instinct.»] *ink.*

97–1 Hamilton: Could not be traced; possibly a member of the Athenaeum Club.
101–1 Hume 1826, 1:232–35; 4:121–26, 'Of the Reason of Animals' vol. 1, pp. 232–35, and vol. 4, 'Of the Reason of Animals'.
101–2 Hume 1826, 2:419–548, 'Dialogues Concerning Natural Religion.'

I suspect the endless round of doubts & scepticisms might be solved by considering the origin of reason. as gradually developed. see Hume on Sceptical Philosophy.[3]

Hume has written "Natural Hist. of Religion" on its origin in Human mind.—[4] Andrew Smith says hen doves & the *female* chamaeleon court the males by odd gestures.

102 In one of the six (?) first Vol of Silliman's Journal paper showing that the signs invented for Deaf & dumb school & used between Indian tribes are Many the same.—[1]

103 Philosoph. Transactions Vol 44. 1746–47. Paper. like. Sir Ch. Bell on Expression «First Croonian Lectures by Parsons.» following pages contain remarks worthy of attention p. 15, 25. 40. 61. CD[a person is here said to open mouth in fright because nature intends to lay open all senses: ‹do›[1] Horse prick his ears «& snort clears nostrils» when frightened, does not hair & rabbit depress. them from squatting.— p. 64 closing both eyelids express contempt.[2] p. 76.— children have been tickled into excessive laughter & so into convulsions.—[3] «Paper» must be referred to, if I follow up this subject & a reference to Brun's work.—[4] Shutting eyes in *contempt* opposite action to opening eyes in *fear*

104 The effect of habitual movements in muscles of face, is well seen in shortsighted people.— hence origin of expression—

105 There are some instincts unintelligible, ‹both› in the end gained «& therefore the» cause, and origin being so is not odd.; for instance wild cattle[1] & deer pursuing a wounded one.— porpoises a ditto— it is probably some

102 *pencil.*
103 'Shutting eyes . . . *fear*'] *pencil.*

101–3 Hume 1826, 1:236–347, 'Of the Sceptical and other Systems of Philosophy'.
101–4 Hume 1826, 4:435–513, 'The Natural History of Religion'.
102–1 Akerly 1824:351, 'If we examine the signs employed by the Indians, it will be found that some are peculiar and arise from their savage customs, and are not so universal as sign language in general; but others are natural, and universally applicable, and are the same as those employed in the schools for the deaf and dumb, after the method of the celebrated Abbe Sicard.'
103–1 Parsons 1746:60, 'The reason why the Eyes and Mouth are suddenly open'd in Frights, seems to be, that the Object of Danger may be the better perceived and avoided; as if Nature intended to lay open all the Inlets to the Senses for the Safety of the Animal . . . the mouth . . . that they may hear it . . . [for] there is a Passage from the *Meatus auditorius*, which opens into the Mouth.' See also C. Bell 1824.
103–2 Parsons 1746:64, 'As soon as the Mind suggests a Contempt for Persons or Things, whether deservedly or not, the first Muscles that begin to act are, the *Elevator Labii superiorus proprius Cowperi*, and the *Pyramidalis*, on one Side only . . .'
103–3 Parsons 1746:76, 'A Person playing with a Child tickled him in the Sides very much . . . till the poor Child grew black in the Face, was convulsed all over . . .'
103–4 Le Brun 1701. *Expression*: 247 (note 13) says, 'Le Brun, in his well-known "Conférence sur l'Expression" ("La Physionomie, par Lavater" edit. of 1820. vol. ix. p. 268), remarks that anger is expressed by the clenching of the fists.'
105–1 See also D48, E117 and M142.

secondary one— blood being disagreeable & anything disagreeable being pursued.—

A dog turning round & round is some old instinct ‹perverted› handed down & down.— mem. Nina used to get into hay & make a nest for herself.— the object is to make saucer-shaped depression.—

106 [blank]

107 Does music bear any relation to the period when men. communicated before language was invented,— were musical notes the language of passion & hence does music now excite our feelings.—[1]

How does Social animal recognize «& take pleasure in» other animal, (especiall as in some ‹instinct› «insects» which become in imago state social) by smell or looks. but it does not know its own smell or look, & therefore there must be some instinctive feeling which is pleased by other animals smell & looks.— no doubt it may be attempted to be said that young animal learns parent smell & look so by association receives pleasure. This

108 [blank]

109 will not do for insects. if this view holds good— then man, a socialist, does not know other men by smell, but by looks. hence. some obscure picture of other men. & hence idea of beauty.— the social affections of animal taking man in place of other animals is hostile «is subversive of» to this view, & fowls hatching stones. in some degree is so.— idea of beauty of music are great distinguishing character between man & animals.—

110 [blank]

111 Double consciousness. only extreme step of an ideal argument held in one's own mind, & D^{r}. Hollands story of man in Delirium tremens hearing other man speaks.[1] shows, that consciousness of personnal identity is by no means a necessary part of man's mind.—

At Maer. Pool. I saw many coots & waterhens feeding on grassy bank some way from water, *suddenly*, as if by word of command, they all took flight & flappered across pool to bed of flags I was astonished & having looked round saw at considerable distance a very large

109 this view, & fowls] *ink blob over illeg. letter between* '&' *and* 'fowls'.

107–1 See Erasmus Darwin 1794, 1:155, '. . . the singing of birds, like human music, is an artificial language rather than a natural expression of passion. . . . Our music, like our language, is perhaps entirely constituted of artifical tones, which by habit suggest certain agreeable passions.'

111–1 Holland 1839b:139, 'I have known these illusions of hearing such, in a case of delirium tremens, that the patient held a long and angry colloquy with an imagined person, whom he supposed (there being no illusion of sight) in an adjoining room.'

112 hawk, which are «so» rare « ‹s.› » here,, that probably few had ever before seen one, yet all— flew to bed of flags. hernes are common. not unlike in size in the air at a distance.— How can such an instinct arise?? «it would appear that an instinct long remains, if no steps are taken to eradicate it.—» «Emma says, «her» tame rabbits were not frightened at a dog.— »
The instinct against man is perhaps, as strong as against hawk, but the birds at Maer have learned that he is not dangerous— wild-ducks would have fled equally if man had appeared— though instinct so firmly implanted, birds soon ‹dis› learn to disobey it— I have seen hawk & sparrow in Shrewsbury garden picking from same bone

---

113 A child born on the 1st March was frightened on the 24th of May at Cresselly by the boys making faces at it, so much so that the nurse had to carry it out of the room. nearly 3 months old.[1]

What is absurdity, why does one laugh at it—

---

sensation of disgust with nausea, (when stomach a little disordered) at thought of almost anything ugly. baby— association— pouting child same as anger, lips not compressed sullen, protruded. determined to do nothing. & so manifesting sulleness.

114 [blank]

115e Circumstances having given to the Bee its instinct is not ‹more› less wonderful than man his intellect

---

Lyell has seen a little dog go to the assistance & bite a big dog. which was fast struggling with another large dog his companion. **Descent —Affection &** [. . .]

Monkeys «Ogleby» **seen Zool. Soc— 1838** remember with distress their companions—[1] a «blue» Gibbon. whose companion had **S**[. . .] been dead

112 hawk, which . . . frightened at a dog.—] *written on bottom half of page.*
The instinct . . . same bone] *written on top half of page.*
such an instinct arise??] 'i' *in* 'instinct *over* 'a'.
birds soon] 'birds' *over 2 illeg. letters.*
113 A child . . . months old.] *Emma Darwin's handwriting.*
115 Circumstances. . . his companion.] *crossed blue crayon.*
'Descent— Affection & [. . .]] *two words illeg., added blue crayon.*
Monkeys «Ogleby»] *underlined blue crayon:*
Seen Zool. Soc— 1838] *added blue crayon.*
'S[. . .]]' *two words illeg., added blue crayon.*

113–1 Cresselly, Pembrokeshire: home of John Bartlett Allen. Probably John Darwin, the child of Allen's grandson Henry Allen (Harry) and Jessie Wedgwood.
115–1 W. Ogilby 1839:31, '. . . a new species of Monkey, now living at the Society's Menagerie . . . will probably become morose and saturine as it advances in age and physical development . . .' Darwin probably discussed the behavior of monkeys with Ogilby.

about two months. saw a «black» spider monkey brought it at opposite end of house. & commenced a most lamentable howls & & was not comforted until the Keeper took it ‹her› in his arms & carried to see.—

116e [blank]

117 A Dog «whilst» dreaming, growling. & yelpings. «& twitching paws» which they only do when ‹great› considerably excited, shows their power of imagination— for it will not be allowed they can dream, & not have day-dreams— think well over this;— it shows similarity in mind.— think of Eyton's horses becoming ‹white› with ‹lather› ‹foame› & sweat, when hearing merely hunting horn— association or imagination[1]

---

118 [blank]

119e–120e [not located]

121 Ernest W.[1] playing with Snow.[2] when 2½ years old. was frightened when Snow put a guaze over her head. & came near him, although knowing it was Snow.— Is this part of same feeling which make us think anything ugly— a beau-ideal feeling. Same effect as acting on us— ‹The Baby› «Effie Wedgwood»[3] April 28th 1840 was frightened at wild beasts in Zoolog. Garden

122 [blank]

123e–124e [not located]

125 A child crying. frowning, pouting, «smiling», just as much instinctive as a bull ‹tr› calf, just born butting, or young crocodile snapping.— these I think are better instances of instincts (highly useful as only means of communication) in man, than sucking.— [I assume a child pouts who has never seen others pout][CD]

126 [blank]

127 Goldsmiths Essays No XV,, on sounds of words being expressive, (Vol. 4 of Works)[1]

128–183 [blank]

184 "Adam Smith Moral Sentiments" much on life & character[1]

---

121 Same . . . Garden] *pencil.*
125 *pencil.*
127 *pencil.*
184 "Adam . . . observations] *boxed left and bottom.*

117–1 Eyton: Thomas Campbell Eyton.
121–1 Ernest Wedgwood.
121–2 Snow: Frances Julia (Snow) Wedgwood.
121–3 Katherine Euphemia (Effie) Wedgwood.
127–1 Goldsmith 1806, Essay XV: 185–97. P. 195, 'The words we term *emphatical*, are such as by their sound express the sense they are intended to convey . . .'
184–1 Adam Smith 1808.

"Humes Dissertation on the Passions."[2]

"Hartley" I should think well worth studying—[3]

"Thomas Brown" on Association[4] worthy of close study.— full of practical observations

[184] Ourang do not move eyebrows.— or skin of head,— «scarcely able St.— » Cyanocephalus, macacus. *Cercopithecus*? very much., «Keeper says some of the monkeys move ‹its› «the» ears but ‹not› Chimpaze. does not gradation towards man.» Macacus especially pulls back skin of whole forehead 2 ears.— emotions of every kind.— «[Are monkeys ‹are› right-handed??]CD» Cyanocephalus, Macacus, Niger. Cercopithecus make labial st st. S. American monkeys. pull back skin from head very little Does blood go in ‹body› face in pashion.?— cry?

INSIDE BACK COVER Do people of weak intellects easily fall into *habits*
Get facts about instincts of mongrel dogs
Do blubbering children, if of. convulsive tendency easily fall into convulsions

---

A carrier pidgeon carried & turned round & round in fainting state would it then know its direction.—

In slight convulsions. are the muscles of the face first affected?— Can shivering & trembling be considered convulsive.—

---

is convulsion. are involuntary movement of voluntary muscles— if so what is trembling palsy?

BACK COVER

Expressions
N

[184] Ourang ... cry? *this passage written in grey ink on a slip of paper pinned to page 184.*
«Keeper ... towards man.» *brown ink.*
whole forehead 2] '2' *over* '&'.
«[Are monkeys ‹are› right-handed??]CD» *pencil.*
Does blood ... cry?—] *verso slip of paper.*

184–2 Hume 1800, 2:175–211, 'A Dissertation on the Passions'.
184–3 Hartley 1834.
184–4 T. Brown 1820a:123–26; 148–49; 181–84; 187–88; 191–92; 204; 224–30; 231–39; 251, considers 'Association'. See T. Brown 1820b, 2:210, 'On Mr. Hume's Classification of the Associating or Suggesting Principles'; p. 333, 'Reasons for preferring the term Suggestion to the phrase Association of Ideas'; and p. 245, 'Refutation of Dr. Hartley's theory of Association'. NB., Mackintosh 1837:336, 'I very early read Brown's *Observations on the Zoonomia* of Dr. Darwin, the perhaps unmatched work of a boy in the eighteenth year of his age ...' In Erasmus Darwin 1794–96, 1:49–53, 441–51, and vol. 2, 1796, pp. 413–536, are discussions of association. See T. Brown 1798, for a 560 page review of Erasmus Darwin's *Zoonomia*, 1794–96.

Old & Useless Notes

Introduction by Sandra Herbert & Paul H. Barrett

Transcribed and edited by Paul H. Barrett

These items (DAR 91:4–55) are not a notebook but a collection of notes on miscellaneous sheets of paper that Darwin grouped together as '[Old & USELESS notes about the moral sense & some metaphysical points written about the year 1837 & Earlier —]CD.' Darwin added this title to the notes some years after they were written, almost certainly when writing the *Descent of Man* (1871). The title is misleading in that it reflects the dismissive attitude of an author filing away notes no longer useful, and the date '1837 & Earlier' is in error. The dated entries extend from 6 September 1838 (OUN25) to March 1840 (OUN33). Other dated entries include 2 October 1838 (OUN29), 13 January 1839 (OUN31), 14 January 1839 (OUN32) and 5 May 1839 (OUN42). Of the undated entries about half refer to authors cited at the close of Notebook C or in D, M, and N, all of which date from 1838–1839. Therefore, while some notes in this collection are possibly earlier than 1838, the bulk of the notes date from 1838–1840.

Old & Useless Notes consists of 42 sheets of paper, contains 26 abstracts, and counting folded sheets and verso sides totals 112 pages, 76 are written upon. A table of sheet sizes and watermarks is listed below. Darwin's practice of writing research notes on the right side of a sheet leaving the left margin and, or, the left facing page blank for future annotations is evident in the notes. Some notes Darwin paginated separately, in which case his numbers are treated as part of the text. Numbers in the margin were added by Cambridge University Library staff and refer to folios in DAR 91:4–55. The last folio, not numbered, is here given the number 55 for index purposes. Darwin wrote in pencil and in ink, OUN25–30 being the only section definitively in grey ink. Darwin's brown crayoned '25' on OUN38^{v} refers to one of his portfolios.

The subject matter of Old & Useless Notes is essentially the same as that of Notebooks M and N, the origin of language, intelligence, expression, instincts, perception and the moral sense being common to each. From the overlap in chronology and content we can assume that Darwin wrote the so-called Old & Useless Notes while reading books and journals at the *Athenaeum*, at Maer or elsewhere and, not having the 'Expression' Notebooks M and N to hand, or wishing to write at greater length used whatever sheets of paper were readily available, including the official *Athenaeum* stationery. Names of family members appear in OUN, for example, Hensleigh Wedgwood's opinions are cited in M58 and his own notes appear in OUN39–41. Darwin's broad and well-grounded education is again evident from the literature cited. Works of 26 different authors ranging from philosophers, political economists and historians to biologists are abstracted. Darwin's notes on Mackintosh's *Dissertation on the Progress of Ethical Philosophy* fill folios 42–55 and DAR 91:11b contains a portion of his notes on Macculloch, the bulk of which appear separately in this volume.

OUN Sheets

DAR 91 FOLIO	PAPER SIZE (mm)	WATERMARK
4 (part of wrapping)	65 × 201	none
5	226 × 190	[I? 1?] SMITH 1838
6	180 × 160	W. TUCKER
7	125 × 204	18 [??]
8	232 × 201	none
8a	102 × 200	none
8b	114 × 66	none
9	242 × 204	W WARREN 1837
10	108 × 177	37
11	114 × 187	yes; insufficient for description
11b	112 × 184	same watermark as 11
12	186 × c 111	H & Son
13	c 74 × c 105	MAN 33
14	c 80 × 99	none
15	52 × 191	none
16	127–155* × 200	Warren 1837
17	111 × 180	TMAN 38[?]
18–21	183 × 111	none
22–24	180 × 113–114	none
25–30	201 × 160	none
31	205 × 162	none
32	105 × 183	none
33	200 × c 166	none
34–38	313 × 187	35, 37 none 34, 36, 38 cipher, but cannot read
39	227 + 62† × 183	none
40–41	227 × 183	none
42–52	232 × 188	none
53–54	188 × 114–117	none
55	113 × 181	none

* Ragged, uneven torn edge
† Added, paste-on piece

4 [Old & USELESS notes about the moral sense & some metaphysical points written about the year 1837 & Earlier—][CD]

5 in Athenæum

"Smart— Beginning of a new School of Metaphysic,"— give my doctrines about origin of language— & effect of reason. reason could not have existed without it.—[1] quotes Ld Mondobbo.—[2] language commenced in whole sentences.—[3] signs— ? were signs originally muscical!!!??—

At least it appears all speculations of the origin of language.— must presume it originates slowly— if these speculations are utterly valueless— then argument fails— if they have, then language was progressive.—

5[v] We cannot doubt that language is an altering element, we see words invented— we see their origin in names of People.— Sound of words— argument of original formation.— declension &c often show traces of origin.—

6 Mayo Philosophy of Living. p. 264.[1] "Architecture is a fine amplification of two ideas in nature; a developement of the thoughts expressed in Fingals cave, & in the arched & leafy forests"
Very good!.

7 I grant that the thrill, which runs throug every fibre, when one behold the last rays of & & or grand chorus are utterly inexplicable— I cannot ‹admit› think reason sufficient to give up my theory— Viewing from eminence. the wide expanse, of county, netted with edges & crowded with towns & thoroughfares, I grant that man, from the effects of heredetary knowledge, has produced almost →

7[v] greater changes in the polity of Nature than any other animal—

8 Aimé Martin de l'Education des Mères Vol. I. p. 198.— "Moralité, raison, beau ideal, infini conscience; voilà l'homme separe de la matiere et du temps! voila les facultes, q'il possede seul sur la terre. J'ai trouve son âme" &c—[1]

5–1 Smart 1839:3–5, 'As to the question, whether speech was or was not, in the first instance, revealed to man, we shall not meddle with it: we do not propose to inquire how the first man came to speak, but whether language is not a necessary effect of reason, as well as its necessary instrument, growing out of those powers originally bestowed on man, and essential to their further development.'

5–2 Monboddo 1773–92. Darwin's spelling 'Mondobbo'.

5–3 Smart 1839:26–27, note, ' "It may be asked," says Lord Monboddo, "what words were first invented. My answer is, that if by words are meant what are commonly called parts of speech, no words at all were first invented; but the first articulate sounds that were formed denoted whole sentences; and those sentences expressed some appetite, desire, or inclination, relating either to the individual, or to the common business which I suppose must have been carrying on by a herd of savages before language was invented. And in this way, I believe language continued, perhaps for many ages, before names were invented." '

6–1 H. Mayo 1838:264 (see N31).

8–1 L. Martin 1837, 1, bk. 2, chap. 8:198, 'Moralité, raison, beau idéal, infini, conscience; voilà l'homme séparé de la matière et du temps! voilà les faculté qu'il possède seul sur la terre. J'ai trouvé son âme, et dans son âme la source morale de l'être humain, c'est-à-dire la nécessité d'une autre vie!'

— Confesses these faculties of soul, treating of infinites not definable.—[2] Has little Chapter on each faculty of Soul.—[3] (1) ‹Conscience› «Moral Sentiments» imperative sense of duty—[4] which makes struggle in man.— two souls in one body— (2) Beau ideal, refers chiefly to moral, beau desires conscience & love.—[5] [With regard to ordinary Beau ideal, Mem. Negro, beau,— Jeffrey denies all Beau—[6] How does Hen determine which most beautiful cock, which best singer— Remember.— avarice a compounded passion gained in life time]CD 3. The Infinite, — *lives by hopes*, looks

8v to eternity.[1] (4) Reason, some transcendental kind—[2] (5) Conscience, not clear—[3] Then these last heads. of separation between soul of man. & intellect of beasts, not clear.— ⁽does not Mackintosh[4] make great difference between moral sense & conscience? we admire what is right by one & are *ordered* to do it by other.—

I suspect conscience, an heredetary compound passion. like avarice.—[5]

Is there not something analogous to imperiousness of Conscience: in Maternal instinct *domineering* over love of Master and sport &c &c — The Bitch does not so act, because maternal instinct gives most pleasure. but because most imperious.—

8a It would indeed be wonderful, if, mind of animal was not closely allied to that of men, when the *five senses* were the same—[1] In its action— emotions—

8v I suspect . . . heredetary] 'et' *over* 'ar'.
Is there not . . . most imperious.—] *small handwriting, lower left corner.*
8a L Aimé] 'e' *accent crossed.*

8–2 L. Martin 1837, 1, bk. 2, chap. 9:201–3, 'De l'instinct de l'homme, et de l'impossibilité de définir les facultés de l'âme'.
8–3 L. Martin 1837, 1, bk. 2, chap. 1:157–62, 'De L'étude des facultés de l'âme'.
8–4 L. Martin 1837, 1, bk. 2, chap. 10:205, 'Il en résulte que le sentiment moral est indépendant de notre intelligence, et qu'il nous commande impérativement ce qu'il faut faire pour être heureux!'
8–5 L. Martin 1837, 1, bk. 2, chap. 11:207, 'Le type du beau est immuable, êternel; il existe, car nous en avons la conscience et l'amour: la conscience pour nous incline à sa recherche, l'amour pour nous rendre dignes de le contempler.'
8–6 Jeffrey 1811b:2, (in a review of Alison 1811) '. . . we can never ascertain what is beauty, without having clear notions of the state of mind which it produces . . . and . . . it is utterly impossible to ascertain what is the nature of the effect produced by beauty on the mind, till we can decide what are the common properties that are found in all the objects which produce it.'
8v–1 L. Martin 1837. 1, bk. 2, chap. 12:208–9, 'Ainsi le passé meurt, le présent s'evanouit, et l'avenir n'est qu'une espérance! Une espérance! Ô mortel, voilà ta grandeur! . . . tu espéres une vie qui ne doit pas finir.'
8v–2 L. Martin 1837, 1, bk. 2, chap. 13:211–12, 'La raison est le sentiment du vrai; c'est une révélation de la sagesse et de l'ordre. Tantôt elle plonge dans le monde des vérités transendantes . . .'
8v–3 L. Martin 1837, 1, bk. 2, chap. 14:216, 'L'homme n'est pas toujours innocent quand sa conscience l'absout! il n'est pas toujours coupable quand sa conscience l'accuse!'
8v–4 See extensive discussion of Mackintosh's ideas in Darwin's abstract of Mackintosh 1837, OUN42–55.
8v–5 See L. Martin 1837, 1, bk. 2, chap. 13:214, 'L'entendement est une puisance composée, par conséquant variable: ses facultés sont à la fois spirituelles et animales; elles comprennenet les sensations et les passions . . . Mais la raison est une puissance simple . . .'
8a–1 L. Martin, 1837, 1, bk. 2, chap. 5:173, 'Voyez ce chein qui repose à mes pieds: les nerfs de son cerveau se projettent aux organes des cinq sens . . .'

p 176 & 177 good passage in French on what dog dreams, awakes— does when Master takes Hat de l'education des Mères par L Aimé Martin[2]

8b ———

Leroy Lettres. Philosophique sur l'intelligence des Animaux—[1]

———

& Le Parfait Chasseur, par Desgraviers, un Vol 8vo[2]

———

Keratry— Inductions morales et physiologiques[3]

———

The first of these books I daresay good.

———

9 1. Sensation is the ‹conse› ordering contraction (that is the only evidence. when consciousness is absent) in fibres united with nervous filaments.— ⸮plants? yes by distinct mechanism
2. Sensation of higher order. where the sensation is conveyed over whole body (which it may be in first case. as when the excised heart is pricked) and certain action. (only evidence. when not consciousness) are produced in consequence having some relation to the primary sensation.— man moving leg when asleep— «or habitual actions» perhaps polypi— (so that lower animals are sleeping higher animals & not plants as supposed by Buffon)[1] Consciousness is *sensation No. 2.* with memory added to it, man in sleep not conscious, nor child— Evidence of consciousness, ‹t› movements «⸮» anterior to any direct sensation, in order to avaoid it— beetles feigning death upon seeing an object.— are Planariæ conscious.—

———

Consciousness bears some relation to time & memory

8b Leroy . . . Animaux] *not in Darwin's handwriting.*
& Le Parfait . . . Vol 8vo] 'Desgraviers' *alternate reading* 'Durivier'.
Keratry . . . physiologiques] *not in Darwin's handwriting.*
9 1. Sensation . . . nervous filaments.—] '⸮plants?' *boxed.*
Evidence . . . movements] '⸮' *circled.*

8a–2 L. Martin 1837, 1, bk. 2, chap. 5:176–77, 'Voilà mon chien qui vient de s'endormir au coin de mon feu: son sommeil est agité, il a un songe, et dans ce songe il poursuit sa proie, il attaque son ennemi, il le voit, il l'entend, il le dévore; il a des sensations, des passions and des idées. Je l'appelle, je le tire de ses visions; il redevient calme. Je prends mon chapeau, il s'élance, saute, me regarde, m'étudie, se traîne à mes pieds, court à ma porte, se rejouit ou s'attriste, suivant la volonté que j'exprime. . . . voilà un animal qui pense, qui veut, qui se ressouvient, qui combine. Il y a des moments où je suis tenté de lui croire une âme: car enfin je trouve dans son intelligence les phénomenès qui sont dans la mienne . . .'
8b–1 LeRoy 1802.
8b–2 Desgraviers 1810.
8b–3 Keratry 1841.
9–1 Buffon 1785, 2:7, 'If the sensation of an oyster, for example, differs in degree only from that of a dog, why do we not ascribe the same sensation to vegetables, though in a degree still inferior? This dinstinction, therefore, between the animal and vegetable, is neither sufficiently general nor decided.'
P. 8: '[Thus] there is no absolute and essential distinction between the animal and vegetable kingdoms; but . . . nature proceeds by imperceptible degrees from the most perfect to the most imperfect animal, and from that to the vegetable: Hence the fresh water polypus may be regarded as the last of animals, and the first of plants.'

10 Reynolds X discourse very curious as showing "the perfection of this science of *abstract* form" is the source of part of the highest enjoyment in mutilated statues[1]

10ᵛ In Elliotson's Physiology much about sleep— Nerves.— Volition &c[1]

11 Reynold XIII Discourse (p 115) a very good passage. about actions & decisions bein the result of sagacity, or intuition. when individual cannot give reason, though he feels he is right—[1] it is because each decision &c is made up of many partial results, & the impressions on them are ‹all› remembered, when the meaning or reasons are forgotten. Our happiness &c, our well-being depends upon the "habitual reason".— This power of the mind, faintly approaches to instinct

11ᵛ How strange it, that Nature should have so little to do with art (p 128) R. compares a view taken by a camera obscura &c a Poussin.—[1]

How are my ideas of a general notion of everything applicable to the high idea «p. 131.» in Tragic acting—[2] CD [My idea. would make the mind have mysterious & *sublime* ideas independent of the senses & experience

11b p. 142 "Upon the whole it seems."— "that the object of «all» art is the realizing and embodying, what never existed but in the imagination".—[1]

Macculloch Vol I. p. 115. Attributes of Deity. on Belief.— you belief things you can give no proof for, & one often replies "what you say is perfectly true, but you do not convince me.—"[2] Belief allied to instinct.—

10–1 Reynolds 1798, 2:17, '... what artist ever looked at the Torso without feeling a warmth of enthusiasm, as from the highest efforts of poetry? From whence does this proceed? What is there in this fragment that produces this effect, but the perfection of this science of abstract form?

'A mind elevated to the contemplation of excellence perceives in this defaced and sheltered fragment, *disjecti membra poetae*, the traces of superlative genius, the reliques of a work on which succeeding ages can only gaze with inadequate admiration.'

10ᵛ–1 Elliotson 1840, chap. 27:598–698, 'Sleep'.

11–1 Reynolds 1798, 2:114, 'There is in the commerce of life, as in Art, a sagacity which is far from being contradictory to right reason, and is superior to any occasional exercise of that faculty; which supercedes it; and does not wait for the slow progress of deduction, but goes at once, by what appears a kind of intuition, to the conclusion. A man endowed with his faculty, feels and acknowledges the truth, though it is not always in his power, perhaps, to give a reason for it; because he cannot recollect ... all the materials that gave birth to his opinion ...'

11ᵛ–1 Reynolds 1798, 2:127, '... the comparatively inferior branches of art ... take their rank and degree in proportion as the artist departs more, or less, from common nature ...' P. 128, 'If we suppose a view of nature represented with all the truth of the *camera obscura*, and the same scene represented by a great Artist, how little and mean will the one appear in comparison of the other, where no superiority is supposed from the choice of the subject ... Like Nicolas Poussin, [the same Artist] transports us to the environs of ancient Rome, with all the objects which a literary education makes so precious and interesting to man ...'

11ᵛ–2 Reynolds 1798, 2:131, '... the best stage-representation [e.g., Hamlet] appears even more un-natural to a person ... who is supposed never to have seen a play before, than it does to those who have had a habit of allowing for those necessary deviations from nature which the Art requires.'

11b–1 Reynolds 1798, 2:142, 'Upon the whole, it seems to me, that the object and the intention of all the Arts is to supply the natural imperfection of things, and often to gratify the mind be realising and embodying what never existed but in the imagination.'

11b–2 Macculloch 1837, 1:115.

11b^{v} ———

p. 134. a painted must not a *actors*, or a scene in garden.— yet both beautiful![1] p. 136. Says Architecture does not come under imitative art[2] [my view says yes. ‹old› mass of rock—]^{CD} or poetry, ^{CD}[my thery says yes. imitating song — two primary sources, sight, & hearing—

12 Staunton Embassy Vol II p. 405.— Speculates on origin of sacrifices. «common to many races»— thinks action towards ‹man› «a king» ‹changed into› is carried on toward deity.— & as king might like cruel pleasure, so sacrifices cruel.—[1] Something wrong here.— Origin is certainly curious. Chinese, S. American. Polynesians Jews, African all sacrifices. How completely men must have personified the deity.—

13 H. Tooke has shown one chief object of language is promptness «of consequence» hence languages become corrupt, & whole classes of words «are abbreviations» he thus derives from nouns & verbs—

═══

so that much of EVERY language shows traces of anterior state??[1]

14 Edinburgh Review Vol 18. (1st Article) on Taste «EXCELLENT». Deficient in not explaining the possibility of ‹handsome› «UGLY healthy» young woman, with good expression— statues not painted— ‹music› very good article— why flower beautiful? ⸢even to children[1]

11b^{v} my thery says yes.] 'yes' *over* 'yet'.

11b^{v}–1 Reynolds 1798, 2:134, '. . . no Art can be engrafted with success on another art . . . If a Painter should endeavour to copy the theatrical pomp and parade of dress, and attitude, instead of that simplicity, which is not a greater beauty in life than it is in Painting, we should condemn such Pictures, as painted in the meanest style. . . . So, also, Gardening, as far as Gardening is an Art . . . is a deviation from nature . . .'

11b^{v}–2 Reynolds 1798, 2:136, '. . . Architecture . . . applies itself, like Musick, (and, I believe, we may add Poetry), directly to the imagination, without the intervention of any kind of imitation.'

12–1 Staunton 1797, 2:405, 'If [the sovereign] should be pleased with gold, the bowels of the earth were ransacked for it; if, as in the excesses to which princes and conquerors in the first ages were supposed to have been prone, he should delight in the riotous and sanquinary pleasures of the table, bloody sacrifices were prepared and offered at his altar.'

13–1 Tooke 1798:27, 'The first aim of *Language* was to *communicate* our thoughts: the second, to do it with *dispatch*.' P. 25, '*Abbreviations* are the *wheels* of language, the *wings* of Mercury.' P. 27, 'Many words are merely abbreviations employed for dispatch . . .' P. 94, 'Abbreviations and Corruption are always busiest with the words which are most frequently in use.' P. 100, '. . . French, Italian, Anglo-saxon, Dutch, German, Danish and Swedish . . . (together with English) are little more than different dialects of one and the same language.' P. 147, '. . . the perpetual accession of new words . . . forbid the deduction of the *whole* language from any one single source.'

14–1 Alison 1811b:10 [in *Edinb. Rev.*], 'The most beautiful object in nature, perhaps, is the countenance of a young and beautiful woman . . . what we admire is not a combination of forms and colours . . . but a collection of signs and tokens of those feelings and affections, which are universally recognized as the proper objects of love and sympathy.'

P. 17, 'The forms and colours that are peculiar to [children], are not necessarily or absolutely beautiful in themselves; for in a grown person, the same forms and colours would be either ludicrous or disgusting.'

P. 18, 'Take, again, for example, the instance of female beauty,— and think what different and inconsistent standards would be fixed for it in the different regions of the world;— in Africa, in Asia, and in Europe;— in Tartary and in Greece;— in Lapland, Patagonia and Circassia. If there was anything absolutely or intrinsically beautiful, in any of the forms thus distinguished, it is inconceivable that men should differ so outrageously in their conceptions of it: If beauty were a real and independent quality, it seems impossible that it should be

15 S. Jenyn's Inquiry into the Origin of Evil. Reviewed by Johnson in the Literary Magazine. 1756— Ceased in 1758— Read the Review or the article.[1]

16 A Planaria must be looked at as animal, with consciousness,, it choosing food— crawling from light.— Yet we can split Planaria into three animals, & this consciousness becomes multiplied with the organisms structure, it looks as if consciousness an effects of sufficient perfection of organization & if consciousness, individuality.—

17 Quotes D. Stewarts System of Emotions.— T. Mayo— Pathology of the Human Mind.[1] Poor.— on insanity.— Prevailing idea. owing to loss of *will*.— chiefly excited by passive emotions.— Cannot quite perceive drift of Book.— Sympathy & affections chiefly fail.— Notices. struggle ‹between› when insanity is coming on «Thinks clearest analogy between dreams & insanity.»[2]

18 D. Stewart on the Sublime[1]

The literal meaning of Sublimity is height. & with the idea of ascension we associate something extraordinary & of great power— —
2 From these & other reasons we apply to God the notion of living in lofty regions.
3 Infinity eternity. darkness, power. being associated with God. these phenomena we (feel & ?) call sublime.—
4 From the association of power &c &c with height, we often apply the term sublime, where there is no real ***sublimity***

16 *Scored left margin, with two large question marks.*
17 «Thinks . . . insanity»] *left margin, connected by line to* 'System of Emotions.—'
18 3 . . . ?] *added pencil.*
3 . . . ')'] *pencil over* ')'.
6 . . . moral excellences] *crossed letter, probably* 't,' *between second* 'c' *and second* 'e'.

distinctly and clearly felt by one set of persons, where another set, altogether as sensitive, could see nothing but its opposite . . .'

Pp. 18–19, 'The style of dress and architecture in every nation, if not adopted from mere want of skill, or penury of materials, always appears beautiful to the natives, and somewhat monstrous and absurd to foreigners . . . The fact is still more striking, perhaps, in the case of Music . . .'

15–1 S. Jenyns 1757a & 1757b.

17–1 T. Mayo 1839:4, '. . . much of my reasoning . . . flows similarly to that by which the operations of sleep and dreaming are explained by Mr. Dugald Stewart. These states have, in truth, always appeared to me to possess a striking affinity to that of the insane,— with this important difference, that there is a constant readiness in the mind to be roused out of sleeping or dreaming state, but no such readiness to emerge out of insanity, when once incurred. And again, that in sleep every voluntary action is suspended; whereas in madness, the will acts with considerable force, though in a more limited extent, than in the sane state.'

17–2 See Abercrombie 1838 (Darwin's copy):258, '. . . in dreaming and insanity, [the voluntary effort] is suspended, and the mind is left entirely under the influence of the chain of thoughts which happens to be present, without being able either to vary or dismiss it.' Darwin's marginal notation, 'they ought not to be classed together, one like reality of the thoughts or absence of doubt in one case being owing to the ‹weakness› absence of contending impressions, & in insanity *opposed* to many present impressions.' And at the top of the page Darwin wrote, 'No, a *vivid* thought «neither pleasant, nor painful, but merely vivid» cannot be dismissed even by the strongest will,— is insanity an unhealthy vividness of thought.'

18–1 Stewart 1829, 4:265–317. 'Essay Second. On the Sublime.' Darwin here outlines the main principles discussed by Stewart in this essay.

18v 5 The emotions of terror & wonder so often concomitant with sublime. adds not a little to the effect: as when we look at the vast ocean from any height.— 6 That the superiority & "inward glorrying, which height. by its accompanying & associated sensations so often gives, when excited by other means, as moral excellences, brings to our recollection the original cause of these feelings & thus we apply to them the metaphorical term sublime

19 7 So that in this Essay. D. Stewart does not attempt «by one common principle» to explain the various causes of those sensations, which we call metaphorically sublime, but that it is through a complicated series of associations that we apply to such emotions. this same term.—

19v Hence it appears, that when certain causes, as great height, eternity, &c &c. produces an inward pride & glorying. (often however accompanied with terror & wonderment) ‹which› «this» emotion, from the associations before mentioned. we call sublime.—
It appears to me, that we may often trace the source of this "inward glorying" to the greatness of an object itself or to the ideas excited & associated with it. as the idea of Deity. with vastness of Eternity. which superiority we transfer to ourselves in the same manner as we are acted on by sympathy.

20 D. Stewart on taste
The object of this essay is to show how taste is gained how it originates, & by what means it becomes an almost instantaneous perception.—[1] Taste has been supposed by some to consist of "an exquisite susceptibility from **Blair** receiving pleasures from beauties of nature & art."[2] But as we often see people who are susceptible of pleasure from these causes who are not men of taste & the reverse of this. taste

20v evidently does not consist of this. but rather in the power of discriminating & respecting good from bad.
And it is manifestly from this fact & the instantaneousness of the result, that

19 7 So that . . . metaphorically] '2' *above* 'call'; 'l' *above* 'metaphorically'.
this same term] 'this' *over* 'thes'.
20 susceptibility from] 'from' *over* 'G' and *2–3 letters illeg.*
beauties of nature] 'ies' *over* 'y'.
"an exquisite . . . pleasures] '**Blair**' *or possibly* '**Blain**' *in margin.*
reverse of this. taste] 'this' *over* 'tis'.

20–1 Stewart 1829, 4:318, 'Essay Third. On Taste.' Includes taste as one of the 'intellectual processes, which, by often passing through the mind, come at length to be carried on with a rapidity that eludes all our efforts to remark it; giving to many of our judgments, which are really the result of thought and reflection, the appearance of instantaneous and intuitive perceptions. The most remarkable instance [of these] . . . are commonly called the *acquired perceptions of sight* . . .'
20–2 Stewart 1829, 4:327, '[Taste] is said to consist in "a power of receiving pleasure from the beauties of nature and of art." '

the term taste is metaphorically applied to this mental power[1] Although taste must necessarily be acquired by a long series of experiments & observations. & yet, like in vision, it becomes

21 so instantaneous. that we cannot ever perceive the various operations which the mind undergoes in gaining the result.—[1]

22 Lessings Laocoon. 2d Lect— The object of art., sculpture & painting, is beauty.—[1] which he thinks is a better definition than Winkleman's. who says it is simplicity with grandeur of character.—[2] Hence Lessings shows expression of pain cannot be represented. But what is beauty?— it is an ideal standard, by which real objects are judged; & how obtained.— implanted in our bosoms.— how comes it there?

23 Laocoon p. 75 "The beauties developed in a work of art are not approved by the eye itself, but by the imagination through the medium of the eye"; he will allow the secondary pleasures of harmonious colours &c &c surely to be added.[1]

---

24 Lessings Laocoon p. 125— says *new* subjects are not fit for painter or sculpture, but rather subjects which we know,[1] it is therefore the embodying of a floating idea.— as statue of beauty, is of the "beau ideal", my *instinctive* impression

25 1) September 6th. 1838
Every action whatever is the effect of a motive.— [— must be so, analyse [(a)] ones feelings when wagging one's finger— one feels it in passion, love— jealousy— «as» effect of bodily organisms— one knows it, when one wishes to do some action (as jump off a bridge to save another) & yet dare not — one *could* do it, but other motives prevent the action see Abercrombie conclusive

22 Laocoon,.] *over* 'Laoocon'.
25 must be so, analyse[(a)] '[(a)]one well . . . chance)' *added as a footnote, verso page 25.*
N.B. *Items numbered 25 through 28 were written on right-hand pages, and the left-hand pages were reserved for marginal notations, viz., footnotes.*

20v–1 Stewart 1829:332, '"The feeling," [Voltaire] observes, "by which we distinguish beauties and defects in the arts, is prompt in its discernment, and anticipates reflection, like the sensations of the tongue and palate. Both kinds of Taste, too, enjoy, with a voluptuous satisfaction, what is good; and reject what is bad, with an emotion of disgust. Accordingly," he adds, "this metaphorical application of the word *taste*, is common to all known languages."'

21–1 Stewart 1829. 4:325, 'The fact seems to be . . . "the mind, when once it has felt the pleasure, has little inclination to retrace the steps by which it arrived at it." It is owing to this, that Taste has been so generally ranked among our original faculties; and that so little attention has hitherto been given to the process by which it is formed.'

22–1 Lessing 1836:19, '. . . among the ancients, Beauty was the supreme law of the arts of design.'

22–2 Lessing 1836:1, 'The general characteristics of the Grecian masterpieces in Painting and Sculpture, are held by Winkelmann to consist in a noble simplicity, and a majestic composure, both of attitude and expression.'

23–1 Lessing 1836:75–76.

24–1 Lessing 1836:122, 'The poet, indeed, enjoys a great advantage who treats a story, or a character already known . . . same advantage is possessed by the painter . . .'

remarks p. 205 & 206.]CD[1]
Motives are units in the universe.[2]
[Effect of heredetary constitution,— education under the influence of others— varied capability of receiving impressions— *accidental* (so called like chance) circumstances.
As man hearing Bible for first time, & great effect being produced.— the wax was soft,— the condition of mind which leads to motion being inclined that way]CD one sees this law in man in somnambulism or insanity.[3] free will (as generally used) is not there present, but he acts from motives, nearly as usual

(a) one well feels how many actions are not determined by what is called free will,

25–1 In his copy of Abercrombie 1838, Darwin wrote in the margins, p. 199, 'A man may wish to jump from a bridge to save another, but absolutely will not let him.— makes the muscles fail, & the heart sink—'

Pp. 202–3, 'Yes, but what determines his *consideration*— his own previous conduct— & what has determined that? & so on— *Heredetary* character & education— & *chance* (aspect of his will) circumstances.

'‹Change of character possible from change of organization.›

'What has given these *desires* & *conduct*

'‹Then why does not act of insanity give shame??›

'According to all this ones disgust at villain ‹ought to be› is nothing more than disgust at some one under foul disease, & pity accompanies both. Pity ought to «banish» disgust. P → P For wickedness is no more a man's fault than bodily disease!! (Animals do persecute the sick as if were their fault). If this doctrine were believed— pretty world we should be in!— But it could not be believed excepting by intellectual people— if I believed it— it would make not one difference in my life, for I feel more virtue more happiness— Believers would ‹will› only marry good women & pay detail attention to education & so put their children in way of being happy.

'It is yet right to punish criminals for public good. All this delusion of free will would necessarily follow from man feeling power of action.

'View no more unreasonable than that there should be *sick* & therefore unhappy men.

'What humility this view teaches.

'A man ‹reading› hearing bible by chance becomes good. This «is» effect of accident with his state of desire (neither by themselves sufficient) effect of birth & other accidents: may be congratulated, but deserves no credit.

'When opposed desires are absolutely equal which is possibility, may free-will then decide.— but it must be decided by habit or wish & these all originate as before.'

At the bottom of pages 206 and 207, '‹A man may put himself in the way of above accidents. but desire to do so arises as before; & knowledge that the effect will be good, arises as before. education & mental disposition—

'One feels how many actions, not determined by will, passion— when the motive power feeble & complicated & opposed we say free will (or chance)›'

25–2 Abercrombie 1838:204–5, 'Of physical relations all that we know is the fact of their uniformity; and it would appear equally philosophical to apply the same term to mental phenomena. On this principle, therefore, we should say,— that the tendency of moral causes or motives is not necessary, but uniform; and that on this depends all our confidence in the uniformity of human character, and in the power of truths, motives, or arguments, to produce particular results on human conduct. "To suppose the mind possessed of a power of determining, apart from all this influence of moral causes or motives, would be to overthrow this confidence, and to reduce our whole calculations on human character to conjecture and uncertainty."' (Darwin in his copy of Abercrombie, pencilled in the added quotation marks above and drew a marginal line beside the passage. The same is true for the following.) '"A power of determining without any reason, appears to be not only unphilosophical, but, in point of fact, inapplicable to any conceivable case."'

25–3 Abercrombie 1838:259, '[Somnambulism] differs from dreaming, in the senses being, to a certain degree, awake to external things; though that power is suspended, by which the mental impressions are corrected by the influence of the external world. . . . the remarkable difference between [the somnambulist] and the maniac is, that the somnambulist can be roused from his vision, and then the whole is dissipated.' Darwin made a marginal stroke beside this passage, and wrote, 'There is some sophistry here: insane man has perfect *consciousness*— somnambulism has not.'

but by strong invariable passions— when these passions, weak, opposed & complicated one calls them free will—the chance of mechanical phenomena.— (Mem: M. Le Comte case of Philosophy, & savage calling laws of nature chance)

26 2) difference is from imperfect condition of mind all motives do not. come into play.—
†It may be urged how often one try to persuade person to change line of conduct. as being better & making him happier.— he agrees & yet does not.— because motive power not in proper state.— When the admonition succeeds who does not recognize an accidental spark falling on prepared materials.
From *contingencies* a mans character may change— because motive power changes with organization
The general delusion about free will obvious.— because man has power of action, & he can seldom analyse his motives (originally mostly INSTINCTIVE, & therefore now great effort of reason to discover them: this is important explanation) he thinks they have none.—
Effects.— One must view a wrecked man, like a sickly one(P)— We cannot help loathing a diseased offensive object, so we view wickedness.— it would however be more proper to pity than to hate & be

†A man may put himself in the way of Contingencies.—but his desire to do arises from motives.—& his knowledge that it is good for him effect of Education & mental capabilities.—
'(P) Animals do attack the weak & sickly as we do the wicked.—we ought to pity & assist & educate by putting contingencies in the way to aid motive power.—if incorrigably bad nothing will cure him'

27 3) disgusted. with them. Yet it is right to punish criminals; but solely to *deter* others.— It is not more strange that there should be necessary. wickedness than disease.
This view should teach one profound humility, one deserves no credit for anything. (yet one takes it for beauty & good temper), nor ought one to blame others.— This view will not do harm, because no one can be really *fully* convinced of its truth. except man who has thought very much, & he will know his happiness lays in doing good & being perfect, & therefore will not be tempted, from knowing every thing he does is independent of himself to do harm.—

Believer in these views will pay great attention to Education.—

---

28 4) These views are directly opposed & inexplicable if we suppose that the sins of a man, are under his control, & that a future life is a reward or retribution.— it may be a consequence but nothing further.—

26 It may be urged . . . prepared materials.] '+A man may . . . capabilities.—' *footnote opposite page.* Effects . . . sickly one (P)—] '(P) Animals do attack . . . cure him' *footnote opposite page.*

29 October ‹8› 2^{d}. 1838
Those emotions which are strongest in man, are common to other animals & therefore to progenitor far back, (anger ‹to› at the very beginning, & therefore most deeply impressed). shame perhaps an exception. (does it originate in a doubting feel between conscience & impulse) but shame «we alas know» is far easier *conquered* than the deeper & worser feelings. These bad feelings no doubt *orginally* necessary revenge was justice.— No checks were necessary to the vice of intemperance, circumstances made

29^{v} the check.— to licentiousness jealousy, & every one being married to keep up population. with the existences of so many positive checks.— (This is encroaching on views in second volume of Malthus)[1]. Adam Smith[2] also talks of the necessity of these passions, but refers (I believe) to present day & not to ruder state of Society.— Civilization is now altering these instinctive passions—, which being unnecessary we call vicious.— (jealousy in a dog no one calls vice). on same principle that Malthus had shown incontinence to be a vice & especially in the female

30 October ^{d}. 1838 **Perhaps insist??**
Two classes of moralists: one says our rule of life is what *will* produce the greatest happiness.— The other says we have a moral sense.— But my view ‹says› unites both «& shows them to be almost identical» + What *has* produced the greatest good «or rather what was necessary for good at all» *is* the «instinctive» moral senses: (& this alone explains why our moral sense points ‹is› to revenge). In judging of ‹our ha› of the rule of happiness we must look *far forward* «& to the *general* action» — certainly because it is the result of what has *generally* been best for our good *far back*.— (much further than we can look forward: hence our ‹[. . .]› rule may sometimes be hard to tell) ++ Society could not go on except for the moral sense, any more than a hive of Bees without their instincts.— **Gives art to when I say How social instincts generated?**

30^{v} The origin of the social instinct «in man & *animals*» must be separately considered.— The difference between civilized man & savage,— is that former is endeavoring to change that part of the moral sense which experience (education is the experience of others) shows does not tend to

29^{v} female] 'female' *over* '—'.
30 **Perhaps insist??**] *added blue pencil.*
But my view . . . moral senses:] '++' *added left margin.*
hence our ‹[. . .]›] '[. . .]' *3–4 letters crossed, illeg.*
Gives art . . . generated?] *added blue pencil, left margin.*
30^{v} The difference . . . to greatest good.—] *scored blue pencil left margin.*

29^{v}–1 Malthus 1826, 2 (bk 4, chap. 1):255–69, 'Of moral restraint, and our obligation to practise this virtue'; chap. 2:270–82, 'Of the effects which would result to society from the prevalence of moral restraint.'

29^{v}–2 Adam Smith 1808, 1 (Sec 2):42–75, 'Of the degrees of the different passions which are consistent with propriety.'

greatest good.— Therefore rule of happiness is to certain degree ‹of› right.— The change ‹of› our moral sense, is strictly analogous to change of instinct amongst animals.—

31 Jan 13th. 1839
My father received a letter from Mr Roberts— «a person he had long known & directed many letters to»— could not ‹remember› «read» Christian name; fancied it looked like. W. but concluded it could not be so.—Looked at a direction book, but could not find out— Directed his letter, & I observed he had written Wilson & pointed it out; he was astonished, & said how very odd.— —could not think what had put Wilson into his head.— remembered, that he had. looked in direction book under head of Wilson, referred to Robert & found his Christian name was *Wilson*!!— How curious an inward. unconscious memory.—

32 Jan 14th. 1839.—
My father says he has heard of many cases of ideots knowing things, which are often repeated in a wonderful manner.— as the hour of the day &c— All habits must conduce to their health & comforts.— Both ideots, old People & those of weak intellects.—

33 Westminster Review. March 1840
p. 267— says the great division amongst metaphysicians— the school of Locke, Bentham, & Hartley, &. the school of Kant. to Coleridge,[1] is regarding the sources of knowledge.— whether ‹we th there› "anything can be ‹any› «the» object of «our» knowledge except our experience".—[2] is this not almost a question whether we have any instincts, or rather the amount of our instincts— surely in animals according to usual definition, there is much knowledge without experience. so there *may* be in men— which the reviewer seems to doubt.

34 [RHC] 1) Effects of Life in the abstract is matter united by certain laws different from those., that govern in the inorganic world; life itself being, the

31 Wilson & pointed] '&' *over* '—'.
32 All habits must] 'must' *alternate reading* 'much'.
34 *Items 34 through 38 were written in two columns down the page, the left being narrower than the right, and used for explanatory notes, footnotes, etc. These are marked* [RHC] = *right hand column*; [LHC] = *left hand column.*
Effects of Life in the] 'the' *over* 'an'.
During growth . . . tissue ‹[. . .]›] '‹. . .›' *2–3 letters crossed, illeg.*

33-1 [Mill] 1840:264, 'Every consistent scheme of philosophy requires, as its starting point, a theory respecting the sources of human knowledge . . . The prevailing theory in the eighteenth century . . . was that proclaimed by Locke, and attributed to Aristotle— that all our knowledge consists of generalizations from experience . . . From this doctrine Coleridge with . . . Kant . . . strongly dissents . . . He distinguishes in the human intellect two faculties . . . Understanding and Reason. The former faculty judges of phenomena, or the appearance of things, and forms generalizations from these: to the latter it belongs, by direct intuition, to perceive things, and recognize truths, not cognizable by our senses.' The metaphysical school of Locke, Hartley, and Bentham is contrasted with that of Coleridge throughout the article.
33-2 [Mill] 1840:267, 'We see no ground for believing that anything can be the object of our knowledge except our experience, and what can be inferred from our experience by the analogies of experience itself . . .'

capability of such matter obeying a certain & peculiar system of movements.[1]
different from inorganic movements.—
See Lamarck for this definition given in full.—[2]
[LHC] ¿Has any vegetable or animal *matter* been formed by the union of *simple* non-organic matter, without action of vital laws—
According to the individual forms of living beings, matter is united in different modification, peculiarities of external form impressed, & different laws of movements.
[LHC] Hence there are two great ‹worlds, inor› systems of laws «in the world» the organic & inorganic— The inorganic are probably one principle for connect of electricity chemical attraction, heat & gravity is probable.—[3]
And the Organic laws probably have some unknown relation to them—
[RHC] In the simplest forms of living beings namely «*one individual*» vegetables, the vital laws act definitely (‹like› «as» chemical laws,) as long as certain contingencies are present, (contingencies as heat light &c).
[LHC] This is true as long as movement of sensitive plant can be shewn to be direct physical effect of touch & not irritability, which at least shows a local will, though perhaps not conscious sensation.
[RHC] During growth ‹extres› tissue ‹[. . .]› unites matter into certain form;[4] invariable, as long as not modified by external accidents, & in such cases modifications bear fixed relation to such accidents.
But such tissue ‹must› bears relation to whole, that is enough must be present to be able to exist as individual.—

34–1 Kirby 1835, 1:xli, '"We have seen," says [Lamarck], "that the life which we remark in certain bodies, in some sort resembled nature, insomuch that it is not a being, but an order of things animated by movements; which has also its power, its faculties, and which exercises them necessarily while it exists."* (**Anim. sans Vertèbr.* i. 321.)'

34–2 Lamarck 1830, 1:403, '. . . on peut donc embrasser ce qui la *constitue* essentiellement dans la définition suivante.

'La vie, dans les parties d'un corps qui la possède, est un ordre et un état de choses qui y permettent les mouvemens organiques; et ces mouvemens, qui constituent la vie active, résultent de l'action d'une cause stimulante qui les excite.'

34–3 Kirby 1835, 1:xxiii–xxiv, 'Body [Lamarck] observes, being essentially constituted of cellular tissue, this tissue is in some sort the matrix, from the modification of which by the fluids put in motion by the stimulus of desire, membranes, fibres, vascular canals, and divers organs, gradually appear . . . and thus progressively new parts and organs are formed, and more and more perfect organizations produced; and thus, by consequence, in the lapse of ages, a monad becomes a man!!!'

34–4 Kirby 1835, 1:xli–xlii, 'Speaking of the imponderable incoercible fluids, and specifying heat, electricity, the magnetic fluid, etc., to which he [Lamarck] is inclined to add light, he says, it is certain that without them, or certain of them, the phenomenon of life could not be produced in any body . . . [but] neither caloric nor electricity, though essential concomitants of life, form its essence.'

35 [RHC] 2) In animals, growth of body precisely same as in plants, but as animals bear relation to less simple bodies, and to more extended space, such powers of relation required to be extended.

Hence a sensorium, which receives communication from without, & gives wondrous power of willing.[1] These +*willings* are common to every animal instinctive and unavoidable.— +**Can the word willing be used without consciousness, for it is not evident, what animals have consciousness.**
These willings have relation to external contingencies, as much as growth of tissue and are subject to accident; the sexual willing comes on period of year as much as inflorescence.—
[LHC] I here omit the case (if such there are) of animals enjoying only movements such as sensitive plants. (But I include irritability for that require will in part. ⸮Why more so than movement of sap. or sunflower to sun? ∵ I should think there. was direct «physical» effects of more or less turgid vessels; effect of heat, light or shade.)

Joining two difficulties into one common one always satisfactory, though not adding to positive knowledge. lessening amount of ignorance

---

[RHC] The radicle of plants absorb by physical laws of endosmic & exosmic juices. arms of polypus, show either local or general will, & stomach likewise «does».
[LHC] ⸮ in Corallina are not two kinds of *life* vegetable and animal strictly united?

[RHC] It is easy to conceive such movements & choice, & obedience to certain stimulants without conscience in the lower animals, as in stomach, intestines & heart of man.
[LHC] ⸮How near in structure is the ganglionic system of lower animals & sympathetic of man[2]

---

35 Hence a sensorium . . . power of willing. These +*willings*] '+**Can the word** . . . **consciousness**' *added as interlinear notation, circled and connected by line to '+willings'.*
It is easy . . . heart] 'heart' *over 3 unidentified letters.*
Where pain and pleasure . . . consciousness???] *brace left;* '?' *lower right corner of page.*
⸮Can insects . . . have?] *brace right.*

35–1 Kirby 1835, 1:xxviii–xxix, '"Every action of an intelligent individual, whether it be a movement or a thought . . . is necessarily preceded by a want of that which has power to excite such action [according to Lamarck]. This want felt immediately moves the internal sentiment, and in that same instant, that sentiment directs the disposable portion of the nervous fluid . . ."'
35–2 In this discussion, Darwin, with his own interpretation, parallels a similar treatment in Kirby 1835, 1:150–51, eg., p. 150. 'Lamarck indeed regards them [Infusoria] as having no volition, as taking their food by absorption like plants; as being without any mouth, or internal organ; in a word as transparent gelatinous masses, whose motions are determined not by their will, but by the action of the medium in which they move.' P. 151, 'Admitting that the observations of Spallanzani just stated record facts, it appears clearly to follow from them that these animals *have* volition, and therefore cannot properly be denominated *apathetic*, or insensible. The fact that they almost all have a mouth and a digestive system; many of them eyes, and some rudiments of a nervous one, implies a degree; more or less, of sensation . . .'

[RHC] ¶How does consciousness commence; where other senses come into play, when relation is kept up with distant object. where many such objects are present, & where will directs ‹to› other parts of body. to do such.—

All this can take place & man not conscious as in sleep; or in sleep is man momentarily conscious, but is memory gone?—
Where pain & pleasure is felt there must be consciousness??? ?
[LHC] ¶Can insects live with no more consciousness than our intestines have?

36 [RHC] 5) Kirby thinks that ‹all› there is one one instinct to all animals modified according to species.[1] This I suppose he deduces from the ends in each case being the same, & the means very similar.— It does not appear more than saying that the thinking principle is the same in all animals.
[LHC] «3)» Eyton told me that his retriever Sailor he has seen push a hare through the bar of a gate before him, & then jump over the gate & bring it. ——— Agrees with ONE animal

---

[RHC] Kirby extends instinct to plants,[2] but surely instincts imply willing, therefore word misplaced

The meaning of Words, must be made out
Reason
Will
Consciousness

Definite instincts being acquired, is a most important argument, to show that they result from organization of brain; «[LHC] not used by Kirby» (:analogy:— as races are formed or modification of external form. so modifications of brain) As in animals no prejudices about souls, we see particular trains of thoughts as fear of man,— crows fear gun,— pointers method of standing,— method of attacking peccari— —retriever— produced as soon as brain developed, and as I have said, no soul superadded,[3] so

36–1 Kirby 1835, 2:247, 'That the same action should unfold such an infinite variety of forms in one case and instincts in the other is equally astounding and equally difficult to explain.— Compare the sunflower and the hive-bee, the compound flowers of the one, and the aggregate combs of the other . . .'
36–2 Kirby 1835, 2:246, '. . . as the most remarkable instincts of animals are those connected with the propagation of the species, so the analogue of these instincts in plants is the developement of these parts peculiarly connected with the production of the seed . . .' Pp. 247–48, 'Again, as all plants have their appropriate fruitification, so they have other peculiarities connected with their situation, nutriment, and mode of life, corresponding in some measure with these instincts that belong to other parts of an animal's economy. Some with a climbing or voluble stem, constantly turn one way, and some as constantly turn another. . . . others close their leaves in the night, and seem to go to sleep; others shew a remarkable degree of irritability when touched . . .'
36–3 Kirby 1835, 1:xxviii, '[Lamarck] admits [man] to be the most perfect of animals, but instead of a son of God, the root of his genealogical tree, according to him, is an animalcule, a creature without sense or voluntary motion, or internal or external organs . . . no wonder therefore that he considers his intellectual powers, not as indicating a spiritual substance derived from heaven though resident in his body, but merely as a result of his organization* (**N. Dict. D'Hist. Nat.* xvi. Artic. Intelligence, 344, comp. Ibid. Artic. Idéa, 78, 80.), and ascribes to him in the place of a soul, a certain *interior sentiment* . . .' See B232, 'The soul by consent of all is superadded . . .'

37 [RHC] 6) thought, however unintelligible it may be, seems as much function of organ, as bile of liver.— ⸮ is the attraction of carbon.[1] hydrogen ‹&c› in certain definite proportions,, (different from what takes place out of bodies) really less wonderful than thoughts—[2] One organic body likes one ‹m› kind more than another— What is matter? the whole a mystery.—
[LHC] This Materialism does not tend to Atheism.[3] inutility of so high a mind without further end just same argument. without indeed we are step towards some final end.— production of higher animals— perhaps, say attribute of such *higher* animals may be looking back, ∴ therefore consciousness, therefore reward in good life

---

[RHC] Instinct appear like heredetary memory; but first memory in many cases cannot be acquired by experience for child sucking.— And is it more wonderful that memory should be transmitted from generation.; than from hour to hour in ‹man:› individual—

[LHC] Perhaps even the most complicated instinct. might be analysed into steps, as species change.— Must be so if Lamarck's theory true

---

[RHC] Acquired instincts analogous «(& replace)» to experience gained by man in lifetime

---

Heredetary memory not so wonderful as at first appears, & no too great advantage.; for superiority of memory does not depend on its length.: Many animals (as horses) very long & good memories— but on its multiplicity & the comparison of ideas.—
As man has so very few (in adult life) instincts.— this loss is compensated by vast power of memory, reason &. & many general instincts, as love of virtue, of association, parental affection— The very existence of mankind requires these instincts,: though very weak so as to be overcome easily by reason.— Conscience is one of these instinctive feelings.
[LHC] As sexual instinct comes on late in life, man almost alone in this case can perceive instinct. boy takes delight in mammæ before any reason had told

37–1 Kirby 1835, 1:xxiv–xxv, 'When, indeed, one reads the above account [progression of molecules to monad to man] of the mode by which, according to [Lamarck's] hypothesis, the first vegetable and animal forms were produced, we can scarcely help thinking that we have before us a receipt for making the organized beings at the foot of the scale in either class— a mass of irritable matter formed by *attraction*, and a *repulsive* principle to introduce into it and form a cellular tissue, are the only ingredients necessary. Mix them, and you have an animal which begins to absorb fluid, and move about as a monad or a vibrio, multiplies itself by scissions or germes . . .'

37–2 Kirby 1835, 1:xxix, '. . . Lamarck sees nothing in the universe but bodies, whence he confounds sensations with intellect.'

37–3 Kirby 1835, 1:xxvii, 'Lamark's great error, and that of many others of his compatriots, is materialism; he seems to have no faith in any thing but *body*, attributing every thing to a physical, and scarcely any thing to a metaphysical cause.' P. xxxiv, 'From [Lamarck's] statements . . . he appears to admit the existence of a Deity . . .'

him this distinctive mark, it is downright instinct, leading to touch a particular organ.—

36v I think Pincher shows surprise, walking home one day met him, with Mark riding instantly followed, me and for five minutes every now and then howled.— Now I don't think this only pleasure; for it was different way of showing it, nor was there any cause, & if surprise was felt.— analyse feelings. Mr Wynne says, that beyond doubt *courage* is heredetary in fowls & not effect of feeling of individual force in any individual.— His Malay breed «of fowl» totally different habits from Europæan. begin to prowl about in the evening «seldom leave their perch till evening» crow different.— *Heredetary* effect of former tropical climate

analogous to inflorescence of Tropical plants when imported & plants sleeping

---

good show acquirement or obliteration of instincts

But habits acquired even by ‹children› «plants»!

38 [RHC] 7) As definite instincts modified by heredetary;— so succession so perhaps general ones.— Parental feelings weakened in Otahiati; fear of death in Hindoo population.— Slightly modified in many countries, hence national character, love of country, of association &c stronger in some than others— Hence superiority of Christian over Heathen race.— But as no great modification in brain would probably take place without corresponding change in «external» man; and as all men nearly same species, so general instincts nearly same; which same argument probably applies to particular instincts of animals. even in wild state; certainly to the domesticated.—

[LHC] NB. Two dogs having very different instinct always obtain peculiarities of external configuration.

[RHC] General— Instincts, certainly appear a sort of acquired memory. a permanent secretion of thought, (or under contingencies of stimulants of certain kinds such secretion)

---

or an association of pleasures with certain actions performed by your parents, conscience

This «X» memory especially «the» general kind taking pleasure in virtue because acquired in past ages; seems to indicate that when we ‹return› turn

36v **analogous . . . «plants»!**] *added grey ink?; also brace added to right of* 'Mr Wynne . . . climate'.
38 'NB. Two dogs . . . configuration.] *double scored, pencil.*
or an association . . . conscience] *added grey ink?;* **association**, *underlined.*
contingencies of stimulants] 'of' *over* 'or'.
This «X» memory . . . in others] *left- and right-hand column format dropped*

into angels. this imperfect memory may become perfect & we may look back to definite action or to our conscious selves.— **Such memory may go back to animals which were changed into man ∴ they meet their reward!**

× Perhaps should hardly be called memory; you cannot call the frame of mind which makes music pleasant, a memory; yet that frame is enhanced by memory of what has been heard; so love of virtue enhanced by this heredetary kind of memory.—

The difference between heredetary memory & individual secretion of thought, may be no more «difference» than sexual intercourse in plants is involuntary, in man voluntary: ? False,— secretion in both involuntary, ‹application in› «ejection only has» will: there must be cases of secretion being some time governed by will in some animals, involuntary in others.

39 [1)] Why may it not be said that thought perceptions will, consciousness memory &c. have the same relation to a living body (especially the cerebral portions of it) that attraction has to ordinary matter.
The relation of attraction to ordinary matter is that which action bears to the agent. Matter is by a metaphor said to attract; & hence if thought &c bore the same relation to the brain that attraction does to matter, it might with equal propriety be said that the **living** brain perceived, thought, remembered &c. **Well the heart is said to feel**
Now this would certainly be a startling expression, & so foreign to the use of ordinary language that the onus probandi might fairly be laid with those who would support the propriety of the expression. They would do well to ask themselves the converse of the ‹expr› question above stated, **because there are living bodies without these faculties** & indeed until we know what answer they would give in support of their view it is impossible to shew satisfactorily it's erroneousness. **it is a point of indifference**

39^{v} 2) In the absence of such a guide we can only ‹shew› point out the mode «of perceptive action» by which we come to conceive of matter as attracting & shew that the groundwork ‹of this› is entirely wanting by which thought or memory. might be in like manner attributed to the brain. There are two modes of perceptive action by which bodily action is made known to us, revealing respectively what are called it's subjective & objective aspect. The

Such memory . . . meet their reward!)] *circled.*
× Perhaps should . . . kind of memory.—] '×' *connected by marginal line to* «'×'» *before* 'memory'.
The difference between . . . False,] 'False' *circled.*
39 *Entries 39, 40, and 41 are in Hensleigh Wedgwood's handwriting, except where otherwise indicated.*
'living' *Darwin's handwriting, pencil.*
'Well the heart is said to feel' *added Darwin's handwriting, left margin, pencil.*
'because there are living bodies without these faculties' *added Darwin's handwriting, pencil.*
'it is point of indifference' *Darwin's handwriting, pencil,* **'point'** *underlined.*
39^{v} **'How do the senses affect us, except by internal consciousness'** *Darwin's handwriting, pencil.*
'external' underlined pencil; **'what'** *added Darwin's handwriting, pencil.*
'force' underlined pencil.

subjective aspect of bodily action is revealed to us by the effort it costs us to exert force or by internal consciousness; the objective, by our *external* **what** senses in the way in which we apprehend the *force* of inamimate bodies. How we identify the two aspects as different phases of the same object of thought is a question which ought to be clearly comprehended by anyone who wishes to fully understand this subject, but the answer to it would require a considerable degree of attention. **How do the senses affect us, except by internal consciousness**

40 3) We must endeavour to do without it as well as we can.
‹The objective aspect of› bodily action **as recognised by our external senses** consists in the *manifestation of force* **i.e. movement**? capable of being traced to the body of the individual to whom the action is attributed; force (be it remembered) being a phenomenon apprehended by the same faculty with matter & being necessarily exhibited in & by matter.
The phenomena of gravity considered in themselves consist in a force manifested in every particle of matter directed towards every other particle; but FORCE, ‹objectively› considered, **by our external senses** is a phenomenon the essence of whose existence consists in its communication to other matter in the course of its DIRECTION, & thus when we apprehend force in inanimate matter we feel dissatisfied until we can point out **How can force be recognized by our external senses—only movement can.—**

40v 4) the source from which it arises.
But coming round to the ‹subjective› aspect of action **as known by the exertion of our own power & consciousness of it** we are conscious that we ourselves can originate in any point an opposition of forces balancing each other & moving in opposite directions. We are satisfied therefore, if we can trace any force in inanimate matter up to the action of some animated agent Now the phenomena of gravity are manifestly the same as if every particle of matter were an animated being pulling every other particle by invisible strings & as on this supposition the forces manifested would be ‹sat› fundamentally accounted for, we prefer this metaphorical mode of stating the fact to the mere statement of the ‹force exhibited in every› phenomena actually apprehensible by sense.

40 The objective aspect of] *crossed, pencil.* 'B' *over* 'b' *of* 'bodily' *pencil*; **'as recognized by our external senses'** *Darwin's handwriting, pencil.*
'*manifestation of force*' *underlined pencil*; **'i.e. movement?'** *Darwin's handwriting, pencil.*
The phenomena of gravity . . . but FORCE] 'FORCE' *double underlined pencil*
objectively] *crossed pencil*; **'by our external senses'** *Darwin's handwriting, pencil.*
'DIRECTION' *double underlined pencil.*
'How can force be recognized by our external senses— only movement can.—' *Darwin's handwriting, pencil.*

40v 'subjective' *crossed pencil.*
'as known by the exertion of our own power & consciousness of it' *Darwin's handwriting, pencil.*

41 5) There is nothing analogous to this in the relation of thought, perceptions, memory &c. either to our bodily frame or the cerebral portion of it Thoughts, perception &c. are modes of subjective action— they are known only by internal consciousness, & have no objective aspect. If thought bore the same relation to the brain that force does to the bodily frame, they could be perceived by the faculty by which the brain is perceived but they are known by courses of action quite independent of each other. A person might be quite familiar with thought & yet be ignorant of the existence of the brain. We cannot perceive the thought **attraction of sulphuric acid for metal** of another person at all, we can only infer it from his **its** behaviour.
Thought is only known *subjectively*?—? the brain only objectively. **We do not know attraction objectively**

41v 6) The reason why thought &c. should imply «**X**» the existence of something in addition to matter is because our knowledge of matter is quite insufficient to account for the phenomena of thought. (The objects of thought have no reference to place. [We see a particle move one to another, & ‹the› (or conceive it) & that is all we know of attraction, but we cannot see an atom think: they are as incongruous as *blue* & *weight*; all that can be said that thought & organization run in a parallel series: if blueness & weight always went together. & as a thing grew blue it «uniquely» grew heavier yet it could not be said that the blueness caused the weight, anymore that weight, the blueness, still less between ‹action› «things» so different as action thought & organization: But if the weight never came untill the blueness had a certain intensity (& the experiment was varied) then might it now be said, that blueness caused weight, because both due to some common cause:— The argument reduces itself to what is cause & effect: it merely is «invariable» priority of one to other: no not only thus, for if day was first, we should not think night an effect.]CD **Cause and effect has relation to forces & mentality because effort is felt**

42 [LHC] 1) May 5th. 1839.— Maer

Mackintosh Ethical Philosophy

---

41 they are known only . . . that force does to] '?' *added with scoring, pencil left margin.*
'attraction of sulphuric acid for metal *Darwin's handwriting, pencil.*
'its' *Darwin's handwriting, pencil.*
'subjectively' underlined and followed by '?—?', *pencil.*
'We do not know attraction objectively' *Darwin's handwriting, pencil.*

41v '«X»' *and score added pencil, erased.*
The objects of thought] '(' *added before* 'The', *pencil.*
[We see . . . night an effect.]CD *Darwin's handwriting.*
'Cause and effect has relation to forces & mentally because effort is felt' *Darwin's handwriting, pencil.*

42 *Items 42 through 52 written in two columns, the left, narrower side being used for marginal notations and footnotes. These are marked* [RHC] *right hand column;* [LHC] *left hand column.*
«some force» . . . hurting them.—] *double score blue pencil left margin.*

[RHC] *On the Moral Sense*

Looking at Man, as a Naturalist would at any other mammiferous animal, it may be concluded that he has parental, conjugal and social instincts, and perhaps others.—

[LHC] ——————
p. 113. Mackintosh Grotius has argued nearly so[1]

——————

[RHC] The history of every race of man shows this, if we judge him by his habits, as ‹if› another animal. These instincts consist of a feeling of love ‹and sympathy› «or benevolence» to the object in question. Without regarding their origin, we see in other animals they consist in such active sympathy that the individual forgets itself, & aids & defends & acts for others at its own expense.— Moreover ‹the› any action in accordance to an instinct gives great pleasure, & ‹an› «such» actions being prevented by ‹necessity› «some force» give pain: for instance either protecting sheep or hurting them.— Therefore in man we should expect that acts of benevolence towards fellow ‹living› creatures, or of kindness to wife

43 [RHC] 2) and children would give him pleasure, without any regard to his own interest. likewise if such actions were prevented by force he would feel pain. [.By a very slight change in association if others injured these objects, without his being able to prevent it, he would likewise feel pain.— If he saw another man ‹say go› «acting in» accordance to his instincts, ‹he would know that many experienced pleasure,› & by association he would feel part of that pleasure, which the acter received.— If either man did not obey his instincts from interference of passion. he would feel pain, which would generally be anger, as he would be tempted to interfere, but with respect to himself it would be remorse as will be presently shown.— This then is moral approbation, as far as it goes.].[CD] But should he prevented by some passion or appetite, what would be the result? In a dog we see a struggle between its appetite, or love of exercise & its love of its puppies: the latter generally soon conquers, & the dog

43 [.By a very slight . . . as far as it goes.].[CD]] *scored left.*
which the acter received.—] ‘r’ *in* ‘acter’ *over* ‘d’.
obey his instincts] *first* ‘n’ *in* ‘instincts’ *over unidentified letter.*
as far as it goes.] ‘goes’ *over 3–4 unidentified letters.*

42–1 Mackintosh 1837:113, ‘To this [the opinion that man as well as other animals prefer their own interest to every other object] Grotius answered, that even inferior animals under the powerful though transient impulse of parental love, prefer their young to their own safety or life; that gleams of compassion, and, he might have addeed, of gratitude and indignation, appear in the human infant long before the age of moral discipline; that man at the period of maturity is a social animal, who delights in the society of his fellow-creatures for its own sake, independently of the help and accomodation which it yields. . . .’ Darwin has a marginal line beside this passage in his personal copy.

44 [RHC] 3) probably thinks no more of it.— Not so man, from his memory & ‹pow› mental capacity of calling up past sensations, he will be forced to reflect on his choice: an appetite gratified gives only short pleasure. passion in its nature is only temporary, & we do not afterwards think of it.— Whatever the cause of this may be, everyone must know, how soon the pleasure from good dinner, or from a blow struck in passion fades away, so that when man afterwards thinks why was such an instinct not followed for a pleasure now though so trifling he feels remorse.— He reasons on it & determines to act more wisely other time, for he knows that the instinct. (or conscience) is always present (which is indeed, often felt at very time it is disobeyed) & is sure guide.— Hence conscience is improved by attending & reasoning on its action, & on the results following our conduct.— If the temptation to disobey the conscience is extremely great

[LHC] The cause perhaps lies in its frequency & in its consisting in desire gratified & therefore as soon as desire is fullfilled, pleasure forgotten.

45 [RHC] 4) as starvation, or fear of death, one makes allowance & either excuses the «non-» following of ones conscience. & palliates the offence; one always admire the habit formed by «obediance to instinct» ‹conscience›., or rather the strengthened instinct, even when our reason tells— + us the action was superfluous, as one man trying to save another in desperation.— This shows, that our feeling, that the instinct *ought* to be followed is a consequence of that being part of our nature, & its effects lasting, whilst passions although equally natural leave effects not lasting. By association one gains the rule, that the passions & appetites should «almost» always be sacrificed to the instincts.— — One does not feel it wrong in very young child to be in passion, any more than in an animal.— which shows that. it is owing to some ‹subsequent› power (reason) obtained by age, which should show the child, which of its instincts are best to be followed.— Yet even at this time, malevolence,, when not urged to it by passion, shows a bad child.— Hence there are certain instincts pointing out lines of conduct to other men,

46 [RHC] 5) which are natural (& which «when present» give pleasure.) & which man *ought* to follow— it is his duty to do so.— So we say a pointer *ought* to stand a ‹spaniels› «housedog's» duty is to watch the house.— it is part of ‹duty› their nature.— When a pointer spring his bird. one says for shame (& the «old» dog really feels ashamed?) not so puppy, we ‹do› try to teach him & strengthen his instincts.— so man *ought* to follow certain lines of conduct, ‹although› even when tempted not to do so, by other natural appetites.— he

44 probably thinks . . . afterwards think of it.—] *scored blue pencil left.*
45 conscience. & palliates . . . strengthened instinct, even when] *scored blue pencil left.*
being part of our nature,] 'of' *over* '='.
One does not feel it wrong . . . animal.— which shows] *scored blue pencil left.*
46 So we say . . . & strengthen his instincts.—] *scored blue pencil left.*
feels ashamed?)] '?)' *over* '.—'.

is *monster*, or unnatural if malevolent, or hates his children without some passion.— If his passions strong & his instincts weak. he will have many struggles, & experience only will teach him, that the instinctive feeling in its nature being always present. & his passion shortlived, it is to his interest to follow the former; & likewise ‹that the› then receive the moral approbation of his fellow men.—

47 [RHC] 6) Hence man must have a feeling, that he *ought* to follow certain lines of conduct, & he must soon *necessarily* learn that it is his interest to follow it. even when opposed by some natural passion.— (a)
[LHC] The conscience rebukes malevolent feelings, as much as actions, therefore Sir J. M. talks too much about the contiguity to will.[1]
(a) The origin of passions too strong for our present interest receive simple explanations from origin of man.—
[RHC] By interest I do not mean any calculated pleasure, but the satisfaction of the mind, which is «much» formed by past recollections.— Hence he has the right & wrong in his mind.— Now we know it is easy by association to give «almost» any taste to a young person. or it is accidentally acquired from some trifling circumstance.— Thus a child may be taught to think almost anything nasty. (‹& accidentally› «by odd association» comes to this conclusion, not owing to peculiarity of organ of taste, for when grown up often conquers it). It will be only rarely that it thinks that nasty, which the *natural* tastes say is good. yet horseflesh show that even this is possible.— So that as there nice & nasty in taste, & right & wrong in action, so a child may be taught, or will acquire from seeing conduct of others, the feeling that *almost* (rarely if opposed to *natural* instincts) any action is either right or wrong.—

48 [RHC] 7) Hence, what parents think will be good for the child on the long run, & for themselves & others, (as the parents are instinctively benevolent) they will teach to be wrong or right; this teaching may be curiously modified

47 By interest I do not . . . satisfaction of the mind; which] *scored blue pencil right margin.*
it is easy by] 'by' *over* 'to'.
child may be taught. . . . to this conclusion,] *scored blue pencil left margin.*

47–1 Mackintosh 1837:199, 'But volitions and actions are not themselves the end, or last object in view, of any other desire or aversion. Nothing stands between the moral sentiments and their object. They are, as it were, in contact with the will . . . Conscience may forbid the will to contribute to the gratification of a desire. No *desire* ever forbids *will* to obey *conscience*.' [Darwin's italics.] P. 201, '. . . man becomes happier, more excellent, more estimable, more venerable, in proportion as conscience acquires a power of banishing malevolent passions . . .' See also P. 198, 'The truth seems to be, that the moral sentiments in their mature state, are *a class of feelings which have no other object but the mental dispositions leading to voluntary action, and the voluntary actions which flow from these dispositions*.' Darwin drew a line beside this passage, and made the following marginal notation, 'How can cowardice, or avarice, or unfeelingness be said to be dispositions leading to action, yet conscience rebukes. a man, who allows another to drown, without «trying to» save his life.'

by circumstances of country, so will the conscience in these cases.— Those instructions, which the child sees uniformly performed by the teachers & all around him, will be paramount,— hence the law of honour. & the etiquettes of Society.—
[LHC] Sir J. M. gives different explanation of law of honour from Paley[1]
[RHC] Anyone, who will reflect must feel, how like to injured conscience, is the feeling of any custom of society broken..— & how far more ‹feelin› acute the feeling really is.— All these associated «habitual» feelings. become like the instinctive, ones, ‹which either lead to actions or not, as feeling of cowardice› «This is not connected with sense» instantaneous so declaring it is right or wrong.— «[just as in tastes of the mouth]CD»
[LHC] My theory of instincts, or heredetary habits fully explains the cementation of habits into instincts.
[RHC] Feelings of the mind, whether leading to action or not, are the parts of our nature, ‹sub› subject to their instincts & associations.— often feelings which do not lead to action are repressed thus avarice. &c &c.—

49 [RHC] 8) in the beginning I mentioned only three instincts.— I am far from saying there are not more, or that the three are as simple as I have said.—
[LHC] instinctive fear of death: of *hoarding* . . Ld. Kames, which Sir. J. says is so ridiculous.[1]
[RHC] the social instinct may be combined with feeling towards one as a leader,— the conjugal feeling may be directed towards one or more.— It will be hard to discover this, for the different races of man may have different instincts, as we see in dogs & pidgeons.— But as man is animal at head of series in which «special» instincts decrease, I should think they were very few & general in their nature.— So that we have some, it is sufficient to give rise to the feeling of right & wrong.— on which «almost» any other might be grafted.—

Origin of the instincts

Hartley, (according to Sir J) explains our love of another, as pleasure arising from association from having received benefits from this person.—
[LHC] p. 254. &c &c
[RHC] But the love is instinctive, & how does it apply to mother loving child, from whom, she has never received any benefit.— Yet I think there is much truth in doctrine, for

48 Feelings of . . . avarice. &c &c.—] *brace left.*
49 [LHC] instinctive fear . . . ridiculous.] *brace right.*

48–1 Darwin's source and meaning of the 'law of honour' is obscure. However, see Manier 1978:169.
49–1 Mackintosh 1837:255, note, 'A very ingenious man, Lord Kames, whose works had a great effect in rousing the mind of his contemporaries and countrymen, has indeed fancied that there is a "hoarding instinct" in man and other animals. But such conclusions are not so much objects of confutation, as ludicrous proofs of the absurdity of the premises which lead to them.'

50 [RHC] 9) We can thus explain love of place.— although here we have not received pleasure *from* the place, but merely *in* the place.[1] & yet place calls up pleasure.—

[LHC] the instinct of sociability & sociability, doubtless grow together

---

[RHC] This feeling seems to vary in races of man. & certainly in «species of» animals, in which case it undoubtedly is instinctive. But does not Hartley explanation apply perfectly to origin of these instincts.— the having received pleasure from some *one* «person» in early infancy, during many generations giving love of mother; the having received some advantages from man. during many generations giving the social feelings.—

[LHC] According to my theory, all instincts demand some explanation
[RHC] Although I cannot pretend to say how far & minutely our instincts extend, yet as they are acquired by social animals, living under certain conditions, in this world, they ‹will conform to the law,› «can only be such, as are consistent» with social animals, that in which have a beneficial tendency, (not to any one individual but to the whole past race).— ‹no one› doubts) « ‹I cannot› »
[LHC] *On the Law of Utility* Nothing but that which has beneficial tendency through ‹all› «many» ages. could be acquired, & we are certain from our reason, that all which ‹has› (as we must admit) has been acquired, does possess the beneficial tendency

51 [RHC] 10) that the instincts of bees & beavers «& deer» have ‹been formed› a beneficial tendency to them, as ‹*social*› animals of peculiar ‹kinds› «social feelings», & living under certain conditions; by my theory they have been formed by the circumstances, which have led to the peculiarities, & hence ‹must have› «only that which» had a beneficial tendency during past races could become instinctive.—
[LHC] ˣIt is probably That becomes instinctive, which is repeated under many generations. (& under unknown conditions) (for pig will not so readily attain *instinct* of pointing as a dog.— also age has much influence.)— & only that which is beneficial to race, will have reoccurred'.

50–1 Mackintosh 1837:257, 'It is easy to perceive how the complacency inspired by a benefit, may be transferred to a benefactor, thence to all beneficient beings and acts. The well-chosen instance of the nurse familiarly exemplifies the manner in which the child transfers his complacency from the gratification of his senses to the cause of it, and thus learns an affection for her who is the source of his enjoyment. With this simple process concur, in the case of a tender nurse, and far more of a mother, a thousand acts of relief and endearment, of which the complacency is fixed on the person from whom they flow, and in some degree extended by association to all who resemble that person.' Darwin has a double line beside this passage, and he wrote in the margin, 'common to animals hence Love of Place.— X! X will not explain love of parent to child—except heredetary—'

NB. Until, it can be shewn, what things easiest become instinctive, this part of argument fails, or rather is weak.—
[RHC] Better simply put it, beneficial tendency in every instinct to the species in which it occurs. [or, more correctly «in which it» has been so in some past time, hence passions]CD «although perhaps useful at present to some extent.» Hence this is the law of our instinctive feelings of right & wrong,— education, of parents strives* to same end.— & general actions of community must frequently teach same end.— Hence this becomes the law of right & wrong, though, that part, which is acquired by association from education & imitation, has often been perverted from want of reason.— Hence as Eugenius says, slow growth of rule of right.—[1]
[LHC] *for it strives to give conduct beneficial to all the children, « ‹then› each himself» & parents, & hence to nearly all the world.—

As conditions change, from civilization, education changes, & probably likewise instincts, for the same law effects both.—

‹such› changes «in accordance to beneficial tendency» will most readily affect. the instincts, for they are in accordance with it. thus a dog may be trained to hunt one pig sooner than other, rather than change hunting instinct.

50v *Our tastes in mouth by my theory are due to ‹habit› heredetary habit (& modified & associated during lifetime). so is our moral taste
p. 152. Reason never can lead to action.—[1]
p. 164. Ld. Shatsbury under term of *Reflex Senses* seems to have ‹compared› «perceived» the comparison between our instinctive feelings & our short lived Passions'[2]

50v *This page continues the left margin column of 51.*

51–1 Eugenius 1838:13, 'Tardily and gradually, no doubt, do the principles of moral truth emerge into view, even among the sagest and most virtuous of the heathen.' N.B. Eugenius is not mentioned in Mackintosh, 1837.

50v–1 Mackintosh 1837:152–53, 'Reason, as reason, can never be a motive to action. It is only when we superadd to such a being [viz., one capable of reason but incapable of receiving pleasure or pain] sensibility, or the capacity of emotion or sentiment (or what in corporeal cases is called sensation), of desire and aversion, that we introduce him into the world of action. We then clearly discern, that when the conclusions of a process of reasoning presents to his mind an object of desire, or the means of obtaining it, a motive of action begins to operate; and reason may then, but not till then, have a powerful though indirect influence on conduct. Let any argument to dissuade a man from immorality be employed, and the issue of it will always appear to be an appeal to a feeling. You prove that drunkenness will probably ruin health. No position founded on experience is more certain. Most persons with whom you reason must be as much convinced of it as you are.' Scored by Darwin.

50v–2 Mackintosh 1837:164–65, '. . . goodness consists in the prevalence of love for the system of which we are a part, over the passions pointing to our individual welfare; a proposition which somewhat confounds the motives of right acts with their tendency, and seems to favour the melting of all particular affections into general benevolence, because the tendency of these affections is to general good. The next, and certainly the most original, as well as important, is, that there are certain affections of the mind which, being contemplated by the mind itself through what he [Earl of Shaftesbury] calls *a reflex sense*, become the objects of love, or the contrary, according to their

State broadly in child or animal it is equally proper to obey anger as benevolence (but not cool malevolence). it is only after reason comes into play that anger can be said to be wrong.—. for then only is it perceived, that our passions are too strong for our instincts. to gain long-lived good, ie happiness— yet this system not *selfish.— explained by principles if Mackintosh.—*

p. 262. Some good remarks, on analogy of pleasure of imagination «the utility part being blended & lost» & moral sense.— My theory explains both, perhaps, by habit—[3]

52 [LHC] 11) Whewells preface.
[RHC] It appears that Sir. J. & others think there is distinct faculty, of conscience.—[1] I believe that certain feelings & actions are implanted in us. & that doing them gives pleasure & being prevented uneasiness, & that this is the feeling of right & wrong.— so far it has *independent* existence. & is supreme. because it is «a» part of our nature, ‹not› which regulates our feelings steadily & not like our appetites & passion, which receive enjoyment from gratification & hence are forgotten— only so far do I admit its *supremacy* p. 37. Whewells gives Mackintosh's theory: the remarks about "contact with

nature. So approved and loved, they constitute *virtue* or *merit*, as distinguished from mere *goodness*, of which there are traces in animals who do not appear to *reflect* on the state of their own minds, and who seem, therefore, destitute of what he elsewhere calls *a moral sense*.' From 'reflex sense' to 'goodness' scored by Darwin.

50^v–3 Mackintosh 1837:261, 'The sentiment of *Moral Approbation*, formed by association out of antecedent affections, may become so perfectly independent of them, that we are no longer conscious of the means by which it was formed, and never can in practice repeat, though we may in theory perceive, the process by which it was generated. It is in that mature and sound state of our nature that our emotions at the view of *Right* and *Wrong* are ascribed to *Conscience*.' Darwin drew a vertical marginal line beside this passage, and wrote, 'rather instinctive' in the margin. Pp. 262–64, 'The pleasures (so called) of Imagination appear, at least in most cases, to originate in association. [N.B., this passage is in the section on Hartley.] But it is not till the original cause of the gratification is obliterated from the mind, that they acquire their proper character. Order and proportion may be at first chosen for their convenience: it is not until they are admired for their own sake that they become objects of taste. Though all the proportions for which a horse is valued may be indications of speed, safety, strength, and health, it is not the less true that they only can be said to admire the animal for his beauty, who leave such considerations out of the account while they admire. The pleasure of contemplation in these particulars of nature and art becomes universal and immediate, being entirely detached from all regard to individual beings. It contemplates neither use nor interest. In this important particular the pleasures of imagination agree with the moral sentiments. Hence, the application of the same language to both in ancient and modern times. . . . But the essential distinction still remains. The purest moral taste contemplates these qualities only with *quiescent* delight or reverence. It has no further view;— it points towards no action. Conscience, on the contrary, containing in it a pleasure in the prospect of doing right, and an ardent desire to act well, having for its sole object the dispositions and acts of voluntary agents, is not, like moral taste, satisfied with passive contemplation, but constantly tends to act on the will and conduct of the man. Moral taste may aid it, may be absorbed into it, and usually contributes its part to the formation of the moral faculty; but it is distinct from that faculty . . .'

52–1 Mackintosh 1837:41 (Whewell's preface includes 1–46.), '[Mackintosh replies] all the separate objects which conscience approves, the social affections, the decisions of justice, the maxims of enlightened prudence, tend to the happiness of some part of the species, and that thus the general rules of conscience must agree with the rules of the general happiness.' In the margin beside this statement Darwin wrote, 'but why the separate parts?'

will"[2] is unintelligible to me.— conscience regulates feelings, as of cowardice.— the whole appears to me rather rigmarole.— He does not say anything about any principles *born* in us.— Great difference with my theory.— see p. 349.— remark on this point.—[3]
[LHC] p. 194. «&c &c» Butler's view given on conscience: I cannot admit it.— see notes to it by me. .'[4]

'p. 333 «& p. 377» some remarks. showing that instinct cannot be said to guide *will.* as bird building nest, but supplies it— instinctive feelings will doubtless lead to similar actions which in prior ‹races› generations led to their formation.—

‹————›

'N.B. If feeling or emotion rises from heredetary action on body.— This feeling, when instinctive will lead to action.— the passion rising from weariness leads to striking blows.—'[5]

52–2 Mackintosh 1837:38, 'According to [Mackintosh], the moral faculty consists of a class of desires and affections which have dispositions and volitions for their sole object . . . the moral sentiments are *in contact with the will* . . .' P. 36, 'Man's soul at first, says Professor Sedgwick, is one unvaried blank, till it has received the impressions of external experience.'
52–3 Mackintosh 1837:349–50, '. . . the formation of . . . affections is acknowledged to belong to a time of which there *is no remembrance* [Darwin's italic; and, a score beside this passage]; —an objection fatal to every theory of any mental function,— subversive, for example, of Berkeley's discovery of acquired visual perception, and most strangely inconsistent in the mouth of a philosopher whose numerous simplifications of mental theory are and must be founded on occurrences which precede experience.' Darwin has a score by this phrase, and in the margin wrote, 'so in birds it is'.
52–4 Mackintosh 1837:194, 'This natural supremacy [of man in nature] belongs to the faculty which surveys, approves, or disapproves the several affections of our minds and actions of our lives. As self-love is superior to the private passions, so conscience is superior to the whole of man. Passion implies nothing but an inclination to follow it; and in that respect passions differ only in force. But no notion can be formed of the principle of reflection, or conscience, which does not comprehend judgment, direction, superintendency. "Authority over all other principles of action is a constituent part of the idea of conscience, and cannot be separated from it." ' NB. The quotation marks were added by Darwin in pencil, and in the margin beside this statement he wrote, 'if so, my theory goes.— in child one sees pain & pleasure struggling' This passage is in the Section on Butler.
52–5 Mackintosh, 1837, p. 333, '[Mr. Dugald Stewart] considers the appearance of moral sentiment at an early age, before the general tendency of actions could be ascertained, as a decisive objection to the origin of these sentiments in association,— an objection which assumes that if utility be the criterion of morality, associations with utility must be the mode by which the moral sentiments are formed, which no skilful advocate of the theory of association will ever allow. That the main, if not sole, object of conscience is to govern our voluntary exertions, is manifest. But how could it perform this great function if it did not *impel the will*? and how could it have the latter effect as a mere act of reason, or indeed in any respect otherwise than as it is made up of *emotions*, by which alone its grand aim could in any degree be attained? Judgment and reason are therefore preparatory to conscience, not properly a part of it.' Darwin has drawn a marginal line beside this passage, and has written in the margin, 'can the instinct of bird building nest be said to imply will.—' And at the bottom of the page he wrote, 'yet emotions are results— are trains of thought ‹firmly› long associated with action' N.B., Darwin underlined 'impel the will' and 'emotions.' Pp. 376–77, 'But it may still be reasonably asked, why these useful qualities [pursuit of truth and knowledge for their own sake, without regard for power or fame] are morally improved, and how they become capable of being combined with those public and disinterested sentiments which principally constitute conscience? The answer is, because they are entirely conversant with volitions and voluntary actions, and in that respect resemble the other constituents of conscience, with which they are thereby fitted to mingle and coalesce.' Darwin drew a marginal line beside this passage, and wrote, 'nonsense— similar association may be made with actions, involuntary, as———[CD's blanks] & etiquettes of society broken unconsciously.—' And beside the follow-

52v p. 224.— Hume's Inquiry— good abstract of Butler & arguments of beneficial tendency of affections.—[1]

If ever I write on these subjects consult ‹following› pages. ‹p. 231› marked in my Mackintosh[2]

53 1) *Mackintosh's Ethnical Philosophy*
p. 6— "The pleasure which results when the object is attained (the gratification of one's offspring) is not the aim of the agent, for it does not enter into his contemplation.—"[1] Now Eugenius would contend against this—[2] but the pleasure a dog has in obeying its instinct,— as young pointer to point— clearly shows this is true.
p. 13. Affections cannot be analysed into "power" &c &c &c— & if termed "selfish", must be subclassed as "disinterested"[3]
p. 14. It is *allowed*, that we have conception of moral obligation «when grown up???» & the question is, whether this can be resolved into some operation of intellectual faculties—[4] Will Eugenius allow this moral obligation?

ing passage, Darwin has a marginal line and a large question mark and two large exclamation marks: 'All those sentiments of which the final object is a state of the will, become thus intimately and inseparably blended; and of that perfect state of solution . . . the result is *Conscience*— the judge and arbiter of human conduct; which though it does not supercede *ordinary motives* of virtuous feelings and habits, which are the ordinary motives of good actions, yet exercises a lawful authority even over them, and ought to blend with them.'

52v–1 Mackintosh, 1837, p. 224, '. . . all the qualities and actions of the mind which are generally approved by mankind agree in the circumstance of being useful to society. In the proof (scarcely necessary) that benevolent affections and actions have that tendency, [Hume] asserts the real existence of these affections with unusual warmth; and he well abridges some of the most forcible arguments of Butler* (**Inquiry*, sect. ii, part i., especially the concluding paragraphs; those which precede being more his own.) whom it is remarkable that he does not mention.' Darwin has a marginal line beside this passage.

52v–2 Mackintosh 1837:230–31, '[Hume's] general doctrine is, that an interest in the wellbeing of others, implanted by nature, which he calls *Sympathy* in his *Treatise of Human Nature*, and much less happily *Benevolence* in his subsequent *Inquiry*,* (**Essays and Treatises*, vol. ii), prompts us to be pleased with all generally beneficial actions . . . though he truly represents our approbation, in others, of qualities useful to the individual, as a proof of benevolence, he makes no attempt to explain our moral approbation of such virtues as temperance and fortitude in ourselves. He entirely overlooks that consciousness of the *rightful supremacy of the moral faculty* over every other principle of human action, without an explanation of which, ethical theory is wanting in one of its vital organs.'

53–1 Mackintosh 1837:6, 'The pleasure which results when the object is attained is not the aim of the agent, for it does not enter into his contemplation; nor could we derive pleasure from the attainment of such objects except the desire had previously existed.' Note that the parenthetical phrase in CD's notes is Darwin's.

53–2 Eugenius 1838:17, 'The contenders for the progress of morality by means of the gradual progress of reason, will, I think, all be pleased with that statement which places the discovery of moral and physical truth on the same foundation.'

53–3 Mackintosh 1837:13, 'If the love of a parent be a compound of love of power and similar ingredients, will it not follow, that if we expect to gain power by sacrificing a parent rather than by serving him, it is *consistent with the nature* of our best affections that we should do so? And is not this a conclusion too monstrous to be accepted by any moralist? The benevolent and family affections, and the desire of power, appear, then, to differ in some other way than in being modifications of the same elements; and, even if we choose . . . to call the latter class of principles selfish, the former must be arranged in a different group, which we cannot designate better than by calling them disinterested.'

53–4 Mackintosh 1837:14, 'It is allowed on all sides that we have a conception of moral obligation; and the question is, Whether this conception can be resolved into some operation of the intellectual faculties, as the perception of general utility; or whether, on the contrary, it is incapable of being thus resolved, and must properly be ascribed to a separate faculty.' The insertion is Darwin's.

53v [2] [The improvement of the instinct of a shepherd dog, is strictly analogous to education of child,— causing many actions to be considered right & wrong,— to be associated with the approving or disapproving instinct— which were not originally, if the shepherd dog had no instinct to commence with scarcely possible to teach it— all *dogs* might be taught, but not *cat*, that is not act by gusto, though by fear it might be partly made.]CD
p. 21. "Why ought I to keep my word"[1]— gives the problem, of ethics— [my answer would be to all such cases— either, that from the necessities «& good» of society such conduct is instinctive in me (& as a consequence, but not cause gives me

54 [3] pleasure) or that I have been taught or habituated to associatical, the emotions of this instinct, with that line of conduct, & if taught rightly, it will be for the general good, that is, the same cause, which gives the instinct.—]CD
p. 22. says affections, desires, & moral sense all different.—
P. 22. Butler & Mackintosh characterize the moral sense, by its "supremacy",—[1] I make its supremacy, solely due to greater duration of impression of social instincts, than other passions, or instincts.— is this good?—

I should think some parts of the emotive part of man, may be quite artificial, as avarice love of *gold*.— love of fame— Yes Hartley explains this & Mackintosh shows the change produced.—

54v 4) p 38 Conscience checks the *wish* to ‹other› outward gratification, whilst ‹the› no desire of gratification will check the consciences desire for virtue.—[1] [I expect there is some fallacy here.— at least point of «false» honour will stop all wish to gratify ‹it›— anything contrary to it]CD NB. the very end of conscience is stop to wishes of passion &c. whilst the passions have no relation I think this ‹boshes›«nonsense»— My theory of durableness will explain it.—

53v such conduct is instinctive] 'instinctive' *alternate reading* 'instructive'.
54v the passions have no relation] *long rule under* 'no'.

53v–1 Mackintosh 1837:21, '[Paley] reduces moral obligation to two elements— external restraint, and the command of a superior. This attempt at an analysis of morality is singularly futile . . . external constraint annihilates the morality of the act, and the reference to a superior presupposes moral obligation . . . If Paley had stated his question . . . "Why *ought* I to keep my word?" he would have had before him a problem more to the purpose of moral philosophy, and one to which his answer would have been palpably inapplicable.'

54–1 Mackintosh 1837:22, 'Thus, as we separate the affections from the desires, we distinguish the moral sense, or conscience, from both. Butler, and Mackintosh with him, express the relation of conscience to the other principles of action, by ascribing to it a *supremacy*, or a right of command.'

54v–1 Mackintosh 1837:38, 'The conscience requires virtuous acts and dispositions to action; and by such requisition it can check and control any desires of external objects; but no desire of any outward gratification can prevent the conscience from demanding a virtuous direction of the will; and this mental relation explains and justifies, Mackintosh conceives, that attribution of supremacy and command to the conscience on which moral writers have often insisted.'

Would not the maternal affections (in a dog. & therefore not ‹instinct› «conscience») equally ‹prefe› destroy all wish of outward gratification,— see what cases Mackintosh gives & try it.—

55v p. 241

---

(1) Any action by habit may be thought wrong.— & conscience will imperiously say so, & produce shame & remorse—[1] [Thus pungency of one's feelings for indecency— preposterously so, for Marquesans think only of prepuce, crepitando,]CD & *if passions makes one break these artifical rules, get remorse*— ((hence desires do not intervene between this kind of conscience & the will, though «this» conscience does between the desires & will?)) (2) It is other question what it is desirable to be taught,— all are agreed general utility (3) It is other question whether any thing is taught instinctively; I say yes, & my explanation agrees. with last head.— (4) It is other question, how the feeling of ought, shame. right & wrong comes into mind in first case— seeing how shame is accompanied by blushing, bears some relation to others

55 5) if so, it is perhaps deviation from the *instinctive*. right & wrong.— (animals excepting domesticated ones have no right & wrong except instinctive ones) Perhaps my theory of greater permanence of social instincts explains the feeling of right & wrong.— arrived at first ‹rationally› by feeling— reasoned on, steps forgotten, habit formed,— & such habits carried on to other feelings, such as temperance, acquired by education.— CD[In similar manner our *desires* become fixed to ambition. money, books &c &c.— ‹]› the "secondary passion" of Hutcheson unfolded by D. Hartley.—[1]

55v *preceeds 55 because the sheet was incorrectly mounted.*
prepuce, crepitando . . . intervene between this] *triple scored left margin, pencil.*
if passion makes . . . get remorse—] *underlined pencil.*

55v–1 Mackintosh 1837:240, 'For it is certain that in many, nay in most cases of moral approbation, the adult man approves the action or disposition merely *as right*, and with a distinct consciousness that no process of sympathy intervenes between the approval and its object.' Darwin made a marginal line beside this passage, and at the bottom of the page wrote, 'My whole question with the breaking mere rule of etiquette.'

55–1 Mackintosh 1837:251–52, in Section on David Hartley, '[Mr. Gay] blames, perhaps justly, that most ingenious man [Hutcheson], for assuming that these sentiments and affections are implanted, and partake of the nature of instincts . . . he [Gay] well exemplifies the power of association in forming the love of money, of fame, of power, etc; but he still treats these effects of association as aberrations and infirmities, the fruits of our forgetfulness and short-sightedness, and not at all as the great process employed to sow and rear the most important principles of a social and moral nature.

'This precious mine may therefore be truly said to have been opened by Hartley; for he who did such superabundant justice to the hints of Gay, would assuredly not have witheld the like tribute from Hutcheson, had he observed the happy expression of "secondary passions" . . .'

Darwin's Abstract of John Macculloch 1837
Proofs and Illustrations of the Attributes of God

Transcribed and edited by Paul H. Barrett

After the *Beagle* voyage it was Darwin's practice to make marginal annotations in books which he owned. For books not his own, if important, he wrote abstracts on separate sheets of paper. The following manuscripts are abstracts of one volume of a 3-volume work by John Macculloch (1773–1835), *Proofs and Illustrations of the Attributes of God from the Facts and Laws of the Physical Universe, being the Foundation of Natural and Revealed Religion* (1837). Macculloch was already known to Darwin for his books on geology and especially for his article on the 'Parallel Roads of Glen Roy', that Darwin in the summer of 1838 had set out personally to investigate.

Darwin's abstracts of *Attributes* survive in three locations in the Darwin archive. The main body of the abstract, which cites eighteen pages of volume 1, is found in DAR 71:53–59. Another section, which abstracts six pages of volume 1, is in DAR 205.5:28–29, 167. In addition a fragment abstracting three pages of volume 1 is item 11b from Old & Useless Notes in DAR 91. DAR 71 contains many of Darwin's book abstracts and DAR 205.5 contains classification folios.

DAR 71:53–58 consists of three folded sheets of paper; all but 53 and 57 embossed with the Athenaeum seal and bearing a [. . .] Smith & Son 1838 watermark. DAR 71:59 is also on Athenaeum paper but bears a [. . .]tman 1838 watermark. Of the two classification sheets one, DAR 205.5:28–29, is identical to the Athenaeum paper except no embossed seal is evident. The second classification sheet, DAR 205.5, 167 is on plain paper, but its contents carry on the theme of these abstracts and it has pin holes which match DAR 205.5:28. The OUN sheet is very similar to the Athenaeum sheets, but is one-half their size and its watermark is not the same (see p. 000), nor does it have the Athenaeum seal.

All the Macculloch notes are in brown ink or pencil. While the order of writing is uncertain, the several bold and assertive references to Malthus with a number of detailed examples of natural selection in operation, along with a critical stance towards natural theology, suggest the notes were written after Darwin's 28 September 1838 reading of Malthus, and, the bulk of them, probably late in 1838.

The marginal numbers in the following transcriptions are to Cambridge archive folios. Darwin's page numbers, where they exist, are also given.

Darwin's Abstract of John Macculloch 1837
Proofs and Illustrations of the Attributes of God

DAR 71:
53r Macculloch. Attribs of Deity. Vol: I[1]

it will be better always to refer to the author if I use these facts
p. 280. adduces provision of seeds for transportation through the air.— cocoa nut by water «fucus for adhesion[2]».— as examples of design.— perhaps they are so.— but the coral rock might have been uninhabited as the Alpine pinnacles.—[3] One thing must be admitted there would not be these plants, if there was not some provision for transportation:— But I do not want to deny laws.— The whole universe is full of adaptations.— but these are, I believe, only direct consequences of still higher laws.— I do not «then» believe the pappus of ‹th› any one seed. (all have not it) was DIRECTLY *created.* for transportation. it follows from some more general law.— [that the laws of propagation, were created with reference to successive developement I admit, but the admission is probably from ignorance]CD

Who would ever have thought that the intestines of a thrush were means sufficient to ensure propagation of Misseltoe?—[4]

53r *top of page:* '(12)' *added brown crayon to indicate Abstract No. 12 in Darwin's collection of book abstracts (DAR 71).*
it will . . . these facts] *added in circle.*
all have] *over* 'or one'.

53r–1 Macculloch 1837,1.
53r–2 Macculloch 1837, 1:280, '. . . how were the seeds of the fuci to root themselves amid the waves? . . . They are surrounded by a mucilage which water cannot dissolve, and which enables them to adhere to whatever solid body they touch; . . . Let chemistry name another mucilage, a substance which water cannot dissolve, though apparently already in solution in water, and then ask if this extraordinary secretion was not designed for the special end attained, and whether also it does not afford an example of that Power which has only to will, that it may produce what it desires, even by means the most improbable.'
53r–3 Macculloch 1837, 1:279–80, 'If the floating of seeds through water is a contrivance which, like the action of the winds, appears too much akin to what we carelessly term accident, to deserve notice, yet thus chiefly are the naked coral rocks of the great Pacific Ocean clothed with vegetation, and rendered fit for the habitation of man. Are we entitled to give the name of accident to that cause, or combination of causes, by which so great an end is produced—even though metaphysics, and religion equally, did not show that there can be no accident to the Creator and Governor of all things? The buoyancy of cocoa nut, the resisting investments, and the vitality of seeds, were not necessities: but there can be no accident when the end in question is thus attained, and when, without it, all those previous and wonderful contrivances by which these islands are created in the ocean would have been fruitless: while we can even believe that the important cocoa palm was created a maritime plant for this very purpose.'
53r–4 Macculloch 1837, 1:281–82, 'If there is, at first sight, a similar appearance of accident in dispersing the seeds of plants through the digestive organs of animals, the intention is here also rendered evident, by a still more complicated system of contrivance; . . . The fruit is the food of the animal; but the seed is protected from the action of the digestive powers by its investments; as it is also empowered to defy the animal chemistry by its vitality, even appearing to be thus quickened for its peculiar destination.

'Thus alone, it is thought, is the mistletoe propagated . . .'

53v do p. 284. it is hard on my theory of gain of small advantages thus to explain the curling of the valves of the broom.— or the springing of other seeds.—[1] But are we certain that these are necessary adaptations.— May they not be accidental? We have good reason to know that they would not be detrimental accidents, & domesticated variations show us accidents may become heredetary [produce some peculiarity in seed vessel]CD if man takes care they are not detrimental.—

NB. One limit to the transmission of abortive organs will be as long as they are not detrimental.—

---

p. 285 the seed-pod of a desert plant (Anastatica) is rolled along, & splits when it comes to a damp. place.—[2]
Kolreuter mentions some hybrid, whose flower great tendency to break off

54r p. 292. Mac. has long rigmarole about plants being created to arrest mud &c at deltas.—[1] Now my theory makes all organic beings perfectly adapted to all situations, where in accordance to certain laws they can live.— Hence the mistake they are created for them. If we once venture to say plants created to ‹arrest› «prevent» the valuable soil in its seaward course,— we sink into such contemptible queries, as why should the earth have drifted; why should plants require earth, why not created to live on alpine pinnacle? if we once to presume that God «created plants to» arrests earth, (like a Dutchman plants them to stop the moving sand) we ‹do› lower the creator to the standard of one his weak creations.—[2]
All such facts are merely relations of one general law. the plants were no more created to arrest the earth, than the earth revolves to form rain to wash down earth, from the mountains upheaved by volcanic force, for these Marsh plants. All flow from some grand & simple laws.—

54v 4 «Study Cuviers Anatomie Comparé[1]»
p 308. Traces the gradation of skeleton in Vertebrates & constantly alludes

53v **Kolreuter . . . break off**] *added pencil.*

53v–1 Macculloch 1837, 1:284, 'The last division of contrivances for dispersing the seeds of plants, is founded on that most inexplicable property of matter, elasticity, so largely used throughout all Creation: . . . Under this principle, the seed-vessel, or some part connected with it, is provided with a latent spring, to be brought into action as soon as the seeds are fit for dispersion, and not before. . . . This alone is an ample proof of design; because it is a train long laid, and implying foresight.'

53v–2 Macculloch 1837, 1:285, 'Thus also it is with the rose of Jericho (anastatica), where the seed-vessel is rolled along the sands by the winds, until, meeting with a moist spot, it opens and parts with its seeds in that only place amid the parched plain where provision has been made for their vegetation. Can anything have been neglected, where calculations so minute as this exist?'

54r–1 See Macculloch 1837, 1:291–93.

54r–2 Macculloch 1837, 1:295, 'Here alone we see the proofs of design: but no doubt can remain when we examine the long roots of these plants, tenacious as they are numerous, and intersecting the sand in every direction, as if the root, rather than the plant, was the contrivance and the object, and thus rendering it a firm mass. Man even takes advantage of them for the same end; and, when he would have invented them had it been in his power, he cannot doubt that for this purpose they were invented and ordained.'

54v–1 G. Cuvier 1829–30.

«(& at p. 312)» to the abortive bones.[2]
He explains it ‹"By› saying "It is the determination to adhere to a plan once adopted; & it is from these very circumstances, that we become satisfied respecting an original thought, or design, pursued to its utmost exhaustion, & till it must be abandoned for another".—[3]

What bosch!! Put it to case of man.

‹&› The ‹design› determination of a God-head.— the designs of an omnipotent creator, exhausted & abandoned. Such is Man's philosophy. when he argues about his Creator!

---

p. 309. says the ribs in Draco support the flying membrane?!!—[4] that the phalanges have separate movements in the Holocentrus ruber (a fish)[5]

Man has abortive muscles to his ears.— p. 313[6]

Many other good cases — p. do]CD ‹Mac. remarks all Mammifers originally land—animals. as›

55^{r} 5
p. 314. Mac. remarks all ‹land› Mammiferous animals originally terrestrial.— for we find even in Cetaceæ traces of hind extremities.—[1] How are we to explain this.— Did reptiles first inhabit seas.— Were they then killed out «by the increase cold», & did mammifers then take their place? Would they not first occupy the Poles? Is this origin of Polar attributes of the Cetaceæ.— How came Bats also.? before birds? They are ancient.— Are Cetaceæ found in Paris Basin?.—

NB) The explanation of types of structure in classes— as resulting from the *will* of the deity, to create animals on certain plans.— is no explanation— *it*

54^{v} **What bosch . . . of man**] *added opposite quoted passage.*

54^{v}–2 For example, Macculloch 1837, 1:308, 'And in the same animal [the horse], the coffin bone, representing the last phalanges of the fingers, bears even the lateral marks of those, though, for all its offices, this bone is a single one: marking that determination to adhere to a plan, which will appear throughout the whole of this slender examination.'
54^{v}–3 Macculloch 1837, 1:306–7.
54^{v}–4 Macculloch 1837, 1:309, '. . . in the flying lizards some of [the ribs] are extended straight, so as to be the base of wings.'
54^{v}–5 Macculloch 1837, 1:309, '. . . in the Holocentrus ruber, the rows of phalanges become as independent as fingers.'
54^{v}–6 Macculloch 1837, 1:313, 'There is no purpose in the external muscles of the human ear, though they were practised and exerted to the utmost: but they conform to that general plan under which they are useful in quadrupeds.'
55^{r}–1 Macculloch 1837, 1:314–15, 'The unexpected nature of the Cetaceous fishes seems also best explained on the same principles: it is the terminal point of a plan commencing in the quadrupeds as land animals, and gradually traced to a perpetual residence in the water, through intermediate stages of construction or variation. The seal forms the first great remove from the pure terrestrial quadrupeds, though I need not now pursue the anatomical differences: while the Manati and the Dugong carry down this exhaustive plan, till it terminates in the Whale; a fish in the more obvious sense, yet breathing air, warm-blooded, giving suck, and retaining hands, such as they are, for uses similar to those in many of the quadrupeds.'

has not the character of a physical law, «& is therefore utterly useless— it foretells nothing» because we know nothing of the will of the Deity. how it acts & whether constant or inconstant like that of Man.— the cause given we know not the effect

55v [blank]

56r 6
p. 412. Macculloch explains the shortness of life (peculiar to each species) owing to the *growing size of the world*? & the physical changes it was to undergo «animals feeding on each other &c &c».— «(Causing death to some, &c &c)»[1] These are reasons, just as liability to accidents & any other cause.— (& my theory [ALL PARTS OF ONE GREAT SYSTEM. C.D]CD [All this does not explain *death*, but *reproduction*]CD though such a scheme. would require constant miracles.—
p. 420 thinks the great fecundity of *germs* is to afford support to other beings.—[2] *true*, (& the doctrine of checks & my theory)

56v Macculloch. Attrib. Vol I.
p. 330. Mentions the many cases, as in Papilionaceous flower, where such care seems to be taken that the anthers should not be exposed to weather.—[1] this is against my theory of frequent intermarriage.—

56v p. 330 . . . intermarriage.—] *scored ink and 'X' in margin*; 'a' *added and crossed beside* 'X'.
living on the] 'on' *over* 'in'.
Do races . . . seedlings] *scored ink*, 'X' *in left margin.*
Macculloch says . . . world.—] *pencil.*

56r–1 Macculloch 1837, 1:412–13, 'I may now ask in what manner the Creator could have peopled such an earth with an original and undying creation of animals; to omit, at present, the similar and almost superfluous question which relates to plants. The hypothesis of immortal organization excludes the system of increase: and thus, if we consider merely the simple enlargement of the earth, it must have gradually become deficient in replenishment. The total system becomes defective: the earth is a provision for life, since we can conceive no purpose in a vacant globe; but, in such a case, there is a provision without an object.
'. . . Species beyond numbering were originally created to occupy every climate and every soil, every conceivable point of the world: and that all might be filled, their sizes, forms, constitutions, and inclinations, have been varied accordingly. . . . But, as I have just shown, the new lands differ from the original ones, the climates are changed, . . . the character of the vegetation is entirely altered, and the food that maintained myriads has vanished. The constitutions and the inclinations of the animals must be altered to meet those changes: or rather, the Creator should never have produced that variety; in reforming one part of His plan we must condemn the whole. But even this would not suffice.'

56r–2 Macculloch 1837, 1:420, 'The germs of plants are produced in myriads, without any design that they should grow into representatives of the original: this excessive fecundity is designed for the support of animals, and the mortality of the former is the life of the latter.'

56v–1 Macculloch 1837, 1:329–30, 'Botanists have often noticed certain contrivances for protecting the stamina of flowers; and adding to these some others, intended to insure the desired consequences in the fertility of the seed, the whole may be viewed as constituting one of the plans in question. The anthera was, in the first place, to be protected from rain and wind, lest its pollen should be lost; and the means adopted are varied in such a manner, that the peculiar forms of the flowers seem directed to this sole end.
'. . . In the papilionaceous flowers, the securities are multiplied beyond all apparent need or use; since there are three distinct contrivances bearing on the same object. The stamina surround the pistil in contact, with an equivalent effect to that which they have in the Syngenesia; and again, the whole are generally so inclosed by the keel

A plant is in the same predicament as a group of bisexual animals living on the borders of a country favourable to change.—

It might be concluded that Plants would be subject to extreme variation as long as crossing with other varieties was prevented

---

Do races of peas become intermixed & gardner have hybrid seedlings}
p. 333. Macculloch. brings forward. the impregnation of Diœcious Plants by foreign agency— as insects, as wonderful case of adaptation.![2] There would not have been any Diœcious plants, had there been no insects. The right inference is, there were insects «⸮when were Palms formed?» as soon as Diœcious Plants were formed.
Macculloch says, life, forms a broken, recurrent series, whilst the habitation «or world» simple series.—[3]
My theory shows life equally simple series, & therefore trace of beginning in organic world.—

57r Macculloch. Attrib. of Deity. Vol I
p. 232. gives Woodpecker as instance of beautiful adaptation.—[1] & then Chamelion, which feeding on same food, differs in every respect, except «in» quick movements. (sliminess instead of barbs)—[2] In all these cases it should be remembered, that animals could not exist without these adaptations. —fossil forms show such losses.—

Consider *ground* Woodpecker stiff tailed cormorant: pain & disease in world & yet talk of perfection

57r *top of page:* '12' *added brown crayon in circle, see textual note on 53r above.*

and wings, that no injury can reach them. But, beyond all this, the erect breadth and partial bending in the middle of the vexillum render it a perfect wind vane, so that not even the wind can reach what seemed already sufficiently protected.'

56v–2 Macculloch 1837, 1:333.

56v–3 Macculloch 1837, 1 (chap. 5):128–46, 'On the creation and the progressive changes of the earth.' For example, see 1:133, 'But I do not rest this argument on the retrograding simplicity of the Earth alone, as an uninhabited globe. The analogy which occurs in the gradual improvement and multiplication, or, reversely, in the as gradual simplification and diminution, retrogressively, of the forms and numbers of Life, presents not only all that support which analogy ought to do in the works of the Deity, but that further assistance which arises from finding that the two creations, Life, and the Habitation of Life, run a parallel and dependent course; singularly entangled, or acting on each other, alternately, in the nature of cause and effect.'

57r–1 Macculloch 1837, 1:232, 'The tongue of the woodpecker . . . departs from the general principle by which this organ in birds has been constructed, because this family has been destined to feed on insects which it must extract out of deep holes: and it is therefore a sort of spear, provided with barbs.'

57r–2 Macculloch 1837, 1:232–33, 'Similar food, under different circumstances, was destined for the chameleon: and he who should compare the sluggishness and awkward construction of the animal with the activity of its food, would determine that it could never succeed in securing a prey. But the Creator of all things is never at a loss. . . . the activity of the tongue is a match for that of the food . . .'

57^{v} Get instances of adaptations in varieties.— greyhound to hare.— waterdog hair to water— bull dog to bulls.— primrose to ‹open fields› banks— cowslip to ‹banks› **fields**— these are adaptations just as much as Woodpecker. —only we here see means— but not in the other. ǂAll Bridgewater Treatises. are reduced simply statement of productiveness, & laws of adaptationǂ

p. 234. The non-absorbing Camel's stomach is puzzler—[1]

p. do says *inconvenience* would have arisen had «‹not› some» some insects ‹not› not been provided. «with proboscis»[2]
«as bee & butterfly» inconvenience.! *extinction*, utter *extinction*! let him study Malthus & Decandoelle.—[3]

58^{r} The Final cause of innumerable eggs is explained by Malthus.— [is it anomaly in me to talk of Final causes: consider this!—]CD consider these barren Virgins[1]

p. 235. talks of the long spinous processes in Giraffe &c, as adaptations to long necks— why they may as well say, «long» neck is adapted to long necks.—
p. 236. Marsupial bones especial adaptation, to «young».—[2] good God & yet Mails have them. What trash
p. 237. Gives as Summary of adaptations Horny point to chickens beak, to break egg. shells—[3] why chicken could not have lived had it not been so.— let egg shells grow harder. so must those with weak beaks be sifted away.—

57^{v} hare] pencil 're' *over* 'ir'.
‹banks›] *crossed pencil*; **fields** *pencil*.

57^{v}–1 Macculloch 1837, 1:234, 'The stomach of the camel offers another of those special contrivances, where the purpose, and the means of attaining that, are so perfectly adapted, that the design has been universally admitted. It was created to live in a land of little water; and thence is it not only patient of thirst to a degree which appears almost miraculous when compared with other animals, but is furnished with the means of carrying water for further exigencies. This contrivance consists in certain appendages to one of the stomachs: and the mere mechanism, or superfluity of structure, is sufficiently remarkable under the present argument; though it is nothing when compared to that breach of a universal law, without which this would have been unavailing. Every similar cavity has an actively absorbent surface; and water, in particular, would disappear in a short time in every analogous one that we know. But it is ordered that the water receptacles of the camel shall not be absorbent, or shall not at least absorb water rapidly; and thus is the perfection of this design evinced.'

57^{v}–3 See D134.

57^{v}–2 Macculloch 1837, 1:234.

58^{r}–1 Bacon 1665, bk iii, V used in Whewell 1833:355–56.

58^{r}–2 Macculloch 1837, 1:236, 'the marsupial bones in the opposum race offer another instance of the same nature. . . . But it being granted that the young required the protection which the pouch affords, the invention is perfect, as it is also one of those pure inventions for an unusual end, which excludes all chance, and even all hypotheses which unite a sort of limited casualty to a general intended plan; since it is an exception, and not a deviation or modification departing from such a fundamental design.'

58^{r}–3 Macculloch 1837, 1:237.

58v 4
& the species, like 10,000 others, perish. & who will dare to say that this is an infringement on the wisdom or Providence. when **whole** rocks nay very mountains are formed of such dead & extinct forms.— **the exuviæ of the dead & extinct**

———

The analogy between the works of art «or intellect» such as hinge, & hinge of shell, works of laws of organization is remarkable— what is intellect, but organization, with mysterious consciousness superadded
This is similar idea, to cells of bee, corresponding to ‹every› «one or any»— brain making structure, instead of parts of body.— Now we know what instinct is— consider this

———

I look at every adaptation, as the surviving one of ten, thousand trials.— each step being perfect «or nearly so (except no in is[d]) although having heredetary superfluities Man could exist without Mammæ.» to the then existing conditions.— An adaptation made by intellect this process is shortened, but yet analogous, no savage ever made a perfect hinge.— reason, & not death rejects the imperfect attempts.

59 In the «Bee» Mollusca the nervous system is endowed with the knowledge of trying a hundred schemes of structure, in the *course of ages* «step by step».— in Man, the nervous system, gains that knowledge, before hand. & can in idea (with consciousness.) ‹th› form these schemes.—

———

I see no reason, why structure of brain should not be born. with tendency to make animal perform some action.— as well as gain it. by habit.— New theory of instinct, returning to Kirby's view.—

DAR 205.5:
28r Macculloch. Attributes of Deity Vol I.
p. 251— stomach hump, kinds of foot. power of closing nostril, foot, sack. power of endurance &c &c[1] **Camels**? all good cases of corelations.— [There

58v **whole**] *added pencil.*
the exuvæ . . . extinct] *added pencil.*
superfluities] *alternate reading* 'organization'.
59 *small sheet of unwatermarked letter-writing paper, embossed with Athenaeum seal; formerly pinned to the main batch of paper.*
New theory . . . view.—] *double scored left margin.*
28r **Camels?**] *added pencil.*
hand in . . . cricket] *underlined brown crayon.*
Pincers . . . *woodcuts*] *circled ink.*
stones . . . Aphysia] *boxed ink.*
analogy . . . respects] *added in a mixture of pencil and ink*; **Ornithorhynchus** *underlined.*
Harvest mouse] *pencil.*

28r–1 Macculloch 1837, 1:250–51, 'The foot of the camel, reversely, is a broad, elastic, and soft cushion, perfectly adapted to those sands which every other peculiarity in its construction shows to have been its

must have been deserts in the *old* world!]CD
p. 252 analogy of *hand in mole, & mole cricket*[2] & rodents (?)
p. 251. all animals run by hind legs— Kangaroo. only a caricature;[3]
Penguin.— Pincers in Scorpion & Crust in Squilla. & Mantis. CD *woodcuts*
stones swallowed by birds & by Aphysia. C.D
p. 258. «grinding» teeth in ‹stomach of› sun-fish, in mouth of swine & in stomach of lobsters— **analogy in Flamingo & Duck, Ornithorhyncus «externally». Petrel & Whale in some respects**
Chamælion like power in Octopus & Chamælion.— C.D. Sucking feet in Frog. *Walrus*. Fly. Gecko &c. Prehensile tail. in Monkeys & Marsupials. **Harvest mouse** & (Chamælion?) C.D. Spines in Hedge Hog & Echidna.. & Aphrodites C.D. Endless cases.—

28v Macculloch p. 260 intimates canines no special use to Man.[1] Applicable to Bell's sneering-theory.—[2]
p. 263. This kind of doctrine runs through Macculloch, the bills of the Grallæ ‹are› «have been made» long «(as adapted to)» because their food lies deep.—[3] I say it is «as» simple consequence they become long. not at once, but by steps. of which we have manifold traces in the several genera of Grallæ

———

Suppose six puppies are born «& it so chances, that one out of every hundred litters is born with long legs» & in the Malthusian rush for life, only two of them live to breed, if circumstances determine that, the long legged one shall rather oftener than any other one. survive. in ten thousand years the long legged race will get the upper hand. though continually dragged back to old

28v any other one.] *alternate reading* 'every other one'.
fertility of Man] *alternate reading* 'fertility of Mother'.
women are same:] *alternate reading* 'women age same'.

intended dwelling-place: while the union of all those circumstances forms so perfect a design in itself, under the intended destination of this animal, that I must notice the whole, before proceeding with the organs now under examination. The stomach I described in the last chapter; but with this provision, there is a singular endurance of thirst, and also of hunger: while, for this also, there is an analogous provision in the hump, which is an internal store of food, and is gradually absorbed to supply the wants of the system. While it is willing, moreover, or inclined to feed on the thorny plants of the desert which scarcely any other animal will touch, it is provided against injury from them, by a tough cartilaginous mouth: as a power of closing the nostrils against sand, with an analogous provision in the eyes for evading its annoyance, complete a design, so perfect in all its parts, that no perversion of understanding can overlook it, or doubt the intention.'

28r–2 Macculloch 1837, 1:252, 'It is an unexpected extension of the mechanism of the mole's hand, to find it adopted in so very different a department of creation as it is in its application to the mole cricket; . . .'

28r–3 Macculloch 1837, 1:251.

28v–1 Macculloch 1837, 1:260, 'Man, it has been remarked, possesses all the three varieties of teeth . . . but the pointed ones seem to belong to that analogy of structure which pervades whole races of different animals very widely, though the parts are of no use.'

28v–2 C. Bell 1824:62–64. Here Bell discusses the muscles of the lips of carnivorous animals which he says (p. 62) 'are so directed as to raise the lip from the canine teeth.' And (p. 63) 'The snarling muscles take their origin from the margin of the orbit of the eye, and from the upper jaw . . . This action of snarling is quite peculiar to the ferocious and carnivorous animals.'

28v–3 Macculloch 1837, 1:263, 'The food . . . [of the] Gralæ consists of worms or larvæ which reside deep in the earth, and would have been unattainable by the bills of the preceding birds.'

type by intermarrying with ordinary race.— «There is no way of eliminating the evils of old age, after breeding season, or gaining adaptations, but for youth most necessary: the fertility of Man in old age keeps woman alive: for Man & woman are same: fertility of either sex determines life:.»

167r «With respect to whether Galapagos beings are species. it should be remembered that Naturalists are prone, fortunately, to take their ideas, which are arbitrary & empirical, from their own Faunas, which in this case is only true criterion.—[1] Hence it is highly unphilosophical to assert, that they are not species, until their breeding together has been tried.—[2]

———

With respect to the six puppies, if a hare was introduced, or ‹a spe› became more numerous. (from death of its destroyer), or other cause, the long legged race would prevail, even if have afforded only 10th part before & now formed eighth part.— or if other prey diminished, total number of dogs. would diminish, whilst the long legged variety would prevail.— Not separately: NB. These views quite exclude the idea of domesticated animals changing.—

167v From these views we can deduce why small islands. should possess many peculiar species. for as long as physical change is in progress or is, present with respect to new arrivers, the small body of species would far more easily be changed.— Hence the Galapagos Islds are explained.[1] On distinct Creation, how anomalous, that the smallest newest, & most wretched isld should possess species to themselves.— Probably no case in world like Galapagos. no hurricanes.— islds never joined, nature & climate very different, from adjoining coast. Admirable explanation is thus offered.— From these views, one would infer that Mollusca would offer few species, or rather be very slowly changed & vertebrata much so.— so far true, but do not fish offer a most striking anomaly to this. Have they wide ranges? Agassiz has shewn that they most widely differ[2]»

167r «With . . . differ»] *This passage is on a separate sheet of cream paper, with faint horizontal rules on 167r. Matching pin holes with folio 28 show that the sheet was at one time attached to the Macculloch Abstract. Note that the passage can also be read in the order 167v, 167r.*

167r–1 See C137, 'The simple expression of such a naturalist "splitting up his species **& genera** very finely" show how arbitrary & optional operation it is.—'

167r–2 See C161, 'My definition ‹in wild› of species. has nothing to do with hybridity . . .'

167v–1 See B7, 'according to this view animals, on separate islands, ought to become different if kept long enough.— «apart, with slightly differen circumstances.—» Now Galapagos Tortoises, Mocking birds; . . .' Between B7 and Mac167 Darwin's 'views' of how the formation of Galapagos species is 'explained' had changed. They now comprehended the effect of 'slightly differen circumstances' and the 'Malthusian rush for life'. In Mac167v we have a clear crystalization of natural selection, the origin of species, and the singularity of the Galapagos case.

167v–2 Agassiz 1833–43, 1:xxv, 'Placés par leur organisation au dessus des Rayonnés, des Mollusques et des Articulés, ils [fish] présentent des particularités de structure plus variées et sujettes à des différenciations plus nombreuses; aussi remarque-t-on chez eux, dans des limites géologiques plus étroites, des différences plus considérables que chez les animaux inférieurs.'

29^{r} 3
A very wide range must be destructive to species, when physical changes are in progress; (on the same principles that islands are favourable,) because it must take so long to change species— yet this is contradicted by continents ‹bri› abounding with species— there will be a balance, continents have been split up.— who can decide their limits.—

To show how little we understand of the Physiological relations of animals. equatorial countries are supposed favourable to terrestrial Mammifers— Marine ones «of large size» ‹to› are best nourished by arctic regions— Whales. «Narwhal» Polar bear. Walrus, great *Seals* of Antarctic seas. (on other hand Spermaceti Whale & Manatee.— Naturalists must be cautious.— ‹some others›: study these facts read Lacépède on Cetacea & Geographical Distrib of larger Seals— Are Porpoises numerous in cold Oceans I think not.— Does this bear on, the absence of their remains in the Wealden?

29^{v} In the strongly separated Arctic genera, there is evidence of antiquity & extinction of such forms— these views will bear on geology—

There is an analogy between fang of snake, (jaw of spider?) sting of bee, sting of nettle.— Are there any other analogies— prickly plants or animals— Exudation of fetid «& acrid» secretion in Mollusca. insects «Carabids & Staphylini» & Mammalia.
The eye being formed in Mollusca, Articulata, & Vertebrata, & Planaria, & light affecting plants. in insects the end is gained by some very different method. in pedunculated eye of Chamelion. **crabs** Crabs & Mollusca we have analogues

---

The stillness p. 276) of flight of Owl remarkable, [gained by very different process from Bats. CD]^{CD}. «Macculloch says no other bird could catch mouse by night[1]»
Sailing lizards. squirrels & Opossums «& fish»: *flying* lizards.—Mammalia. C.D.—

29^{r} *Seals*] 'S' *over* 's'.
29^{v} The eye . . . plants] *brace in margin.*
Chamelion.**crabs**] '**crabs**' *pencil.*
stillness] 't' *emmended pencil.*
p. 276)] *vertical stroke following.*

29^{v}–1 Macculloch 1837, 1:275–76, 'The most remarkable variation, however, occurs in the owl; as, united to the other peculiarities of this animal, it forms another of those completed designs which I have already noticed in the chameleon, the camel, and the mole. There is a peculiar laxity in the feathers, which, with the mode of managing its wings, confers on it that power of inaudible flight . . . it is by no means easy to discover what is the variation through which this silence of flight is attained. . . . nor is there any other bird in creation which could have surprised the quick-hearing and active shrew-mouse during the stillness of the night.'

Table of Location of Excised Pages

References to location of pages in the Darwin Archive, Cambridge University Library include the number of the volume in which the pages appear and, where applicable, the folio number of the volume. DAR 42:35 thus refers to folio 35 of volume 42 of the collection. If the section of the archive in which the page is found is devoted to a particular subject, that fact is noted under 'Subject' with subject headings written in Darwin's hand being given in quotation marks.

For Notebooks B–E, the bulk of the excised pages are currently in a single volume DAR 208. These pages were formerly distributed in the subject portfolios Darwin maintained for *Natural Selection* and the *Origin* now catalogued as DAR 205. The excised pages were removed to DAR 208 by the De Beer editorial team. We have been able to reconstruct the original locations of these pages because most bear brown crayon classification numbers that correlate with the contents of the extant portfolios.

† Indicates which of the two MS pages of an excised sheet is marked with a brown crayon classification number.

* Indicates the presence of pin holes, that mark Darwin's use of straight pins to attach the excised sheet to another MS.

‡ Indicates pencil classification numbers for divisions of DAR 205.7 (Hybridism portfolio): 1c = Hybridism in animals, 1b = Prepotency in animal hybrids.

} Indicates that the pages were pinned together.

Red Notebook

Page	DAR Location	Subject
5–6	42:35	'Scraps Cleavage'
7–8	40:4	geology
13–14*frag	42:90	'Gravel, Valleys Denudation &c &c'
15–16	42:90	'Gravel, Valleys Denudation &c &c'
19–20	42:24	'Scraps Cleavage'
33–34* }	40:6	geology
35–36* }	40:2	geology
37–38* }	42:84	'Gravel, Valleys Denudation &c &c'
39–40* }	42:84	'Gravel, Valleys Denudation &c &c'
43–44	40:2	geology
45–46	42:80	'Gravel, Valleys Denudation &c &c'
53–54 frag	42:57	'Scraps to end of Pampas Chapter'
55–56 frag	42:56	'Scraps to end of Pampas Chapter'
65–66	42:24	'Scraps Cleavage'
67–68	42:84	'Gravel, Valleys Denudation &c &c'
87–88 frag	40:5	geology
93–94* }	42:90	'Gravel, Valleys Denudation &c &c'
95–96* }	42:90	'Gravel, Valleys Denudation &c &c'
97–98* }	42:90	'Gravel, Valleys Denudation &c &c'
99–100	40:3	geology
105–106	40:3	geology
113–114	42:58	'Scraps to end of Pampas Chapter'
134–135	42:56	'Scraps to end of Pampas Chapter'
136–137	40:4	geology
140–141	5:86	Original notes for glacier paper
142–143	40:1	geology
144–145	42:57	'Scraps to end of Pampas Chapter'
157–158	42:67	'Scraps to end of Pampas Chapter'
163–164	42:28	'Scraps Cleavage'
165–166* }	42:28	'Scraps Cleavage'
167–168* }	42:28	'Scraps Cleavage'
169–170* }	42:29	'Scraps Cleavage'
171–170*[172] }	42:29	'Scraps Cleavage'
174–175* }	42:29	'Scraps Cleavage'
176–177* }	42:30	'Scraps Cleavage'

Notebook A

Page	DAR Location	Subject
1–2	42:81	'Gravel, Valleys Denudation &c &c'
13–14	208	
15–16	42:75	'Gravel, Valleys Denudation &c &c'
25–26	42:75	'Gravel, Valleys Denudation &c &c'
27–28	208	
33–34	42:60	'Scraps to end of Pampas Chapter'
37–38	42:75	'Gravel, Valleys Denudation &c &c'
39–40	42:61	'Scraps to end of Pampas Chapter'
41–42 top frag	42:25	'Scraps Cleavage'
bottom frag	40:1	geology
43–44*	42:66	'Scraps to end of Pampas Chapter'
45–46* }	42:25	'Scraps Cleavage'
47–48* }	42:25	'Scraps Cleavage'
49–50* }	42:26	'Scraps Cleavage'
51–52* }	42:26	'Scraps Cleavage'
53–54* }	42:26	'Scraps Cleavage'

Page	DAR Location	Subject
67–68	42:36	'Scraps Cleavage'
69–70	42:91	'Gravel, Valleys Denudation &c &c'
71–72*	42:87	'Gravel, Valleys Denudation &c &c'
73–74*	42:87	'Gravel, Valleys Denudation &c &c'
75–76*	42:87	'Gravel, Valleys Denudation &c &c'
81–82 frag	40:1	geology
83–84	42:27	'Scraps Cleavage'
87–88 frag	42:75	'Gravel, Valleys Denudation &c &c'
97–98	42:91	'Gravel, Valleys Denudation &c &c'
99–100*	39:2	geology
101–102 frag	42:59	'Scraps to end of Pampas Chapter'
103–104 frag	40:1	geology
109–110	42:87	'Gravel, Valleys Denudation &c &c'
111–112	42:27	'Scraps Cleavage'
115–116 top frag	42:86	'Gravel, Valleys Denudation &c &c'
bottom frag	40:2	geology
117–118	42:69v	'Scraps to end of Pampas Chapter'
119–120	42:76	'Gravel, Valleys Denudation &c &c'
127–128	42:76	'Gravel, Valleys Denudation &c &c'
133–134	42:91	'Gravel, Valleys Denudation &c &c'
135–136 frag	42:76	'Gravel, Valleys Denudation &c &c'
141–142	42:77	'Gravel, Valleys Denudation &c &c'
143–144*	42:64	'Scraps to end of Pampas Chapter'
147–148	42:27	'Scraps Cleavage'
153–154	42:62	'Scraps to end of Pampas Chapter'

Glen Roy Notebook

Page	DAR Location	Subject
64–65	replaced in MS formerly in DAR 205	—

Notebook B

Page	DAR Location Present	Former	Brown Crayon Portfolio	Number	Subject
29–30†	208	205.7	17 1c‡		Hybridism
51–52†	208	205.7	17 [1]c‡		Hybridism
55–56†	208	205.3	19		Island endemism: animals
†69–70†	208	205.5	7[?] 11	[B69] [B70]	*NS* Chap. 7: Divergence? Divergence
75–76†	208	205.4	20		Island endemism: plants
†107–108	208	205.4	20		Island endemism: plants
123–124†	208	205.2	18		Migration
125–126†	208	205.5	11		Divergence
151–152	208	—	—		—
153–154†	208	205.3	19		Island endemism: animals

	DAR Location		Brown Crayon		
Page	Present	Former	Portfolio	Number	Subject
159–160†	208	205.3	19		Island endemism: animals
165–166	208	—	—		—
†173–174	208	205.4	20		Island endemism: plants
†177–178	208	205.3	19		Island endemism: animals
187–188	5(ser. 2):67	—	—		Early geological notes: notes for glacier paper
†189–190†	208	205.7	17 1c‡		Hybridism
197–198	208	—	—		—
†199–200	208	205.3	19		Island endemism: animals
†201–202	208	205.3	19		Island endemism: animals
†209–210	208	205.3	19		Island endemism: animals
233–234†	208	205.3	19		Island endemism: animals
†249–250	208	205.3	19		Island endemism: animals
†255–256	208	205.3	19		Island endemism: animals

Notebook C

	DAR Location		Brown Crayon		
Page	Present	Former	Portfolio	Number	Subject
†13–14	208	205.3	19		Island endemism: animals
17–18†	208	205.3	19		Island endemism: animals
†23–24	208	205.3	18		Migration
†25–26	208	205.3	19		Island endemism: animals
†27–28	208	205.3	19		Island endemism: animals
39–40†	208	205.2	18		Migration
†41–42	208	205.3	19		Island endemism: animals
†47–48	208	—	3[?]		*1842 Sketch* Chap. 3: Marked varieties in nature?
49–50†	208	205.3	19		Island endemism: animals
		205.11	23		Instinct: change in habits
†71–72	208	—	5		*Origin* Chap. 5: Laws of variation
		205.5	11		Divergence
91–92†	208	205.3	19		Island endemism: animals
†93–94	208	205.4	20	[C93]	Island endemism: plants
		205.2	Ch. 2 Dogs	[C94]	
†95–96	208	205.3	19		Island endemism: animals
101–102†	208	205.4	20		Island endemism: plants
105–106	208	—	—		
†107–108	208	205.5	11		Divergence

Page	DAR Location Present	Former	Brown Crayon Portfolio	Number	Subject
†109–110†	208	205.3	19	[C109]	Island endemism: animals
		205.7	11 Book	[C110]	Divergence
†111–112*	208	205.5	11		Divergence
113–114	208	205.5	pinned to C111–112		
*115–116	208	205.5	pinned to C111–112		
141–142†	208	205.5	11		Divergence
147–148†	208	205.4	20		Island endemism: plants
159–160†	45:45	—	7?		Notes for *NS* Chap. 4: Variation under nature
		205.11	23		Instinct: change in habits
†161–162	208	205.5	11		Divergence
†183–184†	208	205.3	19	[C183]	Island endemism: animals
		205.4	20	[C184]	Island endemism: plants
†185–186	208	205.5	11		Divergence
†205–206†	208	205.3	19	[C205]	Island endemism: animals
		205.5	11	[C206]	Divergence
209–210†	208	205.7	‹5› 17 1b‡		Hybridism: prepotency
†213–214	208	205.2	1 18		Migration
†215–216†	208	205.1	10	[C215]	Rudimentary & abortive organs
		205.3	19	[C216]	Island endemism: animals
221–222†	208	205.5	11		Divergence
†225–226	208	205.3	19		Island endemism: animals
†227–228	208	205.3	19		Island endemism: animals
		205.2	18		Migration
237–238†	208	205.4	20		Island endemism: plants
†239–240	208	205.4	20		Island endemism: plants
†241–242	208	205.4	20		Island endemism: plants
†249–250†	208	205.3	19	[C249]	Island endemism: animals
		205.4	20	[C250]	Island endemism: plants
†251–252	208	205.3	19		Island endemism: animals
†253–254†	11.1:14^{v}	—	9[?]	[C253]	*NS* Chap. 7. Laws of variation: acclimatisation
		205.11	23	[C254]	Instinct: change in habits
†257–258	208	205.9	21		Palaeontology: extinction

Notebook D

Page	DAR Location Present	Former	Brown Crayon Portfolio	Number	Subject
†5–6	208	205.7	Ch IX 1[b?]‡		Hybridism: prepotency

Page	DAR Location Present	DAR Location Former	Brown Crayon Portfolio	Number	Subject
†7–8*	208	205.7	16 1b**		Hybridism: prepotency
11–12†	208	—	3[?]		*1842 Sketch* Chap. 3: marked varieties in nature?
†29–30	208	205.5	11		Divergence
†31–32*	208	205.3 205.7	19 17		Island endemism: animals Hybridism
33–34*	208	—	pinned to D31–32		
47–48*†	46.1:25	—	Ch. 3		Notes for *NS* Chap. 5 = *Origin* Chap. 3: Struggle for existence
53–54†	208	205.1	10		Rudimentary & abortive organs
†55–56	208	205.5	11		Divergence
†61–62	208	205.3	19		Island endemism: animals
63–64†	208	205.3	19		Island endemism: animals
†73–74	208	205.7	17 1c‡		Hybridism
†85–86	69:5	—	17		*Descent* 1871 scraps Hybridism
†87–88*	208	205.7	17 1b‡		Hybridism: prepotency
†89–90*	208	205.7	17 1c‡	[D89 top]	Hybridism
		205.7	16	[D89 bot]	Hybridism: prepotency
†95–96*	84.2:34	—	N 12	[blue crayon]	*Descent* 1871 notes: 'Scraps Birds. Sex Selection'
101–102	208	205.5	7[?]		Divergence: *NS* Chap. 7?
†103–104*	84.2:35	—	12		*Descent* 1871 notes: 'Scraps Birds. Sex Selection'
†105–106	208	205.7	17 1c‡		Hybridism
†113–114*	83:70	—	‹5›	[D113]	*Descent* 1871 notes: 'Scraps Used Sex Selection Mammalia'
			12 11 'Chapt I also Latent Character'	[D114]	Divergence
†133–134	208	205.3	19		Island endemism: animals
135–136†	208	205.7	16 1b‡		Hybridism: prepotency
†147–148†	84.2:36	—	12	[D147]	*Descent* 1871 notes: 'Scraps Birds. Sex Selection'

Page	DAR Location Present	Former	Brown Crayon Portfolio	Number	Subject
			'Ch 6 Sexual Selection'		
		205.11	23	[D148]	Instinct: change in habits
†151–152	208	205.3	19		Island endemism: animals
†159–160	208	205.7	17 1c‡		Hybridism
†173–174	208	205.7	17 1c‡		Hybridism

Notebook E

Page	DAR Location Present	Former	Brown Crayon Portfolio	Number	Subject
5–6†	208	205.9	22		Palaeontology: extinction
9–10†	208	205.3	19		Island endemism: animals
11–12*†	208	205.3	19		Island endemism: animals
†13–14*	208	205.3	19 pinned to D12		Island endemism: animals
†19–20	208	205.3	19		Island endemism: animals
†21–22	208	205.3	19		Island endemism: animals
†25–26	208	205.5	11		Divergence
†35–36	208	205.7	17 1c‡	[E35 top]	Hybridism
		205.3	19	[E35 bot]	Island endemism: animals
41–42†	208	205.3	19		Island endemism: animals
55–56†	208	205.11	5 over 3		*Origin* Chap. 5: Laws of variation
			23		Instinct: change in habits
85–86	50(ser. 5):45	—	—		Notes on alpine, polar & glacier distribution
†87–88†	208	205.9	22 21	[E87] [E88]	Palaeontology: extinction
†91–92	208	205.5	11		Divergence
103–104†	208	205.4	20		Island endemism: plants
†115–116	50(ser. 5):46	—	—		Notes on alpine, polar & glacier distribution
119–120*†	208	205.9	21		Palaeontology: extinction
121–122*	208	205.9	pinned to E121–122		
†123–124	208	205.7	17 1c‡		Hybridism
125–126†	208	205.9	22		Palaeontology: extinction
129–130†	208	205.1	10		Rudimentary & abortive organs
†139–140	208	205.2	18		Migration
165–166*†	208	205.3	18 19	[E166 top]	Migration Island endemism: animals
		205.9	22	[E166 bot]	Palaeontology: extinction

Page	DAR Location Present	DAR Location Former	Brown Crayon Portfolio	Brown Crayon Number	Subject
167–168*	208	205.9	pinned to E166 bot		
†169–170	208	205.7	17 1c‡		Hybridism
†173–174	208	205.3	19 *bis*		Island endemism: animals
175–176	208	205.3	19		Island endemism: animals
179–180†	208	205.3	19		Island endemism: animals
181–182†	208	205.3	19	[E182 bot]	Island endemism: animals

Torn Apart

Page	DAR Location Present	DAR Location Former	Brown Crayon Portfolio	Brown Crayon Number	Subject
1–2	208	—	—		
Frag 1	205.3:61	—	—		
Frag 2	208	—	—		
9	208	—	—		
†13*	208	205.7	1b‡	[T13 top]	Hybridism: prepotency
		205.7	1c‡	[T13 bot]	Hybridism
†15*	208	205.7	1c‡ pinned to T13 & 205.7:209	[T15 top]	Hybridism
†19	208	205.9	22		Palaeontology: extinction
†25–26	208	205.6?	Embryology		
37	47:64	—	Ch. 6		*NS* Chap. 7, *Origin* Chap. 5: Laws of variation. Many notes marked 'Ch. 6' and then 'Ch. 5'
39	208	—	—		
†41	208	205.9	21		Palaeontology: extinction
			5		*Origin* Chap. 5: Laws of variation
Frag 3	208	205.7	—		
Frag 4	208	—	—		
Frag 5	208	—	—		
51	208	—	‹5›	[T51 top]	*Origin* Chap. 5: Laws of variation
		205.2	18		Migration
	205.4:5	—	—	[T51 bot]	Island endemism: plants
55	208	205.5	11		Divergence
57	208	—	—		
59	208	—	—		
63	208	—	—		
65	208	—	—		
Frag 6	208	205.9	—		Palaeontology: extinction
79–80	208	—	—		
81	208	205.9	22		Palaeontology: extinction

Page	DAR Location Present	Former	Brown Crayon Portfolio	Number	Subject
89	208	—	—		
91	208	—	—		
93–94	49:22	—	—		Dichogamy: 'Scraps about Plants. All Finally used'
95–96	208	—	—		—
99–100	109:16	—	—		?
103–104	77:62c	—	—		Notes for *Effects Crossing*: 'Used Scraps. Chiefly Calculations'
105–106	49:144	—	—		Dichogamy: 'Scraps about Plants. All Finally used'
111–112	46.2(ser. 3):46	—	—		'Habits of Bees'
119	208	—	—		
135	46.2(ser. 3):5	—	—		'Habits of Bees'
151	208	pinned to 1 July 1842 letter from R. W. Darwin			
153	46.2(ser. 3):6	—	—		'Habits of Bees'
175	49:131	—	—		Dichogamy: 'Scraps about Plants. All Finally used'
177–178	46.2(ser. 3):29	—	—		'Habits of Bees'
Frag 7	208	205.7	—		Hybridism
Frag 8	208	—	—		
Frag 9	208	—	—		
Frag 10	208	—	—		
Frag 11	208	205.5	—		Divergence

Summer 1842

Page	DAR Location Present	Former	Brown Crayon Portfolio	Number	Subject
1	76:2	—	—		'Old observations on Diptera sucking Flowers'
2	46.2 (ser. 3):23	—	—		'Habits of Bees'
3	208	—	—		—
4	208	205.9	—		Palaeontology: extinction
5	208	—	—		—
6	208	—	—		—
7	208	—	—		—

Edinburgh Notebook

Page	DAR Location Present	DAR Location Former	Brown Crayon Portfolio Number	Subject
1 frag	205.2:30	—	18	Migration
8 frag	replaced in MS	—	—	—
9 frag	replaced in MS	205.2	18	Migration
12 frag	replaced in MS	205.2	18	Migration
13 frag	replaced in MS	—	—	—

Notebook M

Page	DAR Location Present	DAR Location Former	Brown Crayon Portfolio Number	Subject
61–62*	British Museum (Natural History) Manuscripts Collection, MSS DAR:5			—
63–64*	British Museum (Natural History) Manuscripts Collection, MSS DAR:5a		pinned to M61–62	—
83–84	53.1(ser. 2):18		—	Notes for *Expression*
101–102	53.1(ser. 2):20		—	Notes for *Expression*
131–132*	88:23		—	Notes for *Descent* 1874 'Moral Sense'
139–140	87:83		—	Notes for *Descent* 1874 'Mental Powers New Ch. 2'
145–146	53.1(ser. 1):1		—	Notes for *Expression*
153–154	53.1(ser. 2):29		—	Notes for *Expression*

Notebook N

Page	DAR Location Present	DAR Location Former	Brown Crayon Portfolio Number	Subject
19–20	87:84		—	Notes for *Descent* 1874 'Mental Powers New Ch. 2'
21–22	53.1(ser. 2):37		—	Notes for *Expression*
43–44	53.1(ser. 2):28		—	Notes for *Expression*
75–76	53.1(ser. 2):20a		—	Notes for *Expression*
87–88	87:85		—	Notes for *Descent* 1874 'Mental Powers New Ch. 2'
115–116	88:85		—	Notes for *Descent* 1874 'Mammals'

Bibliography

Priority of reference has been given to editions of works in Darwin's Library and to works he abstracted. Works with Darwin's marginalia are principally located in the Darwin Archive at Cambridge University Library. Darwin's collection of abstracts are principally in volumes DAR 71 to DAR 74 of the Darwin Archive. For the period of these *Notebooks* Darwin prepared abstracts almost exclusively of works he did not own. In this bibliography, '*' means the copy or the abstract is in the Darwin Library at CUL; in the case of an abstract, the DAR Catalogue number is given. Works are indexed to the notebook pages where they are mentioned; works mentioned in editors' introductions are not indexed. 'IBC' means inside back cover; 'IFC' means inside front cover. Lengthy book titles are shortened; periodical abbreviation style follows that of the *List of Serial Publications in the British Museum (Natural History) Library*, 1968.

Short Titles of Darwin Works and Editions

Calendar — *A Calendar of the correspondence of Charles Darwin, 1821–1882.* Eds. Frederick Burkhardt, Sydney Smith, David Kohn, William Montgomery and Stephen V. Pocock. New York, London. 1985 C107, 176 QEIFC, 9

Correspondence — *The Correspondence of Charles Darwin.* Vol. 1. Eds. Frederick Burkhardt, Sydney Smith; David Kohn, William Montgomery and Stephen V. Pocock. Cambridge 1985; Vol. 2, Eds. Frederick Burkhardt, Sydney Smith; Janet Browne, David Kohn, William Montgomery, Stephen V. Pocock, Charlotte Bowman, Anne Secord. Cambridge 1986. Vol. 3, *idem* 1987. RNIFC, 129, 178 A9, 146 B8, 136, 139, 141, 176 C100, 107, 223 DIFC, 9, 21, 25, 151, 180 E15, 58, 103, 110, 113, 141, 143, 151, 183 T151, 177 S1, 5 QE3, 5, [6^{v}], 7, 8, 10–19 Zed8 M142 N4, 61

CP — *The collected papers of Charles Darwin.* 2 vols. Ed. Paul H. Barrett. Chicago 1977. E16 N71, 74

CR — *The structure and distribution of coral reefs. Being the first part of the geology of the voyage of the 'Beagle', under the command of Capt. FitzRoy, R.N. during the years 1832–1836.* London 1842.

Crossing — *The effects of cross and self fertilisation in the vegetable kingdom.* 1876 D159 E48

Descent — *The descent of man, and selection in relation to sex.* 2 vols. London 1871. C244 E49 M132, 139 N115 [2nd ed. 1874]

Diary — *Charles Darwin's diary of the voyage of H.M.S. 'Beagle'.* Ed. Nora Barlow. Cambridge, New York 1933. RNIFC, 32 M131 QE21..

Expression — *The expression of the emotions in man and animals.* London 1872. N41, 103

GSA — *Geological observations on South America. Being the third part of the geology of the voyage of the 'Beagle', under the command of Capt. FitzRoy, R.N. during the years 1832 to 1836.* London 1846. RN38, 49, 100, 134, 137, 143, 178 A3, 6, 39, 44, 68, 94, 100, 111, 117, 131, 144 GR57

JR — *Journal of researches into the geology and natural history of the various countries visited by H.M.S. 'Beagle', under the command of Captain FitzRoy, R.N. from 1832 to 1836.* London 1839. RN17, 51, 56, 65, 77, 86, 99, 102, 125, 126, 127, 129, 130, 134, 135, 139, 141, 142, 144, 155, 177, 178 A9, 10, 36, 57, 88, 133, 141 GR13, 85, 109 B2, 23, 33, 38, 53, 60, 61, 62, 91, 143, 223, 231 C23, 73 D39, 96 E32, 37, 47, 66 Zed17, 131 M69, 131

JR 1845 — *Journal of researches* . . . 2d ed. London 1845. E12, 100 ZEd13

LL — The life and letters of Charles Darwin, including an autobiographical chapter. 3 vols. Ed. Francis Darwin. London 1887.

Living Cirripedia — *A monograph of the sub-class Cirripedia, with figures of all the species.* (Vol. 1) *The Lepadidae; or pedunculated Cirripedes.* London 1851. (Vol. 2) *The Balanidae (or Sessile Cirripedes); the Verrucidae, &c.* London 1854. D157

ML — *More letters of Charles Darwin. A record of his work in a series of hitherto unpublished letters.* 2 vols. Eds. Francis Darwin and A. C. Seward. Cambridge 1903. M29, 69 N47

Natural Selection — *Charles Darwin's natural selection, being the second part of his big species book written from 1856 to 1858.* Ed. Robert C. Stauffer. Cambridge 1975. BIFC C25, 47, 160, 163, 210, 253 DIFC, 48, 73, 87, 89, 99, 100, 105, 113, 159, 173 E16, 48, 53, 56, 123, 127, 181 T2, 15

Origin — *On the origin of species by natural selection, or the preservation of favoured races in the struggle for life.* London 1859. C25, 177 D60, 100, 148, 159 E6, 17, 55, 59, 63, 114 M142

Power of Movement — *The power of movement in plants.* London 1880.

Variation — *The variation of animals and plants under domestication.* 2 vols. London 1868. BIFC, 167, 176 C120, 121, 210 DIFC, 87, 99, 100 E15 M142 QEIFC, 18

VI — *Geological observations on volcanic islands, visited during the voyage of H.M.S. 'Beagle', together with some brief notices on the geology of Australia and the Cape of Good Hope. Being the second part of the voyage of the 'Beagle', under the command of Capt. FitzRoy, R.N., during the years 1832 to 1836.* London 1844. RN71, 120, 150, 159 A40, 94 N26

Zoology — *The zoology of the voyage of H.M.S. 'Beagle' under the command of Captain FitzRoy, R.N., during the years 1832 to 1836. Published with the approval of the Lords Commissioners of Her Majesty's Treasury.* Edited and superintended by Charles Darwin . . . naturalist to the expedition. 5 pts. London 1839–43. C23

Fossil Mammalia — R. Owen, with a geological introduction by C. Darwin, 1838–40 (Pt 1). RN113, 129 B223, 231 C72 E105 T63

Mammalia — G. R. Waterhouse, with a notice of their habits and ranges by C. Darwin, 1839 (Pt 2). B250 C23, 29, 35, 36, 116 D99 E24

Birds — J. Gould, with a notice of their habits and ranges by C. Darwin, and with an anatomical appendix by T. C. Eyton, 1841 (Pt 3). RN127, 130 C68, 69, 71, 88, 105, 163 D96 E56 ZEd14, 16, 17

Fish — L. Jenyns, 1842 (Pt 4). C20

Reptiles — T. Bell, 1843 (Pt 5). C18, 36, 54, 71 E56

1842 Sketch — *Charles Darwin's sketch of 1842.* Cambridge 1909. Ed. Francis Darwin. In C. Darwin and A. R. Wallace, *Evolution by natural selection, with a foreword by Sir Gavin de Beer*, pp. 39–88. Cambridge 1958. D37 E22, 50, 63, 118

1844 Essay — *Charles Darwin's essay of 1844.* Cambridge 1909. Ed. Francis Darwin. In C. Darwin and A. R. Wallace, *Evolution by natural selection, with a foreword by Sir Gavin de Beer*, pp. 89–254. Cambridge 1958. C176 D21, 37 E24

Bibliography

Abel, Clarke
1818 *Narrative of a journey in the interior of China . . . in the years 1816 and 1817.* London. N94

Abercrombie, John
*1838 *Inquiries concerning the intellectual powers and the investigation of truth.* 8th ed. London. C61,270 D21,56 M8,11,15,63,78,114,115,141 N74 OUN17,25

Acosta, José de
1600 *Histoire naturelle et moralle des Indes, tant Orientalles qu'Occidentalles.* 2d Fr. ed. Paris. RN116

Agassiz, Louis Jean Rodolphe
1833–43 *Recherches sur les poissons fossiles.* 5 vols & atlas of 5 vols. Neuchâtel. D133 E60 Mac167^{v}
1838 Prodromus of a monograph of the Radiata and Echinodermata. *Ann. Mag. Nat. Hist.* 1:30–43, 297–307, 440–49. C103,104
1840 Discovery of the former existence of glaciers in Scotland, especially in the highlands. *Scotsman* 7 Oct. 1840:3
1842 The glacial theory and its recent progress. *Edinb. New Phil. J.* 33:217–83.

Aimé-Martin, Louis
1837 *De l'éducation des mères de famille ou, de la civilisation du genre humain par les femmes.* 2 vols. Brussels. OUN8,8^{v},8a

Akerly, Samuel
1824 Observations on the language of signs. *Am. J. Sci.* 8:348–58. N102

Alexander, James Edward
1830 Notice regarding the Salt Lake Inder, in Asiatic Russia. *Edinb. New Phil. J.* 8:18–20. RN127

*1832 Notes of two expeditions up the Essequibo and Mazaroony Rivers in 1830 and 1831. *J. R. Geogr. Soc.* 2:65–72. A30

Alexander, John H.
1835 See Ducatel, J.T. and J.H. Alexander

Alison, Archibald
1811a *Essays on the nature and principles of taste.* 2 vols. Edinburgh.
1811b 'Review of 1811a.' *Edinb. Rev.* 18:1–46. OUN8, 14

Allan, Thomas
1831 On a mass of native iron from the Desert of Atacama in Peru. *Trans. R. Soc. Edinb.* 11:223–28. RN156

Allardyce, James
1836 On the granitic formation, and direction of the primary mountain chains, of Southern India. *Madras J. Lit. Sci.* 4:327–35. RN57

Anderson, James
1799–1803 *Recreations in agriculture, natural-history, arts, and miscellaneous literature.* 6 vols. London. QE22

Angelis, Pedro de
*1836–37 *Coleccion de obras y documentos relativos a la historia antigua y moderna de las provincias del Rio de la Plata.* 6 vols. Buenos Aires. See review Cooley 1837. RN143 A88
1837a Review of 1836–37. *Athenaeum* (496) 29 April: 300–03. RN143
1837b Review of 1836–37. *J. R. Geogr. Soc.* 7, 1837:351–68. A88

Anonymous
1810–29 Siberia. *Encyclopaedia Londinensis.* 24 vols. London. Vol. 23, 1828:177–79. A7
1824 A list of the donations to the library, to the collections of maps, plans, sections and models; and to the cabinet of minerals. *Trans. Geol. Soc. Lond.* 2d ser. 1:425–39. A97
1827 Irish furze, broom, and yew. *Edinb. New Phil. J.* 2:207. B107
1833–38 Report of the committee appointed to examine and report on the state of the museums and library. *Proc. Geol. Soc. Lond.* 2 (17 February 1837): 460–65. A63
1835a On the pleasures and advantages of the study of natural history. *Edinb. J. Nat. Hist.* 1(2):5–6. B190
1835b Notice of the Transactions of the Geological Society of Pennsylvania, Pt I. *Am. J. Sci.* 27:347–55. RN142
1835c Earthquake at sea. *The Times.* 28 March. RN33
1835d Earthquake at sea. *Carmarthen J.* 3 April. RN33
1836 Corals. *Chambers' Edinb. J.* 5:151. A26
*1837a Chronique. *Institut* 5 (222) December:404. B165
1837b Notice respecting *Rhèa Darwinii* Gould. *Mag. Nat. Hist.* 1:504. RN127
1838 Account of the great annual show of the Highland and Agricultural Society of Scotland at Dumfries. *Q. J. Agric. Edinb.* 8:365–77. GR1 D43,44
1839a French scientific expedition in Northern Europe. *Athenaeum* (603) 18 May:377–78. E161
1839b 'Memorandum on breeds of cattle and sheep.' *Athenaeum* (607) 15 June:451. E173
1839c On the geographical distribution of insects. *Edinb. New Phil. J.* 27:94–111,333–51. ZEd18
1854 Commission of lunacy on Capt. Childe. A lover of the queen. *Shrewsbury Chron.* 28 July :6. M10
n.d. *The British aviary, and bird breeder's companion; containing copious directions for propagating the breed of canaries: also, goldfinch and linnet mules.* London. B217 C248,275

Aoust, Pierre-Théodore Virlet d'
1833 See Le Puillon de Boblaye, E. and P.-T. Virlet d'Aoust

Aubuisson de Voisins, Jean François d'
*1819 *Traité de géognosie.* 2 vols. Strasbourg. RNIFC

Audubon, John James
*1831–49 *Ornithological biography, or an account of the habits of the birds of the United States of America.* 5 vols. Edinburgh. C253,266 N45

Azara, Félix de
*1801 *Essais sur l'histoire naturelle des quadrupèdes de la Province du Paraguay.* 2 vols. Paris. C41,276 E113
*1809 *Voyages dans l'Amérique méridionale, par don Félix de Azara, commissaire et commandant des limites espagnoles dans le Paraguay depuis 1781 jusqu'en 1801.* 4 vols & atlas. Paris. RN126 C105,276 ZEd1,2,3,12
1838 *The natural history of the quadrupeds of Paraguay and the river La Plata.* Only vol. 1 published. Edinburgh. C276

Babbage, Charles
1838 *The ninth Bridgewater treatise. A fragment.* 2d ed. London. RN32 E59

Babington, Charles Cardale
*1838 A notice, with the results, of a botanical expedition to Guernsey and Jersey, in July and August 1837. *Mag. Zool. Bot.* 2:397–99. C192

Bachman, John
1836 On the migration of the birds of North America. *Am. J. Sci.* 30:81–100. C254,IBC
1837a Observations on the different species of hares (genus *Lepus*) inhabiting the United States and Canada. *J. Acad. Nat. Sci. Philad.* 7:282–361. C251
1837b Some remarks on the genus Sorex, with a monograph of the North American species. *J. Acad. Nat. Sci. Philad.* 7:362–402. C251

1838 Monograph of the species of squirrel inhabiting North America. *Proc. Zool. Soc. Lond.* 6:85–103. D29

Bacon, Francis
1665 *Francisci Baconi . . . opera omnia.* Frankfort. Mac58v

Baer, Karl Ernst von
*1837a Aurochs du Caucase. *Institut* 5 (218) August:260–61. B136
1837b Bericht ueber die neuesten Entdeckungen an der Küste von Nowaja-Semlja. *Bull. scient. Acad. Sci. St. Petersb.* II: col. 137–72. D24
1837c Ueber das Klima von Nowaja-Semlja und die mittlere Temperatur insbesondere. *Bull. scient. Acad. Sci. St. Petersb.* II: col. 2225–38. D24
1837d Ueber den jährlichen Gang der Temperatur in Nowaja-Semlja. *Bull. scient. Acad. Sci. St. Petersb.* II:col. 242–54. D24
1837e Ueber den täglichen Gang der Temperatur in Nowaja-Semlja. *Bull. scient. Acad. Sci. St. Petersb.* II: col. 289–300. D24
*1838a On the ground ice or frozen soil of Siberia. *J. R. Geogr. Soc.* 8:210–13. A117 C101
*1838b Recent intelligence upon the frozen ground in Siberia. *J. R. Geogr. Soc.* 8:401–6. A135

Baird, William
*1837 The natural history of the British Entomostraca. *Mag. Zool. Bot.* 1:35–41,309–33,514–26. C235
*1838 The natural history of the British Entomostraca. *Mag. Zool. Bot.* 2:132–44,400–12. C162,235 ZEd 17

Ball, John
1835 Remarks upon the geology, and physical features of the country west of the Rocky Mountains, with miscellaneous facts. *Am. J. Sci.* 28:1–16. RN142

Banks, Joseph, Sr.
1812 Some hints respecting the proper mode of inuring tender plants to our climate. *Trans. R. Hort. Soc.* 1:21–25. E111

Barclay, John
*1822 *An inquiry into the opinions, ancient and modern, concerning life and organization.* Edinburgh. C166

Barker-Webb, Philip and Sabin Berthelot
*1836–50 *Histoire naturelle des Iles Canaries.* 8 vols in 3. Paris.
[* Vol. 3, pt 1 only]. C184 D74
*1837 [Review of 1836–50, livraisons i, vi. [1836]]. *Mag. Zool. Bot.* 1:470–82. C184

Barlow, Emma Nora
1945 *Charles Darwin and the voyage of the 'Beagle'.* London. M1
1958 *The autobiography of Charles Darwin 1809–1882.* London, New York. GR4
1963 Darwin's ornithological notes. *Bull. Br. Mus. (Nat. Hist.) Hist. Ser.* 2:201–78. RN130 C82,105
1967 *Darwin and Henslow: the growth of an idea.* London. RN178

Barrett, Paul H.
1960 A transcription of Darwin's first notebook on 'Transmutation of Species'. *Bull. Mus. Comp. Zool.* 122:247–96.
1973 Darwin's 'gigantic blunder'. *J. Geol. Educ.* 21: 19–28.
1974 The Sedgwick-Darwin geologic tour of North Wales. *Proc. Am. Phil. Soc.* 118:146–64. RN93
1980 *Metaphysics, materialism, & the evolution of mind. Early writings of Charles Darwin.* Chicago B139, 141

Barrett, Paul. H., Donald J. Weinshank and T. Gottleber
1981 *A concordance to Darwin's Origin of Species, first edition.* Ithaca.

Barrett, Paul H., Donald J. Weinshank, Paul Ruhlen, Stephan J. Ozminski and Barbara Newell-Berghage
1986 *A concordance to Darwin's The Expression of the Emotions in Man and Animals.* Ithaca.

Barrett, Paul H., Donald J. Weinshank, Paul Ruhlen and Stephan J. Ozminski
1987 (in press) *A concordance to Darwin's The Descent of Man and Selection in Relation to Sex.* Ithaca.

Barrington, Daines
1774 Experiments and observations on the singing of birds. *Phil. Trans. R. Soc.* 63:249–91. C178

Barrow, John Jr.
1835 *A visit to Iceland, by way of Tronyem, in the 'Flower of Yarrow' Yacht, in the summer of 1834.* London. RN159

Barton, John
1827 *A lecture on the geography of plants.* London. C40 E15

Bartram, William
1791 *Travels through North and South Carolina, Georgia, East and West Florida.* Philadelphia. A120 C269 E160

Bauer, Franz Andreas
1830–38 *Illustrations of orchidaceous plants with notes and prefatory remarks by John Lindley.* London. QE21..

Beatson, Alexander
1816 *Tracts relative to the Island of St. Helena . . . Illustrated with views engraved by Mr. William Daniell, from the drawings of Samuel Davis, Esq.* London. B151, 156,173 E119

Beaufort, Francis
1817 *Karamania; or, a brief description of the south coast of Asia-Minor . . . in the years 1811 & 1812*. London. RN181

Beechey, Frederick William
1831 *Narrative of a voyage to the Pacific and Beering's Strait . . . in the years 1825, 26, 27, 28*. 2 vols. London. A132–33
*1832 *Narrative of a voyage to the Pacific and Beering's Strait . . . in the years 1825, 26, 27, 28*. Philadelphia. RN15,45,97
1839 *The zoology of Captain Beechey's voyage . . . in His Majesty's Ship 'Blossom' . . . in the years 1825, 26, 27, and 28*. London. See also Sowerby (Elder) 1839b. E98,99 T2
1843 *A voyage of discovery towards the north pole . . . 1818*. London. A133

Beer, Gillian
1985 Darwin's reading and the fictions of development. In Kohn 1985.

Beer, William and Johann Heinrich Madler
*1838 Survey of the surface of the moon. *Edinb. New Phil. J.* 25:38–69. A104–05

Bell, Charles
1806 *Essays on the anatomy of expression in painting*. London. C243
1824 *Essays on the anatomy and philosophy of expression*. 2d ed. London. N41,103 Mac28^{v}
1833 *The hand. Its mechanism and vital endowments as evincing design. The Bridgewater treatises on the power wisdom and goodness of god as manifested in the creation*. Treatise 4. 2d ed. London. C269 E157,158 N89,90

Bell, Thomas
*1834 Characters of a new genus of freshwater tortoise (Cyclemys). *Proc. Zool. Soc. Lond.* 2:17. B279
*1836 Cheiroptera. In Todd, 1836–59, Vol. 1:594–600. D113
*1837 *A history of British quadrupeds, including the Cetacea*. London. [*abst DAR 71:116–24.] B7,31,183, 219 C177 D179 E91,103
1843 See *Zoology: Reptiles*

Bellinsgauzen, Faddei Faddeevich
1831 Dvukratnyia izyskaniia v Iuahnom Ledovitom Okeane i plavanie vokrug svieta, v prodolzhenii 1819,20 i 21 godov, sovershennyia na shliupakh Vostok i Mirnom, pod nachal'stvom Kapitana Bellinsgauzena, Komandira Shliupa Vostoka. Shliupom Mirnym nachal'stvoval Leitenant Lazarev. 2 vols & atlas. St Petersburg. RN181

Bennett, E.T.
1835 On the Chinchillidae, a family of herbivorous *Rodentia*, and on a new genus referrible to it. *Trans. Zool. Soc. Lond.* 1:35–64. ZEd4

Bennett, Frederick Debell
*1837 Extracts from the journal of a voyage round the globe in the years 1833–36. *J. R. Geogr. Soc.* 7:211–29. A88 C93–94

Bennett, George
1834 *Wanderings in New South Wales, Batavia, Pedir Coast, Singapore, and China . . . during 1832, 1833, and 1834*. 2 vols. London. E167 N94

Bentham, George
1832–36 *Labiatarum genera et species: or a description of the genera and species of plants of the order Labiatae; with their general history, characters, affinities, and geographical distribution*. London. E100,104

Benza, P.J.
1835 Geological sketch of the Neilgherries. *J. Asiat. Soc. Beng.* 4:413–38. A145

Berendt, G.C.
*1839 Dr Berendt's investigations on amber. *Edinb. New Phil. J.* 27:211,12. ZEd18

Berghaus, Heinrich Karl Wilhelm
1832–43 *Atlas von Asia*. Gotha. A142

Berkeley, Miles Joseph
*1838 On the fructification of the pileate and clavate tribes of hymenomycetous fungi. *Ann. Nat. Hist.* 1:81–100. C95

Berthelot, Sabin. See Barker-Webb, P. and S. Berthelot.

Bevan, Edward
*1827 *The honey-bee; its natural history, physiology and management*. London. C266,268

Bewick, Thomas
1797–1804 *History of British birds*. 2 vols. Newcastle. D101

Bibron, Gabriel
1834–54 See Duméril, A.M.C. and G. Bibron

Bicheno, James Ebenezer
*1827 [1826] On systems and methods in natural history. *Trans. Linn. Soc. Lond.* 15:479–96. *Also Phil. Mag.* 3, 1828:213–19, 265–71. [For responses, see Fleming 1829 and MacLeay 1829b.] C42, 155,156,276

Bird, Golding
1837 Notice of his paper 'On the crystallization of metals by galvanic influence' presented to a meeting of the British Association. *Athenaeum* (516) 16 September:670. A29

Bird, James
*1834 Observations on the manners of the inhabitants who occupy the southern coast of Arabia and shores of the Red Sea; with remarks on the ancient and modern geography of that quarter, and the route, through the desert, from Kosir to Keneh. *J. R. Geogr. Soc.* 4:192–206. RN148

Bischoff, Carl Gustav Christoph
*1836–38 On the cause of the temperature of hot and thermal springs; and on the bearings of this subject as connected with the general question regarding the internal temperature of the earth. *Edinb. New Phil. J.* 20:329–76; 23:330–98; 24:132–64, 252–300. Vol. 23, 1837. [*Vol. 24 only.] A132,133,151

Blackwall, John
1829 Facts relating to the natural history of the cuckoo. *Zool. J. Lond.* 4:294–300. C67

Blainville, Henri-Marie Ducrotay de
1834 See Geoffroy Saint-Hilaire, I. and H.-M. D. de Blainville.
*1837 Ossemens de quadrumanes. *Institut* 5 (216) 28 June:205 B94
*1838a Chauve-Souris. *Institut* 6 (223) January:6–8. B200
*1838b Mammifères insectivores. *Institut* 6 (231) 31 May:174. C227
*1838c Dépôt d'ossemements fossiles de Sansan. *Institut* 6 (238) 19 July:230. C257
*1838d Doutes sur le prétendu Didelphe fossile de Stonefield. *Institut* 6 (243) 23 August:275. D62

Blakeway, John Brickdale
1831 *The Sheriffs of Shropshire.* Shrewsbury. M42,44

Blomefield, Leonard. See Jenyns [Blomefield], L.

Blume, Carel Lodewijk
1824a Ueber die Vegetation des Berges Gedee auf der Insel Java. *Flora, Jena* 7:289–304. QE17
1824b See Reinwardt C.G.C., C.L. Blume, and C.G.D. Nees von Esenbeck.
1827–28 *Enumeratio plantarum Javae et insularum adjacentium minus cognitarium vel novarum ex herbariis Reinwardtii, Kuhlii, Hasseltii et Blumii.* Lugduni Batavorum. C17,268

Blumenbach, Johann Friedrich
[1792] *An essay on generation.* Transl. by Alexander Crichton. C204,269

Blyth, Edward
*1835 An attempt to classify the 'varieties' of animals with observations on the marked seasonal and other changes which naturally take place in various British species, and which do not constitute varieties. *Mag. Nat. Hist.* 8:40–53. C70
*1836a Observations on the various seasonal and other external changes which regularly take place in birds, more particularly in those which occur in Britain; with remarks on their great importance in indicating the true affinities of species; and upon the natural system of arrangement. *Mag. Nat. Hist.* 9:393–409. C222 E124,125
*1836b Further remarks on the affinities of the feathered race; and upon the nature of specific distinctions. *Mag. Nat. Hist.* 9:505–14. [Continuation of 1836a.] C149,222
*1837a On the psychological distinctions between man and all other animals; and the consequent diversity of human influence over the inferior ranks of creation, from any mutual and reciprocal influence exercised among the latter. *Mag. Nat. Hist.* 1:1–9, 77–85, 131–41. C198,199
*1837b On the counterfeiting of death, as a means to escape from danger, in the fox and other animals. *Mag. Nat. Hist.* 1:566–74. C197
1838a On the geographical distribution of birds. *Athenaeum* (537) 10 February:107. B199 C36
*1838b Remarks on the plumage and progressive changes of the crossbill and linnet. *Proc. Zool. Soc. Lond.* 6:115. D95
1841 An amended list of the species of the genus Ovis. *Ann. Nat. Hist.* 7:248–61 TFRAG6

Boblaye, Emile Le Puillon de. See Le Puillon de Boblaye, E.

Boece, ———
1526 *Scotorum historiae.* Paris. D48

Boerhaave, Hermann
1725 Preface. In Marsilli, L.F. *Histoire physique de la mer.* Amsterdam. QE21..

Bolingbroke, Henry
1807 *A voyage to the Demerary, containing a statistical account of the settlements there, and of those on the Essequebo, the Berbice, and other contiguous rivers of Guyana.* London. RN124

Bollaert, William
1833–38 Description of the insulated masses of silver found in the mines of Huantaxaya, in the province of Tarapaca, Peru. *Proc. Geol. Soc. Lond.* 2 (31 January 1838):598–99. A44

Bonaparte, Charles Lucien Jules Laurent, Prince de Canino
*1838a *A geographical and comparative list of the birds of Europe and North America.* London. C67
1838b Review of 1838a. *Ann. Nat. Hist.* 1:318–19. C225

Bongard, Heinrich Gustav
1838 Végétation des îles de Bonin-Sima. *Institut* 6 (232) 7 June:184. C241

Bonpland, Aimé. See Humboldt, F.H.A. von and A. Bonpland.

Boott, Francis
1845–47 Descriptions of six new North American Carices. *Boston J. Nat. Hist.* 5:112–16. QE21..

Bory de Saint-Vincent, Jean Baptiste
1804 *Voyage dans les quatre principales îles des mers d'Afrique . . . (1801 et 1802) . . . sur la corvette 'Le Naturaliste'. 3 vols & atlas. Paris.* RN71 B152 C18,268
1822–31 *Dictionnaire classique d'histoire naturelle.* 17 vols incl. atlas. Paris. Vol. 7, 1825; vol. 12, 1827. RN130 ZEd5

Boswell, James
1799 *The life of Samuel Johnson.* 3d ed. 4 vols. Oxford. C269

Boteler, Thomas
1835 *Narrative of a voyage of discovery to Africa and Arabia, performed in HMS 'Leven' and 'Barracouta', from 1821 to 1826, under the command of Capt. F. W. Owen, R.N.* 2 vols. London. TFRAG1

Bougainville, Louis Antoine de
1772 *Voyage autour du monde, par le frégate du roi 'la Boudeuse' et la flûte 'l'Étoile' en 1766, 1767, 1768 & 1769.* 2d ed. 2 vols. Paris. RN177 ZEd5

Boussingault, Jean Baptiste Joseph
1834–35 Sur les tremblemens de terre des Andes. *Bull. Soc. Géol. Fr.* 6:52–57. RN158

Bowers, Fredson
1976 Transcription of manuscripts: the record of variants. *Studies in Bibliography* 29:212–64.

Bowman, John Eddowes
*1837 On the longevity of the yew, as ascertained from actual sections of its trunk; and on the origin of its frequent occurrence in churchyards. *Mag. Nat. Hist.* 1:28–35,85–90. QE9

Brewster, David
1837 A treatise on magnetism. Forming the article under that head in the seventh edition of the *Encyclopaedia Britannica.* Edinburgh. A21
1838 *Cours de philosophie positive.* Par M. Auguste Comte. 2 tom. 8vo. Paris: 1830–5. *Edinb. Rev.* 67:271–308. D152 M70,72,81,135 N12
1842 Magnetism. In *Encyclopaedia Britannica.* 7th ed. 21 vols. 13:685–774. Edinburgh. A21

Broderip, William John
1827 Observations on the jaw of a fossil mammiferous animal, found in the Stonesfield slate. *Zool. J. Lond.* 3:408–12. B87 E60
1834 Table of the situations and depths at which recent genera of marine and estuary shells have been observed. In De la Beche 1834, Appendix: 399–408. ZEd8

Brongniart, Adolphe Théodore
*1837 Végétaux fossiles. *Institut* 5 (220) October:318–21. B150

Bronn, Heinrich Georg
1837–38 [1838] Sur l'âge géologique des terrains tertiaires du bassin de Mayence. *Bull. Soc. Géol. Fr.* 9:23. C16

Brougham, Henry Peter
*1839 *Dissertations on subjects of science connected with natural theology: being the concluding volumes of the new edition of Paley's work.* 2 vols. London. C266 N62

Brown, Robert
1814 General remarks, geographical and systematical, on the botany of Terra Australis. In Flinders vol. 2, 1814, Appendix 3:533–613. B76,77 C102, 237–240,268 QE21..
1818 Observations, systematical and geographical, on Professor Christian Smith's collection of plants from the vicinity of the River Congo. In Tuckey, 1818, Appendix V:420. C224,248,251,268,275
*1832 General view of the botany of the vicinity of Swan River. *J. R. Geogr. Soc.* 1:17–21. B93 C102
1833 On the organs and mode of fecundation in Orchideae and Asclepiadeae. *Trans. Linn. Soc. Lond.* 16:685–745. C237 E133,165 QE[5]a, 15,21..

Brown, Thomas
1798 *Observations on the Zoonomia of Erasmus Darwin, M.D.* Edinburgh. N184
1820a *Sketch of a system of the philosophy of the human mind.* Part First. *Comprehending the physiology of the mind.* Edinburgh. N184
1820b *Lectures on the philosophy of the human mind.* 4 vols. Edinburgh. N184

Browne, Janet
1985 Darwin and the expression of the emotions. In Kohn 1985.

Browne, Thomas
1835–36 *Sir Thomas Browne's works, including his life and correspondence.* Simon Wilkins, ed. 4 vols. London. C270 D54 M126

Browne, William George
1799 *Travels in Africa, Egypt and Syria.* London. C269

Brun, Conrad Malte. See Malte-Brun, C.

Brun, Charles Le. See Le Brun, C.

Brush, Stephen G.
1979 Nineteenth-century debates about the inside of the earth: solid, liquid or gas? *Ann. Sci.* 36:225–54.

Buch, Leopold von
*1813 *Travels through Norway and Lapland, during the years 1806, 1807, and 1808*. Transl. by John Black. Notes and illustrations by Robert Jameson. London. A97 C178,275 E153
*1836 *Description physique des îles Canaries, suivie d'une indication des principaux volcans du globe*. With atlas. Paris. RN137,150 A39,40,42 B156–159,164 D69
1838 Coquilles fossiles de l'Amérique du Sud. *Institut* 6 (242) 16 August:270–71. A115 D68

Buckland, William
1823 *Reliquiae diluvianae; or, observations on the organic remains contained in caves, fissures, and diluvial gravel, and on other geological phenomena, attesting the action of an universal deluge*. London. A36 E44,45,60
*1836 *Geology and mineralogy considered with reference to natural theology*. 2 vols. London. [*abst DAR 71:125–27.] B149

Buffon, Georges Louis Leclerc, Comte de
1762 *The natural history of the horse. To which is added, that of the ass, bull . . . and swine. With . . . full directions for breeding . . . and improving, those useful creatures*. London. C268 D112
1785 *Natural history, general and particular*. 2d ed. 9 vols. London. D40,112 OUN9

Bulkeley, John and John Cummins
1743 *A voyage to the South-Seas, in the years 1740–1. Containing a faithful narrative of the loss of his Majesty's ship the 'Wager'*. London. RN17

Burchell, William J.
*1822–24 *Travels in the interior of southern Africa*. 2 vols. London. RN86,134

Burckhardt, John Lewis
1829 *Travels in Arabia, comprehending an account of those territories in Hedjaz which the Mohammedans regard as sacred*. London. C221

Burgess, T.H.
*1839 *On the physiology or mechanism of blushing; illustrative of the influence of mental emotion on the capillary circulation; with a general view of the sympathies and the organic relations of those structures with which they seem to be connected*. London. C265 M144

Burke, Edmund
1823 *Philosophical inquiry into the origin of our ideas of the sublime and beautiful, with an introductory discourse concerning taste, and several other additions*. London. M51,88,89,108 N10,57

Burke, John Bernard
1925 *A genealogical and heraldic history of the landed gentry*. 14th ed. London. M1

Burnes, Alexander
1832 Some account of the salt mines of the Punjab. *J. Asiat. Soc. Beng.* 1:145–48. A143

Burnet, James, Lord Monboddo
1773–92 *On the origin and progress of language*. 6 vols. Edinburgh. OUN5

Burney, James
1803–17 *A chronological history of the discoveries in the South Sea or Pacific Ocean*. 5 vols. Vol. 4, 1816. London. A144

Bury, Charles William, Lord Tullamore
1802 Analysis of turf ashes. *Trans. R. Ir. Acad.* 8:135–39. A126

C., A. E.
1915–18 *The family of Corbet*. 2 vols. M1

Caldcleugh, Alexander
1833–38 Some observations on the elevation of the strata on the coast of Chili. *Proc. Geol. Soc. Lond.* 2 (4 January 1837):444–46. A88

Campbell, James
*1834 Geographical memoir of Melville Island and Port Essington, on the Cobourg Peninsula, Northern Australia; with some observations on the settlements which have been established on the north coast of New Holland. *J. R. Geogr. Soc.* 4:129–81. A30 B132

Camper, Peter
1779 Account of the organs of speech of the orang outang. *Phil. Trans. R. Soc.* 69:139–59. T119

Candolle, Augustin Pyramus de
[1810] Note sur les Georgina (Dahlia *Cav.* et *Hort. Par.*) *Annls Mus. Hist. Nat. Paris* 15:307–16. E13
*1820 Géographie botanique. In Cuvier, F.G. *Dictionnaire des sciences naturelles*, 18:359–422. B13,156, 280,IBC C268
1821 *Elements of the philosophy of plants*. Edinburgh. C268
1824 Memoir on the different species, races, and varieties of the genus Brassica, (Cabbage) and of the genera allied to it, which are cultivated in Europe. *Trans. R. Hort. Soc.* 5:1–43. QE3
1832 *Physiologie végétale*. 3 vols. Paris. E184

Cannon, Walter F. (Susan Faye)
1961 The impact of uniformitarianism: Two letters from John Herschel to Charles Lyell, 1836–1837. *Proc. Am. Phil. Soc.* 105:301–14. RN32,115 A46,104,114,118

Cantor, T.
1837 Notice of a skull fragment of a gigantic fossil Batrachian. *J. Asiat. Soc. Beng.* 6:538–41 T1

Carlile, H.
1838 Teratologie: difformites du cerveau. *Institut* 6 (251) 18 October:340. N33

Carlyle, Thomas
1837 *The French revolution: a history*. 3 vols. London. C269

Carmichael, Dugald
1818 Some account of the Island of Tristan da Cunha and of its natural productions. *Trans. Linn. Soc. Lond.* 12:483–513. B192,193,195 ZEd8

Carne, Joseph
1822 On the relative age of the veins of Cornwall. *Trans. R. Geol. Soc. Cornwall* 2:49–128. RN20

Carus, C.G.
1837 The kingdoms of nature, their life and affinity. *Scient. Mem.* 1:223–54. C102–104,210,275

Cary, Byron
1834 Note on a large specimen of the Gallapagos tortoise. *Proc. Zool. Soc. Lond.* 2:113. ZEd 6

Cautley, Proby T.
1833–38 An extract of a letter, dated Saharumpore, 18th November, 1836. *Proc. Geol. Soc. Lond.* 2 (3 May 1837):544–45. B 126,250
1838 Note on a fossil ruminant genus allied to a Giraffidae in the Sewalik hills. *J. Asiat. Soc. Beng.* 7:658–60. T1

Cautley, Proby T. and Hugh Falconer
1833–38 Notice on the remains of a fossil monkey from the tertiary strata of the Sewalik Hills in the north of Hindostan. *Proc. Geol. Soc. Lond.* 2 (14 June 1837):568–69. B126
1838 See Falconer, H. and P.T. Cautley

Chambers, Robert
1848 *Ancient sea-margins, as memorials of changes in the relative level of sea and land.* Edinburgh, London.

Chamisso, Adelbert von
1821 *Remarks and opinions of the naturalist of the expedition.* In Kotzebue 1821, 2:351–433; 3:1–318. A15 B91,124,234 C28,38 ZEd6

Chapin, A.B.
1835 Junction of trap and sandstone, Wallingford, Conn. *Am. J. Sci.* 27:104–12. RN142

Charlton, ———
1838 [Communication by Dr. Charlton on a hybrid Tetrao. Notice of the eighth Br. Ass. Advmt. Sci. meeting, Durham.] *Athenaeum* (567) 8 September: 654. D73

Charpentier, Johann von
1836 Account of one of the most important results of the investigations of M. Venetz, regarding the present and earlier condition of the glaciers of the canton Vallais. With later additions by the author. *Edinb. New Phil. J.* 21:210–20. A131

Chasseboeuf, Volney de. See Volney, C.F.C. de

Chevalier, Yves Eugène
1837a Observations relatives à la géologie et minéralogie faites dans le voyage de circum-navigation de la Bonite. *C. R. Hebd. Séanc. Acad. Sci. Paris* 5:720–22. A39
1837b Géologie: Observations faites par les officiers de la Bonite. *Institut* 5 (222) 20 November:405. A39
1844 *Voyage autour du monde exécuté pendant les années 1836 et 1837 sur la corvette 'a Bonite' Géologie et minéralogie.* Paris. A39

Children, J.G. and N.A. Vigors
1826 Zoology. In Denham, Clapperton and Oudney 1826 [Northern and Central Africa], Appendix XXI. C41

Chladni, Ernst Florenz Friedrich
1809 Supplément au catalogue des météores, à la suite desquels des pierres ou des masses de fer sont tombées. *J. Mines Paris.* 26:79–80. RN172

Christie, Alexander Turnbull
1827 Geognostical structure of the country around Darwar. *Edinb. New Phil. J.* 2:194–95. A25
1828–29 Sketches of the meteorology, geology, agriculture, botany, and zoology of the southern Mahratta country. *Edinb. New Phil. J.* 5:292–303; 6:98–120; 7:49–65. A27

Christol, Jules de
1829 Cavernes à ossements renfermant de débris humains. *Bull. Sci. Nat. Géol.* 18:101–02. E35
1835 Recherches sur les caractères des grandes espèces de rhinocéros fossiles. *Annls Sci. Nat.* 4:44–111. QE16

Christ's College, Cambridge
1909 *Darwin centenary. The portraits, prints, and writings of Charles Robert Darwin. Exhibited at Christ's College, Cambridge, 1909.* Cambridge.

Clapperton, Hugh. See Denham, D.N., H. Clapperton and W. Oudney

Clarke, Edward Daniel
1816–24 *Travels in various countries of Europe Asia and Africa.* 4th ed. 11 vols. London. CIBC

Clarke, W.B.
*1834 A few words on cats. *Mag. Nat. Hist.* 7:139–41. B178
*1838 Observations on the *Caprimulgus europaeus* (night-jar). *Mag. Zool. Bot.* 2:158–63. C186

Cleghorn, James
1787 The history of an ovarium, wherein were found teeth, hair, and bones. *Trans. R. Ir. Acad.* 1:73–89. D164

Clemençon, ———
*1837 Sur le district des Diamans du Brésil. *Institut* 5 (221) November:366–67. A42

Clift, William
1831 Report by Mr Clift, of the College of Surgeons, London, in regard to the fossil bones found in the caves and bone-breccias of New Holland. *Edinb. New Phil. J.* 10:394–95. C131

Cline, Henry
1805 *On the form of animals.* London. B145
*1829 *Observations on the breeding and form of domestic animals.* London. [*Pamphlet 166] B145 C269, IBC E123,124

Coleridge, Samuel Taylor
1829 *The poetical works of Coleridge, Shelley, and Keats.* Paris. M88
1840 Review of the works of Samuel Taylor Coleridge [by John Stuart Mill]. *Lond. Westm. Rev.* 33, 1840:257–302. OUN33

Comte, Isidore Auguste Marie François Xavier
1830–35 *Cours de philosophie positive.* 2 vols. Paris. See review, Brewster 1838. D152 E88 M70,72, 81,135 N12 OUN25

Conrad, Timothy Abbott
1835 Observations on the Tertiary strata of the Atlantic Coast. *Am. J. Sci.* 28:104–11,280–82. RN142

Conybeare, William Daniel
1833 Report on the progress, actual state, and ulterior prospects of geological science. *Rep. Br. Ass. Advmt Sci.* (Meetings at York, 1831, and Oxford, 1832) 1(a):364–414. RN142

Conybeare, William Daniel and William Phillips
*1822 *Outlines of the geology of England and Wales.* [Only pt 1 published.] London. RN22,23,46,47

Cook, James
1777 *A voyage towards the South Pole, and round the world . . . in . . . the 'Resolution' and 'Adventure' in the years 1772, 1773, 1774, and 1775.* 2 vols. London. RN140
*1784 *A voyage to the Pacific Ocean . . . in . . . the 'Resolution' and 'Discovery' in the years 1776, 1777, 1778, 1779, and 1780.* 3 vols & atlas. Vol. 3 by James King. London. RN140,141
1815 *An account of a voyage towards the South pole and round the world, performed in His Majesty' ships the 'Resolution' and 'Adventure' in the years 1772, 3, 4, and 5.* In Kerr 1811–24, Vol. 14, 1815. A81

[Cooley, W.D.]
1837 Review of Angelis 1836–37. *Edinb. Rev.* 65:87–109. RN143 A88

Cooper, ———
1828 See Mitchill, J.A., ——— Smith and ——— Cooper

Cordier, Pierre-Louis-Antoine
1834 Rapport sur la partie géologique du voyage de M. D'Orbigny dans l'Amérique méridionale. In Rapport sur les résultats scientifiques du voyage de M. Alcide d'Orbigny dans l'Amérique du Sud, pendant les années 1826, 1827, 1828, 1829, 1830, 1831, 1832 et 1833. *Nouv. Ann. Mus. Hist. Nat. Paris* 3:84–115. A99
1838 Géologie. *Institut* 6 (228) 10 May:150–51. A102

Cortès, ——— and A.M. de Jonnès
1810 Mémoire sur la géologie des Antilles. *J. Phys. Chim. Hist. Nat.* 70:129–34. RN90,91

Coulter, Thomas
*1835 Notes on Upper California. *J. R. Geogr. Soc.* 5:59–70. B235

Courcier, Alexandre
1838 See Montbrun, ——— de and A. Courcier.

Courrejolles, F.
1802 Observations qui prouvent la nécessité d'observer et de méditer longtemps, avant de rien prononcer en physique en général, et en particulier sur la cause des tremblements de terre. *J. Phys. Chim. Hist. Nat.* 54:103–15. A18,19

Courtet de l'Isle, Victor
1838 *La science politique fondée sur la science de l'homme, ou étude des races humaines sous le rapport philosophique, historique et social.* Paris. TFRAG5
1839 Review of 1838. 'Morgan' *Athenaeum* (624) 12 October:771–73. TFRAG5

Cousin, Victor
1828 *Cours de philosophie. Introduction à l'histoire de la philosophie.* Paris. N49

Cowper, William
1784 Unnoticed properties of that animal the hare. *Gentleman's Mag.* 54:412–14. E117
1836–37 *The works of William Cowper. With a life of the author,* by Robert Southey. 15 vols. London. N94

Coxe, William
1817 *A view of the cultivation of fruit trees.* Philadelphia. C265

Crabbe, George
1834 *The poetical works. With his letters and journals, and his life, by his son.* 8 vols. London. C269

Crawford, ———
1827 Account of Mr Crawford's mission to Ava. *Edinb. New Phil. J.* 3:359–70. B119

Crawfurd, John
1820 *History of the Indian Archipelago. Containing an account of the manners, arts, languages, religions, institutions, and commerce of its inhabitants.* 3 vols. Edinburgh. CIBC

Cummins, John
1743 See Bulkeley, J. Cummins

Cuvier, Frédéric Georges
1816–30 *Dictionnaire des sciences naturelles . . . par plusiers professeurs du jardin du roi et des principales écoles de Paris.* 60 vols. Strasbourg, Paris. [*1820, vol. 18 & plates] BIBC
1822 Instinct. In Cuvier, F.G. 1816–30 *Dictionnaire des sciences naturelles* 23:528–44. C268
1827–28 Essay on the domestication of mammiferous animals, with some introductory considerations on the various states in which we may study their actions. *Edinb. New Phil. J.* 3:303–18; 4:45–60, 292–97. B118 C165

Cuvier, Georges Léopold Chrétien Frédéric Dagobert
1803 Sur le Pennatula cynomorium (*Alcyonium epipetrum*, Gmelin) et sur les coraux en général. *Bull. Soc. Philom. Paris.* 3:133. ZEd4
1804 Sur l'Ibis des anciens Egyptiens. *Annls Mus. Hist. Nat. Paris* 4:116–35. E142
1821–24 *Recherches sur les ossemens fossiles.* 2d ed. 5 vols. Vol. 1, 1821; Vol. 5, pt 1, 1823. Paris. A6,10 B52 D67 E23,55 T63
1825 Nature. In Cuvier, F.G. 1816–30 *Dictionnaire des sciences naturelles.* B112
1827 *Essay on the theory of the earth. With geological illustrations, by Professor Jameson.* 5th ed. Edinburgh B53
1827–35 See Griffith, E.
*1829–30 *Le règne animal distribué d'après son organisation, pour servir de base à l'histoire naturelle des animaux et d'introduction à l'anatomie comparée.* 5 vols. Paris. B79 E9,63 Mac54v
1834–36 *Recherches sur les ossemens fossiles* 4th ed. Paris. B53

Cuvier, Georges Léopold Chrétien Frédéric Dagobert and Achille Valenciennes
1828–49 *Histoire naturelle des poissons.* 22 vols. Paris. B112

d'Alton, Eduard
1833 Uber die von dem verstorbenen Herrn Sellow aus der Banda oriental mitgebrachten fossilen Panzertragmente und die dazu gehörigen Knochen-Uberreste. *Abh. Akad. Wiss. Berlin.* 369–424. A151

Dampier, William
1698–1703 *A new voyage round the world.* 3rd ed. 3 vols. London. Vol. 2, 1699; vol. 3, 1703. RN8–10,15 C266 E182

Darby, William
1816 *A geographical description of the State of Louisiana.* Philadelphia. RN84

Darwin, Bernard
1955 *The world that Fred made.* London.

Darwin, Charles Robert
1833–38a On certain areas of elevation and subsidence in the Pacific and Indian oceans, as deduced from the study of coral formations. *Proc. Geol. Soc. Lond.* 2 (31 May 1837):552–54. *CP* 1:46–49. RN15 E22
1833–38b On the connection of certain volcanic phaenomena, and on the formation of mountain-chains and volcanos, as the effects of continental elevations. *Proc. Geol. Soc. Lond.* 2 (7 March 1838):654–60. *CP* 1:53–86. A79
*1839 Observations on the parallel roads of Glen Roy, and of other parts of Lochaber in Scotland, with an attempt to prove that they are of marine origin. *Phil. Trans. R. Soc.*, 129:39–81. *CP* 1:89–137 A133 M90
1842 On the distribution of the erratic boulders and on the contemporaneous unstratified deposits of South America. *Trans. Geol. Soc. Lond.* 2d ser. 6:425–31. *CP* 1:145–63. A88
1843 Double-flowers—their origin. *Gard. Chron.* (36), 9 Sept.:628 *CP* 1:175–77. E184
*1844a Observations on the structure and propagation of the genus *Sagitta*. *Ann. Mag. Nat. Hist.* 8:1–6. *CP* 1:177–82. RN174 ZEd5
1844b Brief descriptions of several terrestrial Planariae, and of some remarkable marine species, with an account of their habits. *Ann. Nat. Hist.* 14:241–51. *CP* 1:182–93. C214
1846 On the geology of the Falkland Islands. *Q. J. Geol. Soc. Lond.* 2 (pt 1):267–74. *CP* 1:203–12. A97
*1870 Note on the habits of the Pampas Woodpecker (Colaptes campestris). *Proc. Zool. Soc. Lond.* (1807):705–6. *CP* 2:161–62. N71
1877 A biographical sketch of an infant. *Mind.* 2:285–94. *CP* 2:191–200. MIBC
1879 [Preliminary notice] In Krause, E. L. *Erasmus Darwin.* London. M25
1958 *The autobiography of Charles Darwin 1809–1882.* Ed. Nora Barlow. London, New York.
1962 *Coral islands. Atoll Research Bull.* No. 88. Washington. E22
1980a *The Red Notebook of Charles Darwin.* Ed. Sandra Herbert. London.

1980b *Metaphysics, materialism, & the evolution of mind*. Ed. Paul H. Barrett. Chicago.

Darwin, Erasmus
*1791–90 *The botanic garden, a poem, in two parts*. Pt I. *The economy of vegetation*. Pt 2. *The loves of the plants*. London. Pt 1 is 1791 1st ed.; Pt 2 is 1790 2d ed. [Darwin's Library also contains a 1789 1st ed. of Pt 2 published in Lichfield.] RN172
*1794–96 *Zoonomia: or, the laws of organic life*. 2 vols. London. B1,2,4,21 C67,243 D19,57,70,178 M7,9,11,18,39,41,45,57,58,63,71,103,146,147,149, 153 N39,47,50,107,184
*1803 *The temple of nature; or the origin of society: A poem with philosophical notes*. London B2 D37

Darwin, George Howard
1907–16 *Scientific papers*. 5 vols. [Vol. 5, Supplementary volume, containing biographical memoirs by Sir Francis Darwin and Professor E. W. Brown.] Cambridge.

Darwin, Francis
1902 *Charles Darwin: his life told in an autobiographical chapter, and in a selected series of his published letters*. London. QE22

Darwin, Robert Waring
*1810 *Principia botanica; or a concise and easy introduction to the sexual botany of Linnaeus*. 3d ed. Newark, England C269

Daubeny, Charles G.
*1826 *A description of active and extinct volcanos*. London. RN38,43,44,46,47,51,58,89–91
1832 Remarks on thermal springs, and their connexion with volcanos. *Edinb. New Phil. J.* 12:49–78. A124
1836 Report on the present state of our knowledge with respect to mineral and thermal waters. *Rep. Br. Ass. Advmt Sci.* 5:1–95. A124
1838 On the geology and thermal springs of North America. *Athenaeum* (567) 8 September:652. A118, 124 M131

Davy, Humphry
*1824 On the corrosion of copper sheeting by sea water, and on methods of preventing this effect; and on their application to ships of war and other ships. *Phil. Trans. R. Soc.* 114:151–58. RN95
1830 *Consolations in travel or the last days of a philosopher*. Philadelphia. C269 N36,62

de Beer, Gavin Rylands
1959 Darwin's Journal. *Bull. Br. Mus. (Nat. Hist.) Hist. Ser.* 2:1–21.
1960 Darwin's notebooks on transmutation of species. Pt I. First Notebook (July 1837–February 1838); Pt II. Second Notebook (February to July 1838); Pt III. Third Notebook (July 15th 1838–October 2nd 1838): Pt IV. Fourth Notebook (October 1838–10 July 1839) *Bull. Br. Mus. (Nat. Hist.) Hist. Ser.* 2 (2,3,4,5):23–183.

de Beer, Gavin Rylands and M.J. Rowlands
1961 Darwin's Notebooks on Transmutation of Species. Addenda and Corrigenda. *Bull. Br. Mus. (Nat. Hist.) Hist. Ser.* 2(6):185–200.

de Beer, Gavin Rylands, M.J. Rowlands, and B.M. Skramovsky
1967 Darwin's Notebooks on Transmutation of Species. Pt VI Pages excised by Darwin. *Bull. Br. Mus. (Nat. Hist.) Hist. Ser.* 3 (5):129–176. E120

Debenham, Frank
1945 *The voyage of Captain Bellingshausen to the Antarctic seas 1819–1821*. 2 vols. London. RN181

Dekay, J.E.
1828 On a fossil ox from the Mississippi. *Edinb. New Phil. J.* 5:326–27. B125

De la Beche, Henry Thomas
1831 *A geological manual*. London. RN19,20,67 A147
*1834 *Researches in theoretical geology*. London. A64,72–75,83 B201 ZEd8

Delalande, Jean Marie
1822 *Précis d'un voyage au Cap de Bonne-Espérance*. Paris. C18

Delta
1835 Thoughts on the geographical distribution of insects. *Ent. Mag.* 2:44–54;280–86. B91

Deluc, Jean-André
1810–11 *Geological travels in some parts of France, Switzerland, and Germany*. 3 vols. London. RN181

Deluc, Jean-André (nephew of the above)
1826 Mémoire sur le phénomène des grandes pierres primitives alpines, distribuées par groupes dans le bassin du Lac de Genève et dans la vallée de l'Arve; et en particulier des groupes qui sont entièrement composés de granites, suivi de conjectures sur la cause qui les a ainsi distribués. *Mém. Soc. Phys. Hist. Nat. Genève* 3 (2):139–200. A20

Denham, Dixon N., Hugh Clapperton and Walter Oudney
1826 *Narrative of travels and discoveries in northern and central Africa in the years 1822, 1823, and 1824*. London. Appendix XXI. Zoology: Children and Vigors 1826. Paris. C41
1828 Review of 1826. *Zool. J. Lond.* 3:452–54. C41

Desgraviers, Augustin-Claude Leconte
1810 *Le parfait chasseur, traité général de toutes les chasses*. Paris. OUN8b

Deshayes, Gérard Paul
1833–34a [Response to M. Geoffroy Saint-Hilaire.] *Bull. Soc. Géol. Fr.* 4:99. B178
1833–34b [Fossils from Java.] *Bull. Soc. Géol. Fr.* 4:212–24. B179
1836–37 [Tertiary fossil shells from the Crimea.] *Bull. Soc. Géol. Fr.* 8:215–18. B54

Desjardins, Julien
*1832 Extract from 'Analyse des travaux de la Société d'Histoire Naturelle de l'Ile Maurice, pendant la 2de année.' *Proc. Zool. Soc. Lond.* 2:111. B160, 166

Demarest, Gaëtan Anselm
1825 Abstract of a 'Memoir on a new genus of the order Rodentia, named Capromys.' *Zool. J. Lond.* 1:81–89. C40

Desmond, Adrian J.
1985 The making of institutional zoology in London, 1822–1836. *Hist. Sci.* 23:153–85,223–50.

Desnoyers, Jules Pierre François Stanislas
1831–32 Sur les ossements humains des cavernes. *Bull. Soc. Géol. Fr.* 2:126–33. E35

Dick, Thomas Lauder
1823 On the parallel roads of Lochaber. *Trans. R. Soc. Edinb.* 9:1–64. A102–3, GR35,41,45,78

Dillwyn, Lewis Weston
*1823 On fossil shells. *Phil. Trans. R. Soc.* 113:393–99. C40
1824 Analytical notice of 1823. *Zool. J. Lond.* 1:120–21. C40

Disraeli, Isaac
1835 *Curiosities of literature.* London. D35

Dobrizhoffer, Martin
1822 *An account of the Abipones, an equestrian people of Paraguay.* 3 vols. London. C13,276 ZEd10

Don, David
1817–20 Descriptions of several new or rare native plants, found in Scotland, chiefly by the late George Don, of Forfar. *Mem. Wernerian Nat. Hist. Soc.* 3:294–305. QE10
1830 Description of the new genera and species of the class Compositae belonging to the floras of Peru, Mexico and Chile. *Trans. Linn. Soc. Lond.* 16:169–303. B193
1833 On the Coniferae at present growing in Australia. *Edinb. New Phil. J.* 14:158–59. B187
1841 Descriptions of two new genera (*Cryptomeria* and *Athrotaxis*) of the natural family of plants called Coniferae. *Trans. Linn. Soc. Lond.* 18:163–80. TFRAG8

Dubreuil, Joseph-Marie (?)
1837 Études anatomiques de têtes ayant appartenu à des individus de races humaines diverses. *Annls Sci. Nat.* 7:254. QE16

Ducatel, Julius T.
1837 Outlines of the physical geography of Maryland, embracing its prominent geological features. *Trans. Maryland Acad. Sci. Lit.* 1:24–54. A130

Ducatel, Julius T. and John H. Alexander
1835 Report on a projected geological and topographical survey of the State of Maryland. *Am. J. Sci.* 27:1–38. RN142

Dufrénoy, Ours Pierre Armand Petit and Jean Baptiste Elie de Beaumont
*1830–38 *Mémoires pour servir à une description géologique de la France.* 4 vols. Paris. Vol. 3, 1836 [Vols 3 & 4 only.] A127 C270

Dujardin, Felix
*1838 [Sur les oeufs du Taenia.] *Institut* 6 (240) 2 August:249–50. D49

Dumas, J.A.
1821, 1832 See Prévost, J.L. and J.A. Dumas

Duméril, Andre Marie Constant and Gabriel Bibron
1834–54 *Erpétologie générale ou histoire naturelle complète des reptiles.* 9 vols & atlas. [*Suites à Buffon.*] Paris. B234,251 C18,IBC D40,179 ZEd12,19

Dumont d'Urville, Jules-Sébastian-César
1825 De la distribution des fougères sur la surface du globe terrestre. *Annls Sci. Nat.* 6:51–73. C16
1826 Flore des Malouines. *Mém. Soc. Linn. Paris* 4:573–621. C24
1830–35 *Voyage de la corvette 'l'Astrolabe'. . . pendant les années 1826–1827–1828–1829.* 14 vols & atlas. Paris. B31 C13,16,276 ZEd10. See also Quoy and Gaimard 1830–35.

Duperrey, Louis Isidore
1825–30 *Voyage autour du monde . . . sur la corvette de sa 'Majesté', 'La Coquille' pendant les années 1822, 1823, 1824 et 1825,* 6 vols. and Atlas. Paris. RN62, 101,102 B31,54,249 C16–28,276 E42 ZEd5,7,11,13

D'Urville, Jules-Sebastian-César Dumont. See Dumont d'Urville, J.-S.-C.

Dutrochet, Henri
1832 Mémoire sur les organes aérifères des végétaux, et sur l'usage de l'air que contiennent ces organes. *Annls Sci. Nat.* 25:242–59. C268

Duvernoy, George Louis
1835 Fragments d'histoire naturelle sysématique et physiologique sur les Musaraignes. *Mém. Soc. Mus. Hist. Nat. Strasbourg,* vol. 2. B31 C183

1838 Notice of Duvernoy 1835. *Institut* 6 (226) April: 111–12. C183

Dwight, Timothy
1821–22 *Travels in New England and New York*. 4 vols. New Haven. C221

Earl, George Windsor
1837 *The Eastern Seas, or voyages and adventures in the Indian Archipelago, in 1832–33–34*. London. C270 E18,19,182

Edwards, Henri-Milne. See Milne-Edwards, H.

Edwards, William Frédéric
1832 *On the influence of physical agents on life*. London. C269 E133,134,184

Ehrenberg, Christian Gottfried
1834 Recherches sur le développement et la durée de la vie des animaux infusoires, suivies d'une description comparative de leurs différens organes. *Annls Sci. Nat.* 1:199–232,266–81. C143,146
*1837a Remarks on the real occurrence of fossil infusoria, and their extensive diffusion. *Scient. Mem.* 1:400–07. A93 C146,275
*1837b Further notices of fossil infusoria. *Scient. Mem.* 1:407–13. A84 C146,275
*1837c On the origin of organic matter from simple perceptible matter, and on organic molecules and atoms; together with some remarks on the power of vision of the human eye. *Scient. Mem.* 1:555–83. RNIFC B19,22
*1837d Infusoires fossiles du tripoli d'Oran. *Institut* 5 (220) October:330–31. B150 C146
1838a Infusoires. *Institut* 6 (224) February:62. C15,95
1838b [Microscopical inquiries relative to the organization of the lower animals.] *Athenaeum* (567) 8 September:653. D130,172
*1839 Communication respecting fossil and recent infusoria made to the British Association at Newcastle. *Ann. Nat. Hist.* 2:121–24. D167

Eichwald, Carl Eduard von
1834–38 *Reise auf dem Caspischen Meere und in den Caucasus. Unternommen in 1825–1826*. 2 vols. Stuttgart, Tubingen. E85
1838a Einige Bemerkungen über das kaspische Meer. *Arch. Naturgesch.* 4:97–112. D151
1838b Sur des ossements fossiles de mammifères trouvés en Pologne. *Institut* 6 (256) 22 November:384. E58
1838c [Sur la faune de la mer Caspienne.] *Institut* 6 (259) 13 December:412. E85
*1839 Remarks on the Caspian Sea. *Ann. Nat. Hist.* 2:135–36. D151

Elie de Beaumont, Jean Baptiste Armand Léonce
*1838 *Recherches sur la structure et sur l'origine du Mont Etna*. In Dufrénoy, O.P.A.P., and J.P. Elie de Beaumont 1830–38, 4 (1838):1–226. C270

Elliotson, John
1840 *Human physiology*. 5th ed. London. OUN10v

Ellis, William
1831 *Polynesian researches, during a residence of nearly eight years in the Society and Sandwich Islands*. 2d ed. 4 vols. London. B196,220

Endlicher, Stephano Friedrich Ladislaus
*1833 *Prodromus Florae Norfolkicae*. Vienna. C238
1836 Bemerkungen über die Flora der Südseeinseln. *Annln Wien. Mus. Naturg.* 1:129–90. C266

Erman, Adolf Georg
*1838a Letter on the frozen soil of Siberia. *J. R. Geogr. Soc.* 8:212–13. A117 C101
1838b Letter on frozen soil in Siberia. *Athenaeum* (546) 14 April:274. A90 C101

Escholtz, Franz
1830 Zoological appendix. In Kotzebue 1830, vol. 2.

Esenbeck. See Reinwardt, C.G.C., C.L. Blume, and C.G.D. Nees von Esenbeck

Eugenius. See Herries, I.M. (?)

Evelyn, John
1664 *Sylva, or a discourse of forest-trees, and the propagation of timber in His Majesty's Dominions*. London. C269
1827 See Mitchell, J.

Everest, Robert
1832 Note on Indian saline deposits. *J. Asiat. Soc. Beng.* 1:149–50. A143

Eyton, Thomas Campbell
1835 Some account of a hybrid bird, between the cock pheasant (*Phasianus Colchicus*, Linn.) and the grey hen (*Tetrao tetrix*, Ej.). *Proc. Zool. Soc. Lond.* 3:62–63. B189
*1837a Some remarks upon the theory of hybridity. *Mag. Nat. Hist.* 1:357–59. B30,138,139,212 DIFC,25, 26, E169
*1837b Notice of some osteological peculiarities in different skeletons of the genus *Sus*. *Proc. Zool. Soc. Lond.* 5:23–24. B162,212 C124,125
1838a *A monograph on the Anatidae*. London. D25,26
1838b Irish Hare (*Lepus Hibernicus*, Yarrell). *Mag. Zool. Bot.* 2:283–84. B7 D61

F.K.P.
1855 Five-toed Spanish. *Poultry Chronicle* 2:355 DIFC

Falconer, Hugh and Proby T. Cautley
1837 See Cautley, P.T. and H. Falconer

1838 Sur de nouvelles espèces fossiles de l'ordre des Quadrumanes. *Institut* 6 (223) January:37–39. B250

Falconer, William
1781 *Remarks on the influence of climate, situation, nature of country, population, nature of food, and way of life on the disposition and temper, manners and behaviour, intellects, laws and customs, form of government, and religion of mankind.* London. B272 C268
1791 *A dissertation on the influence of the passions upon disorders of the body.* 2d ed. London. N93

Falkner, Thomas
1774 *A description of Patagonia, and the adjoining parts of South America.* London. RN67 C13,276

Ferguson, William
1823 On the nature and history of the marsh poison. *Trans. R. Soc. Edinb.* 9:273–98. D24

Fernandez de Oviedo y Valdes, Gonzalo
1547 *La hystoria general de las Indias.* Salamanca. C42

[Ferrier, James]
1838a An introduction to the philosophy of consciousness. *Blackwood's Edinb. Mag.* 43:187–201,437–52, 784–91. C267 M18,78,155
1838b *Expedition to Novaïa Zemlia and Lapland.* By M. Baer.—Bulletin Scientifique, &c. St Petersburg. *Athenaeum* (560) 21 July:505–08. D24

Férussac, André-Etienne-Just-Paschal-Joseph-François D'Audebard de and G.P. Deshayes
1819–51 *Histoire naturelle, générale et particulière des mollusques terrestres et fluviatiles.* 2 vols & atlas of 2 vols. Paris. C245

Fitton, William Henry
1827 An account of some geological specimens. In King, P.P. 1827, vol. 2:566–629. RN6,38,101

FitzRoy, Robert
1836 Sketch of the surveying voyages of H.M.S. 'Adventure' and 'Beagle', 1825–36. *J. R. Geogr. Soc.* 6:311–42. A79
1837 Extracts from the diary of an attempt to ascend the River Santa Cruz, in Patagonia, with the boats of his Majesty's sloop 'Beagle's' *J. R. Geogr. Soc.* 7:114–26. A88
*1839 *Narrative of the surveying voyages of His Majesty's Ships 'Adventure' and 'Beagle' between the years 1826 and 1836, describing their examination of the southern shores of South America, and the 'Beagle's' circumnavigation of the globe.* 3 vols. & Appendix. London.
Vol. 1: *Proceedings of the first expedition, 1826–1830, under the Command of Captain P. Parker King, R.N., F.R.S.* C269
Vol. 2: *Proceedings of the second expedition, 1831–36, under the Command of Captain Robert Fitz-Roy, R.N.* [& Appendix]. RN12,17,35,99 A141 C269 E166
Vol. 3: *Journal and remarks. 1832–36.* By Charles Darwin, Esq., M.A., Sec. Geol. Soc. See *JR.*

Fleming, John
*1822 *The philosophy of zoology.* 2 vols. Edinburgh. B272 C149,166,268 QE1^{v},11,15
*1828 *A history of British animals.* Edinburgh, London. E102
1829 *On systems and methods in natural history.* By J.E. Bicheno, Esq. 1829. [unsigned response to Bicheno 1827.] *Q. Rev.* 41:302–27. C42,155–57,276

Flinders, Matthew
1814 *A voyage to Terra Australis . . . in the years 1801, 1802, and 1803, in . . . The 'Investigator', and subsequently in the armed vessel 'Porpoise' and 'Cumberland' schooner. With an account of the shipwreck of the 'Porpoise' arrival of the 'Cumberland' at Mauritius, and imprisonment of the Commander during six years and a half in that island.* 2 vols & atlas. London. B76,77 C102,237, 238,240,268 QE21

Flourens, Marie-Jean-Piere
1835–37 Review of Cuvier, G.L.C.F.D. 1834–36. *J. Savants* 1837:237–50. B52,53
*1854 *De la longévité humaine et de la quantité de la vie sur la globe.* Paris.

Flourens, Marie-Jean-Pierre and Marcel de Serres
1838 Métis du moufflon et du mouton. *Institut* 6 (251) 18 October:338. E35

Forbes, Edward
*1839 On the land and freshwater mollusca of Algiers and Bougia. *Ann. Nat. Hist.* 2 (10):250–55. E90, 91

Forchhammer, G.
1828 On the chalk formation of Denmark. *Edinb. J. Sci.* 9:56–68. GR109

Forrest, Thomas
1780 *A voyage to New Guinea, and the Moluccas, from Balambangan . . . in the 'Tartar Galley' during the years 1774, 1775, and 1776.* 2d ed. London. C17,23,268 E173,175,181,182

Forster, Johann Reinhold
*1778 *Observations made during a voyage round the world.* London. C54

Fournet, Joseph-Jean-Baptiste-Xavier
*1837 Filons de l'Arbresle. *Institut* 5 (218) August:246–49. A15,31

Fox, Robert Were
*1830 On the electro-magnetic properties of metalliferous veins in the mines of Cornwall. *Phil. Trans. R. Soc.* 120:399–414. RN20
1837 Substance of a communication on the temperature of some mines in Cornwall and Devonshire, made by Robert Were Fox, to the Royal Geological Society of Cornwall at their last annual meeting. *Phil. Mag.* 11:520–23 A76
1838 Sur la température des mines du Cornwall et du Devonshire. *Institut* 6 (223) January:39–40. A76

Franklin, John
1823 *Narrative of a journey to the shores of the Polar Sea, in the years 1819, 20, 21, and 22.* London. M103

Freeman, Richard B.
1977 *The works of Charles Darwin. An annotated bibliographical handlist.* 2d ed. Folkestone, Hamden. QE3
1978 *Charles Darwin. A companion.* Folkestone, Hamden.

Freeman, Richard B. and P.J. Gautrey
1969 Darwin's *Questions about the breeding of animals*, with a note on *Queries about expression. Soc. Biblphy Nat. Hist. J.* 5:220–25. QE3,5,15,19

French, John Oliver
1825 An inquiry respecting the true nature of instinct, and of the mental distinction between brute animals and man; introductory to a series of essays, explanatory of the various faculties and actions of the former, which have been considered to result from a degree of moral feeling, and of intellect. *Zool. J. Lond.* 1:1–32,153–73,346–67. C240

Frere, John
1800 Account of flint weapons discovered at Hoxne in Suffolk. *Archaeologia* 13:204–5. ZEd20

Freycinet, Louis Claude de Saulses de
1807–16 See Péron, F.
1824–44 *Voyage autour du monde, entrepris . . . sur les corvettes de S. M.* 'l'Uranie' et 'la Physicienne' . . . 1817, 1818, 1819 et 1820. 7 vols & 4 vols of plates. *Zoologie.* [See Quoy and Gaimard 1824.] Paris. B31,54,234 D74 ZEd3

Fries, Bengt Frederik
*1839a On the genus *Syngnathus. Ann. Nat. Hist.* 2:96–105. D169 E57 T25,26
*1839b Metamorphosis observed in *Syngnathus lumbriciformis. Ann. Nat. Hist.* 2:225. E90
*1839c Metamorphosis observed in the small pipe-fish (*Syngnathus lumbriciformis*). *Ann. Nat. Hist.* 2:451–55. T25

Fries, Elias Magnus
1825 *Systema orbis vegetabilis.* Lund. B46 CIBC
1836 Entwurf zu einer neuen Beantwortung der Frage: Welche Gewächse sind die vollkommensten? *Flora, Jena* 19:1–16. B200

Froude, Richard Hurrell
1838a *Remains of the late Rev. Richard Hurrell Froude, M.A., Fellow of Oriel College, Oxford.* London. 2 vols. M60
1838b Review of 1838a. *Edinb. Rev.* 67:525–35. M60

Fuller, Edward
*1836 Notice of the rearing of two hybrids from a barn-door hen, having a cross from a pheasant and a pheasant-cock. *Proc. Zool. Soc. Lond.* 4:84,85. C164

Gaertner, Joseph
1801–7 *De fructibus et seminibus plantarum.* 3 vols. Leipzig. QE7,11[15v]

Gaertner, Karl Friederich von
1805 *Supplementum carpologiae; seu, continuati operis Josephi Gaertner.* Vol. 3. In Gaertner, J. 1801–07. QE7,11[15v]
1836 Einige Bemerkungen über die Befruchtung der Gewächse, und die Bastard- Erzeugung im Pflanzenreich. *Flora, Jena* 19:177–92. QE7

Gaimard, Paul Joseph
1838–51 *Voyage en Islande et au Groënland exécté pendant les années 1835 et 1836 sur la corvette 'la Recherche'. 7 vols & atlas in 4 vols. Paris.* B256
See Quoy, J.R.C. and P.J. Gaimard

Gallesio, Giorgio
*1816 *Teoria della riproduzione vegetale.* Pisa. [*abst DAR 71:95–111.] E183 QE16

Gardiner, William
1832 *The music of nature; or, an attempt to prove that what is passionate and pleasing in the art of singing, speaking, and performing upon musical instruments, is derived from the sounds of the animated world.* London. C269 N64–66,94

Garnot, Prosper
1826 Remarques sur la zoologie des Iles Malouines, faites pendant le voyage autour du monde de la corvette 'la Coquille' exécuté en 1822, 1823, 1824, et 1825. *Annls Sci. Nat.* 7:39–59. B31,54
1826–30 See Lesson, R.P. and P. Garnot

Gaudichaud-Beaupré, Charles.
1826–30 Botanique. In Freycinet, L.C. *Voyage autour du monde, fait par ordre du roi, sur les corvettes de S. M. 'l'Uranie' et 'la Physicenne' pendant les années 1817, 1818, 1819 et 1820.* 1 vol. & atlas. Paris. D74

Gautrey, P.J. See R.B. Freeman and P.J. Gautrey 1969.

Gawler, J. Bellenden
1812 Allium Cepa (β). Common onion. The bulb-bearing variety, or tree onion. *Curtis's Bot. Mag.* 36, Pl. 1469. E165

Gay, Claude
1833 Aperçu sur les recherches d'histoire naturelle faites dans l'Amérique du sud, et principalement dans le Chili, pendant les années 1830 et 1831. *Annls Sci. Nat.* 28:369–93. RN65
1838 Terrains du Chili. *Institut* 6 (236) 5 July:216–17. A103

Gebhard, John
1835 On the geology and mineralogy of Schoharie, N.Y. *Am. J. Sci.* 28:172–77. RN142

Geoffroy Saint-Hilaire, Etienne
1818–22 *Philosophie anatomique.* 2 vols. Paris. B114 E89
1830 *Principes de philosophie zoologique.* Paris. B110–14,IBC D69 QE17
1833–34 [Fossiles mammifères en Auvergne.] *Bull. Géol. Soc. Fr.* 4:89–90. B178
1837a Des changements à la surface de la terre, qui paraissent dépendre originairement et nécessairement de la variation pré-existente incessante, lente et successive des milieux ambiants divers et consecutifs du globe terrestre. *C. R. Hebd. Séanc. Acad. Sci. Paris* 5:183–94. B135
1837b Singe fossile de Sansan. *Institut* 5 (218) August: 242–44 B94,133,135 E70
1837c Sur la singularité et la haute portée en philosophie naturelle de l'existence d'une espèce de singe trouvée à l'état fossile dans le midi de la France. *C. R. Hebd. Séanc. Acad. Sci. Paris* 5:35–44. B133

Geoffroy Saint-Hilaire, Isidore
*1832–36 *Histoire générale et particulière des anomalies de l'organisation chez l'homme et les animaux, ouvrage comprenant des recherches sur les caractères, la classification, l'influence physiologique et pathologique, les rapports généraux, les lois et les causes des monstruosités, des variétés et des vices de conformation, ou traité de tératologie.* 3 vols & atlas. Paris. C267 D57 QE17
1838a Nouveaux genres d'oiseaux Madagascar. *Institut* 6 (226) April:127–28. C205
1838b Rapport sur un mémoire de M. Alcide d'Orbigny, intitulé: Sur la distribution géographique des oiseaux passeeaux dans l'Amérique méridionale. *C. R. Hebd. Séanc. Acad. Sci. Paris* 6:190–94. C183
*1841 *Essais de zoologie générale, ou mémoires et notices sur la zoologie générale, l'anthropologie, et l'histoire de la science.* Paris. QE3,17

Geoffroy Saint-Hilaire, Isidore and Henri-Marie Ducrotay de Blainville
1834 Partie zoologique. In Rapport sur les résultats scientifiques du voyage de M. Alcide d'Orbigny dans l'Amérique du Sud, pendant les années 1826, 1827, 1828, 1829, 1830, 1831, 1832 et 1833. *Nouv. Ann. Mus. Hist. Nat. Paris* 3:84–115. C183

Geological Society of London
1846 *Catalogue of the books and maps in the library of the Geological Society of London.* London. RN157

Gervais, P.
*1837 [Une note sur les animaux mammifères des Antilles.] *Institut* 5 (218) August:253–54. E42

Gibbon, Edward
1830 *Memoirs of the life and writings of Edward Gibbon.* 2 vols. London. C270

Goebel, K.
1909 The biology of flowers. In *Darwin and modern science.* A.C. Seward, ed. Cambridge. D175

Goldsmith, Oliver
1806 *The miscellaneous works of Oliver Goldsmith.* 5 vols. London. N127

Gore, Catherine Grace Frances
1838 *The rose fancier's manual.* London. C267

Gould, John
1832–37 *The birds of Europe.* 5 vols. London. B241 C82,111,125 ZEd15
*1837a Observations on the raptorial birds in Mr. Darwin's collection, with characters of the new species. *Proc. Zool. Soc. Lond.* 5:9–11. RN130 D96 QE10a
*1837b On a new Rhea (Rhea Darwinii) from Mr. Darwin's collection. *Proc. Zool. Soc. Lond.* 5:35–36. RN127
*1837c Characters of a large number of new species of Australian birds. *Proc. Zool. Soc. Lond.* 5:138–57. QE10a
*1837d Observations on some species of the genus Motacilla of Linnaeus. *Mag. Nat. Hist.* 1:459–61. B138
1837e Observations on the raptorial birds of Australia and the adjacent islands. *Proc. Zool. Soc. Lond.* 5:96–100. D96
1837f Three species of the genus Orpheus, from the Galapagos, in the collection of Mr. Darwin. *Proc. Zool. Soc. Lond.* 5:27. RN130
[1837–38] *A synopsis of the birds of Australia, and the adjacent islands.* London. B50 C239 QE10a
1838a *A monograph of the Trogonidae, or family of trogons.* London C40
1838b Proceedings of learned societies: Zoological Society, October 10, 1837. *Ann. Nat. Hist.* 1:401–6. D61
*1838–41 See *Zoology*: *Birds.*
*1839 Letter from Van Diemen's Land, accompanied with descriptions of some new Australian birds. *Proc. Zool. Soc. Lond.* 7:139–45. QE10a
*1840a On some new species of Australian birds. *Proc. Zool. Soc. Lond.* 8:113–15. QE10a

*1840b Descriptions of new birds from Australia. *Proc. Zool. Soc. Lond.* 8:147–51,159–65,169–78. QE10a
*1845 Descriptions of a new Trogon and seven new birds from Australia. *Proc. Zool. Soc. Lond.* 13:18–20. QE10a, [11v]

Graah, Wilhelm August
1837 *Narrative of an expedition to the East Coast of Greenland.* London. E109

Grant, C.W.
1840 Memoir to illustrate a geological map of Cutch. *Trans. Geol. Soc. Lond.* 2d ser. 5:289-329. D68

Gray, John Edward
1834 Some observations on the economy of molluscous animals, and on the structure of their shells. *Phil Trans. R. Soc.* 123:771-819. B9
*1835 Remarks on the difficulty of distinguishing certain genera of testaceous mollusca by their shells alone, and on the anomalies in regard to habitation observd in certain species. *Phil. Trans. R. Soc.* 125:301–10. B9
*1838a On the two species of Echidna. *Ann. Nat. Hist.* 1:335. C225
1838b See Gunn, R. and J.E. Gray
*1839 Note on the wild cattle of Chillingham Park. *Ann. Nat. Hist.* 2:284. QE17
*1846a On a species of hippopotamus from Sierra Leone. *Ann. Mag. Nat. Hist.* 18:136. QE17
*1846b On the arrangement of the hollow-horned ruminants (Bovidae). *Ann. Mag. Nat. Hist.* 18:227-33. QE17

Griffith, Edward
1827–35 *The animal kingdom arranged in conformity with its organization, by the Baron Cuvier . . . with additional descriptions of all the species.* 16 vols. London. C228

Gruber, Howard E.
1981 *Darwin on man: a psychological study of scientific creativity.* 2d ed. Chicago.

Gruber, Howard E. and Paul H. Barrett
1974 *Darwin on man.* New York, London. B139,141

Gubbins, Charles
1838 Mode of manufacture of the Salumba salt of Upper India. *J. Asiat. Soc. Beng.* 7:363–64. A143

Guillemin, Jean-Baptiste-Antoine
*1836–37 Zephyritis Taitensis—énumération des plantes découvertes par les voyageurs dans les Iles de la Société, principalement dans celle de Taïti. *Anls Sci. Nat. (Bot)* 2d ser. 6:297–30; 7:177–92,241–55,349–70. [*abst of vol. 6:297–320 DAR 72.93–94] B198

Gunn, Ronald and John Edward Gray
*1838 Notices accompanying a collection of quadrupeds and fish from Van Diemen's Land. With notes and descriptions of the new species. *Ann. Nat. Hist.* 1:101–11. C95

Guyon, ———
1838 [Un cas tératologique analogue à celui de l'enfant bi-corps de Prunay- sous-Ablis.] *Institut* 6 (260) 20 December:414. E88

Hall, James
1812 Account of a series of experiments, shewing the effects of compression in modifying the action of heat. *Trans. R. Soc. Edinb.* 6:71–186. A77,111
1815 On the revolutions of the earth's surface. *Trans. R. Soc. Edinb.* 7:139–67 (pt 1), 169–211 (pt 2). A36
1826 On the consolidation of the strata of the earth. *Trans. R. Soc. Edinb.* 10:314–29. A155

Hamilton, William John
1842 *Researches in Asia Minor, Pontus, and Armenia with some account of their antiquities and geology.* 2 vols. London. C214

Hamilton, William John, and Hugh Edwin Strickland
1833–38 An account of a tertiary deposit near Lixouri, in the island of Cephalonia. *Proc. Geol. Soc. Lond.* 2 (3 May 1837):545–46. C55

Hammond, Graham
1837 Bank of seventy fathoms off Staten Island. *Naut. Mag.* 6:822. A102

Hancock, John
1838 On the *Falco Islandicus* of authors. [Notice of 8th meeting of the Br. Ass. Advmt Sci.] *Athenaeum* (565) 25 August:613 B256

Harbottle, Thomas Benfield
1897 *Dictionary of quotations (classical).* London. M18

Hardwicke, Thomas
1828 On the Bos gour of India. *Zool. J. Lond.* 3:231–33. C41

Harlan, Richard
1835 *Medical and physical researches: or original memoirs in medicine, surgery, physiology, geology, zoology, and comparative anatomy.* Philadelphia. C266 S3r

Harmer, Frederic
1909 See W. G. Ridewood.

Hartley, David
1834 *Observations on man, his frame, his duty, and his expectations.* 6th ed. 2 parts. London. N184

Harvey, J.B.
1836 Letter referring to a collection of marine productions, including a specimen of *Capros aper*, Lacep., and a new species of *Tubularia* (*T. gracilis*, Harv.), collected on the south coast of Devonshire, and presented by the writer to the Society. *Proc. Zool. Soc. Lond.* 4:54–55. ZEd6

Hasselt, Jan Conrad von
1827–28 See Blume, C.L.

Hausmann, Johann Friedrich Ludwig
*1838 On metallurgical phenomena as illustrative of geology. *Edinb. New Phil. J.* 24:65–85. A149

Haydn, Joseph
1817 *The life of Haydn. Followed by the life of Mozart.* London. C269 N66

Heber, Reginald
1828 *Narrative of a journey through the upper provinces of India, from Calcutta to Bombay, 1824–1825.* 2 vols. London. C235 E178

Heberden, William (The Younger)
1801 *Observations on the increase and decrease of different diseases, and particularly of the plague.* London. C266

Hedges, William
*1820 Account of experiments on the production of blue instead of red flowers on the *Hydrangea Hortensis*; with some notes on the propagation and management of the plant. *Trans. R. Hort. Soc.* 3:173–77. [*abst DAR 74:61.] E150

Heineken, C.
1835 [1830] Observations on the *Fringilla canaria, Sylvia atricapilla*, and other birds of Madeira. *Zool. J. Lond.* 5:70–79. C160

Helms, Anthony Zachariah
1807 *Travels from Buenos Ayres by Potosi, to Lima.* 2d ed. London. RN155 A16

Henslow, John Stevens
*1821–22 Geological description of Anglesea. *Trans. Camb. Phil. Soc.* 1:359–452. RN5,7 A25,52
*1830a On the specific identity of the primrose, oxlip, cowslip, and polyanthus. *Mag. Nat. Hist.* 3:406–09. C194 E16
*1830b On the specific identity of *Anagallis arvénsis* and *coerulea*. *Mag. Nat. Hist.* 3:537–38. C192,194
*1832 On variations in the cotyledons and primordial leaves of the sycamore (*Acer pseudo-platanus*, L.). *Mag. Nat. Hist.* 5:346–47. C192
*1837 *The principles of descriptive & physiological botany.* London. B12,13 D162 QE10
*1838 Florula Keelingensis. An account of the native plants of the Keeling Islands. *Ann. Nat. Hist.* 1:337–47. B234 C16,100 QE7 [15v]

Herbert, Sandra
1968 *The logic of Darwin's discovery.* Ph.D. diss. Brandeis University.
1971 Darwin, Malthus and selection. *J. Hist. Biol.* 4:209–17.
1974–77 The place of man in the development of Darwin's theory of transmutation. Pt I. *J. Hist. Biol.* 7:217–58. Pt II. *J. Hist. Biol.* 10:155–227.
1980 *The Red Notebook of Charles Darwin.* London. RN130
1985 Darwin the young geologist. In Kohn 1985.

Herbert, William
1820 Instructions for the treatment of the *Amaryllis longifolia*, as a hardy aquatic, with some observations on the production of hybrid plants, and the treatment of the bulbs to the genera *Crinum* and *Amaryllis*. *Trans. R. Hort. Soc.* 3:187–96. B180,191
1822 On the production of hybrid vegetables; with the result of many experiments made in the investigation of the subject. In a letter to the Secretary. *Trans. R. Hort. Soc.* 4:15–50. B180,191 E143 QE2,[6v],7
*1837 *Amaryllidaceae; preceded by an attempt to arrange the monocotyledonous orders, and followed by a treatise on cross-bred vegetables, and supplement.* London. B180, 191 C125,219,265,269 D66,91 E103,107, 111–113,133,141,143,144 QE3,11,[13v],14

Herries, Isabelle M. (?) Eugenius [pseudonym]
1838 *Observations on the foundation of morals; suggested by Professor Whewell's sermons on the same subject.* C269 OUN51,53

Herschel, John Frederick William
*1831 *A preliminary discourse on the study of natural philosophy.* London. C269 D71 N49,60
1833 *A treatise on astronomy.* London. A105,113,114,118
1833–38 Extracts from a letter from Sir John F.W. Herschel to C. Lyell, Esq., postmarked Fredhausen, Cape of Good Hope, 20th February, 1836. *Proc. Geol. Soc. Lond.* 2 (17 May 1837):548–50. RN32 A46 E59
1838 Extracts from letters of Sir John Herschel. In Babbage 1838:225–47. RN32 E59

Hilhouse, William
*1834a Journal of a voyage up the Massaroony in 1831. *J. R. Geogr. Soc.* 4:25–40. A89 B131
*1834b Memoir on the Warow land of British Guiana. *J. R. Geogr. Soc.* 4:321–33. A87 B151

Hill, M.
*1838 Habits of the blue titmouse (*Parus coerulus*). *Ann. Nat. Hist.* 1:158–59. C96

Hindmarsh, L.
1838 See Tankerville, C.A.B. and L. Hindmarsh
*1839a On the wild cattle of Chillingham Park. *Ann. Nat. Hist.* 2:274–84 D48 E91 M142
1839b On the wild cattle of Chillingham Park. *Rep. Br. Ass. Advmt Sci.* (Meeting at Newcastle, 1838) 7:100–4. D48 E91

Hitchcock, Edward
1835 *Report on the geology, mineralogy, botany, and zoology of Massachusetts.* 2d ed. Amherst, Massachusetts. A131

Hodge, M.J.S.
1982 Darwin and the laws of the animate part of the terrestrial system (1835–1837): on the Lyellian origins of his zoonomical explanatory program. *Stud. Hist. Biol.* 7:1–106.

Hodge, M.J.S. and David Kohn
1985 The immediate origins of natural selection. In Kohn 1985.

Hodgson, Bryan Houghton
1831 Contributions in natural history. *Gleanings in Sci.* 3:320–24. T9
1832 On the Mammalia of Nepal. *J. Asiat. Soc. Beng.* 1:335–49. E178,179
1838 Notice of a classified catalogue of Nepalese Mammalia, read at the Linnean Society 20 February 1838. *Ann. Nat. Hist.* 1:152–54. C96

Hoff, Karl E.A. von
*1837 Remarques sur l'origine des pierres météoriques plus spécialement sur les opinions de M. Berzélius à ce sujet. *Institut* 5 (219) September: 297–300. A32

Hoffman, Friedrich
1838 *Geschichte der Geognosie, und Schilderung der vulkanischen Erscheinungen.* Berlin. RN165

Holland, Henry
1811 Preliminary dissertation on the history and literature of Iceland. In Mackenzie, 1811, pp. 1–70. RN159
*1839a *Medical notes and reflections.* London. E183 M127 N111 QE1,16
1839b *Medical notes and reflections.* Philadelphia, New Orleans. N111

Holman, James
1834 *A voyage round the world.* [1827–1832] 4 vols. London. B124

Home, Everard
*1802 Description of the anatomy of the *Ornithorhynchus hystrix. Phil. Trans. R. Soc.* 92:348–64. C225

Home, Henry, Lord Kames
1774 *Sketches of the history of man.* 2 vols. London. C257,275 OUN49

Hooker, Joseph Dalton
*1847–60 *The Botany of the Antarctic voyage of H.M. Discovery Ships 'Erebus' and 'Terror' in the years 1839–1843, under the Command of Captain Sir James Clark Ross.* 6 vols in 3 parts, vol. of plates. London. I. 1847 *Flora Antarctica*, 2 pts; II 1853 *Flora Novae-Zelandiae*, 2 pts; III 1860 *Flora Tasmania*, 2 pts, 4 plates. QE21..
1847a An enumeration of the plants of the Galapagos Archipelago; with descriptions of those which are new. *Trans. Linn. Soc. Lond. (Bot.)* 20:163–233. E100
1847b On the vegetation of the Galapagos Archipelago as compared with that of some other tropical islands and of the continent of America. *Trans. Linn. Soc. Lond. (Bot.)* 20:235–62. E100

Hooker, William Jackson
1811 List of Icelandic plants. Appendix 2 in Mackenzie 1811. B159 C184

Hope, Frederick William
1836 Observations on insects producing silk, and on the possibility of rearing silk crops in England. *Trans. Ent. Soc. Lond.* 1:123–27. QE12
*1837–40 On insects and their larvae occasionally found in the human body. *Trans. Ent. Soc. Lond.* 2:256–71. C233
1839 Remarks on the modern classification of insects. *Rep. Br. Ass. Advmt Sci.* 8 (2) 1838:113. E11

Hopkins, William
1835 [1836] Researches in physical geology. *Trans. Camb. Phil. Soc.* 6 (1):1–84. [Darwin's offprint dated 1835; journal dated 1836.] A103,107,145
1836 *An abstract of a memoir on physical geology.* Cambridge. A65,107,145

Horner, Leonard
1836 On the occurrence of the Megalichthys in a bed of cannel coal in the West of Fifeshire, with observations on the supposed lacustrine limestone at Burdiehouse. *Edinb. New Phil. J.* 20:309–20. QE17

Horsburgh, James
1836 *India Directory.* 4th ed. 2 vols in 1. London. E177

Horsfield, Thomas
1822 Systematic arrangement and description of birds from the island of Java. *Trans. Linn. Soc. Lond.* 13:133–200. D40
1824 *Zoological researches in Java, and the neighbouring islands.* London. B164 C17,268
1826 Description of the *Helarctos euryspilus*; exhibiting in the bear from the Island of Borneo, the type of a subgenus of *Ursus. Zool. J. Lond.* 2:221–34. C41
1827 See also Vigors, N.A. and T. Horsfield
1828 Notice of a new genus of Mammalia, found in Sumatra by Sir T. Stamford Raffles. *Zool. J. Lond.* 3:246–49. C41

Houlton, Joseph
1831 Protraction of vegetable life in a dry state. *Arc. Sci. Art* 4:160. C92

Humboldt, Friedrich Wilhelm Heinrich Alexander von
*1811 *Political essay on the Kingdom of New Spain.* Transl. by John Black. 4 vols. London. [* vols 1 & 2 New

York] RN160,161,163–171,175–77 B32 C268 M38
1814–34 *Atlas géographique et physique des régions équinoxiales du nouveau continent.* Paris. RN157
1817 *De distributione geographica plantarum secundum coeli temperiem et altitudinem montium, prolegomena.* Lutetiae Parisiorum. B157,159 D69
*1831 *Fragmens de géologie et de climatologie asiatiques.* 2 vols. Paris. RN72–73 A15
*1837 Volcans des montagnes de Quito. *Institut* 5 (209) 10 May:155–56. A26
1838a Description géognostique et physique des volcans du plateau de Quito. *Bull. Soc. Géol. Fr.* 9:24–25. A81
1838b Humboldt on El Dorado. *Athenaeum* (539) 24 February:137–38. A76
1838c Mémoire sur le plateau de Bogota. *Institut* 6 (242) 16 August:268–70. A115

Humboldt, Friedrich Wilhelm Heinrich Alexander von and Aimé Bonpland
1805 *Essai sur la géographie des plantes; accompagné d'un tableau physique des régions équinoxiales. In Voyage aux régions équinoxiales du nouveau continent.* Part V. Paris. B92
1811–33 *Recueil d'observations de zoologie et d'anatomie comparée.* 2 vols. Paris. Vol. 1, 1811; vol. 2, 1833. Deuxième partie in *Voyage de Humboldt et Bonpland,* 1805–1834, 23 vols. Paris. ZEd6
*1819–29 *Personal narrative of travels to the equinoctial regions of the New Continent, during the years 1799–1804.* Transl. by Helen Maria Williams. Vols. 1–2 1822, 3d ed. London. vol. 3 1822, 2d ed.; vol. 4, 1819; vol. 5, 1821; vol. 7, 1829, 1st ed. 7 vols. RN24,43,70,91,100 B142 D69 E182 QE[15^{v}] M38

Hume, David
1750 *Philosophical essays concerning human understanding.* 2d ed. London. C267 M155
1800 *Essays and treatises on several subjects.* 2 vols. Edinburgh. N184
1807 See Ritchie, T.E.
1817 *Essays and treatises on several subjects.* 2 vols. Edinburgh. M104
1826 *The philosophical works of David Hume.* 4 vols. Edinburgh. M104 N101

Hunt, Thomas Carew
*1845 A description of the island of St. Michael (Azores). *J. R. Geogr. Soc.* 15:268–96. QE18

Hunter, John
1793 *An historical journal of the transactions at Port Jackson and Norfolk Island . . . since the publication of Phillip's voyage.* London. RN126

Hunter, John
1780 Account of an extraordinary pheasant. *Phil. Trans. R. Soc.* 70 (2):527–35. C215
1786 *Observations on certain parts of the animal oeconomy.* London. B51 C215,267
1792 *Observations on certain parts of the animal oeconomy.* 2d ed. London. B51 C267
1835–37 *The works of John Hunter, F.R.S. with notes.* Ed. James F. Palmer. 4 vols & atlas. London. C270 QE21..
1837 *Observations on certain parts of the animal oeconomy . . . with notes by Richard Owen.* [Vol. 4 of 1835–37 but published as a separate work.] London. B51, 161 C215,267,270 D57,67,112–16,127,152, 154,156–59,161,163,173–75 E26,60 M44, 147 N43

Hutton, James
1795 *Theory of the earth, with proofs and illustrations.* In Four Pts. 2 vols. Edinburgh. RN112 D21

Hutton, T.
1832 Notes in natural history. *J. Asiat. Soc. Beng.* 1:554–58. D21 E177
1833 On the nest of the tailor bird. *J. Asiat. Soc. Beng.* 2:502–05. E181 T2

Imrie, Lieutenant-Colonel
1818 A geological account of the southern district of Stirlingshire, commonly called the Campsie Hills, with a few remarks relative to the two prevailing theories as to geology, and some examples given illustrative of these remarks. *Mem. Wernerian Nat. Hist. Soc.* 2:24–50. A36

Isabelle, Arsene
1835 *Voyage à Buenos-Ayres et à Porto-Alègre, par la Banda-Oriental, les missions d'Uruguay et la province de Rio-Grande-do-Sul (de 1830 à 1834).* Havre. RN158

Jamieson, Thomas F.
1863 On the parallel roads of Glen Roy, and their place in the history of the glacial period. *Q. J. Geol. Soc. Lond.* 19:235–59.

Jeffrey, Francis
1811a *Essays on the nature and principles of taste.* By Archibald Alison. 2 vols. Edinburgh. OUN8
1811b Review of 1811a. *Edinb. Rev.* 18:1–46. OUN8
1835 Letters to the editor. In Mackintosh vol. 2, 1835 chap. 8:490–507. C218

Jenner, Edward
*1824 Some observations on the migration of birds; with an introductory letter to Sir Humphry Davy. *Phil. Trans. R. Soc.* 114:11–44. C40 ZEd12
1825 Some observations on the migration of birds. *Zool. J. Lond.* 1:125–29. ZEd12

Jenyns [Blomefield], Leonard
1835 *A manual of British vertebrate animals*. Cambridge. B7,31 E102
1837 Some remarks on the study of zoology, and on the present state of the science. *Mag. Zool. Bot.* 1:1–31. ZEd1
*1838a On the dentition and other characters of the British shrews, with reference to M. Duvernoy's recent researches into the structure of this genus of animals. *Mag. Zool. Bot.* 2:24–42. B31
*1838b Further remarks on the British shrews including the distinguishing characters of two species previously confounded. *Ann. Nat. Hist.* 1:417–27. B31 C268
1839a On certain species of Sorex. *Rep. Br. Ass. Advmt Sci.* (Meeting at Newcastle, 1838) 7:104. B31
1839b Additional note on the British shrews. *Ann. Mag. Nat. Hist.* 2:43. B31
1842 See *Zoology: Fish*

Jenyns, Soame
1757a *A free inquiry into the nature and origin of evil. In six letters to—.* OUN15
1757b Review of 1757a. *Lit. Mag., Univl Rev.* 2,171–75,251–53,301–6. OUN15

Johnson, James Rawlins
1822 Observations on the genus Planaria. *Phil. Trans. R. Soc.* 111:437–47. ZEd1
1825 Further observations on Planariae. *Phil. Trans. R. Soc.* 115:247–56. ZEd1

Johnson, Samuel
1799 See Boswell, J.

Johnston, George
*1838a The British Ariciadae. *Mag. Zool. Bot.* 2:63–73. B126 C185
*1838b The natural history of British zoophytes. *Mag. Zool. Bot.* 2:319–40. C162 ZEd17,18

Johnston, James Finlay Weir
1840 [On some varieties of peat.] *Athenaeum* (677) 17 October:823. QE[23v]

Jones, Thomas Rymer
1836–39 Gastropoda. In Todd vol. 2 1836–59:377–404. D166

Jonnès, A.M. de
1810 See Cortès, ——— and A.M. de Jonnès

Jourdan, Antoine Jacques Louis
*1837 Mammifères nouveaux. *Institut* 5 (221) November: 351. D64

Juan, George and Antonio de Ulloa
*1806 *A voyage to South America*. 4th ed. 2 vols. London. [* Vol. 1 only] RN75,105,106,125,177 ZEd5

Kames, Lord. See Home, Henry, Lord Kames.

Keilhau, Baltazar Mathias
*1838 Theory of granite, and the other massive rocks; together with that of crystalline slate. *Edinb. New Phil. J.* 24:387–403; 25:80–101,263–272 A99

Keir, Thomas
*1837 Remarks upon the Apteryx. *Proc. Zool. Soc.* 5:24. B162

Kendal, Edward
*1832 Account of the Island of Deception, one of the New Shetland Isles. Extracted from the private Journal of Lieutenant Kendal, R.N., embarked on board his Majesty's sloop 'Chanticleer', Captain Forster, on a scientific voyage. *J. R. Geogr. Soc.* 1:62–66. RN138,139

Kennedy, Alexander
1823 Account of a non-descript worm (the *Ascaris pellucidus*) found in the eyes of horses in India. *Trans. R. Soc. Edinb.* 9:107–11. C227

Kenyon, John
1838 *Poems: for the most part occasional*. London. M141

Keratry, Auguste Hilarion de
1841 *Inductions morales et physiologiques*. 3d ed. Paris. OUN8b

Kerr, Robert
1811–24 *A general history and collection of voyages and travels, arranged in systematic order*. 18 vols. Edinburgh, London. Vols 14–15, 1815: *An account of a voyage towards the South pole and round the world, performed in His Majesty's ships the 'Resolution' and 'Adventure', in the years 1772, 3, 4, and 5*, by James Cook. A81 E12,13

King, James
1784 See Cook, J.

King, Philip Parker
1827 *Narrative of a survey of the intertropical and western coasts of Australia performed between the years 1818 and 1822*. 2 vols. London. RN6,38,101 C213
1828–29 Extracts from a letter addressed by Capt. Philip Parker King, R.N., F.R.S. and L.S., to N.A.Vigors, Esq., 'On the animals of the Straits of Magellan'. *Zool. J. Lond.* 3:422–32; 4:91–105. C61,99 ZEd12
1831 Characters of new species of birds from the Straits of Magellan. *Proc. Zool. Soc. Lond.* 1:14–16,29–30. ZEd12
1839 *Proceedings of the first expedition, 1826–30*. See FitzRoy 1839, vol. 1. A141 C269

Kirby, William
*1802 *Monographia apum Angliae; or, an attempt to divide into their natural genera and families, such species of the*

Linnean genus Apis as have been discovered in England, with descriptions and observations. 2 vols. Ipswich. B2

1835 *On the history habits and instincts of animals. On the power wisdom and goodness of god as manifested in the creation. The Bridgewater treatises*. Treatise VII. 2 vols. London. B141–43 M98 OUN34–37 QE[5a]

Kirby, William and William Spence

*1818–26 *An introduction to entomology, or, elements of the natural history of insects*. 3d ed. London. D40,41

Kneeland, S.S.

1860 [Barking of the wild dogs in Southern Africa.] *Proc. Amer. Acad. Sci.* 4:426–28. N44

Knight, Thomas Andrew

*1799 An account of some experiments on the fecundation of vegetables. *Phil. Trans. R. Soc.* 89:195–204. E152,162

*1809 On the comparative influence of male and female parents on their offspring. *Phil. Trans. R. Soc.* 99:392–99. QE11

1815 Introductory remarks relative to the objects which the Horticulture Society have in view. *Trans. R. Hort. Soc.* 1 (2d ed.) 1:1–7. E77

1818a On the want of permanence of character in varieties of fruit, when propagated by grafts and buds. *Trans. R. Hort. Soc.* 2 (2d ed.):160–161. E77,148,153,162

*1818b Upon the advantages of propagating from the roots of old ungrafted fruit trees. *Trans. R. Hort. Soc.* 2 (2d ed.):252–54. [*abst DAR 74:60.] E153,154,184

1822 Observations on hybrids. *Trans. R. Hort. Soc.* 4:367–73. QE1v,7,11

1824 An account of some mule plants. *Trans. R. Hort. Soc.* 5:292–96. QE1v

*1837 On the hereditary instinctive propensities of animals. *Phil. Trans. R. Soc.* 127:365–69. QE11

Kohn, David

1980 Theories to work by: rejected theories, reproduction, and Darwin's path to natural selection. *Stud. Hist. Biol.* 4:67–170.

1985 *The Darwinian heritage*. Princeton.

Kohn, David, Sydney Smith and Robert C. Stauffer

1982 New light on *The Foundation of the Origin of Species*: A reconstruction of the archival record. *J. Hist. Biol.* 15:419–42.

Kolff, Dirk Hendrick

1840 *Voyages of the Dutch brig of war 'Dourga', through . . . the Moluccan archipelago . . . during 1825 & 1826*. London. E180,181

Kölreuter, Joseph Gottlieb

*1761–66 *Vorlaufige Nachricht von einigen das Geschlecht der Pflanzen betreffenden Versuchen und Beobachtungen*. 4 pts in 1 vol. Leipzig. QE[5]a

Kotzebue, Otto von

1821 *A voyage of discovery, into the South Sea and Beering's Straits . . . 1815–1818 . . . in the Ship 'Rurick'. Remarks. . . .* [See Chamisso 1821.] 3 vols. London. RN181 A15 B91,124,234 C28,269 E20–22 ZEd6

1830 *A new voyage round the world, in the years 1823, 24, 25, and 26*. 2 vols. London. C269 E19,21

Krause, Ernst Ludwig

*1879 *Erasmus Darwin. With a preliminary notice by Charles Darwin*. Leipzig. (* 1880 edition) M25

Krauss, Christian Ferdinand Friederich

1837 *Beitrag zur Kenntniss der Corallinen und Zoophyten der Südsee*. Stuttgart. ZEd8

1843 *Die süd-afrikanischen Crustaceen. Eine Zusammenstellung aller bekannten Malacostraca, Bemerkungen über deren Lebensweise und geographische Verbreitung*. Stuttgart. ZEd8

Kuhl, Heinrich

1827–28 See Blume, C.L.

Labillardière, Jacques Julien Houton de

*[1800] *Relation du voyage à la recherche de 'La Pérouse' . . . pendant les années 1791, 1792, et pendant la 1ère et la 2ème année de la République française*. 2 vols & atlas. Paris. RN5,12

La Condamine, Charles Marie de

1747 *A succinct abridgment of a voyage made within the inland parts of South America*. London. RN56

Laizer, Louis, Comte de and Parieu, ——— de

1838a Description et determination d'une mâchoire appartenant a un Mammifère jusqu'a present inconnu. *Institut* 6 (260) 20 December:419–20. E88

1838b Notice of 1838a. *C. R. Hebd. Séanc. Acad. Sci. Paris*. 7:442. E88

Lamarck, Jean Baptiste Pierre Antoine de Monet de

[1802] *Hydrogéologie*. Paris. B178 C119

*1815–22 *Histoire naturelle des animaux sans vertèbres*. 7 vols. Paris. B178

[1822–24] *Genera of shells*. Transl. by J.G. Children. London. E46

*1830 *Philosophie zoologique*. New ed. 2 vols. Paris. [* Vol. 1 only, text identical to 1809 ed.] B6,9,16,21, 129,178,214,216 C63,168,269 D69 E145, 148,150,159,160 N19,51,90,91 OUN34,37

Landor, Walter Savage

1824–29 *Imaginary conversations of literary men and statesmen*. 5 vols. London. C270

Lang, John Dunmore
1834a *View of the origin and migrations of the Polynesian nation*. London. C221
1834b *An historical and statistical account of New South Wales*. London. C275

La Pérouse, Jean François de Galaup
1798–99 *A voyage round the world performed in the years 1785, 1786, 1787, and 1788, by the 'Boussole' and 'Astrolabe'*. 2 vols & atlas. London. Vol. 2 (1799). RN15

Lardner, Dionysius (see also Phillips, John)
1837–39 *Cabinet of Cyclopaedia. Natural History*. 2 vols. London. A54–57,147 B167–168,170,173 C269 TFRAG9

Lartet, Edward
1839 Fossils. *Athenaeum* (620) 14 Sept.:708. B133 TFRAG4

La Salle, Jean-Anne-Amédée de
1851 *Voyage autour du monde exécuté pendant les années 1836 et 1837 sur la corvette 'la Bonite'. Relation du voyage*. Vol. 2. Paris A39

Lasch, W.
1829 Beitrag zur Kenntniss der Varietäten und Bastardformen einheimischer Gewächse. *Linnaea* 4:405–34. C265

Latham, John
1821–28 *A general history of birds*. 10 vols. Winchester. C68

Latreille, Pierre André
1819 *Mémoires sur divers sujets de l'histoire naturelle, des insectes, de géographie ancienne et de chronologie*. Paris. C21,268

Lauder, Thomas Dick. See Dick, T.L.

Lavater, Johann Caspar
1804 *Essays on physiognomy*. 2d ed. 3 vols (Vol. 3 in 2 pts). London. C270 D164,165 M145 N6,9,10
1820 *L'art de connaître les hommes par la physionomie*. Nouvelle édition. 10 vols. Paris. C270 N10,103

[Lawrence, John]
1820 *The sportsman's repository; comprising a series of highly-finished engravings representing the horse and the dog*. By John Scott [pseudonym]. London. C266

Lawrence, William
*1822 *Lectures on physiology, zoology, and the natural history of man*. London C204,268

Le Brun, Charles
*1701 *The conference of Monsieur Le Brun . . . upon expression*. London. N103

Lecoq, Henri
1833–34 Explications qu'il donne sur l'action érosive des eaux. Réponse à la question qui lui est adressée sur les abaissemens et les élévations successives des eaux du lac Leman d'Auvergne. *Bull. Soc. Géol. Fr.* 4:33–37 A44

Le Couteur, John
*1836 *On the varieties, properties and classification of wheat*. Jersey and London. [*n.d. Jersey] QE11,15 M155

Le Puillon de Boblaye, Emile and Pierre-Théodore Virlet d'Aoust
1833 Géologie et minéralogie. In *Expédition scientifique de Morée*. Vol. 2 (2). 1831–38. Paris. C55

LeRoy, Charles Georges
1802 *Lettres philosophiques sur l'intelligence et la perfectibilité des animaux, avec quelques lettres sur l'homme*. Nouvelle édition. Paris. OUN8b

Lessing, Gotthold Ephraim
1836 *Laocoon; or the limits of poetry and painting*. London. C266,269 OUN22–24

Lesson, René Primevère
1827 *Manuel de mammalogie, ou histoire naturelle des mammifères*. Paris. B31 C23
1830 In Lesson, R.P. and P. Garnot 1826–30.

Lesson, René Primevère and Prosper Garnot
1826–30 *Zoologie*. In Duperrey, 1825–1830 [Coquille], vol. 5. 2 vols. in 4 pts: 1826, Lesson and Garnot, vol. 1 (pt 1); 1828, Lesson and Garnot, vol. 1 (pt 2); 1830, Lesson, vol. 2 (pt 1); 1830, Lesson, vol. 2 (pt 2). RN62,101,102 B31,54,220,234,249 C16–29,276 E42 ZEd5,7,11,13

Levaillant. See Vaillant, François Le.

Lewis, R.H.
1836 Case of maternal attendance on the larva by an insect of the tribe of Terebrantia, belonging to the genus *Perga*, observed at Hobarton, Tasmania. *Trans. Ent. Soc. Lond.* 1:232–34. M64

Liebig, Justus von
1840 Abstract. New chemical views relative to agriculture and physiology. In Report on the applications of organic chemistry in agriculture and physiology. *Athenaeum* (675) 3 October:773–74. QE[23v]

Limoges, Camille
1970 *La sélection naturelle*. Paris.

Lindley, John
1826 Some account of the spherical and numerical system of nature of M. Elias Fries. *Phil. Mag.* 68:81–91. B46

*1830 *An introduction to the natural system of botany.* London. [*1836 *A natural system of botany*, 2d ed. London.] BIBC C268

*1839 *School botany.* London. [*1856 *School botany and vegetable physiology.* London.] QE10

Lindsay, H.H.

1832 See Franklin, J., J. Gould and H.H. Lindsay.

Linnaeus, Carl

1749–90 *Amoenitates academicae.* 10 vols. Holmiae, Lipsiae B155

1775 *The elements of botany.* London. QE8

1781 *Select dissertations from the amoenitates academicae.* London. B155

1786 *A dissertation on the sexes of plants.* London. QE12

*1789–96 *Systema naturae per regna tria naturae.* Ed. 13a (bound in 10 vols). Lugduni. C157

Lisiansky, Urey

1814 *A voyage round the world, in the years 1803, 4, 5, & 6 . . . in the ship, 'Neva'. London.* C270

Lisle, Edward

1757 *Observations in husbandry.* London. C275

Lister, Joseph Jackson

*1834 Some observations on the structure and functions of tubular and cellular Polypi, and of Ascidiae. *Phil. Trans. R. Soc.* 124:365–88. RN24

Litchfield, Henrietta Emma

1915 *Emma Darwin. A century of family letters. 1792–1896.* 2 vols. London, New York. M41 N61

Litke, Fedor Petrovich

1835–36 *Voyage autour du monde . . . sur la corvette 'le Séniavine', dans les années 1826, 1827, 1828, et 1829.* 3 vols & atlas. Paris. C241,269 ZEd19 N22

Lockhart, John Gibson

1833 *The history of Napoleon Buonaparte.* 2 vols. New York. C269

1837–38 *Memoirs of the life of Sir Walter Scott.* 7 vols. Edinburgh. C269,270 M126,129 N19

Loddiges, Conrad

*1823 *Catalogue of plants, in the collection of Conrad Loddiges & sons.* 13th ed. London. [*abst DAR 71:89–94.] D118,128

Lord, P.B.

1838 Some account of a visit to the plain of Koh-i-Daman, the mining district of Ghorband, and the pass of Hindu Kush, with a few general observations respecting the structure and conformation of the country from the Indus to Kabul. *J. Asiat. Soc. Beng.* 7:521–37. A145

Lowe, Richard Thomas

*1833a Letter from Rev. R. T. Lowe read at meeting of Aug. 27, 1833. *Proc. Zool. Soc. Lond.* 1:102. QE18

*1833b Primitiae faunae et florae Maderae et Portus-Sancti; sive species quaedam novae vel hactenus minus rite cognitae animalium et plantarum in his insulis degentium breviter descriptae. *Trans. Camb. Phil. Soc.* 4:1–70. [*1831 article date] A72 B204

Lund, Peter Wilhelm

1838 Blik på Brasiliens dyreverden för sidste jordomvaeltning. *Overs. K. danske Vidensk. Selsk. Forh.* pp. 7–8. T63,79

*1840 *View of the fauna of Brazil, previous to the last geological revolution.* [transl.] *Mag. Nat. Hist.* (n.s.) 4:1–8, 49–57, 105–12, 153–61, 207–13, 251–59, 307–17,373–89. T63,79

Lyell Charles

1827 [Review of] *Memoir on the geology of central France; including the volcanic formations of Auvergne, the Velay, and the Vivarais, with a volume of maps and plates,* by G.P. Scrope, F.R.S., F.G.S. Murray, London, 1827. *Q. Rev.* 36:437–83. N19

*1830–33 *Principles of geology, being an attempt to explain the former changes of the earth's surface, by reference to causes now in operation.* 3 vols. London. Vol. 1, 1830; vol. 2, 1832; vol. 3, 1833. RNIFC,44,52, 57,60,61,63,65,67,68,70,79,82–84,88,100 A11, 95 B10,115,116,153,155–157,172 C53,106,168 D60 E4,167

1835a On the proofs of a gradual rising of the land in certain parts of Sweden. *Phil. Trans R. Soc.* 125:1–38. A111,113

1835b *Principles of geology: being an inquiry how far the former changes of the earth's surface are referable to causes now in operation.* 4 vols. London. RN158

*1837 *Principles of geology: being an inquiry how far the former changes of the earth's surface are referable to causes now in operation.* 5th ed. 4 vols. London. RN115 A7,11,79,85,116,121 GR109 B6,10–13,23, 59, 63, 69, 81–82, 87, 91, 96, 115, 153, 155, 157, 170,172,200,201,202,249 C39,53,137,168,270 D21,39,60,104,134 E4,26,35,38,65,105,109 M128

*1838a *Elements of geology.* London. RNIFC A86,87,89, 94,115,120,121 C270 D60,133,134 E66,105

1838b On certain phaenomena connected with the junction of granitic and transition rocks, near Christiania in Norway. *Rep. Br. Ass. Advmt Sci.* 7 (2):67–69. A68

1840a On the cretaceous and tertiary strata of the Danish islands of Seeland and Moen. *Trans. Geol. Soc. Lond.* 2d ser. 5:243–58. GR109 T89

1840b On the boulder formation, or drift and associated freshwater deposits composing the mud-cliffs of eastern Norfolk. *Phil. Mag.* 16 (104) May:345–80. T89

McClelland, John
1833–38 On the geology of Upper Assam. *Proc. Geol. Soc. Lond.* 2 (14 June 1837):566–68. A28 B153

Macculloch, John
1817 On the parallel roads of Glen Roy. *Trans. Geol. Soc. Lond.* 2d ser. 4:314–92. GR35,63,65,69,70,78.
1824 Hints on the possibility of changing the residence of certain fishes from salt water to fresh. *Q. J. Lit. Sci. Arts Lond.* 17:209–31. B54
1837 *Proofs and illustrations of the attributes of god, from the facts and laws of the physical universe; being the foundation of natural and revealed religion.* 3 vols. London. [*abst DAR 71:53–59 205.5:28–29,167.] N35 OUN11b

Macdonald, William
*1839 [Verbal communication on the osseous structure of fishes.] *Ann. Nat. Hist.* 2:69–70. D62

Macgillivray, William
*1839a *A history of the British birds.* Vol. 2. *Cantatores, songsters.* London. [1837–52.5 vols.] TFRAG3
1839b Review of 1839a. *Athenaeum.* (616) 10 Aug.:604–6. TFRAG3

Mackenzie, George Steuart
1811 *Travels in the Island of Iceland, during the summer of the year MDCCCX.* Edinburgh. RN90,159 A39 B159,255,256 C276 ZEd9

Mackintosh, James
1835 *Memoirs of the life of the Right Honourable Sir James Mackintosh.* 2 vols. London. C218
*1837 *Dissertation on the progress of ethical philosophy.* Preface, pp. 1–46, by William Whewell. 2d ed. Edinburgh. C267,269 M75,151 N89,92,184 OUN8v,42–55

MacLeay, William Sharp
*1819–21 *Horae Entomologicae.* 1 vol. in 2pts. London. [*book: CUL Rare Books, abst DAR 71:128–38.] B2,8,46,129,IBC C49,103,108,119,139,157,202,211,213,216,218,275 D59 ZEd18
1825 Remarks on the identity of certain general laws which have been lately observed to regulate the natural distribution of insects and fungi. *Trans. Linn. Soc. Lond.* 14:46–68. C157,158
1829a Notes on the genus *Capromys* of Desmarest. *Zool. J. Lond.* 4:269–78. C42 E42
1829b A letter to J.E. Bicheno, Esq., F.R.S.; in examination of his 'Paper on Systems and Methods', in the Linnean Transactions. *Zool. J. Lond.* 4:401–15. C42
1830a *On the dying struggle of the dichotomous system.* Proof copy, 35 pages. C149,157,158,276 D59
1830b On the dying struggle of the dichotomous system. In a letter to N.A. Vigors. *Phil. Mag.* 7:431–45; 8:53–57,134–40,200–7. C149,155,157,158,276
1838 Invertebratae. In Smith, Andrew 1838–49, vol. [5]. Also *Illustrations of the Annulosa of South Africa.* London. D50–52,107

Madler, Johann Heinrich
1838 See Beer, W. and J.H. Madler

Malaspina, Alessandro
1885 *Viaje politico-cientifico alrededor del mundo por las corbetas Descubierta y Atrevida al mando de los capitanes de navio D. Alejandro Malaspina y Don José de Bustamante y Guerra desde 1789 a 1794.* Madrid. RN181

Malcolmson, John Grant
1833 Note on saline deposits in Hyderabad. *J. Asiat. Soc. Beng.* 2:77–79. A143
1833–38 On the fossils of the eastern portion of the great basaltic district of India. *Proc. Geol. Soc. Lond.* 2 (16 December 1837):579–84. A41
1836 Notes explanatory of a collection of geological specimens from the country between Hyderabad and Nagpur. *J. Asiat. Soc. Beng.* 5:96–122. A144

Mallet, Robert
1838 Roches de trapp. *Institut* 6 (253) 1 November: 359–60. A135

Malte-Brun, Conrad
1822–33 *Universal geography.* 10 vols [including Index, vol. 10]. Edinburgh. Vol. 2 1822; vol. 5, 1825. A6,16 E19

Malthus, Thomas Robert
*1826 *An essay on the principle of population.* 6th ed. 2 vols. London. C266,270 D134,135 E3 N10 OUN29v Mac57v

Mancuniensis
1832 Cats without tails in the Isle of Man. *Mag. Nat. Hist.* 5:717. B178

Manier, Edward
1978 *The young Darwin and his cultural circle.* Dordrecht, Boston. OUN48

Mantell, Gideon Algernon
*1837 On the bones of birds discovered in the strata of Tilgate Forest, in Sussex. *Trans. Geol. Soc. Lond.* 2d ser. 5:175–77. C163

Marsden, William
1783 *The history of Sumatra.* London. E21
1811 *The history of Sumatra.* 3d ed. London. RN90 ZEd6

Marsilli, Louis Ferdinand
1725 *Histoire physique de la mer.* Preface by H. Boerhaave. Amsterdam. QE21..

Martens, Martin
1838 On hybridity in ferns. *Athenaeum* (539) 24 February: 154. B235 D162
1839a On hybridity occurring in ferns. [Transl. by W.H. White.] *Proc. R. Bot. Soc. Lond.* 1:55–57. QE12
*1839b Hybridity in ferns. *Ann. Nat. Hist.* 2:236. E90

Martin, Louis, Aimé-
1837 See Aimé-Martin, Louis

Martin, William Charles Linnaeus
*1837 Description of a new bat (*Rhinolophus landeri*) from Fernando Po, and a new hedgehog (*Erinaceus concolor*) from Trebizond. *Proc. Zool. Soc. Lond.* 5:101–03. B209

Martineau, Harriet
1838 *How to observe. Morals and manners.* London. C270 M75,76,142,151

Martius, Carl Friedrich Philipp von
1824 See Spix, J.B. von and C.F.P. von Martius

Mathison, Gilbert Farquhar
1825 *Narrative of a visit to Brazil, Chile, Peru, and the Sandwich Islands, during the years 1821 and 1822.* London. RN181

Mawe, John
*1825 *Travels in the gold and diamond districts of Brazil.* A new ed. London. RN181

Mayo, Herbert
1837 *The philosophy of living.* London. C270 M126
1838 *The philosophy of living.* 2d ed. London. D49 M102,103,110,111,114,115,116 N31 OUN6

Mayo, Thomas
1839 *Elements of the pathology of the human mind.* Philadelphia. C266,269 OUN17

Meckel, Johann Friedrich
1808–12 *Beyträge zur vergleichenden Anatomie.* 2 vols in 3. Leipzig. E89
1831 On varieties in the animal kingdom, depending on procreation between individuals of different species. *Gleanings in Science* 3:81–84. TFRAG2

Meyen, Franz Julius Ferdinand
1833 Beiträge zur Zoologie, gesammelt auf einer Reise um die Erde. Zweite Abhandlung. Säugethiere. *Nova Acta Physico Med.* 16:549–610. ZEd6
*1834 *Beiträge zur Zoologie, gesammelt auf einer Reise um die* Erde, und W. Erichson's und H. Burmeister's Beschreibungen und Abbildungen . . . Insekten. Breslau, Bonn. ZEd9,10
1834–35 *Reise um die Erde . . . in den Jahren 1830, 1831 und 1832.* 2 vols. Breslau, Bonn. A87
*1836 Analysis of 1834–35. *J. R. Geogr. Soc.* 6:365–77. A87
1838a Rapport sur les progrès de la physiologie végétale pendant l'année 1836. *Institut* 6 (223) January:23–24. B200
1838b Animaux spermatiques des végétaux. *Institut* 6 (245) 6 September:291 E100,123

Michell, John
*1760 Conjectures concerning the cause, and observations on the phaenomena of earthquakes; particularly of that great earthquake of the first of November 1755, which proved so fatal to the city of Lisbon, and whose effects were felt as far as Africa, and more or less throughout almost all Europe. *Phil. Trans. R. Soc. Lond.* 51:566–634. RN80

Miers, John
*1826 *Travels in Chile and La Plata.* 2 vols. London. RN56,100
1841 On some new Brazilian plants allied to the natural order Burmanniaceae, as well as some Indian species *Gonyanthes candida, G. Nepalensis, G. Wallichii, G. pusilla*, etc.). *Trans. Linn. Soc. Lond.* 18:535–56. S6

[Mill, John Stuart]
1840 Review of the works of Samuel Taylor Coleridge. *Lond. Westm. Rev.* 33:257–302. OUN33

Milne-Home, David
1849 On the parallel roads of Lochaber, with remarks on the change of relative levels of sea and land in Scotland, and on the detrital deposits in that country. *Trans. R. Soc. Edinb.* 16:395–418.

Milne-Edwards, Henri
1833 Mémoire sur l'organisation de la bouche chez les crustacés suceurs. *Annls Sci. Nat.* 28:78–86. B112
*1834–40 *Histoire naturelle des crustacés.* 3 vols & atlas. (*Suites à Buffon.*) Paris. CIBC D52
*1837 Classification naturelle des polypes. *Institut* 5 (212) 31 May:178–79. ZEd7
1838a Sur les Polipes du genre des *Tubulipores. Institut* 6 (225) March:75. ZEd14
1838b Sur les Crisies, Hornères et plusieurs autres polypes vivants ou fossiles dont l'organisation est analogue à celle des *Tubulipores. Institut* 6 (227) 3 May:138–39. ZEd18
1838c Géographie zoologique: Crustacés. *Institut* 6 (245) 6 September:290–91. E100
1838d [Un mémoire sur le mécanisme de la respiration chez les Crustacés.] *Institut.* 6 (250) 11 October: 329–30. E25

Mitchell, James
1827 *Dendrologia; or, a treatise of forest trees with Evelyn's Sylva revised.* London. C269

Mitchell, Thomas Livingstone
1838 *Three expeditions into the interior of eastern Australia.* 2 vols. London. A92 C130,131,159,270 D71, 75,99,112 M132

Mitchill, J.A., ——— Smith and ——— Cooper
1828 Discovery of a fossil walrus or sea-horse in Virginia. *Edinb. New Phil. J.* 5:325–26. B125

Mitchill, Samuel Latham
1826 Facts and considerations showing that the two-headed snakes of North America and other parts of the world, are not individuals of a distinct race, but universally monsters. *Am. J. Sci.* 10:48–53. C27

Mitscherlich, Eilhert
1830 On artificial crystals of oxide of iron. *Edinb. J. Nat. Geogr. Sci.* 2:302. RN165

Mittre, H.
1839 New crustacea. *Athenaeum* (585) 12 January:36. E90

Mohl, Hugo von
1836 *Erläuterung und Vertheidigung meiner Ansicht von der Structur der Pflanzen-Substanz.* Tübingen. B200

Molina, Juan Ignacio
*1788–95 *Compendio de la historia geografica, natural y civil del reyno de Chile.* 2 vols. Madrid. RNIFC,63 C42,276 ZEd9

Monboddo, James Burnet
1773–92 See Burnet, J. Lord Monboddo

Mons, Jean Baptiste van
*1820 Substance of a memoir on the cultivation and variation of Brussels sprouts. *Trans. R. Hort. Soc.* 3:197–200. [*abst DAR 74:61–62.] E141
1835–36 *Arbres fruitiers.* 2 vols. Louvain. QE13

Montagu, George
1804 Observations on some species of British quadrupeds, birds, and fishes. *Trans. Linn. Soc. Lond.* 7:274–290. C46,267
1831 *Ornithological dictionary of British birds.* 2d ed. London. C46,267

Montaigne, Michel Eyquem de
1588 [1873–75] *Les essais de Montaigne réimprimes sur l'édition originale de 1588.* 4 vols. Paris. M89,104

Montbrun, ——— de and Alexandre Courcier
1838 Présence de l'ichthyosaure dans la craie. *Institut* 6 (226) April:136. C205

Monteith, William
*1833 [1834] Journal of a tour through Azerdbijan and the shores of the Caspian. *J. R. Geogr. Soc.* 3:1–58. B131

Montgomery, William
1985 Charles Darwin's thought on expressive mechanisms in evolution. In Zivins 1985.

Moor, J.H.
1837 *Notices of the Indian Archipelago, and adjacent countries.* Singapore, London. E174–176

Moore, ———
1833–38 Extracts from two letters 'On the earthquake in Syria in January last;' addressed by Mr. Moore, his Majesty's Consul-General at Beyrout, to Viscount Palmerston. *Proc. Geol. Soc. Lond.* 2 (5 April 1837):540–41. A28

[Moore, John]
1765 *A treatise on domestic pigeons.* London. C269 D100

Moreton, Lord
1821 See Morton, G.

[Morgan]
1839 Review of Courtet de l'Isle, W. 1838. *Athenaeum* 624, 12 October:771–73. TFRAG5

Morris, John and Daniel Sharpe
*1846 Description of eight species of brachiopodous shells from the Palaeozoic rocks of the Falkland Islands. *Q. J. Geol. Soc. Lond.* 2:274–78. RN144

Morton, George
*1821 A communication of a singular fact in natural history. *Phil. Trans. R. Soc.* 110:20–22. B181, 197,219 D9,41,113,152,165,168,171–73,176 E79

Morton, Samuel George
1835 Notice of the fossil teeth of fishes of the United States, the discovery of the Gault in Alabama, and a proposed division of the American Cretaceous group. *Am. J. Sci.* 28:276–78. RN142

Mozart, Wolfgang Amadeus
1817 See Haydn, J.

Mulhall, Michael G.
1878 *The English in South America.* London. RN155

Müller, Johannes
*1838–42 *Elements of physiology.* 2 vols. 2d ed. London. Vol. 1, 1838. E57,83,85,127

Murchison, Roderick Impey
1829 Supplementary remarks on the strata of the Oolitic series, and the rocks associated with them, in the counties of Sutherland and Ross, and in the Hebrides. *Trans. Geol. Soc. Lond.* 2d ser. 2:353–68. A36
1833–38 The gravel and alluvia of S. Wales and Siluria as distinguished from a northern drift covering Lancashire, Cheshire, N. Salop, and parts of Worcester and Gloucester. *Proc. Geol. Soc. Lond.* 2 (3 February 1836):230–336. A131

1835 On a fossil fox found at Oeningen near Constance; with an account of the deposit in which it was imbedded. *Trans. Geol. Soc. Lond.* 2d ser. 3:277–90. B279

*1839a *The Silurian System.* London. [* 2 parts in 3 vols] RN142 A36,131

*1839b See Sedgwick, A. and R.I. Murchison

Murray, John

1828 Uber das Licht und die leuchtende Materie der *Lampyris noctiluca. Z. Organ. Physik.* 2:94–103. ZEd4

1845 *Strictures on morphology: its unwarrantable assumptions, and atheistical tendency.* London. QE17

Necker, Albertine Adrienne

1839–43 See Saussure, A.-A. de

Nees von Esenbeck, C.G.D.

1824 See Reinwardt, C.G.C., C.L. Blume and C.G.D. Nees von Esenbeck

Neill, Patrick

1830 Horticulture. In Brewster, D. [ed.] *The Edinburgh Encyclopaedia*, 4th ed. 11:177–315. C265

Newbold, Thomas John

1836 Sketch of the State of Muar, Malay Peninsula. *J. Asiat. Soc. Beng.* 5:561–67. [Reprinted in Moor, J.H. 1837:73–76.] E176

Newman, Edward

1832 *Sphinx vespiformis: an essay.* London. B46

Newport, George

1836 On the predaceous habits of the common wasp, *Vespa vulgaris* Linn. *Trans. Ent. Soc. Lond.* 1:228-29. M63

Nichol, John Pringle

1838 Remarques sur les explications que l'examen de la surface de la lune peut présenter pour certains phénomènes géologiques. *Institut* 6 (258) 6 December:400. A135

Nilsson, Sven

1817–21 *Ornithologia Svecica.* 2 vols. Copenhagen. D105

1835 *Skandinavisk fauna.* 2 vols in 1. Copenhagen. D105

Nind, Scott

1832 Description of the natives of King George's Sound (Swan River Colony) and adjoining country. *J. R. Geogr. Soc.* 1:21–51. C131

Ogilby, John

1671 *America.* London. E167

Ogilby, William

1838a Rongeurs australasiens. *Institut* 6 (224) February: 67. C15,46,95

*1838b [A new species of Muntjac deer.] *Proc. Zool. Soc. Lond.* 6:105. D29

*1839 On a new species of monkey (*Papio melnotus*). *Proc. Zool. Soc. Lond.* 7:31. N115

1841 Notice of certain Australian quadrupeds belonging to the order Rodentia. *Trans. Linn. Soc. Lond.* 18:121–32. C46

Orbigny, Alcide Charles Victor Dessalines d'

1835–47 *Voyage dans l'Amérique méridionale . . . pendant les années 1826, 1827, 1828, 1829, 1830, 1831, 1832 et 1833.* 7 vols & atlas in 2 vols. Paris. Vol. 2, 1839–43; vol. 3, 1842; vol. 5, pt 3, 1835–43. RN127,174 A99 C50,56,57,67,99,129,183 ZEd5,6,14

1837a Sur la distribution géographique des Passereaux, dans l'Amérique méridionale. *Institut* 5 (221) November:347–48. ZEd6

*1837b Observations on the raptores of South America. Translated from 'Voyages dans l'Amérique méridionale'. *Mag. Zool. Bot.* 1:347–59. ZEd7

1838 L'homme américain (de l'Amérique méridionale), considéré sous ses rapports physiologiques et moraux. *C. R. Hebd. Séanc. Acad. Sci. Paris* 7:568–70. D136 E79

Orbigny, Charles d'

1836a Existence d'un étage de calcaire marin particulier au-dessous du terrain tertiaire du bassin de Paris et d'une assise, également nouvelle dépendante de l'argile plastique; découverte d'ossements fossiles dans ce dernier étage. *C. R. Hebd. Séanc. Acad. Sci. Paris* 3:228–34. C178

1836b Observations sur des ossemens fossiles découverts dans une assise nouvelle dépendante de l'argile plastique du bassin de Paris. *Annls Sci. Nat.* (Seconde Série-Zoologie) 6:126–28. C178

Osler, Edward

1826 On the burrowing and boring marine animals. *Phil. Trans. R. Soc.* 116:342–71. ZEd1

Ospovat, Dov

1981 *The development of Darwin's theory.* Cambridge.

Oudney, Walter

1826 See Denham, D.N., H. Clapperton and W. Oudney

Ovington, John

*1696 *A voyage to Suratt in the year 1689.* London. C227, 275

Owen, Richard

1833–38 Description of the cranium of the *Toxodon Platensis*, a gigantic extinct mammiferous species, referrible by its dentition to the Rodentia, but with affinities to the Pachydermata and the herbivorous Cetacea. *Proc. Geol. Soc. Lond.* 2 (19 April 1837):541–42. A5

1835a *Syllabus of an elementary course of lectures on comparative anatomy*. London. C33
1835b On the anatomy of *Distoma clavatum*, Rud. *Trans. Zool. Soc. Lond* 1:381–84. ZEd13
1835c Description of a new species of tape-worm, *Taenia lamelligera*, Owen. *Trans. Zool. Soc. Lond.* 1:385–86. ZEd13
1835d Remarks on the Entozoa, and on the structural differences existing among them: including suggestions for their distribution into other classes. *Trans. Zool. Soc. Lond.* 1:387–94. C47–49,213 ZEd13
*1836 Notes on the anatomy of the wombat, *Phascolomys wombat*, Per. *Proc. Zool. Soc. Lond.* 4:49–53. C239
*1836–39 Entozoa. In Todd, 1836–59:111–12. C266
1837 Notes. In Hunter, John 1837. B161 C267
1838 On the anatomy of the Apteryx (*Apteryx Australis* Shaw). *Proc. Zool. Soc. Lond.* 6:48–51,71–72,105–10. D29,30
*1838–40 See *Zoology*: *Fossil mammalia*.
1838–42 Description of a tooth and part of the skeleton of the Glyptodon, a large quadruped of the Edentate order, to which belongs the tessellated bony armour figured by Mr. Clift in his memoir on the remains of the Megatherium, brought to England by Sir Woodbine Parish, F.G.S. *Proc. Geol. Soc. Lond.* 3 (27 March 1839):108–113. A9
*1839a [Notes on the Dugong.] *Ann. Nat. Hist.* 2:300–7. E87,91,92
1839b On the bone of an unknown Struthious bird of New Zealand. *Proc. Zool. Soc. Lond.* 7:169–71. T65
1839c Megatheriidae. *The penny cyclopaedia of the society for the diffusion of useful knowledge* (1833–43, 27 vols) 15:65–73. S3[r]
1840a A description of a specimen of the *Plesiosaurus macrocephalus*, Conybeare, in the collection of Viscount Cole. *Trans. Geol. Soc. Lond.* 2d ser. 5:515–35. C72
1841a On the anatomy of the southern Apteryx (*Apteryx australis*, Shaw). *Trans. Zool. Soc. Lond.* 2:257–301 [Communicated 10 April 1838]. C3 D35
1841b Description of the Lepidosiren annectens. *Trans. Linn. Soc. Lond.* 18:327–61. [Read 1839.] E168

Owen, William Fitzwilliam
*1832 Some remarks relative to the geography of the Maldiva Islands. *J. R. Geogr. Soc.* 2:81–92. A128
1833 *Narrative of voyages to explore the shores of Africa, Arabia, and Madagascar; performed in H.M. Ships 'Leven' and 'Barracouta'. 2 vols. London.* RN64,65 B190

Paley, William
1839 *The principles of moral and political philosophy*. 2 vols in 1. New York. M132

Pallas, Peter Simon
1802–3 *Travels through the southern Provinces of the Russian Empire, in the years 1793 and 1794*. 2 vols. London. RN127 A6

Palmer, Henry R.
1834 Observations on the motions of shingle beaches. *Phil. Trans. R. Soc.* 124:567–76. A83

Parent-Duchâtelet, Alexandre-Jean-Baptiste
1836 *De la prostitution dans la ville de Paris, considérée sous le rapport de l'hygiène publique, de la morale et de L'administration*. 2 vols. Paris. C266

Parieu, ——— de
1838 See Laizer, L. and ——— de Parieu

Parish, Woodbine
*1834 Notice as to the supposed identity of the large mass of meteoric iron now in the British Museum, with the celebrated Otumpa Iron described by Rubin de Celis in the Philosophical Transactions for 1786. *Phil. Trans. R. Soc.* 124:53–54. RN172
1838 *Buenos Ayres, and the Provinces of the Rio de la Plata*. London. [re-issued 1839] RN142, 143, 155–157 A8

Park, Mungo
1800 *Travels in the interior districts of Africa . . . in the years 1795, 1796, and 1797*. 4th ed. London. C269

Park, Thomas
1828 Mr. Thomas Park's journey into the interior of Africa. *Edin. New Phil. J.* 4:410. A27 B121

Parker, Samuel
1838 Review of his journal of an exploring tour beyond the Rocky Mountains. *Athenaeum* (575) 3 November:790–93. A132

Parrot, Georg Friedrich
1831 Considérations sur la température du globe terrestre. *Mém. Acad. Sci. St.-Pétersb.* (6th ser.) 1:501–62. A134,137–40

Parsons, James
1746 [–1748] Human physiognomy explain'd: in the Crounian lectures on muscular motion. *Phil. Trans. R. Soc.* 44 (Supplement): 1–82. N103

Paxton, Joseph
1838 *A practical treatise on the cultivation of the Dahlia*. London. C267

Pennant, Thomas
1773 *Genera of birds*. Edinburgh. D101

Pentland, Joseph Barclay
1835 On the ancient inhabitants of the Andes. *Rep. Br. Ass. Advmt Sci.* (Meeting at Edinburgh, 1834) 3:623–24. B208,209

Pernéty, Antoine Joseph
*1769 *Journal historique d'un voyage fait aux îles Malouines en 1763 & 1764*. 2 vols. Berlin. RN102 C276

Péron, François
1804 Sur quelques faits zoologiques applicables à la théorie du globe, lu à la classe des sciences physiques et mathematiques, de l'Institut national. *J. Phys. Chim. Hist. Nat.* 59:463–79. A58 B177
1807–16 *Voyage de découvertes aux terres australes . . . sur les corvettes 'le Géographe' 'le Naturaliste' et la goëlette 'le Casuarina' pendant les années 1800, 1801, 1802, 1803 et 1804*; continué par M. Louis Freycinet . . . 2 vols & atlas. Paris. Vol. 1, 1807; Vol. 2, 1806; atlas 1811 RN118–20 B152,234 C268
1824 *Voyage de découvertes aux terres australes*. 2d ed. 4 vols. Paris. C21

Phillip, Arthur
1789 *The voyage of Governor Phillip to Botany Bay*. London. RN126

Phillips, John
*1837–39 *Treatise on geology*. In Lardner, D., ed. *The cabinet cyclopaedia. Natural history*. 2 vols. London. A54–57,147 B167–68,170,173 C269 TFRAG9

Phillips, William
1822 See Conybeare, W.D. and W. Phllips.
*1823 *An elementary introduction to the knowledge of mineralogy*. 3d ed. London. RN89

Pictet, M.F.G.
*1839 On the writings of Goethe relative to natural history. *Ann. Nat. Hist.* 2:313–22. E89,92,96

Plato
1977 *Phaedo*. Transl. G.M.A. Grube. Indianapolis. M128

Playfair, John
*1802 *Illustrations of the Huttonian theory of the earth*. Edinburgh. RN110,112,161

Plinius, Secundus Caius .
1601 *The historie of the world. Commonly called the naturall historie of C. Plinius Secundus*. 2 vols. London. C269

Poeppig, Eduard Friedrich von
*1836 Analysis of Poeppig's Travels in Chile, Peru, and on the Amazon River, in the years 1827–32. *J. R. Geogr. Soc.* 6:381–85. A37

Prévost, Constant
1835 Notes sur l'île Julia, pour servir à l'histoire de la formation des montagnes volcaniques. *Mém. Soc. Géol. Fr.* 2:91–124. C270
1836–37 Exemple remarquable de cristallisation en pyramide dans une marne jaune: idées théoriques qu'il émet à ce sujet. *Bull. Soc. Géol. Fr.* 8:320–22. A70
1838 Développement de la théorie du synchronisme des formations. Réponses aux objections. *Bull. Soc. Géol. Fr.* 9:90-95. A101

Prévost, J.L. and J.A. Dumas
1821 Essai sur les animalcules spermatiques de divers animaux. *Mém. Soc. Phys. Nat. Genève*. 1:180–207. E184
1832 On electricity. Appendix. In W.F. Edwards 1832:285–306. E184

Prichard, James Cowles
1826 *Researches into the physical history of mankind*. 2d ed. London. B91
*1836–47 *Researches into the physical history of mankind*. 3d ed. 2 vols. London. [*book at CUL, abst DAR 71:143–145.] B91,136 C204,268
1840 On the extinction of human races. *Edinb. New Phil. J.* 28:166–70. T81

Prinsep, James
1833a Occurrence of the bones of man in the fossil state. *J. Asiat. Soc. Beng.* 2:632–35. E182
1833b Fall of fish fom the sky. *J. Asiat. Soc. Beng.* 2:650–52. E182

Prudhoe, Lord
1835 Extracts from private memoranda kept by Lord Prudhoe on a journey from Cairo to Sennar, in 1829, describing the Peninsula at Sennar. *J. R. Geogr. Soc.* 5:38–58. C91

['Proteus']
1834 The Bahama Islands. *United Service J. Naval Military Mag.* 3:215–26. RN27

Quetelet, Lambert Adolphe Jacques
*1835a *Sur l'homme et le développement de ses facultés*. 2 vols. Paris C268 D152
1835b Review of 1835a. *Athenaeum* (406) 8 August:593–95; (407) 15 August:611–13; (409) 29 August: 658–61. D152

Quoy, Jean-René Constant and Paul Joseph Gaimard
1824 *Zoologie*. In Freycinet 1824–44 B31,54,234 D74 ZEd3
1827 Observations zoologiques faites à bord de 'l'Astrolabe', en Mai 1826, dans le Détroit de Gibraltar. *Annls Sci. Nat.* 10:5–21,172–93,225–39. RN174
1830–35 *Zoologie*. In Dumont d'Urville 1830–35 ['Astrolabe'], vols. 8–11. B54 C13–15,28,276 E42 ZEd10

Rachootin, Stan P.
1985 Owen and Darwin reading a fossil: *Macrauchenia* in a bony light. In Kohn 1985.

Rackett, Thomas
1815 Observations on *Cancer salinus*. *Trans. Linn. Soc. Lond.* 11:205–6. RN127

Raffles, Thomas Stamford Bingley
1830 *The history of Java.* 2d ed. 2 vols. London. C17, 268,IBC

Ramond de Carbonnières, Louis-Elizabeth
1815 On the vegetation of high mountains. *Trans. R. Hort. Soc.*, (2d ed.) 1 (Appendix):15–23. E114–116

Rang, Paul-Charles-Alexandre-Léonard
*1829 *Manuel de l'histoire naturelle des mollusques et de leurs coquilles, ayant pour base de classification celle de M. le baron Cuvier.* Paris. C23

Raverat, Gwen
1960 *Period piece.* London. M41

Ray, John
*1692 *The wisdom of god manifested in the works of the creation.* 2d ed. 2 pts in 1 vol. London. C248,270, 275

Rees, Abraham
1819 Siberia. In Rees, A. *Cyclopaedia.* 39 vols. London. Vol. 32. A7

Reinwardt, Caspar Georg Carl, C.L. Blume and C.G.D. Nees von Esenbeck
1824 Hepaticae javanicae, editae conjunctis studiis et opera. *Nova Acta Physico-Med.* 12:181–238; 409–18. C17,268

Rengger, Johann Rudolph
*1830 *Naturgeschichte der Saeugethiere von Paraguay.* Basel. C267,270 T81 ZEd11

Reynolds, Joshua
1798 *The works of Sir Joshua Reynolds.* 2d ed. 3 vols. London. M32 N26,32 OUN10,11,11v,11b, 11bv
1831 *Discourses delivered to the students of the Royal Academy.* London. C269

Richards, Robert J.
1979 Influence of sensationalist tradition on early theories of the evolution of behavior. *J. Hist Ideas* 40:85–105.

Richardson, John
*1829–37 *Fauna Boreali-Americana; or the zoology of the northern parts of British America.* 3 vols (in 4 pts). London. [*Pts 1–3] B91 C69,268 E52
1831 See Swainson, W. and J. Richardson
*1837 Report on North American zoology. *Rep. Br. Ass. Advmt Sci.* (Meeting at Bristol, 1836) 5:121–224. [*1836 *Philos. Tracts*, 1:18] B154
1839 Notice of a few simple observations which it is desirable to make on the frozen soil of British North America. [Article by John Washington] *Athenaeum* (586) 19 January:52–53. A136

Ridewood, William George
1909 *British Museum (Natural History) Special Guide No. 4. Memorials of Charles Darwin: a collection of manuscripts, portraits, medals, books and natural history specimens to commemorate the centenary of his birth and the fiftieth anniversary of the publication of 'The Origin of Species'.* Preface by Sidney Frederic Harmer. London.

Riley, John
1839 Reply to M. Marten's paper on the hybridity of ferns. *Proc. R. Bot. Soc. Lond.* 1:60–62. QE12

Ritchie, Catherine
1932 [The Gorringer catalogue of Charles Darwin's manuscripts.] University Library, Cambridge.

Ritchie, Thomas Edward
1807 *An account of the life and writings of David Hume, Esq.* London. C270

Robert, Eugene
*1837 Observations faites dans la deuxième expedition de la corvette 'la Recherche'. *Institut* 5 (214) 14 June:192–93. A20

Rogers, Henry D.
1835a On the Falls of Niagara and the reasonings of some authors respecting them. *Am. J. Sci.* 27:326–35. RN142
1835b Report on the geology of North America. *Rep. Br. Ass. Advmt Sci.* (Meeting at Edinburgh, 1834) 3:1–66. A56,57 B174
1861 On the origin of the parallel roads of Lochaber (Glen Roy), Scotland. In *The Royal Institution Library of Science. Earth Sciences*, 1971, ed. S.K. Runcorn, 3 vols. Vol. 1:200–4. London.

Rose, Gustav
1835 Mémoire sur les roches qu'on nomme grunstein et grunstein porphyrique. *Annls Mines* 8:3–32. A84

Ross, James Clark
1847–60 See Hooker, J.D.

Ross, John
1819 *A voyage of discovery, made under the orders of the Admiralty, in His Majesty's Ships 'Isabella' and 'Alexander', for the purpose of exploring Baffin's Bay, and inquiring into the probability of a North-West Passage.* London. RN114,181

Roussin, Albin-Reine
1826 *Le Pilote du Brésil, ou description des côtes de l'Amérique méridionale . . . exécutée en 1819 et 1820 sur la corvette 'la Bayadère' et le brig 'le Favori'.* Paris. RN22,91

Roxburgh, William
1816 An alphabetical list of plants, seen by Dr. Roxburgh growing on the Island of St. Helena, in

1813–14. Appendix [I], pp. 295–326, in Beatson, 1816. B173

Royle, John Forbes
*1835 Illustrations of the botany and other branches of the natural history of the Himalayan Mountains, and of the flora of Cashmere. *J. R. Geogr. Soc.* 5:361–65. B151,236,272 C268
*1839 *Illustrations of the botany and other branches of the natural history of the Himalayan Mountains, and of the flora of Cashmere.* 2 vols. London. [*abst DAR 71:20–25.] QE12
*1840 *Essay on the productive resources of India.* London. [*abst. DAR 71:26–28] QE12

Rudwick, Martin, J.S.
1974–75 Darwin and Glen Roy: a 'great failure' in scientific method? *Studies Hist. Phil. Sci.* 5:97–185.
1982 Charles Darwin in London: The integration of public and private science. *Isis* 73:186:206.

Rüppell, Wilhelm Peter Eduard Simon
1826–28 *Atlas zu der Reise im nördlichen Afrika.* 5 vols. Frankfurt. C42
1829 Review of 1826–28. *Zool. J. Lond.* 4:385–94. C42 ZEd12

Rutherford, Henry William
1908 *Catalogue of the library of Charles Darwin.* Cambridge. E145

Sabine, Edward
1840 *Narrative of an expedition to the Polar Sea in the years 1820–23.* London. QE17

Sabine, Joseph
*1820 Observations on, and account of, the species and varieties of the genus *Dahlia*; with instructions for their cultivation and treatment. *Trans. R. Hort. Soc. Lond.* 3:217–43. [*abst DAR 74:62.] E13, 142
*1812 Observations on the different species of Dahlia. *Trans. R. Hort. Soc.* 1:84–98 [*abst DAR 74:56] E113

Saint-Hilaire, Auguste de [Augustin François-César-Prouvensal de]
*1841 *Leçons de botanique comprenant principalement la morphologie végétale.* Paris. QE17

Saint-Pierre, Jacques-Henri Bernardin de
1773 *Voyage à l'Isle de France, à l'Isle de Bourbon, au cap de Bonne- Espérance, &c. Avec des observations nouvelles sur la nature & sur les hommes, par un officier du roi.* 2 vols in 1. Neuchâtel. B190,234,255 C276 ZEd4

Salisbury, Richard Anthony
1807 An account of a storm of salt, which fell in January, 1803. *Trans. Linn. Soc. Lond.* 8:286–90. A1

Saussure, Albertine Adrienne de
1839–43 *Progressive education; or, considerations on the course of life.* London. 3 vols. C265 N74

Savigny, Marie-Jules-César Lelorgne de
1816 *Mémoires sur les animaux sans vertèbres.* 2 pts in 1 vol. Paris. C216

Schiede, Christian Julius Wilhelm
1825 *De plantis hybridis spontenatis.* Kassel. C265

Schlegel, Hermann
1834 *Monographie van het geslacht Zonurus.* Hoeven en Vriese, Tijdschrift 1:203–21. QE17
1836 Beschreibung von *Zonurus microlepidotus*, Cuv., und *Zonurus Novae Guineae*, *Archiv. Naturgesch.* 2:101–3. QE17

Schmidtmeyer, Peter
1824 *Travels into Chile ... in the years 1820 and 1821.* London. A100

Schomburgk, Robert Hermann
*1836 Report of an expedition into the interior of British Guayana, in 1835–6. *J. R. Geogr. Soc.* 6:224–84. A87
*1837 Diary of an ascent of the River Berbice in British Guayana in 1836–7. *J. R. Geogr. Soc.* 7:302–50. A87 C94

Schouw, Joakim Frederick
1823 *Grundzüge einer allgemeinen Pflanzengeographie ... Aus dem Danischen übersetzt.* Berlin. B156 QE17

Schweber, Silvan S.
1977 The origin of the *Origin* revisited. *J. Hist. Biol.* 10:229–316.

Scoresby, Jr., William
1820 *An account of the Arctic regions, with a history and description of the northern whale-fishery.* 2 vols. Edinburgh. RN181

Scott, John [pseudonym]
1820 See [Lawrence, John]

Scott, Walter
1815 *Guy Mannering; or, the astrologer.* 3 vols. Edinburgh. M8,59
1829 *Waverley novels.* 48 vols. London. N46
1837–38 See Lockhart, J.G.

Scrope, George Julius Poulett
*1825 *Considerations on volcanos.* London. RN77 A34
1827 *Memoir on the geology of central France; including the volcanic formations of Auvergne, the Velay, and the*

Vivarais, with a volume of maps and plates. London. See review, Lyell 1827. N19
Review of 1827. *Q. Rev.* 36:437–83. N19
1829 Notice on the geology of the Ponza Isles. *Trans. Geol. Soc. Lond.* 2d ser. 2:195–236. A43,51

Seale, Robert F.
1834 *The geognosy of the island St. Helena*. London.

Sebright, John Saunders
*1809 *The art of improving the breeds of domestic animals*. London. C133,275 D179
*1836 *Observations upon the instinct of animals*. London. B250 C134,165,275 N63

Sedgwick, Adam
1835 On the geological relations and internal structure of magnesian limestone, and the lower portions of the new red sandstone series in their range through Nottinghamshire, Derbyshire, Yorkshire, and Durham, to the southern extremity of Northumberland. *Trans. Geol. Soc. Lond.* 2d ser. 3:37–124. A45,49

Sedgwick, Adam and Roderick Impey Murchison
*1839 Classification of the older stratified rocks of Devonshire and Cornwall. *Phil. Mag.* 14:241–60, 317,354–58. [*Pamphlet 66] E126

Selby, Prideaux John
*1837–38 Observations on the importance of a local fauna;—exemplified in the fauna of Twizel. *Mag. Zool. Bot.* 1:421–24; 2:389–97. C160

Serres, Antoine-Etienne-Renaud-Augustin
*1837 Anatomie des mollusques. *Institut* 5 (221) November:370–71. B163

Serres, Marcel, [Pierre Marcel Toussaint de]
1823 Observations sur les ossements humains découverts dans les crevasses des terrains secondaires, et en particulier sur ceux que l'on observe dans le caverne de Durfort. *Bibltque Univ. Genève* 23:277–95. E35
*1837 De la présence du fer sulfuré sublimé dans les calcaires tertiaires des environs de Montpellier. *Institut* 5 (220) October:331. A32 B151
1838a *Essai sur les cavernes à ossements et sur les causes qui les y ont accumulés*. 3d ed. Paris. C267
1838b See Flourens, M.J.P. and M. Serres

Seward, A.C.
1909 *Darwin and modern science*. Cambridge. D175

Sharpe, D.
1846 See Morris, J. and D. Sharpe

Shaw, John
1836 An account of some experiments and observations on the parr, and on the ova of the salmon, proving the parr to be the young of the salmon. *Edinb. New Phil. J.* 21:99–110. QE18
*1838 Experiments on the development and growth of the fry of the salmon from the exclusion of the ovum to the age of seven months. *Edinb. New Phil. J.* 24:165–76. QE18
1840 Account of experimental observations on the development and growth of salmon-fry, from the exclusion of the ova to the age of two years. *Trans. R. Soc. Edinb.* 14:547–66. QE18

Shepard, Charles U.
1835 On the strontianite of Schoharie, (N.Y.) with a notice of the limestone cavern in the same place. *Am. J. Sci.* 27:363–70. RN142

Sherborn, Charles Davies and B.B. Woodward
1901 Notes on the dates of publication of the natural history portions of some French voyages. *Ann. Mag. Nat. Hist.* 7th ser. 7:388–92. RN127

Shuttleworth, Nina Louisa Kay
1910 *A life of Sir Woodbine Parish, K.C.H., F.R.S., 1796–1882*. London. RN155

Siebold, Phillip Franz von
1833–50 *Fauna Japonica; sive, descriptio animalium, quae in itinere per Japoniam, jussu et auspiciis superiorum, qui summum in India Batava imperium tenent, suscepto, annis 1823–1830 collegit, notis, observationibus et adumbrationibus illustravit Ph. Fr. de Siebold, conjunctis studiis C.J. Temminck et H. Schlegel pro vertebratis, atque W. de Haan pro invertebratis elaborata*. 6 vols in 4. Luduni Batavorum. C225 E170

Simpson, George Gayland
1980 *Splendid isolation*. New Haven, London. RN129

Sloan, Phillip R.
1985 Darwin's invertebrate program, 1826–1836: preconditions for transformism. In Kohn 1985.
1986 Darwin, vital matter, and the transformation of species. *J. Hist. Biol.* 19:369–445.

Smart, Benjamin Humphrey
1839 *Beginnings of a new school of metaphysics: three essays in one volume: Outline of sematology.—MDCCCXXXI. Sequel to sematology.—MDCCCXXXVII*. London. OUN5

Smee, Walter
1835 Some account of the maneless lion of Guzerat. *Trans. Zool. Soc. Lond.* 1:165–74. C47

Smellie, William
*1790–99 *The philosophy of natural history*. 2 vols. Edinburgh. B272 C268 D56

Smith, ———
1828 See Mitchill, J.A., ——— Smith and ——— Cooper.

Smith, Adam
1795 *Essays on philosophical subjects. By the late Adam Smith . . . To which is prefixed, an account of the life and writings of the author, by Dugald Stewart F.R.S.E.* London. M108,155
1808 *The theory of moral sentiments.* 11th ed. 2 vols. Edinburgh. N184 OUN29[v]
1829 See Stewart, D. 1829

Smith, Andrew
1831 Observations relative to the origins and history of the Bushmen. *Phil. Mag.* 9:119–27,197–200,339–42,419–23. B32 QE16
1836a *Report of the expedition for exploring Central Africa, from the Cape of Good Hope, June 23, 1834. Cape Town.* [Location: British Museum (Natural History) Zoology Dept. Library]. B33,69
1836b Report of the expedition for exploring Central Africa. *J. R. Geogr. Soc.* 6:394–413. A94
1837 *A catalogue of the South African Museum now exhibiting in the Egyptian Hall Piccadilly. London.* [Location: British Museum (Natural History) Zoology Dept. Library] B67
*1838–49 *Illustrations of the zoology of South Africa . . . collected . . . in the years 1834, 1835, and 1836.* 5 pts. London. [*4 vols] B33,67 C41 D50–52, 107 QE16

Smith, Christen
1818 Professor Smith's Journal. In Tuckey 1818. C250,275

Smith, James [of Jordanhill]
1838 On the last changes in the relative levels of the land and sea in the British Islands. *Edinb. New Phil. J.* 25:378–94. A111 E161
1838–42 On the geology of the island of Madeira. *Proc. Geol. Soc. Lond.* 3 (6 January 1841):351–55. QE1

Smith, James Edward
1826 *A grammar of botany, illustrative of artificial, as well as natural, classification, with an explanation of Jussieu's system.* London. QE[15[v]]

Smith, Samuel Stanhope
1810 *An essay on the causes of the variety of complexion and figure in the human species . . . Also, strictures on Lord Kaim's* [Kames] *discourse on the original diversity of mankind.* 2d ed. New Brunswick. C275

Smith, Sydney
1960 The origin of the *Origin*, as discerned from Charles Darwin's notebooks and his annotations in the books he read between 1837 and 1842. *Advancement Sci.* 16:391–401.
1968 The Darwin collection at Cambridge, with one example of its use: Charles Darwin and *Cirripedes. Actes du Xe Cong. Int. d'Hist. Sci.* (1965) pt. 5:96–100.

Somerville, Mary
*1834 *On the connexion of the physical sciences.* London. C219

Southey, Robert
1836–37 See Cowper, W.

Sowerby, George Brettingham (Elder)
1812 An account of a new Scarabaeus, discovered by Mr. Neale, and observations on two other rare insects. *Trans. Ent. Soc. Lond.*, 1:246–247. QE1
*1833 Characters of new species of Mollusca and Conchifera, collected by Mr. Cuming. *Proc. Zool. Soc. Lond.* 1:16–22,34–38,52–56,70–74,82–85,134–39. QEIFC,1
1839a On certain monstrosities of the genus Encrinus. *Rep. Br. Ass. Advmt Sci.* (1838) 7:115–16. QE1
1839b *The zoology of Captain Beechey's voyage.* London. E98,99

Sowerby, George Brettingham (Junior)
*1840 On some new species of the genus Cardium, chiefly from the collection of H. Cuming, Esq. *Proc. Zool. Soc. Lond.* 8:105–11. QE1

Sowerby, James de Carle
1806–38 *The British miscellany.* 2 vols. London. QE1
1812–46 *The mineral conchology of Great Britain.* 7 vols. London. E121
1820–25[–34] *The genera of recent and fossil shells, for the use of students in conchology and geology.* Commenced by James Sowerby, and continued by George Brettingham Sowerby. 2 vols. London. C244–246

Spallanzani, Lazzaro
1769 *An essay on animal reproduction.* London. C269 E90,148 QE1[v],10

Spence, William
1836 Observations on a mode practised in Italy of excluding the common house-fly from apartments. *Trans. Ent. Soc. Lond.* 1:1–7. D22

Spencer, Herbert
1863–65 *The principles of biology.* 2 vols. London.

Spenser, Edmund
1589 *The faerie queene disposed into twelve bookes fashioning XII. moral virtues.* London. [1923 reprint] M84

Spicer, John W. G.
1854 Note on hybrid gallinaceous birds. *Zoologist* 12:4294–96. B189

Spix, Johann Baptiste von
1815 *Cephalogenesis.* Monachii E89

Spix, Johann Baptist von and Carl Friedrich Philipp von Martius
*1824 *Travels in Brazil, in the years 1817–20.* 2 vols in l. London. C67 ZEd20

Sprengel, Christian Konrad
1793 *Das entdeckte geheimniss der natur in bau und in der befruchtung der blumen*. Berlin E165

Spry, Henry Harpur
1832 Note on Indian saline deposits. *J. Asiat. Soc. Beng.* 1:503. A143

Stanley, Edward
1835 *A familiar history of birds: their nature, habits and instincts*. 2 vols in 1. London. C269 D103 E136,138–140 QE19

Stark, John
*1838 On the food of the vendace, herring, and salmon. *Ann. Nat. Hist.* 1:74–75. C95

Stauffer, Robert C.
1975 *Charles Darwin's natural selection. Being the second part of his big species book written from 1856 to 1858*. Cambridge.

Staunton, George Leonard
1797 *An authentic account of an embassy from the King of Great Britain to the Emperor of China*. 2 vols & atlas. London. C269 OUN12

Staunton, George Thomas
1821 See Tu-Li-Shin

Stephen, James
1838 Remains of the Rev. Richard Hurrell Froude, M.A., Fellow of Oriel College, Oxford, 2 vols. 8vo. London. *Edinb. Rev.* 67:525–35. M60

Stephenson, J.
1833 On the manufacture of saltpetre, as practised by the natives of Tirhut. *J. Asiat. Soc. Beng.* 2:23–27. A143
1834a On the efflorescence of khari nun, or sulphate of soda, as found native in the soil of Tirhut and Sarun, in the province of Behar. *J. Asiat. Soc. Beng.* 3:188–89. A143
1834b On the saline nature of the soil of Ghazipoor, and manufacture of common salt, as practised by the natives of the villages of Tuttulapoor, Ratouly, Sahory, Chilar and Becompoor. *J. Asiat. Soc. Beng.* 3:36–39. A143
1836 Specimens of the soil and salt from the Samar, or Sambhur lake salt-works. *J. Asiat. Soc. Beng.* 5:798–806. A143

Sternberg, Caspar von (Count)
1827 On the distribution of living and fossil plants. *Edinb. New Phil. J.* 3:190–92. B117

Stewart, Dugald
1795 See Smith, A.
1829 *The works*. 7 vols. Vol. 7: *Account of the life and writings of Adam Smith, Account of the life and writings of Thomas Reid*. Cambridge, Mass. C267 M155 OUN17–21

Strachan, Charles
*1820 Account and description of the different varieties of the onion. *Trans. R. Hort Soc.* 3:369–79. [*abst. DAR 74.63] E165

Straus-Durckheim, Hercule Eugène
1819–20 Mémoire sur les Daphnia, de la classe des Crustacés. *Mém. Mus. Hist. Nat. Paris* 5:380–425, 6:149–62. C162,235

Strickland, Hugh Edwin
1837 [1838] See Hamilton, W.J. and H.E. Strickland

Sturt, Charles
1833 *Two expeditions into the interior of southern Australia during the years 1828, 1829, 1830, and 1831*. 2 vols. London. RN126

Sulloway, Frank J.
1982 Darwin's conversion: The '*Beagle*' voyage and its aftermath. *J. Hist. Biol.* 15:325–96.

Swainson, William
*1835 *A treatise on the geography and classification of animals*. London. B8,33,46,67,92,276 C73,74,170,185 D26
*1836–37 *On the natural history and classification of birds. Lardner's Cabinet Cyclopaedia*. 2 vols. London. C112,113
*1837 *Dr Lardner's Cabinet Cyclopaedia, Natural History*. 1. *On the geography and classification of animals*. By W. Swainson, Esq. . . . [Review of 1835.] *Mag. Zool. Bot.* 1:545–66. B67,92,276 C170,185

Swainson, William and John Richardson
1831 *The birds*. In Richardson 1829–37 [Fauna-Boreali], vol. 2. C69,73

Sweet, Robert
1830 *Sweet's Hortus Britannicus: or, a catalogue of plants indigenous, or cultivated in the gardens of Great Britain*. 2d ed. London. C265

Swisher, Charles N.
1967 Charles Darwin on the origins of behavior. *Bull. Hist. Med.* 41:24–43.

Sykes, William Henry
1835 Description of the wild dog of the Western Ghats. *Trans. R. Asiat. Soc.* 3:405–09. QE12
1836a Descriptions of new species of Indian ants. *Trans. Ent. Soc. Lond.* 1:99–107. MIFC,62
1836b Observations upon the habits of *Copris midas*. *Trans. Ent. Soc. Lond.* 1:130–32. ZEd19
*1838a Observations on the *Canis jubatus* and some skins of the *Felis pardina*. *Proc. Zool. Soc. Lond.* 6:111–13. QE12

*1838b On the identity of the wild ass of Cutch and the Indus, with the Dziggetai (*Equus hemionas* of Pallas). *Ann. Nat. Hist.* 1:322–27. C225

1838c On the Calandra Lark. *Proc. Zool. Soc. Lond.* 6:113–14. D95

Tankerville, Charles August Bennet and L. Hindmarsh.

1838 On the wild cattle of Chillingham Park. *Athenaeum* (565) 25 August:611–12. D48

Tanselle, G. Thomas

1978 The editing of historical documents. *Studies in Bibliography* 31:1–56.

Taylor, Richard

1837–53 *Scientific memoirs.* London. C275

Telfair, Charles

1826–33 [Letter from Mr. Telfair] *Proc. Geol. Soc. Lond.* 1 (1 May 1833):479. C216

1833 Letter on the history of a living specimen of the Indian tortoise (*Testudo Indica*, Linn.), presented by General Sir Charles Colville. *Proc. Zool. Soc. Lond.* 1:81. ZEd6

Temminck, Coenraad Jacob

*1813–15 *Histoire naturelle générale des pigeons et des Gallinacés.* 3 vols. Amsterdam, Paris. [*abst DAR 71:6–19.] C36,IBC D101

1817 *Observations sur la classification méthodique des oiseaux.* Amsterdam, Paris. D25

1826 Sur le genre chat, ou Felis. [Review of 1827–41, 1 (1827), 4th monograph]. *Zool. J. Lond.* 2:526–34. C41

1827–41 *Monographies de mammalogie, ou description de quelques genres de mammifères, dont les espèces ont été observées dans les différens musées de l'Europe.* 2 vols. Paris, Amsterdam. B249 C41

1833 *Discours préliminaire. Coup-d'oeil sur la faune des îles de la Sonde et de l'empire du Japon.* In Siebold, 1833–50. CIBC E170

*1838 Zoology of Java. *Ann. Nat. Hist.* 1:335. C225

Temple, Edmond

1830 *Travels in various parts of Peru, including a year's residence in Potosi.* 2 vols. London. RN125,157,158

Templeton, John

1802 On the naturalization of plants. *Trans. R. Ir. Acad.* 8:111–29. A126

Thompson, John Vaughan

1835a Discovery of the metamorphosis in the second type of Cirripedes, viz., the Lepades, completing the natural history of these singular animals, and confirming their affinity with the Crustacea. *Phil. Trans. R. Soc.* 126:355–58. ZEd1

1835b On the double metamorphosis in the Decapodous Crustacea, exemplified in *Cancer maenas*, Linn. *Phil. Trans. R. Soc.* 126:359–62. ZEd1

Thompson, William

*1837a Notes relating to the natural history of Ireland, with a description of a new genus of fishes (*Echiodon*). *Proc. Zool. Soc. Lond.* 5:52–63. B166

*1837b On hybrids produced in a wild state between the black-grouse (*Tetrao tetrix*), and common pheasant (*Phasianus colchicus*). *Mag. Zool. Bot.* 1:450–53. B189 C184,228

*1838a Contributions to the natural history of Ireland. No. 2. On the birds of the order Raptores. *Mag. Zool. Bot.* 2:42–57. C185

*1838b Contributions to the natural history of Ireland. No 4. On the birds of the order Insessores. *Mag. Zool. Bot.* 2:427–40. C189

*1838c Contributions to the natural history of Ireland. No. 6. On the birds of the order Insessores. *Ann. Nat. Hist.* 1:181–95. C189

1838d On the Irish hare (*Lepus Hibernicus*). *Trans. R. I. Acad.* 18:260–71. B7 D61

*1839a On the Irish hare. *Ann. Nat. Hist.* 2:70–71. D61

*1839b On fishes; containing a notice of one species new to the British, and of others to the Irish fauna. *Ann. Nat. Hist.* 2:266–73. E91

Thomson, James Jr.

1848 On the parallel roads of Lochaber. *Edinb. New Phil. J.* 45:49–61.

Tickell, Samuel Richards

1833 List of birds, collected in the jungles of Borabhum and Dholbhum. *J. Asiat. Soc. Bengal* 2:569–83. E181

Todd, Robert Bentley

*1836–59 *The cyclopaedia of anatomy and physiology.* 5 vols in 6. London. C266 D113,166

Tooke, John Horne

1798–1805 *ΕΠΕΑ ΠΤΕΡΟΕΝΤΑ, Or the diversions of Purley*, 2d ed. 2 pts. London. N65 OUN13

Totten, Joseph G.

1835 Descriptions of some shells, belonging to the coast of New England. *Am. J. Sci.* 28:347–53. RN142

Tournal, Paul (?)

1833 Considérations générales sur le phénomène des cavernes à ossemens. *Annls Chim.* 52:161–81. E35,182

Treviranus, Gottfried Reinhold

1802–22 *Biologie, oder philosophie der lebenden Natur für Naturforscher und Aerzte.* 6 vols. Göttingen. C104

Tuckey, James Kingston

1818 *Narrative of an expedition to explore the river Zaire, usually called the Congo, in South Afrca, in 1816.* London. B154 C224,248–251,268,275

Tu-Li-Shin
1821 *Narrative of the Chinese Embassy to the Kahn of the Tourgouth Tartars, in the years 1712, 13, 14, and 15.* [Transl. by George Thomas Staunton.] London. C269

Tullamore, Lord. See Bury, Charles William, Lord Tullamore

Turnbull, Alexander See Christie, A.T.

Turner, Edward
1837 *Elements of chemistry*. London. A84

Turner, Samuel
1800 *An account of an embassy to the court of the Teshoo Lama, in Tibet*. London. C266

Turpin, Pierre-Jean-Francois
[1827] *Organographie végétale. Observations sur quelques végétaux microscopiques, et sur le rôle important que leurs analogues jouent dans la formation et l'accroissement du tissu cellulaire*. Paris. E165
1838 [Milk globules turn into a vegetable.] *Athenaeum* (553) 2 June:396. C224

Ulloa, Antonio de
1792 *Noticias Americanas: entretenimientos fisico-históricos sobre la América meridional*. 2d ed. Madrid. RN125
1806 See Juan G. and A. de Ulloa

Unanue, Jose Hipolito
1815 *Observaciones sobre el clima de Lima*. 2d ed. Madrid. A153

Ure, Andrew
*1823 *A dictionary of chemistry*. 2d ed. London. A22–24

Urville, Jules-Sébastian César Dumont d'. See Dumont d'Urville, J-S-C.

Vaillant, François Le
1790 *Travels from the Cape of Good-Hope, into the interior parts of Africa*. London. 2 vols. C129

Valenciennes, Achille
1828–49 See Cuvier, G.L.C.F.D. and A. Valenciennes.
1838a Ossements fossiles de Stonesfield. *Institut* 6(246) 13 September:297–98. D62
1838b Poissons des îles Canaries. *Institut* 6 (251) 18 October:338. E35
1838c Rhinocéros fossile. *Institut* 6 (258) 6 December: 394. E72
1838d Considérations générales sur l'ichthyologie de l'Atlantique. *C.R. Hebd. Séanc. Acad. Sci. Paris* 7:717–22. E35

Van-Hasselt
1827–28 See Blume, C.L.

Vigors, Nicholas Aylward
1826a Sketches in ornithology: or observations on the leading affinities of some of the more extensive groups of birds. 2. On a group of Psittacidae known to the ancients. 3. On a new genus of Falconidae (*Gampsonyx*). *Zool. J. Lond.* 2:37–70. C40
1826b See Children, J.G. and N.A. Vigors.
1828 Sketches in ornithology, &c. &c. On some species of birds from Cuba. *Zool. J. Lond.* 3:432–48. C61

Vigors, Nicholas Aylward and Thomas Horsfield
1827 A description of the Australian birds in the collection of the Linnean Society; with an attempt at arranging them according to their natural affinities. *Trans. Linn. Soc. Lond.* 15:170–331. D40,61

Virey, Julien-Joseph
1816–19 Métamorphose. In *Nouveau dictionnaire d'histoire naturelle*. 20(1818):347–76. 36 vols. Paris. D41
1835 *Philosophie de l'histoire naturelle, ou phénomènes de l'organisation des animaux et des vegetaux*. Paris. C267

Virlet d'Aoust, Pierre-Théodore
1833 See Le Puillon de Boblaye, E. and P.-T. Virlet d'Aoust

Volney, Constantin François Chasseboeuf de
1787a *Travels through Syria and Egypt, in the years 1783, 1784, and 1785*. London. 2 vols. CIBC
1787b *Voyage en Syrie et Egypte pendant les années 1783, 1784 et 1785*. 2d ed. 2 vols. Paris. RN8,52

Vrolik, Gerardus
1837 Notice sur les dents incisives et le nombre des côtes du Rhinocéros africain. *Annls Sci. Nat.* 7:20–26. QE16

Wagner, Franz Auguste
*1838 Observations on rabies or madness in dogs, oxen, horses, pigs, and sheep. *Edinb. New Phil. J.* 24:353–60. E36–37

Wahlenberg, Georg (Goran)
1812 Einiges zur physikalischen Erdbeschreibung von Lappland, und über die Gesetze, nach welchen die Pflanzen verbreitet sind. Transl. from Swedish. *Annln Phys.* 41:233–325. C178

Waitz, Theodor
1859–72 *Anthropologie der naturvölker*. 6 vols. Leipzig. N22

Walker, Alexander
*1838 *Intermarriage; or, the mode in which, and the causes why, beauty, health and intellect, result from certain unions, and deformity, disease and insanity, from others; demonstrated by delineations of the structure and forms, and descriptions of the functions and capacities which each parent, in every pair, bestows on children, in conformity with certain natural laws, and by an account of corresponding effects in the breeding of animals*. London. C265 E113,169 QE4

1839 Review of 1838. *Brit. For. Med. Rev.* 7:370–85. C265

Walker, Francis
1838 Descriptions of some Chalcidites discovered by C. Darwin, Esq. *Entom. Mag.* 5:469–77. C107

Wallich, Nathaniel
1830–32 *Plantae Asiaticae Rariores.* 3 vols. London. S6

Warlow, ———
1832 Catalogue of Indian birds. *J. Asiat. Soc. Beng.* 1:261–64, 313–23. E177

Wartmann, Elie
1846 Memoir on Daltonism [or colour blindness]. *Scient. Mem.* 4:156–87. QE16

[Washington, John] See Richardson, J.

Waterhouse, George Robert
*1837 Characters of new species of the genus *Mus*, from the collection of Mr. Darwin. *Proc. Zool. Soc. Lond.* 5:15–21,27–32. B250 C29 E24
1837–40 Descriptions of some new species of exotic insects. *Trans. Ent. Soc. Lond.* 2 (3) [read 5 December 1836]:188–96. B46,56,57,79 C107, 108
1838 [Marsupial skins from Van Diemen's Land] *Proc. Zool. Soc. Lond.* 6:105 D29
*1839a See *Zoology: Mammalia.* B250 C23,29,35,36,116 D99 E24
*1839b Observations on the Rodentia, with a view to point out the groups, as indicated by the structure of the crania, in this order of mammals. *Mag. Nat. Hist.* 3:90–96,184–88,274–79,593–600. C29 E169
*1839c On the geographical distribution of the Rodentia. *Proc. Zool. Soc. Lond.* 7:172–74. C36,116
*1845 Descriptions of coleopterous insects collected by Charles Darwin, Esq., in the Galapagos Islands. *Ann. Mag. Nat. Hist.* 16:19–40. E170

Waterton, Charles
*1833 The habits of the carrion crow. *Mag. Nat. Hist.* 6:208–14. N1
*1838 *Essays on natural history, chiefly ornithology.* London. [*1838 2d ed.] C270 D147,148 N1

Watson, Hewett Cottrell
1835 *Remarks on the geographical distribution of British plants.* 2d ed. London. [*1843 ed. at Down, abst of pt 1 of 1843 ed. DAR 71:153,160–61] GR108 C267 M155

Waugh, Francis Gledstanes
[1888] *Members of the Athenaeum Club from its foundation.* [London?] D151

Webb, Philip Barker-. See Barker-Webb, P. and S. Berthelot.

Webster, John W.
1821 *A description of the Island of St. Michael, comprising an account of its geological structure; with remarks on the other Azores or Western Islands.* Boston. RN126

Webster, Thomas
1836 On the changes of temperature consequent on any change in the density of elastic fluids, considered especially with reference to steam. *Trans. Inst. Civil Eng.* 1:219–26. A55

Webster, William Henry Bayley
1834 *Narrative of a voyage to the southern Atlantic Ocean, in the years 1828, 29, 30, performed in H.M. Sloop 'Chanticleer'.* 2 vols. London. RN33,125,128

Wedgwood, Hensleigh
1866 *On the origin of language.* London. N31,39

Wedgwood, John
1812 Observations on the culture of the Dahlia in the northern parts of Great Britain, &c. In a letter to Richard Anthony Salisbury. *Trans. R. Hort. Soc.* 1:113–15. E13

Weismann, August
1893 'Telegony' In *Oxford English Dictionary.* B181

Weiss, Christian Samuel
1827 Uber das südliche Ende des Gebirgszuges von Brasilien in der Provinz S. Pedro do Sul und der Banda oriental oder dem Staate von Monte Video; nach den Sammlungen des Herrn Fr. Sellow. [1827–28] *Abh. Akad. Wiss. Berlin.* 217–93. A151

Weissenborn, W.
*1838a On spontaneous generation. *Mag. Nat. Hist.* 2:369–81,621–24. QE[5]a,17
*1838b On the transformation of oats into rye. *Mag. Nat. Hist.* 2:670–72. QE[15^v]
*1838c Transmission of experience in birds, in the form of instinctive knowledge. *Mag. Nat. Hist.* 2:50–53. B199

Wells, Algernon
1834 *On animal instinct.* Colchester. C269 N68–72

Wellsted, James Raymond
*1835 Memoir on the Island of Socotra. *J. R. Geogr. Soc.* 5:129–229. C92
*1836 Observations on the coast of Arabia between Ras Mohammed and Jiddah. *J. R. Geogr. Soc.* 6:51–96. B48 C93

Westwood, John Obadiah
1835 On the supposed existence of metamorphoses in the Crustacea. *Phil. Trans. R. Soc.* 126:311–28. ZEd1
1836 On the earwig. *Trans. Ent. Soc. Lond.* 1:157–63. M63

1837 On Diopsis, a genus of dipterous insects, with descriptions of twenty-one species. *Trans. Linn. Soc. Lond.* 17:283–313. B139

*1838 Notes upon subaquatic insects, with the description of a new genus of British Staphylinidae. *Mag. Zool. Bot.* 2:124–32. C186 ZEd17

1841 On the family Fulgoridae, with a monograph of the genus *Fulgora*, Linn. *Trans. Linn. Soc.* 18:133–54. TFRAG7

Whewell, William

1833 *Astronomy and general physics considered with reference to natural theology. The Bridgewater treatises on the power, wisdom, and goodness of god as manifested in the creation.* Treatise III. 2d ed. London. C72,91 D49 Mac58r

*1837 *History of the inductive sciences, from the earliest to the present times.* 3 vols. London. C269 D26,71 E57,60,69,70,83,89,97 N14

1833–38 Address to the Geological Society, delivered at the anniversary, on the 16th of February, 1838. *Proc. Geol. Soc. Lond.* 2 (16 February 1838):624–49. C55,62

1838–42 Address delivered at the anniversary meeting of the Geological Society of London, on the 15th February 1839; and the announcement of the award of the Wollaston Medal and Donation Fund for the same year. London. [* Pamplet 42.] Also in *Proc. Geol. Soc. Lond.* 3 (15 February 1839):7–44. E128

Whitby, Mary Anne Theresa

1848 *A manual for rearing silkworms in England.* London QEIFC

White, Charles

1799a On the Dhanésa; or, Indian Buceros, communicated by Lieut. Fraser. *Asiat. Reschs* 4:119–28. QE10

1799b *An account of the regular gradation in man, and in different animals and vegetables; and from the former to the latter.* London. B272 C268

White, Gilbert

*1825 *The natural history and antiquities of Selborne.* 2 vols. London. C248,254,275

Whitehead, G. Kenneth

1953 *The ancient white cattle of Britain and their descendants.* London. M142

Wiegmann, Arend Friedrich

*1828 Ueber die Bastarderzeugung im Pflanzenreiche. *Arch. Ges. Naturlehre* 15:204–09. [* Braunschweig] C265 QE[13v]

*1838 On the genus Procyon, with a description of two new species. *Ann. Nat. Hist.* 1:132–35. C96

Wilbraham, Roger

*1818 Report of the fruit committee. *Trans. R. Hort. Soc.* 2:58–63. [*abst DAR 74:58.] E141

Wilkinson, John Gardner

*1820 *Remarks addressed to Sir J. Sebright.* London. C133,275

1837 *Manners and customs of the ancient Egyptians.* 3 vols. London. C269 E142,145

Williams, John

1837 *A narrative of missionary enterprises in the South Sea Islands.* London. RN12 B220 E10 TFRAG10

Wilson, Alexander

1832 *American ornithology.* 3 vols. London, Edinburgh. C67–69,112,267 ZEd14

Wilson, James

1842 Notice of the occurrence in Scotland of the *Tetrao medius*, shewing that supposed species to be a hybrid. *Proc. R. Soc. Edinb.* 1(1842–44):395. D73

Wilson, Leonard G.

1972 *Charles Lyell: The years to 1841. The revolution in geology.* New Haven and London. RN129 A9

Woodward, Samuel Peckworth

*1851–56 *A manual of the mollusca; or, rudimentary treatise of recent and fossil shells.* 3 pts. London. E120

Wordsworth, William

1802 *Lyrical ballads.* 2 vols. London M40 N57

Yarrell, William

*1827 On the change in the plumage of some hen-pheasants. *Phil. Trans. R. Soc.* 117:268–75. C215, 267 D114,154

*1830–31 On the specific identity of the Gardenian and night herons (*Ardea gardeni* and *Nycticorax*). *Proc. Zool. Soc. Lond.* 1:27. D95

1833 Characters of the Irish Hare, a new species of the genus *Lepus*, Linn. *Proc. Zool. Soc. Lond.* 1:88. B7

1836 *A history of British fishes.* 2 vols. London. T26

*1843 *A history of British birds.* 3 vols. London. [*1839, vol. 1; 1845, abst of 2d ed. DAR 71:166–79] C121 D96

Youatt, William

1831 *The horse; with a treatise on draught. Library of useful knowledge.* London. C267,IBC D179

*1834 *Cattle: their breeds. Library of useful knowledge.* London. C267,IBC D179

*1837 *Sheep: their breeds, management, and diseases. Library of useful knowledge.* London. C267,IBC D104

Young, R.M.

1985 Malthus and the evolutionists: the common context of biological and social theory. In R.M. Young, *Darwin's metaphor: nature's place in Victorian culture*, pp. 23–55. Cambridge.

Zivins, Gail (Ed.)

1985 *The development of expressive behavior: biology-environment interactions.* Orlando.

Biographical Index

Abberly, *c.* 1841; Robert Waring Darwin's gardener at Shrewsbury. EIBC T151 QE8, 11, 14, ‹15› M49

Allen, John Bartlett, 1733–1803; of Cresselly, Pembrokeshire, father of Elizabeth Allen who married Josiah Wedgwood II. N113

D'Arblay, Alexander Charles Louis, *c.* 1795–1837; student at Christ's College, Cambridge; mathematician, clergyman. M99

Aristotle, 384–322 B.C.; philosopher. C267

Arrowsmith, John, 1790–1873; geographer and mapmaker; a founding member of the Royal Geographical Society, 1830. A2

Ash, John, 1723–1798; physician, practised in Birmingham where he suffered temporary mental illness; FRS (1787). M14, 19, 80, 117

Babbage, Charles, 1792–1871; mathematician, designed early mechanical computers, member of Cambridge University network; FRS (1816). A55

Babington, Charles Cardale, 1808–1895; botanist, entomologist, archaeologist; Professor of Botany, Cambridge University, 1861–95; FRS (1851). C192, 195 M79 QEIFC, 10a

Bachman, John, 1790–1874; American clergyman and naturalist; collaborator of J. J. Audubon. C251–53, 255 D31–34, 103

Baldwin, Jack; possibly son of Andrew Baldwin of Coreby, Shropshire. M11

Basket, Fuegia, 1821–1883?; native Fuegian, returned from England to Tierra del Fuego on board H.M.S. *Beagle*, 1833. N15

Bauer, Franz Andreas, 1758–1840; botanical artist at Kew Gardens; FRS (1821). C237 QE21..

Beaufort, Francis, 1774–1857; naval officer, retired as Rear-Admiral in 1846; hydrographer to the Admiralty (1829–55); one of the founders of the Royal Astronomical Society and of the Royal Geographical Society; FRS (1814). A55

Beck, Henrick Henricksen, 1799–1863; Danish zoologist and conchologist. A85 B153, 202 C84, 137 D133 E59, 86, 101

Beechey, Frederick William, 1796–1856; naval officer and geographer; FRS (1824). A132, 133

Bell, J.; of Oxford Street; bred bloodhounds. B183, 184

Bell, Thomas, 1792–1880; dental surgeon at Guy's Hospital, London, 1817–61; Professor of Zoology, King's College, London, 1836; described the reptiles from the *Beagle* voyage. FRS (1828). C18 QE16

Bennett, John Joseph, 1801–1876; botanist, assistant to Robert Brown at the British Museum; FRS (1841). QE[15^{v}]

Bentham, George, 1800–1884; botanist; Honorary Secretary of the Horticultural Society, 1829–40; FRS (1862). E104 QE17, 21..

Berwick [William Noel Hill, 3d Baron Berwick] d. 1842; diplomat. D14 M25

Bewick, Thomas, 1753–1828; wood engraver and naturalist. E91

Bibron, Gabriel, 1806–1848; French herpetologist who described the reptiles collected by Alcide d'Orbigny in South America. C18, 54

Blyth, Edward, 1810–1873; naturalist; curator of the Museum of the Royal Asiatic Society of Bengal, 1841–62. D29, 30, 33, 95, 96 QEIFC, 19

Boerhaave, Hermann, 1668–1738; chemist, botanist; held professorships in botany, chemistry, and medicine at the University of Leiden; FRS (1730). QE21..

Bollaert, William, 1807–1873; chemist, South American traveller and mine assistant. A39, 44

Boott, Francis, 1792–1863; physician and botanist; secretary, Linnean Society 1832–39. QEIFC, 21..

Bowerbank, James Scott, 1797–1877; geologist with special interest in London clay fossils; FRS (1842). A93

Brayley, Edward William, 1802–1870; writer on science, chemist, geologist; FRS (1854). A41

Brehm, Christian Ludwig, 1787–1864; German clergyman and ornithologist. QE18

Broderip, William John, 1789–1859; magistrate and naturalist; FRS (1828). D55–57

Brown, Robert, 1773–1858; botanist; supervised the botanical collection at the British Museum (1827–58) where he advised Darwin on the latter's specimens from the *Beagle* voyage; FRS (1811). RN161 C148, 237, 239, 266 QEIFC, 1^{v}, [5]a, 15, 21..

Brown, William, 1777–1857; Admiral in the navy of Buenos Aires. RN155

Bunbury, Charles James Fox, 1809–1886; botanist and palaeobotanist; collected plants in South America, 1833–34, and in South Africa, 1838–39; FRS (1851). S5, 6 QE21..

Burchell, William J., 1782–1863; naturalist who collected in southern Africa (1811–15) and South America (1825–29). RN134

Butler, Joseph, 1692–1752; Bishop of Durham. OUN52, 54

Button, Jemmy; native Fuegian who returned to Tierra del Fuego from England on board H.M.S. *Beagle* in 1833. N15

Bynoe, Benjamin, 1804–1865; naval surgeon, 1825–63; assistant and later acting surgeon aboard H.M.S. *Beagle*, 1831–36. RN141 M79, 142

Campbell, Thomas, 1777–1844; poet, prominent in literary and educational circles in London. A26

Candolle, Augustin Pyramus de, 1778–1841; Swiss botanist; foreign member, Royal Society (1822). Mac57ᵛ

Carlyle, Thomas, 1795–1881; essayist and historian. B255

Charlesworth, Edward, 1813–1893; naturalist and palaeontologist; editor, *Magazine of Natural History*. QE[5]a, [15ᵛ]

Clive, Edward, 1st Earl of Powis, 1754–1839; Lord Lieutenant of Shropshire, 1804–1839; his seat, Powis Castle, is in the vicinity of Shrewsbury. B140

Clive, Miss; presumably a member of one of the branches of the Clive family, which also included Edward Clive. N61

Comte, Isidore Auguste Marie François Xavier, 1798–1857; mathematician and philosopher of science; founder of positivism. OUN25

Conybeare, William Daniel, 1787–1857; geologist; FRS (1819). E182

Corbet, Dryden Robert, 1805–1859; of Sundorne Castle, Shropshire. M1, 9, 10, 29, 42, 44, 156

Covington, Syms, 1816?–1861; Darwin's servant and assistant 1833–39; emigrated to Australia in 1839. M112

Cuming, Hugh, 1791–1865; naturalist; collected shells and living orchids in the Pacific, on the coast of Chile, and in the Philippine Islands; returned to England in 1839. QEIFC ZEd13

Darwin, Caroline Sarah. See Wedgwood, Caroline Sarah.

Darwin, Emily Catherine, 1810–1866; Darwin's sister, became Charles Langton's second wife in 1863. M28, 33, 83

Darwin, Emma, 1808–1896; daughter of Elizabeth and Josiah Wedgwood II; Darwin's wife and cousin, married 29 January 1839. D21, 176, 177 E125 N61, 73, 112

Darwin, Erasmus, 1731–1802; physician, botanist, poet; Darwin's grandfather; FRS (1761). EIBC M13, 15, 131, IBC

Darwin, Erasmus Alvey, 1804–1881; Charles Darwin's elder brother; medical degree Edinburgh 1828; qualified but never practised as a physician; well connected in London literary circles and knowledgeable in the natural sciences, particularly chemistry. RN115 A118, 119 C183, 266 D10, 63 E125 M28, 83, 111, 128 N33

Darwin, Francis, 1848–1925; Darwin's son, botanist; edited Darwin's letters; FRS (1882).

Darwin, Francis Sacheveral, 1786–1859; Darwin's half-uncle, son of Erasmus Darwin by his second marriage. B136

Darwin, Marianne. See Parker, Marianne.

Darwin, Robert Waring, 1766–1848; Darwin's father; physician with a large practice in Shrewsbury, resided at The Mount; married Susannah, daughter of Josiah Wedgwood I, in 1796; FRS (1788). D1–3, 151, 180 E67, IBC T151, 176 QEIFC, 4, 6, 7 M1, 2, 7, 9, 10, 15, 16, 18, 21, 22, 24–26, 29, 35, 42-44, 50, 55, 78, 83, 90, 156, IBC N1, 39, 53, 60, 63 OUN31, 32

Darwin, Susan Elizabeth ('Granny'), 1803–1866; Darwin's sister, lived at The Mount, Shrewsbury, until her death. M33

Darwin, Violetta, 1783–1874; Darwin's aunt; married Samuel Tertius Galton. B175

Deshayes, Gérard Paul, 1797–1875; French palaeontologist. A85 E182

Dewar, possibly A. Dewar of Fife; Scottish botanist. M78

Diard, Pierre Medard, 1794–1863; French naturalist, travelled in Indochina. C17

Dick, Thomas Lauder, 1784–1848; geologist and writer. GR78

Don, David, 1800–1841; Professor of Botany, King's College, London, 1836–1841; Linnean Society librarian, 1822–41. B79, 187, 188, 191, 192, 193 E100 ZEd8, QE10

Don George, 1798–1856; botanist; collected plants for the Horticultural Society of London, 1821–23, in Brazil, West Indies, Sierra Leone and São Tomé. B79

Donn, probably James, 1758–1813; botanist; curator of the Cambridge Botanic Garden, 1790–1813. ZEd8

Douglas, Gavin, *c.* 1474–1522; poet, Latin scholar. N39

Drinkwater, Richard, 1806–1853; woolstapler of Shrewsbury, Mayor of Shrewsbury. D42 M129

Duchesne, Henri-Gabriel, 1739–1822; ecclesiastical administrator and archivist; author of *Manuel du naturaliste* (1770). Or Antoine-Nicholas Duchesne, 1747–1827, author of *Manuel de botanique* (1764). E183 QE1

Duvaucel, Alfred, 1793–1824; French naturalist, travelled to East Indies and India. C17

Eyton, Thomas Campbell, 1809–1880; Shropshire naturalist, close friend of Darwin since childhood. B248 DIFC, 4, 26 E168, 169 QEIFC, 9, 18, 22 N117 OUN36

Falconer, Hugh, 1808–1865; palaeontologist and botanist; specialist in Indian materials; FRS (1845). QE17

FitzRoy, Robert, 1805–1865; naval officer, hydrographer, meteorologist; Commander and from July 1835 Captain of H.M.S. *Beagle*; FRS (1851). RN16, 35, 67, 135, 137 E42 M43, 60

Fleming, John, 1785–1857; zoologist, geologist, clergyman. QE11

Ford, Richard Sutton, 1785–1850?; farmer of Newstead, near Trentham, Staffordshire; agent to the Fitzherbert estate at Swynnerton, Staffordshire. QE3

Foster, Henry, 1796–1831; naval officer, astronomer; FRS (1824). RN138

Fox, William Darwin, 1805–1880; clergyman, Darwin's second cousin and close friend at Cambridge and after, shared an enthusiasm for entomology and natural history generally. B83, 141, 159, 176, 177, 182–84 D5–15, 70 M49, 97, 128

Fraser, Louis, *fl.* 1866; naturalist, Curator of the Zoological Society until 1841; naturalist to the Niger expedition, 1841–42. C107

Galton, John Howard, 1794–1862; of Hadzor House, Worcestershire; High Sheriff of Worcestershire, 1834; married Isabella Strutt in 1819; brother of Samuel Tertius Galton. B175, 183

Galton, Samuel Tertius, 1783–1844; married Violetta Darwin. B175

Gillies, John, 1792–1834; physician and naturalist, resided in Mendoza in 1820s. A38

Gloger, Constantin Wilhelm Lambert, 1803–1859; zoologist and ornithologist. QE18

Gould, John, 1804–1881; ornithologist; described the birds Darwin collected on the *Beagle* expedition; FRS (1843). RN127, 130 B50, 171, 249 C15, 71, 80–82, 88, 109, 110, 113, 114, 143, 144, 165, 189, 239 D61, 102 QEIFC, 10a, 17

Gowen, James Robert, ?–1862; hybridized rhododendrons, amaryllis. QEIFC, 12

Granny. See Darwin, Susan Elizabeth.

Gray, John Edward, 1800–1875; naturalist; Assistant Zoological Keeper at the British Museum, 1824; Keeper, 1840–75. FRS (1832). B9 QEIFC, 12, 17

Greenough, George Bellas, 1778–1855; geologist; a founder of the Geological Society of London; FRS (1807). E173

Haller, Victor Albrecht von, 1708–1777; Swiss anatomist, physiologist and botanist. N93

Hamilton, William John, 1805–1867; geologist; succeeded Darwin as Secretary of the Geological Society; FRS (1855). C214

Harrison, Mrs; Darwin family acquaintance. M1

Hartley, David, 1705–1757; physician, philosopher; FRS (1736). OUN33, 49, 50

Henry, Samuel P., 1800–1852; ship's captain whom Darwin met at Tahiti. RN12

Henslow, John Stevens, 1796–1861; clergyman, botanist, and mineralogist; Professor of Mineralogy at Cambridge University, 1822–27; Professor of Botany, 1825–61; instructed Darwin in natural history at Cambridge University and arranged for him to be offered the position of naturalist aboard H.M.S. *Beagle*. RN127 A49, 97 B68, 198, 230, 234 C16, 92, 100, 192, 265 E100, 129, 143, 162, 165 QEIFC, 2, 7, 11, 12, 15, [15^{v}] N74, 76

Herbert, William, 1778–1847; naturalist, horticulturist, classical scholar, linguist, clergyman; studied plant hydridization; rector of Spofforth, Yorkshire, 1814–40; Dean of Manchester, 1840–47. B91 D25 E103, 106, 110, 141, 143, 151, 183

Heron, Robert, 1765–1854; politician, naturalist; collected a menagerie; MP for Peterborough, 1820–47. DIFC QE18

Herschel, John Frederick William, 1792–1871; astronomer, chemist, philosopher of science; carried out astronomical observations at the Cape of Good Hope, 1834–38; FRS (1813). RN32, 115 A46, 104 D151 E59

Herschell, William, 1738–1822; astronomer. FRS (1781). A121 D151

Holland, Henry, 1788–1873; fashionable London physician, essayist, distant relative of Charles Darwin through the Wedgwoods; FRS (1815). RN159 E183 QEIFC, 16

Home, Henry, Lord Kames, 1696–1782; Scottish judge, author of *Sketches of the History of Man*. OUN49

Hooker, Joseph Dalton, 1817–1911; botanist; Darwin's close friend and frequent correspondent; FRS (1847). QEIFC, 1, [9v], 10a, 17

Hope, Frederick William, 1797–1862; entomologist and clergyman; FRS (1834). B248 C233 E11

Hopkins, William, 1793–1866; Cambridge University mathematician and geologist. FRS (1837). A137

Horner, Leonard, 1785–1864; geologist; father-in-law of Charles Lyell; FRS (1813). QE17

Horsfield, Thomas, 1773–1859; naturalist; born Bethlehem, Pennsylvania; served Dutch and British in East Indies, 1799–1819; FRS (1828). B164 QEIFC, 12

Hubberly. See Abberley

Humboldt, Friedrich Wilhelm Heinrich Alexander von, 1769–1859; German naturalist, traveller, natural philosopher and geographer whose scientific travels in South America, 1799–1804, inspired Darwin to emulation; foreign member Royal Society (1815). RN157 E44, 45

Hunt, ———; Keeper at the Zoological Gardens of the Zoological Society of London in Regent's Park. M138.

Hunter, John (1728–1793); surgeon and anatomist; his collection of zoological specimens formed the basis of the Hunterian Museum of the Royal College of Surgeons; FRS (1767). B51, 161 C215, 267, 270 D57, 67, 112–6, 152, 156–59, 161, 163, 173–75 E26 QE21.. M44, 147 N43

Hutton, James, 1726–1797; Scottish natural philosopher and geologist. RN112, 161 A77 C119

Jenyns [Blomefield], Leonard, 1800–1893; naturalist and clergyman; described Darwin's collection of fish from the *Beagle* voyage; adopted the name Blomefield in 1871; a friend of Darwin's at Cambridge. B217 C20

Jones, R.; Shropshire or Staffordshire clergyman. D41–43, 104 M155

Jones, Thomas, 1775–1852; a founder of the Astronomical Society and member of the Athenaeum Club; FRS (1835). D151

Kames, Lord. See Henry Home.

Kant, Immanuel, 1724–1804; philosopher. E57 OUN33

Keilhau, Baltazar Mathias, 1797–1858; Norwegian stratigrapher, petrologist, and geologist. A68

King, Philip Parker, 1793–1856; naval officer and hydrographer; Commander of the *Adventure* and *Beagle* on the first surveying expedition to South America, 1826–30; FRS (1824). C61, 99 ZEd12–13

Kirby, William, 1759–1850; clergyman and entomologist; FRS (1818). Mac59

Knight, Thomas Andrew, 1759–1838; horticulturist; President of the Horticultural Society, 1811–38; FRS (1805). E130, 148, QE11, 21..

Kuhl, Heinrich, 1797–1821; naturalist; collected in Dutch East Indies. C17

Lamarck, Jean Baptiste Pierre Antoine de Monet de, 1744–1829; naturalist; held botanical posts at the Jardin du Roi, 1788–93; Professor of Zoology, Muséum d'Histoire Naturelle, 1793; major exponent of transformism. B9, 214 C119, 123, 157, 158 OUN37

Le Brun, Charles, 1618–1690; French artist. N103

Leighton, Francis Knyvett, 1772–1834; army officer; a close friend of Robert Waring Darwin; Mayor of Shrewsbury, 1834. M1

Leighton, William Allport, 1805–1889; Shropshire botanist, clergyman, and antiquary. EIBC

Leschenault de la Tour, Jean Baptiste, 1773–1826; botanist, travelled extensively in Australia, East Indies, India, Africa and South America. C17

Lindley, John, 1799–1865; botanist and horticulturist; Editor of *Gardeners' Chronicle* from 1841; FRS (1828). QE21..

Locke, John, 1632–1704; philosopher; FRS (1668). M84 OUN33

Lonsdale, William, 1794–1871; geologist; served the Geological Society of London in various capacities including librarian. C175–177 E6, 119–121, 126 M97

Lowe, Richard Thomas, 1802–1874; clergyman and naturalist; specialist on Madeira. QE18

Lyell, Charles, 1797–1875; uniformitarian geologist; Charles Darwin's friend and chief scientific mentor in the late 1830s; FRS (1826). RNIFC, 115, 129, 137, 145 A7, 14, 46, 68, 85, 86, 153 B153, 170, 272 C39, 74, 84, 119, 135, 153, 176, 178, 266 E6, 59, 65, 86, 100, 101, 134 S3r ZEd10 N115

MacLeay, William Sharpe, 1792–1865; zoologist and diplomat; known for quinary system of taxonomy. B154, 179, 180 C55, 56, 153, 213, 249 D153 E22, 23

Macphee, Donald; presumably a resident of Scotland in the vicinity of Glen Roy. GR102

Malaspina, Alessandra, 1754–1809; Spanish navigator of Italian birth. RN181

Malcolmson, John Grant, 1802–1844; geologist, surgeon; served as Secretary to the Medical Board, Madras; FRS (1840). A142, 146

Malthus, Thomas Robert, 1766–1834; clergyman and political economist; argued the relation between population growth and food supply; FRS (1818). Mac28v, 57v

Mark; Dr Robert Darwin's coachman at Shrewsbury. D3 QE‹15› OUN36v

Marsh, Arthur Cuthbert, 1786–1849; banker; husband of Anne Marsh-Caldwell the novelist. E108

Martin, William Charles Linnaeus, 1798–1864; Superintendent of the Museum of the Zoological Society, 1830–38. B165

Mayo, Herbert, 1796–1852; physiologist, anatomist; FRS (1828). M110

Miller, Alexander; Superintendent, Zoological Society Gardens, London (1829–52). D87 T2 QE4, 20 M137

Miquel, Friedrich Anton Wilhelm, 1811–71; prolific author of botanical works. QE17

Mitchell, Thomas Livingston, 1792–1855; Australian explorer and Surveyor-General, New South Wales, 1828–55. A92 C130, 131, 159, 213

Monro, Alexander III, 1773–1859; anatomist; physician; Darwin attended Monro's lectures at Edinburgh University. M144

Murchison, Roderick Impey, 1792–1871; geologist, founder of the Silurian system; FRS (1826). RN142 A36

Nancy; Darwin's childhood nurse who was kept on as a servant of the Darwin family at The Mount, Shrewsbury. M53

Ogilby, William, 1808–1873; Irish barrister and zoologist; studied fossil mammals from the Stonesfield Slate of England. B163 D29, 30, 34

Orbigny, Alcide Charles Victor Dessalines d', 1802–1857; French palaeontologist who explored and collected in South America, 1826–34. C18, 129, 165

Owen, Richard, 1804–1892; comparative anatomist and palaeontologist; Hunterian Professor at the Royal College of Surgeons, 1836–56; described the *Beagle* fossil mammal specimens; FRS (1834). RN113, 127, 129 B19, 113, 161, 163, 231 C3, 72, 131, 266 D29, 35, 113 E136 QEIFC, 17

Parish, Woodbine, 1796–1882; diplomat, geographer; represented the British Government in Buenos Ayres 1823–32; FRS (1824). RN126, 142, 143, 155, 157 A38

Parker, Marianne, 1798–1858; Darwin's eldest sister, married Henry Parker in 1824. N37

Peacock, George, 1791–1858; mathematician; Lowndean Professor of Geometry and Astronomy at Cambridge University, 1837; Dean of Ely, 1839–58; FRS (1818). M100

Phillips, John, 1800–1874; geologist and palaeontologist; Professor of Geology, King's College London, 1834–40; palaeontologist to the Geological Survey, 1840–44; FRS (1834). E121, 122

Plato, 427–348 BC; philosopher. E76 M128

Power, Mrs; presumably a resident of Port Louis, Mauritius. RN17

Powis, Earl of. See Clive, Edward.

Priestley, Joseph, 1733–1804; chemist and Unitarian theologian FRS (1766). M42

Richardson, John, 1787–1865; Arctic explorer and naturalist; FRS (1825). C129 E52

Roberts, Wilson; a correspondent of Darwin's father. OUN 31

Rosales, Francisco Javier, d. 1875; Chilean Chargé d'Affaires to Paris from 1836–53. RN155

Rosales, Vincente Perez, 1807–1886; author, colonization agent for the Chilean Government in Europe. RN155

Roussin, Albin-Reine, 1781–1854; French naval commander and later admiral; hydrographer. RN22

Roy, Nead; presumably a resident of Scotland in the vicinity of Glen Roy. GR102

Royle, John Forbes, 1799–1858; surgeon and naturalist; surgeon in East India Company's service; superintendent of garden at Saharunpore, 1823–31; FRS (1837). B242, 273 QEIFC, 12

Rozales. See Rosales, Francisco Javier, and Rosales, Vincente Perez. RN155

Sagaret, Augustin, 1763–1862, agronomist. QE3

Schomburgk, Robert Hermann, 1804–1865; traveller; explored British Guiana; FRS (1859). Or Richard Schomburgk, 1811–1890; botanist. QEIFC

Schouw, Joakim Frederik, 1789–1852; botanist. C266

Sebright, John Saunders, 1767–1846; Whig politician and agriculturist; published on animal breeding. C120 D108

Sedgwick, Adam, 1785–1873; geologist, clergyman; Woodwardian Professor of Geology, Cambridge University; gave Darwin early field training in geology; FRS (1821). RN93

Selby, Prideaux John, 1788–1867; ornithologist; cofounder, *Magazine of Zoology and Botany*, 1837. D102

Sellow, Friedrich, 1789–1831; Prussian naturalist, and particulary botanist, who collected in Brazil and Uruguay, 1814–31. A151

Silliman, Benjamin, 1779–1864; American chemist, geologist, and mineralogist. C267

Smith, Andrew, 1797–1872; zoologist; army surgeon stationed in South Africa, 1821–37; FRS (1857). RN32 A34, 94 B32–34, 68, 83, 167, 180, 182, 233, 256 C35, 129, 150, 227, 228 D136, 137, 139, 168 Tfrag11 QEIFC, 16 N101

Smith, James ('Smith of Jordanhill'), 1782–1867; geologist and man of letters; FRS (1830). QEIFC

Smith, Sydney, 1771–1845; essayist. E108

Socrates, 470–399 BC; philosopher. E76

Sorrell, Thomas, *c.* 1797–?; boatswain on voyage of H.M.S. *Beagle*. RN99

Sowerby, George Brettingham, 1788–1854; conchologist and artist. E45, 76

Sowerby, James de Carle, 1787–1871; fossil conchologist who examined some of Darwin's specimens from the *Beagle* voyage. RN144 C245, 246 E120

Spenser, Edmund, 1552?–1599; poet, author of 'Faerie Queene' quoted in Hensleigh Wedgwood, *On the Origin of Language* (1866). N39

Stephens, James Francis, 1792–1852; entomologist and zoologist. B256

Stokes, John Lort, 1812–1885; naval officer; served in the *Beagle* as midshipman, 1826–31; mate and assistant surveyor, 1831–37; later admiral. N26

Strutt, Mr; cousin of John Howard Galton. D9

Sulivan, Bartholomew James, 1810–1890; naval officer and hydrographer; Lieutenant on voyage of H.M.S. *Beagle*, 1831–36; surveyed the Falkland Islands, 1838–46. D48 QE11v

Sykes, William Henry, 1790–1872; naturalist and soldier; served in the India military service of the East India Company; FRS (1834). D99 QEIFC, 12

Taylor, Richard, 1781–1858; printer, naturalist, and editor, *Scientific Memoirs*, 1837–52; established *Annals of Natural History* in 1838. C104, 275 QE16

Thompson, Henry Stephen (later Meysey Thompson), 1809–1874; agriculturist and a founder of the Royal Agricultural Society, 1838. C231

Tollet, George, 1767–1855; of Betley Hall, Staffordshire; Justice of the Peace of Staffordshire; agricultural reformer, friend of the Wedgwoods and Darwins. E15 QE19

Turner, Edward, 1798–1837; chemist; FRS (1830). RN156

Van Voorst, John; publisher of natural history works, London. QE18

Voltaire, Francois-Marie Arouet de, 1694–1778; French author and philosopher; FRS (1743). M145

Waterhouse, George Robert, 1810–1888; naturalist; curator, Zoological Society of London, 1836–43; described mammals and entomological specimens from the *Beagle* voyage. A14 B71, 162, 166, 249, 250, 256, IBC C22, 36, 95, 107, 116, 162, IBC D29, 61 E170 T55, 57, 59 ZEd9 M98

Webster, Thomas, 1810–1875; barrister; author of works on the physical sciences; Darwin's contemporary at Cambridge University; FRS (1847). A55

Wedgwood, Caroline Sarah, 1800–1888; Charles's elder sister, married her cousin Josiah Wedgwood III on 1 August 1837. RN125 M56

Wedgwood, Elizabeth (Bessy), 1764–1846; eldest daughter of John Bartlett Allen, wife of Josiah Wedgwood II whom she married in 1792, and mother of Emma, Darwin's wife. M8, 12, 26, 79

Wedgwood, Emma. See Darwin, Emma.

Wedgwood, Ernest Hensleigh, 1838–1898; son of Hensleigh and Francis Wedgwood. N37, 121

Wedgwood, Frances (Fanny), 1800–1889; daughter of James and Catherine Mackintosh; married Hensleigh Wedgwood in 1832. M53

Wedgwood, Frances Julia (Snow), 1833–1915; writer; daughter of Hensleigh and Frances Wedgwood. N121

Wedgwood, Hensleigh, 1803–1891; philologist and barrister; B.A., Christ's College, Cambridge and Fellow 1829–30; son of Elizabeth (Bessy) and Josiah Wedgwood II; married Frances Mackintosh in 1832. C165, 244 E144 T51 ZEd20 M58, 61, 96 N37, 39, 81

Wedgwood, John, 1766–1844; banker and horticulturist; a founder of the Horticultural Society of London, 1804; son of Sarah and Josiah Wedgwood I. E13–17, 98

Wedgwood, John Allen (Allen), 1796–1882; son of John and Louisa Jane Wedgwood; Vicar of Maer, Staffordshire, 1825–63. S1r M79

Wedgwood, Katherine Euphemia (Effie), 1839–?; daughter of Hensleigh and Frances Wedgwood. N121

Wedgwood, Sarah Elizabeth (Sarah), 1778–1856; youngest daughter of Sarah and Josiah Wedgwood I; Darwin's aunt. M83

Wedgwood, Sarah Elizabeth (Elizabeth), 1793–1880; eldest daughter of Elizabeth (Bessy) and Josiah Wedgwood II. E144 S1[v]

Wellsted, James Raymond, 1805–1842; Lieutenant in the Indian navy; traveller and surveyor; FRS (1837). A133

Whitby, Mary Anne Theresa, 1784–1850; landowner, artist, silk producer. QEIFC

Willis, M.; hairdresser in Great Marlborough Street when Darwin lived at no. 36. C232 D24, 163, 173

Wilson. See Roberts, Wilson.

Wombwell, George, 1778–1850; owner of travelling menageries. QE20

Woodward, Samuel Peckworth, 1821–1865; naturalist and geologist; Sub-curator, Geological Society, 1839–45. E120

Wordsworth, William, 1770–1850; poet. M40 N57

Worsley, Thomas, 1797–1885; Master, Downing College, Cambridge. N4, 74

Wynne, Mr; not positively identified, but often mentioned by Darwin especially in connection with breeding of animals; resident of North Wales or Shropshire; several families of Wynne lived in and near Shrewsbury in the 1830–40 period. B139, 141 C106, 120, 133, 134 M25 OUN36[v] QE18

York Minster; native Fuegian who returned from England to Tierra del Fuego in 1833 aboard the H.M.S. *Beagle*. C244 M135

Yarrell, William, 1784–1856; zoologist, bookseller, newspaper agent; author of standard works on British birds and fishes. B138, 140, 171 CIFC, 4, 68, 71, 120, 121, 202 DIFC, 8, 29, 34, 72, 95, 99, 100, 102–4, 108, 152, 153, 155, 168 E35, 110, 112, 141, 168, 169, 174 QEIFC, 4, 18 M138

Subject Index

Aberrant species
  See Species, aberrant
Abhandlungen der Königlichen Akademie der Wissenschaften zu Berlin A151
Aborigines
  See Australia
Abortive organs C215–16
  bones QE17
  callosities QE20
  function B84, D59, 172, E57, MAC58^{v}
  in birds B70, 162, 251, C25, 57, 114, 161, 215–16, D29–30, 59, 60, 140
  inheritance of D132, T37, MAC53^{v}
  in insects B84, C186, 215, 216, D116, 172, MAC58^{v}
  in mankind B84, 99, C215, D132, 158, 169, 172, E57, 66–7, 89, 147, MAC54^{v}, 58^{v}
  in plants C267, D162, E129–30, T99, QE13, 14, 21
  in reproductive system D157, 158–9, 162, E129–30, T99, QE13, 14
  larvae of animals with D57
  likelihood of their disappearance D132
  origin B99, D162, 166, E57
Abrolhos, Brazil RN10, C100
Acalepha D156, E164
Acanthosoma M64
Acclimatisation C253, 265, E111–12
  See also Adaptation; Domestication; Selection, artificial
Acrita C48–9
Aconcagua, Chile
  deposition of rivers GR30
Acting N24, 121, OUN11^{v}
Adaptation B46, 55, C57, 72, 84, 140–1, 208, 236, E91–2, 157, T37, MAC58^{v}–59, S7^{r}–7^{v}
  and design MAC53^{r}–53^{v}, 54^{r}
  and extinction B38, 90
  and generation B3, 4, 78
  and monstrosities C65, 83–7, D107
  and the origin of species RN133, B64, C64, D135
  as a trifling character B69–70, 194
  burrowing marine animals ZEd1
  fine balance of C73, E9
  law of B21, 45
  leads to simplified organisms as well as more complex ones E95–7
  mechanism B227, D135
  necessary MAC57^{r}, 57^{v}
  not perfect B45, 115, C174, D167
  of Annelidae B143
  of eggs of Taenia D49
  of fish B113, 143, E134, T57, MAC29^{v}
  of fossil animals B99, 205–6
  of seeds E137
  of seeds for transportation D74, MAC53^{r}, QE[5]a, 6, 10, [15^{v}]
  of tradesmen M44–5
  of varieties B236, C4, D107, MAC57^{r}–57^{v}
  overdevelopment of C109
  perfect E57, MAC54^{r}
  prevented by crossing B210–11, C60
  to desert conditions A5, C88
  to external conditions of existence B24, 45, 55, 113, 205–6, 235–6, C72, 84, 160, T37
  to nocturnal way of life C186
  to snow C252
  webbed feet of shrews C252
  See also Cause, final; Colour; Conditions of existence; Domestication; Economy of nature; Monsters; Selection, artificial; Selection, natural; Selection, sexual; Variation; Varieties
Affinity
  a real relationship B162, 243
  caused by adaptation to similar conditions B24, 43
  Darwin's theory explains B228, C62–3
  distinguished from analogy C115, 139–40, 151, D26
  gives appearance of circular relationships between organisms B139–40
  indicated by the generative organs E91–2
  shown by structure not habits B162, C81–2, 86–7
  See also Analogy; Classification; Quinarian system
Africa C269, D50
  animals B133, 157, C21, 40, 46, 233, 249, D72, E60, QE17
  climate C38, E39–41
  fossil mammals B72, C132, D25, T79
  monstrous forms common in E88
  native species found in other countries B192, 218, 242, C183–4, D25
  plants B194
  relationship of animals with those of the East Indies B242, C38, E39
  relationship of animals with those of Europe C37, 233, E72
  relationship of animals with those of India B242, C46
  shells A20, B157, C246, E41, 90
  See also Africa, West; Africa, South; Egypt; Fernando Po; Madagascar; Negro
Africa, South
  animals RN113, B62, 67, 94, 249, C18, 116, ZEd14
  geology RN32, B83
  organisms have not arrived from the north B77
  relationship of animals with those of South America C18, 116
  zebra B62, 72
  See also Africa; Cape of Good Hope
Africa, West B190, 223, C224, 246, 249, E90–1, 167
  See also Africa; Cape Verde Islands
Agate A53
Agouti B80
Aisa caespitosa E165
Albatross D47
Albinoes B119, C85, E176
Albite A66
Alcedo C26
Algae
  See Seaweeds
Alleghanies, USA
  wren C255
Alpine plants
  Darwin's theory of origin of B195, C85, 168, 191, E104–5, 115–16, 165, TFRAG6, QE21..

Alpine plants (*cont.*)
structure reflects conditions at different altitudes E72
See also Antarctica; Arctic
Alps
erratic blocks A20, 110
fossils C178
geology RN100
ibex TFRAG6
rivers A116
Amaryllidaceae B191, E111
Amazon, river basin
geology RN100
Amber
insects in ZEd18
Amboina (Amboyna), Moluccas C13, 14, 18, 24
Ambylrhynchus E168, ZEd19
America
animals A3, B66, C46, 221, D103, E42, TFRAG6, M85–7, ZEd14
birds B47, 98, C36, 69, 254, 256, 267
climate C38, E37–41
fossils A117, B72, 223, C37, 247, E11, 105
geology RN15, A5, C38, E38
plants B194, 218, C224, E166
relationship between fauna of North and South B66, 98, C37, 67, 105, 132, 183, D39, E37, 42
See also America, North; America, South
America, North B174, C269, D31
alluvial plains of Mississippi river A10
animals A57, B134, C40, 69, 175, 251, 266, D31
animals compared to those of Asia B65, 242
animals compared to those of Europe B47, C36, 225, 246, 247, 256, D96, 102, E100, 109, S3[r]–3[v]
birds B98, 199, C36, 67, 105, 225, 255, 256
climate C37, E42
fossils B125, 172, 174, C37, 247, E32, 100
fossils compared to those of Europe C246, 247, E100, S3[r]–3[v]
geology C247, M131
lakes B251
plants C265, E116, QE21..
sandbars A130
shells A37, B172, 174, C246, D39, E100, 109
America, South
animals B94, 250, C18, 42, 245, 252, 257, D38, 61, 99, E13, 32, T81', ZEd9, 12, 13
animals compared with those of Australia B137, C15, 37, 116, D72, E41
animals compared with those of South Africa C18, 116, 257
birds B13, C50, 96, 157, 253, 256, TFRAG10, ZEd6, 7, 16
climate C168, E37, 38
different mineralogical character of rocks to those of Europe RN125
fish C183, E35
fossil mammals RN85, B13, 53–4, 60, 62, 63, 133, 135, 137, C37, 39, 167, 247, D72, E40, T79
geographic limits of birds and animals RN128
horses RN85, 113, A3, B62, 63, 72, 106, 133, 222, N67
insects C197, D111
marsupials D73, E41
monkeys M143, N183
relationship of animals and plants to those of Galapagos Islands C54, 209, E100
representative species in B8, 13
shells RN36, 106, A33, 39, 72, 102, 112, 144, B95, C244, 245, 246, D39, ZEd8, 13, 15
See also America; America, North; Cordillera; Fossils; Geographical distribution; Geology; Patagonia; Species; Tierra del Fuego
American Journal of Science RN142, C27, 254, IBC, N102
Ammocoetus E97
Ammonites E121
Amphidesmas C245
Anagallis C194
Analogy (biological)
a divalent power C60, 61
a guide to species classification C127–30
arguments against a theory of B110
between adaptations in different species B79, 87, C107–8, 111–15, 143, 208, D132, MAC28[r], 29[v], 58[v]
characteristics are recently acquired C202
distinguished from affinity B57–8, 139, C115, 139–40, 151, D26
exhibited between aberrant species C61
similar to degradation B129
See also Adaptation; Affinity; Species, aberrant
Analogy (philosophical)
Darwin's theory rests on C176–7, E128
Anapsis T106
Anastatica MAC53[v]
Anatidae D25, 26
Anatifera ZEd5
Anatomy
See Morphology
Ancestor B112, D179, E117, T1
classification reveals B206, 207, C52, 186, MAC29[v]
Darwin's theory does not require going back to first stock D21
of aquatic mammals E92
of birds C52, 206–8
of fish B88
of mammals B87, 88
of mankind B119, C178, 243, D137, M123, N10, 27, 41
of reptiles B88
of vertebrates E89
of zebra B72
Ancestor, common B14, 35, 43, 204, 232
of armadillo and megatherium B54
of birds and mammals B26, 42, 112, 263
of fish and mammals B97
of insects B58
of mankind C154
of pig and tapir B86–7
See also Mankind; Progression; Transmutation; Tree of Life
Anchusa T111, QE[11[v]]
Andes
See Cordillera
Andite RN138
Anemone QE[15[v]]
Anger M69, 70, 84, 94
akin to insanity M13–14
associated with tiredness M51, 131
Darwin's feelings of M51
existence in ancestor of man and animals OUN29, 51
facial expression C212, M84, 147, 149, 152–3, N6, 46, 113
in dogs C212, E36, M147

in monkeys M94, 107, N94
movement of blood during N14
nature of M12, 14, 51, 122, 124, 148, N8, OUN43, 51
sensation of N66
Anglesea
geology RN5–7, A97
Animal magnetism
See Hypnosis
Animals
can live only on matter already organized C104
cannot blush N15, 52, 60
dependence on plants B108–9
feel emotion just like mankind B231
feel pleasure in music N64
gradation between human beings and C55, 222–3, 244, D22
gradation between plants and A180, B43, 73, 110, OUN35
have no notion of beauty C71
mode of generation different to that of plants D167
vary in same manner as plants B210
Annales des Sciences naturelles RN65, 174, B31, 54, 112, C16, 143, 146, 178, 268, QE16
Annales des Sciences naturelles. Botanique B198
Annales du Muséum National d'Histoire Naturelle. Paris E13, 142
Annals and Magazine of Natural History RN174, D101, ZEd5, QE17. See also *Annals of Natural History*; *Magazine of Natural History*
Annals of Natural History B31, 234, C16, 95, 96, 100, 188, 214, 225, 268, D48, 61, 62, 151, 167, 169, E87, 89, 90, 91, 92, 96, T4, 25, M142, QE7, [15^v], 17
Annelidae B126, 143, D117, 156
Annulosa D55
Anoplotherium B250
Antarctica
currents QE17
fish C28
large marine mammals MAC29^r
living organisms at great depth RN114
origin of plants B195
vegetation RN125
Anteaters B135, C257
Antelopes B14, C14, 93
Anticlines RN59, 175, A66, 107
Antiquity
Aristotle C267
Greek philosophy M135
ideas of beauty N27
intellect of Greeks N47
Plato and Socrates E76
Temple of Serapis A95
See also Egypt
Antilles, West Indies E42
Ants
MIFC, 62–3
Aphids B181, D40, 41, 175, QE1^v
Aphrodite MAC28^r
Aphysia MAC28^r
Apion radiolum M63
Apples D178, E14–15, 17, 110, QE13, 21
death of golden pippen trees B63, 72, D165
fertilization E15, 110, QE1^v
golden pippen B83, 230, D70
grafting E17
reversion to crab B229, 230
Ribston pippen B83, 229–30, D70
Apricots QE23
Apteryx
abortive bones B162, QE17
feathers C3
structure B70, 162, 251, C3, 51, 215, D29–30, 59, E135
taxonomic relationships B162, C74, 207, 208, T65
Aquilegia QE1
Arabia C92–3
Araucaria B187
Archaeology
arrow heads N26
Archipelagos
See Islands
Architecture
imitates nature OUN11b^v
ugliness of new buildings N30–1
amplifies ideas in nature OUN6
See also Scenery
Arctic A8, B153, 157, 221, C36, 84, 116, D133, E166, MAC29^r, 29^v, ZEd6, QE17
plants A7, B247, C224, D24, E109, 115, 151–2, QE21..
See also Alpine plants; Antarctica; Greenland; Iceland; Lapland; Norway; Siberia; Spitzbergen; Sweden
Argentina
See Bahia Blanca; Buenos Aires; Córdoba; La Plata; Mendoza; Pampas; Port Desire; Tucumán; Ventana
Arica, Chile
geology A99
Arion ZEd13
Aristolochia QE14
Aristotle C267
Armadilloes A9, B20, 54, 69–70, 99, TFRAG4
Aroe (Aru) Island, Indonesia C13
Art
enjoyment of OUN10, 20, 23
not based on nature OUN11^v, 11b, 11b^v
object of OUN11b, 22
See also Acting; Architecture; Beauty; Music; Painting; Poetry; Photography; Singing; Sublimity
Articulata B27, 43, 208, 225, 252, D49
Artificial Selection
See Selection, artificial
Arum QE14
Ascension Island RN79, 129, B31
animals D65, E35, ZEd13
calcareous encrustations on rocks RN93
geology RN17, 42, 45, 60, 72, 90, 107, A40, 60
plants RN79
Ascidia C49
Asclepias E133, 165, QE14, 21..
Ash tree E17
Ash, volcanic RN67, 72
See also Dust
Asia B65, 136, 142, C116, 214, E11, 20, 41, M152
Asia Minor
domestic animals C214
Asparagus T177, QE14
Ass D66, N75
cross breeding with horse B219, D113, QE20
cross breeding with zebra T2, QE20

Ass (*cont.*)
propensity for hybridization E75, 103
tame D112
wild C225, D112, M153
Association of ideas GR12, C50, 172, 235, D111, M10, 29, 46, 62, 78, 87, 111, 127, 139, N16, 20, 41, 50, 54, 89, 107, 184, OUN18–19, 48–9
and origin of instincts N107–9, OUN38, 42–7, 49–51
cultural values taught by OUN47–8, 51
Asteriae RN22–3
Astronomy A24, B101, C123, 157, D36, E49, N5, 36
See also Earth; Gravity; Moon; Science; Sun
Atacama desert, Chile RN156
Atavism B58, 59, C166, D180, E183, QE1
effect on offspring C66, D112
in domestic animals C30, D3, 5–8, 13, 15, 148
in garden vegetables E15–16
in human families D8, 42, 151
in plants E141
reduces changes in species B75, C65
reversion of hybrid offspring B229, C30, 33, D3, 5–8, 13, 15, 148
reversion to youthful characteristics D30
similar to the production of monsters C59
See also Crossing & Hybridization; Hybrids; Inheritance; Recapitulation
Atheism M74, OUN37
See also Creation; God; Materialism; Religion
Athenaeum RN143, A29, 76, 88, 90, 118, 124, 132, 136, B133, 199, 235, 256, C36, 101, 224, 268, D24, 48, 73, 130, 152, 162, 172, E90, 161, 173, TFRAG3, FRAG4, FRAG5, M131, QE[23v]
Athenaeum Club M81, N97
Atmosphere A7, B46, D128
See also Climate; Rainfall; Winds
Auckland Island RN138
Auk B28, 144, D29, ZEd15
Auroch B136
Australia
aborigines C83, 244, D111, 112, E47, 64, M137
abundance of species B77
animals B15, C13, 15, 36, 38, 95, 102, 239, D72–3, E41, ZEd9
animals and plants that are also found elsewhere B187, C116, 224, 240, D73, E46
beetles E11
birds C15, 19, 27, 67, 80–1, 102, 239, D61, 75
dogs C17, 46, E26, ZEd4
droughts RN126
fossils B173, C130–1, 257
geographic distribution of animals in C27, 102
geographic distribution of plants in B76, 93, C102, 239, D25
geology RN6, 15, 101, 118–19, A92, B152, C130, 213
origin of fauna B15, 137, E136
ornithorhynchus B89, 97, 162, C87, D51, 115, E66, MAC28r
plants B187, C224, 240, QE21..
present-day fauna analogous to that of Oolitic period A55, B173
relationship of organisms with those of America C15, 37, 116
relationship of organisms with those of the East Indies C13, 19, C213
relationship of organisms with those of New Guinea C14–15
relationship of organisms with those of New Zealand B50, C19, 20, 239
relationship of organisms with those of Tasmania B50, 152, 264, C95, 130, 225, 239, D61
shells C244
See also Marsupials
Auvergne, France
geology RN38, A40
Avarice OUN8, 8v, 48, 54
Avestruz
See Rhea
Azalea T93, 111, 112, QE[5]a
Azores A40, QE18
geology RN107, 165
resemble Galapagos Islands E44
Axolotl E133

Baboons E18, 167
ancestor of mankind C243, D137, M123
behaviour C212, D136, M105–6, 137
expressions M105–6
Badgers E18
Bahamas RN27
Bahia Blanca, Argentina
geology RN67, 93–4, 113, A142, 146
shells A33, ZEd8, 15
Balancement
See Morphology
Balance of Nature
See Economy of Nature
Balanidae RN24, B113
See also Cirripedia
Bananas QE2, 12, 21..
Banda Oriental, Uruguay A68
Bantams C120–1, 231, D101, QE18, 20
See also Breeders; Crossing & Hybridization; Fowl
Barbaroussa C14
Barnacles
See Cirripedia
Basalt RN150, 153, A31, 32, 39
Batopilas, Mexico
mines RN168
Batrachians C18, E157
See also Frogs
Bats
flight D113, N72, MAC29v
fossil B200, T19, MAC55r
geographic distribution B104, C25
origin of structure is a great difficulty for Darwin's Theory T55, 57
vampire E10
Beagle channel RN148, A115, GR73, 109, E115
Beagle specimens
comparison with Henslow's mineralogical specimens from Anglesea A97
fish C20
fossils E99, T63
fossil wood RN178
insects C107, ZEd17
Planariae B143
Sagitella ZEd10
Sagitta RN174, ZEd5
variation in Darwin's mice E24
See also Fossils; Galapagos Islands; Rhea

Beagle voyage
  Darwin's assessment of his character during M79
  dates of sailing from England RNIFC
  gales RNIFC
  soundings taken during RN16, 19
  Woolich docks AIFC, 112
Beans
  Darwin's experiments on QE10a, 11, [11^{v}], 14
  fertilization T93, 103, 104
  permanence of varieties T151
Bears B65, 164, 222, C41, 46, D60, E18
  fossil T1
  mating in captivity QE20
  See also Zoological Society Gardens
Beauty
  and ideas of ugliness N121
  and taste OUN14, 20–4
  idea of, in animals C71, N64, 109
  instinctive perception of M32, 37, N109, OUN20–1
  origin of ideas of M71, N19, 26–9, OUN22–4
  origin of pleasure in M36–41, OUN6–7
  the object of the arts OUN22
  the object of flowers QE[5]a
  See also Pleasure; Selection, sexual
Beavers T89
Bees B208, C266, 268
  aid in pollination of plants E14, T93–4, 103–4, 105–6, 111–12, 135, 153, 177–8, S2^{r}–2^{v}, QE[5]a, [6^{v}], 8, 9, 10
  cells in honeycomb N70, 77, 81, MAC58^{v}
  Darwin's experiments with QE[5]a, 9, 22
  instincts B74, 208, C221, N3, 69, 70, 72, 81, 115
  races of QE15
  sexes C167, 215, 221, 235, D158, 172
  sting MAC29^{v}
Bees, neuter C215, 221, 235, D116, 172
Beetles C88, 108, 223, N36, ZEd18
  feign death OUN9
  wings B84, C86–7, 107, E147
Beetles, dung C197, ZEd19
Behaviour
  animals shamming death C197, OUN9
  birds assist in feeding young of other parents C67, 96
  birds return to same nest C40
  continuity between mankind and animals C154
  defence of nest by birds N1
  different dispositions of individual animals E117
  diverge between actions and feelings in mankind M60
  hoarding C199
  nest-building in birds C105, 189, N68
  of animals at Zoological Society gardens M106, 137–9, 141, N97–8, 138
  of baboons C212, D136–9
  of birds B105, C68, 113, 114, 124, 143, 159, 161, 185, 245, D33, 75, 101, 102, 105, E138, 140, 174, M141, 147, 152–3, N1
  of cats C186, D155, M49, 146, 147, 152, N64
  of cattle D12, 48, M71, 97, N125
  of cuckoo C67
  of deer D99, M76, N105
  of diving birds C160–1
  of dogs C212, 243, E36, M23, 56, 84, 118, 144, 152, N1, 7, 64, 90, 115, OUN36^{v}, 38
  of domestic fowl B140, C61, 70, D163, N109, OUN36^{v}
  of elephants C199, 235, D155, M144, N20
  of foxes M147, N69
  of hares N103
  of horses C163, D10, M71, 85, 129, 144, 146–7, N7, 75, 90, 103, 117
  of hyaena D139, M71, 153
  of insects C197, M63, 64, 97, 99, OUN35–6
  of jackal N44
  of monkeys C212, D136–7, E18, M28, 106, 137, 138, 142, 145, 156, N2, 14, 115, 183
  of orangutans C235, D138, M85, 107, 129, 137–40, N13, 94
  of pigs B255
  of rabbits C154, N103, 112
  of tigers M152
  of wild and tame horses together C41
  of wolves M147
  of zebra M146, N75
  polygamous in geese E174
  sexual B176–7, C51, 135, 235, D10–11, 90, 99, 137–9, 155, 177, M106, N41, 53, 59, QE19
  suckling as an innate form of N79
  temperament of hybrids D180
  value for classification C82, 165, D102
  See also Courage; Expressions; Fear; Habits; Instincts
Benchuca ZEd3
Ben Erin, Scotland GR55–6
Berberis QE5
Berbice, British Guiana
  geology of A87
Beroides ZEd7, 10
Bible
  New Testament M121
  Old Testament C219, D104
Biology
  definition of C104
  See also Classification; Crossing & Hybridization; Economy of nature; Fertilization; Generation & Reproduction; Inheritance; Natural History; Nervous system; Organization; Science
Bird of Paradise C109, C144, E147, E181
Birds B26, 81, 199, D29–30, MAC28^{r}, 28^{v}, ZEd6, 16
  ability to learn B4, C255, D32, M31, 32, 58, N11, 112
  adapted to elements other than air B27, 45, 113, C111–15
  aid in dispersal of seeds B193
  ancestry B26, 42, 112, 263, C52, 206–8
  as predators E138–40
  behaviour B105, C67, 68, 113, 124, 143, 159, 161, 185, 254, D75, 102, E140, M141
  classification B27, C71, 82, 88, 149, 205, D35
  close species of C46, 206, 267, D29, 31
  colour of plumage B98, 217, CIFC, 3, 68–9, 71, 80, 84, 88, 109, 144, 163–4, 209, D47, 92, 97, 100–1, 114, 147–8, 154, 160, E147, 160
  communication C68, M97
  eggs C4, D31, 148, 154, E158, 160, 168, T58FRAG I
  feathers an ancient feature C51–2
  fossil C58, 163, 207, D133, S3^{v}
  flightless B251, C25, 57, 114, 161, D29, 30, 60, 140
  geographic distribution B67, 81, 98, 104–5, 124, 199, C50, 55, 88, 94, 253, D26, 31, 180, E101–2, ZEd15, 17
  great variety of adaptations in C82, 141, 160

Birds (*cont.*)
instincts B4, 105, 151, 218, C51, 67, 68, 69, 96, T2, N4–5
naturalized E114
nest C40, M49, 126, N1, 68
nest-building behaviour C105, 189, T2, N1, 68
raptores C227, D114, ZEd7
rudimentary wings B70, D29–30, 59
soft plumage C186
tameness of wild B4, C50, 189, E174
taxonomic distinctness B42–3, D117
taxonomic type C52, 88
web feet C73, 115, 156, 173, 199
wingless B251
young C149, D95–6, 147, 148, ZEd7
See also Apteryx; Birdsong; Crossing & Hybridization; Ducks; Fowl; Geese; Hybrids; Migration; Penguins; Pheasants; Pigeons; Rhea
Birdsong
and sexual differences C178, D103, 113–14, E106, E138, 160
birds' enjoyment of M109, N18
comparison of, between different countries C209, 255–6, M31, ZEd19, 20
duration D103, E138
learnt D102–3, E108, M8, 31–2, 58
man's pleasure in M39
variety of M97
Bison D65
Blackbirds C149, 255, D96, 103, M49
cross-breeding with thrush TFRAG3
Black cock C184, 228, D73, 105–6
Black Sea D151
Blacksmiths M45, N42
Blackwood's Magazine C267, M18, 78, 135, 155
Blattae C107
Blood
of allied species QE4
See also Blushing; Crossing & Hybridization; Inheritance
Bloodhound B175–8, 182, 183–4, D179
Bluebell T106
Blushing M144
absence in animals N15, 52, 60
and morality N53, 54
Darwin's theory of N25, 60
in Negroes N15
in the dark N53
movement of blood during N51, 184
nature of N51–5
See also Conscience; Morality
Bogotá (Santa Fé de Bogotá), Colombia A115
Bolivia RN106, 152, 166, A2, B209, C56, E64
Bombs (geology) RN74, A135
Bonin Islands, Pacific Ocean
flora C241
Borneo C17, 27, 41, 46, E18, 20, 174, 177, 182
Bourou (Buru) Island, Moluccas
animals C14, 24, 27
Brain B74, C33, 76
action of M18, 20, 53, 62, 81, N42, 45, 91
relationship between structure and animal and human behaviour B226, C33, 166, 172–3, 226, D94, M57, N42–3, 89, OUN36ᵛ–37, MAC59
the organ of mind GR126, C166, 211, M46, 80–1, 131, N38–9, 43, 89, OUN37, 38, 39–39ᵛ, 41
water on D3
See also Consciousness; Insanity; Instincts; Intellect; Memory; Mind; Nervous system; Phrenology
Branching
See Tree of life
Brassicas E14, 16, 141, 146
Brassica
See also Cabbages
Brazil RN37, 134, A79, E90
animals TFRAG6, QE3, 4
Araucaria B187
birds ZEd17, 20
geology RN33, 63, 93–4, 100, 158, A42
soundings off coast RN15–16, 91, 98, A130
Breccia RN42, 110, 121, 170, A69, C131
Breeders
Bantam Clubs C231
bird fanciers DIFC, 34, 103
Darwin's trust in their knowledge D89
Darwin's questions for QE3
of wild English birds QE19
prizes QE18
See also Horticulturists
Breeding
See Crossing & Hybridization; Cultivation; Domestication; Hybrids; Selection, artificial
Breeding-in
See In-Breeding
Bridgewater Treatises B141–3, C91, 269, E59, 157, 158, N89, 90
Brine
organisms in lakes of RN127, 128
Britain
See England; Isle of Man; Scotland; Shropshire; Wales
British Association for the Advancement of Science
Reports RN142, A56, 57, 68, 124, B31, 154, 174, 208, 209, 256, C266, D48, 73, E11, 91, M142, QE1
Broccoli E14, T105
Broom B107, T103, 178, MAC53ᵛ
Butchers QE[9ᵛ]
Buccinum C145, D134
Budding
See Generation
Buenos Aires, Argentina A37, 117, C22
Buffaloes B132, D180, E19, 21
Bulletin de la Société géologique de France RN158, A44, 70, 81, 101, B54, 178, 179, C16
Bullfinches B217, C70, E150
Bullrushes E137, QE[15ᵛ]
Bulls
See Cattle
Buntings B193, D31
Bustards B198–9, 218
Butterflies D22, 62, 63, T57FRAG11, QE10

Cabbages T105, QE2, 10
experiments on breeding of QE13, 14
hybridization QE1ᵛ, 3, 5, [5]a, 8
variation in E146, QE11ᵛ
Cacti B180
Cairn tan leer peek (Càrn Leac), Scotland GR90

Calandria RN130
Calceolaria E103
Caledonian canal, Scotland
  terraces at mouth of GR22–3
California, Gulf of RN148
Callao, Peru RN95, 105
Caltha palustris T91
Cambrian system
  geographic extent E137
Cambridge E129, M82, 99, N4, 19
Camelidae
  species do not breed together ZEd17
Camels RN113, A5, B244, C46, E103, 157, T1, MAC57^{v}, 28^{r}, QE20
Camouflage
  See Adaptation; Colour
Campanula T153, QE[11^{v}]
Campanularia ZEd9
Campbell Island C19
Canaries QE18
Canary Islands B157, 220, C193, E90–1, QE10a
  plants B156, C184, 193, 240, 250, 275
  See also Tenerife
Cancer (disease) D2, N53
Capercailzie D73, 105
Cape of Good Hope
  animals RN134, B14–15, 135, 233, C28, 42, 45, E12, 173, QE16
  geology RN15, A94
  Hottentots B32, 189, 233, QE16
  plants C91, 250, E14, QE7, 10a
Cape Possession RN138
Cape Tres Montes, Chile
  dikes at RN7
  geology RN88
Cape Verde Islands RN99, A20, B154, C224, 249
  See also St Jago
Capromys C40, 42
Carabidae B47, 56
Caracara RN130, B55, D75, 96, ZEd7, 14
Carbon A116
Carbonate of lime
  in Plas Newydd dike RN7
Carbonate of soda
  formation of A84
  See also Soda
Carbonic acid (carbon dioxide) A80, 176, QE25
Carboniferous period B167, C216, 247, E164
  fossils C216, 247, D133, E164
  plants B150, C58, D133
Cardium CIBC, QE1
Cardoon C35
Carex QE[15^{v}], 21
Caribbean Islands
  See West Indies
Caricaridae ZEd14
Carimon Island, Indonesia
  animals E177
Caroline Islands, Pacific Ocean
  animals C22, 25, 27, E21
  inhabitants E22, N22–3
Carrots C195, E149, QE1^{v}, [11^{v}]
Caspian Sea B131, D151, E85
Cassowary D29, 154
Castration
  effects of B2, D76, 85–7, 99, 156, 162
Caterpillars C173, D40, 41, 62, 63, 170, QE5
Catorce, Mexico
  mines RN175
Cats
  behaviour C186, D155, M49, 146, 147, 152, N64
  deformities of D108, 112, E19, 182
  eye colour and deafness D11
  hair B190, C214, D15
  heterogenous offspring B250, D5–6, 15
  intellect M49, N72
  length of intestine QE3
  mummified Egyptian species B6, 16
  shorttailed B163, C249, E113
  tailless B178, C175, M25, QE22
  tortoise-shell B250, QE3
  varieties B148, 181, 250
Cattle B154, C157, 159, QE[11^{v}], 17
  African B167
  behaviour D12, 48, M71, 97, N125
  Caledonian C185, D48
  Caspian B131
  Chillingham D48, E91, 117, M142, QE10
  colour C29, 92, D42, 104
  cross-breeding of B140, 233, C121, 231, D3, 48, 100, 152, 155, 177, DIBC, E75, QE10
  cross-breeding with buffalo B132
  cross-breeding with yak QE12
  diseases E36
  East Indies E175
  fossil E92
  hornless B255, E23, 113
  horns S7^{r}, QE19, 21
  humped B131, 155, 190, 217, 255, 273, C56, D65, E12
  Indian B136, 139, 155, 217, 234, 273–4, C121, D65
  short-horn D42–3
  size B255, C92
  wild C267, D48, 101, N105
  See also Oxen
Cauliflowers E98
Cause C204, D36, N49
  idea of M135, 151
  nature of connection with effects B148, N12–13, OUN39–41^{v}, 53
  origin of idea of N60
Cause, finel B5, 49, C236, D114, 135, 167, E48–9, M154, MAC58^{r}
  absence of D114, E146–7
  See also Creation; God
Caves A65, C130, E66
  Proteus anguiformis E134
Cavia ZEd6
Cayenne Island, French Guiana A20
Celebes
  animals B249, C13, 14, E175
Cephalopoda B149, C40, D133
  classificatory position C186, D151, E96
  fewness of species B168, E135
  sex D156–7

Ceram, Indonesia
  cassowary C27
Ceratophytes A15, ZEd6
Cercopithecus C249, D137, E19, N183
Cervus
  See Deer
Cetacea E92, 134
  fossil B206, 172, T19, MAC55^{r}
  origin E92
  size D151, E59
  traces of hind limbs MAC55^{r}
Ceylon A39, B220
  elephant C46, E179
Chaffinch E138
Chalk
  purity RN120–1
  solubility RN30–1
Chalk deposits
  organisms in RN27, 42, 61, C205
Chambers Edinburgh Journal A26
Chamelion N101, MAC28^{r}, 29^{v}, 57^{r}
Chance B147, 221, C73, E68, 163
  of having successors B41, 134, 146–7, 148, C61, D112, E137
  law of B55, OUN25
  origin of man E68–9
  substituted for free will M27, 31, 74, 126, OUN25
  See also Selection, Natural; Will, free
Change
  in species by adaptation to external conditions B16, 20, 21, 22–3, 61, 63, 102, 214, 217, 235, 245, 246, 252, D69, 173, E4–6, 50, T41
  law of B228, C164
  sudden differences not part of nature's plan B239, D69, E113
  See also Adaptation; Gradation; Progression; Selection, natural; Species; Transmutation; Uniformitarianism; Variation
Channel Islands C192, 193, M155
Chemistry
  See Carbon; Carbonate of lime, Carbonate of soda; Carbonic acid; Chlorite; Mercury; Salt; Soda; Solubility; Specific gravity; Sublimation; Sulphur
Chesil bank
  movement of shingle on RN67
Chetah QE12
Chickens
  See Fowl
Chile C276, E91
  animals ZEd2
  Araucaria B187
  birds B51, C56, 126
  earthquakes RN75, 142, 144, 154–5, 157–8, A137, M16
  fish C28
  geographic distribution of coal RN34, 36
  geology A103, GR24
  geology of the valleys A87, 109
  human inhabitants M16, 69
  introduced species of birds B51
  rain and springs in RN31
  shells RN36, 106, A39, 72, 102, 144, T80, ZEd13
  See also Aconcagua; Arica; Atacama desert; Cape Tres Montes; Cobija; Chiloé Island; Chonos archipelago; Concepción; Copiapó; Coquimbo; Cordillera; Guantajaya; Iquique; Maipo; Obstruction Sound; Portezuelo; Valparaiso
Chiloé Island, Chile
  animals B7, M131, ZEd19
  birds RN130, C126
  geology RN28, 42, 69, 87, 160, 165
Chimborazo, Ecuador A81
Chimpanzee
  facial expression N76, 88, 183
China C17, 102, 223, 228, 269, E18, 123, OUN12
  dogs C228, E18
  geese B139, D9, 13, 180, E169
  language N31
  pigs D33, E123
Chinchillidae ZEd4
Chionis B55, C113
Chlorite RN7
Chonos archipelago, Chile
  geology RN7, 36, 63
Chrysomela B56
Chubut river, Patagonia RN67
Cirripedia D55, 156, E60, 71
  Balanidae RN24, B113
  remarkable sexual system E60
Cistus T111
Civet cats C92, 178
Civilization
  development of B4, C72, 217, 226, M87, N65, OUN30^{v}
  effect on mankind E47, M136, OUN29^{v}, 51
  original nature of D54, N65
Clarkia QE21..
Classification
  absence of gaps in tree of life B79, C96, 145, D52, E68, TFRAG8
  an activity based on experience C70, 205
  and the fossil record C110, 156
  artificiality of Linnaeus' system E109
  Darwin's respect for good describers E52
  Darwin's theory explains C126–30, 138, 139, 202
  describers of species must have theoretical ends in view E51–2
  difficulty of discriminating between groups of organisms B9, 229, C54–5, 70, 96, 137, 145, 205, E51–3, 68, TFRAG8, QE10a
  doubtful value of bones as a guide to B95, E23
  impact of increased number of specimens B9, 79, C200, E11
  natural C155, 156, 205
  origin of gaps in tree of life B35–44, 71, 79, 113, C157
  purpose of C158, MAC54^{v}
  relative 'highness' or 'lowness' of organisms B29, 73, 200, 204–5, 252, D116–17
  reveals ancestral form of organisms B206, 207, C3, 52, 186, MAC29^{v}
  value of colour as a guide to C71, E22
  value of constant characters B213, C149, 158, E22
  value of geographical distribution as a guide to C45, 54, 88
  value of habits as a guide to C82, 165, 226, D102, 115
  value of mode of generation as a guide to C33, 47, 149, 162, 195, 213, E91–2
  See also Ancestor; Ancestor, common; Genera; Gradation; Perfection; Quinarian system; Transitional forms; Tree of life; Type; Species; Varieties
Clay A20, 70, 111, B131, C178
Clay slate A94, 95–6, B121

Cleavage A60, 70, 86
Cleavage
and concretions RN5, A46–9, 53
and force of gravity A49, 51–2, 60, 62
and heat of the earth RN41, A108
and lamination of rocks RN5, A45–9, 53, 60
and magnetism A62
and metamorphism A94–6
and veins RN21
an illustration of the symmetrical structure of earth RN38, 41
Darwin's account of RN101, 102, A62
of sand in Scotland GR15
Climate
and internal geology of the earth RN41, E109
dry season RN126
effect of changes in former RN41, B243, 245, C33, 37, 38, 99, D105–6, 115, E37–41, 109, 161
effect on geographic distribution of organisms RN128–9, B243, 245, C147, D115
effects of monsoon B11
effect on parasites D91
onset of freezing weather on Shetland RN139
origin of Arctic A8
possible cause of extinction of organisms C37, 99
sources of information on meteorology RN8
the origin of impurities in chalk RN121
uniformity B19, 246, C38, 99, E37–40
See also Conditions of existence; Currents; Erosion; Rainfall; Sea; Waves; Winds
Close Species
See Species, close
Clover T103, 178, QE11
Clubs
See Athenaeum Club; Breeders; Societies and institutions
Coal RN34, 36, A107, 115
formation of RN28, 34–6
fossil animals in RN36, D133
fossil plants in D133, T19
Coal period
See Carboniferous period
Coastlines
shelving RN97–8
Coati C21, ZEd11
Cobalt RN172
Cobija, Chile
elevation A39
shells A102
Cocos Islands
See Keeling Islands
Coipu ZEd2
Coleoptera
See Beetles
Colobes B94
Colour C13, D43, 104, T9
albinoes B119, C85, E176
black hair in mankind D24
chamelion N60, MAC28r
effect of, on plants QE5
effect of surroundings on C inside front cover, 68–9, 70, 214, 252, D140, E43
inheritance C inside front cover, 29, 88, 120–1, D101, 127, E31
innate sense of N88
metallic D147, 160
of birds B98, 217, C inside front cover, 3, 68–9, 70–1, 80, 88, 109, 144, 163–4, 209, D47, 92, 97, 100–1, 114, 147–8, 154, 160, E147, 160
of horses C2, E9, 175
of plants T95–6, QE3
of rabbits C29, E9
of young deer D103, E31
octopus N60, MAC28r
origin M109
tortoise-shell in one sex only QE3
value for classification C71, E22
See also Man; Negro; Selection, sexual
Colour blindness QE16
Columbia river, Oregon A132
Columbine T94
Comparative Anatomy
See Morphology
Competition
Darwin's interest in C73
dreadful war of organic beings D134–5, E114–15
See also Conditions of existence; Population; Selection, natural; Selection, sexual
Compositae B193, 200, E100
Comptes rendus hebdomadaires des Séances de l'Académie des Sciences A39, B133, 135, C178, 183, D136, E79, 88
Concepción, Chile
animals C21, ZEd11
earthquake RN75, 154–5
elevation RN106, 115–16, 149
geology RN7, 102
Concholepas C244
Concretions (geology) RN5, 23, 66, 72, 121, 160, 165, A46–9, 53, 54, 84
Conditions of existence GR24, B112, C33, 57, E52, 113
effect of changes in B2–3, 7, 17, 38, 103, C140, 200, D23–4, E43, 71, 72, E146–7, 163, T17
effects of enclosing common land C160
effect of unfavourable B126, C65, 151, D37
quantity of organisms related to B12, C147
unknown causes of change in B17
See also Adaptation; Climate; Economy of nature
Condor ZEd6, 10
Confervae RN5, B200, N49
Conglomerates RN19, 23, 100, 151, 152
Conifers B149, 150, 202, C58, D133, 174, TFRAG8
Conscience N34, OUN8–8v
acquisition of N26–8
an instinctive feeling OUN37, 38
in animals N2–3, 46–7
nature of M60–1, 75–6, 119–20, 124–5, N3–4, 46–7, 99–100, OUN44–5, 52, 55v–55
purpose OUN54v
See also Blushing; Morality; Shame; Will, free
Consciousness N10, 111, OUN16, 35
absence of recollection of unpleasant incidents M115, 125
and drunkenness M79
depends on memory M103, OUN9
existence of double M78, 83, 110, 115–17, 156, N43, 61, 111
in animals M155, OUN35–6
See also Memory; Sleep; Will, free

Contempt
  facial expression N7–8, 9
Continents
  influence on fauna and flora of neighbouring islands B10, 11, 30, 31, 116, 138, 156, 160, 221, 227, 278, C6, 106, 129, 176, 184, D74
  former extensions of A1, B133, 224, C6, 130, 247, D139
  process of formation A104, B11–12, 72, 221–2, 224, 227, 241, C80, D34, 140
  See also Geographic distribution; Islands
Coots B193, N111
Copiapó, Chile RN50, 155
Copper RN161, 167
Coquimbo, Chile
  geology RN42, 61, 102
  shells A72
Coral
  See Coral polyps
Coral islets
  arrival of plants on B81
Corallinas E56, 151, 155, OUN35
Corallines ZEd7, 8
Coral of life
  See Tree of life
Coral polyps
  anatomy ZEd10
  germination RN24, E164
  rapid growth ZEd4
  survival in violent surfs RN93–4
  See also Polyps
Coral reefs RN69, A26, 80, 83, B224
  Darwin's theory of RN117, E22
  volcanic eruptions form foundations for RN91
Coral rocks
  in North Wales RN93
Cordillera RN136, 157, A115
  barrier to geographical distribution of plants RN52
  bones at great height in E44–5
  craters of A38
  different species of mice on either side of C252
  dikes RN7, A42, 51, 103
  elevation and subsidence RN22, 64, 137, 149, A79, 127
  faults RN37
  formation RN87, 131, 137, A107
  fossils D68
  geology of Portillo pass RN44, 150, 151
  geology of Uspallata pass RN152, 154, A120
  gold deposits RN176, A100
  lakes RN155
  lava RN87, 143
  metamorphosis of rocks of A16
  rocks RN99, 137, 143, A20, 38, 66, 115, E137
  salt deposits A16
  shells RN36, 106, A144, ZEd13
  terraces GR30
  volcanic activity RN10–11, 32, 144, A38
  volcanic bombs RN73
  water action in valleys A70–1, GR30
  See also Chile; Chimborazo; Cotopaxi
Córdoba, Argentina
  earthquake RN142, 144, 157–8
  mountains A100
Cormorant MAC57[r]
Coronella QE[5]a
Corry (Coire), Scotland GR63, 84, 85
Corsica
  moufflon E35
Costorphine Hills, Edinburgh
  grooves in rocks RN69
Cotopaxi, Ecuador
  geological structure RN43
Cotton plant QE10
Courage B140, C120, D85, N44, OUN36[v]
Cowpox C174, E36
Cows
  See Cattle
Cowslip
  and primrose E16, 113, 141
  experiments on QE1[v], 5
  wild C35, 194
Crab apples
  See Apples
Crabs D129–30, 151, 156, MAC29[v]
Crag B202, 243, 245, E166–7
Cranes B242
Craters RN58, 59, 107, 110, A2, 38, 40, B192
Craters of elevation RN139, 144
Creation RN127–8, B84, 115, 116, 243, C146, 240, D25, M136, N36
  an assumption B104
  an expression of ignorance C106
  by miracles C55, 156, MAC56[r]
  Darwin's theory does not require B227, D22
  difficulties incurred by belief in D115
  difficulties of Darwin's theory may lead to belief in C64
  double QE17
  likelihood of having occurred D19
  of animals all at once D19
  of island species RN127, B30, 98, 100, 103, 160, MAC167[v]
  See also Cause, final; God; Selection, natural; Species; Transmutation
Creator
  See God; Religion
Cretaceous period C216, E60
Criminality
  See Evil
Crocodiles B234, 250, C18, 28, E21
Crocodiles
  fossil C39
  instincts N71, 125
Crocuses QE[6[v]]
Crossing & Hybridization B59, 273, D25, 105, E70–1, 123, QE20
  and perfect adaptation B210, CIFC, E124, N44
  a poor test of species B240
  as the origin of species C151
  between asses and horses B219, D113, T2, QE20
  between asses and zebras T2, QE20
  between birds in a wild state D72, 73, TFRAG3
  between breeds of canaries QE18
  between breeds of cats B250, D5–6, 15
  between breeds of cattle B140, 233, C121, 231, D3, 100, 152, 155, 177, IBC
  between breeds of cucumbers QE8, 14
  between breeds of dogs B140, 155, 181–4, C1, 120, 189, 228,

D7, 63, 180, IBC, N44, QE4
between breeds of domestic fowl C120–1, D101
between breeds of ducks D180
between breeds of geese D180
between breeds of horses B184, CIFC, 2
between breeds of pigeons CIFC, 2–4, QE18
between breeds of sheep GR25–6
between dog and fox C187, D7, N44
between dog and jackal D90, IBC, N44
between dog and hairless dog of Africa D89
between dog and wild dogs D7–8
between dog and wolf B51, 140, 163, D7, 173, N44
between domestic and wild animals B132, C41, 185, D7–8, 112
between horses and zebras QE20
between lion and tiger C228, DIFC, 8
between races of hares E184
between races of mankind B32, 33–4, D136
between races of plants B75, E123, 143–4, QE1v, 3, 5, [5]a, 8, 21
between sheep and moufflon E35
between species of birds B141, C4, 68, 164, 184, D25, 26, 31–4, 87–8, 92, 103, 164, T13, 15, QE22
between species of Camelidae B244, ZEd17
between species of crow where their geographic ranges meet E101–2
between tame and wild ducks D148
between trout and salmon QE18
Darwin's experiments on QE1v, 4
Darwin's questions regarding QE3
Darwin's theory of frequent MAC56v
Darwin's theory of generation and D177–9
depends on amount that parent varies E103
depends on instincts in animals E143–4
desirability of crossing the breed QE10
effect of back-crossing with parents C2
implications of large number of races for GR1–2
leguminosae difficult to cross QE7
no difference between that of species and of varieties E141–2
of a variety with its parent species B203, E106
of ferns B230, C52
of wild animals in captivity QE18, 20
of wild dogs C94
origin of garden plants E14–16
permanence of qualities after C231, E63
possibility of between two new varieties C135
repugnance to B24, 33, 34, 59, 93, 120, 158, 189, 209, 241, C30, 34
See also Breeders; Fertility; Hybrids; In-breeding; Inheritance; Selection, artificial
Crows C80, 177, E106
behaviour B218, D75
geographical range C67, E101–2
Crozet Islands RN138
Crustacea RN127, D55, 133, E71, 90, 97, 100, TFRAG4, ZEd1, 3, 8
mouth parts B112, 143
respiration E25
Crying
See Expressions
Cryptogamia B46, 150, 159, 229, C95, 240, 244
reproduction D177, E100, 123, 164
See also Seaweeds
Crysomelidae B56, 58, C108
Crystallization
activity of atoms during A96, 97
and laminar structure of rocks A46–7, 53
arrangement of crystals A56
in clay A111
law of A46, 49
Crystals
in igneous rocks RN88, 123
arranged in planes RN21, 115
Ctenomys C35
Cuba RN24, C42, 61, E42
Cuckoos C113, 142
behaviour C67
Cuculionidae C108
Cucumbers QE8, 14
Cultivated plants
See Cultivation
Cultivation E13–17, QE[15v]
affects transmission of variations T95–6
origin of double flowers QE[5]a, 6
See also Dahlias; Domestication; Fruit trees; Horticulturists; Plants; Roses
Cuscus B249, C23, 24
Cuttlefish RN10, E134
Currents A82–3, 124, 138–9
Cyanocephalus C13
behaviour D136, N2, 14, 183
facial expressions M105, N183
Cyanoglossum T177, QE14
Cyclops ZEd3
Cyclostoma C245
Cyrena C127

Dahlias C267, E13, 113, 142, M63
Daisy QE6
Daltonism
See Colour blindness
Dandelion QE[15v]
Darwin, Charles
anger M51, 70
appreciation of Kensington Gardens M41
assessment of his character during *Beagle* voyage M79
belief in materialism C166, M57
believes he takes after R. W. Darwin M83
capacity for inventive thought M34–5, 90–1
childhood memories M29, 41
considers new buildings ugly N30–1
Caroline Darwin provides a reference for RN125
E. A. Darwin provides information on dogs D63
E. A. Darwin provides information on horses C183, D10
E. A. Darwin recommends books to C266
distracted by association of ideas D111
dreams M111, 143–4, N33–4
first memory C242
handwriting like Erasmus Darwin's EIBC, M83
intends to keep tumbling pigeons QE4a
memory of Fitzwilliam Museum, Cambridge M82
recollections of Robert FitzRoy's behaviour M43, 60
respect for good describers of species E52
sand-walk at Down House QE22

Darwin, Charles (*cont.*)
taste for music M28, 51, N45
tea chest leads to change in habits C217
visits Windsor Park E31–2
Hensleigh and Fanny Wedgwood provide information on infant behaviour M58
See also *Beagle* specimens; *Beagle* voyage; Cambridge; Maer; Shrewsbury
Darwin — Difficulties of species theory B4, 16, 235, 248
adults instincts evident in young offspring T95–6
age of the world E155–6
constancy of plants in absence of crossing E163–4
crossing of hermaphrodites E70–1
crossing of mosses E123
Darwin's attitude to C145, D71, OUN35
Darwin's theory must account for both reason and instinct C199
Darwin to summarize argument against E55
dreadful war of organic beings C114
existence of fertile hybrids in plants B240
existence of sudden changes not part of nature's plan E113
extinction of South American fossil quadrupeds D72
few changes in species during historical times E4–6, 17
how the first eye is formed C175, D21
lack of diversity in fossils of oldest formations E137
likelihood of civilizing savages E47
likelihood of observable change in species in thick geological strata E6, 17
long lifespan of some species B170
may lead to belief in creation C64
migration of turtles E138
necessity that every animal should cross E150–1
not possible to show details of hybridity C151
no traces of passage between birds with and without webbed feet C156
of believing in doctrine of slow movements C74–5
of explaining preservation of colour C71
of fixed organization being modified B4
of thinking of Plato and Socrates as descendants of animals E76
of tracing change of species to species C64
origin of instincts OUN51
origin of intermediate structures T55, 57
presented by cases where there cannot be gradation C124
rarity of gradation in anatomical structures C222
separation of the sexes D159
See also Classification; Crossing & Hybridization; Transitional forms
Darwin—Experiments
comparison of bones of wild and tame ducks QE3, ‹15›
microscopic examination of plant anatomy T100
observations at Maer T91, 93, 99, 105–6, 135, S2^{r}
on association of ideas M91
on beans QE10a, 11, [11^{v}], 14
on bees QE[5]a, 9, 22
on crossing and hybridization QE1^{v}, 4
on crossing Indian fowl with common D180
on cowslips QE5
on fertilization QE1^{v}
on habits C217
on land shells in salt water B248
on lizards B248
on peas QE10a, 11, [13^{v}]–14, 22
on perception N75
on plant reversion B230
on soaking seeds B125
on strawberries E15
on zoophytes A180
to be performed at Maer QEIFC
to be performed at Shrewsbury EIBC, QEIFC
Darwin—Health
feelings of fear in the night M54
headaches M81, 89–90
weak stomach M34, N113
Darwin—Personal communications
with Abberly (Hubberly?) E184, T151, M49, QE8, 11, 14, ‹15›
with Fanny Allen N54
with Pedro de Angelis A117
with John Arrowsmith A2
with C. C. Babington C192, 195, M79, QEIFC, 10a
with John Bachman C251, 253, D31, 103
with Francis Beaufort A55
with Thomas Bell T13, 15, QE16
with George Bentham QE21
with Gabriel Bibron C54
with Edward Blyth D29, 33, 95, QEIFC, 20
with William Bollaert A44
with Francis Boott QEIFC, 21..
with Edward Brayley A41
with Mark Briggs D3, QE‹15›
with Robert Brown RN155, C237, 238, 239, QEIFC, 21..
with C. J. F. Bunbury S5–6, QE21..
with Benjamin Bynoe RN141, M79, 142
with Thomas Carlyle B255
with D. R. Corbet QE‹15›
with Hugh Cuming ZEd13, QEIFC
with Caroline Darwin (Wedgwood) M56
with Catherine Darwin M28, 33
with E. A. Darwin RN115, A118, 119, C183, 266, D10, 63, M128
with Emma Darwin E125, N61, 73, 112
with Sir F. S. Darwin B136
with R. W. Darwin D1–4, T175, M1–2, 9–11, 13–18, 20–2, 24–6, 35, 42, 43, 44, 50, 55, 78, 156, IBC, N1, 39, 53, 60–1, 63, OUN31, 32, QEIFC, 4, 6, 7, 16
with Susan Darwin M33
with David Don B79, 187, 191–3, QE10
with Edwards QE12, 13
with Thomas Eyton B248, DIFC, 4, E168–9, OUN36^{v}, QEIFC, 9, 18, 22
with Hugh Falconer QE17
with Robert FitzRoy RN67, A141, E42, M43, 60
with R. S. Ford QE3
with W. D. Fox B83, 141, 176–7, 183–4, D5–15, 70, M49, 128
with Louis Frazer C107
with J. H. Galton B175
with John Gould B171, C71, 80–2, 109, 110, 113–14, 143, 144, 189, D102, QEIFC, 10a, [11^{v}]
with Mr Gowen QEIFC, 12
with J. E. Gray QEIFC, 12, 17
with W. J. Hamilton C214
with J. S. Henslow B68, 198, 230, C16, 265, E129–30, T99, N74, 76, QEIFC, 2, 3, 7–12, 15–15^{v}

with William Herbert QE[6[v]], 12
with Sir R. Heron QE18
with Henry Holland RN159, E183, N111, QEIFC, 1, 16
with J. D. Hooker QEIFC, 1, [9[v]], 10a, 17
with F. W. Hope B248, E11, QE12
with Thomas Horsfield C268, QEIFC, 12
with Mr Hunt M138
with Leonard Jenyns B217, C19
with Rev. R. Jones M99, 155
with W. A. Leighton E184
with John Lindley QE21..
with Conrad Loddiges D118
with William Lonsdale C175–6, 177, E119–21, M97
with Charles Lyell RN145, A14, 68, E59, 86, 100–1, N115, ZEd10
with J. G. Malcolmson A142, 146
with W. C. L. Martin B165
with Herbert Mayo M110
with Alexander Miller T2, M137
with Major Mitchell C159, 213
with W. S. MacLeay B179–80
with R. I. Murchison RN142, A36
with Nancy M53
with William Ogilby B163, N115
with Richard Owen B161, C3, 72, 131, D29–30, 113, QEIFC, 17
with Woodbine Parish RN142, 143, 155, 157, A38
with Marianne Parker N37
with John Phillips E121
with Mrs Power at Port Louis RN17
with John Riley QE12
with J. F. Royle B242, 273, QEIFC, 12
with Signor Rozales RN155
with R. H. Schomburgk QEIFC
with shepherds in Wales GR11, 25, 125, D43, 47, E123
with Andrew Smith A34, B83, 233–4, D136–9, QEIFC, [15[v]], 16
with James Smith of Jordan Hill QEIFC, 1
with Thomas Sorrell RN99
with G. B. Sowerby C246, E76, 98–9, 120, QEIFC, 1
with G. B. Sowerby jr. RN144
with B. J. Sulivan QE[11[v]]
with Col. Sykes D99, QEIFC, 12
with Henry Thompson C231
with George Tollett QEIFC, 19
with G. R. Waterhouse A14, B71, 162, 166, 249, 250, 256, C107, 162, T55, 57, 59, M98
with Allen Wedgwood M79, S1[r]
with Elizabeth (Bessy) Wedgwood M12, 26, 79
with Ernest Hensleigh Wedgwood N37
with Fanny Wedgwood M53
with Henry Allen (Harry) Wedgwood N113
with Hensleigh Wedgwood C165, 244, E144, T51, M58, 61, 96, N37, 39, 74, 81, 121, ZEd20
with John Wedgwood E13–17, 98
with Sarah Elizabeth Wedgwood E144, S1[v]
with M. A. T. Whitby QEIFC
with Mr Willis C232, D24, 163, 173
with S. P. Woodward E120
with Thomas Worsley N4, 74
with Mr Wynne B139, 141, OUN36[v], QE18
with William Yarrell B138, 140, C1–4, D72, 99, 102, 108, 152, 153, E112, 174, M138, QEIFC, 18
with keepers at the Zoological Society Gardens D89, N97–8, 184, QE20
Darwin—Species theory
and geographic distribution RN127, B164, C29, 69, 77, D23
and origin of language N39, OUN5
and natural selection DIFC, 134–5, 175, MAC54[r]
and spontaneous generation D132, E160
antiquity of Darwin's views C267
arguments in favour of B44, 261
based on three principles E58
boldness of D26
contains no absolute tendency to progression N47
Darwin fears opposition on subject of classification C202
Darwin's belief in advancement of science C123
Darwin's belief in truth of E118
Darwin's description of C62–3
Darwin's list of books examined with reference to C276IBC, D179
Darwin's methodology D117
Darwin's summary of facts relating to his theory of generation D176–8, 174–5 sic
does not go back to first stock of all animals D21
does not require idea of creation B104, 227, D22
effect on classification B228, C126–30, 138, 139, 145, 202, 206, D52–3, E23, 42, 51
established on the idea that all animals basically hermaphrodite D161, 162
explained metaphorically as a human history book E6
explains gradation between forms B226
explains instincts D26
explains mules C135
explains organization C70
explains origin of man M84
explains the fossil record C57
explains the separation of the sexes D159, 162
first speculations on RN127–30
gives one great final cause to nature B49, 226–7, E48–9,
gives new meaning and scope to natural history B47, 224–9, C164, E53–4
grandeur of D36–8
implications for crossing between varieties C30, E106
most hypothetical part of C30
not about origins of life C58
not original D69
of inherited habits N63
of morality OUN30
of sensitive plants N50
of the origin of sublime ideas OUN11[v]
possibility of proof of C177, E51, 69
reason is not sufficient to make Darwin give up OUN7
relationship with Lamarck's theory B214, N91
requires crossing between individuals E164
requires existence of progression in fossil record E60
rests on analogy C176–7
shows life as a simple series MAC56[v]
the beauty of it E71
Elizabeth and Hensleigh Wedgwood's opinion of E144
Hensleigh Wedgwood objects to T51
whole fabric totters if granted C76–7
Dasypus A117, ZEd6
Dasyurus C130

Deception Island, Antarctic RN138, 139
Deer C14, 22, 24, 46, E18, 44, 173, 175, 177, 181, M41
  behaviour M76, N105
  colour D30, 103, E31, 42
  crossing of breed by keepers QE10
  males C61, D99
  young D30, T9
Deity
  See Creation; God; Religion
Deluge
  no fossil evidence for sudden extinctions D39
Demerara, British Guiana
  earthquake RN124
  subsidence A118, B131
Denmark A115
Denudation
  See Erosion
Deserts RN156, B235–6, C68
  adaptations to conditions of A5, B55, C88
  characteristic species of RN52
Design
  See Adaptation; Cause, final; Creation; God;
Development D57, 112, 132, 176, 178, E123, 158
  variations appear in foetus before gender emerges T95
  See also Growth; Metamorphosis; Monsters; Progression; Recapitulation; Transmutation
Devonian period
  absence of plant fossils S3^{v}
Didelphis B87, 133, 219, C13, D62, 73, E60, 128, ZEd9
  See also Stonesfield Slate
Digestive system
  grinding mechanisms MAC28^{r}
  non-absorbing stomachs in camels MAC57^{v}, 28^{r}
  pigeons secrete milk from stomach D132, E57
  stomach B213, D94, M13
Dikes RN7, 71, A42, 51, 65, 83, 89, 103, 147
  and faults RN37
  and formation of veins A116
  and mountain axis A58–9, 65
  and movements of the earth's crust RN49, A63, 65, 108, 127
  frequent in granitic countries RN56
  in superficial strata RN58
  laminated structure A42, 51
  not always points of volcanic eruption RN59
  position in relation to other rocks RN59, A65, 67, 86, 87
  rocks in RN7, 63, 78–9, A35, 42, 61
Dingo D127, N97
Dinornis
  relationship with Apteryx T65
Diorite A20, 84
Dip RN71, 87, A50, 52, GR6
Diseases RN178, D166, E67, 178, N43, 53
  apoplexy D1
  asthma QE16
  cancer D1, 2, N53
  caused by in-breeding B145, C175, E45
  changes in frequency C266, D2
  common to man and animals C174, EIBC, 36
  convulsions NIBC
  distemper in dogs D163
  epidemics C64, E3
  epilepsy M7, 20, 21, 26
  erysipelus D1
  gout QE16
  haemorrhage QE16
  heart M54
  hydrocele QE16
  hydrophobia RN126, 177
  infectious B119, C174, E36, M142
  inflammatory M43
  inherited B148, C65, D18, 172, M157
  pleurisy D3
  St. Vitus' dance N38
  transmission through females QE16
  worm fever D3–4
  See also Darwin—Health; Malaria
Divergence
  amount of difference between forms CIFC
  crossing prevents B210e–211, 212
  See also Selection, natural; Tree of life
Dodo B251, C25
Dogs B163, 219, D9, 44, S7^{r}, QE1
  and man C165, E36, 37, M23–4, 97, N46
  anger C212, E36, M147
  barking C42, 159, M84, 92, 94, 96, N94
  beagle D42
  behaviour C243, E36, M23, 56, 84, 118, 144, 152, N1, 7, 64, 90, 115, OUN36^{v}, 38
  bloodhounds B175–8, 182, 183–4, D179
  bulldog B217, QE‹15›
  Chinese C228, E18, QE22
  conscience N3, M24, 148
  contempt N7
  courage D85
  cross-breeding with fox C189, D7, N44
  cross-breeding with jackal C210, D90, IBC, N44
  cross-breeding with wild dogs D7–8
  cross-breeding with wolf B51, 140, 163, D7, 173, N44
  crossing between breeds of B140, 155, 181, 183, 184, C1, 120, 189, 228, D7, 63, 180, IBC, N44, QE4
  descent B51, 278, E47, 88
  diseases RN126, D163, E36
  dreams M102, N117, OUN8a
  ears C210, M146, N97
  Egyptian B6, E142
  Eskimo B140, 155, C1, D7, QE4
  fear M152, 153
  geographic distribution B177, C25, 46, 131, E18
  greyhounds B159, 171, 184, 217, C120, 214, D20–1, E75, 118, 145, N43, QE‹15›
  hairless C21, 228
  hare-coursing C120, E26, 75, N44, MAC57^{v}
  harrier D41–2
  hunting C94, 213, M132
  in-breeding B176
  instincts B165, C134, N1, 17, 43–4, 47, 105, 185
  intellect C210, E46, 47, M15, 97, 98, 128, N43
  intermediate forms between breeds B155, 198, 217, 219, D9, E69, 88, M102
  kinds of B198, 262, C1
  memory N72, 90
  monsters D108, E113
  natural selection in MAC28^{v}, 167^{r}
  Newfoundland N45

physiological differences between breeds of QE4
reproduction B183–4, C228, 232, D115, 138, 155, 163, inside back cover, M71
sense of smell M71, 109, N87
spaniel B184, C228, D9, N43
staghound D42
tailless QE20, 22
terrier C94, 210
teleogony in B32, 184, D9, 172, M25
toothless dogs in Brazil QE3, 4
training of C134, D21, M132
variation in QE19, 22
yawning M85, 95
Dogs, Australian B122, C1, D66
behaviour B250, M128
burrowing B165
colour E26
domestication E103, 149
half-breeds C189, D180
introduction C46
hunting C213
voice N94
Dogs, Shepherd D47
colour GR2, D44
cross-breeding with other breeds B183
instinct GR12, M101, N17
young GR26
Dogs, Wild C94, D7–8, 127, ZEd4
Domestication
absence of breeding during C135, D172, E174
breeds of animals resemble races of mankind B51
changes in animals during B145, 190, C176
characteristics compared with wild animals C134, D20–1, E71, N112
characteristics of animals running wild RN133, B31, 240
effects of, on silkworm QE12
great number of breeds D23–4, E113
instincts of animals during B197, E174, OUN38
interaction with wild animals B132, C41, 185, D7–8
loss of fertility during B120, 123, C219, DIFC, 103, 108, QE12
of animals on Pacific Islands C54
provincial breeds D44
tendency to revert to oldest race C30
value of introducing new breeds C159
variation in animals during B100, 262, C219
See also Cats; Cattle; Crossing & Hybridization; Cultivation; Dogs; Ducks; Fowl; Geese; Goats; Horses; In-breeding; Selection, artificial; Species, introduced; Tameness
Domingo (Santo Domingo) Island, West Indies A18
Doves D72, E169, M88, N101
Draco MAC54v
Dreams
compared with insanity M21, OUN17v
images during M102–3, 111–12, 130
of dogs M102, N117, OUN8a
nature of M111–15
recollected in waking thoughts M102
restful for the mind M90
See also Sleep
Drift deposits RN114, A36, 57
Ducks C82, D89, 96, 101, 180, MAC28r, QE20
crossing between kinds of B141, C228, D25, 26, 33, 148
Darwin to examine bones of QE3, 15
loggerhead B251
Muscovy C228, D9, 32–3, 101
wild C134, D33, 103, E174
Ducks, penguin D32, 87–8, 91
cross-breeding with ducks T13
eggs QE4
skeleton QE20
Ducks, pintail D inside front cover, 25, 26, 89
Dugong C131, E91–2
Dust
blown far out to sea A25
See also Ash, volcanic

Eagles B55, D61, E138, 139, 140
Eagles, hawk C143
Eap Island (Carolines) E21, 22
Ear
wax C174
Earth
age of B225
central heat of RN41, A76, 77, 78, 79, 90–1, 108, 121
cooling of central heat of A77–9, 121–6, B137, 247
fluid nature of core RN57–8, 111–12, A106–8, 113, 126, 140
history B224
metallic core A21
movements of fluid core RN41, 48–9, 111–12, A85
origin A104
perturbation of axis of A24
shape of A113–14, 118, 121
symmetry RN123
variation of compass measurements A85
Earth, crust of RN41, 123, A64, 123, 128
conduction of heat A139
cracking A107
crustal movements A92, B80
Darwin's hypothesis of coral reefs and movements of RN117
equilibrium of land and sea A129
temperature A121–5, 134, 137–40, 151
thickness RN154, A77–9, 114, 133, 136, 147
thinness RN131, 154
See also Elevation; Elevation & Subsidence; Subsidence
Earthquakes RN75, 112, 116
absence at Mendoza RN158
accompanied by rain RN76–7
action of sea waves during RN80–3
and collection of sediment GR53
at Acapulco RN177
at Calabria RN75
at Concepción RN75, 154–5
at Córdoba RN142, 144, 157–8
at Demerara RN124
at Pasto RN144
at Quito RN158
come from several directions RN177
connections with volcanoes RN90, A76
effect of passage of the moon A137, 153
first movements of RN80–1, 82, A18
geographic extent of RN70, 116, A28
in Antarctica RN17
in Sumatra RN90

Earthquakes (*cont.*)
  mankind's fear of M16
  metamorphic rocks A106
  movement of earth's crust during RN154, A80, D140
  on Melville Island A30
  on Radack and Ralix Islands RN101
  on St. Helena RN125
  on shores of Pacific Ocean RN157
  submarine RN33
Earwig M64
East Indies
  See Indonesia
Echidna C225
  spines B79, 87, C208, D132, T17, MAC28^{r}
Echinites RN22–3
Echinoderms C103, D156
Economy of nature B104, D115
  absence of gaps in E43
  effect of introduced species on E114–15
  gaps in D135
  number of living beings in existence C146–7, D134–5, E85, 95–7, 108
  physiological relations of animals MAC29^{r}
  relative numbers of organisms B12, C147, E136, 140, MAC56^{r}
  See also Competition; Conditions of existence; Divergence
Edentata A57, B99, C36
  fossil A6, B94, 173, 223, C37, E40
  geographic distribution A5, 6, B94–5, 106, 133, 223, C36, 38, D68, 73
Edinburgh New Philosophical Journal RN127, A25, 27, 99, 104, 105, 111, 124, 131, 132–3, 149, 151, B107, 117, 118, 119, 121, 125, 187, C131, 165, E36–7, T81, QE17, 18
Edinburgh Review RN143, A88, D152, M60, 70, 72, 81, 135, N12, OUN8, 14, 25
Education
  ability of animals to learn C134, D21, M132, N70
  ability of birds to learn B4, C255, D32, M31, 32, 58, N11, 112
  analogous to improvement of instincts OUN53^{v}
  instinct to learn in children N74–5
  leads to improvement of mankind C220, OUN27
  See also Civilization
Egypt
  ancient animals domesticated C153, ZEd12
  germination of ancient bulb from C92
  present-day animals have not changed from ancient times B6, 16, D106–7, E142
  wheat QE[11^{v}]
Electricity RN6, 115, A29, B17
Elephants RN134, B233, D172
  behaviour C199, 235, D155, M144, N20
  descent C167, D22, E65
  extinction in South America B62–3
  fossil C132, D25, E65
  geographic distribution B80, 233, C17, 46, D139, E20, 31, 60, 173, 176, 179
  intellect C196, 210, M98, 128, 131
Elevation A13, 34, 82–3, 101, 104–5
  absence of signs of RN115–16
  accompanied by mountain formation A62
  accompanied by volcanoes RN117, 137, A34, 62
  and earthquakes RN75, 124
  and formation of salt pans A5
  and origin of dikes A63, 127
  and terraces RN60
  at Stockholm A113
  caused by movements of fluid rock RN61, 83
  cracks in the earth's crust after RN83, A65, 112
  Darwin's effort of thought to trace evidence of M90–1
  effect on geographic distribution of species B82, C191
  effect on subterranean temperatures A77–9
  erosion of cliffs after RN39–40
  evidence from shells at heights above sea level RN37, 106, A4, 20, 39, E86
  lines of RN137
  of Chile A39, 109
  of freshwater lakes RN69
  of land near Red Sea RN148
  of sea bottom RN106
  of Siberia A7, 8, 11
  of Sumatra and Java B241
  rate of RN69, 83, 113, 117, 135, 149, A13, 109, 118
  recent RN60–1, 140, A4
  regularity of slope of valleys after RN69
  sea waves obliterate signs of RN115–16
  See also Elevation & Subsidence; Glen Roy; Mountain building; Subsidence; Terraces
Elevation & Subsidence RN77, A8, 85, 113, 118
  areas not equal in extent C220
  Cordillera marks line between areas of A79
  Darwin's wish to be cautious when describing A134
  Darwin's theory of A101
  former A11
  law of equilibrium applied to the Earth's crust A114
  See also Earth; Elevation; Subsidence
Elk, Irish B80, C257
Elms QE7, [11^{v}]
Emberiza brasiliensis B193, ZEd8
Emberiza melanodera ZEd11
Embryology
  See Development; Monsters; Recapitulation
Emotion
  explained by origin of man OUN47
  hereditary OUN8, 8^{v}
  materialistic view of origin M57
  nature of N57, 66, 68, 89, 92, 93, 97
  origin in muscular exertion M51–5, 57, 70
  origin in pain and pleasure M101
  passions analogous to pleasures of association GR12
  passions are the origin of evil M122–3
  strongest passions are common to man and other animals OUN29–29^{v}
  See also Anger; Avarice; Contempt; Courage; Expressions; Fear; Instinct; Love; Modesty; Pain; Pleasure; Pride; Shame; Surprise; Suspicion; Sympathy
Emu C27, D29, M132
  See also Cassowary
Encrinites RN22–3
England B220, C223, ZEd16
  animals B65, 138, 191, 280, C29, 83, 159, 177, 189, 256, D105, E45, 102, 138, 162, M31
  geology RN45, 46, 47, 51, A1, 115, B80, D139, E121
  great changes in fauna from previous geological period C132
  plants D180, E184
  relationship of species to those of Europe B80, 138, 220, C123

relationship of species to those of Ireland B7, 166, 191, 221, 262, D16
relationship of species to those of North America C246, D96
Entomological Magazine B91, C107
Entomostraca C162, 235, D40, 41
Entozoa C266, D156
Eocene period B30, 88, 200, 245, C36, E40, 105, 117, 167
Epidote RN6
Epilepsy M7, 20, 21, 26
Erosion
action on rocks RN28–30, 33, 105–6, 153, A69, 103, 145
by rivers RN108–9, A70, 72–5
by wave action RN38–41, 94, 95, 139, 144, A39, 98
great power of water action RN108–9, 151, A69, 70, GR30
of alluvial deposits by rivers GR20, 27–8, 29
of cliffs RN38–41, 67, 100, A59
of river valleys A12, 70–1, 72–5
of Salisbury Crags GR10–11
subaqueous A82–3, 92
Erratic blocks RN155, A36, 87, E37, 38
dispersal RN114, A56, 70, 71, 131
geographic distribution RN32, 155, A56–7
height of final position of A110
in Alps A20, 110
indicate former climate of whole world E37–8
in drift deposits RN114
in Glen Collarig, Scotland GR40
in Glen Roy, Scotland GR76, 82, 84, 85
in Meall Doire, Scotland GR32
on Ben Erin, Scotland GR55–6
on Càrn Leac, Scotland GR90–1
shape RN20
size A131
Eskimo dog
See Dogs, Eskimo
Essequibo, British Guiana
rocks of A30
Ethics
See Morality
Etna, Mount RN57, 137
Eucalyptus C239, M64
Euphorbia C250
Europe
absence of endemic genera C39
ancient climate E37–40, 42
animals B47, 220, 247, C37, 225, 233, E72
cryptogamia same as those in Australia C240
fossil mammals D25, TFRAG9, 79
geology RN18, 22, 73, A3, B167
Evil
See Sin
Evolution
See Darwin—Species theory; Species; Transformism; Transmutation; Tree of life
Evolutionism
See Gradation; Progression
Experiments
See Darwin—Experiments
Expressions
bird behaviour considered as instances of N1
chimpanzee N76, 88, 183, 184
common to all races of mankind N22–3, 35
common to man and animals C154, 243, D22, M93–4, 97
crying C243, M89, 94, 107, N20, 56–7, 59, 125, 182, NIBC
Darwin's attempts to classify in monkeys M142
difficult to conceal M93
frowning E125, M58, 95–6, IBC, N58
imitation produces corresponding emotion N10
innate recognition of M145
inheritance of D18, M107
intimately connected with mind M151
nodding N37
of passion M146
of respect N32
origin N104
reveal affinities between animals M137, 146–7, N13–14
sighing N73, 92, 93
shrugging N22–4, 73
sneering E125, MAC28v
understood by young children N37
use of nose and nostrils in M71, 153, N9, 45, 103
yawning M84, 93, 95–6
See also Anger; Avarice; Blushing; Contempt; Emotion; Fear; Laughter; Pain; Pleasure; Shame; Sneering
Extinction B19, 53, 245, C168, D25, E43
and number of species B21, T19, MAC58v
and origin of species B211, C234, D58, E48, E122
caused by non-adaptation to conditions B38–9, 61–2, 64, C153, MAC57r–57v
causes of RN85, 129, 133, 134, A9, B135, 148, C37, 39, 64, 99, 110, D49, T81
explains existence of taxonomic categories B27, 29, 37, C167–8, 169, 200, 203, 217
marsupials almost destroyed by D73
of great reptiles B53
of man B169
of species like that of individuals RN133, B22, 153
species never reappear after E105
taxonomic groups that are few in number are remnants of B39, C186
the foundation of geology E87
Eye
analogues of, in different groups of organisms MAC29v
blinking D111, M157
chamelion MAC29v
colour C204, D11
crab MAC29v
Darwin avoids having to explain first formation of D21
formation of C175, D21
in expression of emotions M70, 79, 105, 156, N9, 103
in Negroes C204
parasites of C227
primitive D56
variation in C158
See also Colour blindness; Darwin—Difficulties of species theory; Eyesight
Eyebrows
in expression of emotions M95, 106, 143, N6, 183
frowning E125, M58, 95–6, IBC, N58
Eyesight
and locomotion D56
perception M9, 61, 130, N82

Faculties B74, C198, M30, 108, N36, 46, OUN8, 40–1, 52–3

Faculties (*cont.*)
nature of N69–72, 76–82, 92, 94
See also Instincts; Intellect; Mind; Perception; Soul
Falcon C185
Falkland Islands
animals B219, C21, 23, ZEd3, 5
animals not subject to change B240–1
animals would change if conditions did C168, 169
birds C160, E56, ZEd5, 9, 11, QE[11^{v}]
bones on A58, 141
cleavage RN102, A52
fauna similar to that of South America E91
geology RN98, 102, 140, 142, 144, A66, 97, E137
plants B173, C24
rabbits B31, 54, C29, ZEd5
Silurian fossils RN142, 144
Faults (geology) RN37, 87, 97
Fear
an involuntary wish to avoid danger M52–5, 57
associated with ideas of the sublime OUN18–19^{v}
behaviour during M153
bodily responses to M54, 57, 148, 150, N7, 56, 59, 93, 97, 103
expression of M131, 149, N7, 103
in animals B231, C120, 197, D22, M24, 52, 140, 152–3, N6, 7, 25, 98
in birds B4, 198, N112, OUN36
in children N89, 113, 121
of earthquakes M16
of death B231, C197, M89, 101, 104, OUN38, 45, 49, 89
See also Courage; Instincts; Tameness
Fecundation
See Fertilization
Fecundity C143, E56
See also Competition; Selection, natural
Feldspar RN150, 176, A35, 46, 48, 61, 66
Fennel QE[11^{v}]
Fernando Noronha Island B56, T59
Fernando Po, Gulf of Guinea B209, 220, C183
Ferns
at The Mount, Shrewsbury M156
geographic distribution B156, 193, C16
hybrid B230, 235, C52, D157, 162, E90, M156, QE12
the first seeds to arrive on islands B193
Fertility C125, 152, DIFC, E75, MAC28^{v}
absence in young animals D40
affected by cross-breeding B120, DIFC, 134, 164, T13, QE4
between slightly altered species B209
gradation of C122
increase of B211, QE16
infertility as a test of species B122, E24
loss of C52, 178, D103, QE15
loss of in domestic animals B120, 123, C219, DIFC, 103, 108, QE12
loss of in hybrid offspring C52, D31, 112, S1^{r}
of crosses between hybrid offspring and parent C33
of hybrid offspring B2, C33, D9–10, 14, 15–19, 26, 70, E143, T2, QE18, 20, 22
of offspring of crosses between recent varieties C34
of plants E183, 184, QE15
of second generation of cross-bred animals D10
vigour of propagating powers D88
See also In-breeding; Fecundity
Fertilization D173, QE4a
aided by insects in plants E14, 90, T93–4, 103–5, 111–12, 135, 153, 177–8, MAC56^{v}, S2^{r}–2^{v}
artificial fecundation of moths D180
Darwin's experiments on QE1^{v}
difficulties prevent cross-impregnation QE[15^{v}]
drifting of pollen in wind E44, 123, QE[5]a
in invertebrates D156–7
of plants E13–15, 133, 143, MAC56^{v}, QE2, 6, 7, 8, 9, 12–14, 21.., 21
of plants by pollen from different species E133
of plants by pollen from same species E110, 143, 144
of varieties of apple trees E15, 110, QE1^{v}
pollen B96
semen D132, 171, 173, E70–1
small amount of spermatic fluid can effect E90
spermatic animalcules in mosses E100
Festuca E165
Figs QE6, 10, 14, [15^{v}]
Final Cause
See Cause, final
Finches B47, C70, ZEd16
bull C70
gold C249, D25, 95, E138
green D25, 95
Firola ZEd5
Fish B25, 162, 255, C20–1, 58, M98
adaptations of B113, 143, E134, T57, MAC29^{v}
agents of geographic dispersal of seeds A14, C245
Beagle specimens C20
descent B26, 88, 97, 170, C58, D49, 117, 133, E96
extinct species of B149
fresh-water D34, 151, E85, 91
hybrids E32, 79, QE18
instincts D173
migratory ZEd9
rain of E182
reproduction D114, 169, 180
viviparous T25
wide geographic range C20, 21, MAC167^{v}
FitzWilliam Museum, Cambridge
Darwin's memories of M82
Flamingo D47, MAC28^{r}
Flax B68, 275, E183, QE15
Fleas D91, N1, ZEd3
Flint
origin of A48
Fluorspar RN167
Flustra C49
Flycatchers
adaptations of B137, C61
pipra C253, 255
Folkestone, Kent
fossils E119–20
Florida
keys of A19
Forres, Scotland
shells at GR47
Fort Augustus, Scotland GR85, 106–7
marine shells at GR106–7
Fort William, Scotland GR23
Fossilization RN23, 165

absence of decomposition of buried body in Shetland RN139
casts of shells in calcareous rocks RN42, 61
of bones RN22, 68, B172
of wood RN28
See also Petrification
Fossil plants B117, 149, 202, C58, D133, T19
See also Coal; Fossil wood
Fossil record B15, 202, 205, C40, 137, E99, T1
Darwin's intention to study C77
each epoch characterized by great group of fossils D151
evidence for transmutation in E120–1, 126–7
extinction in B19–20, 72, 106, C205, E48, 128
fills up gaps between existing species C110, 156
first appearance of insects in B207–8
first appearance of Man in B207, 208
imperfect nature of B86–8, 239, 244, E66, T63
increase in complexity not related to passage of time T19
numerical relations between different deposits E166–7
transitional forms in C110, 257, E88, 119–20, 126–7
See also Extinction; Fossils; Gradation; Progression; Transmutation
Fossils RN113, 165, B88, 95, 99, C246–7, 257, E58, 166
absence of MAC29r
Darwin's specimens E99
distribution of bones by sharks RN8–9, 12
footprints in New Red Sandstone A109
in Australia B173, C130–1, 257
in Europe C184, D25, 39, TFRAG9, 79
in India B126, 250, C39, T79
in Java B179, CIBC
in North America B125, 172, 174, C37, 246–7, E32, 100, S3r–3v
insects in amber ZEd18
in South America RN85, B13, 53–4, 60, 62, 63, 133, 135, 137, C37, 39, 132, 167, 247, D68, 72, E40, T79
organic origin of A4
preservation is accidental GR13
relationship with modern species A28, B14, 99, 167–8, 174, 200, 202, 228, C77, 146, 246–7, E41, 92
scarcity of RN73
See also Apteryx; Didelphis; Dinornis; Extinction; Fossil plants; Fossil wood; Macrauchenia; Macrotherium; Mammoth; Mastodon; Megalonyx; Megatherium; Mylodon; Palaeotherium; Plesiosaurus; Shells; Sivatherium; Toxodon; Trilobite
Fossil wood RN45–6, 64, 161, 178
See also Coal; Fossil plants
Fowl
aboriginal species D66
behaviour B140, C61, 70, D163, N109, OUN36v
bones QE17
chickens C120, 236, N74
cross-breeding with pheasants B52, C164, D85, 87, 164
crossing between different breeds of QE4a, 20
Dorking QE20
effect of castration D156
hybrids D33, 153
instincts of chicks C51, E158, N4–5, 11
Malay C240, D180
old varieties of B181, C120, D66
plumage C138, 215, D101, 138, 154, 172, S7v
Polish C120, D163
purpose of horny point on beak MAC58r
secondary sexual characteristics C215, D113, 114, 147, 153–4, QE18
See also Bantams
Foxes B279
behaviour M147, N69
black B255
breeding in captivity QE18, 20
cross-breeding with dogs C189, D7, N44
cubs E140
facial expression M147
geographical distribution of B65, 104, 255–6, C223
in North America B65, 222, C45
intellect E46
in South America B222
Magellanic GR2, D44
on Chiloé B7
on Falkland Islands B7, 219
Foxglove QE6
France B133, C123, E121
geology RN38, 68, A40, C270
lakes RN69, A44
plants B157
See also Paris basin
Fraxinella T111–12, 135
Friendly (Tonga) Islands, Pacific Ocean
animals C27
Frigate bird C143
Fringilla D22
Frogs B255, D173, 180, E90, T1, N18, MAC28r
tadpoles E133
See also Batrachians
Frowning
See Expressions
Fruit trees
cultivation of C265, E162, QE21
degeneration D161
grafting E17, E153
propagation B1, D70, 178, E148, 153
run wild B83
varieties B83
See also Apples
Fuegians C79, M142, 153
absence of religious feelings C244
respect for human artefacts RN177
Fulgoridae TFRAG7
Fulica chloropus B193, ZEd8
Fungus B229
See also Mushrooms
Furnarius RN130, ZEd16
Furze
See Gorse

Galápagos Islands B70, 103, 157, 173, 193, 227, C54, D53, E32, 44, 104, 170, ZEd19
amphibia RN55
animal remains in strata being formed today RN55
arbitrary nature of species on MAC167r, QE10a
birds B7, 55, 70, 98, 100, 103, C145, 209
geographic dispersal on A41, B98, 100, 220, D65
geology RN69, 72, 165, 177, A32, B227
insects E170

Galápagos Islands (*cont.*)
land shells ZEd8
lizards RN55
mouse B220, C29, D65
ocean currents RN55
plants B157, 173, 193, C184, 194, E100, 104, QE[15v]
shrews B221
South American character of species B11, 98, 103, 160, C54, 209, E100, 170
species a result of geological elevation E44
species peculiar to separate islets B7, D53, E32, 100, ZEd8
tortoise RN55, B7, E12, ZEd12
volcanic structure of RN31, 38, 43, 110
Galeum saxatile T178
Galiopithecus C17, E19
Gastrobranchus C28
Gecko C28, D129, MAC28r
Geese B55, C40
behaviour E174, M153
Brent E168
Canada D9, 87
Chinese B139, D9, 13, 180, E169
hybrids DIFC, 9, 12–13, 25, 32, 87, E169, QE22
instincts D180
upland B55
Generation RN132, B1–5, 14, 26, 99, 111, 114–17, D19
a modification of law of growth E148
analogy between production by gemmation and by seed E162
and variation B3–5, 78, 219, C59, 84, 162, D177, 179, E48–51
asexual budding B1, D19, E164, 184, QE1v, 2
asexual individuals followed by sexual C235
comparison between sexual and asexual D68, 128–31, E77–8
Darwin's summary of facts relating to D176–8, 174–5
Darwin's views on B63–4, 204, 227, 228, C66, 162, IBC
development of two sexes C236, D41, 156–7, E60, 89, MAC56
differences between that of animals and plants D167
gemmation considered as artificial division D130–1, 172
importance of B2, 5, 230
interruption of gemmation by sexual reproduction E150–2, 153–5
length of gestation D155
likened to teleogony D168–9
mother's growth checked during D153
of plants by cuttings QE15
of zoophytes D180
overwintering eggs C162
possibility of degeneration during D161
production of seeds in plants E162, 163, 164–5, QE1v
separate roles of male and female in D178, E17
value as a guide to classification C33, 47, 149, 162, 195, 213, E91–2
See also Crossing & Hybridization; Fertility; Inheritance; Parthenogenesis; Regeneration; Spontaneous generation
Generation, spontaneous
See Spontaneous generation
Genera
and geographic distribution B13, 261, 273, C109
and origin of aberrant species B28
closely allied B229
Darwin's theory gives new meaning to C129
die together B29
origins of B21, 36, C138, 203
relative numbers of in Australia B77
relative numbers of species in B28, 39, 149
resemblance of species within B57, 79, 85
Geographic dispersal B102, 105, 115, 158, C240
adaptation of seeds for transportation D74, MAC53r, QE[5]a, 6, 10, [15v]
animals on drifting ice E166
bones carried by shark RN8–9
carcasses drifting at sea RN10, A88, C94
effect of winds A25, B11
experiments to determine effect of salt-water on seeds B125, C148, QE[5]a
means by which animals arrive on islands C29, D65, E21
means by which plants arrive on islands C100
of coconuts C148
of live animals by birds A14, QE19
of ova of shells C245
plants floating in the sea C249
stomach contents of animals C95, D180, ZEd3
transport of seeds A25, B124–5, 192–4, 234, C184
transport of seeds by birds T39, QE[5]a, 6
transport of seeds by fish A14, C245
transport of seeds by winds B11, 209, QE[5]a
See also Species, introduced
Geographic distribution B14, 138, 156, C80, D31
absence of endemic genera in Europe B200, C39
climate not a primary cause of D115
correlations between that of man, animals and plants D25
Darwin's theory of B164, C29, 69, 77, D23
deserts B235–6
division of the earth into regions B135, D38
each species has its own limits RN128, 130, C253
effect of climate RN128–9, B243, 245, C147, D115
effect of continents on species of neighbouring islands B10–11, 30, 116, 221–2, 278, C106
effect of geographic barriers RN52, 128, B103, 142, 209, C119, 251, D31, 61, E115
effect of isolation on species B15, 33, 75, 155, MAC167v, 29r
effect of previous configuration of the earth A11, B12, 95, 224, C247, D68, E100–1
endemic species at southern points of continents in Southern hemisphere C37
endemic species on continents B12–13
existence of two sexes limits extent of range C54–5
extent of range of mundane genera QE10a, 17
extent of range of species in endemic genera QE10a
great extent of range of fossil pachyderms C46
idea of creation does not help explain B243, C106
idea of 'propagation' of species helps explain B12–13, 212, 227
in Arctic regions B221, C36, 116
latitude more important than longitude for wide-ranging species C245
number of shells decrease with depth of ocean E122
of birds B67, 81, 98, 104–5, 124, 199, C50, 55, 88, 94, 253, D26, 31, 180, E101–2, ZEd15, 17
range of dioecious plants C245
of Edentata A5, 6, B94–5, 106, 133, 223, C36, 38, D68, 73
of fossils B200, C36
of insects according to heat of the earth ZEd18
of mammals B31, 67, 81, 91, 115, 249, C41, ZEd11
of man D25

of plants B76, 81, 93, 280, C102, 239, D25, QE17
of shells B204, C20, 21, 244
of species in topographic space parallels changes in species over time B17, 200, T41
on islands B7, 160, C18–19
on mountains B210, 235–6, C191, E105
range of species compared to genera C35
ranges of birds RN129, B137, C50, 99, ZEd12, 17
relationship of species in north and south hemispheres B66, C22, 28, C7
scarcity of animals outside their own area QE16
substantiates a belief in extreme antiquity of animals D59
tropics supposedly favourable to terrestrial mammals MAC29^{r}
value to systematists C45, 54, 88
wide range not favourable for preservation of adaptations C60
wide ranges of fish C20, 21, MAC167^{v}, ZEd9
wide-ranging species A11, B100, 104–5, 134, 187–8, 223, 245, C25, 116, E11
See also Alpine plants; Geographic dispersal; Islands; Species, introduced; Species, representative
Geological Society of London
collections in A63, 97
See also *Proceedings of the Geological Society of London; Quarterly Journal of the Geological Society*
Geological time
See Time
Geology
chronology is a sublime discovery E4
conducive to understanding species E5
considered as a book with missing pages D60
Darwin's aesthetic appreciation of M40
Darwin's conclusions applicable to whole world RN18
Darwin's difficulty of pursuing close trains of thought in M34, 90
effect of Darwin's theory on E87
Europe the birth place of the science RN73
See also Cleavage; Coral reefs; Dikes; Earth; Earthquakes; Elevation; Elevation & Subsidence; Erratic blocks; Faults; Fossils; Glen Roy; Laminated rocks; Metamorphism; Mountain building; Strata; Stratigraphy; Subsidence; Terraces; Time; Uniformitarianism; Veins; Volcanoes
Geranium T94, QE14
Germ
first B228, E83
Germs B26, MAC56^{r}
Germany
volcanoes RN68
Gibbons N115
Gilolo Island, Indonesia
animals E181
Giraffes B233, 234, N75
fossil T1
neck vertebrae MAC58^{r}
Glaciation
See Ice Age
Glaciers RN63, 135, B188, 195
grooved rocks A36
Glen Bought, Scotland GR87–9
Glen Bright, Scotland GR92
Glencoe, Scotland GR21–2, 23
Glen Collarig (Caol Lairig), Scotland GR50, 52, 57, 67
lake-theory improbable GR45
terraces GR37–46
Glen Fintec (Fintaig), Scotland GR54
Glengarry, Scotland GR101
Glen Guoy, Scotland GR58, 79
Glen Roy, Scotland GR35, 49, 50, 63
and absence of cracks after elevation A112
as evidence for equilibrium of land and sea A129
equal movements in roads of A107
Darwin's belief terraces are marine in origin GR64–5, 71–2, 80–1
Darwin's belief terraces are marine or lacustrine in origin GR50–1, 88
errors of previous observers in the area GR78
geology GR62–85
river-theory not applicable GR47–8, 50
terraces correspond on either side of valley GR29–30
terraces on east side of valley GR49–53
transported material deposited on terraces GR67, 68–70, 74–6
Glen Tarf, Scotland GR88–9
river GR103–4, 124
Glen Turrit, Scotland GR85, 125
terraces GR62–3, 86
Glowworms C221, N27, ZEd4
reproduction D40–1
rudimentary wings C215
Gnats
swarming ZEd18
Gneiss RN150, A54, 97, 105
Gnu B67
Goats TFRAG6
hair B48, C93, 214
resemble sheep B48, D66, E35
Goatsucker C112–13
God B216, C196
as the creator B101, 114, C184, D37, 54, 72, M1, 69
design B45, E23
mankind likes to think origin is godlike C155–6
man's estimate of C244, MAC54^{r}, 54^{v}
operates through laws of nature B101, 193–4, E3, M69–70, 135–6, 154, N12
origin of idea of M135–7, 151, N4, 11, 35
personification of OUN12
See also Bible; Cause, final; Creation; Religion
Godwit C209
Gold deposits RN163–4, 176, A100
Gold mines
in Cordillera RN176
in Mexico RN163–4
Golden pippen
death B63
longevity D165
production of crab apples B230
reproduction B72, 83, D70
Goldfinch D25
Gorse B107, E152–3, QE2
Gory (Gorée) Island, Africa A20
Gradation B44, 112, C222, D69
absence of a perfect chain of animals B209, C167–8, 216–7, E25, T81
a perfect chain of animals C161, 205
between animals and man C55, 222–3, 244, D22

Gradation (*cont.*)
between Apteryx and birds C207, 208
between taxonomic units B85, 224, 225
Darwin's interest in B226, C76
demonstrates accommodation to different external conditions B235–6, E72
exhibited in birds C80–1, 96, 157, TFRAG10, FRAG11, ZEd16
in fertility of hybrids T2
in the separation of the sexes D156–9
vertebrae exhibit passage from form to form T55, 57, MAC54v
See also Perfection; Progression; Transitional forms
Gradualism RN108–9, C74–5, E99
Grafting A180, E17, 77–8, 153, 162, QE11
Graham Island, Mediterranean Sea B192
Granite RN100, 120, A17, 81, 99, 147
decomposition RN33, 153, A145
gradation into other rocks RN88–9, A35, 149
previous extent of RN110
structure of RN79, 88, 100, A25, 61
Grasses
viviparous QE2
Grauwacke RN99, 164
Gravity
and cleavage of rocks A49, 51–2, 60, 62
law of B196, C166, M41, OUN40v, 41v
Grebes C160, 222
Greece C55
See also Antiquity
Greenland
alpine plants E115
geology RN159, A39, E100–1, 109
Greenstone RN88–9, 163, 170, A84, 110
Greyhounds
See Dogs
Greywacke
See Grauwacke
Groundsel QE6
Grouse GR126
black D72
cross-breeding with pheasant B189, D33, 106, E106
red D105
three species of D25
See also Capercailzie
Growth
modification of matter OUN34–5
reproduction acts as check to D153
See also Development; Regeneration
Guanaca ZEd6
Guanaco A9, 34, D38
Guanaxuato, Mexico
geology of RN170–1, 175
Guantajaya, Chile
salt deposits A44
Guava A88, C73, 93–4
Guayana (Guiana)
geology A87, 89
wild dogs C94
Guayaquil, Ecuador RN177, A102, C99
Guillemot ZEd15
Guinea fowl D114, 154
cross-breeding with pea hen D31
Gulls C164, D47, E168
Guzerat (Gujarat), India
maneless lion C47
Gypsum RN44, A38

Habits C125, 235, 240, D48, M75, N17, 29, OUN36v
become hereditary C105, N40, 42, 63, 67
become instinctive C171, N91, OUN48
become unconscious M104–5, N51
connected with mind C210, M83, 104–5, N51
Darwin's view of C199, 217
dissimilar in close species C125
easily acquired by animals M82
easily acquired by children C212, OUN36v
effects of change in on animal structure C63, 81–2, 85–7, 124, 160–1, 163, D147
men and animals exist by C232, M2
persist in different surroundings C105
repetition leads to C190, 217
retention of childhood N62
the origin of facial expressions C243, M52, 54, 107, 150
the origin of mankind's ideas of taste N26–30
value for classification C82, 165, 226, D102, 115
See also Behaviour; Education
Halimeda E155
Hamster E169
Handwriting
inheritance of E89, 184, IBC, M83
Hares D61, E139, N103
chased by dogs E75, N1, 44
colour C84, 252, E9
geographic distribution C22, 251, E13
kinds of B7, 80, 221, 262, C23, 251, D61, E184
tame E117
Hares, Irish B7, 80, 221, 262, D61, E184
Harpalus B256
Hawks A14, C111–12, 143, QE19
classification C111, 157, 222, D61
numbers of C111, D135, N112
pellets QE[5]a, 19
predator of other birds D135, N112
Health
See Darwin—Health; Diseases; Insanity; Parasites
Heart
action during emotion M52, 54–5, 57
Heartsease S2r–2v, QE2, 13, 21
Heat (geological)
conduction through rocks RN6, A139
See also Earth
Heaths QE7, 10a
Hebrides
animals C27
Heckla, Mount A39
Hedgehog
spines B87, T17, MAC28r
Helix QE10a
Hemiptera C107–8, 186
Hens
See Fowl
Heredity
See Inheritance

Hermaphrodites C245, E78
classification of D156
freemartins in cattle D152
in flowers C167
mutual coupling of B96, 232, E70, 155, 159
the basic form of animals D154–5, 162, 172, 174, E57, 80
structure D114, 158–9, 161–2, 174
See also Castration; Secondary sexual characteristics
Hermaphroditism
animals not subject to variation C169
a step in the scale of nature C167
correlated with liquid semen E70–1, 124, 155
Darwin's theory of C167, D154, 156–9, 161, E70, 80
different kinds of E80
Hermoso, M. RN69, 113
Herons A14, D95
Hervey (Cook) Islands, Pacific Ocean E10
Heteromera B56–7, 58, C108, E170
Himalayas B151, 242, C203
animals at great heights in E44–5
fossils T1
geology A2
shells D68
species from other regions in C36, 223–4
Hippopotamus RN8–9, B233, 242, D25, 52
fossil C216, D25, S3[r]
geographical distribution B234, 242, D25, E60
Hollyhocks E144, 183, M63, QE2, 14
Holocentrus ruber MAC54[v]
Homology
of flipper E25
See also Analogy (biological)
Hops QE9
Hornblende RN176, A15
Horns B233, D42, 101, S7[r], QE19, 21
absence B145, 255, E23, 113, 123
as secondary sexual characteristics D99, 172, E113
inheritance B145, D42, E123
time of emergence QE19
Horses B35, 72, 273, C inside front cover, D10, 66
behaviour C163, D10, M71, 85, 129, 144, 146–7, N7, 75, 90, 103, 117
breeds of B184, E112
cross-breeding between different breeds of B184, C inside front cover, 2
cross-breeding with ass B219, D113, T2, QE20
cross-breeding with zebra QE20
diseases E36
extinction in South America RN85, 113, B63, 222
fossil B62, 72, 106, E44
gaits C163, E127, N67
geographic distribution B35, 62, 133, 222, E175
insanity M15
in South America A3, B62, 72, 106, 133, N67
instincts D4, M101
intellect M71, 108, 131, 141
jumping M129, QE22
memory D7, M129, N72, OUN37
parasites B142, C227
persistence of good qualities in offspring of the racehorse Eclipse E112–3
size B106, E175, T2
species of C177, 186
structure D159, QE17, 20, 22
variation in D66, E103
wild C41, E175
Horticulturists QEIFC, 18, 21
Darwin's questions for seedsmen QE9
seedsmen supply country landowners E98
selection of seeds QE8
Hottentots B32, 189, 233, QE16
Hudson's Bay
terebratula from A37
Humming birds B278, C111–13, D118
Hungary
geology RN171
Hyacinth QE[5]a
Hyaena M71, 153
cross-breeding with jackal N43
fossil T1
geographical distribution C42, 92, D22, 25, 139, E60, N6
related to dogs E88
Hybrids B140, C219, 228, DIFC, 8, 19, 113, 159, 160, 180, E103
abortive genital organs C219
analogous to new species CIFC, D14, 73, E106
between different species of birds B189, C249, D9, 31, 32, 34, 86, 160, 163–4, IBC, T13
between varieties C34, 122
breeding among C164, 184
can live in the wild D103
compared with mutilated animals D13, 15–19
cross-bred sheep not so hardy as parents GR26
Darwin's theory explains mules C135
ducks B141, D180
evidence for in fossil record E120
ferns B230, 235, C52, D157, 162, E90, M156, QE12
fertility of B2, C33, D9–10, 14, 15–19, 26, 70, E143, T2, QE18, 20, 22
first offspring most like parents C228
fish E32, 79, QE18
geese D inside front cover, 9, 12–13, 25, 32, 87, E169, QE22
gradation of fertility in C122, E107, QE11
health D163, QE21
heterogeneity of siblings B250, D5–6, 15, 33, 104, 106, 180
infertile B10, D14–16, 17, 19, 31–2, 85–6, 112, QE4
infertility of sibling crosses D134
instincts GR126
intermediate between parents in character B34, 120, 219, D7, 31–2, N43–4
likelihood of being hermaphrodites D159
mongrel loses all traces of parentage TFRAG5
nature of instincts in D33, M101, N43–4, IBC
nature of offspring of two half-bred animals QE3, 4
not exactly intermediate in character B5, D8
number of generations that can be reproduced QE12
occur naturally in the wild QE5, [5]a, [6[v]]
peas MAC56[v]
pheasants QE19
plants D70, E103, 143, T175, S1[r]–1[v], QE1[v], 3, 8, [15[v]]
reproductive behaviour D90, 177, QE19
reptiles E79
resemble either parent C133, D154
resemble one parent only B203, C164, 189, D3, 5–6, 7–8, 13,

Hybrids
- resemble one parent only (*cont.*) 15, 88, 90, 92–4, QE3
- tendency to revert to parents' characteristics B33, 203, 229, DIFC
- uniformity of siblings D89, QE2

Hydra D177, N13, ZEd17
Hydrangea QE9
Hydromys D115
Hymenoptera M98–9
Hypnosis C211

Ibex TFRAG6
Ibis B6, D47
Ice RN141, 153
- and subterranean isothermal lines A125, 135
- freezing of the sea bottom RN115

Ice Age
- See Drift deposits; Erratic blocks; Glaciers; Glen Roy; Terraces

Icebergs RN99, 114, QE17
Iceland B159, 220
- absence of endemic species B188, 255–6
- birds ZEd9, 19
- flora B247
- geology RN90, 159, A39, 120, B188
- hornless cattle B255

Ichneumon C173
Ichthyosaurus C205, E157
Icterus E160
Igneous rocks
- amygdaloid A60
- rate of cooling RN123
- See also Basalt; Granite; Lava; Mountain formation

Iguana ZEd3
Imago D40, N107
In-Breeding B176, D48
- and loss of passion D163
- effects of B6, 7, 13, C133, D85–7, E10
- in pigeons D88, 104
- leads to degeneration in dogs D180, QE18
- leads to diseased offspring C175, E45, 67
- leads to infertility C232–3, D103
- repugnance to D177

India B220, C227
- animals B94, C26, 39, 40, 46, 47, 227, D60, E175, 181, M152
- cattle B131, 136, 139, 155, 217, 234, 273–4, C121, D65
- comparison of organisms with those of Africa B242, C46, 224
- fall of fish in E182
- fertility of animals QE12
- fertility of man QE12
- forms intermediate between Africa and America B223
- fossils A28, B126, 250, C39, T79
- geological formation of Coromandel Coast A146
- geology A27, 28, 145
- Hindu fear of death OUN38
- parrots QE1
- races of mankind C91
- salt deposits A143
- See also Himalayas
- Indians, location of habitations RN125

Indian rubber RN178
Indonesia
- absence of quadrupeds in B219
- American species in C17
- ancient climate E39
- animals B80–1, 242, C17, 24, 46, 50, E20–1, 170, 173–6, 177, 180
- Australian species in C19
- elevation and subsidence of E45
- floating marine confervae RN5
- relationship of animals with those of Africa B242, C38, E39
- volcanoes RN44, 90
- See also Borneo; Celebes; Ceram; Gilolo Island; Java; Lao Choo; Malaya; Moluccas Islands; New Guinea; Philippines; Sumatra; Timor

Induction (philosophy) E51, N14, 16
Indus, river
- geology RN68

Infusoria C15
- classification B229, C104
- fossil B150, C146, E128
- numbers of C143
- persistence of simplest forms RNIFC

Inheritance B47, 118, 145, C1–4, D19
- analogous in grafting and sexual generation E77–8
- changes not acquired by parent can be handed down to offspring D127
- characteristics that appear only in adult plants T95–6
- Darwin's theory leads to study of B228
- Darwin's view that characters are inherited at a corresponding age and sex T95–6
- effect of father on character of offspring GR1, D9
- effect of mother's imagination on foetus D104, 171
- extent to which young resemble parents indicates age of species D148
- influence of parent on offspring D92–4, 172, 176, E17, 26, 110, 127
- in twins D131, 152, 175
- likened to teleogony D168–9, E79
- of accidental variations B239, C83–7, 221, D19, E148–9, QE16, 17
- of adult characteristics by young offspring $S7^r$–7^v
- of behavioural traits D164–5, M156, OUN8, 8^v
- of colour B70, CIFC
- of deformities C65, 66, 83–4, D14–18, 108
- of diseases D18, QE16
- of facial resemblances E108
- of family resemblances C219, D35, 151, M1–2, 42, 44
- offspring intermediate between parents in character GR25, B33–4, 183–4, 219, C3, T15
- offspring of black and white parents B32, C228
- offspring resemble either parent D180
- offspring resemble father B32, 203, C3, 121, D44, 47, 165, 171–2, 176, 178, 180, E123, QE21..
- offspring resemble mother GR25, 125, B32, 203, C3, 228, D5, 43, 88, 165, 171, 176, 180, E35, 78, M1
- of grandparent's characteristics D42
- of long-established traits D16–17, 18, E106
- of mutilations C83, 232
- of non-deleterious variations T37, $MAC53^v$
- old breeds have greatest effect on characteristics of offspring CIFC, 1–3
- perpetuation of cultivated varieties of plants by budding D128–131

- persistence of peculiarities C133, D172, E91
- reversion B33, 203, 229, DIFC
- the same laws are consistent throughout nature E36
- through the female line B32, D147, 165, 168, E108, QE16
- See also Atavism; Colour; Mutilations; Teleogony; Yarrell's Law

Injury
- accidental B4
- muscular response to D166, M52, N56
- recollection of C172
- See also Mutilation; Regeneration

Insanity C226, M13, 14, 16, 17, 20, 26
- a loss of will OUN17^{v}, 25
- and memory OUN32
- characteristics of M21
- compared with dreaming M21, OUN17^{v}
- compared with sleeping M25, 35, 127
- consciousness of M14–15, 17, 19, 80, 116, 127
- cures for M18, 42
- delirium N111
- feelings during M14
- forgetfulness of M15, 35, 78, 116
- gradation between sanity and M13
- ideas return to those of childhood M21
- in animals M15
- in civilized countries C226
- intellect impaired by paralytic stroke M9–11
- likened to double consciousness M116–17
- limited to particular ideas M18
- mania M13, 26, 94, 114, 127
- moral causes M16

Insects B207–8, 218
- abortive organs B84, C186, 215, 216, D172
- acrid secretions MAC29^{v}
- act in pollination of flowers E60, 90, 165, T93–4, 103–6, 111–12, 135, 153, 177–8, MAC56^{v}, QE[5]a, 6^{v}, 9, [9]a, 11, 14
- as food for other species B108, C105, 112
- as parasites B108, C174, M63, ZEd3
- behaviour C197, M64, 97, 99, OUN35–6
- butterflies D22, 62, 63, T57FRAG11, QE10
- caterpillars C173, D40, 41, 62, 63, 170
- classification B116–17, 208, C33, 107–8, 162, M98
- classification by Quinarian system B27, 56–7
- Darwin to examine collection of B47
- eye MAC29^{v}
- food B218, D25
- geographic distribution C21, E19
- in amber ZEd18
- instincts B74, 207, 218, C210, 221, M98–9, N3, 69, 70, 77, 81, 107–9, 115
- locomotion C107, D55
- metamorphosis B200, C162
- origin of B108, 171, 206
- reproduction D40–1, 76, 116, E71, 80, 114, 177, M64, N69
- structure B206, C114, 218, 236, D159, MAC57^{v}
- submarine ZEd17
- swarming ZEd18
- variation B280, QE1
- See also Bees; Beetles; Carabidae; Hemiptera; Heteromera; Instincts, social; Longicornes

Instinct B3, 197
- abhorrence of perverse N99–100
- acquired C76, N11, OUN36^{v}
- acquisition of C134, M101
- acts in concert with conscience OUN43–7
- a material property of the brain N42–4
- animal migration not explained by Darwin's theory of N72, 77–80
- awakening of dormant C235
- blending of QE4a, 22
- can be stopped by intervention of another instinct N1–2
- changes of under domestication B197, E174, OUN38
- continuity between those of mankind and animals C196
- Darwin to study B227, 228, C76
- distinction between reason and B34, C198
- effect of crossing on GR126, C30, D180, N44, QE22
- existence of some that are unintelligible N105
- hereditary B198–9, C134, MAC59, OUN36^{v}
- impulse to perform N67
- innate C70, 235, E158, M126, N11
- likened to an unconscious memory C171, M7, 110, OUN11, 37, 38
- loss of C173
- maternal D154, M64, OUN8^{v}, 38
- modifications of C159, T2, N69, OUN36^{v}, 37
- nature of C236, N33, 48, 68–73, 77–82, 112
- nature of in animals B34, C212, M59, N29, 68–9, OUN36^{v}
- number of OUN42, 49
- of birds B4, 105, 151, 218, C51, 67, 68, 69, 70, 96, 134, 199, T2, N4–5, 71, 112
- of dogs B165, C134, N1, 17, 43–4, 47, 105, 185
- of insects B74, 207, 218, C210, 221, M98–9, N3, 69, 70, 77, 81, 107–9, 115
- of horses D4, M101
- of plants N48, OUN36^{v}
- of mankind B6, 18, 93, 161, M31, 58–9, 96, 108–9, 122, 124, 126, N72, 90, 99, 125, OUN26, 37, 49, 116
- of savages C150, 172, D4
- originate in habits C232, OUN48
- pleasure felt in fulfilling GR12
- relationship with physical structure C51, N71, OUN38
- sexual B161, D138–9, 173, OUN37, 50^{v}
- value for classification B9, M98
- See also Association of ideas; Emotions; Habits

Instincts, social
- as the origin of moral feeling M132, 150–1, N2–3, OUN53–4
- common to man and animals OUN42
- conjugal OUN49
- feelings towards a leader OUN49
- increase with increase in sociability OUN50
- in insects N107–9
- in mankind M76
- origin N107–9, OUN30^{v}, 50–1
- pleasure felt in fulfilling OUN42–4
- the most beautiful of moral sentiments E49
- See also Civilization; Education; Love

Institut A15, 20, 26, 31, 32, 39, 42, 76, 102, 103, 115, 135, B94, 133, 135, 136, 150, 151, 163, 165, 200, 250, C15, 95, 146, 183, 205, 227, 241, 257, D49, 62, 64, 68, E25, 35, 42, 58, 70, 72, 88, 100, 123, N33, ZEd6, 7, 14, 18

Intellect B215, 244, 252
- absence of improvement in N47
- advantage of considering as gradually developed N101

Intellect (*cont.*)
and natural selection in human beings E63–5
civilizations inherit powers of reasoning C72
difference between that of mankind and animals B214, E46, N62
nature of B74, C198, E46, N48, 62, 92, OUN5
nature of weak N40, 60–1, IBC
of animals B34, C196, E46, M49, 62–3, 64, 128, 131, 141, N62, 70, 90, 117, OUN8^{v}, 8a, 36^{v}
origin in laws of organization MAC58^{v}–59
powers exhibited by chess playing M99–101
powers of mental arithmetic M92, 99–100
progressive improvement in D49, N62
reason originates in acquisition of memory N18, 21
See also Insanity; Mind
Intermarriage
See Crossing & Hybridization
Inverness, Scotland GR124
Inverorum, Scotland GR19, 21
birch wood GR24
Iquique, Chile
elevation of A39
fresh water at RN31
shells A39
Ireland A126, C192
number of doubtful species on QE10a
species B107, E91, QE10
Iris T111, QE[5]a, 7
Iron RN156, 160, A32
oxidation RN165, 172
Iron ore RN160, 161, 165
Iron oxide RN165, 167, 168, 172
Ironstone A93, 94
Irritability
of plants OUN34, 35
See also Nervous system
Ischia, Italy RN90
Islands B30, 188
and creation of species B30, 160, MAC167^{v}
gales of wind between would blend species E44
geological origin of B80, 221–2
large mammals not found on B115
number of species on B158, 160, 221
numerous doubtful species on QE10a
origin of new species on RN127, B7, C25, E170
peculiar species on B12, 67, 156, MAC167^{v}
plants B81, 192, C100, E104–5
relationship of organisms to those of neighbouring continent B10, 11, 30, 31, 116, 138, 156, 160, 221, 227, 278, C6, 106, 129, 176, 184, D74, E91
See also Galápagos Islands; Geographic distribution; Isolation; Volcanic islands
Islas de Sonda, Indonesia
animals E170
Isle of France
See Mauritius
Isle of Man B163, C193
Isolation B20, 155, C53
compared with time as an agent of change in species E135
constancy of species during B6, 7
effect in the origin of species B7, 15, 17, 24, 209, D23
effect on domesticated races D23–4, 44
local varieties formed slowly C59
species formation favoured by MAC167^{v}, 29^{r}
See also Islands; Species
Italy
earthquake in Calabria RN75

Jackal C92, N44
cross-breeding with dog C210, D90,IBC, N44
cross-breeding with hyaena N43
in Zoological Society of London Gardens D7, IBC
Jaguar C63, D140, E97
geographical distribution B196, 222–3
Jamaica C42
Japan C36, 203, 225, 241
Java B241, C213, T51
alpine flora QE17
animals C14, 26, 28, E175
birds B164, C83, D40
fossils B179, CIBC
rhinoceros B67, 82
volcanoes RN137
Jay C68, E138
Jersey
wheat M155
Journal de physique, de chimie et d'histoire naturelle RN91, A18–19
Journal des mines. RN172
Journal of the Asiatic Society of Bengal A143, 144, 145, D21, E176, 177, 178, 179, 181, 182, T1, 2
Journal of the Royal Geographical Society RN138, 139, 143, 148, A30, 37, 79, 87, 88, 89, 94, 117, 128, 135, B48, 93, 131, 132, 151, 235, 236, 272, C92, 93, 94, 101, 102, 131, 268, QE18
Joy
See Pleasure
Juan Fernandez B156, 157, 173, 193, 278, D74
birds C106
elevation RN80
fauna similar to that of Chile B11, E91
plants C100, 106
Jura
erratic boulders in A110

Kangaroo C13–14, M132, MAC28^{r}
Keeling Islands RN129
animals E177
plants B157, 234, C100, 184, QE15^{v}
Kelp RN52, 93–4, 140–1
Kerguelen Land RN140–1, B173
Darwin's interest in RN138
Kilfinnan, Scotland GR124
Kingfishers
adaptations of B55, C114
behaviour C114, E138
geographical distribution C26
great ZEd10
plumage B70, D147, 148
Kordofan, Sudan C42

Labiatæ E104
Laburnum QE13
Lake Constance, Germany
fossils E11

Lakes
formed by geographical barriers A71
fossils in lacustrine formations T89
let out in steps A44
shores of A98–9, 102–3
subterranean RN128
terraced alluvial deposits A71
unequal elevation of RN69
See also Glen Roy
Laminated rocks A45, 49, 115
and cleavage RN5, A45–9, 53, 60
origin of laminae RN23, 45
Language B244, N31, 48, 65, 88
and signs N102, OUN5
as a distinction between man and animals C154, M58, 96–7, 153
children can understand before they can talk N13, 64–5
diversity of C53
grows and alters over time N65, OUN5–5v
origin N60, OUN5, 13
origin in animal sounds N20, 31, 39
origin in singing N18
Lao Choo, Indonesia
horse E175, T2
Lapland
alpine plants E115
flora QE21..
insects swarming ZEd18
La Plata, Argentina
absence of fossil shells T80
animals ZEd2
fossils A5
Larks C113, 256, D95, 102, TFRAG11
Larkspur T94, 111, 153, QE[5]a
Las Tres Marias Islands, Pacific Ocean
animals E13
Lathyrus B187
Lathyrus pratensis T104
Laughter
caused by tickling N103
follows smiling in infant development N58
in monkeys M145
involuntary N6
origin in animal expressions M84, 92, 96
Lava RN17, 87, 99, 143
bubbles in RN90, 159
flowing uphill RN159
fresh appearance of old RN17
Law
all living matter governed by B22, 43, N97, OUN34
as the will of a superior being M55r, 69–70
of adaptation B21, 45
of balancing of organs B114, E157, QE17
of change B228, C164
of generation B101–2, E148, MAC53r
of hybridity B236, C30
of organization T17
of population E3
of progression E70
of repugnance to crossing in the wild B5, 33
of the inorganic world OUN34
of unused organs being absorbed D166
of variation E70
science based on D36–7, 67, N36
See also Gravity; Science
Lead RN161, 167, 169, 178
Leeches ZEd6
Leeks T106
Leeward Islands C42
Lemna E162–4
Lemurs B249, E159, N74
Leopards E144, 174, M152
hunting TFRAG6
Lepidosiren E168
Leptosiphon densifolium QE[5]a
Letter Finlay (Leiter Finlay), Scotland GR100
Lice D91, M156
Life
and fecundity C143
Darwin believes useless to speculate on origin of C58
defined as matter united by certain laws OUN34
exists at great depths of ocean RN19, 114, 140–1
final cause of B5
intimate relation with laws of chemical combination C102
number of living beings in existence C146–7, D134–5, E85, 95–7, 108
original germ of B228, E83
origin of B35, E83
principle of C210–11
See also Monads
Lifespan
causes for relative differences in MAC56r
determined by fertility MAC28v
duration RN inside front cover, B23, 63, 64, 72, 81, 223
duration connected to gaps in fossil record B35
duration connected to perfection of species B29
duration in mammals B223, C234
effect if endless B5
individuals of one species die together B73
prolonged by budding D165
shortness of B2, 23, 27
Lily B180, T93, 94
Lima, Peru
animals ZEd13
earthquake RN75
subsidence A153
Lime RN121, A41
Limestone RN7, 22–3, 93, 171, 176, A119, D133, E99
absence in intertropical regions RN41
animal origin of RN22–3
formed in shallow water RN19
fossils in RN36–7, 61, D133
silver veins in RN164
solubility RN29–30
Limestone, artificial A77
Linaria T177
Linnet D95, ZEd11
Linum QE15
Linum flavum QE14
Lions D140, N20, QE16
breeding in captivity QE20
cross-breeding with tiger C228, DIFC, 8
fear N98
maneless C47

Lithospermum QE[11v]
Lizards B56, 149, 248
flying MAC29v
fossil B205, 206, 208, C57–8
geographical distribution E21, TFRAG10
sailing MAC29v
tail regeneration D129
Llamas B35
extinction of fossil species RN129
Lobelia QE21..
Lobster MAC28r
Loch Dochart, Scotland GR27
buttresses of alluvium GR13–18
Darwin's search for marine remains in GR17
marine origin of alluvium GR17–18
Loch Lochy, Scotland GR98–100, 105, 122, 124, A115
Loch Ness, Scotland GR102, 105
Loch Oich, Scotland GR92, 98, 105, 122, 123
Loch Spey, Scotland GR78, 79–80
Loch Tulla, Scotland GR19–20
marine origin of alluvium GR20
Locomotion D56, E113, N48, 92, MAC28r
Longicornes B56
Lotus corniculatus T104
Louse C233–4
Love M101, N9, 59
affection between animals N115
Luminescence ZEd10
Lupins T103, 104, 111, QE11
Lychnis B275, E129, QE1v, [9v]
Lycopodium D133

Macquarie Island, Antarctic Ocean C19
parrots C61, 99
Macrauchenia RN113, 129, A9, B231, C132
Macrotherium C257
Madagascar B220, E90–1
animals B190, C26, 205, 216
plants B187
soundings off coast RN98
Madeira B220
birds C160, QE[11v], 18
rabbit on Porto Santo QE18
shell deposits A72, QE1
Madrepores ZEd10
Maer, Staffordshire C249, 257, QE6, 8, 13–[13v]
Darwin's observations of waterfowl at N111–12
Darwin's observations on dogs at M24
Darwin's observations on plants at T91, 93, 99, 105–6, 135, S2r
Magazine of Natural History B30, 138, 139, 178, 199, 212, C29, 70, 149, 192, 194, 197, 198, 199, 222, D inside front cover, 25, 26, E124, 169, N1, QE[5]a, 9, [15v], 17
See also *Annals and Magazine of Natural History*
Magazine of Zoology and Botany B7, 31, 67, 92, 126, 189, 276, C160, 162, 170, 184, 185, 186, 189, 192, 228, 235, ZEd1, 7, 17, 18
Magellan, Straits of
birds ZEd12
submarine channels RN86–7
Magnesia A119
Magnetism
and flight of bees QE[5]a
connection with cleavage A62
Magnolia QE[5]a
Magpie D147
Mahé island, Seychelles
Cocos do Mar C148
Maipo, Rio, Chile RN20
Maize C40
Malaria E178
resistance of Negro to D24
Malaya
animals C17, D60, E176
family of albinoes in E176
Maldonado, Uruguay
birds C163, ZEd16
Maldonado
limestone RN7
Mammals
brain B226
carnivorous B223
classification B27, 42, C130
extinction B40, D72
fossil B88, 206, 223, 278, C57, D25, E58
geographic distribution A11, B15, 81, 95, 115, 169, 223, 234, D31, 115, E20, 40
hair D132
lifespan B23, C234
mammae D62, 132, 155
mammae in males B99, C215, D61, 132, 169, 172, E57, 147
origin B15, 40, 88, 89, 112, 196, 202, 204–6, 220, 263, D39, 132, 174, MAC54v–55r
web feet in C57, 252, E63
Mammoth A3
Mangaia Island, Pacific Ocean
bats E10
Manila, Philippines
animals E19
Mankind B49, 119, 248, C77, 171, 196, 223, D49, E64, M102, 138, 151, N49, 69, 81, QE4
acquisition of reason B244
albinism B119, E176
castes E166
common ancestor of races of C154, 217
crossing between races B32, 33, 34, 68, 179–80, 181, 189, D136, QE16
Darwin's theory explains origin C77–9
differences between races C204, 217, M75–6, N3–4
diseases C174, D1–4, E178
extinction of B169, E63–5, T81
future blending of races of D38–9
future of B49, 226–7, 228
geographical distribution D25, 65, E31, 35, 65, 66
geological history of B207, E65, 182
habits C163, 198–9, M82, N47, 100
hair B119, C178, D24, E124
imagination D155, M30
inheritance in B119, CIFC, QE16
instinct B6, 18, 93, 161, M31, 58–9, 96, 108–9, 122, 124, 126, N72, 90, 99, 125, OUN26, 37, 49, 116
national characters of M85–7
nipples B84, 99
non-miraculous origin C55, E49

  origin B78, 207, 231, 244, C53, 55, 156, 204, 234, D162, 170, 174, E66–8, 89, M84, N49
  originates from animals C74, 154–5, 196–7, 223, D61, E47, 65, 68–9, 134, OUN8^{v}
  prehistoric arrow heads in Suffolk ZEd20
  produces civilization B4, D54, E47, OUN7–7^{v}
  reproduction B6, 93, D57, 68, 99, 162, 175, M104, N59
  races not all equally related C138–9, 140–2
  races of B68–9, 152, C91, TFRAG5
  rudimentary organs B84, 99, C215, D132, 158, 169, 172, E57, 66–7, 89, 147, MAC54^{v}, 58
  sense of smell M109, N109
  sexual instincts M71, N41, OUN37
  similarities between races D71, OUN38, 42
  single origin of all races C174, 234
  species of B34, 169, 209, OUN38
  struggle between races of E63–5
  transmission of mental characteristics C163
  tribes of B33, 34
  varieties of B215, 262, D25
  See also Civilization; Disease; Hottentots; Insanity; Instincts, social; Intellect; Language; Memory; Mind; Negro; Soul
Mantis MAC28^{r}
Maoris
  See New Zealand
Marble A19
Marianas Islands, Pacific Ocean
  animals C14, E42
  cultivation of foreign plants on E21
Marsupials
  ancestor of mammals C36, D132
  bones MAC58^{r}
  brain C47
  classification B141, 242, C47, D72
  diversification in Australia D72–3
  fossil B87, T19
  geographic distribution B15, 106, C36, 38, D68, 73, E41
  prehensile tail MAC28^{r}
  rudimentary organs D132, E57
  without pouch T25
  young D41
Massachusetts AIBC
Mastodon B53, E65, S3^{r}
  extinction B62
  geographic distribution A3, 5, B80, 135, C46, E32, 173
  nature of country when alive RN85–6
Materialism C166, M19, 57
  does not tend to atheism OUN37
  See also Atheism; Metaphysics; Mind
Matter OUN34, 37
  operates through laws of attraction OUN39–41^{v}
Mauritius B219, 234
  animals B160, 190, 234, C20, 22, 25, 28, 183, ZEd1, 12
  geology RN71–2, B222
  introduced species B255, C22
  plants C265
  relationship with Madagascar B166, 220
Meall Doire, Scotland
  geology GR32
Mediterranean Sea
  coral reefs A26
Megalonyx B53
Megatherium A117, B12, E32, S3^{r}
  ancestry B20, 54
  extinction RN85, B20
  relationship with armadillo B20, 54, 69–70
Melegethes T105, 106
Melons QE[11^{v}], 14
  experiment in hybridization QE1^{v}
Melville Island, Arctic Ocean A30, C224
Mémoires de la Société de physique et d'histoire naturelle de Genève A20, E184
Memoirs of the Wernerian Natural History Society A36, C166, QE10
Memory C173, N18, OUN37, 44
  analogous to instinct N46, 63
  and insanity OUN32
  as a repetition of sensations in brain M61–2
  dormant M7
  for poetry and music M8
  hereditary OUN37–8
  impaired by age M12, 21, 22, 26, 55, 78, 79, N60–1
  impaired by illness M43–4, N43
  impaired by paralytic stroke M9–11
  in animals D4, 12, N72–3, 90
  in birds C50, 255
  nature of M28, 29, 46, 50, 110–11
  recollection after long interval C172
  the origin of plant sensitivity N50
  unconscious workings of C241, OUN31
  vividness of first thoughts C242
  weak M2
Mendoza, Argentina
  former extension of sea to A71
  level plains RN155
  tranquility during earthquake at Quito RN158
Menyanthes QE13
Mephites ZEd17
Mercurialis QE[9^{v}]
Mercury RN178, A32
Mesites bird C205
Metals RN145
  at core of Earth A21
  chemical activity of A29, 119, 149
  in volcanoes A116
  magnetic properties RN162
  origin of veins of RN163–9
  powers of conduction of heat A121–2, 124
  See also Copper; Gold deposits; Iron; Iron ore; Lead; Mercury; Nickel; Silver deposits
Metamorphism (geology) A16, 105–8
  metamorphic rocks RN6, A29, 48, 94–6, B201, C101
  nature of RN21, A16–17, 91
  near surface of earth RN156, C101
  of clay slates A94, 95–6
  of shales RN63
  of slate A15
Metamorphosis E154
  difference between parent forms and young D55, 57
  not akin to transformations of species E158
  not wonderful D62–3
  of crustacea ZEd1
  of fish E90
  of insects B200, C162, D62, 63, M63
  of plants B200

Metamorphosis (*cont.*)
of tadpoles E133–4
of vertebrae E89
Metamorphosis, theory of
See Morphology
Metaphysics D54
Darwin's theory contributes to M84
Darwin's theory leads to study of B228
experience rests on instincts OUN33
foundations of N5
legitimacy of arguing from man to animals N49
no natural starting place for reasoning about human nature C218
Meteorolite A21
Meteorology
See Climate
Meteors RN172, A22–4
magnetic properties RN161–2
Methodology
See Science
Mexico
difficulty of extraction of nickel RN160
earthquakes at Acapulco RN177
elevation of land A114
Gulf fouled by sand bars A130
gold RN163–4, 176
limestone RN171, 176
metalliferous ores RN160, 161, 163–4, 166–9
meteors RN172
offspring of interracial crosses in B32
opinions of miners at Potosi RN106
porphyry RN163–4, 170–1, 176
rocks RN164, 170, 176, A26
silver RN162, 163–4, 166, 168–9, 176
slate RN170, 171, 176
Mica
hardness of Chilean formations GR24
Mice B100, 104, C22, D22
and mankind D65, ZEd12
as prey C185, N37, 44
classification B250, C29, D115
geographic distribution B65, 82, 250, C24, 35, 45, 252
numbers E140
on Galapagos Islands B220, C29, D65
Mice, Harvest MAC28[r]
Midwifery
accidents during QE16
Migration
birds follow narrow bands of C253–4
birds return to same place for many years M32
different kinds of C253–5
geographic extent of QE19
of birds B100, 104, 217, 242, C36, 68, 159, 160, 253–5, IBC, D4, N72, 76–81, IBC, ZEd12, QE[11[v]], 19
of birds with full stomachs C40
of fish ZEd9
of turtles E138
role of sun N76
Millepora ZEd3, 5, 10
Milvulus B137, C111, 112, 124
Mimosa B151, E184, EIBC
Darwin's experiments on QE5
Mind B207–8, 228, C198, 218
a gap between man and animals B2, 208, C244
a thinking principle C210–11
connected with organization M26–7, N5
continuity of between man and animals C222–3, OUN8a
differences in between the races of mankind C196, E46–7
effect on reproduction B3, 197, 219
effect on the body N52–3
enlarged powers of D118
sequence of trains of thought M34–5, 80, 90–1
the product of the action of the brain C166, 211, M46, 61–2, 101, 131, N38–9, 43, OUN37, 38, 39–39[v], 41
thoughts become habitual M46, OUN11
See also Association of ideas; Consciousness; Faculties; Insanity; Intellect; Memory; Perception; Phrenology; Will, free
Mindanao, Philippines
animals E175
Mineralogy RN125, A32, 111, B201
See Basalt; Chalk; Chlorite; Clay; Clay slate; Crystalization; Crystals; Diorite; Epidote; Gneiss; Granite; Grauwacke; Greenstone; Igneous rocks; Ironstone; Lava; Lime; Limestone; Mercury; Meteorolite; Meteors; Obsidian; Olivine; Pearlstones; Pitchstone; Porphyry; Pumice; Sandstone; Selenite; Serpentine; Shales; Slate; Syenite; Tosca; Trachyte; Trap rock; Tufa; Volcanic rocks
Minerals
See Cobalt; Copper; Feldspar; Fluorspar; Gold deposits; Gypsum; Hornblende; Iron; Iron ore; Lead; Magnesia; Mica; Nickel; Pyrites; Quartz; Silica; Silver deposits; Soda
Mining
for gold RN163–4, 176
for iron RN160, 161, 165
for silver RN162, 163–4, 166, 167–9, 176
opinions of miners RN106
Miocene period B30, 59, 202, 243, 245, 278
Mites B208, 229
Moa Island, Indonesia
absence of wild animals E180
Mocking birds B7
Modesty
absence in animals N25
facial expression N25
shyness N17
See also Shame
Mojo Islands, Indonesia
behaviour of birds at missions of C50
Molluscs B113, 223, D157, N51, MAC29[v]
Darwin's intention to experiment on B248
eye C158, MAC29[v]
fossil B167, 168, C40, E32, 85, 164
hermaphroditism B96, D158, 162, E155, 164
hibernation RN65
lifespan of species B252, C234
organization B97, E60
ova C245, D180
reproduction B96, 232, D156, E90
See also Shells
Moluccas Islands, Indonesia
animals C13, 17, 27, 28
kingfisher resembles European form C26
monkeys E180–1

See also Amboina; Bourou
Monads B18–19, 22, 29, 35, 78, C206
Monkeys B133, 135, 214
anger M94, 107, N94
behaviour C212, D136–7, E18, M28, 106, 138, 156, N115
expressions D22, M106–7, 142–3, N183
extent of geographic range B135, 242
fossil B94, 126, 133, 250, T79
geographic distribution B154, 164, 250, C22, 46, 92, 183, 249, E18, 19, 177, 180–1, M138
parasites of M156
prehensile tail MAC28[r]
recognize sexes in other animals DIFC, 137–8, M138, 156
sexual characteristics C204, D138, NIFC
the ancestor of man B169, 214–15, C74, 204, E68–9, M128
voice C79, M94, 97, 107, 109, 137, N18
See also Cercopithecus
Monkeys, spider M107, 143, N115
Monoceros E98
Monocotyledons
flowers D177
in coal formations B150
Monotremata B168
Monsters B111, 114, 161, 190, E45, 88
and artificial selection C52, D107
and disease B119
classification of D161
correlation of deafness with eye colour D11
Darwin thinks worth studying C226
deformities of foetus D57, 67, 164, M25
effect of inorganic substances A180
effect on pregnancy B181, D14, M25
hereditary C4, 59, D14–18, M25
hermaphrodite D174
in domestic animals B175–7, 190, D108, 112, E19, 182
in mankind B119, 181–2, M25
in plants A180, C267, QE2, 6, 7, 10, 14
law of B161, D66–7, 112
likened to adaptations B230, C4, 65, 66, 85, D107
organs of lower animals appear in B112
origin B161, D57, 107, 175, E9, T17
prevented by infertility of hybrid crosses C52
reproduction B119, D108, 112
resemble other species QE1
some varieties resemble C4
supernumerary fingers C83, 84, D14, 15, M25
See also Abortive organs; Mutilation
Montevideo, Uruguay
geology RN56, 87, 144
Moon
mountains A104–5, 135
Morality OUN27, 30[v], 44, 53–4
absence of universal sense of M75–7, N3
associated with ideas of the sublime OUN18–19
ethics C267, 269, N89
instinct and conscience act together OUN43–7, 51
origin M132, 150–1, OUN30, 43, 47
origin of chastity N10
See also Conscience; Sin
Mormodes QE21..
Morocco E90
Morphology
change in number of vertebrae C124–5
changes in organs as related to their functions C56
development of multiplicity of parts in animals C103
effect of external circumstances on C33, E146–7
law of balancement of organs B114, E157, QE17
persistence of features through whole classificatory family C109
repetition of organs in invertebrates C48–9, E45
skull composed of metamorphosed vertebrae E89
some organs more fixed in form than others C33, 139
spiral structure of echinoderms C103
See also Affinity; Analogy; Metamorphosis; Type
Moruffetes
different kinds of ZEd2
Mosquitoes C174, D111, ZEd18
Mosses
hybrid E inside front cover, M156
sexual reproduction E100, 123
Moths B2, C215, D170, 180, E79
Moufflon E35
The Mount, Shrewsbury (home of R. W. Darwin)
See Shrewsbury
Mountain building RN118–19, A64, 68, 104–5, D140, M69
action of lateral pressure in A66
and central heat of the Earth A107
and elevation of landmasses A62
caused by movements in fluid nucleus of Earth RN48–9, A65
effects of subsidence on A127, 128
injection of rock during A31, 62
origin of parallel lines of elevation RN101
thickness of strata A66
Mountain chains
axis of A58
direction of A42, B249, M131
effect on geographic distribution RN52, B142, 209, C251, D31, 61
geographic extent RN155
cold summits of equatorial A7
escarpments and line of sea coast A55
of coral limestone A18
See also Alpine plants; Chimborazo; Etna; Volcanoes
Mules
See Hybrids
Mullein, white QE[8[v]]
Mummies, Egyptian C92, D39, E142
Muscicapa C109, 254
Mushrooms C240, E184
Music N64, 66
and origin of language N107, OUN5
appreciation of distinguishes between man and animals N109
a primitive feeling N19
birds' pleasure in M109
Darwin's taste for M28, 51, N45
effect of memory on playing N61
feelings excited by M33, 39, N107
orangutan's pleasure in M156, N64
origin of pleasure in M33, 36, 39, 88, N18, 19, 45, 107, OUN38
pianoforte M51, N91
relationship with poetry M33, N88
sexual basis of pleasure in N41, 107
Mutilations C232, D15, 174

Mutilations (*cont.*)
not inherited C65, 83, 232, D18, 112, 172
See also Injury; Regeneration
Mylodon S3[r]
Myothera B13

Natural History
Darwin's work will create a new system of B47
Natural selection
See Selection, natural
Natural theology
See Creation; God; Science
Nature
never extravagant C86–7
architecture mimics scenes of OUN6
See also Conditions of existence; Economy of nature
Nautilus C58, D134
Navigators (Samoan) Islands, Pacific Ocean
animals E10, TFRAG10
Nectarines QE21
Negro B71, M86–7
appearance B34, C204, D24
a separate species of mankind B231, D39
blushing N15
compared with Brazilians and Indians M85–7
cross-breeding with other races B32, 34, 179, D38–9
diminishing numbers of D38
first appearance in fossil record B208, C204
ideas of beauty M32, N27, OUN8
in ancient drawings and writings C219
parasites B142
prejudices against C154–5
reasoning powers N33
resistance to malaria D24
secondary sexual characteristics QE18
See also Slavery
Nepal
animals C96, E178
Nervous system
action during emotion M53, 54–5
and development D118
and origin of facial expressions M150
as the basis for behaviour D118
endowed with knowledge MAC59
relationship between that of animals and man OUN35
sympathetic nerves C236, 242, D166–7, M150
Nettles MAC29[v], QE[11[v]], 14, [15[v]]
New Britain, Pacific Ocean
animals C17
volcanoes A40
New Caledonia, Pacific Ocean
animals C17, 20
races of mankind D25
Newfoundland
geese C40
New Guinea B220, C27
animals C13–14, 15, 18, 27, 28
fauna allied to that of Australia C14–15
fauna allied to that of the East Indies C15
New Holland
See Australia
New Ireland, Pacific Ocean
animals C13, 15, 23, 28, D73
New South Wales
See Australia
Newspapers and periodicals
See *Annales des sciences naturelles*; *Annals and Magazine of Natural History*; *Annals of Natural History*; *Athenaeum*; *Blackwood's Magazine*; *Chambers' Edinburgh Journal*; *Edinburgh New Philosophical Journal*; *Edinburgh Review*; *Entomological Magazine*; *Institut*; *Journal des Mines*; *Journal de physique*; *Magazine of Natural History*; *Philosophical Magazine*; *Quarterly Journal of Agriculture*; *Quarterly Review*; *Scientific Memoirs*; *The Times*; *Zoological Journal*
New York C175
New Zealand B219
animals RN129, C25, 29, IBC
Apterix in C74
birds C15, 19, QE10a, [11[v]]
character of flora RN62, 102, C239, QE10a
Dinornis in T65
doubtful species in B50, QE10a
races of mankind D25
species similar to those in Australia B50, C20, 239
volcanic rocks RN102
Nickel RN160, 172
Nightingales C160, 254, D102
Nipples C257, DIFC
See also Abortive organs; Mammals; Mankind
Nodules RN5
Nomenclature
See Classification
Norfolk Crag
fossil mammals T89
Norfolk Island, Pacific Ocean C20
plants B187, C238–9, M38
guava C93
impact of geology on fauna QE21..
North America
See America, North
Norway
alpine plants E115
geology A68, 86, 97
gorse E152, 153, QE2
minerals RN168
Nova Zemlia Island, Arctic Ocean
plants D24

Oats
transmutation into rye QE15[v]
Obsidian RN43, 45, 121
Obstruction sound, Chile
force of tides RN141
Oceans
See Sea
Octopus N60
colour MAC28[r]
Oenothera QE[6[v]], 9
Olivine RN156, A32
Onions E165, QE2
Oolitic period A55, B173, 227, E121
Opetiorhyncus C126
Ophrys QE21..

Opossum B80, 95, MAC29v
Oranges, horned QE7, 10, 14
Orangutan C79, D61, T119, M153, N94
  a close relative of man C79
  behaviour C235, D138, M85, 107, 129
  behaviour of 'Jenny' at the Zoological Gardens M137–40, N13, 94
  behaviour of 'Tommy' at the Zoological Gardens M138–9, N13
  facial expressions M129, N88, 183
  pleasure in music M156
  sense of smell M85, 156
  sense of touch N64
Orchids QE[5]a, 7
  fecundation QE15, 21..
  plants intermediate between S6
Organization B26, 60, 69–70, 73, C47–9, 76
  absence of progression in E60
  and transitional forms B87
  animals tend to multiply parts and improve in B19, 78, D49
  as a criterion of species B198
  complexity of structure related to sexual reproduction E154–5
  constant in shells B97
  Darwin's views on B227, 227–8, C70, E150
  determines the kind of variation which takes place T17
  of insects and molluscs B113
  limits of form B47
  loss of organs leads to extinction D58
  See also Development; Digestive system; Growth; Life; Locomotion; Monsters; Nervous system; Perfection
Orinoco, river basin
  geology RN100
Orioles E160
Ornithogalum QE12
Ornithorhynchus B89, 97, 162, D115, MAC28r
  classification C87, D51, E66
  fossil E66
  structure B97, 162
  the ancestor of mammals B89
Orpheus B37–8, 51, ZEd16
  on Galapagos Islands B103, C145
Osorno, Chile
  lakes RN155
  volcano in lake A35
Osteopora platycephalus
  identity of A57
Ostrich, African
  allied to Rhea D29
Ostrich, South American
  See Rhea
Otaheite
  See Tahiti
Otters B216–17, C57, 178
Oualan (Ualan) Island, Carolines
  animals C25, 27
Ouzel C189
  ring C160
  rock C254
  water C51
Owls B82, C222, MAC29v, ZEd12
  barn D102
  short-eared E140
Oxalis QE2
Oxen B142, 190, E75, 80
  deterioration of domestic breeds C159
  fossil B125
  geographic distribution B190, C14, E174
  wild C14
  with hump B190
  See also Cattle
Oxyurus ZEd16
Oysters RN134, N14
  cross-fertilization between E159
  living above reach of tide RN134
  organization M72–3
  reproductive organs D156
  variation in QE10a

Pachydermata
  aquatic E92
  classification B59, C186, E92
  extinction of fossil species B59, 149, C37, 178
  geographic distribution B223, D68
  intellect C196, M131
Pacific Islands
  absence of metals RN145
  animals B220, C18, 24, 27, E10, 13
  birds C15
  destruction of children by inhabitants M76
  domesticated animals C54, T81
  epidemics C64
  flora C16, 241, E22
  See also Caroline Islands; Marianas Islands; Navigators; New Britain; New Caledonia; New Ireland; Norfolk Island; Pelew Islands; Radack Islands; Society Islands; Tahiti
Pacific Ocean
  former continent of B11–12
  great extent of RN73
  shores profoundly deep RN97
Pain
  facial expression N45, OUN22
Painting
  Darwin's appreciation of M82
  facial expressions in N7
  FitzWilliam Museum M82
  limits of C266
Palaeotherium B53, C39
Palms RN128, C239, E164, MAC56v
Pampas, Argentina
  geology A147
  predominance of thistle C73
Panama E167
Paradoxurus B80, D61, 64
Parakeet C14
Parameles C14
Paraguay
  fleas ZEd3
  introduction of animals ZEd2
  mammals ZEd11
Parasites
  absence on tigers in zoological gardens D91
  different forms on races of man B142
  different forms on related animals B142, 252, C233

Parasites (*cont.*)
different species have different hosts C233, 234
do not live outside their own country D4
effect of climate on B142, 248, C233, D4
leave a diseased person D3–4
relationship with host C233
Paris
Museum E173
Paris basin
fossils A28, B222, 223, C39, 178, E40, 72, 105, 166–7, MAC55r
Parnassia QE13
Parrots
classification C205
claws C113
desert C99
geographical distribution B67, C40, 61, 99, QE1
green E173
ground B55, C99
Parsley QE[11v]
Parthenogenesis
experiment to test incidence in plants QE1v
Partridges B51, M109, N5
Passion flower QE14
Passions
See Emotion
Patagonia A79
absence of fossil bones RN45
absence of fossil shells T80
animals C13, ZEd9
birds C56, ZEd16
desert animals and plants RN52
elevation RN46–7, 83
erratic boulders in A36, 110
former seaward extension of A1
geology RN36, 62, 86, 100, 114, A112
shells A112, C99
transport of pebbles by sea RN67
See also Port St Julian; Santa Cruz river
Patella QE10a
Payta, Peru
geology RN102, A102
hairless dogs C21
Peaches QE21
Peacocks QE20
behaviour D114, M147, N1
cross-breeding with guinea fowl D31
tail C109, D32, 114, N64
Peas E17
Pearlstones RN89
Peas T104, 151
Darwin's experiments on QE10a, 11, [13v]–14, 22
hybrid seedlings MAC56v
impregnation QE21..
'late' varieties of QE2
Peat
preservation of animal and plant substances in QE[23v]
Pebbles RN50, 56, 67
motion in beds of RN84
Peccari B134
Pecten D56
Pelew Islands, Pacific Ocean
animals C17, 28, E21
Pelican D47
Penguins B25, 251, C161, D29
cross-breeding with duck D32, 87–8, 91
replace auk in Southern hemisphere ZEd15
wings C161, D59
Pennatula ZEd4
Pentstemon QE[6v]
Perception
analogous to action of force on matter OUN39–41v
distinguished from emotion N89
See also Emotion; Eyesight; Intellect; Mind; Nervous System; Smell; Taste; Touch
Perfection
connected to duration of life B29
correlated with multiplicity of parts C103
defined as being able to reproduce D55
See also Progression
Periwinkle E165
Pernambuco RN93, 94, 100
Peru
geographic range of shells C99
geology RN89
Indians' use of arrows RN156
silver mines RN166, 167, 168
See also Callao; Lima; Payta
Petise
See Rhea
Petrels B28, 144, MAC28r, ZEd8, 15
Petrification
effect of carbonic acid in QE[23v]
Peuquenes, Cordillera
geology RN151
Phalangista B249, C13, 225
Phalerope D96
Pheasants C113, QE19
behaviour D33, 105, N1
cross-breeding among breeds of D85, 104, 159, 160
cross-breeding with black game C184, 228, D105–6
cross-breeding with common hen C164, D85, 87, 164, QE[11v]
cross-breeding with grouse B189, D33, E106
plumage C267, D104, 160
sexual characteristics D87
silver B52, D104, 160
white C164
Philippines A142, D64
animals C26, 27, 245, E19, 175, 182
Philosophical Magazine A76, B46, C149, 155, 157, 158, 276, E126, T89
Philosophical Transactions of the Royal Society of London RN20, 24, 80, 95, 172, A111, 113, B9, 181, 197, 219, C40, 178, 215, 225, 267, D9, 114, 152, 154, 165, 168, 171–3, 176, E79, 152, 162, T119, M90, N3, 103, ZEd1, QE11
Phlox T177, 178
Phocea B206
Phonolite RN21, 170
Photography OUN11v
Phrenology
mental exercise alters form of head M30–1, 57
Phryganea A44
Physiognomy
See Expressions
Pichincha, Ecuador A81

Pigeons
  behaviour D101, N1
  breeding C1FC, 2–4, QE18
  classification C205
  Darwin's intention to study B180
  effects of in-breeding D88, 104
  inheritance of adult characteristics in young offspring S7r
  instinct M62, N72, 76, 78–9
  kinds of B181, E136, QE10a
  male sexual characteristics D132, E57
  plumage C1FC, 3, 51, D100–1, 148, E146
  structure C121, 205, D101, E146
  teleogony in QE1v
  voice D101
Pigeons, Cape ZEd3
Pigeons, carrier M62, N72, 76, 79, 185
Pigeons, owl D103
Pigeons, pouter C121, D101, 112, E118
Pigeons, rock C3, 71, D100, 101
Pigeons, tumbler E127, QE4a, 20
Pigeons, wood N1, QE1v
Pigeon-fancy
  prize birds D100
Pigs B162, 219, C23
  bones B162, C124, 202, 228, E157
  breeds of B162, 198
  breeds with undivided foot D1FC, QE18
  Chinese D33, 180, E123
  crossing of different breeds QE20
  diseases E36
  feet B217, C inside front cover
  geographic distribution B256, C21, E18, M131
  hybrids D42, 180
  instinct N66
  intellect B255, M131
  relationship with tapirs B86–7
  young B190, 219, C3, T9
Pineapple QE21..
Pineaster B151
Pinks QE21..
Pintail duck See Duck, pintail
Pipe-fish D169, E57, 90, T25
Pippen, golden See Apples
Pitchstone RN43, 63, 121
Planariae ZEd1, 5, 10
  Beagle specimens B143
  consciousness OUN9, 16
  Darwin to write paper on ZEd13
  eye MAC29v
  parasitic ZEd13
  regeneration B1, D129
  reproduction E70
  terrestrial C214
Plants B46, 76, OUN35, QE27
  abortive organs B59, C267, D162, E129–30, T99–100, QE13, 14, 21
  adaptations to ensure fertilization C237–8, QE[5]a
  and carbonic acid gas B109
  annual B2, D153, 165, 176, E1FC, 184
  characters of dioecious QE15
  colours T95–6, QE3, 5
  Compositae the most perfect B193, 200, E100
  conversion of annual into biennial D153
  crossing among B75, E123, 143–4, QE1v, 3, 5, [5]a, 8, 21
  Darwin's questions and experiments on D180, M156, QE1v–2, 5–[15v], 19, 22, 23, [24v], [23v]
  dependence of animals on B108–9
  difference from animals B43, 210, 214, D167
  effect of hot-house on QE5, 11
  fecundity E56
  flowering period C91, QE13
  germination of ancient bulb C92
  inheritance of double flowers T95–6
  instincts N48, OUN36v
  intermediate between animal and inorganic realms C104
  irritability C241–2
  habitual actions C236, M156
  memory N50
  mode of generation B1, 2, 96, 197, D167
  mode of generation in monoecious C167, D157, 175, E80, 184
  nature of variation in B210, 239, E13–17
  originally hermaphrodite E109
  origin of monoecious QE1v, 2, [9v], 15
  origin of separation between monoecious and dioecious forms D175
  races of B68, 275, QE[11v]
  repugnance to crossing B189, E143–4
  sensitive D21, N50, OUN9, 34, 35
  sensitivity to light D21, E43, MAC29v
  sexuality D156–7, 162, E129–30
  sleep E inside front cover, 184, N81, OUN36v, QE5, [5]a
  tendency of hybrids to reversion B33
  viviparous E163, 165
  See also Alpine plants; Conifers; Confervae; Cryptogamia; Cultivation; Ferns; Fertilization; Fossil plants; Fruit trees; Geographic dispersal; Geographic distribution; Horticulturists; Kelp; Monocotyledons; Mosses; Palms; Seeds; Trees
Pleasure
  and conscience M119–20, 124–5, N46–7
  facial expressions M146–7, N45
  felt by animals C163, M156, N64.
  muscular movements of smile N6, 8–9
  weeping during M89
  See also Laughter; Music; Poetry
Pleasure-pain principle
  and nature of emotions N68
  and nature of happiness M118–27
Plesiosaurus C72
Poetry C266
  Darwin's theory explains B11
  description of emotion in N57
  imitates song OUN11bv
  memory of M8, 39, 50
  onomatopia N31, 39, 127
  origin of pleasure in M33, 39, 88
Poland E58
Pollination
  See Fertilization
Polymorphism
  on volcanic islands B152
Polynesia
  See Pacific Islands
Polyps B1, C49, D68, 156, M72, ZEd13, 17–18
  See also Coral polyps; Hydra; Zoophytes

Poplars QE9
Poppies T111, S1r–1v
Population
decrease of B235
Malthusian checks to size of OUN29v
Malthus' law of E3
Porphyry RN68, 86, 87, 88, 140, 163–4, 166, 170–1, 176, A141
Porpoises E47, M142, N105
absence of fossil remains MAC29r
Port Desire, Argentina
erosion at RN139
geology RN86, 114
Portezuelo, Chile A100
Portillo Pass, Cordillera
geology RN44, 150, 151
Port Phillip, Australia A92
Port St. Julian, Patagonia
erosion at RN139
fossil guanaco A34
slowness of elevation RN113
Potatoes D170, T104
Potosi, Bolivia RN106, 166
Preparis Isles, Bay of Bengal
animals E177
Pride M20, 24, 77, 93
facial expression N7, 8
Primroses C35, 194
change into cowslips E16, 113, 141
crossing between sports of QE21
Darwin's experiments on QE1v, 5
Prince Edward Islands RN138
Probability
See Chance
Proceedings of the Geological Society of London A5, 28, 41, 44, 46, 63, 79, 131, B126, 153, 250, C55, 62, 216, E22, QE1
Proceedings of the Zoological Society of London RN127, 130, B7, 160, 162, 166, 189, 209, 212, 250, 279, C29, 36, 116, 124, 125, 164, D29, 95, E24, 128, N71, 115, ZEd6, QEIFC, 1, 10a, [11v], 12, 18
Progression B18, 88, 108, 113, 205, C157, E70
absence of B44, C210, E60
animal development shows a B49, 113, D49
comparison between that of living world and of the earth MAC56v
inward developing power of E159
no necessary tendency towards greater complexity E95–7
Proof
See Darwin—Species theory; Science
Propagation
See Generation
Prostitution C266
Proteaceae B77
Proteus E133–4
Ptarmigan C84, D101
cross-breeding with black game D72, 105–6
Puffins C86–7, ZEd15
Pumas D140, M152
Pumice RN38, 42, 43, 67, A38, 63
Pyrenees, Spain E114–6, TFRAG6
Pyrites RN169
Pyrocephalus ZEd15
Quagga B181, C145, D113, 172
Quarterly Journal of Agriculture D43, 44, GR1
Quarterly Journal of the Geological Society of London RN144, A97
Quarterly Review C42, 155–7, 276, N19
Quartz RN99, 150, 164, 176, A3, 61, 94, 106, E137
Quillota, Chile
geology RN150
Quinarian system B27–8, 46, 57, 112, 126, 129–30, C150, 222, D50–3, 58–9
analogies between groups C111–15, 143, 202, 158
based on adaptations to the three elements B23–4, 57, 112, 263, C218
Darwin's opinion of C74, 170, 185, D62
Darwin's rejection of C73–4
Darwin's theory explains B46
explains affinities C139
evaluation of taxonomic characters in C149, 158, D50–2
inosculation of species RN130, B8
osculant groups B126, C202–3
plants do not conform to B46
possibility of a quaternary arrangement C95, D62
used to classify birds C111–15, 143, 227
used to classify insects B56–7
Quito, Ecuador RN177
earthquake RN158

Rabbits
appear to resemble hares E9
as prey E139
at Porto Santo QE18
behaviour C154, N103, 112
colour C29, E9
ears D101, N103
instinct C134
on Falkland Islands B31, 54, 122, C29, ZEd5
varieties of B181
Rabies E36
Radack Islands, Pacific Ocean
animals B157, C28
earthquakes RN101
Rainfall RN108, MAC54
Rafflesia QE[9v]
Ralix Islands, Pacific Ocean RN101
Ranunculus T178, QE1, 6
Rats B55, 166, C24, E139
arrive first on island RN129
Reason
See Intellect
Recapitulation B163, C48–9
affinities shown by larvae C162
foetus resembles lower forms of life B163, C149, 162, D170, E89
individuals pass through whole series of forms D179
man passes through a caterpillar state D170
young birds resemble each other C149
young birds resemble earlier stage of existence E125
Red Sea
hills which make noises RN50–1
Redpole D102
Regeneration B210
comparable to absorption of useless organs D166–7
Darwin to experiment on D165

gemmation considered as a kind of D130–1, 172
healing of wounds D129–31
Religion
an inherited memory from animal state OUN36[v]–37, 38
as the origin of idea of sublime OUN18–19[v]
history of M69
origin of ideas of M135–7, 151, N11–12, 20, 101
origin of sacrifices OUN12
spirituality has a materialistic origin M19
superiority of Christianity OUN38
views of Australian aborigines M137
See also Atheism; Bible; Creation; God; Soul; Will, free
Representative species
See Species, representative
Reproduction
See Generation
Reproductive organs
an uncertain guide to classification C47
development of D158–9, 162
ovaries D158, 164, 172
testes D158, 172
Reptiles B86, 113, 226, 251, C213
classification B86, C54, E168
fossil B88, 170, C39, D133, E96
geographic distribution B113, 251, C25, D40, ZEd13
hybrid E79
in the fossil record B53, 202, 205, D151
structural relationship with birds D35
structural relationship with fishes D133, E168
Reversion
See Inheritance
Rhea RN127, 130, B105
allied to ostrich D29
bones D35
feet E157
geographic distribution B242, 251, C207–9
poorly adapted B37
species of RN153, B13, 16, C126
wings B251, D60
Rhinoceros C45, QE16, 17
African E60
East Indian B82, 234, D37, E176
fossil C132, E72
geographic distribution B67, 82, 233–4, D37, E60, 174, 176
Indian B241
nature of country inhabited RN85–6, 134
species of B67, 82, 233, 241, C45
Rhododendrons E114–6, QE13
fertilization by insects T93, 111, 112
Rhubarb T105, 106, N87
Ribston pippin
See Apples
Rio de Janeiro, Brazil
animals E90
argument for elevation of RN37
Rio Negro, Patagonia A99
Rivers A55, 96, 116
and elevation of land A109
and subsidence of land A76
annual rise in water levels RN108
as origin of Scottish terraces GR47–8, 59–61, 63–5, 71–2, 81
changing course in Australia A92–3
deposition of sand GR16
erosion of valleys A12, 70–1, 72–5
fossils in estuary formations T89
inosculation of tributaries of A76
origin of Serpentine course of A73
transport of sediment RN108–9
See also Springs
Rocky Mountains, USA
separate close species of birds and mammals D31
wren C255
Rodentia A5, B141, C46, 131, 196
Rongeur C15
Rooks B4, 199
Rosa QE10a
Roses D118, QE2
culture C267, D128
degeneration D161
Rotifers D156
Royal Geographical Society
See *Journal of the Royal Geographical Society*
Ruscus E129
Rye
transmutation from oats QE[15[v]]

Sagitella ZEd10
Sagitta RN174, ZEd5
St Helena B157, 173, 193, D74
absence of quadrupeds RN62
craters RN110
earthquakes RN125
erosion of cliffs RN38–41
fish E35
gecko C28
geology RN43, 63, 84, 107
native mouse RN79
plants RN62, B151, 156, C100, 184, E119
St Jago, Cape Verde Islands
birds C68, 163
geology RN99, 107
monkey B154, C249
plants C250
shells C224
St Marc, Haiti
mountains of A18–19
St Pauls Rocks C108
St Peter and St Pauls
insects E19
Salamanders C183, T1
Salamandria aquatica ZEd19
Salisbury Crags, Edinburgh GR2
geology A110, 116, GR3–7, 9–11
veins GR3–4
Saliva N38, 41
Salmon C183, 212, N67
Salpa ZEd5
Salt
taste for M117
solubility of A109
Salta RN155, 157
Salt deposits
efflorescence in A10
formation of A1, 5, 32, 55, 68, 109

Salt deposits (*cont.*)
  in India A143
  in South America A16, 44, 115
Salt lakes A6, 93
Salt-water
  adaptation of fresh-water fish to D151, E85
  Crustacea in RN127
  effect of soaking seeds in B125, C148, QE[5]a
  fish D34
  lizards in B248
  putrefaction of ZEd17
  shells in B54, 201, 248
Saman (Samar) Island, Philippines
  animals E19
Samoyed
  women C257
Sand
  beaches A83
  graduates into gravel RN91
  movement of makes noise RN50, 51
Sand bars A130
  See also Silt
Sandpipers B100
Sandstone RN42, 107, 150, A96
  hypothetical origin of RN30
  shells in RN27
  Silurian character in Falkland Islands RN142
Sandstone, New Red T80, S3[r]–3[v]
Sandstone, Old Red C58, E137
Sandstone, Red RN125, A16, 17
Sandwich Islands C28
  Australian species on B187
  plant species peculiar to separate islets E104
  shells QE1
Santa Cruz river, Patagonia
  degradation of basalt RN107
  fresh-water springs RN140
  geology A12, 13, 88
  terraces in valley of A12
Santa Maria A79
Santos, Brazil
  oysters RN134
Sapajou B94
Saurians
  See Lizards
Savages B4, C150, N40, 66, QE1
  facial expressions N22–3
  primitive culture of C79
Saxifrage C224
Scabies T177
Scandinavia
  See Lapland; Norway; Sweden
Scarabadae B56
Scenery
  and buildings N30–1
  origin of pleasure in M36–41
Science
  definition of N14, 15
  Darwin's belief in advancement of C123
  difficulty of framing methodology E5
  methodology B194
  natural theology N12
  opposition of clerics to N18
  prediction is the aim of D67
  use of analogy in E128
  See also Cause; Cause, final; Chance; Induction; Law; Materialism; Metaphysics
Scientific Memoirs C102–4, 210, 275
Scincus C18, 28
Scolopax ZEd11
Scorpion MAC28[r]
Scotland
  absence of fossil shells T80
  ancient volcanic activity RN159
  erratic boulders in A110
  fossil shells similar to those of North America and Greenland E100–1
  metamorphic rocks A115
  See also Glen Roy
Sea
  action on coastlines RN38–41, 93–5, A82
  freezing of ocean floor RN115
  modelling power of GR30
  movement accompanying earthquakes RN80–1
  temperature of ocean floor A138–9, 151
  transport of coastal matter RN97–8
  separates sand from finer matter RN28
  See also Erosion; Geographic dispersal; Waves
Seals B172, 206, C61, 131, E92, N46
Seashells
  See Shells
Sea-water
  See Salt-water
Seaweeds RN140–1, C245, M38
  See also Confervae; Kelp
Secondary Sexual Characteristics D154–5, 161
  absence in fish D114
  effect of in-breeding on D153
  of crustacea D55
  of females D76, 114
  of hybrid birds D85–6
  transmission of male QE18
  See also Castration
Secretary birds C114, ZEd14
Seeds
  capsules QE21
  comparison between buds and seeds of plants D128–31
  experiment to determine function of QE1[v]
  number of DIFC
  rarity of production in Lemna E162, 163
  vary in rich soil B3
  viviparous grasses E164–5
  See also Geographic dispersal
Seedsmen
  See Horticulturists
Selection, artificial
  different forms of domestic animals created by different races of man C141
  effect on silkworms QE12
  extent of differences made by QE18
  for horns in sheep and cattle S7[r]
  formation of new species by B244
  formation of new varieties by B34, CIFC, 133, D20–1
  for milking cows QE19

improvement of breed by picking C231
inheritance of characters acquired by QE19
in primitive societies QE16
last litters considered most valuable D163
man's interest in perpetuating domestic breeds B155
of domestic animals D20–1
of horses CIFC
of secondary sexual characteristics in fowl QE18
picking offspring CIFC, 17
practised judgement in E63
races of domestic animals made by isolation D23–4, 44
See also Cultivation; Domestication; In-Breeding
Selection, natural
accounts for long legged dogs MAC28^{v}, 167^{r}
acts in formation of instincts N42–3
a force like a hundred thousand wedges D135
analogy with artificial selection D20–1, E63, 71, 118, MAC167^{r}
a result of law of population E3, 9
Darwin first thought of D inside front cover
Darwin's experiments on Aquilegia for QE1
explains acclimatization of plants E111–2
explains adaptations in chickens MAC58^{r}–58^{v}
late marriage is a form of N67
natural checks on population size D134–5
only a small part of the grand mystery of nature E145
See also Chance; Darwin—Species theory; Divergence; Population
Selection, sexual C61
animals have no notion of beauty C71
choice based on beauty OUN8
contest between the sexes of deer QE10
females' love for the victor D99
male characteristics considered as a departure from specific type D113–14, 147
rivalry of male birds C178, D103, 113, 114, E106
See also Beauty; Birdsong; Secondary Sexual Characteristics
Selenite RN169
Senegal A20, M138
Septaria A47, 60
Seraphis
Temple of A95
Serpentine RN170
Sertularia ZEd10
Sexes, origin of separate
See Hermaphroditism
Sexual selection
See Selection, sexual
Shales RN36, 63
Shame M144
a feeling different to fear M148, 156
a recent acquisition OUN29
facial expressions of M149, 150, N58
felt by animals M23–4, N25
origin of OUN55^{v}–55
See also Conscience; Modesty
Sharks RN8–9, BC, B205
Sheep CIFC, 267, N44
big-tail B233, C214, E12
black faced GR11, 25, 125
colour C93, D104
cross-breeding with moufflon E35
differences in lambs of different breeds QE19
diseases E36
geographic distribution B48, 233, C93, 175, E173
horns S7^{r}, QE19
kinds of B48, C221, 227, D44, 65, E123, QE19
of Arabia resemble goats B48
Shells A16, 115, B137, 157, 204, C99, 169, N51
arctic D133
as evidence for elevation of land RN37, 106, A4, 20, 39, E86
as evidence for subsidence of land A153
classification B9, 110, 207, C127
fresh-water B54, 201, C245
hinge MAC58^{v}
in Africa A20, B157, 204, C246, E41, 90
in Britain AIFC
in calcareous rocks RN27, 42, 61
in Chile RN36, 106, A39, 72, 102, 144, T80, ZEd13
indicate passage of geological time T79–80
in eastern South America A33, 112, ZEd8, 15
in Europe D39, E100
in fossil record A85, B167, 172, 207, C110, 220, D133, E32, 92
in India B153, D68
in North America B172, 174, D39, E100
in Scandinavia A111, E86
in South America B95, C244, 245, 246, D39
land A72, B9, 204, 284, C184, E76
life span of species of B252, C234
long preservation of T1
marine RN84, 143, A72, B54, 201
north and south limits of distribution ZEd13
number of species of E122
on Barrier Island, New Zealand CIBC
on Cape Verde Islands C224
on Juan Fernandez E166
on Madeira A72, QE1
organization remains constant over long periods B97
origin of closely allied species of B229
preservation of original social groups in fossiliferous rocks RN65
proportions of B167–8
relationship between coal distribution and RN36
replacement of one species by another in geographic distribution ZEd15
tropical C21, E59
variation in C82, QE1
See also Molluscs
Shepherds E140
talk to Darwin D43, 47, E123
Shepherd dogs
See Dogs, Shepherd
Shetlands, south RN138, A63
ancient body recovered on RN139
Shiant Island, Scotland
birds E140
Shoals RN15–16, 91
Shrews B31, 61, T frag 4
Shrewsbury, 'The Mount' GR109
Shrewsbury,
books at C267
Darwin's experiments on plants at QE14–‹15›
Darwin's observations of wild birds at N112
dogs N2, 41, 105

Shrewsbury (*cont.*)
hothouse M49, 156
Nina M26, 56
Pincher M70
tailless cats at M25
vegetable garden experiments T151, 176
Shropshire, UK
geology A1, 115, GR109
See also Shrewsbury
Siam
animals C14, E175
Siberia
animals RN85, B60, 62, E169, ZEd19
frozen soil A90–1, 117–18, 124, 136, C101
plants A7, D24, E116
recent elevation of A7, 8, 11
Sicily
change of species in C168
migration of fauna to E105
number of species on B160
Sierra Leone QE17
Silene T177
Silex A25
Silica A176
Silkworms QEIFC, 5, 12
Silliman's Journal
See *American Journal of Science*
Silt
in coastal channels A59
Silurian system B95, 170, D37, 133, 134
Silver deposits RN162, 163–4, 166, 167–9, 176
Sin OUN12, 15
a necessary passion OUN28–29[v]
criminality deserves pity OUN26–8
the origin of evil passions N10–11
See also Avarice
Singing
in animals N18
in mankind M55, 78, N18
origin of pleasure in M33, 39
See also Birdsong
Sivatherium T1
Skunk ZEd9
Slate RN5, 6, 150, A42, B87, 121, E11
in Mexico RN170, 171, 176
primitive RN166
Slate, clay
See Clay slate
Slavery M87
debases mankind B231, C154–5
slave holders B231
trade in slaves D38
Sleep B11, C268, M45, 46, 56, OUN10[v]
and sensation OUN9
consciousness during M112–14
duration D49
in plants E inside front cover, 184, N81, OUN36[v], QE5, [5]a
forgetfulness after C172
function EIFC, M114
like insanity M114, 127
lower animals are 'sleeping' forms of higher animals OUN9
nature of OUN35
See also Dreams; Somnambulism
Sleeptalking M43, 102
Sleepwalking C171, 211
Sloths B20
Smell
and power of association M82, N107
of man C174
of plants D21, M109
Smell, sense of
and sexuality M71, 85, N19, 41
in animals M71, N74
in apes M85, 139, 156
in dogs M71, 109, N87
in horses D10, N7
in hyaenas M71
in insects N107
in mankind M109, N109
in monkeys D137, M109
relationship to sense of taste M71
the same in man and animals N64
Smile
See Pleasure
Snails D166, M63, 97
See also Molluscs
Snakes B233, C27, E10, T55
fangs T57, MAC29[v]
Snap-dragon QE6
Sneering M95, 96
Snipe QE[11[v]]
Snow
adaptations to C252, E109
Snow-line B195, E48, 109
Societies and Institutions
See *Abhandlungen der Königlichen Akademie der Wissenschaften zu Berlin*; *Annales du Muséum Nationale d'Histoire Naturelle*; Athenaeum Club; Breeders; British Association for the Advancement of Science; *Bulletin de la Société géologique de France*; *Comptes rendus hebdomadaires des séances de l'Académie des Sciences*; *Journal of the Asiatic Society of Bengal*; *Journal of the Royal Geographical Society*; *Mémoires de la Société de physique et d'histoire naturelle de Genève*; *Memoirs of the Wernerian Natural History Society*; *Philosophical Transactions of the Royal Society of London*; *Proceedings of the Geological Society of London*; *Proceedings of the Zoological Society of London*; *Quarterly Journal of the Geological Society of London*; Statistical Society; *Transactions of the Cambridge Philosophical Society*; *Transactions of the Entomological Society of London*; *Transactions of the Geological Society of London*; *Transactions of the Linnean Society of London*; *Transactions of the (Royal) Horticultural Society*; *Transactions of the Royal Irish Academy*; *Transactions of the Royal Society of Edinburgh*; *Transactions of the Zoological Society of London*; Zoological Society gardens.
Society Islands, Pacific Ocean C24, E10
See also Tahiti
Socotra Island, Arabian Sea
animals C92
Soda A41, 126
See also Carbonate of soda
Solanum QE21..
Solfateras RN162, 165
Solomon Islands, Pacific Ocean A40

Solor Island, Indonesia C13
Solubility RN74, 172, A61, 68
Somerset, UK
  monstrous cockles E45
Somnambulism M156
  and consciousness M110, 114–15
  free will not present OUN25
Sooloo (Sulu) Islands, Indonesia
  animals E173, 176
Sorex C183
Sorrel QE[9v]
Soul
  faculties of OUN8–8v
  immortality of E76
  pre-existence of M128
  separates man from animals B232, OUN36v
  See also Will, free
South America
  See America, South
South Sea Islands
  See Pacific Islands
Spain
  deterioration of human intellect in N47
  ibex T frag 4
  limits to geographic range of rhododendron E114–16
Sparrows D95, N112
Spean, Scotland
  alluvial deposits near GR26–9, 32–5
  Darwin's belief that deposits are marine in origin GR28–9, 31–2
Species B9, 158, 280, C42
  absence of gradation between B209
  age of determined by extent to which young resemble parents D147–8
  analogous to an individual B22, 39, 63–4, E83–4
  balanced number of B21, 37
  capacity for making new species B37
  constant B5
  Darwin's definition of C161
  Darwin's theory abolishes specific names C126–30
  Darwin's theory explains persistance of form C60
  definition B122, 212, 213, C152, 161, E24, MAC167r, 167v
  disposition to deviate from nature D161
  distinct B171, 212, 213, 225
  distinctions between B9, 57–8, C26
  duration of existence B22–3, 29, 35, 38, 170
  effect of changing environment on B8, 210
  effect of geographic isolation on B209, C21, 43, 53–4, E135
  effect of geological changes on B82, C25, 80
  formation of new B24, 51, 78, 101–2, 161, 189
  gradation between B228
  increase in total number of B29, 202, C146
  inequality in numbers of B225
  inhabit different localities B261, 275
  inosculating B8
  like buds of plants B73
  made up of innumerable variations E57
  must generate or die B72
  new appearance of is the mystery of mysteries E59
  origin similar to first cross between hybrids D14
  permanent in plants C192–5
  progress in B18
  not a fixed entity C152
  rate of change in B8, 18, C16, 234, E48–51
  relationship with genus B28, 51, 57, 245
  retrograde development in D57
  small number that change B39
  See also Darwin—Species theory
Species, aberrant B43, 126, D11
  classification of C73, 207–8, E168
  nature of C87, 144
  origin B28, C144, 185
Species, close B8, 50, 51, 65, 69, 100, 280, C46, E11, ZEd12
  and hybridity C151
  come from different localities C80
  Darwin's theory will explain C176
  Darwin to investigate geographic distribution of QE10a
  geographic range B153
  habits dissimilar C125
  in same country RN153
  of birds C46, 206, 267, D29, 31
  origin B229
  value for Darwin's theory E51–2
  variation B241
Species, introduced
  attacked by different predators M63
  birds in Chile B51
  cattle on Madagascar B255
  dog in Australia C46
  domestic animals in Sierra Leone QE17
  elephants on island of Sooloo E173, 176
  guava on Tahiti A88, C73, 93–4
  in gardens E114–5
  in Iceland B188, 255–6
  in Paraguay ZEd2
  on Mauritius B255, C22
  on South Sea Islands C54, E21
  on Tasmania B263
  plants on Marianne Islands E21
  rabbit on Falkland Islands C29, ZEd5
  rats in New Zealand RN129
  relationship between plants and insects B218
  snails in Mauritius ZEd1
  without agency of man QE7
Species, representative
  among birds B47, 199, C27, 50, 56, D29, 31, 96, ZEd11
  among butterflies D22
  among insects B56–7
  change in B8
  effect of crossing on E51
  formed after long separation B15
  geographical barriers lead to formation of B103, D31
  in Australia C27, D29
  in East Indies C19, 41
  in North America C251–2, D31
  inosculating B8
  in South America B8, C50, 56
  of clenomys C35
  replace each other in natural economy B15, 67, 144, C35, 144, 150, 225, ZEd14, 15
Specific gravity
  of artificial limestones A77
Speedwell T111
Sperm E90, 155, 184

Sperm (*cont.*)
  in mosses E100, 123
Spiders C197, D156, M99
  poisonous MAC29^{v}
Spitzbergen RN159, ZEd18
Spondylus D56
Sponges C49, D55, 156
Spontaneous generation C49, 224, E160, OUN34
  Darwin's views on C102, D132
  not improbable C102
  See also Generation; Life
Sports
  See Inheritance; Monsters; Variation
Springs RN31, 127
  Darwin's observations on RN140
  in frozen soil A124, 125, 136
  under the ocean floor A123, 128, 132, 133
Springs, hot A108, 124, 132–3, C62
Sprouts
  See Brassica
Squilla MAC28^{r}
Squirrels C199, D22, E177, MAC29^{v}
  colour C252, D12
Stalactites A47, 53
Staphylinidae C108, T105, 106
Starlings D147
Statistical Society C268
Sterility
  See Fertility
Stoats E139
Stocks QE11, 15
Stomach
  See Digestive system
Stonehenge N77
Stonesfield Slate
  marsupial fossils in B87, 219, D62, E11, 60
  See also Didelphis
Strata A13, 58, 59, 65, 155
Stratigraphy A101
Strawberries B230, E15, QE11^{v}
Stromboli RN137
Struthionidae D29–30, E135
Sturnus magellanicus ZEd11
Sublimation (geology) A32
Sublimity N59
  origin of ideas of OUN18–19^{v}
Subsidence RN87, 100, A8, 39, 89, 92, 118, 132, 153, B131
  and volcanic activity RN60, A76, 80
  effect on climate E109
  effect on forests A131, 141
  effect on fossil record S4
  effect on topography RN51, 140, A76
  evidence for RN140, A76, 95, B222
  fluid matter moves away from underneath land RN77
  indicated by silicified trees RN152, 154
  leads to formation of new species B82, 222
  of Arctic A8
  of Greenland A39
  of mountain chains A127, 128
  See also Elevation & Subsidence
Subularia QE12
Sugar C217, N87
Sugar-cane QE12
Suicide N99
Sulphur RN78, A32, 116
Sulphuric acid C84
Sumatra B187, 241, T51
  animals C17, 26, 41, 46, E21, 175, ZEd6
  bears B164, C41
  earthquakes RN90
  Priaman volcano RN90
  rhinoceros in B67, 82
Sun E125, N36, 78
  and migratory instincts of birds N76
  protection from light of M64
Sunda Islands, Indonesia
  species different to those of Southern Asia E20–1
Sunfish MAC28^{r}
Superfoetation
  Darwin's belief in D176
  one impregnation suffices D41
  See also Teleogony
Superstition C244, M137, N20
  See also Religion
Surprise
  in animals OUN36^{v}
Sus americana
  fossil S3^{r}
Suspicion M20
  facial expression N7
Swallows
  feed on insects C112
  migrations C253
  numbers of E136
  other birds have same role as B137, C111–12
Swans
  behaviour M147, 152–3
  black E174, M141, 153
Sweden A76, 113, D105
Sweet peas
  Darwin's experiment on QE22
Swifts C40
Sycamores QE[15^{v}]
Syenite RN138, 170, A20
Sympathy M88–9, 108
  absence in animals M142
Synallaxis C15, 69, 71, 207, ZEd16
Syria RN52
Systematics
  See Classification

Taenia C49, D49
Tahiti D74
  animals C28
  bananas QE2, 21..
  fish C20
  guava on A88, C73, 93–4
  parental feelings of islanders OUN38
  plants B198
  shells ZEd13
Tailor birds T2
Tameness
  absence in wild birds B4, C50, 189, E174
  and breeding behaviour QE1, 16

effect of crossing with wild animals on D7
inheritance of B136, C165, D71
lost in cross-bred ducks D148
of Australian dogs E103
of ducks E174
of hares E117
of wolf E103
Tapirs
fossil E105
geographic distribution B242, C17, E18, 176, 182
structural relationship with pig B86–7
Tartary C102
Tasmania B219, 220, C27
absence of dog B177, C131
animals B187, 263, 264, E11
animals different to those of Australia B50, C95, 225
animals identical to those of Australia C239, D30, 61
doubtful species in QE10a
fish C28
former connection with Australia B152, C130
fossils C30–1, 257
geological resemblance of Hobart to Tierra del Fuego RN21
rocks RN120, E137
Taste, sense of OUN50[v]
an acquired habit N29
the same in man and animals N64
Tatous ZEd2, 11, 12
Taxonomy
See Classification
Tea plant D117
Teal D96
Teazle
See Thistles
Teeth QE16, 17
canines of no use to mankind MAC28[v]
Teleogony D6, 9, 113, 152, T13, QE1[v]
absence of explanation for D173
applied to temperament D164–5
applies to crosses between species as well as individuals of same species E79
contrasted with superfoetation D172
Darwin's belief in D176
effect of male on character of offspring D9
effect on fixing characteristics in the blood D168–9
experiment to test incidence in plants QE1[v]
facts against D9
in dogs B32, 184, D9, 163, 172, M25
Lord Morton's mare B197
not applicable to animals which have eggs impregnated externally E79
obscures study of transmission of parents' qualities E35
Tenerife, Canary Islands B158, 193
absence of polymorphous plants E182
fish E35
geological structure of peak RN43
Tenioptera C88
Tenrecs C22
Teratology
See Monsters
Terebrantia M64
Terebratula A37, B19–20, D134
Terraces A12
absence in Loch Tulla valley GR20
absence of GR41
absence of fossil shells on T80
as proof of slow erosion A12
at Glen Roy, Scotland GR62–85
at mouth of Caledonian canal GR22–3
compared with stepped streams of lava RN99
different from platforms formed by sea erosion RN40–1
indicate successive elevations RN60
lake required to deposit GR29–30, 35
lips level with shelves GR46, 78, 98
marine origin GR27–9, 31–2
preservation by peat and heather GR101
preservation of RN63, GR66
seen most easily on steep earthy slope GR38–9
supernumerary shelves GR63
vestiges in Loch Dochart GR13–14
Theodicy
See Sin
The Times RN33
Thistles C73, T177, QE[15[v]]
Thought
See Mind
Thrips T105
Thrushes B193, C105, D31, ZEd11
and mistletoe MAC53[r]
cross-breeding with blackbirds TFRAG3
missel C189
mocking C209, 255–6
song C255–6
Thyme QE10, 11
abortive stamens T99, QE13, 14
Ticks E177
Tierra del Fuego
absence of birdsong M31
absence of earthquakes RN17
absence of hot springs RN17
animals B187, C99, ZEd10, 17, 18, 19, QE7
distinctness of tribes in B33
plants B221, QE7
religious views of inhabitants M135, 137
rocks RN21, 88–9, 99, A141
savages E47
soundings off coast of RN140
topography a result of subsidence RN140
See also Beagle Channel; Magellan, straits of
Tigers B196, E97, E144, M41
behaviour M152
cross-breeding with lion DIFC
Tilgate beds C163
Time
an element in change of species E99, 142
an element in defining species C3, D33, 106–7, 139–40
children have loose idea of N61
compared with isolation as an agent of change in species E135
complexity of organic world not related to T19
fossil record indicates enormous periods of T79–80
geological C224–5
geological indicators of passage of RN107, E87
great antiquity of world E125
length that pigeon varieties have existed for D100
slowness of changes in species C17

Time (*cont.*)
slow rate of great changes C17, D167
Timor, Indonesia
animals E170, 173, 180
Australian forms of birds C83
connected to Australia by shallow seas C213
plants C239
Tongatabou (Tonga or Amsterdam) Island
birds C15
Tortoises B250, C157, ZEd6
absence of fossils on Isle of France ZEd12
fossil T1
fresh-water B279
geographic distribution B234, 279, C18, E12
on Galapagos Islands RN55, B7, E12, ZEd12
Tosca RN84, A142
Touch, sense of N64, 75
Touraine beds E166–7
Toxodon A5, C132, 167, E23, 168
Trachyte RN38, 43, 150, A43
action of steam on RN165
a primitive rock RN44
first production of RN78
origin of A21, 35
Transactions of the Cambridge Philosophical Society RN5, 7, A25, 52, 72, 103, 107, 145, B204
Transactions of the Entomological Society of London B46, 56, 57, 79, C107, 108, 233, D22, MIFC, 62, 63, 64, ZEd19, QE1, 12
Transactions of the Geological Society of London A36, 45, 49, 88, 97, B279, C72, 163, D68, GR35, 63, 65, 69, 70, 78, 109
Transactions of the Linnean Society of London RN127, A1, B139, 192, 193, 195, 218, C42, 46, 155, 156, 157, 158, 237, 276, D40, 61, E100, 133, 165, 168, TFRAG7, ZEd8, S6, QE[5]a, 15, 21..
Transactions of the (Royal) Horticultural Society B180, 191, E13, 77, 111, 114–16, 141, 142, 143, 148, 150, 153, 154, 162, 165, 184, QE1^{v}, 2, 3, [6^{v}], 7, 11
Transactions of the Royal Irish Academy A126, B7, D61, 164
Transactions of the Royal Society of Edinburgh RN156, A36, 77, 102–3, 111, 155, C227, D24, GR35, 41, 45, 78
Transactions of the Zoological Society of London C47–9, 213, D29, 30, 35, T65, ZEd4, 13
Transformism B110–12
changes brought about by the 'willing' of organisms B21, 219, 227, N12–13
Darwin's belief that the theory is absurd B216
Lamarck's theory of 'willing' OUN34–5
theory of not applicable to plants C63
theory of precedes that of Darwin B178
See also Transmutation
Transitional forms B86
absence of B9, 53, 217, C201
atavism reduces number of B59
between orchids and other plants S6
between primrose and cowslip QE5
Darwin's theory explains B226
difficulty of tracing B25, 26, C64
extinction of C203, D52, T57
in cattle B154
in dogs B155, 198, 217, 219, D9, E69, 88, M102
in reptiles B86
in some genera of South American birds C96
intermediate species have perfect organs C206
in the fossil record C257, E88, 119–20, 126–7
nature full of B87, C73
not expected B154
often exist between two similar groups of birds C107, 110
origin of intermediate structures is a great difficulty for Darwin's theory T55, 57
reasons for absence of gradation B209
See also Darwin—Difficulties of Species theory; Gradation
Transitional rocks A27
Transmutation
Darwin's theory different to that of Lamarck B214
Darwin's theory of B227
evidence for C73, E120–1
facts induce towards a theory of E51
fossil record presents little evidence for E126–7
isolation plays a part in E122
of oats into rye QE[15]^{v}
of wheat QE11–[11^{v}]
passage of genus into genus B79
passage of organ into organ B112, 143, 225–6, 228
passage of species into species B178, 224, C76, 156, 222, E92
the foundation of geology E87
See also Change; Darwin—Species theory: Extinction; Gradation; Progression; Selection, natural; Species; Tree of life
Trap rocks A106, GR10
Tree of life C139, 151–2, 155, D59
agreement of Quinarian system with B57, D59
a representation of organized beings B21, 25, 26
branching B19, 21, 23–5, 28, 113–14, 263
branching is irregular B21
classification is linear in lower classes B27
dying of branches explains gaps in nature B20, 27, 29, 35–44, 71, 79, 113, C157
number of primary divisions in D52
similarity of animals from one branch B35
species like buds on B72
stems adapted for three elements B23–4
Trees
considered as compound animals M41
injured by salt A1
sporting in nursery gardens QE3
See also Conifers; Fossil wood; Fruit trees; Palms
Trematodes D156
Trigonia C244, E120
Trilobite D55, 134
Tristan da Cunha B218, 220, D74
birds ZEd8
Darwin's interest in RN138
glaciation on B195
plants B192, C100
Troglodytis furnarius ZEd11
Trogon C40, 109–10
Trout S5
Tubilipores ZEd14
Tubularia ZEd6
Tucumán, Argentina RN155, 157
Tucuman Mountains, Argentina
condor ZEd10
Tufa RN74, 86
Tulips E151, 162
Turbo CIFC, D134

Turdus B193, C105
Turdus falklandii ZEd11
Turdus Guayanensis B193, ZEd8
Turdus Magellanicus ZEd17
Turkey-cock M18, 93, 147, N14
Turnips E14, 98
Turritella D134
Turtles C70, E138, N71
Tyndrum, Scotland GR19
  flat topped hill at GR14–15
Type B14, 30, 44, 112, 142
  adaptations as changes in typical structure C72
  changes from original C208
  Darwin's theory explains B225, D21
  Darwin to study unity of C76
  degrees of closeness between C206
  exceptions to law of B133
  formation of distinct B15
  in birds C52, 88, 111–15, 143
  law of B133, 154
  mammalian B59
  of great quadrupeds B60
  originates from will of the deity MAC55r
  relationships of between different countries B75
  typical forms are most like ancient B37, 207, E108
  See also Ancestor; Ancestor, common; Atavism; Inheritance; Morphology
Tyrannidae C111, 124
Tyrannulae ZEd15

Uniformitarianism
  and climate B246, 247
  doctrine of B201
  slow changes in operation B209, D167
  slow changes in rock constituents B201
  slow insensible changes in species E4–6
  unequal rate of physical changes C153
  See also Gradualism
Urtica QE[9v]
Uruguay
  birds C163, ZEd16
  geology RN7, 56, 87, 144, A68
Uspallata Pass, Cordillera
  no trace of subsidence except for silicified trees RN152, 154, A120

Vaginulus ZEd13
Valerian QE[9v]
Valleys A87, 131
  effect of tides on submersed RN141
  erosion of RN109, A12, 70
  formation of transverse RN141
Valparaiso, Chile RN5, 88, GR30
Valvata B229
Van Diemen's Land
  See Tasmania
Variation B6, 7, 148, 241, 236
  absence of limits to B43
  absent in simplest organisms E163
  accidental MAC53v
  classification schemes are based on absence of D51
  complex animals more subject to C169
  correlations of C192, E51, 53–5, T95, MAC28r
  cumulative E57, 70
  determined by laws of organization E50, T17
  differences between those of seedling plants and cuttings QE5
  during asexual reproduction C237
  endless changes prevented by crossing of two sexes E49–51
  extremes of, in thorax of insects TFRAG5
  fixity of character prevents E141–2
  great in birds owing to variety of stations inhabited C82
  impossibility of discovering origin of peculiarities D100
  in abortive parts of plants QE[6v]
  in animals which have recently acquired their peculiarities GR31
  in cultivated plants D118, 128–30, E13–17, QE1v, 3, 5, 6, 7, [15v]
  in domestic animals B100, E142, 168
  in personal character of animals N98
  likelihood of after crossing is prevented MAC56v
  likelihood of a law of D112
  limits to D104, E136
  of individuals in one country more than in another QE1, 10a
  of individuals like that of species E12
  oldest organ least likely to undergo C149, E23
  possibility of degeneration after several generations D161
  similarity to mutilations C232
  some forms of may be analogous to specific character of other species in the genus D65–7
  some species vary more than others C53
  under new conditions of existence B3, T81
  See also Inheritance
Varieties B83, 96, 158, 171, C1–4, MAC57v
  cannot be counteracted by man C106
  change in B64, 85, QE17
  comparison between wild and domestic D20–1
  Darwin's theory of origin of E118
  difference between races and E72, 75
  grow together under same conditions QE[15v]
  in plants B83, 107, C192–5, QE2
  length of time that pigeon varieties have been known D100
  length of time to become fixed D42–3
  man's interest in perpetuating B155
  old varieties prone to extinction D49
  permanence B68, 123, 130, C136
  permanent in plants C192–5
  possibility of being raised by seed E183
  produced per saltum B8, 278
  production of B38, 125, 191, CIFC, 59, 73, 195
  production of plant varieties compared with that of animals B210, QE20
  resemble species B159, D107
  result of cross-breeding between B203, C30, 34, D15–16, QE8
  reversion to parent stock B180–1, 203
  run wild B83
  two classes of C3–4, 106
  See also Yarrell's Law
Veins A147
  and dikes A42, 61, 116
  a proof of elevation of land A13
  chemical activity in RN78, 165
  cleavage coincides with line of RN21
  direction RN106
  formation A47, 60, 149

Veins (*cont.*)
in Chiloé RN165
in granite A25
in Mexico RN171, 175–6
metallic RN20, 74, 165, 168
origin RN66
quartz RN7, 87
structure RN66, 88, 165
Vellellae ZEd13
Ventana, Mount, Argentina A3, GR13
Verbena M139, QE[11^{v}]
Veronica T178
Vertebrates B43, 108, 110, E89
Vesuvius RN63, 137
Viburnum
sterile form of QE[6^{v}]
Vinca QE[9^{v}], 10
Vinchuca ZEd3
Violets QE21..
Viscacha ZEd2
Volcanic bombs RN74
Volcanic gases RN74
emission of sulphuric vapours RN44, 78, 137, A116
Volcanic islands RN127, QE10a
plants on A2, B193–4
formation of new species on B17, 103, 152, 192–4
Volcanic rocks
absence of lime B201
distinguished from metamorphic rocks A29
olivine RN156
origins of RN79
variety of ejected matter RN41
See also Lava; Pumice; Tufa
Volcanoes RN75, A15, 34, 40, 67, 89, 113, 134, 138–40
absence in subsiding areas A80
activity prevented by accumulation of sedimentary rocks RN32
aerial RN68
and surface heat of the earth A123
and thickness of earth's crust A78
and warmth of ocean A138
as evidence for a central core of molten rock RN57–9, A108
built up by series of lava coatings RN57–8
chemical activity within RN78, 111–12, A116
effect of water on RN79
eruption accompanied by elevation of land RN117, 137, A34, 62
geographic distribution RN70, 73, 90, 137, 144, 158, A39, 40
in lakes A35, 142
origin of volcanoes of Cordillera RN10–11
origins in dislocation of strata RN10–11, 146–7
nature of eruptions RN44, 57, 70, 79, 146–7
submarine RN68, 74, 76, 146, A127, 138
See also Cordillera; Craters; Craters of elevation; Etna; Metamorphic rocks; Stromboli; Vesuvius
Voluta ZEd8
Vultures C67

Wagtails C160, D102
Wales
geology of Anglesea RN5–7, A97
limestone RN93
trout S5
Wallflowers QE1^{v}
Walnut QE10a
Walrus E92, MAC28^{r}
Warbler, grasshopper C160
Wasps D116, M63
Water A61, 68, 123
Water action
See Currents; Erosion; Lakes; Rivers; Springs; Waves
Waves
action of A59
erosion of coastal cliffs RN38–41, 67, 100, A59
caused by earthquakes RN80–3
See also Sea
Wealden beds RN152, A55, B172, D133, E55, MAC29^{r}
Weathering
See Erosion
Weeping
See Expressions
West Indies A85, B32, C42, E42
geology RN27, 124, A85
See also Cuba
Whales B162, 229, C157, E92, MAC28^{r}, QE17
Wheat D135, M155
transformations in QE11–[11^{v}]
Whitethorn E17
Widgeons B141, D25
Widow birds C109, 144
Will
arises from fixed laws of organization M69–70
governs secretion of body fluids OUN39
Will, free M118, 126
actions are effects of motives OUN25–8
belief in material origin of tends to aetheism M72–3
distinguished from power of imagination N94
existence of M26–7, N49
in animals M72–3
origin in instincts OUN26, 52
synonymous with chance M27, 30–1
Willing of animals
See Transformism
Willows B151, C156, E130, QE8, 9
Winds A41, 131, E44, 123
Windsor Park
animals E31–2
Wolves E47, 102, ZEd5
behaviour M147
cross-breeding with dogs B51, 140, 163, D7, 173, N44
intellect E46
same species as dogs B51
tame E103, N97
Wombat C239
Women
and transmission of characteristics C163
education C220
exhibit ancestral type C178
genitals C204
Wood
conversion into coaly matter RN28
See also Coal; Conifers; Fossil Wood; Trees
Woodlouse TFRAG4
Woodpeckers C61, 113, MAC57^{r}
adaptations B55, N71

tail C14, C111
Woodpeckers, ground C57, 64, 114, MAC57^r–57^v
habitat B55
tail C82
Woolich AIFC
geology of A112
Worms, parasitic B108, D3–4, M42, ZEd13
Wrens B213, C160, 255
Wrens, willow B241, C125, 177, D102

Yaguarundi ZEd11
Yak QE12
Yarrell's Law D7
and colours of different breeds of birds C2–4, 68, 71
and hybridity C30, 33–4
and reversion D8
anomaly in D88, 89, 91–4
evidence about old varieties illusory C121
evidence against E35, 169
experimental test of QE4
partial truth of E112
statement of C1–3
Yawning M84, 93, 95–6
Yews RN178, B107, QE9, 13, 14

Zebra C145, D66
behaviour M146, N75
cross-breeding with ass T2, QE20
cross-breeding with horse D113, T2, QE20
geographic distribution B62, 72, 234
origin B72
Zizania E111
Zoological Journal B87, C40, 41, 42, 61, 67, 99, 160, 240, E42, 60, ZEd12
Zoological Society Gardens
animals' affection for keepers N2, 14, 115
animals' behaviour towards keepers M137–9, 141
baboon behaviour C212
baboons M105–6
bantams QE18, 20
bears QE20
birds C164, 210, D163, QE20
camels QE20
Darwin's childhood memories of M29
Darwin's observations of animal expressions at M105–8, 129
Darwin's questions for keepers at QEIFC, 20
dogs D90, 127
fertile hybrid between ass, zebra and pony T2
foxes QE20
hybrid birds D32, 159–60, 180, IBC
jackal D7, D inside back cover
Jenny the orangutan M137–40, N13, 94
keepers' views on animal behaviour M106, 137–8, N97–8, 138
keepers' views on cross-breeds D91
lions N98, QE20
monkey behaviour M137, 142, 145
monkeys M106–7
orangutan M107, 129
orangutan behaviour M85, 129
panther N6
pigs QE20
propagation rare in QE4a
rhinoceros QE20
the Hensleigh Wedgwoods' visit N121
Tommy the orangutan M138–9, N13
wolf E103, N97
zebra T2, QE20
Zoological Society of London
meetings at C107, D26, 95
See also *Proceedings of the Zoological Society of London*; *Transactions of the Zoological Society of London*
Zoophytes RN132, A180, DIFC, N49
See also Polyps
Zorilla RN128, ZEd2
Zostera QE[15^v], 21..